QUANTUM INFORMATION SCIENCE

Quantum Information Science

Riccardo Manenti Mario Motta

OXFORD
UNIVERSITY PRESS

Great Clarendon Street, Oxford, OX2 6DP,
United Kingdom

Oxford University Press is a department of the University of Oxford.
It furthers the University's objective of excellence in research, scholarship,
and education by publishing worldwide. Oxford is a registered trade mark of
Oxford University Press in the UK and in certain other countries

Published in the United States of America by Oxford University Press
198 Madison Avenue, New York, NY 10016, United States of America

British Library Cataloguing in Publication Data
Data available

Library of Congress Control Number: 2023938489

ISBN 978–0–19–878748–8

DOI: 10.1093/oso/9780198787488.001.0001

Printed and bound by
CPI Group (UK) Ltd, Croydon, CR0 4YY

Preface

About one hundred years ago, several scientists, including Planck, Bohr, De Broglie, Schrödinger, Pauli, Dirac, and Heisenberg, made significant contributions to the discovery of a new physical theory called quantum mechanics. Initially developed to address specific problems, such as the black body radiation and atomic absorption spectra, quantum mechanics gradually extended to other physical phenomena, correctly predicting the collective behavior of interacting particles. The success of quantum mechanics allowed us to construct electronic devices of immense practical relevance, including the transistor, underpinning semiconductor-based computers, and the laser, an essential component in optical communication.

Although modern computers contain some quantum components, their computational processes are still governed by classical laws and formalized by a mathematical model known as the Turing machine [1]. For this reason, we call them "classical computers". For the past forty years, universities and research centers have been investigating new types of computational models based on the principles of quantum mechanics to determine whether it is possible to build a machine more powerful than Turing machines.

The objective of this textbook is to introduce readers to the basic concepts of quantum computing, quantum information, and quantum engineering. Since this research field evolves rapidly and has many specialized ramifications, we decided to present the essential notions necessary to build solid fundamentals in the field.

Basic concepts

A classical computer contains billions of transistors. These are electronic devices that can be in two different states usually denoted 0 and 1. Whenever we write a text on a classical computer, it is encoded into a string of 0's and 1's, i.e. a binary string. Mathematical operations are carried out by transforming binary strings into other strings by applying a set of logical operations, such as the OR and AND gates.

Building a quantum computer does not simply mean miniaturizing transistors but rather replacing them with a new fundamental component called a qubit. A qubit is a quantum system with two states, usually denoted $|0\rangle$ and $|1\rangle$. Unlike a classical system, a qubit can be in a linear combination of $|0\rangle$ and $|1\rangle$ called a superposition state. Quantum computers can execute classical logic gates and purely quantum mechanical operations such as the Hadamard and CPHASE gates. Leveraging this broader set of transformations one can prepare a register of N qubits into a superposition of 2^N states. Some of these states, known as entangled states, have statistical properties that cannot be mimicked by a classical distribution of states.

The creation of entangled states in a large vector space and their manipulation with destructive and constructive interference underpins the power of quantum computers. However, the manipulation of an exponential number of states does not imply that quantum computers can solve any mathematical problem with an exponential speed-up with respect to classical machines by testing all possible solutions in parallel. Indeed, the measurement of a register of qubits collapses the quantum state into one of the 2^N states. A quantum algorithm must therefore coordinate the interference between the components of the superposition state so that the correct solution is measured with high probability. The class of problems that a quantum computer can solve in polynomial time with a small error probability is known as BQP [2]. The relation between this class and the class P (the complexity class containing all problems that a deterministic machine can solve in polynomial time) is still unknown [3], but it is widely believed that BQP contains P.

A timeline of quantum computation

Quantum computation has its roots in the seminal works by Szilard, von Neumann, and Landauer [4, 5, 6]. These scientists connected the thermodynamic concept of entropy to the theory of information and showed that applying non-reversible binary operations to one or more bits leads to a dissipation of energy, a phenomenon known as Landauer's inequality. Inspired by these works, Bennet introduced the notion of reversible computation as a possible strategy to circumvent Landauer's limit [7, 8]. Bennet's intuition was later developed by Toffoli, Fredkin, and Benioff [9, 10].

This set of works led Feynman to publish a series of papers between 1982–1985 in which he proposed to use quantum devices to simulate the time evolution of quantum systems [11, 12]. His intuition was based on the observation that quantum computing extends the idea of reversible computing since the unitary evolution generated by a Hamiltonian is a more general version of reversible dynamics.

A significant contribution to the development of quantum computers is due to Deutsch, who in 1985 introduced the quantum Turing machine, a computational model leveraging the superposition principle and entanglement to execute a computation [13]. In 1989, Deutsch introduced the circuit model of quantum computing and the concept of a universal set of quantum gates [14]. In collaboration with Jozsa, he formulated the first problem that a quantum algorithm can solve with fewer oracle queries than a classical algorithm [15].

The field of quantum computing blossomed in the 1990s with the discovery of several quantum algorithms with a lower query complexity than their classical counterparts. Among these, we must mention the Bernstein–Vazirani algorithm (1993) [16], Simon's algorithm (1994) [17], Shor's algorithm for prime factorization (1994) [18], and Grover's algorithm for unstructured database search (1996) [19]. These algorithms are known as early quantum algorithms.

In the last two decades, researchers have developed various quantum algorithms to tackle complex problems. These include:

- Algorithms to demonstrate that quantum computers can sample some probability distributions much faster than classical computers. Important examples are the random circuit simulations problem and boson sampling [20, 21].

- Quantum algorithms for optimization problems. One of the goals of these algorithms is to find the global minimum of a complex cost function.
- Quantum algorithms for machine learning. This field focuses on constructing mathematical models that capture trends and properties of a data set. This is an emerging research field beyond the scope of this textbook.
- Quantum algorithms for the simulation of complex systems based on an existing physical theory. An important problem in quantum simulation is the calculation of the eigenstates of a Hamiltonian, which has plenty of applications in chemistry (e.g. the analysis of chemical reaction rates), in life science (e.g. drug discovery), and materials science (e.g. high-temperature superconductivity). A variety of heuristic quantum algorithms have been developed to approximately solve this problem, including the adiabatic simulation (2000) [22], the quantum approximate optimization algorithm (2014) [23], and the variational quantum eigensolver (2015) [24]. Another important problem is the simulation of the time evolution generated by a Hamiltonian. For a broad family of Hamiltonians, this problem lies in the BQP class. Important examples of quantum algorithms for the simulation of time evolution are the Trotterization (1996) [25, 26], the Taylor series (2014) [27], and the qubitization method (2016) [28]. These algorithms let us study the dynamical properties of quantum systems efficiently.

This textbook will focus on early quantum algorithms and quantum algorithms for the simulation of many-body quantum systems.

Building a quantum computer

At the beginning of the 1950s, it was still unknown which physical platform would have been used to build a large-scale classical processor. Different technologies were explored, including mechanical and electromechanical devices, until the community converged on integrated circuits. Similarly, in the context of quantum computing, multiple technologies are under consideration by universities, research centers, and private companies. They include:

- Solid-state devices: in this technology, qubits are built out of artificial structures embedded in a solid-state device. Some examples are superconducting circuits and quantum dots.
- Quantum computers based on individual atoms, ions, or molecules: in this technology, the qubits are pairs of energy levels of atoms, ions, or molecules that are kept at a fixed position by force fields. Some examples are trapped ions, Rydberg atoms in optical tweezers, ultracold atoms in optical lattices, and cold molecules.
- Photonic quantum computers: this technology uses photons to encode quantum information. Photons propagate along waveguides in integrated circuits and are manipulated with non-linear optical elements.

In this book, we focus on superconducting circuits, a technology that has attracted great interest in recent years.

Content

This book is divided into four parts:

- In the first part, titled "Foundations", we present fundamental concepts of mathematics, computational theory, linear algebra, and quantum mechanics. A good understanding of quantum information science cannot disregard these notions.
- In the second part, titled "Modern quantum mechanics", we present more advanced tools of quantum mechanics to understand how noise affects the functioning of a real quantum computer. Indeed, a quantum computer can be considered an open quantum system constantly interacting with the environment. This interaction is unavoidable and might destroy the fragile quantum effects, thereby introducing errors in the computation.
- In the third part, titled "Applications", we explain the concept of entanglement and some quantum algorithms. We start with early quantum algorithms and then discuss quantum algorithms for time evolution and Hamiltonian eigenstates simulation.
- In the fourth part, titled "Quantum engineering", we explain the physics of superconducting devices.

The book contains more than 200 figures and more than 100 exercises with solutions. At the end of each chapter, we include a summary and a section with further reading to broaden the reader's knowledge. The book's final part contains some appendices, the bibliography, and the index. The bibliography includes the references cited in the text. We apologize in advance to any researcher whose work has been inadvertently omitted. The book might contain typos as well as mathematical imprecisions. If you find any, please send us your comments at **qis.manenti.motta@gmail.com**. We are open to suggestions to improve the textbook.

For reasons of space, the book does not discuss quantum error correction, recent results about entanglement theory (e.g. the area law and tensor networks), quantum cryptography, quantum communication, quantum sensing, quantum machine learning, and other approaches for constructing quantum computers. We hope that some of these topics will be covered in future editions.

Target audience

This book is aimed at students who recently finished a bachelor of science, master's students, graduate students, and researchers that want to deepen their understanding of quantum information science. We expect our readers to be familiar with mathematics, particularly calculus and the rudimentary notions of quantum mechanics. We believe that readers passionate about quantum information will develop (with patient study) a good understanding of the fundamentals of this field.

Acknowledgments

We would like to thank C.J. Winkleblack, E. Vitali, R. Renzas, P. Bartesaghi, E. Chen, D. Berry, L. Lin, A. M. Childs, J. Shee, D. Maslov, R. Smith, A. Nersisyan, N. Alidoust, M. J. O'Rourke, J.-P. Paquette, and W. Swope for helpful feedback about the manuscript. It goes without saying that no responsibility should be attributed to any of the above regarding possible mistakes in the book, for which the authors alone are to blame. We thank Sonke Adlung and Giulia Lipparini, currently working at the Oxford University Press, for their constant support and encouragement throughout these (many) years.

About the authors

This book was completed while Riccardo Manenti was a Ph.D. student at the University of Oxford under the supervision of Dr. Peter Leek and a senior quantum engineer at Rigetti computing, and while Mario Motta was a Ph.D. student at the University of Milan under the supervision of Prof. Davide Galli, a postdoctoral researcher at the College of William and Mary under Prof. Shiwei Zhang, a postdoctoral researcher at the California Institute of Technology under Prof. Garnet K.-L. Chan and a research staff member at the IBM Almaden Research Center.

To our readers

Contents

PART II MODERN QUANTUM MECHANICS

PART III APPLICATIONS

PART I
FOUNDATIONS

1
Mathematical tools

Solving a decision problem
means answering yes or no
to a question.

One of the main goals of quantum information science is to develop efficient techniques to solve complex mathematical problems. Examples of such problems are decision problems of the form: "does a graph contain a complete subgraph with k nodes?" or "does a graph contain a Hamiltonian cycle?" Intuitively, solving a **decision problem** means answering yes or no to a question [29, 30]. These types of problems can often be solved using algorithms: finite sequences of computational steps performed by a human or a machine [31, 32, 33, 34, 35]. Over the past several decades, researchers have developed multiple computational models to group mathematical problems into classes according to the time and space required to solve them [36, 37, 38].

After introducing some mathematical tools, we will present some decision problems, including ways to classify their resource requirements. A rigorous theory of computational complexity allows us to make precise statements about the formulation and solution of problems using computational devices. Additionally, it enables us to characterize and appreciate the impact of quantum computers in solving complex problems that are currently beyond the reach of classical computers. We encourage readers to revisit these sections as necessary.

1.1 Basic notions

In this section, we introduce some important mathematical ideas that will be widely used throughout this textbook, to provide readers from different backgrounds with a good understanding of the notation and terminology. First, we will review sets, sequences, relations, and functions; then, we will learn the main properties of graphs. These concepts will be crucial to understand some important decision problems presented in the second part of the chapter.

1.1.1 Sets and sequences

A **set** is a collection of objects. The objects contained in a set are called its elements. A set is usually named by uppercase letters with its elements contained within curly brackets, like

$$A = \{1, 2, 3\} . \tag{1.1}$$

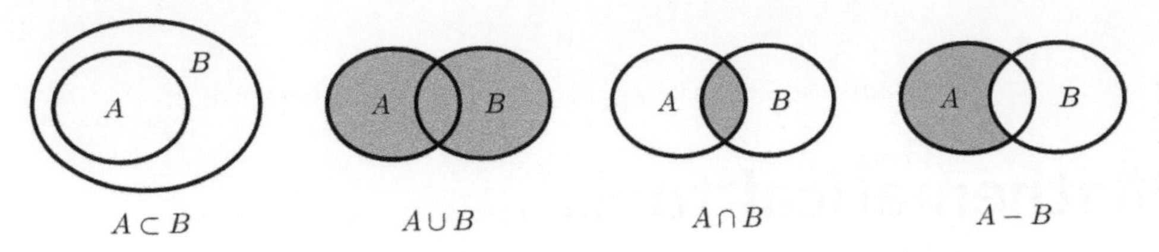

Fig. 1.1 **Sets.** From left to right: inclusion, union, intersection, and subtraction operations.

Here, 1 belongs to A, which is denoted $1 \in A$, while 4 does not belong to A, which is denoted $4 \notin A$. The sets $A_2 = \{2, 3, 1\}$ and $A_3 = \{1, 1, 2, 3, 3, 3\}$ are equivalent to A. In other words, neither the order nor the multiplicity of the elements matters in a set. On the other hand, a **multiset** is a set in which the multiplicity of each element is relevant, but not the order. Multisets are also indicated with uppercase letters and curly brackets. Some examples are: $B_1 = \{1, 2, 3\}$, $B_2 = \{2, 3, 1\}$ and $B_3 = \{1, 1, 2, 3\}$. The multisets B_1 and B_2 are equal, but B_1 is not equal to B_3.

The **empty set** does not contain any elements and is denoted by the symbol $\varnothing$. A set with only one element is called a **singleton**. For singletons, we will sometimes omit the curly brackets. A finite (infinite) set contains a finite (infinite) number of elements. The **cardinality** of a set is denoted as $|A|$ and, for a finite set, it indicates the number of elements (for example $|A| = 3$). Typical examples of infinite sets are $\mathbb{N}$, $\mathbb{Z}$, $\mathbb{Q}$, $\mathbb{R}$, and $\mathbb{C}$: the sets of all natural, integer, rational, real, and complex numbers respectively. The set of all real numbers between a and b inclusive is indicated with square brackets $[a, b]$. We will use the notation (a, b) to indicate all real numbers larger than a but smaller than b. To indicate that a set contains only non-negative numbers, we will use the subscript "≥ 0". For example, $\mathbb{R}_{\geq 0}$ indicates the set of non-negative real numbers.

A convenient way to express a set is to specify a property shared by all of its elements. For instance, the set of even numbers can be written as

$$A = \{m \in \mathbb{Z} \; : \; m = 2n \text{ where } n \in \mathbb{Z}\}, \tag{1.2}$$

which should be read as "the set A contains all integers m such that $m = 2n$ where n is an integer". Another commonly used notation to describe a set is

$$B = \{b_1, \ldots, b_n\} = \{b_j\}_{j=1}^{n} \, . \tag{1.3}$$

For example, $\{2j + 1\}_{j=1}^{3} = \{3, 5, 7\}$.

The phrases "for all", "there exists", and "such that" are very frequent in mathematics. The quantifier "**for all**" is abbreviated with the symbol "$\forall$". The symbol "$\exists$" means "**there exists**". Finally, we abbreviate "**such that**" with the symbol ":". According to these definitions, the statement "for all natural numbers n, there exists a real number r such that r is greater than n" is abbreviated as

$$\forall n \in \mathbb{N} \quad \exists r \in \mathbb{R}: \quad r > n\,. \tag{1.4}$$

Let us list some fundamental definitions and properties regarding sets:

$$\begin{aligned}
A = B &: \ A \text{ contains the same elements as } B, \\
A \neq B &: \ A \text{ and } B \text{ do not contain the same elements}, \\
A \subseteq B &: \ A \text{ is a } \textbf{subset} \text{ of } B \text{ (all elements of } A \text{ are contained in } B), \\
A \subset B &: \ A \text{ is a } \textbf{proper subset} \text{ of } B \text{ (} A \text{ is a subset of } B \text{ and } A \neq B), \\
A \cup B &: \ \textbf{union} \text{ of } A \text{ and } B \text{ (all elements in } A \text{ or } B), \\
A \cap B &: \ \textbf{intersection} \text{ of } A \text{ and } B \text{ (elements in both } A \text{ and } B), \\
A - B &: \ \text{elements that are in } A \text{ but not in } B.
\end{aligned}$$

The last four definitions are illustrated in Fig. 1.1. When the sets A and B do not share any element, we will write $A \cap B = \varnothing$, and we will say that A and B are **disjoint** sets. Lastly, the set $\mathcal{P}(A)$ is defined as the collection of all subsets of A, and it is called the **power set** of A. For example, if $A = \{1, 7\}$, then

$$\mathcal{P}(A) = \{\varnothing, \{1\}, \{7\}, \{1, 7\}\}\,. \tag{1.5}$$

Observe that the power set of A contains both $\varnothing$ and A.

A **sequence** is an ordered list of elements indexed by natural numbers. In this book, we indicate sequences with round brackets,

$$s = (1, 7, -5)\,. \tag{1.6}$$

We will sometimes omit the commas and brackets, especially when dealing with binary strings of the form $(1, 0, 1, 1) = 1011$. Unlike sets, the elements in a sequence are ordered in a precise way and the multiplicity of each element is relevant. For example, the sequences $s_2 = (1, -5, 7)$ and $s_3 = (1, 1, 1, 7, -5)$ are not equal to s. A finite sequence with n elements is called n-**tuple** and a sequence with two elements is usually called an **ordered pair**. A compact notation for sequences is given by

$$s = (a_j)_{j=1}^{n}\,. \tag{1.7}$$

For example, the sequence in eqn (1.6) can be written as $(3 + (-2)^j)_{j=1}^3$. We will sometimes use the notation $A = \{a_j\}_j$ and $s = (a_j)_j$ to indicate that the set A and the sequence s might have a finite or countably infinite number of elements (note the missing superscript). The concept of a countably infinite set will be formalized in Section 1.1.3.

1.1.2 Cartesian product and binary relations

The **Cartesian product** between two sets A and B is the set $A \times B$ of all ordered pairs of the form (a, b) where $a \in A$ and $b \in B$. For example, if $A = \{1, 7\}$ and $B = \{\text{red}, \text{blue}\}$, then

$$A \times B = \{(1, \text{red}), (1, \text{blue}), (7, \text{red}), (7, \text{blue})\}\,. \tag{1.8}$$

The generalization to multiple sets is straightforward: the Cartesian product between the sets $A_1, \ldots, A_n$ is the set

$$A_1 \times \ldots \times A_n = \{(a_1, \ldots, a_n) : a_i \in A_i\}\,. \tag{1.9}$$

If these n sets are copies of the same set, we will write

$$\underbrace{A \times A \times \ldots \times A}_{n \text{ times}} = A^n\,. \tag{1.10}$$

It is easy to verify that for a finite set A, A^n is a finite set with cardinality $|A|^n$. As an example, consider the **Boolean set** $\mathbb{B} = \{0, 1\}$. The Cartesian product $\mathbb{B}^2$ is given by $\mathbb{B}^2 = \{(0,0), (0,1), (1,0), (1,1)\}$. Another example of a Cartesian product is the well-known Cartesian plane $\mathbb{R}^2$.

A **binary relation** R from a set A to a set B is a subset of $A \times B$. For example, if $A = \{1, 7\}$ and $B = \{3, 4\}$, then the set $R = \{(1, 3), (1, 4)\}$ is a relation from A to B. When we write $a\,R\,b$, we mean that the pair $(a, b) \in R$. Typical examples of binary relations between real numbers are: less-than "$<$", greater-than "$>$", and equal-to "$=$". A binary relation on a set A is a subset of $A \times A$. For all $a_i \in A$, a binary relation is:

reflexive: when the relation $a_1\,R\,a_1$ holds,
symmetric: if $a_1\,R\,a_2$ then $a_2\,R\,a_1$,
transitive: if $a_1\,R\,a_2$ and $a_2\,R\,a_3$ then $a_1\,R\,a_3$.

A binary relation that is reflexive, symmetric, and transitive is called an **equivalence relation**. If R is an equivalence relation and $a\,R\,b$ holds, the elements a and b are said to be R-equivalent. A simple example of an equivalence relation between real numbers is "$=$".

Let us now introduce an important mathematical relation. Consider two real numbers a, b, and a real number $n \neq 0$. We say that a and b are **congruent modulo** n if the difference $a - b$ is an integer multiple of n, and we write

$$a = b \pmod{n}. \tag{1.11}$$

Here are some examples of congruence relations:

$$3 = 13 \pmod{5}, \qquad 3 = -2 \pmod{5}, \qquad -12 = 3 \pmod{5}.$$

The congruence relation satisfies some important properties:

- $a = a \pmod{n}$ (reflexive),
- $a = b \pmod{n}$ implies $b = a \pmod{n}$ (symmetric),
- if $a = b \pmod{n}$ and $b = c \pmod{n}$, then $a = c \pmod{n}$ (transitive).

Thus, the congruence relation is an equivalence relation.

The **equivalence class** $[a]_R$ is a set that contains all elements that are equivalent to a with respect to a binary relation R. As an example, consider the equivalence class $[1]_{\text{mod}\,3}$ that contains all integers x that are congruent to 1 modulo 3,

$$[1]_{\text{mod}\,3} = \{\ldots, -5, -2, +1, +4, +7, \ldots\} = \{x \in \mathbb{Z} \,:\, x = 1 \pmod 3\}\,. \tag{1.12}$$

Since $[1]_{\text{mod}\,3} = [4]_{\text{mod}\,3} = [7]_{\text{mod}\,3} = \ldots$, the element 1 can be considered to be the representative element of this equivalent class. It can be shown that two equivalence classes $[a]_R$ and $[b]_R$ on a set A are either identical or disjoint.

1.1.3 Functions

A **function** (or map or transformation) f between two sets A and B is a relation that associates each element in A with exactly one element in B. It is written as

$$\begin{aligned} f: \quad A &\to B \\ a &\mapsto b \end{aligned}$$

The set A is called the **domain** of the function and B is called the **codomain**. If the function f associates an element a with an element b, we write $f(a) = b$ and we say that a is mapped onto b. Element a is called the **argument** of the function and b is called the **image** of a (note that by definition an element $a \in A$ has exactly one image in B). The image of A, denoted as $f(A)$, contains all of the images of the elements in the domain and is, therefore, a subset of B. In mathematical terms, this can be expressed as $f(A) \subseteq B$. We define the **support** of a real-valued function f as the subset of the domain containing those elements which are not mapped onto zero. Three different types of functions are shown in Fig. 1.2:

injective functions: if $a_1 \neq a_2$, then $f(a_1) \neq f(a_2)$,
surjective functions: for any $b \in B$, there exists $a \in A$ such that $f(a) = b$,
bijective functions: functions that are both injective and surjective.

If a function f is bijective, it is possible to define the **inverse function** $f^{-1} : B \to A$ that satisfies $f^{-1}(f(a)) = a$ for all $a \in A$, and we say that f is **invertible**. If a function is bijective, the domain and codomain have the same cardinality. In particular, if there exists a bijective function that maps elements of a set A to the set of natural numbers, then we say that the set A is **countably infinite**.

In arithmetic, the division of an integer a by a natural number n always produces a **quotient** $q \in \mathbb{Z}$ and a **remainder** r,

$$a = qn + r, \tag{1.13}$$

where $0 \leq r < n$. This is called **Euclidean division**. The function $a \,\%\, n$ returns the remainder r of the Euclidean division of a by n. The remainder satisfies some important properties:

- $(a \,\%\, n) \,\%\, n = a \,\%\, n$,
- $(a + b) \,\%\, n = [(a \,\%\, n) + (b \,\%\, n)] \,\%\, n$,

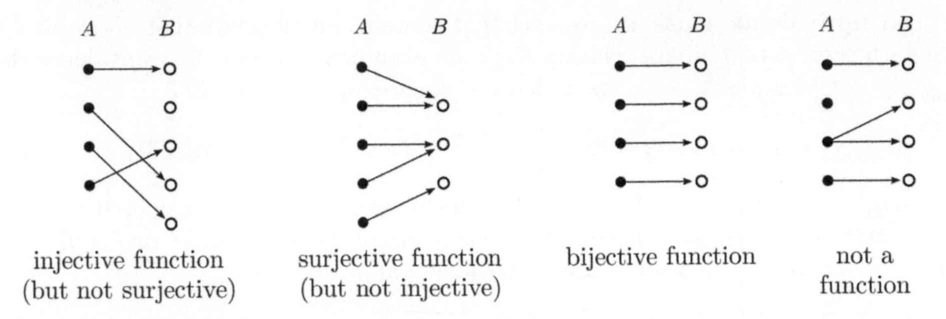

Fig. 1.2 Functions. From left to right: schematic representation of an injective, a surjective, and a bijective function, and of a relation that is not a function. In the first three cases, A is the domain, and B is the function's codomain.

- $(a \cdot b) \,\%\, n = [(a \,\%\, n) \cdot (b \,\%\, n)] \,\%\, n$.

It is worth mentioning that in some textbooks about quantum information, the symbol "%" is denoted "mod".

When presenting logic circuits, we will deal with functions whose domain and codomain are finite sets. These types of transformations can be conveniently described with tables. For example, consider the sum modulo two, the transformation $\oplus : \mathbb{B}^2 \to \mathbb{B}$ defined as $x_1 \oplus x_0 = (x_1 + x_0) \,\%\, 2$. This map is described in the following table

x_1	x_0		$x_1 \oplus x_0$
0	0		0
0	1	$\mapsto$	1
1	0		1
1	1		0

We shall see more examples of finite functions in Section 1.1.7 when we introduce Boolean functions.

In calculus, real-valued functions play a crucial role. These are functions that assign a real number to members of the domain. Important examples of real functions $f : \mathbb{R} \to \mathbb{R}$ are **polynomials** defined as

$$p(x) = \sum_{i=0}^{n} c_i x^i = c_0 x^0 + c_1 x^1 + \ldots + c_n x^n \,, \tag{1.14}$$

where $x \in \mathbb{R}$, $c_i \in \mathbb{R}$, and $n \in \mathbb{N}$. The degree of a non-zero polynomial is n, the greatest power[1] of the variable x. For example, the polynomial $p_1(x) = 2x^7 + 1$ has degree 7. Polynomials of degree 0 are called **constant functions**, of degree 1 are called **linear functions**, and of degree 2 are called **quadratic functions**.

An **exponential function** is a real function of the form $f(x) = a^x$, where $a > 0$. An important example of an exponential function is e^x, where $e \approx 2.71828$ is Napier's constant. The **logarithmic** function $\log_a x$ is the inverse function of the exponential,

[1]By convention, the degree of the polynomial $p(x) = 0$ is $-\infty$.

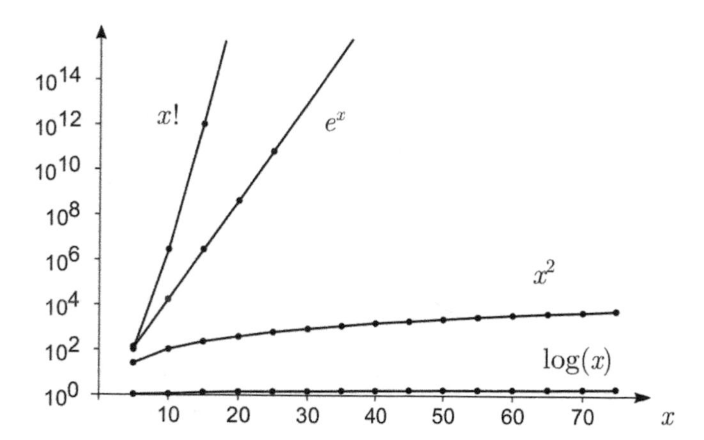

Fig. 1.3 Real-valued functions. A graphical representation of some functions $f : \mathbb{N} \to \mathbb{R}$. It is evident that $\log(x) < x^2 < e^x < x!$ for $x \geq 10$. Note that the y-axis is in log scale.

---

i.e. $\log_a(a^x) = x$, where a and x are positive. When the base of the logarithm is Napier's constant, we will write $\log x$ or $\ln x$. The **factorial** of a natural number x is defined as $x! = x \cdot (x-1) \cdot (x-2) \cdot \ldots \cdot 1$. It can be shown that, for $k > 1$, there exists a number x_0 such that for all natural numbers $x > x_0$, we have $(\log_a(x))^k < x^k < e^x < x!$. These relations are illustrated in Fig. 1.3.

When we introduce decidable languages, we will make use of the **characteristic function** $\chi_C : A \to \{0,1\}$. This transformation maps an element $x \in A$ onto 1 if $x \in C$ and onto 0 otherwise,

$$\chi_C(x) = \begin{cases} 1 & \text{if } x \in C \\ 0 & \text{if } x \notin C \end{cases} \qquad \text{where } C \subseteq A. \tag{1.15}$$

For example, the function $\chi_{[a,b]}(x) : \mathbb{R} \to \{0,1\}$ assigns 1 to all real numbers in the interval $[a,b]$ and 0 to all numbers outside this interval.

Example 1.1 The exponential function can be expressed as the sum of an infinite number of polynomials

$$e^x = \sum_{n=0}^{\infty} \frac{x^n}{n!} = 1 + x + \frac{x^2}{2} + \frac{x^3}{6} + \ldots$$

This is an example of a Taylor series.

1.1.4 Big-O notation

In complexity theory, a fundamental equivalence relation between functions is the **big-O relation** [39, 40]. Consider two functions $f : \mathbb{N} \to \mathbb{R}_{>0}$ and $g : \mathbb{N} \to \mathbb{R}_{>0}$. If for a positive number c there exists a natural number n_0 such that for all $n > n_0$,

Class	Name	Example	
$O(1)$	constant	6	subpolynomial
$O(\log n)$	logarithmic	$3+2\log n$	
$O(\log^c n)$	polylogarithmic	$\log^5 n - \log^4 n$	
$O(n^k)$	polynomial	$5n^7+2n^2-1$	
$O(e^n)$	exponential	$2^n - n^2$	superpolynomial
$O(n!)$	factorial	$(2n)!$	

Table 1.1 Important O equivalence classes, where $c > 1$ and $k > 0$.

$$f(n) \le cg(n), \tag{1.16}$$

then we write $f \in O(g(n))$ and we say "f is big-O of g". With the notation $O(g(n))$ we will indicate the set of functions that do not grow faster than $g(n)$ for large n. Table 1.1 lists the most important classes of functions categorized according to big-O notation. Observe that functions that only differ by a multiplicative constant belong to the same class (for instance, $f_1(n) = 1000n^2$ and $f_2(n) = 2n^2$ both belong to $O(n^2)$). With the term **subpolynomial**, we will indicate the set of functions that grow slower than any polynomial of degree 1. Similarly, with the term **superpolynomial**, we refer to all functions that grow faster than any polynomial. Lastly, we will often use the notation $f = \text{poly}(n)$ to indicate that the quantity f grows polynomially with n.

There exist several related notations expressing the asymptotic behavior of real-valued functions. These notations are indicated with the letters o, Ω, and ω, and their meaning is summarized in the following table:

notation	asymptotic behavior for $n \to +\infty$
$f \in o(g)$	$f(n)/g(n) = 0$
$f \in O(g)$	$f(n) \le cg(n)$
$f \in \Omega(g)$	$f(n) \ge cg(n)$
$f \in \omega(g)$	$f(n)/g(n) = +\infty$

Little-o notation, $f \in o(g)$, means that the asymptotic growth of f is strictly less than the growth of g. Hence, little-o notation is a stronger statement than big-O notation. Indeed, if $f_1 \in o(f_2)$, then it always holds that $f_1 \in O(f_2)$. **Little-ω notation** can be considered to be the opposite of little-o notation, i.e. if $f_1 \in o(f_2)$ then $f_2 \in \omega(f_1)$. We note that since $\lim_{n\to\infty} f(n)/f(n) = 1$, it is evident that $f(n) \notin o(f(n))$ and $f(n) \notin \omega(f(n))$. The following examples clarify these notations.

Example 1.2 Consider the functions $f_1(n) = 7n^2 + 4$ and $f_2(n) = 2n^3$. If we set $n_0 = 4$ and $c = 1$, we see that $|f_1(n)| \le c|f_2(n)|$ for all $n > n_0$ and thus $f_1 \in O(f_2)$.

Example 1.3 It holds that:

$$\begin{array}{rrr} \sqrt{n} \in o(n) & n \in \omega(\sqrt{n}) & 5n^3 \in \Omega(2n^3) \\ n \in o(n \log n) & n \log(n) \in \omega(n) & n! \in \Omega(2^n) \\ n \log n \in o(n^2) & n^2 \in \omega(n \log n) & 7n^2 \in \Omega(n^2) \\ n^2 \in o(2^n) & 2^n \in \omega(n^2) & n \in \Omega(6)\,. \end{array}$$

1.1.5 Probability

Chapter 4 will show that in quantum mechanics, the process of measurement is inherently random. This means that the outcome of a measurement, associated with an event E, cannot be predicted with certainty. However, if the measurement is repeated many times under identical conditions, the frequency with which event E occurs will converge to a well-defined limit. This limit is what we define as the probability of event E. A probability close to 0 means that event E is unlikely to occur, while a probability close to 1 means that event E is likely to occur. In summary, probability is a measure of the likelihood of observing a particular outcome in a series of repeated measurements.

Consider a finite set Ω that contains all **possible outcomes** of a measurement and consider its power set $\mathcal{P}(\Omega)$. The sets in $\mathcal{P}(\Omega)$ are called the **possible events** and are usually indicated with the letter E. A probability measure is a function that associates each event with a number in the interval $[0, 1]$,

$$\begin{array}{rccc} p: & \mathcal{P}(\Omega) & \to & [0,1] \\ & E & \mapsto & p(E)\,. \end{array}$$

For p to be a probability, it must satisfy:

p1. $p(E) \geq 0$ (positivity),

p2. $p(\Omega) = 1$ (unit measure),

p3. $p\Big(\bigcup_i E_i\Big) = \sum_i p(E_i)$ for disjoint sets E_i (countable additivity).

In property **p3**, the sum is either finite or countably infinite.

We define the **conditional probability** $p(E_2 \mid E_1) = p(E_2 \cap E_1)/p(E_1)$ as the probability that event E_2 occurs given that event E_1 has already happened. Two events E_1 and E_2 are said to be **independent** if $p(E_1 \cap E_2) = p(E_1)p(E_2)$. When Ω is a finite set, the probability associated with the elements in Ω can be collected in a **probability distribution** $\mathbf{p} = (p_1, \ldots, p_n)$. This is a sequence of real numbers $p_i \geq 0$ such that $\sum_i p_i = 1$.

Example 1.4 Consider the rolling of a die. The set of outcomes is Ω ={1, 2, 3, 4, 5, 6} and the set of all possible events is

$$\mathcal{P}(\Omega) = \{\varnothing, \{1\}, \{2\}, \ldots, \{1,2,3,4,5,6\}\}\,. \tag{1.17}$$

This set contains 64 events. The probability that the outcome of a measurement is a number between 1 and 6 is $p(1,2,3,4,5,6) = p(\Omega) = 1$. For a fair die, the probability that the event $E_5 = \{5\}$ occurs is $p(E_5) = 1/6$. Furthermore, the probability that the outcome is an even number is one-half: $p(E_{\text{even}}) = 1/2$ where $E_{\text{even}} = \{2, 4, 6\}$.

Example 1.5 Let us consider two consecutive die rolls. In this case, the set of possible outcomes is $\Omega \times \Omega$ and the set of all possible events is $\mathcal{P}(\Omega \times \Omega)$. The events $E_{\text{A}} =$ "the outcome of the first roll is 1" and $E_{\text{B}} =$"the outcome of the second roll is 6" are denoted as

$$E_{\text{A}} = \{1\} \times \Omega = \{(1,1), (1,2), \ldots, (1,6)\}, \tag{1.18}$$
$$E_{\text{B}} = \Omega \times \{6\} = \{(1,6), (2,6), \ldots, (6,6)\}. \tag{1.19}$$

The probability that the outcome of the second measurement is 6 given that the outcome of the first measurement is 1 is $p(E_{\text{B}} \mid E_{\text{A}}) = p(E_{\text{A}} \cap E_{\text{B}})/p(E_{\text{A}})$. Since $E_{\text{A}} \cap E_{\text{B}} = (1,6)$ and $p(E_{\text{A}} \cap E_{\text{B}}) = 1/36$, then $p(E_{\text{B}} \mid E_{\text{A}}) = 1/6$.

1.1.6 Binary system

A binary digit, or **bit**, is an element of a set with cardinality 2. We will indicate a single bit with a lower case letter x and a finite sequence of bits with a bold letter $\mathbf{x}$. A bit is usually taken from the **Boolean set** $\mathbb{B} = \{0, 1\}$ [41, 42]. The **complement** of a bit will be denoted with the symbol $\overline{x}$ and it is defined by the expression $x + \overline{x} = 1$ (here, the symbol $+$ indicates the usual sum between integers; for instance, if $x = 0$, then $\overline{x} = 1$). Any natural number $a < 2^n$ can be expressed using n binary digits. This can be done by first writing a as a power series,

$$a = \sum_{i=0}^{n-1} x_i 2^i = x_{n-1}2^{n-1} + \ldots + x_1 2^1 + x_0 2^0 , \tag{1.20}$$

where $x_i \in \mathbb{B}$ are the coefficients of the expansion. The **binary representation** of a is the sequence of coefficients in eqn (1.20). For example, the binary representation of 4 is $\mathbf{x} = x_2 x_1 x_0 = 100_2$, where the subscript 2 refers to the binary representation. It is also possible to express a rational number $0 \leq q < 2^n$ in binary representation. This can be done by first expressing q as a power series,

$$q = \sum_{i=-\infty}^{\infty} x_i 2^i = \ldots + x_1 2^1 + x_0 2^0 + x_{-1} 2^{-1} + \ldots , \tag{1.21}$$

where again $x_i \in \mathbb{B}$. The binary representation of q is the sequence of coefficients in eqn (1.21). For example, the binary representation of 1.375 is $1.375 = 1 \cdot 2^0 + 0 \cdot 2^{-1} + 1 \cdot 2^{-2} + 1 \cdot 2^{-3} = 1.011_2$. Note that the binary representation of a rational number might require an infinite number of bits. This is the case for 0.1, whose binary representation is $0.0\overline{0011}_2$.

We will conclude this section with an interesting observation. The number of *decimal* digits required to represent[2] a natural number N is $n_{\text{d}} = \lfloor \log_{10} N \rfloor + 1$. Similarly,

[2]The floor function $\lfloor x \rfloor$ returns the largest integer less than or equal to $x \in \mathbb{R}$, e.g. $\lfloor 3.412 \rfloor = 3$ and $\lfloor -7.2 \rfloor = -8$.

				AND $\wedge$ ↓					XOR $\oplus$ ↓	OR $\vee$ ↓							NAND $\uparrow$ ↓	
x_1	x_0		f_0	f_1	f_2	f_3	f_4	f_5	f_6	f_7	f_8	f_9	f_{10}	f_{11}	f_{12}	f_{13}	f_{14}	f_{15}
0	0		0	0	0	0	0	0	0	0	1	1	1	1	1	1	1	1
0	1	$\mapsto$	0	0	0	0	1	1	1	1	0	0	0	0	1	1	1	1
1	0		0	0	1	1	0	0	1	1	0	0	1	1	0	0	1	1
1	1		0	1	0	1	0	1	0	1	0	1	0	1	0	1	0	1

Table 1.2 Complete list of two-bit operations $f_i : \mathbb{B}^2 \to \mathbb{B}$.

the number of *binary* digits required to represent N is $n_\mathrm{b} = \lfloor \log_2 N \rfloor + 1$. For example, the number 721 has three decimal digits while its binary representation requires 10 bits. It can be shown that for $N \gg 1$, the binary representation requires $\log_2 10$ times the number of digits of the decimal representation and thus $n_\mathrm{b} \approx n_\mathrm{d} \cdot 3.32$.

1.1.7 Boolean functions

A **Boolean function** is a transformation $f : \mathbb{B}^n \to \mathbb{B}$ that maps a binary string into a bit. The simplest Boolean function is the single-bit **identity**, defined as $\mathbb{1} : \mathbb{B} \to \mathbb{B}$ where $\mathbb{1}(x) = x$. Another single-bit operation is the **negation** $\mathsf{NOT} : \mathbb{B} \to \mathbb{B}$ that gives back the complement of the input bit. These functions can be expressed with the following tables

x		$\mathbb{1}(x)$
0	$\mapsto$	0
1		1

x		$\mathsf{NOT}(x)$
0	$\mapsto$	1
1		0

Boolean functions that map pairs of bits into a bit are called two-bit operations, and they are all listed in Table 1.2. The most important ones are the **conjunction** AND $\wedge$, the **disjunction** OR $\vee$, the sum modulo 2 XOR $\oplus$, and the operation NAND $\uparrow$. Their definitions are given in Table 1.2 and their circuit representations are illustrated in Fig. 1.4. In these graphical representations, we will follow the standard convention that each line corresponds to one bit, and any sequence of logic gates must be read from the left (input) to the right (output). Lastly, we mention the function $\mathsf{COPY} : \mathbb{B} \to \mathbb{B}^2$ which is defined as $\mathsf{COPY}(x) = (x, x)$. This transformation copies an input bit into two output bits.

A **minterm** is a Boolean function that returns 1 for only one input string of the domain and 0 for all others. An example of a minterm is the transformation $m_{011} : \mathbb{B}^3 \to \mathbb{B}$ defined as

$$m_{011}(\mathbf{x}) = \begin{cases} 1 & \text{if } \mathbf{x} = 011 \\ 0 & \text{if } \mathbf{x} \neq 011\,. \end{cases} \tag{1.22}$$

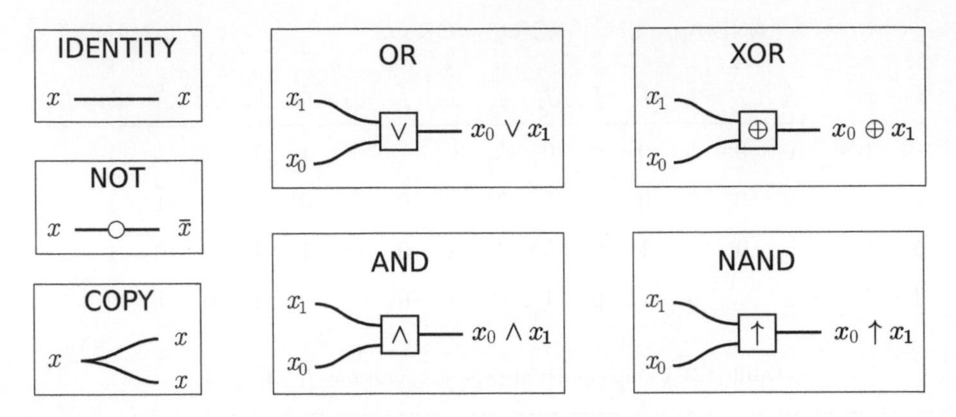

Fig. 1.4 Logic circuits. Circuits representing the COPY operation, one-bit, and two-bit operations.

A minterm can be expressed in terms of AND and NOT operations only. For example, the minterm m_{011} is given by

$$m_{011}(\mathbf{x}) = \overline{x}_2 \wedge x_1 \wedge x_0 \,. \tag{1.23}$$

The reader can easily verify that this expression is equal to 1 only when $x_2 = 0$, $x_1 = 1$, $x_0 = 1$. In general, a minterm $m_{\mathbf{c}} : \mathbb{B}^n \to \mathbb{B}$ with $\mathbf{c} \in \mathbb{B}^n$ takes the form

$$m_{\mathbf{c}}(\mathbf{x}) = y_{n-1} \wedge \ldots \wedge y_0, \quad \text{where} \quad y_i = \begin{cases} \overline{x}_i & \text{if } c_i = 0 \\ x_i & \text{if } c_i = 1 \,. \end{cases} \tag{1.24}$$

Minterms can be used to express any function $f : \mathbb{B}^n \to \mathbb{B}^m$ with AND, OR, NOT, and COPY operations only [43, 44, 45]. For an intuitive idea of this concept, consider the function $f : \mathbb{B}^3 \to \mathbb{B}^2$ defined by the table

$\mathbf{x}$				$f(\mathbf{x})$	
x_2	x_1	x_0		$f_1(\mathbf{x})$	$f_0(\mathbf{x})$
0	0	0		1	0
0	0	1		0	1
0	1	0		0	1
0	1	1	$\mapsto$	0	0
1	0	0		0	0
1	0	1		1	0
1	1	0		0	1
1	1	1		0	0

The outputs of this transformation are two bits, $f_0(\mathbf{x})$ and $f_1(\mathbf{x})$. These Boolean functions can be written as

$$
\begin{aligned}
f_0(\mathbf{x}) &= m_{001}(\mathbf{x}) \vee m_{010}(\mathbf{x}) \vee m_{110}(\mathbf{x}) \\
&= (\overline{x}_2 \wedge \overline{x}_1 \wedge x_0) \vee (\overline{x}_2 \wedge x_1 \wedge \overline{x}_0) \vee (x_2 \wedge x_1 \wedge \overline{x}_0)\,,
\end{aligned}
$$

$$
\begin{aligned}
f_1(\mathbf{x}) &= m_{000}(\mathbf{x}) \vee m_{101}(\mathbf{x}) \\
&= (\overline{x}_2 \wedge \overline{x}_1 \wedge \overline{x}_0) \vee (x_2 \wedge \overline{x}_1 \wedge x_0)\,.
\end{aligned}
$$

This shows that $f : \mathbb{B}^3 \to \mathbb{B}^2$ can be expressed in terms of {AND, OR, NOT, COPY}. This can be easily generalized.

Theorem 1.1 Any transformation $f : \mathbb{B}^n \to \mathbb{B}^m$ can be expressed with AND, OR, COPY, and NOT. Thus, the set {AND, OR, COPY, NOT} is universal for classical computation.

Proof The output of the function $f : \mathbb{B}^n \to \mathbb{B}^m$ is the sequence $(f_1(\mathbf{x}), \ldots, f_m(\mathbf{x}))$ where f_i are Boolean functions of the form $f_i : \mathbb{B}^n \to \mathbb{B}$. We just need to show that each of these functions can be expressed in terms of {AND, OR, COPY, NOT}. The function $f_i(\mathbf{x})$ assigns 1 to k input strings $\mathbf{c}_j$, where $0 \leq k \leq 2^n$. For these input strings only, define

$$
m_{\mathbf{c}_j}(\mathbf{x}) = \begin{cases} 1 & \text{if } \mathbf{x} = \mathbf{c}_j \\ 0 & \text{if } \mathbf{x} \neq \mathbf{c}_j\,. \end{cases} \tag{1.25}
$$

Then, the function $f_i(\mathbf{x})$ takes the form

$$
f_i(\mathbf{x}) = m_{\mathbf{c}_1}(\mathbf{x}) \vee \ldots \vee m_{\mathbf{c}_k}(\mathbf{x})\,. \tag{1.26}
$$

This shows that $f_i\,(\mathbf{x})$ is a conjunction of minterms, each expressed with NOT and AND operations only. The COPY operation is required because the minterms need k copies of the input string. □

A **universal set** is a set of operations that can be used to build any function in a class. We have just demonstrated that {AND, OR, NOT, COPY} is universal for functions that transform binary strings. The set {NAND, COPY} is also universal. This is because the operations AND, OR, and NOT can be expressed in terms of NAND and COPY only, as illustrated in Fig. 1.5. If a machine can implement the operations NAND and COPY, it can transform a binary string into any other binary string.

1.1.8 Landauer's principle and reversible gates

The two-bit operations presented in Section 1.1.7 are not reversible: given one output bit, it is impossible to uniquely determine the value of the two input bits. When performing these operations, a bit of information is irreparably lost. According to **Landauer's principle** [4, 6, 46, 47], thermodynamics arguments indicate that whenever a bit of information is erased, the amount of energy dissipated into the environment is greater than

$$
E \geq k_{\mathrm{B}} T \ln 2\,, \tag{1.27}
$$

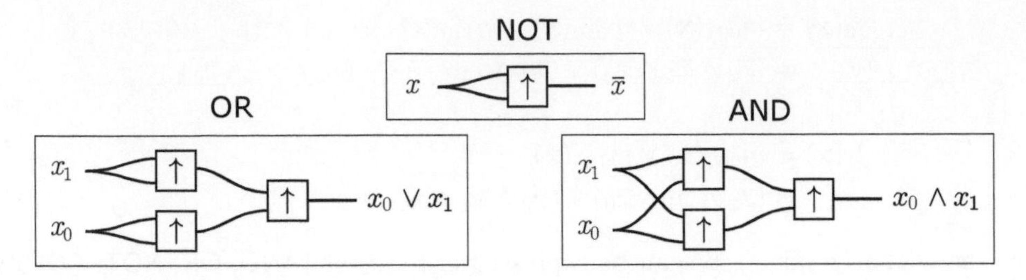

Fig. 1.5 Relations between logical operations. Circuits representing AND, OR, and NOT in terms of NAND and COPY gates. For example, the NOT gate is equivalent to NAND(COPY(x)).

where k_B is the Boltzmann constant and T is the temperature of the machine. If the machine operates at room temperature, then $T \approx 300$ K and therefore $E \gtrsim 0.02$ eV. Two-bit irreversible gates running in modern computers dissipate an amount of energy much greater than 0.02 eV, and so (1.27) has to be considered as a lower bound. If a machine were to operate reversible operations only, bits of information would never be erased during the computation and therefore no energy is dissipated. As we shall see in Chapter 5, an *ideal* quantum computation is intrinsically reversible.

It is always possible to construct a reversible transformation from an irreversible transformation $f : \mathbb{B}^n \to \mathbb{B}^m$. One way this can be done is by converting f into a bijective function $\tilde{f} : \mathbb{B}^{n+m} \to \mathbb{B}^{n+m}$ of the form

$$\tilde{f}(\mathbf{x}, \mathbf{y}) = (\mathbf{x}, \, (\mathbf{y} + f(\mathbf{x})) \,\%\, 2^m) \,, \tag{1.28}$$

where $\mathbf{x}$ is a n-bit string, and $\mathbf{y}$ and $f(\mathbf{x})$ are m-bit strings[3].

As an example, consider the AND gate,

a			$f(\mathbf{a})$
0	0		0
0	1	$\mapsto$	0
1	0		0
1	1		1

A reversible version of this gate is given by the function $\tilde{f}$ defined as

[3] Here, the binary strings $\mathbf{x}$ and $\mathbf{y}$ should be regarded as integers expressed in binary representation and in this sense the % operation (which gives the remainder of a division) is correctly defined.

$\mathbf{x}$			CNOT($\mathbf{x}$)	
x_1	x_0		x_1'	x_0'
0	0		0	0
0	1	$\mapsto$	0	1
1	0		1	1
1	1		1	0

$\mathbf{x}$			SWAP($\mathbf{x}$)	
x_1	x_0		x_1'	x_0'
0	0		0	0
0	1	$\mapsto$	1	0
1	0		0	1
1	1		1	1

Table 1.3 Truth tables for the CNOT and SWAP gates.

$\mathbf{a}$				$\tilde{f}(\mathbf{a})$		
a_2	a_1	a_0		a_2'	a_1'	a_0'
0	0	0		0	0	0
0	0	1		0	0	1
0	1	0		0	1	0
0	1	1	$\mapsto$	0	1	1
1	0	0		1	0	0
1	0	1		1	0	1
1	1	0		1	1	1
1	1	1		1	1	0

Here, the role of $\mathbf{x}$ in eqn (1.28) is played by (a_2, a_1) and the role of $\mathbf{y}$ is played by a_0. Then, $(a_2', a_1') = (a_2, a_1)$, and $a_0' = (a_0 + \mathsf{AND}(a_2, a_1)) \,\%\, 2$ as in (1.28). It is evident that $\tilde{f}$ is a three-bit reversible function since there is a one-to-one correspondence between input and output strings.

Reversible operations are bijective functions that map n-bit strings onto n-bit strings. When $n = 1$, we have the identity and the NOT gate. For $n = 2$, there is a total of $2^n! = 24$ reversible operations that map two input bits into two output bits. The most important one is the controlled-NOT operation defined as $\mathsf{CNOT}(x_1, x_0) = (x_1, x_0 \oplus x_1)$. Its truth table is shown in Table 1.3. Here, bit x_1 acts as a control and its value is not changed by this gate. The bit x_0 is the target bit and its value is flipped if the control bit is set to one. This gate is reversible, since two applications of the CNOT gate are equal to the identity

$$\mathsf{CNOT}(\mathsf{CNOT}(x_1, x_0)) = \mathsf{CNOT}(x_1, x_0 \oplus x_1) = (x_1, x_0 \oplus x_1 \oplus x_1) = (x_1, x_0)\,. \quad (1.29)$$

This relation also indicates that the inverse of this gate is the CNOT itself.

Another important two-bit reversible gate is the SWAP, whose truth table is shown in Table 1.3. This gate simply swaps the two input bits, $\mathsf{SWAP}(x_1, x_0) = (x_0, x_1)$. Two applications of the SWAP gate give the identity, meaning that the inverse of this operation is the SWAP gate itself. The circuits representing the CNOT and the SWAP gates are shown in Fig. 1.6.

It is possible to demonstrate that one-bit and two-bit *reversible* gates are not sufficient for universal classical computation [29]. In other words, one-bit and two-bit reversible operations are not sufficient to build all functions of the form $f : \mathbb{B}^n \to \mathbb{B}^m$ and a three-bit reversible gate, such as the **Toffoli gate**, is needed [9]. The Toffoli

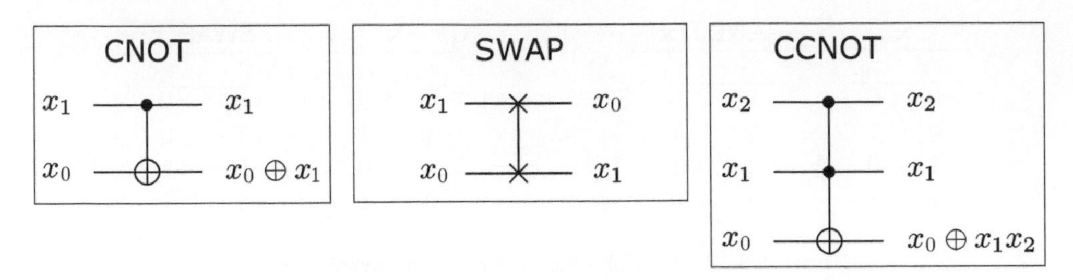

Fig. 1.6 Two-bit and three-bit operations. Circuits representing the CNOT, SWAP, and Toffoli gates. The symbol • indicates that a specific operation is applied to the target bit when the control bit is 1. The symbol $\oplus$ indicates the sum modulo 2. Finally, two crosses $\times$ connected by a segment indicate a swap operation between the corresponding bits.

gate (or controlled-controlled-NOT) is a three-bit reversible gate. This operation flips a target bit if two control bits are set to 1; its definition is $\text{CCNOT}(x_2, x_1, x_0) = (x_2, x_1, x_0 \oplus x_1 x_2)$, where $x_1 x_2$ indicates the usual multiplication between integers. Figure 1.6 shows the circuit associated with the CCNOT operation. The Toffoli gate is an important reversible gate because it can be used to carry out any classical computation.

Theorem 1.2 The Toffoli gate CCNOT is a universal gate for classical computation.

Proof To prove that the Toffoli gate can transform any binary string into any other binary string, we just need to show that it can be used to generate the COPY and NAND operations (as seen in the previous section, these two gates are universal). As shown in the left panel of Fig. 1.7, if we set the initial value of one control bit to 1, the Toffoli gate operates like a CNOT on the remaining bits, $\text{CCNOT}(1, x_1, x_0) = (1, \text{CNOT}(x_1, x_0))$. In turn, the CNOT can be used to generate the COPY. This can be done by setting the initial value of the target bit to 0 so that $\text{CNOT}(x_1, 0) = (x_1, x_1) = \text{COPY}(x_1)$ as illustrated in the middle panel of Fig. 1.7. Thus, the Toffoli gate generates the COPY gate. Let us show how to use the Toffoli gate to generate the NAND gate. This can be done by setting the initial value of x_0 to 1, such that $\text{CCNOT}(x_2, x_1, 1) = (x_2, x_1, 1 \oplus x_1 x_2) = (x_2, x_1, \text{NAND}(x_2, x_1))$. This relation is illustrated in the right panel of Fig. 1.7. □

Another important three-bit reversible operation is the **Fredkin gate**. This gate swaps two target bits only when the control bit is set to one. Similarly to the Toffoli gate, the Fredkin gate is universal for classical computation [9].

1.1.9 Conjunctive normal form

Boolean functions in conjunctive normal form are important to introduce some fundamental problems in complexity theory, such as the satisfiability problem. To define them, we first need to introduce a few tools. A **literal** is a Boolean variable x or its complement $\overline{x}$. A **clause** is defined as a disjunction of literals. An example of a clause

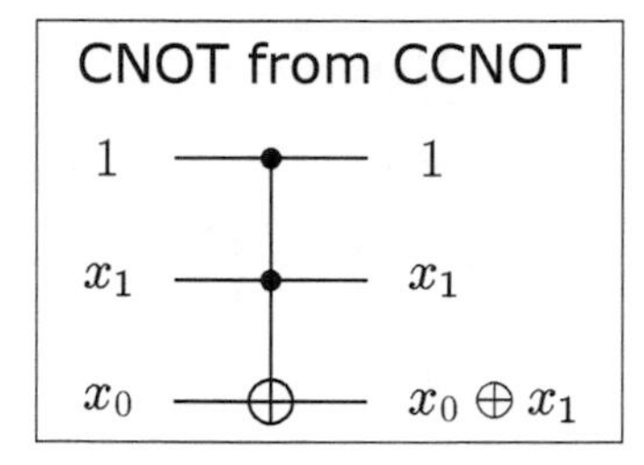

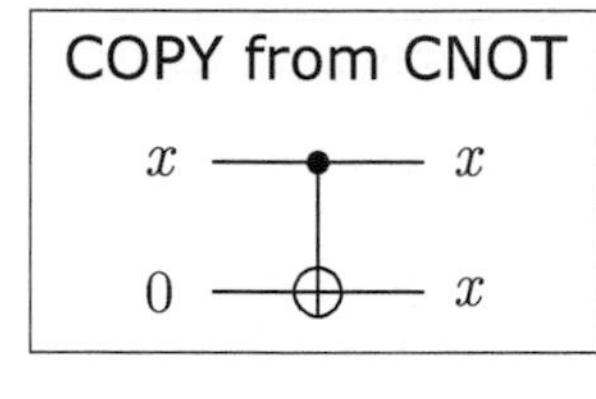

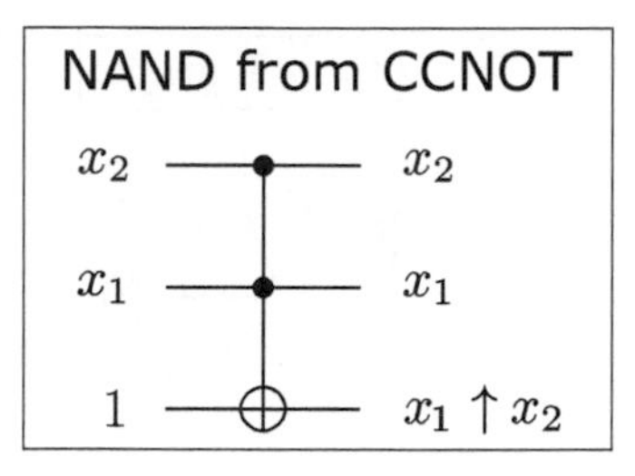

Fig. 1.7 Relations between logical operations. The CCNOT generates the CNOT (left panel). The CNOT generates the COPY (middle panel). The CCNOT generates the NAND gate (right panel).

is $x_1 \vee \overline{x}_3 \vee x_6$. A Boolean function $f : \mathbb{B}^n \to \mathbb{B}$ is in **conjunctive normal form** (CNF) if it is expressed as a conjunction of one or more clauses, as in

$$f(\mathbf{x}) = \underbrace{(x_1 \vee x_2)}_{1^{\text{st}}\text{ clause}} \wedge \underbrace{(x_2 \vee \overline{x}_3 \vee \overline{x}_4)}_{2^{\text{nd}}\text{ clause}}. \tag{1.30}$$

If each clause has exactly k literals, we say it is in kCNF. For example, the transformation

$$f_{2\text{CNF}}(\mathbf{x}) = (\overline{x}_5 \vee x_2) \wedge (x_4 \vee x_3) \wedge (x_5 \vee x_1) \wedge (x_2 \vee \overline{x}_3) \tag{1.31}$$

is a Boolean function in 2CNF because each clause contains two literals. Similarly, the function

$$f_{3\text{CNF}}(\mathbf{x}) = (x_7 \vee x_4 \vee \overline{x}_1) \wedge (x_3 \vee \overline{x}_4 \vee \overline{x}_6) \tag{1.32}$$

is a Boolean function in 3CNF since each clause contains three literals. The conjunctive normal form will be important when we introduce the satisfiability problem in Section 1.2.4.

1.1.10 Graphs

Graphs are crucial for the description of automata, Turing machines, and quantum circuits, and they play an essential role in many fields of science, including mathematics, physics, computer science, and engineering [48, 49, 50, 51, 52]. Figure 1.8 shows some examples of undirected graphs, i.e. a set of nodes possibly connected by some edges. More formally, a **graph** $G = (V, E)$ is a pair of sets where V contains a finite number of nodes and $E \subseteq V \times V$ is a multiset of edges connecting them. For **undirected graphs**, edges are unordered pairs of nodes. Two nodes connected by an edge are called **adjacent** nodes and the **degree** of a node is the number of edges connected to that node. If a node has degree one, it is called an **endpoint**. The definitions of the graphs shown in Fig. 1.8 are:

simple graphs: pairs of nodes are connected by a maximum of one edge,
complete graphs: simple graphs where each node is connected to all others,

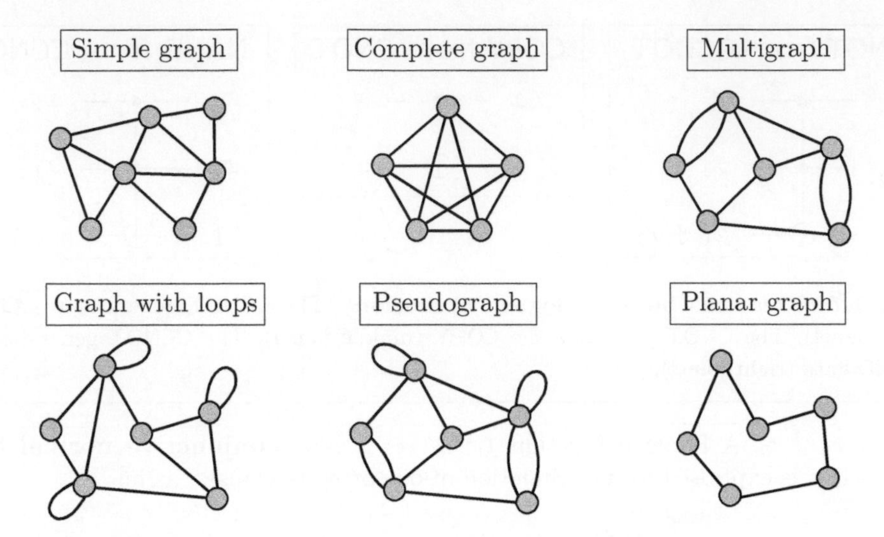

Fig. 1.8 Categories of graphs. The complete graph shown in this figure is an example of a non-planar graph.

multigraphs: pairs of nodes may be connected by more than one edge,
graphs with loops: nodes can be connected to themselves by one or more edges,
pseudographs: graphs that are both multigraphs and graphs with loops,
planar graphs: graphs that can be drawn in a plane without edges crossing.

A graph $G_1 = (V_1, E_1)$ is a **subgraph** of $G_2 = (V_2, E_2)$ if the nodes and edges of G_1 are a subset of the nodes and edges of G_2. A subgraph with k nodes that is complete is called a k-**clique**.

Many problems in complexity theory are concerned with finding a particular **walk** within a graph. A walk w is defined as a sequence of nodes and edges

$$\mathsf{w} = (v_0, e_1, v_1, \ldots, e_n, v_n)\,, \tag{1.33}$$

where edge e_i connects node v_{i-1} to node v_i. The **length** of a walk $\ell(\mathsf{w})$ is simply its number of edges. The **distance** between two nodes $d(v_1, v_2)$ is the length of the shortest walk connecting v_1 to v_2. There exist different types of walks:

cycles: a closed walk with at least three distinct nodes,
trails: a walk with distinct edges,
paths: a walk with distinct edges and nodes.

These concepts can be used to introduce additional types of graphs, shown in Fig. 1.9:

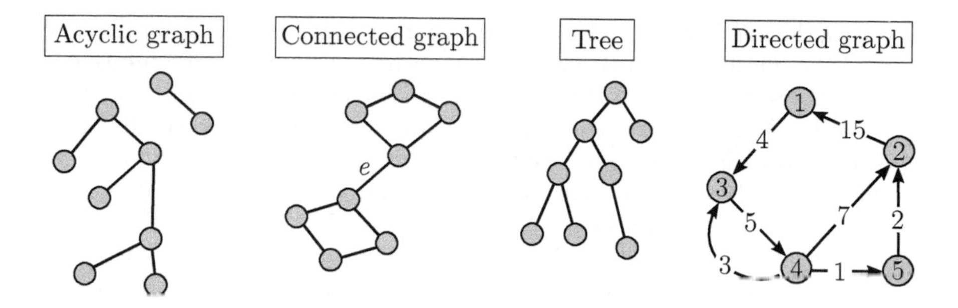

Fig. 1.9 Examples of graphs. The edge labeled e in the connected graph is a bridge.

---

acyclic graphs: graphs not containing cycles,
connected graphs: any pair of nodes can be connected by a walk,
trees: connected graphs without cycles.

If an edge is removed from a connected graph and the resultant graph is not connected, that edge is called a **bridge**. In a tree, there is often one distinguished node, called the **root**, from which several branches spread out and terminate with an endpoint, called a **leaf**. We define a **branch** as the shortest path connecting the root to a specific leaf. The concept of branches will be useful when we introduce probabilistic Turing machines in Section 2.2.7.

If the edges connecting adjacent nodes are replaced by arrows, we obtain a **directed graph**. Directed graphs are crucial in describing automata, Turing machines, and logic circuits. The properties and definitions presented for graphs can easily be generalized to directed graphs. A labeled directed graph is a graph whose arrows are labeled with symbols. The mathematical description of the labeled directed graph shown in Fig. 1.9d is given by $G = \{V, E, w\}$, where

$$V = (1, 2, 3, 4, 5) \,, \qquad E = \{(1,3), (2,1), (3,4), (4,2), (4,3), (4,5), (5,2)\} \,, \tag{1.34}$$

and $w : E \to L$ is a function that assigns a label $l \in L$ to each arrow. If w assigns non-negative real numbers only, it is called a weight function and the corresponding graph is called a **weighted graph**. For weighted graphs, the length of a walk can be defined as the sum of the weights assigned to its edges, $\ell(\mathsf{w}) = w(e_1) + \ldots + w(e_n)$. We now have all the necessary tools to introduce some decision problems.

1.2 Decision problems

Decision problems are mathematical problems whose solution is either *yes* or *no*. A computation aims to solve a decision problem by exploiting two physical resources: **time** and **space**. The time required by an algorithm to solve a problem is the number of elementary steps that the machine has to perform before terminating a computation (in most cases, the number of steps increases with input size). On the other hand, space is the size of the physical support used by the machine during the computation[4].

The running time of an algorithm usually depends on the input size. For instance, a machine will take more time to multiply two numbers with 200 digits than two numbers with two digits. Here, the input size can be considered the number of digits of the multiplying factors. In general, the **input size** n is the number of bits needed to encode the input of the problem. This is a reasonable choice since most of the mathematical problems encountered in computer science can be encoded using binary strings. When we say that an algorithm solves a problem in **polynomial time**, we mean that the number of elementary steps $t(n)$ executed by the machine to reach a final state scales polynomially with input size, i.e. $t(n) \in O(n^k)$ for some real number $k \geq 1$.

In complexity theory, decision problems can be informally divided into two categories: **easy** problems and **difficult** problems. If a classical computer[5] can solve a decision problem X with resources of time and space that scale polynomially with input size, we say that X is easy and that the algorithm can **efficiently** solve it. On the contrary, if a classical machine needs resources of time *or* space that scale super polynomially with input size, we say that the problem is difficult and that the algorithm does not solve it efficiently. Examples of difficult problems include those that require exponential time, i.e. $t(n) \in O(2^n)$. Different algorithms can solve a given problem; some may be more efficient than others. An optimal algorithm minimizes the time and space used by the computation. When we say that a problem is difficult, we mean that the best classical algorithm requires super polynomial resources to solve it. Throughout history, some problems considered difficult were subsequently shown to be easy.

The reader might argue that the distinction between easy and difficult problems is somewhat arbitrary. We will illustrate the effectiveness of this distinction with an example. Consider a problem with input size n and suppose that an algorithm A takes n^2 steps to solve it, while algorithm B takes 2^n steps. If $n = 1000$, algorithm A will take 1 million steps to reach the solution. Is this a big number? Well, if compared to 2^{1000}, the number of steps taken by algorithm B, it is certainly not. To better appreciate the enormous difference between these two quantities, suppose that each elementary step performed by the machine takes 1 microsecond. Algorithm A will output the solution in 1 second, while algorithm B will take more time than the age of the universe. This example suggests that algorithms that solve problems with polynomial resources are usually useful from a human perspective; algorithms that need super polynomial resources often are not. There is a more fundamental reason for the distinction between polynomial and super-polynomial resources: two deterministic machines take the same number of steps (up to a polynomial-time overhead) to solve a computational problem. For this reason, a decision problem is either easy or difficult regardless of the deterministic model considered.

[4]These concepts will be formalized in the next chapter once we introduce some computational models.

[5]Hereafter, the terms "classical machine" and "classical computer" will intend a deterministic Turing machine. The definition of such a machine will be given in Chapter 2.

We now give the reader an intuitive understanding of three important complexity classes: P, NP, and NPC. We will define these classes formally in the next chapter when introducing Turing machines. The class P is the collection of all decision problems that a deterministic Turing machine can solve with a number of steps that is polynomial in the input size. This class roughly corresponds to the kind of problems a classical machine can solve efficiently; addition and multiplication are members of this class. Another interesting class of decision problems, called NP, are those whose *affirmative* answer can be verified by a deterministic Turing machine in a polynomial number of steps. An example of a problem whose solution is easy to check is the RSA problem[6]. This mathematical problem can be formulated as follows: given $N = p_1 \cdot p_2$ where p_1 and p_2 are prime numbers, find p_1 and p_2. For $N \gg 1$, this is a difficult problem to solve. However, if we were provided the values of p_1 and p_2, we could quickly check that their product equals N. This indicates that the RSA problem is in NP.

We highlight that if a problem is in NP, it does not necessarily mean that it is difficult. For example, the addition of two integers is an easy problem (thus, it is in P) and it is also in NP (if someone provides the solution to this problem, we can easily verify it). In general, it holds that $\mathsf{P} \subseteq \mathsf{NP}$. This follows from the fact that if a problem is in P, it is possible to verify a putative solution by quickly computing the solution itself.

In complexity theory, many decision problems are difficult to solve, but their solution is straightforward to verify. This observation naturally leads to the following question: is finding a solution to specific problems intrinsically more difficult than verifying their solution? In other words, is the set $\mathsf{P} \subset \mathsf{NP}$ or is $\mathsf{P} = \mathsf{NP}$? A significant number of computer scientists believe that $\mathsf{P} \subset \mathsf{NP}$, i.e. that there exists a collection of problems that are intrinsically difficult to solve with a deterministic Turing machine, but whose solution is very easy to verify. This belief partly originates from the lack of actual progress—despite many attempts—in finding classical algorithms that efficiently solve some difficult NP problems. However, this does not rule out the possibility that future research will find such algorithms. The so-called P versus NP problem of determining whether $\mathsf{P} \subset \mathsf{NP}$ or $\mathsf{P} = \mathsf{NP}$ is arguably the most important open question in complexity theory (see Fig. 1.10 for a schematic representation of this problem).

It is sometimes useful to reformulate a decision problem in terms of a different problem. As we shall see in Section 2.5.4, the problem of satisfying a Boolean function in 3CNF can be formulated in terms of the coloring of a graph. The process of mapping a problem into another is called **reduction**. It is possible to show that all NP problems can be efficiently formulated in terms of a subset of NP problems called NPC (or NP **complete**). This implies that if one day someone finds an algorithm that efficiently solves one NPC problem, then it would be possible to solve all NP problems efficiently; on these lines, NPC problems can be considered the most difficult problems in the class NP. We now introduce some mathematical concepts before listing the most important decision problems we will encounter throughout this book.

[6] The RSA problem is named after R. Rivest, A. Shamir, L. Adleman, the inventors of a cryptographic protocol based on the factorization of large semiprimes.

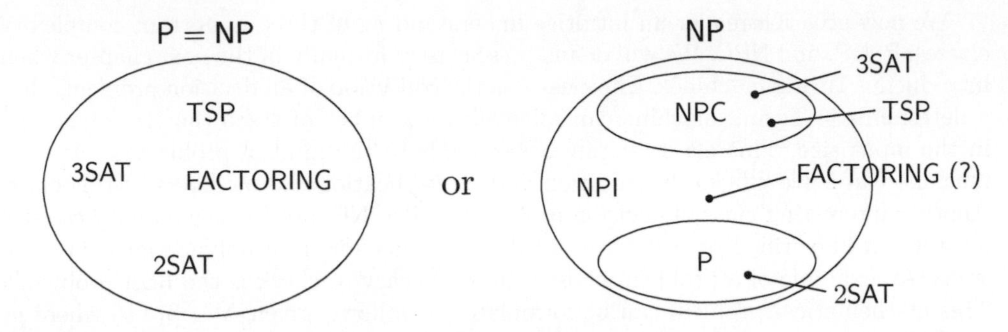

Fig. 1.10 The P vs. NP problem. The diagram on the left shows the case P = NP. To demonstrate that P = NP, it would be sufficient to find a classical algorithm that solves an NPC problem in polynomial time. If instead P ⊂ NP, then the set NP includes two disjoint sets, P and NPC, as shown in the right diagram. The decision problems not contained in these two subsets are usually called NPI. FACTORING is believed to be one of them. The problems shown in this figure are formally defined in the main text.

1.2.1 Alphabets

Alphabets are essential for discussing decision problems, automata, and Turing machines [53]. We define an **alphabet** as a finite set of symbols. Some examples of alphabets are

$$\mathbb{B} = \{0, 1\},$$
$$\Sigma_{\text{Eng}} = \{\text{a, b, c, d, e, f, g, h, i, j, k, l, m, n, o, p, q, r, s, t, u, v, w, x, y, z}\},$$

where $\mathbb{B}$ is the Boolean alphabet and Σ_{Eng} is the English alphabet. Alphabets may contain a special symbol "␣" called **blank**. This symbol plays a key role in Turing machines.

A **string x** over an alphabet Σ is a finite sequence of symbols taken from Σ. The empty string is denoted with the letter ε. The length of a string $|\mathbf{x}|$ is the number of symbols in $\mathbf{x}$ (by definition $|\varepsilon| = 0$). A generic string $\mathbf{x}$ of length n over an alphabet Σ can thus be written as

$$\mathbf{x} = x_1 x_2 \dots x_{n-1} x_n , \tag{1.35}$$

where x_1 is the first symbol and x_n is the n-th symbol. The concatenation of two strings $\mathbf{x}$ and $\mathbf{y}$ is defined as a new string $\mathbf{z} = \mathbf{x} \circ \mathbf{y} = x_1 \dots x_n y_1 \dots y_m$ that has $|\mathbf{z}| = n + m$ symbols. The set of all strings of length n over the alphabet Σ will be denoted as

$$\Sigma^n = \{\mathbf{x} = x_1 \dots x_n : x_i \in \Sigma\} . \tag{1.36}$$

The set Σ^n is clearly a finite set with $|\Sigma|^n$ elements. The **Kleene closure** Σ^* is defined as the set of all possible strings that can be constructed from the alphabet Σ [53, 54],

$$\Sigma^* = \bigcup_{n=0}^{\infty} \Sigma^n . \tag{1.37}$$

For instance, the Kleene closure of $\mathbb{B}$ is

$$\mathbb{B}^* = \{\varepsilon, 0, 1, 00, 01, 10, 11, 000, 001, 011, \dots\} . \tag{1.38}$$

Note that the set $\mathbb{B}^*$ is countable, i.e. there exists a bijective function $f : \mathbb{B}^* \to \mathbb{N}$.

In light of the above definitions, we can introduce the concept of language. A **language** L is a collection of strings constructed from an alphabet Σ. In other words, a language is a subset of Σ^*. For example, the English language L_{Eng} can be considered to be a subset of Σ^*_{Eng}. Another example of a language is the set of binary strings with the first bit equal to zero,

$$L = \{\mathbf{x} \in \mathbb{B}^* : x_1 = 0\} . \tag{1.39}$$

As we shall see in Section 2.1.1, languages are a crucial concept in computability theory.

We will conclude this section by presenting the **encoding** of a string. Formally, the encoding of a string is a bijective map of the form

$$\begin{array}{rccc} \langle\,\cdot\,\rangle : & \Sigma & \to & \mathbb{B}^n \\ & x & \mapsto & \langle x \rangle \end{array}$$

where $n = \lfloor \log_2 |\Sigma| \rfloor + 1$ is the number of bits required to uniquely specify a symbol in the alphabet. As an example, consider the English alphabet. Its symbols can be encoded into binary strings with n digits, where $n = \lfloor \log_2 |\Sigma_{\text{Eng}}| \rfloor + 1 = 5$. The encoding of the English letters into five-digit bit strings can be performed in numerical order

⟨a⟩	00001
⟨b⟩	00010
⟨c⟩	00011
⟨d⟩	00100
⟨e⟩	00101
⟨f⟩	00110
⟨g⟩	00111
⟨h⟩	01000
⟨i⟩	01001
⟨j⟩	01010
⟨k⟩	01011
⟨l⟩	01100
⟨m⟩	01101
⟨n⟩	01110
⟨o⟩	01111
⟨p⟩	10000
⟨q⟩	10001
⟨r⟩	10010
⟨s⟩	10011
⟨t⟩	10100
⟨u⟩	10101
⟨v⟩	10110
⟨w⟩	10111
⟨x⟩	11000
⟨y⟩	11001
⟨z⟩	11010

where the notation $\langle a \rangle$ indicates the binary encoding of a. Here, the order of the bits is irrelevant as long as there is a one-to-one correspondence between symbols and encodings. Modern computers use a similar way to encode alphabets into binary strings with eight bits (ASCII) or with 16 bits (UNICODE). Any alphabet can be represented with the Boolean alphabet and any text can be encoded into a sequence of bits. For instance, the binary encoding of the word "hello" according to the table above is

$$\langle \text{hello} \rangle = \underbrace{01000}_{\text{h}}\underbrace{00101}_{\text{e}}\underbrace{01100}_{\text{l}}\underbrace{01100}_{\text{l}}\underbrace{01111}_{\text{o}} . \tag{1.40}$$

Similarly, other mathematical objects, like functions and graphs, can be encoded into a binary string, although with some exceptions, as we shall see in Section 2.6. We now have all the necessary tools for presenting a list of decision problems that we will discuss in more detail in the following chapters.

1.2.2 Primality

Primality is an important decision problem in number theory. A **prime number** is a natural number greater than one that can be divided (without any remainder) only by one and itself. Some examples of prime numbers are 2, 7, and 11. The fundamental theorem of arithmetic states that for any natural number $N > 1$ there exist some prime numbers $a_1, \ldots, a_r$, called **factors**, and some natural numbers $k_1, \ldots, k_r$ such that $N = a_1^{k_1} \ldots a_r^{k_r}$. If a number is not prime, it is called **composite**. If a composite number is a product of two prime numbers, it is called **semiprime**; 6 and 21 are examples of semiprimes.

The primality problem can be formulated as follows: given a natural number $N > 1$, is N prime? In a more formal way, PRIMES can be defined as a language containing all binary strings associated with prime numbers

$$\begin{aligned} \mathsf{PRIMES} &= \{\langle N \rangle \in \mathbb{B}^* : N \text{ is prime}\} \\ &= \{10_2, 11_2, 101_2, 111_2, 1011_2, 1101_2, \ldots\}, \end{aligned}$$

where the notation $\langle x \rangle$ indicates the binary encoding of x (see Section 1.2.1). Given a natural number N, does $\langle N \rangle$ belong to the set PRIMES? This problem has two possible answers, either yes or no. A classical computer can solve the primality problem in polynomial time and therefore PRIMES $\in$ P. This was demonstrated for the first time in 2002 by Agrawal, Kayal, and Saxene [55].

1.2.3 Factoring

Factoring is arguably one of the most fascinating problems in number theory: given a natural number $N > 1$ and a number $k_N < N$, does N have a factor smaller than k_N? This problem can be formulated in terms of languages. The language FACTORING contains all binary strings encoding natural numbers with a factor smaller than k_N,

$$\mathsf{FACTORING} = \{\langle N \rangle \in \mathbb{B}^* : N \text{ has a factor smaller than } k_N\}. \tag{1.41}$$

All known classical algorithms take superpolynomial time to solve FACTORING and it is widely suspected (but not proven) that it is unsolvable in polynomial time by a classical computer. This fact is exploited in the RSA cryptosystem [56] in which two large prime numbers are multiplied together to produce a public key that can be used by anyone to generate an encrypted message. It is important to mention that the factoring problem is neither known nor believed to be NP complete.

Trial division is the simplest algorithm to solve FACTORING. This brute force method divides the number N by all primes between 1 and $\sqrt{N}$. In the worst case, $N = p^2$ where p is prime and $O(\sqrt{N})$ divisions must be calculated before finding a factor. Since each division can be executed in $O(\log^2 N)$ steps, trial division has time complexity $t(N) \in O(\sqrt{N} \log^2 N)$. Recall that the size of a problem is the number of bits needed to encode the input of the problem, $n = \log_2 N$. Thus, the time complexity of trial division is clearly super polynomial, $t(n) \in O(2^{n/2} n^2)$. More sophisticated classical algorithms have been developed for solving the factoring problem, but they all take super polynomial time [57, 58, 59, 60]. It is important to note that in 1994, Peter

Shor theoretically demonstrated that a quantum computer could solve FACTORING in polynomial time [18]. To this day, no one has built a fully-functional quantum computer with a sufficient number of qubits to factor a non-trivial semiprime using Shor's algorithm. The typical semiprimes used in the RSA protocol have 1024 bits and recent studies indicate that $10^5 - 10^6$ noisy qubits are needed to factor a 1024–bit semiprime on a physical quantum computer [61, 62]. At the time of writing, the largest digital quantum computer on the planet has less than 10000 qubits.

1.2.4 Satisfiability

The **satisfiability** of a Boolean formula is one of the most important NP complete problems. This problem can be stated as follows: given a Boolean function $f_{k\text{CNF}}$ in conjunctive normal form[7], is it possible to find a string $\mathbf{x}$ such that $f_{k\text{CNF}}(\mathbf{x}) = 1$? More formally, this problem requires us to understand if the input function $f_{k\text{CNF}}$ belongs to the collection of functions

$$k\mathsf{SAT} = \{\langle f_{k\text{CNF}}\rangle \in \mathbb{B}^* : \exists \mathbf{x} \text{ such that } f_{k\text{CNF}}(\mathbf{x}) = 1\}. \tag{1.42}$$

Solving this problem is easy when $k = 2$ and difficult when $k \geq 3$. It is quite intuitive that kSAT is an NP problem: given a string $\mathbf{x}$, it is very easy to check whether $f_{k\text{CNF}}(\mathbf{x}) = 1$. It can be shown that any problem in NP can be efficiently mapped into 3SAT, implying that 3SAT $\in$ NPC [29].

1.2.5 Eulerian cycles

An **Eulerian cycle** is a closed trail that traverses all of the edges of a graph exactly once (see Fig. 1.11 for an example of an Eulerian cycle). Given a connected graph $G = (V, E)$, does G contain an Eulerian cycle? This decision problem can be formulated in terms of a language

$$\mathsf{ECYCLE} = \{\langle G\rangle \in \mathbb{B}^* : G \text{ contains an Eulerian cycle}\}. \tag{1.43}$$

Here, the input size is the number of nodes of the graph, $n = |V|$. This is a typical problem in P and Euler's theorem [48], stated below, offers a simple way to solve it.

Theorem 1.3 (Euler's theorem) A connected graph G contains an Eulerian cycle if and only if each node has even degree.

A systematic way to draw an Eulerian cycle in a graph G, in which all nodes have even degree, is given by **Fleury's algorithm** [63]:

1. Start from an arbitrary node $v_1 \in V$ and set the current node to v_1 and current trail to (v_1).
2. Select an edge $e_1 = (v_1, v_2)$ that is not a bridge. If there is no alternative, select a bridge.
3. Add (e_1, v_2) to the current trail (v_1) and set the current node to v_2.

[7] See Section 1.1.9 for a definition of conjunctive normal form.

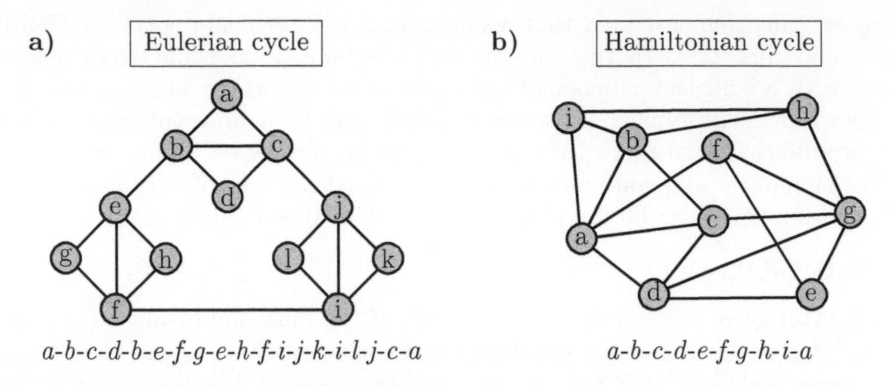

Fig. 1.11 Eulerian and Hamiltonian cycles. a) A graph containing an Eulerian cycle. In this example, each node has either degree 2 or 4. **b)** A graph containing a Hamiltonian cycle.

4. Remove e_1 from the graph and remove any isolated nodes.
5. Repeat steps 2–4 until all edges have been traversed. The final trail $(v_1, e_1, v_2, \ldots, v_N)$ is an Eulerian cycle.

The reader may test this algorithm with the graph shown in Fig. 1.11.

1.2.6 Hamiltonian cycles

A **Hamiltonian cycle** is a closed path that traverses all nodes of a graph exactly once (see Fig. 1.11b). All graphs G that contain a Hamiltonian cycle can be grouped together in a language

$$\mathsf{HAMCYCLE} = \{\langle G \rangle \in \mathbb{B}^* : G \text{ has a Hamiltonian cycle}\}\,. \tag{1.44}$$

Deciding whether a graph contains a Hamiltonian cycle remains a difficult problem for classical computers even after decades of research. Note that this problem is an NP problem: we can easily verify whether a closed walk w is a Hamiltonian cycle.

Let us consider the complement set of HAMCYCLE. This language can be defined as

$$\overline{\mathsf{HAMCYCLE}} = \{\langle G \rangle \in \mathbb{B}^* : G \text{ does not have a Hamiltonian cycle}\}\,. \tag{1.45}$$

The affirmative answer to this problem is "yes, G does not contain a Hamiltonian cycle". However, verifying this statement is not a trivial task. One approach is to check all the closed paths in G and ensure that none of them is Hamiltonian. Unfortunately, this method requires a superpolynomial number of steps. $\overline{\mathsf{HAMCYCLE}}$ is an example of a decision problem that may not belong to NP [29, 64].

1.2.7 The traveling salesman problem

The **traveling salesman problem** (TSP) is one of the most important problems in graph theory and it finds ubiquitous applications. Suppose that a salesman would like

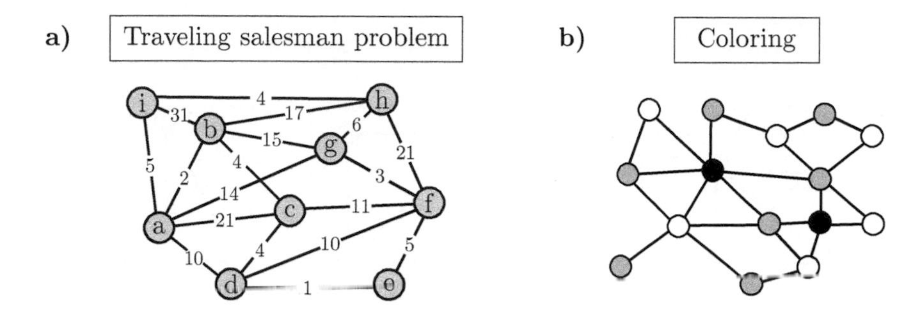

Fig. 1.12 TSP and coloring problem. a) An undirected weighted graph. The Hamiltonian cycle *a-b-c-d-e-f-g-h-i-a* is the one with minimal length. **b)** An example of a simple connected graph colored with three colors.

---

to visit different cities. The TSP consists in finding a closed path that visits all of the cities exactly once and minimizes the distance traveled by the salesman. This problem can be formulated in terms of graphs. First, each city can be associated with a node of a weighted graph (see Fig. 1.12a). The weights assigned to each edge represent the distance between pairs of cities. The length of a closed path $\mathsf{w} = (v_0, e_1, \ldots e_k, v_k)$ through these cities is given by $\ell(\mathsf{w}) = w(e_1) + \ldots + w(e_k)$. We define the language TSP as

$$\mathsf{TSP} = \{\langle G \rangle \in \mathbb{B}^* : G \text{ has a Hamiltonian cycle with length } \ell(\mathsf{w}) < r_n\}. \tag{1.46}$$

The TSP boils down to understanding whether a weighted graph contains a Hamiltonian cycle whose length is shorter than a real number r_n. One might argue that this problem can easily be solved by computing $\ell(\mathsf{w})$ for all Hamiltonian cycles in G and then picking the one with minimal length. Unfortunately, this approach is inefficient because the number of Hamiltonian cycles increases in a factorial way with the number of nodes in a highly connected graph. The TSP is a difficult NP problem. In addition, it can be shown that TSP is NPC, i.e. any problem in NP can be efficiently mapped into the traveling salesman problem [65].

1.2.8 Coloring

Coloring a graph consists in assigning colors to its nodes such that adjacent nodes do not share the same color. We can define 3COLOR as the set of all graphs that can be colored with three colors

$$\mathsf{3COLOR} = \{\langle G \rangle \in \mathbb{B}^* : G \text{ is colorable with three colors}\}. \tag{1.47}$$

Figure 1.12b shows a 3-coloring of a graph. It can be shown that 3COLOR is NPC.

1.2.9 Clique

Consider a graph G with n nodes and suppose that we want to find a k_n-clique, i.e. a complete subgraph with k_n nodes. The CLIQUE language is the set of graphs containing

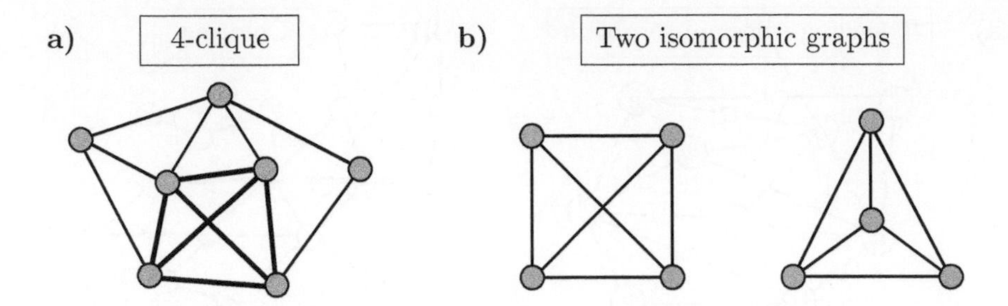

Fig. 1.13 Clique and isomorphic graphs. a) A graph with a 4-clique indicated with a thick line. **b)** Two isomorphic graphs.

at least one k_n-clique,

$$\mathsf{CLIQUE} = \{\langle G \rangle \in \mathbb{B}^* : G \text{ is an undirected graph with a } k_n\text{-clique}\}. \tag{1.48}$$

For large values of k_n, understanding whether G contains a complete subgraph with k_n nodes is a difficult NP problem. Figure 1.13a shows an example of a 4-clique in a simple graph. In Section 2.5.4, we will show how to convert a 3SAT problem into a CLIQUE problem.

1.2.10 Graph isomorphism

Consider two simple graphs $G_1 = (V_1, E_1)$ and $G_2 = (V_2, E_2)$. These graphs are **isomorphic** if there exists a bijective function $f : V_1 \to V_2$ such that two vertices v_i, v_j are adjacent in G_1 if and only if $f(v_i)$ and $f(v_j)$ are adjacent in G_2. The set of all pairs of isomorphic graphs forms a language,

$$\mathsf{GI} = \{\langle (G_1, G_2) \rangle \in \mathbb{B}^* : G_1,\ G_2 \text{ are isomorphic simple graphs}\}. \tag{1.49}$$

Figure 1.13b shows two isomorphic graphs. Determining whether two graphs are isomorphic can be a difficult problem [66, 67]. However, polynomial-time algorithms have been discovered for planar graphs.

1.2.11 MAXCUT

Consider a simple graph $G = (V, E)$ and a function $w(v_i, v_j)$ that assigns a real number to each edge of the graph. The vertices of G can be divided into two disjoint sets, A and A^c, such that their union is $A \cup A^c = V$; the set A is called a **cut** of the graph. We define the weight of a cut $w(A)$ as the sum of the weights of the edges connecting A to A^c,

$$w(A) = \sum_{v_i \in A} \sum_{v_j \in A^c} w(v_i, v_j),$$

where $w(v_i, v_j) = 0$ if $(v_i, v_j) \notin E$. It is evident that $w(A) = w(A^c)$. In Fig. 1.14 we show two different cuts of the same graph (for simplicity, we assumed that all edges in

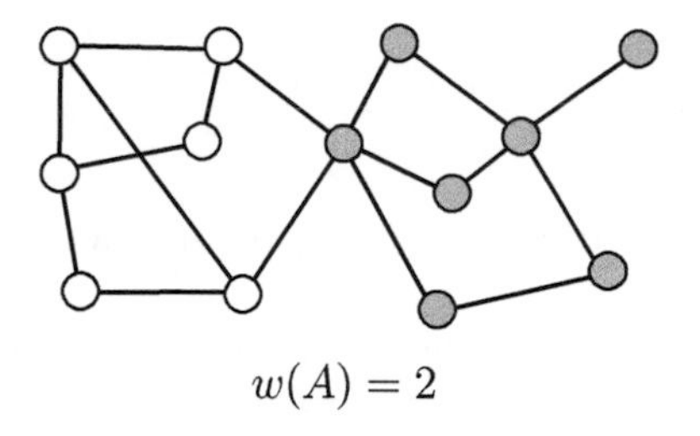

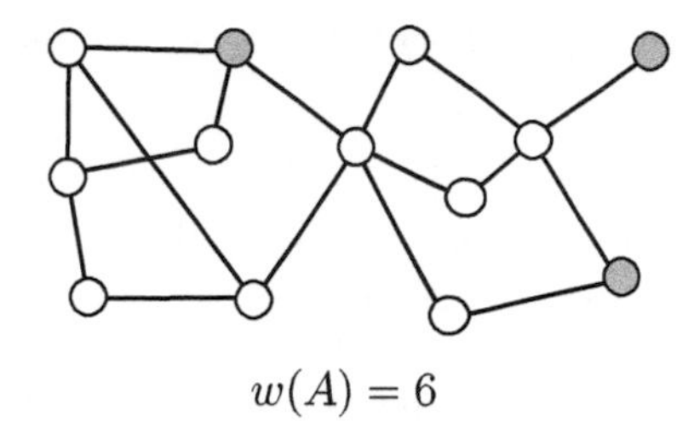

Fig. 1.14 MAXCUT. Two cuts of the same graph. The set A is indicated with white nodes, and the set A^c with grey nodes. In the left panel, A and A^c are connected with two edges, and in the right panel with six edges. Assuming that each edge has a weight equal to one, the cut on the left has weight two, the one on the right has weight six.

this graph have weight equal to one). The set of all graphs containing a cut of weight greater than a certain threshold w_n defines a language

$$\mathsf{MAXCUT} = \{\langle G \rangle \in \mathbb{B}^* : G \text{ is a simple graph with a cut } A \text{ where } w(A) > w_n\}.$$

Given a non-planar graph G, determining whether it contains a cut A of weight $w(A) \geq w_n$ is a difficult NP problem [68]. Several problems in engineering and statistical mechanics can be formulated in terms of MAXCUT. This problem has thus been of continued interest and has developed extensive literature [68, 65].

1.2.12 Groups

In this section, we will introduce some notions of group theory and we will then present the hidden subgroup problem, an important decision problem relevant in the context of quantum algorithms.

A group is a pair $(G, \cdot)$ where G is a non-empty set and $\cdot$ is the group operation

$$\begin{aligned} \cdot : \quad G \times G &\rightarrow \quad G \\ (a_1, a_2) &\mapsto \quad a_1 \cdot a_2 \end{aligned}$$

For any $a_i \in G$, the group operation must satisfy:

G1. $a_1 \cdot a_2 \in G$ (closure),

G2. $a_1 \cdot (a_2 \cdot a_3) = (a_1 \cdot a_2) \cdot a_3$ (associativity),

G3. $\exists e : a \cdot e = e \cdot a = a$ (unit element),

G4. $\exists a^{-1} : a \cdot a^{-1} = a^{-1} \cdot a = e$ (inverse element).

If $(G, \cdot)$ satisfies **G1**, **G2**, and **G3** but not necessarily **G4**, then it is called a **semigroup**. A group $(G, \cdot)$ is called **finite** if G contains a finite number of elements and the number $|G|$ is called the **order** of the group. We will often denote a group $(G, \cdot)$ simply as G.

A subset of G can generate the elements of G. More precisely, consider a subset $S = \{s_1, \ldots, s_k\} \subseteq G$. We say that S **generates** G if all elements in G can be expressed as a finite product of elements in S and we write

$$G = \langle s_1, \ldots, s_k \rangle.$$

A group obtained by repeatedly applying the group operation to just one element is called a **cyclic group**.

Consider a group G and a subset $H \subseteq G$. If the operation $\cdot : H \times H \to H$ is a valid group operation, then H is a **subgroup** of G. For instance, the set $H = \langle a_1, \ldots, a_m \rangle$ with $a_i \in G$ is a subgroup of G.

Example 1.6 An example of a group is $(\mathbb{Z}, +)$ where $+$ is the usual addition between integers, the unit element is 0, and the inverse element of $a \in \mathbb{Z}$ is $-a$. A generating set of $\mathbb{Z}$ is the set $S = \{-1, 1\}$.

Example 1.7 An example of a group is $(\mathbb{R} - \{0\}, \cdot)$ where $\cdot$ is the usual product between real numbers, the unit element is 1 and the inverse element of a is $1/a$.

Example 1.8 An example of a finite group is $(\mathbb{B}^n, \oplus)$ where the group operation $\oplus$ is defined by $\mathbf{x} \oplus \mathbf{y} = (x_1 \oplus y_1, \ldots, x_n \oplus y_n)$. For example, the sum between $\mathbf{x} = (1,1,0)$ and $\mathbf{y} = (1,0,1)$ is given by $\mathbf{x} \oplus \mathbf{y} = (0,1,1)$. The group $|\mathbb{B}^n|$ contains 2^n elements, the unit element is $(0, \ldots, 0)$ and the inverse of an element $\mathbf{x} = (x_1, \ldots, x_n)$ is $\mathbf{x}^{-1} = (x_1 \oplus 1, \ldots, x_n \oplus 1)$. The group $\mathbb{B}^n$ is generated by the set $S = \{\mathbf{e}_1, \ldots, \mathbf{e}_n\}$ where $\mathbf{e}_i = (\delta_{i1}, \ldots, \delta_{in})$. An interesting subgroup of $\mathbb{B}^n$ is $G_k = \{\mathbf{x} \in \mathbb{B}^n : \mathbf{x} = (x_1, \ldots, x_k, 0, \ldots, 0)\}$.

1.2.13 The hidden subgroup problem

We need two final ingredients before introducing the hidden subgroup problem (HSP). Consider a group G, a subgroup H, and an element $g \in G$. The **left coset** of H is the set

$$g \cdot H = \{g \cdot h : h \in H\}.$$

The **right coset** $H \cdot g$ is defined in a similar way: $H \cdot g = \{h \cdot g : h \in H\}$.

Consider a group G, a finite set $A = (a_1, \ldots, a_m)$ and a function $f : G \to A$. We say that f **hides** a subgroup $H \subseteq G$ if for all $g_i \in G$, we have

$$f(g_1) = f(g_2) \qquad \Leftrightarrow \qquad g_1 \cdot H = g_2 \cdot H.$$

In other words, the left cosets $g_1 \cdot H$ and $g_2 \cdot H$ are different if and only if the function f assigns different values to g_1 and g_2. We are finally ready to formulate the **hidden subgroup problem**:

Given We are given a group G and a function $f : G \to A$. The function f hides a subgroup $H \subseteq G$.

Goal By evaluating the function f on elements of G, find the subgroup H.

Many interesting computational problems, such as FACTORING and the Deutsch problem, are special cases of the hidden subgroup problem [69]. As an example, consider the group $(\{0,1\},\oplus)$ and the set $A=\{0,1\}$. The function $f:\{0,1\}\to A$ defined as $f(x)=x$ hides the subgroup $H=\{0\}$. Indeed, for $g_1=0$, we have $f(g_1)=0$ and the left coset $g_1\oplus H$ is equal to $\{0\}$. On the other hand, for $g_2=0$, we have $f(g_2)=1$ and the left coset $g_2\oplus H$ is $\{1\}$. Since $f(g_1)\neq f(g_2)$, the left cosets $g_1\oplus H$ and $g_2\oplus H$ are different as required.

Further reading

We recommend readers unfamiliar with computational complexity theory start with the textbook by M. Sipser [29]. Another excellent reference covering similar topics is the textbook by J. Savage [70]. A comprehensive monograph on classical algorithms is "The Art of Computer Programming" by D. Knuth [32]. For an introduction to group theory, we recommend the textbook by P. Ramond [71]. Readers interested in the physical applications of group theory can find more information in Refs. [72, 73, 74]. Probability is the cornerstone of various fields of mathematics and physics. Its foundations are covered in a variety of textbooks, including [75, 76]. Graphs are essential for the mathematical description of random walks, automata, and Turing machines. References [52, 51] cover foundations and applications of graph theory.

Summary

Sets and relations

$A = \{a_1, \ldots, a_n\} = \{a_i\}_{i=1}^n$	Common notation for finite sets
$A = B$	A and B contain the same elements
$A \neq B$	A and B do not contain the same elements
$A \subseteq B$	A is a subset of B
$A \subset B$	A is a proper subset of B
$A \cup B$	Union of A and B
$A \cap B$	Intersection of A and B
$A - B$	All elements that are in A but not in B
$\mathcal{P}(A)$	Power set of A
$A \times B$	Cartesian product
$A^n = A \times \ldots \times A$	Cartesian product of A with itself n times
$a = b \pmod n$, with $n > 0$	The division $(a-b)/n$ has no remainder

Functions

$f : A \to B$	Function from A to B
$p(x) = c_0 x^0 + \ldots + c_n x^n$	Polynomial of degree n (assuming $c_n \neq 0$)
$f \in O(g(n))$	f does not grow faster than g
$p : \mathcal{P}(\Omega) \to [0,1]$	Probability

Boolean functions

$\mathbb{1}(x) = x$	Identity
$\mathsf{NOT}(x) = \overline{x}$	NOT gate
$\mathsf{COPY}(x) = (x, x)$	COPY gate
$\mathsf{AND}(x_1, x_0) = x_1 \wedge x_0$	AND gate
$\mathsf{OR}(x_1, x_0) = x_1 \vee x_0$	OR gate
$\mathsf{XOR}(x_1, x_0) = x_1 \oplus x_0$	XOR gate (or sum modulo two)
$\mathsf{NAND}(x_1, x_0) = \overline{x_1 \wedge x_0}$	NAND gate

All functions $f : \mathbb{B}^n \to \mathbb{B}^m$ can be decomposed into $\{\mathsf{NAND}, \mathsf{COPY}\}$.

Erasing a bit of information dissipates an amount of energy $\geq k_{\mathrm{B}} T \ln 2$.

$\mathsf{CNOT}(x_1, x_0) = (x_1, x_0 \oplus x_1)$	CNOT gate
$\mathsf{SWAP}(x_1, x_0) = (x_0, x_1)$	SWAP gate
$\mathsf{CCNOT}(x_2, x_1, x_0) = (x_2, x_1, x_0 \oplus x_1 x_2)$	Toffoli gate
$f(\mathbf{x}) = (x_1 \vee x_2) \wedge (x_2 \vee \overline{x}_3) \wedge (\overline{x}_1 \vee x_2)$	This function is in 2CNF form

Graphs

$G = (V, E)$	Common notation for graphs
Simple graphs	Pairs of nodes are connected by a maximum of one edge
Complete graphs	Simple graph where each node is connected to all others
Multigraphs	Pairs of nodes may be connected by more than one edge
Graphs with loops	Nodes can be connected to themselves
Pseudographs	Both multigraphs and graphs with loops
Planar graphs	Can be drawn in a plane without edges crossing
$\mathsf{w} = (v_0, e_1, v_1, \ldots, e_n, v_n)$	A walk is a sequence of nodes and edges
$\ell(\mathsf{w})$	Length of a walk
$d(v_1, v_2)$	Distance between two nodes
Cycles	A closed walk with at least three distinct nodes
Trails	A walk with distinct edges
Paths	A walk with distinct edges and nodes
Acyclic graphs	Graphs not containing cycles
Connected graphs	Any pair of nodes can be connected by a walk
Trees	Connected graphs without cycles

Alphabets and languages

Σ	An alphabet is a finite non-empty set
$\mathbf{x} = x_1 \ldots x_n$	A string is a sequence of symbols taken Σ
$\mathbf{z} = \mathbf{x} \circ \mathbf{y}$	The string $\mathbf{z}$ is the concatenation of $\mathbf{x}$ and $\mathbf{y}$
Σ^n	All strings with n symbols taken from Σ
Σ^*	Strings of any length composed of symbols taken from Σ
L_Σ	A language is a subset of Σ^*
$\langle \cdot \rangle : \Sigma \to \mathbb{B}^n$	Encoding of a symbol into a binary string

---

Decision problems

$\mathsf{PRIMES} = \{\langle N \rangle \in \mathbb{B}^* : N \text{ is prime}\}$
$\mathsf{FACTORING} = \{\langle N \rangle \in \mathbb{B}^* : N \text{ has a factor smaller than } k_N\}$
$\mathsf{kSAT} = \{\langle f_{k\text{CNF}} \rangle \in \mathbb{B}^* : \text{there exists } \mathbf{x} \text{ such that } f_{k\text{CNF}}(\mathbf{x}) = 1\}$
$\mathsf{ECYCLE} = \{\langle G \rangle \in \mathbb{B}^* : G \text{ contains an Eulerian cycle}\}$
$\mathsf{HAMCYCLE} = \{\langle G \rangle \in \mathbb{B}^* : G \text{ contains an Hamiltonian cycle}\}$
$\mathsf{TSP} = \{\langle G \rangle \in \mathbb{B}^* : G \text{ has a Hamiltonian cycle with length } \ell(\mathsf{w}) < r_n\}$
$\mathsf{3COLOR} = \{\langle G \rangle \in \mathbb{B}^* : G \text{ is colorable with three colors}\}$
$\mathsf{CLIQUE} = \{\langle G \rangle \in \mathbb{B}^* : G \text{ is an undirected graph with a } k\text{-clique}\}$
$\mathsf{GI} = \{\langle G_1, G_2 \rangle \in \mathbb{B}^* : G_1,\ G_2 \text{ are isomorphic simple graphs}\}$
$\mathsf{MAXCUT} = \{\langle G \rangle \in \mathbb{B}^* : G \text{ is a simple graph with a cut } A \text{ where } w(A) > w_n\}$
$\mathsf{HSP} = \{\langle f, G \rangle \in \mathbb{B}^* : f : G \to A \text{ hides } H \subseteq G\}$

Exercises

Exercise 1.1 Factorize 16445 using trial division.

Exercise 1.2 Write the function f_5 presented in Table 1.2 in conjunctive normal form.

Exercise 1.3 Find all binary strings (if any) such that the function

$$f(x_1, x_0) = (x_1 \vee x_0) \wedge (\overline{x}_1 \vee \overline{x}_0)$$

outputs one.

Exercise 1.4 Consider the graphs $G_1 = (V_1, E_1)$ and $G_2 = (V_2, E_2)$, where

$$V_1 = \{0, 1, 2, 3\}, \qquad E_1 = \{(0,1), (0,3), (1,2), (1,3), (2,3)\},$$

and

$$V_2 = \{0, 1, 2, 3, 4\}, \qquad E_2 = \{(0,1), (0,2), (1,2), (2,3), (2,4), (3,4)\}.$$

For each graph, find an Eulerian cycle (if any).

Exercise 1.5 Show that the number of edges in a complete undirected graph is $n(n-1)/2$ where n is the number of nodes in the graph.

Solutions

Solution 1.1 Since $\sqrt{16445} \approx 128.23$, the factors of 16445 are smaller than 128. List the prime numbers smaller than 128, namely 2, 3, 5, 7, 11, 13, and so on. Dividing 16445 by these numbers one by one, we find $16445 = 5 \cdot 11 \cdot 13 \cdot 23$.

Solution 1.2 The Boolean function f_5 assigns one to strings 01 and 11 only. Thus, $f_5(x_1, x_0) = m_{01}(x_1, x_0) \vee m_{11}(x_1, x_0) = (\overline{x}_1 \wedge x_0) \vee (x_1 \wedge x_0)$.

Solution 1.3 $x_1 = 1$, $x_0 = 0$ and $x_1 = 0$, $x_0 = 1$.

Solution 1.4 Graph G_1 does not have any Eulerian cycle because nodes 1 and 3 have an odd degree (see Euler's theorem presented in Section 1.2.5). On the contrary, all of the nodes of G_2 have even degree. An Eulerian cycle for this graph can be found using Fleury's algorithm:

Step	Trail
1	(0)
2	(0, (0,1), 1)
3	(0, (0,1), 1, (1,2), 2)
4	(0, (0,1), 1, (1,2), 2, (2,4), 4)
5	(0, (0,1), 1, (1,2), 2, (2,4), 4, (4,3), 3)
6	(0, (0,1), 1, (1,2), 2, (2,4), 4, (4,3), 3, (3,2), 2)
7	(0, (0,1), 1, (1,2), 2, (2,4), 4, (4,3), 3, (3,2), 2, (2,0), 0)

Solution 1.5 Each node is connected to $n-1$ edges. Thus, the total number of edges is $n(n-1)$. Since the graph is undirected, this number should be divided by two.

2
Computational models

In our current state of knowledge,
it is widely accepted (but not rigorously proven)
that quantum computers are more powerful
than classical computers.

In theoretical computer science, a computational model is a model that describes how an input string is processed to generate an output. One of the simplest computational models is the Turing machine, a machine that can edit the information stored on a tape. Although a Turing machine is made of simple elements, it can perform very complex tasks using two resources: time (the number of movements along the tape) and space (the number of tape cells scanned during the computation). Algorithms are a collection of instructions that direct the activity of the machine. These programs are classified into certain complexity classes according to how much time and space they require to perform a computational task.

In this chapter, we first introduce the definition of an automaton, a machine that can perform very limited tasks. This discussion will lead us to the presentation of Turing machines. We will present different flavors of Turing machines, including deterministic, probabilistic, and multi-tape Turing machines. In each case, we will explain the advantages of these computational models. In Section 2.2.5, we present some complexity classes, i.e. collections of problems that a specific machine can solve with similar resources. The most important complexity classes presented in this chapter are P and NP. The class P contains all decision problems that a deterministic Turing machine can efficiently solve. Instead, the class NP contains all the decision problems whose affirmative solution can be verified by a deterministic machine with a number of steps that scales polynomially with input size. In Section 2.3, we introduce the circuit model, a flexible computational model important to quantum information science. In Section 2.4 we present the Church–Turing thesis, which states that if a problem can be solved by a physical machine, then it can also be solved by a Turing machine. In Section 2.5 we will present non-deterministic machines. Even if these machines cannot be built in the real world, they are very important from an abstract point of view to examine the limitations of practical machines. The last part of this chapter is dedicated to quantum Turing machines and the quantum circuit model. We will study the complexity class BQP, a class of problems that can be solved by a quantum computer with polynomial resources and a small error probability. Finally, we introduce QMA,

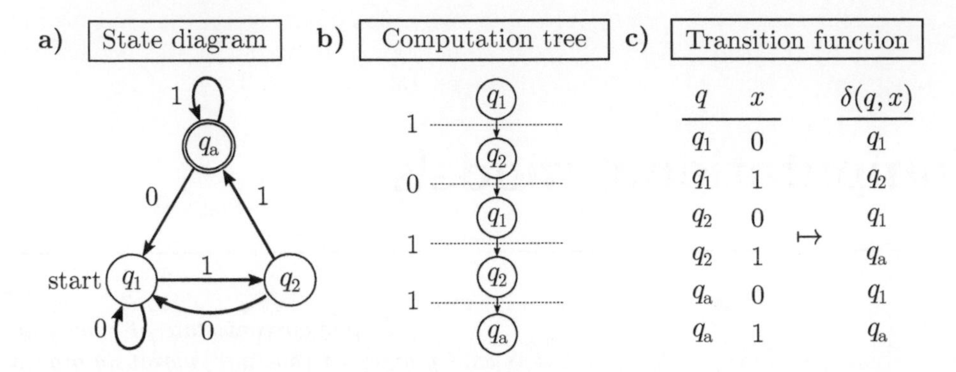

Fig. 2.1 Automaton A_1. a) State diagram of automaton A_1. **b)** Computation steps that A_1 executes when the input string is 1011. **c)** Transition function of automaton A_1.

the quantum version of the complexity class NP. We present the k-LOCAL problem and we show that this problem is QMA complete.

2.1 Deterministic automata

An **automaton** is a machine that processes information following a predetermined set of instructions. An elevator is an example of an automaton from our daily life: this machine can be in one of several states (floors) and its movements depend on the information provided by a user (by pressing a specific floor button). More generally, we can define a **deterministic finite-state automaton** (DFA) as a machine composed of several parts: an **alphabet** Σ that indicates the symbols the automaton can deal with; a finite number of **states** Q (including the **initial state** of the machine q_1); a collection of **accept states** q_a; and a **transition function** δ, which specifies a set of instructions that the automaton must follow during its operations.

As an example, consider an automaton A_1 with alphabet $\mathbb{B}$ and suppose that such a machine operates on binary strings $\mathbf{x} = x_1 \ldots x_n$ where x_i are the input symbols. The operations and states of this automaton can be described with a labeled directed graph known as the **state diagram** (see Fig. 2.1a). As this graph suggests, A_1 has three states: q_1, q_2, and q_a. The machine's initial state, q_1, is indicated with the word "start". The accept state, q_a, has a double circle around it. One or more arrows connect pairs of states indicating possible transitions between them and each transition is labeled with a symbol taken from the Boolean alphabet. Once all of the symbols of the input string have been processed, the computation either stops in an accept state, q_a, in which case the output of the computation is "**accept**", or stops in a state that is not an accept state, in which case the output of the computation is "**reject**".

Let us investigate how automaton A_1 operates. Suppose that this machine, which is initially in state q_1, receives the input string $\mathbf{x} = x_1x_2x_3x_4 = 1011$. Since the first input symbol is $x_1 = 1$, the machine follows the edge labeled 1 connecting state q_1 to state q_2 and the new machine state becomes q_2. The next symbol of the input string is $x_2 = 0$: the machine accordingly moves from q_2 to q_1 following the edge labeled 0. Since the last two input symbols are $x_3 = 1$ and $x_4 = 1$, the state of the automaton first changes from q_1 to q_2 and then from q_2 to q_a. As no other symbols are left in the input string, the machine stops in state q_a. Since this is an accept state, the output of this procedure is "accept". We can list these operations as follows:

$$q_1 \xrightarrow{1} q_2 \xrightarrow{0} q_1 \xrightarrow{1} q_2 \xrightarrow{1} q_a\,. \tag{2.1}$$

This sequence of steps, illustrated in Fig. 2.1b by a tree with a single branch, summarizes the computation performed by the automaton A_1 on input 1011.

We could investigate the operations of A_1 on different input strings. In doing so, we would discover that, regardless of the input string length, the automaton will always take $n = |\mathbf{x}|$ steps to terminate the computation. We would also discover that the output will be "accept" only when the last two symbols of the string are $x_n = x_{n-1} = 1$. For this reason, we say that this machine **recognizes** the language that contains all binary strings ending with 11. Formally:

$$L(A_1) = \{\mathbf{x} \in \mathbb{B}^* : \ |\mathbf{x}| = n \geq 2 \ \text{ and } \ x_{n-1} = x_n = 1\}\,. \tag{2.2}$$

This is called the language of automaton A_1.

We are ready to give a more formal definition of a deterministic finite-state automaton, namely a machine with five elements:

1. an alphabet Σ,
2. a non-empty finite set of states Q,
3. a transition function $\delta : Q \times \Sigma \to Q$,
4. an initial state $q_1 \in Q$,
5. a set of accept states $A \subseteq Q$.

The only element that still needs to be discussed is the **transition function** δ. This function indicates how the state of the automaton changes according to its current state and the input symbol received. The table in Fig. 2.1c shows the transition function for automaton A_1. This function encapsulates the same information provided by the state diagram. In light of these definitions, the formal description of A_1 is $(\mathbb{B}, Q = \{q_1, q_2, q_a\}, \delta, q_1, \{q_a\})$.

We will conclude this section with two observations. First, an automaton does not have a memory to which it can write notes. This constitutes the main difference between automata and Turing machines. Secondly, the state diagram of an automaton is a labeled directed graph where the edges leaving a node have distinct symbols. As we shall see in Section 2.5.1, this constitutes the main difference between deterministic and non-deterministic automata (in the latter, the same symbol can be assigned to more than an edge leaving a node).

2.1.1 Regular languages

In the previous section, we discovered that automaton A_1 accepts the string 1011, i.e. the final state of the machine on this input is q_a. We have also guessed that any binary string ending with two ones is accepted by A_1 and that the collection of such strings is the language recognized by this machine. We will now formalize these two observations. A string $\mathbf{x} = x_1 \dots x_n$ is said to be **accepted** by an automaton (Σ,

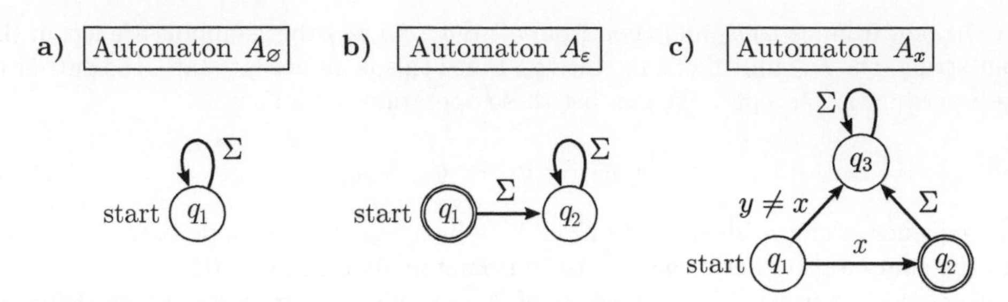

Fig. 2.2 Examples of automata. Panels **a**), **b**), and **c**) show the state diagrams of automata $A_\varnothing$, A_ε, and A_x respectively. The symbol Σ means any symbol in the automaton alphabet. The state diagram of automaton $A_\varnothing$ does not contain any accept state. In the state diagram of automaton A_ε (A_x), the accept state is q_1 (q_2).

Q, δ, q_1, A), if there exists a sequence of states $(p_1, \ldots, p_{n+1})$ in Q that satisfy the conditions:

1. p_1 is the initial state q_1,
2. $\delta(p_i, x_i) = p_{i+1}$ for all $i \in \{1, \ldots, n\}$,
3. p_{n+1} is an accept state.

For instance, the strings 10011 and 01011 are both accepted by A_1. This definition leads to the concept of a regular language. A language L is said to be **regular** if there exists a deterministic finite-state automaton A such that:

$$\text{if } \mathbf{x} \in L, \text{then } A \text{ terminates in an accept state,}$$
$$\text{if } \mathbf{x} \notin L, \text{then } A \text{ does not terminate in an accept state.}$$

Does there exist a language that is not regular? In other words, does there exist a language that cannot be recognized by any deterministic finite-state automaton? Yes, and here is an example:

$$L = \{0^m 1^m : m \geq 0\} = \{\varepsilon, 01, 0011, 000111, \ldots\} . \tag{2.3}$$

This language contains binary strings with a sequence of zeros followed by a sequence of ones of the same length. There does not exist any deterministic automaton with a *finite* number of states that can count the number of 0's of the strings in L and check that their number matches the number of 1's for *any* natural number n (in principle n could be greater than the number of states of the automaton). This epitomizes the limitations of DFA. In Section 2.2.6, we will see that a Turing machine can decide this language.

Example 2.1 Consider the automaton $A_\varnothing = \{\Sigma, \{q_1\}, \delta, q_1, \varnothing\}$ with state diagram shown in Fig. 2.2a. This automaton does not accept any language, $L(A_\varnothing) = \varnothing$.

Example 2.2 Consider the automaton $A_\varepsilon = \{\Sigma, \{q_1, q_2\}, \delta, q_1, \{q_1\}\}$ with state diagram shown in Fig. 2.2b. This automaton accepts the empty string only, $L(A_\varepsilon) = \{\varepsilon\}$.

Example 2.3 Consider the automaton $A_x = \{\Sigma, \{q_1, q_2, q_3\}, \delta, q_1, \{q_2\}\}$ with state diagram shown in Fig. 2.2c. This automaton accepts the language that has only one symbol x, $L(A_x) = \{x\}$.

Example 2.4 The union of two regular languages $L = L_1 \cup L_2$ is a regular language. This can be shown as follows: since L_1 and L_2 are regular, there exist two finite automata $M_1 = \{\Sigma, Q_1, \delta_1, q_{\text{in}1}, A_1\}$ and $M_2 = \{\Sigma, Q_2, \delta_2, q_{\text{in}2}, A_2\}$ that recognize these two languages. From these two machines, one can construct an automaton M that recognizes the language $L_1 \cup L_2$,

$$M = \{\Sigma, Q_1 \times Q_2, \delta, (q_{\text{in}1}, q_{\text{in}2}), A\}. \tag{2.4}$$

Here, the accept states are defined as $A = (A_1 \times Q_2) \cup (Q_1 \times A_2)$. In other words, the accept states of M are the states (q_1, q_2) where either $q_1 \in A_1$ or $q_2 \in A_2$. If the transition function is $\delta((q_1, q_2), x) = \{\delta_1(q_1, x), \delta_2(q_2, x)\}$, the machine M recognizes $L_1 \cup L_2$ since $\delta((q_1, q_2), x) \in A$ if and only if $\delta(q_1, x) \in A_1$ or $\delta(q_2, x) \in A_2$.

Example 2.5 Consider a regular language L. The complement of this language, $\overline{L} = \{x \in \Sigma^* : x \notin L\}$, is regular. This can be shown as follows: since L is regular, there exists an automaton $M = \{\Sigma, Q, \delta, q_1, A\}$ that recognizes L. Now consider the DFA $\overline{M} = \{\Sigma, Q, \delta, q_1, \overline{A}\}$, where $\overline{A} = Q - A$. This machine recognizes $\overline{L}$.

Example 2.6 The intersection of two regular languages $L_1 \cap L_2$ is a regular language. This is because $L_1 \cap L_2 = \overline{\overline{L_1} \cup \overline{L_2}}$ and (as seen in the previous two examples) regular languages constructed from the union operation and the complement operation are regular.

2.2 Turing machines and complexity classes

In the following sections, we give a formal definition of a Turing machine and discuss an example of a deterministic Turing machine with a single tape. We present the halting problem, a decision problem that cannot be decided by a Turing machine. We introduce the main complexity classes for deterministic machines and their relations. We introduce multi-tape Turing machines and explain the main advantages with respect to single-tape machines. Lastly, we present probabilistic Turing machines and the complexity classes BPP and PP.

2.2.1 Deterministic Turing machines

A Turing machine is a mathematical model describing an abstract machine capable of simulating computational algorithms implemented in real-world computers [1, 53]. Turing machines are more powerful than automata: not only can they accept and reject input strings, but they can also store the output of a computation on a tape with an

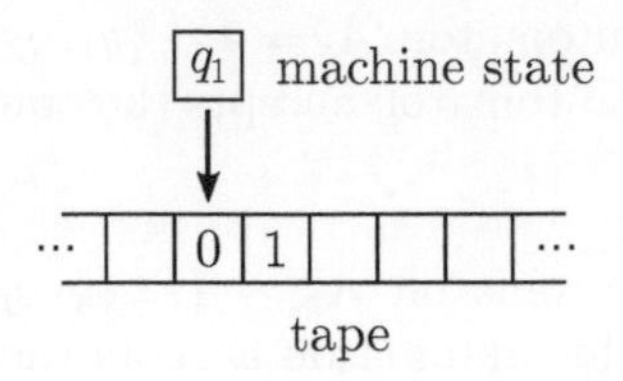

Fig. 2.3 Example of a single-tape Turing machine. The content of the tape is 01. The head is in state q_1 and points to the first cell.

---

infinite number of cells. As done in the previous section on automata, we will first give a simple description of a Turing machine (TM) and then outline a more formal definition.

A deterministic finite-state Turing machine (DTM) consists of several parts, as illustrated in Fig. 2.3. First, it has a **tape** divided into **cells**. The tape is infinite and it extends in both directions. Each cell can be indexed with an integer and contains only one symbol at a time. The input string $\mathbf{x}$ is recorded onto the first n cells and it is preceded and followed by an infinite number of blank symbols. A TM also has a **head** that is assigned to read, write, and erase symbols from the tape. The head can move left and right and at the beginning of the computation it points to the first input symbol. A finite-state unit controls the operations of the head. Lastly, a **transition function** indicates the instructions that the machine has to follow during the computation.

More formally, a **deterministic finite-state Turing machine** is a sequence of seven elements:

1. an alphabet Σ that does not contain the blank symbol $\sqcup$,
2. a tape alphabet Γ such that $\sqcup \in \Gamma$ and $\Sigma \subset \Gamma$,
3. a non-empty finite set of states Q,
4. a transition function $\delta : Q \times \Gamma \to Q \times \Gamma \times \{\mathrm{L}, \mathrm{R}, \mathrm{N}\}$,
5. an initial state $q_1 \in Q$,
6. an accept state[1] $q_\mathrm{a} \in Q$,
7. a reject state $q_\mathrm{r} \in Q$, where $q_\mathrm{r} \neq q_\mathrm{a}$.

The computation of a DTM is divided into several steps: at the beginning of the computation, the input string $\mathbf{x} \in \Sigma^*$ is recorded into the first n cells of the tape (the remaining cells are left blank), the head points to the first symbol x_1 and the state of the machine is the initial state q_1 as shown in Fig. 2.3. Depending on the current machine state q and the symbol x the head is reading, the transition function outputs a new machine state q', a new symbol x', and a shift (which can be left L, right R, or "no movement" N). The symbol x is replaced by x', the machine state changes from q to q', and the head moves either left or right, or it does not move at all. This procedure stops as soon as the state of the machine either reaches an accept state q_a or a reject

[1]Requiring that a Turing machine has only one accept state does not lead to any loss of generality. Even if the accept states are more than one, we can in principle add a final step to the computation and group all of the accept states into one.

state q_r. The output of the computation, denoted by $T(\mathbf{x})$, is the string remaining on the right side of the head once the machine has reached either an accept state or a reject state.

During a computation, a DTM continues to change the machine state, the content of the tape, and the position of the head. The collection of these three pieces of information is called the **TM configuration**. A TM configuration can be concisely written with the notation $\mathsf{c} = \mathbf{x}^q\mathbf{y}$, which means that the current state of the machine is q, the content of the tape is the concatenation $\mathbf{x} \circ \mathbf{y}$ and the head points to the first symbol of the string $\mathbf{y}$. For example, the machine

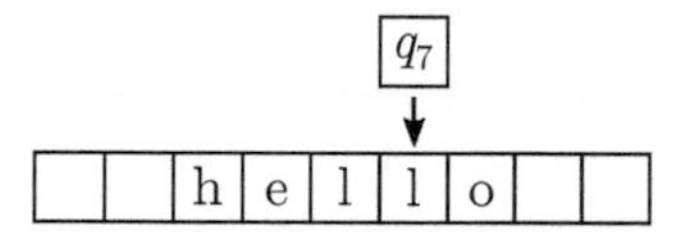

is in the configuration $\mathsf{c} = \text{hel}^{q_7}\text{lo}$. If a TM enters into an accept state q_a, the computation terminates in the configuration $\mathsf{c}_a = \mathbf{z}^{q_a}\mathbf{y}$ and the string $\mathbf{y}$ is the output of the computation[2].

It is worth pointing out that the configuration of a TM can also be specified with a triplet $\mathsf{c} = (q, \mathbf{x}, z)$, where q is the state of the machine, $\mathbf{x}$ is the content of the tape, and $z \in \mathbb{Z}$ is the index of the cell the head is pointing to. In our example, $\mathsf{c} = (q_7, \text{hello}, 4)$. This alternative way to represent a TM configuration will be convenient when we introduce quantum Turing machines in Section 2.6.

2.2.2 Decidable languages and computable functions

As seen in Section 2.1.1, automata can only recognize a limited class of languages, the so-called regular languages. The need to recognize non-regular languages prompted us to introduce Turing machines to overcome the memory limitations of automata. In this section, we formalize the concept of a computation performed by a Turing machine.

An input string $\mathbf{x}$ is **accepted** by a Turing machine if there exists a sequence of configurations $(\mathsf{c}_1, \ldots, \mathsf{c}_N)$ that satisfy the following conditions:

1. $\mathsf{c}_1 = {}^{q_1}\mathbf{x}$ is the initial configuration of the machine,
2. configuration c_i leads to c_{i+1} for all $1 \leq i < N$,
3. c_N is an accept configuration.

A language L is **Turing recognizable** if there exists a Turing machine that accepts all strings in L and rejects all strings not in L. A string can be rejected in two ways: either the machine enters into a reject state or loops forever (i.e. it never reaches a halting configuration). This latter situation is not ideal: it is difficult to understand whether a machine is looping or carrying out a very long computation. For this reason, it is often useful to introduce **deciders**, i.e. deterministic Turing machines that on any input string either enter an accept state or a reject state (they never loop). A language L is **decidable** if there exists a decider T such that:

[2]Note that the string $\mathbf{z}$ might be the empty string ε.

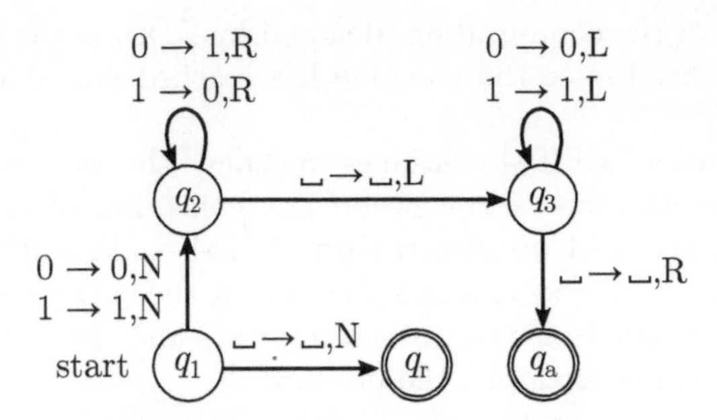

Fig. 2.4 State diagram of a Turing machine. State diagram of the Turing machine T_1 computing the bit-by-bit complement of the input string. The state q_1 deals with the initialization of the computation (basically, it rejects the empty string). State q_2 takes care of the main part of the computation by substituting the symbol 0 with 1 and vice versa. State q_3 brings the computation to a final state by moving the head to the leftmost symbol on the tape. The label "$0 \to 1, \text{R}$" means that the head replaces 0 with 1 and moves to the right. The label "$0 \to 0, \text{N}$" means that the head replaces 0 with 0 and does not move.

---

if $\mathbf{x} \in L$, then T ends in an accept state,
if $\mathbf{x} \notin L$, then T ends in a reject state.

The process of accepting or rejecting an input string is what we call a **computation**.

One can offer an alternative definition of decidable languages based on the notion of **computable functions**. Consider an alphabet Σ such that $\sqcup \notin \Sigma$. We say that a function $f : \Sigma^* \to \Sigma^*$ is computable by a Turing machine if for any $\mathbf{x} \in \Sigma^*$ there exists a sequence of configurations $(\mathsf{c}_1, \ldots, \mathsf{c}_N)$ such that:

1. $\mathsf{c}_1 = {}^{q_1}\mathbf{x}$ is the initial configuration of the machine,
2. the configuration c_i leads to c_{i+1} for all $1 \le i < N$,
3. the last configuration is $\mathsf{c}_N = \mathbf{z}^{q_\mathrm{a}}\mathbf{y}$, where q_a is an accept state, $\mathbf{y} = f(\mathbf{x})$ and $\mathbf{z}$ is an arbitrary string.

The definitions of decidable languages and computable functions are equivalent. Indeed, given a language L, one can always define a characteristic function $\chi_L : \Sigma^* \to \mathbb{B}$ of the form

$$\chi_L(\mathbf{x}) = \begin{cases} 1 & \text{if } \mathbf{x} \in L \\ 0 & \text{if } \mathbf{x} \notin L. \end{cases} \tag{2.5}$$

A language L is decidable if its associated characteristic function $\chi_L(\mathbf{x})$ is computable. Similarly, given a Boolean function $f : \Sigma^* \to \mathbb{B}$, it is always possible to define a language

$$L_f = \{\mathbf{x} \in \mathbb{B}^* : f(\mathbf{x}) = 1\}. \tag{2.6}$$

A Boolean function f is computable if L_f is decidable. Examples of decidable languages are $L(A_1)$ of eqn (2.2) and the language presented in eqn (2.3).

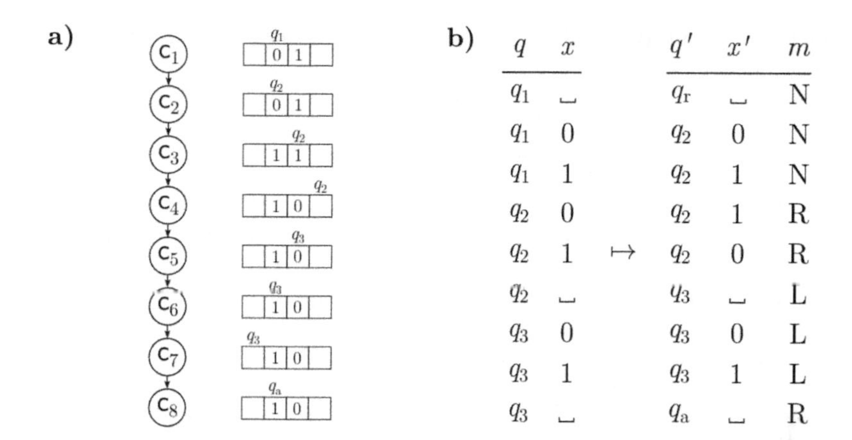

q	x		q'	x'	m
q_1	$\sqcup$		q_r	$\sqcup$	N
q_1	0		q_2	0	N
q_1	1		q_2	1	N
q_2	0		q_2	1	R
q_2	1	$\mapsto$	q_2	0	R
q_2	$\sqcup$		q_3	$\sqcup$	L
q_3	0		q_3	0	L
q_3	1		q_3	1	L
q_3	$\sqcup$		q_a	$\sqcup$	R

Fig. 2.5 The Turing machine $\mathbf{T_1}$. a) Computation tree of the Turing machine T_1 executing the bitwise complement of the string 01. **b)** Transition function of the Turing machine T_1: q is the current machine state, x is the symbol read by the head, q' is the new machine state, x' is the new symbol written on the tape and m indicates the next movement of the head.

2.2.3 Deterministic Turing machine for bit-by-bit complement

Let us present an example of a deterministic Turing machine T_1 that transforms each bit of a binary string into its complement (simply speaking, it replaces 0's with 1's and vice versa). The transition function of T_1 is represented by the state diagram shown in Fig. 2.4. We can verify that T_1 operates as expected. Suppose that the input string is $\mathbf{x} = 01$. The computation starts with the TM in state q_1 and the head points to the first input symbol $x_1 = 0$ (see Fig. 2.3). The state diagram shows an edge connecting q_1 to q_2 labeled "$0 \to 0, \text{N}$". Thus, the machine state changes from q_1 to q_2, the head replaces 0 with 0, and the position of the head does not change. The symbol read by the head is again $x_1 = 0$; the state diagram shows a self-loop connecting state q_2 with itself. Hence, this time the state of the machine does not change. However, the input symbol is replaced with 1 and the head moves to the right. In the new configuration, the head is pointing to the second symbol $x_2 = 1$ and the state of the machine is still q_2. The state diagram instructs the head to replace 1 with 0 and to move to the right. One can verify that after some steps the TM reaches the final configuration ${}^{q_a}10$. A more convenient way to describe this procedure is to list all the configurations the machine goes through:

$$ {}^{q_1}01 \to {}^{q_2}01 \to 1\,{}^{q_2}1 \to 10\,{}^{q_2} \to 1\,{}^{q_3}0 \to {}^{q_3}10 \to {}^{q_3}\sqcup 10 \to {}^{q_a}10\,. \tag{2.7} $$

These configurations are represented by a tree with a single branch in Fig. 2.5a. The formal definition of T_1 is given by $T_1 = (\mathbb{B},\ \{0, 1, \sqcup\},\ \{q_1, q_2, q_3, q_a, q_r\},\ \delta,\ q_1,\ q_a,\ q_r)$. For completeness, we report the transition function of this machine in Fig. 2.5b.

We conclude this section with some observations. First of all, a Turing machine accepts exactly one language, but the same language can be accepted by more than one Turing machine. Secondly, the Turing machine T_1 is not doing anything other than computing the function

$$f_1 : \begin{array}{ccc} \mathbb{B}^* & \to & \mathbb{B}^* \\ \mathbf{x} & \mapsto & \overline{\mathbf{x}} \, , \end{array}$$

where $\overline{\mathbf{x}}$ is the bit-by-bit complement of the input string $\mathbf{x}$ (in our example $f_1(01) = 10$). Hence, we say that f_1 can be computed by T_1. Finally, this Turing machine will take approximately $2n$ steps to get to the final result and it will scan approximately n distinct tape cells, where n is the length of the input string.

2.2.4 The halting problem

Are all languages decidable? No, and this can be shown in two ways: one way consists in demonstrating that the set of all TMs is countably infinite, and so is the set of decidable languages $\mathcal{L}_{\text{dec}}$, whereas the set of all languages $\mathcal{L}$ is uncountable[3]. This implies that the set $\mathcal{L} - \mathcal{L}_{\text{dec}}$ is not empty, i.e. there exists a language that is not decidable. This argument is purely theoretical and does not illustrate an example of a non-decidable language. A second approach is to come up with a language that is not decidable. An example of such a language is the HALTING problem. This problem can be formulated as follows. Consider a deterministic Turing machine T running on an input $\mathbf{x}$. Will the machine T stop its computation after a *finite number* of steps? In 1936, Alan Turing proved that there does not exist a classical algorithm that solves the HALTING problem for all possible pairs of input strings and deterministic Turing machines [1, 53].

To define the HALTING problem in a more rigorous way, let us first observe that a Turing machine is described by a 7-tuple of finite elements. All of this information can be encoded into a binary string $\mathbf{a}_T = \langle T \rangle$ (see Section 1.2.1 for some examples of binary encodings). The collection of binary strings encoding all possible Turing machines will be denoted with the letter C and clearly $C \subset \mathbb{B}^*$. The HALTING problem consists in deciding if a given pair of binary strings $(\mathbf{a}_T, \mathbf{x})$ belongs to the language[4]

$$L_{\mathsf{HALTING}} = \{(\mathbf{a}_T, \mathbf{x}) : \mathbf{a}_T \in C, \ \mathbf{x} \in \mathbb{B}^*, \text{ and } T(\mathbf{x}) \text{ terminates}\}.$$

If this language is decidable, then there exists a decider T_{H} that takes two inputs (the encoding of a Turing machine $\mathbf{a}_T$ and a string $\mathbf{x}$) and outputs 1 if the computation $T(\mathbf{x})$ terminates and outputs 0 if it loops forever,

$$T_{\mathrm{H}}(\mathbf{a}_T, \mathbf{x}) = \begin{cases} 1 & T(\mathbf{x}) \text{ halts} \\ 0 & T(\mathbf{x}) \text{ does not halt.} \end{cases}$$

In other words, the machine T_{H} is capable of examining the blueprint of a Turing machine T and deciding whether the computation $T(\mathbf{x})$ will halt or not. We now show

[3]The set of all languages over an alphabet Σ is $\mathcal{L} = \mathcal{P}(\Sigma^*)$, i.e. the power set of the countably infinite set Σ^*. The power set of a countably infinite set is uncountable.

[4]Here, we are assuming that the alphabet of the Turing machine is the Boolean alphabet. This is general enough since any alphabet can be encoded into a collection of binary strings.

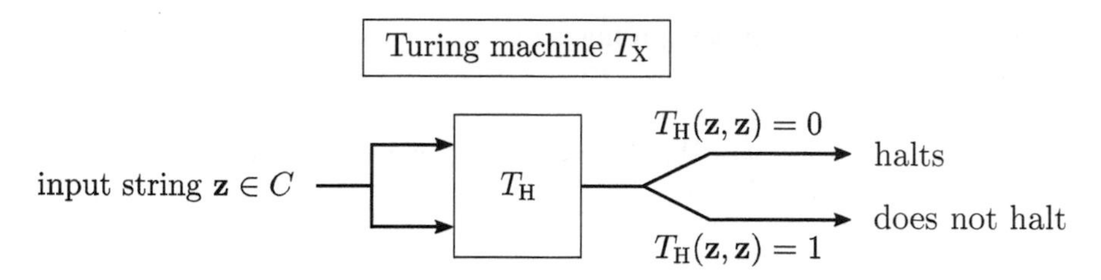

Fig. 2.6 The halting problem. A schematic of the Turing machine T_X. This machine takes an input string $\mathbf{z}$ and runs the subroutine $T_H(\mathbf{z}, \mathbf{z})$. The output is either 0 or 1.

that L_{HALTING} is not decidable by proving that T_H does not exist.

Theorem 2.1 L_{HALTING} is not decidable.

Proof Suppose that T_H exists. Therefore, we can create a new Turing machine T_X (as shown in Fig. 2.6) that uses T_H as a subroutine. The machine T_X takes as input a binary string $\mathbf{z}$ and operates as follows:

$$T_X(\mathbf{z}) = \begin{cases} \text{halts} & \text{if } \mathbf{z} = \mathbf{a}_T \text{ and } T_H(\mathbf{z}, \mathbf{z}) = 0 \\ \text{does not halt} & \text{if } \mathbf{z} = \mathbf{a}_T \text{ and } T_H(\mathbf{z}, \mathbf{z}) = 1. \end{cases} \quad (2.8)$$

Figure 2.6 shows a schematic of the operations performed by T_X. The last part of the proof analyzes what happens when the input of the Turing machine T_X is its own encoding. Consider the computation $T_X(\mathbf{a}_{T_X})$ and suppose that $T_H(\mathbf{a}_{T_X}, \mathbf{a}_{T_X}) = 0$; then the machine T_H indicates that the computation $T_X(\mathbf{a}_{T_X})$ does not halt, but this is in contrast with the definition (2.8). Suppose instead that $T_H(\mathbf{a}_{T_X}, \mathbf{a}_{T_X}) = 1$; then T_H suggests that the computation $T_X(\mathbf{a}_{T_X})$ terminates, but again this is in contrast with the definition of T_X expressed by eqn (2.8). The machine T_H is supposed to output the right answer for all Turing machines and input strings. This shows that T_H cannot exist and thus L_{HALTING} is not decidable. □

2.2.5 Complexity classes

Complexity theory focuses on the **resources** required to solve a problem. With the term resources, we primarily mean the time and space needed by a machine to carry out a computation. In mathematical terms, we define the **running time** $t(n)$ as the number of steps taken by a deterministic Turing machine to reach a final state on any input string of length n (the final state can either be an accept state or a reject state). Typically, $t(n)$ increases as n increases[5]. The **deterministic time class** $\mathsf{TIME}(t(n))$ is the collection of all languages that a deterministic Turing machine can decide in

[5]For example, the Turing machine T_1 presented in Section 2.2.3 flips the bits of the input string in approximately $t_{T_1}(n) = 2n$ steps on any input string. For this reason, we say that T_1 runs in linear time.

$O(t(n))$ steps[6]. Time classes are grouped in several categories, some of which are listed below:

- P is the class of languages that a DTM can decide in a number of steps that is polynomial in the length of the input strings:

$$\mathsf{P} = \bigcup_{k=1}^{\infty} \mathsf{TIME}(n^k) \, . \tag{2.9}$$

  This set roughly corresponds to the kind of decision problems a deterministic machine can solve efficiently. An example of a problem in P is 2SAT, presented in Section 1.2.

- EXPTIME is the class of languages that a DTM can decide in a number of steps that increases exponentially with the input size n:

$$\mathsf{EXPTIME} = \bigcup_{k=1}^{\infty} \mathsf{TIME}(2^{n^k}) \, . \tag{2.10}$$

  This class contains easy problems as well as problems that are not efficiently solved by deterministic machines (important examples are HAMCYCLE, TSP, 3SAT, and 3COLOR presented in Section 1.2).

We define the **space** function $s\,(n)$ as the maximum number of distinct tape cells[7] scanned by a DTM on any input string of length n. For example, the Turing machine presented in Section 2.2.3 approximately scans $s_{T_1}\,(n) = n$ cells before reaching a final state on any input string. For this reason, we say that T_1 uses linear space. The **space complexity class** SPACE$(s(n))$ is the collection of all languages that can be decided by a DTM scanning $O(s(n))$ tape cells before reaching a final state on any input string of length n. The most important ones are:

- PSPACE is the class of languages that can be decided by a DTM scanning a number of tape cells polynomial in the input size n:

$$\mathsf{PSPACE} = \bigcup_{k=1}^{\infty} \mathsf{SPACE}(n^k). \tag{2.11}$$

  This class corresponds to the decision problems that can be solved by a DTM using polynomial space and any amount of time. ECYCLE, presented in Section 1.2.5, is an example of a language in PSPACE.

- EXPSPACE is the class of languages that can be decided by a single-tape DTM scanning a number of tape cells exponential in the input size n:

[6] See Section 1.1.4 for more details about the big-O notation.

[7] With the word "distinct", we mean that if a tape cell is scanned twice during a computation, it still counts as one.

$$\mathsf{EXPSPACE} = \bigcup_{k=1}^{\infty} \mathsf{SPACE}(2^{n^k}). \tag{2.12}$$

It is natural to wonder if it is possible to define a class of languages that can be decided by a DTM using logarithmic space. It turns out that this is not possible with single-tape Turing machines: in most cases, the machine has to read the entire input string, which already takes $O(n)$ scans. To introduce the sublinear complexity class L, we first need to introduce multi-tape Turing machines, a variant of Turing machines (see Section 2.2.6).

Complexity classes are related to each other by important inclusion relations. It is quite intuitive that if a DTM can decide a language in polynomial time, then it can also decide it in exponential time, implying that $\mathsf{P} \subseteq \mathsf{EXPTIME}$. Furthermore, the **time hierarchy theorem** states that $\mathsf{P} \subset \mathsf{EXPTIME}$ [77]. It is intuitive to see that $\mathsf{P} \subseteq \mathsf{PSPACE}$ and $\mathsf{EXPTIME} \subseteq \mathsf{EXPSPACE}$ (during a computation, the same tape cell can be scanned many times, while the number of steps taken by the machine continuously increases). Finally, it can be shown that $\mathsf{PSPACE} \subseteq \mathsf{EXPTIME}$ [77]. Collecting these results together, we arrive at

$$\mathsf{P} \subseteq \mathsf{PSPACE} \subseteq \mathsf{EXPTIME} \subseteq \mathsf{EXPSPACE}\,. \tag{2.13}$$

The number of complexity classes is remarkably high. In this chapter, we will only mention the most relevant ones and refer the interested reader to Ref. [30] for more information.

The alphabet adopted by a machine does not significantly affect its computation speed. More precisely, if a DTM decides a language in $t(n)$ steps using an alphabet Σ with more than two symbols, then it is possible to build a DTM that decides the same language in $O(t^k(n))$ steps using the Boolean alphabet, where $k > 1$. However, the encoding of the input string might play an important role. Consider, for example, the language containing all binary strings ending with 1: $L = \{\mathbf{x} \in \mathbb{B}^* : x_n = 1\}$. It is easy to build a Turing machine that decides this language in $O(n)$ steps: this machine just needs to scroll to the end of the input string and check that the last bit is equal to 1. If the input string is initially recorded in reverse order, however, the machine can check the first input symbol and decide L in one step. Luckily, the time complexity of the most interesting decision problems does not strongly depend on the encoding of the input string.

2.2.6 Multi-tape Turing machines

A **multi-tape deterministic finite-state Turing machine** (MDTM) is a variant of the standard TM, the main difference being that multi-tape machines can use $k > 1$ tapes (see Fig. 2.7). The first tape is called the **input tape**, whereas the additional tapes are usually called **work tapes**. These extra tapes can be exploited to speed up the computation and can be regarded as scratch pads on which the machine can annotate some information during its operations. Each tape is equipped with a head that can move left and right. At the beginning of the computation, the input string is recorded on the input tape, the other tapes are left blank and the k heads point to

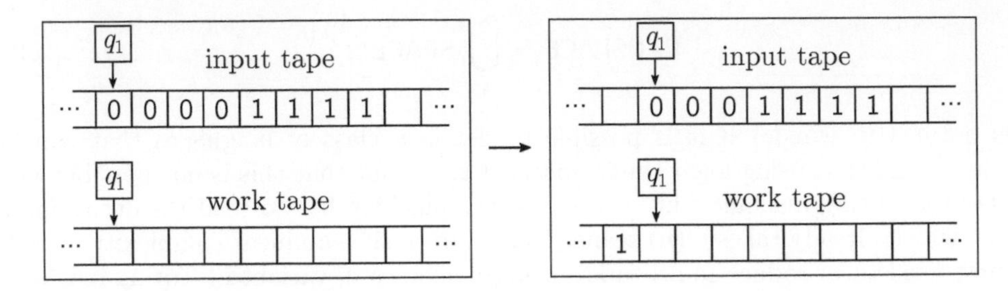

Fig. 2.7 A multi-tape Turing machine. A schematic of a multi-tape Turing machine that decides language $L = \{0^m 1^m : m \geq 0\} = \{\varepsilon, 01, 0011, 000111, \ldots\}$. The left panel shows the initial configuration of the machine with input string 00001111. The right panel shows a configuration of the machine during the computation. This multi-tape Turing machine has only one work tape.

the first cell of the respective tape. The output is stored on the first tape at the end of the computation.

More formally, a MDTM is a sequence $T = \{\Sigma, \Gamma, Q, \delta, q_1, q_\mathrm{a}, q_\mathrm{r}\}$, where the transition function is defined as

$$\delta : Q \times \Gamma^k \rightarrow Q \times \Gamma^k \times \{\mathrm{L}, \mathrm{R}, \mathrm{N}\}^k . \tag{2.14}$$

In multi-tape machines, the input of the transition function is the state q of the machine, and the symbols $(x_1, \ldots, x_k)$ read by each head. The output is the new state of the machine q', the symbols $(x'_1, \ldots, x'_k)$ each head has to overwrite, and the directions $(m_1, \ldots, m_k)$ each head has to take.

The functioning of a multi-tape Turing machine can be better understood with a concrete example. Consider the non-regular language

$$L = \{0^n 1^n : n \geq 0\} = \{\varepsilon, 01, 0011, 000111, \ldots\} . \tag{2.15}$$

While, as explained in Section 2.1.1, a DFA cannot recognize this language, a Turing machine with two tapes can do it. The operations of such a machine on a binary string can be summarized as follows:

1. The first head moves to the right until it reads the first 1; during this procedure, every time that the first head reads 0, it replaces this symbol with ␣ and moves right. At the same time, the second head writes 1 on the work tape and moves right.
2. As soon as the first head reads 1, the two heads simultaneously stop. The first head now starts moving to the right and the second head starts moving to the left. The machine checks that the two heads are reading the same symbol. If this is not the case, the computation stops in a reject state.
3. If the two heads reach a blank symbol at the same time, the computation stops in an accept state.

The transition function of this multi-tape machine is described by the following table:

current state	symbols read		new state	symbols overwritten	movements of the heads
q_1	$(\sqcup, \sqcup)$		q_a	$(\sqcup, \sqcup)$	(N, N)
q_1	$(0, \sqcup)$		q_2	$(0, \sqcup)$	(N, N)
q_1	$(1, \sqcup)$		q_r	$(1, \sqcup)$	(N, N)
q_2	$(0, \sqcup)$		q_2	$(\sqcup, 1)$	(R, R)
q_2	$(1, \sqcup)$	$\mapsto$	q_3	$(1, \sqcup)$	(N, L)
q_3	$(0, 1)$		q_r	$(0, 1)$	(N, N)
q_3	$(\sqcup, 1)$		q_r	$(\sqcup, 1)$	(N, N)
q_3	$(1, \sqcup)$		q_r	$(1, \sqcup)$	(N, N)
q_3	$(1, 1)$		q_3	$(\sqcup, \sqcup)$	(R, L)
q_3	$(\sqcup, \sqcup)$		q_a	$(\sqcup, \sqcup)$	(N, N)

As an example, suppose that the input string is $\mathbf{x} = 00001111$ as shown in Fig. 2.7. Then, the input tape and the work tape will go through the configurations

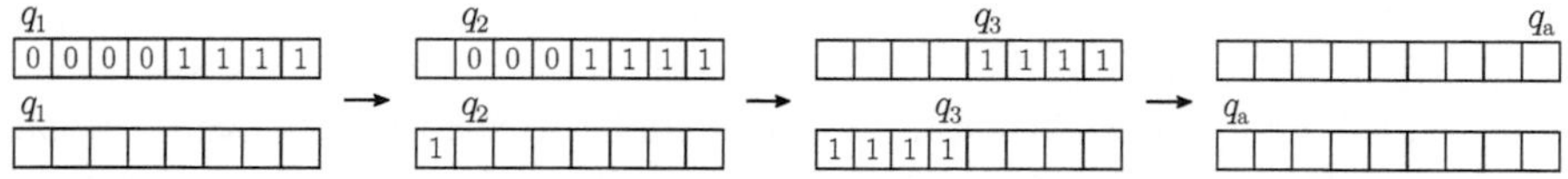

q_1: 0 0 0 0 1 1 1 1 / q_1: (blank) $\rightarrow$ q_2: $\sqcup$ 0 0 0 1 1 1 1 / q_2: 1 $\rightarrow$ q_3: 1 1 1 1 / q_3: 1 1 1 1 $\rightarrow$ q_a: (blank) / q_a: (blank)

In the worst case, the first head moves to the right n times, where n is the number of symbols in the input string. Hence, the running time of this computation is linear. How much space does this machine use? For multi-tape Turing machines, we define space as the number of cells scanned on the k work tapes during the computation. In our example, the second head examines approximately $n/2$ cells before reaching an accept state. Therefore, the space used is linear.

Can we do any better than this? Can we decide this language using a multi-tape Turing machine that scans less than $n/2$ work cells? Yes, we can. In principle, the second head can annotate the number of 0's read by the first head in *binary* representation. In the example above, the second head could write 100 (the binary representation of four) instead of four 1's. This would save some work cells. Adopting this strategy, the number of work cells scanned by the machine becomes $\approx \log_2(n/2)$ instead of $n/2$. This argument indicates that the language defined in (2.15) belongs to L, the set of languages that can be decided by a MDTM scanning a logarithmic number of work cells, $\mathsf{L} = \mathsf{SPACE}\{O(\log(n))\}$.

Can we decide this language using a Turing machine with one tape instead of two? This is indeed possible, but the number of steps taken by the machine is higher. A single-tape Turing machine can decide this language in the following way:

1. The head starts from the leftmost symbol recorded on the input tape. If it is 0, it overwrites $\sqcup$, moves to the end of the string, and checks that the rightmost symbol is 1. If this is not the case, the computation stops in a reject state.
2. If the rightmost symbol is 1, the head overwrites $\sqcup$ and moves to the left until it reaches a blank symbol.
3. The machine checks the symbol on the right side of the blank symbol. If this is also a blank symbol, the computation stops in an accept state. If there is a 1,

the computation stops in a reject state. If it finds a 0, the machine reiterates the entire procedure from the first step.

This single-tape machine takes many more steps compared to the multi-tape machine presented before. Since the head has to move back and forth n times, the number of steps executed scales quadratically[8]. This simple example shows that a multi-tape TM can offer a speed-up over single-tape TMs. In general, it can be shown that if a multi-tape Turing machine decides a language in $t(n)$ steps, it is always possible to build a single-tape Turing machine that decides the same language in $O(t^2(n))$ steps [29].

Does there exist a language that can be decided by a single-tape TM, but not by a multi-tape TM? No. If a language can be decided by a single-tape TM, it is always possible to build a multi-tape TM that decides the same language and *vice versa* [29]. For this reason, we say that multi-tape TMs are **equivalent** to single-tape ones even if they can solve some problems with a quadratic speed-up over single-tape machines.

2.2.7 Probabilistic Turing machines

Given a machine state and an input symbol (q_1, x_1), the transition function of a deterministic Turing machine will always output the same triplet (q_2, x_2, m_2) where q_2 is the new machine state, x_2 is the new symbol that the head has to overwrite on the tape and m_2 is the movement that the head has to take. We call this property **determinism**. Dropping this deterministic feature of TM can speed up the computation for some problems at the cost of reaching the wrong solution in some instances. In a **probabilistic finite-state Turing machine** (PTM), given a machine state and an input symbol (q_0, x_0), the transition function assigns a probability to each triplet (x_1, q_1, m_1), ..., (x_N, q_N, m_N). It is not possible to predict with certainty which step the machine will take based on the machine state and the symbol read by the head. In a PTM, the transition function is defined as

$$\delta : Q \times \Gamma \times Q \times \Gamma \times \{\mathrm{L}, \mathrm{R}, \mathrm{N}\} \to [0,1] \,. \tag{2.16}$$

If a PTM is in a configuration $\mathsf{c} = \mathbf{x}^q\mathbf{y}$, the machine will move to a new configuration c' with probability $\delta(q, y_1, q', y_1', m)$ where q' is the new machine state, y_1' is the symbol the head will overwrite and m is the movement the head will take.

For example, suppose that a PTM receives $\mathbf{x}$ as an input string. As illustrated by Fig. 2.8, the machine will move from one configuration to another with a certain probability. The main novelty of PTMs is that the machine can randomly move from one configuration to another and the computation tree has several branches (contrary to deterministic machines whose computation tree has a single branch). The branches are divided into accepting and rejecting. A branch is an **accepting branch** b_a (or a **rejecting branch** b_r) if the last node of the branch is an accept state q_a (or a reject state q_r). The probability $p(b)$ associated with a branch b is the product of the probabilities from the first edge to the last edge since the crossings of the various edges in b are statistically independent events. If we start a PTM in the initial configuration ${}^{q_1}\mathbf{x}$ and repeat the computation sampling the transition function several times, the

[8] Note that it is possible to design a single-tape Turing machine that decides this language in $O(n \log n)$ steps, see Ref. [29].

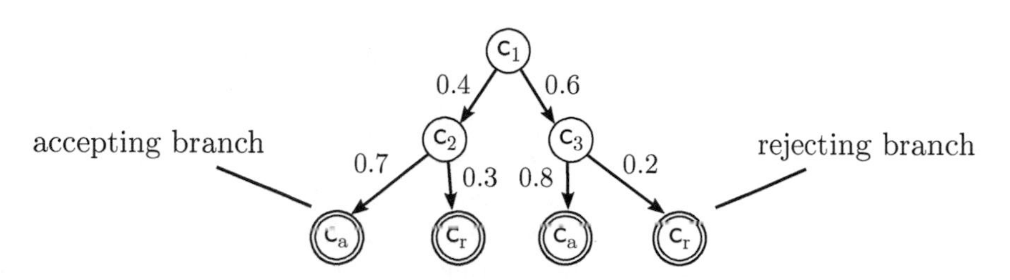

Fig. 2.8 Computation tree of a probabilistic Turing machine. Each node of this graph represents a configuration of the PTM. The number on each arrow indicates the probability of transitioning from one configuration to another. This machine accepts the input string with probability $\sum_{b_a} p(b_a) = 0.4 \cdot 0.7 + 0.6 \cdot 0.8 = 0.76$ and rejects it with probability $1 - \sum_{b_a} p(b_a) = 0.24$.

output will sometimes be an accept state and other times will be a reject state. It is not possible to predict a priori in which state the computation will terminate. We thus define the probability that a PTM accepts an input string $\mathbf{x}$ as

$$p(\, q_a \mid \mathbf{x}\,) = \sum_{b_a} p(b_a) \qquad \text{where } b_a \text{ are accepting branches.} \tag{2.17}$$

The probability that a PTM rejects an input string is given by $p(\, q_r \mid \mathbf{x}\,) = 1 - p(\, q_a \mid \mathbf{x}\,)$. We say that a probabilistic Turing machine T **decides** a language L if these two conditions are satisfied:

$$\text{if } \mathbf{x} \in L, \text{ then } T \text{ accepts } \mathbf{x} \text{ with high probability } p(\, q_a \mid \mathbf{x}\,) \geq 2/3,$$
$$\text{if } \mathbf{x} \notin L, \text{ then } T \text{ rejects } \mathbf{x} \text{ with high probability } p(\, q_r \mid \mathbf{x}\,) \geq 2/3.$$

The number $2/3$ is merely conventional. Indeed, it can be shown that if p_1 is a fixed constant in the interval $0.5 < p_1 < 1$ and a probabilistic Turing machine T_1 decides a language with success probability p_1 in polynomial time, then it is possible[9] to build a probabilistic Turing machine T_2 that decides the same language in polynomial time with success probability $p_2 > p_1$. This is the so-called **amplification lemma** [29]. The running time of a probabilistic Turing machine is the length of the longest path in the computation tree. The number of repetitions determines the accuracy of the measured success probability.

The class of languages that can be decided by a PTM in $O(t(n))$ steps is called $\mathsf{BPTIME}(t(n))$. We are ready to introduce the **bounded-error probabilistic polynomial** time class BPP, an important class in complexity theory. This class contains all languages that can be decided by a PTM with a number of steps polynomial in the input size and a success probability greater than $2/3$,

[9]The important point here is that the error probability is a fixed quantity that does not depend on the input size.

$$\mathsf{BPP} = \bigcup_k \mathsf{BPTIME}(n^k). \tag{2.18}$$

The class BPP is the probabilistic version of the deterministic class P. It is intuitive that $\mathsf{P} \subseteq \mathsf{BPP}$, since DTMs are special cases of probabilistic Turing machines.

An interesting complexity class closely related to BPP is PP. A language $L \in \mathsf{PP}$, if a PTM efficiently accepts strings in L with a probability greater than $1/2$ and rejects strings not in L with a probability smaller or equal to $1/2$. In principle, the probability of acceptance can scale as $1/2 + 1/2^n$, where n is the size of the input string. Since the probabilities of accepting or rejecting a string can be exponentially close, problems in PP lack efficient error amplification: one might need to repeat the computation an exponential number of times to gather enough statistics and decide whether the string is accepted or not. Since the condition of acceptance has been relaxed, it is evident that $\mathsf{BPP} \subseteq \mathsf{PP}$.

Is there a language that can be decided by a probabilistic Turing machine, but cannot be decided by a deterministic Turing machine? No. If a language can be decided by a PTM, it is always possible to build a DTM that decides the same language and *vice versa*. Thus, PTM and DTM are equivalent. However, there are some decision problems that a PTM can solve in polynomial time, but it is still unclear if a DTM can solve them in polynomial time. An interesting case is PRIMES, presented in Section 1.2.2. For many years this problem was known to be in BPP, but not in P. Only in 2002, did Agrawal, Kayal, and Saxena discover a deterministic algorithm that solves PRIMES in polynomial time [55].

2.3 Circuit model

The Turing machine is one of many computational models that can be used to solve a computational problem. Another important computational model is the **circuit model**. Circuits are graphical representations of **straight-line program**. These programs only contain assignment statements and do not contain loops, branches, or conditional statements. A straight-line program can be decomposed into input steps, computation steps, and output steps:

(s READ x): in this input step, the program reads the input symbol x. Here, s indicates the step number of the computation.

(s OP $s_1 \dots s_n$): in this computation step, the program performs the operation OP on the variables $s_1 \dots s_n$ elaborated in previous steps. Here, s still indicates the step number of the computation.

(s OUTPUT s_i) : in this output step, the program outputs a variable s_i; again s is the step number.

Each part of the program must be executed in a precise order (input steps must precede output steps, for instance). Let us present a simple example of a straight-line program called **adder** (this example will clarify how this computational model works in practice). The adder performs the sum modulo 2 of two binary digits. Consider the computation

step number	operation	in/output
(1	READ	x_0)
(2	READ	x_1)
(3	XOR	1 2)
(4	AND	1 2)
(5	OUTPUT	3)
(6	OUTPUT	4).

In the first two steps, the program reads the input bits; in the third and fourth step, the program operates a XOR and an AND on the two input bits; finally, the program outputs two bits in steps 5 and 6. The circuit representing this straight-line program is illustrated in Fig. 2.9a. Each line represents one bit and the sequence of gates must be read from left (input) to right (output). The reader can check that if the input of the circuit is $x_0 = 0$ and $x_1 = 1$, then the sum is $s = 1$ and the carry is $c = 0$.

A **circuit** C is a **directed acyclic graph** (DAG) representing a straight-line program. If the alphabet used by the program is $\mathbb{B}$ and the program executes Boolean operations only, the circuit is called a **logic circuit**. Each circuit node (apart from input and output nodes) is labeled with an operation. The **size** of a circuit, SIZE(C), is the number of nodes in the circuit, excluding input and output nodes. The **depth** of a circuit, DEPTH(C), is the length of the longest path connecting an input node to an output node. A **basis** of a circuit is the set of gates used in the circuit and the standard basis is {AND, OR, NOT, COPY}.

In Section 1.1.7, we demonstrated that any function $f : \mathbb{B}^n \to \mathbb{B}^m$ can be expressed with a combination of {AND, OR, COPY, NOT}. It can be shown that these functions can be computed by a logic circuit [70]. We will not prove this result rigorously, limiting ourselves only to the presentation of a simple example. Consider the function $f : \mathbb{B}^2 \to \mathbb{B}^2$ defined as

$\mathbf{x}$			$f(\mathbf{x})$	
x_1	x_0		x'_1	x'_0
0	0		0	1
0	1	$\mapsto$	0	1
1	0		1	0
1	1		1	0

This function can be expressed with minterms,

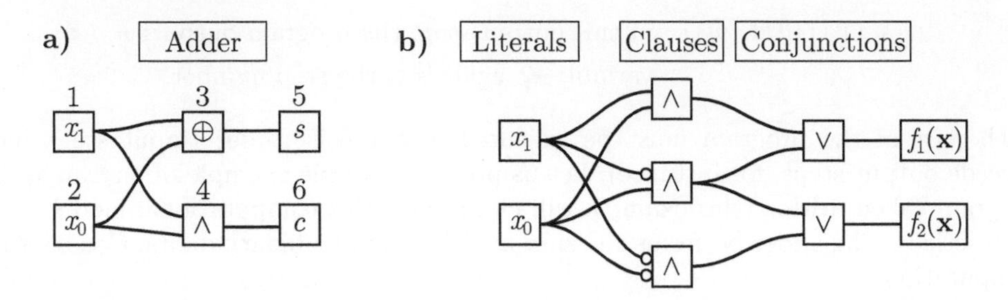

Fig. 2.9 Circuits. a) A directed acyclic graph with labels representing a straight-line program called adder. In this circuit, each number associated with a node indicates the corresponding step of the straight-line program. This circuit has $\mathsf{SIZE}(C) = 2$ and $\mathsf{DEPTH}(C) = 2$. Observe that the COPY and NOT operations are not usually considered in the estimate of SIZE and DEPTH. **b)** A logic circuit with two inputs and two outputs that computes the function in eqn (2.19). This circuit has $\mathsf{SIZE}(C) = 5$ and $\mathsf{DEPTH}(C) = 3$.

$$\begin{aligned} f(\mathbf{x}) = (x_1', x_0') &= (\, m_{10}(\mathbf{x}) \vee m_{11}(\mathbf{x}),\ m_{00}(\mathbf{x}) \vee m_{01}(\mathbf{x})\,) \\ &= (\,(x_1 \wedge \overline{x}_0) \vee (x_1 \wedge x_0)\,,\ (\overline{x}_1 \wedge \overline{x}_0) \vee (\overline{x}_1 \wedge x_0)\,). \end{aligned} \tag{2.19}$$

The logic circuit that computes this function is shown in Fig. 2.9b. This example suggests that given a function $f : \mathbb{B}^n \to \mathbb{B}^m$, it is always possible to construct a logic circuit with a sufficient number of input and output wires that computes f.

Unlike Turing machines, a circuit can only process input strings of a certain length because the number of input wires in a circuit is fixed. To overcome this limitation and construct a computational model based on logic circuits that resembles the infinite length of a Turing machine tape, it is necessary to introduce the concept of a circuit family. A **circuit family** is an infinite sequence of circuits $(C_0, C_1, C_2, \ldots)$ where the circuit C_n has n input wires and a finite number m_n of output wires. The output string will be indicated with the notation $C_n(\mathbf{x})$ where $\mathbf{x}$ is an input string with n bits. Now consider a circuit family $(C_i)_i$, a string $\mathbf{x}$ with s symbols and a number $n > s$. If the output string $C_n(\mathbf{x})$ (suitably padded with a sequence of 0's) is equal to $C_s(\mathbf{x})$, then we say that the circuit family is **consistent**.

The requirements discussed above are not restrictive enough to build a physical computational model; a consistent circuit family can in principle be used to solve any kind of problem, even the halting problem[10]! To address this issue, it is necessary to add a constraint. We say that a circuit family $(C_i)_i$ is **uniform** if there exists a DTM that takes the binary representation of n and outputs a binary description of C_n for all circuits in the family. In other words, a circuit family is uniform if there exists a DTM capable of describing the functioning of each circuit in the family, including the number of input and output wires, the logic gates, and the interconnections between them. This description provides all the information required to build the actual circuit.

[10]Consider the halting language restricted to n bit strings, $\mathsf{HALTING} \subseteq \mathbb{B}^n$. The associated Boolean function $\chi_{\mathsf{HALTING}}(\mathbf{x})$ transforms an n bit string into a bit. We showed that these types of functions can be computed by a circuit C_n. Thus, a circuit family $(C_i)_i$ can solve the halting problem.

Uniform and consistent circuit families can be used to precisely define what it means for a circuit to decide a language. A uniform and consistent circuit family $(C_i)_i$ **decides** a language $L \subseteq \mathbb{B}^*$ when

$$C_n(\mathbf{x}) = \begin{cases} 1 & \text{if } \mathbf{x} \in L \text{ and } |\mathbf{x}| = n \\ 0 & \text{if } \mathbf{x} \notin L \text{ and } |\mathbf{x}| = n. \end{cases}$$

The **size complexity** of a circuit family $f(n)$ is a function that returns the number of gates of each circuit in the family $(C_i)_i$,

$$f(n) = \mathsf{SIZE}(C_n). \tag{2.20}$$

The circuits in a family usually include gates taken from a universal set such as $\{\mathsf{NOT}, \mathsf{AND}, \mathsf{OR}, \mathsf{COPY}\}$.

The following theorem outlines a connection between the time required by a deterministic TM to decide a language and the number of gates required by a circuit family to decide the same language.

Theorem 2.2 If a DTM decides a language in $t(n)$ steps where $t(n) \geq n$, then there exists a uniform and consistent circuit family $(C_i)_i$ with size complexity $f(n) \in O(t^2(n))$ that decides the same language.

The proof of this theorem can be found in Ref. [29]. It can also be shown that a language can be decided by a deterministic Turing machine if and only if it can be decided by a uniform and consistent circuit family. Thus, these two computational models are equivalent.

2.4 Church–Turing thesis

In the previous sections, we understood that a function is computable by a single-tape Turing machine if and only if it is computable by a multi-tape TM (or a probabilistic TM) and vice versa. This is valid for circuits too: a function is computable by a TM if and only if it is computable by a uniform and consistent circuit family. Throughout history, many other computational models have been proposed and investigated, such as Markov algorithms, μ-recursive functions, and random access machines [78, 79, 80]. It has been demonstrated that they are all equivalent to a deterministic Turing machine, meaning that all these models can compute the same class of functions. This observation led to the **Church–Turing thesis** [81, 1, 82]:

Any physical computational model is equivalent
to a deterministic Turing machine.

The Church–Turing thesis claims that if a problem cannot be decided by a Turing machine (such as the HALTING problem), then it cannot be decided by any other physical machine. Some decision problems are intrinsically unsolvable regardless of the computational model adopted. It can be demonstrated that any deterministic computational model takes the same number of steps (up to a polynomial) to solve a given decision problem. For this reason, deterministic computational models are said to be **polynomially equivalent**.

2.4.1 The extended Church–Turing thesis

The Church–Turing thesis claims that if a Turing machine cannot compute a function, it cannot be computed by any other physical machine. This thesis, however, does not assert anything about the computation speed. Is it possible to build a machine that decides a language much faster than a deterministic Turing machine? As discussed in Section 2.2.7, probabilistic Turing machines sometimes offer a speed-up over deterministic ones. In the last century, many other computational models have been developed and many of them do not provide a systematic exponential speed-up over probabilistic machines. This observation is the basis of the **extended Church–Turing thesis** [83]:

Any physical computational model can be efficiently simulated by a probabilistic Turing machine.

Here, the key word is *physical.* In fact, it is possible to imagine some computational models that can outperform a PTM. Consider for example a deterministic machine that executes the first computation step in 1 second, the second computation step in 0.1 seconds, the third computation step in 0.01 seconds and so on. This machine can potentially offer an exponential speed-up over a probabilistic machine, since each step would take less time as the computation progresses. Unfortunately, this type of machine is not physical at all. After 100 computation steps, this hypothetical machine would execute each step in less than 10^{-100} seconds. Elementary physical arguments suggest that such a machine cannot be constructed (it would require an impractically large amount of energy to resolve such short time intervals).

This is only part of the story though. At the end of the last century, a group of scientists advanced the hypothesis that a quantum computer, i.e. a Turing machine governed by the laws of quantum physics, can in principle solve some decision problems with an exponential speed-up with respect to PTMs [11, 13, 25, 16]. There are some preliminary signs that PTMs cannot efficiently simulate quantum machines and the extended Church–Turing thesis is not valid [20, 84, 85]. We will present quantum Turing machines in Section 2.6.

2.5 Non-deterministic machines

Non-deterministic machines are profoundly different from deterministic machines. A non-deterministic machine differs from a DTM in that the machine can move from one configuration to multiple configurations at the same time. Another way of thinking about non-deterministic machines is to imagine a probabilistic Turing machine in which the machine always guesses the right path through the computation tree. Clearly, this interpretation precludes the possibility of building a physical non-deterministic machine and one often reads that they are “magical” [86] or “hypothetical” [87]. Nevertheless, they are important from a theoretical point of view to examine the limitations of physical machines. We first explain non-deterministic automata and

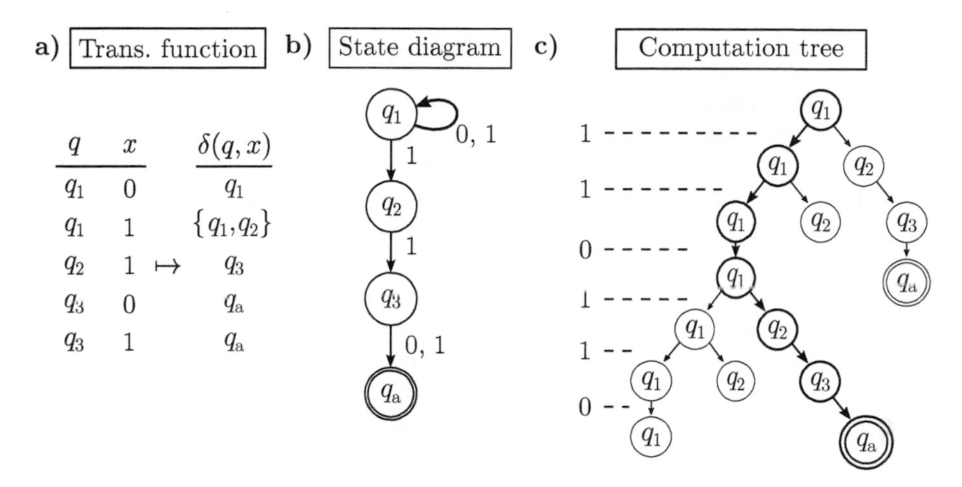

Fig. 2.10 The non-deterministic finite-state automaton A. **a)** The transition function of an NFA and **b)** its associated state diagram. Observe that the same label may be assigned to two edges leaving a node. **c)** Computation tree of NFA A processing the input string 110110. An accepting branch is indicated with a thick line.

present a simple example. We will then introduce non-deterministic Turing machines in Section 2.5.2.

2.5.1 Non-deterministic automata

Non-deterministic finite-state automata (NFA) are very similar to DFA. They have a set of states, an alphabet, an initial state, and a set of accept states. The main difference between these two machines lies in the transition function. In Section 2.1, we learned that DFAs carry out a computation in a systematic way: the machine moves from one state to another according to a predetermined set of rules. This process is deterministic, i.e. repeating this procedure always leads to the same result. For NFA, the situation is slightly different: given a machine state and an input symbol, the transition function outputs **one or more** possible states. This allows an NFA to explore different computation branches at the same time[11]. The transition function of a non-deterministic machine is given by

$$\delta : Q \times \Sigma_\varepsilon \to \mathcal{P}(Q) \tag{2.21}$$

where the symbol Σ_ε indicates that the machine's alphabet includes the empty string ε. Observe that the codomain of the transition function is $\mathcal{P}(Q)$, the power set of the machine states. In short, the formal definition of NFA is a sequence of five elements $\{\Sigma, Q, \delta, q_1, A\}$, where δ is now given by (2.21).

As an example, consider a non-deterministic machine A that decides the language

[11]This behavior is different from that of a probabilistic Turing machine. The computation performed by a PTM follows only one branch of the computation tree at a time.

$$L = \{\mathbf{x} \in \mathbb{B}^* : |\mathbf{x}| \geq 3 \text{ and } x_{n-2} = x_{n-1} = 1\} . \tag{2.22}$$

This language contains all binary strings that have the bits x_{n-2} and x_{n-1} equal to 1. The transition function and state diagram of A are illustrated in Fig. 2.10a–b. Suppose that the input string is $\mathbf{x} = 110110$. The automaton starts in state q_1 and the first input symbol is $x_1 = 1$. Which step will the automaton execute? The state diagram shows that there are two edges labeled 1 connected to state q_1, and one can interpret this situation by imagining that *the machine splits into two copies of itself.* One copy of the machine will move to state q_2 and another copy will stay in q_1. This behavior is represented by the computation tree illustrated in Fig. 2.10c. The second symbol is $x_2 = 1$: the machine will move from q_2 to q_3 and *at the same time* the copy of the machine that remained in state q_1 will split again into two parts. The machine now receives the remaining input symbols and the computation follows the steps shown in the computation tree of Fig. 2.10c. At the bottom of the graph, one branch ends with a node labeled q_a. This is an accept state. Hence, the string $\mathbf{x} = 110110$ is accepted by A_4.

In general, an input string is **accepted** by a non-deterministic automaton if there exists *at least one* accept branch in the computation tree that ends with an accept leaf once the *entire* input string has been processed[12]. An NFA rejects a string $\mathbf{x}$ if *every* path through the computation tree ends with a reject leaf. A language L is recognized by an NFA A when the following conditions are satisfied:

$$\begin{aligned} &\text{if } \mathbf{x} \in L, \text{ then } A \text{ accepts } \mathbf{x}, \\ &\text{if } \mathbf{x} \notin L, \text{ then } A \text{ rejects } \mathbf{x}. \end{aligned}$$

It is quite natural to wonder if there exists a language that can be recognized by a non-deterministic automaton, but not by a deterministic one. This never happens; these two machines are equivalent. However, NFA can potentially recognize a language with an exponential speed-up with respect to DFA since NFA can explore multiple paths at the same time.

2.5.2 Non-deterministic Turing machines

As the reader might expect, **non-deterministic finite-state Turing machines** (NTM) are the non-deterministic variant of DTM. Non-deterministic Turing machines still have an alphabet Σ, a tape alphabet Γ, an initial state, a reject state, and an accept state. The main difference is in the transition function δ: for single-tape NTM, this function is defined as

$$\delta : Q \times \Gamma \to \mathcal{P}(Q \times \Gamma \times \{\text{L}, \text{R}, \text{N}\}) . \tag{2.23}$$

[12] This point marks a significant difference between DFAs and NFAs: in a DFA the automaton can always process the entire string and end in a state, but in an NFA it is possible that the automaton "stalls" along a computation path before processing the entire string, rendering the end state of that path meaningless. For example, the computation tree in Fig. 2.10c contains a branch that ends with an accepting node before the entire string has been processed; this branch is not a valid accepting branch.

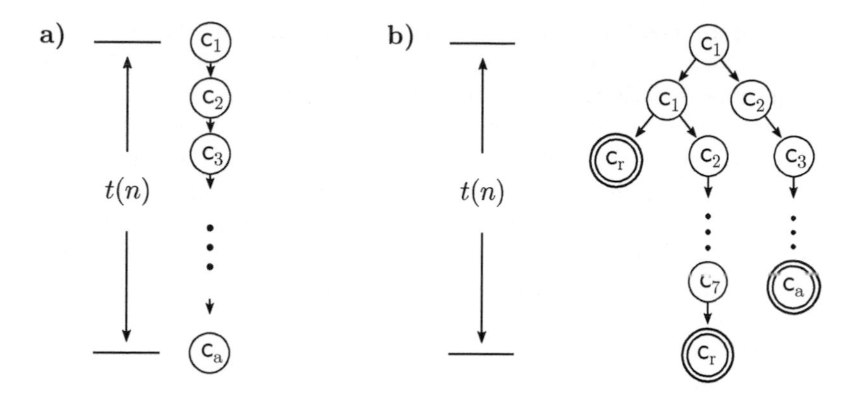

Fig. 2.11 The running time t(n). a) Computation tree of a DTM. **b)** Computation tree of an NTM. For an NTM the longest branch determines the running time of the computation. Since this computation tree has at least one accept leaf, the machine accepts the input string.

More precisely, a non-deterministic Turing machine is a 7-tuple T comprising:

1. an alphabet Σ that does not contain the blank symbol $\sqcup$,
2. a tape alphabet Γ such that $\sqcup \in \Gamma$ and $\Sigma \subset \Gamma$,
3. a non-empty finite set of states Q,
4. a transition function $\delta : Q \times \Gamma \to \mathcal{P}(Q \times \Gamma \times \{\mathrm{L}, \mathrm{R}, \mathrm{N}\})$,
5. an initial state $q_1 \in Q$,
6. an accept state $q_\mathrm{a} \in Q$,
7. a reject state $q_\mathrm{r} \in Q$, where $q_\mathrm{r} \neq q_\mathrm{a}$.

The computation tree of a non-deterministic Turing machine has more than one branch. If, given an input string $\mathbf{x} \in \Sigma^*$, there is at least one branch that ends in an accept state, the machine accepts the input string; if this is not the case, the input string is rejected. NTMs whose branches accept or reject an input string (without entering into an infinite loop) are called **deciders**. A language L is **decided** by an NTM when the following two conditions are satisfied:

if $\mathbf{x} \in L$, the computation tree has at least one accepting branch,
if $\mathbf{x} \notin L$, the computation tree does not have any accepting branch.

It can be shown that if a language can be decided by an NTM, it can also be decided by a DTM and vice versa.

When an NTM receives an input string $\mathbf{x}$ with n symbols, the various branches b of its computation tree will accept or reject it in a number of steps $t(b, n)$. The **running time** of an NTM is the function $t(n) = \max_b t(b, n)$, where the maximum is evaluated over all branches. Figure 2.11 compares the running time of a DTM with the running time of an NTM. In the latter case, the computation tree might have several rejecting branches and one or more accepting branches; we emphasize again that the longest branch determines the running time of the computation. The collection of all languages L that can be decided by an NTM in $O(t(n))$ steps is denoted $\mathsf{NTIME}(t(n))$. This is the non-deterministic version of the complexity class $\mathsf{TIME}(t(n))$.

It can be shown that if a single-tape NTM solves a decision problem in $O(t(n))$ steps, it is always possible to build a single-tape DTM that solves the same problem in $2^{O(t(n))}$ steps [29]; this suggests that NTMs can offer an exponential speed-up over deterministic machines. However, just like non-deterministic automata, non-deterministic Turing machines are not a physical computational model (a real machine cannot split into multiple copies during a computation!).

2.5.3 NP problems

Many problems in arithmetic and linear algebra belong to the complexity class P. This is the class of decision problems that a classical computer can solve with polynomial resources. Another important class of decision problems is the class NP. This class contains all languages whose *affirmative* solution can be verified in polynomial time by a DTM with the aid of an appropriate *certificate*.

To better understand this definition, we now present two simple examples. Consider 3COLOR, introduced in Section 1.2.8. In this problem, one has to determine if the nodes of a simple graph $G = (V, E)$ can be colored with three colors so that adjacent vertices do not share the same color. Thus, the goal is to find a function $f : V \to \{\text{green}, \text{red}, \text{blue}\}$ such that

$$v_1 \text{ and } v_2 \text{ are connected} \qquad \Longleftrightarrow \qquad f(v_1) \neq f(v_2). \tag{2.24}$$

In general, the number of steps required by a classical computer to find the correct coloring scales exponentially with the number of nodes of the graph, and therefore 3COLOR $\in$ EXPTIME. However, verifying that a given function f is a valid coloring is quite simple: one just needs to examine all edges (there are no more than n^2 where $n = |V|$) and check that $f(v_i) \neq f(v_j)$ for all adjacent nodes. If this is the case, then the graph belongs to the language 3COLOR and f is a valid certificate of the problem.

Similarly, in the 3SAT problem, one has to determine whether a Boolean function φ expressed in conjunctive normal form is satisfiable, i.e. there exists a binary string $\mathbf{x}$ such that $\varphi(\mathbf{x}) = 1$. While 3SAT is a difficult problem, verifying that a given string $\tilde{\mathbf{x}}$ satisfies $\varphi(\tilde{\mathbf{x}}) = 1$ is easy: one just has to evaluate $\varphi(\tilde{\mathbf{x}})$ and see if it is equal to one. If this is the case, φ belongs to the language 3SAT and $\tilde{\mathbf{x}}$ is a valid certificate of the problem. These two decision problems are important examples of NP problems.

The **NP class** is the collection of languages whose affirmative solution can be checked by a deterministic verifier in polynomial time. A **verifier** is a Turing machine T_{V} that exploits a binary string $\mathbf{c}$ (called the **certificate**) to check if a string $\mathbf{x}$ belongs to a language. More formally, a language $L \in$ NP if there exists a DTM T_V running in polynomial time such that:

1. $\forall \mathbf{x} \in L$, there exists a certificate $\mathbf{c} \in \Sigma^*$ such that[13] $T_{\text{V}}(\mathbf{x}, \mathbf{c}) = 1$,
2. $\forall \mathbf{x} \notin L$ and certificates $\mathbf{c} \in \Sigma^*$, $T_{\text{V}}(\mathbf{x}, \mathbf{c}) = 0$.

[13]With the notation $T_{\text{V}}(\mathbf{x}, \mathbf{c})$ we mean that the computation of the verifier starts with the strings $\mathbf{x}$ and $\mathbf{c}$ recorded on the tape and properly separated by a special symbol such as an asterisk.

The length of the certificate must be polynomial in input size, $|\mathbf{c}| = \text{poly}(|\mathbf{x}|)$. Almost all of the decision problems presented in Chapter 1 belong to the class NP.

Let us outline an alternative definition for the class NP. First, recall the definition of the non-deterministic time class $\mathsf{NTIME}(t(n))$, the class containing all languages L that can be decided by a non-deterministic Turing machine in $O(t(n))$ steps. The **class** NP can be defined as the set of languages decidable by an NTM with a number of steps polynomial in the input size,

$$\mathsf{NP} = \bigcup_{k=1}^{\infty} \mathsf{NTIME}(n^k)\,. \tag{2.25}$$

The two definitions of the NP class presented in this section are equivalent: the affirmative solution to a decision problem L can be verified by a DTM in polynomial time if and only if an NTM can decide L in polynomial time [29].

2.5.4 Reducibility

Creating connections between languages is important to characterize the resources required to solve some decision problems. These connections are provided by bijective functions that map strings of a language into strings of a different language. In a more formal way, we say that a language $L_1 \subseteq \Sigma^*$ is **mapping reducible** to $L_2 \subseteq \Sigma^*$, and we write $L_1 \leq_{\text{m}} L_2$, if there exists a computable function $f : \Sigma^* \to \Sigma^*$ that maps strings in L_1 into strings in L_2 and maps strings that are not in L_1 into strings that are not in L_2:

$$\begin{aligned} \mathbf{x} \in L_1 \quad &\overset{f}{\mapsto} \quad \mathbf{x}' \in L_2, \\ \mathbf{x} \notin L_1 \quad &\overset{f}{\mapsto} \quad \mathbf{x}' \notin L_2. \end{aligned}$$

The function f is called a **reduction**. Figure 2.12 illustrates a reduction between two languages. If such a function exists, checking that $\mathbf{x}$ belongs to L_1 is equivalent to verifying that $f(\mathbf{x})$ belongs to L_2. If f can be computed in polynomial time by a DTM, we say that L_1 is **polynomial time reducible** to L_2 and we write

$$L_1 \leq_{\text{p}} L_2. \tag{2.26}$$

The reduction f is called an **efficient mapping**. It can be shown that if L_2 is decidable and $L_1 \leq_{\text{m}} L_2$, then L_1 is decidable too.

An efficient reduction can be exploited to solve a decision problem by mapping it into a different problem that we already know how to solve. Suppose that we know how to decide a language L_2 in polynomial time with a DTM and suppose $L_1 \leq_{\text{p}} L_2$. This implies that deciding whether $\mathbf{x} \in L_1$ can be done by first transforming this string into $f(\mathbf{x}) = \mathbf{x}'$ and then checking if $\mathbf{x}'$ belongs to L_2. Since $L_2 \in \mathsf{P}$ and the implementation of an efficient reduction requires a polynomial number of steps, the language L_1 can be decided in polynomial time too. To summarize, it is evident that

$$\text{if } L_2 \in \mathsf{P} \text{ and } L_1 \leq_{\text{p}} L_2 \quad \Rightarrow \quad L_1 \in \mathsf{P}\,. \tag{2.27}$$

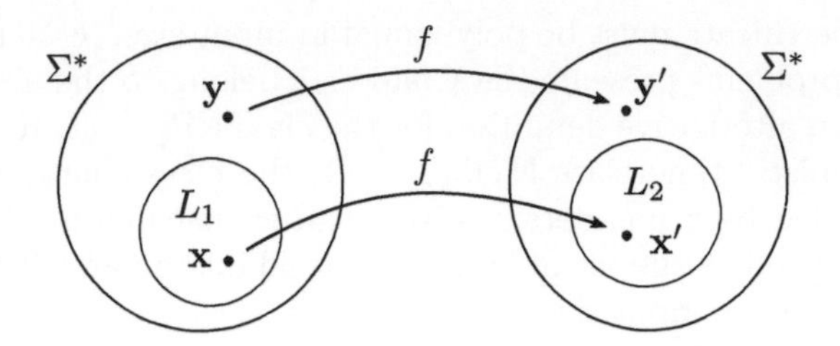

Fig. 2.12 Reducibility. Reduction f between two languages L_1 and L_2 defined over the same alphabet Σ. If the function f can be computed by a classical machine in polynomial time, we write $L_1 \leq_{\mathrm{p}} L_2$.

In complexity theory, reductions are essential to relate decision problems within the same class.

Example 2.7 Suppose we would like to know if there exists a string $\mathbf{x}$ such that the function

$$\varphi(\mathbf{x}) = \underbrace{(x_1 \vee x_1 \vee x_2)}_{1^{\text{st}}\text{ clause}} \wedge \underbrace{(\overline{x}_1 \vee \overline{x}_2 \vee \overline{x}_2)}_{2^{\text{nd}}\text{ clause}} \wedge \underbrace{(\overline{x}_1 \vee x_2 \vee x_2)}_{3^{\text{rd}}\text{ clause}} \tag{2.28}$$

is equal to 1. Let us show that this 3SAT problem can be mapped into CLIQUE. This can be done by first converting our satisfiability problem into a connected graph, as shown in Fig. 2.13. The three clauses are grouped together and a segment connects each literal to all others with some constraints: a literal cannot be connected to its negation and literals in the same clause are not connected. This prescription generates the connected graph shown in Fig. 2.13. Does this graph contain a complete subgraph with three nodes? We shall see that answering this question provides a solution to our original satisfiability problem. It is easy to see that the nodes $(x_2, \overline{x}_1, x_2)$ are connected by a complete subgraph, the one bolded in Fig. 2.13. If we now assign 1 to these literals, we obtain $(x_2 = 1, \overline{x}_1 = 1, x_2 = 1)$ which corresponds to the assignment $x_1 = 0$, $x_2 = 1$. Interestingly, this is also the solution to eqn (2.28),

$$\varphi(\mathbf{x}) = (0 \vee 0 \vee 1) \wedge (1 \vee 0 \vee 0) \wedge (1 \vee 0 \vee 0) = 1 \wedge 1 \wedge 1 = 1\,. \tag{2.29}$$

One can verify that this procedure holds for Boolean functions with any number of clauses, and it suggests that any 3SAT instance can be mapped into a CLIQUE problem.

2.5.5 NP complete problems

The concept of reduction is crucial to understanding NP **complete problems** (NPC). NPC is a subset of problems in NP to which *all* NP problems can be reduced in polynomial time. More formally, a language $L \in \mathsf{NPC}$ if:

1. $L \in \mathsf{NP}$,
2. $\forall L_1 \in \mathsf{NP}$, $L_1 \leq_{\mathrm{p}} L$.

NPC problems can be considered the most difficult problems in the class NP. If an efficient algorithm for a NPC problem were found, it could be adapted to solve all other NP problems efficiently. The existence of such an algorithm would prove that

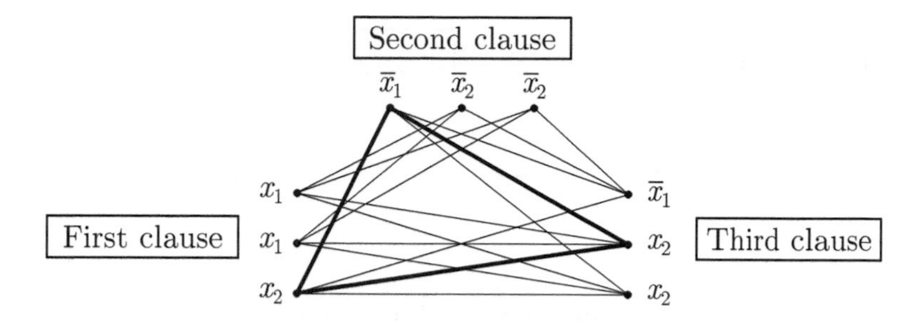

Fig. 2.13 An example of a reduction. A connected graph mapping the 3SAT problem of eqn (2.28) into an instance of the CLIQUE problem. The complete subgraph in bold is one solution of the CLIQUE problem.

P = NP. At the time of writing, a deterministic algorithm that efficiently solves an NPC problem has never been found.

It is quite intuitive that if a NPC language L_1 is polynomial-time reducible to L_2, then L_2 is also NPC. In mathematical terms

$$\text{if } L_1 \in \mathsf{NPC} \text{ and } L_1 \leq_{\mathrm{p}} L_2 \quad \Rightarrow \quad L_2 \in \mathsf{NPC}\,. \tag{2.30}$$

The first language to be identified as NPC was SAT, a result known as the Cook–Levin theorem [88]. Richard Karp subsequently identified a list of twenty-one NPC problems [65].

How does the NP class relate to other complexity classes? It is evident that $\mathsf{P} \subseteq \mathsf{NP}$, because a DTM can simply ignore the certificate for a P problem and solve it in polynomial time. It is conjectured (though not proven) that $\mathsf{P} \subset \mathsf{NP}$. If this is the case, then according to Ladner's theorem [89] there exists at least one NP problem that does not belong to either P or NPC. These types of problems are called NPI (or NP **intermediate**). FACTORING and some versions of GI are believed to be NPI. Furthermore, $\mathsf{NP} \subseteq \mathsf{PSPACE}$, meaning that any problem in NP can be solved by a deterministic machine that uses polynomial space and an unlimited amount of time. This follows from the fact that a deterministic machine can solve NP problems by trying all possible solutions. This brute-force approach takes considerable time, but a limited amount of space[14]. Lastly, the relation between NP and BPP (the class of languages that can be decided in polynomial time by a probabilistic machine with a small error) is unknown. Collecting these results together, we arrive at the relations

$$\mathsf{P} \subseteq \mathsf{NP} \subseteq \mathsf{PSPACE} \subseteq \mathsf{EXPTIME} \subseteq \mathsf{EXPSPACE}. \tag{2.31}$$

We will encounter more complexity classes in Section 2.7.1 once we introduce quantum Turing machines.

[14]For instance, a deterministic machine can solve 3SAT by writing a binary string $\mathbf{x}_1$ on a tape and checking if $f_{\mathsf{3CNF}}(\mathbf{x}_1) = 1$. If this is not the case, the machine can erase the string $\mathbf{x}_1$ and reuse the same cells for another binary string $\mathbf{x}_2$. This procedure takes a lot of time, but only $O(n)$ space, indicating that 3SAT (like any other NP language) is in PSPACE.

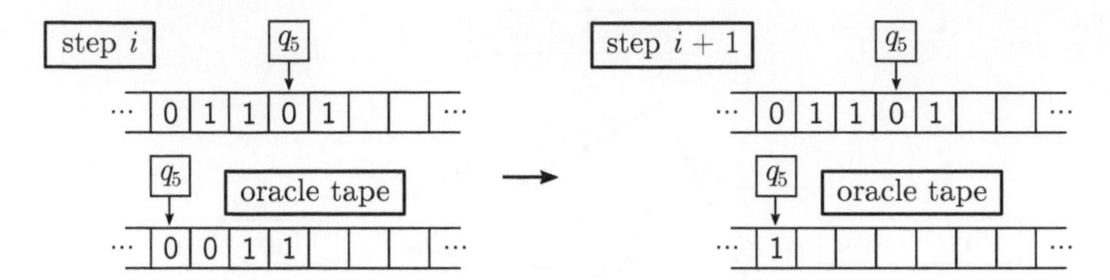

Fig. 2.14 An oracle Turing machine. Example of a Turing machine with an oracle for the language $L = \{0^n 1^n : n \geq 0\}$ of eqn (2.3). In one step, the oracle correctly states that the string 0011 belongs to this language.

2.5.6 Oracle Turing machines

Oracles are an important concept in classical and quantum complexity theory. Oracles are black boxes that an algorithm can query to obtain the solution to a problem in a single step. The internal structure of these black boxes is irrelevant; the important thing is that they always give the correct answer and can be queried an unlimited number of times. As we shall see in Chapter 10, early quantum algorithms were primarily concerned with minimizing the number of queries of an oracle within a computational program.

An **oracle Turing machine** (OTM) [90, 91] is a deterministic, probabilistic, or non-deterministic Turing machine with two tapes. The first tape is the usual input tape on which the input and output of the computation are stored. The second tape is an **oracle tape** with an associated **oracle function**[15] χ_L. This tape is provided with a head that can move in both directions, as shown in Fig. 2.14. During the computation, the machine can write a binary string $\mathbf{x}$ on the oracle tape and ask the oracle whether this string belongs to a language L. The machine takes n steps to write a string with n symbols on the oracle tape. In a single step, the oracle reads the string and outputs 1 if $\mathbf{x} \in L$ and 0 otherwise. The computation then proceeds taking into account the oracle's response.

We will use the notation T^L to indicate a Turing machine T that has access to an oracle for the language L. For these types of machines, the function **time** $t(n)$ indicates the number of steps taken by the computation, where one oracle consultation counts as one step. On the other hand, the function **space** $s(n)$ indicates the number of cells scanned on the input tape. As mentioned before, it is not important to specify the internal structure of the oracle and it is not even required that this is a physical machine [90].

Let us introduce the complexity classes associated with oracle Turing machines. For pedagogical reasons, first consider a deterministic machine T^{3SAT}, i.e. a deterministic Turing machine with an oracle for the 3SAT problem. The set of all languages that this machine can decide in polynomial time is denoted as $\mathsf{P}^{\mathsf{3SAT}}$. This notation can be easily extended as follows:

[15]It is not required for this function to be computable.

- P^{NP} is the class of problems decided in polynomial time by a deterministic TM that has access to an oracle that can solve any NP problem in one step.
- $\mathsf{EXPTIME}^{\mathsf{NP}}$ is the class of problems decided in exponential time by a deterministic TM that has access to an oracle that can solve any NP problem in one step.
- $\mathsf{NP}^{\mathsf{NP}}$ is the class of problems decided in polynomial time by a non-deterministic TM that has access to an oracle that can solve any NP problem in one step.

Oracle machines will be relevant when we discuss quantum algorithms in Chapter 10.

As explained in Section 1.2.6, the language

$$\overline{\mathsf{HAMCYCLE}} = \{\langle G \rangle \in \mathbb{B}^* : G \text{ does not have a Hamiltonian cycle}\} \qquad (2.32)$$

may not be in NP. However, a deterministic machine T^{HAMCYCLE} can easily decide this language. The encoding of the graph is fed into the oracle. If the oracle's output is 1, then the graph has a Hamiltonian cycle implying that $G \notin \overline{\mathsf{HAMCYCLE}}$. If, instead, the output of the oracle is 0, $G \in \overline{\mathsf{HAMCYCLE}}$.

We will now present the formal definition of a quantum Turing machine based on the work of Bernstein and Vazirani [16]. We decided to present quantum Turing machines in this chapter so as not to interrupt the discussion on computational models. However, we recognize that the arguments presented in the next sections require a thorough understanding of quantum mechanics. We therefore advise our readers to read Chapters 3, 4, and 5 and come back to this part only when they are familiar with the necessary quantum theory.

2.6 Quantum Turing machines

The computation executed by a deterministic machine can be thought of as a sequence of mechanical operations governed by the laws of classical physics. Nowadays, we know that the behavior of matter is described by quantum mechanics, a more fundamental theory of nature. It is natural to wonder if a "quantum machine" (a variant of a Turing machine that performs a computation under the constraints of quantum theory) is more powerful than a deterministic one. This quantum Turing machine (QTM) consists of the same components as a standard DTM: an alphabet, a tape, and a set of states. Its transition function, however, can exploit the superposition principle and transform an initial configuration c into a superposition of configurations $\alpha_1\mathsf{c}_1 + \ldots + \alpha_n\mathsf{c}_n$. The main novelty here is that the configurations appearing in the state $\alpha_1\mathsf{c}_1 + \ldots + \alpha_n\mathsf{c}_n$ coexist at the same time, and the coefficients α_i are probability amplitudes satisfying the normalization condition $\sum_i |\alpha_i|^2 = 1$. This feature seems to suggest that a QTM is equivalent to a non-deterministic Turing machine since multiple configurations can be explored at the same time. A big caveat is the fact that a measurement of the superposition state $\alpha_1\mathsf{c}_1 + \ldots + \alpha_n\mathsf{c}_n$ produces a single configuration c_i with probability $|\alpha_i|^2$ and therefore not all the information contained in this superposition can be accessed with a single measurement. This observation outlines a clear distinction between QTMs and NTMs. It is conjectured (but not proven) that quantum Turing machines are not as powerful as non-deterministic Turing machines. To put it another way, it is widely accepted that quantum computers will never solve NPC problems in polynomial time [92]. The good news is that quantum computers can simulate the time evolution of a broad class of quantum systems with an exponential speed-up over classical algorithms. This point has important implications in many scientific fields, including quantum chemistry.

A **quantum Turing machine** is a variant of a Turing machine in which the evolution from one configuration to another is described by a unitary operator [13]. A quantum Turing machine is therefore a 7-tuple $(\Sigma, \Gamma, Q, \delta, q_1, q_\text{a}, q_\text{r})$ where the quantum transition function δ is now given by[16]

$$\delta : \quad Q \times \Sigma \times Q \times \Sigma \times \{\text{L}, N, \text{R}\} \quad \rightarrow \quad \tilde{\mathbb{C}}.$$

This function assigns a complex number to a transition from one configuration to another as shown in Fig. 2.15. Recall that a configuration of a Turing machine can be indicated with a vector $\mathsf{c} = (q, \mathbf{x}, z)$, where $q \in Q$ is the state of the machine, $\mathbf{x} \in \Gamma^*$ is the content of the tape, and $z \in \mathbb{Z}$ is the index of the cell the head is pointing to. To keep our notation consistent with quantum mechanics, we will indicate a configuration $\mathsf{c} \in Q \times \Gamma^* \times \mathbb{Z}$ with a ket $|\mathsf{c}\rangle = |q, \mathbf{x}, z\rangle$. According to this notation, the function δ assigns a complex number α to the transition from $|q, \mathbf{x}, z\rangle$ to $|q', \mathbf{x}', z + m\rangle$ where $m = -1$ if the head moves to the left, $m = +1$ if the head moves to the right, and $m = 0$ if the head doesn't move[17]. We denote with the letter H the set of states that can be written as a linear combination of configurations with a *finite* number of terms,

$$H = \big\{|\psi\rangle : |\psi\rangle = \sum_{i=1}^{n} \alpha_i |\mathsf{c}_i\rangle, \ n \in \mathbb{N} \text{ and } \sum_{i=1}^{n} |\alpha_i|^2 = 1\big\}.$$

The set H (along with the usual addition between vectors and scalar product) forms a vector space containing *finite* linear combinations. Hence, this vector space is not complete. With some caution [16], H can be converted into a Hilbert space with the usual inner product between complex vectors such that $\{|\mathsf{c}_i\rangle\}_i$ is an orthonormal basis.

The computation of a quantum Turing machine is divided into several parts. At the beginning of the computation, the machine is prepared in the configuration $|\mathsf{c}_1\rangle = |q_1, \mathbf{x}, 1\rangle$, where $\mathbf{x}$ is the input string of the problem. The transition function governs the evolution of the machine. Since in quantum mechanics the time evolution is described by a unitary transformation, it is reasonable to express the action of the transition function with a unitary operator $\hat{U}_\delta$,

[16]The symbol $\tilde{\mathbb{C}}$ requires a clarification. A real number x is computable if there exists a DTM which takes as input a string with n symbols, such as $(1, \ldots, 1)$, and outputs in polynomial time a number y that approximates x with high precision (i.e. $|x - y| \leq 2^{-n}$). A complex number is computable if the real and imaginary parts are computable. The symbol $\tilde{\mathbb{C}}$ indicates the set of all computable complex numbers. Interestingly, there exist some numbers that are not computable. The restriction of the quantum transition function to computable numbers is required to avoid some problematic cases.

[17]It can be shown that a QTM whose head can only move {L,R} is as powerful as a QTM whose head can move {L,N,R}.

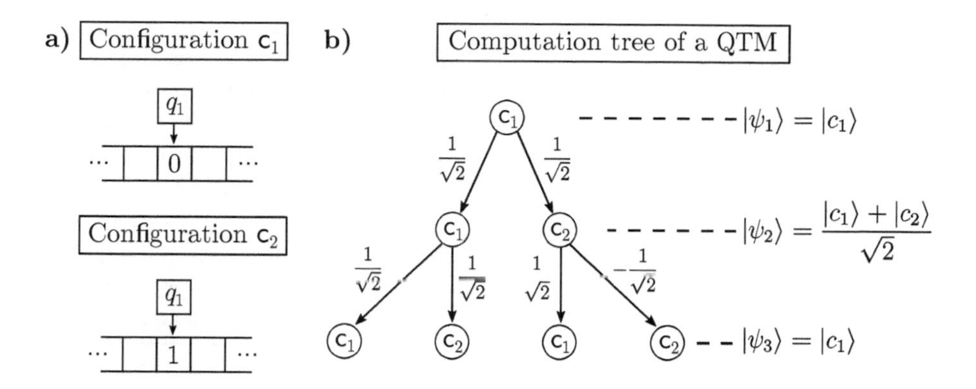

Fig. 2.15 A quantum Turing machine. a) Configurations of a quantum Turing machine with a single tape. **b)** Computation tree of a QTM. The transition function transforms the initial configuration of the machine from $|\psi_1\rangle = |\mathsf{c}_1\rangle$ to a superposition state $|\psi_2\rangle = (|\mathsf{c}_1\rangle + |\mathsf{c}_2\rangle)/\sqrt{2}$. In the second computation step, the transition function applies the same transformation to the state $|\psi_2\rangle$. Some machine configurations interfere destructively and others constructively. The resulting machine configuration is $|\psi_3\rangle = |\mathsf{c}_1\rangle$. This computation is equivalent to applying two consecutive Hadamard gates to the state $|0\rangle$.

$$\hat{U}_\delta|\mathsf{c}\rangle = \sum_i \alpha_i|\mathsf{c}_i\rangle.$$

After t computation steps, the Turing machine will be in a superposition state $|\psi_t\rangle = (\hat{U}_\delta)^t|\mathsf{c}_1\rangle$. The computation ends with a measurement of the state $|\psi_t\rangle$ in the canonical basis. This measurement will collapse $|\psi_t\rangle$ into a configuration $|\mathsf{c}_j\rangle$ with probability $p(\mathsf{c}_j \mid \psi_t) = |\langle\mathsf{c}_j|\psi_t\rangle|^2$.

Defining what it means for a QTM to accept a string can be tricky, as a QTM might reach a superposition of configurations, in which some are accept configurations $|\mathsf{c}_\mathrm{a}\rangle$, while others are not. To overcome this issue, we say that a QTM accepts (or rejects) an input string $\mathbf{x}$ with a running time t if the state of the machine after t computation steps is a superposition of only accept (or reject) configurations. A **polynomial QTM** is a quantum Turing machine that halts on every input $\mathbf{x}$ in polynomial time.

Example 2.8 The quantum Turing machine shown in Fig. 2.15b starts in the initial configuration $|\mathsf{c}_1\rangle = |q_1, 0, 1\rangle$. The transition function transforms this configuration into a superposition state $(|\mathsf{c}_1\rangle + |\mathsf{c}_2\rangle)/\sqrt{2}$, where $|\mathsf{c}_2\rangle = |q_1, 1, 1\rangle$. A measurement of this superposition state will give either $|\mathsf{c}_1\rangle$ or $|\mathsf{c}_2\rangle$ with 50% probability. As illustrated in Fig. 2.15b, if δ is applied again to $(|\mathsf{c}_1\rangle + |\mathsf{c}_2\rangle)/\sqrt{2}$, we obtain $|\mathsf{c}_1\rangle$. A measurement of this state will give $|\mathsf{c}_1\rangle$ with 100% probability. We note that this computation is equivalent to applying two consecutive Hadamard gates to a qubit prepared in state $|0\rangle$.

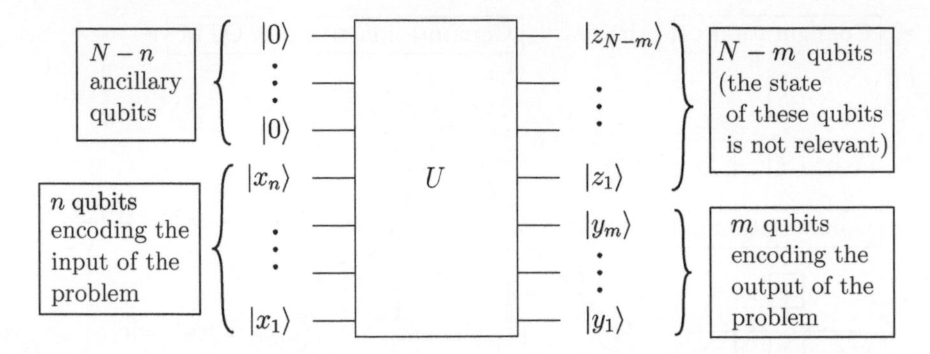

Fig. 2.16 A quantum circuit. A quantum circuit implementing the unitary $\hat{U}$ that transforms an input state $|\mathbf{0x}\rangle$ into $\hat{U}|\mathbf{0x}\rangle$. In this particular example, the output of the computation is a product state $|\mathbf{zy}\rangle$. In general, the output $\hat{U}|\mathbf{0x}\rangle$ will be a superposition state.

2.7 Quantum algorithms

The QTM formalism can be rather abstract and not so easy to handle. The quantum circuit model, first described by David Deutsch in 1989 [14], provided a more practical alternative and soon became the preferred computational model for studying quantum algorithms and their efficiency. The transition from QTM to quantum circuits was made possible by a significant discovery by Andrew Yao in 1993 [93]: he proved that QTMs and uniform quantum circuit families are equivalent up to a polynomial overhead. The interested reader is referred to Refs. [93, 16, 94] for additional information about this equivalence. We will limit our discussion to an explanation of uniform quantum circuits and how they can be used to solve decision problems. Our presentation will follow the same lines as Ref. [2].

A **quantum circuit** U is a directed acyclic graph with N input wires and N output wires as shown in Fig. 2.16. Each wire represents a qubit and each vertex (apart from input and output nodes) represents a **gate**, i.e. a unitary operator acting on a set of qubits. The input of the circuit is a binary string $\mathbf{x}$ of length $n \leq N$ encoding the input of the problem, which can be written in Dirac notation as

$$|\mathbf{0}\rangle_{N-n}|\mathbf{x}\rangle_n \;=\; |0\rangle \otimes \ldots \otimes |0\rangle \otimes |x_n\rangle \otimes \ldots \otimes |x_1\rangle,$$

where $|\mathbf{0}\rangle_{N-n}$ indicates a collection of $N-n$ qubits initialized in $|0\rangle$. These additional qubits are usually called **ancilla qubits**. They can be exploited to reduce the number of gates required by the computation. Each gate of the circuit can be decomposed into unitary gates acting on one or two qubits (see Section 5.5). A sequence of unitary operations transforms the initial state into[18] $\hat{U}|\mathbf{0}\,\mathbf{x}\rangle$ as shown in Fig. 2.16. The computation ends with a measurement in the computational basis. This operation will collapse the bottom m qubits onto $|\mathbf{y}\rangle_m$ with probability[19]

[18]The notation $|\mathbf{0x}\rangle_N$ indicates a product state of the form $|0\rangle \otimes \ldots \otimes |0\rangle \otimes |x_n\rangle \otimes \ldots \otimes |x_1\rangle$.
[19]As usual, the notation $|\mathbf{y}\rangle_m$ should be interpreted as $|\mathbf{y}\rangle_m = |y_m\rangle \otimes \ldots \otimes |y_1\rangle$.

$$\boxed{p(\mathbf{y} \mid \mathbf{x}) = \sum_{\mathbf{z} \in \mathbb{B}^{N-m}} |\langle \mathbf{z}\,\mathbf{y}| \hat{U} |\mathbf{0}\,\mathbf{x}\rangle|^2.} \tag{2.33}$$

Here, $\mathbf{z}$ is a subset of the final string that does not encode the output of the computation and can be ignored. For this reason, eqn (2.33) includes a sum over the states $|\mathbf{z}\rangle_{N-m}$ as we are interested only in one particular $|\mathbf{y}\rangle_m$. We say that a quantum circuit $\hat{U}$ **computes** a function $f : \mathbb{B}^n \to \mathbb{B}^m$, if for any input $|\mathbf{x}\rangle$ the quantum circuit produces a state $\hat{U}|\mathbf{0}\rangle|\mathbf{x}\rangle$ whose measurement gives $|f(\mathbf{x})\rangle$ with high probability,

$$p(f(\mathbf{x}) \mid \mathbf{x}) = \sum_{\mathbf{z} \in \mathbb{B}^{N-m}} |\langle \mathbf{z}\, f(\mathbf{x})| \hat{U} |\mathbf{0}\,\mathbf{x}\rangle|^2 \geq 1 - \epsilon, \tag{2.34}$$

where $0 \leq \epsilon < 1/2$ and the number of qubits N is greater than n and m. In other words, a quantum circuit computes a function when the output state has a high overlap with the state $|\mathbf{z}\, f(\mathbf{x})\rangle$ where $f(\mathbf{x})$ is the image of the input string and $\mathbf{z}$ is an arbitrary string.

A quantum circuit with N input wires can only handle strings with no more than N symbols. To construct a circuit model that can process bit strings of any length, it is necessary to introduce the concept of a **quantum circuit family**. This is an infinite sequence of quantum circuits $(\hat{U}_i)_i$ where the circuit $\hat{U}_n$ has N input and N output wires, and in general $N \geq n$. A circuit family $(\hat{U}_i)_i$ is **uniform** if there exists a DTM[20] that outputs a complete description of $\hat{U}_n$ in polynomial time for any n (this description includes the number of input and output wires, the gates executed, and their interconnections). We will mainly deal with **polynomial-size quantum circuits**, namely circuit families where each circuit $\hat{U}_n$ contains a polynomial number of qubits and gates taken from a universal set of single and two-qubit gates.

We can finally give a precise definition of a quantum algorithm in terms of quantum circuits. A uniform family of quantum circuits $(\hat{U}_i)_i$ **computes** a function[21] $f : \mathbb{B}^* \to \mathbb{B}^*$ if for all binary strings $\mathbf{x} \in \mathbb{B}^*$ the circuits in the family output $f(\mathbf{x})$ with high probability

$$p(f(\mathbf{x}) \mid \mathbf{x}) = \sum_{\mathbf{z} \in \mathbb{B}^{N-m}} |\langle \mathbf{z}\, f(\mathbf{x})| \hat{U}_n |\mathbf{0}\,\mathbf{x}\rangle|^2 \geq 1 - \epsilon, \tag{2.35}$$

where $0 \leq \epsilon < 1/2$, $n = |\mathbf{x}|$ and $N > n$. In short, a **quantum algorithm** is a uniform family of quantum circuits that output $f(\mathbf{x})$ with high probability. A **quantum computer** is a physical machine that implements quantum circuits.

2.7.1 BQP and its relations to other complexity classes

The class BQP (which stands for **bounded-error quantum polynomial time**) is the most important complexity class in quantum computing. BQP is the class of languages a quantum computer can decide in polynomial time with a small error probability.

[20] We could also use a QTM to describe this family of circuits. This definition would not change.

[21] We have only considered functions defined on the Boolean alphabet; this does not cause any loss of generality since the Boolean alphabet can be used to encode any other alphabet.

More formally, a language $L \in \mathsf{BQP}$ if there exists a uniform family of *polynomial-size* quantum circuits $(U_i)_i$ such that:

$$\text{if } \mathbf{x} \in L \text{ and } |\mathbf{x}| = n, \text{ then the measurement of } \hat{U}_n|\mathbf{0x}\rangle \text{ gives } 1$$
$$\text{with high probability, } p(1 \mid \mathbf{x}) \geq 1 - \epsilon,$$

$$\text{if } \mathbf{x} \notin L \text{ and } |\mathbf{x}| = n, \text{ then the measurement of } \hat{U}_n|\mathbf{0x}\rangle \text{ gives } 0$$
$$\text{with high probability, } p(0 \mid \mathbf{x}) \geq 1 - \epsilon.$$

In this definition, the probabilities $p(0 \mid \mathbf{x})$ and $p(1 \mid \mathbf{x})$ are defined as

$$p(k \mid \mathbf{x}) = \sum_{\mathbf{z} \in \mathbb{B}^{N-1}} |\langle \mathbf{z}\, k| \, \hat{U}_n \, |\mathbf{0}\, \mathbf{x}\rangle|^2, \tag{2.36}$$

where $k \in \{0, 1\}$. The value of ϵ is not important as long as it is a fixed quantity in the interval $[0, 1/2)$. If this is the case, the correct answer can be determined by repeating the experiment multiple times and selecting the most frequently observed outcome.

The class BQP corresponds to the type of problems a quantum computer can easily solve with a small error probability. This class contains FACTORING, HSP, ECYCLE, and all problems in BPP. In 1997, Bennett, Bernstein, Brassard, and Vazirani gave the first evidence that NP $\not\subseteq$ BQP with respect to an oracle [95]. It is widely believed (but not proven) that NPC problems are not in BQP, meaning that a quantum computer will never solve important decision problems (such as 3SAT) in polynomial time [92, 95].

The quantum complexity class BQP has some interesting relations with classical complexity classes. It is intuitive that BPP $\subseteq$ BQP, since a quantum Turing machine can efficiently simulate the behavior of a probabilistic Turing machine. A less intuitive result is that BQP $\subseteq$ PSPACE. To show this, we demonstrate that a deterministic Turing machine can calculate the probability (2.36) using polynomial space and an unlimited amount of time. Consider the probability that a circuit $\hat{U}_n$ with N wires outputs 1 on input $\mathbf{x}$,

$$p(1 \mid \mathbf{x}) = \sum_{\mathbf{z} \in \mathbb{B}^{N-1}} |\langle \mathbf{z}\, 1| \, \hat{U}_n \, |\mathbf{0}\, \mathbf{x}\rangle|^2. \tag{2.37}$$

The unitary $\hat{U}_n$ can be decomposed as $\hat{U}_n = \hat{V}_{p(n)} \ldots \hat{V}_1$ where $p(n)$ is a polynomial and $\hat{V}_i$ are unitary operators acting on one or two qubits. Introducing multiple completeness relations, $\mathbb{1} = \sum_{\mathbf{w}_i} |\mathbf{w}_i\rangle\langle\mathbf{w}_i|$, the matrix element in (2.37) can be written as

$$\begin{aligned}&\langle \mathbf{z}\, 1| \, \hat{U}_n \, |\mathbf{0}\, \mathbf{x}\rangle = \langle \mathbf{z}\, 1| \, \hat{V}_{p(n)} \ldots \hat{V}_1 \, |\mathbf{0}\, \mathbf{x}\rangle = \\ &= \sum_{\mathbf{w}_{p(n)-1} \in \mathbb{B}^N} \cdots \sum_{\mathbf{w}_1 \in \mathbb{B}^N} \langle \mathbf{z}\, 1| \, \hat{V}_{p(n)} |\mathbf{w}_{p(n)-1}\rangle \langle \mathbf{w}_{p(n)-1}| \hat{V}_{p(n)-1} |\mathbf{w}_{p(n)-2}\rangle \cdot \ldots \cdot \langle \mathbf{w}_1 | \hat{V}_1 | \mathbf{0}\, \mathbf{x}\rangle.\end{aligned} \tag{2.38}$$

This sum contains an exponential number of elements of the form $\langle \mathbf{w}_{i+1}|\hat{V}_{i+1}|\mathbf{w}_i\rangle$. Each of these elements can be easily calculated with a deterministic machine (recall that $\hat{V}_i$

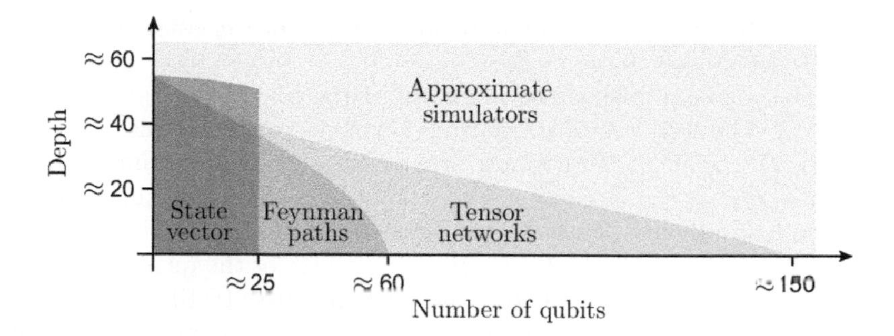

Fig. 2.17 Simulating quantum circuits with classical computers. The boundaries of the gray regions are not well defined because the classical simulation depends on many factors.

are unitary matrices acting on two qubits or less). To compute the sum in (2.38), a classical machine can calculate the first term, add it to a cumulative sum, erase it from its memory to free up some space, and use the same cells to compute the following term. This procedure is repeated until the entire sum has been estimated. This process takes an exponential number of steps but only polynomial space. Hence, $\mathsf{BQP} \subseteq \mathsf{PSPACE}$ as anticipated. A more formal proof based on quantum Turing machines can be found in the work of Ethan Bernstein and Umesh Vazirani [16]. Since $\mathsf{PSPACE} \subseteq \mathsf{EXPTIME}$, this also means that $\mathsf{BQP} \subseteq \mathsf{EXPTIME}$. This is consistent with the simple intuition that a deterministic machine can simulate a quantum computer by calculating the evolution $\hat{U}|\mathbf{0}\,\mathbf{x}\rangle$ in exponential time. Collecting these results together, we have

$$\mathsf{P} \subseteq \mathsf{BPP} \subseteq \mathsf{BQP} \subseteq \mathsf{PSPACE} \subseteq \mathsf{EXPTIME}. \tag{2.39}$$

Does there exist a language that can be decided by a quantum computer and not by a classical computer? No. A quantum computer can decide the same languages as a deterministic Turing machine. This means that quantum computers do not violate the Church–Turing thesis. Can quantum computers solve some decision problems faster than classical computers? At the time of writing, even if the majority of the scientific community believes that quantum computers are more powerful than classical machines, a formal mathematical proof that $\mathsf{BPP} \subset \mathsf{BQP}$ is still missing. Though, to be fair, we do not even have a rigorous proof that BPP is strictly contained in PSPACE. If one day the inclusion $\mathsf{BPP} \subset \mathsf{BQP}$ is finally demonstrated, then $\mathsf{BPP} \subset \mathsf{PSPACE}$ and this would be an extraordinary result in complexity theory.

In the last two decades, we collected some encouraging signs that quantum computers outperform probabilistic Turing machines by developing multiple quantum algorithms that offer an exponential speed-up with respect to the classical ones. For this reason, many scientists believe that $\mathsf{BPP} \subset \mathsf{BQP}$. We also have preliminary *experimental* evidence that quantum computers are more powerful than classical computers – albeit on mathematical problems with no practical applications [20, 84]. The competition between classical and quantum computing is fascinating. The boundary between what can be efficiently solved by a quantum computer and not by a classical machine is a blurry region due to the constant improvement of both classical and quantum algorithms.

The main methods used to simulate a quantum circuit on a classical computer are shown in Fig. 2.17. The state vector simulation, Feynman path integrals [16], and tensor networks [96, 97] can simulate quantum circuits exactly when the number of qubits and circuit depth are not too large. It is important to mention that the gray regions illustrated in the figure should be taken with a grain of salt: what can be simulated by a classical computer strongly depends on the connectivity of the quantum hardware, the fidelity of single and two-qubit gates, the CPU adopted to run the simulation, and the targeted quantum circuit. Approximate simulators (i.e. classical algorithms that approximate the output of a quantum circuit) can stretch the boundaries of what is simulatable by classical machines. However, unlike the methods mentioned before, they do not simulate the quantum circuit exactly. For more information about the subject, see Refs [98, 99, 100].

2.8 QMA problems

An important quantum complexity class is QMA (which stands for quantum Merlin Arthur), the quantum version of the complexity class NP. Before diving into the definition of QMA, it's worth clarifying a few concepts about NP. This class is defined as the set of languages whose affirmative solution can be verified in polynomial time by a deterministic Turing machine with the aid of a certificate. In this definition, two parties are implicitly assumed: an all-knowing entity (usually called Merlin) that can provide certificates on demand and a human (typically named Arthur) that owns a deterministic machine T_V and wants to verify if a string $\mathbf{x}$ belongs to a specific language. To accomplish this task, Arthur asks Merlin to provide a suitable certificate $\mathbf{c}$. Once Arthur has received the certificate, he can run the computation $T_V(\mathbf{x}, \mathbf{c})$ and discover in polynomial time whether $\mathbf{x} \in L$.

If we slightly change this protocol and allow Arthur to operate a *probabilistic* Turing machine instead of a deterministic one (see Table 2.1), we obtain a new complexity class called MA. A language $L \in \mathbb{B}^*$ is in MA if there exists a probabilistic Turing machine T_V running in polynomial time such that:

$$\text{if } \mathbf{x} \in L, \text{ then } \exists \mathbf{c}_\mathbf{x} \in \mathbb{B}^* \text{ such that } T_V(\mathbf{x}, \mathbf{c}_\mathbf{x}) = 1 \text{ with probability } p > 1 - \epsilon,$$
$$\text{if } \mathbf{x} \notin L, \text{ then } \forall \mathbf{c} \in \mathbb{B}^* \text{ we have } T_V(\mathbf{x}, \mathbf{c}) = 0 \text{ with probability } p > 1 - \epsilon.$$

The parameter ϵ is a constant in the interval $[0, 1/2)$ and the length of the string $\mathbf{c}$ is polynomial in input size. It is intuitive that NP $\subseteq$ MA, since any probabilistic verifier can simply run the same algorithm that a deterministic verifier would use to check if $\mathbf{x} \in L$.

We can finally introduce the complexity class QMA. QMA contains all languages whose affirmative solution can be verified by a *quantum computer* in polynomial time and can be considered the quantum counterpart of the class NP. In a more formal way, a language $L \in$ QMA if there exists a uniform family of polynomial-size quantum circuits $(\hat{U}_i)_i$ such that:

Class	Merlin prepares:	Arthur operates:
NP	a binary string $\mathbf{c}$	a DTM T_V to compute $T_V(\mathbf{x}, \mathbf{c})$
MA	a binary string $\mathbf{c}$	a PTM T_V to compute $T_V(\mathbf{x}, \mathbf{c})$
QMA	multiple copies of $\|\psi\rangle$	a quantum computer to compute $\hat{U}\|\psi\rangle\|\mathbf{x}\rangle$

Table 2.1 Merlin and Arthur. Comparison between the complexity classes NP, MA and QMA. The main difference between these classes is the machine used by Arthur to check whether the string $\mathbf{x}$ belongs to a language or not with the aid of a certificate provided by Merlin.

$$\text{if } x \in L \text{ and } |\mathbf{x}| = n, \text{then there exists a certificate } |\mathbf{c_x}\rangle \text{ such that}$$
$$\text{the measurement of } \hat{U}_n|\mathbf{x}\rangle|c_\mathbf{x}\rangle \text{ gives 1 with high probability, } p(1) \geq 1 - \epsilon,$$

$$\text{if } x \notin L \text{ and } |\mathbf{x}| = n, \text{then for all certificates } |c\rangle$$
$$\text{the measurement of } \hat{U}_n|\mathbf{x}\rangle|c\rangle \text{ gives 0 with high probability, } p(0) \geq 1 - \epsilon.$$

As usual, $\epsilon \in [0, 1/2)$ and the certificate must be constructed with a polynomial number of qubits. The probabilities $p(0)$ and $p(1)$ are defined in the standard way,

$$p(k) = \sum_{\mathbf{z} \in \mathbb{B}^{N-1}} |\langle \mathbf{z}k| \, \hat{U}_n \, |c\mathbf{x}\rangle|^2 \qquad \text{where } k \in \{0, 1\}.$$

The states $|\mathbf{z}\rangle$ are $N-1$ qubit states that do not encode any useful information and can be ignored. The certificate $|\mathbf{c}\rangle$ prepared by Merlin can be an arbitrary superposition state. As explained in Section 5.6, a quantum computer might require an exponential number of gates to prepare such a superposition state. However, this is not a problem since we have assumed that Merlin has infinite power and can easily prepare any certificate for Arthur.

It is quite intuitive that NP $\subseteq$ MA $\subseteq$ QMA, because the machine used by Arthur becomes more and more powerful with each inclusion. A less intuitive result is that QMA $\subseteq$ PP. This was demonstrated by Kitaev and Watrous [2, 101]. The complexity classes presented in this chapter are related to each other as shown below:

```
        NP → MA
      ↗         ↘
P                  QMA → PP → PSPACE → EXPTIME
      ↘         ↗
        BPP → BQP
```

In this diagram, each arrow represents an inclusion symbol $\subseteq$. A rigorous proof of these inclusions is out of the scope of this textbook. The interested reader can find more information in the lecture notes by Scott Aaronson on quantum complexity theory [102].

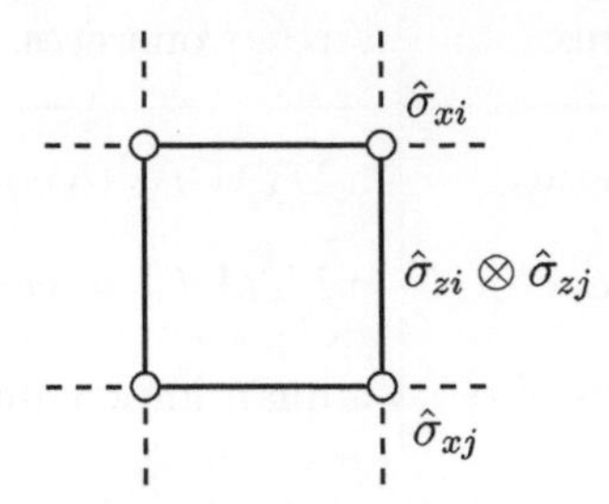

Fig. 2.18 Ising model. A square grid of spins. The white dots indicate lattice sites, while the black segments connect nearest neighbors.

The main takeaway is that quantum computers are at least as powerful as classical computers (BPP $\subseteq$ BQP) and at most exponentially faster (BQP $\subseteq$ EXPTIME).

2.8.1 k-local Hamiltonians

This section presents some examples of k-LOCAL Hamiltonians. k-LOCAL Hamiltonians are an interesting type of Hamiltonians encountered in many fields of physics. In the next section, we will show that deciding whether the lowest eigenvalue of a k-LOCAL Hamiltonian is greater than a certain threshold is a QMA problem.

A Hamiltonian acting on n qubits is said to be k-LOCAL if it can be decomposed as

$$\boxed{\hat{H} = \sum_{i=1}^{r} c_i \hat{H}_i,} \tag{2.40}$$

where $\hat{H}_i$ are positive operators acting on no more than k qubits, the coefficients c_i are positive numbers and $r = \text{poly}(n)$. Note that if the operators $\hat{H}_i$ are Hermitian but not positive, it is often possible to redefine them so that $\hat{H}$ can be expressed as a sum of positive operators.

Example 2.9 Consider the lattice shown in Fig. 2.18 where each node represents a spin. Let us assume that their interactions are governed by the Hamiltonian of the **Ising model**

$$\hat{H} = \sum_{i=1}^{n} B_i \hat{\sigma}_{xi} + \sum_{\langle i,j \rangle} J_{ij} \hat{\sigma}_{zi} \hat{\sigma}_{zj}. \tag{2.41}$$

The notation $\sum_{\langle i,j \rangle}$ indicates summation over nearest neighbors. This Hamiltonian is local since it contains Pauli operators acting on no more than two qubits. However, it is not expressed as a sum of positive operators because the Pauli operators are not positive (their eigenvalues are $+1$ and -1). One can express $\hat{H}$ as a sum of positive operators by introducing the positive operators

$$\begin{array}{llcll} \hat{P}_x \equiv \frac{\mathbb{1}+\hat{\sigma}_x}{2}, & \hat{P}_z \equiv \frac{\mathbb{1}+\hat{\sigma}_x}{2}, & \Rightarrow & \hat{\sigma}_x = 2\hat{P}_x - \mathbb{1}, & \hat{\sigma}_x = 2\hat{P}_z - \mathbb{1}, \\ \hat{Q}_x \equiv \frac{\mathbb{1}-\hat{\sigma}_x}{2}, & \hat{Q}_z \equiv \frac{\mathbb{1}-\hat{\sigma}_x}{2}, & & \hat{\sigma}_x = \mathbb{1} - 2\hat{Q}_x, & \hat{\sigma}_x = \mathbb{1} - 2\hat{Q}_z. \end{array}$$

Substituting these expressions into eqn (2.41) and dropping some constant terms, one obtains a k-LOCAL Hamiltonian.

2.8.2 k-LOCAL is in QMA

The k-LOCAL problem consists in deciding whether the lowest eigenvalue of a k-local Hamiltonian is lower than a certain threshold. To formulate a rigorous definition of this problem, it is convenient to introduce the binary encoding of a k-local Hamiltonian. First, each operator $\hat{H}_i$ in eqn (2.40) can be divided by its largest eigenvalue η_i. This leads to

$$\hat{H} = \sum_{i=1}^{r} c_i \eta_i \frac{\hat{H}_i}{\eta_i} = \sum_{i=1}^{r} c_i \eta_i \hat{h}_i,$$

where we introduced the positive operators $\hat{h}_i = \hat{H}_i/\eta_i$ with eigenvalues in the interval $[0, 1]$. Secondly, we can divide $\hat{H}$ by the positive constant $\gamma = \sum_i c_i \eta_i$ and obtain

$$\hat{H} = \gamma \sum_{i=1}^{r} \frac{c_i \eta_i}{\gamma} \hat{h}_i = \gamma \sum_{i=1}^{r} p_i \hat{h}_i, \tag{2.42}$$

where $(p_1, \ldots, p_r)$ is a probability distribution[22]. Recall that each operator $\hat{h}_i$ acts on a set of qubits S_i with cardinality $|S_i| \leq k$. Hence, we can decompose each operator $\hat{h}_i$ in the Pauli basis using $4^{|S_i|}$ elements

$$\hat{H} = \gamma \sum_{i=1}^{r} p_i \sum_{j=1}^{4^{|S_i|}} A_{ij} \hat{P}_j,$$

where A_{ij} are the coefficients of the expansion and $\hat{P}_j$ are tensor products of Pauli operators. Finally, the Hamiltonian $\hat{H}$ can be encoded into a binary string

$$\boxed{\mathbf{x}_{\hat{H}} = \langle\, n, r, \gamma, (p_1, S_1, \mathbf{A}_1), \ldots, (p_r, S_r, \mathbf{A}_r) \,\rangle,} \tag{2.43}$$

where n indicates the number of qubits in the system, r is the number of terms in the Hamiltonian, γ is a positive number, $(p_1, \ldots, p_r)$ is a probability distribution, S_i indicates the qubits on which the i-th term of the Hamiltonian is acting, $\mathbf{A}_i = (A_{i,1}, \ldots, A_{i,4^{|S_i|}})$ and the brackets $\langle\ \rangle$ indicate a binary encoding. The length of this string grows polynomially with n.

Let us provide a formal definition of the k-LOCAL problem. We are given a k-local Hamiltonian acting on n qubits. The Hamiltonian is the sum of a polynomial number of positive operators $\hat{H}_i$ with norm not greater than one[23]. We are also given two numbers a and b satisfying $0 \leq a < b \leq 1$. The difference between them is not too

[22] By definition $\sum_i p_i = \sum_i c_i \eta_i / \gamma = \sum_i c_i \eta_i / \sum_i c_i \eta_i = 1$.

[23] Assuming that the norm of the operators $\hat{H}_i$ is not greater than one is general enough. Indeed, any k-local Hamiltonian can be expressed as $\hat{H} = \gamma \sum_i p_i \hat{h}_i$, see eqn (2.42). The positive operators $\hat{h}_i$ have norm $\|\hat{h}_i\| \leq 1$ and the constant γ can be factored out.

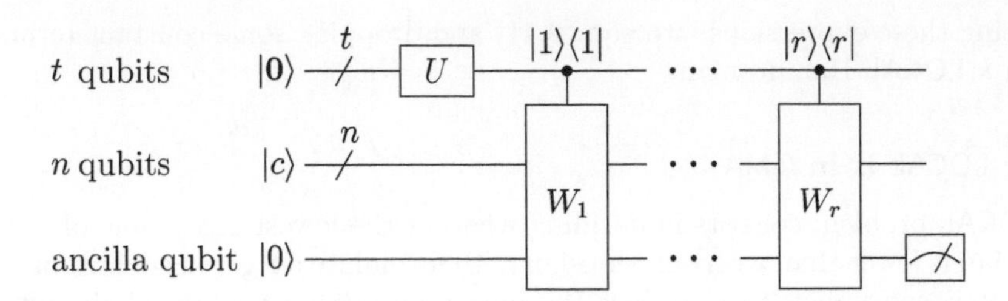

Fig. 2.19 k-LOCAL is in QMA. The quantum circuit used by Arthur to decide if the lowest eigenvalue of $\hat{H}$ is smaller than a or larger than b. The input state $|c\rangle$ created by Merlin can be a complicated superposition state of n qubits. The measurement of the bottom qubit will statistically tell Arthur whether $\mathbf{x}_{\hat{H}} \in L_{k\text{-LOCAL}}(a,b)$. Arthur might need to run this computation several times to support his conclusion with high confidence.

small, $b - a \geq 1/\text{poly}(n)$. The k-LOCAL **Hamiltonian problem** consists in deciding whether the lowest eigenvalue $\tilde{\lambda}$ is smaller than a or larger than b. Provided that $\mathbf{x}_{\hat{H}}$ is the binary encoding of the Hamiltonian, we can express this problem in terms of a language

$$L_{k\text{-LOCAL}}(a,b) = \{\mathbf{x}_{\hat{H}} \in \mathbb{B}^* : \hat{H} \text{ is } k\text{-LOCAL and } \tilde{\lambda} \leq a < b\}.$$

One must calculate the lowest eigenvalue of $\hat{H}$ to decide this language.

Let us show that $L_{k\text{-LOCAL}}$ is in QMA.

Theorem 2.3 $L_{k\text{-LOCAL}} \in \text{QMA}$.

Proof The proof of this theorem boils down to constructing a polynomial-size quantum circuit that Arthur can operate to verify that the certificate $|c\rangle$ provided by Merlin indicates that the Hamiltonian has an eigenvalue lower than a with high probability. To this end, Arthur can run a quantum circuit that outputs 1 with probability

$$p(1) = 1 - \frac{\langle c|\hat{H}|c\rangle}{r},$$

where r is the number of terms in the Hamiltonian. Suppose that Merlin gives Arthur the eigenvector $|c\rangle$ associated with the lowest eigenvalue $\tilde{\lambda}$. If $\tilde{\lambda} \leq a$, the quantum circuit executed by Arthur will give outcome 1 with probability

$$p(1) = 1 - \frac{\langle c|\hat{H}|c\rangle}{r} = 1 - \frac{\tilde{\lambda}}{r} \geq 1 - \frac{a}{r}.$$

If instead $\mathbf{x}_{\hat{H}} \notin L_{k\text{-LOCAL}}$, all eigenvalues of the Hamiltonian are larger than b and therefore the quantum circuit operated by Arthur will give outcome 1 with probability

$$p(1) = 1 - \frac{\langle c|\hat{H}|c\rangle}{r} \leq 1 - \frac{b}{r}$$

for all states $|c\rangle$ provided by Merlin. Arthur must run the quantum circuit multiple times to gather enough statistics and check whether $p(1) \geq 1 - a/r$ or $p(1) \leq 1 - b/r$. The rest of the proof explains how to build a quantum circuit with a polynomial number of gates that outputs the probability $p(1) = 1 - \langle c|\hat{H}|c\rangle/r$. For simplicity, we first consider only one term of the Hamiltonian $\hat{H} = \sum_i c_i \hat{H}_i$ with spectral decomposition

$$\hat{H}_i = \sum_{l=1}^{2^{|S_i|}} \lambda_l |E_l\rangle\langle E_l|,$$

where $\{|E_l\rangle\}_l$ is an orthonormal basis of eigenvectors. Suppose that Merlin provides a certificate $|c\rangle = \sum_l c_l |E_l\rangle$. Arthur will apply a unitary $\hat{W}_i$ to the state $|c\rangle|0\rangle$, where $|0\rangle$ is the initial state of an ancilla qubit. The state of the ancilla qubit at the end of the computation will tell Arthur whether $\mathbf{x}_{\hat{H}} \in L_{k\text{-LOCAL}}$. The transformation $\hat{W}_i$ is defined as

$$\hat{W}_i = |E_1\rangle\langle E_1| \otimes \mathsf{Y}_{\theta_1} + \ldots + |E_{2^{|S_i|}}\rangle\langle E_{2^{|S_i|}}| \otimes \mathsf{Y}_{\theta_{2^{|S_i|}}},$$

where Y_θ is a rotation about the y axis of the Bloch sphere by an angle θ and $\theta_l = 2\arccos(\sqrt{\lambda_l})$. This unitary transforms a state $|E_l\rangle|0\rangle$ into

$$\hat{W}_i|E_l\rangle|0\rangle = |E_l\rangle \otimes \left(\sqrt{\lambda_l}|0\rangle + \sqrt{1-\lambda_l}|1\rangle\right).$$

The final measurement will reveal the ancilla qubit in state $|1\rangle$ with probability

$$\begin{aligned}
p_i(1) &= \mathrm{Tr}\left[\left(\hat{W}_i(|c\rangle\langle c| \otimes |0\rangle\langle 0|)\hat{W}_i^\dagger\right)(\mathbb{1} \otimes |1\rangle\langle 1|)\right] = \langle c0|\hat{W}_i^\dagger\,(\mathbb{1} \otimes |1\rangle\langle 1|)\,\hat{W}_i|c0\rangle \\
&= \sum_{j=1}^{2^{|S_i|}} \sum_{l=1}^{2^{|S_i|}} c_j^* c_l \langle E_j 0|\hat{W}_i^\dagger\,(\mathbb{1} \otimes |1\rangle\langle 1|)\,\hat{W}_i|E_l 0\rangle \\
&= \sum_{j=1}^{2^{|S_i|}} \sum_{l=1}^{2^{|S_i|}} c_j^* c_l \sqrt{1-\lambda_j}\sqrt{1-\lambda_l}\langle E_j 1|E_l 1\rangle = \sum_{l=1}^{2^{|S_i|}} |c_l|^2(1-\lambda_l) = 1 - \langle c|\hat{H}_i|c\rangle.
\end{aligned}$$

Let us now consider a Hamiltonian with r terms. The general circuit in Fig. 2.19 applies $\hat{W}_i$ onto $|c\rangle|0\rangle$ when the $t = \lceil \log_2 r \rceil$ qubits in the top part of the circuit are in state $|i\rangle$. These qubits are initially prepared in a superposition

$$\hat{U}|\mathbf{0}\rangle = \frac{1}{\sqrt{r}} \sum_{i=0}^{r-1} |i\rangle.$$

The final measurement will reveal the bottom qubit in state $|1\rangle$ with probability

$$p(1) = \frac{1}{r}\sum_{i=1}^{r} p_i(1) = \frac{1}{r}\sum_{i=1}^{r}\left(1 - \langle c|\hat{H}_i|c\rangle\right) = 1 - \frac{\langle c|\hat{H}|c\rangle}{r},$$

as required. Note that the controlled operations $\sum_i |i\rangle\langle i| \otimes \hat{W}_i$ can be decomposed with a polynomial number of single and two-qubit gates [103]. □

2.8.3 k-LOCAL is QMA complete

In this section, we show that k-LOCAL is QMA complete, i.e. that all problems in QMA can be efficiently mapped into a k-LOCAL problem. Let us start by considering a generic language $L \in \mathsf{QMA}$ and a binary string $\mathbf{x} \in \mathbb{B}^n$. If $\mathbf{x} \in L$, there exists at least one certificate $|c_\mathbf{x}\rangle$ and a quantum circuit $\hat{U}$ both with polynomial size such that the measurement of $\hat{U}|\mathbf{x}\rangle|c_\mathbf{x}\rangle$ produces 1 with high probability: $p(1) \geq 1 - \epsilon$ where ϵ is a constant in $[0, 1/2)$. On the contrary, if $\mathbf{x} \notin L$, for all certificates $|c\rangle$ the measurement of $\hat{U}|\mathbf{x}\rangle|c\rangle$ produces 0 with high probability: $p(0) \geq 1 - \epsilon$. We will assume that the certificate is an arbitrary superposition state of m qubits and the quantum register is initialized in $|\mathbf{0}c\rangle$ where $|\mathbf{0}\rangle$ is a n qubit state. The quantum circuit $\hat{U}$ operates on $n+m$ qubits and can be decomposed into a polynomial number of single and two-qubit gates,

$$\hat{U} = \hat{U}_T \dots \hat{U}_1.$$

The operator U_1 transforms $|\mathbf{0}c\rangle \mapsto |\mathbf{x}c\rangle$ with some NOT gates. The state of the quantum register goes through the steps,

$$|\psi_0\rangle = |\mathbf{0}c\rangle, \qquad |\psi_1\rangle = \hat{U}_1|\psi_0\rangle, \qquad \dots \qquad |\psi_T\rangle = \hat{U}_T|\psi_{T-1}\rangle.$$

We can demonstrate that determining whether $\mathbf{x}$ is a member of L with the aid of a certificate $|c\rangle$ is equivalent to solving a k-LOCAL problem [2]. To do this, we must construct a k-LOCAL Hamiltonian that satisfies two conditions:

1. If $p(1) \geq 1 - \epsilon$, then $\hat{H}$ has an eigenvalue lower than ϵ.
2. If $p(0) \geq 1 - \epsilon$, then all eigenvalues of $\hat{H}$ are greater than $3/4 - \epsilon$.

The main idea behind this demonstration is to create a time-independent Hamiltonian $\hat{H}$ that mimics the evolution induced by $\hat{U}$. This way, we can use the ground state energy of $\hat{H}$ to determine whether $\mathbf{x}$ is in L.

First, $\hat{H}$ acts not only on $n+m$ qubits, but also on a third register with $l = \lceil \log_2 T \rceil$ qubits. The third register is nothing but a counter: at each step, the content of the counter is increased by one,

$$\underset{\text{step 0}}{|0\dots 00\rangle} \rightarrow \underset{\text{step 1}}{|0\dots 01\rangle} \rightarrow \underset{\text{step 2}}{|0\dots 10\rangle} \rightarrow \dots \rightarrow \underset{\text{step T}}{|T\rangle}.$$

The Hamiltonian that simulates the transitions operated by the quantum circuit $\hat{U}$ is the sum of three terms,

$$\hat{H} = \hat{H}_\mathrm{A} + \hat{H}_\mathrm{B} + \hat{H}_\mathrm{C}.$$

The term $\hat{H}_\mathrm{A}$ checks that the three registers are initialized correctly. It does that by assigning a "penalty" when the registers are not in the state $|\mathbf{0}, c, \mathbf{0}\rangle$,

$$\hat{H}_\mathrm{A} = \sum_{s=1}^{n} |1\rangle\langle 1|_s \otimes \mathbb{1}_m \otimes |\mathbf{0}\rangle\langle \mathbf{0}|.$$

In this expression, the operators $|1\rangle\langle 1|_s$ are defined as

$$\begin{aligned}
|1\rangle\langle 1|_1 &= \mathbb{1} \otimes \mathbb{1} \otimes \ldots \otimes \mathbb{1} \otimes |1\rangle\langle 1|, \\
|1\rangle\langle 1|_2 &= \mathbb{1} \otimes \mathbb{1} \otimes \ldots \otimes |1\rangle\langle 1| \otimes \mathbb{1}, \\
&\vdots \\
|1\rangle\langle 1|_n &= |1\rangle\langle 1| \otimes \mathbb{1} \otimes \ldots \otimes \mathbb{1} \otimes \mathbb{1}.
\end{aligned}$$

Similarly, $\hat{H}_{\mathrm{C}}$ assigns a penalty when the first qubit ends up in 0 at the end of the computation,

$$\hat{H}_{\mathrm{C}} = \underbrace{|0\rangle\langle 0| \otimes \mathbb{1} \otimes \ldots \otimes \mathbb{1}}_{n+m \text{ qubits}} \otimes \underbrace{|T\rangle\langle T|}_{l \text{ qubits}}.$$

Finally, $\hat{H}_{\mathrm{B}}$ checks that the computation follows the same steps performed by the verifier $\hat{U}$,

$$\hat{H}_{\mathrm{B}} = \sum_{r=1}^{T} \hat{H}_r,$$

where

$$\begin{aligned}
\hat{H}_r = &\frac{1}{2}\mathbb{1}_{n+m} \otimes |r\rangle\langle r| + \frac{1}{2}\mathbb{1}_{n+m} \otimes |r-1\rangle\langle r-1| \\
&- \frac{1}{2}\hat{U}_r \otimes |r\rangle\langle r-1| - \frac{1}{2}\hat{U}_r^\dagger \otimes |r-1\rangle\langle r|.
\end{aligned}$$

Let us assume that $\mathbf{x} \in L$, i.e. there exists at least one certificate $|c_{\mathbf{x}}\rangle$ such that $p(1) \geq 1 - \epsilon$. This implies that $\hat{H}$ has at least one eigenvalue lower than ϵ. To show this, we need to calculate the expectation value of $\langle\psi|\hat{H}|\psi\rangle$ where $|\psi\rangle$ is a quantum state defined as

$$|\psi\rangle = \frac{1}{\sqrt{T+1}} \sum_{j=0}^{T} (\hat{U}_j \ldots \hat{U}_1 |\mathbf{0}\rangle|c_{\mathbf{x}}\rangle)|j\rangle = \frac{1}{\sqrt{T+1}} \sum_{j=0}^{T} |\psi_j\rangle|j\rangle,$$

The term $\langle\psi|\hat{H}_{\mathrm{A}}|\psi\rangle$ is zero,

$$\begin{aligned}
\langle\psi|\hat{H}_{\mathrm{A}}|\psi\rangle &= \sum_{i,j=0}^{T} \langle\psi_i, i|\Big(\sum_{s=1}^{n} |1\rangle\langle 1|_s \otimes \mathbb{1}_m \otimes |\mathbf{0}\rangle\langle\mathbf{0}|\Big)|\psi_j, j\rangle \\
&= \langle\psi_0|\Big(\sum_{s=1}^{n} |1\rangle\langle 1|_s \otimes \mathbb{1}_m\Big)|\psi_0\rangle = \langle 0c|\Big(\sum_{s=1}^{n} |1\rangle\langle 1|_s \otimes \mathbb{1}_m\Big)|\mathbf{0}c\rangle = 0.
\end{aligned}$$

The term $\langle\psi|\hat{H}_{\mathrm{B}}|\psi\rangle$ is zero as well,

$$\langle\psi|\hat{H}_{\mathrm{B}}|\psi\rangle = \frac{1}{2(T+1)}\sum_{i,j,r=0}^{T}\langle\psi_i,i|\Big(\mathbb{1}_{n+m}\otimes|r\rangle\langle r| + \mathbb{1}_{n+m}\otimes|r-1\rangle\langle r-1|$$

$$-\frac{1}{2}\hat{U}_r\otimes|r\rangle\langle r-1| - \frac{1}{2}\hat{U}_r^\dagger\otimes|r-1\rangle\langle r|\Big)|\psi_j,j\rangle$$

$$= \frac{1}{2(T+1)}\sum_{r=0}^{T}\langle\psi_r|\psi_r\rangle + \langle\psi_{r-1}|\psi_{r-1}\rangle - \langle\psi_r|\hat{U}_r|\psi_{r-1}\rangle - \langle\psi_{r-1}|\hat{U}_r^\dagger|\psi_r\rangle$$

$$= \frac{1}{2(T+1)}\sum_{r=0}^{T}\langle\psi_r|\psi_r\rangle + \langle\psi_{r-1}|\psi_{r-1}\rangle - \langle\psi_r|\psi_r\rangle - \langle\psi_{r-1}|\psi_{r-1}\rangle = 0,$$

where we used $\hat{U}_r|\psi_{r-1}\rangle$ and $\hat{U}_r^\dagger|\psi_r\rangle = |\psi_{r-1}\rangle$. Lastly,

$$\langle\psi|\hat{H}_{\mathrm{C}}|\psi\rangle = \frac{1}{T+1}\sum_{i,j=0}^{T}\langle\psi_i,i|\Big(|0\rangle\langle 0|\otimes\mathbb{1}_{n+m-1}\otimes|T\rangle\langle T|\Big)|\psi_j,j\rangle$$

$$= \frac{1}{T+1}\langle\psi_T|\Big(|0\rangle\langle 0|\otimes\mathbb{1}_{n+m-1}\Big)|\psi_T\rangle$$

$$= \frac{1}{T+1}\langle 0c|\hat{U}^\dagger\Big(|0\rangle\langle 0|\otimes\mathbb{1}_{n+m-1}\Big)\hat{U}|0c\rangle$$

$$= \frac{p(0)}{T+1} = \frac{1-p(1)}{T+1} \leq \frac{\epsilon}{T+1} < \epsilon.$$

Putting all together, $\langle\psi|\hat{H}|\psi\rangle < \epsilon$. This means that if $\mathbf{x} \in L$, $\hat{H}$ must have at least one eigenvalue lower than ϵ. In summary, we showed that the operations performed by the verifier to check whether a string belongs to a QMA language are equivalent to verifying that a local Hamiltonian has at least one eigenvalue lower than a certain threshold. We should demonstrate that if $\mathbf{x} \notin L$, then all of the eigenvalues of $\hat{H}$ are greater than $3/4 - \epsilon$. This part of the proof is a bit technical and interested readers can refer to Refs [2, 104]. □

It is widely believed (but not rigorously proven) that QMA complete problems are not in BQP, meaning that a quantum computer cannot solve QMA complete problems with polynomial resources and bounded errors.

Further reading

The theory of automata and Turing machines is explained very well in the textbooks by M. Sipser and J. Savage [29, 70]. Cook, Karp, and Levin developed the theory of NP complete problems in the 1970s [88, 65]. Quantum complexity theory can be further explored by reading the notes by Aaronson, Bouland, and Schaeffer [102], the comprehensive paper by Bernstein and Vazirani [16, 105] and the textbook by Kitaev, Shen, and Vyalyi [2]. Another helpful source is the survey by Watrous [101] in which the author presents the group non-membership problem (an important QMA complete problem), quantum interactive proofs, and the class QIP not presented in this chapter. More information about QMA problems can be found in Ref. [104]. The theory of quantum Turing machines is explained in Refs [16, 94]. The equivalence between quantum Turing machines and quantum circuits was demonstrated for the first time by Yao [93].

Summary

Deterministic finite-state automaton (DFA)

Σ	An alphabet is a finite set of symbols
$L \subseteq \Sigma^*$	Language
Q	A non-empty set of states
$A \subseteq Q$	Set of accept states
$\delta : Q \times \Sigma \to Q$	Transition function for a finite-state automaton
$M = (\Sigma, Q, \delta, q_1, A)$	Definition of finite-state automaton
$L(M)$	Language recognized by automaton M

The union and intersection of two regular languages is a regular language.
The complement of a regular language is a regular language.

Deterministic Turing machine (DTM)

Σ	Alphabet
Γ	Tape alphabet, $\Sigma \subset \Gamma$ and $\text{␣} \in \Gamma$
$\delta : Q \times \Gamma \to Q \times \Gamma \times \{\mathrm{L}, \mathrm{R}, \mathrm{N}\}$	Transition function for a deterministic TM
$T = \{\Sigma, \Gamma, Q, \delta, q_1, q_\mathrm{a}, q_\mathrm{r}\}$	Deterministic Turing machine (DTM)
$\mathsf{c} = \mathbf{x}^q\mathbf{y}$	Configuration of a TM
$\mathsf{c} = (q, \mathbf{x}, z)$	Alternative notation for a TM configuration

$f : \Sigma^* \to \mathbb{B}$ is computable if $L_f = \{\mathbf{x} \in \mathbb{B}^* : f(\mathbf{x}) = 1\}$ is decidable.

$\mathcal{L}$	Set of all languages
$\mathcal{L}_\mathrm{dec}$	Set of decidable languages
$\mathcal{L} - \mathcal{L}_\mathrm{dec}$	This set is not empty
$\mathbf{a}_T = \langle T \rangle$	Binary encoding of a DTM
$C \subset \mathbb{B}^*$	Set of binary strings encoding all possible DTM

$L_{\mathsf{HALTING}} = \{(\mathbf{a}_T, \mathbf{x}) : \mathbf{a}_T \in C, \mathbf{x} \in \mathbb{B}^*, T(\mathbf{x}) \text{ terminates}\}$ is not decidable by a DTM.

$t(n)$	Steps required to decide a string of length n
$\mathsf{TIME}(t(n))$	Languages decided by a DTM in $O(t(n))$ steps
P	Languages decided by a DTM in polynomial time
EXPTIME	Languages decided by a DTM in exponential time
$\mathsf{SPACE}(t(n))$	Languages decided by a DTM scanning $O(t(n))$ tape cells
PSPACE	Languages decided by a DTM using polynomial space
EXPSPACE	Languages decided by a DTM using exponential space

$\mathsf{P} \subseteq \mathsf{PSPACE} \subseteq \mathsf{EXPTIME} \subseteq \mathsf{EXPSPACE}$

Multi-tape deterministic Turing machine (MDTM)

k	Number of tapes
$\delta : Q \times \Gamma^k \to Q \times \Gamma^k \times \{\mathrm{L}, \mathrm{R}, \mathrm{N}\}^k$	Transition function for an MDTM
$T = \{\Sigma, \Gamma, Q, \delta, q_1, q_\mathrm{a}, q_\mathrm{r}\}$	Definition of an MDTM

If an MDTM decides L in $t(n)$ steps, a DTM can decide L in $O(t^2(n))$ steps.

Probabilistic Turing machine (PTM)

$\delta : Q \times \Gamma \times Q \times \Gamma \times \{\mathrm{L}, \mathrm{R}, \mathrm{N}\} \to [0,1]$	Transition function for a PTM
$T = \{\Sigma, \Gamma, Q, \delta, q_1, q_\mathrm{a}, q_\mathrm{r}\}$	Definition of a PTM
$b_\mathrm{a}, b_\mathrm{r}$	Accepting and rejecting branches
$p(q_\mathrm{a} \mid \mathbf{x}) = \sum_{b_\mathrm{a}} p(b_\mathrm{a})$	Probability that a PTM accepts $\mathbf{x}$
$1 - p(q_\mathrm{a} \mid \mathbf{x})$	Probability that a PTM rejects $\mathbf{x}$

A PTM decides L if it returns the correct answer more than 66% of the time.

$\mathsf{BPTIME}(t(n))$	Languages decided by a PTM in $t(n)$ steps
$\mathsf{BPP} = \bigcup_k \mathsf{BPTIME}(n^k)$	Languages decided by a PTM in polynomial time

$\mathsf{P} \subseteq \mathsf{BPP} \subseteq \mathsf{PSPACE} \subseteq \mathsf{EXPTIME} \subseteq \mathsf{EXPSPACE}$

Circuit model

DAG	Directed acyclic graph
C	A circuit is a DAG with n input wires and m_n output wires
$(C_0, C_1, C_2, \ldots)$	Circuit family
$C_n(\mathbf{x})$	The output of a circuit acting on a string $\mathbf{x}$ with $\lvert\mathbf{x}\rvert = n$
$f(n) : \mathbb{N} \to \mathbb{N}$	Size complexity (number of gates in the circuit C_n)

If $|\mathbf{x}| = s$, $n > s$ and $C_n(\mathbf{x}) = C_s(\mathbf{x})$, then the circuit family $(C_i)_i$ is consistent.
$(C_i)_i$ is uniform if a DTM can describe any C_n in polynomial time.

If a DTM decides L in $t(n) > n$ steps, then $(C_i)_i$ decides L with $f(n) \in O(t^2(n))$.

(C–T thesis) Any physical computational model is equivalent to a DTM.

(Extended C–T thesis) Any computation performed by a physical machine can be efficiently simulated by a PTM.

Non-deterministic finite-state Turing machine (NTM)

$\delta : Q \times \Gamma \to \mathcal{P}(Q \times \Gamma \times \{\mathrm{L}, \mathrm{R}, \mathrm{N}\})$	Transition function of an NTM
$T = \{\Sigma, \Gamma, Q, \delta, q_1, q_{\mathrm{a}}, q_{\mathrm{r}}\}$	Definition of an NTM
b_{a}, b_{r}	Accepting and rejecting branches
$t(b, n)$	Length of a branch b when the input has length n
$t(n) = \max_{b_{\mathrm{a}}} t(b, n)$	Running time of an NTM
$\mathsf{NTIME}(t(n))$	Languages decided by an NTM in $O(t(n))$ steps

If an NTM decides L in $t(n)$ steps, a DTM can decide L in $2^{O(t(n))}$ steps.
A problem is in NP if a candidate solution can be efficiently recognized as correct.

$\mathsf{NP} = \bigcup_k \mathsf{NTIME}(n^k)$	Alternative definition of NP
$L_1 \leq_p L_2$	L_1 can be mapped into L_2 efficiently

If $L_2 \in \mathsf{P}$ and $L_1 \leq_{\mathrm{p}} L_2 \quad \Rightarrow \quad L_1 \in \mathsf{P}$
L is NP complete if $L \in \mathsf{NP}$ and $L_1 \leq_{\mathrm{p}} L$ for all $L \in \mathsf{NP}$.

$\mathsf{P} \subseteq \mathsf{NP} \subseteq \mathsf{PSPACE} \subseteq \mathsf{EXPTIME} \subseteq \mathsf{EXPSPACE}$

Quantum computers (QC)

U	A quantum circuit is a DAG with N inputs and N outputs
$\vert\mathbf{0}\rangle\vert\mathbf{x}\rangle$	Input of the quantum circuit
$\hat{U}\vert\mathbf{0}\rangle\vert\mathbf{x}\rangle$	Output of the quantum circuit
$p(\mathbf{y} \mid \mathbf{x})$	Probability that a measurement of $U\vert\mathbf{0}\rangle\vert\mathbf{x}\rangle$ gives $\mathbf{y}$
BQP	Decision problems decided by a QC in polynomial time with a small error probability

$\mathsf{P} \subseteq \mathsf{BPP} \subseteq \mathsf{BQP} \subseteq \mathsf{PSPACE} \subseteq \mathsf{EXPTIME}$

A problem is QMA if its affirmative solution can be efficiently verified by a QC.

S_i	Set of qubits with cardinality $\leq k$
$\hat{H} = \sum_i c_i \hat{H}_i[S_i]$	k-local Hamiltonian ($\hat{H}_i$ are positive operators and $c_i > 0$)
$\tilde{\lambda}$	Lowest eigenvalue of $\hat{H}$

$L_{k-\mathsf{LOCAL}}(a, b) = \{\mathbf{x}_{\hat{H}} \in \mathbb{B}^* : \hat{H} \text{ is } k\text{-local and } \tilde{\lambda} < a < b\}$
$L_{k-\mathsf{LOCAL}}$ is QMA complete

Exercises

Exercise 2.1 Design a DFA that recognizes the language $L = \{\mathbf{x} \in \mathbb{B}^* : x_n = x_{n-1} = x_{n-2} = 1\}$. This language contains all binary strings whose last three digits are 1.

Exercise 2.2 Design a DFA that recognizes binary strings that contain 101.

Exercise 2.3 Design a deterministic TM that converts a binary string into a string of zeros of the same length.

Exercise 2.4 Design a Turing machine with two tapes that takes two binary strings and checks whether they are identical or not. **Hint:** Design a multi-tape TM similar to the one presented in Section 2.2.6.

Exercise 2.5 Consider the 3-SAT problem

$$\varphi(\mathbf{x}) = (x_1 \vee x_2 \vee x_4) \wedge (x_1 \vee \overline{x}_3 \vee \overline{x}_4) \wedge (\overline{x}_2 \vee \overline{x}_3 \vee x_4)\,.$$

Map this problem into a CLIQUE problem and find a solution.

Solutions

Solution 2.1

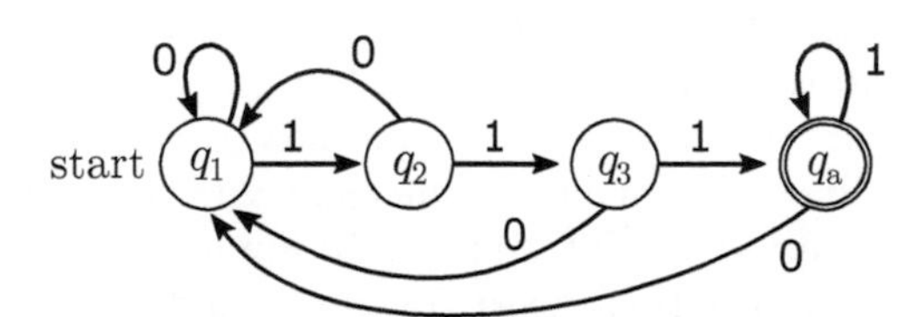

Solution 2.2

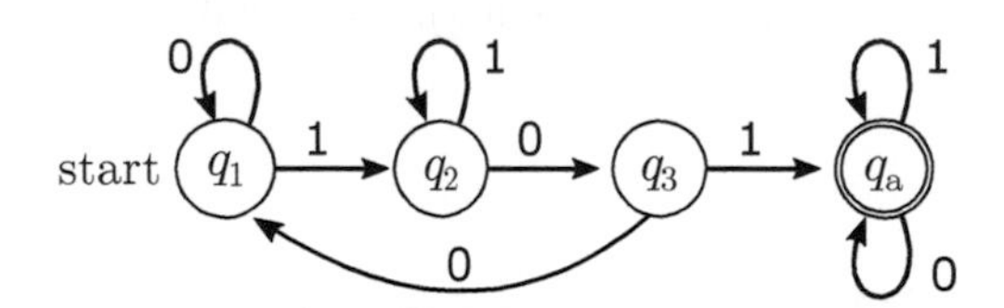

Solution 2.3

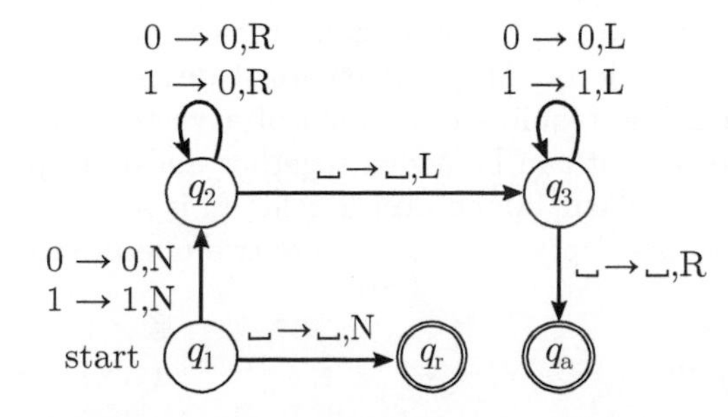

Solution 2.5 One possible solution is $(x_1 = 1,\ x_2 = 1,\ x_3 = 1,\ x_4 = 1)$.

3
Linear algebra

Eigenvalues and eigenvectors
are very important in the field
of quantum mechanics.

Quantum information theory strongly relies on linear algebra. In this chapter, we present an overview of this topic focusing on finite-dimensional vector spaces. As most readers are, to various degrees, familiar with linear algebra, the purpose of this presentation is to refresh the concepts relevant to quantum computing using the notation common to the field. Linear algebra is a subject of fundamental importance across many scientific disciplines, and it is discussed in detail in many excellent textbooks of pure and physical mathematics [106, 107, 108, 109, 110, 111, 112].

This chapter is divided into two parts. In the first part, we introduce vector spaces. We present the notions of linear dependence, dimensionality, inner product, and distance. These concepts will be used to introduce finite-dimensional Hilbert spaces, a key ingredient for quantum information theory. In the second part of the chapter, we introduce linear operators. We describe the main classes of linear operators and show their mutual relations. This mathematical framework will be crucial for discussing quantum circuits in Chapter 5.

3.1 Vector spaces

In many fields of physics, the states of a system obey the superposition principle: if s_1, s_2 are possible states of the system and c is a number different from zero, then $s_1 + s_2$ and $c\,s_1$ are possible states too. Important examples include electromagnetic fields and states of quantum objects. The mathematical description of a system obeying the superposition principle requires the notion of a vector space, i.e. a collection of objects, called **vectors**, that can be added together and multiplied by a number. In more rigorous terms, a **vector space** over a field[1] $\mathbb{K}$ is a sequence of three elements $(V, +, \cdot)$ where V is a non-empty set and $+$, $\cdot$ are two operations

$$\begin{array}{rccc} +: & V \times V & \to & V \\ & (\mathbf{v}_1, \mathbf{v}_2) & \mapsto & \mathbf{v}_1 + \mathbf{v}_2 \end{array} \qquad \begin{array}{rccc} \cdot: & \mathbb{K} \times V & \to & V \\ & (c, \mathbf{v}) & \mapsto & c \cdot \mathbf{v} \end{array} \tag{3.1}$$

[1] In this textbook, the field $\mathbb{K}$ will be either $\mathbb{R}$, $\mathbb{C}$, or the Boolean set $\mathbb{B} = \{0, 1\}$.

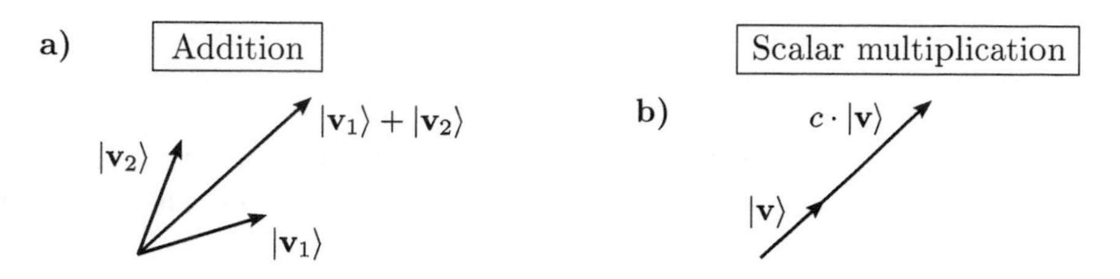

Fig. 3.1 Addition and scalar multiplication. a) A graphical representation of the sum between two vectors in $\mathbb{R}^2$. **b)** Scalar multiplication of a vector $|\mathbf{v}\rangle \in \mathbb{R}^2$ by a number $c > 1$.

called **addition** and **scalar multiplication**. For any vector $\mathbf{v}_i \in V$ and number $c_i \in \mathbb{K}$, these operations must satisfy eight requirements:

V1. $(\mathbf{v}_1 + \mathbf{v}_2) + \mathbf{v}_3 = \mathbf{v}_1 + (\mathbf{v}_2 + \mathbf{v}_3)$ (associativity of addition),

V2. $\mathbf{v}_1 + \mathbf{v}_2 = \mathbf{v}_2 + \mathbf{v}_1$ (commutativity of addition),

V3. $\exists\, \underline{\mathbf{0}} \in V\colon\ \underline{\mathbf{0}} + \mathbf{v} = \mathbf{v}$ (existence of additive zero),

V4. $\exists\, (-\mathbf{v}) \in V\colon\ \mathbf{v} + (-\mathbf{v}) = \underline{\mathbf{0}}$ (existence of additive inverse),

V5. $1 \cdot \mathbf{v} = \mathbf{v}$, where $1 \in \mathbb{K}$ (multiplicative identity),

V6. $(c_1 + c_2) \cdot \mathbf{v} = c_1 \cdot \mathbf{v} + c_2 \cdot \mathbf{v}$ (distributivity of addition),

V7. $c \cdot (\mathbf{v}_1 + \mathbf{v}_2) = c \cdot \mathbf{v}_1 + c \cdot \mathbf{v}_2$ (distributivity of multiplication),

V8. $(c_1 c_2) \cdot \mathbf{v} = c_1 \cdot (c_2 \cdot \mathbf{v})$ (compatibility).

The operations of addition and scalar multiplication are schematically illustrated in Fig. 3.1. We will mostly work with numbers taken from the complex field $\mathbb{C}$, and we will often use the well-known **Dirac notation** (also called bra-ket notation). According to this notation, vectors are indicated with "kets" $|\mathbf{v}\rangle$. We will often denote the vector space $(V, +, \cdot)$ simply with the letter V and we will omit the symbol $\cdot$ when referring to scalar multiplication. When we say that "$\mathbf{v}$ is a non-null vector" we mean that $\mathbf{v}$ is not the zero vector[2] $\underline{\mathbf{0}}$. Complex numbers $c = x + iy$ will sometimes be called **scalars** and their complex conjugate will be indicated with an asterisk[3], $c^* = x - iy$. Finally, the vector $c \cdot |\mathbf{v}\rangle$ will sometimes be written as $|c\mathbf{v}\rangle$. We now present some important examples of vector spaces.

Example 3.1 An important vector space in quantum information theory is $\mathbb{C}^n$, which is defined as

$$\boxed{\mathbb{C}^n = \{|\mathbf{v}\rangle = (v_1, \ldots, v_n) : \ v_i \in \mathbb{C}\},}$$

with addition and scalar multiplication defined as

[2]Here, the underline helps to distinguish the number 0 from the zero vector $\underline{\mathbf{0}}$.

[3]In some textbooks, the complex conjugate is denoted as $\bar{c} = x - iy$.

$$|\mathbf{v}\rangle + |\mathbf{w}\rangle = (v_1 + w_1, \ldots, v_n + w_n)$$
$$c \cdot |\mathbf{v}\rangle = (cv_1, \ldots, cv_n).$$

The vectors in this space are finite sequences of complex numbers and they are usually called $\mathbb{C}^n$ vectors. The numbers $v_1, \ldots, v_n$ are the components of $|\mathbf{v}\rangle$. The components of a $\mathbb{C}^n$ vector can be conveniently arranged in a column,

$$|\mathbf{v}\rangle = \begin{bmatrix} v_1 \\ \vdots \\ v_n \end{bmatrix}.$$

When the components v_j are restricted to real numbers only, the vector space is denoted $\mathbb{R}^n$ and it is called the **Euclidean space**.

Example 3.2 Another important vector space is $\mathrm{M}_{m\times n}(\mathbb{C})$, the set of all complex matrices with m rows and n columns,

$$\mathrm{M}_{m\times n}(\mathbb{C}) = \left\{ A = \begin{bmatrix} A_{11} & \ldots & A_{1n} \\ \vdots & \ddots & \vdots \\ A_{m1} & \ldots & A_{mn} \end{bmatrix} : A_{ij} \in \mathbb{C} \right\},$$

with addition and scalar multiplication defined as

$$A + B = \begin{bmatrix} A_{11} + B_{11} & \ldots & A_{1n} + B_{1n} \\ \vdots & \ddots & \vdots \\ A_{m1} + B_{m1} & \ldots & A_{mn} + B_{mn} \end{bmatrix},$$
$$c \cdot A = \begin{bmatrix} cA_{11} & \ldots & cA_{1n} \\ \vdots & \ddots & \vdots \\ cA_{m1} & \ldots & cA_{mn} \end{bmatrix}.$$

If $m = n$, the set $\mathrm{M}_{n\times n}(\mathbb{C})$ is usually indicated as $\mathrm{M}_n(\mathbb{C})$.

Example 3.3 The vector space $\mathbb{B}^n$ over the field $\mathbb{B} = \{0, 1\}$ is defined as the set of all binary strings of length n,

$$\mathbb{B}^n = \{\mathbf{x} = (x_1 \ldots x_n) : x_i \in \mathbb{B}\},$$

with addition and scalar multiplication defined by

$$\mathbf{x} + \mathbf{y} = (x_1 \oplus y_1, \ldots, x_n \oplus y_n)$$
$$c \cdot \mathbf{x} = (cx_1, \ldots, cx_n),$$

where $c \in \{0, 1\}$. This vector space is usually called the **Hamming space**. Here, the symbol $\oplus$ indicates the sum modulo 2 (also called the XOR gate) introduced in Section 1.1.3.

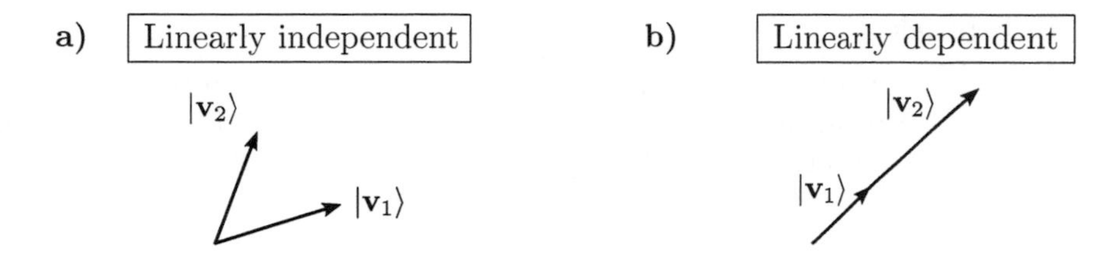

Fig. 3.2 Linear combinations. a) An example of two linearly independent vectors in $\mathbb{R}^2$. **b)** An example of two linearly dependent vectors in $\mathbb{R}^2$.

3.1.1 Linear independence and dimensionality

Addition and scalar multiplication, the fundamental operations on a vector space, can be used to define the concept of **linear combination**. A vector $|\mathbf{v}\rangle$ is a linear combination of n vectors $|\mathbf{v}_1\rangle \dots |\mathbf{v}_n\rangle$, if there exist some numbers $c_1, \dots, c_n \in \mathbb{C}$ such that $|\mathbf{v}\rangle = c_1|\mathbf{v}_1\rangle + \dots + c_n|\mathbf{v}_n\rangle$. A **convex combination** is a linear combination $\sum_i c_i|\mathbf{v}_i\rangle$ in which the coefficients c_i are non-negative real numbers that sum up to one. In the next chapter, we will see that a quantum system in a superposition state is described by a linear combination of vectors.

Linear combinations allow us to introduce the notion of linear independence, a key concept in linear algebra. The vectors $|\mathbf{v}_1\rangle, \dots, |\mathbf{v}_n\rangle$ are **linearly independent** if the linear combination $c_1|\mathbf{v}_1\rangle + \dots + c_n|\mathbf{v}_n\rangle$ is equal to the zero vector only when all the coefficients c_i are zero (see Fig. 3.2). If there exist non-zero coefficients such that $c_1|\mathbf{v}_1\rangle + \dots + c_n|\mathbf{v}_n\rangle = \underline{\mathbf{0}}$, these vectors are said to be linearly dependent.

The **dimension** of a vector space, $\dim(V)$, is the maximum number of linearly independent vectors in V. A vector space is said to be n-dimensional if $\dim(V) = n$. A set of n linearly independent vectors in a n-dimensional vector space is called a **basis**[4]. A vector space has **finite dimension** if $\dim(V) < +\infty$. An important example of a n-dimensional vector space is $\mathbb{C}^n$. In the rest of the chapter, we will focus on finite-dimensional spaces, where many technicalities specific to infinite-dimensional spaces are absent.

Example 3.4 Consider the vectors

$$\underline{\mathbf{0}} = \begin{bmatrix} 0 \\ 0 \end{bmatrix}, \qquad |\mathbf{v}_1\rangle = \begin{bmatrix} 1 \\ 0 \end{bmatrix}, \qquad |\mathbf{v}_2\rangle = \begin{bmatrix} 1 \\ 2 \end{bmatrix}, \qquad |\mathbf{v}_3\rangle = \begin{bmatrix} 2 \\ 4 \end{bmatrix}.$$

The vectors $|\mathbf{v}_1\rangle$ and $|\mathbf{v}_2\rangle$ are linearly independent, because $c_1|\mathbf{v}_1\rangle + c_2|\mathbf{v}_2\rangle = \underline{\mathbf{0}}$ if and only if $c_1 = c_2 = 0$. The vectors $|\mathbf{v}_2\rangle$ and $|\mathbf{v}_3\rangle$ are linearly dependent because $c_2|\mathbf{v}_2\rangle + c_3|\mathbf{v}_3\rangle = \underline{\mathbf{0}}$ is satisfied with $c_2 = 2$ and $c_3 = -1$. Similarly, the vectors $\underline{\mathbf{0}}$ and $|\mathbf{v}_1\rangle$ are linearly dependent because $c_0\underline{\mathbf{0}} + c_1|\mathbf{v}_1\rangle = \underline{\mathbf{0}}$ is satisfied with $c_0 = 7$ and $c_1 = 0$.

[4]In this chapter, we will use the letter n to indicate the dimension of a vector space. In the following chapters, we will predominantly use the letter d.

3.1.2 Subspaces

A subset of vectors in a vector space might form a vector space. A **subspace** W of a vector space V is a non-empty subset $W \subseteq V$ closed under both addition and scalar multiplication. This means that for all $|\mathbf{w}_i\rangle \in W$ and $c \in \mathbb{C}$ we must have:

$$\begin{aligned}|\mathbf{w}_1\rangle + |\mathbf{w}_2\rangle &\in W, \\ c \cdot |\mathbf{w}\rangle &\in W.\end{aligned}$$

A simple example of a subspace of $\mathbb{R}^2$ is the vector space $\mathbb{R}$. An important example of a subspace is the linear span, a vector space obtained by the linear combinations of a set of vectors. More formally, the **linear span** of the vectors $|\mathbf{v}_1\rangle, \ldots, |\mathbf{v}_n\rangle$ is the set

$$\text{span}(|\mathbf{v}_1\rangle, \ldots, |\mathbf{v}_n\rangle) = \{|\mathbf{v}\rangle \in V : \ |\mathbf{v}\rangle = c_1|\mathbf{v}_1\rangle + \ldots + c_n|\mathbf{v}_n\rangle, \ c_i \in \mathbb{C}\} . \tag{3.2}$$

A set of n independent vectors generates a vector space V if $\text{span}(|\mathbf{v}_1\rangle, \ldots, |\mathbf{v}_n\rangle) = V$.

A subset $W \subseteq V$ is **convex** if all convex combinations of vectors in W are still in W. An interesting example of a convex subset in quantum mechanics is the set of separable quantum states, as we shall see in Section 9.8.

Example 3.5 Consider the vector space $\mathbb{B}^3$ formed by all binary strings with three bits. The vector space $\mathbb{B}^2$ is a subspace of $\mathbb{B}^3$. The vector space $\mathbb{B}^2$ is generated by the vectors $|\mathbf{e}_1\rangle = (0, 1)$ and $|\mathbf{e}_2\rangle = (1, 0)$, i.e. $\text{span}(|\mathbf{e}_1\rangle, |\mathbf{e}_2\rangle) = \mathbb{B}^2$.

3.1.3 Inner product

The inner product is a function that associates each pair of vectors with a number. This function allows us to define the length of a vector, the distance between two vectors, and the angle between two vectors. Furthermore, it can be used to determine if two vectors are orthogonal.

The **inner product** is a function denoted as

$$\begin{array}{rccc}\langle \cdot | \cdot \rangle : & V \times V & \to & \mathbb{C} \\ & (\mathbf{v}_1, \mathbf{v}_2) & \mapsto & c = \langle \mathbf{v}_1 | \mathbf{v}_2 \rangle.\end{array} \tag{3.3}$$

In simple terms, the inner product maps a pair of vectors into a complex number. For any $\mathbf{v}_i \in V$ and $c_i \in \mathbb{C}$, the inner product must satisfy the following requirements:

I1. $\langle \mathbf{v}|\mathbf{v}\rangle \geq 0$, and $\langle \mathbf{v}|\mathbf{v}\rangle = 0 \Leftrightarrow |\mathbf{v}\rangle = \underline{\mathbf{0}}$ (positive-definiteness),

I2. $\langle \mathbf{v}|c_1\mathbf{v}_1 + c_2\mathbf{v}_2\rangle = c_1\langle \mathbf{v}|\mathbf{v}_1\rangle + c_2\langle \mathbf{v}|\mathbf{v}_2\rangle$ (linearity),

I3. $\langle \mathbf{v}_1|\mathbf{v}_2\rangle = \langle \mathbf{v}_2|\mathbf{v}_1\rangle^*$ (conjugate symmetry).

A vector space equipped with an inner product $(V, \langle \cdot | \cdot \rangle)$ is called an **inner product space**.

A **linear form** ϕ is a linear function that maps vectors into complex numbers,

$$\begin{array}{rccc}\phi : & V & \to & \mathbb{C} \\ & \mathbf{v} & \mapsto & \phi(\mathbf{v}).\end{array} \tag{3.4}$$

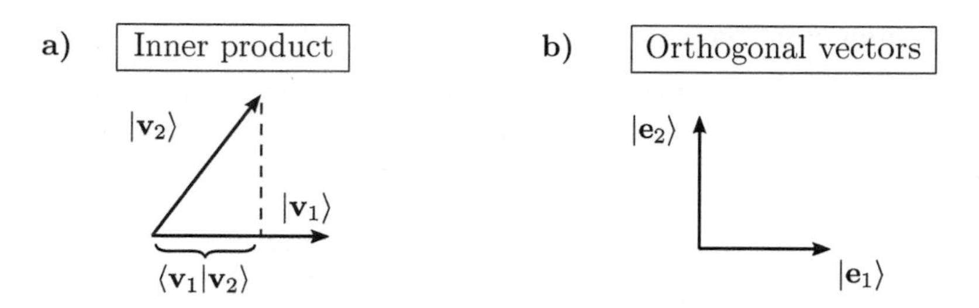

Fig. 3.3 The inner product. a) A graphical representation of the inner product between two unit vectors in $\mathbb{R}^2$. **b)** Two orthonormal vectors in $\mathbb{R}^2$.

In an inner product space $(V, \langle\cdot|\cdot\rangle)$, the function

$$\begin{array}{rccc} \langle\mathbf{w}|\cdot\rangle: & V & \to & \mathbb{C} \\ & \mathbf{v} & \mapsto & \langle\mathbf{w}|\mathbf{v}\rangle \end{array} \tag{3.5}$$

is a linear form, usually called a **bra**. When the bra $\langle\mathbf{w}|$ operates on a vector $|\mathbf{v}\rangle$, it returns the complex number $\langle\mathbf{w}|\mathbf{v}\rangle$. A collection of linear forms composes a vector space $(V^*, +, \cdot)$ called the **dual space**. The bra $\langle\mathbf{w}|$ is usually referred to as the **dual vector** of $|\mathbf{w}\rangle$.

In quantum information theory, the inner product between two vectors $|\mathbf{v}\rangle = (v_1, \ldots, v_n)$, and $|\mathbf{w}\rangle = (w_1, \ldots, w_n)$ is defined as

$$\boxed{\langle\mathbf{w}|\mathbf{v}\rangle = w_1^* v_1 + \ldots + w_n^* v_n = \sum_{j=1}^{n} w_j^* v_j.} \tag{3.6}$$

This is the inner product between $\mathbb{C}^n$ vectors that will be used in this textbook and it can be interpreted as the usual dot product between a row vector and a column vector,

$$\langle\mathbf{w}|\mathbf{v}\rangle = \begin{bmatrix} w_1^* & \ldots & w_n^* \end{bmatrix} \begin{bmatrix} v_1 \\ \vdots \\ v_n \end{bmatrix} = \sum_{j=1}^{n} w_j^* v_j. \tag{3.7}$$

Figure 3.3a illustrates the inner product between two unit vectors in $\mathbb{R}^2$. In the Euclidean space, the absolute value of the inner product can be interpreted as the length of the projection of one vector onto the other.

Example 3.6 Consider the vectors $|\mathbf{v}\rangle = (1, 2)$ and $|\mathbf{w}\rangle = (3, -4i)$. Their inner product is given by

$$\langle\mathbf{w}|\mathbf{v}\rangle = \sum_{j=1}^{2} w_j^* v_j = \begin{bmatrix} 3 & 4i \end{bmatrix} \begin{bmatrix} 1 \\ 2 \end{bmatrix} = 3 + 8i. \tag{3.8}$$

3.1.4 Orthonormal bases

In the Euclidean space $\mathbb{R}^2$, two orthogonal vectors can be visualized as two arrows perpendicular to one another as shown in Fig. 3.3b. In general, two vectors are **orthogonal** when $\langle \mathbf{v}_1 | \mathbf{v}_2 \rangle = 0$. Similarly, a set of independent vectors $\{|\mathbf{e}_1\rangle, \ldots, |\mathbf{e}_n\rangle\}$ is **orthonormal** when

$$\boxed{\langle \mathbf{e}_i | \mathbf{e}_j \rangle = \delta_{ij} = \begin{cases} 0 & \text{when } i \neq j \\ 1 & \text{when } i = j \end{cases}.} \tag{3.9}$$

Here, we introduced the **Kronecker delta** δ_{ij}, a very useful function in linear algebra. In the vector space $\mathbb{C}^n$, the orthonormal basis

$$B = \{|\mathbf{e}_1\rangle = (1, 0, \ldots, 0), \ldots, |\mathbf{e}_n\rangle = (0, \ldots, 0, 1)\} = \{|\mathbf{e}_i\rangle\}_{i=1}^n \tag{3.10}$$

is called the **canonical basis**. To simplify the notation, we will often write this basis as $B = \{|1\rangle, |2\rangle, \ldots, |n\rangle\} = \{|i\rangle\}_{i=1}^n$. In quantum information, the canonical basis of the vector space $\mathbb{C}^2$ is usually written as $B = \{|0\rangle, |1\rangle\}$ where

$$|0\rangle = \begin{bmatrix} 1 \\ 0 \end{bmatrix}, \qquad |1\rangle = \begin{bmatrix} 0 \\ 1 \end{bmatrix}.$$

This basis is called the **computational basis**.

Example 3.7 Orthonormal bases can be used to calculate the inner product between two vectors in a simple way. Consider an orthonormal basis $B = \{|\mathbf{e}_1\rangle, |\mathbf{e}_2\rangle\}$ of $\mathbb{C}^2$. The inner product between the vectors

$$|\mathbf{v}\rangle = 1|\mathbf{e}_1\rangle + 2|\mathbf{e}_2\rangle, \qquad |\mathbf{w}\rangle = 3|\mathbf{e}_1\rangle - 4i|\mathbf{e}_2\rangle \tag{3.11}$$

can be obtained by constructing the dual vector $\langle \mathbf{w}| = 3\langle \mathbf{e}_1| + 4i\langle \mathbf{e}_2|$ and calculating

$$\begin{aligned} \langle \mathbf{w}|\mathbf{v}\rangle &= (3\langle \mathbf{e}_1| + 4i\langle \mathbf{e}_2|)\,(1|\mathbf{e}_1\rangle + 2|\mathbf{e}_2\rangle) \\ &= 3\underbrace{\langle \mathbf{e}_1|\mathbf{e}_1\rangle}_{1} + 3 \cdot 2\underbrace{\langle \mathbf{e}_1|\mathbf{e}_2\rangle}_{0} + 4i \cdot 2\underbrace{\langle \mathbf{e}_2|\mathbf{e}_1\rangle}_{0} + 8i\underbrace{\langle \mathbf{e}_2|\mathbf{e}_2\rangle}_{1} = 3 + 8i. \end{aligned}$$

Here, $\langle \mathbf{e}_1|\mathbf{e}_1\rangle = \langle \mathbf{e}_2|\mathbf{e}_2\rangle = 1$ and $\langle \mathbf{e}_1|\mathbf{e}_2\rangle = \langle \mathbf{e}_2|\mathbf{e}_1\rangle = 0$ since B is orthonormal.

3.1.5 Norm

The **norm** is a function that quantifies the length of a vector (see Fig. 3.4a for a representation of the norm of a vector in $\mathbb{R}^2$). This function is defined as

$$\begin{aligned} \|\cdot\| : \quad V &\to \quad \mathbb{R} \\ \mathbf{v} &\mapsto \quad \|\mathbf{v}\|. \end{aligned} \tag{3.12}$$

For any $\mathbf{v}_i \in V$ and $c_i \in \mathbb{C}$, the norm must satisfy:

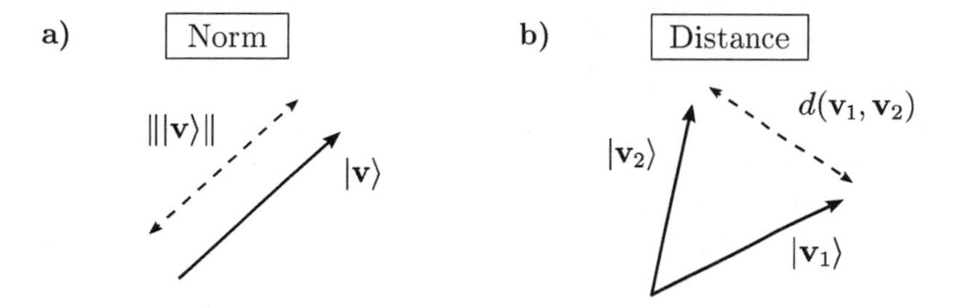

Fig. 3.4 Norms and distances. a) Norm of a vector in $\mathbb{R}^2$. **b)** Distance between two vectors in $\mathbb{R}^2$.

N1. $\|\mathbf{v}\| \geq 0$, and $\|\mathbf{v}\| = 0 \iff \mathbf{v} = \underline{\mathbf{0}}$ (positive-definiteness),

N2. $\|c\mathbf{v}\| = |c| \, \|\mathbf{v}\|$ (homogeneity),

N3. $\|\mathbf{v}_1 + \mathbf{v}_2\| \leq \|\mathbf{v}_1\| + \|\mathbf{v}_2\|$ (triangle inequality).

Vectors that satisfy $\|\mathbf{v}\| = 1$ are called **normalized vectors** (or unit vectors). Given a vector $\mathbf{v} \neq \underline{\mathbf{0}}$, the corresponding normalized vector is $c|\mathbf{v}\rangle$ where $c = 1/\,\|\mathbf{v}\|$. A vector space equipped with a norm $(V, \|\cdot\|)$ is called a **normed space**. Note that an inner product always induces a norm through the map $\sqrt{\langle \mathbf{v}|\mathbf{v}\rangle} = \|\mathbf{v}\|$.

The norm of a $\mathbb{C}^n$ vector can be defined using the inner product presented in Section 3.1.3,

$$\boxed{\|\mathbf{v}\| = \sqrt{\langle \mathbf{v}|\mathbf{v}\rangle} = \sqrt{\sum_{i=1}^{n} v_i^* v_i} = \sqrt{\sum_{i=1}^{n} |v_i|^2}\,.} \tag{3.13}$$

This is the norm we will adopt hereafter for the vector space $\mathbb{C}^n$ (unless otherwise specified). We stress once more that an inner product can always be used to define a norm in a vector space.

Example 3.8 The norm of the vector $|\mathbf{v}\rangle = 2i|\mathbf{e}_1\rangle + (5 + 7i)|\mathbf{e}_2\rangle$ is given by

$$\begin{aligned}\|\mathbf{v}\| = \sqrt{\langle \mathbf{v}|\mathbf{v}\rangle} &= [(-2i\langle \mathbf{e}_1| + (5 - 7i)\langle \mathbf{e}_2|) \cdot (2i|\mathbf{e}_1\rangle + (5 + 7i)|\mathbf{e}_2\rangle)]^{1/2} \\ &= [(-2i)(2i) + (5 - 7i)(5 + 7i)]^{1/2} = \sqrt{78}.\end{aligned}$$

The vector $|\mathbf{v}\rangle/\sqrt{78}$ is a normalized vector.

3.1.6 Fidelity

The **fidelity** (or Uhlmann fidelity) between two normalized vectors is the absolute value of their inner product,

$$\begin{array}{rccc} F: & V \times V & \to & [0,1] \\ & (\mathbf{v}_1, \mathbf{v}_2) & \mapsto & |\langle \mathbf{v}_1 | \mathbf{v}_2 \rangle|. \end{array} \tag{3.14}$$

The fidelity provides a quantitative measure of the similarity of two vectors: if two vectors are identical, the fidelity is equal to one. If two vectors are orthogonal, the fidelity is equal to zero. The fidelity is an important function in quantum information theory as it can be used to estimate the performance of a quantum protocol. In some textbooks, the fidelity is defined as $|\langle \mathbf{v}_1 | \mathbf{v}_2 \rangle|^2$. This quantity is sometimes called the **Jozsa fidelity**.

3.1.7 Distance

The distance between two vectors on a plane can be interpreted as the length of the vector connecting their ends when their origins coincide (see Fig. 3.4**b** for a graphical example). In general, the **distance** between two vectors is defined as

$$\begin{array}{rccc} d: & V \times V & \to & \mathbb{R} \\ & (\mathbf{v}_1, \mathbf{v}_2) & \mapsto & d(\mathbf{v}_1, \mathbf{v}_2). \end{array} \tag{3.15}$$

For any $\mathbf{v}_i \in V$, the function d must satisfy[5]:

D1. $d(\mathbf{v}_1, \mathbf{v}_2) \geq 0$ (positive-definiteness),

D2. $d(\mathbf{v}_1, \mathbf{v}_2) = 0 \iff \mathbf{v}_1 = \mathbf{v}_2$ (non-degenerate),

D3. $d(\mathbf{v}_1, \mathbf{v}_2) = d(\mathbf{v}_2, \mathbf{v}_1)$ (symmetry),

D4. $d(\mathbf{v}_1, \mathbf{v}_3) \leq d(\mathbf{v}_1, \mathbf{v}_2) + d(\mathbf{v}_2, \mathbf{v}_3)$ (triangle inequality).

A **metric vector space** (V, d) is a vector space V endowed with a distance d.

An inner product can always be used to define a distance. This follows from the fact that the function $d(\mathbf{v}_1, \mathbf{v}_2) = \sqrt{\langle \mathbf{v}_1 - \mathbf{v}_2 | \mathbf{v}_1 - \mathbf{v}_2 \rangle}$ is a valid distance. Thus, the distance between two $\mathbb{C}^n$ vectors can be defined as

$$\boxed{d(\mathbf{v}, \mathbf{w}) = \sqrt{\langle \mathbf{v} - \mathbf{w} | \mathbf{v} - \mathbf{w} \rangle} = \sqrt{\sum_{i=1}^{n} |v_i - w_i|^2}.}$$

This is the most important distance for the vector space $\mathbb{C}^n$.

Example 3.9 Let us calculate the distance between the vectors $|\mathbf{v}\rangle = (0, 2i, -2)$, and $|\mathbf{w}\rangle = (3, 1, 2)$ in the vector space $\mathbb{C}^3$. Since $|\mathbf{v} - \mathbf{w}\rangle = (-3, 2i - 1, -4)$, then

$$d(\mathbf{v}, \mathbf{w}) = \sqrt{\langle \mathbf{v} - \mathbf{w} | \mathbf{v} - \mathbf{w} \rangle} = \sqrt{|-3|^2 + |2i - 1|^2 + |-4|^2} = \sqrt{30}.$$

[5] Property **D1** could be dropped, because it can be derived by combining the other three requirements.

Example 3.10 Let us calculate the distance between the vectors $|\mathbf{v}\rangle = (1,0,0)$, and $|\mathbf{w}\rangle = (1,0,1)$ in the vector space $\mathbb{B}^3$. First, we calculate the vector $|\mathbf{v}\rangle - |\mathbf{w}\rangle = (1\oplus 1, 0\oplus 0, 0\oplus 1) = (0,0,1)$. The distance between two binary strings can be defined as the number of 1's in the string $|\mathbf{v}\rangle - |\mathbf{w}\rangle$. In our example, $d(\mathbf{v},\mathbf{w}) = 1$. This is called the Hamming distance.

3.1.8 Cauchy–Schwarz inequality

The **Cauchy–Schwarz inequality** is a very important relation between vectors. This inequality states that for any two vectors $\mathbf{v}$, $\mathbf{w}$ in an inner product space

$$\boxed{|\langle\mathbf{w}|\mathbf{v}\rangle|^2 \leq \langle\mathbf{w}|\mathbf{w}\rangle\langle\mathbf{v}|\mathbf{v}\rangle = \|\mathbf{w}\|^2 \cdot \|\mathbf{v}\|^2\,.} \tag{3.16}$$

By taking the square root, this inequality can also be written as $|\langle\mathbf{w}|\mathbf{v}\rangle| \leq \|\mathbf{w}\| \cdot \|\mathbf{v}\|$. The equality holds only if $\mathbf{v}$ and $\mathbf{w}$ are linearly dependent. The Cauchy–Schwarz inequality has a simple geometric interpretation in the Euclidean space: the absolute value of the overlap between two vectors is not greater than the product between their lengths. We will now outline a proof of this relation.

The Cauchy–Schwarz inequality (3.16) clearly holds for $|\mathbf{w}\rangle = \underline{\mathbf{0}}$. Hence, we will assume $|\mathbf{w}\rangle \neq \underline{\mathbf{0}}$. Consider the vector $|\mathbf{y}\rangle = |\mathbf{v}\rangle - c|\mathbf{w}\rangle$. By definition of inner product

$$\begin{aligned} 0 \leq \langle\mathbf{y}|\mathbf{y}\rangle &= (\langle\mathbf{v}| - c^*\langle\mathbf{w}|)\,(|\mathbf{v}\rangle - c|\mathbf{w}\rangle) \\ &= \langle\mathbf{v}|\mathbf{v}\rangle - c\langle\mathbf{v}|\mathbf{w}\rangle - c^*\langle\mathbf{w}|\mathbf{v}\rangle + cc^*\langle\mathbf{w}|\mathbf{w}\rangle. \end{aligned} \tag{3.17}$$

No assumption was made about the complex number c. We can set $c = \langle\mathbf{w}|\mathbf{v}\rangle/\langle\mathbf{w}|\mathbf{w}\rangle$ and therefore $c^* = \langle\mathbf{v}|\mathbf{w}\rangle/\langle\mathbf{w}|\mathbf{w}\rangle$. Substituting these expressions into (3.17), we find

$$0 \leq \langle\mathbf{v}|\mathbf{v}\rangle - \frac{\langle\mathbf{w}|\mathbf{v}\rangle}{\langle\mathbf{w}|\mathbf{w}\rangle}\langle\mathbf{v}|\mathbf{w}\rangle\,. \tag{3.18}$$

Multiply both members by $\langle\mathbf{w}|\mathbf{w}\rangle$ and obtain $\langle\mathbf{w}|\mathbf{v}\rangle\langle\mathbf{v}|\mathbf{w}\rangle \leq \langle\mathbf{w}|\mathbf{w}\rangle\langle\mathbf{v}|\mathbf{v}\rangle$, as required.

3.1.9 Gram–Schmidt decomposition

The **Gram–Schmidt decomposition** is a systematic procedure that converts any basis into an orthonormal basis. More formally, this decomposition creates a set B of orthonormal vectors starting from a generic set A of linearly independent vectors such that $\text{span}\,(B) = \text{span}\,(A)$. Figure 3.5 shows the steps of the Gram–Schmidt decomposition for two linearly independent vectors in $\mathbb{R}^2$.

Let us systematically construct a set of orthonormal vectors B starting from A. The first vector of the set B can be defined as $|\mathbf{w}_1\rangle = |\mathbf{v}_1\rangle$, where $|\mathbf{v}_1\rangle \neq \mathbf{0}$ by hypothesis. The second vector $|\mathbf{w}_2\rangle$ can be constructed by removing from $|\mathbf{v}_2\rangle$ the component collinear to $|\mathbf{w}_1\rangle$,

$$|\mathbf{w}_2\rangle = |\mathbf{v}_2\rangle - \frac{\langle\mathbf{w}_1|\mathbf{v}_2\rangle}{\langle\mathbf{w}_1|\mathbf{w}_1\rangle}|\mathbf{w}_1\rangle\,. \tag{3.19}$$

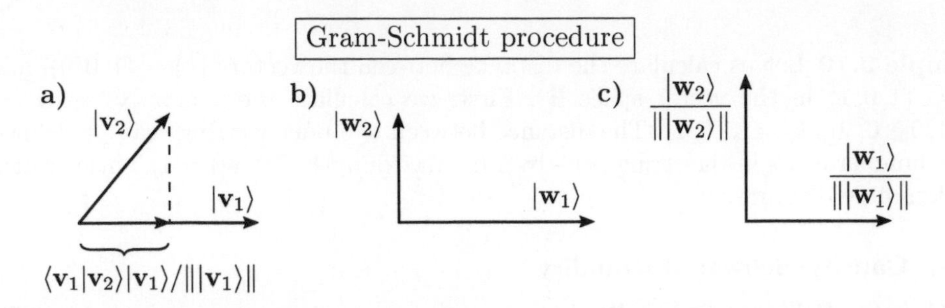

Fig. 3.5 The Gram–Schmidt procedure in $\mathbb{R}^2$. a) We start from two linearly independent vectors $A = \{|\mathbf{v}_1\rangle, |\mathbf{v}_2\rangle\}$. **b)** We set $|\mathbf{w}_1\rangle = |\mathbf{v}_1\rangle$ and we construct the vector $|\mathbf{w}_2\rangle = |\mathbf{v}_2\rangle - \langle\mathbf{v}_1|\mathbf{v}_2\rangle|\mathbf{v}_1\rangle/\||\mathbf{v}_1\rangle\|$. The vectors $|\mathbf{w}_1\rangle$, $|\mathbf{w}_2\rangle$ are orthogonal by definition. **c)** In the last step, we normalize the vectors $|\mathbf{w}_1\rangle$, $|\mathbf{w}_2\rangle$ and obtain the orthonormal basis $B = \{|\mathbf{w}_1\rangle/\||\mathbf{w}_1\rangle\|, |\mathbf{w}_2\rangle/\||\mathbf{w}_2\rangle\|\}$.

The vector $|\mathbf{w}_3\rangle$ can be defined in a similar way. At the i-th step, we have

$$\boxed{|\mathbf{w}_i\rangle = |\mathbf{v}_i\rangle - \sum_{k=1}^{i-1} \frac{\langle\mathbf{w}_k|\mathbf{v}_i\rangle}{\langle\mathbf{w}_k|\mathbf{w}_k\rangle}|\mathbf{w}_k\rangle.} \tag{3.20}$$

Therefore, $B = \{|\mathbf{w}_1\rangle/\,\||\mathbf{w}_1\rangle\|\,, \ldots, |\mathbf{w}_n\rangle/\,\||\mathbf{w}_n\rangle\|\}$ is a set of orthonormal vectors. This set also satisfies span (B) = span (A). As mentioned before, this procedure can be used to transform any basis into an orthonormal basis.

3.1.10 Cauchy sequences and Hilbert spaces

Cauchy sequences are crucial to introduce the notions of complete spaces and Hilbert spaces. A sequence of vectors $(\mathbf{v}_1, \mathbf{v}_2, \mathbf{v}_3, \ldots)$ in a metric vector space (V, d) is called **Cauchy** if

$$\forall\epsilon > 0, \quad \exists N > 0: \quad \forall n, m > N \quad d(\mathbf{v}_n, \mathbf{v}_m) < \epsilon\,. \tag{3.21}$$

In other words, a sequence whose vectors become arbitrarily close to one another as the sequence progresses is called Cauchy. The limit of a Cauchy sequence does not need to belong to the vector space V.

A metric vector space (V, d) is **complete** if the limit of all Cauchy sequences in V belongs to V. An inner product space $(V, \langle\cdot|\cdot\rangle)$ is a **Hilbert space** if it is complete with respect to the distance induced by the inner product (i.e. all Cauchy sequences converge in V). A Hilbert space is **separable** if and only if it has a basis with a finite or countably infinite number of elements. From here on, we will only consider separable Hilbert spaces and we will indicate them with the letter H. Hilbert spaces are the essence of quantum mechanics.

Example 3.11 A simple example of Cauchy sequence in $\mathbb{R}$ is $(1/n)_{n=1}^{\infty}$. This sequence converges to zero.

Example 3.12 The inner product space $(\mathbb{C}^n, \langle \cdot | \cdot \rangle)$ with the inner product defined by $\langle \mathbf{w}|\mathbf{v}\rangle = \sum_i w_i^* v_i$ is a Hilbert space. This is the most important Hilbert space in quantum information.

3.2 Linear operators

Physical theories are often concerned with the time evolution of a physical system, which usually entails transforming one state into another. These operations typically preserve the superposition principle. This requirement naturally leads us to introduce the notion of a linear operator.

An **operator** $\hat{A}$ is a function between two vector spaces[6]

$$\begin{aligned} \hat{A}: \quad & V && \rightarrow && W \\ & |\mathbf{v}\rangle && \mapsto && \hat{A}|\mathbf{v}\rangle \end{aligned} \tag{3.22}$$

where in general $\dim(V) \neq \dim(W)$. The space V is the domain of $\hat{A}$, and W is the codomain. An operator is called **linear** if for all $|\mathbf{v}_i\rangle \in V$ and $c_i \in \mathbb{C}$,

$$\boxed{\hat{A}\left(c_1|\mathbf{v}_1\rangle + c_2|\mathbf{v}_2\rangle\right) = c_1(\hat{A}|\mathbf{v}_1\rangle) + c_2(\hat{A}|\mathbf{v}_2\rangle).} \tag{3.23}$$

A linear operator is determined by its action on the basis vectors.

The set of all linear operators is denoted as $\mathrm{L}\,(V, W)$. If the domain and codomain are the same, the set $\mathrm{L}\,(V, V)$ is usually denoted as $\mathrm{L}\,(V)$ and the operators in this set are called **endomorphisms**. The vector $\hat{A}|\mathbf{v}\rangle$ will sometimes be written as $|\hat{A}\mathbf{v}\rangle$. Similarly, the bra corresponding to the vector $\hat{A}|\mathbf{v}\rangle$ will sometimes be written[7] as $\langle \hat{A}\mathbf{v}|$. A bijective function between two vector spaces is called an **isomorphism**. If there exists an isomorphism between two vector spaces, they are said to be **isomorphic**.

The **composition** of two endomorphisms $\hat{A}, \hat{B} \in \mathrm{L}(V)$ is the operator $\hat{C} = \hat{A}\hat{B}$. The application of $\hat{C}$ on a generic vector is performed as

$$\hat{C}|\mathbf{v}\rangle = (\hat{A}\hat{B})|\mathbf{v}\rangle = \hat{A}(\hat{B}|\mathbf{v}\rangle)\,, \tag{3.24}$$

i.e. applying the operator $\hat{C}$ is equivalent to applying $\hat{B}$ to $|\mathbf{v}\rangle$ and then $\hat{A}$ to the vector $\hat{B}|\mathbf{v}\rangle$. The operator $\hat{A}^m = \hat{A}\ldots\hat{A}$ is the composition of $\hat{A}$ with itself m times. In a vector space V, the **identity operator** is defined by the relation $\mathbb{1}|\mathbf{v}\rangle = |\mathbf{v}\rangle$ for any $\mathbf{v} \in V$. We will sometimes use the notation $\mathbb{1}_V$ to highlight that the identity operator acts on vectors in V. The composition $\hat{A}\hat{B}$ might differ from $\hat{B}\hat{A}$. We will come back to this point in the next section.

Example 3.13 Important examples of operators are linear maps $\hat{A} : \mathbb{C}^n \rightarrow \mathbb{C}^n$ that transform $\mathbb{C}^n$ vectors into $\mathbb{C}^n$ vectors. From a geometrical point of view, these transformations can be visualized as operations that rotate, contract, stretch, and project

[6] The hat ^ distinguishes operators from complex numbers. It will be omitted in some cases to simplify the notation.

[7] For example, if $|\hat{A}\mathbf{v}\rangle = 7|\mathbf{e}_1\rangle + 2i|\mathbf{e}_2\rangle$, then $\langle \hat{A}\mathbf{v}| = 7\langle \mathbf{e}_1| - 2i\langle \mathbf{e}_2|$.

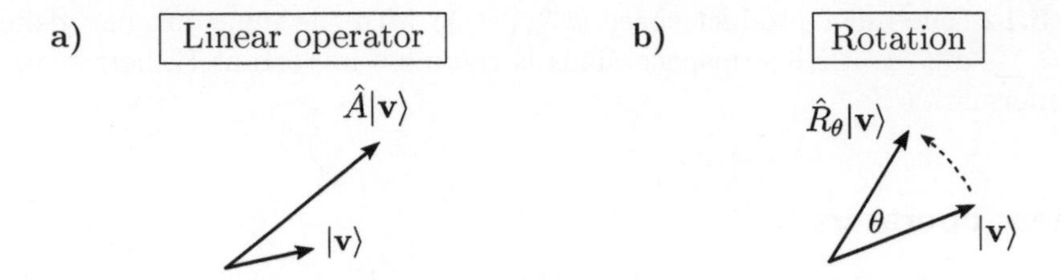

Fig. 3.6 Operators. a) Representation of a vector $|\mathbf{v}\rangle$ in $\mathbb{R}^2$ and the transformed vector $\hat{A}|\mathbf{v}\rangle$. The operator $\hat{A}$ stretches and rotates the initial vector. **b)** Graphical representation of the rotation $\hat{R}_\theta$ acting on a two-dimensional vector.

arrows (see Fig. 3.6a for a graphical representation of an operator acting on a vector in $\mathbb{R}^2$). Their application to a vector $|\mathbf{v}\rangle = (v_1, \dots, v_n)$ can be expressed with a matrix

$$\hat{A}|\mathbf{v}\rangle = \begin{bmatrix} A_{11} & \dots & A_{1n} \\ \vdots & \ddots & \vdots \\ A_{n1} & \dots & A_{nn} \end{bmatrix} \begin{bmatrix} v_1 \\ \vdots \\ v_n \end{bmatrix} = \begin{bmatrix} A_{11}v_1 + \ldots + A_{1n}v_n \\ \vdots \\ A_{n1}v_1 + \ldots + A_{nn}v_n \end{bmatrix}. \tag{3.25}$$

The composition of two linear operators $\hat{A}, \hat{B} \in \mathrm{L}(\mathbb{C}^n)$ is given by the usual matrix multiplication

$$\hat{A}\hat{B} = \begin{bmatrix} A_{11}B_{11} + \ldots + A_{1n}B_{n1} & \dots & A_{11}B_{1n} + \ldots + A_{1n}B_{nn} \\ \vdots & \ddots & \vdots \\ A_{n1}B_{11} + \ldots + A_{nn}B_{n1} & \dots & A_{n1}B_{1n} + \ldots + A_{nn}B_{nn} \end{bmatrix}. \tag{3.26}$$

As we shall see in Section 3.2.3, it is always possible to associate a matrix with a linear operator.

Example 3.14 An interesting example of a linear operator in the Euclidean plane $\mathbb{R}^2$ is the rotation $\hat{R}_\theta$,

$$\hat{R}_\theta = \begin{bmatrix} \cos\theta & -\sin\theta \\ \sin\theta & \cos\theta \end{bmatrix}.$$

This operator rotates vectors counterclockwise by an angle θ. Figure 3.6b shows an example of such a rotation.

3.2.1 Commutator and anticommutator

Unlike ordinary products between real numbers, operator composition is not always commutative. This means that the operator $\hat{A}\hat{B}$ might act in a different way from the operator $\hat{B}\hat{A}$. This observation naturally leads to the definition of the **commutator** between two operators,

$$\boxed{[\hat{A}, \hat{B}] = \hat{A}\hat{B} - \hat{B}\hat{A}.} \tag{3.27}$$

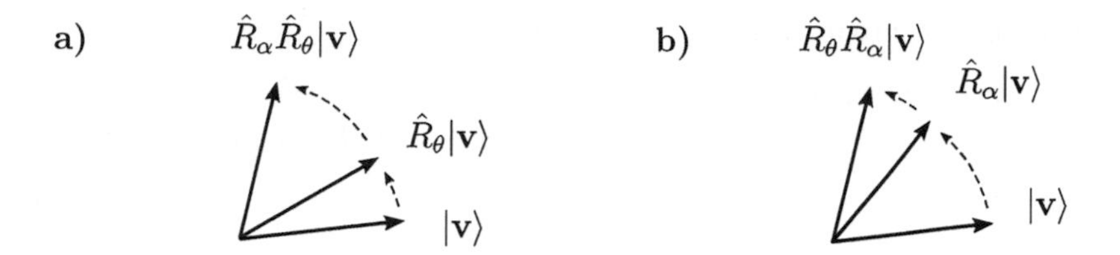

Fig. 3.7 Rotations in the Euclidean plane. a) A rotation $\hat{R}_\theta$ of the vector $|\mathbf{v}\rangle$ followed by a rotation $\hat{R}_\alpha$. **b)** A rotation $\hat{R}_\alpha$ of the vector $|\mathbf{v}\rangle$ followed by a rotation $\hat{R}_\theta$. The combination of these two rotations can be performed in any order and the result remains unchanged.

Two operators are said to commute if $[\hat{A}, \hat{B}] = \underline{\mathbf{0}}$, with $\underline{\mathbf{0}}$ denoting the zero operator. Similarly, the **anticommutator** between two operators is defined as[8]

$$\boxed{\{\hat{A}, \hat{B}\} = \hat{A}\hat{B} + \hat{B}\hat{A}.} \tag{3.28}$$

The commutator and anticommutator satisfy the properties listed below:

Commutator	Anticommutator
C1. $[\hat{A}, \mathbb{1}] = \underline{\mathbf{0}}$	**AC1.** $\{\hat{A}, \mathbb{1}\} = 2\hat{A}$
C2. $[\hat{A}, \hat{A}^n] = \underline{\mathbf{0}}$	**AC2.** $\{\hat{A}, \hat{A}^n\} = 2\hat{A}^{n+1}$
C3. $[\hat{A}, \hat{B}] = -[\hat{B}, \hat{A}]$	**AC3.** $\{\hat{A}, \hat{B}\} = \{\hat{B}, \hat{A}\}$
C4. $[\hat{A}, \hat{B} + \hat{C}] = [\hat{A}, \hat{B}] + [\hat{A}, \hat{C}]$	**AC4.** $\{\hat{A}, \hat{B} + \hat{C}\} = \{\hat{A}, \hat{B}\} + \{\hat{A}, \hat{C}\}$
C5. $[\hat{A}\hat{B}, \hat{C}] = \hat{A}[\hat{B}, \hat{C}] + [\hat{A}, \hat{C}]\hat{B}$	**AC5.** $\{\hat{A}\hat{B}, \hat{C}\} = \hat{A}\{\hat{B}, \hat{C}\} - \{\hat{A}, \hat{C}\}\hat{B}$

where $n \geq 1$. These relations can be proven by explicitly expanding commutators and anticommutators according to their definition.

Example 3.15 Consider the operators

$$\hat{R}_\theta = \begin{bmatrix} \cos\theta & -\sin\theta \\ \sin\theta & \cos\theta \end{bmatrix}, \qquad \hat{R}_\alpha = \begin{bmatrix} \cos\alpha & -\sin\alpha \\ \sin\alpha & \cos\alpha \end{bmatrix}.$$

These operators rotate vectors in the Euclidean plane counterclockwise by an angle θ and α, respectively. With a simple matrix multiplication, it is possible to show that they commute, $\hat{R}_\theta\hat{R}_\alpha = \hat{R}_\alpha\hat{R}_\theta$. This is consistent with the fact that rotations in $\mathbb{R}^2$ can be executed in any order and the result remains unchanged (see Fig. 3.7a–b for a graphical representation).

[8]In some texts, the anticommutator is denoted $[\hat{A}, \hat{B}]_+$.

3.2.2 Bounded operators and operator space

Most of the operators presented in this textbook are bounded. A linear operator $\hat{A} : V \to V$ is **bounded** if there exists a real number $M > 0$ such that for any $|\mathbf{v}\rangle \in V$,

$$\|\hat{A}|\mathbf{v}\rangle\| \le M \, \||\mathbf{v}\rangle\| \, . \tag{3.29}$$

The smallest M that satisfies this requirement is called the **norm of the operator** $\hat{A}$ and it is indicated with the notation $\|\hat{A}\|_{\mathrm{o}}$. The operator norm can be interpreted as the largest value by which $\hat{A}$ stretches an element of the vector space[9]

$$\boxed{\|\hat{A}\|_{\mathrm{o}} = \sup_{\|\mathbf{v}\|=1} \|\hat{A}|\mathbf{v}\rangle\|.} \tag{3.30}$$

The set of all bounded operators is indicated as $\mathrm{B}(V)$. It can be shown that if a vector space V is finite dimensional, then $\mathrm{L}(V) = \mathrm{B}(V)$ [113]. For readers already familiar with the concept of eigenvalue, we anticipate that the operator norm of a normal operator $\hat{A}$ is nothing but the absolute value of the eigenvalue with maximum magnitude.

Operators can be considered vectors themselves. A collection of linear operators forms a vector space $(\mathrm{L}(V, W), +, \cdot)$ with the operations of addition and scalar multiplication defined as

$$(\hat{A} + \hat{B})|\mathbf{v}\rangle = \hat{A}|\mathbf{v}\rangle + \hat{B}|\mathbf{v}\rangle, \tag{3.31}$$

$$(c \cdot \hat{A})|\mathbf{v}\rangle = c(\hat{A}|\mathbf{v}\rangle). \tag{3.32}$$

When the domain and codomain are identical, the vector space $\mathrm{L}(V, V)$ is usually denoted as $\mathrm{L}(V)$. A linear operator $\mathcal{L} : \mathrm{L}(V) \to \mathrm{L}(V)$ that transforms an operator into another is called a **superoperator**. The transposition and the dephasing map, presented in Sections 7.5–7.4, are examples of superoperators.

Example 3.16 An important example of a vector space whose elements are linear operators is $\mathrm{L}(\mathbb{C}^n)$, the space of linear operators transforming $\mathbb{C}^n$ vectors into $\mathbb{C}^n$ vectors. In this space, the addition and scalar multiplication are defined as

$$\hat{A} + \hat{B} = \begin{bmatrix} A_{11} & \dots & A_{1n} \\ \vdots & \ddots & \vdots \\ A_{n1} & \dots & A_{nn} \end{bmatrix} + \begin{bmatrix} B_{11} & \dots & B_{1n} \\ \vdots & \ddots & \vdots \\ B_{n1} & \dots & B_{nn} \end{bmatrix}, \tag{3.33}$$

$$c \cdot \hat{A} = \begin{bmatrix} cA_{11} & \dots & cA_{1n} \\ \vdots & \ddots & \vdots \\ cA_{n1} & \dots & cA_{nn} \end{bmatrix}. \tag{3.34}$$

[9] The notation "sup" indicates the supremum of a set, not to be confused with the maximum of a set. For example, the supremum of the set $S = [0, 1) \subset \mathbb{R}$ is 1, whereas $\max(S)$ does not exist.

As we shall see in Section 3.2.14, the vector space $\mathrm{L}(\mathbb{C}^n)$ can be endowed with an inner product, such as the Hilbert–Schmidt inner product.

Example 3.17 Consider the operator $A = \begin{bmatrix} 0 & 2 \\ 3 & -1 \end{bmatrix}$. In Example 3.22, we will show that this operator has eigenvalues -3 and 2. The norm of this operator is $\|\hat{A}\|_{\mathrm{o}} = 3$, since -3 is the eigenvalue with the largest absolute value.

3.2.3 Projectors and completeness relation

Projectors are important operators in quantum theory as they describe a class of measurement processes. Formally, a **projector** is an endomorphism $\hat{P} : V \to V$ that satisfies

$$\boxed{\hat{P}^2 = \hat{P}.} \tag{3.35}$$

From a geometrical point of view, $\hat{P}_{\mathbf{e}}$ projects vectors along the direction $|\mathbf{e}\rangle$, where $|\mathbf{e}\rangle$ is a normalized vector. The application of $\hat{P}_{\mathbf{e}}$ on a generic vector can be expressed as

$$\hat{P}_{\mathbf{e}}|\mathbf{v}\rangle = \Big(|\mathbf{e}\rangle\langle\mathbf{e}|\Big)\ |\mathbf{v}\rangle = |\mathbf{e}\rangle\underbrace{\langle\mathbf{e}|\mathbf{v}\rangle}_{c} = c|\mathbf{e}\rangle\,. \tag{3.36}$$

This is the beauty of Dirac notation: inner products can be moved to either side of a vector in an elegant way and the expression still holds. The relation that defines a projector $\hat{P}^2 = \hat{P}$ indicates that when a projector is applied to a vector twice, the same result is obtained, i.e.

$$\hat{P}_{\mathbf{e}}^2|\mathbf{v}\rangle = |\mathbf{e}\rangle\underbrace{\langle\mathbf{e}|\mathbf{e}\rangle}_{1}\langle\mathbf{e}|\ |\mathbf{v}\rangle = |\mathbf{e}\rangle\langle\mathbf{e}|\ |\mathbf{v}\rangle = \hat{P}_{\mathbf{e}}|\mathbf{v}\rangle\,. \tag{3.37}$$

Figure 3.8 shows the action of a projector on a two-dimensional vector. In general, given a set of orthonormal vectors $|\mathbf{e}_1\rangle, \ldots, |\mathbf{e}_k\rangle$, a projector $\hat{P}_W$ onto the subspace $W = \mathrm{span}\,(|\mathbf{e}_1\rangle, \ldots, |\mathbf{e}_k\rangle)$ is given by

$$\boxed{\hat{P}_W = \sum_{i=1}^{k} |\mathbf{e}_i\rangle\langle\mathbf{e}_i|.} \tag{3.38}$$

Note that if $|\mathbf{v}\rangle \in W$, then $\hat{P}_W|\mathbf{v}\rangle = |\mathbf{v}\rangle$. If instead $|\mathbf{v}\rangle$ is orthogonal to this subspace, then $\hat{P}_W|\mathbf{v}\rangle = \underline{\mathbf{0}}$. Projectors along orthonormal vectors satisfy the relation $\hat{P}_{\mathbf{e}_i}\hat{P}_{\mathbf{e}_j} = \hat{P}_{\mathbf{e}_i}\delta_{ij}$. The importance of projectors in quantum theory will be evident once we introduce projective measurements in Section 4.3.3.

In finite-dimensional vector spaces, projectors are expressed by square matrices. For example, consider the application of $\hat{P}_{\mathbf{e}_1} = |\mathbf{e}_1\rangle\langle\mathbf{e}_1|$ on a generic vector,

$$\begin{aligned}\hat{P}_{\mathbf{e}_1}|\mathbf{v}\rangle &= \hat{P}_{\mathbf{e}_1}\left(v_1|\mathbf{e}_1\rangle + \ldots + v_n|\mathbf{e}_n\rangle\right) \\ &= |\mathbf{e}_1\rangle\langle\mathbf{e}_1|\left(v_1|\mathbf{e}_1\rangle + \ldots + v_n|\mathbf{e}_n\rangle\right) \;=\; v_1|\mathbf{e}_1\rangle,\end{aligned}$$

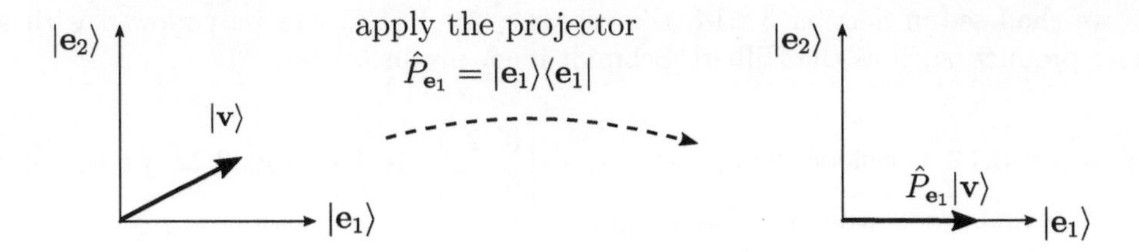

Fig. 3.8 Projectors. A representation of the projector $\hat{P}_{\mathbf{e}_1} = |\mathbf{e}_1\rangle\langle\mathbf{e}_1|$ acting on a vector $|\mathbf{v}\rangle \in \mathbb{R}^2$.

where we used $\langle\mathbf{e}_i|\mathbf{e}_j\rangle = \delta_{ij}$. This relation can be written in matrix form as

$$\hat{P}_{\mathbf{e}_1}|\mathbf{v}\rangle = \begin{bmatrix} 1 & & & \\ & 0 & & \\ & & \ddots & \\ & & & 0 \end{bmatrix} \begin{bmatrix} v_1 \\ v_2 \\ \vdots \\ v_n \end{bmatrix} = \begin{bmatrix} v_1 \\ 0 \\ \vdots \\ 0 \end{bmatrix} = v_1|\mathbf{e}_1\rangle\,. \tag{3.39}$$

This clearly shows that the matrix associated with a projector along a basis vector is diagonal with all entries equal to zero apart from one. The sum of all projectors in an n-dimensional vector space must be equal to the identity

$$\boxed{\sum_{i=1}^{n} \hat{P}_{\mathbf{e}_i} = \sum_{i=1}^{n} |\mathbf{e}_i\rangle\langle\mathbf{e}_i| = \begin{bmatrix} 1 & & \\ & \ddots & \\ & & 1 \end{bmatrix} = \mathbb{1}.} \tag{3.40}$$

This is called the **completeness relation**. The completeness relation $\mathbb{1} = \sum_{i=1}^{n} |\mathbf{e}_i\rangle\langle\mathbf{e}_i|$ is a powerful tool that will be used many times throughout this textbook. This relation can be used to decompose a vector in an orthonormal basis

$$|\mathbf{v}\rangle = \mathbb{1}|\mathbf{v}\rangle = \sum_{i=1}^{n} |\mathbf{e}_i\rangle\underbrace{\langle\mathbf{e}_i|\mathbf{v}\rangle}_{v_i} = \sum_{i=1}^{n} v_i|\mathbf{e}_i\rangle\,. \tag{3.41}$$

It can also be used to decompose a linear operator

$$\hat{A} = \mathbb{1}\hat{A}\mathbb{1} = \sum_{i,j=1}^{n} |\mathbf{e}_i\rangle\underbrace{\langle\mathbf{e}_i|\hat{A}|\mathbf{e}_j\rangle}_{A_{ij}}\langle\mathbf{e}_j| = \sum_{i,j=1}^{n} A_{ij}|\mathbf{e}_i\rangle\langle\mathbf{e}_j|\,, \tag{3.42}$$

where we defined the matrix $A_{ij} = \langle\mathbf{e}_i|\hat{A}|\mathbf{e}_j\rangle$. From eqns (3.41, 3.42), the action of an operator on a generic vector can be written as

$$\hat{A}|\mathbf{v}\rangle = \Big(\sum_{i,j=1}^{n} A_{ij}|\mathbf{e}_i\rangle\langle\mathbf{e}_j|\Big)\sum_{k=1}^{n} v_k|\mathbf{e}_k\rangle = \sum_{i,j,k=1}^{n} A_{ij}v_k|\mathbf{e}_i\rangle\langle\mathbf{e}_j|\mathbf{e}_k\rangle$$

$$= \sum_{i,j}^{n} A_{ij}v_j|\mathbf{e}_i\rangle = \sum_{i}^{n} b_i|\mathbf{e}_i\rangle \;=\; |\mathbf{b}\rangle,$$

where in the last step we introduced the vector $|\mathbf{b}\rangle$ with components $b_i = \sum_{j=1}^{n} A_{ij}v_j$. The matrix form of this expression is

$$\hat{A}|\mathbf{v}\rangle = \begin{bmatrix} A_{11} & \cdots & A_{1n} \\ \vdots & \ddots & \vdots \\ A_{n1} & \cdots & A_{nn} \end{bmatrix}\begin{bmatrix} v_1 \\ \vdots \\ v_n \end{bmatrix} = \begin{bmatrix} b_1 \\ \vdots \\ b_n \end{bmatrix} = |\mathbf{b}\rangle\,. \tag{3.43}$$

In conclusion, any operator $\hat{A} : V \to V$ can be represented by a square matrix, usually called the **associated matrix** to $\hat{A}$ in the basis B. These calculations also suggest that there exists a one-to-one correspondence between the space of linear operators $\mathrm{L}(V)$ and the space of complex matrices $M_n(\mathbb{C})$. Hence, these two vector spaces are isomorphic.

We conclude this section by introducing some important matrix operations. The **transpose of a matrix** A is a matrix A^{T} whose entries are given by $A^{\mathrm{T}}_{ij} = A_{ji}$. The **conjugate** of A is the matrix A^* whose entries are the complex conjugate of the entries of A. Finally, the **conjugate transpose** of a matrix A is the matrix $A^\dagger$ whose entries are given by: $A^\dagger_{ij} = \big(A^{\mathrm{T}}_{ij}\big)^* = A^*_{ji}$. These operations will be useful when we introduce adjoint and Hermitian operators.

Example 3.18 The operators $P_1 = \begin{bmatrix} 1 & 0 \\ 0 & 0 \end{bmatrix}$, $P_2 = \begin{bmatrix} 1/2 & 1/2 \\ 1/2 & 1/2 \end{bmatrix}$ are both projectors. To show this, compute P^2 and check that $P^2 = P$.

Example 3.19 Consider the matrix

$$A = \begin{bmatrix} 1 & 5i & 7 \\ 0 & -i & 2 \\ 2 & -1 & 3 \end{bmatrix}.$$

Then:

$$A^{\mathrm{T}} = \begin{bmatrix} 1 & 0 & 2 \\ 5i & -i & -1 \\ 7 & 2 & 3 \end{bmatrix}, \qquad A^* = \begin{bmatrix} 1 & -5i & 7 \\ 0 & +i & 2 \\ 2 & -1 & 3 \end{bmatrix}, \qquad A^\dagger = \begin{bmatrix} 1 & 0 & 2 \\ -5i & i & -1 \\ 7 & 2 & 3 \end{bmatrix}.$$

3.2.4 Pauli operators

Pauli operators are linear operators in $\mathrm{L}\left(\mathbb{C}^2\right)$. These operators are crucial in quantum information theory as they represent important single-qubit gates. Their definition and matrix representation in the computational basis[10] $B = \{|0\rangle, |1\rangle\}$ is

$$\mathbb{1} = |0\rangle\langle 0| + |1\rangle\langle 1| = \begin{bmatrix} 1 & 0 \\ 0 & 1 \end{bmatrix}, \qquad \hat{\sigma}_x = |0\rangle\langle 1| + |1\rangle\langle 0| = \begin{bmatrix} 0 & 1 \\ 1 & 0 \end{bmatrix},$$
$$\hat{\sigma}_y = -i|0\rangle\langle 1| + i|1\rangle\langle 0| = \begin{bmatrix} 0 & -i \\ i & 0 \end{bmatrix}, \qquad \hat{\sigma}_z = |0\rangle\langle 0| - |1\rangle\langle 1| = \begin{bmatrix} 1 & 0 \\ 0 & -1 \end{bmatrix}.$$

We denote Pauli operators as $\hat{\sigma}_0 = \mathbb{1}$, $\hat{\sigma}_x = \hat{\sigma}_1 = \hat{X}$, $\hat{\sigma}_y = \hat{\sigma}_2 = \hat{Y}$, $\hat{\sigma}_z = \hat{\sigma}_3 = \hat{Z}$ and we will use these notations interchangeably. These operators satisfy some important properties:

PO1. $\hat{\sigma}_0^2 = \hat{\sigma}_1^2 = \hat{\sigma}_2^2 = \hat{\sigma}_3^2 = \mathbb{1}$ (reversibility),

PO2. $\hat{\sigma}_a\hat{\sigma}_b = \delta_{ab}\mathbb{1} + i\epsilon_{abc}\hat{\sigma}_c$ (composition),

PO3. $[\hat{\sigma}_a, \hat{\sigma}_b] = 2i\epsilon_{abc}\hat{\sigma}_c$ (commutator),

PO4. $\{\hat{\sigma}_a, \hat{\sigma}_b\} = 2\delta_{ab}\mathbb{1}$ (anticommutator),

where a, b, $c \in \{1, 2, 3\}$ and the **Levi–Civita symbol** ϵ_{abc} is defined as

$$\epsilon_{abc} = \begin{cases} -1 & \text{if } abc \in \{321, 213, 132\} \\ 0 & \text{if } a = b \text{ or } a = c \text{ or } b = c \\ 1 & \text{if } abc \in \{123, 231, 312\}\ . \end{cases} \tag{3.44}$$

The first and third lines in this definition include odd and even permutations of the sequence $(1, 2, 3)$, respectively.

The **raising and lowering operators**, usually denoted by $\hat{\sigma}^+$ and $\hat{\sigma}^-$, are defined as linear combinations of Pauli operators

$$\hat{\sigma}^+ = \frac{1}{2}(\hat{\sigma}_x - i\hat{\sigma}_y) = |1\rangle\langle 0| = \begin{bmatrix} 0 & 0 \\ 1 & 0 \end{bmatrix},$$
$$\hat{\sigma}^- = \frac{1}{2}(\hat{\sigma}_x + i\hat{\sigma}_y) = |0\rangle\langle 1| = \begin{bmatrix} 0 & 1 \\ 0 & 0 \end{bmatrix}.$$

Their name is related to the way they act on the basis vectors. For instance, the application of $\hat{\sigma}^+$ onto $|0\rangle$ leads to $\hat{\sigma}^+|0\rangle = |1\rangle$. Similarly, $\hat{\sigma}^-|1\rangle = |0\rangle$. When we introduce the Hamiltonian of a qubit in Section 4.4, it will be useful to remember that $\hat{\sigma}^+\hat{\sigma}^- = |1\rangle\langle 1|$ and $\hat{\sigma}^-\hat{\sigma}^+ = |0\rangle\langle 0|$.

[10] As a reminder, $|0\rangle$, $|1\rangle$ are two orthonormal vectors in $\mathbb{C}^2$. In the previous sections, they were indicated with the symbols $|\mathbf{e}_1\rangle$ and $|\mathbf{e}_2\rangle$, respectively.

3.2.5 Inverse operator

In linear algebra, a linear operator $\hat{A} \in \mathrm{L}(V)$ is said to be **invertible** if there exists an operator $\hat{A}^{-1}$, called the **inverse operator**, such that

$$\hat{A}\hat{A}^{-1} = \hat{A}^{-1}\hat{A} = \mathbb{1}\,. \tag{3.45}$$

If $\hat{A}^{-1}$ is the inverse operator of $\hat{A}$, then $\hat{A}|\mathbf{v}\rangle = |\mathbf{w}\rangle$ implies $|\mathbf{v}\rangle = \hat{A}^{-1}|\mathbf{w}\rangle$.

In a finite-dimensional space, an operator is invertible if and only if the determinant of the associated matrix is different from zero [114]. There exist multiple equivalent ways of defining the determinant of a matrix. The **determinant** of an $n \times n$ complex matrix A with entries A_{ij} is a complex number given by

$$\boxed{\det(A) = \sum_{\sigma \in S_n} (-1)^\sigma \prod_{i=1}^{n} A_{i\,\sigma(i)}.} \tag{3.46}$$

Here, S_n is a set containing all permutations σ of the sequence $(1, 2, \ldots, n)$. A **permutation** σ is one possible reordering of a sequence of numbers and $(-1)^\sigma$ is the **signature** of the permutation, a quantity that is $+1$ when the permutation σ can be obtained from $(1, \ldots, n)$ with an even number of transpositions of adjacent numbers, and -1 whenever it can be achieved with an odd number of such interchanges. The notation $\sigma(i)$ indicates the i-th entry of the permutation σ. For example, the set S_3 contains a total of six permutations

$$\begin{aligned} S_3 &= \{\sigma_1, \sigma_2, \sigma_3, \sigma_4, \sigma_5, \sigma_6\} \\ &= \{\underbrace{(1,2,3)}_{+1}, \underbrace{(1,3,2)}_{-1}, \underbrace{(2,1,3)}_{-1}, \underbrace{(2,3,1)}_{+1}, \underbrace{(3,1,2)}_{+1}, \underbrace{(3,2,1)}_{-1}\}, \end{aligned}$$

where the number under the permutation indicates its signature. According to our notation, $\sigma_2(2) = 3$ and $\sigma_6(3) = 1$. Thus, the determinant of a 3×3 matrix A is given by:

$$\begin{aligned} \det \begin{bmatrix} A_{11} & A_{12} & A_{13} \\ A_{21} & A_{22} & A_{23} \\ A_{31} & A_{32} & A_{33} \end{bmatrix} &= \sum_{\sigma \in S_3} (-1)^\sigma \prod_{i=1}^{3} A_{i\sigma(i)} \\ &= (-1)^{\sigma_1} \prod_{i=1}^{3} A_{i\sigma_1(i)} + (-1)^{\sigma_2} \prod_{i=1}^{3} A_{i\sigma_2(i)} + (-1)^{\sigma_3} \prod_{i=1}^{3} A_{i\sigma_3(i)} \\ &\quad + (-1)^{\sigma_4} \prod_{i=1}^{3} A_{i\sigma_4(i)} + (-1)^{\sigma_5} \prod_{i=1}^{3} A_{i\sigma_5(i)} + (-1)^{\sigma_6} \prod_{i=1}^{3} A_{i\sigma_6(i)} \\ &= A_{11}A_{22}A_{33} - A_{11}A_{23}A_{32} - A_{12}A_{21}A_{33} \\ &\quad + A_{12}A_{23}A_{31} + A_{13}A_{21}A_{32} - A_{13}A_{22}A_{31}. \end{aligned} \tag{3.47}$$

Similarly, it can be shown that the determinant of a 2×2 matrix is

$$\det \begin{bmatrix} A_{11} & A_{12} \\ A_{21} & A_{22} \end{bmatrix} = A_{11}A_{22} - A_{12}A_{21}.$$

We stress once more that a square matrix is invertible if and only if its determinant is not zero. With some matrix multiplications, one can verify that the inverse of a 2×2 invertible matrix is given by:

$$A^{-1} = \frac{1}{\det(A)} \begin{bmatrix} A_{22} & -A_{12} \\ -A_{21} & A_{11} \end{bmatrix}.$$

The set of **invertible operators** is indicated as $\mathrm{GL}(V)$. An important observation is that projectors (apart from the identity operator) are not invertible. This has profound consequences since projectors describe an important class of measurement processes in quantum mechanics.

Example 3.20 In the Euclidean space, $\mathbb{R}^2$, the operator $\hat{R}_\theta$ performs a counterclockwise rotation of a vector by an angle θ,

$$\hat{R}_\theta = \begin{bmatrix} \cos\theta & -\sin\theta \\ \sin\theta & \cos\theta \end{bmatrix}.$$

It is straightforward to show that $\det(R_\theta) = 1$. Hence, $\hat{R}_\theta$ is invertible. This is consistent with the fact that rotations are invertible geometric operations. The inverse operator is $\hat{R}_\theta^{-1} = \hat{R}_{-\theta}$. The reader can verify this with a matrix multiplication: $\hat{R}_{-\theta}\hat{R}_\theta = \mathbb{1}$.

3.2.6 Eigenvalues and eigenvectors

Consider a vector space V and a linear operator $\hat{A} : V \to V$. A non-null vector $|\mathbf{a}\rangle$ is called an **eigenvector** of $\hat{A}$ with **eigenvalue** $a \in \mathbb{C}$ if

$$\boxed{\hat{A}|\mathbf{a}\rangle = a|\mathbf{a}\rangle.} \tag{3.48}$$

When the operator $\hat{A}$ acts on an eigenvector $|\mathbf{a}\rangle$, it simply multiplies this vector by a complex number. Figure 3.9 shows the transformation of an eigenvector in the vector space $\mathbb{R}^2$. Eigenvalues and eigenvectors are crucial in quantum mechanics because in quantum theory the result of a measurement is the eigenvalue of a specific operator. Right after the measurement, the state of the system is described by the eigenvector associated with that particular eigenvalue.

An operator can have multiple eigenvalues. The problem of finding the eigenvalues of an operator is called the **eigenvalue problem**. Let us show how to solve this problem in a systematic way. Equation (3.48) can be written as

$$(\hat{A} - a\mathbb{1})|\mathbf{a}\rangle = \underline{\mathbf{0}}.$$

The operator $\hat{A} - a\mathbb{1}$ maps the vector $|\mathbf{a}\rangle$ into the null vector. Thus, this operator is not invertible, i.e. the determinant of the matrix $A - a\mathbb{1}$ is zero,

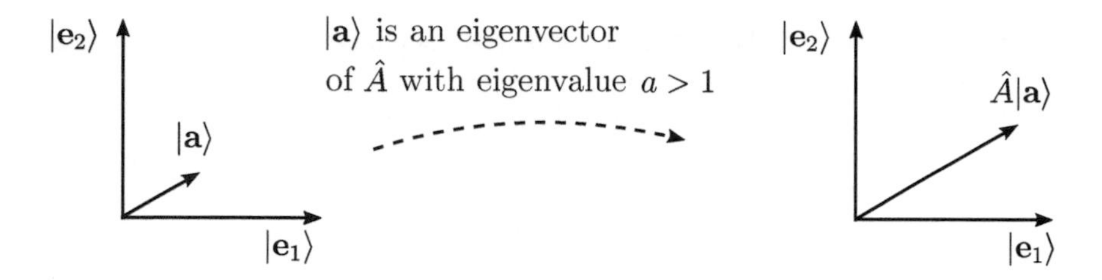

Fig. 3.9 Eigenvalues and eigenvectors. The application of an operator $\hat{A}$ onto an eigenvector $|\mathbf{a}\rangle$ with positive eigenvalue a. In this example, the operator stretches the length of the eigenvector. The action of $\hat{A}$ on a generic vector might be very different.

$$\boxed{\det(A - a\mathbb{1}) = 0.}$$

This is called the **characteristic equation**. This equation yields a polynomial in the variable a of degree n. The roots of this polynomial are the eigenvalues of $\hat{A}$.

The number of eigenvalues of an operator is equal to the dimension of the vector space (some eigenvalues might be equal to each other). The collection of the eigenvalues associated with an operator is called the **spectrum** of the operator and this multiset is denoted $\sigma(\hat{A})$. The number of times an eigenvalue appears in the spectrum is called the **multiplicity** of the eigenvalue. If the multiplicity of an eigenvalue is greater than one, it is said to be **degenerate**. It can be shown that the eigenvalues of an operator do not depend on the choice of basis.

Example 3.21 Consider the Pauli operator $\hat{\sigma}_x = \begin{bmatrix} 0 & 1 \\ 1 & 0 \end{bmatrix}$. The characteristic equation for this operator is

$$\det \begin{bmatrix} -a & 1 \\ 1 & -a \end{bmatrix} = a^2 - 1 = 0.$$

The solutions of this equation are the eigenvalues of $\hat{\sigma}_x$: $a_1 = 1$, $a_2 = -1$. Thus, the spectrum of this operator is $\sigma(\hat{\sigma}_x) = \{1, -1\}$. The normalized eigenvectors associated with these eigenvalues are $|\mathbf{a}_1\rangle = (1/\sqrt{2}, 1/\sqrt{2})$ and $|\mathbf{a}_2\rangle = (1/\sqrt{2}, -1/\sqrt{2})$.

Example 3.22 Consider the operator $A = \begin{bmatrix} 0 & 2 \\ 3 & -1 \end{bmatrix}$. The characteristic equation for this operator is $\det(A) = a + a^2 - 6$. The roots of this polynomial are $a_1 = -3$ and $a_2 = 2$. Thus, the eigenvalues of $\hat{A}$ are $\{-3, 2\}$.

3.2.7 Adjoint operator

The concept of an adjoint operator is necessary to introduce several classes of operators, such as normal, Hermitian, and unitary operators. The **adjoint** of an operator $\hat{A} \in \mathrm{L}(V)$ is an operator $\hat{A}^\dagger \in \mathrm{L}(V)$ satisfying

$$\boxed{\langle \mathbf{w} | \hat{A}^\dagger \mathbf{v} \rangle = \langle \hat{A} \mathbf{w} | \mathbf{v} \rangle,} \tag{3.49}$$

or equivalently $\langle \mathbf{w}|\hat{A}^\dagger \mathbf{v}\rangle = \langle \mathbf{v}|\hat{A}\mathbf{w}\rangle^*$. The operator $\hat{A}^\dagger$ is called the **adjoint operator** of $\hat{A}$. It can be shown that the adjoint operator is unique.

In a finite-dimensional space, it is possible to construct the associated matrix to the adjoint operator in a simple way. The associated matrix to $\hat{A}$ has entries

$$A_{ij} = \langle \mathbf{e}_i|\hat{A}|\mathbf{e}_j\rangle = \langle \mathbf{e}_i|\hat{A}\mathbf{e}_j\rangle . \tag{3.50}$$

Therefore, the entries of the associated matrix to $\hat{A}^\dagger$ are

$$\boxed{A^\dagger_{ij} = \langle \mathbf{e}_i|\hat{A}^\dagger \mathbf{e}_j\rangle = \langle \hat{A}\mathbf{e}_i|\mathbf{e}_j\rangle = \big(\langle \mathbf{e}_j|A\mathbf{e}_i\rangle\big)^* = A^*_{ji}.} \tag{3.51}$$

This shows that the matrix associated with the adjoint operator is the conjugate transpose of A.

The adjunction operation satisfies the properties listed below:

AD1. $(c\hat{A})^\dagger = c^*\hat{A}^\dagger$,

AD2. $(\hat{A}^\dagger)^\dagger = \hat{A}$,

AD3. $(\hat{A}+\hat{B})^\dagger = \hat{A}^\dagger + \hat{B}^\dagger$,

AD4. $(\hat{A}\hat{B})^\dagger = \hat{B}^\dagger\hat{A}^\dagger$.

These properties are useful in various calculations and can be proved in the following way:

Proof

Property **AD1** follows from $(cA)^\dagger_{ij} = (cA)^*_{ji} = c^*A^*_{ji} = c^*A^\dagger_{ij}$. □

Property **AD2** follows from $(A^\dagger_{ij})^\dagger = \big(A^*_{ji}\big)^\dagger = A_{ij}$. □

Property **AD3** follows from

$$(A+B)^\dagger_{ij} = (A+B)^*_{ji} = A^*_{ji} + B^*_{ji} = A^\dagger_{ij} + B^\dagger_{ij}.$$ □

To prove property **AD4**, write the explicit form of the operators $(\hat{A}\hat{B})^\dagger$ and $\hat{B}^\dagger\hat{A}^\dagger$,

$$(AB)^\dagger_{ij} = (AB)^*_{ji} = \sum_{k=1}^n A^*_{jk}B^*_{ki},$$

$$\big(B^\dagger A^\dagger\big)_{ij} = \sum_{k=1}^n B^\dagger_{ik}A^\dagger_{kj} = \sum_{k=1}^n B^*_{ki}A^*_{jk} = \sum_{k=1}^n A^*_{jk}B^*_{ki}\,.$$

Since these two expressions are identical, $(\hat{A}\hat{B})^\dagger = \hat{B}^\dagger\hat{A}^\dagger$ as required. □

Example 3.23 Consider the matrix $A = \begin{bmatrix} 7 & 2i \\ 0 & i \end{bmatrix}$. Its adjoint is $A^\dagger = \begin{bmatrix} 7 & 0 \\ -2i & -i \end{bmatrix}$.

Example 3.24 The Pauli operators are equal to their adjoint, $\hat{\sigma}^\dagger_i = \hat{\sigma}_i$.

3.2.8 Normal operators and the spectral theorem

Normal operators are an important subset of linear operators. An operator $\hat{A} \in \mathrm{L}(V)$ is said to be normal if

$$\boxed{\hat{A}\hat{A}^\dagger = \hat{A}^\dagger \hat{A},} \tag{3.52}$$

or equivalently if it commutes with its adjoint $[\hat{A}, \hat{A}^\dagger] = \underline{\mathbf{0}}$. Most of the operators we discuss in this textbook (including Hermitian, unitary, and positive operators) are normal. The set of all normal operators is indicated as $\mathrm{N}(V)$. A normal operator can be expressed as a linear combination of projectors where the coefficients of the decomposition are the operator's eigenvalues. This important result of linear algebra is called the **spectral theorem**.

Theorem 3.1 (Spectral theorem) Consider a Hilbert space V with dimension $n = \dim(V)$ and a normal operator $\hat{A}$. Then, there exists a multiset of eigenvalues $\{a_1, \ldots, a_n\}$ and orthonormal eigenvectors $\{|\mathbf{a}_1\rangle, \ldots, |\mathbf{a}_n\rangle\}$ such that

$$\hat{A} = \sum_{i=1}^{n} a_i |\mathbf{a}_i\rangle\langle\mathbf{a}_i|. \tag{3.53}$$

This is called the **spectral decomposition** of $\hat{A}$. The operators $\hat{P}_i = |\mathbf{a}_i\rangle\langle\mathbf{a}_i|$ are orthogonal projectors and the eigenvectors $\{|\mathbf{a}_i\rangle\}_i$ form an orthonormal basis of the vector space.

The proof of the spectral theorem can be found in various textbooks [114, 113]. As stated by the theorem, the eigenvectors of a normal operator $\hat{A}$ form an orthonormal basis. In this basis, the associated matrix is diagonal and the elements on the diagonal are the eigenvalues of the operator

$$A = \begin{bmatrix} a_1 & & & \\ & a_2 & & \\ & & \ddots & \\ & & & a_n \end{bmatrix}.$$

This is called the **diagonal representation** of a normal operator. In short, an operator is diagonalizable with respect to an orthonormal basis of eigenvectors if and only if it is normal. There exist numerous efficient algorithms for diagonalizing a normal operator, see Ref. [115] for more information.

Let us present the connection between commuting normal operators and their eigenvectors.

Theorem 3.2 (Simultaneous diagonalization) Consider a vector space V of dimension n. Two normal operators $\hat{A}, \hat{B} \in \mathrm{N}(V)$ share a common basis of orthonormal eigenvectors if and only if $[\hat{A}, \hat{B}] = \underline{\mathbf{0}}$.

Proof ($\rightarrow$) Suppose there exists an orthonormal basis $\{|i\rangle\}_{i=1}^{n}$ such that $\hat{A}|i\rangle = a_i|i\rangle$ and $\hat{B}|i\rangle = b_i|i\rangle$. Then, for any state $|\mathbf{v}\rangle \in V$, we have

$$\hat{A}\hat{B}|\mathbf{v}\rangle = \sum_{i=1}^{n} \hat{A}\hat{B}|i\rangle\langle i|\mathbf{v}\rangle = \textstyle\sum_{i=1}^{n} a_i b_i |i\rangle\langle i|\mathbf{v}\rangle = \sum_{i=1}^{n} b_i a_i |i\rangle\langle i|\mathbf{v}\rangle$$
$$= \textstyle\sum_{i=1}^{n} \hat{B}\hat{A}|i\rangle\langle i|\mathbf{v}\rangle = \hat{B}\hat{A}|\mathbf{v}\rangle$$

and therefore $[\hat{A}, \hat{B}] = \underline{\mathbf{0}}$.

Proof ($\leftarrow$) To show the converse, suppose that $[\hat{A}, \hat{B}] = \underline{\mathbf{0}}$. Since $\hat{A}$ is normal, there exists an orthonormal basis of eigenvectors $\{|i\rangle\}_i$ of $\hat{A}$. Let us show that $\{|i\rangle\}_i$ are also eigenvectors of $\hat{B}$. First of all, we have

$$\hat{B}|i\rangle = \sum_{j=1}^{n} |j\rangle\langle j|\hat{B}|i\rangle = \sum_{j=1}^{n} \langle j|\hat{B}|i\rangle|j\rangle, \tag{3.54}$$

where we used the completeness relation $\sum_{j=1}^{n} |j\rangle\langle j| = \mathbb{1}$. Thus

$$[\hat{A}, \hat{B}]|i\rangle = \hat{A}\hat{B}|i\rangle - \hat{B}\hat{A}|i\rangle = \hat{A}\Big[\sum_{j=1}^{n} \langle j|\hat{B}|i\rangle|j\rangle\Big] - \hat{B}a_i|i\rangle$$
$$= \sum_{j=1}^{n} \langle j|\hat{B}|i\rangle a_j|j\rangle - \sum_{j=1}^{n} \langle j|\hat{B}|i\rangle a_i|j\rangle = \sum_{j=1}^{n} \langle j|\hat{B}|i\rangle(a_j - a_i)|j\rangle = \underline{\mathbf{0}},$$

where the last step is valid because we assumed $[\hat{A}, \hat{B}] = \underline{\mathbf{0}}$. If the eigenvalues are not degenerate, then $a_i \neq a_j$ for $i \neq j$. This implies that $\langle j|\hat{B}|i\rangle = 0$ for $i \neq j$. Let us define $\beta_j = \langle j|\hat{B}|j\rangle$. Therefore, $\langle j|\hat{B}|i\rangle = \beta_j\delta_{ij}$. Substituting this expression into (3.54), we obtain

$$\hat{B}|i\rangle = \beta_i|i\rangle.$$

This implies that $\{|i\rangle\}_i$ is an orthonormal basis of eigenvectors for $\hat{B}$. This proof can be modified to account for the case in which the eigenvalues are degenerate. □

3.2.9 Functions of normal operators

In quantum theory, we will often compute **functions of normal operators**. Suppose that $\hat{A}$ is a normal operator and f is a complex analytic function. Since $\hat{A}$ is normal, according to the spectral theorem, it can be expressed as a sum of projectors $\hat{A} = \sum_i a_i|\mathbf{a}_i\rangle\langle\mathbf{a}_i|$ where $\{|\mathbf{a}_i\rangle\}_i$ is an orthonormal basis. We define the operator $f(\hat{A})$ as

$$\boxed{f(\hat{A}) = \sum_{i=1}^{n} f(a_i)\,|\mathbf{a}_i\rangle\langle\mathbf{a}_i|.} \tag{3.55}$$

The function of a normal operator can also be defined with a Taylor expansion

$$f(\hat{A}) = \sum_{n=0}^{\infty} \frac{f^{(n)}(0)}{n!} \hat{A}^n. \tag{3.56}$$

Here, $f^{(n)}(0)$ indicates the n-th derivative of f evaluated in zero. The definitions (3.55) and (3.56) are equivalent. Note that the operator $f(\hat{A})$ is diagonal in the basis $\{|\mathbf{a}_i\rangle\}_i$.

An important function of a normal operator is the exponential $e^{\hat{A}}$. The reader might be confused seeing a number to the power of an operator. An example will clarify this notation. Consider the exponential $e^{-i\alpha\hat{\sigma}_z}$ where $\hat{\sigma}_z$ is a Pauli operator and $\alpha \in \mathbb{R}$. Since $\hat{\sigma}_z$ has two eigenvalues ($+1$ and -1) and two eigenvectors ($|0\rangle$ and $|1\rangle$), according to (3.55) the function $e^{-i\alpha\hat{\sigma}_z}$ can be expressed as

$$e^{-i\alpha\hat{\sigma}_z} = e^{-i\alpha}|0\rangle\langle 0| + e^{i\alpha}|1\rangle\langle 1|. \tag{3.57}$$

This equation can also be derived using a Taylor expansion. From (3.56), we have

$$\begin{aligned} e^{-i\alpha\hat{\sigma}_z} &= \sum_{n=0}^{\infty} \frac{(-i\alpha\hat{\sigma}_z)^n}{n!} = \mathbb{1} + \frac{-i\alpha\hat{\sigma}_z}{1!} + \frac{-\alpha^2\mathbb{1}}{2!} + \frac{i\alpha^3\hat{\sigma}_z}{3!} + \dots \\ &= \left[1 + \frac{-\alpha^2}{2!} + \dots\right]\mathbb{1} - i\left[\frac{\alpha}{1!} - \frac{\alpha^3}{3!} + \dots\right]\hat{\sigma}_z \\ &= \cos(\alpha)\,\mathbb{1} - i\sin(\alpha)\,\hat{\sigma}_z. \end{aligned} \tag{3.58}$$

The reader can verify that eqn (3.57) and (3.58) are identical[11]. These formulas will be useful when we introduce single-qubit rotations in Section 5.3.

The exponential of a normal operator has multiple applications in quantum mechanics. This function can be used, for example, to calculate the **similarity transformation** $e^{\alpha\hat{A}}\hat{B}e^{-\alpha\hat{A}}$ where α is a complex number and $\hat{A}$, $\hat{B}$ are two normal operators. The operator $e^{\alpha\hat{A}}\hat{B}e^{-\alpha\hat{A}}$ can be expanded with a series using the **Baker–Campbell–Hausdorff formula** [116, 117],

$$\boxed{e^{\alpha\hat{A}}\hat{B}e^{-\alpha\hat{A}} = \hat{B} + \alpha[\hat{A},\hat{B}] + \frac{\alpha^2}{2!}[\hat{A},[\hat{A},\hat{B}]] + \frac{\alpha^3}{3!}[\hat{A},[\hat{A},[\hat{A},\hat{B}]]] + \dots} \tag{3.59}$$

Equation (3.59) will be used multiple times in this textbook.

Example 3.25 Let us calculate the similarity transformation $e^{\alpha\hat{A}}\hat{B}e^{-\alpha\hat{A}}$ with $\hat{A} = \hat{\sigma}_z$ and $\hat{B} = \hat{\sigma}^-$. Note that

$$[\hat{A},\hat{B}] = [\hat{\sigma}_z, \hat{\sigma}^-] = 2\hat{\sigma}^- \tag{3.60}$$

$$[\hat{A},[\hat{A},\hat{B}]] = [\hat{\sigma}_z,[\hat{\sigma}_z,\hat{\sigma}^-]] = 2[\hat{\sigma}_z,\hat{\sigma}^-] = 4\hat{\sigma}^-. \tag{3.61}$$

Substituting these expressions into (3.59), we arrive at

[11]Recall that $\mathbb{1} = |0\rangle\langle 0| + |1\rangle\langle 1|$ and $\hat{\sigma}_z = |0\rangle\langle 0| - |1\rangle\langle 1|$.

$$e^{\alpha\hat{\sigma}_z}\hat{\sigma}^- e^{-\alpha\hat{\sigma}_z} = \hat{\sigma}^- \sum_{n=0}^{\infty} \frac{(2\alpha)^n}{n!} = \hat{\sigma}^- e^{2\alpha} . \tag{3.62}$$

With similar calculations, it can be shown that $e^{\alpha\hat{\sigma}_z}\hat{\sigma}^+ e^{-\alpha\hat{\sigma}_z} = \hat{\sigma}^+ e^{-2\alpha}$.

Example 3.26 Consider a similarity transformation involving single-qubit operators

$$e^{i\frac{\theta}{2}\vec{n}\cdot\vec{\sigma}}\,\vec{a}\cdot\vec{\sigma}\,e^{-i\frac{\theta}{2}\vec{n}\cdot\vec{\sigma}}. \tag{3.63}$$

Here, $\vec{n} = (n_x, n_y, n_z)$ and $\vec{a} = (a_x, a_y, a_z)$ are unit vectors, and $\vec{\sigma} = (\hat{\sigma}_x, \hat{\sigma}_y, \hat{\sigma}_z)$ is the Pauli vector. With the notation $\vec{n}\cdot\vec{\sigma}$ we mean $n_x\hat{\sigma}_x + n_y\hat{\sigma}_y + n_z\hat{\sigma}_z$. Equation (3.63) can be simplified using the Baker–Campbell–Hausdorff formula. It can be shown that

$$\boxed{e^{i\frac{\theta}{2}\vec{n}\cdot\vec{\sigma}}\vec{a}\cdot\vec{\sigma}e^{-i\frac{\theta}{2}\vec{n}\cdot\vec{\sigma}} = [\vec{n}\,(\vec{n}\cdot\vec{a}) + \cos\theta\,(\vec{a} - \vec{n}\,(\vec{n}\cdot\vec{a})) + \sin\theta\,(\vec{n}\times\vec{a})]\cdot\vec{\sigma}.} \tag{3.64}$$

This identity will be used multiple times when dealing with single-qubit rotations in Section 5.3.

Example 3.27 Consider the similarity transformation $e^{i\hat{\sigma}_z\theta/2}\hat{\sigma}_x e^{-i\hat{\sigma}_z\theta/2}$. This expression is equivalent to (3.64) with $\vec{n} = (0,0,1)$ and $\vec{a} = (1,0,0)$. Thus

$$e^{i\hat{\sigma}_z\theta/2}\hat{\sigma}_x e^{-i\hat{\sigma}_z\theta/2} = \hat{\sigma}_x\cos\theta - \hat{\sigma}_y\sin\theta\,. \tag{3.65}$$

With similar calculations, it can be shown that $e^{-i\hat{\sigma}_z\theta/2}\hat{\sigma}_x e^{i\hat{\sigma}_z\theta/2} = \hat{\sigma}_x\cos\theta + \hat{\sigma}_y\sin\theta$.

Example 3.28 An important function of a normal operator $\hat{A}$ is its absolute value $|\hat{A}| \equiv \sqrt{\hat{A}^\dagger\hat{A}}$. Using the spectral decomposition $\hat{A} = \sum_i = a_i|\mathbf{a}_i\rangle\langle\mathbf{a}_i|$, we can write

$$|\hat{A}|^2 = \hat{A}^\dagger\hat{A} = \Big(\sum_i a_i^*|\mathbf{a}_i\rangle\langle\mathbf{a}_i|\Big)\Big(\sum_j a_j|\mathbf{a}_j\rangle\langle\mathbf{a}_j|\Big) = \sum_i |a_i|^2|\mathbf{a}_i\rangle\langle\mathbf{a}_i|, \tag{3.66}$$

and therefore $|\hat{A}| = \sum_i |a_i||\mathbf{a}_i\rangle\langle\mathbf{a}_i|$. This clearly shows that the eigenvalues of $|\hat{A}|$ are non-negative real numbers, i.e. $|\hat{A}|$ is a positive operator.

3.2.10 Hermitian operators

Hermitian operators are important examples of normal operators. An operator $\hat{A} \in \mathrm{L}(V)$ is **Hermitian** (or self-adjoint) if it is equal to its adjoint,

$$\boxed{\hat{A} = \hat{A}^\dagger.} \tag{3.67}$$

In a finite-dimensional space, this definition is equivalent to requiring that the associated matrix is equal to its conjugate transpose, $A_{ij} = A_{ji}^*$. Therefore, the diagonal elements of a Hermitian matrix are real. Hermitian operators play a key role in quantum mechanics since they represent physical observables, such as position and momentum.

Hermitian operators are normal. Hence, they can be diagonalized and written in the form $\hat{A} = \sum_i a_i|\mathbf{a}_i\rangle\langle\mathbf{a}_i|$, where $|\mathbf{a}_i\rangle$ is the eigenvector associated with the eigenvalue a_i and $\hat{P}_i = |\mathbf{a}_i\rangle\langle\mathbf{a}_i|$ are the projectors associated with the eigenvectors $|\mathbf{a}_i\rangle$. Since $\hat{A} = \hat{A}^\dagger$, the eigenvalues are real numbers. The set of all Hermitian operators will be indicated as $\mathrm{Her}(V)$.

3.2.11 Unitary operators

Unitary operators describe the time evolution of physical systems in quantum mechanics and quantum gates in quantum computing. A linear operator $\hat{A} \in \mathrm{L}(V)$ is **unitary** if it satisfies

$$\boxed{\hat{A}^\dagger \hat{A} = \hat{A}\hat{A}^\dagger = \mathbb{1},} \tag{3.68}$$

or equivalently $\hat{A}^\dagger = \hat{A}^{-1}$. Unitary operators are usually indicated with the letters $\hat{U}$ and $\hat{V}$. Note that the product $\hat{C} = \hat{U}\hat{V}$ of two unitary operators is still unitary, i.e.

$$\hat{C}^\dagger \hat{C} = (\hat{U}\hat{V})^\dagger(\hat{U}\hat{V}) = \hat{V}^\dagger \underbrace{\hat{U}^\dagger \hat{U}}_{\mathbb{1}} \hat{V} = \hat{V}^\dagger \hat{V} = \mathbb{1}\,.$$

Unitary operators preserve the inner product between vectors

$$\langle \hat{U}\mathbf{w}|\hat{U}\mathbf{v}\rangle = \langle \mathbf{w}|\underbrace{\hat{U}^\dagger \hat{U}}_{\mathbb{1}}|\mathbf{v}\rangle = \langle \mathbf{w}|\mathbf{v}\rangle\,,$$

and therefore also their lengths

$$\langle \hat{U}\mathbf{v}|\hat{U}\mathbf{v}\rangle = \langle \mathbf{v}|\hat{U}^\dagger \hat{U}|\mathbf{v}\rangle = \langle \mathbf{v}|\mathbf{v}\rangle = \||\mathbf{v}\rangle\|^2\,.$$

Consequently, a unitary operator transforms an orthonormal basis into an orthonormal basis. The set of all unitary operators is denoted $\mathrm{U}(V)$. This set is a subset of $\mathrm{GL}(V)$, the set of invertible operators. These concepts are fundamental in linear algebra and have important applications in quantum mechanics.

It is straightforward to show that all of the eigenvalues of a unitary operator are complex numbers with modulus one. Assuming that $|\mathbf{v}\rangle$ is an eigenvector of $\hat{U}$ with eigenvalue λ,

$$\langle \mathbf{v}|\mathbf{v}\rangle = \langle \mathbf{v}|\hat{U}^\dagger \hat{U}|\mathbf{v}\rangle = \langle \mathbf{v}|\lambda^* \lambda|\mathbf{v}\rangle = |\lambda|^2 \langle \mathbf{v}|\mathbf{v}\rangle.$$

By dividing both members by $\langle \mathbf{v}|\mathbf{v}\rangle$, we obtain $|\lambda|^2 = 1$ as anticipated.

Since a unitary operator transforms an orthonormal basis into another orthonormal basis, a unitary transformation can be considered to be a **change of basis**. To better understand this point, consider two orthonormal bases $B = \{|\mathbf{e}_i\rangle\}_i$, $B' = \{|\mathbf{e}'_i\rangle\}_i$, and the operator $\hat{U}$ that expresses the basis B' in terms of the basis B, $|\mathbf{e}'_i\rangle = \hat{U}|\mathbf{e}_i\rangle$. This equation suggests that the operator $\hat{U}$ has the form $\hat{U} = \sum_k |\mathbf{e}'_k\rangle\langle\mathbf{e}_k|$ and the entries of the associated matrix are

$$U_{ij} = \langle \mathbf{e}_i|\Big(\sum_{k=1}^{n} |\mathbf{e}'_k\rangle\langle\mathbf{e}_k|\Big)|\mathbf{e}_j\rangle = \langle \mathbf{e}_i|\mathbf{e}'_j\rangle\,. \tag{3.69}$$

The entries of U are given by the inner product between the vectors of the two bases.

Unitary operators are related to Hermitian operators in an elegant way.

Theorem 3.3 Any unitary operator $\hat{U}$ acting on a finite-dimensional space V can be expressed as

$$\boxed{\hat{U} = e^{i\hat{A}},} \tag{3.70}$$

where $\hat{A}$ is a Hermitian operator.

Proof Since $\hat{U}$ is unitary, its spectral decomposition is $\hat{U} = \sum_{j=1}^{n} \lambda_j |\mathbf{e}_j\rangle\langle\mathbf{e}_j|$ where the eigenvalues λ_j satisfy $|\lambda_j| = 1$. This means that the complex numbers λ_j can be written as $\lambda_j = e^{ia_j}$ where $a_j \in [0, 2\pi)$. Now consider the Hermitian operator $\hat{A} = \sum_{j=1}^{n} a_j |\mathbf{e}_j\rangle\langle\mathbf{e}_j|$, and note that

$$\hat{U} = \sum_{j=1}^{n} \lambda_j |\mathbf{e}_j\rangle\langle\mathbf{e}_j| = \sum_{j=1}^{n} e^{ia_j} |\mathbf{e}_j\rangle\langle\mathbf{e}_j| = e^{i\hat{A}}.$$

When a set of unitary operators is expressed as $\hat{U}_t = e^{it\hat{A}}$, we say that $\hat{A}$ generates the transformation $\hat{U}_t$. □

Example 3.29 In the vector space $\mathbb{C}^2$, a generic unitary operator is described by a 2×2 matrix

$$U = \begin{bmatrix} a & b \\ c & d \end{bmatrix}.$$

For unitary matrices, the products $U^\dagger U$ and $UU^\dagger$ must be equal to the identity

$$U^\dagger U = \begin{bmatrix} |a|^2 + |c|^2 & a^*b + c^*d \\ ab^* + cd^* & |b|^2 + |d|^2 \end{bmatrix} = \begin{bmatrix} 1 & 0 \\ 0 & 1 \end{bmatrix} = \begin{bmatrix} |a|^2 + |b|^2 & ac^* + bd^* \\ a^*c + b^*d & |c|^2 + |d|^2 \end{bmatrix} = UU^\dagger.$$

Since the diagonal elements are equal to 1, we have $|a|^2 + |c|^2 = |a|^2 + |b|^2 = 1$ and also $|b|^2 + |d|^2 = |c|^2 + |d|^2 = 1$. Therefore, these parameters can be written as

$$a = \cos\frac{\theta}{2}e^{iA}, \qquad b = \sin\frac{\theta}{2}e^{iB}, \qquad c = \sin\frac{\theta}{2}e^{iC}, \qquad d = \cos\frac{\theta}{2}e^{iD}.$$

Furthermore, since the off-diagonal elements are zero, we have $B = A + D - C + \pi$. Thus, a generic unitary transformation in the Hilbert space $\mathbb{C}^2$ is parametrized by four real parameters,

$$\begin{aligned} \hat{U} &= \begin{bmatrix} \cos\frac{\theta}{2}e^{iA} & -\sin\frac{\theta}{2}e^{i(A+D-C)} \\ \sin\frac{\theta}{2}e^{iC} & \cos\frac{\theta}{2}e^{iD} \end{bmatrix} \\ &= e^{i\delta} \begin{bmatrix} \cos\frac{\theta}{2}e^{-\frac{i}{2}(\alpha+\beta)} & -\sin\frac{\theta}{2}e^{-\frac{i}{2}(\alpha-\beta)} \\ \sin\frac{\theta}{2}e^{\frac{i}{2}(\alpha-\beta)} & \cos\frac{\theta}{2}e^{\frac{i}{2}(\alpha+\beta)} \end{bmatrix}, \end{aligned} \tag{3.71}$$

L: Linear
B: Bounded
N: Normal $[\hat{A}, \hat{A}^\dagger] = \underline{0}$
U: Unitary $\hat{A}^\dagger = \hat{A}^{-1}$
Her: Hermitian $\hat{A}^\dagger = \hat{A}$
Pos: Positive $\hat{A} \geq 0$
D: Density operators

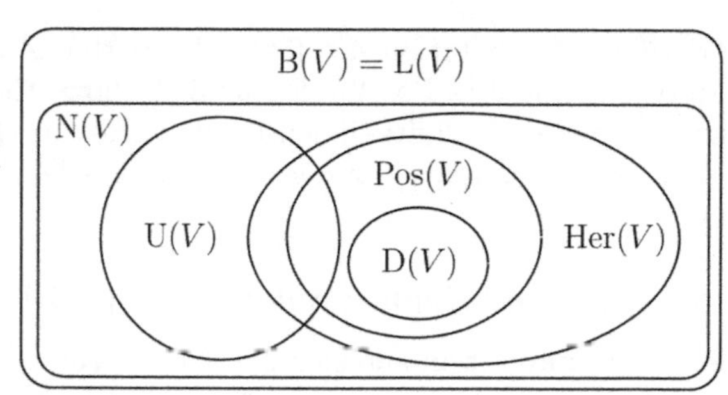

Fig. 3.10 Linear operators. Relation between linear operators for finite-dimensional spaces.

---

where in the last step we introduced $\delta = (D + A)/2$, $\alpha = C - A$, and $\beta = D - C$ for convenience. This expression will be useful in Section 5.3.1 when we show that any unitary transformation acting on a qubit state can be decomposed in a sequence of single-qubit rotations.

Example 3.30 Consider two orthonormal bases in the vector space $\mathbb{C}^2$, $B = \{|0\rangle, |1\rangle\}$ and $B' = \{\frac{|0\rangle+|1\rangle}{\sqrt{2}}, \frac{|0\rangle-|1\rangle}{\sqrt{2}}\}$. The unitary matrix implementing the change of basis from B to B' is given by

$$\hat{U} = \begin{pmatrix} \langle 0| \left(\frac{|0\rangle+|1\rangle}{\sqrt{2}}\right) & \langle 0| \left(\frac{|0\rangle-|1\rangle}{\sqrt{2}}\right) \\ \langle 1| \left(\frac{|0\rangle+|1\rangle}{\sqrt{2}}\right) & \langle 1| \left(\frac{|0\rangle-|1\rangle}{\sqrt{2}}\right) \end{pmatrix} = \frac{1}{\sqrt{2}} \begin{pmatrix} 1 & 1 \\ 1 & -1 \end{pmatrix}. \tag{3.72}$$

This unitary is called the Hadamard gate. We will study this unitary in more detail in Section 5.3.2.

3.2.12 Positive operators

Positive operators are an important class of linear operators. An operator $\hat{A} \in \mathrm{L}(V)$ is **positive** if, for all $|\mathbf{v}\rangle \in V$,

$$\boxed{\langle \mathbf{v}|\hat{A}|\mathbf{v}\rangle \geq 0.} \tag{3.73}$$

Positive operators are Hermitian operators with non-negative eigenvalues. Let us prove this statement. From the definition (3.73), we know that $\langle \mathbf{v}|\hat{A}|\mathbf{v}\rangle$ is a non-negative real number, and therefore $\langle \mathbf{v}|\hat{A}|\mathbf{v}\rangle = \langle \mathbf{v}|\hat{A}|\mathbf{v}\rangle^*$. Thus,

$$\langle \mathbf{v}|\hat{A}|\mathbf{v}\rangle - \langle \mathbf{v}|\hat{A}|\mathbf{v}\rangle^* = \langle \mathbf{v}|\hat{A}|\mathbf{v}\rangle - \langle \mathbf{v}|\hat{A}^\dagger|\mathbf{v}\rangle = \langle \mathbf{v}|(\hat{A} - \hat{A}^\dagger)|\mathbf{v}\rangle = 0. \tag{3.74}$$

The operator $\hat{A} - \hat{A}^\dagger$ is normal and the spectral theorem implies that $\hat{A} - \hat{A}^\dagger = \sum_k \alpha_k |k\rangle\langle k|$. Equation (3.74) suggests that $\alpha_k = 0$ for all k and therefore $\hat{A} - \hat{A}^\dagger = \underline{0}$. This means that $\hat{A}$ is Hermitian.

We will use the notation $\hat{A} \geq 0$ to indicate that $\hat{A}$ is positive. The set of all positive operators is denoted as Pos(V). Figure 3.10 shows the relations between some classes of linear operators in finite-dimensional spaces. In this figure, the only set we have not yet presented is the set of density operators D(V). This set will be introduced in Section 3.2.14.

3.2.13 Polar and singular value decomposition

The polar decomposition and singular value decomposition are two important ways of decomposing a linear operator, and they will be utilized frequently in this textbook.

Theorem 3.4 (Polar decomposition) Consider a Hilbert space V. Any linear operator $\hat{A} : V \to V$ can be expressed as a product of two operators:

$$\boxed{\hat{A} = \hat{L}\hat{U},}$$

or equivalently $\hat{A} = \hat{U}\hat{R}$. Here, $\hat{L}$ and $\hat{R}$ are positive operators and $\hat{U}$ is unitary. The expression $\hat{A} = \hat{L}\hat{U}$ ($\hat{A} = \hat{U}\hat{R}$) is called left (right) **polar decomposition** of $\hat{A}$.

Proof Suppose $\hat{A}$ is invertible. Then, the operator $\hat{R} = \sqrt{\hat{A}^\dagger \hat{A}}$ is positive, Hermitian and invertible (see Example 3.28). If we define $\hat{U} = \hat{A}\hat{R}^{-1}$, then the operator $\hat{U}^\dagger$ is given by

$$\hat{U}^\dagger = \left(\hat{R}^{-1}\right)^\dagger \hat{A}^\dagger = \left[\left(\sqrt{\hat{A}^\dagger \hat{A}}\right)^{-1}\right]^\dagger \hat{A}^\dagger = \left(\sqrt{\hat{A}^\dagger \hat{A}}\right)^{-1} \hat{A}^\dagger.$$

It is simple to show that $\hat{U}$ is unitary:

$$\hat{U}^\dagger \hat{U} = \left(\sqrt{\hat{A}^\dagger \hat{A}}\right)^{-1} \hat{A}^\dagger \hat{A} \left(\sqrt{\hat{A}^\dagger \hat{A}}\right)^{-1} = \mathbb{1}\,. \tag{3.75}$$

Furthermore, the operator $\hat{A}$ is equal to the product $\hat{U}\hat{R}$,

$$\hat{U}\hat{R} = \hat{A}\left(\sqrt{\hat{A}^\dagger \hat{A}}\right)^{-1} \sqrt{\hat{A}^\dagger \hat{A}} = \hat{A}\,. \tag{3.76}$$

Similarly, the left polar decomposition of $\hat{A}$ is $\hat{A} = \hat{U}\hat{R} = \hat{U}\hat{R}\hat{U}^\dagger\hat{U} = \hat{L}\hat{U}$ where we introduced $\hat{L} = \hat{U}\hat{R}\hat{U}^\dagger$.

If $\hat{A}$ is not invertible, this proof changes slightly. Define $\hat{R} = \sqrt{\hat{A}^\dagger \hat{A}}$. This operator is positive and Hermitian by definition, and its spectral decomposition is $\hat{R} = \sum_i r_i |\mathbf{r}_i\rangle\langle\mathbf{r}_i|$ where $r_i \geq 0$ are real eigenvalues and $\{|\mathbf{r}_i\rangle\}_i$ is an orthonormal basis of the Hilbert space. Some eigenvalues r_i will be non-zero. For these eigenvalues, define the vectors

$$|\mathbf{e}_i\rangle = \frac{\hat{A}}{r_i}|\mathbf{r}_i\rangle\,. \tag{3.77}$$

It is simple to show that these vectors are orthonormal:

$$\langle \mathbf{e}_j | \mathbf{e}_i \rangle = \langle \mathbf{r}_j | \frac{\hat{A}^\dagger \hat{A}}{r_i r_j} | \mathbf{r}_i \rangle = \langle \mathbf{r}_j | \frac{\hat{R}^2}{r_i r_j} | \mathbf{r}_i \rangle = \delta_{ij} \,. \tag{3.78}$$

Starting from the vectors $|\mathbf{e}_i\rangle$ and using the Gram–Schmidt procedure, it is possible to construct an orthonormal basis $\{|\mathbf{e}_i\rangle\}_i$ of V. Finally, define the operator $\hat{U} = \sum_k |\mathbf{e}_k\rangle\langle \mathbf{r}_k|$. This operator is unitary, since $\hat{U}^\dagger \hat{U} = \sum_{jk} |\mathbf{e}_k\rangle\langle \mathbf{r}_k|\mathbf{r}_j\rangle\langle \mathbf{e}_j| = \mathbb{1}$. Let us show that $\hat{U}\hat{R} = \hat{A}$ by calculating the action of $\hat{U}\hat{R}$ on a generic element $|\mathbf{r}_i\rangle$ of the orthonormal basis

$$\hat{U}\hat{R}|\mathbf{r}_i\rangle = \Big(\sum_{kj} r_j |\mathbf{e}_k\rangle\langle \mathbf{r}_k|\mathbf{r}_j\rangle\langle \mathbf{r}_j|\Big)|\mathbf{r}_i\rangle = \sum_k r_k |\mathbf{e}_k\rangle\langle \mathbf{r}_k|\mathbf{r}_i\rangle = r_i|\mathbf{e}_i\rangle = \hat{A}|\mathbf{r}_i\rangle \,. \tag{3.79}$$

The left polar decomposition is $\hat{A} = \hat{L}\hat{U}$ with $\hat{L} = \hat{U}\hat{R}\hat{U}^\dagger$. □

Let us now discuss the **singular value decomposition** (SVD), an important decomposition of a complex matrix in terms of unitary and diagonal matrices. We will focus on square matrices since this is the most relevant case for future discussions. A similar result holds for rectangular matrices.

Theorem 3.5 (Singular value decomposition) Any endomorphism $\hat{A}$ can be expressed as a product of three operators

$$\boxed{\hat{A} = \hat{U}\hat{D}\hat{V},}$$

where $\hat{U}$ and $\hat{V}$ are unitary operators and $\hat{D}$ is a diagonal operator. In matrix form

$$A = \begin{bmatrix} U_{11} & \cdots & U_{1n} \\ \vdots & \ddots & \vdots \\ U_{n1} & \cdots & U_{nn} \end{bmatrix} \begin{bmatrix} D_{11} & \cdots & 0 \\ \vdots & \ddots & \vdots \\ 0 & \cdots & D_{nn} \end{bmatrix} \begin{bmatrix} V_{11} & \cdots & V_{1n} \\ \vdots & \ddots & \vdots \\ V_{n1} & \cdots & V_{nn} \end{bmatrix},$$

where n is the dimension of the vector space.

Proof Exploiting the polar decomposition, we can express the operator $\hat{A}$ as $\hat{A} = \hat{C}\hat{R}$, where $\hat{C}$ is unitary and $\hat{R}$ is positive. Since $\hat{R}$ is positive, it can be written as $\hat{R} = \hat{M}\hat{D}\hat{M}^\dagger$, where $\hat{M}$ is unitary and $\hat{D}$ is a diagonal matrix with non-negative real elements. Thus, $\hat{A} = \hat{C}\hat{M}\hat{D}\hat{M}^\dagger$. Now define $\hat{U} = \hat{C}\hat{M}$ and $\hat{V} = \hat{M}^\dagger$. From these definitions, it follows that $\hat{A} = \hat{U}\hat{D}\hat{V}$. □

3.2.14 Trace and the Hilbert–Schmidt inner product

The trace function is one of the most important operations in quantum information theory. The **trace** is a function that maps an endomorphism into a complex number

$$\begin{aligned} \mathrm{Tr}: \quad \mathrm{L}(V) &\rightarrow \quad \mathbb{C} \\ \hat{A} &\mapsto \quad \mathrm{Tr}[\hat{A}]. \end{aligned} \tag{3.80}$$

The complex number $\mathrm{Tr}[\hat{A}]$ is equal to the sum of the diagonal elements of the matrix associated with $\hat{A}$,

$$\boxed{\mathrm{Tr}[\hat{A}] = \sum_{i=1}^{n} \langle \mathbf{e}_i | \hat{A} | \mathbf{e}_i \rangle = \sum_{i=1}^{n} A_{ii}.} \tag{3.81}$$

The trace function has some important properties that are listed below.

T1. $\mathrm{Tr}[\hat{A} + \hat{B}] = \mathrm{Tr}[\hat{A}] + \mathrm{Tr}[\hat{B}]$ (linearity),

T2. $\mathrm{Tr}[c\hat{A}] = c\mathrm{Tr}[\hat{A}]$ (scalar multiplication),

T3. $\mathrm{Tr}[\hat{A}\hat{B}\hat{C}] = \mathrm{Tr}[\hat{C}\hat{A}\hat{B}]$ (cyclicity),

T4. $\mathrm{Tr}[\hat{U}\hat{A}\hat{U}^\dagger] = \mathrm{Tr}[\hat{A}]$ (unitary invariance),

T5. $\mathrm{Tr}[\hat{A}|\mathbf{a}\rangle\langle\mathbf{b}|] = \langle\mathbf{b}|\hat{A}|\mathbf{a}\rangle$ (matrix entry),

where $\hat{U}$ is a unitary operator. Property **T4** indicates that the trace is independent of the choice of basis. The set of positive operators with trace equal to 1 is denoted as

$$\mathrm{D}(V) = \left\{ \hat{A} \in \mathrm{L}(V) : \hat{A} \geq 0 \text{ and } \mathrm{Tr}[\hat{A}] = 1 \right\}. \tag{3.82}$$

The elements in $\mathrm{D}(V)$ are often indicated with the letters ρ, σ, and τ and are called **density operators**. For these operators, the hat $\hat{}$ is usually omitted.

The **Hilbert–Schmidt inner product** is an important example of an inner product between linear operators based on the trace,

$$\langle \cdot | \cdot \rangle_2 : \quad \mathrm{B}(V) \times \mathrm{B}(V) \quad \rightarrow \quad \mathbb{C}. \tag{3.83}$$

Its definition is

$$\boxed{\langle \hat{A} | \hat{B} \rangle_2 = \mathrm{Tr}[\hat{A}^\dagger \hat{B}].}$$

Here, subscript 2 distinguishes this inner product from other inner products between linear operators. The operator space $\mathrm{B}(V)$ endowed with the Hilbert–Schmidt inner product $\langle \cdot | \cdot \rangle_2$ is a Hilbert space usually called the **Hilbert–Schmidt space**. Let us show that the inner product $\langle \cdot | \cdot \rangle_2$ satisfies all necessary requirements presented in Section 3.1.3.

Proof

1. The Hilbert–Schmidt inner product is positive-definite because $\hat{A}^\dagger \hat{A}$ is positive

$$\langle \hat{A} | \hat{A} \rangle_2 = \mathrm{Tr}\left[A^\dagger A \right] = \sum_{i=1}^{n} \langle i | \hat{A}^\dagger \hat{A} | i \rangle \geq 0. \tag{3.84}$$

2. The Hilbert–Schmidt inner product is linear with respect to the second argument because the trace is a linear operation

$$\langle \hat{A}|c_1\hat{B}_1 + c_2\hat{B}_2\rangle_2 = c_1\mathrm{Tr}[\hat{A}^\dagger\hat{B}_1] + c_2\mathrm{Tr}[\hat{A}^\dagger\hat{B}_2] = c_1\langle\hat{A}|\hat{B}_1\rangle_2 + c_2\langle A|\hat{B}_2\rangle_2. \tag{3.85}$$

3. The third requirement is satisfied by the Hilbert–Schmidt inner product because

$$\langle\hat{A}|\hat{B}\rangle_2 = \mathrm{Tr}[\hat{A}^\dagger\hat{B}] = \sum_{i=1}^{n}\langle i|\hat{A}^\dagger\hat{B}|i\rangle = \sum_{i=1}^{n}\langle i|\hat{B}^\dagger\hat{A}|i\rangle^* = \mathrm{Tr}[\hat{B}^\dagger\hat{A}]^* = \langle\hat{B}|\hat{A}\rangle_2^*.$$

This concludes the proof. □

The Hilbert–Schmidt inner product naturally induces a norm for operators in $\mathrm{B}(V)$,

$$\|\hat{A}\|_2 = \sqrt{\langle\hat{A}|\hat{A}\rangle_2} = \sqrt{\mathrm{Tr}[\hat{A}^\dagger\hat{A}]}. \tag{3.86}$$

This is called the **Hilbert–Schmidt norm**. Furthermore, the Hilbert–Schmidt inner product induces a distance between linear operators, called the **Hilbert–Schmidt distance**

$$D_2(\hat{A},\hat{B}) = \frac{1}{2}\|\hat{A}-\hat{B}\|_2 = \frac{1}{2}\sqrt{\mathrm{Tr}[(\hat{A}-\hat{B})^\dagger(\hat{A}-\hat{B})]}. \tag{3.87}$$

It is worth noting that this is not the only distance that can be defined between linear operators.

3.2.15 Tensor spaces

A quantum system composed of two or more subsystems is described by a **tensor product space**. Consider two Hilbert spaces H_A and H_B of dimension n_A and n_B, respectively. We say that $H = H_\mathrm{A} \otimes H_\mathrm{B}$ is the tensor product of H_A and H_B if it is possible to associate a vector $|\mathbf{v}\rangle \otimes |\mathbf{w}\rangle$ in H with each pair of vectors $|\mathbf{v}\rangle \in H_\mathrm{A}$, and $|\mathbf{w}\rangle \in H_\mathrm{B}$. The operation

$$\begin{array}{cccc} \otimes: & H_\mathrm{A}\times H_\mathrm{B} & \to & H_\mathrm{A}\otimes H_\mathrm{B} \\ & (\mathbf{v},\mathbf{w}) & \mapsto & \mathbf{v}\otimes\mathbf{w} \end{array} \tag{3.88}$$

must satisfy the following requirements for all $|\mathbf{v}\rangle \in H_\mathrm{A}$, $|\mathbf{w}\rangle \in H_\mathrm{B}$ and $c \in \mathbb{C}$:

TP1. $c\,(|\mathbf{v}\rangle \otimes |\mathbf{w}\rangle) = (c|\mathbf{v}\rangle) \otimes |\mathbf{w}\rangle = |\mathbf{v}\rangle \otimes (c|\mathbf{w}\rangle)$,

TP2. $(|\mathbf{v}_1\rangle + |\mathbf{v}_2\rangle) \otimes |\mathbf{w}\rangle = |\mathbf{v}_1\rangle \otimes |\mathbf{w}\rangle + |\mathbf{v}_2\rangle \otimes |\mathbf{w}\rangle$,

TP3. $|\mathbf{v}\rangle \otimes (|\mathbf{w}_1\rangle + |\mathbf{w}_2\rangle) = |\mathbf{v}\rangle \otimes |\mathbf{w}_1\rangle + |\mathbf{v}\rangle \otimes |\mathbf{w}_2\rangle$.

The vector $|\mathbf{v}\rangle \otimes |\mathbf{w}\rangle$ is called the **tensor product** of $|\mathbf{v}\rangle$ and $|\mathbf{w}\rangle$. The following notations will be used interchangeably,

$$\boxed{|\mathbf{v}\rangle \otimes |\mathbf{w}\rangle = |\mathbf{v}\rangle|\mathbf{w}\rangle = |\mathbf{v},\mathbf{w}\rangle = |\mathbf{vw}\rangle = |\mathbf{vw}\rangle_\mathrm{AB}\,.} \tag{3.89}$$

It is possible to construct a tensor product between multiple Hilbert spaces. This is important since a collection of n distinguishable particles in quantum mechanics is represented by a unit vector in the tensor product space $H_1 \otimes \ldots \otimes H_n$.

When a quantum system is described by a vector $|\mathbf{x}\rangle$ belonging to the tensor space $H_\mathrm{A} \otimes H_\mathrm{B}$, the vector $|\mathbf{x}\rangle$ is called a bipartite state. If $B_\mathrm{A} = \{|i\rangle\}_{i=0}^{n_\mathrm{A}}$ and $B_\mathrm{B} = \{|j\rangle\}_{j=0}^{n_\mathrm{B}}$ are two orthonormal bases for H_A and H_B, the set

$$B = \{|i\rangle \otimes |j\rangle\}_{i,j}$$

with $1 \leq i \leq n_\mathrm{A}$ and $1 \leq j \leq n_\mathrm{B}$ is an orthonormal basis of $H_\mathrm{A} \otimes H_\mathrm{B}$. This space has dimension $n_\mathrm{A} n_\mathrm{B}$. In this basis, a bipartite state can be expressed as

$$|\mathbf{x}\rangle = \sum_{i=1}^{n_\mathrm{A}} \sum_{j=1}^{n_\mathrm{B}} c_{ij} |i\rangle \otimes |j\rangle, \tag{3.90}$$

where $c_{ij} = \langle ij|\mathbf{x}\rangle$. The inner product between this vector and another vector $|\mathbf{y}\rangle = \sum_{ij} d_{ij}|i\rangle|j\rangle$ is given by

$$\begin{aligned}
\langle \mathbf{y}|\mathbf{x}\rangle &= \Big[\sum_{k=1}^{n_\mathrm{A}} \sum_{l=1}^{n_\mathrm{B}} d_{kl}^* \langle k| \otimes \langle l|\Big] \Big[\sum_{i=1}^{n_\mathrm{A}} \sum_{j=1}^{n_\mathrm{B}} c_{ij}|i\rangle \otimes |j\rangle\Big] \\
&= \sum_{i,k=1}^{n_\mathrm{A}} \sum_{j,l=1}^{n_\mathrm{B}} d_{kl}^* c_{ij}\, \delta_{ki}\delta_{lj} = \sum_{i=1}^{n_\mathrm{A}} \sum_{j=1}^{n_\mathrm{B}} d_{ij}^* c_{ij}.
\end{aligned}$$

If $\hat{A}$ and $\hat{B}$ are linear operators acting on H_A and H_B, then the action of the operator $\hat{A} \otimes \hat{B}$ on a generic vector $|\mathbf{x}\rangle \in H_\mathrm{A} \otimes H_\mathrm{B}$ reads

$$\hat{A} \otimes \hat{B}|\mathbf{x}\rangle = \hat{A} \otimes \hat{B}\Big(\sum_{ij} c_{ij}|i\rangle \otimes |j\rangle\Big) = \sum_{ij} c_{ij} \left(\hat{A}|i\rangle \otimes \hat{B}|j\rangle\right). \tag{3.91}$$

The matrix representation of $\hat{A} \otimes \hat{B}$ in the basis $\{|i\rangle|j\rangle\}_{i=1,j=1}^{n_\mathrm{A},n_\mathrm{B}}$ is given by

$$\hat{A} \otimes \hat{B} = \begin{bmatrix} A_{11}B & \cdots & A_{1n_\mathrm{A}}B \\ \vdots & \ddots & \vdots \\ A_{n_\mathrm{A}1}B & \cdots & A_{n_\mathrm{A}n_\mathrm{A}}B \end{bmatrix}, \tag{3.92}$$

where $A_{ij}B$ are $n_\mathrm{B} \times n_\mathrm{B}$ submatrices. In physical terms, the tensor notation $\hat{A}|\mathbf{v}\rangle \otimes \hat{B}|\mathbf{w}\rangle$ means that the operator $\hat{A}$ ($\hat{B}$) is changing the state of the subsystem A (B).

Example 3.31 Consider the two-qubit space $\mathbb{C}^2 \otimes \mathbb{C}^2$. An orthonormal basis for this space is $\{|00\rangle, |01\rangle, |10\rangle, |11\rangle\}$ and a generic vector can be expressed as $|\mathbf{x}\rangle = c_{00}|00\rangle + c_{01}|01\rangle + c_{10}|10\rangle + c_{11}|11\rangle$.

Example 3.32 Consider the operator $\hat{\sigma}_x \otimes \hat{\sigma}_z$. Using (3.92), its matrix representation is given by

$$\hat{\sigma}_x \otimes \hat{\sigma}_z = \begin{bmatrix} 0 & 1 \\ 1 & 0 \end{bmatrix} \otimes \hat{\sigma}_z = \begin{bmatrix} 0 & \hat{\sigma}_z \\ \hat{\sigma}_z & 0 \end{bmatrix} = \left[\begin{array}{cc|cc} 0 & 0 & 1 & 0 \\ 0 & 0 & 0 & -1 \\ \hline 1 & 0 & 0 & 0 \\ 0 & -1 & 0 & 0 \end{array}\right]. \tag{3.93}$$

The action of $\hat{\sigma}_x \otimes \hat{\sigma}_z$ on a vector $|\mathbf{x}\rangle = \sum_{ij} c_{ij}|i\rangle|j\rangle$ is

$$\hat{\sigma}_x \otimes \hat{\sigma}_z|\mathbf{x}\rangle = \begin{bmatrix} 0 & 0 & 1 & 0 \\ 0 & 0 & 0 & -1 \\ 1 & 0 & 0 & 0 \\ 0 & -1 & 0 & 0 \end{bmatrix} \begin{bmatrix} c_{00} \\ c_{01} \\ c_{10} \\ c_{11} \end{bmatrix} = \begin{bmatrix} c_{10} \\ -c_{11} \\ c_{00} \\ -c_{01} \end{bmatrix}. \tag{3.94}$$

Example 3.33 The **Bell states** are important examples of two-qubit states. These vectors belong to the Hilbert space $\mathbb{C}^2 \otimes \mathbb{C}^2$ and they are defined as

$$|\beta_{00}\rangle = \frac{|00\rangle + |11\rangle}{\sqrt{2}}, \qquad |\beta_{10}\rangle = \frac{|00\rangle - |11\rangle}{\sqrt{2}},$$
$$|\beta_{01}\rangle = \frac{|01\rangle + |10\rangle}{\sqrt{2}}, \qquad |\beta_{11}\rangle = \frac{|01\rangle - |10\rangle}{\sqrt{2}}.$$

As we shall see in Section 9.1, these are examples of maximally entangled states. In other textbooks, these states are sometimes indicated with the notation $|\Phi^+\rangle = |\beta_{00}\rangle$, $|\Phi^-\rangle = |\beta_{01}\rangle$, $|\Psi^+\rangle = |\beta_{10}\rangle$, $|\Psi^-\rangle = |\beta_{11}\rangle$.

3.2.16 The Pauli basis and the Pauli group

A convenient basis of the operator space $\mathrm{L}(\mathbb{C}^2 \otimes \ldots \otimes \mathbb{C}^2)$ is the **Pauli basis**, which consists of tensor products of n single-qubit Pauli operators. The Pauli basis contains 4^n operators and is defined as

$$\boxed{B = \{\hat{A}_1 \otimes \ldots \otimes \hat{A}_n : \hat{A}_i \in \{\mathbb{1}, \hat{\sigma}_x, \hat{\sigma}_y, \hat{\sigma}_z\}\}.} \tag{3.95}$$

The operators in this basis are unitary and Hermitian. They will be indicated with the letter $\hat{P}$ and will be called **collective Pauli operators** or simply Pauli operators. The Pauli basis is orthogonal[12] with respect to the Hilbert–Schmidt inner product $\langle\hat{A}|\hat{B}\rangle_2 = \mathrm{Tr}[A^\dagger B]$. Indeed, $\langle\hat{P}_i|\hat{P}_j\rangle_2 = \mathrm{Tr}[\hat{P}_i\hat{P}_j] = \delta_{ij}d$ where $d = 2^n$.

Any linear operator $\hat{O}$ acting on a set of n qubits can be decomposed in the Pauli basis as

$$\hat{O} = \sum_{i=1}^{4^n} c_i \hat{P}_i, \tag{3.96}$$

where the coefficients of the expansion are given by the inner product $c_i = \langle\hat{P}_i|\hat{O}\rangle_2/d$.

[12]Note that with a slight modification, the Pauli basis can be normalized: $\tilde{B} = \{\frac{1}{\sqrt{d}}\hat{A}_1 \otimes \ldots \otimes \hat{A}_n : \hat{A}_i \in \{\mathbb{1}, \hat{\sigma}_x, \hat{\sigma}_y, \hat{\sigma}_z\}\}$ so that $\langle\hat{P}_i|\hat{P}_j\rangle_2 = \delta_{ij}$ for all $\hat{P}_i \in \tilde{B}$.

When two Pauli operators are multiplied together they generate another Pauli operator (up to a constant factor): $\hat{P}_i\hat{P}_j = \alpha\hat{P}_k$, where $\alpha \in \{1, i, -1, -i\}$. This suggests that a suitable collection of Pauli operators can form a group. In particular, the n-qubit **Pauli group** $\mathcal{P}_n$ consists of tensor products of n single-Pauli operators together with a multiplicative factor of ± 1 or $\pm i$,

$$\boxed{\mathcal{P}_n = \{i^k \hat{A}_1 \otimes \ldots \otimes \hat{A}_n : \ \hat{A}_i \in \{\mathbb{1}, \hat{\sigma}_x, \hat{\sigma}_y, \hat{\sigma}_z\} \text{ and } 0 \leq k \leq 3\}.}$$

The Pauli group $\mathcal{P}_n$ contains 4^{n+1} elements and they will be indicated with the letter $\hat{P}$. The unit element of this group is clearly $\mathbb{1} \otimes \ldots \otimes \mathbb{1}$ and the inverse of $\hat{P}$ is $\hat{P}^\dagger$.

Example 3.34 The smallest Pauli group is $\mathcal{P}_1$, which contains 16 elements

$$\mathcal{P}_1 = \{\pm\mathbb{1},\ \pm i\mathbb{1},\ \pm\hat{\sigma}_x,\ \pm i\hat{\sigma}_x,\ \pm\hat{\sigma}_y,\ \pm i\hat{\sigma}_y,\ \pm\hat{\sigma}_z,\ \pm i\hat{\sigma}_z\}. \tag{3.97}$$

Another interesting example of a Pauli group is $\mathcal{P}_2$

$$\mathcal{P}_2 = \{\pm\,\hat{A}_1 \otimes \hat{A}_2,\ \ \pm i\,\hat{A}_1 \otimes \hat{A}_2\},$$

where as usual $\hat{A}_1,\ \hat{A}_2 \in \{\mathbb{1}, \hat{\sigma}_x, \hat{\sigma}_y, \hat{\sigma}_z\}$. The Pauli group $\mathcal{P}_2$ contains 64 elements.

Further reading

A good understanding of linear algebra is vital for quantum information theory. A comprehensive introduction is "Linear Algebra" by Lang [114]. The connection between linear algebra and quantum mechanics can be found in multiple textbooks, including "Mathematical Methods of Quantum Optics" by Puri [117] and "Geometry of quantum states" by Bengtsson and Zyczkowski [111]. For more information about functional analysis and infinite-dimensional Hilbert spaces, we recommend the monograph by Reed and Simon [113]. Numerical linear algebra is important for simulating quantum systems on classical and quantum processors. The following references cover basic topics on numerical linear algebra [118, 119, 120]. Computational packages available to the public can be found in Refs [121, 122, 123].

Summary

Vector spaces

V	A non-empty set of elements called vectors
$+ : V \to V$	Addition between two vectors
$\cdot : \mathbb{K} \times V \to V$	Scalar multiplication
$(V, +, \cdot)$	Vector space
$\mathbb{C}^n = \{\lvert\mathbf{v}\rangle = (v_1, \cdots, v_n) : v_i \in \mathbb{C}\}$	Complex vector space
$\underline{\mathbf{0}} = (0, \ldots, 0)$	Null vector in $\mathbb{C}^n$
$\mathbb{B}^n = \{\mathbf{x} = x_1, \cdots, x_n : x_i \in \mathbb{B}\}$	Hamming space
$\lvert\mathbf{v}\rangle = c_1\lvert\mathbf{v}_1\rangle + \ldots + c_n\lvert\mathbf{v}_n\rangle$	Linear combination
$\sum_i c_i\lvert\mathbf{v}_i\rangle = \underline{\mathbf{0}} \;\Leftrightarrow\; c_i = 0$	Independent vectors $\{\lvert\mathbf{v}_i\rangle\}_i$
$\lvert\mathbf{v}\rangle = \sum_i c_i\lvert\mathbf{v}_i\rangle$, $\sum_i c_i = 1$	Convex combination
$\text{span}(\lvert\mathbf{v}_1\rangle, \ldots, \lvert\mathbf{v}_n\rangle)$	Linear span
$\langle \cdot \vert \cdot \rangle : V \times V \to \mathbb{C}$	Inner product
$\langle \mathbf{w} \vert \cdot \rangle : V \to \mathbb{C}$	Bra
$\langle \mathbf{w} \vert \mathbf{v} \rangle = \sum_i w_i^* v_i$	Inner product between two $\mathbb{C}^n$ vectors
$\langle \mathbf{v}_1 \vert \mathbf{v}_2 \rangle = 0$	$\mathbf{v}_1$ and $\mathbf{v}_2$ are orthogonal
$\langle \mathbf{e}_i \vert \mathbf{e}_j \rangle = \delta_{ij}$	The basis $\{\lvert\mathbf{e}_1\rangle, \ldots, \lvert\mathbf{e}_n\rangle\}$ is orthonormal
$B = \{\lvert 1\rangle, \lvert 2\rangle \ldots, \lvert n\rangle\}$	Canonical basis in $\mathbb{C}^n$
$B = \{\lvert 0\rangle, \lvert 1\rangle\}$	Computational basis in $\mathbb{C}^2$
$\lVert \cdot \rVert : V \to \mathbb{R}$	Norm
$\mathbf{w} = \mathbf{v}/\lVert\mathbf{v}\rVert$ where $\mathbf{v} \neq \underline{\mathbf{0}}$	$\mathbf{w}$ is a normalized vector
$\lVert\mathbf{v}\rVert = \sqrt{\sum_i \lvert v_i\rvert^2}$	Norm of a $\mathbb{C}^n$ vector
$F(\mathbf{v}_1, \mathbf{v}_2) = \lvert\langle \mathbf{v}_1 \vert \mathbf{v}_2 \rangle\rvert$	Uhlmann fidelity
$F_{\text{J}}(\mathbf{v}_1, \mathbf{v}_2) = \lvert\langle \mathbf{v}_1 \vert \mathbf{v}_2 \rangle\rvert^2$	Jozsa fidelity
$d : V \times V \to \mathbb{R}$	Distance
$d(\mathbf{v}, \mathbf{w}) = \sqrt{\sum_i \lvert v_i - w_i\rvert^2}$	Distance between two $\mathbb{C}^n$ vectors
$\lvert\langle \mathbf{w} \vert \mathbf{v} \rangle\rvert \leq \lVert\mathbf{w}\rVert \cdot \lVert\mathbf{v}\rVert$	Cauchy–Schwarz inequality

The metric vector space (V, d) is complete if all Cauchy sequences in V converge in V.

An inner product space $(V, \langle \cdot \vert \cdot \rangle)$ that is complete is called a Hilbert space.

Linear operators

$\hat{A} : V \to W$	Linear operator, $\hat{A} \in \mathrm{L}(V, W)$
$\hat{A} : V \to V$	Endomorphism, $\hat{A} \in \mathrm{L}(V)$
$\exists M > 0 : \ \|\hat{A}\|\mathbf{v}\rangle\| \leq M \| \|\mathbf{v}\rangle\|$	Bounded operator, $\hat{A} \in \mathrm{B}(V)$
$\hat{P}^2 = \hat{P}$	Projector, $\hat{P} \in \mathrm{Proj}(V)$
$\hat{A}^{-1}$	Inverse operator
$\det(A) \neq 0$	Invertible operator, $\hat{A} \in \mathrm{GL}(V)$
$\hat{A}^\dagger$	Adjoint operator of $\hat{A}$
$[\hat{A}, \hat{A}^\dagger] = \underline{\mathbf{0}}$	Normal operator, $\hat{A} \in \mathrm{N}(V)$
$\hat{A} = \hat{A}^\dagger$	Hermitian operator, $\hat{A} \in \mathrm{Her}(V)$
$\hat{U}^\dagger = \hat{U}^{-1}$	Unitary operator, $\hat{U} \in \mathrm{U}(V)$
$\langle \mathbf{v} \| \hat{A} \| \mathbf{v} \rangle \geq 0$ for all $\|\mathbf{v}\rangle$	Positive operator, $\hat{A} \in \mathrm{Pos}(V)$
$\hat{A} \geq 0$ and $\mathrm{Tr}[\hat{A}] = 1$	Density operator, $\hat{A} \in \mathrm{D}(V)$

$$\mathrm{D}(V) \subseteq \mathrm{Pos}(V) \subseteq \mathrm{Her}(V) \subseteq \mathrm{N}(V) \subseteq \mathrm{B}(V) \subseteq \mathrm{L}(V)$$

$\hat{C} = \hat{B}\hat{A}$	Composition of two operators
$\mathbb{1}\|\mathbf{v}\rangle = \|\mathbf{v}\rangle$	Identity operator
$[\hat{A}, \hat{B}] = \hat{A}\hat{B} - \hat{B}\hat{A}$	Commutator
$\{\hat{A}, \hat{B}\} = \hat{A}\hat{B} + \hat{B}\hat{A}$	Anticommutator
$\|\hat{A}\|_{\mathrm{o}} = \sup_{\|\mathrm{v}\|=1} \|\hat{A}\|\mathbf{v}\rangle\|$	Norm of an operator
$\sum_i \|\mathbf{e}_i\rangle\langle \mathbf{e}_i\| = \mathbb{1}$	Completeness relation
$(A_{ij})^{\mathrm{T}} = A_{ji}$	Tranpose
$(A_{ij})^\dagger = A_{ji}^*$	Conjugate tranpose
$\hat{\sigma}_x, \hat{\sigma}_y, \hat{\sigma}_z$	Pauli operators
$\hat{\sigma}^+ = \|1\rangle\langle 0\|, \hat{\sigma}^- = \|0\rangle\langle 1\|$	Raising and lowering operators
$\det(A) = \sum_{\sigma \in S_n} (-1)^\sigma \prod_i A_{i\sigma(i)}$	Determinant of a matrix A
$\hat{A}\|\mathbf{a}\rangle = a\|\mathbf{a}\rangle \qquad (\|\mathbf{a}\rangle \neq \underline{\mathbf{0}})$	$\|\mathbf{a}\rangle$ is an eigenvector of A with eigenvalue a
$\det(A - a\mathbb{1}) = 0$	Characteristic equation
$f(\hat{A}) = \sum_i f(a_i)\|\mathbf{a}_i\rangle\langle \mathbf{a}_i\|$	Function of a normal operator $\hat{A}$
$\|\hat{A}\| = \sqrt{\hat{A}^\dagger \hat{A}} = \sum_i \|a_i\|\|\mathbf{a}_i\rangle\langle \mathbf{a}_i\|$	Absolute value of an operator $\hat{A} \in \mathrm{N}(V)$
$\mathrm{Tr}[\hat{A}] = \sum_i \langle i\|\hat{A}\|i\rangle$	Trace of an operator
$\langle \hat{A}\|\hat{B}\rangle = \mathrm{Tr}[\hat{A}^\dagger \hat{B}]$	Hilbert–Schmidt inner product
$\|\hat{A}\|_2 = \sqrt{\mathrm{Tr}[\hat{A}^\dagger \hat{A}]}$	Hilbert–Schmidt norm
$D_2(\hat{A}, \hat{B}) = \sqrt{\mathrm{Tr}[(\hat{A} - \hat{B})^\dagger (\hat{A} - \hat{B})]}$	Hilbert–Schmidt distance
$H_{\mathrm{A}} \otimes H_{\mathrm{B}}$	Tensor space

Important theorems

(Spectral) Any normal operator $\hat{A}$ can be expressed as $\hat{A} = \sum_i a_i |\mathbf{a}_i\rangle\langle\mathbf{a}_i|$.

$\hat{A}$ and $\hat{B}$ can be diagonalized w.r.t. to a common basis if and only if $[\hat{A}, \hat{B}] = \underline{\mathbf{0}}$.

If $\hat{U}$ is unitary, then $\hat{U} = e^{i\hat{A}}$ where $\hat{A}$ is Hermitian.

(Polar decomp.) Any endomorphism $\hat{A} = \hat{L}\hat{U}$, where $\hat{L} \geq 0$ and $\hat{U}$ is unitary.

(SVD) Any square matrix $A = UDV$, where $U, V \in \mathrm{U}(V)$ and $D \geq 0$ is diagonal.

Exercises

Exercise 3.1 Show that the vectors $|\mathbf{v}_1\rangle = (1, 0)$, $|\mathbf{v}_2\rangle = (1, 1)$ generate $\mathbb{R}^2$.

Exercise 3.2 Show that the space generated by the vectors $|\mathbf{v}_1\rangle = (1, 2)$, $|\mathbf{v}_2\rangle = (-1, -2)$ is one dimensional.

Exercise 3.3 Calculate the inner product between $|\mathbf{v}\rangle = (5, -2+3i, -4)$ and $|\mathbf{w}\rangle = (1, 2, -4i)$.

Exercise 3.4 Calculate the norm of the vector $|\mathbf{v}\rangle = -|\mathbf{e}_1\rangle + 2|\mathbf{e}_2\rangle - 7i|\mathbf{e}_3\rangle$ and find the corresponding normalized vector.

Exercise 3.5 Consider the vector space $\mathbb{C}^2$. Show that the distance between the vectors $|\mathbf{v}_1\rangle = (1, i)$ and $|\mathbf{v}_2\rangle = (-i, -1)$ is $d(\mathbf{v}_1, \mathbf{v}_2) = 2$.

Exercise 3.6 Show that the determinant of the matrix

$$A = \begin{bmatrix} i & 2 & 4 \\ 2i & 3 & -4 \\ 0 & 4 & 1 \end{bmatrix}$$

is $\det(A) = 47i$.

Exercise 3.7 Consider the linearly independent vectors $|\mathbf{v}_1\rangle = (1, 2, 0)$, $|\mathbf{v}_2\rangle = (1, 1, 0)$, $|\mathbf{v}_3\rangle = (0, 2, -1)$. Convert them into a set of orthonormal vectors using the Gram–Schmidt procedure.

Exercise 3.8 Consider the operators $A = \begin{bmatrix} 1 & i \\ -i & 0 \end{bmatrix}$ and $B = \begin{bmatrix} 0 & 1 \\ 1 & 2 \end{bmatrix}$. Using the inner product $\langle A|B\rangle_2 = \mathrm{Tr}[A^\dagger B]$, calculate $\|A\|_2$, $\|B\|_2$, $\langle A|B\rangle_2$, and $D_2(A, B)$. Lastly, calculate the operator norms $\|A\|_o$, $\|B\|_o$ and the operator distance $D_o(A, B) = \|A - B\|_o$.

Solutions

Solution 3.1 The vector space $\mathbb{R}^2$ has dimension 2 and is generated by two independent vectors. The vectors $|\mathbf{v}_1\rangle$ and $|\mathbf{v}_2\rangle$ are independent because the equation $c_1|\mathbf{v}_1\rangle + c_2|\mathbf{v}_2\rangle = 0$ is satisfied only when $c_1 = c_2 = 0$.

Solution 3.2 The vector space generated by $|\mathbf{v}_1\rangle$ and $|\mathbf{v}_2\rangle$ is one dimensional because these vectors are linearly dependent. Indeed, the relation $c_1|\mathbf{v}_1\rangle + c_2|\mathbf{v}_2\rangle = 0$ is satisfied with $c_1 = 1$ and $c_2 = -1$.

Solution 3.3 $\langle \mathbf{w}|\mathbf{v}\rangle = 1 \cdot 5 + 2 \cdot (-2+3i) + (4i) \cdot (-4) = 1 - 10i$.

Solution 3.4 $\||\mathbf{v}\rangle\| = \sqrt{\langle \mathbf{v}|\mathbf{v}\rangle} = \sqrt{1+2+49} = \sqrt{52}$. The normalized vector is $|\mathbf{v}\rangle/\||\mathbf{v}\rangle\| = -1/\sqrt{52}\,|\mathbf{e}_1\rangle + 2/\sqrt{52}\,|\mathbf{e}_2\rangle - 7i/\sqrt{52}\,|\mathbf{e}_3\rangle$.

Solution 3.5 Since $|\mathbf{v}_1\rangle - |\mathbf{v}_2\rangle = (1+i, i+1)$, then $d(\mathbf{v}_1, \mathbf{v}_2) = \sqrt{|1+i|^2 + |1+i|^2} = 2$.

Solution 3.6 Using eqn (3.47), we have

$$\det(A) = i \cdot 3 \cdot 1 - i \cdot (-4) \cdot 4 - 2 \cdot 2i \cdot 1 + 2 \cdot (-4) \cdot 0 + 4 \cdot 2i \cdot 4 - 4 \cdot 3 \cdot 0 = 47i.$$

Solution 3.7 Using the Gram–Schmidt procedure presented in Section 3.1.9, we find

$$|\mathbf{w}_1\rangle = |\mathbf{v}_1\rangle = \begin{bmatrix} 1 \\ 2 \\ 0 \end{bmatrix},$$

$$|\mathbf{w}_2\rangle = |\mathbf{v}_2\rangle - \frac{\langle \mathbf{w}_1|\mathbf{v}_2\rangle}{\langle \mathbf{w}_1|\mathbf{w}_1\rangle}|\mathbf{w}_1\rangle = \begin{bmatrix} 2/5 \\ -1/5 \\ 0 \end{bmatrix},$$

$$|\mathbf{w}_3\rangle = |\mathbf{v}_3\rangle - \frac{\langle \mathbf{w}_1|\mathbf{v}_3\rangle}{\langle \mathbf{w}_1|\mathbf{w}_1\rangle}|\mathbf{w}_1\rangle - \frac{\langle \mathbf{w}_2|\mathbf{v}_3\rangle}{\langle \mathbf{w}_2|\mathbf{w}_2\rangle}|\mathbf{w}_2\rangle = \begin{bmatrix} 0 \\ 0 \\ -1 \end{bmatrix}.$$

Thus, $|\tilde{\mathbf{w}}_1\rangle = \frac{1}{\sqrt{5}}(1,2,0)$, $|\tilde{\mathbf{w}}_2\rangle = \frac{1}{\sqrt{5}}(2,-1,0)$, $|\tilde{\mathbf{w}}_3\rangle = (0,0,-1)$ is an orthonormal basis.

Solution 3.8 Since

$$\|A\|_2^2 = \langle A|A\rangle_2^2 = \mathrm{Tr}[A^\dagger A] = \mathrm{Tr}\begin{bmatrix} 1 & i \\ -i & 0 \end{bmatrix}\begin{bmatrix} 1 & i \\ -i & 0 \end{bmatrix} = \mathrm{Tr}\begin{bmatrix} 2 & i \\ -i & 1 \end{bmatrix} = 3,$$

$$\|B\|_2^2 = \langle B|B\rangle_2^2 = \mathrm{Tr}[B^\dagger B] = \mathrm{Tr}\begin{bmatrix} 0 & 1 \\ 1 & 2 \end{bmatrix}\begin{bmatrix} 0 & 1 \\ 1 & 2 \end{bmatrix} = \mathrm{Tr}\begin{bmatrix} 1 & 2 \\ 2 & 5 \end{bmatrix} = 6,$$

we have $\|A\|_2 = \sqrt{3}$, and $\|B\|_2 = \sqrt{6}$. The Hilbert–Schmidt inner product between A and B is

$$\langle A|B\rangle_2 = \mathrm{Tr}[A^\dagger B] = \mathrm{Tr}\begin{bmatrix} 1 & i \\ -i & 0 \end{bmatrix}\begin{bmatrix} 0 & 1 \\ 1 & 2 \end{bmatrix} = \mathrm{Tr}\begin{bmatrix} i & 1+2i \\ 0 & -i \end{bmatrix} = 0.$$

In other words, A and B are orthogonal. Since

$$C = A - B = \begin{bmatrix} 1 & i-1 \\ -i-1 & 2 \end{bmatrix},$$

the Hilbert–Schmidt distance between A and B is $D_2(A,B) = \frac{1}{2}\sqrt{\mathrm{Tr}[C^\dagger C]} = 3/2$.

Both A and B are Hermitian. The eigenvalues of A are $(1 \pm \sqrt{5})/2$, which means that its operator norm is given by $\|A\|_o = (1+\sqrt{5})/2$. Similarly, the eigenvalues of B are $1 \pm \sqrt{2}$. The operator norm is the modulus of the maximum magnitude eigenvalue, $\|B\|_o = 1 + \sqrt{2}$. The operator distance between A and B is $D_o = \|A - B\|_o = (1+\sqrt{17})/2$ because the eigenvalues of $A - B$ are $(-1 \pm \sqrt{17})/2$.

4 Quantum mechanics

The postulates of quantum mechanics
introduce a radical departure from classical physics.

Quantum mechanics is a physical theory that was developed at the beginning of the twentieth century. This theory aims to describe and predict the behavior of matter at the fundamental level. When it was first proposed, the principal intent was to explain black-body radiation and the physics of the hydrogen atom in a non-relativistic context. Over the years, numerous phenomena, such as atomic emission spectra, have been convincingly explained by quantum mechanics. The growing number of experimental results in favor of the new theory indicated its predictive power in a wide range of scientific fields.

In this chapter, we will present the theoretical foundations of quantum mechanics. Excellent textbooks which give a detailed and thorough presentation of this topic already exist [5, 124, 125, 126, 127, 128]. As such, we will limit ourselves to presenting the concepts necessary to understand the content of this textbook. For simplicity, we will consider systems with finite degrees of freedom since these are the systems mainly involved in quantum computation protocols.

Our presentation of quantum mechanics will follow the **von Neumann postulates.** We will discuss them through the steps of a typical quantum experiment: the preparation of the system, its time evolution, and the final measurement (see Fig. 4.1 for their graphical representation).

Regarding state preparation, a fundamental concept of quantum mechanics is that a system well isolated from the environment, such as an electron or a particle of light, can be described by a unit vector in a Hilbert space. This vector is usually called the quantum state of the system and its time evolution is governed by a linear differential equation called the Schrödinger equation. The solution of this equation shows that the state of a quantum system at future times is obtained by applying a unitary transformation. Physical quantities that can be measured, such as the position or the energy of a particle, are called observables. In quantum mechanics, observables are described by Hermitian operators, which are linear operators with real eigenvalues. The outcome of a measurement is one of the eigenvalues of the observable and the state of the system after the measurement lies in the eigenspace associated with the eigenvalue. Unlike the time evolution operation, the measurement process is an irreversible operation.

After presenting the von Neumann postulates, we will dwell on two concrete examples of quantum systems: the qubit and the quantum harmonic oscillator. We will study the interaction between them in two different regimes.

4.1 Postulates of quantum mechanics

Without further ado, let us start with the first postulate.

Postulate 4.1 The **pure state** of a quantum system is completely described by a non-null vector $|\psi\rangle$ in a complex separable Hilbert space H and is called a **quantum state**. Vectors differing by a multiplicative factor represent the same quantum state. The Hilbert space of a composite system is the tensor product space $H_1 \otimes \ldots \otimes H_n$, where H_i is the Hilbert space of the i-th individual subsystem.

The first postulate introduces a radical change from classical physics because it implies that if $|\psi_1\rangle$ and $|\psi_2\rangle$ are two physical states[1], so is any linear combination $c_1|\psi_1\rangle+c_2|\psi_2\rangle$. This **superposition principle** is necessary to explain the phenomenology of multiple experiments including the double-slit experiment [129, 130]. The first postulate also implicitly claims that a pure state is described by a ray in a Hilbert space. A **ray** S is a collection of non-zero vectors differing by just a constant factor

$$\boxed{S(|\psi\rangle) = \{|v\rangle \in H : |v\rangle = c|\psi\rangle \text{ where } c \neq 0\}.} \tag{4.1}$$

Any vector in S can be chosen to describe the state of the system. A unit vector is usually picked as representative of this set. This choice is justified by the probability distribution produced by a measurement, as we shall see later. It is worth noting that the global phase η of a quantum state can be chosen freely, meaning that $|\psi\rangle \equiv e^{i\eta}|\psi\rangle$.

Postulate 4.2 The unperturbed **time evolution** from time t_0 to time t of a quantum state $|\psi(t_0)\rangle$ is described by a unitary operator $\hat{U}(t, t_0)$ according to

$$\boxed{|\psi(t)\rangle = \hat{U}(t, t_0)\,|\psi(t_0)\rangle.} \tag{4.2}$$

The operator $\hat{U}(t, t_0)$ is called the **time evolution operator**.

This axiom has several consequences. First of all, if $t = t_0$, the operator $\hat{U}(t_0, t_0)$ must be equal to the identity. Secondly, since $\hat{U}(t, t_0)$ is unitary, the reversed time evolution from t to t_0 is described by the operator

$$\hat{U}(t_0, t) = \hat{U}^{-1}(t, t_0) = \hat{U}^{\dagger}(t, t_0)\,. \tag{4.3}$$

The composition of two consecutive time evolutions takes the form

[1] To simplify the notation, in this chapter and the following ones, the symbol inside a ket will no longer be indicated with bold letters.

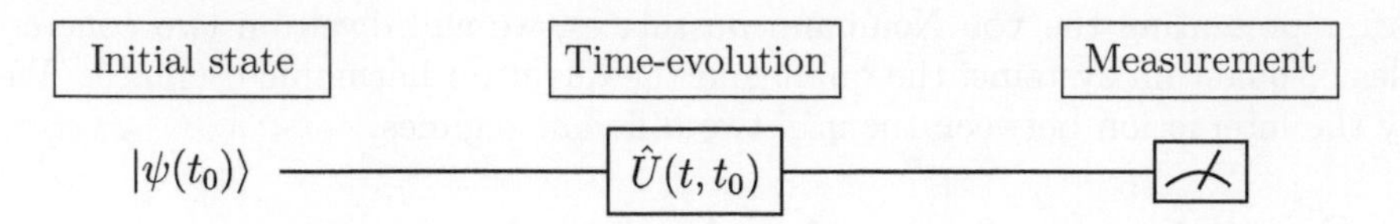

Fig. 4.1 A quantum experiment. Steps of a typical experiment in quantum physics. A quantum system is initially in state $|\psi(t_0)\rangle$. The time evolution is described by a unitary operator $\hat{U}(t, t_0)$. The state of the system at $t > t_0$ is given by $\hat{U}(t, t_0)|\psi(t_0)\rangle$. The experiment ends with a measurement of the quantum system (indicated with a meter in this schematic).

$$\hat{U}(t_2, t_0) = \hat{U}(t_2, t_1)\, \hat{U}(t_1, t_0) \,, \tag{4.4}$$

where $t_0 < t_1 < t_2$. If time-dependent forces are not applied to the system, its dynamics are invariant under time translations (in other words, the time evolution operator only depends on the difference $t - t_0$). Lastly, this postulate implies that the normalization of a quantum state does not change over time

$$\langle \psi(t) | \psi(t) \rangle = \langle \psi(t_0) | \underbrace{\hat{U}^\dagger(t, t_0)\, \hat{U}(t, t_0)}_{\mathbb{1}} | \psi(t_0) \rangle = \langle \psi(t_0) | \psi(t_0) \rangle = 1 \,. \tag{4.5}$$

Let us move on to the next postulate.

Postulate 4.3 Any **physical observable** A is associated with a Hermitian operator $\hat{A} \in \mathrm{L}(H)$ acting on the Hilbert space H of the quantum system.

Examples of physical observables are the position $\mathbf{x}$, the linear momentum $\mathbf{p}$, the angular momentum $\mathbf{L}$, the energy E, and the spin $\mathbf{S}$. Many operators representing observables in quantum mechanics are the generators of a symmetry transformation. The discussion of this beautiful correspondence is out of the scope of this book. The reader can find more information in Refs [127, 128]. Hereafter, we will often denote an observable with the associated Hermitian operator since this postulate establishes a one-to-one correspondence between them.

What are the possible outcomes of a measurement? The fourth postulate answers this question.

Postulate 4.4 The only **possible outcomes of a measurement** of an observable A are the eigenvalues of the corresponding Hermitian operator $\hat{A}$.

It is now clear why Postulate 4.3 requires observables to be associated with Hermitian operators: the eigenvalues of these operators are real numbers and, according to Postulate 4.4, the only possible measurement outcomes (which must be real quantities) are the eigenvalues of these operators. If the eigenvalues of an observable are discrete, such as the energy of a particle in a potential well, the measurement outcomes will take on *discrete* values. This phenomenon captures the *quantum* nature of physical systems.

The set of eigenvectors $\{|a_j\rangle\}_j$ of an observable $\hat{A}$ forms an orthonormal basis of the Hilbert space and a generic quantum state can be decomposed in this basis as[2]

$$|\psi\rangle = \sum_j |a_j\rangle\underbrace{\langle a_j|\psi\rangle}_{c_j} = \sum_j c_j|a_j\rangle, \tag{4.6}$$

where c_j are complex numbers usually called the **probability amplitudes**. If a state $|\psi\rangle$ is a linear combination of orthonormal vectors as in eqn (4.6), it is called a **superposition state**. The decomposition of a quantum state in an orthonormal basis is convenient for calculating the probability of a measurement outcome.

Postulate 4.5 When a measurement of an observable A is performed on a pure state $|\psi\rangle$, the **probability** of obtaining the measurement outcome a_i is

$$p(a_i) = \langle\psi|\hat{P}_i|\psi\rangle, \tag{4.7}$$

where $\hat{P}_i$ is the projector on the eigenspace associated with the eigenvalue a_i. This is called the **Born rule**.

The Born rule quantifies the probability of measuring an eigenvalue a_i of an observable A. Suppose that a quantum system is prepared in a state $|\psi\rangle = \sum_i c_i|a_i\rangle$ where $|a_i\rangle$ are the eigenvectors of the observable. Let us assume for simplicity that the eigenvalues a_i are non-degenerate, i.e. the associated eigenspaces are one-dimensional. Then $\hat{P}_i = |a_i\rangle\langle a_i|$ and the probability of observing a_i is given by

$$p\,(a_i) = \langle\psi|\hat{P}_i|\psi\rangle = \langle\psi|a_i\rangle\langle a_i|\psi\rangle = \left|\langle a_i|\psi\rangle\right|^2 = |c_i|^2. \tag{4.8}$$

Thus, the probability of measuring an eigenvalue a_i is the absolute square of the coefficient c_i. As a consequence, if a quantum state is an eigenstate $|a_i\rangle$ of an observable, a measurement will produce a specific outcome with certainty. Lastly, the probability of obtaining any result must be equal to 1, i.e.

$$\sum_i p\,(a_i) = \sum_i \langle\psi|\hat{P}_i|\psi\rangle = \langle\psi|\psi\rangle = 1\,, \tag{4.9}$$

where we used the completeness relation $\sum_i \hat{P}_i = \mathbb{1}$. The last step is valid because the quantum state is normalized.

Postulate 4.6 Immediately after the measurement of an observable A with outcome a_i, the quantum system is described by the state[3]

[2] Recall the completeness relation $\sum_j |a_j\rangle\langle a_j| = \mathbb{1}$.

[3] Here, the denominator is required for the correct normalization of the state.

$$\boxed{|\psi_i\rangle = \frac{\hat{P}_i|\psi\rangle}{\sqrt{\langle\psi|\hat{P}_i|\psi\rangle}},} \tag{4.10}$$

where $\hat{P}_i$ is the projector on the eigenspace associated with the eigenvalue a_i.

This postulate, also known as the **collapse of the wavefunction**, is the most controversial one[4]. It is motivated by empirical evidence: if the measurement of an observable yields outcome a_i and this observable is immediately measured again, the result of the second measurement will be a_i with 100% probability. This peculiar evolution is well captured by the transformation $|\psi\rangle \mapsto \hat{P}_i|\psi\rangle/\|\hat{P}_i|\psi\rangle\|$. In the particular case where the eigenvalue a_i is non-degenerate, the state of the system immediately after the measurement is $|a_i\rangle$. If a quantum state is prepared in a superposition of an observable's eigenstates, it is not possible to infer a priori in which eigenstate the state will collapse.

Example 4.1 The vector space $\mathbb{C}^2$ with the inner product $\langle w|v\rangle = \sum_{i=0}^{1} w_i^* v_i$ is a two-dimensional Hilbert space. Vectors in this space can be written as $|\psi\rangle = c_0|\mathbf{e}_0\rangle + c_1|\mathbf{e}_1\rangle$ where $|\mathbf{e}_0\rangle$ and $|\mathbf{e}_1\rangle$ are two orthonormal vectors. According to Postulate 4.1, the vectors $|\psi\rangle$ and $-7|\psi\rangle$ describe the same quantum state. To simplify the notation, the basis $\{|\mathbf{e}_0\rangle, |\mathbf{e}_1\rangle\}$ is usually indicated as $\{|0\rangle, |1\rangle\}$.

Example 4.2 Consider the Lebesgue space $\mathrm{L}^2(\mathbb{R})$ of square-integrable functions ψ satisfying

$$\int_{-\infty}^{\infty} |\psi(x)|^2 dx < \infty.$$

The space $\mathrm{L}^2(\mathbb{R})$ together with the inner product $\langle\phi|\psi\rangle = \int_{-\infty}^{+\infty} \phi^*(x)\psi(x)dx$ forms a Hilbert space.

Example 4.3 Suppose a quantum system is in the state $|\psi(0)\rangle = c_0|0\rangle + c_1|1\rangle$ where $\{|0\rangle, |1\rangle\}$ is an orthonormal basis of $\mathbb{C}^2$. Consider the time evolution operator $\hat{U}(t) = e^{i\hat{\sigma}_z t}$ where t is a real number. This operator is unitary because $e^{i\hat{\sigma}_z t}e^{-i\hat{\sigma}_z t} = \mathbb{1}$. The time evolution of the state $|\psi(0)\rangle$ is given by

$$\begin{aligned}
|\psi(t)\rangle = \hat{U}(t)|\psi(0)\rangle &= c_0 e^{i\hat{\sigma}_z t}|0\rangle + c_1 e^{i\hat{\sigma}_z t}|1\rangle \\
&= c_0 \sum_{n=0}^{\infty} \frac{(i\hat{\sigma}_z t)^n}{n!}|0\rangle + c_1 \sum_{n=0}^{\infty} \frac{(i\hat{\sigma}_z t)^n}{n!}|1\rangle \\
&= c_0 \sum_{n=0}^{\infty} \frac{(it)^n}{n!}|0\rangle + c_1 \sum_{n=0}^{\infty} \frac{(-it)^n}{n!}|1\rangle = c_0 e^{it}|0\rangle + c_1 e^{-it}|1\rangle,
\end{aligned}$$

where we used $\hat{\sigma}_z|0\rangle = |0\rangle$ and $\hat{\sigma}_z|1\rangle = -|1\rangle$.

[4] The interested reader can find more information about the interpretation of this postulate in Chapter 3 of Ref. [131].

Example 4.4 Consider the quantum state $|\psi\rangle = c_0|0\rangle + c_1|1\rangle \in \mathbb{C}^2$ and suppose that a measurement of the Pauli operator $\hat{\sigma}_z$ is performed. This operator is Hermitian because $\hat{\sigma}_z^\dagger = \hat{\sigma}_z$. This observable has two eigenvalues and two eigenvectors. The eigenvector $|0\rangle$ has eigenvalue $+1$, while the eigenvector $|1\rangle$ has eigenvalue -1. The projectors associated with these eigenvalues are: $\hat{P}_0 = |0\rangle\langle 0|$ and $\hat{P}_1 = |1\rangle\langle 1|$. The probability of measuring the eigenvalue $+1$ is given by: $p(1) = \langle\psi|\hat{P}_0|\psi\rangle = \langle\psi|0\rangle\langle 0|\psi\rangle = |c_0|^2$. If the measurement outcome is 1, the state after the measurement is $|0\rangle$.

4.1.1 No-cloning theorem

An important consequence of the unitary evolution of quantum states is the **no-cloning theorem** formulated by Wootters and Zurek in 1982 [132]. This theorem states that there does not exist any quantum machine that can copy an *arbitrary* quantum state onto another. This implies that it is not possible to define a quantum version of the COPY gate introduced in Section 1.1.8. This has some important implications for quantum computation and quantum error correction.

Theorem 4.1 (No-cloning) Consider two quantum systems prepared in the states $|a\rangle \in H_{\rm A}$ and $|b\rangle \in H_{\rm B}$, where $\dim(H_{\rm A}) = \dim(H_{\rm B})$. There does not exist any unitary operator $\hat{U}$ acting on the composite system $H_{\rm A} \otimes H_{\rm B}$ such that *for all* $|a\rangle \in H_{\rm A}$,

$$\hat{U}|a\rangle|b\rangle = |a\rangle|a\rangle\,. \tag{4.11}$$

Proof Suppose by contradiction that such a machine does exist. Then, we would be able to copy arbitrary states $|a_1\rangle$ and $|a_2\rangle$,

$$\hat{U}|a_1\rangle|b\rangle = |a_1\rangle|a_1\rangle, \tag{4.12}$$

$$\hat{U}|a_2\rangle|b\rangle = |a_2\rangle|a_2\rangle\,. \tag{4.13}$$

Since the scalar product between two vectors is unique, the scalar product between the left members must be equal to the scalar product between the right members,

$$\langle a_1 b|\hat{U}^\dagger\hat{U}|a_2 b\rangle = \langle a_1|a_2\rangle\langle a_1|a_2\rangle\,. \tag{4.14}$$

Since $\hat{U}^\dagger\hat{U} = \mathbb{1}$, this expression can be written in the form

$$z = \langle a_1|a_2\rangle = \langle a_1|a_2\rangle\langle a_1|a_2\rangle = z^2, \tag{4.15}$$

where we have introduced the complex number $z = \langle a_1|a_2\rangle$. The relation $z = z^2$ can be satisfied in only two cases: when $z = 0$ (which implies that $|a_1\rangle$ and $|a_2\rangle$ are orthogonal) and when $z = 1$ (which implies that $|a_1\rangle$ and $|a_2\rangle$ are identical). In conclusion, this quantum machine can only copy either orthogonal or identical states. Thus, it cannot make a perfect copy of an arbitrary superposition state. It can be shown that this theorem also holds when $\hat{U}$ is replaced by a quantum map, an interesting type of transformation that will be presented in Chapter 7. □

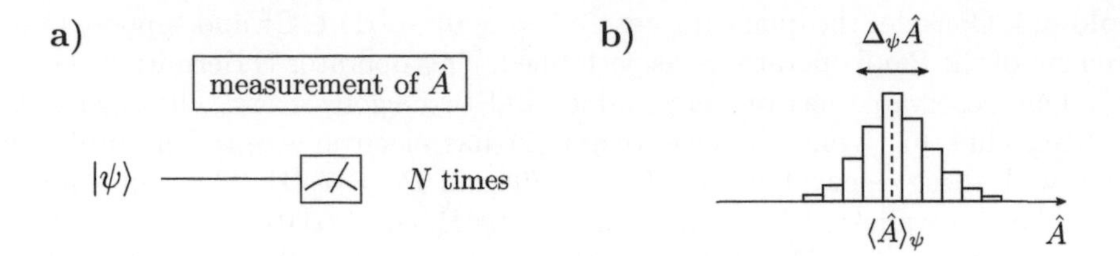

Fig. 4.2 Expectation value of an observable. a) A quantum system is prepared in a superposition state $|\psi\rangle = \sum_i c_i|a_i\rangle$ and an observable $\hat{A}$ is measured N times. The outcomes are ordered in a sequence $(x_1, \ldots x_N)$. From this sample, we can extract the sample average $\bar{A}_N$ and the sample standard deviation s_N. **b)** The measurement outcomes $(x_i)_i$ are grouped in a histogram. When $N \gg 1$, the sample average A_N converges to the real number $\langle \hat{A} \rangle_\psi$ indicated with a dashed line in the figure. Similarly, the sample standard deviation s_N converges to $\Delta_\psi \hat{A}$.

4.2 Expectation value of an observable

The measurement of a quantum system can be studied in two different scenarios. On the one hand, when a quantum system is prepared in an eigenstate of an observable $\hat{A}$, such as $|a\rangle$, a measurement will collapse the system in the state $|a\rangle$ and the measurement result will be the associated eigenvalue a. On the other hand, when a quantum system is prepared in a *superposition* of multiple eigenstates, $|\psi\rangle = \sum_i c_i|a_i\rangle$, and the observable $\hat{A}$ is measured, it is not possible to predict a priori the measurement outcome. Although, if the system can be prepared in the state $|\psi\rangle$ multiple times, one can analyze the distribution of the measurement results. The goal of this section is to show that the average of this distribution is the real number $\mu = \langle\psi|\hat{A}|\psi\rangle$ and its variance is given by $\sigma^2 = \langle\psi|\hat{A}^2|\psi\rangle - \langle\psi|\hat{A}|\psi\rangle^2$.

Suppose a quantum system is prepared in a quantum state $|\psi\rangle = \sum_i c_i|a_i\rangle$ and a measurement of an observable $\hat{A}$ is performed (see Fig. 4.2a). This measurement will yield an eigenvalue a_i with probability $p(a_i \mid \psi) = \langle\psi|\hat{P}_i|\psi\rangle$. If we repeat this measurement N times, we will obtain a sequence of real numbers $(x_1, \ldots, x_N)$. From this set of numbers, we can extract the **sample average**

$$\boxed{\mu_N = \frac{1}{N}\sum_{j=1}^{N} x_j.} \tag{4.16}$$

The sample average converges almost surely to the number $\langle\psi|\hat{A}|\psi\rangle$ for $N \to +\infty$. Let us show this by using the strong law of large numbers,

$$\lim_{N\to\infty} \mu_N = \lim_{N\to\infty} \frac{1}{N}\sum_{j=1}^{N} x_j = \sum_{i=1}^{d} a_i\, p(a_i \mid \psi) =$$

$$= \sum_{i=1}^{d} a_i \langle\psi|\hat{P}_i|\psi\rangle = \langle\psi|\hat{A}|\psi\rangle \equiv \langle\hat{A}\rangle_\psi,$$

where d is the dimension of the Hilbert space, we used the decomposition $\hat{A} = \sum_i a_i \hat{P}_i$, and we introduced the short-hand notation $\langle\hat{A}\rangle_\psi = \langle\psi|\hat{A}|\psi\rangle$. The quantity $\langle\hat{A}\rangle_\psi$ is called the **expectation value** of the observable $\hat{A}$ in state ψ. This quantity is an unbiased estimator for the average of the distribution. It is worth mentioning that the expectation value $\langle\psi|\hat{A}|\psi\rangle$ cannot be smaller (greater) than the lowest (highest) eigenvalue of $\hat{A}$.

From the finite sequence of real numbers $(x_1, \ldots, x_N)$, one can also extract the **sample standard deviation** σ_N, a real number that quantifies the dispersion of the measured values,

$$\boxed{\sigma_N = \sqrt{\frac{1}{N-1}\sum_{j=1}^{N}(x_j - \bar{A}_N)^2}.} \tag{4.17}$$

The sample standard deviation converges almost surely to $\sqrt{\langle\psi|\hat{A}^2|\psi\rangle - \langle\psi|\hat{A}|\psi\rangle^2}$ when the number of experiments $N \to +\infty$. This can be demonstrated in a few steps,

$$\begin{aligned}
\lim_{N\to\infty} \sigma_N &= \lim_{N\to\infty} \sqrt{\frac{1}{N-1}\sum_{j=1}^{N} x_j^2 - 2x_j\bar{A} + \bar{A}^2} \\
&= \sqrt{\sum_{i=1}^{d} a_i^2\, p(a_i \mid \psi) - 2a_i\, p(a_i \mid \psi) \cdot \langle\hat{A}\rangle_\psi + \langle\hat{A}\rangle_\psi^2} \\
&= \sqrt{\langle\hat{A}^2\rangle_\psi - 2\langle\hat{A}\rangle_\psi\langle\hat{A}\rangle_\psi + \langle\hat{A}\rangle_\psi^2} = \sqrt{\langle\hat{A}^2\rangle_\psi - \langle\hat{A}\rangle_\psi^2} \equiv \Delta_\psi\hat{A},
\end{aligned}$$

where in the second step we leveraged the strong law of large numbers and in the final step we defined the **standard deviation** $\Delta_\psi\hat{A} = \sqrt{\langle\hat{A}^2\rangle_\psi - \langle\hat{A}\rangle_\psi^2}$. This parameter best represents the dispersion of the measured values and is sometimes called the **uncertainty** of the observable $\hat{A}$ in state $|\psi\rangle$. The concepts of expectation value and standard deviation are illustrated in Fig. 4.2b. These quantities will often be denoted with the short-hand notation $\langle\hat{A}\rangle$ and $\Delta\hat{A}$, respectively.

Example 4.5 Consider a quantum state $|\psi\rangle = c_0|0\rangle + c_1|1\rangle$ where $\{|0\rangle, |1\rangle\}$ is an orthonormal basis of $\mathbb{C}^2$. The expectation value of the observable $\hat{\sigma}_z$ is

$$\begin{aligned}
\langle\psi|\hat{\sigma}_z|\psi\rangle &= (c_0^*\langle 0| + c_1^*\langle 1|)\,(c_0\hat{\sigma}_z|0\rangle + c_1\hat{\sigma}_z|1\rangle) \\
&= (c_0^*\langle 0| + c_1^*\langle 1|)\,(c_0|0\rangle - c_1|1\rangle) \\
&= c_0^*c_0 - c_1^*c_1 = |c_0|^2 - |c_1|^2.
\end{aligned}$$

The expectation value of $\hat{\sigma}_z^2$ is $\langle\psi|\hat{\sigma}_z^2|\psi\rangle = \langle\psi|\mathbb{1}|\psi\rangle = 1$. Therefore, the standard

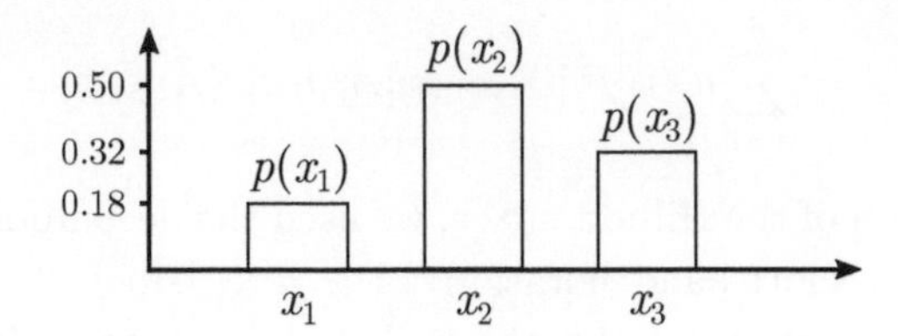

Fig. 4.3 Probability distribution. Probability distribution of the discretized position operator $\hat{x} = \sum_{i=1}^{3} x_i |x_i\rangle\langle x_i|$ for the quantum state (4.18).

deviation of the Pauli operator $\hat{\sigma}_z$ in the state $|\psi\rangle$ is given by $\Delta\hat{\sigma}_z = \sqrt{\langle\hat{\sigma}_z^2\rangle - \langle\hat{\sigma}_z\rangle^2} = \sqrt{1-(|c_0|^2-|c_1|^2)^2}$. When the quantum state is $|\psi\rangle = |0\rangle$, we have $c_0 = 1$ and $c_1 = 0$. Thus, the expectation value is 1 with a standard deviation of 0. When the quantum state is the superposition state $|\psi\rangle = \frac{1}{\sqrt{2}}(|0\rangle + |1\rangle)$, we have $c_0 = c_1 = 1/\sqrt{2}$. In this case, the expectation value is 0 with a standard deviation of 1. This example shows that the expectation value of an observable does not need to be one of the eigenvalues.

Example 4.6 Consider a particle in the superposition state

$$|\psi\rangle = \frac{1}{\sqrt{50}}\left(3|x_1\rangle + 5|x_2\rangle + 4|x_3\rangle\right), \tag{4.18}$$

where $x_1 = 5\ \mu\text{m}$, $x_2 = 6\ \mu\text{m}$, $x_3 = 7\ \mu\text{m}$ are eigenvalues of the position operator. The probability of finding the particle in states $|x_1\rangle$, $|x_2\rangle$, $|x_3\rangle$ is given by the probability distribution $\mathbf{p} = (p(x_1),\ p(x_2),\ p(x_3))$ illustrated in Fig. 4.3. From this plot, it is evident that the expectation value of the position is close to x_2. More quantitatively,

$$\langle\psi|\hat{x}|\psi\rangle = \sum_{i=1}^{3} p(x_i)x_i = \frac{1}{50}\left(9x_1 + 25x_2 + 16x_3\right) = 6.14\ \mu\text{m}. \tag{4.19}$$

To calculate the uncertainty related to the position of the particle, we need to compute the expectation value of the operator $\hat{x}^2$:

$$\langle\psi|\hat{x}^2|\psi\rangle = \sum_{i=1}^{3} p(x_i)x_i^2 = \frac{1}{50}\left(9x_1^2 + 25x_2^2 + 16x_3^2\right) = 38.18\ \mu\text{m}^2. \tag{4.20}$$

Therefore, $\Delta_\psi\hat{x} = \sqrt{\langle\hat{x}^2\rangle - \langle\hat{x}\rangle^2} = 0.69\ \mu\text{m}$.

4.2.1 Heisenberg uncertainty principle

The Heisenberg uncertainty principle is one of the most striking aspects of quantum mechanics. Unfortunately, this fascinating principle is sometimes misinterpreted. To avoid any possible confusion and give the reader a clear explanation of this principle, we start our discussion with a simple example. Suppose that a particle is prepared multiple times in the quantum state $|\psi_1\rangle$ and suppose that the position of the particle is measured each time. The outcomes of these experiments can be collected in a sequence $(x_1, \ldots, x_N)$ from which we can extract the sample average $\bar{x}_{\psi_1} = 5\,\text{m}$ and the sample standard deviation $\Delta_{\psi_1} x = 4.1 \times 10^{-7}$ m. Now suppose that a particle is prepared multiple times in the same state $|\psi_1\rangle$ and each time the momentum of the particle is measured. This set of experiments will produce a sequence of outcomes $(p_1, \ldots, p_N)$ from which we can extract the sample average $\bar{p}_{\psi_1} = 6 \times 10^{-27}\,\text{kg m/s}$ and the sample standard deviation $\Delta_{\psi_1} p = 2.6 \times 10^{-27}$ kg m/s. We can now compute the product between $\Delta_{\psi_1} x$ and $\Delta_{\psi_1} p$ and obtain:

$$\Delta_{\psi_1} x \cdot \Delta_{\psi_1} p = 1.05 \cdot 10^{-33}\ [\text{Js}].$$

This quantity is very small, but it is legitimate to wonder if it is possible to prepare the system in a different quantum state $|\psi_2\rangle$ such that the *product* between the standard deviations $\Delta_{\psi_2} x \cdot \Delta_{\psi_2} p$ is lower. The answer is yes, but not indefinitely. If we prepare and measure the position of a particle in a quantum state $|\psi\rangle$ N times, repeat the same procedure for the momentum, and compute the sample standard deviations from the measured values, they will always satisfy the relation

$$\boxed{\Delta_\psi x \cdot \Delta_\psi p \geq \hbar/2,} \tag{4.21}$$

where $\hbar = 1.054 \times 10^{-34}$Js. This is the **Heisenberg uncertainty principle**. This principle states that it is not possible to prepare a system in a quantum state such that the product between the standard deviations of the position and the momentum estimated from many experiments is less than $\hbar/2$. While we can definitely prepare a particle in a quantum state $|\psi'\rangle$ such that $\Delta_{\psi'} x$ is arbitrarily small, this will inevitably lead to a much larger $\Delta_{\psi'} p$ such that the inequality (4.21) is still valid. This peculiar phenomenon follows from the fact that in quantum mechanics the commutator between the position and momentum is proportional to the identity [128]

$$[\hat{x}, \hat{p}] = i\hbar\mathbb{1}.$$

Let us show that the product between the standard deviations of two observables is intrinsically related to their commutation relation.

Theorem 4.2 Consider two Hermitian operators $\hat{A}$ and $\hat{B}$. Then,

$$\boxed{\Delta_\psi \hat{A} \cdot \Delta_\psi \hat{B} \geq \frac{|\langle\psi|\,[\hat{A}, \hat{B}]\,|\psi\rangle|}{2}.} \tag{4.22}$$

This is called the **Heisenberg uncertainty relation**.

Proof Consider the operators $\hat{P} = \hat{A} - \langle\psi|\hat{A}|\psi\rangle\mathbb{1}$ and $\hat{Q} = \hat{B} - \langle\psi|\hat{B}|\psi\rangle\mathbb{1}$. The

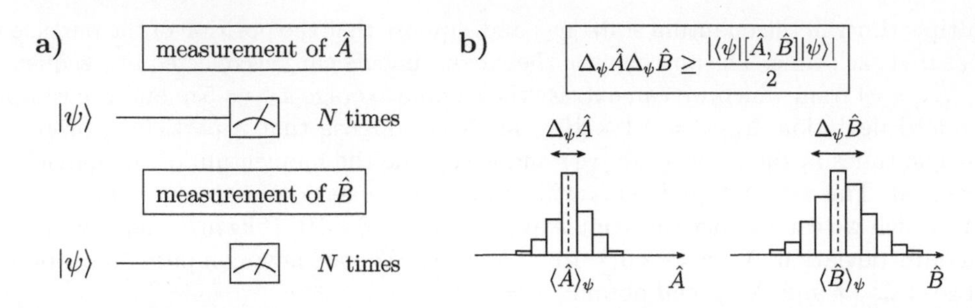

Fig. 4.4 Heisenberg uncertainty relation. a) A quantum system is prepared in state $|\psi\rangle$ and the observables $\hat{A}$ and $\hat{B}$ are measured N times. The outcomes are collected in two sequences $(a_{(1)}, \ldots, a_{(N)})$ and $(b_{(1)}, \ldots, b_{(N)})$. One can extract the sample averages and sample standard deviations from these sequences. In the limit $N \to +\infty$, these quantities converge to the expectation values $\langle\hat{A}\rangle_\psi$ and $\langle\hat{B}\rangle_\psi$, and to the uncertainties $\Delta_\psi \hat{A}$ and $\Delta_\psi \hat{B}$. **b)** The measurement outcomes are grouped in a histogram. The expectation values are indicated with dashed lines. The product between the standard deviations obeys the Heisenberg uncertainty relation.

---

expectation values of the products $\hat{P}\hat{Q}$ and $\hat{Q}\hat{P}$ are the complex numbers[5] $\langle\psi|\hat{P}\hat{Q}|\psi\rangle = a + ib$, and $\langle\psi|\hat{Q}\hat{P}|\psi\rangle = a - ib$. From these relations, it follows that

$$\langle\psi|\,[\hat{P},\hat{Q}]\,|\psi\rangle = \langle\psi|\hat{P}\hat{Q}|\psi\rangle - \langle\psi|\hat{Q}\hat{P}|\psi\rangle = 2ib, \tag{4.23}$$

$$\langle\psi|\,\{\hat{P},\hat{Q}\}\,|\psi\rangle = \langle\psi|\hat{P}\hat{Q}|\psi\rangle + \langle\psi|\hat{Q}\hat{P}|\psi\rangle = 2a\,. \tag{4.24}$$

From (4.23), we have $|\langle\psi|[\hat{P},\hat{Q}]|\psi\rangle|^2 = 4b^2$. By summing (4.23) and (4.24) together, we obtain $|\langle\psi|\hat{P}\hat{Q}|\psi\rangle|^2 = a^2 + b^2$. By substituting these two identities into the inequality $4b^2 \leq 4\left(a^2 + b^2\right)$ valid for all real numbers, we arrive at

$$|\langle\psi|\,[\hat{P},\hat{Q}]\,|\psi\rangle|^2 \leq 4|\langle\psi|\hat{P}\hat{Q}|\psi\rangle|^2 \leq 4\langle\psi|\hat{P}^2|\psi\rangle\langle\psi|\hat{Q}^2|\psi\rangle\,, \tag{4.25}$$

where in the last step we used the Cauchy–Schwarz inequality[6]. One can verify that $[\hat{P},\hat{Q}] = [\hat{A},\hat{B}]$. In addition[7], $\langle\psi|\hat{P}^2|\psi\rangle = (\Delta\hat{A})^2$ and similarly $\langle\psi|\hat{Q}^2|\psi\rangle = (\Delta\hat{B})^2$. By substituting these three identities into (4.25), we finally have

$$\frac{|\langle\psi|[\hat{A},\hat{B}]|\psi\rangle|^2}{4} \leq (\Delta\hat{A})^2\,(\Delta\hat{B})^2\,. \tag{4.26}$$

The square root of this inequality gives the Heisenberg uncertainty relation. Note that if the two operators commute $[\hat{A},\hat{B}] = \underline{\mathbf{0}}$, from Theorem 3.2 there exists a common

[5] Even if $\hat{P}$ and $\hat{Q}$ are Hermitian, their product might not be Hermitian. That is why the expectation value of the product $\hat{P}\hat{Q}$ is a complex number.

[6] Define the vectors $|v\rangle = \hat{Q}|\psi\rangle$ and $|w\rangle = \hat{P}|\psi\rangle$. The Cauchy–Schwarz inequality (3.16) states that $|\langle\psi|\hat{P}\hat{Q}|\psi\rangle|^2 = |\langle w|v\rangle|^2 \leq \langle v|v\rangle\langle w|w\rangle = \langle\psi|\hat{P}\hat{P}|\psi\rangle\langle\psi|\hat{Q}\hat{Q}|\psi\rangle$.

[7] Indeed, $\langle\hat{P}^2\rangle = \langle(\hat{A} - \langle\hat{A}\rangle\mathbb{1})^2\rangle = \langle\hat{A}^2\rangle - \langle\hat{A}\rangle^2 = (\Delta\hat{A})^2$.

basis of eigenvectors $\{|v_i\rangle\}_i$ such that $\Delta_{v_i}\hat{A} = \Delta_{v_i}\hat{B} = 0$. □

To summarize, the Heisenberg uncertainty relation provides a lower limit for the product between the standard deviations of two observables extracted from a sequence of measurements executed on the same quantum state. Figure 4.4 shows a concise representation of this concept.

Example 4.7 Consider a qubit in state $|\psi\rangle = |0\rangle$. Measurements of the operator $\hat{\sigma}_z$ will give $+1$ with 100% probability. The outcomes of 10 measurements will clearly be $(1, 1, 1, 1, 1, 1, 1, 1, 1, 1)$ with sample average $\bar{\sigma}_z = +1$ and sample standard deviation $\Delta\sigma_z = 0$. If we now prepare the system in the same state and measure the operator $\hat{\sigma}_x$ multiple times, we will obtain $+1$ and -1 with equal probability. The outcome of eight measurements might be $(1, 1, -1, 1, 1, -1, -1, 1)$ with sample average $\bar{\sigma}_x = 0.25$ and sample standard deviation $\Delta\sigma_x = 1.035$. Thus, we have $\Delta\sigma_z \cdot \Delta\sigma_x = 0$, which is consistent with the Heisenberg uncertainty relation. Indeed, one can verify that

$$\Delta_\psi\hat{\sigma}_z\Delta_\psi\hat{\sigma}_x \geq \frac{|\langle 0|[\hat{\sigma}_x, \hat{\sigma}_z]|0\rangle|}{2} = \frac{|\langle 0|2i\hat{\sigma}_y|0\rangle|}{2} = 0.$$

In an actual experiment, the quantum state must be measured multiple times in order to get an estimate of the standard deviation. As an exercise, the reader can calculate the lower limit for the product $\Delta_\psi\hat{\sigma}_z\Delta_\psi\hat{\sigma}_y$ using eqn (4.22) when $|\psi\rangle = |0\rangle$.

4.3 Evolution of quantum systems and projective measurements

4.3.1 The Schrödinger equation

In quantum mechanics, the time evolution of an isolated quantum system is described by a unitary operator. This section explores the relationship between the time evolution operator $\hat{U}(t, t_0)$ and the Hamiltonian of the system. This discussion will lead to the Schrödinger equation. To derive this equation, we draw an analogy with the classical case. A more formal derivation can be found in Ref. [128].

In classical mechanics, the evolution of an observable A, such as the position or the momentum, is governed by **Poisson's equation**. For simplicity, we assume that the system under consideration is not subject to time-dependent forces. The time derivative of a generic observable A that depends on the position x, the momentum p, and time t is given by Poisson's equation

$$\frac{dA}{dt} = \{A, H\} = \frac{\partial A}{\partial x}\frac{\partial H}{\partial p} - \frac{\partial A}{\partial p}\frac{\partial H}{\partial x}, \tag{4.27}$$

where H is the classical Hamiltonian of the system and $\{\cdot, \cdot\}$ is the Poisson bracket. It is natural to expect a similar relation to hold in quantum mechanics.

According to Postulate 4.2, the time evolution of a quantum system is described by a unitary operator. Let us assume for simplicity that the system is invariant under time translation, i.e. the Hamiltonian is time-independent. Therefore, the time evolution operator only depends on the difference $t - t_0$ and can be expressed as

$$\hat{U}(t,t_0) = e^{i(t-t_0)\hat{B}}, \tag{4.28}$$

where $\hat{B}$ is a Hermitian operator (see Section 3.2.11). For an infinitesimal time evolution, we have

$$\hat{U}(t_0+\epsilon,t_0) = e^{i\epsilon\hat{B}} = \mathbb{1} + i\epsilon\hat{B} + O\left(\epsilon^2\right), \tag{4.29}$$

where we performed a Taylor expansion and grouped small terms in $O(\varepsilon^2)$. Using the short-hand notation $\langle\psi(t)|\hat{A}|\psi(t)\rangle = \langle\hat{A}\rangle_t$, the expectation value of an observable after an infinitesimal time interval is given by

$$\begin{aligned}\langle\hat{A}\rangle_{t_0+\epsilon} &= \langle\hat{U}^\dagger(t_0+\epsilon,t_0)\,\hat{A}\,\hat{U}(t_0+\epsilon,t_0)\rangle_{t_0} \\ &= \langle(\mathbb{1} - i\epsilon\hat{B} + O(\epsilon^2))\hat{A}(\mathbb{1} + i\epsilon\hat{B} + O(\epsilon^2))\rangle_{t_0} \\ &= \langle\hat{A}\rangle_{t_0} - i\epsilon\langle\hat{B}\hat{A} - \hat{A}\hat{B}\rangle_{t_0} + O(\epsilon^2).\end{aligned}$$

Rearranging this equation and taking the limit $\epsilon \to 0^+$, we find

$$\frac{d\langle\hat{A}\rangle_{t_0}}{dt} = \lim_{\epsilon\to 0^+}\frac{\langle\hat{A}\rangle_{t_0+\epsilon} - \langle\hat{A}\rangle_{t_0}}{\epsilon} = -i\langle\hat{B}\hat{A} - \hat{A}\hat{B}\rangle_{t_0} = i\langle[\hat{A},\hat{B}]\rangle_{t_0}. \tag{4.30}$$

The comparison between (4.30) and (4.27) suggests that the operator $\hat{B}$ must be proportional to the Hamiltonian. More precisely, $\hat{B} = -\hat{H}/\hbar$, where the minus sign is conventional and the parameter $\hbar$ (pronounced "hbar") must have the units of Js. This parameter is related to the **Planck constant** $h = 6.626 \cdot 10^{-34}$ Js by the relation $\hbar = h/2\pi$. In conclusion, from (4.28) the time evolution of a quantum state is

$$|\psi(t)\rangle = \hat{U}(t,t_0)\,|\psi(t_0)\rangle = e^{\frac{1}{i\hbar}(t-t_0)\hat{H}}|\psi(t_0)\rangle. \tag{4.31}$$

By taking the time derivative, we arrive at the well-known **Schrödinger equation**,

$$\boxed{i\hbar\frac{d}{dt}|\psi(t)\rangle = \hat{H}|\psi(t)\rangle.} \tag{4.32}$$

If a system is subject to forces that depend on time, the derivation presented in this section changes slightly and the Hamiltonian appearing in the Schrödinger equation becomes time-dependent. Solving the Schrödinger equation is one of the most important tasks in quantum mechanics as its solution provides the state of the system at future times.

The Hamiltonian is a Hermitian operator, and its eigenvalues and eigenvectors are defined by the equation $\hat{H}|E\rangle = E|E\rangle$. This relation is usually called the **time-independent Schrödinger equation**. This eigenvalue problem provides the energy levels of the system and it will play an important role in Sections 4.5 and 4.7 when we derive the energy levels of the harmonic oscillator and the Jaynes–Cummings Hamiltonian.

4.3.2 The Schrödinger, Heisenberg, and interaction pictures

In many physical situations, the Hamiltonian of a system is the sum of two terms,

$$\boxed{\hat{H} = \hat{H}_0 + \hat{V},} \tag{4.33}$$

where $\hat{H}_0$ is a simple Hamiltonian with eigenstates and eigenvalues that can be derived analytically, while the Hamiltonian $\hat{V}$ is usually a small contribution that makes the global Hamiltonian analytically intractable.

To simplify calculations, it is often convenient to perform a change of basis and introduce a new **frame** (often called a "picture"). Suppose that a time-dependent change of basis $\hat{R}$ is applied to the state vectors. The states in the new basis are given by $|\psi'(t)\rangle = \hat{R}|\psi(t)\rangle$. Since $\hat{R}$ is invertible, we have

$$|\psi(t)\rangle = \hat{R}^{-1}|\psi'(t)\rangle. \tag{4.34}$$

Substituting this expression into the Schrödinger equation (4.32) leads to

$$i\hbar\Big(\frac{d}{dt}\hat{R}^{-1}\Big)|\psi'(t)\rangle + i\hbar\hat{R}^{-1}\Big(\frac{d}{dt}|\psi'(t)\rangle\Big) = (\hat{H}_0 + \hat{V})\hat{R}^{-1}|\psi'(t)\rangle\,. \tag{4.35}$$

By applying $\hat{R}$ on the left and rearranging some terms, we obtain the Schrödinger equation in the new frame

$$i\hbar\frac{d}{dt}|\psi'(t)\rangle = \Big[\hat{R}\left(\hat{H}_0 + \hat{V}\right)\hat{R}^{-1} - i\hbar\hat{R}\Big(\frac{d}{dt}\hat{R}^{-1}\Big)\Big]|\psi'(t)\rangle\,. \tag{4.36}$$

By comparing this expression with the Schödinger equation in the old basis (4.32), we conclude that the Hamiltonian in the new frame is

$$\boxed{\hat{H}' = \hat{R}(\hat{H}_0 + \hat{V})\hat{R}^{-1} - i\hbar\hat{R}\Big(\frac{d}{dt}\hat{R}^{-1}\Big).} \tag{4.37}$$

If the transformation is time-independent, then $\frac{d}{dt}\hat{R}^{-1} = \underline{0}$ and $\hat{H}'$ reduces to $\hat{H}' = \hat{R}(\hat{H}_0 + \hat{V})\hat{R}^{-1}$. We now present three important pictures.

The Schrödinger picture. For $\hat{R} = \mathbb{1}$, there is no actual change of basis. This frame is called the Schrödinger picture. As indicated by eqn (4.31), quantum states evolve as $|\psi(t)\rangle_{\mathrm{S}} = e^{\frac{1}{i\hbar}\hat{H}t}|\psi(0)\rangle_{\mathrm{S}}$. The subscript S stands for Schrödinger. In the Schrödinger picture, states evolve in time, but operators are time-independent.

The Heisenberg picture. For $\hat{R}(t) = e^{-\frac{1}{i\hbar}\hat{H}t}$, the new frame is called the Heisenberg picture. In this frame, quantum states do not evolve in time: $|\psi(t)\rangle_{\mathrm{H}} = |\psi(0)\rangle$. On the other hand, operators are time-dependent: $\hat{A}_{\mathrm{H}}(t) = \hat{R}^\dagger(t)\hat{A}\hat{R}(t)$. The expectation value of an observable in the Schrödinger and the Heisenberg picture must be the same at any time

$$\langle\psi(0)|\hat{A}_{\rm H}(t)|\psi(0)\rangle = \langle\psi(0)|\hat{R}^\dagger(t)\,\hat{A}\,\hat{R}(t)|\psi(0)\rangle = {}_{\rm S}\langle\psi(t)|\hat{A}|\psi(t)\rangle_{\rm S}.$$

Here, the subscript H stands for Heisenberg. At $t = 0$, the kets and operators in the Heisenberg picture coincide with those in the Schrödinger picture.

The interaction picture. For $\hat{R}(t) = e^{-\frac{1}{i\hbar}\hat{H}_0 t}$, the new frame is called the interaction picture. The quantum states in this basis are given by

$$|\psi'(t)\rangle = e^{-\frac{1}{i\hbar}\hat{H}_0 t}|\psi(t)\rangle_{\rm S}\,. \tag{4.38}$$

From eqn (4.37), the Hamiltonian in the interaction picture is

$$\begin{aligned}\hat{H}'(t) &= e^{-\frac{1}{i\hbar}\hat{H}_0 t}(\hat{H}_0 + \hat{V})e^{\frac{1}{i\hbar}\hat{H}_0 t} - i\hbar e^{-\frac{1}{i\hbar}\hat{H}_0 t}\left(\frac{d}{dt}e^{\frac{1}{i\hbar}\hat{H}_0 t}\right)\\ &= e^{-\frac{1}{i\hbar}\hat{H}_0 t}\hat{V}e^{\frac{1}{i\hbar}\hat{H}_0 t} \;=\; \hat{R}(t)\hat{V}\hat{R}^{-1}(t).\end{aligned} \tag{4.39}$$

The interaction picture is very useful for studying physical phenomena in which two subsystems interact with one another, particularly when the contribution of $\hat{H}_0$ to the time evolution is trivial and only the contribution from $\hat{V}$ is relevant for the dynamics. At $t = 0$, the kets and operators in the interaction picture coincide with those in the Schrödinger picture.

4.3.3 Projective measurements (PVM)

The last part of a quantum experiment is the measurement of an observable. This procedure generates an outcome a_i according to a probability distribution. The collection of all the possible results is usually denoted with the symbol $\Omega = \{a_i\}_i$. Immediately after the measurement, the state of the system is described by the vector $\hat{P}_i|\psi\rangle/\|\hat{P}_i|\psi\rangle\|$ where $\hat{P}_i$ is the projector associated with a_i. The correspondence between measurement outcomes and projectors can be formalized by introducing the projection-valued measure (PVM).

A **projection-valued measure** (or projective measurement) is a function that maps measurement outcomes into projectors acting on a Hilbert space H,

$$\begin{aligned} P:\quad \Omega &\;\to\; \text{Proj}\,(H)\\ a_i &\;\mapsto\; \hat{P}_i.\end{aligned} \tag{4.40}$$

The function P must satisfy two requirements:

PVM1. $\hat{P}_i\hat{P}_j = \delta_{ij}\hat{P}_i$ (idempotency),
PVM2. $\sum_i \hat{P}_i = \mathbb{1}$ (completeness).

The first property indicates that the projectors $\hat{P}_i$ are mutually orthogonal (this is consistent with the fact that the eigenspaces of an observable are orthogonal). The second requirement is the usual completeness relation. In this textbook, we will often use the term PVM to denote a set of orthogonal projectors $\{\hat{P}_i\}_i$ satisfying the requirements listed above.

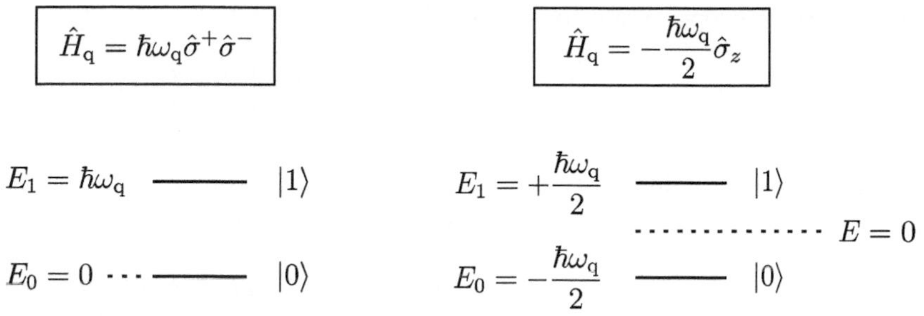

$$\hat{H}_\mathrm{q} = \hbar\omega_\mathrm{q}\hat{\sigma}^+\hat{\sigma}^- \qquad\qquad \hat{H}_\mathrm{q} = -\frac{\hbar\omega_\mathrm{q}}{2}\hat{\sigma}_z$$

$$E_1 = \hbar\omega_\mathrm{q} \quad |1\rangle \qquad\qquad E_1 = +\frac{\hbar\omega_\mathrm{q}}{2} \quad |1\rangle$$

$$E = 0$$

$$E_0 = 0 \quad |0\rangle \qquad\qquad E_0 = -\frac{\hbar\omega_\mathrm{q}}{2} \quad |0\rangle$$

Fig. 4.5 The energy levels of a qubit. a) The zero-point of the potential energy is set to E_0. **b**) The zero-point of the potential energy is set to $(E_1 + E_0)/2$. The Hamiltonians $\hat{H}_\mathrm{q} = \hbar\omega_\mathrm{q}\hat{\sigma}^+\hat{\sigma}^-$ and $\hat{H}_\mathrm{q} = -\hbar\omega_\mathrm{q}\hat{\sigma}_z/2$ are equivalent.

4.4 The Hamiltonian of a qubit

In the final part of this chapter, we present some concrete examples of quantum systems and their interactions. We will first introduce the Hamiltonian of a qubit. We will then present the quantum harmonic oscillator. This model accurately describes a plethora of physical systems, including electromagnetic cavities and mechanical resonators in the quantum regime. Lastly, we will study the interaction between a qubit and a harmonic oscillator. We will introduce the Jaynes–Cummings Hamiltonian and we will derive its energy levels. This treatment will be useful in Section 14.7 when we introduce circuit quantum electrodynamics.

A qubit is a well-isolated quantum system whose Hilbert space contains a subspace isomorphic to $\mathbb{C}^2$. The basis vectors of this space are denoted as $|0\rangle$ and $|1\rangle$. Suppose that the energies associated with these two states are E_0 and E_1 where $E_0 < E_1$. In the energy eigenbasis, the **Hamiltonian of a qubit** is given by

$$\hat{H}_\mathrm{q} = E_0|0\rangle\langle 0| + E_1|1\rangle\langle 1| = \begin{bmatrix} E_0 & 0 \\ 0 & E_1 \end{bmatrix}. \tag{4.41}$$

Here, E_0 is the energy of the **ground state** $|0\rangle$ and E_1 is the energy of the **excited state** $|1\rangle$. For convenience, we can subtract[8] a constant term $E_0\mathbb{1} = E_0(|0\rangle\langle 0| + |1\rangle\langle 1|)$ from the Hamiltonian and obtain

$$\hat{H}_\mathrm{q} = (E_1 - E_0)|1\rangle\langle 1| = \hbar\omega_\mathrm{q}|1\rangle\langle 1| = \hbar\omega_\mathrm{q}\hat{\sigma}^+\hat{\sigma}^- = \begin{bmatrix} 0 & 0 \\ 0 & \hbar\omega_\mathrm{q} \end{bmatrix}. \tag{4.42}$$

where $\omega_\mathrm{q} = (E_1 - E_0)/\hbar$ is the **qubit transition frequency** and we used[9] $\hat{\sigma}^+\hat{\sigma}^- = |1\rangle\langle 1|$.

To simplify calculations, it is sometimes convenient to write the Hamiltonian of a qubit in terms of the Pauli operator $\hat{\sigma}_z$. This can be done by adding to the Hamilto-

[8]We can add or subtract terms proportional to the identity without affecting the dynamics of the system. In physics, only the difference between energy levels is relevant, not their absolute value.

[9]Recall that the lowering and raising operators are defined as $\hat{\sigma}^+ = |1\rangle\langle 0|$ and $\hat{\sigma}^- = |0\rangle\langle 1|$.

nian (4.42) a term proportional to the identity,

$$\hat{H}_{\rm q} = \hbar\omega_{\rm q}|1\rangle\langle 1| - \frac{\hbar\omega_{\rm q}}{2}\mathbb{1} = \begin{bmatrix} -\hbar\omega_{\rm q}/2 & 0 \\ 0 & \hbar\omega_{\rm q}/2 \end{bmatrix} = -\frac{\hbar\omega_{\rm q}}{2}\hat{\sigma}_z. \tag{4.43}$$

Equations (4.42) and (4.43) express the same Hamiltonian but in two different forms (only differences between energy levels matter for physical purposes and the choice of zero-point is arbitrary). This concept is illustrated in Fig. 4.5.

The Hamiltonian (4.43) has two eigenvalues, $\pm\hbar\omega_{\rm q}/2$. The lowest one is associated with the ground state $|0\rangle$, the highest one is associated with the excited state $|1\rangle$. It is worth mentioning that in some sections of this textbook, the states $|0\rangle$ and $|1\rangle$ will be denoted as $|{\rm g}\rangle$ and $|{\rm e}\rangle$, respectively. This notation makes it possible to distinguish the eigenstates of a qubit from the eigenvectors of the quantum harmonic oscillator.

4.5 Quantum harmonic oscillator

The quantum harmonic oscillator (QHO) is the quantum analog of the classical harmonic oscillator. This system models the behavior of a particle subject to a quadratic potential. The importance of the QHO stems from the fact that it describes a variety of systems, since many potentials with an equilibrium point can be accurately approximated with a quadratic function close to that point.

A harmonic oscillator can be imagined as a particle with mass m attached to an elastic spring. The mass performs small oscillations about the equilibrium point, as shown in Fig. 4.6. The Hamiltonian of a **quantum harmonic oscillator** is the sum of the kinetic energy of the particle and the potential energy,

$$\boxed{\hat{H}_{\rm r} = \frac{\hat{p}^2}{2m} + \frac{1}{2}m\omega_{\rm r}^2\hat{x}^2,} \tag{4.44}$$

where $\omega_{\rm r}$ is the angular frequency, and $\hat{x}$ and $\hat{p}$ are the position and momentum of the particle. Since the function $V(\hat{x}) = \frac{1}{2}m\omega_{\rm r}\hat{x}^2$ is quadratic in position, this potential is said to be **harmonic**.

Let us derive the energy levels E_n of the quantum harmonic oscillator. In quantum mechanics, the momentum operator acting on a state in the position eigenbasis takes the form [128]

$$\langle x|\hat{p}|\psi\rangle = -i\hbar\frac{d}{dx}\psi(x).$$

Using this relation, the time-independent Schrödinger equation $\hat{H}_{\rm r}\psi_n(x) = E_n\psi_n(x)$ becomes

$$-\frac{\hbar^2}{2m}\frac{d^2\psi_n(x)}{dx^2} + \frac{1}{2}m\omega_{\rm r}^2x^2\psi_n(x) = E_n\psi_n(x), \tag{4.45}$$

where E_n are the energy levels of the system associated with the eigenfunctions $\psi_n(x)$ and n is an integer. It can be shown that the solution of this differential equation is:

$$E_n = \hbar\omega_{\rm r}\left(n + \frac{1}{2}\right), \tag{4.46}$$

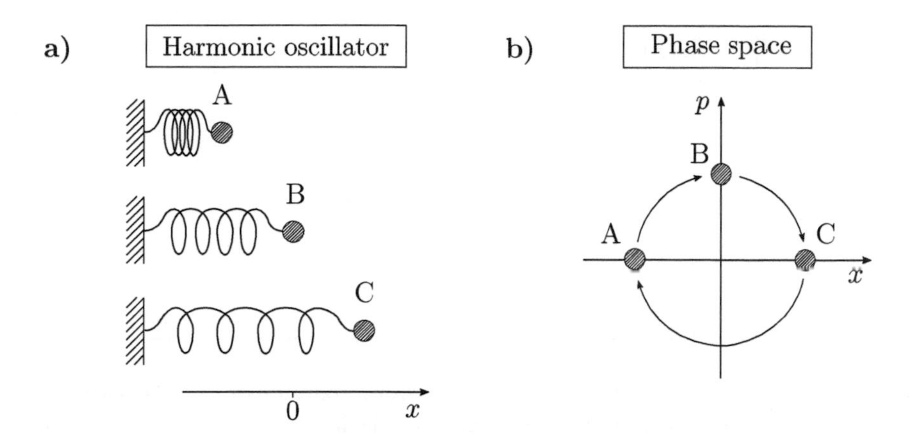

Fig. 4.6 The harmonic oscillator and the phase space. a) A schematic representation of a harmonic oscillator. In this diagram, the mass starts at point A and moves toward the equilibrium point in B, where the position of the oscillator is zero and the momentum is maximum. The mass reaches point C, where the kinetic energy is completely converted into potential energy. This motion carries on cyclically. **b)** The motion of a classical oscillator in the phase space. In this diagram, we are assuming $m = 1$ and $\omega_{\mathrm{r}} = 1$. The x coordinate is the position of the oscillator and the y coordinate represents its momentum. As the mass oscillates back and forth, it draws continuous circles in the phase space. If the spring had some friction, the particle's motion in the phase space would look like a spiral ending at the origin.

$$\psi_n(x) = \langle x|n\rangle = \frac{1}{\sqrt{2^n n!}}\sqrt{\frac{\lambda}{\sqrt{\pi}}}\,H_n(\lambda x)\,e^{-\frac{1}{2}\lambda^2 x^2}, \tag{4.47}$$

where $n \geq 0$, $\lambda = \sqrt{m\omega_{\mathrm{r}}/\hbar}$ is the inverse of a distance, and $H_n(\lambda x)$ are the Hermite polynomials [131]. Equation (4.46) indicates that the energy levels of a QHO are equally spaced by $\hbar\omega_{\mathrm{r}}$. This quantity is called a **quantum of energy**.

The eigenfunctions of the harmonic oscillator belong to the Hilbert space $\mathrm{L}^2(\mathbb{R})$. The first three eigenfunctions are:

Eigenenergies	Eigenfunctions
$E_2 = \frac{5}{2}\hbar\omega_{\mathrm{r}},$	$\psi_2(x) = \sqrt{\frac{\lambda}{8\sqrt{\pi}}}\,(4\lambda^2x^2 - 2)\,e^{-\frac{1}{2}\lambda^2x^2},$
$E_1 = \frac{3}{2}\hbar\omega_{\mathrm{r}},$	$\psi_1(x) = \sqrt{\frac{\lambda}{2\sqrt{\pi}}}\,2\lambda x\,e^{-\frac{1}{2}\lambda^2x^2},$
$E_0 = \frac{1}{2}\hbar\omega_{\mathrm{r}},$	$\psi_0(x) = \sqrt{\frac{\lambda}{\sqrt{\pi}}}\,e^{-\frac{1}{2}\lambda^2x^2},$ (4.48)

and are illustrated in Fig. 4.7. Equation (4.48) indicates that the quantum state with

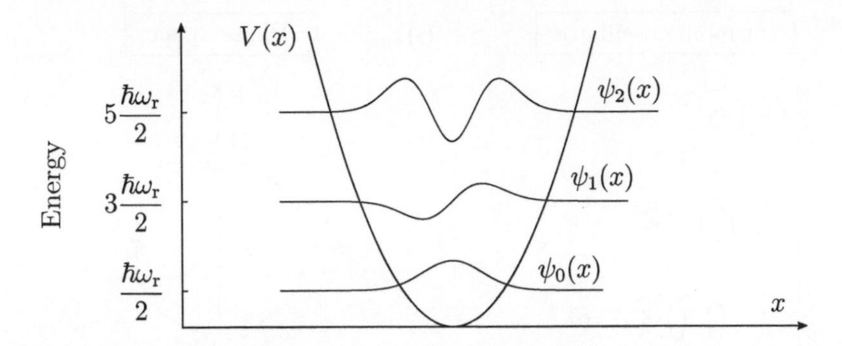

Fig. 4.7 The energy levels of a quantum harmonic oscillator. A schematic representation of the quadratic potential $V(\hat{x}) = \frac{1}{2}m\omega_r^2\hat{x}^2$ of the harmonic oscillator and the first three eigenfunctions. The wavefunctions are displaced in the vertical direction for clarity. Note that the even (odd) eigenfunctions are symmetric (antisymmetric). The associated eigenenergies are equally spaced by $\hbar\omega_r$.

the lowest energy is a Gaussian centered at the origin. This state is usually called the **vacuum state**. When a harmonic oscillator is in the vacuum state, the position's expectation value is zero and does not change over time like in the classical case. When a harmonic oscillator is populated with n excitations, its state is described by the wavefunction $\psi_n(x)$. The probability of finding the particle in a small interval $[x, x+dx]$ is given by $|\psi_n(x)|^2dx$. The total probability $\int_{-\infty}^{\infty}|\psi_n(x)|^2dx = 1$ because the wavefunctions are normalized.

For pedagogical reasons, it is useful to derive the eigenvalues and eigenvectors of the QHO Hamiltonian with a different method. This method is based on two non-Hermitian operators called the **annihilation and creation** operators,

$$\boxed{\hat{a} = \sqrt{\frac{m\omega_r}{2\hbar}}\left(\hat{x} + i\frac{\hat{p}}{m\omega_r}\right),} \tag{4.49}$$

$$\boxed{\hat{a}^\dagger = \sqrt{\frac{m\omega_r}{2\hbar}}\left(\hat{x} - i\frac{\hat{p}}{m\omega_r}\right).} \tag{4.50}$$

From these definitions, it follows that the position and momentum are given by

$$\hat{x} = \sqrt{\frac{\hbar}{2m\omega_r}}(\hat{a} + \hat{a}^\dagger), \tag{4.51}$$

$$\hat{p} = -i\sqrt{\frac{\hbar m\omega_r}{2}}(\hat{a} - \hat{a}^\dagger). \tag{4.52}$$

Since $[\hat{x}, \hat{p}] = i\hbar\mathbb{1}$, one can verify that $[\hat{a}, \hat{a}^\dagger] = \mathbb{1}$. By substituting expressions (4.51, 4.52) into (4.44), we obtain

$$\boxed{\hat{H}_{\mathrm{r}} = \hbar\omega_{\mathrm{r}} \left(\hat{a}^\dagger \hat{a} + \frac{\mathbb{1}}{2} \right) = \hbar\omega_{\mathrm{r}} \left(\hat{n} + \frac{\mathbb{1}}{2} \right),} \tag{4.53}$$

where we used $[\hat{a}, \hat{a}^\dagger] = \mathbb{1}$ and in the last step we introduced the **number operator** $\hat{n} = \hat{a}^\dagger \hat{a}$ with eigenvalues and eigenvectors defined by $\hat{n}|n\rangle = n|n\rangle$. The eigenvectors of the number operator are called the **number states**. Equation (4.53) shows that the Hamiltonian $\hat{H}_{\mathrm{r}}$ and the number operator share the same eigenvectors.

Let us study the number operator and its eigenvectors in more detail. Using the commutation relation $[\hat{a}, \hat{a}^\dagger] = \mathbb{1}$ and properties **C1–C5** presented in Section 3.2.1, we have

$$[\hat{n}, \hat{a}] = -\hat{a}, \tag{4.54}$$

$$\left[\hat{n}, \hat{a}^\dagger\right] = \hat{a}^\dagger. \tag{4.55}$$

Now consider a generic eigenvector $|n\rangle$. From (4.54), it follows that

$$\hat{n}\, \hat{a}|n\rangle = (\hat{a}\hat{n} - \hat{a})\, |n\rangle = (\hat{a}n - \hat{a})\, |n\rangle = (n-1)\, \hat{a}|n\rangle. \tag{4.56}$$

Similarly, from eqn (4.55), we have

$$\hat{n}\, \hat{a}^\dagger|n\rangle = \left(\hat{a}^\dagger\hat{n} + \hat{a}^\dagger\right) |n\rangle = \left(\hat{a}^\dagger n + \hat{a}^\dagger\right) |n\rangle = (n+1)\, \hat{a}^\dagger|n\rangle. \tag{4.57}$$

On the one hand, eqn (4.57) shows that the state $\hat{a}^\dagger|n\rangle$ is an eigenvector of $\hat{n}$ with eigenvalue $n+1$. On the other hand, eqn (4.56) indicates that the state $\hat{a}|n\rangle$ is an eigenvector of $\hat{n}$ with eigenvalue $n-1$. By repeatedly applying the operator $\hat{a}$ to the vector $|n\rangle$, it seems that it would be possible to construct an infinite sequence of states with eigenvalues $n-1$, $n-2$, $n-3\ldots$ At some point, one of these eigenvalues might be negative. This contradicts the fact that the eigenvalues of the number operator are non-negative integers

$$n = \langle n|\hat{n}|n\rangle = \langle n|\hat{a}^\dagger\hat{a}|n\rangle = \langle z|z\rangle \geq 0,$$

where we defined $\hat{a}|n\rangle = |z\rangle$. This contradiction can only be resolved if we impose the existence of a state $|0\rangle$ such that $a|0\rangle = \underline{\mathbf{0}}$. The state $|0\rangle$ is the lowest-energy state of the quantum harmonic oscillator and is called the **vacuum state**. The infinite set of number states $\{|n\rangle\}_{n=0}^{\infty}$ forms an orthonormal basis of the Hilbert space. Their representation in the position eigenbasis $\langle x|n\rangle = \psi_n(x)$ was presented in eqn (4.47).

It is important to calculate the action of the creation and destruction operators on the number states, namely $\hat{a}|n\rangle$ and $\hat{a}^\dagger|n\rangle$. From (4.56), we have

$$\hat{n}\, \hat{a}|n\rangle = (n-1)\, \hat{a}|n\rangle,$$

$$\hat{n}|n-1\rangle = (n-1)\, |n-1\rangle.$$

These expressions indicate that the states $\hat{a}|n\rangle$ and $|n-1\rangle$ are both eigenstates of the operator $\hat{n}$ with the same eigenvalue. Hence, they must be proportional, i.e. $\hat{a}|n\rangle = c|n-1\rangle$. The constant factor c can be calculated from

$$1 = \langle n|n\rangle = \frac{1}{n}\langle n|\hat{n}|n\rangle = \frac{1}{n}\langle n|\hat{a}^\dagger\hat{a}|n\rangle = \frac{|c|^2}{n} \qquad \Rightarrow \qquad c = \sqrt{n},$$

where we took c to be real without loss of generality. A similar argument shows that $\hat{a}^\dagger|n\rangle = \sqrt{n+1}|n+1\rangle$. Thus, we arrive at the relations

$$\boxed{\hat{a}|n\rangle = \sqrt{n}|n-1\rangle,} \tag{4.58}$$

$$\boxed{\hat{a}^\dagger|n\rangle = \sqrt{n+1}|n+1\rangle.} \tag{4.59}$$

These two equations are very important and can be used to calculate the expectation value of the position and momentum when the quantum harmonic oscillator is prepared in an arbitrary quantum state. We will explain this point in the next section.

To summarize, the energy levels of a quantum harmonic oscillator are given by $E_n = \hbar\omega_\mathrm{r}(n+1/2)$ and the eigenstates of the Hamiltonian are $\{|n\rangle\}_{n=0}^{\infty}$. These states are called number states. When the quantum oscillator is in a state $|n\rangle$, the oscillator is populated with n excitations. Depending on the oscillator, these excitations are called photons, phonons, magnons, or other particles. Each excitation carries an amount of energy $\hbar\omega_\mathrm{r}$. In the position eigenbasis, the eigenfunctions are $\psi_n(x)$ and the probability of observing the particle around x is given by $|\psi_n(x)|^2dx$. The raising operator $\hat{a}^\dagger$ creates a photon of frequency ω_r in the oscillator. The lowering operator $\hat{a}$ reduces the number of excitations in the oscillator by one.

4.6 Coherent states

In classical mechanics, the position and momentum of a harmonic oscillator follow a sinusoidal behavior in time. The equations of motion of a classical oscillator are given by

$$x(t) = x_0\cos(\omega_\mathrm{r}t) + \frac{p_0}{m\omega_\mathrm{r}}\sin(\omega_\mathrm{r}t), \tag{4.60}$$

$$p(t) = p_0\cos(\omega_\mathrm{r}t) - m\omega_\mathrm{r}x_0\sin(\omega_\mathrm{r}t), \tag{4.61}$$

where x_0 and p_0 are the position and momentum of the particle at $t = 0$. These equations are plotted in Fig. 4.8b for a classical oscillator that starts from a standstill.

When a quantum harmonic oscillator is in a number state $|\psi\rangle = |n\rangle$, the expectation value of the position and momentum does not change over time and is always zero. This behavior is significantly different from the classical case and indicates that number states are highly non-classical states. To see this, suppose that a harmonic oscillator is in the state $|n\rangle$ at $t = 0$. The expectation value of the position at $t > 0$ is

$$\begin{aligned}\langle n(t)|\,\hat{x}\,|n(t)\rangle &= \langle n|\hat{U}^\dagger(t,0)\,\hat{x}\,\hat{U}(t,0)|n\rangle \\ &= \langle n|e^{-\frac{1}{i\hbar}E_nt}\,\hat{x}\,e^{\frac{1}{i\hbar}E_nt}|n\rangle \;=\; \langle n|\hat{x}|n\rangle \\ &= \sqrt{\frac{\hbar}{2m\omega_\mathrm{r}}}(\langle n|\hat{a}|n\rangle + \langle n|\hat{a}^\dagger|n\rangle) = 0,\end{aligned} \tag{4.62}$$

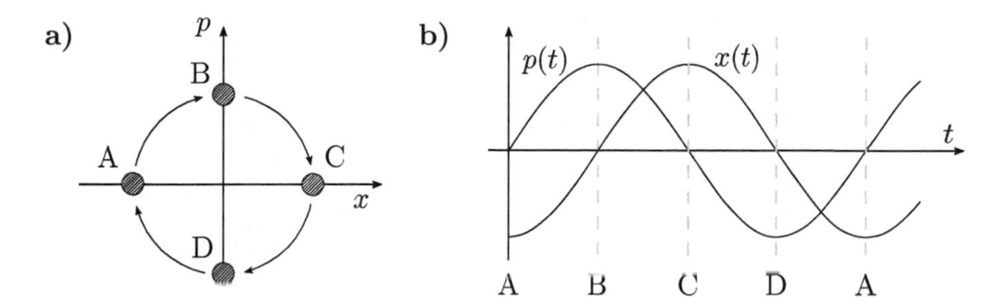

Fig. 4.8 Classical motion of an oscillator. a) Motion of a classical oscillator in the phase space. In this diagram, we are assuming $m = \omega_{\rm r} = 1$. As the mass swings back and forth, it draws a continuous circle in the phase space. At points A and C, the particle's speed is zero, while at points B and D it is maximized. **b)** Plot of the equations of motion of a classical oscillator. The position and the momentum show a sinusoidal behavior with a phase difference of $\pi/2$.

---

where we used[10] eqns (4.51, 4.58, 4.59). In a similar way, it can be shown that $\langle n(t)|\hat{p}|n(t)\rangle = \langle n|\hat{p}|n\rangle = 0$. As anticipated, the expectation value of the position and momentum do not change in time when the system is prepared in a number state, an eigenstate of the Hamiltonian (preparing a quantum harmonic oscillator in a number state is not trivial: in the field of quantum optics, this can be done by exploiting a process known as parametric down conversion [133]).

What is the quantum state of a harmonic oscillator whose time evolution most resembles the motion of a classical oscillator? When a quantum harmonic oscillator is prepared in a **coherent state**, the position and momentum evolve in time with a typical sinusoidal behavior. Formally, a coherent state $|\alpha\rangle$ is defined as the eigenstate of the annihilation operator,

$$\boxed{\hat{a}|\alpha\rangle = \alpha|\alpha\rangle.} \tag{4.63}$$

The eigenvector $|\alpha\rangle$ is normalized and the eigenvalue α is a complex number because the annihilation operator is not Hermitian. A coherent state can be decomposed in the orthonormal basis $\{|n\rangle\}_{n=0}^{\infty}$. In Exercise 4.1, we derive the coefficients of the expansion $|\alpha\rangle = \sum_n c_n|n\rangle$ and we find that

$$\boxed{|\alpha\rangle = e^{-\frac{1}{2}|\alpha|^2}\sum_{n=0}^{\infty}\frac{\alpha^n}{\sqrt{n!}}|n\rangle.}$$

This expression shows that a coherent state is nothing but a superposition of number states. The vacuum state $|0\rangle$ is the only state that is both a number state and a coherent state. The probability of finding n excitations in an oscillator prepared in a coherent state $|\alpha\rangle$ is

[10]This result can also be obtained by calculating $\langle n|\hat{x}|n\rangle = \int_{-\infty}^{\infty}\langle n|\hat{x}|x\rangle\langle x|n\rangle dx = \int_{-\infty}^{\infty} x|\psi_n(x)|^2 dx$. This integral is zero because $x|\psi_n(x)|^2$ are odd functions.

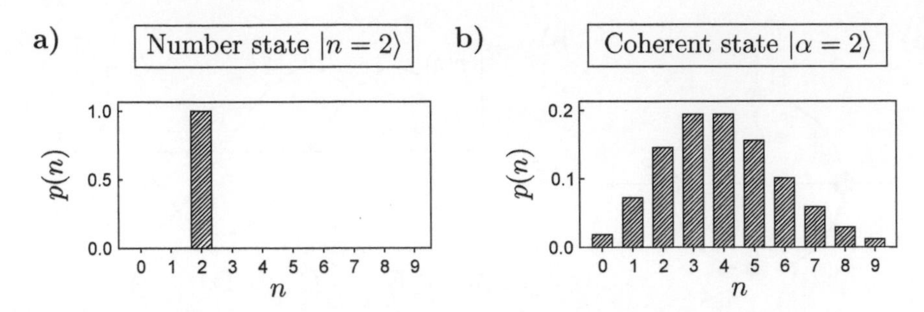

Fig. 4.9 Number state vs. coherent state. a) Histogram of the probability of observing n photons in a quantum harmonic oscillator prepared in the number state $|n=2\rangle$. The probability of measuring two photons is 100%. **b)** Histogram of the probability of measuring n photons in a harmonic oscillator prepared in the coherent state $|\alpha=2\rangle$. The probability distribution $p(n)$ is given by eqn (4.64) with $\alpha=2$. The average of this distribution is $|\alpha|^2=4$, see eqn 4.74.

$$p(n) = |\langle n|\alpha\rangle|^2 = e^{-|\alpha|^2}\frac{|\alpha|^{2n}}{n!}. \tag{4.64}$$

This probability distribution is called a **Poissonian distribution**. In Fig. 4.9, we compare the probability of measuring n photons in a harmonic oscillator when it is prepared in the number state $|n=2\rangle$ and the coherent state $|\alpha=2\rangle$. Creating a coherent state is relatively simple. In the field of superconducting devices, this can be done by sending a resonant drive to a microwave resonator in the quantum regime. This point will be explained in more detail in Section 13.7.

Unlike number states, coherent states are not orthogonal, which means that in general $\langle\alpha_2|\alpha_1\rangle \neq 0$. This follows from the fact that

$$\begin{aligned}
\langle\alpha_2|\alpha_1\rangle &= e^{-\frac{1}{2}|\alpha_1|^2}e^{-\frac{1}{2}|\alpha_2|^2}\sum_{m=0}^{\infty}\sum_{n=0}^{\infty}\langle m|\frac{(\alpha_2^*)^m}{\sqrt{m!}}\frac{(\alpha_1)^n}{\sqrt{n!}}|n\rangle \\
&= e^{-\frac{1}{2}|\alpha_1|^2-\frac{1}{2}|\alpha_2|^2}\sum_{n=0}^{\infty}\frac{(\alpha_2^*\alpha_1)^n}{n!} \\
&= e^{-\frac{1}{2}|\alpha_1|^2-\frac{1}{2}|\alpha_2|^2}\,e^{\alpha_2^*\alpha_1} \;=\; e^{\frac{1}{2}(\alpha_1\alpha_2^*-\alpha_1^*\alpha_2)}e^{-\frac{1}{2}|\alpha_1-\alpha_2|^2}.
\end{aligned} \tag{4.65}$$

The factor $e^{\frac{1}{2}(\alpha_1\alpha_2^*-\alpha_1^*\alpha_2)}$ has modulus 1 and therefore $|\langle\alpha_2|\alpha_1\rangle| = e^{-\frac{1}{2}|\alpha_1-\alpha_2|^2}$. Thus, two generic coherent states are not orthogonal. However, if the distance $|\alpha_1-\alpha_2|$ is large enough, they are nearly orthogonal. For instance, when $|\alpha_1-\alpha_2|=3$, the overlap is relatively small, $|\langle\alpha_2|\alpha_1\rangle| \approx 0.01$.

We have all the ingredients to show that the expectation value of the position and momentum for a quantum harmonic oscillator prepared in a coherent state oscillates like in the classical case. Consider a quantum harmonic oscillator initially prepared in a coherent state $|\alpha\rangle$. Using eqn (4.49), the complex number α can be expressed as

$$\alpha = \langle\alpha|\hat{a}|\alpha\rangle = \sqrt{\frac{m\omega_{\rm r}}{2\hbar}}\langle\alpha|\hat{x}|\alpha\rangle + \frac{i}{\sqrt{2\hbar m\omega_{\rm r}}}\langle\alpha|\hat{p}|\alpha\rangle. \tag{4.66}$$

Therefore, the expectation values of the position and momentum are proportional to the real and imaginary parts of α,

$$x_0 = \langle\alpha|\hat{x}|\alpha\rangle = \sqrt{\frac{2\hbar}{m\omega_{\rm r}}}\,{\rm Re}[\alpha], \tag{4.67}$$

$$p_0 = \langle\alpha|\hat{p}|\alpha\rangle = \sqrt{2\hbar m\omega_{\rm r}}\,{\rm Im}[\alpha]. \tag{4.68}$$

The time evolution of the system is given by[11]

$$\begin{aligned}|\alpha(t)\rangle = \hat{U}(t,t_0)|\alpha\rangle = e^{\frac{1}{i\hbar}\hat{H}_{\rm r}t}|\alpha\rangle &= e^{-\frac{1}{2}|\alpha|^2}\sum_{n=0}^{\infty}\frac{\alpha^n}{\sqrt{n!}}e^{-i\omega_{\rm r}t\hat{n}}|n\rangle \\ &= e^{-\frac{1}{2}|\alpha|^2}\sum_{n=0}^{\infty}\frac{\left(\alpha e^{-i\omega_{\rm r}t}\right)^n}{\sqrt{n!}}|n\rangle \;=\; |\alpha e^{-i\omega_{\rm r}t}\rangle.\end{aligned} \tag{4.69}$$

This equation is simply saying that the complex number $\alpha(t)$ has the same modulus of α but a different phase. The expectation value of the position at $t > 0$ becomes

$$\begin{aligned}x(t) = \langle\alpha(t)|\hat{x}|\alpha(t)\rangle &= \sqrt{\frac{2\hbar}{m\omega_{\rm r}}}\,{\rm Re}[\alpha e^{-i\omega_{\rm r}t}] \\ &= \sqrt{\frac{2\hbar}{m\omega_{\rm r}}}\,{\rm Re}\Big[\sqrt{\frac{m\omega_{\rm r}}{2\hbar}}\Big(x_0 + \frac{ip_0}{m\omega_{\rm r}}\Big)e^{-i\omega_{\rm r}t}\Big] \\ &= x_0\cos(\omega_{\rm r}t) + \frac{p_0}{m\omega_{\rm r}}\sin(\omega_{\rm r}t),\end{aligned} \tag{4.70}$$

where we used (4.66). Similarly, for the momentum we have

$$\begin{aligned}p(t) = \langle\alpha(t)|\hat{p}|\alpha(t)\rangle &= \sqrt{2\hbar m\omega_{\rm r}}\,{\rm Im}[\alpha e^{-i\omega_{\rm r}t}] \\ &= \sqrt{2\hbar m\omega_{\rm r}}\,{\rm Im}\Big[\sqrt{\frac{m\omega_{\rm r}}{2\hbar}}\Big(x_0 + \frac{ip_0}{m\omega_{\rm r}}\Big)e^{-i\omega_{\rm r}t}\Big] \\ &= p_0\cos(\omega_{\rm r}t) - m\omega_{\rm r}x_0\sin(\omega_{\rm r}t).\end{aligned} \tag{4.71}$$

Equations (4.70, 4.71) are equivalent to the classical equations of motion (4.60, 4.61). The motion of a quantum harmonic oscillator in a coherent state resembles the dynamics of the classical case.

When a harmonic oscillator is prepared in a coherent state $|\alpha\rangle$, the probability of measuring the oscillator at a particular position x follows a Gaussian distribution. This is because the wavefunction of a coherence state in the position eigenbasis is given by

[11] Here, we are dropping the vacuum term from the Hamiltonian $\hat{H}_{\rm r} = \hbar\omega_{\rm r}(\hat{a}^\dagger\hat{a} + 1/2)$ without loss of generality.

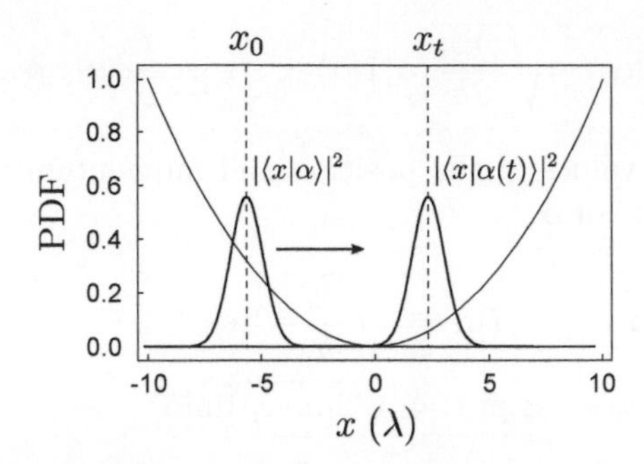

Fig. 4.10 Probability density function for a coherent state. Probability of measuring an oscillator at x when prepared in a coherent state $|\alpha\rangle = |-4\rangle$. From eqn (4.73), the probability distribution $|\langle x|\alpha\rangle|^2$ is a Gaussian centered at $x_0 = \sqrt{2}\text{Re}[\alpha]/\lambda \approx -5.66/\lambda$. At $t > 0$, the oscillator evolves into the coherent state $|\alpha(t)\rangle = |-4e^{-i\omega_r t}\rangle$. The probability distribution $|\langle x|\alpha(t)\rangle|^2$ is a Gaussian with the same width and centered at $x_t = \sqrt{2}\text{Re}[\alpha e^{-i\omega_r t}]/\lambda$. Here, we assumed $\omega_r t = 2$ rad. Clearly, at $t = 2\pi m/\omega_r$ where $m \in \mathbb{N}$, the wavefunction of the particle will be centered again at x_0. The quadratic potential is represented with a solid line in the background.

$$\begin{aligned}\psi_\alpha(x) = \langle x|\alpha\rangle &= e^{-\frac{1}{2}|\alpha|^2}\sum_{n=0}^{\infty}\frac{\alpha^n}{\sqrt{n!}}\langle x|n\rangle \\ &= \sqrt{\frac{\lambda}{\sqrt{\pi}}}e^{-\frac{1}{2}|\alpha|^2}e^{-\frac{1}{2}\lambda^2x^2}\sum_{n=0}^{\infty}\frac{\alpha^n}{n!\sqrt{2^n}}H_n(\lambda x),\end{aligned} \tag{4.72}$$

where we used eqn (4.47) and $\lambda = \sqrt{m\omega_r/\hbar}$. Remember that the generating function of the Hermite polynomials is $\sum_{n=0}^{\infty}\frac{A^n}{n!}H_n(B) = e^{2AB-A^2}$. By using this expression, we can calculate the series in (4.72) and obtain

$$\boxed{\psi_\alpha(x) = |\langle x|\alpha\rangle|^2 = \sqrt{\frac{\lambda}{\sqrt{\pi}}}e^{-\frac{1}{2}(|\alpha|^2-\alpha^2)}e^{-\frac{1}{2}(\lambda x-\sqrt{2}\alpha)^2}.} \tag{4.73}$$

The probability of measuring the oscillator around x is $|\psi_\alpha(x)|^2dx$. The probability density function $|\psi_\alpha(x)|^2$, plotted in Fig. 4.10, is a Gaussian centered at $x_0 = \sqrt{2}\text{Re}[\alpha]/\lambda$. This is consistent with eqn (4.67). If the oscillator is left free to evolve, the system at time $t > 0$ will be in the state $|\alpha e^{-i\omega_r t}\rangle$ and the peak of the probability distribution will move to $x_t = \sqrt{2}\text{Re}[\alpha e^{-i\omega_r t}]/\lambda$.

What is the physical interpretation of α? As shown in eqns (4.67, 4.68), the real part of α is proportional to the expectation value of the position, while its imaginary part is proportional to the expectation value of the momentum. In addition, the modulus squared of α indicates the average number of excitations in the coherent state because

$$\langle\alpha|\hat{n}|\alpha\rangle = \langle\alpha|\hat{a}^\dagger\hat{a}|\alpha\rangle = \alpha^*\alpha = |\alpha|^2. \tag{4.74}$$

Thus, the expectation value of the energy for a coherent state is $\langle\alpha|\hat{H}_\mathrm{r}|\alpha\rangle = \hbar\omega_\mathrm{r}(|\alpha|^2 + 1/2)$.

It is important to mention that for coherent states the product between the standard deviations $\Delta_\alpha\hat{x}$ and $\Delta_\alpha\hat{p}$ reaches the minimum value allowed by the Heisenberg uncertainty principle. For this reason, coherent states are said to be **minimum uncertainty states**. To see this, let us write the Hamiltonian (4.44) in the form

$$\boxed{\hat{H}_\mathrm{r} = \frac{\hat{P}^2}{2} + \frac{\hat{X}^2}{2}.} \tag{4.75}$$

The operators $\hat{X}$ and $\hat{P}$ are called the **quadratures**. These are nothing but rescaled position and momentum operators,

$$\hat{X} = \sqrt{\frac{m\omega_\mathrm{r}}{\hbar}}\hat{x} = \frac{1}{\sqrt{2}}(\hat{a} + \hat{a}^\dagger),$$
$$\hat{P} = \frac{1}{\sqrt{\hbar m\omega_\mathrm{r}}}\hat{p} = -\frac{i}{\sqrt{2}}(\hat{a} - \hat{a}^\dagger).$$

Since $[\hat{x}, \hat{p}] = i\hbar\mathbb{1}$, then $[\hat{X}, \hat{P}] = i\mathbb{1}$ and the Heisenberg uncertainty relation (4.22) becomes

$$\Delta_\psi\hat{X}\cdot\Delta_\psi\hat{P} \geq \frac{|\langle\psi|[\hat{X}, \hat{P}]|\psi\rangle|}{2} = 1/2.$$

Let us derive the uncertainty of the quadrature $\hat{X}$ for a coherent state. We first need to calculate the quantities:

$$\langle\alpha|\hat{X}^2|\alpha\rangle = \frac{1}{2}\langle\alpha|\hat{a}\hat{a} + 2\hat{a}^\dagger\hat{a} + 1 + \hat{a}^\dagger\hat{a}^\dagger|\alpha\rangle = \frac{1}{2}(\alpha^2 + 2|\alpha|^2 + 1 + \alpha^{*2}),$$
$$\langle\alpha|\hat{X}|\alpha\rangle^2 = \frac{1}{2}\langle\alpha|\hat{a} + \hat{a}^\dagger|\alpha\rangle^2 = \frac{1}{2}(\alpha + \alpha^*)^2,$$

where we used $[\hat{a}, \hat{a}^\dagger] = \mathbb{1}$. Therefore,

$$\Delta_\alpha\hat{X} = \sqrt{\langle\alpha|\hat{X}^2|\alpha\rangle - \langle\alpha|\hat{X}|\alpha\rangle^2} = 1/\sqrt{2}.$$

With similar steps, one can verify that $\Delta_\alpha\hat{P} = 1/\sqrt{2}$, which implies that $\Delta_\alpha\hat{X}\cdot\Delta_\alpha\hat{P} = 1/2$. As anticipated, for a coherent state the product between the standard deviations of the position and momentum achieves the limit set by the Heisenberg uncertainty principle.

Coherent states can be depicted in the phase space as shown in Fig. 4.11. The x and y axis of the phase space represent the expectation value of the quadratures $\hat{X}$ and $\hat{P}$, respectively. Recall that the real and imaginary parts of α are proportional to $\langle\hat{X}\rangle$ and $\langle\hat{P}\rangle$. Furthermore, $\Delta_\alpha\hat{X} = \Delta_\alpha\hat{P} = 1/\sqrt{2}$. Thus, we can represent coherent states with a circle with a diameter of $1/\sqrt{2}$ describing a region of uncertainty in the complex plane. The vacuum state is depicted with a circle centered at the origin (see

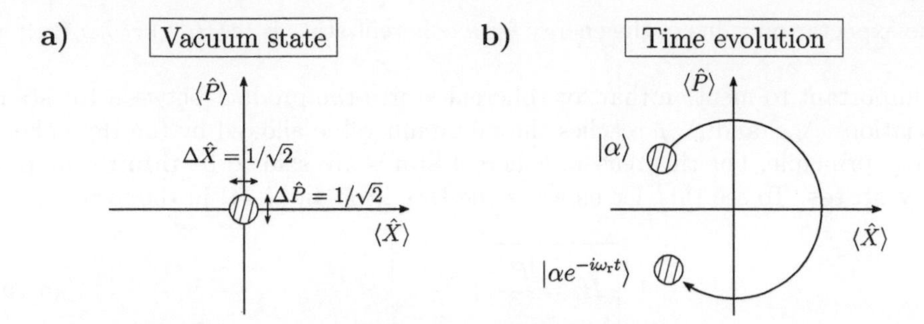

Fig. 4.11 Coherent states in the phase space. a) Illustration of the vacuum state in the phase space. This coherent state is represented by a circle centered at the origin. **b)** Time evolution of a coherent state $\alpha = |\alpha|e^{i\theta}$. This state is represented with a circle at a distance $|\alpha|$ from the origin and rotated by an angle θ with respect to the x axis. The state at time $t > 0$ is rotated clockwise by an angle $\omega_r t$. This evolution is identical to the classical case presented in Fig. 4.6b.

Fig. 4.11a), while a generic coherent state $|\alpha\rangle$ with $\alpha = |\alpha|e^{i\theta}$ is illustrated with a circle at a distance $|\alpha|$ from the origin rotated counter clockwise by an angle θ with respect to the x axis. As seen in eqn (4.69), when a quantum oscillator is prepared in $|\alpha\rangle$ and is left free to evolve, the state at time $t > 0$ is $|\alpha e^{-i\omega_r t}\rangle$. In the phase space, this evolution corresponds to a clockwise rotation by an angle $\omega_r t$ (see Fig. 4.11b). This rotation resembles the dynamics of a classical oscillator in the phase space presented in Fig. 4.6b.

The **displacement operator** is a unitary operator $\hat{D}(\alpha)$ that moves a coherent state from one point of the complex plane to another. This operator is defined as

$$\boxed{\hat{D}(\alpha) = e^{-\frac{1}{2}|\alpha|^2} e^{\alpha\hat{a}^\dagger} e^{-\alpha\hat{a}}.} \tag{4.76}$$

When the displacement operator acts on the vacuum state, it creates a coherent state $|\alpha\rangle$. This can be shown in a few steps

$$\begin{aligned}\hat{D}(\alpha)|0\rangle = e^{-\frac{1}{2}|\alpha|^2} e^{\alpha\hat{a}^\dagger} e^{-\alpha\hat{a}}|0\rangle &= e^{-\frac{1}{2}|\alpha|^2} e^{\alpha\hat{a}^\dagger}|0\rangle \\ &= e^{-\frac{1}{2}|\alpha|^2} \sum_{n=0}^{\infty} \frac{(\alpha)^n}{n!}(\hat{a}^\dagger)^n|0\rangle \\ &= e^{-\frac{1}{2}|\alpha|^2} \sum_{n=0}^{\infty} \frac{(\alpha)^n}{\sqrt{n!}}|n\rangle = |\alpha\rangle,\end{aligned} \tag{4.77}$$

where we used $(\hat{a}^\dagger)^n|0\rangle = \sqrt{n!}|n\rangle$. Figure 4.12a shows the action of the displacement operator $\hat{D}(\alpha)$ on the vacuum state.

The displacement operator can be expressed in other equivalent forms. To show this, it is worth recalling that if two operators $\hat{A}$ and $\hat{B}$ satisfy the relations $[\hat{A},[\hat{A},\hat{B}]] = [\hat{B},[\hat{A},\hat{B}]] = \underline{\mathbf{0}}$, the following Baker–Campbell–Hausdorff formula holds,

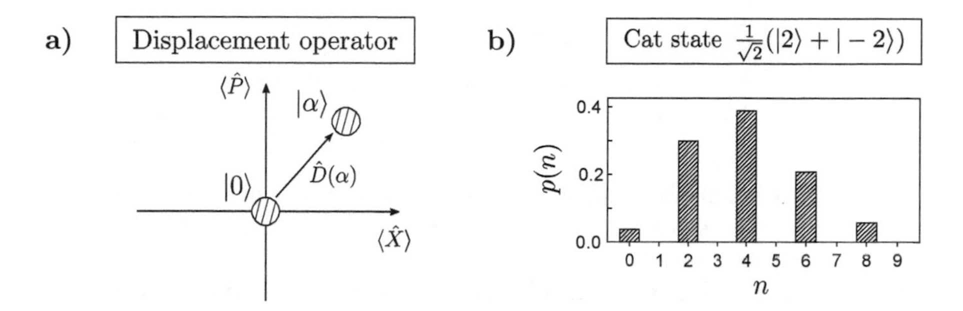

Fig. 4.12 Displacement operators and cat states. a) The displacement operator $\hat{D}(\alpha)$ transforms the vacuum state $|0\rangle$ into the coherent state $|\alpha\rangle$. **b)** Probability of measuring n excitations in a harmonic oscillator prepared in the cat state $\frac{1}{\sqrt{2}}(|2\rangle + |-2\rangle)$. The measurement will reveal only an even number of photons in the system.

$$e^{\hat{A}+\hat{B}} = e^{\frac{1}{2}[\hat{A},\hat{B}]}e^{\hat{B}}e^{\hat{A}} = e^{-\frac{1}{2}[\hat{A},\hat{B}]}e^{\hat{A}}e^{\hat{B}}. \tag{4.78}$$

By setting $\hat{A} = \alpha\hat{a}^\dagger$ and $\hat{B} = -\alpha^*\hat{a}$, we obtain

$$[\hat{A}, [\hat{A}, \hat{B}]] = [\alpha\hat{a}^\dagger, [\alpha\hat{a}^\dagger, -\alpha^*\hat{a}]] = \underline{\mathbf{0}},$$
$$[\hat{B}, [\hat{A}, \hat{B}]] = [-\alpha^*\hat{a}, [\alpha\hat{a}^\dagger, -\alpha^*\hat{a}]] = \underline{\mathbf{0}},$$

where we used $[\hat{a}, \hat{a}^\dagger] = \mathbb{1}$. Since $[\hat{A}, \hat{B}] = [\alpha\hat{a}^\dagger, -\hat{\alpha}^*\hat{a}] = |\alpha|^2$, from eqn (4.78) we have

$$\hat{D}(\alpha) = e^{\alpha\hat{a}^\dagger - \alpha^*\hat{a}} \tag{4.79}$$
$$= e^{\frac{1}{2}|\alpha|^2} e^{-\alpha^*\hat{a}}e^{\alpha\hat{a}^\dagger} \tag{4.80}$$
$$= e^{-\frac{1}{2}|\alpha|^2} e^{\alpha\hat{a}^\dagger} e^{-\alpha^*\hat{a}}. \tag{4.81}$$

These expressions will be useful in Section 13.7 when we show that a microwave resonator in the quantum regime can be prepared in a coherent state by applying an electrical pulse to the system. Let us show that the inverse of the displacement operator $\hat{D}(\alpha)$ is $\hat{D}(-\alpha)$. We can prove this by calculating the product of two generic displacement operators

$$\hat{D}(\beta)\hat{D}(\alpha) = e^{\beta\hat{a}^\dagger - \beta^*\hat{a}}e^{\alpha\hat{a}^\dagger - \alpha^*\hat{a}} = e^{\hat{B}}e^{\hat{A}},$$

where we defined $\hat{B} = \beta\hat{a}^\dagger - \beta^*\hat{a}$ and $\hat{A} = \alpha\hat{a}^\dagger - \alpha^*\hat{a}$. These operators commute with their commutator. Hence, we can use eqn (4.78) and write $e^{\hat{B}}e^{\hat{A}} = e^{-\frac{1}{2}[\hat{A},\hat{B}]}e^{\hat{B}+\hat{A}}$. This leads to

$$\begin{aligned}\hat{D}(\beta)\hat{D}(\alpha) &= e^{-\frac{1}{2}[\hat{A},\hat{B}]}e^{\hat{A}+\hat{B}} \\ &= e^{-\frac{1}{2}(\alpha\beta^*-\alpha^*\beta)}e^{(\alpha+\beta)\hat{a}^\dagger-(\alpha^*+\beta^*)\hat{a}} \\ &= e^{-\frac{1}{2}(\alpha\beta^*-\alpha^*\beta)}\hat{D}(\alpha+\beta). \end{aligned} \tag{4.82}$$

Thus, the combination of two displacement operators is still a displacement operator up to a global phase. By setting $\beta = -\alpha$, eqn (4.82) reduces to $\hat{D}(-\alpha)\hat{D}(\alpha) = \mathbb{1}$ which implies that $\hat{D}^{-1}(\alpha) = \hat{D}(-\alpha)$.

Named after the well-known Schrodinger's cat thought experiment, a **cat state** is a quantum superposition of two macroscopically distinct states. For a quantum harmonic oscillator, a cat state is defined as the superposition of two coherent states of the form[12]

$$\boxed{|\psi_{\alpha,\phi}\rangle = \frac{1}{\sqrt{2}}(|\alpha\rangle + e^{i\phi}|-\alpha\rangle),} \tag{4.83}$$

where $|\alpha| \gg 1$. Cat states play an important role in some implementations of quantum computers based on microwave photons [134, 135, 136]. The cat state $|\psi_{\alpha,0}\rangle$ can be decomposed in the number basis $\{|n\rangle\}_n$ as

$$\begin{aligned}|\psi_{\alpha,0}\rangle = \frac{1}{\sqrt{2}}[|\alpha\rangle + |-\alpha\rangle] &= \frac{1}{\sqrt{2}}e^{-\frac{|\alpha|^2}{2}}\Big[\sum_{n=0}^{\infty}\frac{\alpha^n}{\sqrt{n!}}|n\rangle + \sum_{n=0}^{\infty}\frac{(-\alpha)^n}{\sqrt{n!}}|n\rangle\Big] \\ &= \frac{1}{\sqrt{2}}e^{-\frac{|\alpha|^2}{2}}\Big[2|0\rangle + \frac{2\alpha^2}{\sqrt{2!}}|2\rangle + \frac{2\alpha^4}{\sqrt{4!}}|4\rangle + \frac{2\alpha^6}{\sqrt{6!}}|6\rangle + \dots\Big] \\ &= \sqrt{2}e^{-\frac{|\alpha|^2}{2}}\sum_{n=\text{even}}\frac{\alpha^n}{\sqrt{n!}}|n\rangle.\end{aligned}$$

This decomposition only contains even number states. When an oscillator is prepared in the cat state $|\psi_{\alpha,0}\rangle$, the probability of measuring n excitations is given by $p(n) = |\langle n|\psi_{\alpha,0}\rangle|^2$. This probability distribution is plotted in Fig. 4.12 for the cat state $\frac{1}{\sqrt{2}}[|2\rangle + |-2\rangle]$. With similar calculations, one can verify that the cat state $|\psi_{\alpha,\pi}\rangle = \frac{1}{\sqrt{2}}[|\alpha\rangle - |-\alpha\rangle]$ contains only odd number states. Cat states are highly non-classical states of a harmonic oscillator. In the field of superconducting devices, they can be created by exploiting a dispersive interaction between a microwave resonator and a superconducting qubit. We will return to this point in Section 13.9.

4.7 The Jaynes–Cummings Hamiltonian

The **Jaynes–Cummings Hamiltonian** is a physical model that describes several systems in which an atom, natural or artificial, can be treated as a two-level system interacting with a quantum harmonic oscillator. This model is important in the field

[12]We warn the reader that $|\psi_{\alpha,\phi}\rangle$ is not normalized since $\langle -\alpha|\alpha\rangle = e^{-2|\alpha|^2}$ as seen in eqn (4.65). The normalized version of this state is $|\tilde{\psi}_{\alpha,\phi}\rangle = \frac{1}{N}(|\alpha\rangle + e^{i\phi}|-\alpha\rangle)$ where $N = (2+2e^{-2|\alpha|^2}\cos\phi)^{1/2}$. When $|\alpha| \gg 1$, the state $|\tilde{\psi}_{\alpha,\phi}\rangle$ is approximated by $|\psi_{\alpha,\phi}\rangle$.

of superconducting devices, as it accurately describes the coupling between a superconducting qubit and a microwave resonator in the quantum regime. The interaction between these two systems plays an essential role in all modern approaches to quantum information processing with superconducting circuits [137].

The Jaynes–Cummings Hamiltonian is given by the sum of three terms[13]

$$\hat{H}_{\text{JC}} = \underbrace{-\frac{\hbar\omega_{\text{q}}}{2}\hat{\sigma}_z}_{\text{qubit/atom}} + \underbrace{\hbar\omega_{\text{r}}\hat{a}^\dagger\hat{a}}_{\text{resonator/cavity}} + \underbrace{\hbar g(\hat{\sigma}^+\hat{a} + \hat{\sigma}^-\hat{a}^\dagger)}_{\text{interaction term}}, \tag{4.84}$$

where ω_{q} is the qubit frequency, ω_{r} is the resonator frequency, g is the **coupling strength** between the two systems and $\hat{\sigma}^+ = |\text{e}\rangle\langle\text{g}|$ and $\hat{\sigma}^- = |\text{g}\rangle\langle\text{e}|$ are the lowering and raising operators of the qubit. The term $\hat{\sigma}^+\hat{a}$ describes the physical process in which an excitation is emitted by the cavity and the qubit gets excited. Similarly, the operator $\hat{\sigma}^-\hat{a}^\dagger$ describes the emission of an excitation by the qubit and the creation of an excitation inside the cavity. Depending on the type of cavity, the excitation can be a photon for an optical cavity or a microwave resonator [138, 139], a phonon for a mechanical resonator [140, 141], a magnon for a ferromagnet [142], or other elementary particles. In the field of superconducting devices, the frequency of the qubit is usually in the range $\omega_{\text{q}}/2\pi = 3 - 10$ GHz and the frequency of the resonator $\omega_{\text{r}}/2\pi$ lies in this interval too [20]. The coupling strength between a superconducting qubit and a microwave resonator depends on the application. Its typical value is in the range $|g/2\pi| = 1 - 200$ MHz and can be easily adjusted by varying the capacitance between the two systems.

The Hilbert space of the global system is the tensor product of the Hilbert space of the qubit and that of the harmonic oscillator. The vectors $\{|\text{g}\rangle, |\text{e}\rangle\}$ form an orthonormal basis of the qubit space, while the number states $\{|n\rangle\}_{n=0}^{\infty}$ form an orthonormal basis for the harmonic oscillator. With the notation $|\text{g}\rangle \otimes |n\rangle = |\text{g}, n\rangle$, we mean that the qubit is in the ground state and the oscillator is populated with n excitations. Similarly, the notation $|\text{e}\rangle \otimes |n\rangle = |\text{e}, n\rangle$ indicates that the qubit is in the excited state and the oscillator is in the number state $|n\rangle$.

The interaction term in the Jaynes–Cummings Hamiltonian describes the energy exchange between the qubit and the resonator. In a real experiment, these two systems are also coupled to the environment. This coupling inevitably leads to some loss of information. The qubit **relaxation rate** γ is defined as the rate at which the qubit releases energy into the environment. This parameter is the reciprocal of the relaxation time $T_1 = 1/\gamma$. As regards the resonator, the **photon loss rate** κ quantifies the rate at which the cavity loses energy. This parameter is proportional to the linewidth of the resonator when measured in the frequency domain. In superconducting circuits, the typical value of the qubit relaxation time is $T_1 = 1 - 200\ \mu$s [143], whereas for the resonator $\kappa/2\pi = 0.1 - 10$ MHz.

Let us study the J–C Hamiltonian in three different regimes.

[13]Since the Hilbert space of the global system is the tensor product between the Hilbert space of a qubit and that of a harmonic oscillator, the operators in this Hamiltonian should be written as $\hat{\sigma}_z \otimes \mathbb{1}$, $\mathbb{1} \otimes \hat{a}^\dagger\hat{a}$, $\hat{\sigma}^+ \otimes \hat{a}$, and $\hat{\sigma}^- \otimes \hat{a}^\dagger$. This notation is not practical and we will not adopt it.

4.7.1 Uncoupled systems

For pedagogical reasons, we will start with the case in which the qubit and the harmonic oscillator are not coupled together, i.e. the coupling strength is zero. By substituting $g = 0$ and $\hat{\sigma}_z = -2\hat{\sigma}^+\hat{\sigma}^- + 1$ into the J–C Hamiltonian (4.84), we arrive at

$$\boxed{\hat{H}_{\rm JC} = \hbar\omega_{\rm r}\hat{a}^\dagger\hat{a} + \hbar\omega_{\rm q}\hat{\sigma}^+\hat{\sigma}^-,} \tag{4.85}$$

where we dropped a constant term. The eigenvectors and eigenvalues are[14]:

Eigenvectors	Eigenvalues
$\|{\rm g}, n\rangle$,	$E_{{\rm g},n} = \hbar\omega_{\rm r}n$,
$\|{\rm e}, n\rangle$,	$E_{{\rm e},n} = \hbar\omega_{\rm q} + \hbar\omega_{\rm r}n$,

where $n \geq 0$. When the qubit and the resonator have the same frequency, the **detuning parameter** $\Delta = \omega_{\rm q} - \omega_{\rm r}$ is zero and the eigenstates $|{\rm g}, n\rangle$ and $|{\rm e}, n-1\rangle$ have the same energy. The case $\Delta = 0$ and $g = 0$ is shown in Fig. 4.13a. The energy levels of the J–C Hamiltonian with $\Delta \neq 0$ and $g = 0$ are shown in Fig. 4.14a.

4.7.2 The general case

It is more interesting to study the case in which the coupling strength is non-zero and significantly higher than the loss rates of the qubit and the resonator, $|g| > \gamma, \kappa$. By substituting $\hat{\sigma}_z = -2\sigma^+\sigma^- + 1$ into the J–C Hamiltonian (4.84) and dropping a constant term, we arrive at

$$\hat{H}_{\rm JC} = \hbar\omega_{\rm r}(\hat{a}^\dagger\hat{a} + \hat{\sigma}^+\hat{\sigma}^-) + \hbar\Delta\hat{\sigma}^+\hat{\sigma}^- + \hbar g(\hat{\sigma}^+\hat{a} + \hat{\sigma}^-\hat{a}^\dagger), \tag{4.86}$$

where $\Delta = \omega_{\rm q} - \omega_{\rm r}$ is the usual detuning parameter. The eigenvectors and eigenvalues of this Hamiltonian can be derived analytically. The basis of the Hilbert space consists of an infinite number of vectors $\{|{\rm g}, n\rangle, |{\rm e}, n\rangle\}_{n=0}^{\infty}$. Let us focus on a subspace with a fixed number of excitations. An orthonormal basis of this two-dimensional subspace is given by $\{|{\rm g}, n\rangle, |{\rm e}, n-1\rangle\}$ where $n \geq 1$. It is important to note that the operators in the Hamiltonian (4.86) leave this subspace invariant:

$$\hat{\sigma}^+\hat{a}\,|{\rm g}, n\rangle = \sqrt{n}\,|{\rm e}, n-1\rangle, \qquad \hat{\sigma}^-\hat{a}^\dagger\,|{\rm e}, n-1\rangle = \sqrt{n}\,|{\rm g}, n\rangle.$$

In the subspace spanned by $\{|{\rm g}, n\rangle, |{\rm e}, n-1\rangle\}$, the J–C Hamiltonian can be expressed with a 2×2 matrix

$$\begin{bmatrix} \langle{\rm g}, n|\hat{H}_{\rm JC}|{\rm g}, n\rangle & \langle{\rm g}, n|\hat{H}_{\rm JC}|{\rm e}, n-1\rangle \\ \langle{\rm e}, n-1|\hat{H}_{\rm JC}|{\rm g}, n\rangle & \langle{\rm e}, n-1|\hat{H}_{\rm JC}|{\rm e}, n-1\rangle \end{bmatrix} = \begin{bmatrix} \hbar\omega_{\rm r}n & \hbar g\sqrt{n} \\ \hbar g\sqrt{n} & \hbar\omega_{\rm r}n + \hbar\Delta \end{bmatrix},$$

where we used (4.86). The reader can check that this matrix can be written as

[14]The eigenvalues are defined by $E_{{\rm g},n} = \langle{\rm g}, n|\hat{H}_{\rm JC}|{\rm g}, n\rangle$, and $E_{{\rm e},n} = \langle{\rm e}, n|\hat{H}_{\rm JC}|{\rm e}, n\rangle$. These matrix elements can be calculated using the relations $\langle n|\hat{a}^\dagger\hat{a}|n\rangle = n$, $\langle{\rm g}|\hat{\sigma}^+\hat{\sigma}^-|{\rm g}\rangle = 0$ and $\langle{\rm e}|\hat{\sigma}^+\hat{\sigma}^-|{\rm e}\rangle = 1$.

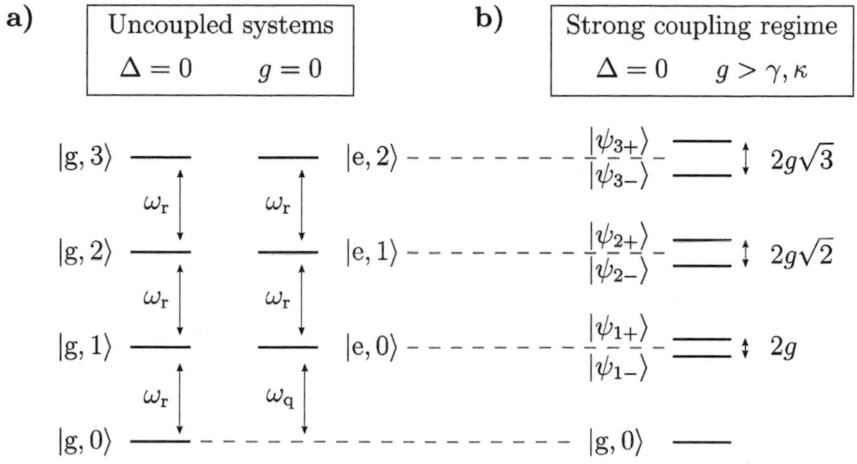

Fig. 4.13 Energy levels of the J–C Hamiltonian a) When the qubit and the resonator have the same frequency but are not coupled to each other ($g = 0$), the eigenstates $|g, n+1\rangle$ and $|e, n\rangle$ have the same transition frequency. **b)** Transition frequencies of a system comprising a qubit coupled to a resonator. In this diagram, we are assuming that the qubit and the resonator are in resonance. The dressed states $|\psi_{n+}\rangle$ and $|\psi_{n-}\rangle$ are entangled states of the qubit and the resonator and their transition frequency is separated by $2g\sqrt{n}$.

$$\hbar\tilde{\omega}_n \mathbb{1} + \hbar g\sqrt{n}\hat{\sigma}_x - \frac{\hbar\Delta}{2}\hat{\sigma}_z, \tag{4.87}$$

where $\tilde{\omega}_n = \omega_r n + \Delta/2$. The diagonalization of this matrix follows the same procedure discussed in Exercise 4.2. The eigenvalues and eigenvectors of the J–C Hamiltonian are given by[15]:

Eigenvectors	Eigenvalues	
$|\psi_{n-}\rangle = \sin\frac{\theta}{2}|g, n\rangle - \cos\frac{\theta}{2}|e, n-1\rangle,$	$E_{n-} = \hbar\tilde{\omega}_n - \hbar\sqrt{g^2 n + \Delta^2/4}$	(4.88)
$|\psi_{n+}\rangle = \cos\frac{\theta}{2}|g, n\rangle + \sin\frac{\theta}{2}|e, n-1\rangle,$	$E_{n+} = \hbar\tilde{\omega}_n + \hbar\sqrt{g^2 n + \Delta^2/4}$	(4.89)

where $n \geq 1$ and $\theta = \arctan(-2g\sqrt{n}/\Delta)$. The eigenstates $|\psi_{n-}\rangle$ and $|\psi_{n+}\rangle$ are called the **dressed states**. When the qubit and the resonator are in resonance ($\Delta = 0$, $\theta = \pi/2$), the composite system is one of the maximally entangled states,

[15]Note that eigenstate of lowest energy is the one in which both the qubit and the resonator are in the ground state, $|g, 0\rangle$. The energy of this state is zero.

$$|\psi_{n-}\rangle = \frac{1}{\sqrt{2}}(|\mathrm{g}, n\rangle - |\mathrm{e}, n-1\rangle),$$
$$|\psi_{n+}\rangle = \frac{1}{\sqrt{2}}(|\mathrm{g}, n\rangle + |\mathrm{e}, n-1\rangle).$$

The excitation is shared between the two systems.

The energy levels E_{n+} and E_{n-} of the J–C Hamiltonian are shown in Fig. 4.13b and in Fig. 4.14b. When $\Delta = 0$, the splitting between the dressed states is proportional to $\sqrt{n}$ [144]. This is different from the splitting between the energy levels of two coupled oscillators, in which the separation does not depend on n (see Exercise 4.4). Note that the derivation of the eigenstates and eigenvalues (4.88, 4.89) is completely general since we did not explicitly use any particular assumption.

3. The dispersive regime

In the dispersive regime, the coupling strength is much smaller than the difference between the qubit frequency and the resonator frequency, $|g/\Delta| \ll 1$. The eigenvalues and eigenvectors of the Hamiltonian are the same as those presented in the previous section. The main difference is that in the dispersive regime we can perform the approximations:

$$\theta \approx \frac{g\sqrt{n}}{-\Delta/2}, \qquad \hbar\sqrt{g^2 n + \Delta^2/4} \approx \frac{\hbar\Delta}{2}\left(1 + \frac{2ng^2}{\Delta}\right).$$

Thus, the eigenvalues and eigenvectors become:

Eigenvectors	Eigenvalues	
$\lvert\mathrm{g}, n\rangle - \frac{g\sqrt{n}}{\Delta}\lvert\mathrm{e}, n-1\rangle,$	$\hbar\omega_\mathrm{r} n - \hbar\chi n,$	(4.90)
$\frac{g\sqrt{n}}{\Delta}\lvert\mathrm{g}, n\rangle + \lvert\mathrm{e}, n-1\rangle,$	$\hbar\omega_\mathrm{q} + \hbar\omega_\mathrm{r}(n-1) + \hbar\chi n,$	(4.91)

where $n \geq 1$ and we introduced the **dispersive shift** $\chi = g^2/\Delta$, a parameter that can be either positive or negative depending on the value of Δ. The energy levels of the J–C Hamiltonian in the dispersive regime are shown in Fig. 4.14b for $\omega_\mathrm{q} < \omega_\mathrm{r}$.

It is instructive to calculate the energy levels of the J–C Hamiltonian in the dispersive regime with a different method. Let us start our analysis from the J–C Hamiltonian,

$$\hat{H}_\mathrm{JC} = \hbar\omega_\mathrm{r}\hat{a}^\dagger\hat{a} - \frac{\hbar\omega_\mathrm{q}}{2}\hat{\sigma}_z + \hbar g(\hat{\sigma}^+\hat{a} + \hat{\sigma}^-\hat{a}^\dagger). \tag{4.92}$$

In Exercise 4.4, we show that there exists a unitary transformation $\hat{U}$ such that $\hat{H}'_\mathrm{JC} = \hat{U}\hat{H}_\mathrm{JC}\hat{U}^\dagger$ is diagonal and takes the form

$$\hat{H}'_\mathrm{JC} = \hbar\omega_\mathrm{r}\hat{a}^\dagger a - \frac{\hbar\omega_\mathrm{q}}{2}\hat{\sigma}_z + \frac{\hbar\Delta}{2}\left(1 - \sqrt{1 + \frac{4g^2}{\Delta^2}\hat{N}_\mathrm{T}}\right)\hat{\sigma}_z, \tag{4.93}$$

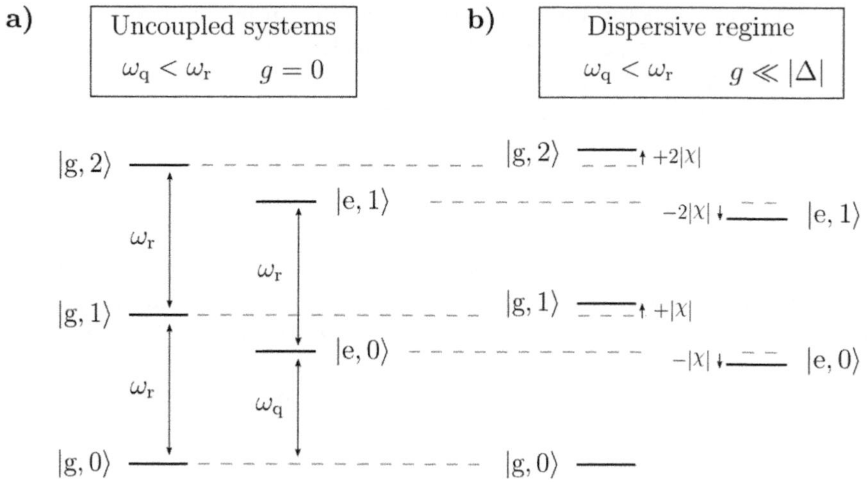

Fig. 4.14 Energy levels of the J–C Hamiltonian a) When the qubit and the resonator are not coupled to each other, the transition frequencies of the states $|g,n\rangle$ and $|e,n\rangle$ are given by $E_{g,n} = \hbar\omega_r n$ and $E_{e,n} = \hbar(\omega_q + \omega_r n)$, respectively. **b)** The energy levels of the J–C Hamiltonian in the dispersive regime. In this diagram, $\omega_q < \omega_r$ which means that Δ and χ are both negative. With respect to the uncoupled case, the energy levels are shifted by multiples of $|\chi|$.

where $\hat{N}_T = \hat{\sigma}^+\hat{\sigma}^- + \hat{a}^\dagger\hat{a}$ is an operator associated with the total number of excitations in the system. This Hamiltonian is diagonal since it only contains the operators $\hat{a}^\dagger\hat{a}$, $\hat{\sigma}^+\hat{\sigma}^-$ and $\hat{\sigma}_z$. The Hamiltonians (4.92, 4.93) are equivalent up to a change of basis.

In the dispersive regime, the Hamiltonian (4.93) can be expressed in a simpler form. Assuming $|g/\Delta| \ll 1$, the term in parentheses in (4.93) can be approximated[16] as $-2g^2\hat{N}_T/\Delta^2$ and therefore

$$\hat{H}'_{JC} = \hbar\omega_r\hat{a}^\dagger\hat{a} - \frac{\hbar\omega_q}{2}\hat{\sigma}_z - \hbar\frac{g^2}{\Delta}(\hat{a}^\dagger\hat{a} + \hat{\sigma}^+\hat{\sigma}^-)\hat{\sigma}_z.$$

By substituting $\hat{\sigma}^+\hat{\sigma}^- = (\mathbb{1} - \hat{\sigma}_z)/2$ and dropping a constant term, we finally obtain the J–C Hamiltonian in the dispersive regime,

$$\boxed{\hat{H}'_{JC} = \hbar(\omega_r - \chi\hat{\sigma}_z)\hat{a}^\dagger\hat{a} - \frac{\hbar}{2}(\omega_q + \chi)\hat{\sigma}_z,} \tag{4.94}$$

where $\chi = g^2/\Delta$ is the usual dispersive shift. The first term of the Hamiltonian shows that the resonator frequency depends on the qubit state: when the qubit is in the ground state, the resonator frequency is $\omega_r - \chi$; if the qubit is excited, the frequency of the resonator shifts to $\omega_r + \chi$. In circuit quantum electrodynamics, this

[16]Recall that $1 - \sqrt{1 + x^2} = -x^2/2 + O(x^4)$ when $x \to 0$.

phenomenon is used to determine the state of the qubit from the frequency response of the resonator. The term $(\omega_\mathrm{q}+\chi)\hat{\sigma}_z$ indicates that the interaction with the resonator shifts the qubit frequency by a constant χ. This phenomenon is known as the **Lamb shift**. The Hamiltonian (4.94) can also be rearranged as

$$\boxed{\hat{H}'_{\mathrm{JC}} = \hbar\omega_\mathrm{r}\hat{a}^\dagger\hat{a} - \frac{\hbar}{2}(\omega_\mathrm{q} + \chi + 2\chi\hat{a}^\dagger\hat{a})\hat{\sigma}_z.} \tag{4.95}$$

This expression shows that the qubit frequency depends on the number of excitations in the cavity: if the resonator is populated with n excitations, the qubit frequency shifts by $2\chi n$. This phenomenon is known as the **AC Stark shift** and has been observed in multiple experiments [145, 146]. We warn the reader that the linear dependence between the AC Stark shift and the number of photons only holds at low photon numbers. When $n > |\Delta^2/4g|$, higher order contributions must be considered [139].

Further reading

Many textbooks about quantum mechanics have been published at various levels. We encourage the reader to start from the masterpiece by Dirac [125]. For a more modern approach, see the textbooks by Sakurai [128] and Weinberg [131]. We also recommend the work by Alonso and Finn [147]. Mathematical, statistical, and foundational aspects of quantum mechanics are thoroughly discussed in the textbooks by Jauch [124], Mackey [148], Holevo [149], and Ludwig [150]. The path integral formulation of quantum mechanics is presented in the textbook by Feynman and Hibbs [151]. See Refs [152, 137] to learn more about coherent states and the Jaynes–Cummings model.

Summary

Postulates of quantum mechanics

$H, \|\psi\rangle \in H$	Hilbert space and quantum state
$\hat{U}(t, t_0)$	Time evolution operator from t_0 to t
$\hat{A} = \hat{A}^\dagger$	Physical observable
$\langle \hat{A} \rangle_\psi = \langle \psi \| \hat{A} \| \psi \rangle$	Expectation value of the observable $\hat{A}$
$\Delta_\psi \hat{A} = \sqrt{\langle \hat{A}^2 \rangle_\psi - \langle \hat{A} \rangle_\psi^2}$	Standard deviation of $\hat{A}$ in the state ψ
$\Delta_\psi \hat{A} \Delta_\psi \hat{B} \geq \|\langle \psi \| [\hat{A}, \hat{B}] \| \psi \rangle\|/2$	Heisenberg uncertainty relation
$\{\hat{P}_i\}_i$, $\hat{P}_i \hat{P}_j = \delta_{ij} \hat{P}_i$, $\sum_i \hat{P}_i = \mathbb{1}$	Projection-valued measure (PVM)
$i\hbar \frac{d}{dt} \|\psi(t)\rangle = \hat{H} \|\psi(t)\rangle$	Schrödinger equation

Qubit

$\|\mathrm{g}\rangle$, $\|\mathrm{e}\rangle$	Ground and excited state
$\hat{H}_\mathrm{q} = -\hbar\omega_\mathrm{q} \hat{\sigma}_z/2$	Hamiltonian of a qubit
$\hat{H}_\mathrm{q} = \hbar\omega_\mathrm{q} \|\mathrm{e}\rangle\langle \mathrm{e}\| = \hbar\omega_\mathrm{q} \hat{\sigma}^+ \hat{\sigma}^-$	Alternative form of $\hat{H}_\mathrm{q}$

Quantum harmonic oscillator

$\hat{H}_\mathrm{r} = \hat{p}^2/2m + m\omega_\mathrm{r}^2 \hat{x}^2/2$	Hamiltonian of a QHO
$\{\|n\rangle\}_{n=0}^{\infty}$	Eigenstates (or number states)
$E_n = \hbar\omega_\mathrm{r}(n + 1/2)$	Eigenvalues
$\lambda = (m\omega_\mathrm{r}/\hbar)^{1/2}$	Inverse characteristic length
$\psi_n(x) = \langle x \| n \rangle = \sqrt{\frac{\lambda}{2^n n! \sqrt{\pi}}} e^{-\lambda^2 x^2} H_n(\lambda x)$	Eigenfunctions in the x basis
$\hat{a} = \sqrt{m\omega_\mathrm{r}/2\hbar}\,(\hat{x} + i\hat{p}/m\omega_\mathrm{r})$	Annihilation operator
$\hat{a}^\dagger = \sqrt{m\omega_\mathrm{r}/2\hbar}\,(\hat{x} - i\hat{p}/m\omega_\mathrm{r})$	Creation operator
$\hat{x} = \sqrt{\hbar/2m\omega_\mathrm{r}}\,(\hat{a} + \hat{a}^\dagger)$	Position operator
$\hat{p} = -i\sqrt{\hbar m\omega_\mathrm{r}/2}\,(\hat{a} - \hat{a}^\dagger)$	Momentum operator
$\hat{H}_\mathrm{r} = \hbar\omega_\mathrm{r}(\hat{a}^\dagger \hat{a} + 1/2)$	Alternative form of $\hat{H}_\mathrm{r}$
$\hat{a}\|n\rangle = \sqrt{n}\|n-1\rangle$	Action of $\hat{a}$ onto $\|n\rangle$
$\hat{a}^\dagger\|n\rangle = \sqrt{n+1}\|n+1\rangle$	Action of $\hat{a}^\dagger$ onto $\|n\rangle$

Representations in quantum mechanics

$\|\psi(t)\rangle_\mathrm{S} = e^{\frac{1}{i\hbar}\hat{H}t}\|\psi_\mathrm{S}(0)\rangle$	$\hat{A}_\mathrm{S}(t) = \hat{A}_\mathrm{S}(0)$	Schrödinger picture
$\|\psi(t)\rangle_\mathrm{H} = \|\psi(0)\rangle_\mathrm{S}$	$\hat{A}_\mathrm{H}(t) = e^{-\frac{1}{i\hbar}\hat{H}t}\hat{A}_\mathrm{S}(0)e^{\frac{1}{i\hbar}\hat{H}t}$	Heisenberg picture
$\|\psi(t)\rangle_\mathrm{I} = e^{\frac{1}{i\hbar}\hat{H}_0 t}\|\psi(0)\rangle_\mathrm{S}$	$\hat{A}_\mathrm{I}(t) = e^{-\frac{1}{i\hbar}\hat{H}_0 t}\hat{A}_\mathrm{S}(0)e^{\frac{1}{i\hbar}\hat{H}_0 t}$	Interaction picture

Jaynes–Cummings Hamiltonian

$\hat{H}_{\rm JC} = -\hbar\omega_{\rm q}\hat{\sigma}_z/2 + \hbar\omega_{\rm r}\hat{a}^\dagger\hat{a} + \hbar g(\hat{\sigma}^+\hat{a} + \hat{\sigma}^- a^\dagger)$	The J–C Hamiltonian
$\hat{H}_{\rm JC} = -\hbar\omega_{\rm q}\hat{\sigma}_z/2 + \hbar\omega_{\rm r}\hat{a}^\dagger\hat{a} + \frac{\hbar\Delta}{2}(1 - \sqrt{1 + \frac{4g^2}{\Delta^2}\hat{N}_{\rm T}})\hat{\sigma}_z$	Diagonal form
$\lvert {\rm g}, n\rangle$, $\lvert {\rm e}, n\rangle$	Eigenstates when $g = 0$
$E_{{\rm g},n} = \hbar\omega_{\rm r}n$, $E_{{\rm e},n} = \hbar\omega_{\rm q} + \hbar\omega_{\rm r}n$	Associated eigenvalues
$\lvert\psi_{n+}\rangle = \cos\frac{\theta}{2}\lvert{\rm g}, n\rangle + \sin\frac{\theta}{2}\lvert{\rm e}, n-1\rangle$	Eigenstate when $g \neq 0$
$\lvert\psi_{n-}\rangle = \sin\frac{\theta}{2}\lvert{\rm g}, n\rangle - \cos\frac{\theta}{2}\lvert{\rm e}, n-1\rangle$	Eigenstate when $g \neq 0$
$E_{n,+} = \hbar\tilde{\omega}_n + \hbar\sqrt{g^2 n + \Delta^2/4}$,	Eigenvalue of ψ_{n+}
$E_{n,-} = \hbar\tilde{\omega}_n - \hbar\sqrt{g^2 n + \Delta^2/4}$,	Eigenvalue of ψ_{n-}
$\kappa, \gamma \ll \lvert g\rvert \ll \lvert\omega_{\rm q} - \omega_{\rm r}\rvert$	Dispersive regime
$\chi = g^2/\Delta$, where $\Delta = \omega_{\rm q} - \omega_{\rm r}$	Dispersive shift
$\hat{H}_{\rm JC} = \hbar(\omega_{\rm r} - \chi\hat{\sigma}_z)\hat{a}^\dagger\hat{a} - \frac{\hbar}{2}(\omega_{\rm q} + \chi)\hat{\sigma}_z$	J–C in the disp. regime
$\hat{H}_{\rm JC} = \hbar\omega_{\rm r}\hat{a}^\dagger\hat{a} - \frac{\hbar}{2}(\omega_{\rm q} + \chi + 2\chi\hat{a}^\dagger\hat{a})\hat{\sigma}_z$	Alternative form

Exercises

Exercise 4.1 Consider the decomposition of a coherent state in the basis of number states, $\lvert\alpha\rangle = \sum_n c_n\lvert n\rangle$. Calculate the coefficients c_n. **Hint:** Calculate the element $\langle n\lvert\hat{a}\rvert\alpha\rangle$.

Exercise 4.2 Consider the Hamiltonian of a qubit

$$\hat{H} = a_x\hat{\sigma}_x + a_y\hat{\sigma}_y + a_z\hat{\sigma}_z,$$

where $\vec{a} = (a_x, a_y, a_z)$ is a real vector. Show that the two eigenvalues are $\lambda_\pm = \sqrt{a_x^2 + a_y^2 + a_z^2}$. Show that the two eigenvectors point in the same and opposite direction of the vector $\vec{a}$.

Exercise 4.3 Consider two coupled quantum harmonic oscillators, $\hat{H} = \hbar\omega_1\hat{a}^\dagger\hat{a} + \hbar\omega_2\hat{b}^\dagger\hat{b} + \hbar g(\hat{a}\hat{b}^\dagger + \hat{a}^\dagger\hat{b})$ where ω_1 and ω_2 are the frequencies of the two oscillators and $g \in \mathbb{R}$ is the coupling strength. Using the unitary transformation $\hat{U} = e^{\alpha(\hat{a}\hat{b}^\dagger - \hat{a}^\dagger\hat{b})}$, find the expression of α so that $\hat{H}' = \hat{U}\hat{H}\hat{U}^\dagger$ is diagonal. Find the energy levels of $\hat{H}'$ and show that on resonance (i.e. when $\omega_1 = \omega_2$), the splitting between the energy levels does not depend on the number of excitations in the system. **Hint:** use the Baker–Campbell–Hausdorff formula.

Exercise 4.4 Show that the JC Hamiltonian (4.84) commutes with the operator $\hat{N} = \hat{a}^\dagger\hat{a} + \hat{\sigma}^+\hat{\sigma}^-$. Then, consider the unitary transformation $\hat{U} = e^{\hat{S}}$ where $\hat{S} = f_{\hat{N}}(\hat{\sigma}^-\hat{a}^\dagger - \hat{\sigma}^+\hat{a})$ (here, $f_{\hat{N}}$ is an analytical function of the operator $\hat{N}$). Find the expression of $f_{\hat{N}}$ so that the Hamiltonian $\hat{H}'_{\rm JC} = \hat{U}\hat{H}_{\rm JC}\hat{U}^\dagger$ is diagonal. **Hint:** Use the Baker–Campbell–Hausdorff formula.

Solutions

Solution 4.1 The decomposition of a coherent state in the number basis is given by

$$|\alpha\rangle = \sum_{n=0}^{\infty} |n\rangle \underbrace{\langle n|\alpha\rangle}_{c_n} = \sum_{n=0}^{\infty} c_n |n\rangle.$$

To calculate the coefficients of this expansion, we note that

$$\langle n|\hat{a}|\alpha\rangle = \alpha\langle n|\alpha\rangle = \alpha c_n. \tag{4.96}$$

Since $\hat{a}^\dagger|n\rangle = \sqrt{n+1}|n+1\rangle$, then $\langle n|\hat{a} = \sqrt{n+1}\langle n+1|$. Hence,

$$\langle n|\hat{a}|\alpha\rangle = \sqrt{n+1}\langle n+1|\alpha\rangle = \sqrt{n+1}c_{n+1}. \tag{4.97}$$

Comparing eqn (4.96) with (4.97), we see that $\sqrt{n+1}c_{n+1} = \alpha c_n$ or equivalently $\sqrt{n}c_n = \alpha c_{n-1}$. This implies that the coefficients c_n are given by

$$c_n = \frac{\alpha}{\sqrt{n}} c_{n-1} = \frac{\alpha}{\sqrt{n}} \frac{\alpha}{\sqrt{n-1}} c_{n-2} = \ldots = \frac{\alpha^n}{\sqrt{n!}} c_0.$$

The last step is to calculate the normalization constant c_0,

$$1 = \langle\alpha|\alpha\rangle = |c_0|^2 \sum_{n=0}^{\infty}\sum_{n'=0}^{\infty} \frac{(\alpha^*)^{n'}\alpha^n}{\sqrt{n'!}\sqrt{n!}}\langle n'|n\rangle = |c_0|^2 \sum_{n=0}^{\infty} \frac{\left(|\alpha|^2\right)^n}{n!} = |c_0|^2 e^{|\alpha|^2},$$

and this leads to $c_0 = e^{-|\alpha|^2/2}$. Putting all together, $c_n = e^{-\frac{|\alpha|^2}{2}} \frac{\alpha^n}{\sqrt{n!}}$.

Solution 4.2 In the Bloch sphere, the vector $\vec{a}$ forms an angle θ with the z axis: $\cos\theta = a_z/|\vec{a}|$ where $|\vec{a}| = \sqrt{a_x^2 + a_y^2 + a_z^2}$. The angle θ can be defined in different equivalent forms:

$$\cos\theta = \frac{a_z}{|\vec{a}|}, \qquad \sin\theta = \frac{\sqrt{a_x^2 + a_y^2}}{|\vec{a}|}, \qquad \tan\theta = \frac{\sqrt{a_x^2 + a_y^2}}{a_z}.$$

In a similar manner, the x and y components define an angle $\phi \equiv \arctan(a_y/a_x)$. The eigenvalues of the Hamiltonian, λ_+ and λ_-, are proportional to the length of $\vec{a}$,

$$\lambda_\pm = \pm\sqrt{a_x^2 + a_y^2 + a_z^2}.$$

With simple matrix multiplications, one can verify that the eigenvector associated with the eigenvalue λ_+ (λ_-) points in the same (opposite) direction of $\vec{a}$. Thus,

$$|v_+\rangle = \cos\frac{\theta}{2}|0\rangle + e^{i\phi}\sin\frac{\theta}{2}|1\rangle, \qquad |v_-\rangle = \sin\frac{\theta}{2}|0\rangle - e^{i\phi}\cos\frac{\theta}{2}|1\rangle.$$

These vectors can also be expressed in a more explicit form as

$$|v_+\rangle = \sqrt{\frac{1 + a_z/|\vec{a}|}{2}}|0\rangle + e^{i\phi}\sqrt{\frac{1 - a_z/|\vec{a}|}{2}}|1\rangle,$$
$$|v_-\rangle = \sqrt{\frac{1 - a_z/|\vec{a}|}{2}}|0\rangle - e^{i\phi}\sqrt{\frac{1 + a_z/|\vec{a}|}{2}}|1\rangle.$$

where we used $\theta = \arccos(a_z/|\vec{a}|)$, and the trigonometric relations $\cos(\frac{1}{2}\arccos x) = \sqrt{(1+x)/2}$ and $\sin(\frac{1}{2}\arccos x) = \sqrt{(1-x)/2}$.

Solution 4.3 We first need to derive the analytical expression of $\hat{H}' = \hat{U}\hat{H}\hat{U}^\dagger$. To this end, let us calculate the action of the unitary $\hat{U}$ on the operator $\hat{a}$. Using eqn (3.59) with $\hat{A} = \hat{a}\hat{b}^\dagger - \hat{a}^\dagger\hat{b}$ and $\hat{B} = \hat{a}$, we obtain

$$\begin{aligned}\hat{U}\hat{a}\hat{U}^\dagger = e^{\alpha(\hat{a}\hat{b}^\dagger - \hat{a}^\dagger\hat{b})}\hat{a}e^{-\alpha(\hat{a}\hat{b}^\dagger - \hat{a}^\dagger\hat{b})} &= \hat{a} + \alpha\hat{b} + \frac{\alpha^2}{2!}(-\hat{a}) + \frac{\alpha^3}{3!}(-\hat{b}) + \frac{\alpha^4}{4!}\hat{a} + O(\alpha^5)\\ &= \hat{a}\Big[1 - \frac{\alpha^2}{2!} + \frac{\alpha^4}{4!} + O(\alpha^6)\Big] + \hat{b}\Big[\alpha - \frac{\alpha^3}{3!} + O(\alpha^5)\Big]\\ &= \hat{a}\cos\alpha + \hat{b}\sin\alpha.\end{aligned}$$

With similar calculations, one can verify that

$$\hat{U}\hat{a}^\dagger\hat{U}^\dagger = \hat{a}^\dagger\cos\alpha + \hat{b}^\dagger\sin\alpha, \qquad \hat{U}\hat{b}\hat{U}^\dagger = \hat{b}\cos\alpha - \hat{a}\sin\alpha, \qquad \hat{U}\hat{b}^\dagger\hat{U}^\dagger = \hat{b}^\dagger\cos\alpha - \hat{a}^\dagger\sin\alpha.$$

Substituting these expressions into $\hat{H}' = \hat{U}\hat{H}\hat{U}^\dagger$, we obtain

$$\begin{aligned}\hat{H}'/\hbar &= \omega_1\hat{U}\hat{a}^\dagger\hat{U}^\dagger\hat{U}\hat{a}\hat{U}^\dagger + \omega_2\hat{U}\hat{b}^\dagger\hat{U}^\dagger\hat{U}\hat{b}\hat{U}^\dagger + g(\hat{U}\hat{a}\hat{U}^\dagger\hat{U}\hat{b}^\dagger\hat{U}^\dagger + \hat{U}\hat{a}^\dagger\hat{U}^\dagger\hat{U}\hat{b}\hat{U}^\dagger)\\ &= (\omega_1\cos^2\alpha + \omega_2\sin^2\alpha - g\sin 2\alpha)\hat{a}^\dagger\hat{a}\\ &\quad + (\omega_1\sin^2\alpha + \omega_2\cos^2\alpha + g\sin 2\alpha)\hat{b}^\dagger\hat{b} + (\tfrac{\Delta}{2}\sin 2\alpha + g\cos 2\alpha)(\hat{a}^\dagger\hat{b} + \hat{a}\hat{b}^\dagger),\end{aligned}$$

where we introduced the detuning $\Delta = \omega_1 - \omega_2$. The Hamiltonian $\hat{H}'$ is diagonal only if the last term $\frac{\Delta}{2}\sin 2\alpha + g\cos 2\alpha$ is zero. This means that the parameter α must be $\alpha = -\frac{1}{2}\arctan(2g/\Delta)$. Using this relation, $\hat{H}'$ becomes

$$\begin{aligned}\hat{H}'/\hbar &= (\omega_1\cos^2\alpha + \omega_2\sin^2\alpha - g\sin 2\alpha)\hat{a}^\dagger\hat{a} + (\omega_1\sin^2\alpha + \omega_2\cos^2\alpha + g\sin 2\alpha)\hat{b}^\dagger\hat{b}\\ &= \Big(\frac{\omega_1+\omega_2}{2} + \frac{\Delta}{2}\cos 2\alpha - g\sin 2\alpha\Big)\hat{a}^\dagger\hat{a} + \Big(\frac{\omega_1+\omega_2}{2} - \frac{\Delta}{2}\cos 2\alpha + g\sin 2\alpha\Big)\hat{b}^\dagger\hat{b}.\end{aligned}$$

Recalling that $\cos(-\arctan(x)) = 1/\sqrt{1+x^2}$ and $\sin(-\arctan(x)) = x/\sqrt{1+x^2}$, we have

$$\begin{aligned}\hat{H}' &= \hbar\underbrace{\Big(\frac{\omega_1+\omega_2}{2} + \frac{1}{2}\sqrt{\Delta^2+4g^2}\Big)}_{\tilde{\omega}_1}\hat{a}^\dagger\hat{a} + \hbar\underbrace{\Big(\frac{\omega_1+\omega_2}{2} - \frac{1}{2}\sqrt{\Delta^2+4g^2}\Big)}_{\tilde{\omega}_2}\hat{b}^\dagger\hat{b}\\ &= \hbar\tilde{\omega}_1\hat{a}^\dagger\hat{a} + \hbar\tilde{\omega}_2\hat{b}^\dagger\hat{b}.\end{aligned}$$

This is the Hamiltonian of two **uncoupled** harmonic oscillators with frequencies

$$\tilde{\omega}_1 = \frac{\omega_1+\omega_2}{2} + \frac{1}{2}\sqrt{\Delta^2+4g^2}, \qquad \tilde{\omega}_2 = \frac{\omega_1+\omega_2}{2} - \frac{1}{2}\sqrt{\Delta^2+4g^2}.$$

The energy levels are given by: $E_{n_1,n_2} = \hbar\tilde{\omega}_1 n_1 + \hbar\tilde{\omega}_2 n_2$. When the two qubits are in resonance ($\Delta = 0$), the splitting between degenerate energy levels is given by $\tilde{\omega}_1 - \tilde{\omega}_2 = 2g$. This splitting does not depend on the number of excitations in the system. This behavior differs from a harmonic oscillator coupled to a qubit in which the separation between adjacent energy levels increases as $\sqrt{n}$.

Solution 4.4 One can verify that $[\hat{N}, \hat{H}_{\rm JC}] = 0$ with an explicit calculation. Since $f_{\hat{N}}$ is an analytical function of $\hat{N}$, this implies that $[f_{\hat{N}}, \hat{H}_{\rm JC}] = 0$. From the Baker–Campbell–Hausdorff formula, we have

$$\hat{H}'_{\rm JC} = e^{\hat{S}}\hat{H}_{\rm JC}e^{-\hat{S}} = \hat{H}_{\rm JC} + [\hat{S}, \hat{H}_{\rm JC}] + \frac{1}{2!}[\hat{S}, [\hat{S}, \hat{H}_{\rm JC}]] + \ldots = \sum_{n=0}^{\infty} \frac{1}{n!}[\hat{S}, \hat{H}_{\rm JC}]_n, \tag{4.98}$$

where in the last step we defined the operation

$$[\hat{S}, \hat{H}_{\rm JC}]_0 = \hat{H}_{\rm JC}, \qquad [\hat{S}, \hat{H}_{\rm JC}]_1 = [\hat{S}, \hat{H}_{\rm JC}], \qquad [\hat{S}, \hat{H}_{\rm JC}]_2 = [\hat{S}, [\hat{S}, \hat{H}_{\rm JC}]], \quad \ldots$$

It is convenient to introduce the operators $\hat{I}_+ = \hat{\sigma}^-\hat{a}^\dagger + \hat{\sigma}^+\hat{a}$ and $\hat{I}_- = \hat{\sigma}^-\hat{a}^\dagger - \hat{\sigma}^+\hat{a}$. With these definitions, $\hat{S} = f_{\hat{N}}\hat{I}_-$ and the JC Hamiltonian can be expressed as

$$\hat{H}_{\rm JC} = \underbrace{\hbar\omega_{\rm r}\hat{a}^\dagger\hat{a} - \frac{\hbar\omega_{\rm q}}{2}\hat{\sigma}_z}_{\hat{H}_0} + \hbar g(\hat{\sigma}^-\hat{a}^\dagger + \hat{\sigma}^+\hat{a}^\dagger) = \hat{H}_0 + \hbar g\hat{I}_+.$$

The operator $\hat{I}_-$ satisfies the relation $[\hat{I}_-, \hat{H}_0] = \hbar\Delta\hat{I}_+$, where we defined $\Delta = \omega_{\rm q} - \omega_{\rm r}$. Thus, eqn (4.98) becomes

$$\begin{aligned}
\hat{H}' &= \sum_{n=0}^{\infty} \frac{1}{n!}[f_{\hat{N}}\hat{I}_-, \hat{H}_0]_n + \sum_{n=0}^{\infty} \frac{1}{n!}[f_{\hat{N}}\hat{I}_-, \hbar g\hat{I}_+]_n \\
&= \sum_{n=0}^{\infty} \frac{1}{n!} f_{\hat{N}}^n[\hat{I}_-, \hat{H}_0]_n + \sum_{n=0}^{\infty} \frac{\hbar g}{n!} f_{\hat{N}}^n[\hat{I}_-, \hat{I}_+]_n \\
&= \hat{H}_0 + \sum_{n=1}^{\infty} \frac{\hbar\Delta}{n!} f_{\hat{N}}^n[\hat{I}_-, \hat{I}_+]_{n-1} + \sum_{n=0}^{\infty} \frac{(n+1)\hbar g}{(n+1)!} f_{\hat{N}}^n[\hat{I}_-, \hat{I}_+]_n,
\end{aligned}$$

where we used the fact that $f_{\hat{N}}$ commutes with $\hat{H}_{\rm JC}$ and $[\hat{I}_-, \hat{H}_0] = \hbar\Delta\hat{I}_+$. Note that the first sum starts from $n = 1$. With a change of variable, we have

$$\begin{aligned}
\hat{H}' &= \hat{H}_0 + \sum_{k=0}^{\infty} \frac{\hbar\Delta}{(k+1)!} f_{\hat{N}}^{k+1}[\hat{I}_-, \hat{I}_+]_k + \sum_{k=0}^{\infty} \frac{(k+1)\hbar g}{(k+1)!} f_{\hat{N}}^k[\hat{I}_-, \hat{I}_+]_k \\
&= \hat{H}_0 + \sum_{k=0}^{\infty} \frac{\hbar\Delta f_{\hat{N}}^{k+1} + (k+1)\hbar g f_{\hat{N}}^k}{(k+1)!}[\hat{I}_-, \hat{I}_+]_k.
\end{aligned} \tag{4.99}$$

Let us calculate the term $[\hat{I}_-, \hat{I}_+]_k$. One can verify that:

$$[\hat{I}_-, \hat{I}_+]_0 = \hat{I}_+, \qquad [\hat{I}_-, \hat{I}_+]_3 = (-4\hat{N})2\hat{N}\hat{\sigma}_z,$$
$$[\hat{I}_-, \hat{I}_+]_1 = 2\hat{N}\hat{\sigma}_z, \qquad [\hat{I}_-, \hat{I}_+]_4 = (-4\hat{N})^2\hat{I}_+,$$
$$[\hat{I}_-, \hat{I}_+]_2 = -4\hat{N}\hat{I}_+, \qquad [\hat{I}_-, \hat{I}_+]_5 = (-4\hat{N})^2 2\hat{N}\sigma_z.$$

A pattern emerges from these nested commutators:

$$[\hat{I}_-, \hat{I}_+]_{2k} = (-4\hat{N})^k \hat{I}_+,$$
$$[\hat{I}_-, \hat{I}_+]_{2k+1} = (-4\hat{N})^k \, 2\hat{N}\hat{\sigma}_z.$$

Using these relations, the sum in eqn (4.99) can be divided into two parts

$$\hat{H}' = \hat{H}_0 + 2\hat{N}\hat{\sigma}_z \sum_{k=0}^{\infty} \frac{\hbar\Delta f_{\hat{N}}^{2k+2} + (2k+2)\hbar g f_{\hat{N}}^{2k+1}}{(2k+2)!}(-4\hat{N})^k$$
$$+ \hat{I}_+ \sum_{k=0}^{\infty} \frac{\hbar\Delta f_{\hat{N}}^{2k+1} + (2k+1)\hbar g f_{\hat{N}}^{2k}}{(2k+1)!}(-4\hat{N})^k. \tag{4.100}$$

The last part of this proof consists in calculating the two sums in eqn (4.100),

$$\hat{H}' = \hat{H}_0 + 2\hat{N}\hat{\sigma}_z \Big[\hbar\Delta \frac{\sin^2(\sqrt{N} f_{\hat{N}})}{2\hat{N}} + \hbar g \frac{\sin(2\sqrt{N} f_{\hat{N}})}{2\sqrt{\hat{N}}}\Big]$$
$$+ \hat{I}_+ \Big[\hbar\Delta \frac{\sin(2\sqrt{N} f_{\hat{N}})}{2\sqrt{\hat{N}}} + \hbar g \cos(2\sqrt{N} f_{\hat{N}})\Big]. \tag{4.101}$$

The Hamiltonian $\hat{H}'$ is diagonal when the last term in eqn (4.101) is zero. This happens when $f_{\hat{N}} = \frac{\arctan(-\frac{g}{\Delta}2\sqrt{\hat{N}})}{2\sqrt{\hat{N}}}$. The first two terms of $\hat{H}'$ lead to eqn (4.93) of the main text.

5

Quantum circuits

Single-qubit rotations and CNOT *gates are universal for quantum computation.*

5.1 Qubits

If a bit is the elementary unit of a classical computation, a quantum bit is the elementary component of a quantum computation. As seen in Chapter 1, a classical bit can be in one of two possible states. In contrast, a **quantum bit** (or qubit) is a well-isolated quantum system whose Hilbert space contains a subspace isomorphic to $\mathbb{C}^2$. Examples of qubits are: a trapped ion, the spin of an electron, two positions of a crystalline defect, the polarization of a photon, and a superconducting circuit. An orthonormal basis of the Hilbert space $\mathbb{C}^2$ is formed by the vectors[1]

$$|0\rangle = \begin{bmatrix} 1 \\ 0 \end{bmatrix}, \qquad |1\rangle = \begin{bmatrix} 0 \\ 1 \end{bmatrix}. \tag{5.1}$$

The states $|0\rangle$ and $|1\rangle$ correspond to the classical states of a bit. A generic **qubit state** can be written as a linear combination of the basis vectors,

$$\boxed{|\psi\rangle = c_0|0\rangle + c_1|1\rangle,} \tag{5.2}$$

where c_i are complex coefficients satisfying $|c_0|^2 + |c_1|^2 = 1$. The quantities $|c_0|^2$ and $|c_1|^2$ indicate the probability of measuring the qubit in $|0\rangle$ and $|1\rangle$, respectively.

Equation (5.2) shows that a qubit state is specified by two complex numbers, i.e. by four real parameters. These parameters are not independent due to two constraints: (1) the normalization condition imposes that $|c_0|^2 + |c_1|^2 = 1$, and (2) the global phase η has no physical significance[2], $e^{i\eta}|\psi\rangle \equiv |\psi\rangle$. These two conditions reduce the number of independent real parameters from four to two. Thus, a generic qubit state is often parametrized as

[1]In Chapter 3, the basis vectors of $\mathbb{C}^2$ were indicated as $|\mathbf{e}_0\rangle$ and $|\mathbf{e}_1\rangle$. In this chapter, we will adopt the notation $|0\rangle$ and $|1\rangle$ for convenience.

[2]Recall from the first postulate of quantum mechanics that a pure state is described by a ray in a Hilbert space.

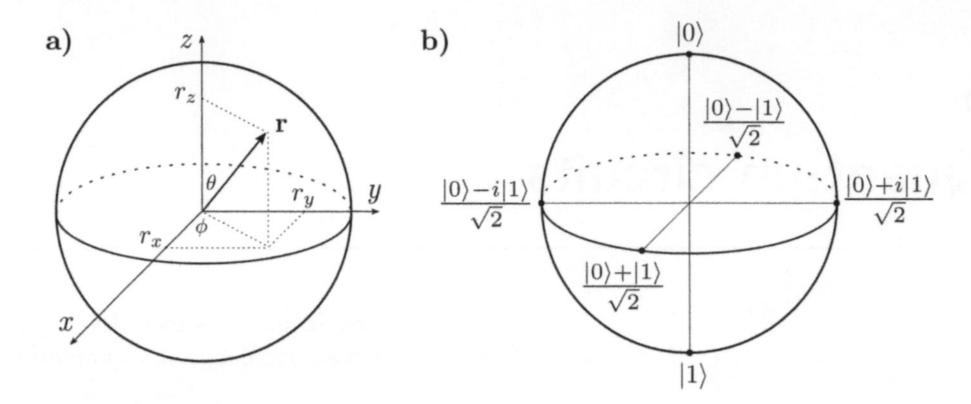

Fig. 5.1 Qubit states. a) A generic qubit state $|\theta, \phi\rangle$ represented by the Bloch vector $\mathbf{r} = (r_x, r_y, r_z)$. The angles θ and ϕ are the polar and azimuth angles, respectively. **b)** The north and south pole of the Bloch sphere correspond to the classical states $|0\rangle$ and $|1\rangle$, respectively. The black dots on the equator indicate four important balanced superposition states.

$$\boxed{|\psi\rangle = \cos\frac{\theta}{2}|0\rangle + e^{i\phi}\sin\frac{\theta}{2}|1\rangle,} \tag{5.3}$$

where $\theta \in [0, \pi]$ and $\phi \in [0, 2\pi)$. The angle ϕ is usually called the **relative phase** of the qubit. For $\theta = 0$ and $\theta = \pi$, we recover the classical states $|0\rangle$ and $|1\rangle$. When $\theta = \pi/2$, the qubit is in a **balanced superposition state**. Unlike a classical bit that can be in only two possible states, a quantum bit is described by two continuous variables and an *infinite* number of states is allowed. For convenience, we will sometimes use the notation $|\theta, \phi\rangle$ to denote a qubit state.

Since two angles completely specify a pure qubit state (see eqn (5.3)), it is possible to establish a one-to-one correspondence between qubit states and points on a unit sphere, usually called the **Bloch sphere** (see Fig. 5.1a). In a spherical coordinate system, θ corresponds to the polar angle and ϕ corresponds to the azimuth angle. The states $|0\rangle$ and $|1\rangle$ are located at the north and south pole of the sphere, respectively. The states on the equator are balanced superposition states. Two important examples of superposition states are

$$|+\rangle = \frac{|0\rangle + |1\rangle}{\sqrt{2}}, \qquad |-\rangle = \frac{|0\rangle - |1\rangle}{\sqrt{2}}, \tag{5.4}$$

which are associated with the spherical coordinates $\{\theta = \frac{\pi}{2}, \phi = 0\}$ and $\{\theta = \frac{\pi}{2}, \phi = \pi\}$, respectively. These vectors lie at the intersection between the equator and the x axis, as shown in Fig. 5.1b. One can verify that these states are eigenvectors of $\hat{\sigma}_x$.

The points on the Bloch sphere can be described by a three-dimensional vector, known as the **Bloch vector r**, with components

$$r_x = \sin\theta\cos\phi, \qquad r_y = \sin\theta\sin\phi, \qquad r_z = \cos\theta\,. \tag{5.5}$$

The relation between the Bloch vector and the coefficients c_0 and c_1 of eqn (5.3) is

$$\sqrt{\frac{1+r_z}{2}} = \cos\left(\frac{\theta}{2}\right) = c_0, \tag{5.6}$$

$$\frac{r_x + ir_y}{\sqrt{2(1+r_z)}} = e^{i\phi}\sin\left(\frac{\theta}{2}\right) = c_1. \tag{5.7}$$

We emphasize that the Bloch vector is only a graphical representation of a qubit state.

In principle, the angle θ could have an infinite decimal expansion. From this observation, it is tempting to conclude that an infinite amount of information could be stored in this angle. There is an important caveat though: to extract the information stored in θ, one needs to measure the qubit state multiple times. As explained in Exercise 5.1, θ can be estimated with a precision of 10^{-k} with a number of measurements that scales as $O(2^k)$. This shows that extracting classical information from a *single* qubit is impractical. For more information about encoding classical information into a quantum computer, see for example Refs [153, 154, 155, 156].

5.2 Quantum circuits and Di Vincenzo criteria

In classical computation, a decision problem can be solved using logic circuits that involve single-bit and two-bit operations (see Section 1.1.7). These circuits can transform binary strings into any other binary string. A quantum computation follows a similar procedure that can be divided into three main parts: 1) the initialization of a set of qubits, 2) their unitary evolution, and 3) a final projective measurement. We will now provide a detailed explanation of each of these parts.

1) In the first part of a quantum computation, a set of n qubits (called the **register**) is initialized in a reliable state, typically $|0\ldots0\rangle$. The state of the register is not described by a n-bit string, but rather by a quantum state in the tensor product space $\mathbb{C}^{\otimes 2n} = \mathbb{C}^2 \otimes \ldots \otimes \mathbb{C}^2$. The vector space of the register has dimension $d = 2^n$ and a convenient orthonormal basis is

$$B = \{|0\ldots0\rangle, |0\ldots1\rangle, \ldots, |1\ldots1\rangle\} = \{|\mathbf{x}\rangle\}_{\mathbf{x}\in\mathbb{B}^n}\ . \tag{5.8}$$

The decomposition of a n-qubit state in this basis is given by[3]

$$\boxed{|\psi\rangle = \sum_{\mathbf{x}\in\mathbb{B}^n} c_{\mathbf{x}}|\mathbf{x}\rangle = c_{0\ldots0}|0\ldots0\rangle + c_{0\ldots1}|0\ldots1\rangle + \ldots + c_{1\ldots1}|1\ldots1\rangle.} \tag{5.9}$$

This expression shows that an n-qubit state is completely specified by 2^n complex coefficients. Recall that a quantum state must satisfy the normalization condition $\langle\psi|\psi\rangle = 1$ and its global phase can take any arbitrary value. These two constraints reduce the number of free complex parameters to $2^n - 1$.

[3]The decomposition $\sum_{\mathbf{x}\in\mathbb{B}^n} c_{\mathbf{x}}|\mathbf{x}\rangle$ is sometimes expressed as $\sum_{i=0}^{2^n-1} c_i|i\rangle$ where it is intended that the integers in this sum should be expressed in binary representation.

It is important to mention that when we write

$$|\psi\rangle = |1\rangle_3|1\rangle_2|0\rangle_1|0\rangle_0 = |1100\rangle \tag{5.10}$$

we mean that qubit numbers 0 and 1 are in state $|0\rangle$, and qubit numbers 2 and 3 are in state $|1\rangle$. The subscript will often be omitted, but the reader should always remember that the qubit indexing starts from the right. We adopt this convention to maintain a simple correspondence with the binary representation. In quantum circuits, such as that shown in Fig. 5.2, qubit 0 is reported at the *bottom* of the circuit.

For example, consider a system of two qubits with Hilbert space $\mathbb{C}^2 \otimes \mathbb{C}^2$. The basis vectors $|00\rangle$, $|01\rangle$, $|10\rangle$, $|11\rangle$ can be conveniently expressed by column vectors

$$|0\rangle \equiv |00\rangle = \begin{bmatrix} 1 \\ 0 \\ 0 \\ 0 \end{bmatrix}, \qquad |1\rangle \equiv |01\rangle = \begin{bmatrix} 0 \\ 1 \\ 0 \\ 0 \end{bmatrix},$$

$$|2\rangle \equiv |10\rangle = \begin{bmatrix} 0 \\ 0 \\ 1 \\ 0 \end{bmatrix}, \qquad |3\rangle \equiv |11\rangle = \begin{bmatrix} 0 \\ 0 \\ 0 \\ 1 \end{bmatrix}.$$

A generic two-qubit state is a linear combination of the basis vectors,

$$|\psi\rangle = \sum_{\mathbf{x}\in\mathbb{B}^2} c_{\mathbf{x}}|\mathbf{x}\rangle = c_{00}|00\rangle + c_{01}|01\rangle + c_{10}|10\rangle + c_{11}|11\rangle. \tag{5.11}$$

This state can also be written as

$$|\psi\rangle = \sum_{i=0}^{2^2-1} c_i|i\rangle = c_0|0\rangle + c_1|1\rangle + c_2|2\rangle + c_3|3\rangle. \tag{5.12}$$

In this equation, we are using the correspondence $|00\rangle = |0\rangle$, $|01\rangle = |1\rangle$, $|10\rangle = |2\rangle$ and $|11\rangle = |3\rangle$. Equations (5.11) and (5.12) are two equivalent ways to express the same state. Similarly, a three-qubit state can be decomposed into a basis of 2^3 orthogonal vectors,

$$|000\rangle = \begin{bmatrix} 1 \\ 0 \\ 0 \\ 0 \\ 0 \\ 0 \\ 0 \\ 0 \end{bmatrix}, \qquad |001\rangle = \begin{bmatrix} 0 \\ 1 \\ 0 \\ 0 \\ 0 \\ 0 \\ 0 \\ 0 \end{bmatrix}, \qquad \ldots, \qquad |111\rangle = \begin{bmatrix} 0 \\ 0 \\ 0 \\ 0 \\ 0 \\ 0 \\ 0 \\ 1 \end{bmatrix}.$$

The length of these vectors increases exponentially with the number of qubits in the register.

2) In the second part of a quantum computation, the state of the register is transformed by a set of **unitary gates**. These gates operate on one or more qubits at a time. In a quantum computer based on superconducting circuits, single and two-qubit gates take approximately $T_{\text{gate}} = 10 - 300$ ns; this timescale has to be compared with the typical coherence time of a superconducting qubit, which lies in the range $T_{\text{coh}} = 10 - 500\,\mu$s. This suggests that the maximum number of gates that can be executed on state-of-the-art superconducting devices is approximately $T_{\text{coh}}/T_{\text{gate}} \approx 1000$ if quantum error correcting protocols are not implemented. Increasing this ratio is one of the main challenges several technological centers are tackling in this decade. This can be done by increasing the coherence time of the qubit and/or reducing the gate time.

Since quantum states differing by a global phase describe the same quantum system, unitary gates differing by an overall phase are equivalent,

$$e^{i\eta}\hat{U} \equiv \hat{U}\,. \tag{5.13}$$

Mathematically, this relation is justified by the fact that $e^{i\eta}\hat{U}|\psi\rangle = \hat{U}e^{i\eta}|\psi\rangle \equiv \hat{U}|\psi\rangle$. This observation will be important when we introduce some two-qubit gates in Section 5.4.

3) The last part of a quantum computation is the **measurement** of the qubits in the computational basis. Each qubit is found either in 0 or 1 and the **output of a quantum computation** is an n-bit string. If before the measurement the register is in a superposition state

$$|\psi\rangle = \sum_{\mathbf{x}\in\mathbb{B}^n} c_{\mathbf{x}}|\mathbf{x}\rangle,$$

the measurement projects $|\psi\rangle$ into a specific state $|\mathbf{y}\rangle$ with probability $|c_{\mathbf{y}}|^2$. If the bit string $\mathbf{y}$ is not the solution to the problem, the computation must be run again. Quantum algorithms must be designed such that at the end of the computation the correct string is measured with the highest probability. It is worth mentioning that it is sometimes necessary to measure the state of the qubits in a basis other than the computational basis. This is done by applying a unitary transformation to the register and then proceeding with a standard measurement. We will show how to do this in Section 5.3.3.

The circuit representing a quantum computation is called a **quantum circuit** (see Fig. 5.2 for an example). In this circuit, each horizontal line corresponds to a qubit, gates are represented by rectangular boxes, projective measurements in the computational basis are indicated with a meter and time flows from left to right.

In the process of building a quantum computer, five requirements must be satisfied:

1. The physical machine must comprise a set of qubits.
2. It must be possible to initialize the qubits in a reliable state, such as $|0\ldots 0\rangle$.
3. The coherence time of the qubits must be longer than the typical gate time.
4. It must be possible to measure the qubits in the computational basis.
5. It must be possible to perform a universal set of quantum gates.

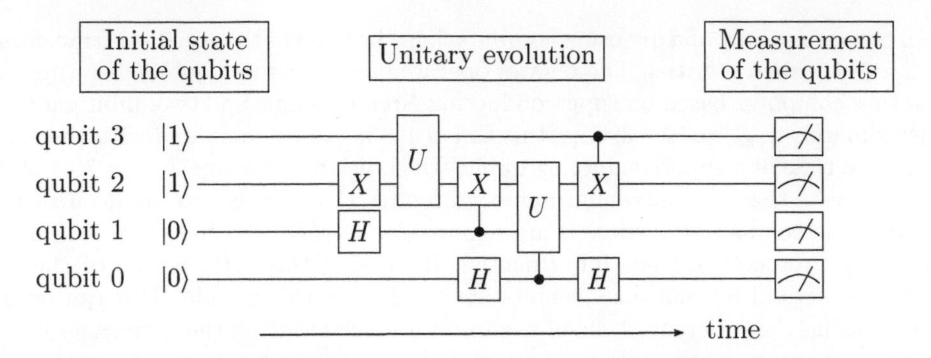

Fig. 5.2 A quantum computation. The quantum register is initialized in $|1100\rangle$. Some unitary transformations are applied to the qubits followed by projective measurements in the computational basis. The vertical lines connecting rectangular boxes to black dots indicate controlled operations.

These five requirements are known as the **Di Vincenzo criteria** [157]. The next sections focus on discussing these criteria in more detail.

5.3 Single-qubit rotations and measurements

Quantum mechanics prescribes that the time evolution of a closed quantum system must be unitary. This means that the only operations that can be applied to a set of qubits (apart from measurement operations) are unitary transformations. These operations are reversible. Thus, according to Landauer's principle presented in Section 1.1.8, an ideal quantum computation does not dissipate any energy.

An important class of unitary transformations on a qubit state are **single-qubit rotations** about an arbitrary axis of the Bloch sphere. For instance, the operations X_α, Y_α, Z_α perform a rotation by an angle α about the x, y, and z axis, respectively. Figure 5.3a shows a graphical representation of these three transformations, whose definitions are

$$\mathsf{X}_\alpha = e^{-i\frac{\alpha}{2}\hat{\sigma}_x}, \qquad \mathsf{Y}_\alpha = e^{-i\frac{\alpha}{2}\hat{\sigma}_y}, \qquad \mathsf{Z}_\alpha = e^{-i\frac{\alpha}{2}\hat{\sigma}_z}. \tag{5.14}$$

As seen in Section 3.2.9, Pauli operators satisfy the relation $\hat{\sigma}_i^2 = \mathbb{1}$ and for this reason we can write[4]

$$\mathsf{X}_\alpha = e^{-i\frac{\alpha}{2}\hat{\sigma}_x} = \cos\left(\frac{\alpha}{2}\right)\mathbb{1} - i\sin\left(\frac{\alpha}{2}\right)\hat{\sigma}_x = \begin{bmatrix} \cos\frac{\alpha}{2} & -i\sin\frac{\alpha}{2} \\ -i\sin\frac{\alpha}{2} & \cos\frac{\alpha}{2} \end{bmatrix}, \tag{5.15}$$

$$\mathsf{Y}_\alpha = e^{-i\frac{\alpha}{2}\hat{\sigma}_y} = \cos\left(\frac{\alpha}{2}\right)\mathbb{1} - i\sin\left(\frac{\alpha}{2}\right)\hat{\sigma}_y = \begin{bmatrix} \cos\frac{\alpha}{2} & -\sin\frac{\alpha}{2} \\ \sin\frac{\alpha}{2} & \cos\frac{\alpha}{2} \end{bmatrix}, \tag{5.16}$$

$$\mathsf{Z}_\alpha = e^{-i\frac{\alpha}{2}\hat{\sigma}_z} = \cos\left(\frac{\alpha}{2}\right)\mathbb{1} - i\sin\left(\frac{\alpha}{2}\right)\hat{\sigma}_z = \begin{bmatrix} e^{-i\frac{\alpha}{2}} & 0 \\ 0 & e^{i\frac{\alpha}{2}} \end{bmatrix}. \tag{5.17}$$

[4]See Section 3.2.9 for more details on how to derive these identities.

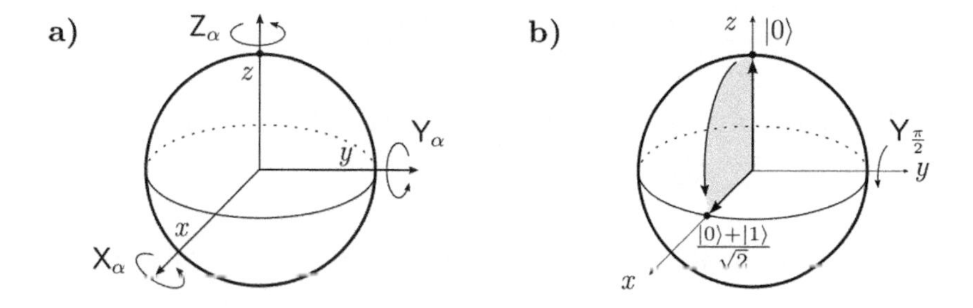

Fig. 5.3 Single-qubit rotations. a) A graphical representation of the single-qubit rotations X_α, Y_α and Z_α. These transformations rotate the Bloch vector by an angle α counterclockwise with respect to a specific axis. **b)** A $\mathsf{Y}_{\frac{\pi}{2}}$ rotation of the state $|0\rangle$ produces the superposition state $[|0\rangle + |1\rangle]/\sqrt{2}$.

Simple relations link π rotations to Pauli operators: $\mathsf{X}_\pi = -i\hat{\sigma}_x$, $\mathsf{Y}_\pi = -i\hat{\sigma}_y$, $\mathsf{Z}_\pi = -i\hat{\sigma}_z$.

A **rotation about an arbitrary axis** $\hat{n}$ of the Bloch sphere by an angle α can be expressed as

$$\mathsf{R}_{\hat{n}}(\alpha) = e^{-i\frac{\alpha}{2}\hat{n}\cdot\vec{\sigma}},$$

where $\hat{n} = (n_x, n_y, n_z)$ is a unit vector and $\vec{\sigma} = (\sigma_x, \sigma_y, \sigma_z)$ is the Pauli vector, a vector composed of three matrices. The quantity $(\hat{n}\cdot\vec{\sigma})^2$ is equal to the identity

$$(\hat{n}\cdot\vec{\sigma})^2 = (n_x\hat{\sigma}_x + n_y\hat{\sigma}_y + n_z\hat{\sigma}_z)(n_x\hat{\sigma}_x + n_y\hat{\sigma}_y + n_z\hat{\sigma}_z) = \mathbb{1},$$

and for this reason the operator $\mathsf{R}_{\hat{n}}(\alpha)$ can be written as[5]

$$\begin{aligned}\mathsf{R}_{\hat{n}}(\alpha) = \sum_{k=0}^{\infty}\frac{(-i\frac{\alpha}{2}\hat{n}\cdot\vec{\sigma})^k}{k!} &= \cos\left(\frac{\alpha}{2}\right)\mathbb{1} - i\sin\left(\frac{\alpha}{2}\right)(n_x\hat{\sigma}_x + n_y\hat{\sigma}_y + n_z\hat{\sigma}_z)\\ &= \begin{bmatrix}\cos\frac{\alpha}{2} - in_z\sin\frac{\alpha}{2} & -i(n_x - in_y)\sin\frac{\alpha}{2}\\ (-in_x + n_y)\sin\frac{\alpha}{2} & \cos\frac{\alpha}{2} + in_z\sin\frac{\alpha}{2}\end{bmatrix}.\end{aligned} \tag{5.18}$$

Let us study some important examples of qubit rotations.

Example 5.1 Suppose that the single-qubit rotation Y_α is applied to a qubit state $|\psi\rangle = (\cos\theta/2, e^{i\phi}\sin\theta/2)$ introduced in eqn (5.3). This transformation yields

$$\mathsf{Y}_\alpha|\psi\rangle = \begin{bmatrix}\cos\frac{\alpha}{2} & -\sin\frac{\alpha}{2}\\ \sin\frac{\alpha}{2} & \cos\frac{\alpha}{2}\end{bmatrix}\begin{bmatrix}\cos\frac{\theta}{2}\\ e^{i\phi}\sin\frac{\theta}{2}\end{bmatrix} = \begin{bmatrix}\cos\frac{\alpha}{2}\cos\frac{\theta}{2} - e^{i\phi}\sin\frac{\alpha}{2}\sin\frac{\theta}{2}\\ \sin\frac{\alpha}{2}\cos\frac{\theta}{2} + e^{i\phi}\sin\frac{\theta}{2}\cos\frac{\alpha}{2}\end{bmatrix}. \tag{5.19}$$

If the initial state is $|0\rangle$ (i.e. $\theta = \phi = 0$) and the rotation angle is $\alpha = \pi/2$, this transformation creates a balanced superposition state

[5] The reader can derive this identity by performing a Taylor expansion as in eqn (3.58).

$$\mathsf{Y}_{\frac{\pi}{2}}|0\rangle = \begin{bmatrix} \cos\frac{\pi}{4} \\ \sin\frac{\pi}{4} \end{bmatrix} = \begin{bmatrix} 1/\sqrt{2} \\ 1/\sqrt{2} \end{bmatrix} = \frac{1}{\sqrt{2}}\left[|0\rangle + |1\rangle\right]. \tag{5.20}$$

Figure 5.3b illustrates a $\mathsf{Y}_{\frac{\pi}{2}}$ rotation on the Bloch sphere. With similar calculations, one can verify that $\mathsf{X}_{\frac{\pi}{2}}|0\rangle = \frac{1}{\sqrt{2}}\left[|0\rangle - i|1\rangle\right]$ and $\mathsf{Z}_{\frac{\pi}{2}}|0\rangle = |0\rangle$.

Example 5.2 Suppose we want to transform a qubit in state $|0\rangle$ into a generic qubit state

$$|\theta, \phi\rangle = \cos\frac{\theta}{2}|0\rangle + e^{i\phi}\sin\frac{\theta}{2}|1\rangle. \tag{5.21}$$

This can be done by applying the unitary transformation $\mathsf{Z}_\phi \mathsf{Y}_\theta$ onto $|0\rangle$. Indeed, with some matrix multiplications, we obtain

$$\mathsf{Z}_\phi \mathsf{Y}_\theta|0\rangle = e^{-i\phi/2}\left[\cos\frac{\theta}{2}|0\rangle + e^{i\phi}\sin\frac{\theta}{2}|1\rangle\right] = e^{-i\phi/2}|\theta, \phi\rangle \equiv |\theta, \phi\rangle, \tag{5.22}$$

where in the last step we dropped a global phase of no physical significance.

Example 5.3 Another interesting example of a single-qubit rotation is provided by the application of X_α to the qubit state $|\theta, -\frac{\pi}{2}\rangle$,

$$\mathsf{X}_\alpha|\theta, -\pi/2\rangle = |\theta + \alpha, -\pi/2\rangle\ . \tag{5.23}$$

Try to visualize this rotation on the Bloch sphere illustrated in Fig. 5.3. Equation (5.23) will be used in Section 5.5 to show that the Hadamard and phase gates are sufficient to construct any unitary transformation in $\mathrm{U}(\mathbb{C}^2)$.

5.3.1 Decomposition of single-qubit unitary operations

In this section, we show that any unitary operator acting on a qubit state can be decomposed into a sequence of single-qubit rotations (up to a global phase). This theorem will be the starting point to prove the universality of single-qubit and CNOT gates later on.

Theorem 5.1 (Euler's decomposition) Any unitary $\hat{U}$ acting on a qubit can be decomposed into a combination of single-qubit rotations plus a global phase,

$$\boxed{\hat{U} = e^{i\delta}\mathsf{Z}_\alpha \mathsf{Y}_\theta \mathsf{Z}_\beta.} \tag{5.24}$$

In addition, there exist $\hat{A}$, $\hat{B}$, $\hat{C}$ such that $\hat{A}\hat{B}\hat{C} = \mathbb{1}$ and

$$\hat{U} = e^{i\delta}\hat{A}\hat{\sigma}_x\hat{B}\hat{\sigma}_x\hat{C}. \tag{5.25}$$

Proof To show that $\hat{U} = e^{i\delta}\mathsf{Z}_\alpha\mathsf{Y}_\theta\mathsf{Z}_\beta$, perform these matrix multiplications one by one and obtain eqn (3.71). Now consider the operators

$$\hat{A} = \mathsf{Z}_\alpha\mathsf{Y}_{\frac{\theta}{2}}, \qquad \hat{B} = \mathsf{Y}_{-\frac{\theta}{2}}\mathsf{Z}_{\frac{-\beta-\alpha}{2}}, \qquad \hat{C} = \mathsf{Z}_{\frac{\beta-\alpha}{2}}. \tag{5.26}$$

By multiplying these matrices together, it is straightforward to show that $\hat{A}\hat{B}\hat{C} = \mathbb{1}$. Furthermore, since $\hat{\sigma}_x^2 = \mathbb{1}$, the transformation $e^{i\delta}\hat{A}\hat{\sigma}_x\hat{B}\hat{\sigma}_x\hat{C}$ is equal to

$$\begin{aligned} e^{i\delta}\hat{A}\hat{\sigma}_x\hat{B}\hat{\sigma}_x\hat{C} &= e^{i\delta}\mathsf{Z}_\alpha\,\underbrace{\mathsf{Y}_{\frac{\theta}{2}}\;\hat{\sigma}_x\mathsf{Y}_{-\frac{\theta}{2}}\hat{\sigma}_x}_{\mathsf{Y}_{\frac{\theta}{2}}}\;\underbrace{\hat{\sigma}_x\mathsf{Z}_{\frac{-\beta-\alpha}{2}}\hat{\sigma}_x}_{\mathsf{Z}_{\frac{\alpha+\beta}{2}}}\;\mathsf{Z}_{\frac{\beta-\alpha}{2}} \\ &= e^{i\delta}\mathsf{Z}_\alpha\;\underbrace{\mathsf{Y}_{\theta/2}\mathsf{Y}_{\theta/2}}_{\mathsf{Y}_\theta}\;\underbrace{\mathsf{Z}_{\frac{\alpha+\beta}{2}}\mathsf{Z}_{\frac{\beta-\alpha}{2}}}_{\mathsf{Z}_\beta} = e^{i\delta}\mathsf{Z}_\alpha\,\mathsf{Y}_\theta\,\mathsf{Z}_\beta = \hat{U}, \end{aligned}$$

as required. These steps can be verified with some matrix multiplications. □

5.3.2 Single-qubit gates

Single-qubit gates are unitary transformations on the Hilbert space $\mathbb{C}^2$. The simplest single-qubit gate is the **identity**. This transformation leaves the qubit state unchanged

$$\begin{aligned} &\text{Dirac notation:} && \mathbb{1}\left[c_0|0\rangle + c_1|1\rangle\right] = c_0|0\rangle + c_1|1\rangle\,, \\ &\text{Matrix form:} && \underbrace{\begin{bmatrix} 1 & 0 \\ 0 & 1 \end{bmatrix}}_{\mathbb{1}} \begin{bmatrix} c_0 \\ c_1 \end{bmatrix} = \begin{bmatrix} c_0 \\ c_1 \end{bmatrix}. \end{aligned} \tag{5.27}$$

Another important single-qubit gate is the NOT operation (or **bit-flip**). The action of the NOT gate on a qubit state is defined as

$$\begin{aligned} &\text{Dirac notation:} && \hat{X}\left[c_0|0\rangle + c_1|1\rangle\right] = c_0|1\rangle + c_1|0\rangle\,, \\ &\text{Matrix form:} && \underbrace{\begin{bmatrix} 0 & 1 \\ 1 & 0 \end{bmatrix}}_{\hat{X}} \begin{bmatrix} c_0 \\ c_1 \end{bmatrix} = \begin{bmatrix} c_1 \\ c_0 \end{bmatrix}. \end{aligned} \tag{5.28}$$

Both the identity and the NOT gates are unitary transformations because $\mathbb{1}\mathbb{1}^\dagger = \mathbb{1}^\dagger\mathbb{1} = \mathbb{1}$ and $\hat{X}\hat{X}^\dagger = \hat{X}^\dagger\hat{X} = \mathbb{1}$. The $\hat{X}$ gate is nothing but the Pauli operator $\hat{\sigma}_x$ and, as seen in eqn (5.15), the $\hat{X}$ gate is proportional to a π rotation about the x axis.

Another important single-qubit gate is the Hadamard gate[6]. The **Hadamard** $\hat{H}$ is a single-qubit operation that transforms $|0\rangle$ into the superposition state $|+\rangle$ and transforms $|1\rangle$ into $|-\rangle$. Its definition is given by

[6]In this textbook, the Hadamard gate is indicated with the same symbol as the Hamiltonian. It will be evident from the context which one we are referring to.

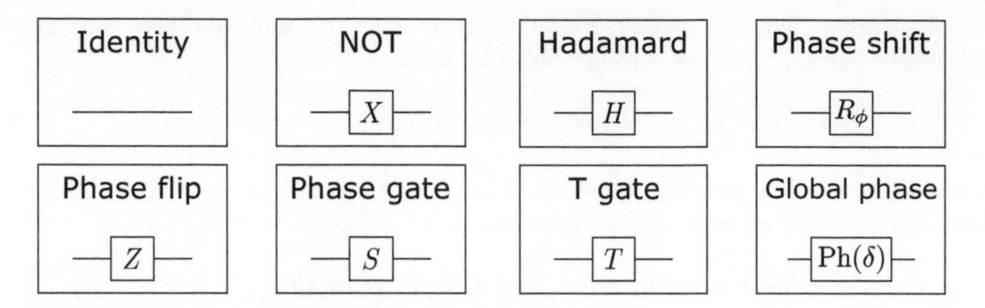

Fig. 5.4 Quantum circuits of single-qubit gates. These gates are the building blocks of a gate-based quantum computation.

$$\text{Dirac notation:} \qquad \hat{H}\left[c_0|0\rangle + c_1|1\rangle\right] = c_0\frac{|0\rangle + |1\rangle}{\sqrt{2}} + c_1\frac{|0\rangle - |1\rangle}{\sqrt{2}},$$

$$\text{Matrix form:} \qquad \underbrace{\frac{1}{\sqrt{2}}\begin{bmatrix}1 & 1\\ 1 & -1\end{bmatrix}}_{\hat{H}}\begin{bmatrix}c_0\\ c_1\end{bmatrix} = \frac{1}{\sqrt{2}}\begin{bmatrix}c_0 + c_1\\ c_0 - c_1\end{bmatrix}. \tag{5.29}$$

With a simple matrix multiplication, one can verify that $\hat{H}\hat{H}^\dagger = \hat{H}^\dagger\hat{H} = \mathbb{1}$. In addition, since $\hat{H}^2 = \mathbb{1}$, the application of two consecutive Hadamard gates leaves the qubit state unchanged. The Hadamard does not have any classical counterpart and it cannot be physically implemented on a classical machine (although it can be simulated). The Hadamard gate can be considered to be (up to a global phase) a π rotation along the bisector $\hat{n} = (1/\sqrt{2}, 0, 1/\sqrt{2})$ between the x and z axes

$$\mathsf{R}_{\hat{n}}(\pi) = e^{-i\frac{\pi}{2}\hat{n}\cdot\vec{\sigma}} = \cos\left(\frac{\pi}{2}\right)\mathbb{1} - i\sin\left(\frac{\pi}{2}\right)(\hat{n}\cdot\vec{\sigma}) = -i\frac{1}{\sqrt{2}}(\hat{\sigma}_x + \hat{\sigma}_z) = -i\hat{H}, \tag{5.30}$$

where we used eqns (5.18, 5.29).

Another gate with no classical counterpart is the **phase shift** $\hat{R}_\phi$. This operation only changes the relative phase of the qubit

$$\text{Dirac notation:} \qquad \hat{R}_\phi\left[c_0|0\rangle + c_1|1\rangle\right] = c_0|0\rangle + c_1 e^{i\phi}|1\rangle,$$

$$\text{Matrix form:} \qquad \underbrace{\begin{bmatrix}1 & 0\\ 0 & e^{i\phi}\end{bmatrix}}_{\hat{R}_\phi}\begin{bmatrix}c_0\\ c_1\end{bmatrix} = \begin{bmatrix}c_0\\ e^{i\phi}c_1\end{bmatrix}.$$

Depending on the value of ϕ, the phase shift generates different single-qubit gates

$$\text{for } \phi = \pi: \qquad \hat{Z} = \begin{bmatrix}1 & 0\\ 0 & -1\end{bmatrix} \qquad \textbf{phase flip},$$

$\mathbb{1} = \|0\rangle\langle 0\| + \|1\rangle\langle 1\|$	Identity
$\hat{X} = \|0\rangle\langle 1\| + \|1\rangle\langle 0\|$	NOT gate
$\hat{H} = \|+\rangle\langle 0\| + \|-\rangle\langle 1\|$	Hadamard gate
$\hat{R}_\phi = \|0\rangle\langle 0\| + e^{i\phi}\|1\rangle\langle 1\|$	Phase shift
$\hat{Z} = \|0\rangle\langle 0\| - \|1\rangle\langle 1\|$	Phase flip
$\hat{S} = \|0\rangle\langle 0\| + i\|1\rangle\langle 1\|$	Phase gate
$\hat{T} = \|0\rangle\langle 0\| + e^{i\pi/4}\|1\rangle\langle 1\|$	T gate
$\text{Ph}(\delta) = e^{i\delta}\|0\rangle\langle 0\| + e^{i\delta}\|1\rangle\langle 1\|$	Global phase

Table 5.1 Single-qubit gates. The single-qubit gates listed here correspond to the circuit elements shown in Fig. 5.4.

$$\text{for } \phi = \frac{\pi}{2}: \qquad \hat{S} = \begin{bmatrix} 1 & 0 \\ 0 & i \end{bmatrix} \qquad \textbf{phase gate},$$

$$\text{for } \phi = \frac{\pi}{4}: \qquad \hat{T} = \begin{bmatrix} 1 & 0 \\ 0 & e^{i\pi/4} \end{bmatrix} \qquad \textbf{T gate}.$$

Observe that $\hat{T}^2 = \hat{S}$ and $\hat{S}^2 = \hat{Z}$.

For later convenience, we introduce a gate denoted as $\text{Ph}(\delta)$. This gate multiplies the qubit state by a global phase

$$\text{Dirac notation:} \qquad \text{Ph}(\delta)\left[c_0|0\rangle + c_1|1\rangle\right] = e^{i\delta}\left[c_0|0\rangle + c_1|1\rangle\right],$$

$$\text{Matrix form:} \qquad \underbrace{\begin{bmatrix} e^{i\delta} & 0 \\ 0 & e^{i\delta} \end{bmatrix}}_{\text{Ph}(\delta)} \begin{bmatrix} c_0 \\ c_1 \end{bmatrix} = \begin{bmatrix} e^{i\delta} c_0 \\ e^{i\delta} c_1 \end{bmatrix}.$$

Clearly, this transformation does not change the qubit state (a measurement cannot detect the global phase). However, this transformation will be useful when we introduce the controlled operation $\text{c}_1\text{Ph}_0(\delta)$ in Section 5.4. The quantum circuits of the single-qubit gates introduced in this section are illustrated in Fig. 5.4 and a concise summary is presented in Table 5.1.

The Hadamard gate and the phase shift are sufficient to implement any rotation of a qubit state in the Bloch sphere.

Theorem 5.2 The Hadamard gate $\hat{H}$ and the phase shift $\hat{R}_\phi$ can transform a qubit state $|\theta_1, \phi_1\rangle$ into any other qubit state $|\theta_2, \phi_2\rangle$.

Proof First, observe that the phase shift $\hat{R}_\alpha$ transforms the qubit state $|\theta, \phi\rangle$ into

$$\hat{R}_\alpha|\theta, \phi\rangle = \cos\frac{\theta}{2}|0\rangle + e^{i(\phi+\alpha)}\sin\frac{\theta}{2}|1\rangle = |\theta, \phi + \alpha\rangle.$$

Furthermore, the operator $\hat{H}\hat{R}_\phi\hat{H}$ is proportional to a rotation about the x axis of the Bloch sphere

$$\hat{H}\hat{R}_\phi\hat{H} = e^{i\phi/2}\begin{bmatrix} \cos\frac{\phi}{2} & -i\sin\frac{\phi}{2} \\ -i\sin\frac{\phi}{2} & \cos\frac{\phi}{2} \end{bmatrix} = e^{i\phi/2}\mathsf{X}_\phi\,, \tag{5.31}$$

where we used (5.15). From eqn (5.23), we also know that $\mathsf{X}_\alpha|\theta_1, -\frac{\pi}{2}\rangle = |\theta_1+\alpha, -\frac{\pi}{2}\rangle$. From these relations, one can show that a combination of Hamadard gates and phase shifts can transform a qubit state $|\theta_1, \phi_1\rangle$ into any other state $|\theta_2, \phi_2\rangle$,

$$\begin{aligned}
\hat{R}_{\frac{\pi}{2}+\phi_2}\,\hat{H}\,\hat{R}_{\theta_2-\theta_1}\,\hat{H}\,\hat{R}_{-\frac{\pi}{2}-\phi_1}\,|\theta_1,\phi_1\rangle &= \hat{R}_{\frac{\pi}{2}+\phi_2}\,\underbrace{\hat{H}\,\hat{R}_{\theta_2-\theta_1}\,\hat{H}}_{e^{i(\theta_2-\theta_1)/2}\mathsf{X}_{\theta_2-\theta_1}}\,|\theta_1,-\tfrac{\pi}{2}\rangle \\
&= e^{i(\theta_2-\theta_1)/2}\,\hat{R}_{\frac{\pi}{2}+\phi_2}\,\mathsf{X}_{\theta_2-\theta_1}|\theta_1,-\tfrac{\pi}{2}\rangle \\
&= e^{i(\theta_2-\theta_1)/2}\,\hat{R}_{\frac{\pi}{2}+\phi_2}\,|\theta_2,-\tfrac{\pi}{2}\rangle \\
&= e^{i(\theta_2-\theta_1)/2}\,|\theta_2,\phi_2\rangle \;\equiv\; |\theta_2,\phi_2\rangle,
\end{aligned}$$

where in the last step we dropped a global phase. □

5.3.3 Measuring a qubit state

The complex coefficients c_0 and c_1 of a qubit state $|\psi\rangle = c_0|0\rangle + c_1|1\rangle$ can be determined with high accuracy, provided that the qubit can be prepared in the same state multiple times (if only one copy of $|\psi\rangle$ is provided, a single measurement will not be enough to accurately determine these coefficients because they might have a long decimal expansion). The set of measurements necessary to determine the coefficients c_0 and c_1 is called **quantum state tomography**.

Since the coefficients c_0 and c_1 are directly related to the Bloch vector (as seen in eqns (5.6, 5.7)), we will focus on the measurement of the Bloch vector components r_x, r_y, and r_z. We will assume that the qubit can be prepared in an unknown quantum state $|\psi\rangle = \cos\frac{\theta}{2}|0\rangle + e^{i\phi}\sin\frac{\theta}{2}|1\rangle$ multiple times.

The r_z component The z component of the Bloch vector can be determined by measuring $|\psi\rangle$ in the computational basis as shown in the top circuit of Fig. 5.5. A simple calculation shows that

$$r_z = \cos\theta = \cos^2\frac{\theta}{2} - \sin^2\frac{\theta}{2} = |\langle 0|\psi\rangle|^2 - |\langle 1|\psi\rangle|^2. \tag{5.32}$$

Thus, the quantity r_z is the difference between the probability of measuring $|0\rangle$ and the probability of measuring $|1\rangle$. In a real experiment, the vector $|\psi\rangle$ will be projected N_{0z} times onto $|0\rangle$ and N_{1z} times onto $|1\rangle$ and r_z is obtained from $r_z = N_{0z}/N - N_{1z}/N$. The r_z component can be determined with arbitrary accuracy as long as a sufficient number of copies of the state $|\psi\rangle$ are available and measurement errors are not too severe.

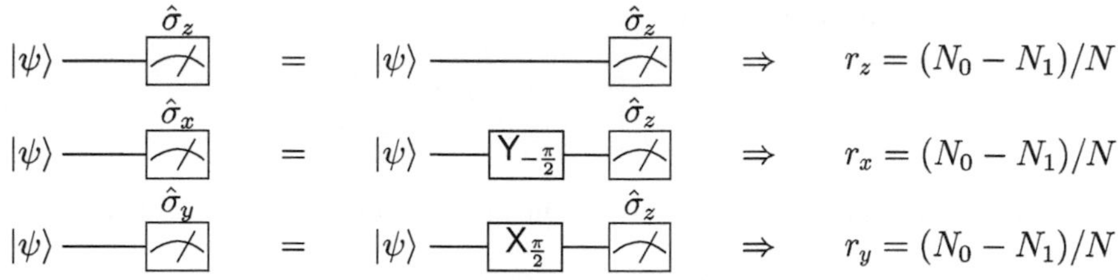

$|\psi\rangle$ — $\hat{\sigma}_z$ = $|\psi\rangle$ — $\hat{\sigma}_z$ $\Rightarrow$ $r_z = (N_0 - N_1)/N$

$|\psi\rangle$ — $\hat{\sigma}_x$ = $|\psi\rangle$ — $\mathsf{Y}_{-\frac{\pi}{2}}$ — $\hat{\sigma}_z$ $\Rightarrow$ $r_x = (N_0 - N_1)/N$

$|\psi\rangle$ — $\hat{\sigma}_y$ = $|\psi\rangle$ — $\mathsf{X}_{\frac{\pi}{2}}$ — $\hat{\sigma}_z$ $\Rightarrow$ $r_y = (N_0 - N_1)/N$

Fig. 5.5 State tomography of a qubit. The z component of the Bloch vector can be estimated by measuring multiple copies of the qubit state in the computational basis. Similarly, the components r_x, r_y can be obtained by applying a single-qubit rotation to $|\psi\rangle$ followed by a measurement in the computational basis. The number of repetitions of these experiments determines the precision of the estimation.

The r_x component One way to estimate the r_x component is to measure the unknown state $|\psi\rangle$ in the eigenbasis of the Pauli operator $\hat{\sigma}_x$, namely $B_x = \{|+\rangle, |-\rangle\}$. Unfortunately, in many experimental situations, a qubit can only be measured in the computational basis $B_z = \{|0\rangle, |1\rangle\}$. To overcome this limitation, one can first apply a unitary transformation to the unknown state and then proceed with a standard measurement in the computational basis, as shown in the second circuit of Fig. 5.5. The single-qubit rotation

$$\mathsf{Y}_{-\frac{\pi}{2}} = \frac{1}{\sqrt{2}} \begin{bmatrix} 1 & 1 \\ -1 & 1 \end{bmatrix} \tag{5.33}$$

transforms the state $|\psi\rangle$ into

$$\mathsf{Y}_{-\frac{\pi}{2}}|\psi\rangle = \frac{1}{\sqrt{2}} \begin{bmatrix} \cos\frac{\theta}{2} + e^{i\phi}\sin\frac{\theta}{2} \\ -\cos\frac{\theta}{2} + e^{i\phi}\sin\frac{\theta}{2} \end{bmatrix} . \tag{5.34}$$

The x component of the Bloch vector is given by

$$\begin{aligned} r_x = \sin\theta\cos\phi &= \left|\frac{1}{\sqrt{2}}\left(\cos\frac{\theta}{2} + e^{i\phi}\sin\frac{\theta}{2}\right)\right|^2 - \left|\frac{1}{\sqrt{2}}\left(-\cos\frac{\theta}{2} + e^{i\phi}\sin\frac{\theta}{2}\right)\right|^2 \\ &= |\langle 0|\mathsf{Y}_{-\frac{\pi}{2}}|\psi\rangle|^2 - |\langle 1|\mathsf{Y}_{-\frac{\pi}{2}}|\psi\rangle|^2, \end{aligned}$$

where we used eqn (5.34). In a real experiment, the state $\mathsf{Y}_{-\frac{\pi}{2}}|\psi\rangle$ is measured in the computational basis N times: we will observe N_{0x} times the state $|0\rangle$ and N_{1x} times the state $|1\rangle$. The r_x component can be estimated from the difference $N_{0x}/N - N_{1x}/N$.

The r_y component Lastly, r_y can be obtained[7] by applying the single-qubit rotation

$$\mathsf{X}_{\frac{\pi}{2}} = \frac{1}{\sqrt{2}} \begin{bmatrix} 1 & -i \\ -i & 1 \end{bmatrix} \tag{5.35}$$

[7]If the qubit state is pure, the measurement of r_y is not strictly necessary since it can be derived from r_x and r_z: $r_y = \sqrt{1 - r_x^2 - r_z^2}$.

to $|\psi\rangle$ and then measuring the state

$$\mathsf{X}_{\frac{\pi}{2}}|\psi\rangle = \frac{1}{\sqrt{2}}\begin{bmatrix} \cos\frac{\theta}{2} - ie^{i\phi}\sin\frac{\theta}{2} \\ -i\cos\frac{\theta}{2} + ie^{i\phi}\sin\frac{\theta}{2} \end{bmatrix} \tag{5.36}$$

in the computational basis. The y component of the Bloch vector is given by

$$\begin{aligned} r_y = \sin\theta\sin\phi &= \left|\frac{1}{\sqrt{2}}\left(\cos\frac{\theta}{2} - ie^{i\phi}\sin\frac{\theta}{2}\right)\right|^2 - \left|\frac{1}{\sqrt{2}}\left(-i\cos\frac{\theta}{2} + e^{i\phi}\sin\frac{\theta}{2}\right)\right|^2 \\ &= |\langle 0|\mathsf{X}_{\frac{\pi}{2}}|\psi\rangle|^2 - |\langle 1|\mathsf{X}_{\frac{\pi}{2}}|\psi\rangle|^2, \end{aligned} \tag{5.37}$$

where we used eqn (5.36). A sequence of N measurements will project the final state $\mathsf{X}_{\frac{\pi}{2}}|\psi\rangle$ onto $|0\rangle$ and $|1\rangle$ N_{0y} times and N_{1y} times, as shown in the third circuit of Fig. 5.5. The r_y component can be estimated from the difference $N_{0y}/N - N_{1y}/N$.

Example 5.4 Suppose that a qubit is in a superposition state $|\psi\rangle = (|0\rangle + |1\rangle)/\sqrt{2}$ and we want to estimate the expectation value $\langle\psi|\hat{\sigma}_z|\psi\rangle$. This can be done by measuring the observable $\hat{\sigma}_z$ multiple times. Suppose that 10 measurements produce the results $\mathbf{x} = (1, 1, -1, 1, 1, -1, -1, 1, 1, -1)$. Therefore, the sample average is $\mu = \sum_i x_i/10 = 0.2$ and the sample standard deviation is $\sigma = \sqrt{\sum_i (x_i - \mu)^2/(10-1)} = 1.0328$. For a large number of experiments, the sample average will converge to $\langle\psi|\hat{\sigma}_z|\psi\rangle = 0$, while the sample standard deviation will converge to $\sqrt{\langle\psi|\hat{\sigma}_z^2|\psi\rangle - \langle\psi|\hat{\sigma}_z|\psi\rangle^2} = 1$.

Example 5.5 Suppose we have a qubit prepared in state $|\psi\rangle$ and we want to measure the expectation value of the non-Hermitian operator $\langle\psi|\hat{\sigma}^+|\psi\rangle$. Although this operator does not correspond to an observable, we can express it as the sum of two Hermitian operators: $\hat{\sigma}^+ = (\hat{\sigma}_x + +i\hat{\sigma}_y)/2$. This means that the expectation value $\langle\psi|\hat{\sigma}^+|\psi\rangle$ can be determined by measuring the expectation values $a = \langle\psi|\hat{\sigma}_x|\psi\rangle$ and $b = \langle\psi|\hat{\sigma}_y|\psi\rangle$. By combining these parameters, we obtain $\langle\psi|\hat{\sigma}^+|\psi\rangle = (a + ib)/2$. Therefore, we can obtain the expectation value of non-Hermitian operators by breaking them down into Hermitian components.

5.4 Two-qubit gates

Single-qubit operations are not sufficient to generate an arbitrary two-qubit state. For instance, the state $|00\rangle$ cannot be transformed into the entangled state $(|00\rangle + |11\rangle)/\sqrt{2}$ with single-qubit operations only. **Two-qubit gates** are unitary transformations acting on the Hilbert space $\mathbb{C}^2 \otimes \mathbb{C}^2$ and are necessary to create entangled states. These types of transformations always involve a physical interaction between two qubits, as opposed to single-qubit gates, which are executed on the subsystems separately.

An important two-qubit gate is the **controlled unitary** operation $\mathsf{c}\hat{U}$ defined as

$$\boxed{\mathsf{c}_1\hat{U}_0 = |0\rangle\langle 0| \otimes \mathbb{1} + |1\rangle\langle 1| \otimes \hat{U}.} \tag{5.38}$$

Here, the letter c stands for control[8]. The notation $\mathsf{c}_1\hat{U}_0$ indicates that the unitary operator $\hat{U}$ acts on qubit 0 only when qubit 1 is set to one

$$\text{Dirac notation:} \qquad \mathsf{c}_1\hat{U}_0|x\rangle|y\rangle = |x\rangle \otimes \hat{U}^x|y\rangle\ ,$$

$$\text{Matrix form:} \qquad \begin{bmatrix} 1 & 0 & 0 & 0 \\ 0 & 1 & 0 & 0 \\ 0 & 0 & u_{00} & u_{01} \\ 0 & 0 & u_{10} & u_{11} \end{bmatrix} \begin{bmatrix} c_{00} \\ c_{01} \\ c_{10} \\ c_{11} \end{bmatrix} = \begin{bmatrix} c_{00} \\ c_{01} \\ u_{00}c_{10} + u_{01}c_{11} \\ u_{10}c_{10} + u_{11}c_{11} \end{bmatrix}. \tag{5.39}$$

Here, u_{ij} are the entries of the matrix U. The term $\hat{U}^x|y\rangle$ means "when $x = 1$, apply $\hat{U}$ to $|y\rangle$; if instead $x = 0$, don't modify $|y\rangle$". In a few steps, it is possible to show that $\mathsf{c}\hat{U}$ is unitary,

$$(\mathsf{c}_1\hat{U}_0)^\dagger \mathsf{c}_1\hat{U}_0|x\rangle|y\rangle = |x\rangle \otimes (\hat{U}^\dagger\hat{U})^x|y\rangle = |x\rangle|y\rangle\ . \tag{5.40}$$

The CNOT gate is an example of a controlled unitary operation: it applies the $\hat{X}$ gate to the target qubit only when the control qubit is set to one[9]

$$\text{Dirac notation:} \qquad \mathsf{c}_1\hat{X}_0|x\rangle|y\rangle = |x\rangle \otimes \hat{X}^x|y\rangle = |x\rangle|x \oplus y\rangle\ ,$$

$$\text{Matrix form:} \qquad \begin{bmatrix} 1 & 0 & 0 & 0 \\ 0 & 1 & 0 & 0 \\ 0 & 0 & 0 & 1 \\ 0 & 0 & 1 & 0 \end{bmatrix} \begin{bmatrix} c_{00} \\ c_{01} \\ c_{10} \\ c_{11} \end{bmatrix} = \begin{bmatrix} c_{00} \\ c_{01} \\ c_{11} \\ c_{10} \end{bmatrix}. \tag{5.41}$$

Another convenient way to express this gate is $\mathsf{c}_1\hat{X}_0 = |0\rangle\langle 0| \otimes \mathbb{1} + |1\rangle\langle 1| \otimes \hat{X}$ where we used eqn (5.38). The CNOT gate combined with the Hadamard gate can be used to generate a **Bell state**, an important example of entangled state

$$\begin{aligned} \mathsf{c}_1\hat{X}_0\hat{H}_1|0\rangle|0\rangle = \mathsf{c}_1\hat{X}_0\frac{|0\rangle + |1\rangle}{\sqrt{2}}|0\rangle &= \frac{1}{\sqrt{2}}\mathsf{c}_1\hat{X}_0[|0\rangle|0\rangle + |1\rangle|0\rangle] \\ &= \frac{1}{\sqrt{2}}[|0\rangle|0\rangle + |1\rangle|1\rangle]\,. \end{aligned}$$

Entangled states will be presented in more detail in Chapter 9.

There exist a few variants of the CNOT gate. For instance, instead of applying the $\hat{X}$ gate when the control qubit is set to one, one could execute this operation when the control qubit is set to zero; in this case, we have $\mathsf{c}_{\bar{1}}\hat{X}_0|x\rangle|y\rangle = |x\rangle|y \oplus \bar{x}\rangle$ (note the bar on the subscript). Alternatively, we can swap the roles of the target qubit and the control qubit. These variations lead to four possible cases

[8] Remember that when we write $|1\rangle|0\rangle$, qubit number 0 is the ket on the right. In a quantum circuit, qubit 0 is indicated with a horizontal line at the bottom of the circuit.

[9] Recall that $x \oplus y = (x + y) \bmod 2$.

$$\mathsf{c}_1\hat{X}_0 = |0\rangle\langle 0| \otimes \mathbb{1} + |1\rangle\langle 1| \otimes \hat{X} = \begin{bmatrix} 1\,0\,0\,0 \\ 0\,1\,0\,0 \\ 0\,0\,0\,1 \\ 0\,0\,1\,0 \end{bmatrix},$$

$$\mathsf{c}_{\overline{1}}\hat{X}_0 = |0\rangle\langle 0| \otimes \hat{X} + |1\rangle\langle 1| \otimes \mathbb{1} = \begin{bmatrix} 0\,1\,0\,0 \\ 1\,0\,0\,0 \\ 0\,0\,1\,0 \\ 0\,0\,0\,1 \end{bmatrix},$$

$$\mathsf{c}_0\hat{X}_1 = \mathbb{1} \otimes |0\rangle\langle 0| + \hat{X} \otimes |1\rangle\langle 1| = \begin{bmatrix} 1\,0\,0\,0 \\ 0\,0\,0\,1 \\ 0\,0\,1\,0 \\ 0\,1\,0\,0 \end{bmatrix},$$

$$\mathsf{c}_{\overline{0}}\hat{X}_1 = \hat{X} \otimes |0\rangle\langle 0| + \mathbb{1} \otimes |1\rangle\langle 1| = \begin{bmatrix} 0\,0\,1\,0 \\ 0\,1\,0\,0 \\ 1\,0\,0\,0 \\ 0\,0\,0\,1 \end{bmatrix}.$$

The quantum circuits associated with these gates are represented in Fig. 5.6a. In these circuits, the black dot indicates that the $\hat{X}$ operation has to be applied to the target qubit when the control qubit is set to $|1\rangle$. Similarly, the white dot indicates that the $\hat{X}$ gate has to be applied to the target qubit when the control qubit is set to $|0\rangle$.

An important controlled operation with no classical counterpart is the CPHASE gate. This gate applies a phase shift $\hat{R}_\phi$ to the target qubit only when the control qubit is set to one. The definition of the CPHASE gate is given by

$$\begin{aligned} &\text{Dirac notation:} && \mathsf{c}_1\hat{R}_{\phi,0}|x\rangle|y\rangle = |x\rangle \otimes \hat{R}_\phi^x|y\rangle = e^{i\phi xy}|x\rangle|y\rangle\,, \\ &\text{Matrix form:} && \begin{bmatrix} 1\,0\,0 & 0 \\ 0\,1\,0 & 0 \\ 0\,0\,1 & 0 \\ 0\,0\,0 & e^{i\phi} \end{bmatrix} \begin{bmatrix} c_{00} \\ c_{01} \\ c_{10} \\ c_{11} \end{bmatrix} = \begin{bmatrix} c_{00} \\ c_{01} \\ c_{10} \\ e^{i\phi}c_{11} \end{bmatrix}. \end{aligned} \tag{5.42}$$

When $\phi = \pi$, the CPHASE gate becomes the CZ gate

$$\begin{aligned} &\text{Dirac notation:} && \mathsf{c}_1\hat{Z}_0|x\rangle|y\rangle = |x\rangle \otimes \hat{Z}^x|y\rangle = (-1)^{xy}|x\rangle|y\rangle\,, \\ &\text{Matrix form:} && \begin{bmatrix} 1\,0\,0 & 0 \\ 0\,1\,0 & 0 \\ 0\,0\,1 & 0 \\ 0\,0\,0 & -1 \end{bmatrix} \begin{bmatrix} c_{00} \\ c_{01} \\ c_{10} \\ c_{11} \end{bmatrix} = \begin{bmatrix} c_{00} \\ c_{01} \\ c_{10} \\ -c_{11} \end{bmatrix}. \end{aligned} \tag{5.43}$$

This gate leaves the states $|00\rangle$, $|01\rangle$, $|10\rangle$ unchanged, while $|11\rangle$ is mapped into $-|11\rangle$. The quantum circuits associated with the CPHASE and the CZ gates are shown in Fig. 5.6b. We note that by definition $\mathsf{c}_1\hat{R}_{\phi,0}$ is equivalent to $\mathsf{c}_0\hat{R}_{\phi,1}$. This concept is illustrated in Fig. 5.7c.

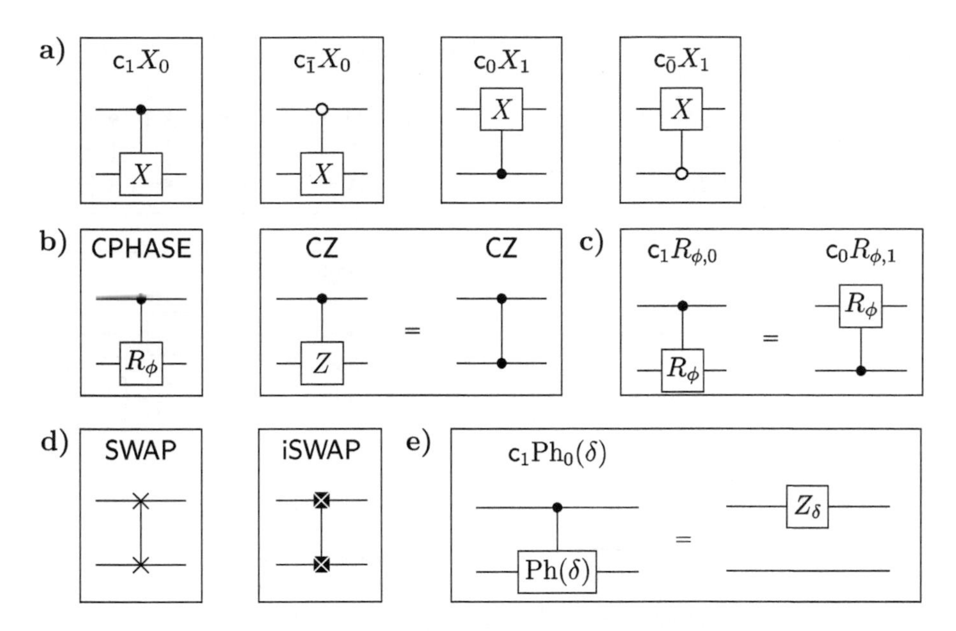

Fig. 5.6 Two-qubit gates. a) In these circuits, qubit number 0 (1) is represented with a horizontal line at the bottom (top) of the circuit. Vertical segments are used to connect control qubits with target qubits. Black dots (white dots) indicate that the $\hat{X}$ gate has to be applied when the control qubit is set to $|1\rangle$ ($|0\rangle$). **b)** Quantum circuits for the CPHASE and CZ gates. Since $c_1\hat{Z}_0 = c_0\hat{Z}_1$, the CZ gate is often represented with two black dots connected by a segment. **c)** The controlled operations $c_1\hat{R}_{\phi,0}$ and $c_0\hat{R}_{\phi,1}$ are equivalent. **d)** Quantum circuits for the SWAP and iSWAP gates. **e)** The gate $c_1\mathrm{Ph}_0(\delta)$ is not an entangling gate since it is equivalent to a single-qubit rotation about the z axis.

The SWAP and the iSWAP are important two-qubit gates. In contrast to the transformations presented so far, these gates are not controlled operations. The SWAP gate simply swaps qubit 0 with qubit 1,

$$\mathsf{SWAP} = |00\rangle\langle 00| + |01\rangle\langle 10| + |10\rangle\langle 01| + |11\rangle\langle 11|. \tag{5.44}$$

The action of this gate on a two-qubit state is given by

$$\begin{aligned} &\text{Dirac notation:} \qquad \mathsf{SWAP}|x\rangle|y\rangle = |y\rangle|x\rangle, \\ &\text{Matrix form:} \qquad \begin{bmatrix} 1\,0\,0\,0 \\ 0\,0\,1\,0 \\ 0\,1\,0\,0 \\ 0\,0\,0\,1 \end{bmatrix} \begin{bmatrix} c_{00} \\ c_{01} \\ c_{10} \\ c_{11} \end{bmatrix} = \begin{bmatrix} c_{00} \\ c_{10} \\ c_{01} \\ c_{11} \end{bmatrix}. \end{aligned} \tag{5.45}$$

Similarly, the iSWAP gate is defined as

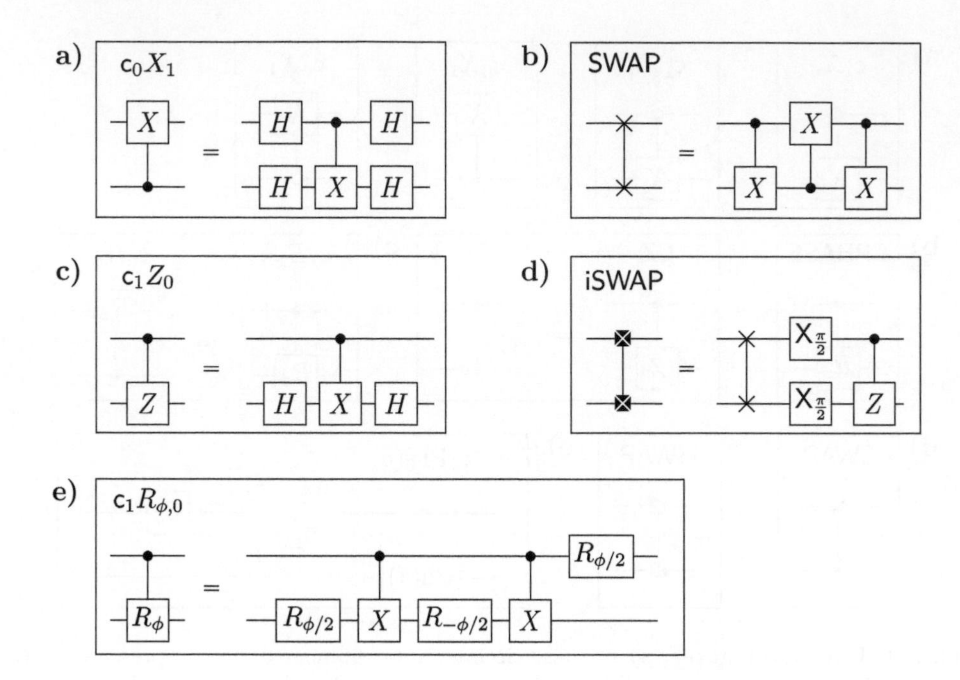

Fig. 5.7 Relations between two-qubit gates. The two-qubit gates $c_0\hat{X}_1$ and $c_1\hat{Z}_0$ are locally equivalent to the CNOT gate.

$$\text{Dirac notation:} \quad \mathsf{iSWAP}|x\rangle|y\rangle = i^{x\oplus y}|y\rangle|x\rangle\,,$$

$$\text{Matrix form:} \quad \begin{bmatrix} 1\,0\,0\,0 \\ 0\,0\,i\,0 \\ 0\,i\,0\,0 \\ 0\,0\,0\,1 \end{bmatrix} \begin{bmatrix} c_{00} \\ c_{01} \\ c_{10} \\ c_{11} \end{bmatrix} = \begin{bmatrix} c_{00} \\ ic_{10} \\ ic_{01} \\ c_{11} \end{bmatrix}. \tag{5.46}$$

The quantum circuits associated with these transformations are shown in Fig. 5.6d. For tunable superconducting qubits, the iSWAP and CZ gates are easier to implement than the CNOT gate on the physical hardware. This is related to the Hamiltonian describing the interaction between two superconducting qubits (see Section 14.8).

A controlled operation that will come up in the next section is the controlled Ph(δ) gate

$$\text{Dirac notation:} \quad c_1\text{Ph}_0(\delta)|x\rangle|y\rangle = |x\rangle e^{i\delta x}|y\rangle\,,$$

$$\text{Matrix form:} \quad \begin{bmatrix} 1\,0\;0\;\;0 \\ 0\,1\;0\;\;0 \\ 0\,0\;e^{i\delta}\;0 \\ 0\,0\;0\;\;e^{i\delta} \end{bmatrix} \begin{bmatrix} c_{00} \\ c_{01} \\ c_{10} \\ c_{11} \end{bmatrix} = \begin{bmatrix} c_{00} \\ c_{01} \\ e^{i\delta}c_{10} \\ e^{i\delta}c_{11} \end{bmatrix}. \tag{5.47}$$

This transformation is equivalent to a single-qubit rotation,

$$\mathsf{c}_1\mathrm{Ph}_0(\delta) = \begin{bmatrix} 1 & 0 & 0 & 0 \\ 0 & 1 & 0 & 0 \\ 0 & 0 & e^{i\delta} & 0 \\ 0 & 0 & 0 & e^{i\delta} \end{bmatrix} = \begin{bmatrix} 1 & 0 \\ 0 & e^{i\delta} \end{bmatrix} \otimes \begin{bmatrix} 1 & 0 \\ 0 & 1 \end{bmatrix} = \hat{R}_\delta \otimes \mathbb{1}$$

$$= \mathrm{Ph}(\delta/2)\mathsf{Z}_\delta \otimes \mathbb{1} \equiv \mathsf{Z}_\delta \otimes \mathbb{1}\,, \tag{5.48}$$

where in the last step we dropped a global phase of no physical significance. This relation is illustrated in Fig. 5.6e and will be used in Section 5.5 to show that any controlled unitary operation on a two-qubit system can be decomposed into single-qubit gates and CNOT gates. Note that $\mathsf{c}_1\mathrm{Ph}_0(\delta)$ is not an entangling gate since it can be decomposed into local operations acting on separate qubits.

A physical quantum computer can implement only a selection of two-qubit gates, such as the CZ and the iSWAP gates. The gates a quantum computer can directly execute are called **native gates**. Other two-qubit gates can be constructed by combining native two-qubit gates with single-qubit rotations. At the time of writing, on state-of-the-art quantum computers single-qubit gates can be performed with a fidelity of 99.9%, while two-qubit gates usually have a lower fidelity [20, 84]. For this reason, quantum compilers always try to minimize the number of two-qubit gates required to run a quantum computation by decomposing large unitary operators with the smallest number of two-qubit gates necessary.

A two-qubit gate A is **locally equivalent** to another two-qubit gate B if A can be generated with a combination of single-qubit and B gates. As shown in Fig. 5.7, the two-qubit gates $\mathsf{c}_0\hat{X}_1$ and $\mathsf{c}_1\hat{Z}_0$ are locally equivalent. In the next section, we will discover that combinations of single-qubit rotations and CNOT gates are sufficient to implement *any* quantum computation. This implies that single-qubit rotations combined with an entangling two-qubit gate, such as the iSWAP gate or the CZ gate, form a universal set for quantum computation.

5.5 Universality

In Chapter 1, we showed that one-bit and two-bit operations are sufficient to convert a binary string into any other string. We have also demonstrated that the Toffoli gate is universal for classical computation. This section aims to demonstrate a similar result for quantum computation [103].

Universality theorem The gates $\{\mathsf{X}_\alpha, \mathsf{Y}_\alpha, \mathsf{Z}_\alpha, \mathsf{CNOT}\}$ are sufficient to decompose any unitary operator acting on n qubits. In the worst case, the number of gates in the decomposition scales as $O(n^3 4^n)$.

This theorem is very important for the construction of quantum computers because it indicates that the experimental efforts should be focused on single and two-qubit gates. Any other unitary operator can be constructed with a combination of these gates. To prove this theorem, we will proceed step by step following Ref. [103]. The proof is somewhat technical and can be skipped on a first reading.

Step 1 Any controlled unitary operator acting on two qubits, $\mathsf{c}_1\hat{U}_0$, can be constructed using the gates $\{\mathsf{X}_\alpha, \mathsf{Y}_\alpha, \mathsf{Z}_\alpha, \mathsf{CNOT}\}$.

Proof First of all, recall from Section 5.3.1 that any unitary operator $\hat{U}$ acting on a single qubit can be expressed as $\hat{U} = \mathrm{Ph}(\delta)\hat{A}\hat{X}\hat{B}\hat{X}\hat{C}$ where

$$\hat{A} = \mathsf{Z}_\alpha \mathsf{Y}_{\frac{\theta}{2}}, \qquad \hat{B} = \mathsf{Y}_{-\frac{\theta}{2}} \mathsf{Z}_{\frac{-\beta-\alpha}{2}}, \qquad \hat{C} = \mathsf{Z}_{\frac{\beta-\alpha}{2}},$$

and $\hat{A}\hat{B}\hat{C} = \mathbb{1}$. A controlled unitary operation on a two-qubit state can be implemented as shown in Fig. 5.8a. If the control qubit is set to 0, the operation $\hat{A}\hat{B}\hat{C} = \mathbb{1}$ is applied to the target qubit. If the control qubit is set to 1, the transformation $\mathrm{Ph}(\delta)\hat{A}\hat{X}\hat{B}\hat{X}\hat{C} = \hat{U}$ is applied to the target qubit. This quantum circuit only uses single-qubit rotations, CNOT gates, and the cPh gate. Recall from eqn (5.48) that $\mathsf{c}_1\mathrm{Ph}_0(\delta) = \mathsf{Z}_\delta \otimes \mathbb{1}$. Thus, any controlled unitary operation $\mathsf{c}_1\hat{U}_0$ acting on two qubits can be implemented with the gates $\{\mathsf{X}_\alpha, \mathsf{Y}_\alpha, \mathsf{Z}_\alpha, \mathsf{CNOT}\}$. □

Step 2 Any controlled unitary operator acting on three qubits $\mathsf{c}_{12}\hat{U}_0$ can be implemented using the gates $\{\mathsf{X}_\alpha, \mathsf{Y}_\alpha, \mathsf{Z}_\alpha, \mathsf{CNOT}\}$. This operation performs a unitary operation on the target qubit (qubit number 0) only when the two control qubits (qubit number 1 and 2) are set to 1.

Proof Figure 5.8b shows how to generate $\mathsf{c}_{12}\hat{U}_0$ using controlled operations acting on two qubits. In this circuit, the unitary operator $\hat{V}$ is defined as $\hat{V}^2 = \hat{U}$. If both the control qubits are set to 0, the target qubit is left unchanged. If only one of the control qubits is set to 1, then the transformation applied to the target qubit is either $\hat{V}^\dagger\hat{V} = \mathbb{1}$ or $\hat{V}\hat{V}^\dagger = \mathbb{1}$. When both the control qubits are set to 1, the operation $\hat{V}\hat{V} = \hat{U}$ is applied to the target qubit. In Step 1, we showed that any controlled operation on two qubits can be implemented using the gates $\{\mathsf{X}_\alpha, \mathsf{Y}_\alpha, \mathsf{Z}_\alpha, \mathsf{CNOT}\}$. Thus, any controlled operation $\mathsf{c}_{12}\hat{U}_0$ acting on three qubits can be constructed using this set of gates. □

The Toffoli gate is a special case of controlled unitary operation $\mathsf{c}_{12}\hat{U}_0$ and corresponds to the case in which $\hat{U} = \hat{X}$. It can be implemented by setting $\hat{V}$ to

$$\hat{V} = \frac{1}{2}\left[(1+i)\,\mathbb{1} + (1-i)\,\hat{X}\right] = \frac{1}{2}\begin{bmatrix} 1+i & 1-i \\ 1-i & 1+i \end{bmatrix}. \tag{5.49}$$

With some matrix multiplications, it is simple to show that $\hat{V}^2 = \hat{X}$. This implementation of the Toffoli gate uses eight CNOT gates. Note that there exists a more efficient decomposition that uses only six CNOT gates as shown in Exercise 5.2.

Step 3 Consider $j+1$ qubits. Any single-qubit gate controlled by j qubits

$$\mathsf{c}_{1\ldots j}\hat{U}_0$$

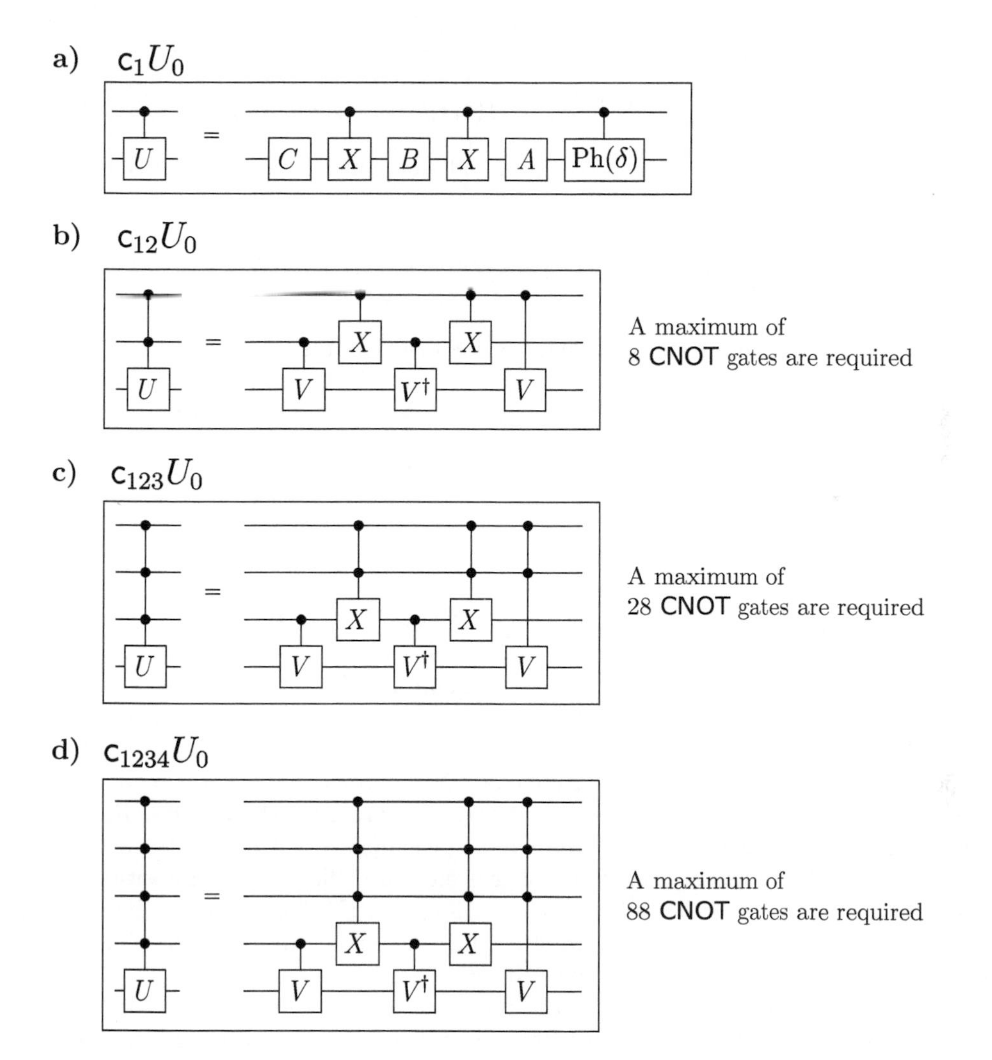

Fig. 5.8 Universality. a) Decomposition of c_1U_0 with single-qubit rotations and CNOT gates. The $c_1\mathrm{Ph}(\delta)$ gate can be expressed as a z rotation on one qubit, $c_1\mathrm{Ph}(\delta) = Z_\delta \otimes \mathbb{1}$. **b-d)** Decomposition of $c_{1...j}U_0$ in controlled unitary operations. Each controlled unitary operator can be implemented with single qubit-rotations and CNOT gates (see text).

can be implemented using the gates $\{X_\alpha, Y_\alpha, Z_\alpha, \mathsf{CNOT}\}$. The operation $c_{1...j}\hat{U}_0$ performs a unitary operation on the target qubit only when all the control qubits from 1 to j are set to 1.

Proof Let us first consider the controlled operations $c_{123}\hat{U}_0$ and $c_{1234}\hat{U}_0$. Both these operations can be decomposed into single-qubit gates and CNOT gates, as shown in Fig. 5.8c–d. A clear pattern is emerging: for $j \geq 3$, the operation $c_{1...j}\hat{U}_0$ can be constructed as[10]

$$\mathsf{c}_{1\ldots j}\hat{U}_0 = \big(\mathsf{c}_{2\ldots j}\hat{V}_0\big)\big(\mathsf{c}_{2\ldots j}\hat{X}_1\big)\big(\mathsf{c}_1\hat{V}_0^\dagger\big)\big(\mathsf{c}_{2\ldots j}\hat{X}_1\big)\big(\mathsf{c}_1\hat{V}_0\big). \tag{5.50}$$

As seen earlier, each member on the right-hand side can be recursively decomposed into CNOT and single-qubit gates. □

The construction of the controlled operations $\mathsf{c}_{1\ldots j}\hat{U}_0$ presented above is not efficient because:

$$\begin{array}{ll}
\mathsf{c}_{12}U_0 & \text{requires no more than } a_2 = 8 \text{ CNOT gates,} \\
\mathsf{c}_{123}U_0 & \text{requires no more than } a_3 = 4 + 3a_2 = 28 \text{ CNOT gates,} \\
\mathsf{c}_{1234}U_0 & \text{requires no more than } a_4 = 4 + 3a_3 = 88 \text{ CNOT gates,} \\
\vdots & \vdots \\
\mathsf{c}_{1\ldots j}U_0 & \text{requires no more than } a_j = 4 + 3a_{j-1} \text{ CNOT gates.}
\end{array}$$

The implementation of $\mathsf{c}_{1\ldots j}\hat{U}_0$ requires a maximum number of CNOT gates that satisfies the recurrence relation $a_j = 4 + 3a_{j-1}$ with $a_2 = 8$ as the initial condition. The solution to this recurrence relation is $a_j = 10 \cdot 3^{j-2} - 2$, indicating that the decomposition of $\mathsf{c}_{1\ldots j}U_0$ requires a number of CNOT gates that scales exponentially with j. However, there exist more efficient ways to construct these gates. Ref. [103] explains how these controlled operations can be decomposed with a quadratic number of CNOT gates. More recent techniques offer even more efficient decompositions (see, for example, Refs [158, 159]). These constructions use $O(n)$ elementary gates, with the best-known lower bound being $2n$ [160]. In conclusion, the gates $\mathsf{c}_{1\ldots j}U_0$ can be implemented with $O(j)$ single and two-qubit gates.

Before discussing the last part of the proof, we need to introduce an important concept. We define a **two-level unitary matrix** (or Givens rotation) as a $d \times d$ unitary matrix that acts non-trivially only on a two-dimensional subspace. As an example, consider the Hilbert space of two qubits, $\mathbb{C}^2 \otimes \mathbb{C}^2$. The following matrices represent all possible two-level unitary matrices acting on a two-qubit state:

$$U_{0,1}(V) = \begin{bmatrix} v_{00} & v_{01} & 0 & 0 \\ v_{10} & v_{11} & 0 & 0 \\ 0 & 0 & 1 & 0 \\ 0 & 0 & 0 & 1 \end{bmatrix}, \qquad U_{0,2}(V) = \begin{bmatrix} v_{00} & 0 & v_{01} & 0 \\ 0 & 1 & 0 & 0 \\ v_{10} & 0 & v_{11} & 0 \\ 0 & 0 & 0 & 1 \end{bmatrix},$$

[10] In eqn (5.50) the rightmost operator $\mathsf{c}_1\hat{V}_0$ corresponds to the leftmost gate in the circuit, as time in the circuit flows from left to right.

$$U_{0,3}(V)=\begin{bmatrix} v_{00} & 0 & 0 & v_{01} \\ 0 & 1 & 0 & 0 \\ 0 & 0 & 1 & 0 \\ v_{10} & 0 & 0 & v_{11} \end{bmatrix}, \qquad U_{1,2}(V)=\begin{bmatrix} 1 & 0 & 0 & 0 \\ 0 & v_{00} & v_{01} & 0 \\ 0 & v_{10} & v_{11} & 0 \\ 0 & 0 & 0 & 1 \end{bmatrix},$$

$$U_{1,3}(V)=\begin{bmatrix} 1 & 0 & 0 & 0 \\ 0 & v_{00} & 0 & v_{01} \\ 0 & 0 & 1 & 0 \\ 0 & v_{10} & 0 & v_{11} \end{bmatrix}, \qquad U_{2,3}(V)=\begin{bmatrix} 1 & 0 & 0 & 0 \\ 0 & 1 & 0 & 0 \\ 0 & 0 & v_{00} & v_{01} \\ 0 & 0 & v_{10} & v_{11} \end{bmatrix}.$$

Here, the submatrix

$$V=\begin{bmatrix} v_{00} & v_{01} \\ v_{10} & v_{11} \end{bmatrix} \tag{5.51}$$

is a generic 2×2 unitary matrix. We can finally add the last piece to the puzzle.

Step 4 Any unitary operator $\hat{U}$ acting on n qubits can be decomposed as $\hat{U} = \hat{A}_1 \dots \hat{A}_k$ where $\hat{A}_i$ are two-level unitary matrices and $k \in O(4^n)$.

Proof Let us present a systematic procedure to find a sequence of two-level unitary matrices $A_1, \dots, A_k$ that satisfy the relation

$$A_k \dots A_1 U = \mathbb{1}, \quad \text{and therefore } U = A_1^\dagger \dots A_k^\dagger. \tag{5.52}$$

For simplicity, we will consider the Hilbert space of two qubits, $\mathbb{C}^2 \otimes \mathbb{C}^2$. In this space, a generic unitary matrix can be written as

$$U=\begin{bmatrix} u_{11} & u_{12} & u_{13} & u_{14} \\ u_{21} & u_{22} & u_{23} & u_{24} \\ u_{31} & u_{32} & u_{33} & u_{34} \\ u_{41} & u_{42} & u_{43} & u_{44} \end{bmatrix}. \tag{5.53}$$

The first part of this procedure consists in defining three two-level unitary matrices A_1, A_2, and A_3 so that the matrix $A_3A_2A_1U$ has the first column equal to zero apart from the top-left entry. Let us start from the definition of A_1,

$$A_1=\begin{bmatrix} v_{11} & v_{12} & 0 & 0 \\ v_{21} & v_{22} & 0 & 0 \\ 0 & 0 & 1 & 0 \\ 0 & 0 & 0 & 1 \end{bmatrix}, \quad \text{with } \begin{bmatrix} v_{11} & v_{12} \\ v_{21} & v_{22} \end{bmatrix} = \frac{1}{\sqrt{|u_{11}|^2+|u_{21}|^2}} \begin{bmatrix} u_{11}^* & u_{12}^* \\ u_{12} & -u_{11} \end{bmatrix}. \tag{5.54}$$

With a matrix multiplication, it is possible to show that

$$A_1U=\begin{bmatrix} u'_{11} & u'_{12} & u'_{13} & u'_{14} \\ 0 & u'_{22} & u'_{23} & u'_{24} \\ u'_{31} & u'_{32} & u'_{33} & u'_{34} \\ u'_{41} & u'_{42} & u'_{43} & u'_{44} \end{bmatrix}. \tag{5.55}$$

Here, the entries u'_{ij} are directly related to the matrix components u_{ij}. We can now define a two-level unitary matrix A_2 such that the product A_2A_1U has another component in the first column equal to zero. The matrix A_2 is defined as

$$A_2 = \begin{bmatrix} v'_{11} & 0 & v'_{12} & 0 \\ 0 & 1 & 0 & 0 \\ v'_{21} & 0 & v'_{22} & 0 \\ 0 & 0 & 0 & 1 \end{bmatrix}, \quad \text{with} \begin{bmatrix} v'_{11} & v'_{12} \\ v'_{21} & v'_{22} \end{bmatrix} = \frac{1}{\sqrt{|u'_{11}|^2 + |u'_{31}|^2}} \begin{bmatrix} u'^*_{11} & u'^*_{13} \\ u'_{13} & -u'_{11} \end{bmatrix}. \tag{5.56}$$

If we now compute the product A_2A_1U, we obtain

$$A_2A_1U = \begin{bmatrix} u''_{11} & u''_{12} & u''_{13} & u''_{14} \\ 0 & u''_{22} & u''_{23} & u''_{24} \\ 0 & u''_{32} & u''_{33} & u''_{34} \\ u''_{41} & u''_{42} & u''_{43} & u''_{44} \end{bmatrix}, \tag{5.57}$$

as required. We now need to find a third two-level unitary matrix A_3 such that the matrix $A_3A_2A_1U$ has the component in the bottom-left corner equal to zero. The matrix A_3 is given by

$$A_3 = \begin{bmatrix} v''_{11} & 0 & 0 & v''_{12} \\ 0 & 1 & 0 & 0 \\ 0 & 0 & 1 & 0 \\ v''_{21} & 0 & 0 & v''_{22} \end{bmatrix}, \quad \text{with} \begin{bmatrix} v''_{11} & v''_{12} \\ v''_{21} & v''_{22} \end{bmatrix} = \frac{1}{\sqrt{|u''_{11}|^2 + |u''_{41}|^2}} \begin{bmatrix} u''^*_{11} & u''^*_{14} \\ u''_{14} & -u''_{11} \end{bmatrix}. \tag{5.58}$$

Finally, we have

$$A_3A_2A_1U = \begin{bmatrix} 1 & u'''_{12} & u'''_{13} & u'''_{14} \\ 0 & u'''_{22} & u'''_{23} & u'''_{24} \\ 0 & u'''_{32} & u'''_{33} & u'''_{34} \\ 0 & u'''_{42} & u'''_{43} & u'''_{44} \end{bmatrix}. \tag{5.59}$$

Since this matrix is unitary (the product of unitary matrices is unitary), the entries u'''_{12}, u'''_{13}, and u'''_{14} are actually zero. We are left with a block-diagonal matrix

$$A_3A_2A_1U = \begin{bmatrix} 1 & 0 & 0 & 0 \\ 0 & u'''_{22} & u''_{23} & u'''_{24} \\ 0 & u'''_{32} & u''_{33} & u'''_{34} \\ 0 & u'''_{42} & u''_{43} & u'''_{44} \end{bmatrix}. \tag{5.60}$$

We repeat this procedure for the 3×3 sub-block. Once this procedure is completed, we will have six two-level unitary matrices such that $A_6A_5A_4A_3A_2A_1U = \mathbb{1}$ and therefore $U = A_1^\dagger A_2^\dagger A_3^\dagger A_4^\dagger A_5^\dagger A_6^\dagger$. The number of two-level unitary matrices needed in this decomposition is equal to half of the off-diagonal entries of the matrix U (in our example, U is a 4×4 matrix and six two-level unitary matrices are necessary for the decomposition).

This procedure can be generalized in a straightforward way to any $d \times d$ unitary matrix where $d = 2^n$. Therefore, any unitary operation can be written in the form

$$\hat{U} = \hat{A}_1^\dagger \dots \hat{A}_k^\dagger, \tag{5.61}$$

where A_1, ..., A_k are two-level unitary matrices and $k = (d^2 - d)/2$. □

This result indicates that if we find a way to implement arbitrary two-level unitary matrices on a quantum computer, we would be able to perform any quantum computation.

Step 5 Any two-level unitary matrix acting on n qubits can be constructed using $\{\mathsf{X}_\alpha, \mathsf{Y}_\alpha, \mathsf{Z}_\alpha, \mathsf{CNOT}\}$. The decomposition requires $O(n^3)$ elementary gates.

Proof For simplicity, let us consider a two-level unitary matrix acting on a quantum register with three qubits

$$U_{1,6}(V) = \begin{bmatrix} 1 & 0 & 0 & 0 & 0 & 0 & 0 & 0 \\ 0 & v_{00} & 0 & 0 & 0 & 0 & v_{01} & 0 \\ 0 & 0 & 1 & 0 & 0 & 0 & 0 & 0 \\ 0 & 0 & 0 & 1 & 0 & 0 & 0 & 0 \\ 0 & 0 & 0 & 0 & 1 & 0 & 0 & 0 \\ 0 & 0 & 0 & 0 & 0 & 1 & 0 & 0 \\ 0 & v_{10} & 0 & 0 & 0 & 0 & v_{11} & 0 \\ 0 & 0 & 0 & 0 & 0 & 0 & 0 & 1 \end{bmatrix}. \tag{5.62}$$

This unitary operator acts non-trivially only on the subspace generated by the states $|1\rangle = |001\rangle$ and $|6\rangle = |110\rangle$. We can implement $U_{1,6}(V)$ using single-qubit and CNOT gates by expressing it as a controlled operation on one target qubit. To do this, we first write the binary representations of 1 and 6: $\mathbf{x} = 001$ and $\mathbf{y} = 110$.

This unitary operator acts non-trivially only on the subspace generated by the states $|1\rangle = |001\rangle$ and $|6\rangle = |110\rangle$. To implement $U_{1,6}(V)$ using single-qubit and CNOT gates, we can express it as a controlled operation on one target qubit. To this end, we start from the binary representation of 1 and 6: $\mathbf{x} = 001$ and $\mathbf{y} = 110$. Then, we connect these two strings with a Gray code. A **Gray code** is a finite sequence of binary strings $(\mathbf{x}_1, \dots, \mathbf{x}_r)$ such that the first string $\mathbf{x}_1 = \mathbf{x}$, the last string $\mathbf{x}_r = \mathbf{y}$, and the strings $\mathbf{x}_i$ and $\mathbf{x}_{i+1}$ are identical apart from one bit. For example, a Gray code connecting 1 and 6 is given by

		qubit 2	qubit 1	qubit 0	
1	=	0	0	1	
		0	0	0	↑ $\hat{W}_1$
		0	1	0	↑ $\hat{W}_2$
6	=	1	1	0	↑ apply $\hat{V}$

This Gray code can be used to construct a sequence of gates that implements $U_{1,6}(V)$ as shown in Fig. 5.9,

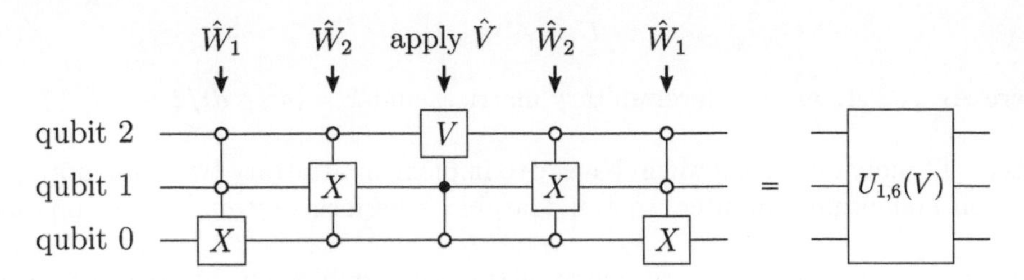

Fig. 5.9 Two-level unitaries. Quantum circuit for the construction of the gate $U_{1,6}(V)$, eqn (5.62), in terms of controlled unitaries. Note that this decomposition only contains single-qubit gates controlled by other qubits.

$$U_{1,6}(V) = \underbrace{(\,\mathsf{c}_{\overline{12}}X_0\,)}_{\hat{W}_1}\underbrace{(\,\mathsf{c}_{\overline{02}}X_1\,)}_{\hat{W}_2}(\,\mathsf{c}_{\overline{0}1}V_2\,)\underbrace{(\,\mathsf{c}_{\overline{02}}X_1\,)}_{\hat{W}_2}\underbrace{(\,\mathsf{c}_{\overline{12}}X_0\,)}_{\hat{W}_1}.$$

The first operator $\hat{W}_1 = \mathsf{c}_{\overline{12}}\hat{X}_0$ maps the state $|1\rangle = |001\rangle$ into $\hat{W}_1|001\rangle = |000\rangle$. The second operator $\hat{W}_2 = \mathsf{c}_{\overline{02}}\hat{X}_1$ maps the state $|000\rangle$ into $\hat{W}_2|000\rangle = |010\rangle$. This state differs from $|6\rangle = |110\rangle$ by only one bit. Hence, the operation V can now be implemented with the controlled operation $\mathsf{c}_{\overline{0}1}V_2$. Finally, the gates $\hat{W}_1$, $\hat{W}_2$ are executed in reverse order. At the end of the circuit, only states $|001\rangle$ and $|110\rangle$ have been affected; all other states are left unchanged. In summary, to perform a unitary operator $\hat{V}$ on the two states $|001\rangle$ and $|110\rangle$, we first operate a sequence of permutations up to a final state that differs from $|110\rangle$ only by one bit. The unitary operation V is then performed on the state of one qubit controlled by the states of all the others. Finally, we undo the permutations.

This scheme can be easily generalized to any two-level unitary matrix $U_{i,j}(V)$. The indices i and j are written in binary representation: $i \to \mathbf{x}$, $j \to \mathbf{y}$. These strings are connected with a Gray code $\{\mathbf{x}_1, \ldots, \mathbf{x}_r\}$ where $\mathbf{x}_1 = \mathbf{x}$ and $\mathbf{x}_r = \mathbf{y}$. The first $r-2$ steps of this code define $r-2$ unitary transformations

$$\hat{W}_k|\mathbf{a}\rangle = \begin{cases} |\mathbf{x}_{k+1}\rangle & \text{if } \mathbf{a} = \mathbf{x}_k \\ |\mathbf{x}_k\rangle & \text{if } \mathbf{a} = \mathbf{x}_{k+1} \\ |\mathbf{a}\rangle & \text{otherwise.} \end{cases} \tag{5.63}$$

The unitary operations $(\hat{W}_1, \ldots, \hat{W}_{r-2})$ are implemented with single-qubit gates controlled by other qubits. Thus, a generic two-level unitary matrix $U_{i,j}(V)$ can be decomposed as

$$U_{i,j}(V) = (W_1 \ldots W_{r-2})Z(V)(W_{r-2}^\dagger \cdots W_1^\dagger)\,, \tag{5.64}$$

where $Z(V)$ is a controlled operation. In conclusion, any two-level unitary matrix acting on n qubits can be decomposed into a product of single-qubit gates controlled by $n-1$ qubits. In Step 3, we showed that such controlled operations can be implemented with the gates $\{\mathsf{X}_\alpha, \mathsf{Y}_\alpha, \mathsf{Z}_\alpha, \mathsf{CNOT}\}$. □

How many controlled operations are required to implement a $2^n \times 2^n$ two-level unitary matrix? Two binary strings of length n can be connected by a Gray code with no more than n steps. This means that $O(n)$ controlled unitary operations $\hat{W}$ are necessary to implement a $2^n \times 2^n$ two-level unitary matrix.

Step 6 Using the results from Steps 1–5, we can finally prove the universality of single-qubit rotations and CNOT gates.

Theorem 5.3 (Universality theorem) A generic unitary transformation $\hat{U}$ acting on n qubits can be implemented using the gates $\{\mathsf{X}_\alpha, \mathsf{Y}_\alpha, \mathsf{Z}_\alpha, \mathsf{CNOT}\}$.

Proof From Step 4, $\hat{U}$ can be decomposed into a product of two-level unitary matrices. From Step 5, each of these two-level matrices can be written as a product of single-qubit gates controlled by $n-1$ qubits. From Step 3, single-qubit gates controlled by $n-1$ qubits can be implemented with the gates $\{\mathsf{X}_\alpha, \mathsf{Y}_\alpha, \mathsf{Z}_\alpha, \mathsf{CNOT}\}$. This completes the proof. □

The universality theorem presented in this section shows that any unitary operator acting on n qubits can be decomposed with $O(n^3 4^n)$ single and two-qubit gates. This decomposition was proposed in 1995 by Barenco et al. [103]. In 1995, Knill reduced the number of CNOT gates[11] to $O(n4^n)$ [161]. In 2004, this result was improved by Vartiainen et al. who discovered a decomposition with $O(4^n)$ CNOT gates [162]. Around the same years, another decomposition known as the NQ decomposition was proposed bringing the number of CNOT gates to $\frac{23}{48}4^n - \frac{3}{2}2^n + \frac{4}{3}$ [163]. The same result was obtained with a different type of decomposition called the QSD decomposition [164]. The takeaway is that a generic unitary operator can be decomposed into $O(4^n)$ elementary gates.

The quantum circuits presented in this textbook are drawn for a fully-connected device, i.e. a quantum computer in which each qubit is connected to any other qubit. In practice, however, some architectures for quantum computing have low connectivity. As the reader might expect, there are different ways to perform a CNOT gate between two qubits that are not directly connected. For example, a CNOT gate between qubits a and c that are both connected to qubit b can be implemented as

$$\mathsf{c}_a\hat{X}_c = \mathsf{c}_b\hat{X}_c\, \mathsf{c}_a\hat{X}_b\, \mathsf{c}_b\hat{X}_c\, \mathsf{c}_a\hat{X}_b.$$

Designing a quantum computer with low connectivity is usually easier, but this comes at the expense of using extra gates in the quantum circuits [165].

The set $\{\mathsf{X}_\alpha, \mathsf{Y}_\alpha, \mathsf{Z}_\alpha, \mathsf{CNOT}\}$ is a well-known universal set for quantum computation, but it is not the only one. This is due to the fact that the CNOT gate can be decomposed into a combination of single and two-qubit gates (see Fig. 5.7). Another interesting set is $\{\hat{H}, \hat{T}, \mathsf{CNOT}\}$, which is composed of discrete gates but still capable of achieving universality. This result is known as the Solovay–Kitaev theorem [166].

[11] Since the fidelity of two-qubit gates is usually worse than that of single-qubit gates, the cost of the decomposition is typically expressed in terms of the number of two-qubit gates required.

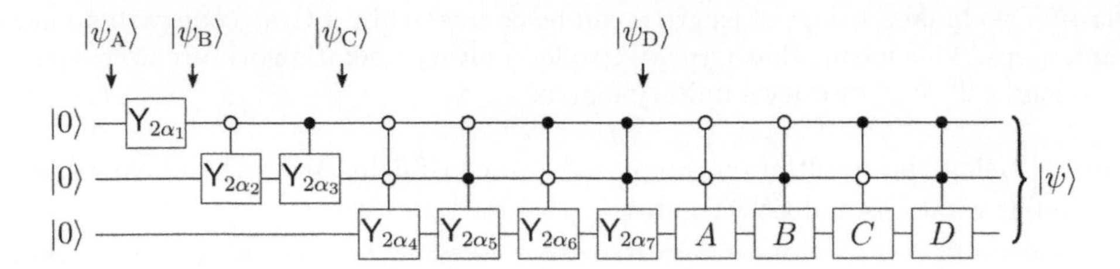

Fig. 5.10 State preparation. A quantum circuit for the preparation of a generic three-qubit state $|\psi\rangle = \sum_{\mathbf{x}} c_{\mathbf{x}} |\mathbf{x}\rangle$. The circuit comprises $2^3 - 1$ gates for fixing the amplitudes $|c_{\mathbf{x}}|$, followed by 2^{3-1} gates for setting the relative phases.

Efficient methods for decomposing a generic unitary $\hat{U}(n)$ into $\hat{H}, \hat{T}$, CNOT gates are presented in Refs [167, 168, 169].

5.6 State preparation

Preparing a register of n qubits in an arbitrary quantum state might require an exponential number of gates. To understand this concept, suppose that the initial state of a quantum computer is $|000\rangle$ and we would like to prepare a generic three-qubit state

$$|\psi\rangle = \sum_{\mathbf{x}\in\mathbb{B}^3} c_{\mathbf{x}}|\mathbf{x}\rangle = c_{000}|000\rangle + c_{001}|001\rangle + \ldots + c_{111}|111\rangle, \tag{5.65}$$

where $c_{\mathbf{x}} = |c_{\mathbf{x}}|e^{i\delta}$ are probability amplitudes. The quantum circuit shown in Fig. 5.10 can be used for this purpose. The first rotation about the y axis transforms the state $|000\rangle$ into

$$|\psi_{\rm B}\rangle = \big(\mathsf{Y}_{2\alpha_1}|0\rangle\big)|00\rangle = \cos\alpha_1|000\rangle + \sin\alpha_1|100\rangle.$$

The application of two controlled rotations leads to

$$\begin{aligned}|\psi_{\rm C}\rangle &= \cos\alpha_1\cos\alpha_2|000\rangle + \cos\alpha_1\sin\alpha_2|010\rangle \\ &\quad + \sin\alpha_1\cos\alpha_3|100\rangle + \sin\alpha_1\sin\alpha_3|110\rangle.\end{aligned}$$

Finally, four more controlled rotations transform this state into

$$\begin{aligned}|\psi_{\rm D}\rangle = \;& \cos\alpha_1\cos\alpha_2\cos\alpha_4|000\rangle + \cos\alpha_1\cos\alpha_2\sin\alpha_4|001\rangle \\ & + \cos\alpha_1\sin\alpha_2\cos\alpha_5|010\rangle + \cos\alpha_1\sin\alpha_2\sin\alpha_5|011\rangle \\ & + \sin\alpha_1\cos\alpha_3\cos\alpha_6|100\rangle + \sin\alpha_1\cos\alpha_3\sin\alpha_6|101\rangle \\ & + \sin\alpha_1\sin\alpha_3\cos\alpha_7|110\rangle + \sin\alpha_1\sin\alpha_3\sin\alpha_7|111\rangle.\end{aligned}$$

This state is equivalent to $|\psi\rangle = \sum_{\mathbf{x}} |c_{\mathbf{x}}||\mathbf{x}\rangle$ when the angles $\{\alpha_1, \ldots, \alpha_7\}$ are chosen such that

$$\cos\alpha_1\cos\alpha_2\cos\alpha_4 = |c_{000}|,$$

$$\cos\alpha_1 \cos\alpha_2 \sin\alpha_4 = |c_{001}|,$$
$$\vdots$$
$$\sin\alpha_1 \sin\alpha_3 \sin\alpha_7 = |c_{111}|.$$

Therefore, $2^3 - 1 = 7$ gates are needed to set the magnitude of the coefficients $c_{\mathbf{x}}$. The second part of this quantum circuit sets the relative phases. The matrices A, B, C, and D are given by

$$A = \begin{bmatrix} e^{i\delta_0} & 0 \\ 0 & e^{i\delta_1} \end{bmatrix}, \qquad B = \begin{bmatrix} e^{i\delta_2} & 0 \\ 0 & e^{i\delta_3} \end{bmatrix},$$
$$C = \begin{bmatrix} e^{i\delta_4} & 0 \\ 0 & e^{i\delta_5} \end{bmatrix}, \qquad D = \begin{bmatrix} e^{i\delta_6} & 0 \\ 0 & e^{i\delta_7} \end{bmatrix}.$$

The operator $\mathsf{c}_{\overline{21}}A_0$ sets the phases for the states $|000\rangle$ and $|001\rangle$. Similarly, the operator $\mathsf{c}_{\overline{2}1}\hat{B}_0$ fixes the phases for the states $|010\rangle$ and $|011\rangle$ and so on. The application of these additional $2^{3-1} = 4$ controlled operations produces the desired state $|\psi\rangle$. In conclusion, the initial state $|000\rangle$ can be transformed into a generic three-qubit state using $(2^3 - 1) + 2^{3-1}$ gates. This simple example suggests that an exponential number of gates is required to prepare an arbitrary quantum state of n qubits. For more information, see Refs [170, 171].

5.6.1 The SWAP test

A ubiquitous task in quantum computing is determining the absolute value of the inner product $|\langle\phi|\psi\rangle|$ between two unknown quantum states $|\psi\rangle$ and $|\phi\rangle$ of n qubits. This task can be accomplished by running the **SWAP test** [172]. The quantum circuit of the SWAP test is shown in Fig. 5.11a. The qubit at the top of the circuit is an ancilla qubit. A Hadamard on the ancilla qubit transforms the initial state $|0\rangle|\psi\rangle|\phi\rangle$ into

$$|\psi_{\rm A}\rangle = \frac{1}{\sqrt{2}}\big(|0\rangle|\psi\rangle|\phi\rangle + |1\rangle|\psi\rangle|\phi\rangle\big).$$

A controlled SWAP and another Hadamard on the ancilla qubit produce the final state

$$|\psi_{\rm C}\rangle = \frac{1}{2}\big(|0\rangle|\phi\rangle|\psi\rangle + |0\rangle|\psi\rangle|\phi\rangle + |1\rangle|\phi\rangle|\psi\rangle - |1\rangle|\psi\rangle|\phi\rangle\big).$$

A measurement of the ancilla qubit will give $|0\rangle$ with probability

$$p(0) = \mathrm{Tr}[(|0\rangle\langle 0| \otimes \mathbb{1} \otimes \mathbb{1})|\psi_{\rm C}\rangle\langle\psi_{\rm C}|] = \langle\psi_{\rm C}|(|0\rangle\langle 0| \otimes \mathbb{1} \otimes \mathbb{1})|\psi_{\rm C}\rangle = \frac{1 + |\langle\phi|\psi\rangle|^2}{2}.$$

Thus, the absolute value $|\langle\phi|\psi\rangle|$ can be estimated from $p(0)$ with the simple relation $|\langle\phi|\psi\rangle| = \sqrt{2p(0) - 1}$. If the input states are two mixed states[12] ρ and σ, the SWAP test returns $\mathrm{Tr}[\rho\sigma]$. In the special case $\rho = \sigma$, the circuit computes the purity of the

[12]Mixed states will be introduced in Section 6.1.

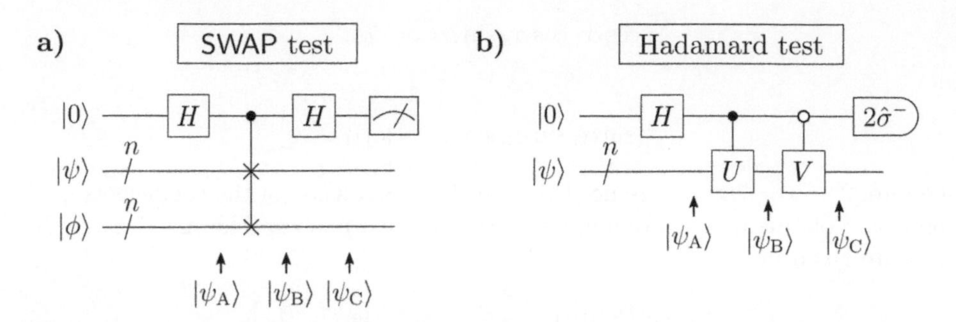

Fig. 5.11 SWAP test and Hadamard test. a) Quantum circuit implementing the SWAP test. $|\psi\rangle$ and $|\phi\rangle$ are two unknown quantum states of n qubits. The probability of measuring the ancilla in $|0\rangle$ is directly related to the quantity $|\langle\phi|\psi\rangle|$. This probability can be accurately estimated only if multiple copies of $|\psi\rangle$ and $|\phi\rangle$ are available. **b)** Quantum circuit implementing the Hadamard test. $|\psi\rangle$ is an unknown quantum state of n qubits. The expectation value of the operator $2\hat{\sigma}^-$ on the ancilla qubit is related to the quantity $\langle\psi|\hat{V}^\dagger\hat{U}|\psi\rangle$. These protocols are common subroutines in quantum simulations.

input state, $\mu = \mathrm{Tr}[\rho^2]$. It is worth mentioning that an algorithm based on a machine-learning approach known as the Bell-Basis Algorithm can estimate the inner product between two quantum states with fewer resources than the SWAP test [173].

5.6.2 The Hadamard test

In many algorithms for quantum simulation, it is required to compute the expectation value $\langle\psi|\hat{V}^\dagger\hat{U}|\psi\rangle$, where $\hat{U}$, $\hat{V}$ are unitary operators and $|\psi\rangle$ is an unknown quantum state of n qubits. The **Hadamard test** is a quantum circuit designed to measure the expectation value $\langle\psi|\hat{V}^\dagger\hat{U}|\psi\rangle$ by using an ancilla qubit and some controlled operations [174]. Note that in the special case $\hat{V} = \mathbb{1}$, the Hadamard test returns the expectation value of a unitary operator, $\langle\psi|\hat{U}|\psi\rangle$.

The quantum circuit of the Hadamard test is shown in Fig. 5.11b. A Hadamard on the ancilla qubit transforms the initial state $|0\rangle|\psi\rangle$ into $|\psi_{\mathrm{A}}\rangle = \frac{1}{\sqrt{2}}(|0\rangle|\psi\rangle + |1\rangle|\psi\rangle)$. Two controlled gates $\mathsf{c}U$ and $\bar{\mathsf{c}}V$ produce the final state

$$|\psi_{\mathrm{C}}\rangle = \frac{1}{\sqrt{2}}(|0\rangle\hat{V}|\psi\rangle + |1\rangle\hat{U}|\psi\rangle).$$

The last step of the protocol boils down to measuring the expectation value of the operator $2\hat{\sigma}^-$ on the ancilla qubit,

$$\begin{aligned}\langle\psi_{\mathrm{C}}|2\hat{\sigma}^- \otimes \mathbb{1}|\psi_{\mathrm{C}}\rangle &= \frac{1}{2}\Big(\langle 0|\langle\psi|\hat{V}^\dagger + \langle 1|\langle\psi|\hat{U}^\dagger\Big)\Big(2|0\rangle\langle 1| \otimes \mathbb{1}\Big)\Big(|0\rangle\hat{V}|\psi\rangle + |1\rangle\hat{U}|\psi\rangle\Big)\\ &= \langle\psi|\hat{V}^\dagger\hat{U}|\psi\rangle.\end{aligned}$$

We note that the operator $2\hat{\sigma}^-$ is not Hermitian and does not correspond to an observable. However, it can be expressed as a sum of two Pauli operators, $2\hat{\sigma}^- =$

$\hat{\sigma}_x + i\hat{\sigma}_y$. This means that the quantity $\langle\psi_C|2\hat{\sigma}^- \otimes \mathbb{1}|\psi_C\rangle$ can be estimated by measuring the expectation value of $\hat{\sigma}_x$ and $\hat{\sigma}_y$ on the ancilla qubit. These measurements will give $a = \langle\psi_C|\hat{\sigma}_x \otimes \mathbb{1}|\psi_C\rangle$ and $b = \langle\psi_C|\hat{\sigma}_y \otimes \mathbb{1}|\psi_C\rangle$, from which one can calculate $\langle\psi_C|2\hat{\sigma}^- \otimes \mathbb{1}|\psi_C\rangle = a + ib$. A variant of the Hadamard test is presented in Exercise 5.4.

5.7 The Clifford group and the Gottesman–Knill theorem

The Clifford group is an important family of quantum gates. The importance of this family comes from the fact that in a circuit containing only Clifford gates the probability distribution of a specific bit string can be calculated efficiently by a classical computer[13]. Hence, they are often used to benchmark quantum protocols since a classical machine can efficiently predict the outcomes of some measurements.

Before presenting the Clifford gates, we will recall the definition of the Pauli group. The n-qubit **Pauli group** consists of all tensor products of n Pauli matrices together with a multiplicative factor of ± 1 or $\pm i$,

$$\mathcal{P}_n = \{i^k \hat{P}_1 \otimes \ldots \otimes \hat{P}_n : \ \hat{P}_i \in \{\mathbb{1}, \hat{X}, \hat{Y}, \hat{Z}\} \text{ and } 0 \leq k \leq 3\}.$$

where $\{\mathbb{1}, \hat{X}, \hat{Y}, \hat{Z}\}$ are the usual Pauli operators. This group contains 4^{n+1} elements. The n-qubit **Clifford group** $\mathcal{C}_n$ is a set of unitary operators that *stabilizes* the Pauli group, i.e. the unitary transformations in $\mathcal{C}_n$ leave the operators of the Pauli group invariant under conjugation

$$\mathcal{C}_n = \{\hat{C} \in \mathrm{U}(H) : \ \hat{C}\hat{P}\hat{C}^\dagger \in \mathcal{P}_n \text{ for all } \hat{P} \in \mathcal{P}_n\} \ / \ \mathrm{U}(1).$$

Here, $/\mathrm{U}(1)$ indicates that Clifford operators differing by a global phase are considered to be identical. It can be shown that $\mathcal{C}_n$ forms a group and the number of its elements increases exponentially with n [175, 176],

$$|\mathcal{C}_n| = \prod_{j=1}^{n} 2 \cdot (4^j - 1) \cdot 4^j = 2^{2n^2 + O(n)}. \tag{5.66}$$

The elements in $\mathcal{C}_n$, called Clifford gates, can be constructed by multiplying together $\hat{H}$, $\hat{S}$, and CNOT gates [176]. This means that Clifford gates alone do not lead to universal quantum computation. This is because the set $\{\hat{H}, \hat{S}, \mathsf{CNOT}\}$ is not universal, and, in order to achieve universality, it is necessary to add the $\hat{T}$ gate to the set.

Example 5.6 The smallest Pauli group $\mathcal{P}_1$ contains 16 elements

$$\mathcal{P}_1 = \{\pm\mathbb{1}, \pm i\mathbb{1}, \pm\hat{X}, \pm i\hat{X}, \pm\hat{Y}, \pm i\hat{Y}, \pm\hat{Z}, \pm i\hat{Z}\}. \tag{5.67}$$

The Clifford group $\mathcal{C}_1$ associated with this set clearly contains the identity, since $\mathbb{1}\hat{P}\mathbb{1} \in \mathcal{P}_1$ for all $\hat{P} \in \mathcal{P}_1$. The set $\mathcal{C}_1$ also contains all operators in the Pauli group $\mathcal{P}_1$ since $\hat{P}'\hat{P}\hat{P}'^\dagger \in \mathcal{P}_1$ for all $\hat{P}, \hat{P}' \in \mathcal{P}_1$. This indicates that $\mathcal{C}_1$ has at least 16 elements. It is

[13] This result is known as the Gottesman–Knill theorem and will be presented at the end of this chapter.

$\mathbb{1}$	H	HS	HS^2	HS^3	HS^2H
S	SH	SHS	SHS^2	SHS^3	SHS^2H
S^2	S^2H	S^2HS	S^2HS^2	S^2HS^3	S^2HS^2H
S^3	S^3H	S^3HS	S^3HS^2	S^3HS^3	S^3HS^2H

Table 5.2 The 24 elements of the Clifford group $\mathcal{C}_1$ written as products of the Hadamard and phase gates.

interesting to observe that the Hadamard gate and the phase gate belong to $\mathcal{C}_1$ too. This can be verified by showing that $\hat{H}\hat{P}\hat{H} \in \mathcal{P}_1$ and $\hat{S}\hat{P}\hat{S} \in \mathcal{P}_1$ for all $\hat{P} \in \mathcal{P}_1$. Table 5.2 lists all the operators in $\mathcal{C}_1$. These gates can be obtained by multiplying the elements $\{\mathbb{1}, \hat{S}, \hat{S}^2, \hat{S}^3\}$ by $\hat{H}$, $\hat{H}\hat{S}$, $\hat{H}\hat{S}^2$, $\hat{H}\hat{S}^3$, and $\hat{H}\hat{S}^2\hat{H}$.

Example 5.7 The Pauli group $\mathcal{P}_2$ has 64 elements

$$\mathcal{P}_2 = \{\pm \hat{P}_1 \otimes \hat{P}_2, \ \ \pm i\, \hat{P}_1 \otimes \hat{P}_2\},$$

where as usual $\hat{P}_1, \hat{P}_2 \in \{\mathbb{1}, \hat{X}, \hat{Y}, \hat{Z}\}$. From eqn (5.66), we know that the associated Clifford group $\mathcal{C}_2$ has 11520 operators. The CNOT is one of them. To see this, we can calculate the action of the CNOT gate on some combinations of X and Z gates:

$$\begin{aligned} \mathsf{CNOT}(\hat{X} \otimes \mathbb{1})\mathsf{CNOT} &= \hat{X} \otimes \hat{X}, & \mathsf{CNOT}(\hat{Z} \otimes \mathbb{1})\mathsf{CNOT} &= \hat{Z} \otimes \mathbb{1}, \\ \mathsf{CNOT}(\mathbb{1} \otimes \hat{X})\mathsf{CNOT} &= \mathbb{1} \otimes \hat{X}, & \mathsf{CNOT}(\mathbb{1} \otimes \hat{Z})\mathsf{CNOT} &= \hat{Z} \otimes \hat{Z}, \\ \mathsf{CNOT}(\hat{X} \otimes \hat{X})\mathsf{CNOT} &= \hat{X} \otimes \mathbb{1}, & \mathsf{CNOT}(\hat{Z} \otimes \hat{Z})\mathsf{CNOT} &= \mathbb{1} \otimes \hat{Z}. \end{aligned}$$

These relations are useful to calculate the action of the CNOT gate on a generic Pauli operator. For example, since $(i\hat{Y}) \otimes (i\hat{Y}) = \hat{Z}\hat{X} \otimes \hat{Z}\hat{X}$, then

$$\begin{aligned} \mathsf{CNOT}(i\hat{Y} \otimes i\hat{Y})\mathsf{CNOT} &= \mathsf{CNOT}(\hat{Z}\hat{X} \otimes \hat{Z}\hat{X})\mathsf{CNOT} \\ &= \mathsf{CNOT}(\hat{Z} \otimes \hat{Z})(\hat{X} \otimes \hat{X})\mathsf{CNOT} \\ &= \mathsf{CNOT}(\hat{Z} \otimes \hat{Z})\mathsf{CNOT}\ \mathsf{CNOT}(\hat{X} \otimes \hat{X})\mathsf{CNOT} \\ &= (\mathbb{1} \otimes \hat{Z})(\hat{X} \otimes \mathbb{1}) = \hat{X} \otimes \hat{Z}. \end{aligned}$$

Thus, the CNOT gate stabilizes the Pauli operator $(i\hat{Y}) \otimes (i\hat{Y})$. The reader can check that for any $\hat{A} \in \mathcal{P}_2$, the operator $\mathsf{CNOT}\ \hat{A}\ \mathsf{CNOT} \in \mathcal{P}_2$.

5.7.1 The Gottesman–Knill theorem

Despite the impressive number of operators in the set $\mathcal{C}_n$, Clifford gates do not form a universal set for quantum computation, and they are not responsible for the origin of the quantum speedup over classical computing, as stated by the following theorem [175, 176].

Theorem 5.4 (Gottesman–Knill) Consider an n-qubit register initialized in $|0\dots0\rangle$ and suppose that a Clifford gate $\hat{C}$ is applied to the register as shown in the figure below:

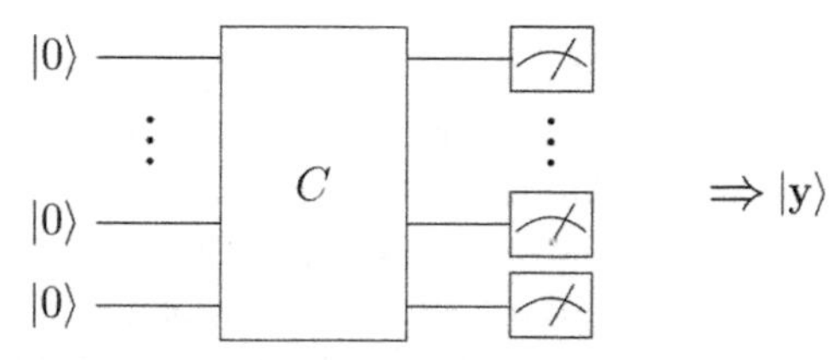

Then, the probability of measuring one specific bit string $|\mathbf{y}\rangle$ at the end of the computation, $p(\mathbf{y}) = |\langle\mathbf{y}|\hat{C}|\mathbf{0}\rangle|^2$, can be efficiently calculated by a classical computer.

Clifford circuits (quantum circuits composed only of Clifford gates) can generate a high degree of entanglement. However, the probability of measuring a specific bit string at the end of a Clifford circuit can be easily simulated by a classical computer. This highlights that while certain types of entangled states are necessary to disallow efficient classical simulation in a quantum computation, they are not sufficient [177].

The Gottesman-Knill theorem has an interesting implication. Suppose a decision problem is solved by a Clifford circuit. If the input string $\mathbf{x}$ belongs to a language L, the final state of qubit number 1 (which stores the output of the circuit) will likely be $|1\rangle$. The Gottesman-Knill theorem states that a classical computer can efficiently calculate the probability of qubit 1 collapsing onto $|1\rangle$. In other words, if a decision problem can be decided by a Clifford circuit with a small error probability, it can also be decided by a probabilistic Turing machine. It can be shown that the problem of simulating a Clifford circuit is complete for the classical complexity class $\oplus\mathsf{L}$ (this class contains all problems that can be mapped to the simulation of a polynomial-size circuit composed entirely of NOT and CNOT gates; therefore $\oplus\mathsf{L} \subseteq \mathsf{P}$) [178].

Further reading

The universality theorem presented in this chapter was proposed in 1995 by Barenco et al. [103]. More efficient decompositions of a generic unitary operator can be found in Refs [161, 162, 179, 180, 163, 164, 181, 182, 165]. The problem of state preparation is discussed in Ref. [171]. A more efficient implementation of the SWAP test is explained in Ref. [173].

Summary

Qubits

$\mathbb{C}^2$	Hilbert space of a qubit
$\lvert\psi\rangle = a\lvert 0\rangle + b\lvert 1\rangle$	Generic qubit state $\lvert\psi\rangle \in \mathbb{C}^2$
$\lvert\psi\rangle = \cos\frac{\theta}{2}\lvert 0\rangle + e^{i\phi}\sin\frac{\theta}{2}\lvert 1\rangle$	Alternative expression
$\lvert\pm\rangle = \frac{1}{\sqrt{2}}[\lvert 0\rangle \pm \lvert 1\rangle]$	Balanced superposition states
$\mathbb{C}^2 \otimes \ldots \otimes \mathbb{C}^2 = \mathbb{C}^{2n}$	Hilbert space of n qubits
$\mathsf{X}_\alpha = \cos\frac{\alpha}{2}\mathbb{1} - i\sin\frac{\alpha}{2}\hat{\sigma}_x$	Rotation about the x axis by an angle α
$\mathsf{Y}_\alpha = \cos\frac{\alpha}{2}\mathbb{1} - i\sin\frac{\alpha}{2}\hat{\sigma}_y$	Rotation about the y axis by an angle α
$\mathsf{Z}_\alpha = \cos\frac{\alpha}{2}\mathbb{1} - i\sin\frac{\alpha}{2}\hat{\sigma}_z$	Rotation about the z axis by an angle α

Any single-qubit unitary operation $\hat{U}$ can be decomposed as $\hat{U} = e^{i\delta}\mathsf{Z}_\alpha\mathsf{Y}_\theta\mathsf{Z}_\beta$.

Single-qubit gates

$\mathbb{1} = \lvert 0\rangle\langle 0\rvert + \lvert 1\rangle\langle 1\rvert$	Identity
$\hat{X} = \lvert 0\rangle\langle 1\rvert + \lvert 1\rangle\langle 0\rvert$	NOT gate
$\hat{H} = \lvert +\rangle\langle 0\rvert + \lvert -\rangle\langle 1\rvert$	Hadamard gate
$\hat{R}_\phi = \lvert 0\rangle\langle 0\rvert + e^{i\phi}\lvert 1\rangle\langle 1\rvert$	Phase shift
$\hat{Z} = \lvert 0\rangle\langle 0\rvert - \lvert 1\rangle\langle 1\rvert$	Phase flip
$\hat{S} = \lvert 0\rangle\langle 0\rvert + i\lvert 1\rangle\langle 1\rvert$	Phase gate
$\hat{T} = \lvert 0\rangle\langle 0\rvert + e^{i\pi/4}\lvert 1\rangle\langle 1\rvert$	$\pi/8$ gate
$\mathrm{Ph}(\delta) = e^{i\delta}\lvert 0\rangle\langle 0\rvert + e^{i\delta}\lvert 1\rangle\langle 1\rvert$	Global phase

The gates $\hat{H}$ and $\hat{R}_\phi$ can transform a qubit state into any other qubit state.

State tomography of a qubit state

$r_x = \lvert\langle 0\rvert\mathsf{Y}_{-\frac{\pi}{2}}\lvert\psi\rangle\rvert^2 - \lvert\langle 1\rvert\mathsf{Y}_{-\frac{\pi}{2}}\lvert\psi\rangle\rvert^2$	r_x component of an unknown state $\lvert\psi\rangle$
$r_y = \lvert\langle 0\rvert\mathsf{X}_{\frac{\pi}{2}}\lvert\psi\rangle\rvert^2 - \lvert\langle 1\rvert\mathsf{X}_{\frac{\pi}{2}}\lvert\psi\rangle\rvert^2$	r_y component of an unknown state $\lvert\psi\rangle$
$r_z = \lvert\langle 0\vert\psi\rangle\rvert^2 - \lvert\langle 1\vert\psi\rangle\rvert^2$	r_z component of an unknown state $\lvert\psi\rangle$

Two-qubit gates

$c_1\hat{U}_0 = \|0\rangle\langle 0\| \otimes \mathbb{1} + \|1\rangle\langle 1\| \otimes \hat{U}$	Controlled unitary operation
$c_1\hat{X}_0 = \|0\rangle\langle 0\| \otimes \mathbb{1} + \|1\rangle\langle 1\| \otimes \hat{X}$	CNOT
$c_{\bar{1}}\hat{X}_0 = \|1\rangle\langle 1\| \otimes \mathbb{1} + \|0\rangle\langle 0\| \otimes \hat{X}$	Variant of the CNOT
$c_1\hat{R}_{\phi,0} = \|0\rangle\langle 0\| \otimes \mathbb{1} + \|1\rangle\langle 1\| \otimes \hat{R}_\phi$	CPHASE
$c_1\hat{Z}_0 = \|0\rangle\langle 0\| \otimes \mathbb{1} + \|1\rangle\langle 1\| \otimes \hat{Z}$	CZ
$\mathsf{SWAP} = \|00\rangle\langle 00\| + \|10\rangle\langle 01\| + \|01\rangle\langle 10\| + \|11\rangle\langle 11\|$	SWAP
$\mathsf{iSWAP} = \|00\rangle\langle 00\| + i\|10\rangle\langle 01\| + i\|01\rangle\langle 10\| + \|11\rangle\langle 11\|$	iSWAP

Universality and Clifford gates

Any unitary operator on n qubits can be decomposed into $\{\mathsf{X}_\alpha, \mathsf{Y}_\alpha, \mathsf{Z}_\alpha, \mathsf{CNOT}\}$.

Preparing a n-qubit state $|\psi\rangle$ might require an exponential number of gates.

$\mathcal{P}_n = \{i^k \hat{P}_1 \otimes \ldots \otimes \hat{P}_n : \ \hat{P}_i \in \{\mathbb{1}, \hat{X}, \hat{Y}, \hat{Z}\} \text{ and } 0 \leq k \leq 3\}.$

$\mathcal{P}_1 = \{\pm\mathbb{1}, \pm i\mathbb{1}, \pm\hat{X}, \pm i\hat{X}, \pm\hat{Y}, \pm i\hat{Y}, \pm\hat{Z}, \pm i\hat{Z}\}.$

$\mathcal{C}_n = \{\hat{C} \in \mathrm{U}(H) : \ \hat{C}\hat{P}\hat{C}^\dagger \in \mathcal{P}_n \text{ for all } \hat{P} \in \mathcal{P}_n\} \,/\, \mathrm{U}(1).$

Any operator in $\mathcal{C}_n$ can be expressed as a combination of $\hat{H}$, $\hat{S}$, and CNOT gates.

The probability of measuring a particular bit string at the end of a Clifford circuit can be efficiently calculated by a classical computer.

Exercises

Exercise 5.1 A qubit is in an *unknown* state $\cos\frac{\theta}{2}|0\rangle + e^{i\phi}\sin\frac{\theta}{2}|1\rangle$ and we would like to determine the angle θ. How many measurements should we execute to estimate θ with a variance smaller than 10^{-k}?

Exercise 5.2 Show that the Toffoli gate can be decomposed using 6 CNOT gates as shown in the figure below

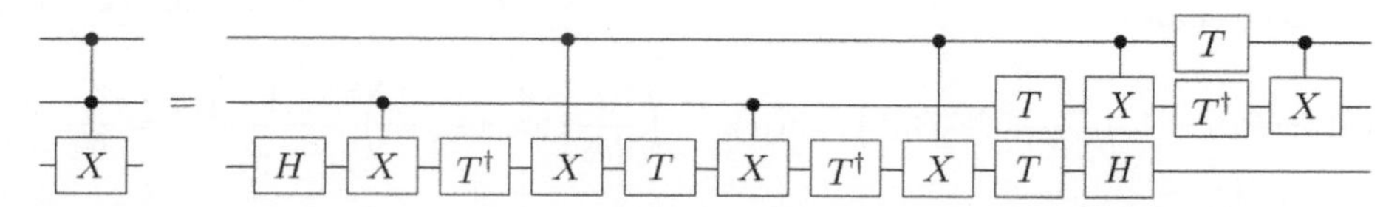

Hint: This decomposition can be checked using Dirac notation or by multiplying some matrices with a piece of software.

Exercise 5.3 As seen in Section 5.6.2, the expectation value $\langle\psi|\hat{V}^\dagger\hat{U}|\psi\rangle$ can be estimated by running a Hadamard test. This expectation value can also be determined using a modified version of the SWAP test. Consider the circuit below:

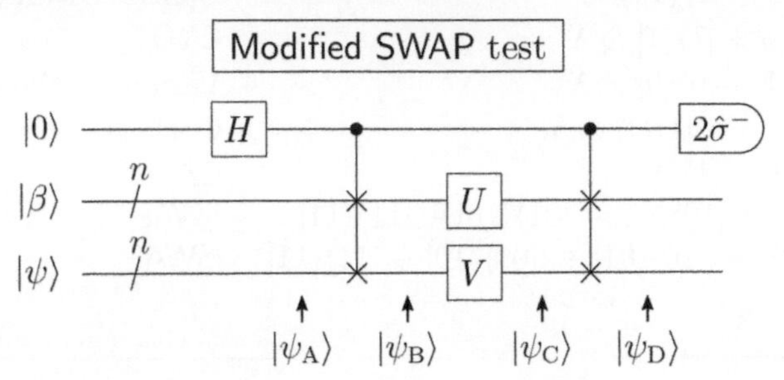

Here, $|\beta\rangle$ is a quantum state that satisfies $\langle\beta|\hat{U}^\dagger\hat{V}|\beta\rangle = 1$. The advantage of this circuit with respect to the Hadamard test is that the unitary operations $\hat{U}$ and $\hat{V}$ are not implemented with controlled operations. Show that $\langle\psi_D|2\hat{\sigma}^- \otimes \mathbb{1} \otimes \mathbb{1}|\psi_D\rangle = \langle\psi|\hat{V}^\dagger\hat{U}|\psi\rangle$.

Exercise 5.4 In some quantum simulations for many-body systems, it is required to estimate expectation values of the form $\langle\psi|\hat{U}^\dagger\hat{A}\hat{U}\hat{B}|\psi\rangle$ where $\hat{A}, \hat{B}, \hat{U}$ are unitary operators and $|\psi\rangle$ is an unknown n-qubit state. A modified version of the Hadamard test can accomplish this task [183]. Consider the circuit below:

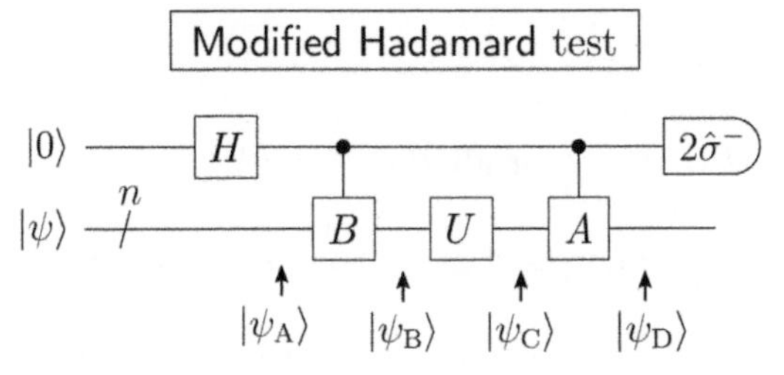

Show that $\langle\psi_D|2\hat{\sigma}^- \otimes \mathbb{1}|\psi_D\rangle = \langle\psi|\hat{U}^\dagger\hat{A}\hat{U}\hat{B}|\psi\rangle$.

Solutions

Solution 5.1 A sequence of measurements in the computational basis is described by a Bernoulli probability distribution. The qubit state will be projected N_0 times onto $|0\rangle$ and N_1 times onto $|1\rangle$, where $N_0 + N_1 = N$. The probability of measuring the qubit in $|0\rangle$ is $P_0 = N_0/N$. For a large number of experiments, the expectation value of this quantity is $\mathrm{E}[P_0] = p_0$ with variance $\mathrm{Var}[P_0] = p_0(1-p_0)/N$ where $p_0 = \cos^2\frac{\theta}{2}$. Our goal is to determine the angle θ. To estimate this parameter, let us introduce the random variable $\Theta(P_0) = 2\arccos\sqrt{P_0}$. The expectation value of this random variable is $E[\Theta(P_0)] = \theta$. The variance is given by

$$\mathrm{Var}[\Theta(P_0)] = \left(\frac{d}{d\mathrm{E}[P_0]}\Theta(E[P_0])\right)^2 \mathrm{Var}[P_0] = \left(\frac{d}{dp_0}2\arccos\sqrt{p_0}\right)^2 \frac{p_0(1-p_0)}{N} = \frac{1}{N}.$$

The angle θ can be estimated with a variance $\mathrm{Var}[\Theta(P_0)] < 10^{-k}$, only if the number of experiments satisfies $N > 10^k$. This number grows exponentially with k.

Solution 5.2 Consider two bits, a, $b \in \{0,1\}$, and a generic qubit state $|\psi\rangle \in \mathbb{C}^2$. From Section 5.3.2, we know that $\hat{X}|a\rangle = \hat{X}|a \oplus 1\rangle$, $\hat{X}^b|a\rangle = \hat{X}|a \oplus b\rangle$ and $\hat{T}|a\rangle = e^{ia\pi/4}|a\rangle$. The action of the circuit on a three-qubit state $|a\rangle|b\rangle|\psi\rangle$ is given by

$$\begin{aligned}
\text{Circuit}|a\rangle|b\rangle|\psi\rangle &= \hat{T}|a\rangle\hat{X}^a\hat{T}^\dagger\hat{X}^a\hat{T}|b\rangle\Big[\underbrace{\hat{H}\hat{T}\hat{X}^a\hat{T}^\dagger\hat{X}^b\hat{T}\hat{X}^a\hat{T}^\dagger\hat{X}^b\hat{H}}_{\hat{C}_{ab}}|\psi\rangle\Big] \\
&= e^{ia\pi/4}|a\rangle e^{ib\pi/4}\hat{X}^a\hat{T}^\dagger|b\oplus a\rangle\hat{C}_{ab}|\psi\rangle \\
&= e^{ia\pi/4}|a\rangle e^{ib\pi/4}e^{-i(a\oplus b)\pi/4}\hat{X}^a|b\oplus a\rangle\hat{C}_{ab}|\psi\rangle \\
&= e^{ia\pi/4}|a\rangle e^{ib\pi/4}e^{-i(a\oplus b)\pi/4}|b\rangle\hat{C}_{ab}|\psi\rangle \\
&= |a\rangle|b\rangle\Big[\underbrace{e^{ia\pi/4}e^{ib\pi/4}e^{-i(a\oplus b)\pi/4}\hat{C}_{ab}}_{\hat{V}_{ab}}|\psi\rangle\Big] = |a\rangle|b\rangle\hat{V}_{ab}|\psi\rangle.
\end{aligned}$$

With some matrix multiplications, one can verify that $\hat{V}_{00} = \hat{V}_{01} = \hat{V}_{10} = \mathbb{1}$ and $\hat{V}_{11} = \hat{X}$.

Solution 5.3 The Hadamard and the controlled SWAP transform the initial state $|0\rangle|\beta\rangle|\psi\rangle$ into

$$|\psi_\text{B}\rangle = \frac{1}{\sqrt{2}}|0\rangle|\beta\rangle|\psi\rangle + \frac{1}{\sqrt{2}}|1\rangle|\psi\rangle|\beta\rangle.$$

Two unitary operators are applied to the register, followed by another controlled SWAP. These transformations yield

$$|\psi_\text{D}\rangle = \frac{1}{\sqrt{2}}|0\rangle\hat{U}|\beta\rangle\hat{V}|\psi\rangle + \frac{1}{\sqrt{2}}|1\rangle V|\beta\rangle U|\psi\rangle.$$

The expectation value of $2\hat{\sigma}^- = 2|0\rangle\langle 1|$ measured on the first qubit is

$$\langle\psi_\text{D}|2\hat{\sigma}^- \otimes \mathbb{1}_n \otimes \mathbb{1}_n|\psi_\text{D}\rangle = \langle\beta|\hat{U}^\dagger\hat{V}|\beta\rangle\langle\psi|\hat{V}^\dagger\hat{U}|\psi\rangle = \langle\psi|\hat{V}^\dagger\hat{U}|\psi\rangle.$$

Solution 5.4 A Hadamard and a controlled operation transform the initial state $|0\rangle|\psi\rangle$ into $|\psi_\text{B}\rangle = \frac{1}{\sqrt{2}}|0\rangle|\psi\rangle + |1\rangle\hat{B}|\psi\rangle$. The unitary operation $\hat{U}$ followed by a controlled unitary operation leads to

$$|\psi_\text{D}\rangle = \frac{1}{\sqrt{2}}|0\rangle\hat{U}|\psi\rangle + |1\rangle\hat{A}\hat{U}\hat{B}|\psi\rangle.$$

The measurement of the operator $2\hat{\sigma}^- = 2|0\rangle\langle 1|$ yields

$$\begin{aligned}
\langle\psi_\text{D}|2\hat{\sigma}^- \otimes \mathbb{1}_n|\psi_\text{D}\rangle &= \frac{1}{2}\Big(\langle 0|\langle\psi|\hat{U}^\dagger + \langle 1|\langle\psi|\hat{B}^\dagger\hat{U}^\dagger\hat{A}^\dagger\Big)\Big(2|0\rangle\langle 1| \otimes \mathbb{1}_n\Big)\Big(|0\rangle\hat{U}|\psi\rangle + |1\rangle\hat{A}\hat{U}\hat{B}|\psi\rangle\Big) \\
&= \langle\psi|\hat{U}^\dagger\hat{A}\hat{U}\hat{B}|\psi\rangle.
\end{aligned}$$

PART II

MODERN QUANTUM MECHANICS

6
Density operators

Density operators
are positive operators
with unit trace.

The first postulate of quantum mechanics states that a quantum system well isolated from the environment is described by a unit vector in a Hilbert space. However, in many practical cases, we do not have complete control over the state of the system and can only assert that the system is in one of many possible states according to a classical probability distribution. In such cases, **density operator formalism** is well suited to describe the state of the system and can be considered an alternative formulation of quantum mechanics with respect to the state vector approach presented in Chapter 4. Density operator formalism is particularly useful for describing subsystems of a composite system and statistical processes in general.

This chapter is devoted to the discussion of density operators. We begin by introducing the definition of a density operator and discussing the unitary evolution of a quantum state. We then explain how to use the density operator approach to describe the measurement of a quantum system in Section 6.3, followed by a presentation of the density operator associated with a qubit state. In Section 6.5, we explore the density operators of a qubit and a harmonic oscillator at thermal equilibrium. The second part of the chapter focuses on on bipartite systems. We show that the quantum state of a subsystem is described by the partial trace of the density operator of the global system. Finally, we explain the concept of purification of a density operator, which elucidates a deep connection between density operators and state vectors.

6.1 Definition and examples

In Chapter 4, we learned that the state of an isolated quantum system can be represented by a unit vector in a complex Hilbert space. This formalism describes an ideal source capable of preparing a quantum system consistently in the same state. In reality, however, sources are not ideal and they often prepare a quantum system in one of many possible states.

To better understand this point, consider a source S_1 that emits one particle per second in the state $|\psi\rangle$ as shown in Fig. 6.1a. Assuming that the state of the particles is consistently the same, this source can be described with the vector $|\psi\rangle$. This quantum system is said to be in a **pure state**. Now consider a second source S_2 that

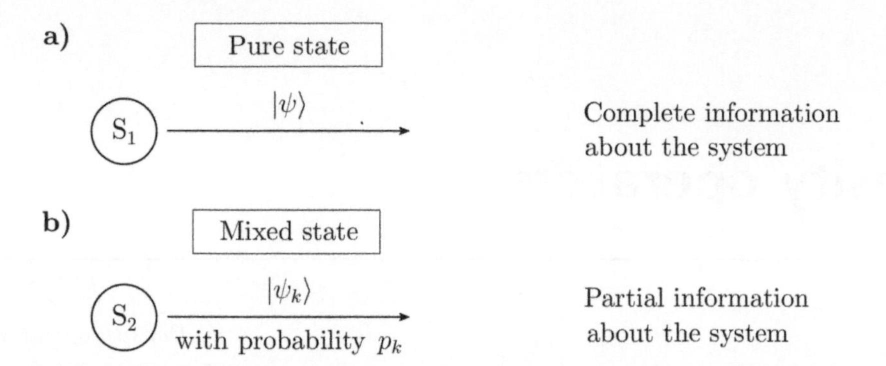

Fig. 6.1 A comparison between two particle sources. a) The source S_1 emits a particle per unit time in the state $|\psi\rangle$. This quantum system is represented by the density operator $\rho = |\psi\rangle\langle\psi|$. This is a pure state. **b)** The source S_2 emits a particle per unit time in a random state $|\psi_k\rangle$ with probability p_k. This quantum system is represented by the density operator $\rho = \sum_k |\psi_k\rangle\langle\psi_k|$. This is a mixed state.

emits one particle per second as illustrated in Fig. 6.1b. The particles emitted by the second source can be in one of the states $\{|\psi_1\rangle, \dots, |\psi_N\rangle\}$ according to a probability distribution $\{p_1, \dots, p_N\}$. Due to the random nature of this process, we cannot predict a priori the state of each particle. The statistical ensemble associated with this source can be concisely expressed with the notation $\{|\psi_k\rangle, p_k\}_k$ and the system is said to be in a **mixed state.**

Suppose that the particles emitted by the source S_2 are subject to a measurement. When the particle is prepared in $|\psi_1\rangle$, multiple measurements of an observable $\hat{A}$ will produce a set of outcomes with average $\langle\psi_1|\hat{A}|\psi_1\rangle$. If the particle is prepared in a different state $|\psi_2\rangle$, the average of the distribution will be $\langle\psi_2|\hat{A}|\psi_2\rangle$ and so on. Therefore, when a quantum system is prepared in a statistical ensemble $\{|\psi_k\rangle, p_k\}_k$, the expectation value of an observable is given by a weighted average[1],

$$\begin{aligned}\langle\hat{A}\rangle = \sum_{k=1}^{N} p_k \langle\psi_k|\hat{A}|\psi_k\rangle &= \sum_{i=1}^{d}\sum_{k=1}^{N} p_k \langle\psi_k|\hat{A}|i\rangle\langle i|\psi_k\rangle = \sum_{i=1}^{d}\sum_{k=1}^{N} p_k \langle i|\psi_k\rangle\langle\psi_k|\hat{A}|i\rangle \\ &= \sum_{i=1}^{d}\langle i|\Big(\underbrace{\sum_{k=1}^{N} p_k|\psi_k\rangle\langle\psi_k|}_{\rho}\Big)\hat{A}|i\rangle = \sum_{i=1}^{d}\langle i|\rho\hat{A}|i\rangle = \mathrm{Tr}[\rho\hat{A}],\end{aligned} \tag{6.1}$$

where we introduced the **density operator**

[1] In the second step, we use the completeness relation $\mathbb{1} = \sum_{i=1}^{d} |i\rangle\langle i|$, where d is the dimension of the Hilbert space.

$$\boxed{\rho = \sum_{k=1}^{N} p_k |\psi_k\rangle\langle\psi_k|.} \tag{6.2}$$

More precisely, any operator expressed as in (6.2) with $p_k \geq 0$ and $\sum_k p_k = 1$ is called a density operator.

Density operators are the most general way of describing quantum systems[2]. Hence, hereafter the terms "quantum state" and "density operator" will be used interchangeably. The matrix associated with a density operator is called a **density matrix** and is defined as

$$\rho_{ij} = \langle i|\rho|j\rangle, \tag{6.3}$$

where $\{|i\rangle\}_{i=1}^{d}$ is an orthonormal basis of the Hilbert space. The diagonal entries of this matrix are called **populations**. They indicate the probability of measuring a basis vector. The off-diagonal elements are called **coherences**, and they signal the presence of a superposition state with respect to the orthonormal basis chosen for the decomposition. In Section 6.1.1, we will demonstrate that density operators are Hermitian; thus, it is always possible to find a basis in which the density matrix is diagonal. *In this basis* the coherences are all equal to zero. Populations and coherences are represented in Fig. 6.2a.

To summarize, when we have complete information about a quantum system, the latter is described by a single vector $|\psi\rangle$. The sum in eqn (6.2) has only one term and the density operator $\rho = |\psi\rangle\langle\psi|$ is a pure state. If the quantum system is prepared in one of many possible states according to a probability distribution, the state of the system is mixed. In this case, the density operator ρ is expressed as a sum of several terms, $\rho = \sum_k p_k |\psi_k\rangle\langle\psi_k|$. A pure state can be easily distinguished from a mixed state because

$$\begin{aligned} \text{for pure states:} \quad & \rho^2 = \rho, \\ \text{and for mixed states:} \quad & \rho^2 \neq \rho. \end{aligned}$$

Since for pure states $\rho^2 = \rho$, pure states are described by projectors.

Example 6.1 Consider a qubit in state $|0\rangle$. The associated density operator $\rho = |0\rangle\langle 0|$ has matrix form

$$\rho = |0\rangle\langle 0| = \begin{bmatrix} 1 & 0 \\ 0 & 0 \end{bmatrix}. \tag{6.4}$$

It is easy to see that $\rho^2 = \rho$ and therefore ρ is a pure state.

Example 6.2 Consider a qubit in a balanced superposition state $|+\rangle = (|0\rangle + |1\rangle)/\sqrt{2}$. The matrix of the corresponding density operator is

[2] This is a consequence of Gleason's theorem [184].

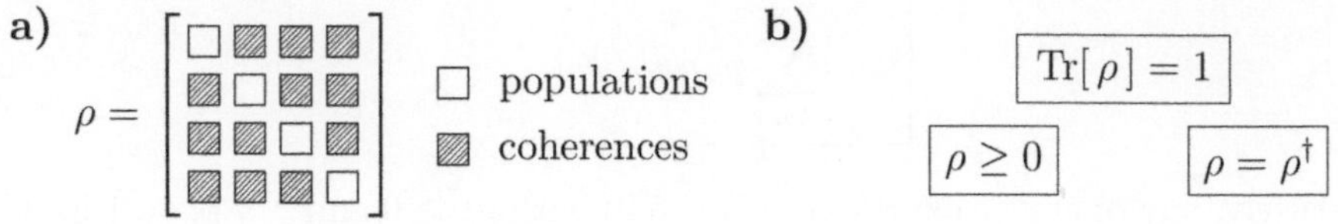

a) $\rho =$ □ populations ▨ coherences

b) $\mathrm{Tr}[\rho] = 1$ $\rho \geq 0$ $\rho = \rho^\dagger$

Fig. 6.2 The density operator. a) The diagonal elements of a density matrix are called the populations, and the off-diagonal elements are called the coherences. **b)** The density matrix is positive and has unit trace. Since it is positive, it is also Hermitian.

$$\begin{aligned}\rho = |+\rangle\langle+| &= \frac{1}{\sqrt{2}}\left[|0\rangle + |1\rangle\right]\left[\langle 0| + \langle 1|\right]\frac{1}{\sqrt{2}} \\ &= \frac{1}{2}\left[|0\rangle\langle 0| + |0\rangle\langle 1| + |1\rangle\langle 0| + |1\rangle\langle 1|\right] = \frac{1}{2}\begin{bmatrix} 1 & 1 \\ 1 & 1 \end{bmatrix}.\end{aligned} \tag{6.5}$$

One can verify that $\rho^2 = \rho$ and therefore ρ is a pure state. This example clearly shows that the off-diagonal elements of a pure state can be non-zero. If the density matrix of a pure state is diagonalized, the elements on the diagonal will all be zero apart from one.

Example 6.3 Consider a source that prepares a qubit in state $|0\rangle$ with 50% probability and state $|1\rangle$ with 50% probability. The density operator associated with this statistical ensemble is

$$\rho = \frac{50}{100}|0\rangle\langle 0| + \frac{50}{100}|1\rangle\langle 1| = \begin{bmatrix} 1/2 & 0 \\ 0 & 1/2 \end{bmatrix} = \frac{\mathbb{1}}{2}. \tag{6.6}$$

This state is called a **completely depolarized state**. It is evident that $\rho^2 \neq \rho$ and therefore ρ is a mixed state. Observe that the state $\frac{1}{2}|+\rangle\langle+| + \frac{1}{2}|-\rangle\langle-|$ is equivalent to (6.6),

$$\begin{aligned}\rho &= \frac{1}{2}|+\rangle\langle+| + \frac{1}{2}|-\rangle\langle-| = \frac{1}{4}\big(|0\rangle + |1\rangle\big)\big(\langle 0| + \langle 1|\big) + \frac{1}{4}\big(|0\rangle - |1\rangle\big)\big(\langle 0| - \langle 1|\big) \\ &= \frac{1}{4}\big(|0\rangle\langle 0| + |0\rangle\langle 1| + |1\rangle\langle 0| + |1\rangle\langle 1| + |0\rangle\langle 0| - |0\rangle\langle 1| - |1\rangle\langle 0| + |1\rangle\langle 1|\big) \\ &= \frac{1}{2}|0\rangle\langle 0| + \frac{1}{2}|1\rangle\langle 1|.\end{aligned}$$

In other words, a source that prepares a qubit in $|0\rangle$ and $|1\rangle$ with the same probability cannot be distinguished from a source that prepares $|+\rangle$ and $|-\rangle$ with the same probability.

6.1.1 Properties of density operators

In this section, we show that density operators are positive operators with unit trace. Consider a density operator $\rho = \sum_k p_k|\psi_k\rangle\langle\psi_k|$ where $p_k \geq 0$ and $\sum_k p_k = 1$. It is straightforward to show that ρ has **unit trace**,

$$\text{Tr}[\rho] = \text{Tr}\Big[\sum_{k=1}^{N} p_k|\psi_k\rangle\langle\psi_k|\Big] = \sum_{k=1}^{N} p_k \text{Tr}\big[|\psi_k\rangle\langle\psi_k|\big] = \sum_{k=1}^{N} p_k\langle\psi_k|\psi_k\rangle = \sum_{k=1}^{N} p_k = 1. \tag{6.7}$$

We can also show that ρ is **positive**. Indeed, for a generic vector $|x\rangle$, we have

$$\langle x|\rho|x\rangle = \sum_{k=1}^{N} p_k\langle x|\psi_k\rangle\langle\psi_k|x\rangle = \sum_{k=1}^{N} p_k|\langle x|\psi_k\rangle|^2 \geq 0\,. \tag{6.8}$$

As seen in Section 3.2.12, all positive operators are **Hermitian** and, according to the spectral theorem, can be expressed in the form

$$\boxed{\rho = \sum_{i=1}^{d} \lambda_i |i\rangle\langle i|,} \tag{6.9}$$

where the sum of the eigenvalues λ_i is equal to 1. These properties are summarized in Fig. 6.2b.

If a linear operator $\hat{A} \in \text{B}(H)$ is positive and has unit trace, then it is a density operator. This follows from the fact that since $\hat{A}$ is positive, it can be diagonalized and expressed in the form $\hat{A} = \sum_i \lambda_i|i\rangle\langle i|$. In addition, since it has unit trace, then $\sum_i \lambda_i = 1$ and we obtain eqn (6.9). We will use the notation $\text{D}(H)$ to indicate the set of all density operators on a Hilbert space H,

$$\text{D}(H) = \{\hat{A} \in \text{B}(H) : \hat{A} \geq 0 \text{ and } \text{Tr}[\hat{A}] = 1\}. \tag{6.10}$$

Figure 6.3 shows the relation between the set of density operators and the sets of positive, Hermitian, and normal operators.

How many parameters are needed to completely specify a density operator? The density matrix is a $d \times d$ complex matrix determined by $2d^2$ real numbers. Since ρ is Hermitian, this property imposes $2\frac{d^2-d}{2} + d$ constraints (the entries in the upper-right triangle are equal to the complex conjugate of the entries in the bottom-left triangle and the diagonal entries must be real), reducing the number of independent parameters to d^2. Taking into account that ρ has unit trace, the number of independent real parameters reduces to $d^2 - 1$. For instance, the Hilbert space of a qubit has dimension $d = 2$ and therefore the qubit density matrix has three independent parameters (the components of the Bloch vector, see Section 6.4).

6.1.2 Purity and von Neumann entropy

The purity and the von Neumann entropy are two important functions that can be used to characterize density operators. The **purity** μ is a real number that quantifies how much a quantum state is close to a pure state,

$$\boxed{\mu(\rho) = \text{Tr}[\rho^2] = \sum_{i=1}^{d} \lambda_i^2,} \tag{6.11}$$

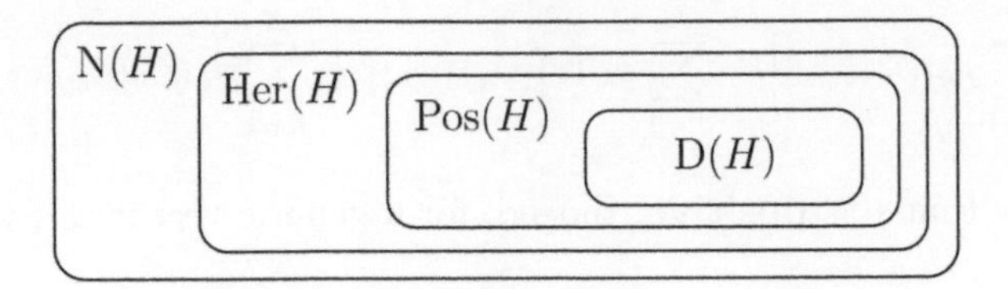

Fig. 6.3 Relations between linear operators. The set of density operators $\mathrm{D}(H)$ is contained in the set of positive operators $\mathrm{Pos}(H)$, Hermitian operators $\mathrm{Her}(H)$, and normal operators $\mathrm{N}(H)$.

where λ_i are the eigenvalues of ρ. The purity lies in the range $1/d \leq \mu(\rho) \leq 1$, where d is the dimension of the Hilbert space. The reader can verify that: 1) $\mu(\rho) = 1/d$ only for the completely depolarized state $\mathbb{1}/d$ and 2) $\mu = 1$ only for pure states. It is sometimes convenient to introduce a slightly modified definition of purity so that its value is bounded between 0 and 1 [20],

$$\tilde{\mu}(\rho) = \frac{d}{d-1}\Big(\mathrm{Tr}[\rho^2] - \frac{1}{d}\Big). \tag{6.12}$$

The definitions (6.11, 6.12) are similar and their use is just a matter of convenience. It is interesting to observe that the purity of a quantum state does not change under a unitary transformation,

$$\mu(\hat{U}\rho\hat{U}^\dagger) = \mathrm{Tr}[\hat{U}\rho^2\hat{U}^\dagger] = \mathrm{Tr}[\rho^2\hat{U}^\dagger\hat{U}] = \mathrm{Tr}[\rho^2] = \mu(\rho). \tag{6.13}$$

This is related to the fact that unitary transformations are associated with reversible time evolutions.

The **von Neumann entropy** is the natural extension of classical entropy to the field of quantum mechanics and quantifies how much a quantum state deviates from a pure state

$$\boxed{S(\rho) = -\mathrm{Tr}[\rho\log_2(\rho)] = -\sum_{i=1}^{d}\lambda_i\log_2(\lambda_i).} \tag{6.14}$$

Since $S(\rho) = S(\hat{U}\rho\hat{U}^\dagger)$, the von Neumann entropy does not depend on the basis in which ρ is decomposed and it only depends on its eigenvalues. Furthermore, $S(\rho)$ is a well-defined quantity, since the eigenvalues $\lambda_i \in [0,1]$ and $\lim_{\lambda\to 0^+}\lambda\cdot\log_2\lambda = 0$. It can be shown that the von Neumann entropy lies in the range $0 \leq S(\rho) \leq \log_2(d)$.

The purity and von Neumann entropy can be used to distinguish mixed states from pure states:

$$\begin{aligned}
&\text{for pure states:} && \rho^2 = \rho, && \mu(\rho) = 1, && S(\rho) = 0,\\
&\text{and for mixed states:} && \rho^2 \neq \rho, && \mu(\rho) < 1, && S(\rho) > 0.
\end{aligned}$$

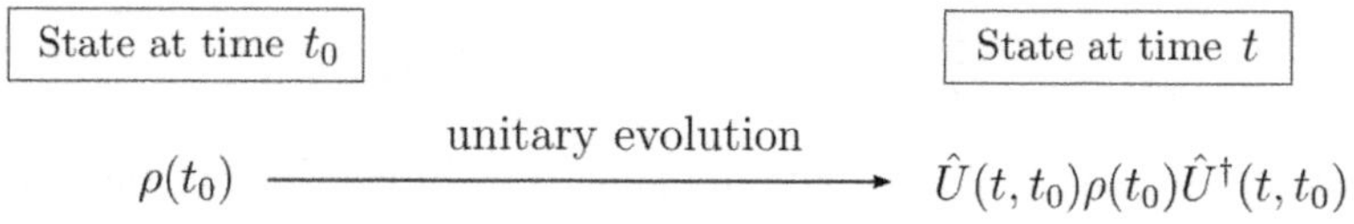

State at time t_0 | State at time t

$$\rho(t_0) \xrightarrow{\text{unitary evolution}} \hat{U}(t,t_0)\rho(t_0)\hat{U}^\dagger(t,t_0)$$

Fig. 6.4 Unitary evolution of a quantum state. When the dynamics of a quantum state $\rho(t_0)$ is governed by a Hamiltonian $\hat{H}$, the state of the system at time t is $\hat{U}(t,t_0)\rho(t_0)\hat{U}^\dagger(t,t_0)$, where $\hat{U}(t,t_0)$ is written in eqn (6.15).

The von Neumann entropy will play an important role in the characterization of entangled states, see Section 9.3.

Example 6.4 A qubit in a completely depolarized state, $\rho = \mathbb{1}/2 = \sum_j \frac{1}{2}|j\rangle\langle j|$, has von Neumann entropy $S(\rho) = -(\frac{1}{2}\log_2\frac{1}{2}) - (\frac{1}{2}\log_2\frac{1}{2}) = 1$ and purity $\mu(\rho) = 1/2^2 + 1/2^2 = 1/2$.

6.2 Unitary evolution of a quantum state

In this section, we investigate the evolution of a density operator in time. The second postulate of quantum mechanics states that the evolution of a quantum state from t_0 to t is described by a unitary operator

$$\hat{U}(t,t_0) = \mathcal{T}\exp\left(\frac{1}{i\hbar}\int_{t_0}^{t}\hat{H}(t')dt'\right), \tag{6.15}$$

where $\hat{H}(t')$ is the Hamiltonian of the system. The action of this operator on an arbitrary quantum state

$$\rho(t_0) = \sum_{k=1}^{N} p_k|\psi_k(t_0)\rangle\langle\psi_k(t_0)|$$

takes the form

$$\rho(t) = \sum_k p_k\hat{U}(t,t_0)\Big(|\psi_k(t_0)\rangle\langle\psi_k(t_0)|\Big)\hat{U}^\dagger(t,t_0) = \hat{U}(t,t_0)\,\rho(t_0)\,\hat{U}^\dagger(t,t_0). \tag{6.16}$$

Figure 6.4 shows the evolution of an isolated quantum system in time. From (6.16), we can easily compute the time derivative of a density operator and obtain the **von Neumann equation**

$$\begin{aligned}\frac{d\rho(t)}{dt} &= \sum_k p_k\left[\left(\frac{d}{dt}|\psi_k(t)\rangle\right)\langle\psi_k(t)| + |\psi_k(t)\rangle\left(\frac{d}{dt}\langle\psi_k(t)|\right)\right]\\ &= \sum_k p_k\left[\frac{1}{i\hbar}\hat{H}|\psi_k(t)\rangle\langle\psi_k(t)| - |\psi_k(t)\rangle\langle\psi_k(t)|\hat{H}\frac{1}{i\hbar}\right] =\end{aligned}$$

$$= \frac{1}{i\hbar}[\hat{H}\rho(t) - \rho(t)\hat{H}] = \frac{1}{i\hbar}[\hat{H}, \rho(t)],$$

where we used the Schrödinger equation, eqn (4.32). In conclusion, the von Neumann equation is

$$\boxed{\frac{d\rho}{dt} = \frac{1}{i\hbar}[\hat{H}, \rho].} \tag{6.17}$$

The von Neumann equation describes the dynamics of an *isolated* quantum system. As we shall see in Chapter 8, quantum systems interacting with the environment can undergo a much broader class of evolution processes.

In many cases, the Hamiltonian of the system is the sum of two terms

$$\boxed{\hat{H} = \hat{H}_0 + \hat{V},}$$

where $\hat{H}_0$ is a tractable Hamiltonian with a known spectrum, while $\hat{H}_0 + \hat{V}$ is more difficult to handle analytically. Since we are mainly interested in the dynamics induced by $\hat{V}$, it is convenient to study the time evolution in the interaction picture. This can be done by performing a time-dependent change of basis, $\hat{R} = e^{-\frac{1}{i\hbar}\hat{H}_0 t}$, so that in the new frame the state of the system becomes

$$\tilde{\rho} = \hat{R}\rho\hat{R}^{-1} = e^{-\frac{1}{i\hbar}\hat{H}_0 t}\rho e^{\frac{1}{i\hbar}\hat{H}_0 t}, \tag{6.18}$$

where the tilde indicates that the corresponding operator is expressed in the interaction picture. The Hamiltonian in the interaction picture is given by

$$\tilde{H} = \hat{R}(\hat{H}_0 + \hat{V})\hat{R}^{-1} - i\hbar\hat{R}\Big(\frac{d}{dt}\hat{R}^{-1}\Big) = \hat{R}\hat{V}\hat{R}^{-1} = \tilde{V}, \tag{6.19}$$

where we used eqn (4.37). Thus, the von Neumann equation reduces to

$$\boxed{\frac{d\tilde{\rho}}{dt} = \frac{1}{i\hbar}[\tilde{V}, \tilde{\rho}].}$$

The relation between $d\tilde{\rho}/dt$ and $d\rho/dt$ can be calculated by taking the time derivative of eqn (6.18),

$$\frac{d\tilde{\rho}}{dt} = -\frac{1}{i\hbar}\hat{H}_0\hat{R}\rho\hat{R}^{-1} + \hat{R}\frac{d\rho}{dt}\hat{R}^{-1} + \frac{1}{i\hbar}\hat{R}\rho\hat{R}^{-1}\hat{H}_0.$$

After multiplying on the left by $\hat{R}^{-1}$ and on the right by $\hat{R}$ and rearranging some terms, we obtain

$$\boxed{\frac{d\rho}{dt} = \frac{1}{i\hbar}[\hat{H}_0, \rho] + \hat{R}^{-1}\frac{d\tilde{\rho}}{dt}\hat{R}.} \tag{6.20}$$

This equation relates the time derivative of the density operator in the Schrödinger picture and the interaction picture. This equation will be useful in Section 8.3 when we derive the master equation of a qubit coupled to a bath of harmonic oscillators.

Example 6.5 Suppose that a qubit is in the initial state

$$\rho(0) = |+\rangle\langle+| = \begin{bmatrix} 1/2 & 1/2 \\ 1/2 & 1/2 \end{bmatrix} \tag{6.21}$$

and its dynamics are governed by the Hamiltonian $\hat{H} = \hbar\omega\hat{\sigma}_z$. The von Neumann equation $i\hbar d\rho/dt = \hat{H}\rho - \rho\hat{H}$ becomes

$$i\hbar\frac{d}{dt}\begin{bmatrix} \rho_{00} & \rho_{01} \\ \rho_{10} & \rho_{11} \end{bmatrix} = \hbar\omega(\hat{\sigma}_z\rho - \rho\hat{\sigma}_z) = \hbar\omega\begin{bmatrix} 0 & 2\rho_{01} \\ -2\rho_{10} & 0 \end{bmatrix}.$$

This expression produces four differential equations:

$$\frac{d\rho_{00}(t)}{dt} = 0, \qquad \frac{d\rho_{01}(t)}{dt} = -i2\omega\rho_{01}(t),$$
$$\frac{d\rho_{10}(t)}{dt} = i2\omega\rho_{10}(t), \qquad \frac{d\rho_{11}(t)}{dt} = 0.$$

Since a density operator is a Hermitian operator with unit trace, $\rho_{10}(t) = \rho_{01}^*(t)$ and $\rho_{11}(t) = 1 - \rho_{00}(t)$. This means that only two of these differential equations are independent. Their solution is

$$\rho_{00}(t) = \frac{1}{2}, \qquad \rho_{01}(t) = \frac{1}{2}e^{-2i\omega t},$$

where we used the initial conditions $\rho_{00}(0) = \rho_{01}(0) = 1/2$ (see eqn 6.21). In conclusion, the populations remain unchanged, while the coherences acquire a dynamical phase. In other words, since the Hamiltonian is proportional to σ_z, the time evolution of the qubit state corresponds to a rotation about the z-axis of the Bloch sphere.

6.3 The measurement of a quantum state

Suppose that a source prepares a quantum system in a statistical ensemble $\{|\psi_k\rangle, p_k\}_{k=1}^N$ and an observable $\hat{A}$ is measured. As stated by the fourth postulate of quantum mechanics presented in Chapter 4, this measurement will produce an eigenvalue a_i. What is the probability of obtaining this measurement outcome? The probability of obtaining a_i, assuming that the system was prepared in $|\psi_k\rangle$, is given by the conditional probability

$$p(a_i \mid \psi_k) = \langle\psi_k|\hat{P}_i|\psi_k\rangle = \text{Tr}[\hat{P}_i|\psi_k\rangle\langle\psi_k|], \tag{6.22}$$

where $\hat{P}_i$ is the projector on the eigenspace of a_i and in the last step we used property **T5** presented in Section 3.2.14. The total probability of obtaining the measurement outcome a_i is the sum of all the possible ways of obtaining this result,

$$\boxed{p(a_i) = \sum_{k=1}^{N} p_k \cdot p(a_i \mid \psi_k) = \sum_{k=1}^{N} p_k \text{Tr}[\hat{P}_i|\psi_k\rangle\langle\psi_k|] = \text{Tr}[\hat{P}_i\rho].} \tag{6.23}$$

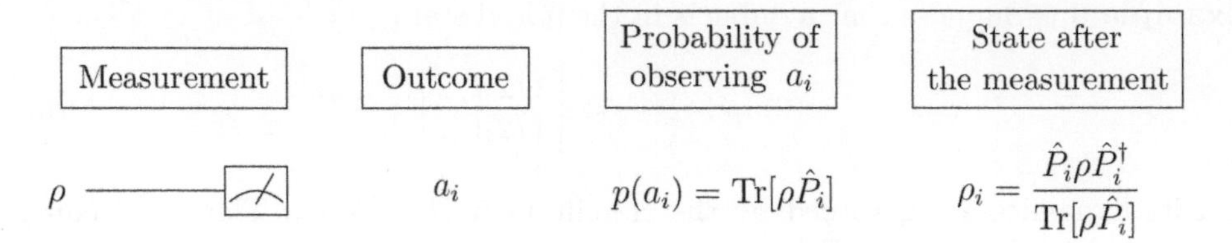

Fig. 6.5 The measurement of a quantum state. The measurement of an observable $\hat{A}$ produces a result a_i. This outcome is observed with probability $p(a_i) = \text{Tr}[\rho \hat{P}_i]$ where $\hat{P}_i$ is the projector onto the eigenspace associated with a_i (if this eigenvalue is non-degenerate $\hat{P}_i = |a_i\rangle\langle a_i|$). The measurement collapses that state of the system into $\rho_i = \hat{P}_i \rho \hat{P}_i^\dagger / \text{Tr}[\hat{P}_i \rho]$.

This is the **Born rule for density operators.** In short, the probability of measuring a_i is given by the trace of the product between the density operator ρ and the projector $\hat{P}_i$.

What about the state of the system after the measurement? The sixth postulate of quantum mechanics tells us that if the measurement of $|\psi_k\rangle$ produces an outcome a_i, then the state of the system after the measurement is[3]

$$|\psi_k^{\text{fin}}\rangle = \frac{\hat{P}_i |\psi_k\rangle}{\sqrt{\langle \psi_k | \hat{P}_i | \psi_k \rangle}} . \tag{6.24}$$

Since the source prepares the quantum system according to a probability distribution, we do not know, with certainty, the state of the system before the measurement. The probability that the system was prepared in $|\psi_k\rangle$ given that the measurement outcome is a_i is the conditional probability $p(\psi_k \mid a_i)$. Therefore, the state of the system after the measurement is a statistical ensemble $\{|\psi_k^{\text{fin}}\rangle, p(\psi_k \mid a_i)\}_k$ described by the density operator

$$\rho_i = \sum_{k=1}^{N} p(\psi_k \mid a_i) \frac{\hat{P}_i |\psi_k\rangle\langle\psi_k| \hat{P}_i}{\langle \psi_k | \hat{P}_i | \psi_k \rangle} . \tag{6.25}$$

Bayes' theorem states that the conditional probability $p(\psi_k \mid a_i)$ satisfies the relation

$$p(\psi_k \mid a_i) = \frac{p_k \cdot p(a_i \mid \psi_k)}{p(a_i)} = \frac{p_k \langle \psi_k | \hat{P}_i | \psi_k \rangle}{p(a_i)} , \tag{6.26}$$

where in the last step we used (6.22). Substituting this relation into (6.25), we obtain

$$\boxed{\rho_i = \frac{\sum_{k=1}^{N} p_k \hat{P}_i |\psi_k\rangle\langle\psi_k| \hat{P}_i}{p(a_i)} = \frac{\hat{P}_i \rho \hat{P}_i}{\text{Tr}[\hat{P}_i \rho]},} \tag{6.27}$$

where in the last step we used eqns (6.2, 6.23). The density operator ρ_i of eqn (6.27) describes the state of the system after a measurement with outcome a_i. The denomi-

[3]If the eigenvalue a_i is not degenerate, $\hat{P}_i = |a_i\rangle\langle a_i|$ and the state after the measurement is $|a_i\rangle$.

nator ensures that the state is correctly normalized. The measurement of a quantum state is illustrated schematically in Fig. 6.5.

Example 6.6 Suppose that a measurement of an observable $\hat{A}$ is performed on a superposition state $|\psi\rangle = \sum_i c_i|a_i\rangle$. Assuming that the eigenvalues a_i are not degenerate, the projector on the eigenspace associated with a_i is given by $\hat{P}_i = |a_i\rangle\langle a_i|$. If the measurement outcome is a_i, the state of the system after the measurement is

$$\rho_i = \frac{\hat{P}_i\rho\hat{P}_i}{\text{Tr}[\hat{P}_i\rho]} = \frac{|a_i\rangle\langle a_i|\psi\rangle\langle\psi|a_i\rangle\langle a_i|}{\text{Tr}[|a_i\rangle\langle a_i|\psi\rangle\langle\psi|]} = \frac{\langle a_i|\psi\rangle\langle\psi|a_i\rangle|a_i\rangle\langle a_i|}{\langle a_i|\psi\rangle\langle\psi|a_i\rangle} = |a_i\rangle\langle a_i|, \tag{6.28}$$

where we used (6.27). The state of the system after the measurement is the eigenstate $|a_i\rangle$ as expected.

Example 6.7 Consider a quantum state defined on the Hilbert state $\mathbb{C}^3$,

$$\rho = \frac{1}{6}|0\rangle\langle 0| + \frac{2}{6}|1\rangle\langle 1| + \frac{3}{6}|2\rangle\langle 2|\,. \tag{6.29}$$

Suppose that a projective measurement on the two subspaces $\{\hat{P}_x = |0\rangle\langle 0| + |1\rangle\langle 1|$, $\hat{P}_y = |2\rangle\langle 2|\}$ is performed. If the measurement outcome is x, the state of the system after the measurement is

$$\rho_x = \frac{\hat{P}_x\rho\hat{P}_x}{\text{Tr}[\hat{P}_x\rho]} = \frac{(|0\rangle\langle 0| + |1\rangle\langle 1|)\,\rho\,(|0\rangle\langle 0| + |1\rangle\langle 1|)}{\text{Tr}[(|0\rangle\langle 0| + |1\rangle\langle 1|)\,\rho]} = \frac{1}{3}|0\rangle\langle 0| + \frac{2}{3}|1\rangle\langle 1|\,. \tag{6.30}$$

In this case, the mixed state ρ turns into the mixed state ρ_x.

6.4 The quantum state of a qubit

Density operators can be used to describe a qubit state. Recall from Section 5.1 that if a qubit state is pure, it can be expressed as

$$|\psi\rangle = \cos\left(\frac{\theta}{2}\right)|0\rangle + e^{i\phi}\sin\left(\frac{\theta}{2}\right)|1\rangle\,. \tag{6.31}$$

The density operator associated with this pure state is

$$\rho = |\psi\rangle\langle\psi| = \begin{bmatrix} \cos^2\frac{\theta}{2} & e^{-i\phi}\cos\frac{\theta}{2}\sin\frac{\theta}{2} \\ e^{i\phi}\cos\frac{\theta}{2}\sin\frac{\theta}{2} & \sin^2\frac{\theta}{2} \end{bmatrix}. \tag{6.32}$$

For any value of $\theta \in [0,\pi]$ and $\phi \in [0,2\pi)$, the density operator satisfies $\rho^2 = \rho$ as expected.

We now derive the density operator associated with a generic qubit state (either pure or mixed). This operator will be associated with a 2×2 positive matrix with unit trace. An orthogonal basis for the space of 2×2 complex matrices is given by the Pauli basis $B = \{\mathbb{1}, \hat{\sigma}_x, \hat{\sigma}_y, \hat{\sigma}_z\}$. Therefore, any qubit density matrix can be decomposed as

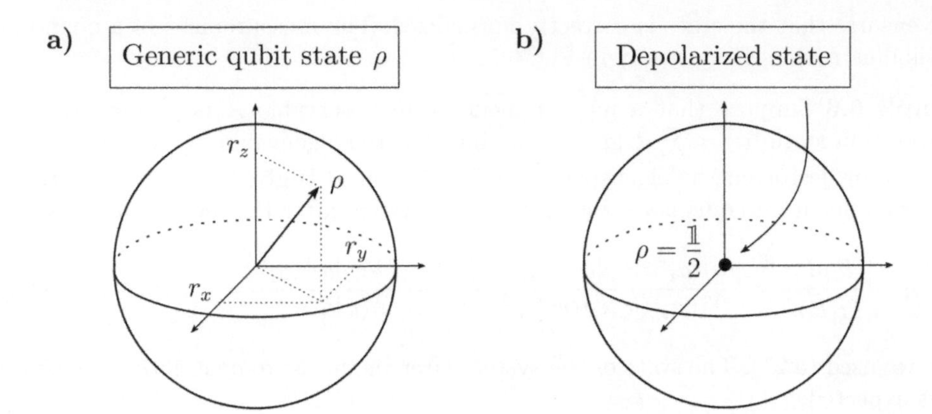

Fig. 6.6 Bloch sphere. a) A representation of a Bloch vector in the Bloch sphere. If the Bloch vector is on the surface of the sphere, the state is pure. If it is inside the sphere, the state is mixed. **b)** The completely depolarized state of a qubit corresponds to the center of the Bloch sphere.

$$\rho = r_{\mathbb{1}}\frac{\mathbb{1}}{2} + r_x\frac{\hat{\sigma}_x}{2} + r_y\frac{\hat{\sigma}_y}{2} + r_z\frac{\hat{\sigma}_z}{2}, \tag{6.33}$$

where $r_{\mathbb{1}}$, r_x, r_y, r_z are real parameters. Since $\mathrm{Tr}\,[\rho] = \mathrm{Tr}\,[\mathbb{1}/2] = 1$ and $\mathrm{Tr}\,[\hat{\sigma}_x] = \mathrm{Tr}\,[\hat{\sigma}_y] = \mathrm{Tr}\,[\hat{\sigma}_z] = 0$, then $r_{\mathbb{1}} = 1$. This means that

$$\boxed{\rho = \frac{1}{2}\left[\mathbb{1} + r_x\hat{\sigma}_x + r_y\hat{\sigma}_y + r_z\hat{\sigma}_z\right] = \frac{1}{2}\begin{bmatrix} 1+r_z & r_x - ir_y \\ r_x + ir_y & 1 - r_z \end{bmatrix},} \tag{6.34}$$

where $\vec{r} = (r_x, r_y, r_z)$ is called the **Bloch vector**. From (6.34), we can derive the explicit expression of the determinant of ρ,

$$\det(\rho) = \frac{1}{4}\left[(1+r_z)(1-r_z) - (r_x - ir_y)(r_x + ir_y)\right] = \frac{1}{4}\left(1 - |\vec{r}|^2\right). \tag{6.35}$$

Since ρ is positive, its determinant is non-negative[4] and thus $0 \le |\vec{r}| \le 1$. This means that the maximum length of the Bloch vector is 1. Equation (6.35) can also be expressed in the form[5]

$$\rho = \begin{bmatrix} \rho_{00} & \rho_{01} \\ \rho_{10} & \rho_{11} \end{bmatrix}, \tag{6.36}$$

where ρ_{00} and ρ_{11} are the populations and ρ_{01}, ρ_{10} are the coherences. Comparing eqn (6.34) with eqn (6.36), we see that the matrix elements ρ_{ij} are related to the coordinates of the Bloch vector,

[4]The determinant of ρ is the product of its eigenvalues, $\det(\rho) = \lambda_1\lambda_2$. Since ρ is positive, the eigenvalues are non-negative and thus $\det(\rho) \ge 0$.

[5]In other texts, the reader will find the equivalent notation $\rho_{00} = \rho_{\mathrm{gg}}$, $\rho_{01} = \rho_{\mathrm{ge}}$, $\rho_{10} = \rho_{\mathrm{eg}}$, $\rho_{11} = \rho_{\mathrm{ee}}$.

$$\rho_{00} = \frac{1+r_z}{2}, \qquad \rho_{01} = \frac{r_x - ir_y}{2}, \qquad \rho_{10} = \frac{r_x + ir_y}{2}, \qquad \rho_{11} = \frac{1-r_z}{2}. \tag{6.37}$$

It is important to remember that the elements ρ_{ij} are not independent: since ρ is Hermitian with unit trace, $\rho_{01} = \rho_{10}^*$ and $\rho_{00} = 1 - \rho_{11}$.

If the qubit state is pure, one of the two eigenvalues of ρ must be equal to zero, which implies that $\det(\rho) = 0$. From (6.35), this happens if and only if $|\vec{r}| = 1$. Thus, the points on the Bloch sphere represent pure states, while points inside the sphere represent mixed states. In particular, the qubit states on the vertical axis of the Bloch sphere are **classical states** that do not possess any particular quantum feature. These states can be considered classical mixtures in which the qubit is prepared in $|0\rangle$ with probability p and in $|1\rangle$ with probability $1-p$. Figure 6.6a shows a representation of a Bloch vector in the Bloch sphere.

We conclude this section by discussing the **completely depolarized state** of a qubit. Suppose that a source prepares a qubit on a random point of the Bloch sphere with equal probability. The density operator associated with this state is

$$\rho = \int_0^{2\pi} \frac{d\phi}{4\pi} \int_0^{\pi} d\theta \sin\theta \left(\cos\frac{\theta}{2}|0\rangle + e^{i\phi}\sin\frac{\theta}{2}|1\rangle\right)\left(\cos\frac{\theta}{2}\langle 0| + e^{i\phi}\sin\frac{\theta}{2}\langle 1|\right).$$

By calculating this double integral, we find

$$\rho = \int_0^{2\pi} \frac{d\phi}{4\pi} \int_0^{\pi} d\theta \sin\theta \begin{bmatrix} \cos^2\frac{\theta}{2} & e^{-i\phi}\cos\frac{\theta}{2}\sin\frac{\theta}{2} \\ e^{-i\phi}\cos\frac{\theta}{2}\sin\frac{\theta}{2} & \sin^2\frac{\theta}{2} \end{bmatrix} = \begin{bmatrix} 1/2 & 0 \\ 0 & 1/2 \end{bmatrix} = \frac{\mathbb{1}}{2}. \tag{6.38}$$

Therefore, the completely depolarized state of a qubit[6] is proportional to the identity. This is a classical state in which the qubit is found 50% of the time in $|0\rangle$ and 50% of the time in $|1\rangle$. Since for a completely depolarized state $r_x = r_y = r_z = 0$, this mixed state corresponds to the center of the Bloch sphere, as shown in Fig. 6.6b.

There is a profound difference between the pure state $|+\rangle = [|0\rangle + |1\rangle]/\sqrt{2}$ and the mixed state $\mathbb{1}/2 = \frac{1}{2}[|0\rangle\langle 0| + |1\rangle\langle 1|]$. If these two states are measured in the computational basis, in both cases we will obtain $|0\rangle$ and $|1\rangle$ with the same probability. However, if the pure state $|+\rangle$ is measured in the basis $B = \{|+\rangle\langle +|, |-\rangle\langle -|\}$, the measurement will collapse the system in $|+\rangle$ with 100% probability. On the contrary, there does not exist any projective measurement that will project the completely depolarized state into a particular state with certainty. The completely depolarized state can be distinguished from the superposition state $|+\rangle$ by measuring the system in a basis other than the computational basis.

The decomposition of a density operator as a sum of pure states is not unique. For example, the completely depolarized state can be expressed as $\frac{1}{2}(|0\rangle\langle 0| + |1\rangle\langle 1|)$ but also as

[6] It is called completely depolarized because the expectation values of the Pauli operators in this state are all equal to zero, $\mathrm{Tr}\left[\frac{1}{2}\hat{\sigma}_x\right] = \mathrm{Tr}\left[\frac{1}{2}\hat{\sigma}_y\right] = \mathrm{Tr}\left[\frac{1}{2}\hat{\sigma}_x\right] = 0$.

$$\frac{1}{2}[|+\rangle\langle+|+|-\rangle\langle-|] = \frac{1}{4}(|0\rangle+|1\rangle)(\langle 0|+\langle 1|) + \frac{1}{4}(|0\rangle-|1\rangle)(\langle 0|-\langle 1|)$$
$$= \frac{1}{2}[|0\rangle\langle 0|+|1\rangle\langle 1|] = \frac{\mathbb{1}}{2}.$$

The reader can easily check that any quantum state of the form[7] $\frac{1}{2}[|0_{\hat{n}}\rangle\langle 0_{\hat{n}}|+|1_{\hat{n}}\rangle\langle 1_{\hat{n}}|]$ is equal to $\mathbb{1}/2$. This example shows that different statistical ensembles might lead to the same density operator.

In the next section, we present two important examples of systems at thermal equilibrium, namely a qubit and a quantum harmonic oscillator. We will show that these two systems can be described using density operator formalism.

6.5 Thermal states

Density operators can accurately describe the state of a quantum system in contact with the environment at a constant temperature. Consider a system (such as a quantum harmonic oscillator or a qubit) governed by a Hamiltonian $\hat{H}_s$ and suppose it is in contact with a thermal bath at a temperature T. Statistical mechanics tells us that the state of the system at thermal equilibrium is the one that maximizes the entropy. Similarly, quantum statistical mechanics (the extension of statistical mechanics to the field of quantum physics) tells us that the state of a quantum system at thermal equilibrium is the one that maximizes the von Neumann entropy $S(\rho)$ with the constraint that the energy of the system, $E = \text{Tr}[\rho\hat{H}_s]$, is constant. The density operator satisfying these two requirements can be found by calculating the maximum of the functional

$$\mathcal{L}(\rho) = S(\rho) - \beta\text{Tr}[\rho\hat{H}_s],$$

where β is a Lagrange multiplier. As shown in Appendix B.2, the density operator that maximizes this functional is given by

$$\boxed{\rho_{\text{th}} = \frac{e^{-\beta\hat{H}_s}}{\text{Tr}[e^{-\beta\hat{H}_s}]} = \frac{e^{-\beta\hat{H}_s}}{Z}.} \tag{6.39}$$

This is called a **thermal state** (or Gibbs state). Here, the parameter $\beta = (k_B T)^{-1}$ is inversely proportional to the temperature of the environment and the real number

$$\boxed{Z = \text{Tr}[e^{-\beta\hat{H}_s}]}$$

is called the **partition function**. The partition function ensures that the thermal state is correctly normalized. We will now study two important systems at thermal equilibrium.

[7] Here, $|0_{\hat{n}}\rangle$ and $|1_{\hat{n}}\rangle$ denote two orthonormal qubit states, such as $|+\rangle$ and $|-\rangle$.

6.5.1 A quantum harmonic oscillator at thermal equilibrium

In this section, we derive the density operator of a quantum harmonic oscillator in contact with a thermal bath. In concrete terms, one can imagine an optical cavity or a microwave resonator at a constant temperature. The Hamiltonian of a quantum harmonic oscillator $\hat{H}_{\text{s}} = \hbar\omega_{\text{r}}\hat{a}^\dagger\hat{a}$ has discrete energy levels $E_n = \hbar\omega_{\text{r}} n$ associated with the eigenstates $|n\rangle$. The parameter ω_{r} is the frequency of the oscillator. The states $\{|n\rangle\}_{n=0}^{\infty}$ are called the number states and they form an orthonormal basis of the Hilbert space. From (6.30), the thermal state of a harmonic oscillator is given by

$$\boxed{\rho_{\text{th}} = \frac{e^{-\beta\hat{H}_{\text{s}}}}{\text{Tr}[e^{-\beta\hat{H}_{\text{s}}}]} = \frac{e^{-\beta\hbar\omega_{\text{r}}\hat{a}^\dagger\hat{a}}}{Z}.}$$

Let us calculate the denominator

$$Z = \text{Tr}[e^{-\beta\hat{H}_{\text{s}}}] = \sum_{n=0}^{\infty}\langle n|e^{-\beta\hbar\omega_{\text{r}}\hat{a}^\dagger\hat{a}}|n\rangle = \sum_{n=0}^{\infty} e^{-\beta\hbar\omega_{\text{r}} n} = \frac{1}{1-e^{-\beta\hbar\omega_{\text{r}}}},$$

where in the last step we used $\sum_{n=0}^{\infty} x^n = 1/(1-x)$ which is valid when $0 < x < 1$. The thermal state ρ_{th} can be written as

$$\rho_{\text{th}} = \sum_{n=0}^{\infty}\sum_{n'=0}^{\infty}|n\rangle\langle n|\rho_{\text{th}}|n'\rangle\langle n'| = \sum_{n=0}^{\infty}\frac{e^{-\beta\hbar\omega_{\text{r}} n}}{Z}|n\rangle\langle n| = \sum_{n=0}^{\infty} p(n)|n\rangle\langle n|, \tag{6.40}$$

where we used the completeness relation $\mathbb{1} = \sum_n |n\rangle\langle n|$ and in the last step we defined the probability distribution

$$p(n) = \frac{e^{-\beta\hbar\omega_{\text{r}} n}}{Z} = (1-e^{-\hbar\omega_{\text{r}}/k_{\text{B}}T})e^{-\hbar\omega_{\text{r}} n/k_{\text{B}}T}.$$

The quantity $p(n)$ is the probability that the harmonic oscillator is populated with n excitations. These excitations are called **photons** for electromagnetic resonators and **phonons** for mechanical resonators. The average number of excitations in a thermal state can be computed as follows[8]

$$\begin{aligned}\langle\hat{n}\rangle = \langle a^\dagger a\rangle = \text{Tr}[\rho\hat{a}^\dagger\hat{a}] &= \sum_{n=0}^{+\infty}\langle n|\hat{n}\frac{e^{-\beta\hat{H}_{\text{s}}}}{Z}|n\rangle = \frac{1}{Z}\sum_{n=0}^{\infty} n e^{-\beta\hbar\omega_{\text{r}} n} \\ &= (1-e^{-\beta\hbar\omega_{\text{r}}})\frac{e^{-\beta\hbar\omega_{\text{r}}}}{(1-e^{-\beta\hbar\omega_{\text{r}}})^2} = \frac{1}{e^{\hbar\omega_{\text{r}}/k_{\text{B}}T}-1}.\end{aligned} \tag{6.41}$$

This is called the **Bose–Einstein distribution.** For high temperatures $k_{\text{B}}T \gg \hbar\omega_{\text{r}}$, one can verify that the average photon number can be approximated by $\langle\hat{n}\rangle = k_{\text{B}}T/\hbar\omega_{\text{r}} - 1/2$. Figure 6.7 shows the probability of observing n photons in a 5 GHz

[8] Recall that $\sum_{n=0}^{\infty} nx^n = \frac{x}{(1-x)^2}$ when $0 < x < 1$.

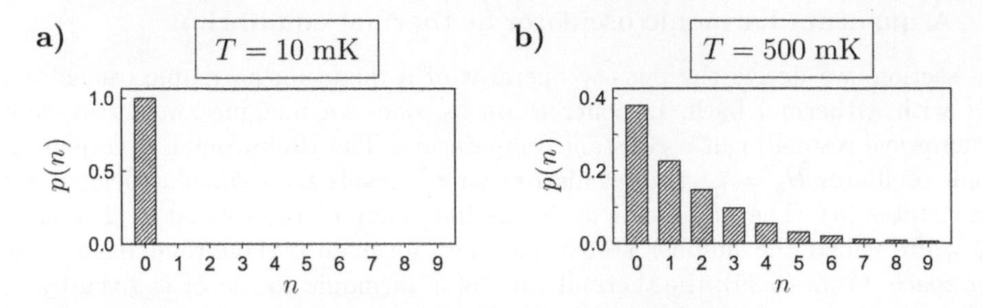

Fig. 6.7 Thermal state of a quantum harmonic oscillator. a) Probability of measuring n excitations in a quantum harmonic oscillator with a frequency of $\omega_{\rm r}/2\pi = 5$ GHz and a temperature $T = 10\,\text{mK}$. Since $10\,\text{mK} \ll \hbar\omega_{\rm r}/k_{\rm B}$, this thermal state almost coincides with the vacuum state. **b)** Same probability distribution at 500 mK. Notice the differences between these histograms and those presented in Fig. 4.9.

---

electromagnetic resonator at 10 mK and at 500 mK. These histograms suggest that a microwave resonator must be cooled down to millikelvin temperatures to suppress thermal excitations. Nowadays, these temperatures can be reached inside dilution refrigerators.

When a harmonic oscillator is prepared in a thermal state, the probability of observing the particle at a position x is a Gaussian centered at the origin. To see this, let us start from the thermal state of a harmonic oscillator per eqn (6.40),

$$\rho_{\rm th} = \frac{1}{Z}\sum_{n=0}^{\infty} e^{-\beta\hbar\omega_{\rm r} n}|n\rangle\langle n| = (1-w)\sum_{n=0}^{\infty} w^n |n\rangle\langle n|,$$

where we introduced $w = e^{-\beta\hbar\omega_{\rm r}}$ for convenience. The probability of observing the particle at the position x is

$$\begin{aligned} p(x) = \text{Tr}[\rho_{\rm th}|x\rangle\langle x|] &= \langle x|\rho_{\rm th}|x\rangle \\ &= (1-w)\sum_{n=0}^{\infty} w^n \langle x|n\rangle\langle n|x\rangle = (1-w)\sum_{n=0}^{\infty} w^n |\psi_n(x)|^2 \\ &= \frac{\lambda}{\sqrt{\pi}}(1-w)e^{-\lambda^2 x^2}\sum_{n=0}^{\infty}\frac{(w/2)^n}{n!}H_n^2(\lambda x), \end{aligned} \tag{6.42}$$

where we used (4.47). Here, $H_n(\lambda x)$ are the Hermite polynomials and $\lambda = \sqrt{m\omega_{\rm r}/\hbar}$ is the inverse of a length. The series in eqn (6.42) is nothing but Mehler's identity[9]. Thus,

$$\boxed{p(x) = \frac{\lambda}{\sqrt{\pi}}\sqrt{\frac{1-w}{1+w}}e^{-\frac{1-w}{1+w}\lambda^2 x^2}.} \tag{6.43}$$

[9]Mehler's identity states that $\sum_{n=0}^{\infty}\frac{(w/2)^n}{n!}H_n(A)H_n(B) = \frac{1}{\sqrt{1-w^2}}\exp\left(\frac{2ABw-(A^2+B^2)w^2}{1-w^2}\right)$.

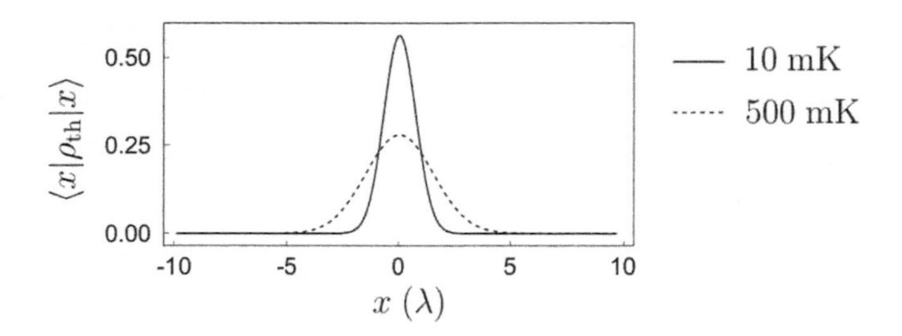

Fig. 6.8 Probability density function for a QHO in a thermal state. Probability density function $\langle x|\rho_{\rm th}|x\rangle$ for a quantum harmonic oscillator (eqn 6.43) in contact with a thermal bath at 10 mK (solid line) and 500 mK (dashed line). In this plot, we assumed $\omega_{\rm r}/2\pi = 5$ GHz.

When the temperature is very low (i.e. when $w \to 0$), this probability distribution reduces to that of the vacuum state $|\psi_0(x)|^2 = \lambda e^{-\lambda^2 x^2}/\sqrt{\pi}$ presented in eqn (4.48).

It is interesting to compare the quantum states of a harmonic oscillator encountered so far. The probability of finding the oscillator at the position x is

$$\text{for number states:} \qquad |\psi_n(x)|^2 = \frac{\lambda}{2^n n!\sqrt{\pi}} H_n^2(\lambda x) e^{-\lambda^2 x^2} \qquad \text{(see eqn 4.47)},$$

$$\text{for thermal states:} \qquad \langle x|\rho_{\rm th}|x\rangle = \frac{\lambda}{\sqrt{\pi}}\sqrt{\frac{1-w}{1+w}} e^{-\frac{1-w}{1+w}\lambda^2 x^2} \qquad \text{(see eqn 6.43)},$$

$$\text{for coherent states:} \qquad |\psi_\alpha(x)|^2 = \frac{\lambda}{\sqrt{\pi}}\left| e^{-\frac{1}{2}(|\alpha|^2-\alpha^2)} e^{-\frac{1}{2}(\lambda x - \sqrt{2}\alpha)}\right|^2 \qquad \text{(see eqn 4.73)}.$$

For number states, the probability density function $|\psi_n(x)|^2$ is an even function centered at the origin (see Fig. 6.8). For thermal states, the probability density function $\langle x|\rho_{\rm th}|x\rangle$ is a Gaussian centered at the origin whose width depends on the temperature of the environment. Finally, for coherent states the probability density function $|\psi_\alpha(x)|^2$ is a Gaussian. Its center depends on α, while its width is determined by λ. The main difference between a thermal state and a coherent state is that the former is a classical mixture, while the latter is a quantum superposition of number states.

6.5.2 A qubit at thermal equilibrium

Another interesting case study is a qubit in contact with a thermal bath. Recall that the Hamiltonian of a qubit is $\hat{H}_{\rm s} = -\hbar\omega_{\rm q}\hat{\sigma}_z/2$ with eigenvectors $\{|0\rangle, |1\rangle\}$ and eigenvalues $\{E_0 = -\hbar\omega_{\rm q}/2, E_1 = \hbar\omega_{\rm q}/2\}$ where $\omega_{\rm q}$ is the transition frequency of the qubit. The associated thermal operator is

$$\boxed{\rho_{\rm th} = \frac{e^{-\beta\hat{H}_{\rm s}}}{\mathrm{Tr}[e^{-\beta\hat{H}_{\rm s}}]} = \frac{e^{\beta\hbar\omega_{\rm q}\hat{\sigma}_z/2}}{Z}.}$$

The partition function is given by

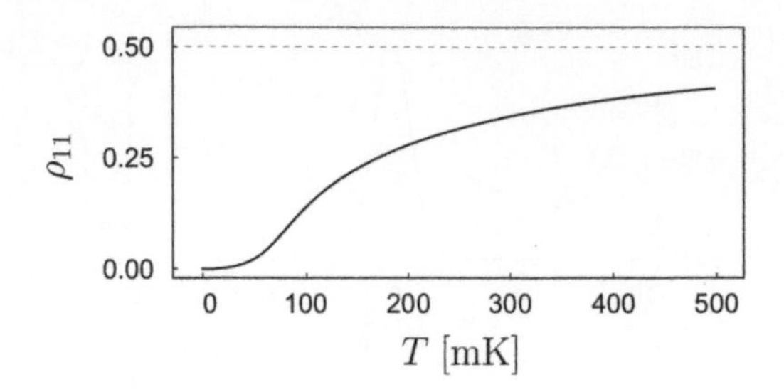

Fig. 6.9 Thermal state of a qubit. Probability of observing a qubit in the excited state $\rho_{11} = \frac{1}{Z}e^{-\hbar\omega_{\mathrm{q}}/2k_{\mathrm{B}}T}$ as a function of the temperature of the environment. In this plot, we assumed $\omega_{\mathrm{q}}/2\pi = 4$ GHz.

$$Z = \mathrm{Tr}[e^{-\beta\hat{H}_{\mathrm{s}}}] = \sum_{i=0}^{1}\langle i|e^{-\beta\hat{H}_{\mathrm{s}}}|i\rangle = e^{-\beta E_0} + e^{-\beta E_1}$$
$$= e^{\beta\hbar\omega_{\mathrm{q}}/2} + e^{-\beta\hbar\omega_{\mathrm{q}}/2} = 2\cosh\left(\frac{\hbar\omega_{\mathrm{q}}}{2k_{\mathrm{B}}T}\right). \tag{6.44}$$

The thermal state of a qubit at temperature T is

$$\rho_{\mathrm{th}} = \sum_{i=0}^{1}\sum_{j=0}^{1}|i\rangle\langle i|\frac{e^{\beta\hbar\omega_{\mathrm{q}}\hat{\sigma}_z/2}}{Z}|j\rangle\langle j|$$
$$= \frac{e^{\beta\hbar\omega_{\mathrm{q}}/2}}{Z}|0\rangle\langle 0| + \frac{e^{-\beta\hbar\omega_{\mathrm{q}}/2}}{Z}|1\rangle\langle 1| = \frac{1}{Z}\begin{bmatrix} e^{\hbar\omega_{\mathrm{q}}/2k_{\mathrm{B}}T} & 0 \\ 0 & e^{-\hbar\omega_{\mathrm{q}}/2k_{\mathrm{B}}T}\end{bmatrix}, \tag{6.45}$$

where in the first step we used the completeness relation. For finite temperatures, the qubit is in a mixed state. The population $\rho_{11} = \frac{1}{Z}e^{-\hbar\omega_{\mathrm{q}}/2k_{\mathrm{B}}T}$ is the probability of observing the qubit in the excited state. This quantity is shown in Fig. 6.9 for a qubit with transition frequency $\omega_{\mathrm{q}}/2\pi = 4\,\mathrm{GHz}$: the qubit will be found in the ground state with high probability only when the temperature of the environment is on the order of a few millikelvin. At high temperatures $k_{\mathrm{B}}T \gg \hbar\omega_{\mathrm{q}}$, the qubit is in the completely depolarized state $\mathbb{1}/2$. This part concludes our discussion on density operators for individual systems. The following sections will focus on bipartite systems.

6.6 Composite systems

Density operator formalism can be used to describe **composite systems**, which are systems comprised by two or more subsystems. When a composite system comprises only two subsystems, such as two electrons or two photons, it is usually called a **bipartite system**. Let us derive the density operator of a bipartite system prepared in a pure state. First of all, recall that a pure state $|\psi\rangle_{\mathrm{AB}} \in H_{\mathrm{A}}\otimes H_{\mathrm{B}}$ can be decomposed as

$$|\psi\rangle_{\mathrm{AB}} = \sum_{i=1}^{d_{\mathrm{A}}}\sum_{j=1}^{d_{\mathrm{B}}} c_{ij}|ij\rangle_{\mathrm{AB}}\,, \tag{6.46}$$

where $d_\mathrm{A} = \dim(H_\mathrm{A})$ and $d_\mathrm{B} = \dim(H_\mathrm{B})$. The corresponding bra is

$$ {}_{\mathrm{AB}}\langle\psi| = \sum_{i=1}^{d_\mathrm{A}}\sum_{j=1}^{d_\mathrm{B}} c_{ij}^* \, {}_{\mathrm{AB}}\langle ij| \,. \tag{6.47}$$

From these expressions, it naturally follows that the density operator associated with the pure state $|\psi\rangle_{\mathrm{AB}}$ is

$$\rho = |\psi\rangle\langle\psi|_{\mathrm{AB}} = \sum_{i,i'=1}^{d_\mathrm{A}}\sum_{j,j'=1}^{d_\mathrm{B}} c_{ij}c_{i'j'}^* |ij\rangle\langle i'j'|_{\mathrm{AB}} = \sum_{i,i'=1}^{d_\mathrm{A}}\sum_{j,j'=1}^{d_\mathrm{B}} \rho_{iji'j'}|ij\rangle\langle i'j'|_{\mathrm{AB}},$$

where we have defined the density matrix

$$\rho_{iji'j'} = \langle ij|\rho|i'j'\rangle = c_{ij}c_{i'j'}^* \,. \tag{6.48}$$

If a source prepares a statistical ensemble of bipartite states $\{|\psi_k\rangle_{\mathrm{AB}}, p_k\}_k$, the associated density operator takes the form $\rho = \sum_k p_k|\psi_k\rangle\langle\psi_k|_{\mathrm{AB}}$. The generalization to composite systems with more than two subsystems is straightforward; in this case, we have $\rho = \sum_k p_k|\psi_k\rangle\langle\psi_k|$ where $|\psi_k\rangle \in H_1 \otimes \ldots \otimes H_m$. In the following, we will mainly focus on bipartite systems with a finite Hilbert space.

6.6.1 Partial trace and reduced density operator

One of the most important and elegant applications of density operator formalism is the description of a subsystem of a composite quantum system. As we shall see in this section, the density operator associated with a subsystem, also known as the reduced density operator, is obtained by tracing over the other subsystems.

Consider two subsystems A and B described by a density operator $\rho \in \mathrm{D}(H_\mathrm{A} \otimes H_\mathrm{B})$. These two subsystems can be two qubits or two particles emitted by a common source. Suppose that a measurement of the observable $\hat{O}$ is performed on subsystem A, while subsystem B is completely ignored. We are interested in the expectation value of such a measurement. According to eqn (6.1), this quantity is given by

$$\langle \hat{O} \otimes \mathbb{1}_\mathrm{B}\rangle_\rho = \mathrm{Tr}_{\mathrm{AB}}[\rho(\hat{O} \otimes \mathbb{1}_\mathrm{B})]. \tag{6.49}$$

In this expression, the trace is calculated over *both* subsystems. We now show that this expectation value can also be calculated as

$$\boxed{\langle \hat{O} \otimes \mathbb{1}_\mathrm{B}\rangle_\rho = \mathrm{Tr}_\mathrm{A}[\rho_\mathrm{A}\hat{O}],} \tag{6.50}$$

where we introduced the **reduced density operator**

$$\boxed{\rho_\mathrm{A} = \mathrm{Tr}_\mathrm{B}[\rho] = \sum_{i=1}^{d_\mathrm{B}} {}_\mathrm{B}\langle i|\rho|i\rangle_\mathrm{B}.} \tag{6.51}$$

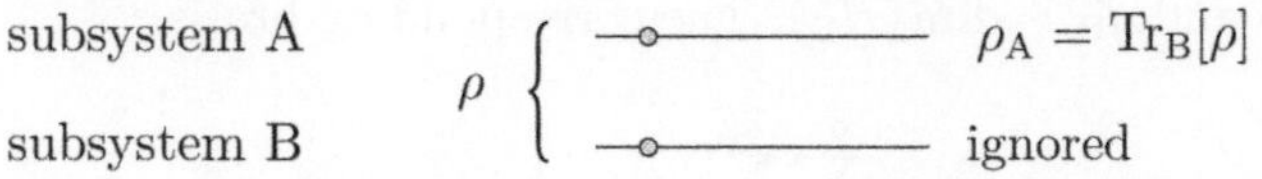

subsystem A $\quad \rho \left\{ \right.$ —o———— $\rho_{\mathrm{A}} = \mathrm{Tr}_{\mathrm{B}}[\rho]$

subsystem B $\quad$ —o———— ignored

Fig. 6.10 The reduced density operator. A bipartite system comprising two subsystems, A and B, is in the initial state ρ. The partial trace averages out subsystem B from the global density matrix. The resultant density matrix $\rho_{\mathrm{A}} = \mathrm{Tr}_{\mathrm{B}}[\,\rho\,]$ describes only A.

The important point here is that in eqn (6.50) the trace is performed with respect to *one* subsystem only. Let us show that eqns (6.49, 6.50) are equivalent,

$$\begin{aligned}
\langle \hat{O} \otimes \mathbb{1} \rangle_\rho &= \mathrm{Tr}_{\mathrm{AB}}[\rho(\hat{O} \otimes \mathbb{1})] = \sum_m \sum_k \langle mk|\rho(\hat{O} \otimes \mathbb{1})|mk\rangle \\
&= \sum_{m,i,i'} \sum_{k,j,j'} \langle mk| \Big(\langle ij|\rho|i'j'\rangle |ij\rangle\langle i'j'| \Big) (\hat{O} \otimes \mathbb{1})|mk\rangle = \sum_{i,i'} \sum_{j,j'} \langle ij|\rho|i'j'\rangle \langle i'j'|(\hat{O} \otimes \mathbb{1})|ij\rangle \\
&= \sum_{i,i'} \sum_j \langle ij|\rho|i'j\rangle \langle i'|\hat{O}|i\rangle = \sum_{i,i'} \langle i| \Big(\sum_j \langle j|\rho|j\rangle \Big) |i'\rangle \langle i'|\hat{O}|i\rangle \\
&= \sum_{i,i'} \langle i|\rho_{\mathrm{A}}|i'\rangle \langle i'|\hat{O}|i\rangle = \sum_i \langle i|\rho_{\mathrm{A}}\hat{O}|i\rangle = \mathrm{Tr}_{\mathrm{A}}[\rho_{\mathrm{A}}\hat{O}].
\end{aligned}$$

In conclusion, when subsystem B is completely ignored, the quantum state describing system A is given by $\rho_{\mathrm{A}} = \mathrm{Tr}_{\mathrm{B}}[\rho]$ as shown in Fig. 6.10. The reduced density operator ρ_{A} can be used to calculate expectation values as we normally do. From a physical point of view, the partial trace is equivalent to averaging out the contributions related to other subsystems in order to focus our attention on the remaining subsystem.

The trace with respect to one subsystem is called the **partial trace**. The partial trace is a function that takes a linear operator defined on a composite Hilbert space and outputs an operator defined on a Hilbert space with lower dimensionality,

$$\begin{aligned}
\mathrm{Tr}_{\mathrm{B}} : \quad \mathrm{B}(H_{\mathrm{A}} \otimes H_{\mathrm{B}}) \quad &\to \quad \mathrm{B}(H_{\mathrm{A}}) \\
\hat{X} \quad &\mapsto \quad \mathrm{Tr}_{\mathrm{B}}[\hat{X}].
\end{aligned}$$

It is simple to demonstrate that if ρ is a density operator, ρ_{A} is still a density operator. Indeed, since $\mathrm{Tr}_{\mathrm{AB}}[\rho] = 1$, then $\mathrm{Tr}_{\mathrm{A}}[\rho_{\mathrm{A}}] = \mathrm{Tr}_{\mathrm{A}}[\mathrm{Tr}_{\mathrm{B}}[\rho]] = 1$. Furthermore, ρ_{A} is a positive operator because for any vector $|\psi\rangle \in H_{\mathrm{A}}$, the element $\langle\psi|\rho_{\mathrm{A}}|\psi\rangle = \mathrm{Tr}_{\mathrm{A}}[\rho_{\mathrm{A}}|\psi\rangle\langle\psi|] = \mathrm{Tr}_{\mathrm{AB}}[\rho(|\psi\rangle\langle\psi| \otimes \mathbb{1}_{\mathrm{B}})] \geq 0$. The partial trace is one of the most fundamental operations in quantum information theory.

Suppose that a bipartite system is prepared in a mixed state. What happens to subsystem A when a measurement is performed on subsystem B? To answer this question, let us consider the circuit shown in Fig. 6.11. A bipartite system is prepared in the quantum state ρ and a measurement of an observable $\hat{X} = \sum_x x\hat{P}_x$ is performed on subsystem B only (here, $\hat{P}_x$ are projectors associated with the eigenvalues x). The probability of measuring an eigenvalue x is given by the Born rule, $p_x = \mathrm{Tr}_{\mathrm{AB}}[\rho(\mathbb{1}_{\mathrm{A}} \otimes \hat{P}_x)]$. From eqn (6.27), the state of subsystem A after the measurement is [185] subsystem A $\qquad \rho_{A,x} = \frac{1}{p_x}\mathrm{Tr}_B[\rho(\mathbb{1}_A \otimes \hat{P}_x)]$

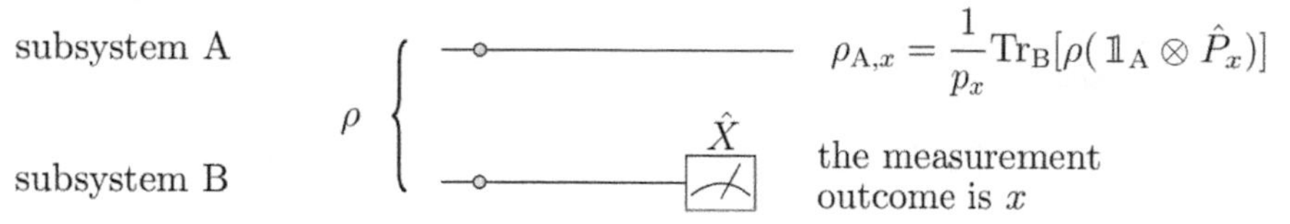

ρ $\qquad \hat{X}$

subsystem B $\qquad$ the measurement outcome is x

Fig. 6.11 Conditional state. A bipartite system is prepared in a quantum state $\rho \in D(H \otimes H)$. An observable $\hat{X}$ is measured on subsystem B. This measurement yields an outcome x with probability $p_x = \mathrm{Tr}_{AB}[\rho(\mathbb{1}_A \otimes \hat{P}_x)]$. The state of subsystem A after the measurement is the conditional state $\rho_{A,x}$.

$$\boxed{\rho_{A,x} = \mathrm{Tr}_B\Big[\frac{1}{p_x}(\mathbb{1}_A \otimes \hat{P}_x)\rho(\mathbb{1}_A \otimes \hat{P}_x)\Big] = \frac{1}{p_x}\mathrm{Tr}_B\big[\rho(\mathbb{1}_A \otimes \hat{P}_x)\big],} \tag{6.52}$$

where we used the cyclic property of the trace and the fact that $\hat{P}_x^2 = \hat{P}_x$. The quantum state $\rho_{A,x}$ is called the **conditional state**. This density operator describes the state of subsystem A when the measurement on subsystem B yields the outcome x. Let us study some concrete examples.

Example 6.8 Consider a source that generates two particles in an entangled state of the polarization, $|\beta_{00}\rangle = \frac{1}{\sqrt{2}}(|00\rangle + |11\rangle)$. The particles move in opposite directions toward two people, whom we will call Alice and Bob as customary in the field. Alice measures the observable $\hat{\sigma}_z$ on the incoming particle. It is reasonable to expect that she will obtain $|0\rangle$ and $|1\rangle$ with equal probability. Let us show this step by step,

$$\begin{aligned}
\rho_A = \mathrm{Tr}_B[|\beta_{00}\rangle\langle\beta_{00}|] &= \frac{1}{2}\mathrm{Tr}_B[|00\rangle\langle 00| + |00\rangle\langle 11| + |11\rangle\langle 00| + |11\rangle\langle 11|] \\
&= \frac{1}{2}\mathrm{Tr}_B[|0\rangle\langle 0| \otimes |0\rangle\langle 0| + |0\rangle\langle 1| \otimes |0\rangle\langle 1| + |1\rangle\langle 0| \otimes |1\rangle\langle 0| + |1\rangle\langle 1| \otimes |1\rangle\langle 1|] \\
&= \frac{1}{2}(|0\rangle\langle 0|_A + |1\rangle\langle 1|_A) \;=\; \frac{\mathbb{1}_A}{2}.
\end{aligned}$$

Thus, the state of Alice's particle is the completely depolarized state and her measurement will produce $|0\rangle$ and $|1\rangle$ with equal probability. With similar calculations, one can verify that the reduced density operator ρ_B is given by $\rho_B = \frac{1}{2}(|0\rangle\langle 0|_B + |1\rangle\langle 1|_B)$.

Example 6.9 Consider a two-qubit state $|\psi\rangle = \frac{1}{\sqrt{2}}(|0\rangle_A|0\rangle_B + |1\rangle_A|1\rangle_B)$ shared by two parties, Alice and Bob. Suppose that Bob measures his qubit in the computational basis and his measurement produces the state $|0\rangle$. Thus, Alice's qubit collapses into the state $|0\rangle$ too. We can show this in a rigorous way by calculating the conditional state using eqn 6.52,

$$\begin{aligned}
\rho_{A,0} &= \frac{1}{p_0}\mathrm{Tr}_B[|\psi\rangle\langle\psi|(\mathbb{1}_A \otimes |0\rangle\langle 0|)] \\
&= \frac{1}{2p_0}\mathrm{Tr}_B[(|00\rangle\langle 00| + |00\rangle\langle 11| + |11\rangle\langle 00| + |11\rangle\langle 11|)(\mathbb{1}_A \otimes |0\rangle\langle 0|)] =
\end{aligned}$$

$$= \frac{1}{2p_0} \text{Tr}_\text{B}[|0\rangle\langle 0| \otimes |0\rangle\langle 0| + |1\rangle\langle 0| \otimes |1\rangle\langle 0|] = |0\rangle\langle 0|,$$

where in the last step we used $p_0 = \text{Tr}_\text{AB}[|\psi\rangle\langle\psi|(\mathbb{1}_\text{A} \otimes |0\rangle\langle 0|)] = 1/2$.

6.6.2 Purification of a density operator

A quantum system in a mixed state can be seen as a part of a larger system that is in a pure state. This procedure of converting a mixed state into a pure state of an enlarged system is called **purification** (see Fig. 6.12). In mathematical terms, given a mixed state ρ on a Hilbert space H_A, it is always possible to add an additional system B with Hilbert space H_B such that the state of the global system $|\psi\rangle \in H_\text{A} \otimes H_\text{B}$ is a pure state and $\rho = \text{Tr}_\text{B}[|\psi\rangle\langle\psi|]$. It is worth pointing out that the additional subsystem does not need to be a real physical system. This technique is often used as a mathematical tool to simplify calculations by working with pure states instead of mixed states.

Theorem 6.1 (Purification) Consider a density operator $\rho \in \text{D}(H_\text{A})$ with spectral decomposition $\rho = \sum_{i=1}^{d_\text{A}} \lambda_i |i\rangle\langle i|_\text{A}$. Then, there exists a Hilbert space H_B and a pure state $|\psi\rangle \in H_\text{A} \otimes H_\text{B}$ such that

$$\rho = \text{Tr}_\text{B}\left[|\psi\rangle\langle\psi|\right] . \tag{6.53}$$

The state $|\psi\rangle$ is called the **purification** of ρ.

Proof Let us set $H_\text{B} = H_\text{A}$. We will now show that the state

$$|\psi\rangle = \sum_{i=1}^{d_\text{A}} \sqrt{\lambda_i}|i\rangle_\text{A}|i\rangle_\text{B} \tag{6.54}$$

is a purification of ρ. Indeed,

$$\begin{aligned}\text{Tr}_\text{B}\left[|\psi\rangle\langle\psi|\right] &= \sum_{i,i'=1}^{d_\text{A}} \sqrt{\lambda_i}\sqrt{\lambda_{i'}}\text{Tr}_\text{B}\left[|i\rangle\langle i'|_\text{A} \otimes |i\rangle\langle i'|_\text{B}\right] \\ &= \sum_{j,i,i'=1}^{d_\text{A}} \sqrt{\lambda_i}\sqrt{\lambda_{i'}}|i\rangle\langle i'|_\text{A}\, \delta_{ji}\delta_{i'j} = \sum_{j=1}^{d_\text{A}} \lambda_j|j\rangle\langle j|_\text{A} = \rho,\end{aligned}$$

as required. We highlight that it is sufficient that $\dim(H_\text{B}) = \dim(H_\text{A})$. □

Purifications are not unique: a density operator has an infinite number of purifications. For example, consider the mixed state $\rho = \frac{1}{2}|0\rangle\langle 0| + \frac{1}{2}|1\rangle\langle 1|$. By calculating the partial trace, one can verify that any two-qubit state of the form $\frac{1}{\sqrt{2}}(|00\rangle + e^{i\phi}|11\rangle)$ is a purification of ρ.

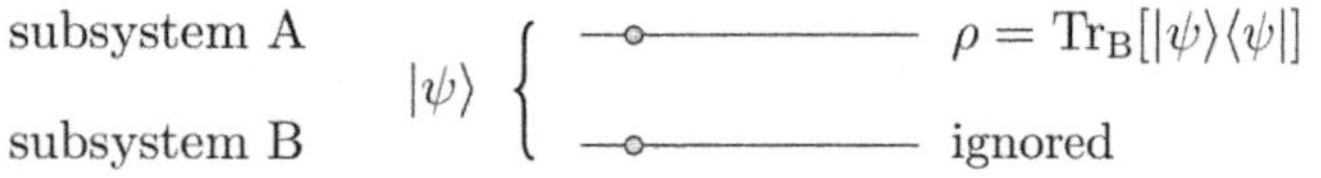

$$\begin{array}{lcll} \text{subsystem A} & \multirow{2}{*}{$|\psi\rangle$} \Big\{ & \multicolumn{1}{l}{\text{—}\circ\text{———}} & \rho = \text{Tr}_{\text{B}}[|\psi\rangle\langle\psi|] \\ \text{subsystem B} & & \text{—}\circ\text{———} & \text{ignored} \end{array}$$

Fig. 6.12 Purification. For any density operator ρ, there exists a pure state $|\psi\rangle$ such that the density operator is the partial trace of that state.

6.7 Quantum state tomography

Quantum state tomography is a set of experiments aimed at measuring all the information contained in an **unknown** quantum state. We already encountered quantum state tomography in Section 5.3.3, when we presented a systematic procedure to determine the coefficients of a qubit state. For pedagogical reasons, we now explain the same procedure with density operator formalism. Let us start from an **unknown** qubit state

$$\rho = \frac{1}{2}\begin{bmatrix} 1+r_z & r_x - ir_y \\ r_x + ir_y & 1-r_z \end{bmatrix}.$$

Our goal is to determine the parameters r_x, r_y, and r_z with high precision. As shown in Fig. 6.13, this can be done by applying a unitary transformation $\hat{U}$ to the qubit state followed by a measurement in the computational basis described by the projectors:

$$\hat{P}_0 = |0\rangle\langle 0| = \begin{bmatrix} 1\,0 \\ 0\,0 \end{bmatrix}, \qquad \hat{P}_1 = |1\rangle\langle 1| = \begin{bmatrix} 0\,0 \\ 0\,1 \end{bmatrix}.$$

To measure the parameter r_x, one can set

$$\hat{U} = \mathsf{Y}_{-\frac{\pi}{2}} = \frac{1}{\sqrt{2}}\begin{bmatrix} 1 & 1 \\ -1 & 1 \end{bmatrix},$$

so that the initial state becomes $\rho' = \mathsf{Y}_{-\frac{\pi}{2}}\rho\mathsf{Y}^{\dagger}_{-\frac{\pi}{2}}$. A measurement in the computational basis projects ρ' onto the ground state with probability

$$\begin{aligned} p_0 = \text{Tr}[\rho'\hat{P}_0] &= \text{Tr}[\mathsf{Y}_{-\frac{\pi}{2}}\rho\mathsf{Y}^{\dagger}_{-\frac{\pi}{2}}\hat{P}_0] = \text{Tr}[\rho\mathsf{Y}^{\dagger}_{-\frac{\pi}{2}}\hat{P}_0\mathsf{Y}_{-\frac{\pi}{2}}] \\ &= \frac{1}{4}\text{Tr}\left[\begin{bmatrix} 1+r_z & r_x - ir_y \\ r_x + ir_y & 1-r_z \end{bmatrix}\begin{bmatrix} 1 & 1 \\ 1 & 1 \end{bmatrix}\right] = \frac{1+r_x}{2}. \end{aligned}$$

One can extract r_x from the real number p_0 with the formula $r_x = 2p_0 - 1$. Similarly, the coefficient r_y can be estimated by setting

$$\hat{U} = \mathsf{X}_{\frac{\pi}{2}} = \frac{1}{\sqrt{2}}\begin{bmatrix} 1 & -i \\ -i & 1 \end{bmatrix}.$$

In this case, the qubit state is projected onto the ground state with probability

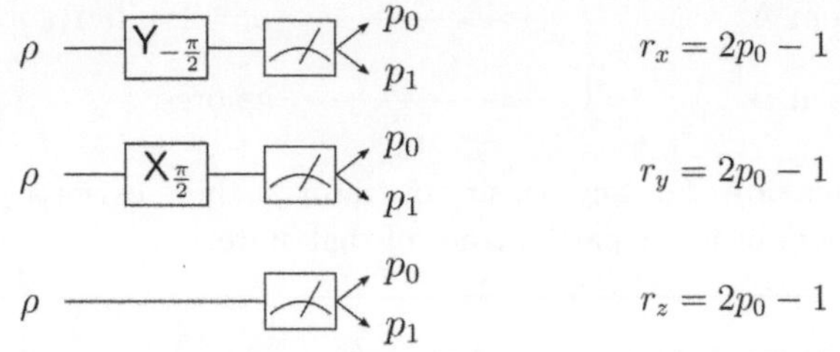

ρ — $\mathsf{Y}_{-\frac{\pi}{2}}$ — measurement — p_0, p_1 $\qquad r_x = 2p_0 - 1$

ρ — $\mathsf{X}_{\frac{\pi}{2}}$ — measurement — p_0, p_1 $\qquad r_y = 2p_0 - 1$

ρ — measurement — p_0, p_1 $\qquad r_z = 2p_0 - 1$

Fig. 6.13 Quantum state tomography. Experiments required to measure the coefficients r_x, r_y, r_z of an unknown qubit state ρ.

$$p_0 = \mathrm{Tr}[\rho' \hat{P}_0] = \mathrm{Tr}[\mathsf{X}_{\frac{\pi}{2}} \rho \mathsf{X}_{\frac{\pi}{2}}^{\dagger} \hat{P}_0] = \mathrm{Tr}[\rho \mathsf{X}_{\frac{\pi}{2}}^{\dagger} \hat{P}_0 \mathsf{X}_{\frac{\pi}{2}}]$$
$$= \frac{1}{4}\mathrm{Tr}\left[\begin{bmatrix} 1+r_z & r_x - r_y \\ r_x + r_y & 1 - r_z \end{bmatrix} \cdot \begin{bmatrix} 1 & -i \\ i & 1 \end{bmatrix}\right] = \frac{1+r_y}{2},$$

and therefore $r_y = 2p_0 - 1$. Lastly, the coefficient r_z is determined by setting $\hat{U} = \mathbb{1}$. A measurement in the computational basis projects the qubit state onto the ground state with probability

$$p_0 = \mathrm{Tr}[\rho \hat{P}_0] = \frac{1}{2}\mathrm{Tr}\begin{bmatrix} 1+r_z & 0 \\ r_x + ir_y & 0 \end{bmatrix} = \frac{1+r_z}{2},$$

and therefore $r_z = 2p_0 - 1$. Each experiment (consisting of a unitary evolution and a measurement) must be repeated multiple times to obtain an accurate estimate of p_0. This can only be done if the unknown state ρ can be prepared on demand. Once the coefficients $\vec{r} = (r_x, r_y, r_z)$ have been determined with the desired accuracy, one can calculate the fidelity F between the reconstructed qubit state and any other qubit state $\vec{s} = (s_x, s_y, s_z)$ with the formula

$$F(\vec{r}, \vec{s}) = \frac{1}{\sqrt{2}}\sqrt{1 + \vec{r}\cdot\vec{s} + \sqrt{\left(1 - |\vec{r}|^2\right)\left(1 - |\vec{s}|^2\right)}}.$$

Let us generalize quantum state tomography to an arbitrary number of qubits. Consider an unknown density operator ρ of n qubits and its decomposition in the Pauli basis[10]

$$\rho = \sum_{i=1}^{d^2} c_i \hat{P}_i,$$

where $c_i = \frac{1}{d}\mathrm{Tr}[\rho \hat{P}_i]$ are real coefficients, $\hat{P}_i$ are tensor products of n Pauli operators and $d = 2^n$ is the dimension of the Hilbert space. The goal of quantum state tomography is to measure the coefficients c_i with high accuracy. To this end, one can

[10] Recall that the Pauli basis $B = \{\hat{\sigma}_{k_1} \otimes \ldots \otimes \hat{\sigma}_{k_n} : k_i \in \{0,1,2,3\}\}$ is a basis of the operator space $\mathrm{B}(H)$ and contains $d^2 = 4^n$ elements.

measure multiple copies of ρ with a set of d^2 binary projectors $\{(\hat{E}_j, \mathbb{1} - \hat{E}_j)\}_j$. The important point is that the operators $\{\hat{E}_j\}_j$ must form a basis of the operator space. A convenient choice is

$$\hat{E}_j = \frac{\mathbb{1} + \hat{P}_j}{2} \qquad \text{where } \hat{P}_j \text{ are Pauli operators.}$$

The operators $\hat{E}_j$ can be decomposed in the Pauli basis as $\hat{E}_j = \sum_k A_{jk}\hat{P}_k$, where the coefficients A_{jk} are known, $A_{jk} = \frac{1}{d}\text{Tr}[\hat{E}_j\hat{P}_k]$. A single measurement projects the unknown quantum state onto the eigenspace associated with $\hat{E}_j$ with probability

$$\begin{aligned} p_j = \text{Tr}[\hat{E}_j\rho] &= \text{Tr}\left[\sum_{k=1}^{d^2} A_{jk}\hat{P}_k \sum_{i=1}^{d^2} c_i\hat{P}_i\right] \\ &= \sum_{k=1}^{d^2}\sum_{i=1}^{d^2} A_{jk}c_i\text{Tr}[\hat{P}_k\hat{P}_i] = d\sum_{k=1}^{d^2}\sum_{i=1}^{d^2} A_{jk}c_i\delta_{ki} = d\sum_{k=1}^{d^2} A_{jk}c_k. \end{aligned}$$

This expression can be written in matrix form as

$$\frac{1}{d}\mathbf{p} = \frac{1}{d}\begin{bmatrix} p_1 \\ \vdots \\ p_{d^2} \end{bmatrix} = \begin{bmatrix} A_{11} & \cdots & A_{1d^2} \\ \vdots & \ddots & \vdots \\ A_{d^21} & \cdots & A_{d^2d^2} \end{bmatrix}\begin{bmatrix} c_1 \\ \vdots \\ c_{d^2} \end{bmatrix} = A\mathbf{c}.$$

The coefficients c_i are given by $\mathbf{c} = \frac{1}{d}A^{-1}\mathbf{p}$ where A^{-1} is the inverse of the matrix A. The set of coefficients c_i gives us enough information to reconstruct the unknown density operator ρ. This procedure works only if the projectors $\hat{E}_j$ are chosen such that the matrix A is invertible (this is the case for $\hat{E}_j = (\mathbb{1} + \hat{P}_j)/2$). Multiple experiments must be run in order to achieve a good estimate of the probabilities p_j. Once the coefficients c_i have been determined, one can calculate the fidelity between the reconstructed state and a density operator σ using $F(\rho, \sigma) = \text{Tr}\left[\sqrt{\sqrt{\rho}\sigma\sqrt{\rho}}\right]$.

Example 6.10 Consider an unknown qubit state

$$\rho = c_1\mathbb{1} + c_2\hat{\sigma}_x + c_3\hat{\sigma}_y + c_4\hat{\sigma}_z. \tag{6.55}$$

To determine the coefficients c_i, one can choose the projectors $\hat{E}_j = (\mathbb{1} + \hat{P}_j)/2$ where $\hat{P}_j$ are Pauli operators:

$$\hat{E}_1 = \mathbb{1}, \qquad \hat{E}_2 = \frac{1}{2}\mathbb{1} + \frac{1}{2}\hat{\sigma}_x,$$
$$\hat{E}_3 = \frac{\mathbb{1}}{2} + \frac{\hat{\sigma}_y}{2}, \qquad \hat{E}_4 = \frac{\mathbb{1}}{2} + \frac{\hat{\sigma}_z}{2}.$$

The decomposition of the operators $\hat{E}_j$ in the Pauli basis is given by $\hat{E}_j = \sum_k A_{jk}\hat{P}_k$. In matrix form, we have

$$\begin{bmatrix}\hat{E}_1\\ \hat{E}_2\\ \hat{E}_3\\ \hat{E}_4\end{bmatrix} = \underbrace{\begin{bmatrix}1&0&0&0\\ \frac{1}{2}&\frac{1}{2}&0&0\\ \frac{1}{2}&0&\frac{1}{2}&0\\ \frac{1}{2}&0&0&\frac{1}{2}\end{bmatrix}}_{A}\begin{bmatrix}\mathbb{1}\\ \hat{\sigma}_x\\ \hat{\sigma}_y\\ \hat{\sigma}_z\end{bmatrix}, \qquad A^{-1} = \begin{bmatrix}1&0&0&0\\ -1&2&0&0\\ -1&0&2&0\\ -1&0&0&2\end{bmatrix}.$$

Each experiment projects the quantum state ρ onto the eigenspace associated with $\hat{E}_j$ with probability p_j. The coefficients $\mathbf{c} = (c_1, c_2, c_3, c_4)$ in (6.55) are given by $\mathbf{c} = \frac{1}{d}A^{-1}\mathbf{p}$ where the vector $\mathbf{p}$ is defined as[11] $\mathbf{p} = (p_1, p_2, p_3, p_4)$. In many practical cases, the qubit state can only be measured in the computational basis. In such cases, one first needs to apply a unitary transformation $\hat{U}$ satisfying $\hat{U}\hat{E}_j\hat{U}^\dagger = |0\rangle\langle 0|$ and then proceed with a measurement in the computational basis.

Further reading

Density operators are discussed in several textbooks. For readers interested in the theoretical description of open quantum systems, we recommend the monograph "The theory of open quantum systems" by Breuer and Petruccione [186]. For readers interested in physical applications, we recommend "Quantum thermodynamics" by Gemmer, Michel, and Mahler [187] and "Statistical mechanics" by Schwabl [188]. Excellent introductory textbooks with a focus on foundational and theoretical aspects are "Quantum Theory: Concepts and Methods" by Peres [109] and "Mathematical Foundations of Quantum Mechanics" by Mackey. The review article by Fano [189] is another excellent introduction to the topic of density operators. For more information about Gleason's theorem, the reader is referred to the original publication [184] and the proof given by Busch [190].

Section 6.7 describes how to implement quantum state tomography for an n-qubit state measuring $4^n - 1$ Pauli operators. The same operation can be implemented by measuring each single qubit in the X, Y, and Z bases, for a total of 3^n measurements, and post-processing the resulting probability distributions. A minimal and optimal way to realize quantum state tomography is achieved by measuring the qubits in a set of 2^n mutually unbiased bases [191, 192].

[11]Note that $p_1 = 1$ since $\hat{E}_1 = \mathbb{1}$.

Summary

Density operators

$\rho = \sum_k p_k \lvert\psi_k\rangle\langle\psi_k\rvert$	Density operator ($p_k \geq 0$, $\sum_k p_k = 1$)
$\langle\hat{A}\rangle_\rho = \mathrm{Tr}[\rho\hat{A}]$	Expectation value of $\hat{A}$
$\rho_{ij} = \langle i\lvert\rho\rvert j\rangle$	Entries of the density matrix
$\rho = \lvert\psi\rangle\langle\psi\rvert,\ \rho^2 = \rho$	Pure state
$\rho^2 \neq \rho$	Mixed state
$\mathrm{Tr}[\rho] = 1,\ \rho \geq 0,\ \rho = \rho^\dagger$	Properties of density operators
$\mathrm{D}(H)$	Set of density operators
$\mu(\rho) = \mathrm{Tr}[\rho^2]$	Purity
$S(\rho) = -\mathrm{Tr}[\rho \log_2 \rho]$	Von Neumann entropy
$S(\rho) = -\lambda_i \sum_i \log_2 \lambda_i$	Von Neumann entropy (λ_i are the eigenvalues of ρ)
$i\hbar\frac{d}{dt}\rho(t) = [\hat{H}, \rho(t)]$	Von Neumann equation
$p(a_i) = \mathrm{Tr}[\hat{P}_i\rho]$	Probability of measuring a_i
$\rho_i = \hat{P}_i\rho\hat{P}_i/\mathrm{Tr}[\hat{P}_i\rho]$	Density operator after a measurement

---

Qubit states

$\rho = (\mathbb{1} + \vec{r}\cdot\vec{\sigma})/2$	Density operator of a qubit
$\vec{r} = (r_x, r_y, r_z)$ where $0 \leq \lvert\vec{r}\rvert \leq 1$	Bloch vector
$\lvert\vec{r}\rvert = 1$	Pure state
$0 \leq \lvert\vec{r}\rvert < 1$	Mixed state
$\rho = \mathbb{1}/2,\ \lvert\vec{r}\rvert = 0$	Completely depolarized state

---

Thermal states

$\rho_{\mathrm{th}} = \frac{1}{Z}e^{-\hat{H}/k_{\mathrm{B}}T}$	Thermal state at temperature T
$Z = \mathrm{Tr}[e^{-\hat{H}/k_{\mathrm{B}}T}]$	Partition function
$\rho_{\mathrm{th}} = (1-w)\sum_{n=0}^{\infty} w^n \lvert n\rangle\langle n\rvert$	Thermal state of a QHO ($w = e^{-\hbar\omega_{\mathrm{r}}/k_{\mathrm{B}}T}$)
$p(x) = \langle x\lvert\rho_{\mathrm{th}}\rvert x\rangle = \frac{\lambda}{\sqrt{\pi}}\frac{\sqrt{1-w}}{\sqrt{1+w}}e^{-\frac{1-w}{1+w}\lambda^2x^2}$	PDF in the position eigenbasis
$\rho_{\mathrm{th}} = \frac{1}{2\cosh(a/2)}\begin{bmatrix} e^{a/2} & 0 \\ 0 & e^{-a/2}\end{bmatrix}$	Thermal state of a qubit ($a = \hbar\omega_{\mathrm{q}}/k_{\mathrm{B}}T$)

Composite systems

$\rho = \sum_{i,i'} \sum_{j,j'} \rho_{iji'j'} \|ij\rangle\langle i'j'\|$	Density operator of a bipartite system
$\mathrm{Tr}_{\mathrm{B}}[\hat{O}] = \sum_i {}_{\mathrm{B}}\langle i\|\hat{O}\|i\rangle_{\mathrm{B}}$	Partial trace
$\rho_{\mathrm{A}} = \mathrm{Tr}_{\mathrm{B}}[\rho]$	Reduced density operator
$\langle \hat{O} \otimes \mathbb{1}_{\mathrm{B}} \rangle_\rho = \mathrm{Tr}_{\mathrm{A}}[\rho_{\mathrm{A}} \hat{O}]$	Expectation value for a subsystem
$\rho = \mathrm{Tr}_{\mathrm{B}}[\|\psi\rangle\langle\psi\|]$, where $\|\psi\rangle \in H_{\mathrm{A}} \otimes H_{\mathrm{B}}$	$\|\psi\rangle$ is a purification of $\rho \in \mathrm{D}(H_{\mathrm{A}})$

	Kets	**Density operators**
vector space	H	$\mathrm{D}(H)$,
quantum state	$\|\psi\rangle$ where $\langle\psi\|\psi\rangle = 1$	ρ where $\rho \geq 0$ and $\mathrm{Tr}[\rho] = 1$,
inner product	$\langle\phi\|\psi\rangle$	$\mathrm{Tr}[\rho_1 \rho_2]$,
composite systems	$\|\psi\rangle_{\mathrm{A}} \otimes \|\psi\rangle_{\mathrm{B}}$	$\rho_{\mathrm{A}} \otimes \rho_{\mathrm{B}}$,
unitary dynamics	$i\hbar \frac{d\|\psi\rangle}{dt} = \hat{H}\|\psi\rangle$	$i\hbar \frac{d\rho}{dt} = [\hat{H}, \rho]$,
measurement	$\|\psi\rangle \mapsto \hat{P}\|\psi\rangle / \\|\hat{P}\|\psi\rangle\\|$	$\rho \mapsto \hat{P}\rho\hat{P}/\mathrm{Tr}[\hat{P}\rho]$,
expectation value	$\langle\hat{A}\rangle_\psi = \langle\psi\|\hat{A}\|\psi\rangle$	$\langle\hat{A}\rangle_\rho = \mathrm{Tr}[\hat{A}\rho]$.

Exercises

Exercise 6.1 Consider a qubit state $\rho = \frac{1}{2}(\mathbb{1} + r_x \sigma_x + r_y \hat{\sigma}_y + r_z \hat{\sigma}_z)$ where $\vec{r} = (r_x, r_y, r_z)$. Show that $r_x = \mathrm{Tr}[\rho\hat{\sigma}_x]$, $r_y = \mathrm{Tr}[\rho\hat{\sigma}_y]$ and $r_z = \mathrm{Tr}[\rho\hat{\sigma}_z]$.

Exercise 6.2 Consider a thermal state of a quantum harmonic oscillator $\rho_{\mathrm{th}} = \sum_n p(n)|n\rangle\langle n|$ where $p(n) = w^n/(1-w)$ and $w = e^{-\hbar\omega_{\mathrm{r}}/k_{\mathrm{B}}T}$. Show that thermal states are stationary states, i.e. $\rho_{\mathrm{th}}(t) = \hat{U}(t,0)\rho_{\mathrm{th}}(0)\hat{U}^\dagger(t,0) = \rho(0)$ where $\hat{U}(t,0) = \exp(\frac{1}{i\hbar}\hat{H}_{\mathrm{r}} t)$ and $\hat{H}_{\mathrm{r}} = \hbar\omega_{\mathrm{r}}\hat{a}^\dagger\hat{a}$.

Exercise 6.3 Prove that $\mathrm{Tr}_{\mathrm{B}}[\hat{A} \otimes \hat{B}] = \hat{A}\mathrm{Tr}[\hat{B}]$ for $\hat{A} \in \mathrm{L}(H_{\mathrm{A}})$ and $\hat{B} \in \mathrm{L}(H_{\mathrm{B}})$.

Exercise 6.4 Consider the operators $\hat{A} \in \mathrm{L}(\hat{H}_{\mathrm{A}})$, $\hat{B} \in \mathrm{L}(\hat{H}_{\mathrm{B}})$, and $\hat{X} \in \mathrm{L}(\hat{H}_{\mathrm{A}} \otimes \hat{H}_{\mathrm{B}})$. Prove that $\mathrm{Tr}_{\mathrm{B}}[\hat{X}(\hat{A} \otimes \hat{B})] = \mathrm{Tr}_{\mathrm{B}}[\hat{X}(\mathbb{1} \otimes \hat{B})]\hat{A}$.

Exercise 6.5 Consider a bipartite state $|\psi\rangle \in H_{\mathrm{A}} \otimes H_{\mathrm{B}}$ where H_{A} and H_{B} are two Hilbert spaces with orthonormal bases $\{|i\rangle_{\mathrm{A}}\}_{i=1}^{d_{\mathrm{A}}}$ and $\{|j\rangle_{\mathrm{B}}\}_{j=1}^{d_{\mathrm{B}}}$ and $d_{\mathrm{A}} \leq d_{\mathrm{B}}$. The state $|\psi\rangle$ can be decomposed as $|\psi\rangle = \sum_{ij} c_{ij}|i\rangle|j\rangle$ where the matrix c_{ij} has the singular value decomposition $c_{ij} = \sum_k u_{ik} d_k v_{kj}$. Here, u_{ik} and v_{kj} define two unitary matrices and d_k are non-negative real numbers. Show that the partial trace of $|\psi\rangle$ is given by $\mathrm{Tr}_{\mathrm{B}}[|\psi\rangle\langle\psi|] = \sum_{l=1}^{d_{\mathrm{A}}} d_l^2 |u_l\rangle\langle u_l|$ where $|u_l\rangle = \sum_i u_{il}|i\rangle$.

Exercise 6.6 Consider a composite system with Hilbert space $H_{\mathrm{A}} \otimes H_{\mathrm{B}}$ and four operators $\hat{A}_1, \hat{A}_2 \in \mathrm{L}(H_{\mathrm{A}})$ and $\hat{B}_1 \otimes \hat{B}_2 \in \mathrm{L}(H_{\mathrm{B}})$. Show that the **partial trace is not a cyclic operation**, i.e. that $\mathrm{Tr}_{\mathrm{B}}[(\hat{A}_1 \otimes \hat{B}_1)(\hat{A}_2 \otimes \hat{B}_2)]$ is not always equal to $\mathrm{Tr}_{\mathrm{B}}[(\hat{A}_2 \otimes \hat{B}_2)(\hat{A}_1 \otimes \hat{B}_1)]$.

Exercise 6.7 Consider a quantum oscillator in a thermal state $\rho_{\text{th}} = (1-w)\sum_n w^n |n\rangle\langle n|$, where $w = e^{-\beta\hbar\omega_{\text{r}}}$. Show that the expectation value of $\hat{n}^2 = \hat{a}^\dagger\hat{a}\hat{a}^\dagger\hat{a}$ is

$$\langle \hat{n}^2 \rangle = \frac{w(1+w)}{(w-1)^2}.$$

Show that at low temperatures, $\langle \hat{n}^2 \rangle \approx \langle \hat{n} \rangle$. **Hint:** Recall from eqn (6.41) that $\langle \hat{n} \rangle = w/(1-w)$ for a thermal state. Calculate $\langle \hat{n}^2 \rangle - \langle \hat{n} \rangle^2$ and $\langle \hat{n} \rangle^2 + \langle \hat{n} \rangle$ and show that they are identical.

Solutions

Solution 6.1 Let us start from

$$\begin{aligned}\text{Tr}[\rho\hat{\sigma}_x] &= \frac{1}{2}\text{Tr}[\mathbb{1} + r_x\hat{\sigma}_x\hat{\sigma}_x + r_y\hat{\sigma}_y\hat{\sigma}_x + r_z\hat{\sigma}_z\hat{\sigma}_x] \\ &= \frac{1}{2}\text{Tr}[\hat{\sigma}_x + r_x\mathbb{1} - ir_y\hat{\sigma}_z + ir_z\hat{\sigma}_y] = \frac{1}{2}\text{Tr}[r_x\mathbb{1}] = \frac{r_x}{2}\text{Tr}[\mathbb{1}] = r_x,\end{aligned}$$

where we used the properties presented in Section 3.2.4 and $\text{Tr}[\hat{\sigma}_i] = 0$. With similar calculations, one can verify that $r_y = \text{Tr}[\rho\hat{\sigma}_y]$ and $r_z = \text{Tr}[\rho\hat{\sigma}_z]$.

Solution 6.2 An explicit calculation leads to

$$\begin{aligned}\hat{U}(t,0)\rho_{\text{th}}(0)\hat{U}^\dagger(t,0) &= \sum_{n=0}^{\infty} p(n)\hat{U}(t,0)|n\rangle\langle n|\hat{U}^\dagger(t,0) = \sum_{n=0}^{\infty} p(n)e^{\frac{1}{i\hbar}\hat{H}_{\text{r}}t}|n\rangle\langle n|e^{-\frac{1}{i\hbar}\hat{H}_{\text{r}}t} \\ &= \sum_{n=0}^{\infty} p(n)e^{\frac{1}{i\hbar}E_n t}|n\rangle\langle n|e^{-\frac{1}{i\hbar}E_n t} = \sum_{n=0}^{\infty} p(n)|n\rangle\langle n| = \rho_{\text{th}}(0),\end{aligned}$$

as required.

Solution 6.3 For two generic vectors $|i\rangle, |j\rangle \in H_{\text{A}}$, we have

$$\langle i|\text{Tr}_{\text{B}}[\hat{A}\otimes\hat{B}]|j\rangle = \sum_k \langle ik|\hat{A}\otimes\hat{B}|jk\rangle = \sum_k \langle i|\hat{A}|j\rangle\langle k|\hat{B}|k\rangle = \langle i|\hat{A}|j\rangle\text{Tr}[\hat{B}].$$

Solution 6.4 Let d_{A} and d_{B} be the dimensions of the Hilbert spaces H_{A} and H_{B}, respectively. For two generic vectors $|i\rangle, |j\rangle \in H_{\text{A}}$, we have

$$\begin{aligned}\langle i|\text{Tr}_{\text{B}}[\hat{X}(\hat{A}\otimes\hat{B})]|j\rangle &= \sum_{m=1}^{d_{\text{B}}} \langle im|\hat{X}(\hat{A}\otimes\hat{B})|jm\rangle \\ &= \sum_{k=1}^{d_{\text{A}}}\sum_{m,n=1}^{d_{\text{B}}} \langle im|\hat{X}|kn\rangle\langle kn|\hat{A}\otimes\hat{B})|jm\rangle \\ &= \sum_{k=1}^{d_{\text{A}}}\sum_{m,n=1}^{d_{\text{B}}} \langle im|\hat{X}|kn\rangle\langle k|\hat{A}|j\rangle\langle n|\hat{B}|m\rangle.\end{aligned}$$

Let us write the matrix element $\langle k|\hat{A}|j\rangle$ as $\sum_{l=1}^{d_A}\langle k|\mathbb{1}|l\rangle\langle l|\hat{A}|j\rangle$. Therefore,

$$\begin{aligned}\langle i|\mathrm{Tr}_B[\hat{X}(\hat{A}\otimes\hat{B})]|j\rangle &= \sum_{k,l=1}^{d_A}\sum_{m,n=1}^{d_B}\langle im|\hat{X}|kn\rangle\langle k|\mathbb{1}|l\rangle\langle l|\hat{A}|j\rangle\langle n|\hat{B}|m\rangle \\ &= \sum_{k,l=1}^{d_A}\sum_{m,n=1}^{d_B}\langle im|\hat{X}|kn\rangle\langle kn|\mathbb{1}\otimes\hat{B}|lm\rangle\langle l|\hat{A}|j\rangle \\ &= \sum_{l=1}^{d_A}\sum_{m=1}^{d_B}\langle im|\hat{X}(\mathbb{1}\otimes\hat{B})|lm\rangle\langle l|\hat{A}|j\rangle = \langle i|\mathrm{Tr}_B[\hat{X}(\mathbb{1}\otimes\hat{B})]\hat{A}|j\rangle.\end{aligned}$$

Solution 6.5 For two generic states $|i\rangle, |i'\rangle \in H_A$, we have

$$\langle i|\mathrm{Tr}_B[|\psi\rangle\langle\psi|]|i'\rangle = \sum_{j=1}^{d_B}\langle ij|\psi\rangle\langle\psi|i'j\rangle = \sum_{j=1}^{d_B}c_{ij}c^*_{i'j}.$$

Now consider the SVD of c_{ij} and $c^*_{i'j}$,

$$\begin{aligned}\langle i'|\mathrm{Tr}_B[|\psi\rangle\langle\psi|]|i'\rangle &= \sum_{j=1}^{d_B}\sum_{k=1}^{d_A}\sum_{l=1}^{d_A}u_{ik}d_k v_{kj}u^*_{i'l}d_l v^*_{lj} \\ &= \sum_{k=1}^{d_A}\sum_{l=1}^{d_A}u_{ik}d_k\delta_{kl}u^*_{i'l}d_l = \sum_{l=1}^{d_A}d_l^2 u_{il}u^*_{i'l} \;=\; \langle i|\left[\sum_{l=1}^{d_A}d_l^2|u_l\rangle\langle u_l|\right]|i'\rangle,\end{aligned}$$

where we used $\sum_k v_{kj}v^*_{lj} = \delta_{kl}$ and in the last step we introduced the vectors $|u_l\rangle = \sum_i u_{il}|i\rangle$.

Solution 6.6 On one hand $\mathrm{Tr}_B[(\hat{A}_1\otimes\hat{B}_1)(\hat{A}_2\otimes\hat{B}_2)] = \hat{A}_1\hat{A}_2\mathrm{Tr}[\hat{B}_1\hat{B}_2]$, on the other hand $\mathrm{Tr}_B[(\hat{A}_2\otimes\hat{B}_2)(\hat{A}_1\otimes\hat{B}_1)] = \hat{A}_2\hat{A}_1\mathrm{Tr}[\hat{B}_2\hat{B}_1]$. These partial traces give the same result only when $\hat{A}_1\hat{A}_2 = \hat{A}_2\hat{A}_1$. Thus, in general $\mathrm{Tr}_B[\hat{O}_1\hat{O}_2] \neq \mathrm{Tr}_B[\hat{O}_2\hat{O}_1]$ where $\hat{O}_1, \hat{O}_2 \in \mathrm{L}(H_A\otimes H_B)$. We remind the reader that the "total" trace is a cyclic operation, $\mathrm{Tr}[\hat{O}_1\hat{O}_2] = \mathrm{Tr}[\hat{O}_2\hat{O}_1]$, as explained in Section 3.2.14.

Solution 6.7 The expectation value $\langle\hat{n}^2\rangle$ can be calculated as

$$\begin{aligned}\langle\hat{n}^2\rangle = \mathrm{Tr}[\rho_{\mathrm{th}}\hat{n}^2] &= (1-w)\sum_{n=0}^{\infty}w^n\mathrm{Tr}[|n\rangle\langle n|\hat{n}^2] = (1-w)\sum_{n=0}^{\infty}w^n\langle n|\hat{n}^2|n\rangle \\ &= (1-w)\sum_{n=0}^{\infty}w^n n^2 \;=\; \frac{w(1+w)}{(w-1)^2},\end{aligned}$$

where we used $\mathrm{Tr}[\hat{A}|\psi\rangle\langle\psi|] = \langle\psi|\hat{A}|\psi\rangle$. From eqn (6.41), we know that $\langle\hat{n}\rangle = w/(1-w)$. Therefore,

$$\langle\hat{n}^2\rangle - \langle\hat{n}\rangle^2 = \frac{w(1+w)}{(w-1)^2} - \frac{w^2}{(1-w)^2} = \frac{w}{(w-1)^2}, \tag{6.56}$$

and also

$$\langle \hat{n} \rangle^2 + \langle \hat{n} \rangle = \frac{w(1+w)}{(w-1)^2} + \frac{w}{1-w} = \frac{w}{(w-1)^2}. \quad (6.57)$$

Since eqn (6.56) and eqn (6.57) are identical, we conclude that

$$\langle \hat{n}^2 \rangle = 2\langle \hat{n} \rangle^2 + \langle \hat{n} \rangle. \quad (6.58)$$

At low temperatures $\langle \hat{n} \rangle \ll 1$ and therefore $\langle \hat{n} \rangle^2 \ll \langle \hat{n} \rangle$. Using this approximation, eqn (6.58) indicates that $\langle \hat{n}^2 \rangle \approx \langle \hat{n} \rangle$ at low temperatures.

7
Quantum maps

Quantum maps can be defined in three equivalent ways.

In an experimental setting, one often observes that a quantum system initially prepared in a superposition state evolves into a mixed classical state due to unavoidable interactions with the environment. The formalism of quantum maps is a general mathematical tool for describing these types of evolutions in a wide range of experiments.

Quantum maps can be defined in three equivalent ways (see Fig. 7.1). One way is to list a set of axioms that a quantum map must satisfy. This approach is a bit theoretical and does not give a physical intuition of these transformations. Another way to define a quantum map is through the Kraus representation, a mathematical representation common to all quantum maps. This approach is particularly useful in calculations but again it lacks physical intuition. Lastly, a quantum map can be interpreted as the partial trace of a unitary transformation acting on the system and the environment. This definition provides a physical intuition of the origin of quantum maps. We will explain these three equivalent definitions throughout this chapter.

We start this chapter with a formal definition of quantum maps. We then present some important examples of quantum maps, including the depolarizing channel, amplitude damping, and dephasing. We briefly present the inversion recovery and the Ramsey experiment, two experimental techniques designed to measure the relaxation time and the dephasing time of a qubit. Section 7.6 shows that any quantum map can be expressed in Kraus form. Section 7.8 presents a simple example of a qubit coupled with two ancilla qubits. This composite system is a toy model describing a qubit interacting with the environment. In Section 7.8.1, we present the general case and show that quantum maps arise when the dynamics of an additional system are averaged out, a result known as the Stinespring's dilation theorem. Lastly, we present the Choi–Jamiolkowski isomorphism, an interesting relation between operator spaces.

7.1 Definition of a quantum map

Quantum mechanics states that the evolution of the universe (taken as a whole) is unitary. When we limit our experiments to a small region, the interactions between the system under consideration and external degrees of freedom might lead to a non-unitary evolution. How should we modify our formalism to describe these dynamics? Quantum maps make up a powerful formalism to model these evolution processes [193,

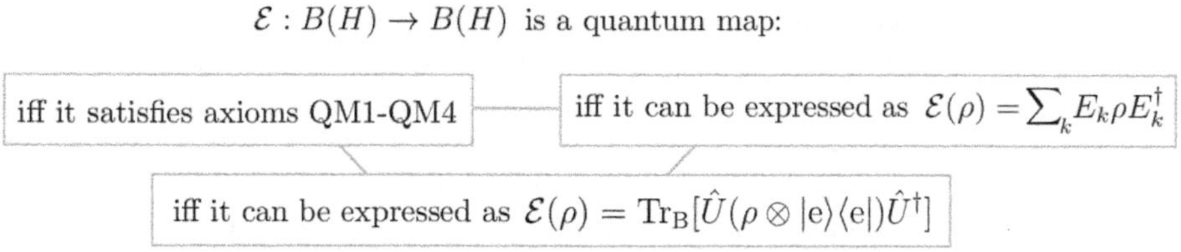

$\mathcal{E} : B(H) \to B(H)$ is a quantum map:

iff it satisfies axioms QM1-QM4

iff it can be expressed as $\mathcal{E}(\rho) = \sum_k E_k \rho E_k^\dagger$

iff it can be expressed as $\mathcal{E}(\rho) = \mathrm{Tr}_\mathrm{B}[\hat{U}(\rho \otimes |\mathrm{e}\rangle\langle\mathrm{e}|)\hat{U}^\dagger]$

Fig. 7.1 Definition of a quantum map. Three equivalent ways to define a quantum map. The operators $\hat{E}_k$ must satisfy $\sum_k \hat{E}_k^\dagger \hat{E}_k = \mathbb{1}$. These three definitions will be discussed one by one in the main text.

194, 195, 196, 197]. They are essential in quantum information theory and can be regarded as the class of physical operations typically encountered in an experiment.

A **quantum map** is a linear transformation $\mathcal{E}$ that maps a quantum state into another quantum state[1],

$$\begin{array}{rccc} \mathcal{E}: & \mathrm{B}(H) & \to & \mathrm{B}(H) \\ & \rho & \mapsto & \mathcal{E}(\rho) \end{array} . \tag{7.1}$$

Quantum maps must satisfy the following requirements for all density operators ρ_i and complex numbers c_i:

QM1.	$\mathcal{E}(c_1\rho_1 + c_2\rho_2) = c_1\mathcal{E}(\rho_1) + c_2\mathcal{E}(\rho_2)$	(linearity),
QM2.	$\mathrm{Tr}[\mathcal{E}(\rho)] = \mathrm{Tr}[\rho]$	(trace preservation),
QM3.	if $\rho \geq 0$, then $\mathcal{E}(\rho) \geq 0$	(positivity),
QM4.	if $\rho \geq 0$, then $(\mathcal{E} \otimes \mathbb{1}_\mathrm{B})(\rho) \geq 0$ for any H_B	(complete positivity).

A function $\mathcal{E}$ satisfying the requirements **QM1–QM4** is called a quantum map, a completely positive trace-preserving (CPTP) map or a quantum channel[2]. A function $\mathcal{E}$ satisfying the requirements **QM1–QM3** only is called a positive map.

The first requirement is quite natural: quantum maps must be linear, a property common to all transformations in quantum mechanics. The second requirement ensures that quantum maps preserve the normalization of the state, while the third one guarantees that quantum maps transform positive operators into positive operators. **QM4** ensures that applying a positive map to only one part of a composite system generates a valid density operator. This requirement seems redundant: if $\mathcal{E}(\rho)$ is positive, one might think that $(\mathcal{E} \otimes \mathbb{1}_\mathrm{B})\,\rho$ is positive too. This is not always the case: there are examples of positive maps that are not completely positive. As we shall see in Section 7.5, the transposition is one of them.

[1]In principle, the domain and the codomain of a quantum map do not need to be the same. However, we will mainly deal with quantum maps transforming density operators from a Hilbert space H into itself.

[2] Note that the requirement **QM3** can be dropped from the list since it is implied by requirement **QM4**: if a map is completely positive, then it is also positive. We included it to emphasize the distinction between positive and completely positive maps.

Are all quantum maps invertible? In other words, given a quantum map $\mathcal{E}$, does there exist *another quantum map* $\mathcal{E}^{-1}$ such that $\mathcal{E}^{-1}(\mathcal{E}(\rho)) = \rho$ for every state ρ? It can be shown that a quantum map is invertible if and only if it is unitary [198], i.e. it can be expressed as $\mathcal{E}(\rho) = \hat{U}\rho\hat{U}^{\dagger}$. From a physical point of view, this suggests that some quantum maps are associated with irreversible processes where some information about the system is irreparably lost. This happens, for example, when the system is coupled to the environment in an uncontrolled way. This phenomenon is known as **decoherence** and can be modeled in an effective way using quantum maps.

We now present some examples of quantum maps acting on a single qubit, in particular the depolarizing channel, amplitude damping, and dephasing. We derive the action of these quantum maps on the Bloch vector and provide some examples to better understand their physical meaning. We warn the reader that in this chapter we will not motivate the microscopic origin of these transformations. We will come back to this point in the next chapter when we study the microscopic derivation of spontaneous emission and dephasing starting from the Hamiltonian of the system and the environment.

7.2 Depolarizing channel

An important example of a quantum map is the **depolarizing channel**. This map can be used to model the depolarization process of a qubit state and is defined as

$$\boxed{\mathcal{E}_{\text{depo}}(\rho) = p\rho + (1-p)\frac{\mathbb{1}}{2}.} \tag{7.2}$$

The depolarizing channel leaves the qubit state unchanged with probability p or maps it into the completely depolarized state with probability $1-p$.

Let us study the action of the depolarizing channel on a Bloch vector. Recall from Section 6.4 that a qubit state can always be written in the form

$$\rho = \frac{\mathbb{1} + r_x\hat{\sigma}_x + r_y\hat{\sigma}_y + r_z\hat{\sigma}_z}{2} = \frac{\mathbb{1} + \vec{r}\cdot\vec{\sigma}}{2}. \tag{7.3}$$

Using (7.2), the action of the depolarizing channel on this state is given by

$$\mathcal{E}_{\text{depo}}(\rho) = p\frac{\mathbb{1} + \vec{r}\cdot\vec{\sigma}}{2} + (1-p)\frac{\mathbb{1}}{2} = \frac{\mathbb{1} + p\,\vec{r}\cdot\vec{\sigma}}{2}. \tag{7.4}$$

Thus, the initial Bloch vector is mapped into

$$\boxed{(r_x, r_y, r_z) \mapsto (pr_x,\ pr_y,\ pr_z).} \tag{7.5}$$

From this expression, it is evident that the depolarizing channel simply reduces the length of the Bloch vector by a factor $p \in [0,1]$. As shown in Fig. 7.2, the effect of the depolarizing channel is to push a qubit state toward the center of the Bloch sphere, the completely depolarized state. An interesting feature of the depolarizing channel is that it is **unital**. A unital map is a transformation that satisfies $\mathcal{E}(\mathbb{1}) = \mathbb{1}$, or, equivalently,

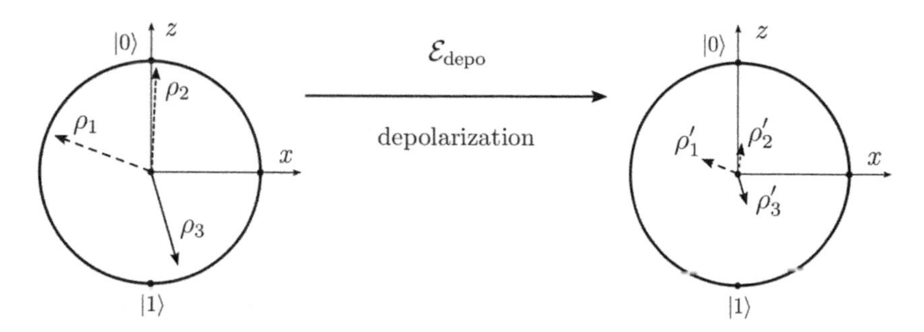

Fig. 7.2 The depolarizing channel. The depolarizing channel pushes a qubit state toward the center of the Bloch sphere. This results in a reduction of the purity of the initial state.

maps the completely depolarized state into itself. The interesting aspect of unital maps is that they never increase the purity of the input state [199, 200, 201, 202].

The depolarizing map can be defined in other equivalent ways, such as

$$\boxed{\mathcal{E}_{\text{depo}}(\rho) = (1 - \mathrm{e_P})\rho + \frac{\mathrm{e_P}}{3}\left(\hat{\sigma}_x\rho\hat{\sigma}_x + \hat{\sigma}_y\rho\hat{\sigma}_y + \hat{\sigma}_z\rho\hat{\sigma}_z\right),} \tag{7.6}$$

where $\mathrm{e_P}$ is a real parameter called the **Pauli error**. This transformation leaves the qubit state unchanged with probability $1 - \mathrm{e_P}$ and it applies a bit flip ($\hat{\sigma}_x$), a phase flip ($\hat{\sigma}_z$), or a combination of the two ($\hat{\sigma}_y$) with equal probability. Let us show that the definitions (7.2, 7.6) are equivalent. The second term in eqn (7.6) can be expressed as[3]

$$\hat{\sigma}_x\rho\hat{\sigma}_x + \hat{\sigma}_y\rho\hat{\sigma}_y + \hat{\sigma}_z\rho\hat{\sigma}_z = 2\mathbb{1} - \rho\,. \tag{7.7}$$

Substituting this identity into (7.6), we arrive at

$$\mathcal{E}_{\text{depo}}(\rho) \;=\; (1 - \mathrm{e_P})\rho + \frac{\mathrm{e_P}}{3}(2\mathbb{1} - \rho) \;=\; \left(1 - \frac{4\mathrm{e_P}}{3}\right)\rho + \frac{4\mathrm{e_P}}{3}\frac{\mathbb{1}}{2}. \tag{7.8}$$

This equation is equivalent to (7.2) with the substitution $p = 1 - 4\mathrm{e_P}/3$.

Example 7.1 The depolarizing channel can be interpreted as a black box that rotates a qubit state about a random axis of the Bloch sphere by a random angle. Figure 7.3 illustrates this concept: each circuit on the left contains a single-qubit rotation about a random axis $\hat{n}$. The vector $\hat{n}$ is distributed in an isotropic way in all directions. This infinite ensemble of quantum circuits describes a statistical process equivalent to a depolarizing channel.

Let us start from a generic qubit rotation[4] $\mathsf{R}_{\hat{n}}(\alpha) = \cos\left(\frac{\alpha}{2}\right)\mathbb{1} - i\sin\left(\frac{\alpha}{2}\right)(\hat{n}\cdot\vec{\sigma})$. Here, $\hat{n} = (n_x, n_y, n_z)$ is a unit vector and $\vec{\sigma} = (\hat{\sigma}_x, \hat{\sigma}_y, \hat{\sigma}_z)$ is the Pauli vector. This rotation transforms a qubit state into

[3] To see this, substitute $\rho = (\mathbb{1} + r_x\hat{\sigma}_x + r_y\hat{\sigma}_y + r_z\hat{\sigma}_z)/2$ into $\hat{\sigma}_x\rho\hat{\sigma}_x + \hat{\sigma}_y\rho\hat{\sigma}_y + \hat{\sigma}_z\rho\hat{\sigma}_z$ and perform some matrix multiplications.

[4] For $\hat{n} = (1, 0, 0)$, this rotation reduces to X_α presented in Section 5.3.

ρ — $\mathsf{R}_{n_1}(\alpha_1)$ —
ρ — $\mathsf{R}_{n_2}(\alpha_2)$ — $=$ ρ — $\mathcal{E}_{\text{depo}}$ —
ρ — $\mathsf{R}_{n_3}(\alpha_3)$ —
⋮

Fig. 7.3 The depolarizing channel. A statistical ensemble of experiments consisting of rotations about a random axis of the Bloch sphere by a random angle is equivalent to a depolarizing channel.

$$\begin{aligned}
&\mathsf{R}_{\hat{n}}(\alpha)\rho\mathsf{R}_{\hat{n}}^{\dagger}(\alpha) = \\
&= \left[\cos\left(\frac{\alpha}{2}\right)\mathbb{1} - i\sin\left(\frac{\alpha}{2}\right)(\hat{n}\cdot\vec{\sigma})\right]\,\rho\,\left[\cos\left(\frac{\alpha}{2}\right)\mathbb{1} + i\sin\left(\frac{\alpha}{2}\right)(\hat{n}\cdot\vec{\sigma})\right] \\
&= \cos^2\left(\frac{\alpha}{2}\right)\rho + \sin^2\left(\frac{\alpha}{2}\right)\,(\hat{n}\cdot\vec{\sigma})\,\rho\,(\hat{n}\cdot\vec{\sigma}) + i\sin\left(\frac{\alpha}{2}\right)\cos\left(\frac{\alpha}{2}\right)\,\left[\rho\,(\hat{n}\cdot\vec{\sigma}) - (\hat{n}\cdot\vec{\sigma})\,\rho\right] \\
&= \cos^2\left(\frac{\alpha}{2}\right)\rho + \sin^2\left(\frac{\alpha}{2}\right)\sum_{i=1}^{3}\sum_{j=1}^{3} n_i n_j\,\hat{\sigma}_i\rho\hat{\sigma}_j + i\sin\left(\frac{\alpha}{2}\right)\cos\left(\frac{\alpha}{2}\right)\sum_{i=1}^{3} n_i(\rho\hat{\sigma}_i - \hat{\sigma}_i\rho).
\end{aligned} \tag{7.9}$$

The rotation angle can take any value in the interval $[0, 2\pi)$ with probability $p(\alpha)$. An infinite ensemble of single-qubit rotations of this form gives rise to a depolarizing channel. To see this, consider the transformation

$$\int_0^{2\pi} d\alpha\; p(\alpha)\frac{1}{4\pi}\int_0^{2\pi} d\phi \int_0^{\pi} d\theta \sin\theta\; \mathsf{R}_{\hat{n}}(\alpha)\,\rho\,\mathsf{R}_{\hat{n}}^{\dagger}(\alpha). \tag{7.10}$$

Recall that[5]

$$\frac{1}{4\pi}\int_0^{2\pi} d\phi \int_0^{\pi} d\theta \sin(\theta)\; n_i n_j = \frac{1}{3}\delta_{ij}, \qquad \frac{1}{4\pi}\int_0^{2\pi} d\phi \int_0^{\pi} d\theta \sin(\theta)\; n_i = 0. \tag{7.11}$$

Using eqns (7.9, 7.11), eqn (7.10) becomes

$$\begin{aligned}
&\int_0^{2\pi} d\alpha\, p(\alpha)\left[\cos^2\left(\frac{\alpha}{2}\right)\rho + \sin^2\left(\frac{\alpha}{2}\right)\sum_{i=1}^{3}\sum_{j=1}^{3}\frac{1}{3}\delta_{ij}\hat{\sigma}_i\rho\hat{\sigma}_j\right] = \\
&= \int_0^{2\pi} d\alpha\, p(\alpha)\cos^2\left(\frac{\alpha}{2}\right)\rho + \frac{1}{3}\int_0^{2\pi} d\alpha\, p(\alpha)\sin^2\left(\frac{\alpha}{2}\right)\sum_{i=1}^{3}\hat{\sigma}_i\rho\hat{\sigma}_i =
\end{aligned}$$

[5]The reader can verify the validity of these equalities by replacing n_i and n_j with $n_x = \sin\theta\cos\phi$, $n_y = \sin\theta\sin\phi$, and $n_z = \cos\theta$.

$$= (1 - e_P)\,\rho + \frac{e_P}{3}\sum_{i=1}^{3} \hat{\sigma}_i \rho \hat{\sigma}_i,$$

where in the last step we defined $e_P = \int_0^{2\pi} d\alpha p(\alpha) \sin^2(\frac{\alpha}{2})$. This is nothing but the depolarizing channel introduced in eqn (7.6). In conclusion, random rotations with an isotropic distribution induce a depolarization of the qubit state. In the context of quantum error correction, undoing a depolarizing error is the most challenging task for an error correcting code.

It is worth highlighting that there exists a big distinction between a deterministic rotation about an axis of the Bloch sphere and a statistical ensemble of rotations about random axes. In the first case, the length of the Bloch vector remains unchanged. In the second case, the Bloch vector gets shorter, as we have demonstrated in this example.

7.3 Amplitude damping

In this section, we present **amplitude damping**, an important example of a quantum map. This is a non-unitary transformation that can be used to model the sudden relaxation of a qubit to the ground state due to spontaneous emission, a physical process in which a quantum of energy is emitted by a qubit in the form of a photon, a phonon, or other elementary particles. This quantum map is defined as

$$\boxed{\mathcal{E}_{\text{damp}}(\rho) = \hat{E}_1 \rho \hat{E}_1^\dagger + \hat{E}_2 \rho \hat{E}_2^\dagger,} \tag{7.12}$$

where the matrices associated with the operators $\hat{E}_1$, $\hat{E}_2$ are given by

$$\hat{E}_1 = \sqrt{p}\hat{\sigma}^- = \begin{bmatrix} 0 & \sqrt{p} \\ 0 & 0 \end{bmatrix}, \tag{7.13}$$

$$\hat{E}_2 = |0\rangle\langle 0| + \sqrt{1-p}|1\rangle\langle 1| = \begin{bmatrix} 1 & 0 \\ 0 & \sqrt{1-p} \end{bmatrix}. \tag{7.14}$$

Here, the parameter $p \in [0, 1)$ can be considered to be the probability that the qubit transitions to the ground state in a short time interval Δt.

Let us consider the action of amplitude damping on a generic qubit state

$$\rho = \frac{1}{2}\begin{bmatrix} 1 + r_z & r_x - ir_y \\ r_x + ir_y & 1 - r_z \end{bmatrix}. \tag{7.15}$$

After substituting (7.13, 7.14, 7.15) into (7.12) and performing some matrix multiplications, we discover that the input state is transformed into

$$\mathcal{E}_{\text{damp}}(\rho) = \frac{1}{2}\begin{bmatrix} 1 + p + (1-p)\,r_z & \sqrt{1-p}\,(r_x - ir_y) \\ \sqrt{1-p}\,(r_x + ir_y) & 1 - p - (1-p)\,r_z \end{bmatrix}. \tag{7.16}$$

Therefore, the initial Bloch vector becomes

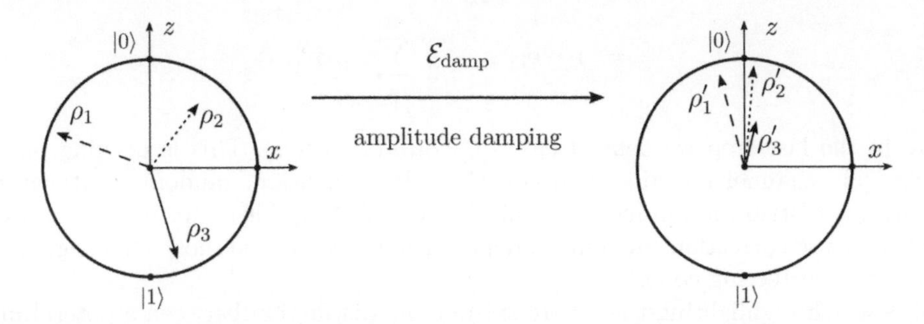

Fig. 7.4 Amplitude damping. Amplitude damping pushes a qubit state toward the north pole of the Bloch sphere. This process affects both longitudinal and transverse components of the Bloch vector.

$$\boxed{(r_x, r_y, r_z) \quad \mapsto \quad (\sqrt{1-p}\, r_x,\ \sqrt{1-p}\, r_y,\ p + (1-p) r_z).} \tag{7.17}$$

For example, the qubit state $|1\rangle\langle 1|$, represented by the Bloch vector $(0, 0, -1)$, is mapped into the Bloch vector $(0, 0, 2p - 1)$.

Figure 7.4 shows how $\mathcal{E}_{\text{damp}}$ acts on some Bloch vectors. This figure shows that the initial state is pushed toward the ground state, the north pole of the Bloch sphere. Amplitude damping is not unital, i.e. $\mathcal{E}_{\text{damp}}(\mathbb{1}/2) \neq \mathbb{1}/2$. Hence, this quantum map can increase the purity of the input state. This is the case for the mixed state $\mathbb{1}/2$, which is mapped into the pure state $|0\rangle\langle 0|$ for $p \to 1$.

By combining (6.37) with (7.16), the action of amplitude damping on a qubit state can also be expressed in terms of populations and coherences,

$$\mathcal{E}_{\text{damp}}(\rho) = \begin{bmatrix} \rho_{00} + p\rho_{11} & \sqrt{1-p}\rho_{01} \\ \sqrt{1-p}\rho_{10} & (1-p)\rho_{11} \end{bmatrix}. \tag{7.18}$$

This expression suggests that ρ_{11} (the probability of finding the qubit in the excited state) is reduced by a factor $1 - p$, whereas the coherence ρ_{01} is reduced by a factor $\sqrt{1-p}$. Since $p \in [0, 1)$, amplitude damping affects the populations more than the coherences.

It is interesting to study the action of amplitude damping on a qubit state several times in a row. From this analysis, an exponential decay of populations and coherences will emerge in a natural way (see Fig. 7.5). Let us start by interpreting the map $\mathcal{E}_{\text{damp}}$ as a discrete change of a qubit state in a short time interval $\Delta t \ll 1$. We can now define the **relaxation rate** γ as the probability that the qubit state jumps to the ground state per unit time, $\gamma = p/\Delta t$. By applying amplitude damping to the initial state n times, we obtain the qubit state after a finite time interval $t = n\Delta t$,

$$\mathcal{E}_{\text{damp},t}(\rho) = \begin{bmatrix} \rho_{00} + [1 - (1-p)^n]\rho_{11} & (1-p)^{n/2}\rho_{01} \\ (1-p)^{n/2}\rho_{10} & (1-p)^n\rho_{11} \end{bmatrix}. \tag{7.19}$$

$$\rho \xrightarrow{\mathcal{E}_{\text{damp}}} \rho(\Delta t) \xrightarrow{\mathcal{E}_{\text{damp}}} \rho(2\Delta t) \xrightarrow{\mathcal{E}_{\text{damp}}} \cdots \xrightarrow{\mathcal{E}_{\text{damp}}} \rho(t) = \begin{bmatrix} 1-\rho_{11}e^{-t/T_1} & \rho_{01}e^{-t/2T_1} \\ \rho_{10}e^{-t/2T_1} & \rho_{11}e^{-t/T_1} \end{bmatrix}$$

Fig. 7.5 Amplitude damping. Repeated applications of amplitude damping lead to an exponential decay of the populations and coherences.

This shows that after n applications the probability of measuring the qubit in the excited state becomes $(1-p)^n \rho_{11}$. In the limit $n \gg 1$, we can express this quantity as

$$\lim_{n\gg 1} (1-p)^n \rho_{11} = \lim_{n\gg 1} \left(1 - \frac{\gamma t}{n}\right)^n \rho_{11} = e^{-\gamma t} \rho_{11} = e^{-t/T_1} \rho_{11}, \tag{7.20}$$

where in the last step we introduced the **relaxation time** $T_1 = 1/\gamma$, the characteristic time in which the qubit abruptly jumps to the ground state. The off-diagonal elements, ρ_{01} and ρ_{10}, decay according to an exponential law too,

$$\lim_{n\gg 1} (1-p)^{n/2} \rho_{01} = \lim_{n\gg 1} \left(1 - \frac{\gamma t}{n}\right)^{n/2} \rho_{01} = e^{-\gamma t/2} \rho_{01} = e^{-t/2T_1} \rho_{01} \,. \tag{7.21}$$

Substituting eqns (7.20, 7.21) into (7.19), we obtain

$$\mathcal{E}_{\text{damp},t}(\rho) = \begin{bmatrix} 1-\rho_{11}e^{-t/T_1} & \rho_{01}e^{-t/2T_1} \\ \rho_{10}e^{-t/2T_1} & \rho_{11}e^{-t/T_1} \end{bmatrix}.$$

Note that the off-diagonal elements statistically decay with a longer characteristic time $2T_1$. In conclusion, amplitude damping causes a suppression of the coherences, a phenomenon known as decoherence.

Amplitude damping projects any initial qubit state onto the ground state after sufficiently long times,

$$\lim_{t\to\infty} \mathcal{E}_{\text{damp},t}(\rho) = \lim_{t\to\infty} \begin{bmatrix} 1-\rho_{11}e^{-t/T_1} & \rho_{01}e^{-t/2T_1} \\ \rho_{10}e^{-t/2T_1} & \rho_{11}e^{-t/T_1} \end{bmatrix} = \begin{bmatrix} 1 & 0 \\ 0 & 0 \end{bmatrix}. \tag{7.22}$$

It turns out that this process is an effective way to reset a qubit into the ground state. If spontaneous emission takes place in a characteristic time T_1, the qubit will statistically be found in the ground state with high probability after a time interval $t \simeq 3T_1$. For quantum devices based on superconducting qubits, $T_1 = 1 - 500$ μs. To speed up this initialization routine, a procedure known as **active reset** can be implemented [203, 204]. The standard version of this procedure consists in measuring the qubit state in the computational basis. If the measurement outcome indicates that the qubit is in the ground state, no further actions have to be taken. If the measurement outcome shows instead that the qubit is in the excited state, a X_π rotation is applied to the qubit. If single-qubit rotations and projective measurements are fast and accurate enough, this procedure can initialize the qubit in the ground state in time intervals much smaller than $3T_1$. Since many quantum algorithms rely on multiple repetitions of a specific circuit, an active reset is often desirable to speed up the initialization routine.

The **generalized amplitude damping** $\mathcal{E}_{\text{g-damp}}$ is a transformation that describes the relaxation of a qubit when in contact with an environment at finite temperature. In contrast to what we discussed before, after sufficiently long times the qubit will not relax to the ground state but to a thermal state ρ_{th} given by eqn (6.45) that we report here for convenience

$$\rho_{\text{th}} = \frac{1}{2\cosh\left(\frac{\hbar\omega_{\text{q}}}{2k_{\text{B}}T}\right)} \begin{bmatrix} e^{\hbar\omega_{\text{q}}/2k_{\text{B}}T} & 0 \\ 0 & e^{-\hbar\omega_{\text{q}}/2k_{\text{B}}T} \end{bmatrix} = \begin{bmatrix} a & 0 \\ 0 & 1-a \end{bmatrix}.$$

Here, ω_{q} is the qubit frequency, $a = e^{\hbar\omega_{\text{q}}/2k_{\text{B}}T}/2\cosh(\frac{\hbar\omega_{\text{q}}}{2k_{\text{B}}T})$ is the probability of finding the qubit in the ground state at thermal equilibrium and T is the temperature of the environment. The action of the generalized amplitude damping on a qubit state is defined as[6]

$$\mathcal{E}_{\text{g-damp}}(\rho) = \hat{E}_1\rho\hat{E}_1^\dagger + \hat{E}_2\rho\hat{E}_2^\dagger + \hat{E}_3\rho\hat{E}_3^\dagger + \hat{E}_4\rho\hat{E}_4^\dagger, \tag{7.23}$$

where

$$E_1 = \sqrt{a}\begin{bmatrix} 0 & \sqrt{p} \\ 0 & 0 \end{bmatrix}, \qquad E_2 = \sqrt{a}\begin{bmatrix} 1 & 0 \\ 0 & \sqrt{1-p} \end{bmatrix},$$

$$E_3 = \sqrt{1-a}\begin{bmatrix} \sqrt{1-p} & 0 \\ 0 & 1 \end{bmatrix}, \qquad E_4 = \sqrt{1-a}\begin{bmatrix} 0 & 0 \\ \sqrt{p} & 0 \end{bmatrix},$$

and the parameter p indicates the probability that the qubit relaxes to the thermal state. The two parameters a and p describe two different features: a is determined by the ratio ω_{q}/T, while p is directly related to the relaxation time of the qubit. It can be shown that after multiple applications, any initial state ρ relaxes to the equilibrium state: $\lim_{n\to+\infty} \mathcal{E}^n_{\text{g-damp}}(\rho) = \rho_{\text{th}}$.

7.3.1 Inversion recovery: measuring T_1

Inversion recovery is an experiment designed to measure the relaxation time of a qubit. As shown in Fig. 7.6a, this experiment applies a π-pulse[7] to a qubit followed by a measurement operation. The π-pulse transforms the initial state $\rho(t_0) = |0\rangle\langle 0|$ into

$$\rho(t_1) = \mathsf{X}_\pi\rho(t_0)\mathsf{X}_\pi^\dagger = \begin{bmatrix} 0 & -i \\ -i & 0 \end{bmatrix}\begin{bmatrix} 1 & 0 \\ 0 & 0 \end{bmatrix}\begin{bmatrix} 0 & i \\ i & 0 \end{bmatrix} = \begin{bmatrix} 0 & 0 \\ 0 & 1 \end{bmatrix} = |1\rangle\langle 1|. \tag{7.24}$$

The qubit then undergoes free evolution for a time interval $t = t_2 - t_1$. In this period of time, the qubit is subjected to spontaneous emission,

[6]Note that if the temperature of the environment satisfies $k_{\text{B}}T \ll \hbar\omega_{\text{q}}$, then this quantum map reduces to $\mathcal{E}_{\text{damp}}$ defined in (7.12).

[7]Here, we are assuming that the relaxation time is much longer than the pulse duration.

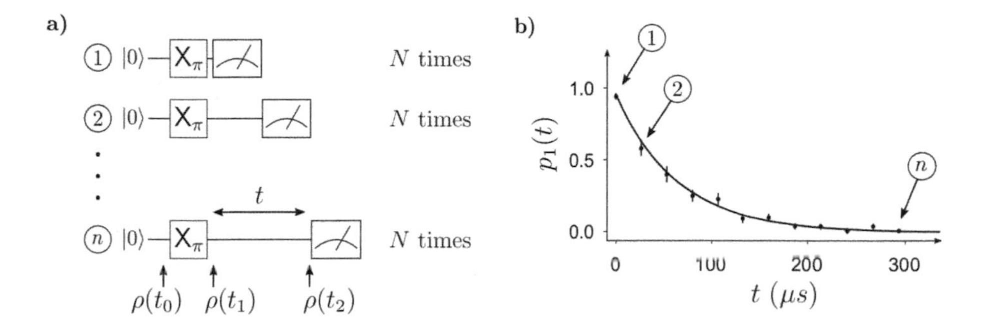

Fig. 7.6 Inversion recovery experiment. a) Scheme of the experiment: a π-pulse is applied to a qubit followed by a projective measurement in the computational basis. The delay time between the π-pulse and the measurement operation is denoted t. Repeated measurements give enough statistics to estimate the probability of finding the qubit in the excited state $p_1 = N_1/N$ as a function of the delay time t. **b)** The relaxation time T_1 is estimated by fitting $p_1(t)$ to an exponential decay. The error bars represent the standard deviation of the mean for a binomial distribution, $\sigma = \sqrt{p_1(1-p_1)/(N-1)}$.

$$\rho(t_2) = \mathcal{E}_{\text{damp}}(\rho(t_1)) = \begin{bmatrix} 1 - e^{-t/T_1} & 0 \\ 0 & e^{-t/T_1} \end{bmatrix}. \tag{7.25}$$

The state $\rho(t_2)$ is measured in the computational basis. Each circuit illustrated in Fig. 7.6a is repeated N times to acquire some statistics: the qubit will be found in the ground state N_0 times and in the excited state N_1 times. This experiment is performed for different values of the delay between the π rotation and measurement pulse. The probability of finding the qubit in the excited state $p_1(t) = N_1(t)/N$ is then plotted as a function of the delay time as shown in Fig. 7.6b. For N sufficiently large, p_1 converges to e^{-t/T_1}. The data points can be fitted with a decay model $A + Be^{-t/C}$ where B is a scaling factor, C is the fit parameter associated with the relaxation time T_1, and A represents the baseline.

7.4 Dephasing

In many experiments, coupling with the environment induces random fluctuations of the *qubit frequency*. Charge and flux noise affecting a superconducting qubit are of this type. These fluctuations lead to a loss of information about the coherence of the qubit state. **Dephasing** (or phase damping, or phase-flip channel) is a quantum channel well suited to describe this process, especially when the fluctuations of the qubit frequency are dominated by white noise. This non-unitary quantum map is defined as

$$\boxed{\mathcal{E}_{\text{deph}}(\rho) = (1-p)\,\rho + p\,\hat{\sigma}_z\rho\hat{\sigma}_z.} \tag{7.26}$$

Here, the parameter $p \in [0, 1)$ can be considered to be the probability that the qubit undergoes a phase flip in a short time interval Δt. This quantum map can be expressed in another equivalent way,

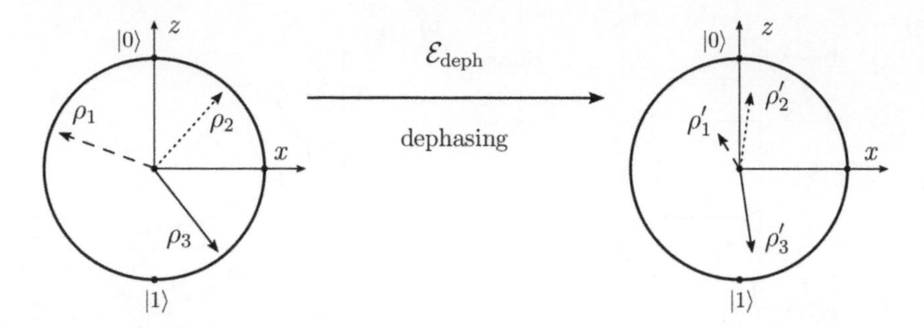

Fig. 7.7 Dephasing. The action of the dephasing channel on some Bloch vectors. The transverse components x and y are reduced in magnitude, whereas the longitudinal component z is left unchanged.

$$\mathcal{E}_{\text{deph}}(\rho) = \hat{E}_1 \rho \hat{E}_1^\dagger + \hat{E}_2 \rho \hat{E}_2^\dagger , \tag{7.27}$$

where the matrices associated with the operators $\hat{E}_1$, $\hat{E}_2$ are given by

$$\hat{E}_1 = \sqrt{1-p}\, \mathbb{1}, \qquad \hat{E}_2 = \sqrt{p}\, \hat{\sigma}_z. \tag{7.28}$$

Dephasing leaves the qubit state unchanged with probability $1-p$, and it applies a phase flip $\hat{\sigma}_z$ with probability p.

Let us investigate the effect of dephasing on a generic qubit state. Substituting expression (7.15) into (7.26), one can verify that the qubit state becomes

$$\mathcal{E}_{\text{deph}}(\rho) = \frac{1}{2} \begin{bmatrix} 1 + r_z & (1-2p)(r_x - ir_y) \\ (1-2p)(r_x + ir_y) & 1 - r_z \end{bmatrix} . \tag{7.29}$$

Therefore, the initial Bloch vector is mapped into

$$\boxed{(r_x, r_y, r_z) \quad \mapsto \quad ((1-2p)\, r_x,\ (1-2p)\, r_y,\ r_z).} \tag{7.30}$$

The z component of the initial vector remains unchanged. Conversely, the x and y components are multiplied by a factor $1-2p$, meaning that their magnitude becomes smaller. Figure 7.7 shows the action of the dephasing channel on three different Bloch vectors. Since $\mathcal{E}_{\text{deph}}(\mathbb{1}) = \mathbb{1}$, dephasing is unital, and never increases the purity of the input state.

The map $\mathcal{E}_{\text{deph}}$ can be interpreted as a discrete change of the qubit state in a short time interval $\Delta t \ll 1$. Let us study the case in which the dephasing channel is applied repeatedly to a qubit state (see Fig. 7.8). The application of this map is better appreciated by looking at its action on populations and coherences,

$$\rho = \begin{bmatrix} \rho_{00} & \rho_{01} \\ \rho_{10} & \rho_{11} \end{bmatrix} \mapsto \rho(\Delta t) = \mathcal{E}_{\text{deph},\Delta t}(\rho) = \begin{bmatrix} \rho_{00} & \rho_{01}(1-2p) \\ \rho_{10}(1-2p) & \rho_{11} \end{bmatrix} . \tag{7.31}$$

This expression shows that the populations do not change, whereas the coherences

$$\rho \xrightarrow{\mathcal{E}_{\text{deph}}} \rho(\Delta t) \xrightarrow{\mathcal{E}_{\text{deph}}} \rho(2\Delta t) \xrightarrow{\mathcal{E}_{\text{deph}}} \cdots \xrightarrow{\mathcal{E}_{\text{deph}}} \rho(t) = \begin{bmatrix} \rho_{00} & \rho_{01}e^{-t/T_\phi} \\ \rho_{10}e^{-t/T_\phi} & \rho_{11} \end{bmatrix}$$

Fig. 7.8 Dephasing. The repeated application of the dephasing channel to a qubit state leads to an exponential decay of the coherences.

---

decrease in magnitude. If a qubit is subjected to a sequence of dephasing channels, the state at a time $t = n\Delta t$ becomes

$$\underbrace{\mathcal{E}_{\text{deph},\Delta t} \circ \ldots \circ \mathcal{E}_{\text{deph},\Delta t}}_{n \text{ times}}(\rho) = \mathcal{E}_{\text{deph},t}(\rho) = \begin{bmatrix} \rho_{00} & \rho_{01}(1-2p)^n \\ \rho_{10}(1-2p)^n & \rho_{11} \end{bmatrix}. \tag{7.32}$$

We can now introduce the dephasing rate as the probability that a phase flip occurs per unit time, $\gamma_\phi \equiv 2p/\Delta t$. After a long time interval ($n \gg 1$), the coherence ρ_{10} becomes

$$\lim_{n\gg 1} \rho_{10}(1-2p)^n = \lim_{n\gg 1} \rho_{10}\left(1 - \frac{\gamma_\phi t}{n}\right)^n = \rho_{10}e^{-\gamma_\phi t} = \rho_{10}e^{-t/T_\phi}, \tag{7.33}$$

where in the last step we introduced the **dephasing time** $T_\phi = 1/\gamma_\phi$, the characteristic time in which the information about the relative phase of the qubit state is lost. Substituting eqn (7.33) into (7.32), we arrive at

$$\mathcal{E}_{\text{deph},t}(\rho) = \begin{bmatrix} \rho_{00} & \rho_{01}e^{-t/T_\phi} \\ \rho_{10}e^{-t/T_\phi} & \rho_{11} \end{bmatrix}.$$

In conclusion, the off-diagonal elements of the density matrix decay exponentially in time, causing the decoherence of the qubit state.

In an actual experiment, a qubit is subjected to spontaneous emission and dephasing *at the same time.* This phenomenon can be described by a combination of dephasing and amplitude damping[8]

$$\mathcal{E}_{\text{damp}}(\mathcal{E}_{\text{deph}}(\rho)) = \begin{bmatrix} 1 - \rho_{11}e^{-t/T_1} & \rho_{01}e^{-t\left(\frac{1}{T_\phi}+\frac{1}{2T_1}\right)} \\ \rho_{10}e^{-t\left(\frac{1}{T_\phi}+\frac{1}{2T_1}\right)} & \rho_{11}e^{-t/T_1} \end{bmatrix}. \tag{7.34}$$

The coherences decay exponentially to zero with a characteristic time

$$\boxed{T_2 = \left(\frac{1}{T_\phi} + \frac{1}{2T_1}\right)^{-1}.} \tag{7.35}$$

The parameter T_2 is called the **coherence time** of the qubit. From (7.35), it is evident that T_2 always lies in the range $0 \leq T_2 \leq 2T_1$. If the dephasing time T_ϕ is much longer

[8]Note that $\mathcal{E}_{\text{damp}}$ and $\mathcal{E}_{\text{deph}}$ commute, i.e. $\mathcal{E}_{\text{damp}} \circ \mathcal{E}_{\text{deph}} = \mathcal{E}_{\text{deph}} \circ \mathcal{E}_{\text{damp}}$.

than the relaxation time, then $T_2 \simeq 2T_1$ and we say that T_2 is $\mathbf{T_1}$**-limited**.

Example 7.2 In Example 7.1, we showed that the depolarizing channel can be interpreted as an infinite ensemble of random single-qubit rotations. Similarly, dephasing can be interpreted as an infinite ensemble of random rotations about the z axis of the Bloch sphere $\mathsf{Z}_\alpha = \cos(\alpha/2)\mathbb{1} - i\sin(\alpha/2)\hat{\sigma}_z$ where $\alpha \in [-\pi, \pi]$ is a random angle with probability distribution $p(\alpha)$. The gate Z_α transforms a qubit state ρ into

$$\mathsf{Z}_\alpha \rho \mathsf{Z}_\alpha^\dagger = \cos^2\left(\frac{\alpha}{2}\right)\rho + i\sin\left(\frac{\alpha}{2}\right)\cos\left(\frac{\alpha}{2}\right)(\rho\hat{\sigma}_z - \hat{\sigma}_z\rho) + \sin^2\left(\frac{\alpha}{2}\right)\hat{\sigma}_z\rho\hat{\sigma}_z. \tag{7.36}$$

Assuming that the probability distribution $p(\alpha)$ is symmetric around zero, an infinite ensemble of random rotations about the z axis is equivalent to the dephasing channel

$$\int_{-\pi}^{\pi} d\alpha\, p(\alpha)\left[\cos^2\left(\frac{\alpha}{2}\right)\rho\right] + \left[i\sin\left(\frac{\alpha}{2}\right)\cos\left(\frac{\alpha}{2}\right)(\rho\hat{\sigma}_z - \hat{\sigma}_z\rho)\right] + \left[\sin^2\left(\frac{\alpha}{2}\right)\hat{\sigma}_z\rho\hat{\sigma}_z\right] =$$
$$= (1-p)\,\rho + p\hat{\sigma}_z\rho\hat{\sigma}_z,$$

where we defined $p = \int_0^{2\pi} d\alpha p(\alpha)\sin^2(\alpha/2)$. This result is consistent with eqn (7.26).

7.4.1 Ramsey experiment: measuring T_2

A **Ramsey experiment** (or free-induction decay) is a sequence of experiments aimed at determining the transverse relaxation time T_2 of the qubit. The qubit is initially prepared in the ground state $\rho(t_0) = |0\rangle\langle 0|$. Two $\pi/2$ rotations separated by a time interval t are applied to the qubit followed by a projective measurement in the computational basis. The two rotations are usually implemented with electrical pulses whose duration is much shorter than the coherence time. After the first $\pi/2$-pulse, the qubit state turns into a balanced superposition state

$$\rho(t_1) = \mathsf{Y}_{\frac{\pi}{2}}\rho(t_0)\mathsf{Y}_{\frac{\pi}{2}}^\dagger = \frac{|0\rangle + |1\rangle}{\sqrt{2}}\frac{\langle 0| + \langle 1|}{\sqrt{2}} = \frac{1}{2}\begin{bmatrix} 1 & 1 \\ 1 & 1 \end{bmatrix}. \tag{7.37}$$

The qubit is subjected to spontaneous emission and dephasing at the same time. After a time interval $t = t_2 - t_1$, the combination of these two processes will transform the qubit state into

$$\rho(t_2) = \mathcal{E}_{\text{damp}}(\mathcal{E}_{\text{deph}}(\rho(t_1))) = \frac{1}{2}\begin{bmatrix} 2 - e^{-t/T_1} & e^{-t/T_2} \\ e^{-t/T_2} & e^{-t/T_1} \end{bmatrix}, \tag{7.38}$$

where we used eqn (7.34) and $T_2 = (1/T_\phi + 1/2T_1)^{-1}$. A second $\pi/2$-pulse is applied to the qubit, yielding

$$\rho(t_3) = \mathsf{Y}_{\frac{\pi}{2}}\rho(t_2)\mathsf{Y}_{\frac{\pi}{2}}^\dagger = \frac{1}{2}\begin{bmatrix} 1 - e^{-t/T_2} & 1 - e^{-t/T_1} \\ 1 - e^{-t/T_1} & 1 + e^{-t/T_2} \end{bmatrix}. \tag{7.39}$$

The qubit is then measured in the computational basis. The probability of observing the qubit in the excited state is given by $\rho_{11}(t_3) = (1 + e^{-t/T_2})/2$. This procedure,

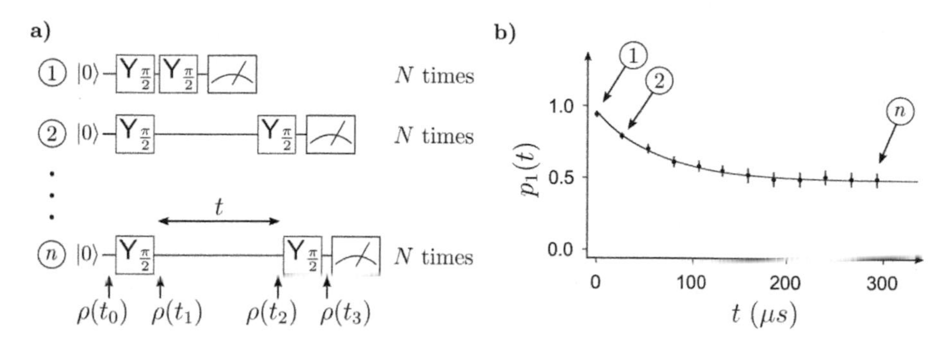

Fig. 7.9 The Ramsey experiment. a) Scheme of the experiment. Two $\pi/2$-pulses are applied to a qubit followed by a projective measurement in the computational basis. The delay time between the two $\pi/2$-pulses is denoted as t. By repeating each experiment N times, we obtain the probability of finding the qubit in the excited state $p_1(t) = N_1/N$ where N_1 is the number of times the qubit is observed in the excited state. **b)** The transverse relaxation time T_2 is obtained by fitting the measured probability $p_1(t)$ to an exponential decay $A + Be^{-t/C}$. The error bars represent the standard deviation of the mean for a binomial distribution $\sigma = \sqrt{p_1(1-p_1)/(N-1)}$.

illustrated in Fig. 7.9, is repeated N times to gather some statistics. The measurements will reveal the qubit in the ground state N_0 times and in the excited state N_1 times. This experiment is performed for different delay times between the two $\pi/2$-pulses and the population of the excited state $p_1(t) = N_1(t)/N$ is plotted as a function of delay time as shown in Fig. 7.9b. The data points are fitted with a decay model $A+Be^{-t/C}$, where the fit parameter C corresponds to the transverse relaxation time T_2. For a superconducting qubit, this quantity is typically in the range $1-200$ μs.

If the electrical pulses implementing the $\pi/2$ rotations have a frequency ω_{d} slightly different from the qubit frequency ω_{q}, the plot in Fig. 7.9b will show some oscillations with frequency $|\omega_{\text{d}}-\omega_{\text{q}}|/2\pi = \delta$. In this case, the fitting function is $A+B\cos(Dt)e^{-t/C}$. Here, D is the fit parameter for the detuning δ between the qubit frequency and the frequency of the pulses. This interferometric experiment turns out to be an excellent way to determine the qubit frequency with high precision [205].

So far, we have explained how to measure the coherence time T_2. What about the dephasing time T_ϕ? This parameter can be determined with different approaches:

1. If $T_2 \ll T_1$, then T_2 is approximately equal to T_ϕ. Thus, T_ϕ can be measured with a Ramsey experiment.

2. If $T_2 \approx T_1$, the dephasing time can be determined by running an inversion recovery experiment followed by a Ramsey experiment. These two experiments give an estimate of T_1 and T_2 from which one can extract the dephasing time from the formula $1/T_\phi = 1/T_2 - 1/2T_1$.

3. If $T_2 \approx T_1$ and both of them fluctuate on a time scale much shorter than the time it takes to measure them[9], it is preferable to estimate T_ϕ using an **interleaved** experiment: the circuits composing the Ramsey experiment and the inversion recovery experiment are executed in a concatenated way to average out the fluctuations [206].

7.5 Transposition

Not all of the superoperators acting on density operators are valid quantum maps. A prime example is the **transposition function**: this function is not a quantum map because it is positive and trace-preserving but not completely positive. Let us explain this point. The transposition T simply transposes the entries of a density operator $\rho = \sum_{i,j} \rho_{ij}|i\rangle\langle j|$,

$$\boxed{\mathrm{T}(\rho) = \sum_{i,j} \rho_{ji}|i\rangle\langle j|.} \tag{7.40}$$

It is simple to show that this function is positive and trace-preserving (see Exercise 7.8). Let us prove that T is not completely positive by showing that the action of $\mathrm{T} \otimes \mathbb{1}$ onto the two-qubit state $|\beta_{00}\rangle = \frac{1}{\sqrt{2}}[|00\rangle + |11\rangle]$ produces a non-positive operator. The state $|\beta_{00}\rangle\langle\beta_{00}|$ can be written as

$$|\beta_{00}\rangle\langle\beta_{00}| = \frac{1}{2}\Big[\,|0\rangle\langle 0| \otimes |0\rangle\langle 0| + |0\rangle\langle 1| \otimes |0\rangle\langle 1| + |1\rangle\langle 0| \otimes |1\rangle\langle 0| + |1\rangle\langle 1| \otimes |1\rangle\langle 1|\,\Big].$$

Therefore,

$$\begin{aligned}(\mathrm{T} \otimes \mathbb{1})|\beta_{00}\rangle\langle\beta_{00}| &= \frac{1}{2}\Big[|0\rangle\langle 0| \otimes |0\rangle\langle 0| + |1\rangle\langle 0| \otimes |0\rangle\langle 1| + |0\rangle\langle 1| \otimes |1\rangle\langle 0| + |1\rangle\langle 1| \otimes |1\rangle\langle 1|\Big] \\ &= \frac{1}{2}\begin{bmatrix} 1 & 0 & 0 & 0 \\ 0 & 0 & 1 & 0 \\ 0 & 1 & 0 & 0 \\ 0 & 0 & 0 & 1 \end{bmatrix} = \rho'.\end{aligned}$$

The eigenvalues of ρ' are $\{-\frac{1}{2}, +\frac{1}{2}\}$. Since one eigenvalue is negative, ρ' is not positive and T is not a quantum map. The partial transposition $\mathbb{1} \otimes \mathrm{T}$ is an important function that can be used to determine if a pure bipartite state is entangled or not, as we shall see in Section 9.8.1.

A simple way to tell whether a linear function $\mathcal{E}$ is a quantum map is by investigating its action on the maximally entangled state $|\Omega\rangle \in \mathbb{C}^d \otimes \mathbb{C}^d$ defined as

$$|\Omega\rangle = \frac{1}{\sqrt{d}} \sum_{i=0}^{d-1} |i\rangle|i\rangle. \tag{7.41}$$

When $d = 2$, this state reduces to the Bell state $|\beta_{00}\rangle = \frac{1}{\sqrt{2}}(|00\rangle + |11\rangle)$. A linear function $\mathcal{E} : \mathbb{C}^d \to \mathbb{C}^d$ is completely positive if and only if it maps the entangled state

[9] A typical inversion recovery experiment consists of $n = 10 - 100$ circuits each repeated $N = 100 - 1000$ times to gather some statistics.

$|\Omega\rangle$ into a positive operator (see Exercise 7.6),

$$(\mathcal{E} \otimes \mathbb{1}_{\mathrm{B}})|\Omega\rangle\langle\Omega| \geq 0. \tag{7.42}$$

In short, one just needs to calculate the eigenvalues of $(\mathcal{E} \otimes \mathbb{1}_{\mathrm{B}})|\Omega\rangle\langle\Omega|$. If one of them is negative, then $\mathcal{E}$ is not a quantum map [207, 208].

7.6 Kraus representation of a quantum map

In the previous sections, we presented some examples of quantum maps. The reader might have noticed that eqns (7.12, 7.23, 7.27) share a very similar form, known as the **Kraus form**. We now present an important theorem that states that any superoperator is a quantum map if and only if it can be expressed in that form.

Theorem 7.1 A superoperator $\mathcal{E} : \mathrm{B}(H) \to \mathrm{B}(H)$ is a quantum map if and only if it can be expressed in the form

$$\boxed{\mathcal{E}(\rho) = \sum_k \hat{E}_k \rho \hat{E}_k^\dagger,} \tag{7.43}$$

where $\sum_k \hat{E}_k^\dagger \hat{E}_k = \mathbb{1}$. Equation (7.43) is called the **Kraus representation** of $\mathcal{E}$ and the operators $\hat{E}_k$ are called the **Kraus operators**.

We highlight that the operators $\hat{E}_k$ do not have to be unitary or Hermitian. From the proof of this theorem, we will discover that any quantum map can be decomposed with no more than d^2 Kraus operators where d is the dimension of the Hilbert space.

Proof ($\leftarrow$) Let us assume that $\sum_k \hat{E}_k^\dagger \hat{E}_k = \mathbb{1}$ and demonstrate that $\mathcal{E}(\rho) = \sum_k \hat{E}_k \rho \hat{E}_k^\dagger$ is a quantum map. To this end, we need to show that this map satisfies the requirements **QM1**–**QM4** presented in Section 7.1.

1. The map $\mathcal{E}$ is linear because

$$\mathcal{E}(\rho_1 + \rho_2) = \sum_k \hat{E}_k (\rho_1 + \rho_2) \hat{E}_k^\dagger = \sum_k (\hat{E}_k \rho_1 \hat{E}_k^\dagger + \hat{E}_k \rho_2 \hat{E}_k^\dagger) = \mathcal{E}(\rho_1) + \mathcal{E}(\rho_2) . \tag{7.44}$$

2. The map $\mathcal{E}$ is trace-preserving because

$$\mathrm{Tr}\big[\mathcal{E}(\rho)\big] = \sum_k \mathrm{Tr}\big[\hat{E}_k \rho \hat{E}_k^\dagger\big] = \sum_k \mathrm{Tr}\big[\rho \hat{E}_k^\dagger \hat{E}_k\big] = \mathrm{Tr}[\rho] . \tag{7.45}$$

3. The map $\mathcal{E}$ is completely positive because for any bipartite state $|\psi\rangle \in H_{\mathrm{A}} \otimes H_{\mathrm{B}}$, the operator $(\mathcal{E} \otimes \mathbb{1}_{\mathrm{B}})\,\rho$ is positive,

$$\langle\psi|\,(\mathcal{E} \otimes \mathbb{1}_{\mathrm{B}})\,\rho|\psi\rangle = \sum_k \langle\psi|(\hat{E}_k \otimes \mathbb{1}_{\mathrm{B}})\,\rho\,(\hat{E}_k^\dagger \otimes \mathbb{1}_{\mathrm{B}})|\psi\rangle = \sum_k \langle\psi_k'|\,\rho\,|\psi_k'\rangle \geq 0,$$

where we introduced $|\psi'_k\rangle = (\hat{E}_k^\dagger \otimes \mathbb{1}_\mathrm{B})|\psi\rangle$. If a map is completely positive, it is positive too. □

Proof ($\rightarrow$) Let us show that any quantum map $\mathcal{E} : \mathrm{B}(H_\mathrm{A}) \rightarrow \mathrm{B}(H_\mathrm{A})$ can be expressed in Kraus form. First, consider an additional Hilbert space H_B with the same dimension as H_A and a density operator[10] ρ_Ω defined on the Hilbert space $H_\mathrm{A} \otimes H_\mathrm{B}$,

$$\rho_\Omega = |\Omega\rangle\langle\Omega| = \frac{1}{d}\sum_{i=1}^{d}\sum_{j=1}^{d}|i\rangle\langle j|_\mathrm{A} \otimes |i\rangle\langle j|_\mathrm{B}, \tag{7.46}$$

where $|\Omega\rangle$ is the maximally entangled state introduced in eqn (7.41) and d is the dimension of H_A. Since $\mathcal{E}$ is completely positive by hypothesis, the operator $\mathcal{E} \otimes \mathbb{1}_\mathrm{B}$ transforms ρ_Ω into a positive operator ρ'_Ω,

$$(\mathcal{E} \otimes \mathbb{1}_\mathrm{B})\,\rho_\Omega = \rho'_\Omega = \sum_{k=1}^{d^2}\lambda_k|v_k\rangle\langle v_k|\,, \tag{7.47}$$

where λ_k and $|v_k\rangle \in H_\mathrm{A} \otimes H_\mathrm{B}$ are the eigenvalues and eigenvectors of ρ'_Ω. Now consider a generic state $|\psi_\mathrm{A}\rangle \in H_\mathrm{A}$ defined as $|\psi_\mathrm{A}\rangle = \sum_{i=1}^{d} c_i|i\rangle_\mathrm{A}$ and consider the state $|\psi_\mathrm{B}\rangle \in H_\mathrm{B}$ given by $|\psi_\mathrm{B}\rangle = \sum_{i=1}^{d} c_i^*|i\rangle_\mathrm{B}$. We can associate an operator $\hat{E}_k \in \mathrm{B}(H_\mathrm{A})$ with each eigenstate $|v_k\rangle$,

$$\langle a|\hat{E}_k|\psi_\mathrm{A}\rangle = \sqrt{\lambda_k d}\,\langle a\psi_\mathrm{B}|v_k\rangle, \tag{7.48}$$

where $|a\rangle$ is any vector of the canonical basis of H_A. Let us demonstrate that the action of the map $\mathcal{E}$ on the state $|\psi\rangle_\mathrm{A}$ can be written as $\mathcal{E}(|\psi\rangle\langle\psi|_\mathrm{A}) = \sum_k \hat{E}_k(|\psi\rangle\langle\psi|_\mathrm{A})\,\hat{E}_k^\dagger$. Consider two generic vectors $|a\rangle$, $|b\rangle \in H_\mathrm{A}$. Then, we have

$$\begin{aligned}
\langle a|\Big(\sum_{k=1}^{d^2}\hat{E}_k|\psi\rangle\langle\psi|_\mathrm{A}\hat{E}_k^\dagger\Big)|b\rangle &= \sum_{k=1}^{d^2} d\,\lambda_k\,\langle a\psi'|v_k\rangle\langle v_k|b\psi'\rangle = d\,\langle a\psi'|\rho_\Omega|b\psi'\rangle \\
&= \sum_{i,j=1}^{d}\,\langle a\psi'|\,(\mathcal{E} \otimes \mathbb{1}_\mathrm{B})\,(|i\rangle\langle j|_\mathrm{A} \otimes |i\rangle\langle j|_\mathrm{B})\,|b\psi'\rangle \\
&= \sum_{i,j=1}^{d}\,\langle a|\mathcal{E}\,(|i\rangle\langle j|_\mathrm{A})\,|b\rangle \cdot \langle\psi'|i\rangle_\mathrm{B} \cdot \langle j|\psi'\rangle_\mathrm{B} \\
&= \langle a|\sum_{i,j=1}^{d}\,c_i c_j^*\,\mathcal{E}\,(|i\rangle\langle j|_\mathrm{A})\,|b\rangle \;=\; \langle a|\mathcal{E}\,(|\psi\rangle\langle\psi|_\mathrm{A})\,|b\rangle.
\end{aligned}$$

We have shown that the action of $\mathcal{E}$ on a pure state $|\psi\rangle_\mathrm{A}$ can be written in Kraus form. Since $\mathcal{E}$ is linear, this result also holds for mixed states and therefore $\mathcal{E}(\rho) =$

[10]For example, if $H_\mathrm{A} = H_\mathrm{B} = \mathbb{C}^2$, the state ρ_Ω is the Bell state $|\beta_{00}\rangle\langle\beta_{00}|$.

$\sum_k \hat{E}_k \rho \hat{E}_k^\dagger$. Finally, we have to demonstrate that $\sum_k \hat{E}_k^\dagger \hat{E}_k = \mathbb{1}_A$. Since $\mathcal{E}$ is trace-preserving, we see that

$$\begin{aligned}\langle j|\mathbb{1}_A|i\rangle &= \text{Tr}\left[|i\rangle\langle j|_A\right] = \text{Tr}\left[\mathcal{E}\left(|i\rangle\langle j|_A\right)\right] = \sum_{k=1}^{d^2} \text{Tr}\left[\hat{E}_k|i\rangle\langle j|_A\hat{E}_k^\dagger\right] \\ &= \sum_{k=1}^{d^2} \text{Tr}\left[\hat{E}_k^\dagger \hat{E}_k|i\rangle\langle j|_A\right] = \sum_{m=1}^{d}\langle m|\sum_{k=1}^{d^2}\hat{E}_k^\dagger\hat{E}_k|i\rangle\langle j|m\rangle = \langle j|\sum_{k=1}^{d^2}\hat{E}_k^\dagger\hat{E}_k|i\rangle.\end{aligned}$$

This proves that $\sum_k \hat{E}_k^\dagger \hat{E}_k = \mathbb{1}_A$. Observe that the number of operators $\hat{E}_k$ is d^2 (one for each eigenvector $|v_k\rangle\rangle$). In conclusion, any quantum map can be expressed in Kraus form with at most d^2 Kraus operators where d is the dimension of the Hilbert space. □

What is the number of real parameters required to specify a quantum map? The Kraus representation provides an easy way to answer this question. Recall that Hermitian matrices are determined by d^2 real numbers. Since a quantum map $\mathcal{E}$ transforms a Hermitian matrix A into a Hermitian matrix A', the map $\mathcal{E}$ is completely specified by $d^2 \times d^2 = d^4$ real parameters. However, not all of them are independent: the condition $\sum_k \hat{E}_k^\dagger \hat{E}_k = \mathbb{1}$ imposes d^2 constraints (one for each entry of the identity matrix) reducing the number of independent parameters to $d^4 - d^2$. For example, a quantum map on single-qubit states is completely specified by $2^4 - 2^2 = 12$ parameters and a quantum map operating on two-qubit states is determined by $4^4 - 4^2 = 240$ parameters.

The Kraus operators associated with a quantum map are not uniquely defined. Consider the operators $\hat{F}_k$ obtained from the Kraus operators $\hat{E}_i$ by means of a unitary matrix V,

$$\hat{F}_k = \sum_i V_{ki}\hat{E}_i. \tag{7.49}$$

The Kraus operators $\hat{F}_k$ describe the same quantum map because

$$\begin{aligned}\sum_k \hat{F}_k \rho \hat{F}_k^\dagger &= \sum_k \sum_{i,j} V_{ki}\hat{E}_i\rho\hat{E}_j^\dagger V_{jk}^\dagger \\ &= \sum_{i,j}\hat{E}_i\rho\hat{E}_j^\dagger\delta_{ij} = \sum_i \hat{E}_i\rho\hat{E}_i^\dagger = \mathcal{E}(\rho).\end{aligned} \tag{7.50}$$

Furthermore

$$\sum_k \hat{F}_k^\dagger \hat{F}_k = \sum_k\sum_{i,j}\hat{E}_i^\dagger V_{ik}^\dagger V_{kj}\hat{E}_j = \sum_{i,j}\delta_{ij}\hat{E}_i^\dagger\hat{E}_j = \sum_i \hat{E}_i^\dagger\hat{E}_i = \mathbb{1}, \tag{7.51}$$

as required. Since the unitary matrix V is arbitrary, there exists an infinite number of Kraus representations associated with a given quantum map.

Example 7.3 In Section 7.4, we defined the dephasing channel as $\mathcal{E}_{\text{deph}}(\rho) = \hat{E}_1\rho\hat{E}_1^\dagger + \hat{E}_2\rho\hat{E}_2^\dagger$ where $\hat{E}_1 = \sqrt{1-p}\mathbb{1}$ and $\hat{E}_1 = \sqrt{p}\hat{\sigma}_z$. This map can be expressed with a different set of Kraus operators $\mathcal{E}_{\text{deph}}(\rho) = \hat{F}_1\rho\hat{F}_1^\dagger + \hat{F}_2\rho\hat{F}_2^\dagger$ where

$$F_1 = \begin{bmatrix} 1 & 0 \\ 0 & 1-2p \end{bmatrix}, \qquad F_2 = \begin{bmatrix} 0 & 0 \\ 0 & \sqrt{1-(1-2p)^2} \end{bmatrix}. \tag{7.52}$$

With a matrix multiplication, one can verify that $\hat{F}_1\hat{F}_1^\dagger + \hat{F}_2\hat{F}_2^\dagger = \mathbb{1}$. These operators are related to the Kraus operators (7.28) by the relations $\hat{F}_1 = V_{11}\hat{E}_1 + V_{12}\hat{E}_2$ and $\hat{F}_2 = V_{21}\hat{E}_1 + V_{22}\hat{E}_2$, where

$$V = \begin{bmatrix} V_{11} & V_{12} \\ V_{21} & V_{22} \end{bmatrix} = \begin{bmatrix} \sqrt{1-p} & \sqrt{p} \\ \sqrt{p} & -\sqrt{1-p} \end{bmatrix}.$$

This matrix is unitary since $VV^\dagger = V^\dagger V = \mathbb{1}$.

7.7 The affine map

It is instructive to study the action of quantum maps on single-qubit states from a geometrical point of view. A quantum map $\mathcal{E}$ transforms a Bloch vector $\mathbf{a}$ into a Bloch vector $\mathbf{b} = M\mathbf{a} + \mathbf{c}$ where M is a 3×3 matrix describing a contraction and/or rotation combined with a displacement $\mathbf{c}$,

$$\begin{bmatrix} b_x \\ b_y \\ b_z \end{bmatrix} = \begin{bmatrix} M_{11} & M_{12} & M_{13} \\ M_{21} & M_{22} & M_{23} \\ M_{31} & M_{32} & M_{33} \end{bmatrix} \begin{bmatrix} a_x \\ a_y \\ a_z \end{bmatrix} + \begin{bmatrix} c_x \\ c_y \\ c_z \end{bmatrix}. \tag{7.53}$$

This is called an **affine map**. The affine map has 12 independent parameters (nine for the matrix M and three for the displacement vector $\mathbf{c}$). This is consistent with the fact that 12 real parameters are required to specify a generic quantum map acting on a qubit state. It can be shown that the matrix M can always be expressed as a composition of a contraction and a rotation (see Exercise 7.2).

Let us show how to derive the matrix M and the displacement vector $\mathbf{c}$ starting from a generic quantum map $\mathcal{E}(\rho) = \sum_k \hat{E}_k \rho \hat{E}_k^\dagger$. The quantum map transforms an initial state ρ_1 into a final state $\rho_2 = \mathcal{E}(\rho_1)$. Their decomposition in the Pauli basis is,

$$\rho_1 = \frac{\mathbb{1}}{d} + \frac{1}{d}\sum_{j=2}^{4^n} a_j \hat{P}_j, \qquad \rho_2 = \frac{\mathbb{1}}{d} + \frac{1}{d}\sum_{j=2}^{4^n} b_j \hat{P}_j, \tag{7.54}$$

where $a_j = \mathrm{Tr}[\rho_1 \hat{P}_j]$ and $b_j = \mathrm{Tr}[\rho_2 \hat{P}_j]$. In eqn (7.54), we intentionally took out the identity from the decomposition. Using the expansion of the Kraus operators in the Pauli basis $\hat{E}_k = \sum_{j=1}^{4^n} c_{jk} \hat{P}_j$, the final state becomes

$$\begin{aligned} \rho_2 &= \sum_{k=1}^{4^n} \hat{E}_k \rho_1 \hat{E}_k^\dagger = \sum_{k=1}^{4^n} \hat{E}_k \Big(\frac{\mathbb{1}}{d} + \frac{1}{d}\sum_{j=2}^{4^n} a_j \hat{P}_j \Big) \hat{E}_k^\dagger \\ &= \frac{1}{d} \underbrace{\sum_{k=1}^{4^n} \hat{E}_k \hat{E}_k^\dagger}_{\hat{T}} + \frac{1}{d} \sum_{j=2}^{4^n} a_j \underbrace{\sum_{k=1}^{4^n} \hat{E}_k \hat{P}_j \hat{E}_k^\dagger}_{\hat{S}_j} = \frac{1}{d}\hat{T} + \frac{1}{d}\sum_{j=2}^{4^n} a_j \hat{S}_j. \end{aligned} \tag{7.55}$$

From this expression, one can calculate the coefficients of the output state,

$$b_\ell = \mathrm{Tr}[\hat{P}_\ell \rho_2] = \mathrm{Tr}\Big[\hat{P}_\ell\Big(\frac{1}{d}\hat{T} + \frac{1}{d}\sum_{j=2}^{4^n} a_j \hat{S}_j\Big)\Big]$$

$$= \underbrace{\frac{1}{d}\mathrm{Tr}[\hat{P}_\ell \hat{T}]}_{c_\ell} + \sum_{j=2}^{4^n} a_j \underbrace{\frac{1}{d}\mathrm{Tr}[\hat{P}_\ell \hat{S}_j]}_{M_{\ell j}} = \sum_{j=2}^{4^n} M_{\ell j} a_j + c_\ell .$$

In conclusion, the matrix M and the displacement vector $\mathbf{c}$ are given by

$$\boxed{M_{\ell j} = \frac{1}{d}\mathrm{Tr}[\hat{P}_\ell \hat{S}_j],} \qquad \boxed{c_\ell = \frac{1}{d}\mathrm{Tr}[\hat{P}_\ell \hat{T}],}$$

where the operators $\hat{S}_j$ and $\hat{T}$ are defined in eqn (7.55).

Example 7.4 The depolarizing channel transforms the Bloch vector (a_x, a_y, a_z) into $(p\,a_x,\ p\,a_y,\ p\,a_z)$ where $p \in [0,1]$ (see eqn 7.5). Therefore, the initial Bloch vector $\mathbf{a}$ becomes $M_{\text{depo}}\mathbf{a} + \mathbf{c}_{\text{depo}}$, where:

$$M_{\text{depo}} = \begin{bmatrix} p & 0 & 0 \\ 0 & p & 0 \\ 0 & 0 & p \end{bmatrix}, \qquad \mathbf{c}_{\text{depo}} = \begin{bmatrix} 0 \\ 0 \\ 0 \end{bmatrix}. \tag{7.56}$$

Example 7.5 Amplitude damping transforms the Bloch vector (a_x, a_y, a_z) into $(a_x \sqrt{1-p},\ a_y\sqrt{1-p},\ p + (1-p)a_z)$ where $p \in [0,1)$ (see eqn 7.17). Therefore, the initial Bloch vector $\mathbf{a}$ is mapped into $M_{\text{damp}}\mathbf{a} + \mathbf{c}_{\text{damp}}$, where:

$$M_{\text{damp}} = \begin{bmatrix} \sqrt{1-p} & 0 & 0 \\ 0 & \sqrt{1-p} & 0 \\ 0 & 0 & 1-p \end{bmatrix}, \qquad \mathbf{c}_{\text{damp}} = \begin{bmatrix} 0 \\ 0 \\ p \end{bmatrix}. \tag{7.57}$$

Example 7.6 A unital quantum map satisfies $\mathcal{E}(\mathbb{1}/2) = \mathbb{1}/2$, which means that it maps the center of the Bloch sphere into itself. Hence, the affine map associated with a unital transformation must have $\mathbf{c} = (0, 0, 0)$.

7.8 Quantum maps from unitary evolutions

In the previous sections, we understood that quantum maps can accurately describe unitary transformations as well as irreversible processes. The latter type of evolution typically originates from the interaction between the system and the **environment**, which can be described as a coupled subsystem not perfectly under control. Let us investigate this phenomenon by considering a simple quantum circuit in which a qubit interacts with two additional qubits, see Fig. 7.10 [209]. In this circuit, ρ is the initial qubit state, while the other two qubits are prepared in the pure state

$$|\mathrm{e}\rangle = \alpha|00\rangle + \beta|01\rangle + \gamma|10\rangle + \delta|11\rangle. \tag{7.58}$$

Throughout this section, these two extra qubits will represent the environment. The initial state of the global system "qubit + environment" is the product state[11]

$$\rho_{\mathrm{tot}} = |\mathrm{e}\rangle\langle\mathrm{e}| \otimes \rho = \begin{bmatrix} |\alpha|^2\rho & \cdot & \cdot & \cdot \\ \cdot & |\beta|^2\rho & \cdot & \cdot \\ \cdot & \cdot & |\gamma|^2\rho & \cdot \\ \cdot & \cdot & \cdot & |\delta|^2\rho \end{bmatrix}. \tag{7.59}$$

Here, the dots indicate matrix elements that are not important for our discussion. Each entry of this matrix is a 2×2 matrix given by the product between a complex number and ρ, the qubit density matrix. As shown in the figure, a unitary operation $\hat{U}$ is applied to the initial state,

$$\begin{aligned} \hat{U} &= |00\rangle\langle 00| \otimes \mathbb{1} + |01\rangle\langle 01| \otimes \hat{\sigma}_x + |10\rangle\langle 10| \otimes \hat{\sigma}_y + |11\rangle\langle 11| \otimes \hat{\sigma}_z \\ &= \begin{bmatrix} \mathbb{1} & 0 & 0 & 0 \\ 0 & \hat{\sigma}_x & 0 & 0 \\ 0 & 0 & \hat{\sigma}_y & 0 \\ 0 & 0 & 0 & \hat{\sigma}_z \end{bmatrix}. \end{aligned}$$

The final state of the global system is

$$\begin{aligned} \rho'_{\mathrm{tot}} = \hat{U}\rho_{\mathrm{tot}}\hat{U}^\dagger &= \begin{bmatrix} |\alpha|^2\rho & \cdot & \cdot & \cdot \\ \cdot & |\beta|^2\hat{\sigma}_x\rho\hat{\sigma}_x & \cdot & \cdot \\ \cdot & \cdot & |\gamma|^2\hat{\sigma}_y\rho\hat{\sigma}_y & \cdot \\ \cdot & \cdot & \cdot & |\delta|^2\hat{\sigma}_z\rho\hat{\sigma}_z \end{bmatrix} \\ &= \left(|00\rangle\langle 00| \otimes |\alpha|^2\rho\right) + \left(|01\rangle\langle 01| \otimes |\beta|^2\hat{\sigma}_x\rho\hat{\sigma}_x\right) \\ &\quad + \left(|10\rangle\langle 10| \otimes |\gamma|^2\hat{\sigma}_y\rho\hat{\sigma}_y\right) + \left(|11\rangle\langle 11| \otimes |\delta|^2\hat{\sigma}_z\rho\hat{\sigma}_z\right) + \ldots \end{aligned}$$

Again, the dots indicate terms that are not needed for our discussion. From this point onwards, the environment will no longer interact with the qubit and hence can be ignored. By tracing over the degrees of freedom of the environment, we obtain a reduced density operator describing the qubit state only,

$$\begin{aligned} \rho' &= \mathrm{Tr}_{\mathrm{e}}[\rho'_{\mathrm{tot}}] = \sum_{i=0}^{1}\sum_{j=0}^{1}\langle ij|\rho'_{\mathrm{tot}}|ij\rangle \\ &= |\alpha|^2\rho + |\beta|^2\hat{\sigma}_x\rho\hat{\sigma}_x + |\gamma|^2\hat{\sigma}_y\rho\hat{\sigma}_y + |\delta|^2\hat{\sigma}_z\rho\hat{\sigma}_z = \sum_k \hat{E}_k\rho\hat{E}_k^\dagger, \end{aligned} \tag{7.60}$$

[11] We could have written the state ρ_{tot} as $\rho \otimes |\mathrm{e}\rangle\langle\mathrm{e}|$; however, we would not have obtained the simple matrix representation (7.59) and the calculations would have been a bit more complicated.

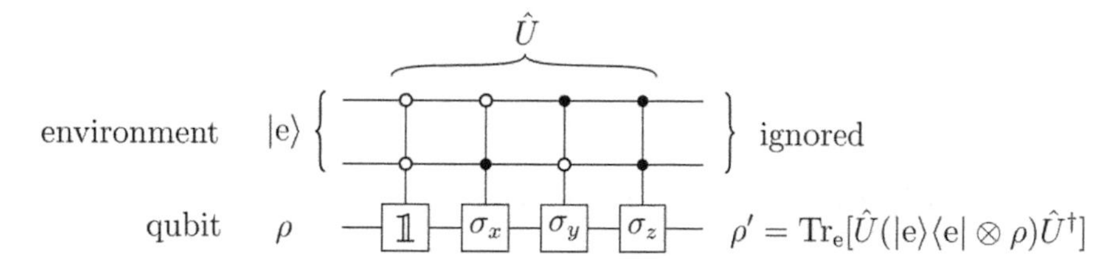

Fig. 7.10 A qubit coupled to the environment. A quantum circuit describing the interaction between a qubit and the environment. In this simple example, the environment is represented by two ancilla qubits.

where in the last step we introduced the Kraus operators

$$\hat{E}_0 = |\alpha|\mathbb{1}, \qquad \hat{E}_2 = |\beta|\hat{\sigma}_x, \qquad \hat{E}_1 = |\gamma|\hat{\sigma}_y, \qquad \hat{E}_1 = |\delta|\hat{\sigma}_z,$$

which satisfy $\sum_k \hat{E}_k^\dagger \hat{E}_k = \mathbb{1}$. Note that the quantum map (7.60) is parametrized by three parameters only because the four parameters $|\alpha|^2$, $|\beta|^2$, $|\gamma|^2$, and $|\delta|^2$ are not independent (they must satisfy the condition $|\alpha|^2 + |\beta|^2 + |\gamma|^2 + |\delta|^2 = 1$). In summary, we have shown that a unitary interaction between a qubit and the environment induces a quantum map on the qubit after tracing over the degrees of freedom of the environment.

The action of the quantum map (7.60) can be visualized on the Bloch sphere. By substituting eqn (6.34) into (7.60) and performing some matrix multiplications, it is possible to show that the initial Bloch vector is mapped into

$$\begin{bmatrix} r_x \\ r_y \\ r_z \end{bmatrix} \quad \mapsto \quad \begin{bmatrix} [1 - 2(|\gamma|^2 + |\delta|^2)]\, r_x \\ [1 - 2(|\delta|^2 + |\beta|^2)]\, r_y \\ [1 - 2(|\beta|^2 + |\gamma|^2)]\, r_z \end{bmatrix}. \tag{7.61}$$

This expression indicates that the quantum map (7.60) transforms the Bloch sphere into an ellipsoid centered at the origin and extending along an axis passing through the origin.

Example 7.7 The depolarizing channel $\mathcal{E}_{\text{depo}} = (1-p)\rho + \frac{p}{3}(\hat{\sigma}_x\rho\hat{\sigma}_x + \hat{\sigma}_y\rho\hat{\sigma}_y + \hat{\sigma}_z\rho\hat{\sigma}_z)$ is equivalent to (7.60) with $|\beta|^2 = |\gamma|^2 = |\delta|^2 = \frac{p}{3}$ and $|\alpha|^2 = (1-p)$.

Example 7.8 The dephasing channel $\mathcal{E}_{\text{deph}} = (1-p)\rho + p\hat{\sigma}_z\rho\hat{\sigma}_z$ is equivalent to (7.60) with $\beta = \gamma = 0$, $|\delta|^2 = p$ and $|\alpha|^2 = 1-p$.

7.8.1 Quantum maps from unitary evolutions: the general case

In the previous section, we learned that a qubit coupled to an external system gives rise to a quantum map describing the time evolution of the qubit alone. We now present an important theorem, proposed by M.-D. Choi and W. F. Stinespring [210, 211, 185], that generalizes this result to any quantum system. This theorem indicates that the origin of irreversible evolutions can be interpreted as an interaction between a quantum system

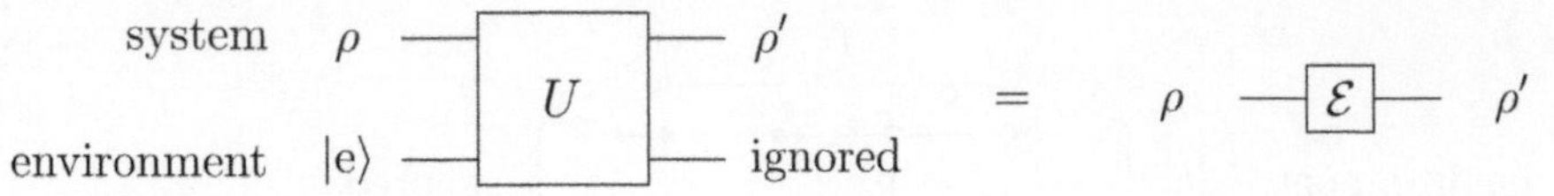

system ρ — U — ρ' = ρ — $\mathcal{E}$ — ρ'
environment $|\mathrm{e}\rangle$ — ignored

Fig. 7.11 Stinespring's dilation theorem. A quantum map $\mathcal{E}$ can always be interpreted as the partial trace of a unitary transformation $\hat{U}$ acting on a factored state $|\mathrm{e}\rangle\langle\mathrm{e}|\otimes\rho$, where $|\mathrm{e}\rangle$ is the initial state of the environment.

and the environment (see Fig. 7.11). This result provides an interesting connection between quantum maps and unitary operators.

Theorem 7.2 (Stinespring's dilation theorem) A superoperator $\mathcal{E}: \mathrm{B}(H_\mathrm{A}) \to \mathrm{B}(H_\mathrm{A})$ is a quantum map if and only if it can be expressed as[12]

$$\mathcal{E}(\rho) = \mathrm{Tr}_\mathrm{B}\left[\hat{U}(\rho \otimes |\mathrm{e}\rangle\langle\mathrm{e}|)\hat{U}^\dagger\right], \tag{7.62}$$

where ρ and $|\mathrm{e}\rangle$ are two quantum states belonging to the Hilbert spaces H_A and H_B, respectively, $\hat{U}$ is a unitary operator acting on the composite system, and the partial trace is performed over H_B.

Proof ($\leftarrow$) We start by showing that the transformation

$$\mathcal{E}(\rho) = \mathrm{Tr}_\mathrm{B}\left[\hat{U}(\rho \otimes |\mathrm{e}\rangle\langle\mathrm{e}|)\hat{U}^\dagger\right] \tag{7.63}$$

is a quantum map acting on subsystem A only. This can be done in two different ways: either we show that (7.63) satisfies requirements **QM1–QM4** or we demonstrate that $\mathcal{E}$ can be written in Kraus form. We will pursue the second approach. Let $\{|k\rangle\}_{k=1}^{d_\mathrm{B}}$ be an orthonormal basis for H_B. By calculating the partial trace on subsystem B, eqn (7.63) becomes

$$\mathcal{E}(\rho) = \sum_{k=1}^{d_\mathrm{B}} \langle k|\,\hat{U}\left(\rho \otimes |\mathrm{e}\rangle\langle\mathrm{e}|\right)\hat{U}^\dagger\,|k\rangle = \sum_{k=1}^{d_\mathrm{B}} \langle k|\hat{U}|\mathrm{e}\rangle\,\rho\,\langle e|\hat{U}^\dagger|k\rangle = \sum_{k=1}^{d_\mathrm{B}} \hat{E}_k\rho\hat{E}_k^\dagger, \tag{7.64}$$

where we defined a total of d_B Kraus operators

$$\hat{E}_k = \langle k|\hat{U}|\mathrm{e}\rangle. \tag{7.65}$$

If $\sum_k \hat{E}_k^\dagger\hat{E}_k = \mathbb{1}_\mathrm{A}$, then $\mathrm{Tr}_\mathrm{B}\left[\hat{U}\left(\rho \otimes |\mathrm{e}\rangle\langle\mathrm{e}|\right)\hat{U}^\dagger\right]$ is a quantum map. This is indeed the case

[12]We assumed that the second subsystem is initially prepared in a pure state $|\mathrm{e}\rangle$. This is general enough because, as explained in Section 6.6.2, it is always possible to perform a purification and convert a mixed state into a pure one.

$$\sum_{k=1}^{d_{\mathrm{B}}} \hat{E}_k^\dagger \hat{E}_k = \sum_{k=1}^{d_{\mathrm{B}}} \langle \mathrm{e}|\hat{U}^\dagger|k\rangle\langle k|\hat{U}|\mathrm{e}\rangle = \langle \mathrm{e}|\hat{U}^\dagger\hat{U}|\mathrm{e}\rangle = \langle \mathrm{e}|\mathbb{1}_{\mathrm{A}} \otimes \mathbb{1}_{\mathrm{B}}|\mathrm{e}\rangle = \mathbb{1}_{\mathrm{A}}.$$

This concludes the first part of the proof. □

Proof ($\rightarrow$) Let us show that any quantum map $\mathcal{E}(\rho)$ acting on a Hilbert space H_{A} can be expressed as in eqn (7.62). Let $\mathcal{E}(\rho) = \sum_k \hat{E}_k \rho \hat{E}_k^\dagger$ be the Kraus form of the quantum map, where $\sum_k \hat{E}_k^\dagger \hat{E}_k = \mathbb{1}_{\mathrm{A}}$. Now consider an additional Hilbert space H_{B}, with orthonormal basis $\{|k\rangle\}_k$, whose dimension is equal to the number of Kraus operators. Consider a pure state $|\mathrm{e}\rangle \in H_{\mathrm{B}}$ and a linear operator $\hat{U} \in \mathrm{L}(H_{\mathrm{A}} \otimes H_{\mathrm{B}})$ such that

$$\hat{U}|\psi\rangle \otimes |\mathrm{e}\rangle = \sum_k (\hat{E}_k|\psi\rangle) \otimes |k\rangle \,. \tag{7.66}$$

It is always possible to transform this operator into a unitary operator acting on $H_{\mathrm{A}} \otimes H_{\mathrm{B}}$. We now prove that $\mathrm{Tr}_{\mathrm{B}}[\hat{U}\rho\otimes|\mathrm{e}\rangle\langle\mathrm{e}|\hat{U}^\dagger] = \mathcal{E}(\rho)$ by using the spectral decomposition $\rho = \sum_j \lambda_j |j\rangle\langle j|$:

$$\begin{aligned}
\mathrm{Tr}_{\mathrm{B}}[\hat{U}\,(\rho \otimes |\mathrm{e}\rangle\langle \mathrm{e}|)\,\hat{U}^\dagger] &= \sum_{j=1}^{d_{\mathrm{A}}} \lambda_j \mathrm{Tr}_{\mathrm{B}}\left[\hat{U}\,(|j\rangle\langle j| \otimes |\mathrm{e}\rangle\langle \mathrm{e}|)\,\hat{U}^\dagger\right] \\
&= \sum_{j=1}^{d_{\mathrm{A}}} \sum_{kk'} \lambda_j \mathrm{Tr}_{\mathrm{B}}\left[\hat{E}_k|j\rangle\langle j|\hat{E}_{k'}^\dagger \otimes |k\rangle\langle k'|\right] \\
&= \sum_{j=1}^{d_{\mathrm{A}}} \sum_{k} \lambda_j \hat{E}_k|j\rangle\langle j|\hat{E}_k^\dagger = \sum_k \hat{E}_k \rho \hat{E}_k^\dagger.
\end{aligned}$$

This concludes the proof. □

As explained at the end of Section 7.6, the Kraus operators associated with a quantum map are not uniquely defined. The quantum circuit illustrated in Fig. 7.12 provides a physical intuition. In this circuit, the interaction between system A and the environment is described by a unitary operator $\hat{U}$. After the interaction, a unitary operation $\hat{V}$ is applied to the environment. It is reasonable to expect that $\hat{V}$ does not affect the future evolution of subsystem A. The final state of system A can be obtained by tracing out the environment,

$$\rho' = \mathrm{Tr}_{\mathrm{B}}[\,(\mathbb{1}_{\mathrm{A}} \otimes \hat{V})\,\hat{U}\,(\rho \otimes |\mathrm{e}\rangle\langle \mathrm{e}|)\,\hat{U}^\dagger(\mathbb{1}_{\mathrm{A}} \otimes \hat{V}^\dagger)\,]. \tag{7.67}$$

This equation gives rise to a quantum map with Kraus operators $\hat{F}_k = \langle k|(\mathbb{1}_{\mathrm{A}} \otimes \hat{V})\hat{U}|\mathrm{e}\rangle$. These operators differ from those defined in eqn (7.65), yet they describe the same dynamics for subsystem A. The relation between these two sets of Kraus operators is

$$\hat{F}_k = \langle k|(\mathbb{1}_{\mathrm{A}} \otimes \hat{V})\hat{U}|\mathrm{e}\rangle = \sum_i (\mathbb{1}_{\mathrm{A}} \otimes \langle k|\hat{V}|i\rangle)\,\underbrace{\langle i|\hat{U}|\mathrm{e}\rangle}_{\hat{E}_i} = \sum_i V_{ki}\hat{E}_i,$$

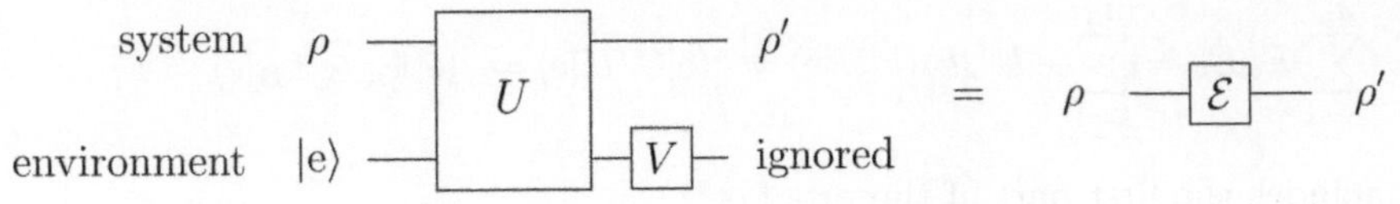

Fig. 7.12 The freedom in the Kraus representation. Applying an an extra unitary operator $\hat{V}$ to the environment does not affect the dynamics of the system.

---

where $V_{ki} = \langle k|\hat{V}|i\rangle$ are the elements of a unitary matrix. Since V is arbitrary, we have an infinite number of Kraus representations for the same quantum map. It is worth mentioning that the Kraus representation can be fixed by expanding the Kraus operators on a particular basis of the operator space. A common choice is the Pauli basis $B = \{\hat{A}_1 \otimes \ldots \otimes \hat{A}_n : \hat{A}_i \in \{\mathbb{1}, \hat{\sigma}_x, \hat{\sigma}_y, \hat{\sigma}_z\}\}$.

Example 7.9 The dephasing channel can be expressed as a partial trace of a unitary operation acting on two qubits. To see this, let us start from the circuit shown below:

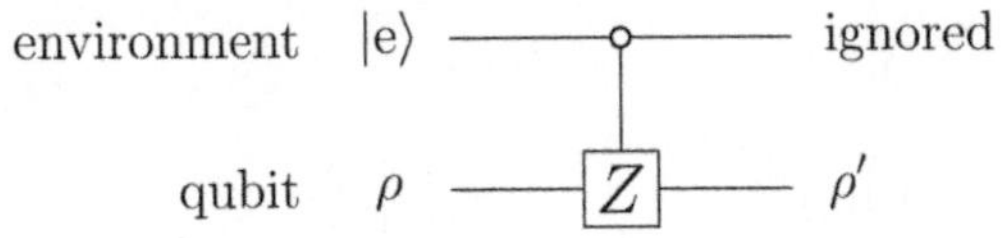

Here, ρ is the initial state of a qubit and $|\mathrm{e}\rangle = \sqrt{p}|0\rangle + \sqrt{1-p}|1\rangle$ is the initial state of a second qubit (representing the environment in this example). At $t = 0$, the state of the two qubits is

$$\rho_{\mathrm{tot}} = |\mathrm{e}\rangle\langle\mathrm{e}| \otimes \rho = \begin{bmatrix} p & \sqrt{p(1-p)} \\ \sqrt{p(1-p)} & 1-p \end{bmatrix} \otimes \rho = \begin{bmatrix} p\rho & \sqrt{p(1-p)}\rho \\ \sqrt{p(1-p)}\rho & (1-p)\rho \end{bmatrix},$$

where $p \in [0,1]$. The controlled unitary $\hat{U} = |0\rangle\langle 0| \otimes \hat{\sigma}_z + |1\rangle\langle 1| \otimes \mathbb{1}$ applies a phase flip to the first qubit only when the environment is in $|0\rangle$,

$$\hat{U} = \begin{bmatrix} 1 & 0 \\ 0 & 0 \end{bmatrix} \otimes \hat{\sigma}_z + \begin{bmatrix} 0 & 0 \\ 0 & 1 \end{bmatrix} \otimes \mathbb{1} = \begin{bmatrix} \hat{\sigma}_z & 0 \\ 0 & \mathbb{1} \end{bmatrix}.$$

The final state of the global system is

$$\hat{U}\rho_{\mathrm{tot}}\hat{U}^\dagger = \begin{bmatrix} p\hat{\sigma}_z\rho\hat{\sigma}_z & \sqrt{p(1-p)}\hat{\sigma}_z\rho \\ \sqrt{p(1-p)}\rho\hat{\sigma}_z & (1-p)\rho \end{bmatrix}.$$

Tracing over the environment, we obtain the final state $\rho' = \mathrm{Tr}_{\mathrm{e}}[\hat{U}\rho_{\mathrm{tot}}\hat{U}^\dagger] = p\hat{\sigma}_z\rho\hat{\sigma}_z + (1-p)\rho$. This is the dephasing channel, eqn (7.2).

Example 7.10 The interaction between a qubit and a harmonic oscillator might lead to some dephasing of the qubit state. Suppose that the interaction between a qubit and a harmonic oscillator is described by the Hamiltonian $\hat{H} = \hbar\chi\hat{\sigma}_z \otimes \hat{x}$, where $\hat{\sigma}_z$ is a Pauli operator, $\hat{x}$ is the position operator, and χ is a coupling constant. This coupling Hamiltonian indicates that the qubit frequency is affected by the movement of the oscillator. Suppose that the global system at $t = 0$ is in the product state $\rho = \rho_A \otimes \rho_B$, where the initial state of the qubit is

$$\rho_A = \sum_{k=0}^{1} \sum_{l=0}^{1} \rho_{kl} |k\rangle\langle l|,$$

and the oscillator starts in the vacuum state $\rho_B = |0\rangle\langle 0|$. The state of the global system at $t > 0$ is given by $\rho(t) = \hat{U}(\rho_A \otimes \rho_B)\hat{U}^\dagger$, where the time evolution operator is $\hat{U} = e^{-i\chi t\hat{\sigma}_z \otimes \hat{x}}$. We are interested in the evolution of the qubit state only

$$\begin{aligned}
\rho_A(t) = \text{Tr}_B[\rho(t)] &= \sum_{k,l=0}^{1} \rho_{kl} \text{Tr}_B[e^{-i\chi t\hat{\sigma}_z \otimes \hat{x}} |k\rangle\langle l| \otimes \rho_B e^{i\chi t\hat{\sigma}_z \otimes \hat{x}}] \\
&= \sum_{k,l=0}^{1} \rho_{kl} |k\rangle\langle l| \underbrace{\text{Tr}_B[e^{-i\chi t(-1)^k \hat{x}} \rho_B e^{i\chi t(-1)^l \hat{x}}]}_{f_{kl}(t)} \\
&= \sum_{k,l=0}^{1} \rho_{kl} f_{kl}(t) |k\rangle\langle l| = \begin{bmatrix} \rho_{00} f_{00}(t) & \rho_{01} f_{01}(t) \\ \rho_{10} f_{10}(t) & \rho_{11} f_{11}(t) \end{bmatrix}. \qquad (7.68)
\end{aligned}$$

Note that $f_{00}(t) = f_{11}(t) = 1$ because the trace satisfies $\text{Tr}[\hat{A}\hat{B}\hat{C}] = \text{Tr}[\hat{B}\hat{C}\hat{A}]$. Let us calculate $f_{01}(t)$ assuming that the resonator is initially in the vacuum state

$$\begin{aligned}
f_{01}(t) = \text{Tr}_B[e^{-i\chi t\hat{x}} \rho_B e^{-i\chi t\hat{x}}] &= \text{Tr}_B[|0\rangle\langle 0| e^{-i2\chi t\hat{x}}] = \int_{-\infty}^{\infty} \langle x|0\rangle\langle 0| e^{-i2\chi t\hat{x}} |x\rangle dx \\
= \int_{-\infty}^{\infty} |\psi_0(x)|^2 e^{-i2\chi tx} dx &= \frac{\lambda}{\sqrt{\pi}} \int_{-\infty}^{\infty} e^{-\lambda^2 x^2} e^{-i2\chi tx} dx = e^{-\chi^2 t^2/\lambda^2},
\end{aligned}$$

where we used (4.48). The qubit state at $t > 0$ is

$$\rho_A(t) = \begin{bmatrix} \rho_{00} & \rho_{01} e^{-t^2/T_\phi^2} \\ \rho_{01} e^{-t^2/T_\phi^2} & \rho_{11} \end{bmatrix},$$

where we introduced the dephasing time $T_\phi = \lambda/|\chi|$. The coherences decay with a Gaussian profile for sufficiently long times.

7.9 Choi–Jamiolkowski isomorphism

We conclude this chapter by presenting an elegant one-to-one correspondence between superoperators from $\text{L}(H)$ to $\text{L}(H)$ and linear operators in $\text{L}(H \otimes H)$, known as the **Choi–Jamiolkowski isomorphism**. As shown in Fig. 7.13, the Choi–Jamiolkowski isomorphism is a function J that maps a superoperator $\Lambda : L(H) \to L(H)$ into an operator in $\text{L}(H \otimes H)$,

$$\text{J}(\Lambda) = (\mathbb{1} \otimes \Lambda)|\Omega\rangle\langle\Omega|. \qquad (7.69)$$

This operator is simply the result of the action of Λ on half of the maximally entangled state $|\Omega\rangle = \frac{1}{\sqrt{d}} \sum_i |i\rangle|i\rangle$ where $d = \dim(H)$. The inverse function J^{-1} maps an operator $\rho \in \mathrm{L}(H \otimes H)$ into a superoperator

$$\mathrm{J}^{-1}[\rho](\hat{A}) = d\,\mathrm{Tr}_{\mathrm{A}}[(\hat{A}^{\mathrm{T}} \otimes \mathbb{1})\rho], \tag{7.70}$$

where T is the transposition and as usual $d = \dim(H)$. It is straightforward to show that eqn (7.70) is the inverse of (7.69). Indeed,

$$\begin{aligned}
\mathrm{J}^{-1}[\mathrm{J}(\Lambda)](\hat{A}) &= d\mathrm{Tr}_{\mathrm{A}}[(\hat{A}^{\mathrm{T}} \otimes \mathbb{1})\mathrm{J}(\Lambda)] = d\mathrm{Tr}_{\mathrm{A}}[(\hat{A}^{\mathrm{T}} \otimes \mathbb{1})(\mathbb{1} \otimes \Lambda)|\Omega\rangle\langle\Omega|] \\
&= \sum_{ij} \mathrm{Tr}_{\mathrm{A}}\left[(\hat{A}^{\mathrm{T}} \otimes \mathbb{1})\left(|i\rangle\langle j| \otimes \Lambda(|i\rangle\langle j|)\right)\right] \\
&= \sum_{ij}\sum_{kl} a_{kl}\mathrm{Tr}_{\mathrm{A}}\left[(|l\rangle\langle k| \otimes \mathbb{1})\left(|i\rangle\langle j| \otimes \Lambda(|i\rangle\langle j|)\right)\right] \\
&= \sum_{ij}\sum_{kl} a_{kl}\mathrm{Tr}_{\mathrm{A}}\left[|i\rangle\langle j|l\rangle\langle k|\right] \Lambda(|i\rangle\langle j|) = \\
&= \sum_{ij}\sum_{kl} a_{kl}\delta_{jl}\delta_{ik}\, \Lambda(|i\rangle\langle j|) = \sum_{kl} a_{kl}\Lambda(|k\rangle\langle l|) = \Lambda(\hat{A}),
\end{aligned} \tag{7.71}$$

where we used the decomposition $\hat{A} = \sum_{kl} a_{kl}|k\rangle\langle l|$. For completeness, we also demonstrate that $\mathrm{J}(\mathrm{J}^{-1}[\rho]) = \rho$. Let us start from

$$\begin{aligned}
\mathrm{J}(\mathrm{J}^{-1}[\rho]) &= (\mathbb{1} \otimes \mathrm{J}^{-1}[\rho])(|\Omega\rangle\langle\Omega|) \\
&= \frac{1}{d}\sum_{ij}(\mathbb{1} \otimes \mathrm{J}^{-1}[\rho])(|i\rangle\langle j| \otimes |i\rangle\langle j|) \\
&= \frac{1}{d}\sum_{ij} |i\rangle\langle j| \otimes \mathrm{J}^{-1}[\rho](|i\rangle\langle j|) = \sum_{ij} |i\rangle\langle j| \otimes \mathrm{Tr}_{\mathrm{A}}[\rho(|j\rangle\langle i| \otimes \mathbb{1})].
\end{aligned} \tag{7.72}$$

Since $\mathrm{Tr}_{\mathrm{A}}[\rho(|j\rangle\langle i| \otimes \mathbb{1})] = \sum_{kl} \rho_{ikjl}|k\rangle\langle l|$, then eqn (7.72) becomes

$$\mathrm{J}(\mathrm{J}^{-1}[\rho]) = \sum_{ij}\sum_{kl} \rho_{ikjl}|i\rangle\langle j| \otimes |k\rangle\langle l| = \rho,$$

as required. In conclusion, J and J^{-1} are mutual inverses.

The Choi–Jamiolkowski isomorphism gives rise to an interesting theorem.

Theorem 7.3 A superoperator Λ is completely positive if and only if $\mathrm{J}(\Lambda)$ is a positive operator,

$$\mathrm{J}(\Lambda) = (\mathbb{1} \otimes \Lambda)|\Omega\rangle\langle\Omega| \geq 0.$$

This theorem is proved in Exercise 7.6.

Provided that Λ is a quantum map, the quantum state $(\mathbb{1} \otimes \Lambda)|\Omega\rangle\langle\Omega|$ has a simple physical interpretation: it is the quantum state obtained by applying the quantum map

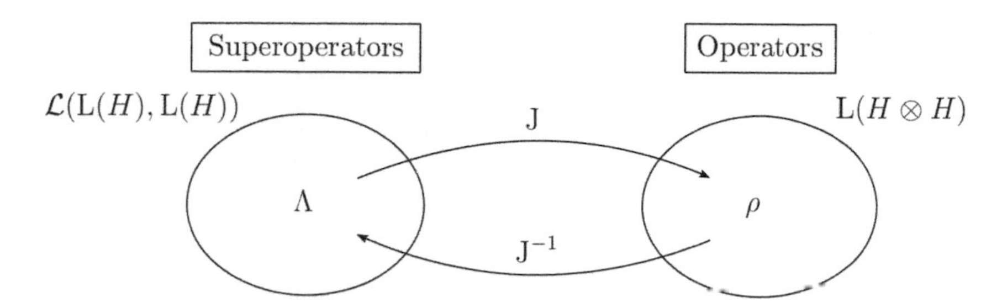

Fig. 7.13 Choi–Jamiolkowski isomorphism. The Choi–Jamiolkowski isomorphism J maps a superoperator $\Lambda : \mathrm{L}(H) \to \mathrm{L}(H)$ into an operator $\rho_\Lambda = (\mathbb{1} \otimes \Lambda)|\Omega\rangle\langle\Omega|$. The inverse function J^{-1} maps a linear operator $\rho \in \mathrm{L}(H \otimes H)$ into a superoperator $\Lambda_\rho(\hat{A}) = d\mathrm{Tr}_\mathrm{A}[\rho(\hat{A}^\mathrm{T} \otimes \mathbb{1})]$. The superoperator Λ is a quantum map if and only if $\mathrm{J}(\Lambda) = \rho_\Lambda$ is positive.

to half of a maximally entangled state. On the other hand, the physical interpretation of $\Lambda_\rho(\hat{A}) = d\mathrm{Tr}_\mathrm{A}[(\hat{A}^\mathrm{T} \otimes \mathbb{1})\rho]$ is not immediate. We will return to this point in Section 9.5.1 when we present the quantum teleportation protocol for mixed states.

Example 7.11 Suppose that $\Lambda = \mathbb{1}$. Then, from eqn (7.69) $J(\Lambda) = J(\mathbb{1}) = \rho = |\Omega\rangle\langle\Omega|$. This means that $J^{-1}[|\Omega\rangle\langle\Omega|]$ is the identity: $J^{-1}[|\Omega\rangle\langle\Omega|](\hat{A}) = \hat{A}$ where we used eqn (7.71).

Example 7.12 Consider the two-qubit state $|00\rangle \in \mathbb{C}^2 \otimes \mathbb{C}^2$. The superoperator associated with this state is $\Lambda(\rho) = 2\mathrm{Tr}_\mathrm{A}[(\rho^\mathrm{T} \otimes \mathbb{1})|00\rangle\langle 00|]$, where ρ is a density operator of a single qubit. The function Λ is not a quantum map because it is not trace-preserving,

$$\mathrm{Tr}[\Lambda(\rho)] = 2\mathrm{Tr}_\mathrm{AB}[(\rho^\mathrm{T} \otimes \mathbb{1})(|0\rangle\langle 0| \otimes |0\rangle\langle 0|)] = 2\rho_{00} \neq \mathrm{Tr}[\rho] = 1.$$

This shows that the superoperator $\mathrm{J}^{-1}[\rho]$ associated with a quantum state ρ is not always a quantum map.

7.10 Quantum process tomography

Quantum process tomography is an experimental procedure aimed at measuring an unknown quantum map. Before explaining the procedure itself, it is necessary to introduce an important mathematical tool, the Pauli transfer matrix (PTM).

7.10.1 The Pauli transfer matrix

For pedagogical reasons, we first present a simple example of a Pauli transfer matrix and will then give a more formal definition. Consider a quantum map $\mathcal{E}(\rho) = \rho'$ acting on a single qubit. The decomposition of ρ and ρ' in the Pauli basis can be expressed with column vectors,

$$R_{\text{depo}} = \begin{bmatrix} \frac{1}{2} & 0 & 0 & 0 \\ 0 & \frac{1-\frac{4}{3}\gamma}{2} & 0 & 0 \\ 0 & 0 & \frac{1-\frac{4}{3}\gamma}{2} & 0 \\ 0 & 0 & 0 & \frac{1-\frac{4}{3}\gamma}{2} \end{bmatrix} \qquad R_{\text{damp}} = \begin{bmatrix} \frac{1}{2} & 0 & 0 & 0 \\ 0 & \frac{\sqrt{1-p}}{2} & 0 & 0 \\ 0 & 0 & \frac{\sqrt{1-p}}{2} & 0 \\ \frac{p}{2} & 0 & 0 & \frac{1-p}{2} \end{bmatrix}$$

Table 7.1 Pauli transfer matrices. Pauli transfer matrices associated with the depolarizing channel and amplitude damping.

$$\rho = c_1 \mathbb{1} + c_2\hat{\sigma}_x + c_3\hat{\sigma}_y + c_4\hat{\sigma}_z = \begin{bmatrix} c_1 \\ c_2 \\ c_3 \\ c_4 \end{bmatrix}, \qquad \rho' = c_1' \mathbb{1} + c_2'\hat{\sigma}_x + c_3'\hat{\sigma}_y + c_4'\hat{\sigma}_z = \begin{bmatrix} c_1' \\ c_2' \\ c_3' \\ c_4' \end{bmatrix}. \tag{7.73}$$

The Pauli transfer matrix associated with $\mathcal{E}$ is the matrix that transforms the vector (c_0, c_1, c_2, c_3) into (c_0', c_1', c_2', c_3'). For instance, the PTM associated with the $\hat{X}$ gate is given by

$$R_X = \begin{bmatrix} 1 & 0 & 0 & 0 \\ 0 & 1 & 0 & 0 \\ 0 & 0 & -1 & 0 \\ 0 & 0 & 0 & -1 \end{bmatrix}$$

The reader can easily check that this matrix correctly maps the ground state $|0\rangle\langle 0| = (1/2, 0, 0, 1/2)$ into the excited state $|1\rangle\langle 1| = (1/2, 0, 0, -1/2)$. In general, the **Pauli transfer matrix** associated with a quantum map $\mathcal{E}$ acting on n qubits is a $d^2 \times d^2$ matrix $R_{\mathcal{E}}$ defined as

$$\boxed{(R_{\mathcal{E}})_{ij} = \frac{1}{d}\text{Tr}[\hat{P}_i\,\mathcal{E}(\hat{P}_j)],} \tag{7.74}$$

where $\hat{P}_i$ are elements of the Pauli basis and $d = 2^n$ is the dimension of the Hilbert space. An interesting property of the Pauli transfer matrix is that the PTM associated with the quantum map $\mathcal{E}_2 \circ \mathcal{E}_1$ is given by the product $R_{\mathcal{E}_2} R_{\mathcal{E}_1}$. In addition, the entries of $(R_{\mathcal{E}})_{ij}$ are real numbers in the interval $[-1, 1]$ and this provides a natural way to represent this matrix with a color plot. Note the matrix $R_{\mathcal{E}}$ associated with a Clifford gate (such as the iSWAP, and CNOT gates) has only one entry different from zero in each column and in each row. This is because Clifford gates $\hat{C}$ map Pauli operators into Pauli operators: $\hat{C}\hat{P}\hat{C}^\dagger = \hat{P}'$. Table 7.1 shows the Pauli transfer matrix associated with the depolarizing channel and amplitude damping.

The Pauli transfer matrices associated with the CZ, iSWAP, and CNOT gates are 16×16 matrices. In the literature, these matrices are usually visualized with a color plot as shown in Fig. 7.14.

7.10.2 The experimental protocol

In the field of quantum information science, quantum process tomography is an experimental procedure that aims at determining an unknown quantum map $\mathcal{E}$ acting

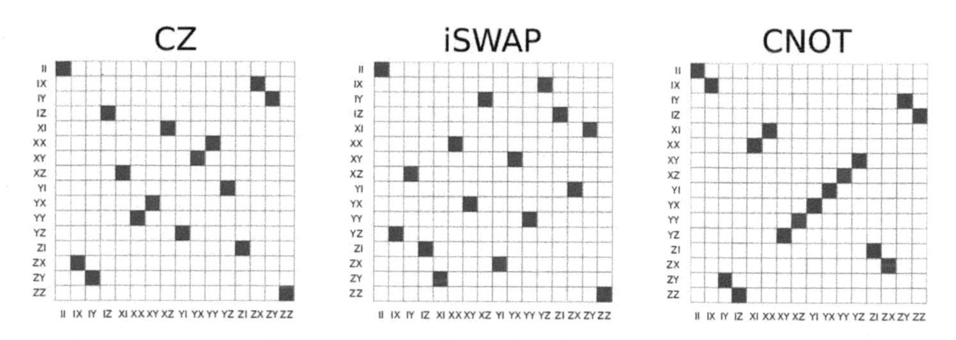

Fig. 7.14 Pauli transfer matrices. Pauli transfer matrices associated with the CZ, iSWAP and CNOT gates. The red, white, blue squares indicate that the corresponding matrix entry is -1, 0, $+1$, respectively.

on n qubits. In practical terms, we are given a black box that implements $\mathcal{E}$ and our goal is to extract all possible information about this quantum map. This can be done by performing a sequence of experiments in which $\mathcal{E}$ operates on a set of qubits prepared in a variety of states ρ_i, followed by a measurement of the PVM $(\hat{E}_j, \mathbb{1} - \hat{E}_j)$, where the set $\{\hat{E}_j\}_j$ forms a basis of the operator space. The experimenter has the freedom to choose the initial states ρ_i and measurement operators $\hat{E}_j$ (see Fig. 7.15a). A convenient choice is

$$\rho_i = \begin{cases} \frac{\mathbb{1}}{d} & \hat{P}_i = \mathbb{1} \\ \frac{\mathbb{1}+\hat{P}_i}{2} & \hat{P}_i \neq \mathbb{1} \end{cases}, \qquad \hat{E}_j = \frac{\mathbb{1} + \hat{P}_j}{2}. \tag{7.75}$$

Each state ρ_i is measured with d^2 PVMs $(\hat{E}_j, \mathbb{1} - \hat{E}_j)$ for a total of d^4 quantum circuits[13]. The measurement collapses the quantum state $\mathcal{E}(\rho_i)$ onto the eigenspace associated with $\hat{E}_j$ with probability

$$\boxed{p_{ij} = \mathrm{Tr}[\mathcal{E}(\rho_i)\hat{E}_j].} \tag{7.76}$$

This is a probability matrix with $d^2 \times d^2$ entries. Since the operators ρ_i and $\hat{E}_j$ can be decomposed in the Pauli basis as

$$\rho_i = \sum_{l=1}^{d^2} c_{il}\hat{P}_l, \qquad \hat{E}_j = \sum_{k=1}^{d^2} A_{jk}\hat{P}_k,$$

eqn (7.76) becomes

[13] In principle, the number of experiments can be reduced from d^4 to $d^4 - d^2$ by exploiting some constraints of the PTM. However, the scaling remains unchanged. In some cases, more sophisticated protocols, such as direct fidelity estimation, can be exploited to reduce the number of experiments and acquire relevant information about the quantum map [212].

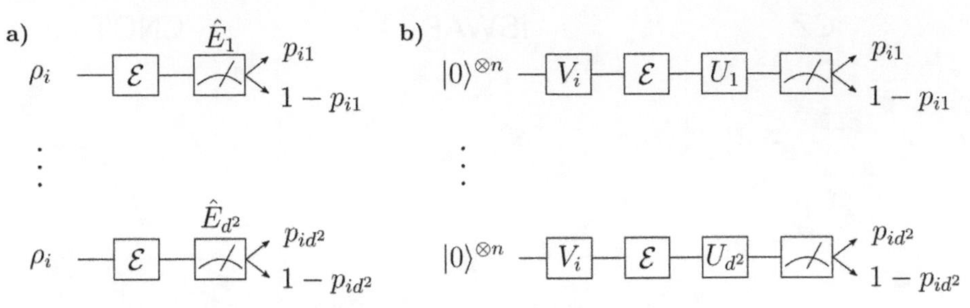

Fig. 7.15 Quantum process tomography. a) Quantum circuits required to construct the probability matrix p_{ij} of eqn (7.76). **b)** If the initial state of the quantum computer is $|0\ldots0\rangle$, suitable single-qubit operations are required to prepare the initial state ρ_i. Similarly, if the quantum states can only be measured in the computational basis, some single-qubit operations are needed to rotate the state before a measurement in σ_z basis.

$$p_{ij} = \mathrm{Tr}[\mathcal{E}(\rho_i)\hat{E}_j] = \sum_{l=1}^{d^2}\sum_{k=1}^{d^2} c_{il}A_{jk}\mathrm{Tr}[\mathcal{E}(\hat{P}_l)\hat{P}_k] = d\sum_{l=1}^{d^2}\sum_{k=1}^{d^2} c_{il}A_{jk}\,(R_{\mathcal{E}})_{kl}\,. \tag{7.77}$$

Using a row-major order method, one can convert the probability matrix p_{ij} and the Pauli transfer matrix $(R_{\mathcal{E}})_{kl}$ into column vectors with d^4 components

$$p_a = p_{ij}, \qquad\qquad r_b = (R_{\mathcal{E}})_{lk}\,,$$

where

$$\begin{aligned} a &= 1 \;\leftrightarrow\; ij = 11, & b &= 1 \;\leftrightarrow\; kl = 11,\\ a &= 2 \;\leftrightarrow\; ij = 12, & b &= 2 \;\leftrightarrow\; kl = 12,\\ &\;\;\vdots & &\;\;\vdots\\ a &= d^2 \;\leftrightarrow\; ij = dd, & b &= d^2 \;\leftrightarrow\; kl = dd. \end{aligned}$$

Similarly, one can combine the matrices c_{il} and A_{jk} into a single matrix with $d^4 \times d^4$ entries,

$$S_{ab} = S_{j+4(i-1),k+4(l-1)} = c_{il}A_{jk}.$$

Substituting these expressions into (7.77), we arrive at

$$\boxed{p_a = d\sum_{b=1}^{d^4} S_{ab}r_b,}$$

which can be written in matrix form as $\mathbf{p} = dS\mathbf{r}$. Provided that ρ_i and E_j are chosen as in eqn (7.75), the matrix S_{ab} is invertible. Therefore, one can estimate the entries of the Pauli transfer matrix $r_b = (R_{\mathcal{E}})_{lk}$ using the expression

$$\mathbf{r} = \frac{1}{d} S^{-1} \mathbf{p}.$$

The vector $\mathbf{r}$ contains all of the information required to reconstruct the PTM $R_{\mathcal{E}}$ and the associated quantum map $\mathcal{E}$. In most cases, we are interested in comparing the quantum map $\mathcal{E}$ with a target unitary map $\mathcal{U}$. This can be done by calculating the average gate fidelity [213],

$$F(\mathcal{E},\mathcal{U}) = \frac{\mathrm{Tr}[R_{\mathcal{U}}^{\mathrm{T}} R_{\mathcal{E}}] + d^2}{d^2(d+1)}, \tag{7.78}$$

where the notation $R_{\mathcal{U}}^{\mathrm{T}}$ indicates the transpose of $R_{\mathcal{U}}$. As usual, if the qubits can only be measured in the computational basis, suitable unitary operations must be implemented before a measurement in the σ_z basis. In addition, if the initial state of the quantum computer is $|0\dots0\rangle$, some unitary operations are required to prepare the states ρ_i. This point constitutes one of the main disadvantages of quantum process tomography: if state preparation and readout operations are not accurate enough, this experimental procedure does not faithfully reconstruct the unknown quantum map. In the literature, these errors are usually called state preparation and measurement errors (SPAM errors).

Example 7.13 We are given a black box $\mathcal{E}$ that operates on a qubit and is supposed to implement an $\hat{X}$ gate, $\mathcal{E}(\rho) = \hat{X}\rho\hat{X}^\dagger$. The associated Pauli transfer matrix is given by

$$R_X = \begin{bmatrix} 1 & 0 & 0 & 0 \\ 0 & 1 & 0 & 0 \\ 0 & 0 & -1 & 0 \\ 0 & 0 & 0 & -1 \end{bmatrix}, \qquad \mathbf{r} = (1,0,0,0,0,1,0,0,0,0,-1,0,0,0,0,-1)\,,$$

where $\mathbf{r}$ is the row-major vectorization of R_X. Our goal is to estimate the entries of the matrix R_X (or equivalently the entries of the vector $\mathbf{r}$) with a sequence of experiments. To this end, the black box $\mathcal{E}$ is applied to the qubit states

$$\begin{bmatrix} \rho_1 \\ \rho_2 \\ \rho_3 \\ \rho_4 \end{bmatrix} = \begin{bmatrix} \mathbb{1}/2 \\ (\mathbb{1}+\hat{\sigma}_x)/2 \\ (\mathbb{1}+\hat{\sigma}_y)/2 \\ (\mathbb{1}+\hat{\sigma}_z)/2 \end{bmatrix}, \qquad c_{il} = \begin{bmatrix} 1/2 & 0 & 0 & 0 \\ 1/2 & 1/2 & 0 & 0 \\ 1/2 & 0 & 1/2 & 0 \\ 1/2 & 0 & 0 & 1/2 \end{bmatrix},$$

where c_{il} are the coefficients of the Pauli decomposition of ρ_i. The measurement is described by the PVM $(\hat{E}_j, \mathbb{1} - \hat{E}_j)$, where

$$\begin{bmatrix} \hat{E}_1 \\ \hat{E}_2 \\ \hat{E}_3 \\ \hat{E}_4 \end{bmatrix} = \begin{bmatrix} \mathbb{1} \\ (\mathbb{1}+\hat{\sigma}_x)/2 \\ (\mathbb{1}+\hat{\sigma}_y)/2 \\ (\mathbb{1}+\hat{\sigma}_z)/2 \end{bmatrix}, \qquad A_{jk} = \begin{bmatrix} 1 & 0 & 0 & 0 \\ 1/2 & 1/2 & 0 & 0 \\ 1/2 & 0 & 1/2 & 0 \\ 1/2 & 0 & 0 & 1/2 \end{bmatrix}.$$

Here, A_{jk} are the coefficients of the Pauli decomposition of the operators $\hat{E}_j$. The measurement projects the states $\mathcal{E}(\rho_i)$ onto the eigenspace associated with $\hat{E}_j$ with probability $p_{ij} = \mathrm{Tr}[\mathcal{E}(\rho_i)\hat{E}_j]$ where

$$p_{ij} = \begin{bmatrix} 1 & \frac{1}{2} & \frac{1}{2} & \frac{1}{2} \\ 1 & 1 & \frac{1}{2} & \frac{1}{2} \\ 1 & \frac{1}{2} & 0 & \frac{1}{2} \\ 1 & \frac{1}{2} & \frac{1}{2} & 0 \end{bmatrix} \qquad \mathbf{p} = \left(1, \frac{1}{2}, \frac{1}{2}, \frac{1}{2}, 1, 1, \frac{1}{2}, \frac{1}{2}, 1, \frac{1}{2}, 0, \frac{1}{2}, 1, \frac{1}{2}, \frac{1}{2}, 0\right).$$

The entries of vector $\mathbf{r}$ can be computed as $\mathbf{r} = \frac{1}{d}S^{-1}\mathbf{p}$, where the 16×16 matrix S is given by $S_{j+4(i-1),k+4(l-1)} = c_{il}A_{jk}$.

Further reading

The key references about quantum maps are the book by Kraus [197], the papers by Hellwig and Kraus [214, 215], and Choi [211]. The connection between Kraus operators and dissipative time-evolutions was discovered by Lindblad [216]. The quantum map formalism has some limitations related to the assumption that the initial state of the system and the environment is a product state. For more information on the subject, see for example the work by Royer [217] and Shaji, Sudarshan [218].

Summary

Quantum maps

$\mathcal{E}$	Quantum map (or CPTP map)
$\mathcal{E}(c_1\rho_1 + c_2\rho_2) = c_1\mathcal{E}(\rho_1) + c_2\mathcal{E}(\rho_2)$	Quantum maps are linear
$\text{Tr}[\mathcal{E}(\rho)] = \text{Tr}[\rho]$	Quantum maps are trace-preserving
If $\rho \geq 0$, then $\mathcal{E}(\rho) \geq 0$	Quantum maps are positive
If $\rho \geq 0$, then $(\mathcal{E} \otimes \mathbb{1}_\text{B})(\rho) \geq 0$	Quantum maps are completely positive

$\mathcal{E}$ is a quantum map if and only if $\mathcal{E}(|\Omega\rangle\langle\Omega|) \geq 0$, where $|\Omega\rangle = \frac{1}{\sqrt{d}}\sum_{i=0}^{d-1}|i\rangle|i\rangle$.

Depolarizing channel

$\mathcal{E}_\text{depo}(\rho) = p\rho + (1-p)\mathbb{1}/2$	Depolarizing channel
$(r_x, r_y, r_z) \mapsto (pr_x, pr_y, pr_z)$	Action of $\mathcal{E}_\text{depo}$ on a Bloch vector
$\mathcal{E}_\text{depo}(\rho) = (1-\text{e}_\text{P})\rho + \frac{\text{e}_\text{P}}{3}\sum_i \hat{\sigma}_i\rho\hat{\sigma}_i$	Equivalent definition of $\mathcal{E}_\text{depo}$

Amplitude damping

$\mathcal{E}_\text{damp}(\rho) = \hat{E}_1\rho\hat{E}_1^\dagger + \hat{E}_2\rho\hat{E}_2^\dagger$	Amplitude damping
$\hat{E}_1 = \begin{bmatrix} 0 & \sqrt{p} \\ 0 & 0 \end{bmatrix}$, $\hat{E}_2 = \begin{bmatrix} 1 & 0 \\ 0 & \sqrt{1-p} \end{bmatrix}$	Kraus operators for $\mathcal{E}_\text{damp}$
$(r_x, r_y, r_z) \mapsto (\sqrt{1-p}r_x, \sqrt{1-p}r_y, p + (1-p)r_z)$	Action of amplitude damping

Dephasing

$\mathcal{E}_\text{deph}(\rho) = (1-p)\rho + p\hat{\sigma}_z\rho\hat{\sigma}_z$	Dephasing
$(r_x, r_y, r_z) \mapsto ((1-2p)r_x, (1-2p)r_y, r_z)$	Action of the dephasing channel
$T_2 = (1/T_\phi + 1/2T_1)^{-1}$	Coherence time

Transposition

$\text{T}(\rho) = \sum_{i,j}\rho_{ji}|i\rangle\langle j|$	Transposition

$\text{T}(\rho)$ is positive, trace-preserving, but not completely positive.

Important results

$\mathcal{E}$ is a quantum map if and only if $\mathcal{E}(\rho) = \sum_k \hat{E}_k\rho\hat{E}_k^\dagger$ where $\sum_k \hat{E}_k^\dagger\hat{E}_k = \mathbb{1}$.
A quantum map $\mathcal{E} : \text{B}(H) \to \text{B}(H)$ is specified by $d^4 - d^2$ real parameters.

$\mathcal{E}$ is a quantum map iff $\mathcal{E}(\rho) = \text{Tr}_\text{B}[\hat{U}(\rho \otimes |\text{e}\rangle\langle\text{e}|)\hat{U}^\dagger]$ where $\hat{U} \in \text{U}(H_\text{A} \otimes H_\text{B})$ and $|\text{e}\rangle \in H_\text{B}$.

Exercises

Exercise 7.1 At the beginning of this chapter, we stated that a quantum map $\mathcal{E}$ is invertible if and only if there exists a *quantum map* $\mathcal{E}^{-1}$ such that $\mathcal{E}^{-1}(\mathcal{E}(\rho)) = \rho$ for any state ρ. Show that the inverse of the depolarizing channel is given by the transformation $\mathcal{D}(\rho) = \rho/p - (1-p)\mathbb{1}/2p$. Prove that $\mathcal{D}$ is not a quantum map by showing that is not positive. This means that the depolarizing channel is not an invertible quantum map.

Exercise 7.2 Show that the affine map M defined in eqn (7.53) can be expressed as the product of a rotation and a contraction.

Exercise 7.3 In Example 7.10, we studied the time evolution of a qubit interacting with a harmonic oscillator. Repeat the derivation assuming that the oscillator is initially prepared in a thermal state $\rho_{\mathrm{B}} = (1-w)\sum_n w^n |n\rangle\langle n|$ where $w = e^{-\hbar\omega_r/k_{\mathrm{B}}T}$. Calculate the partial trace of the global state and show that the coherences of the qubit decay in time.

Exercise 7.4 Show that a quantum map is unital if and only if it does not decrease the von Neumann entropy of the input state

$$\mathcal{E}(\mathbb{1}/d) = \mathbb{1}/d \qquad \Leftrightarrow \qquad S(\mathcal{E}(\rho)) \geq S(\rho).$$

Hint: Use the von Neumann relative entropy $S(\rho \,\|\, \sigma) = \mathrm{Tr}[\rho(\log_2 \rho - \log_2 \sigma)]$ and the fact that $S(\mathcal{E}(\rho) \,\|\, \mathcal{E}(\sigma)) \leq S(\rho \,\|\, \sigma)$ only if $\mathcal{E}$ is a quantum map.

Exercise 7.5 Consider a Hilbert space H of dimension d. Show that any bipartite state $|\psi\rangle \in H \otimes H$ can be expressed as $|\psi\rangle = \mathbb{1} \otimes \hat{A}|\Omega\rangle$, where $\hat{A} = \sqrt{d}\rho_{\mathrm{B}}^{1/2}$, $\rho_{\mathrm{B}} = \mathrm{Tr}_{\mathrm{A}}[|\psi\rangle\langle\psi|]$ and $|\Omega\rangle$ is the entangled state $|\Omega\rangle = \frac{1}{\sqrt{d}}\sum_j |j\rangle|j\rangle$. **Hint:** Use the Schmidt decomposition presented in Section 9.2.

Exercise 7.6 Show that

$$\mathcal{E} : \mathrm{L}(H) \to \mathrm{L}(H) \text{ is CP} \qquad \Leftrightarrow \qquad (\mathcal{E} \otimes \mathbb{1})|\Omega\rangle\langle\Omega| \geq 0,$$

where $|\Omega\rangle \in H \otimes H$ is the entangled state $|\Omega\rangle = \frac{1}{\sqrt{d}}\sum_j |j\rangle|j\rangle$. **Hint:** From Exercise 7.5 we know that any bipartite state $|\psi\rangle \in H \otimes H$ can be written as $|\psi\rangle = \mathbb{1} \otimes \hat{A}|\Omega\rangle$ where $\hat{A}$ is a positive operator.

Exercise 7.7 Derive the affine map associated with the single-qubit rotation X_α presented in Section 5.3.

Exercise 7.8 Show that the transposition map is positive and trace-preserving.

Exercise 7.9 Derive the affine map associated with the dephasing channel.

Solutions

Solution 7.1 It is straightforward to see that $\mathcal{D}$ is the inverse of $\mathcal{E}_{\text{depo}}$,

$$\begin{aligned}\mathcal{D}(\mathcal{E}_{\text{depo}}(\rho)) &= \mathcal{D}\Big(p\rho + (1-p)\frac{\mathbb{1}}{2}\Big) \\ &= \frac{1}{p}\Big(p\rho + (1-p)\frac{\mathbb{1}}{2}\Big) - \frac{(1-p)}{p}\frac{\mathbb{1}}{2} = \rho.\end{aligned}$$

However, $\mathcal{D}$ is not a quantum map because it does not produce a positive operator for all input states. To show this, calculate the action of $\mathcal{D}$ on a generic qubit state $\rho = \sum_j \lambda_j |j\rangle\langle j|$,

$$\begin{aligned}\mathcal{D}(\rho) &= \frac{1}{p}\sum_j \lambda_j |j\rangle\langle j| - \frac{(1-p)}{p}\frac{1}{2}\sum_j |j\rangle\langle j| \\ &= \sum_j \underbrace{\frac{2\lambda_j - (1-p)}{2p}}_{\eta_j} |j\rangle\langle j| = \sum_j \eta_j |j\rangle\langle j|.\end{aligned}$$

The eigenvalues η_j are not all positive if one of the initial eigenvalues λ_j satisfies $2\lambda_j < 1-p$. Hence, $\mathcal{D}$ is not positive and $\mathcal{E}_{\text{depo}}$ is not an invertible quantum map.

Solution 7.2 Let us consider the polar decomposition of M, namely $M = LU$ (see Section 3.2.13). Here, L is a positive matrix and U is unitary. Since the output vector $\mathbf{b}$ must be real, the entries of M are real too. Hence, U is a real matrix associated with a proper rotation.

Solution 7.3 The calculations are the same up to eqn (7.68). Let us calculate $f_{01}(t)$ assuming that ρ_{B} is a thermal state at $t = 0$,

$$\begin{aligned}f_{01}(t) = \text{Tr}_{\text{B}}[\rho_{\text{B}} e^{-i2\chi t\hat{x}}] &= \int_{-\infty}^{\infty} \langle x|\rho_{\text{B}} e^{-i2\chi t\hat{x}}|x\rangle dx = \int_{-\infty}^{\infty} \langle x|\rho_{\text{B}}|x\rangle e^{-i2\chi t x} dx \\ &= \frac{\lambda}{\sqrt{\pi}}\sqrt{\frac{1-w}{1+w}}\int_{-\infty}^{\infty} e^{-\frac{1-w}{1+w}\lambda^2 x^2} e^{-i2\chi t x} dx \\ &= \exp\Big(-\frac{(1+w)\chi^2}{(1-w)\lambda^2}t^2\Big) \;=\; e^{-t^2/T_\phi^2},\end{aligned}$$

where we used eqn (6.43) and we introduced the dephasing time $T_\phi = \sqrt{(1-w)\lambda^2/(1+w)\chi^2}$. The coherences decay with a Gaussian profile for sufficiently long times.

Solution 7.4 We start by showing that if a map is unital, then $S(\mathcal{E}(\rho)) \geq S(\rho)$. To this end, we will make use of the von Neumann relative entropy defined as $S(\rho \,\|\, \sigma) = \text{Tr}(\rho \log_2 \rho) - \text{Tr}(\rho \log_2 \sigma)$. A quantum map never increases the von Neumann relative entropy: $S(\mathcal{E}(\rho) \,\|\, \mathcal{E}(\sigma)) \leq S(\rho \,\|\, \sigma)$ [200]. This property can be used in our favor: by substituting $\sigma = \mathbb{1}/d$ into the left member of the inequality, we obtain

$$\begin{aligned}S(\mathcal{E}(\rho) \,\|\, \mathcal{E}(\mathbb{1}/d)) &= \text{Tr}(\mathcal{E}(\rho)\log_2 \mathcal{E}(\rho)) - \text{Tr}(\mathcal{E}(\rho)\log_2 \mathbb{1}/d) \\ &= -S(\mathcal{E}(\rho)) - \log_2(1/d).\end{aligned}$$

On the other hand, the right member becomes $S(\rho \,\|\, \mathbb{1}/d) = -S(\rho) - \log_2(1/d)$. Combining these two results, we arrive at

$$-S(\mathcal{E}(\rho)) - \log_2(1/d) \leq -S(\rho) - \log_2(1/d),$$

which implies that $S(\mathcal{E}(\rho)) \geq S(\rho)$ as required. Let us show the converse implication. Suppose that $S(\mathcal{E}(\rho)) \geq S(\rho)$. Substituting the completely depolarized state into this inequality, we have $S(\mathcal{E}(\mathbb{1}/d)) \geq S(\mathbb{1}/d) = \log_2 d$. The completely depolarized state is a global maximum for the von Neumann entropy (there does not exist any other state with a higher entropy).

This implies that $S(\mathcal{E}(\mathbb{1}/d)) = S(\mathbb{1}/d)$, i.e. $\mathcal{E}(\mathbb{1}/d) = \mathbb{1}/d$ as required.

Solution 7.5 Any bipartite state can be written as $|\psi\rangle = \sum_j \lambda_j |j\rangle|j\rangle$ where λ_j are non-negative real numbers. This is called the Schmidt decomposition and is presented in Section 9.2. Let us calculate

$$\begin{aligned}\rho_{\mathrm{B}} = \mathrm{Tr}_{\mathrm{A}}\left[|\psi\rangle\langle\psi|\right] &= \mathrm{Tr}_{\mathrm{A}}\Big[\sum_{j,k} \lambda_j\lambda_k|j\rangle|j\rangle\langle k|\langle k|\Big] \\ &= \sum_{j,k} \lambda_j\lambda_k \mathrm{Tr}_{\mathrm{A}}[|j\rangle\langle k| \otimes |j\rangle\langle k|] = \sum_{j,k} \lambda_j\lambda_k\delta_{jk}|j\rangle\langle k| = \sum_k \lambda_k^2|k\rangle\langle k|.\end{aligned}$$

The last step of this proof boils down to showing that the Schmidt decomposition $\sum_j \lambda_j|j\rangle|j\rangle$ is equivalent to $|\psi\rangle = \mathbb{1} \otimes \hat{A}|\Omega\rangle$,

$$\begin{aligned}|\psi\rangle = \mathbb{1} \otimes \hat{A}|\Omega\rangle &= \frac{1}{\sqrt{d}}(\mathbb{1} \otimes \sqrt{d}\rho_{\mathrm{B}}^{1/2})\sum_j |j\rangle|j\rangle \\ &= \sum_j |j\rangle \otimes \sum_k \lambda_k|k\rangle\langle k|j\rangle = \sum_j |j\rangle \otimes \sum_k \lambda_k|k\rangle\delta_{kj} = \sum_j \lambda_j|j\rangle|j\rangle.\end{aligned}$$

Solution 7.6 Suppose that $\mathcal{E}$ is completely positive. Then, $(\mathcal{E} \otimes \mathbb{1})|\Omega\rangle\langle\Omega| \geq 0$ because $|\Omega\rangle\langle\Omega|$ is a positive operator. Let us now consider the converse implication. Suppose that $(\mathcal{E} \otimes \mathbb{1})|\Omega\rangle\langle\Omega| \geq 0$. Since a bipartite state can always be expressed as $|\psi\rangle = (\mathbb{1} \otimes \hat{A})|\Omega\rangle$ where $\hat{A}$ is a positive operator (see Exercise 7.5), then

$$\begin{aligned}(\mathcal{E} \otimes \mathbb{1})|\psi\rangle\langle\psi| &= (\mathcal{E} \otimes \mathbb{1})(\mathbb{1} \otimes \hat{A})|\Omega\rangle\langle\Omega|(\mathbb{1} \otimes \hat{A}) \\ &= (\mathbb{1} \otimes \hat{A})(\mathcal{E} \otimes \mathbb{1})|\Omega\rangle\langle\Omega|(\mathbb{1} \otimes \hat{A}) \geq 0.\end{aligned}$$

The inequality holds because the operators $(\mathcal{E} \otimes \mathbb{1})|\Omega\rangle\langle\Omega|$ and $\mathbb{1} \otimes \hat{A}$ are both positive. We now need to show that $(\mathcal{E} \otimes \mathbb{1})\rho$ is positive for any density operator ρ. Consider the spectral decomposition $\rho = \sum_i p_i|\psi_i\rangle\langle\psi_i|$. Since each state in this decomposition can be written as $|\psi_i\rangle = \mathbb{1} \otimes \hat{A}_i|\Omega\rangle$ where $\hat{A}_i$ are positive operators, then $(\mathcal{E} \otimes \mathbb{1})\rho = \sum_i p_i(\mathcal{E} \otimes \mathbb{1})|\psi_i\rangle\langle\psi_i| \geq 0$.

Solution 7.7 The single-qubit rotation X_α transforms a qubit state ρ into

$$\rho' = \mathsf{X}_\alpha \, \rho \, \mathsf{X}_\alpha^\dagger = \frac{\mathbb{1} + a_x \mathsf{X}_\alpha \hat{\sigma}_x \mathsf{X}_\alpha^\dagger + a_y \mathsf{X}_\alpha \hat{\sigma}_y \mathsf{X}_\alpha^\dagger + a_z \mathsf{X}_\alpha \hat{\sigma}_z \mathsf{X}_\alpha^\dagger}{2}. \tag{7.79}$$

Using eqn (3.64), the numerator can be expressed in a simpler form:

$$\begin{aligned}\mathsf{X}_\alpha \hat{\sigma}_x \mathsf{X}_\alpha^\dagger &= e^{-i\frac{\alpha}{2}\hat{\sigma}_x}\hat{\sigma}_x e^{-i\frac{\alpha}{2}\hat{\sigma}_x} = \hat{\sigma}_x, \\ \mathsf{X}_\alpha \hat{\sigma}_y \mathsf{X}_\alpha^\dagger &= e^{-i\frac{\alpha}{2}\hat{\sigma}_x}\hat{\sigma}_y e^{-i\frac{\alpha}{2}\hat{\sigma}_x} = \cos\alpha\,\hat{\sigma}_y + \sin\alpha\,\hat{\sigma}_z, \\ \mathsf{X}_\alpha \hat{\sigma}_z \mathsf{X}_\alpha^\dagger &= e^{-i\frac{\alpha}{2}\hat{\sigma}_x}\hat{\sigma}_z e^{-i\frac{\alpha}{2}\hat{\sigma}_x} = \cos\alpha\,\hat{\sigma}_z - \sin\alpha\,\hat{\sigma}_y\,.\end{aligned}$$

Substituting these relations into (7.79), we have

$$\rho' = \mathsf{X}_\alpha \, \rho \, \mathsf{X}_\alpha^\dagger = \frac{\mathbb{1} + a_x\hat{\sigma}_x + (a_y\cos\alpha - a_z\sin\alpha)\,\hat{\sigma}_y + (a_y\sin\alpha + a_z\cos\alpha)\,\hat{\sigma}_z}{2}.$$

Thus, the initial Bloch vector $\mathbf{a} = (a_x, a_y, a_z)$ is mapped into $M_\alpha \mathbf{a} + \mathbf{c}_\alpha$ where

$$M_\alpha = \begin{bmatrix} 1 & 0 & 0 \\ 0 & \cos\alpha & -\sin\alpha \\ 0 & \sin\alpha & \cos\alpha \end{bmatrix}, \qquad \mathbf{c}_\alpha = \begin{bmatrix} 0 \\ 0 \\ 0 \end{bmatrix}. \tag{7.80}$$

M_α is the usual rotation matrix about the x axis by an angle α.

Solution 7.8 Recall that any density operator is positive, i.e. for any state $|\psi\rangle = \sum_j c_j |j\rangle$,

$$0 \leq \langle\psi|\rho|\psi\rangle = \sum_{i,j} \rho_{ij} \langle\psi|i\rangle\langle j|\psi\rangle = \sum_{i,j} \rho_{ij} c_i^* c_j. \tag{7.81}$$

Now consider the state $|\psi^*\rangle = \sum_i c_i^* |i\rangle$. Since $\langle\psi^*|\rho|\psi^*\rangle \geq 0$, we have

$$\begin{aligned} 0 \leq \langle\psi^*|\rho|\psi^*\rangle &= \sum_{i,j} c_j c_i^* \langle j|\rho|i\rangle = \sum_{i,j} \langle j|\psi\rangle\langle\psi|i\rangle\langle j|\rho|i\rangle \\ &= \sum_{i,j} \rho_{ji} \langle\psi|i\rangle\langle j|\psi\rangle = \langle\psi|\Big(\sum_{i,j} \rho_{ji} |i\rangle\langle j|\Big)|\psi\rangle \; = \; \langle\psi|\, \mathrm{T}(\rho)\, |\psi\rangle. \end{aligned}$$

This shows that T is positive. The map T is also trace-preserving because

$$\mathrm{Tr}\left[\mathrm{T}(\rho)\right] = \sum_{i,j} \rho_{ji} \mathrm{Tr}[|i\rangle\langle j|] = \sum_{i,j} \rho_{ji}\delta_{ij} = \sum_i \rho_{ii} = \mathrm{Tr}\left[\rho\right]. \tag{7.82}$$

Solution 7.9 The dephasing channel transforms the Bloch vector (a_x, a_y, a_z) into $(a_x(1-2p),\, a_y(1-2p),\, a_z)$ where $p \in [0,1)$ (see eqn 7.30). Therefore, the final Bloch vector can be expressed as $M_{\text{deph}}\mathbf{a} + \mathbf{c}_{\text{deph}}$ where

$$M_{\text{deph}} = \begin{bmatrix} 1-2p & 0 & 0 \\ 0 & 1-2p & 0 \\ 0 & 0 & 1 \end{bmatrix}, \qquad \mathbf{c}_{\text{deph}} = \begin{bmatrix} 0 \\ 0 \\ 0 \end{bmatrix}. \tag{7.83}$$

8
Decoherence

Decoherence is the environment-induced
dynamical suppression of the quantum coherences.

In quantum mechanics, the time evolution of a quantum system is governed by the Schrödinger equation. Its solution provides an accurate description of the dynamics of a **closed** quantum system, i.e. a quantum system that is not affected by uncontrolled interactions with the environment. In practice, a quantum system is never completely isolated from its surroundings. In this sense, all quantum systems are **open**, and in some special cases it is justified to use a closed-system approach as a first approximation.

The dynamics of an open system can be understood with a classical analogy. The system under investigation can be pictured as a hot object in the middle of the ocean. The heat exchange cools the object to the same temperature as the surrounding water. This process will take place in a characteristic time τ. The ocean's temperature will not be significantly affected and can be considered constant during the interaction. In experiments relevant to quantum information processing, the system under investigation is typically a qubit or a quantum harmonic oscillator. The environment is usually modeled as an infinite set of harmonic oscillators at a constant temperature. The interaction between the system and the environment brings the system to an equilibrium state after a sufficient amount of time.

An analytical description of the system-environment dynamics is often impossible due to the complexity and large number of particles in the environment. A common approach in deriving the time evolution of an open system is to first study the evolution of the composite system and then trace over the environmental degrees of freedom. As seen in Section 7.8.1, this inevitably leads to a non-unitary evolution. Calculating the equation of motion for the reduced system, known as the **master equation**, is usually a difficult task and several approximations, valid in many physical contexts, are often taken into account.

The first part of the chapter introduces the Markov master equation in Lindblad form and the Redfield equation. These equations will be helpful in Sections 8.3 and 8.5 to discuss the interaction between a qubit and a thermal bath of harmonic oscillators. This treatment will lead to the introduction of the relaxation time and the dephasing time, two central concepts in quantum information processing. Section 8.4 outlines a correspondence between the relaxation time and the power spectral density. Similarly, Section 8.7 explains the correspondence between the dephasing time and the power

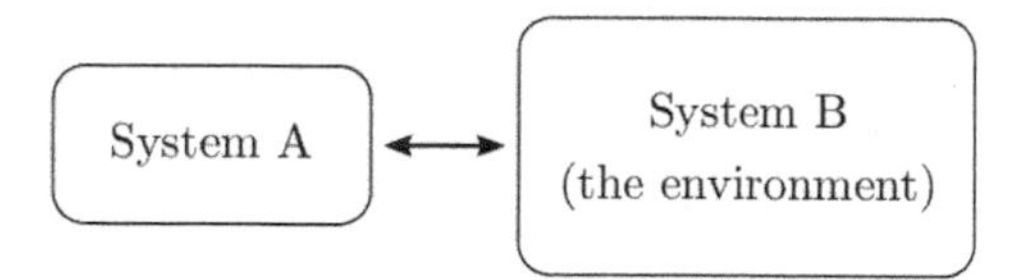

Fig. 8.1 An open quantum system. A schematic showing a quantum system A interacting with an environment B. In this chapter, system A will be a qubit, while the environment will be modeled as an infinite set of harmonic oscillators at a constant temperature. We will use the symbol ρ_{tot} to indicate the state of the "system+environment", while $\rho = \text{Tr}_{\text{B}}[\rho_{\text{tot}}]$ will indicate the state of system A.

spectral density. In Sections 8.8 and 8.9, we present the Hahn-echo experiment and the CPMG experiment, two protocols that provide some information about the noise spectrum affecting a qubit. The chapter ends with a discussion of the driven evolution of a qubit state when subject to spontaneous emission and dephasing.

8.1 Born–Markov master equation

Deriving the time evolution of an open system is often a complex task. Let us start our analysis by considering an open system A interacting with an environment, i.e. another quantum system B (see Fig. 8.1). In many practical cases, system A is composed of a few particles, whereas the environment has an infinite number of constituents. The Hamiltonian of the composite system is given by

$$H = H_{\text{A}} \otimes \mathbb{1} + \mathbb{1} \otimes H_{\text{B}} + V,$$

where H_{A} is the Hamiltonian of the system, H_{B} is the Hamiltonian of the environment and $V \in \text{L}(H_A \otimes H_B)$ is an interaction term that acts on both subspaces. Suppose that at $t = 0$ the system and the environment are in the initial state $\rho_{\text{tot}}(0)$. The state of the composite system at $t > 0$ will be

$$\rho_{\text{tot}}(t) = U(t)\rho_{\text{tot}}(0)U^\dagger(t),$$

where the time evolution operator is defined as $U(t) = \exp(\frac{1}{i\hbar}Ht)$. The state of the system at time $t > 0$ is obtained by tracing over the environment[1],

$$\rho(t) = \text{Tr}_{\text{B}}[U(t)\rho_{\text{tot}}(0)\, U^\dagger(t)].$$

Unfortunately, the state of the environment is very complex and this makes it impossible to calculate this trace directly. However, we can derive a general form of the equation of motion for subsystem A by making some reasonable assumptions.

Assumption 1 (the initial state) Our first assumption is that at $t = 0$ the system and the environment are in a product state $\rho_{\text{tot}}(0) = \rho \otimes \rho_{\text{B}}$, i.e. they are not entangled. We define the **quantum dynamical map** $\mathcal{E}_t$ as the transformation that brings the

[1]See Section 6.6.1 for more details about the reduced density operator.

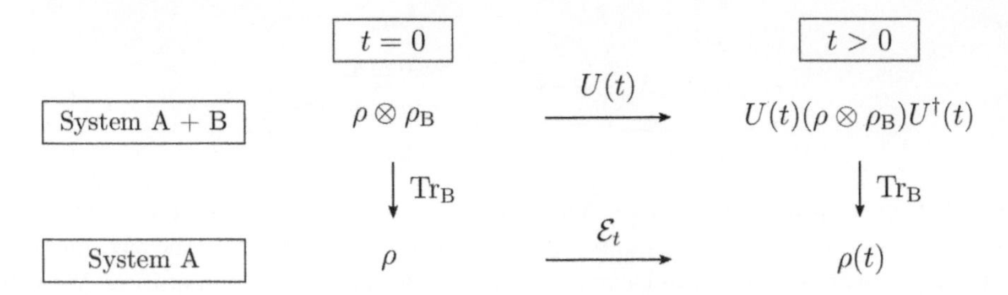

Fig. 8.2 Dynamics of an open system. Diagram showing the relation between the state of the composite system A + B and the state of system A at two different times. The quantum map $\mathcal{E}_t$ acts on subsystem A and transforms the initial state ρ into $\rho(t)$.

state of system A from time zero to $t > 0$ (see Fig. 8.2 for a diagrammatic representation). This map consists of a trace over the degrees of freedom of the environment and is given by

$$\boxed{\mathcal{E}_t(\rho) = \rho(t) = \mathrm{Tr}_B[U(t)(\rho \otimes \rho_B)U^\dagger(t)].} \tag{8.1}$$

This is a quantum map because it can be expressed in Kraus form (see Exercise 8.1),

$$\mathcal{E}_t(\rho) = \sum_{k,l=1}^{d_B} E_{kl}(t)\rho E_{kl}^\dagger(t). \tag{8.2}$$

The function $\mathcal{E}_t$ describes a single dynamical map with a fixed time t. If we allow t to vary, we obtain an infinite set of dynamical maps $\{\mathcal{E}_t : t \geq 0\}$. Clearly, when $t = 0$, we have $\mathcal{E}_0 = \mathbb{1}$.

Assumption 2 (the Born approximation) Our second assumption deals with the state of the environment. In general, the interaction between the system and the environment changes the state of the environment. In practice, the environment returns to equilibrium in a very short time interval compared to the typical variation times of the system. This allows us to approximate the state of the environment with a constant state ρ_B.

Assumption 3 (the Markov approximation) We also assume that the time derivative of the system state only depends on the instantaneous state of the system itself. This approximation is called the **Markov approximation** and is equivalent to requiring that dynamical maps at two different times, $\mathcal{E}_{t_1}$ and $\mathcal{E}_{t_2}$, satisfy the semigroup property

$$\boxed{\mathcal{E}_{t_2}(\mathcal{E}_{t_1}(\rho)) = \mathcal{E}_{t_1+t_2}(\rho).} \tag{8.3}$$

As explained in Section 1.2.12, a semigroup is an algebraic structure in which the inverse element might not exist. Therefore, $\mathcal{E}_t$ does not necessarily have an inverse map $\mathcal{E}_t^{-1}$ such that $\mathcal{E}_t^{-1}\mathcal{E}_t = \mathbb{1}$. This is the typical behavior of quantum maps responsible for the decoherence of a quantum state, such as amplitude damping and dephasing.

A dynamical map satisfying the Markov approximation (8.3) allows us to use Stone's theorem [219] and write $\mathcal{E}_t$ in exponential form

$$\rho(t) = \mathcal{E}_t(\rho) = e^{\mathcal{L}t}(\rho), \tag{8.4}$$

where $\mathcal{L}$ is a superoperator acting on density operators. Since $\mathcal{L}$ does not depend on time, it is straightforward to calculate the time derivative of (8.4) and obtain an equation of motion for system A,

$$\boxed{\frac{d\rho(t)}{dt} = \mathcal{L}(\rho(t)).} \tag{8.5}$$

This is called the **time-homogeneous Markovian master equation**[2]. The term time-homogeneous refers to the fact that the superoperator $\mathcal{L}$ is time-independent. In this chapter, we will mainly focus on these types of master equations.

Let us derive the most general form of a time-homogeneous Markovian master equation following Ref. [186]. First, we assume that the Hilbert space of the system H_{A} has a finite dimension d_{A}. The basis of the operator space is given by $\{F_i\}_i$ where the index $i \in (1, \ldots, d_{\mathrm{A}}^2)$. We assume that this basis is orthonormal with respect to the Hilbert-Schmidt inner product $\langle F_i|F_j\rangle = \mathrm{Tr}[F_i^\dagger F_j] = \delta_{ij}$ and we assume that $F_1 = \mathbb{1}/\sqrt{d_{\mathrm{A}}}$ so that the remaining operators are traceless. The decomposition of the operators $E_{kl}(t)$ in this basis is given by $E_{kl}(t) = \sum_i \langle F_i|E_{kl}(t)\rangle F_i$. By substituting this decomposition into (8.2), we obtain

$$\begin{aligned}\mathcal{E}_t(\rho) &= \sum_{k,l=1}^{d_{\mathrm{B}}} \Big(\sum_{i=1}^{d_{\mathrm{A}}^2} \langle F_i|E_{kl}(t)\rangle F_i\Big)\rho\Big(\sum_{j=1}^{d_{\mathrm{A}}^2} \langle F_j|E_{kl}(t)\rangle^* F_j^\dagger\Big) \\ &= \sum_{i,j=1}^{d_{\mathrm{A}}^2} \underbrace{\sum_{k,l=1}^{d_{\mathrm{B}}} \langle F_i|E_{kl}(t)\rangle\langle F_j|E_{kl}(t)\rangle^* }_{c_{ij}(t)} F_i\rho F_j^\dagger = \sum_{i,j=1}^{d_{\mathrm{A}}^2} c_{ij}(t) F_i \rho F_j^\dagger,\end{aligned} \tag{8.6}$$

where c is a $d_{\mathrm{A}}^2 \times d_{\mathrm{A}}^2$ positive matrix (see Exercise 8.5). Substituting (8.6) into eqn (8.5) leads to

$$\begin{aligned}\frac{d\rho}{dt} &= \mathcal{L}(\rho) = \lim_{\varepsilon\to 0} \frac{\mathcal{E}_\varepsilon(\rho) - \rho}{\varepsilon} = \lim_{\varepsilon\to 0} \frac{1}{\varepsilon} \sum_{i,j=1}^{d_{\mathrm{A}}^2} \left[c_{ij}(\varepsilon) F_i\rho F_j^\dagger - \rho\right] \\ &= \lim_{\varepsilon\to 0} \frac{1}{\varepsilon}\Big[\frac{c_{11}(\varepsilon) - d_{\mathrm{A}}}{d_{\mathrm{A}}}\rho + \sum_{i\geq 2}^{d_{\mathrm{A}}^2}\Big[\frac{c_{i1}(\varepsilon)}{\sqrt{d_{\mathrm{A}}}} F_i\rho + \frac{c_{1i}(\varepsilon)}{\sqrt{d_{\mathrm{A}}}}\rho F_i^\dagger\Big] + \sum_{i,j\geq 2}^{d_{\mathrm{A}}^2} c_{ij}(\varepsilon) F_i\rho F_j^\dagger\Big],\end{aligned} \tag{8.7}$$

[2]The von Neumann equation presented in Section 6.2 is the simplest example of a Markovian master equation: $\frac{d\rho}{dt} = \mathcal{L}(\rho) = \frac{1}{i\hbar}[H,\rho]$. Here, the superoperator $\mathcal{L}$ acts as $\mathcal{L}(A) = \frac{1}{i\hbar}[H, A]$.

where we used $F_1 = \mathbb{1}/\sqrt{d_A}$. To simplify this formula, we introduce the following time-independent parameters:

$$\alpha = \lim_{\varepsilon\to 0} \frac{c_{11}(\varepsilon) - d_A}{\varepsilon d_A},$$

$$\alpha_i = \lim_{\varepsilon\to 0} \frac{c_{i1}(\varepsilon)}{\varepsilon} \qquad \text{where } i \in (2, 3, \ldots, d_A^2),$$

$$\beta_{ij} = \lim_{\varepsilon\to 0} \frac{c_{ij}(\varepsilon)}{\varepsilon} \qquad \text{where } i, j \in (2, 3, \ldots, d_A^2).$$

The parameters β_{ij} define a positive matrix because c is positive. Using these definitions, eqn (8.7) becomes

$$\frac{d\rho}{dt} = \alpha\rho + \sum_{i\geq 2}\Big(\frac{\alpha_i}{\sqrt{d_A}}F_i\rho + \frac{\alpha_i^*}{\sqrt{d_A}}\rho F_i^\dagger\Big) + \sum_{i,j\geq 2}\beta_{ij}F_i\rho F_j^\dagger \tag{8.8}$$

$$= \alpha\rho + F\rho + \rho F^\dagger + \sum_{i,j\geq 2}\beta_{ij}F_i\rho F_j^\dagger, \tag{8.9}$$

where in the last step we defined $F = \sum_{i\geq 2}\alpha_i F_i/\sqrt{d_A}$. To express this equation in a form that resembles the von Neumann equation, we introduce the operators

$$G = \frac{\alpha}{2}\mathbb{1} + \frac{1}{2}(F + F^\dagger), \qquad H = \frac{\hbar}{2i}(F^\dagger - F).$$

With these definitions, eqn (8.9) becomes

$$\frac{d\rho}{dt} = \frac{1}{i\hbar}[H,\rho] + \{G,\rho\} + \sum_{i,j\geq 2}\beta_{ij}F_i\rho F_j^\dagger. \tag{8.10}$$

The operator G can be written in a simpler form. Indeed, note that $\mathrm{Tr}_A[d\rho/dt] = \mathrm{Tr}_A[\mathcal{E}_\varepsilon(\rho)] - \mathrm{Tr}_A[\rho] = 0$, and therefore

$$0 = \mathrm{Tr}_A[d\rho/dt] = \frac{1}{i\hbar}\mathrm{Tr}_A[H\rho - \rho H] + \mathrm{Tr}_A[G\rho + \rho G] + \mathrm{Tr}_A\big[\sum_{i,j\geq 2}\beta_{ij}F_i\rho F_j^\dagger\big]$$

$$= \mathrm{Tr}_A\big[\big(2G + \sum_{i,j\geq 2}\beta_{ij}F_j^\dagger F_i\big)\rho\big].$$

This equation holds for any density operator ρ and thus $G = -\frac{1}{2}\sum_{i,j\geq 2}\beta_{ij}F_j^\dagger F_i$. Inserting this expression into (8.10), we arrive at

$$\frac{d\rho}{dt} = \frac{1}{i\hbar}[H,\rho] + \sum_{i,j=2}^{d_A^2}\beta_{ij}F_i\rho F_j^\dagger - \frac{1}{2}\sum_{i,j=2}^{d_A^2}\beta_{ij}\{F_j^\dagger F_i, \rho\}.$$

Since β is a positive matrix, we can diagonalize it with a change of basis $\beta_{ij} =$

$\sum_k u_{ik}\gamma_k u^*_{kj}$, where the parameters γ_k define a diagonal matrix. Using this expression, we finally obtain the Markovian master equation in **Lindblad form** [216, 220, 221],

$$\boxed{\frac{d\rho}{dt} = \mathcal{L}(\rho) = \frac{1}{i\hbar}[H,\rho] + \sum_{k=2}^{d_{\rm A}^2} \gamma_k \big(A_k\rho A_k^\dagger - \frac{1}{2}\{A_k^\dagger A_k, \rho\}\big),} \tag{8.11}$$

where we introduced the **Lindblad operators** $A_k = \sum_i u_{ik}F_i$. The term proportional to $[H, \rho]$ represents the coherent evolution of the system. Assuming that the operators A_k are dimensionless, the constants γ_k are decay rates with unit 1/s. These parameters are determined by the coupling between the system and the environment. The distinctive feature of time-homogeneous Markovian master equations is that the decay rates γ are time-independent. In literature, the master equation (8.11) is often reported in a more concise form,

$$\boxed{\frac{d\rho}{dt} = \frac{1}{i\hbar}[H,\rho] + \sum_{k=2}^{d^2} \gamma_k \mathcal{D}[A_k](\rho),} \tag{8.12}$$

where the superoperators

$$\mathcal{D}[A_k](\rho) = A_k\rho A_k^\dagger - \frac{1}{2}\{A_k^\dagger A_k, \rho\}$$

are called the **dissipators**. These terms are responsible for the decoherence of the system state, as we shall see in the next sections.

To summarize, a dynamical map $\mathcal{E}_t = \mathrm{Tr}_{\rm B}[U(t)(\rho\otimes\rho_{\rm B})U^\dagger(t)]$ can be seen as a discrete variation of a quantum state from t_0 to t_0+t. Only dynamical maps satisfying the semigroup property $\mathcal{E}_{t_2}(\mathcal{E}_{t_1}(\rho)) = \mathcal{E}_{t_1+t_2}(\rho)$ lead to Markovian master equations. Since some quantum maps do not satisfy the semigroup property, the quantum map formalism presented in the previous chapter is a more general formalism than the one based on Markovian master equations.

In the following sections, we will study some concrete examples of master equations. We will focus on the time evolution of a qubit interacting with the environment and discover that the interaction between the two systems leads to a decay of the off-diagonal elements of the qubit density matrix, a phenomenon known as decoherence.

Example 8.1 Spontaneous emission is a physical phenomenon in which a system transitions from an excited to a lower energy state by emitting a quantum energy into the environment. The Markovian master equation describing the de-excitation of a qubit is

$$\frac{d\rho}{dt} = i\frac{\omega_{\rm q}}{2}[\sigma_z, \rho] + \gamma\mathcal{D}[\sigma^-](\rho),$$

where $\gamma = 1/T_1$ is the relaxation rate of the qubit. We will present the microscopic derivation of this master equation in Section 8.3.

Example 8.2 Dephasing is a physical process that transforms a quantum state ρ into a classical state. For a qubit, this process can be modeled by the Markovian master equation

$$\frac{d\rho}{dt} = i\frac{\omega_{\mathrm{q}}}{2}[\sigma_z, \rho] + \frac{\gamma_\phi}{2}\mathcal{D}[\sigma_z](\rho),$$

where $\gamma_\phi = 1/T_\phi$ is the dephasing rate of the qubit. We will present the microscopic derivation of this master equation in Section 8.5.

8.2 The Redfield equation

In the previous section, we used the Markov approximation to derive the master equation of a system interacting with the environment assuming that the two systems are not entangled initially. For pedagogical reasons, we now study the same problem from a different perspective. We will start from the general case in which the two systems are in an arbitrary state $\rho_{\mathrm{tot}} \in \mathrm{D}(H_{\mathrm{A}} \otimes H_{\mathrm{B}})$ at $t = 0$. By tracing over the environmental degrees of freedom, we will obtain an equation of motion for system A that is often intractable. We will then introduce some approximations that will allow us to derive the Redfield equation, a master equation for system A that can be solved analytically in many practical cases.

We start our analysis from the Hamiltonian[3]

$$\boxed{H = H_{\mathrm{A}} + H_{\mathrm{B}} + V,}$$

where H_{A} is the Hamiltonian of the system, H_{B} is the Hamiltonian of the environment, and $V \in \mathrm{L}(H_{\mathrm{A}} \otimes H_{\mathrm{B}})$ is a small interaction term. It is easier to study the dynamics in the interaction picture. In this frame, the state of the composite system evolves as

$$\tilde{\rho}_{\mathrm{tot}}(t) = e^{-\frac{1}{i\hbar}(H_{\mathrm{A}}+H_{\mathrm{B}})t}\rho_{\mathrm{tot}}(t)e^{\frac{1}{i\hbar}(H_{\mathrm{A}}+H_{\mathrm{B}})t},$$

where the tilde indicates that the corresponding operator is expressed in the interaction picture. The time evolution of the composite system is governed by the von Neumann equation[4]

$$\frac{d\tilde{\rho}_{\mathrm{tot}}}{dt} = \frac{1}{i\hbar}[\tilde{V}(t), \tilde{\rho}_{\mathrm{tot}}(t)], \tag{8.13}$$

where $\tilde{V}(t) = e^{-\frac{1}{i\hbar}(H_{\mathrm{A}}+H_{\mathrm{B}})t}Ve^{\frac{1}{i\hbar}(H_{\mathrm{A}}+H_{\mathrm{B}})t}$ is the coupling term in the interaction picture. The solution of (8.13) can be calculated by integrating the left and right-hand sides

$$\tilde{\rho}_{\mathrm{tot}}(t) = \rho_{\mathrm{tot}}(0) + \frac{1}{i\hbar}\int_0^t [\tilde{V}(t'), \tilde{\rho}_{\mathrm{tot}}(t')]dt'.$$

Substituting this expression into (8.13) leads to

[3]Here, we replaced the terms $H_{\mathrm{A}} \otimes \mathbb{1}$ and $\mathbb{1} \otimes H_{\mathrm{B}}$ with H_{A} and H_{B} for convenience.

[4]See Section 6.2 for a discussion of the von Neumann equation in the interaction picture.

$$\frac{d\tilde{\rho}_{\text{tot}}}{dt} = \frac{1}{i\hbar}[\tilde{V}(t), \rho_{\text{tot}}(0) + \frac{1}{i\hbar}\int_0^t [\tilde{V}(t'), \tilde{\rho}_{\text{tot}}(t')]\,dt']$$
$$= \frac{1}{i\hbar}[\tilde{V}(t), \rho_{\text{tot}}(0)] - \frac{1}{\hbar^2}\int_0^t [\tilde{V}(t), [\tilde{V}(t'), \tilde{\rho}_{\text{tot}}(t')]]\,dt'.$$

We can trace over the environment to obtain the evolution of system A only,

$$\boxed{\frac{d\tilde{\rho}}{dt} = \frac{1}{i\hbar}\text{Tr}_{\text{B}}\big([\tilde{V}(t), \rho_{\text{tot}}(0)]\big) - \frac{1}{\hbar^2}\int_0^t \text{Tr}_{\text{B}}\big([\tilde{V}(t), [\tilde{V}(t'), \tilde{\rho}_{\text{tot}}(t')]]\big)\,dt'.} \tag{8.14}$$

This equation is exact but often unsolvable. Let us make some approximations to simplify it.

Assumption 1 (the initial state) We assume that the system and the environment are not correlated at $t = 0$, i.e. they are in a product state. Therefore, $\rho_{\text{tot}}(0)$ in (8.14) can be replaced with $\rho \otimes \rho_{\text{B}}$.

Assumption 2 (the Born approximation) We assume that the interaction with the environment is small at all times such that the two systems are not strongly correlated during the evolution, $\tilde{\rho}_{\text{tot}}(t') \approx \tilde{\rho}(t') \otimes \tilde{\rho}_{\text{B}}(t')$. We also assume that the state of the environment does not significantly vary. Thus, $\tilde{\rho}_{\text{B}}(t)$ can be replaced with the initial state ρ_{B}. This is called the Born approximation and is discussed in detail in the work of Haake [222]. Under these assumptions, eqn (8.14) becomes

$$\frac{d\tilde{\rho}}{dt} = \frac{1}{i\hbar}\text{Tr}_{\text{B}}\big([\tilde{V}(t), \rho(0) \otimes \rho_{\text{B}}]\big) - \frac{1}{\hbar^2}\int_0^t \text{Tr}_{\text{B}}\big([\tilde{V}(t), [\tilde{V}(t'), \tilde{\rho}(t') \otimes \rho_{\text{B}}]]\big)\,dt'. \tag{8.15}$$

Note that the integral on the right-hand side contains the reduced density operator $\tilde{\rho}(t')$ that depends on the variable of integration t'. This means that the time derivative $d\tilde{\rho}/dt$ depends on the past history of the system.

Assumption 3 (the Markov approximation) In many physical situations, the past evolution of the system does not significantly impact its instantaneous variation. Hence, we will assume that $d\tilde{\rho}/dt$ only depends on the current state of the system. This is called the Markov approximation. This approximation boils down to substituting $\tilde{\rho}(t')$ with $\tilde{\rho}(t)$ in the integral (8.15). Its physical motivation will become clear in the next section when we discuss the auto-correlation function of the environment. Using this approximation, we finally obtain a master equation known as the **Redfield equation** [223],

$$\boxed{\frac{d\tilde{\rho}}{dt} = \frac{1}{i\hbar}\text{Tr}_{\text{B}}\big([\tilde{V}(t), \rho(0) \otimes \rho_{\text{B}}]\big) - \frac{1}{\hbar^2}\int_0^t \text{Tr}_{\text{B}}\big([\tilde{V}(t), [\tilde{V}(t'), \tilde{\rho}(t) \otimes \rho_{\text{B}}]]\big)\,dt'.} \tag{8.16}$$

Here, the only term in the integral that depends on the variable of integration is the interaction Hamiltonian. It is worth pointing out that the first trace in the Redfield equation can always be set to zero by rearranging some terms in the original Hamiltonian (see Exercise 8.4).

To summarize, we have obtained the Redfield equation by assuming that the state of the composite system is a product state at all times, the environment is in a stationary state and the future behavior of the reduced system depends only on its present state. Let us derive the solution of the Redfield equation in a case relevant to quantum information processing.

8.3 Microscopic derivation of amplitude damping

In our first example, we consider a qubit coupled to a thermal bath of harmonic oscillators, as shown in Fig. 8.3. The qubit can be a trapped ion, a superconducting qubit, a quantum dot, or any other type of two-level system. The bath models the environment surrounding the qubit and is described by an infinite number of harmonic oscillators in a thermal state. This is usually called the **spin-boson model**. We will assume that the bath is at a temperature T much smaller than the transition frequency of the qubit, $T \ll \hbar\omega_{\mathrm{q}}/k_{\mathrm{B}}$, so that thermal photons are not energetic enough to excite the qubit. This situation is quite common in quantum information processing where the initialization protocol of the qubits into the ground state usually involves their relaxation to a thermal state determined by the temperature of the environment[5].

Spontaneous emission involves the exchange of energy between the qubit and the environment. When the qubit is in the excited state, the decay mechanism to the ground state involves the instantaneous emission of a quantum of energy, such as a photon or a phonon. This energy will be absorbed by an oscillator in the environment. This process is irreversible, i.e. once the quantum of energy is absorbed by the environment, it is not transferred back to the qubit. The relaxation time T_1 is the average time scale in which this emission occurs. Since this decay mechanism involves a resonant interaction between the qubit and the oscillators, the relaxation time at low temperatures is mainly determined by the coupling between the qubit and the oscillators with frequency $\omega \approx \omega_{\mathrm{q}}$.

Let us model this phenomenon by writing the Hamiltonian of the composite system

$$\boxed{H = \underbrace{-\frac{\hbar\omega_{\mathrm{q}}}{2}\sigma_z}_{\text{qubit } H_{\mathrm{A}}} + \underbrace{\sum_{k=1}^{\infty}\hbar\omega_k b_k^\dagger b_k}_{\text{environment } H_{\mathrm{B}}} + \underbrace{\sum_{k=1}^{\infty}\hbar g_k(\sigma^+ \otimes b_k + \sigma^- \otimes b_k^\dagger)}_{\text{interaction } \hat{V}},} \tag{8.17}$$

where ω_{q} is the transition frequency of the qubit. The environment is modeled with an infinite set of harmonic oscillators: ω_k is the frequency of each oscillator and $b_k^\dagger$, b_k are the corresponding creation and annihilation operators. The term $\sigma^- b_k^\dagger$ describes the emission of a quantum of energy by the qubit and the excitation of the k-th oscillator. The term $\sigma^+ b_k$ describes the opposite phenomenon where the k-th oscillator emits a photon which then excites the qubit. The state of the environment is constant and

[5]If the temperature of the environment is $T \gtrsim \hbar\omega_{\mathrm{q}}/k_{\mathrm{B}}$, the qubit will thermalize into a statistical mixture of $|0\rangle$ and $|1\rangle$ and must be actively reset before it is used in a computation [224].

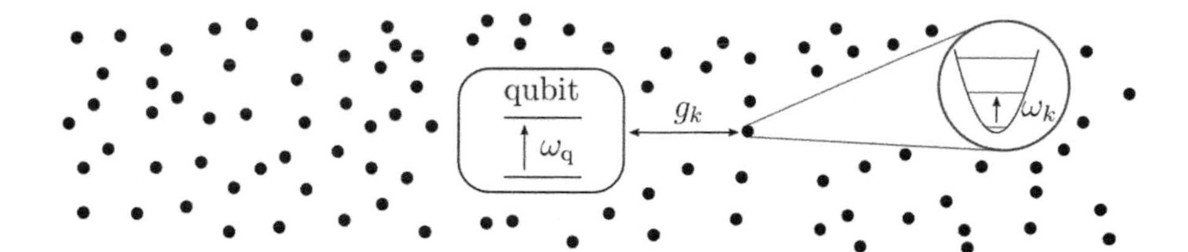

Fig. 8.3 A qubit coupled to a thermal bath of harmonic oscillators. The frequency of the k-th oscillator is ω_k and its coupling to the qubit is quantified by the real number g_k. Only the oscillators with a frequency similar to the qubit frequency determine the longitudinal relaxation time of the qubit, see eqn (8.43).

is given by a tensor product of an infinite number of thermal states, one for each oscillator

$$\rho_{\mathrm{B}} = \rho_1^{(\mathrm{th})} \otimes \rho_2^{(\mathrm{th})} \otimes \rho_3^{(\mathrm{th})} \otimes \ldots$$

where $\rho_k^{(\mathrm{th})} = \frac{1}{Z_k} e^{-\hbar\omega_k b_k^\dagger b_k / k_{\mathrm{B}} T}$ is the thermal state of the k-th oscillator and T is the temperature of the environment[6]. The interaction with the environment is linear[7] and the parameters g_k quantify the coupling strength between the qubit and each oscillator. We assume that the coupling with the environment is small, $|g_k| \ll \omega_{\mathrm{q}}$. The Hamiltonian (8.17) resembles the Jaynes–Cummings Hamiltonian presented in Section 4.7. The main difference is that the number of oscillators coupled to the qubit is now infinite.

Let us derive the equation of motion for the qubit following Refs [225, 226, 227, 228]. We will start from eqn (8.15), which we report here for convenience

$$\frac{d\tilde{\rho}}{dt} = \frac{1}{i\hbar}\mathrm{Tr}_{\mathrm{B}}\big([\tilde{V}(t), \rho \otimes \rho_{\mathrm{B}}]\big) - \frac{1}{\hbar^2}\int_0^t \mathrm{Tr}_{\mathrm{B}}\big([\tilde{V}(t), [\tilde{V}(t'), \tilde{\rho}(t') \otimes \rho_{\mathrm{B}}]]\big)\, dt'. \quad (8.18)$$

First, the Hamiltonian in the interaction picture is given by

$$\tilde{V}(t) = e^{-\frac{1}{i\hbar}(H_{\mathrm{A}}+H_{\mathrm{B}})t} V e^{\frac{1}{i\hbar}(H_{\mathrm{A}}+H_{\mathrm{B}})t} = \sum_{k=1}^{\infty} \hbar g_k \big[\sigma^+(t) \otimes b_k(t) + \sigma^-(t) \otimes b_k^\dagger(t)\big], \quad (8.19)$$

where we used eqn (6.19) and we defined

[6] See Section 6.5.1 for more information about thermal states.

[7] To be precise, the coupling between a qubit and a mode of the electric field is proportional to $\sigma_x(b+b^\dagger) = \sigma^+b+\sigma^-b^\dagger+\sigma^-b+\sigma^+b^\dagger$. Since $g \ll \omega_{\mathrm{q}}$, we can drop the last two terms because they give a small contribution to the dynamics. This approximation is called the rotating wave approximation (see Appendix A.1 for more information).

$$b_k(t) = e^{i\omega_k b_k^\dagger b_k t}\; b_k\; e^{-i\omega_k b_k^\dagger b_k t} = b_k e^{-i\omega_k t}, \tag{8.20}$$

$$b_k^\dagger(t) = e^{i\omega_k b_k^\dagger b_k t}\; b_k^\dagger\; e^{-i\omega_k b_k^\dagger b_k t} = b_k^\dagger e^{i\omega_k t}, \tag{8.21}$$

$$\sigma^-(t) = e^{-i\frac{\omega_q t}{2}\sigma_z}\; \sigma^-\; e^{i\frac{\omega_q t}{2}\sigma_z} = \sigma^- e^{-i\omega_q t}, \tag{8.22}$$

$$\sigma^+(t) = e^{-i\frac{\omega_q t}{2}\sigma_z}\; \sigma^+\; e^{i\frac{\omega_q t}{2}\sigma_z} = \sigma^+ e^{i\omega_q t}. \tag{8.23}$$

These four identities can be derived using the Baker–Campbell–Hausdorff formula. To simplify future calculations, it is useful to write the interaction Hamiltonian (8.19) in a more compact form

$$\tilde{V}(t) = \hbar\big[\sigma^+(t)\otimes\Gamma(t) + \sigma^-(t)\otimes\Gamma^\dagger(t)\big] = \hbar\sum_{i=1}^{2} s_i(t)\otimes\Gamma_i(t), \tag{8.24}$$

where we introduced the operators

$$s_1(t) = \sigma^+(t), \qquad \Gamma_1(t) = \Gamma(t) = \sum_{k=1}^{\infty} g_k b_k e^{-i\omega_k t}, \tag{8.25}$$

$$s_2(t) = \sigma^-(t), \qquad \Gamma_2(t) = \Gamma^\dagger(t) = \sum_{k=1}^{\infty} g_k b_k^\dagger e^{i\omega_k t}. \tag{8.26}$$

The operator Γ ($\Gamma^\dagger$) contains destruction (creation) operators for the modes in the environment. Second, the first trace in (8.18) is zero. This is because the environment is in a thermal state and the interaction Hamiltonian is proportional to b_k and $b_k^\dagger$,

$$\begin{aligned}\mathrm{Tr_B}\big([\tilde{V}(t),\, \rho\otimes\rho_\mathrm{B}]\big) &= \hbar\,\mathrm{Tr_B}\big[\sigma^+(t)\otimes\Gamma(t) + \sigma^-(t)\otimes\Gamma^\dagger(t),\, \rho\otimes\rho_\mathrm{B}\big] \\ &= \hbar\,[\sigma^+(t),\rho]\underbrace{\mathrm{Tr_B}[\rho_\mathrm{B}\Gamma(t)]}_{=0} + \hbar\,[\sigma^-(t),\rho]\underbrace{\mathrm{Tr_B}[\rho_\mathrm{B}\Gamma^\dagger(t)]}_{=0} = 0.\end{aligned}$$

In this calculation, we used the fact that $\mathrm{Tr_B}[\rho_\mathrm{B}\Gamma(t)] = \mathrm{Tr_B}[\rho_\mathrm{B}\Gamma^\dagger(t)] = 0$ because ρ_B is a thermal state and Γ and $\Gamma^\dagger$ contain creation and destruction operators. Thus, eqn (8.18) reduces to

$$\frac{d\tilde{\rho}}{dt} = -\frac{1}{\hbar^2}\int_0^t dt'\mathrm{Tr_B}\big([\tilde{V}(t),[\tilde{V}(t'),\tilde{\rho}(t')\otimes\rho_\mathrm{B}]]\big). \tag{8.27}$$

By substituting (8.24) into (8.27), we obtain

$$\frac{d\tilde{\rho}(t)}{dt} = -\int_0^t dt' \,\mathrm{Tr_B}\Big(\big[\sum_{i=1}^{2} s_i(t) \otimes \Gamma_i(t), \big[\sum_{j=1}^{2} s_j(t') \otimes \Gamma_j(t'), \tilde{\rho}(t') \otimes \rho_\mathrm{B} \big]\big] \Big)$$

$$= \int_0^t dt' \sum_{i,j=1}^{2} \Big\{ \big[s_j(t')\tilde{\rho}(t')s_i(t) - s_i(t)s_j(t')\tilde{\rho}(t') \big] \mathrm{Tr_B}\big[\rho_\mathrm{B}\Gamma_i(t)\Gamma_j(t') \big]$$

$$+ \big[s_i(t)\tilde{\rho}(t')s_j(t') - \tilde{\rho}(t')s_j(t')s_i(t) \big] \mathrm{Tr_B}\big[\rho_\mathrm{B}\Gamma_j(t')\Gamma_i(t) \big] \Big\}. \quad (8.28)$$

The integral (8.28) contains 16 terms because $i, j \in \{1, 2\}$. It can be simplified by noticing that the traces are zero when $i = j$,

$$\mathrm{Tr_B}\big[\rho_\mathrm{B}\Gamma_1(t)\Gamma_1(t') \big] = \sum_{k,l} e^{-i\omega_k t} e^{-i\omega_l t'} g_k g_l \mathrm{Tr_B}[\rho_\mathrm{B} b_k b_l] = 0, \quad (8.29)$$

$$\mathrm{Tr_B}\big[\rho_\mathrm{B}\Gamma_2(t)\Gamma_2(t') \big] = \sum_{k,l} e^{i\omega_k t} e^{i\omega_l t'} g_k g_l \mathrm{Tr_B}[\rho_\mathrm{B} b_k^\dagger b_l^\dagger] = 0, \quad (8.30)$$

$$\mathrm{Tr_B}\big[\rho_\mathrm{B}\Gamma_2(t)\Gamma_1(t') \big] = \sum_{k,l} e^{i\omega_k t} e^{-i\omega_l t'} g_k g_l \mathrm{Tr_B}\big[\rho_\mathrm{B} b_k^\dagger b_l \big] = \sum_{k=1}^{\infty} e^{i\omega_k (t-t')} g_k^2 \bar{n}(\omega_k), \quad (8.31)$$

$$\mathrm{Tr_B}\big[\rho_\mathrm{B}\Gamma_1(t)\Gamma_2(t') \big] = \sum_{k,l} e^{-i\omega_k t} e^{i\omega_l t'} g_k g_l \mathrm{Tr_B}\big[\rho_\mathrm{B} b_k b_l^\dagger \big] = \sum_{k=1}^{\infty} e^{-i\omega_k (t-t')} g_k^2 (\bar{n}(\omega_k) + 1), \quad (8.32)$$

where we used (6.41) and we introduced $\bar{n}(\omega_k) = \mathrm{Tr_B}[\rho_\mathrm{B} b_k^\dagger b_k] = 1/(e^{\hbar\omega_k/k_\mathrm{B}T} - 1)$, the average number of thermal photons in the k-th oscillator. The quantities $\mathrm{Tr_B}[\rho_\mathrm{B}\Gamma_i(t)\Gamma_j(t')]$ are called the **auto-correlation functions** of the environment. The sign and magnitude of the auto-correlation function $\mathrm{Tr}[\rho\hat{A}(t)\hat{A}(t')]$ indicates whether the fluctuations of the observable $\hat{A}$ at two different times are correlated, anti-correlated, or statistically independent. Dropping the terms with $i = j$, eqn (8.28) becomes

$$\frac{d\tilde{\rho}(t)}{dt} = \int_0^t dt' \{\sigma^+\tilde{\rho}(t')\sigma^- - \sigma^-\sigma^+\tilde{\rho}(t')\} e^{-i\omega_\mathrm{q}(t-t')} \mathrm{Tr_B}\big[\rho_\mathrm{B}\Gamma^\dagger(t)\Gamma(t') \big]$$

$$+ \{\sigma^-\tilde{\rho}(t')\sigma^+ - \sigma^+\sigma^-\tilde{\rho}(t')\} e^{i\omega_\mathrm{q}(t-t')} \; \mathrm{Tr_B}\big[\rho_\mathrm{B}\Gamma(t)\Gamma^\dagger(t') \big]$$

$$+ \{\sigma^+\tilde{\rho}(t')\sigma^- - \tilde{\rho}(t')\sigma^-\sigma^+\} e^{i\omega_\mathrm{q}(t-t')} \; \mathrm{Tr_B}\big[\rho_\mathrm{B}\Gamma^\dagger(t')\Gamma(t) \big]$$

$$+ \{\sigma^-\tilde{\rho}(t')\sigma^+ - \tilde{\rho}(t')\sigma^+\sigma^-\} e^{-i\omega_\mathrm{q}(t-t')} \mathrm{Tr_B}\big[\rho_\mathrm{B}\Gamma(t')\Gamma^\dagger(t) \big], \quad (8.33)$$

where we used eqns (8.22, 8.23) and the relations $s_1(t) = \sigma^+(t)$, $s_2(t) = \sigma^-(t)$. Let us perform a change of variable $\{\tau = t - t',\, d\tau = -dt'\}$,

$$\frac{d\tilde{\rho}(t)}{dt} = \int_0^t d\tau \{\sigma^+\tilde{\rho}(t-\tau)\sigma^- - \sigma^-\sigma^+\tilde{\rho}(t-\tau)\} e^{-i\omega_q\tau} \mathrm{Tr_B}\left[\rho_B\Gamma^\dagger(t)\Gamma(t-\tau)\right]$$
$$+ \{\sigma^-\tilde{\rho}(t-\tau)\sigma^+ - \sigma^+\sigma^-\tilde{\rho}(t-\tau)\} e^{i\omega_q\tau}\ \mathrm{Tr_B}\left[\rho_B\Gamma(t)\Gamma^\dagger(t-\tau)\right]$$
$$+ \{\sigma^+\tilde{\rho}(t-\tau)\sigma^- - \tilde{\rho}(t-\tau)\sigma^-\sigma^+\} e^{i\omega_q\tau}\ \mathrm{Tr_B}\left[\rho_B\Gamma^\dagger(t-\tau)\Gamma(t)\right]$$
$$+ \{\sigma^-\tilde{\rho}(t-\tau)\sigma^+ - \tilde{\rho}(t-\tau)\sigma^+\sigma^-\} e^{-i\omega_q\tau} \mathrm{Tr_B}\left[\rho_B\Gamma(t-\tau)\Gamma^\dagger(t)\right]. \quad (8.34)$$

Assuming that the frequencies of the harmonic oscillators are very dense, the sums in eqns (8.31, 8.32) can be replaced by an integral. Hence, ω_k can be substituted by a continuous variable ω, the coupling strengths g_k are replaced by $g(\omega)$ and the average photon number is replaced with $\bar{n}(\omega)$. With these substitutions, the correlation functions of the environment (8.31, 8.32) become

$$\mathrm{Tr_B}\left[\rho_B\Gamma^\dagger(t)\Gamma(t-\tau)\right] = \int_0^\infty d\omega e^{i\omega\tau}\lambda(\omega)g^2(\omega)\bar{n}(\omega), \quad (8.35)$$

$$\mathrm{Tr_B}\left[\rho_B\Gamma(t)\Gamma^\dagger(t-\tau)\right] = \int_0^\infty d\omega e^{-i\omega\tau}\lambda(\omega)g^2(\omega)(\bar{n}(\omega)+1). \quad (8.36)$$

The function $\lambda(\omega)$ is a density of states such that $\lambda(\omega)d\omega$ gives the number of oscillators in the infinitesimal interval $[\omega, \omega+d\omega]$. The next step is to calculate the integrals (8.35, 8.36) and derive the analytical expression of the correlation functions. As explained in Example 8.3, in many practical cases these functions decay to zero on a time scale t_R much smaller than the relaxation time of the qubit. Hence, we can approximate the correlation functions with a Dirac delta, $\mathrm{Tr_B}[\rho_B\Gamma(t)\Gamma^\dagger(t-\tau)] \propto \delta(\tau)$. This allows us to invoke the Markov approximation, replace $\tilde{\rho}(t-\tau)$ with $\tilde{\rho}(t)$ and take the curly brackets in (8.34) out from the integral. The master equation(8.34) finally becomes

$$\frac{d\tilde{\rho}}{dt} = \alpha(\sigma^-\tilde{\rho}\sigma^+ - \sigma^+\sigma^-\tilde{\rho}) + \beta(\sigma^-\tilde{\rho}\sigma^+ - \sigma^+\sigma^-\tilde{\rho} + \sigma^+\tilde{\rho}\sigma^- - \tilde{\rho}\sigma^-\sigma^+)$$
$$+ \alpha^*(\sigma^-\tilde{\rho}\sigma^+ - \tilde{\rho}\sigma^+\sigma^-) + \beta^*(\sigma^+\tilde{\rho}\sigma^- - \sigma^-\sigma^+\tilde{\rho} + \sigma^-\tilde{\rho}\sigma^+ - \tilde{\rho}\sigma^+\sigma^-), \quad (8.37)$$

where we introduced two complex functions,

$$\alpha(t) = \int_0^t d\tau \int_0^\infty d\omega e^{-i(\omega-\omega_q)\tau}\lambda(\omega)g^2(\omega), \quad (8.38)$$

$$\beta(t) = \int_0^t d\tau \int_0^\infty d\omega e^{-i(\omega-\omega_q)\tau}\lambda(\omega)g^2(\omega)\bar{n}(\omega). \quad (8.39)$$

These functions are similar to the correlation functions (8.35, 8.36), the main difference being an additional time integral.

We are interested in the time evolution of the density operator for a duration t comparable to the relaxation time T_1. In many practical cases, the relaxation time is much longer than the time scale in which the environment returns to equilibrium. This implies that the integrands in eqns (8.38, 8.39) are zero almost on the entire integration interval. Hence, we can extend the upper integration limit in (8.38, 8.39) to infinity making the functions α and β time-independent. To derive their analytical form, we will use an identity between distributions

$$\lim_{t\to\infty}\int_0^t d\tau e^{-i(\omega-\omega_{\mathrm{q}})\tau} = \pi\delta(\omega-\omega_{\mathrm{q}}) + i\mathcal{P}\frac{1}{\omega-\omega_{\mathrm{q}}},$$

where δ is the Dirac delta and $\mathcal{P}$ is the Cauchy principal value. The value of α is given by

$$\alpha = \int_0^\infty d\omega\pi\delta(\omega-\omega_{\mathrm{q}})\lambda(\omega)g^2(\omega) + i\underbrace{\mathcal{P}\int_0^\infty \frac{d\omega}{\omega-\omega_{\mathrm{q}}}\lambda(\omega)g^2(\omega)}_{\Delta_1} \tag{8.40}$$

$$= \underbrace{\pi\lambda(\omega_{\mathrm{q}})g^2(\omega_{\mathrm{q}})}_{\gamma/2} + i\Delta_1 = \frac{\gamma}{2} + i\Delta_1, \tag{8.41}$$

while the expression of β is

$$\beta = \int_0^\infty d\omega\pi\delta(\omega-\omega_{\mathrm{q}})\lambda(\omega)g^2(\omega)\bar{n}(\omega) + i\underbrace{\int_0^\infty d\omega\mathcal{P}\frac{1}{\omega-\omega_{\mathrm{q}}}\lambda(\omega)g^2(\omega)\bar{n}(\omega)}_{\Delta_2} \tag{8.42}$$

$$= \underbrace{\pi\lambda(\omega_{\mathrm{q}})g^2(\omega_{\mathrm{q}})}_{\gamma/2}\bar{n}(\omega_{\mathrm{q}}) + i\Delta_2 = \frac{\gamma}{2}\bar{n}_{\mathrm{q}} + i\Delta_2,$$

where $\bar{n}_{\mathrm{q}} = 1/(e^{\hbar\omega_{\mathrm{q}}/k_{\mathrm{B}}T} - 1)$ is the average number of photons in the oscillators with frequency $\omega = \omega_{\mathrm{q}}$ and the **relaxation rate** of the qubit has been defined as

$$\boxed{\gamma = 2\pi\lambda(\omega_{\mathrm{q}})g^2(\omega_{\mathrm{q}}).} \tag{8.43}$$

The product $J(\omega) \equiv 4\lambda(\omega)g^2(\omega)$ is called the **spectral density** of the bath. The important point about eqn 8.43 is that the relaxation rate is determined by the spectral density of the environment evaluated at the qubit frequency. By substituting $\alpha = \gamma/2 + i\Delta_1$ and $\beta = \gamma\bar{n}_{\mathrm{q}}/2 + i\Delta_2$ into eqn (8.37), we have

$$\frac{d\tilde{\rho}}{dt} = i(\Delta_1/2+\Delta_2)[\sigma_z,\tilde{\rho}] + \gamma(\bar{n}_{\mathrm{q}}+1)\left(\sigma^-\tilde{\rho}\sigma^+ - \frac{1}{2}\sigma^+\sigma^-\tilde{\rho} - \frac{1}{2}\tilde{\rho}\sigma^+\sigma^-\right)$$
$$+ \gamma\bar{n}_{\mathrm{q}}\left(\sigma^+\tilde{\rho}\sigma^- - \frac{1}{2}\sigma^-\sigma^+\tilde{\rho} - \frac{1}{2}\tilde{\rho}\sigma^-\sigma^+\right),$$

where we have used $\sigma^+\sigma^- = (\mathbb{1}-\sigma_z)/2$ and $\sigma^-\sigma^+ = (\mathbb{1}+\sigma_z)/2$. Here, $\tilde{\rho}$ is still in the interaction picture. To transform back to the Schrödinger picture, we use (6.20) to obtain

$$\frac{d\rho}{dt} = i\frac{\omega'_q}{2}[\sigma_z, \rho] + \gamma(\bar{n}_q + 1)\left(\sigma^-\rho\sigma^+ - \frac{1}{2}\sigma^+\sigma^-\rho - \rho\sigma^+\sigma^-\right)$$
$$+ \gamma\bar{n}_q\left(\sigma^+\rho\sigma^- - \frac{1}{2}\sigma^-\sigma^+\rho - \frac{1}{2}\rho\sigma^-\sigma^+\right), \quad (8.44)$$

where

$$\omega'_q = \omega_q + \Delta_1 + 2\Delta_2. \quad (8.45)$$

Thus, the interaction with the environment slightly shifts the qubit frequency by an amount $\Delta_1 + 2\Delta_2$. The contribution Δ_1 is called the **Lamb shift**. As seen in (8.40), this term is temperature-independent. The term $2\Delta_2$ is called the **AC Stark shift** and depends on the number of photons in the oscillators, see eqn (8.42). The master equation (8.44) can be expressed in Lindblad form

$$\boxed{\frac{d\rho(t)}{dt} = i\frac{\omega'_q}{2}[\sigma_z, \rho] + \gamma_\downarrow\mathcal{D}[\sigma^-](\rho) + \gamma_\uparrow\mathcal{D}[\sigma^+](\rho),} \quad (8.46)$$

where the rates are defined as:

$$\gamma_\downarrow = \gamma(\bar{n}_q + 1), \qquad \gamma_\uparrow = \gamma\bar{n}_q.$$

The dissipator $\mathcal{D}[\sigma^-]$ describes the de-excitation of the qubit at a rate $\gamma_\downarrow$, while the dissipator $\mathcal{D}[\sigma^+]$ describes the opposite phenomenon in which the bath emits a thermal photon with frequency $\omega = \omega_q$ thereby exciting the qubit at a rate $\gamma_\uparrow$. The **longitudinal relaxation time** is defined as

$$\boxed{T_1 = \frac{1}{\gamma_\downarrow + \gamma_\uparrow} = \frac{1}{\gamma(2\bar{n}_q + 1)}.} \quad (8.47)$$

It is important to mention that in quantum information processing the quantum devices are usually operated at a temperature $T \ll \hbar\omega_q/k_B$. The oscillators in the environment with a frequency $\omega \approx \omega_q$ are in the ground state and the average number of thermal photons in these oscillators is $\bar{n}_q \approx 0$. Therefore, the relaxation rates become $\gamma_\downarrow = \gamma$, $\gamma_\uparrow = 0$ and the relaxation time simplifies to $T_1 = 1/\gamma$. At low temperatures, the equation of motion (8.46) can be written as[8]

$$\boxed{\frac{d\rho(t)}{dt} = \gamma\mathcal{D}[\sigma^-](\rho) = \gamma\left(\sigma^-\rho\sigma^+ - \frac{1}{2}\sigma^+\sigma^-\rho - \frac{1}{2}\rho\sigma^+\sigma^-\right).} \quad (8.48)$$

This is the master equation for a qubit emitting energy into the environment at low temperatures. Let us solve this differential equation and show that the qubit decays to the ground state after sufficiently long times regardless of the initial state. At $t = 0$ the qubit is in a generic state

[8]Here, we are ignoring the trivial evolution related to σ_z. This is equivalent to performing a time-dependent change of basis $R = e^{-\frac{1}{i\hbar}H'_q t}$ where $H'_q = -\hbar\omega'_q\sigma_z/2$.

$$\rho = \rho_{00}|0\rangle\langle 0| + \rho_{01}|0\rangle\langle 1| + \rho_{10}|1\rangle\langle 0| + \rho_{11}|1\rangle\langle 1| = \begin{bmatrix} \rho_{00} & \rho_{01} \\ \rho_{10} & \rho_{11} \end{bmatrix},$$

where ρ_{01}, ρ_{10} are the coherences and ρ_{00}, ρ_{11} are the populations, i.e. the probabilities of measuring the qubit in the ground and excited state. Substituting

$$\sigma^- = \begin{bmatrix} 0 & 1 \\ 0 & 0 \end{bmatrix}, \qquad \sigma^+ = \begin{bmatrix} 0 & 0 \\ 1 & 0 \end{bmatrix}, \qquad \rho = \begin{bmatrix} \rho_{00} & \rho_{01} \\ \rho_{10} & \rho_{11} \end{bmatrix}$$

into eqn (8.48) and performing some matrix multiplications, we arrive at

$$\begin{bmatrix} \dot{\rho}_{00} & \dot{\rho}_{01} \\ \dot{\rho}_{10} & \dot{\rho}_{11} \end{bmatrix} = \begin{bmatrix} \gamma\rho_{11} & -\gamma\rho_{01}/2 \\ -\gamma\rho_{10}/2 & -\gamma\rho_{11} \end{bmatrix}.$$

The solutions of the differential equations

$$\frac{d\rho_{11}}{dt} = -\gamma\rho_{11}, \qquad \frac{d\rho_{01}}{dt} = -\frac{\gamma}{2}\rho_{01} \tag{8.49}$$

are $\rho_{11}(t) = \rho_{11}(0)e^{-\gamma t}$ and $\rho_{01}(t) = \rho_{01}(0)e^{-\gamma t/2}$. In conclusion, the populations decay exponentially with a relaxation time $T_1 = 1/\gamma$, while the coherences relax to zero with a characteristic time $2T_1$.

The relaxation time can be measured with an **inversion recovery** experiment as shown in Fig. 8.4a. In this experiment, the qubit starts in the ground state and is brought into the excited state with a fast X_π pulse. This pulse is followed by a measurement delayed by a time interval t. The experiment is repeated for different delay times. The relaxation time is extracted by measuring the probability of finding the qubit in the excited state and fitting the decay with an exponential curve.

Example 8.3 Let us study the Markov approximation in more detail. This approximation is based on the fact that the relaxation time of the qubit is much longer than the time taken by the environment to return to equilibrium. Suppose that the transition frequency of the qubit is $\omega_{\mathrm{q}}/2\pi = 5$ GHz and the environment is at $T = 10$ mK. This situation is quite common in the field of superconducting devices. Let us calculate the correlation functions of the environment,

$$\mathrm{Tr}_{\mathrm{B}}[\rho_{\mathrm{B}}\Gamma^\dagger(t)\Gamma(t-\tau)] = \int_0^\infty d\omega e^{i\omega\tau}\lambda(\omega)g^2(\omega)\bar{n}(\omega),$$

where $\bar{n}(\omega) = 1/(e^{\hbar\omega/k_{\mathrm{B}}T} - 1)$. For simplicity, we will assume that the spectral density increases linearly with frequency, $\lambda(\omega)g^2(\omega) = C\omega$. Therefore,

$$\mathrm{Tr}_{\mathrm{B}}[\rho_{\mathrm{B}}\Gamma^\dagger(t)\Gamma(t-\tau)] = C\int_0^\infty d\omega e^{i\omega\tau}\frac{\omega}{e^{\hbar\omega/k_{\mathrm{B}}T} - 1} = \frac{C}{t_{\mathrm{R}}^2}\zeta(2, 1 - i\tau/t_{\mathrm{R}}), \tag{8.50}$$

where $\zeta(a, b)$ is the Hurwitz zeta function and $t_{\mathrm{R}} = \hbar/k_{\mathrm{B}}T$ is the characteristic decay time of the correlation functions. Figure 8.4b shows a plot of the real part and imaginary part of (8.50). From the plot, we can see that the correlation functions of the

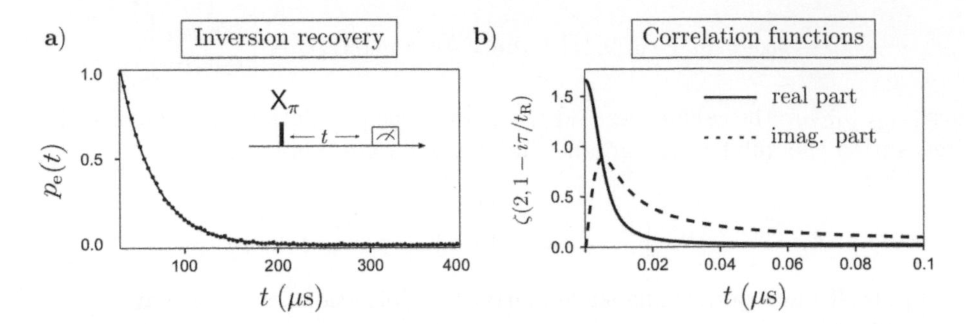

Fig. 8.4 Inversion recovery and correlation functions. a) Probability of measuring a qubit in the excited state $p_e(t) = \rho_{11}(t)$ in an inversion recovery experiment. As shown in the inset, this experiment comprises a short π pulse, followed by a measurement delayed by a time t. The experiment is executed for different delays. Each run is repeated multiple times to acquire some statistics. The longitudinal relaxation time is extracted from an exponential fit. **b)** Plot of real and imaginary part of the correlation function $\mathrm{Tr}_B[\rho_B\Gamma^\dagger(t)\Gamma(t-\tau)]$, eqn (8.50). Here, we are assuming that the temperature of the environment is $T = 10$ mK and the spectral density increases linearly as a function of frequency.

environment decay to zero in a time $t_R = \hbar/k_BT \approx 0.76$ ns. Since the typical relaxation time of a superconducting qubit is $T_1 = 10 - 300$ μs, the Markov approximation $t_R \ll T_1$ is valid.

8.4 Relationship between T_1 and the power spectral density

This section aims to explain the relationship between the relaxation time and the power spectral density of a generic noise source. We start our analysis considering a qubit A interacting with an external system B,

$$H = H_A + H_B + \underbrace{A \otimes F}_{V},$$

where $H_A = -\hbar\omega_q\sigma_z/2$ is the qubit Hamiltonian, H_B is the Hamiltonian of system B, while A and F are two Hermitian operators. Hereafter, the operator F will describe a generic **noise source** coupled to the qubit. The noise source could be a fluctuating magnetic field or a fluctuating charge on some electrodes close to the qubit. The operator A is a qubit operator, such as Pauli operator. When $A = \sigma_z$, we say that the noise source is **longitudinally** coupled to the qubit. This type of coupling leads to fluctuations of the qubit frequency and dephasing. When $A = \sigma_x$ or σ_y, the noise source is **transversally** coupled to the qubit. This type of coupling leads to relaxation processes.

Suppose that the composite system is initially in a product state $|i\rangle\langle i| \otimes \rho_B$ where $|i\rangle$ is an eigenstate of the qubit and ρ_B is a stationary state, $[\rho_B, H_B] = 0$. Our goal is to calculate the probability that system A transitions to a different eigenstate $|f\rangle$ per unit time. This analysis will lead to Fermi's golden rule. It is convenient to study the dynamics in the interaction picture where the Hamiltonian takes the form

$$\tilde{V}(t) = e^{-\frac{1}{i\hbar}(H_{\mathrm{A}}+H_{\mathrm{B}})t} V e^{\frac{1}{i\hbar}(H_{\mathrm{A}}+H_{\mathrm{B}})t} = \tilde{A}(t) \otimes \tilde{F}(t). \tag{8.51}$$

As usual

$$\tilde{A}(t) = e^{-\frac{1}{i\hbar}H_{\mathrm{A}}t} A e^{\frac{1}{i\hbar}H_{\mathrm{A}}t}, \qquad \tilde{F}(t) = e^{-\frac{1}{i\hbar}H_{\mathrm{B}}t} F e^{\frac{1}{i\hbar}H_{\mathrm{B}}t}.$$

We assume that the interaction term is small so that we can truncate the expansion of the time evolution operator to first order

$$\tilde{U}(t) = e^{\frac{1}{i\hbar}\int_0^t \tilde{V}(t')dt'} \approx \mathbb{1} + \frac{1}{i\hbar}\int_0^t \tilde{V}(t')dt'. \tag{8.52}$$

The transition rate from $|i\rangle$ to $|f\rangle$ is the transition probability $P_{i\to f}$ per unit time

$$\begin{aligned}
\gamma_{i\to f} &= \frac{1}{t}P_{i\to f} = \frac{1}{t}\mathrm{Tr}_{\mathrm{AB}}[\tilde{U}(t)(|i\rangle\langle i| \otimes \rho_{\mathrm{B}})\tilde{U}^{\dagger}(t)(|f\rangle\langle f| \otimes \mathbb{1})] \\
&\approx \frac{1}{t}\mathrm{Tr}_{\mathrm{AB}}\Big[\big(\mathbb{1} + \frac{1}{i\hbar}\int_0^t \tilde{V}(t')dt'\big)\big(|i\rangle\langle i| \otimes \rho_{\mathrm{B}}\big)\big(\mathbb{1} - \frac{1}{i\hbar}\int_0^t \tilde{V}(t'')dt''\big)(|f\rangle\langle f| \otimes \mathbb{1})\Big],
\end{aligned}$$

where we used the Born rule (6.23). Since $\langle i|f\rangle = \delta_{if}$, the only term that survives in this trace is the product between the two time integrals. Therefore,

$$\begin{aligned}
\gamma_{i\to f} &\approx \frac{1}{t}\mathrm{Tr}_{\mathrm{AB}}\Big[\big(\frac{1}{i\hbar}\int_0^t \tilde{V}(t')dt'\big)\ \big(|i\rangle\langle i| \otimes \rho_{\mathrm{B}}\big)\ \big(-\frac{1}{i\hbar}\int_0^t \tilde{V}(t'')dt''\big)\big(|f\rangle\langle f| \otimes 1\big)\Big] \\
&= \frac{1}{\hbar^2 t}\int_0^t dt' \int_0^t dt''\mathrm{Tr}_{\mathrm{A}}[(\tilde{A}(t')|i\rangle\langle i|\tilde{A}(t''))|f\rangle\langle f|]\ \mathrm{Tr}_{\mathrm{B}}[\tilde{F}(t')\rho_{\mathrm{B}}\tilde{F}(t'')] \\
&= \frac{1}{\hbar^2 t}\int_0^t dt' \int_0^t dt''\langle f|\tilde{A}(t')|i\rangle\langle i|\tilde{A}(t'')|f\rangle \mathrm{Tr}_{\mathrm{B}}[\tilde{F}(t')\rho_{\mathrm{B}}\tilde{F}(t'')] \\
&= \frac{1}{\hbar^2 t}|\langle f|A|i\rangle|^2 \int_0^t dt' \int_0^t dt'' e^{-i(\omega_i-\omega_f)(t'-t'')}\mathrm{Tr}_{\mathrm{B}}[F(t')\rho_{\mathrm{B}}F(t'')],
\end{aligned} \tag{8.53}$$

where we used $H_{\mathrm{A}}|i\rangle = \hbar\omega_i|i\rangle$ and $H_{\mathrm{A}}|f\rangle = \hbar\omega_f|f\rangle$. The trace in the integral

$$\boxed{G_{FF}(t',t'') \equiv \mathrm{Tr}_{\mathrm{B}}[\tilde{F}(t')\rho_{\mathrm{B}}\tilde{F}(t'')]} \tag{8.54}$$

is the **auto-correlation function**[9] of the noise source F. The auto-correlation function quantifies the correlation between values of an observable at two different times. For stationary states, the auto-correlation function only depends on the time difference $t' - t''$. This is because

[9]If F is a classical variable, the trace is replaced by a time average.

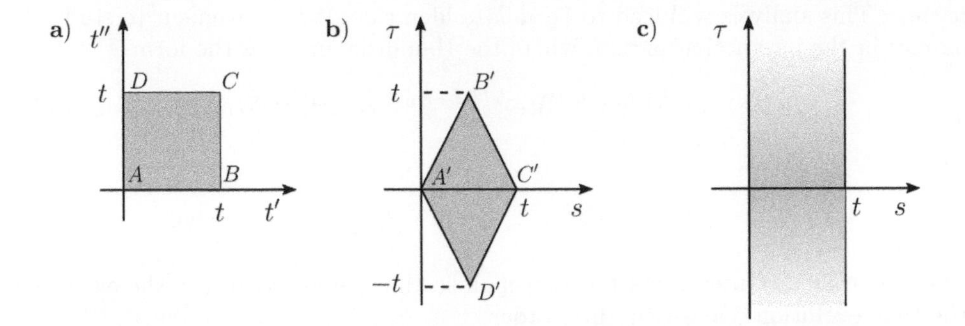

Fig. 8.5 Time integrals. a) The time integrals in eqn (8.55) are defined in the interval $[0,t]\times[0,t]$. **b)** The change of variable $\{t'-t''=\tau,\, s=(t'+t'')/2\}$ modifies the region of integration into a rhombus. **c)** Since the auto-correlation function, $G_{FF}(\tau)=\mathrm{Tr}[F(\tau)\rho_{\mathrm{B}}F(0)]$, usually decays to zero very quickly, the integration in $d\tau$ can be extended to infinity.

$$\begin{aligned} G_{FF}(t',t'') = \mathrm{Tr}_{\mathrm{B}}[\tilde{F}(t')\rho_{\mathrm{B}}\tilde{F}(t'')] &= \mathrm{Tr}_{\mathrm{B}}[e^{-\frac{1}{i\hbar}H_{\mathrm{B}}t'}Fe^{\frac{1}{i\hbar}H_{\mathrm{B}}t'}\rho_{\mathrm{B}}e^{-\frac{1}{i\hbar}H_{\mathrm{B}}t''}Fe^{\frac{1}{i\hbar}H_{\mathrm{B}}t''}] \\ &= \mathrm{Tr}_{\mathrm{B}}[e^{\frac{1}{i\hbar}H_{\mathrm{B}}t''}e^{-\frac{1}{i\hbar}H_{\mathrm{B}}t'}Fe^{\frac{1}{i\hbar}H_{\mathrm{B}}t'}e^{-\frac{1}{i\hbar}H_{\mathrm{B}}t''}\rho_{\mathrm{B}}F] \\ &= \mathrm{Tr}_{\mathrm{B}}[\tilde{F}(t'-t'')\rho_{\mathrm{B}}\tilde{F}(0)] = G_{FF}(t'-t''), \end{aligned}$$

where we used the cyclic property of the trace and the stationarity condition $[\rho_{\mathrm{B}}, H_{\mathrm{B}}] = 0$. Inserting this expression into (8.53) yields

$$\gamma_{i\to f} \approx \frac{1}{\hbar^2 t}|\langle f|A|i\rangle|^2 \int_0^t dt' \int_0^t dt'' e^{-i(\omega_i-\omega_f)(t'-t'')}G_{FF}(t'-t''). \tag{8.55}$$

The final step is to perform a change of variable to simplify the integration: $\{t'-t''=\tau$, $s=(t'+t'')/2\}$. This change of variable, illustrated in Fig. 8.5 on a Cartesian plane, leads to

$$\gamma_{i\to f} \approx \frac{1}{\hbar^2 t}|\langle f|A|i\rangle|^2 \int_0^t ds \int_{-B(s)}^{+B(s)} d\tau e^{-i(\omega_i-\omega_f)\tau}G_{FF}(\tau), \tag{8.56}$$

where the second integration is performed between two lines

$$B(s) = \begin{cases} 2s & 0 \le s \le t/2 \\ -2s+2t & t/2 \le s \le t. \end{cases}$$

Suppose that the auto-correlation function of system B decays to zero in a very short time interval t_{F}. Thus, for $t \gg t_{\mathrm{F}}$, it is reasonable to approximate the integral (8.56) by taking the limit $B(s)\to\infty$. This makes the first integral in (8.56) trivial, $\int_0^t ds = t$. With this approximation, we finally obtain

$$\boxed{\gamma_{i\to f} \approx \frac{1}{\hbar^2}|\langle f|A|i\rangle|^2 S_{FF}(\omega_i - \omega_f),} \tag{8.57}$$

where we defined the **power spectral density** (PSD) $S_{FF}(\omega)$ as the Fourier transform of the auto-correlation function $G_{FF}(\tau)$,

$$\boxed{S_{FF}(\omega) = \int_{-\infty}^{+\infty} d\tau e^{-i\omega\tau} G_{FF}(\tau).} \tag{8.58}$$

The power spectral density characterizes the frequency distribution of the noise power for a stationary noise source. The unit of measure of S_{FF} is $[F]^2/[\mathrm{Hz}]$ where $[F]$ is the unit of measure of F. Using the decomposition of ρ_{B} in the energy eigenbasis and the completeness relation $\mathbb{1} = \sum_l |l\rangle\langle l|$, the power spectral density can be expressed in the form

$$\begin{aligned} S_{FF}(\omega) &= \sum_{k,l} \rho_k \int_{-\infty}^{\infty} d\tau e^{-i\omega\tau} \langle k|F|l\rangle\langle l|F(\tau)|k\rangle \\ &= \sum_{k,l} \rho_k \int_{-\infty}^{\infty} d\tau e^{i(\Omega_l - \Omega_k - \omega)\tau} |\langle k|F|l\rangle|^2 \\ &= 2\pi \sum_{k,l} \rho_k |\langle k|F|l\rangle|^2 \delta(\Omega_l - \Omega_k - \omega), \end{aligned} \tag{8.59}$$

where we used $H_{\mathrm{B}}|k\rangle = \hbar\Omega_k|k\rangle$ and $H_{\mathrm{B}}|l\rangle = \hbar\Omega_l|l\rangle$. Inserting this expression into (8.57), we retrieve the familiar **Fermi's golden rule**,

$$\boxed{\gamma_{i\to f} \approx \frac{2\pi}{\hbar^2} \sum_{k,l} \rho_k |\langle k|F|l\rangle|^2 |\langle f|A|i\rangle|^2 \delta((\Omega_l - \Omega_k) - (\omega_i - \omega_f)).} \tag{8.60}$$

This equation indicates that the decay rate from state $|i\rangle$ to state $|f\rangle$ is given by the sum of all possible transitions that obey the conservation of energy. In other words, the probability of transitioning from $|i, l\rangle$ to $|f, k\rangle$ is non-zero only when $\omega_i - \omega_f = \Omega_l - \Omega_k$. In passing, we note that eqn (8.59) shows that the power spectral density is always positive, in contrast with the auto-correlation function that might be negative.

Let us consider a concrete example. Suppose that the interaction between a qubit and a noise source F is of the form $V = \hbar g \sigma_x \otimes F$ where $g \in \mathbb{R}$ is the coupling strength. Since the noise source is coupled to the $\hat{\sigma}_x$ operator, the coupling is **transversal**. What is the decay rate of the qubit from the excited state to the ground state? The decay rate can be calculated using eqn (8.57) with the substitutions $|i\rangle = |1\rangle$, $|f\rangle = |0\rangle$, $A = \hbar g \sigma_x$ and $\omega_i - \omega_f = \omega_{\mathrm{q}}$. Using these relations, we obtain

$$\boxed{\gamma_\downarrow = \gamma_{1\to 0} = g^2 S_{FF}(\omega_{\mathrm{q}}),} \tag{8.61}$$

where we used $\langle 0|\sigma_x|1\rangle = 1$. Therefore, the relaxation rate is mainly determined by the

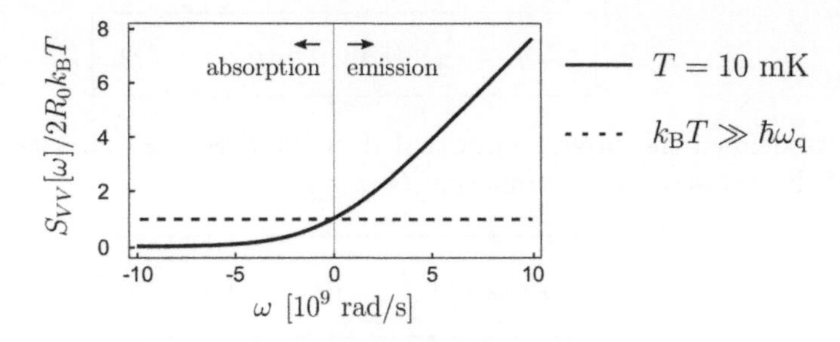

Fig. 8.6 Power spectral density. Power spectral density of the voltage across a resistor at low temperature (solid curve) and high temperature (dashed curve). At high temperatures, the PSD is frequency-independent. At low temperatures, the PSD approaches zero for $\omega < 0$ and increases linearly for $\omega > 0$.

value of the power spectral density at the qubit frequency[10]. Similarly, the probability that the qubit transitions from the ground state to the excited state per unit time is

$$\boxed{\gamma_\uparrow = \gamma_{0\to 1} = g^2 S_{FF}(-\omega_q).} \tag{8.62}$$

This shows that for $\omega > 0$, the power spectral density describes transitions where the qubit emits energy into the environment. In contrast, for $\omega < 0$, it describes transitions where the qubit absorbs energy from the environment. At a finite temperature T the qubit is in a thermal state and the ratio between the population of the excited state and the ground state is $P_e/P_g = e^{-\hbar\omega_q/k_B T}$ (see Section 6.5.2). Hence, at thermal equilibrium the ratio between the power spectral density at $-\omega_q$ and ω_q is given by

$$\frac{S_{FF}(-\omega_q)}{S_{FF}(+\omega_q)} = \frac{\gamma_\uparrow}{\gamma_\downarrow} = \frac{P_e}{P_g} = e^{-\hbar\omega_q/k_B T}. \tag{8.63}$$

This means that the values of $S_{FF}(-\omega_q)$ and $S_{FF}(\omega_q)$ need not be equal, and in fact for low temperatures $S_{FF}(-\omega_q)$ tends to zero. This is consistent with the fact that at low temperatures the environment is in a vacuum state and cannot transfer energy to the qubit.

Example 8.4 An interesting noise source is the fluctuating voltage across a resistor due to the thermal agitation of the charge carriers in the conductor. Even if the applied voltage is zero, the voltage across the resistor varies in time and the amplitude of the fluctuations increases with temperature. This type of noise is called the **Johnson–Nyquist noise**. The power spectral density for the voltage V is given by

[10] The qubit linewidth $\delta\omega_q$ is usually narrow and the power spectral density does not significantly vary in the frequency band $[\omega_q - \delta\omega_q, \omega_q + \delta\omega_q]$. The noise is well behaved and the population of the excited state decays exponentially. There are only a few exceptions to this behavior. For example, in the presence of hot quasiparticles the decay of a superconducting qubit from the excited state to the ground state might be non-exponential [229].

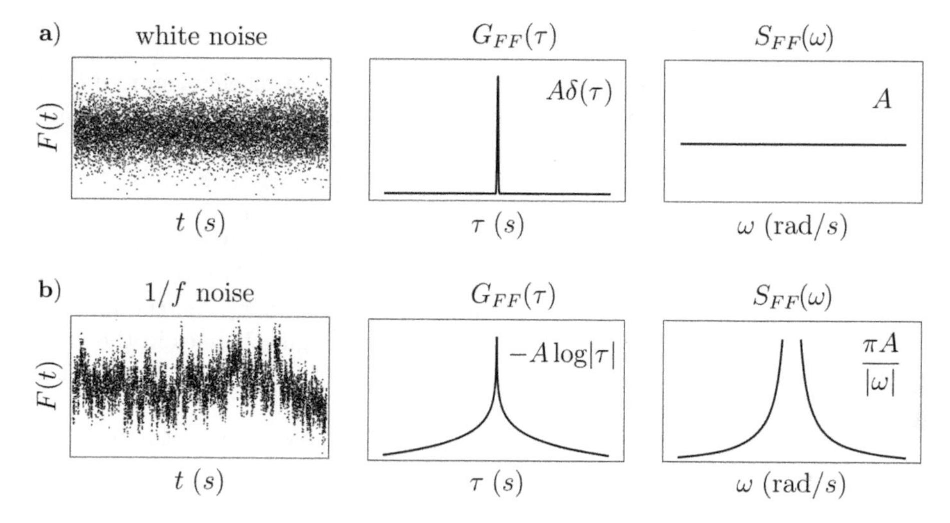

Fig. 8.7 White-noise and $1/f$ noise. a) The quantity $F(t)$ randomly fluctuates in time. The fluctuations do not show any correlation. The auto-correlation function is a Dirac delta, while the power spectral density is a constant. **b)** $1/f$ noise describes processes that slowly fluctuate in time. The power spectral density has a distinctive $1/f$ dependence.

$$S_{VV}(\omega) = \frac{2Z_0\hbar\omega}{1 - e^{-\hbar\omega/k_\mathrm{B}T}}, \tag{8.64}$$

where Z_0 is the resistance. This equation is plotted in Fig. 8.6. Its derivation can be found in Refs [230, 231] and in the original paper by Caldeira and Leggett [232]. Equation (8.64) holds for more complex electrical circuits as long as the quantity Z_0 is replaced by $\mathrm{Re}[Z]$ where Z is the impedance of the circuit. We will use this result in Section 14.6.2 when we calculate the relaxation time of a superconducting qubit due to the capacitive coupling with a control line.

Example 8.5 Consider a noise source F and suppose that $F(t')$ and $F(t'')$ are completely uncorrelated when $t' \neq t''$. This type of noise is called the **white noise.** Since the fluctuations are not correlated, the auto-correlation function $G_{FF}(\tau) = \mathrm{Tr}_\mathrm{B}[F(\tau)\rho_\mathrm{B}F(0)]$ is a Dirac delta. The Fourier transform of a Dirac delta is a constant, meaning that the power spectral density is frequency-independent, as shown in Fig. 8.7a.

Example 8.6 Suppose that the fluctuations of the quantity F are characterized by **pink noise**. The auto-correlation function $G_{FF}(\tau) = \mathrm{Tr}_\mathrm{B}[F(\tau)\rho_\mathrm{B}F(0)]$ decays logarithmically, while the power spectral density has a $1/|\omega|$ dependence (see Fig. 8.7b). Pink noise is often called **$1/f$ noise** due to the analytical form of the power spectral density. In the field of superconducting devices, charge noise and flux noise are characterized by $1/f$ noise.

8.5 Microscopic derivation of pure dephasing

In addition to amplitude damping, there is another important mechanism that leads to decoherence. **Pure dephasing** is a dynamical process that suppresses the off-diagonal elements of the qubit density matrix but, unlike amplitude damping, does not affect the diagonal elements. This mechanism is caused by variations of the qubit frequency due to the coupling with external noise sources and does not involve any energy exchange. We study this process for a qubit coupled to a bath of harmonic oscillators. We assume that the bath is at a temperature $T \ll \hbar\omega_{\mathrm{q}}/k_{\mathrm{B}}$. We also assume that the qubit frequency depends on the position of the oscillators $\hat{x}_k$, i.e. the interaction Hamiltonian is of the form[11] $\hat{\sigma}_z \otimes \hat{x}_k \propto \hat{\sigma}_z \otimes (\hat{b}_k + \hat{b}_k^\dagger)$. Slight changes in the position of the oscillators lead to a small variation of the qubit frequency, thereby dephasing the qubit state.

Let us start our analysis from the Hamiltonian of the composite system

$$H = \underbrace{-\frac{\hbar\omega_{\mathrm{q}}}{2}\sigma_z}_{\text{qubit}} + \underbrace{\sum_{k=1}^{\infty} \hbar\omega_k b_k^\dagger b_k}_{\text{environment}} + \underbrace{\sum_{k=1}^{\infty} \hbar g_k \sigma_z \otimes (b_k + b_k^\dagger)}_{\text{interaction}}. \tag{8.65}$$

Since the oscillators are coupled to the $\hat{\sigma}_z$ operator, we say that the coupling to the environment is **longitudinal**. Suppose that at $t = 0$, the composite system is in a product state $\rho_{\mathrm{tot}}(0) = \rho \otimes |\underline{0}\rangle\langle\underline{0}|$, where $|\underline{0}\rangle = |0\rangle|0\rangle|0\rangle \ldots$ is the vacuum state of an infinite number of harmonic oscillators. Even if the number of oscillators is infinite, this model can be solved without mathematical approximations. In the interaction picture, the Hamiltonian becomes

$$\tilde{H}(t) = \sum_{k=1}^{\infty} \hbar g_k \sigma_z \otimes (\tilde{b}_k(t) + \tilde{b}_k^\dagger(t)), \tag{8.66}$$

where as usual $\tilde{b}_k = b_k e^{-i\omega_k t}$ and $\tilde{b}_k^\dagger = b_k^\dagger e^{i\omega_k t}$. Since $\tilde{H}$ is time-dependent, the time evolution operator of the global system is given by

$$\tilde{U}(t) = \mathcal{T} e^{\frac{1}{i\hbar}\int_0^t \tilde{H}(t')dt'}, \tag{8.67}$$

where $\mathcal{T}$ is the time-ordering operator. Since the commutator of the interaction Hamiltonian $\tilde{H}(t)$ at two different times is proportional to the identity, we can write $\tilde{U}(t)$ in a simpler form (see Exercise 8.6),

$$\tilde{U}(t) = \underbrace{e^{\frac{1}{i\hbar}\int_0^t \tilde{H}(t_1)dt_1}}_{\tilde{V}(t)} e^{i\phi(t)} = \tilde{V}(t)e^{i\phi(t)}.$$

The phase $\phi(t)$ is irrelevant because the state of the composite system at time t is given by

[11] Other forms of the interaction term between the qubit and the environment would lead to the same conclusions, see for example Ref. [228].

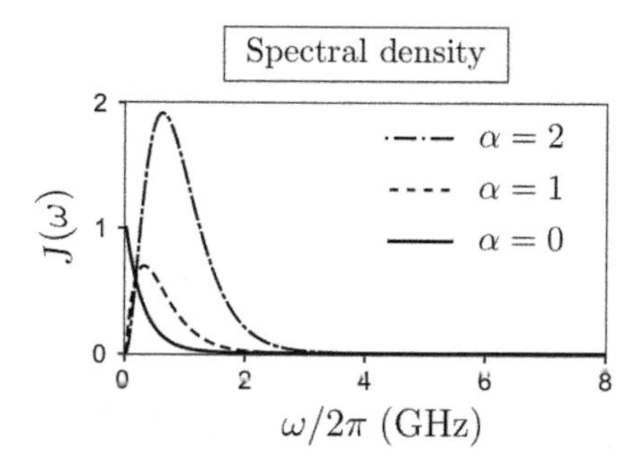

Fig. 8.8 Spectral density. Plot of the spectral density $J(\omega) = A\omega^\alpha e^{-\omega/\Omega}$. In this plot, we have set $A = 1$ and assumed that the cutoff frequency is $\Omega/2\pi = 0.3$ GHz.

$$\tilde{\rho}_{\text{tot}}(t) = \tilde{U}(t)\rho_{\text{tot}}(0)\tilde{U}^\dagger(t) = \tilde{V}(t)\rho_{\text{tot}}(0)\tilde{V}^\dagger(t).$$

Let us calculate the explicit form of the operator $\tilde{V}(t) = \exp(\frac{1}{i\hbar}\int_0^t \tilde{H}(t_1)dt_1)$. The exponent can be expressed as

$$\begin{aligned}
\frac{1}{i\hbar}\int_0^t \tilde{H}(t_1)dt_1 &= -i\sum_{k=1}^{\infty} g_k \int_0^t \sigma_z \otimes (b_k e^{-i\omega_k t_1} + b_k^\dagger e^{i\omega_k t_1})dt_1 \\
&= \sum_{k=1}^{\infty} g_k \sigma_z \otimes \left(\frac{1-e^{i\omega_k t}}{\omega_k} b_k^\dagger - \frac{1-e^{-i\omega_k t}}{\omega_k} b_k\right) \\
&= \sum_k \sigma_z \otimes \left(\alpha_k(t) b_k^\dagger - \alpha_k(t)^* b_k\right),
\end{aligned}$$

where in the last step we defined the parameters $\alpha_k(t) = g_k(1 - e^{i\omega_k t})/\omega_k$. Therefore,

$$\tilde{V}(t) = e^{\sigma_z \otimes \sum_k \alpha_k b_k^\dagger - \alpha_k^* b_k}.$$

This time evolution operator is very similar to the displacement operator presented in Section 4.79. When the qubit is initially prepared in $|0\rangle$, the state of the qubit and the oscillators evolves as

$$\tilde{V}(t)|0\rangle \otimes |\underline{0}\rangle = |0\rangle \otimes \underbrace{e^{\sum_k \alpha_k(t) b_k^\dagger - \alpha_k^*(t) b_k}}_{D_0(\vec{\alpha}_t)} |\underline{0}\rangle = |0\rangle \otimes |\alpha_1(t), \alpha_2(t), \ldots\rangle,$$

where we introduced the displacement operator $D_0(\vec{\alpha}_t)$ and we used eqn (4.77). If the qubit is initially in $|1\rangle$, we obtain

$$V(t)|1\rangle \otimes |\underline{0}\rangle = |1\rangle \otimes \underbrace{e^{-\sum_k \left(\alpha_k(t) b_k^\dagger - \alpha_k(t)^* b_k\right)}}_{D_1(\vec{\alpha}_t)} |\underline{0}\rangle = |1\rangle \otimes | -\alpha_1(t), -\alpha_2(t), \ldots\rangle.$$

The operator $D_1(\vec{\alpha}_t)$ is a product of an infinite number of displacement operators, one for each oscillator. In brief, if the initial state of the qubit is $|x\rangle$ with $x \in \mathbb{B}$, the state of the k-th harmonic oscillator at time t is the coherent state $|(-1)^x \alpha_k(t)\rangle$ where $\alpha_k(t) = g_k(1 - e^{i\omega_k t})/\omega_k$.

Suppose that the initial state of the qubit is an arbitrary state $\rho(0) = \sum_{jk} \rho_{jk}|j\rangle\langle k|$. The qubit state at time t can be calculated by tracing over the degrees of freedom of the environment,

$$\begin{aligned}\tilde{\rho}(t) &= \mathrm{Tr}_\mathrm{B}[\tilde{V}(t)\rho(0) \otimes |\underline{0}\rangle\langle\underline{0}|\tilde{V}^\dagger(t)] \\ &= \sum_{jk} \rho_{jk}|j\rangle\langle k|\mathrm{Tr}_\mathrm{B}[D_j(\vec{\alpha}_t)|\underline{0}\rangle\langle\underline{0}|D_k^\dagger(\vec{\alpha}_t)]. \end{aligned} \tag{8.68}$$

It is easy to see that dephasing does not change the qubit populations (see Exercise 8.7 for a detailed derivation). On the contrary, the coherences decay in time. This can be appreciated by calculating the matrix element $\tilde{\rho}_{01}$ as a function of time,

$$\begin{aligned}\tilde{\rho}_{01}(t) &= \rho_{01}(0)\mathrm{Tr}_\mathrm{B}[D_0(\vec{\alpha}_t)|\underline{0}\rangle\langle\underline{0}|D_1^\dagger(\vec{\alpha}_t)] \\ &= \rho_{01}(0)\mathrm{Tr}_\mathrm{B}[D(\vec{\alpha}_t)|\underline{0}\rangle\langle\underline{0}|D^\dagger(-\vec{\alpha}_t)] \\ &= \rho_{01}(0)\mathrm{Tr}_\mathrm{B}[D(2\vec{\alpha}_t)|\underline{0}\rangle\langle\underline{0}|] \;=\; \rho_{01}(0)\mathrm{Tr}_\mathrm{B}[|2\vec{\alpha}_t\rangle\langle\underline{0}|] \\ &= \rho_{01}(0)\ \langle 0|2\alpha_1(t)\rangle \cdot \langle 0|2\alpha_2(t)\rangle \cdot \langle 0|2\alpha_3(t)\rangle \ldots \\ &= \rho_{01}(0)\prod_{k=1}^{\infty} e^{-|2\alpha_k(t)|^2/2} = \rho_{01}(0)e^{-\sum_k |2\alpha_k(t)|^2/2}, \end{aligned} \tag{8.69}$$

where we used $\mathrm{Tr}[|a\rangle\langle b|] = \langle b|a\rangle$, $D^\dagger(-\alpha) = D(\alpha)$ and $\langle 0|\alpha\rangle = e^{-|\alpha|^2/2}$. Let us calculate the exponent of (8.69),

$$\sum_{k=1}^{\infty} \frac{|2\alpha_k|^2}{2} = \sum_{k=1}^{\infty} 2g_k^2 \frac{(1 - e^{i\omega_k t})}{\omega_k} \frac{(1 - e^{-i\omega_k t})}{\omega_k} = \sum_{k=1}^{\infty} 4g_k^2 \frac{1 - \cos\omega_k t}{\omega_k^2}.$$

We now assume that the oscillator frequencies are so dense that the sum can be replaced by an integral. This allows us to replace the parameters g_k with a continuous function $g(\omega)$ such that

$$\sum_{k=1}^{\infty} 4g_k^2 \frac{1 - \cos\omega_k t}{\omega_k^2} \quad \mapsto \quad \int_0^\infty \underbrace{4\lambda(\omega)g^2(\omega)}_{J(\omega)} \frac{1 - \cos\omega t}{\omega^2} d\omega = \int_0^\infty J(\omega)\frac{1 - \cos\omega t}{\omega^2} d\omega, \tag{8.70}$$

where $\lambda(\omega)$ is the density of oscillators at a particular frequency and $J(\omega)$ is the spectral density. This function contains information on the frequencies of the modes and their coupling to the system and is sufficient to characterize the thermal bath. The analytical form of $J(\omega)$ is usually parametrized as

$$\boxed{J(\omega) = A\omega^\alpha e^{-\omega/\Omega},} \tag{8.71}$$

where A is a constant, whereas Ω [rad/s] is a cutoff frequency above which the coupling with the oscillators decreases. The real parameter α is a characteristic feature of the bath: when $\alpha < 1$, $\alpha = 1$, $\alpha > 1$ we say that the bath has subohmic, ohmic, and superohmic behavior respectively. Figure 8.8 shows the spectral density as a function of frequency for several values of α. In this section, we will assume that the spectral density is constant $J(\omega) = A$, i.e. the environment is characterized by white noise. The integral (8.70) becomes

$$A \int_0^\infty \frac{1 - \cos \omega t}{\omega^2} d\omega = \underbrace{\frac{\pi}{2} A}_{\gamma_\phi} t = \gamma_\phi t. \tag{8.72}$$

The parameter γ_ϕ is the **dephasing rate**. Using this result, eqn (8.69) finally turns into

$$\boxed{\tilde{\rho}_{01}(t) = \rho_{01}(0) e^{-\gamma_\phi t}.}$$

The coherences decay with a characteristic time

$$\boxed{T_\phi = 1/\gamma_\phi.}$$

This is called the **dephasing time**. It is important to highlight that if the spectral density is frequency-dependent, the decay may no longer be exponential (see Exercise 8.10).

The main conclusion is that the dynamics governed by the Hamiltonian (8.65) (which contains a longitudinal coupling between the qubit and the environment) lead to a suppression of the off-diagonal elements of the qubit density matrix, while the populations remain unchanged. If the positions of the thermal oscillators are characterized by white noise, the decay is exponential. The same result can be obtained by assuming that the dynamics of the qubit state are governed by a Markovian master equation of the form

$$\boxed{\frac{d\rho}{dt} = i\frac{\omega_q}{2}[\sigma_z, \rho] + \frac{\gamma_\phi}{2}\mathcal{D}[\sigma_z](\rho) = i\frac{\omega_q}{2}[\sigma_z, \rho] + \frac{\gamma_\phi}{2}(\sigma_z \rho \sigma_z - \rho),}$$

where the parameter γ_ϕ is time-independent. This equation of motion leads to an exponential decay of the coherences as can be seen by studying the dynamics in the interaction picture,

$$\frac{d\tilde{\rho}}{dt} = \frac{\gamma_\phi}{2}(\sigma_z \tilde{\rho} \sigma_z - \tilde{\rho}). \tag{8.73}$$

Substituting $\sigma_z = \begin{bmatrix} 1 & 0 \\ 0 & -1 \end{bmatrix}$ and $\rho = \begin{bmatrix} \tilde{\rho}_{00} & \tilde{\rho}_{01} \\ \tilde{\rho}_{10} & \tilde{\rho}_{11} \end{bmatrix}$ into eqn (8.73) and performing some matrix multiplications, we obtain

$$\begin{bmatrix} \dot{\tilde{\rho}}_{00} & \dot{\tilde{\rho}}_{01} \\ \dot{\tilde{\rho}}_{10} & \dot{\tilde{\rho}}_{11} \end{bmatrix} = \begin{bmatrix} 0 & -\gamma_\phi \tilde{\rho}_{01} \\ -\gamma_\phi \tilde{\rho}_{10} & 0 \end{bmatrix}.$$

The solution of the differential equations

$$\frac{d\tilde{\rho}_{00}}{dt} = 0, \qquad \frac{d\tilde{\rho}_{01}}{dt} = -\gamma_\phi \rho_{01}, \tag{8.74}$$

is $\tilde{\rho}_{00}(t) = \tilde{\rho}_{00}(0)$ and $\tilde{\rho}_{01} = \tilde{\rho}_{01}(0)e^{-\gamma_\phi t}$. As expected, the populations do not change, while the coherences decay exponentially with a decay time $T_\phi = 1/\gamma_\phi$.

8.6 Decoherence under free evolution: the Bloch equations

In a real experiment, a qubit is subject to amplitude damping and dephasing at the same time. The combination of these two processes leads to a faster decay of the coherences. Assuming that the noise sources are weakly coupled to the qubit, the combined master equation reads

$$\boxed{\frac{d\rho}{dt} = \frac{1}{T_1}\mathcal{D}[\sigma^-](\rho) + \frac{1}{2T_\phi}\mathcal{D}[\sigma_z](\rho).} \tag{8.75}$$

Here, we neglected the dissipator $\mathcal{D}[\sigma^+]$ because we are assuming that the temperature of the environment is low compared to the qubit frequency. In addition, we ignored the trivial evolution related to the qubit Hamiltonian by working in the interaction picture. The master equation (8.75) leads to[12]

$$\frac{d\rho_{11}}{dt} = -\frac{1}{T_1}\rho_{11}, \tag{8.76}$$

$$\frac{d\rho_{01}}{dt} = -\Big(\frac{1}{T_\phi} + \frac{1}{2T_1}\Big)\rho_{01}. \tag{8.77}$$

The solution is

$$\rho_{11}(t) = \rho_{11}(0)e^{-t/T_1},$$

$$\rho_{01}(t) = \rho_{01}(0)e^{-(\frac{1}{T_\phi}+\frac{1}{2T_1})t}.$$

The population ρ_{11} decays with a characteristic time T_1. On the other hand, the coherences decay with a characteristic time

$$\boxed{T_2 = \Big(\frac{1}{T_\phi} + \frac{1}{2T_1}\Big)^{-1}.} \tag{8.78}$$

This is called the **transverse relaxation time**. This equation shows that T_2 never exceeds $2T_1$. When $T_\phi \gg T_1$, we have $T_2 \approx 2T_1$ and we say that T_2 is T_1-limited.

The transverse relaxation time T_2 can be measured with a Ramsey experiment as illustrated in Fig. 8.9a. This experiment consists of two $\mathsf{Y}_{\frac{\pi}{2}}$ pulses separated by a delay t. This sequence is repeated for different delays. The probability of measuring

[12]These equations can be derived either with an explicit calculation or by combining eqns (8.49, 8.74).

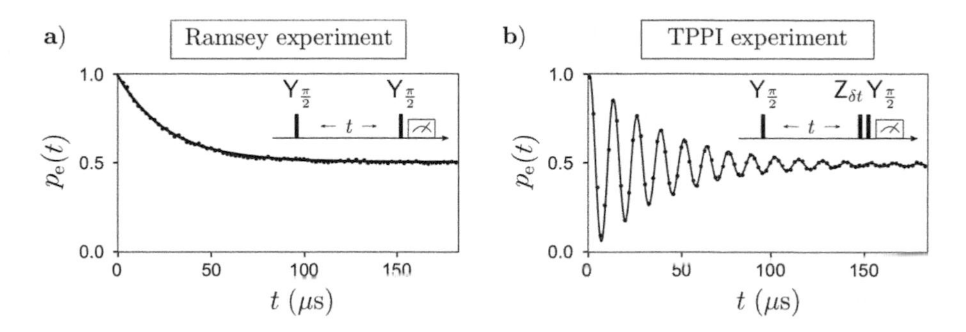

Fig. 8.9 The Ramsey experiment. a) Probability of measuring a qubit in the excited state as a function of the delay time in a Ramsey experiment. As shown in the inset, this experiment is composed of two $\mathsf{Y}_{\frac{\pi}{2}}$ pulses separated by a time interval t. The transverse relaxation time T_2 is extracted from an exponential fit $\frac{1}{2}+\frac{1}{2}e^{-t/T_2}$. **b)** The TPPI protocol is an interferometric experiment comprising two resonant $\mathsf{Y}_{\frac{\pi}{2}}$ pulses separated by a time interval t. The second pulse is preceded by a Z rotation by an angle δt. The parameter δ represents an artificial detuning with respect to the qubit frequency. The value of T_2 is extracted from the fit $\frac{1}{2}+\frac{1}{2}e^{-t/T_2}\cos(\delta t)$. To resolve the oscillations, the sampling rate s [rad/s] must satisfy $s > \delta/2$.

the qubit in the excited state $p_e(t)$ decays in time. The value of T_2 is extracted by fitting $p_e(t)$ to $\frac{1}{2}+\frac{1}{2}e^{-t/T_2}$. For completeness, we mention that the value of T_2 can also be measured with a slightly different interferometric experiment known as the **time-proportional phase incrementation (TPPI) experiment** [233] (see Fig. 8.9b).

When the transverse relaxation time is measured with a Ramsey experiment, the decay constant extracted from the fit is usually indicated with the symbol T_2^* to distinguish it from other relaxation times measured with different pulse sequences, such as the Hahn-echo experiment. The value of T_ϕ can be determined by measuring T_1 and T_2 and then using the formula (8.78).

With some manipulations, eqns (8.76, 8.77) lead to the Bloch equations. The Bloch equations are equations of motion of the Bloch vector $\vec{r} = (r_x, r_y, r_z)$ rather than the elements of the density matrix. To derive them, recall that $\rho_{01} = (r_x - ir_y)/2$ and $\rho_{11} = (1 - r_z)/2$. Therefore,

$$\frac{d\rho_{01}}{dt} = \frac{1}{2}\frac{dr_x}{dt} - \frac{i}{2}\frac{dr_y}{dt} = -\frac{1}{T_2}\Big(\frac{r_x}{2} - i\frac{r_y}{2}\Big),$$
$$\frac{d\rho_{11}}{dt} = -\frac{1}{2}\frac{dr_z}{dt} = -\frac{1}{T_1}\Big(\frac{1}{2} - \frac{r_z}{2}\Big).$$

Equating the real and imaginary parts of these expressions, we obtain the **Bloch equations**:

$$\frac{dr_x}{dt} = -\frac{1}{T_2} r_x, \tag{8.79}$$

$$\frac{dr_y}{dt} = -\frac{1}{T_2} r_y, \tag{8.80}$$

$$\frac{dr_z}{dt} = -\frac{1}{T_1} r_z + \frac{1}{T_1}. \tag{8.81}$$

These equations are equivalent to (8.76, 8.77). The choice of one or the other is only a matter of convenience.

8.7 Relation between T_ϕ and the PSD under free evolution

Pure dephasing is a decoherence process caused by fluctuations of the qubit frequency. To appreciate this point, suppose that a qubit is prepared in a superposition state. The corresponding Bloch vector points along the x axis and remains still in a rotating frame. If external perturbations change the qubit frequency $\omega_q \to \omega_q + \delta\omega_q$, the Bloch vector rotates about the z axis with an angular frequency $\delta\omega_q$. Random fluctuations will result in a different rotation frequency for each experiment. This stochastic process spreads out the qubit state along the equator causing a loss of information about the qubit relative phase.

Let us make this model more concrete by writing the Hamiltonian of a qubit whose frequency changes in time

$$\boxed{\hat{H}_q(t) = -\frac{\hbar(\omega_q + \delta\omega_q(t))}{2}\hat{\sigma}_z,}$$

where ω_q is the qubit frequency and $\delta\omega_q(t)$ are small fluctuations. If the fluctuations are caused by a stationary noise source λ, then to first order

$$\boxed{\delta\omega_q(t) = \frac{\partial\omega_q}{\partial\lambda}\delta\lambda(t) = \frac{1}{\hbar}\frac{\partial H_q}{\partial\lambda}\delta\lambda(t).} \tag{8.82}$$

Here, λ could be a magnetic field or a fluctuating charge close to the electrodes of a superconducting qubit for example. We are interested in the time evolution of the qubit state from zero to t. To simplify our treatment, we will divide this interval into n discrete steps of duration $\Delta t = t/n$. At each step t_i, the random variable $\delta\lambda$ takes a value x_i. We assume that $\delta\lambda$ is a zero-mean random variable with Gaussian distribution. In other words, at a fixed time t_i, the average of $\delta\lambda(t_i)$ over many experiments is zero,

$$\langle\delta\lambda(t_i)\rangle = \int_{-\infty}^{\infty} x_{t_i} p(x_{t_i}) dx_{t_i} = 0,$$

where $p(x_{t_i})$ is a Gaussian with variance σ^2. Figure 8.10 shows two realizations $\mathbf{x}_1$ and $\mathbf{x}_2$ in the time interval $[0, t]$. The probability that a particular realization takes place is $p(\mathbf{x}) = p(\delta\lambda(t_1) = x_1, \ldots, \delta\lambda(t_n) = x_n)$. The fluctuations of λ over many experiments are characterized by the auto-correlation function

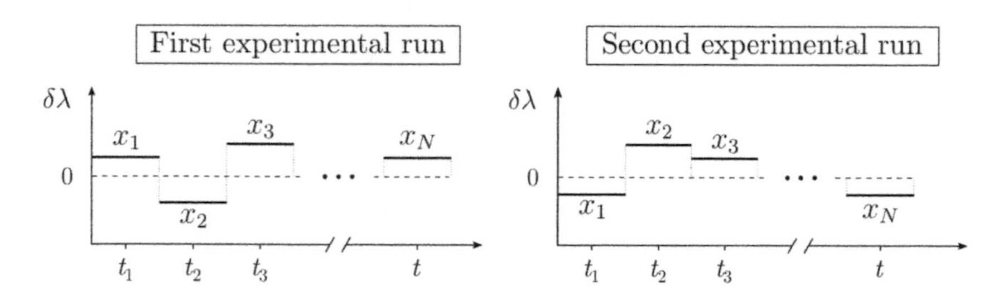

Fig. 8.10 Fluctuations. A noise source λ randomly fluctuates in time. This induces fluctuations of the qubit frequency through eqn (8.82). In our discussion, the time interval $[0, t]$ is discretized in n steps. The vector $\mathbf{x} = (x_1, \ldots, x_n)$ describes a realization of the random variable $\delta\lambda$ in the time interval $[0, t]$. We assume that $\delta\lambda$ at a fixed time t_i has a Gaussian distribution with zero average.

$$\boxed{G_{\lambda\lambda}(t_i - t_j) \equiv \langle \delta\lambda(t_i)\delta\lambda(t_j)\rangle\,.} \tag{8.83}$$

For example, if the fluctuations are completely uncorrelated, the noise source is described by white noise and the auto-correlation function is a Kronecker delta, $G_{\lambda\lambda}(t_i - t_j) = \sigma^2\delta_{ij}$.

Let us assume that the qubit state at the initial time is

$$\rho(0) = \rho_{00}|0\rangle\langle 0| + \rho_{01}|0\rangle\langle 1| + \rho_{10}|1\rangle\langle 0| + \rho_{11}|1\rangle\langle 1| = \begin{bmatrix} \rho_{00} & \rho_{01} \\ \rho_{10} & \rho_{11} \end{bmatrix}.$$

For a particular realization $\mathbf{x}$ (i.e. in a specific experimental run), the time evolution operator from the initial time to the final time t is given by

$$\begin{aligned} \hat{U}_{\mathbf{x}}(t) = \exp\Big[\frac{1}{i\hbar}\sum_{j=1}^{n}\hat{H}_{\mathrm{q},\mathbf{x}}(t_j)\Delta t\Big] &= \exp\Big[\frac{i}{2}\sum_{j=1}^{n}(\omega_{\mathrm{q}} + \delta\omega_{\mathrm{q},\mathbf{x}}(t_j))\Delta t\Big] \\ &= \exp\Big[\frac{i}{2}(\omega_{\mathrm{q}}t + \Phi_{\mathbf{x}}(t))\hat{\sigma}_z\Big], \end{aligned}$$

where we defined

$$\Phi_{\mathbf{x}}(t) = \sum_{j=1}^{n}\delta\omega_{\mathrm{q},\mathbf{x}}(t_j)\Delta t.$$

This quantity is the random phase accumulated due to the small fluctuations of the qubit frequency. The final state of the qubit is

$$\rho_{\mathbf{x}}(t) = \hat{U}_{\mathbf{x}}(t)\rho(0)\hat{U}_{\mathbf{x}}^{\dagger}(t) = e^{\alpha_{\mathbf{x}}\hat{\sigma}_z}\rho(0)e^{-\alpha_{\mathbf{x}}\hat{\sigma}_z},$$

where we introduced the parameter $\alpha_{\mathbf{x}} = \frac{i}{2}(\omega_{\mathrm{q}}t + \Phi_{\mathbf{x}}(t))$ for conciseness. Suppose that multiple experiments are performed. Since the fluctuations are random, $\rho_{\mathbf{x}}(t)$ will be different in each experiment. The final state over many experiments is given by the ensemble average

$$\rho(t) = \int p(\mathbf{x})\rho_{\mathbf{x}}(t)d\mathbf{x} = \int p(\mathbf{x})e^{\alpha_{\mathbf{x}}\hat{\sigma}_z}\rho(0)e^{-\alpha_{\mathbf{x}}\hat{\sigma}_z}d\mathbf{x}$$
$$= \rho_{00}|0\rangle\langle 0| + \rho_{11}|1\rangle\langle 1| + \int p(\mathbf{x})e^{\alpha_{\mathbf{x}}\hat{\sigma}_z}\Big(\rho_{01}|0\rangle\langle 1| + \rho_{10}|1\rangle\langle 0|\Big)e^{-\alpha_{\mathbf{x}}\hat{\sigma}_z}\,d\mathbf{x}.$$

Using the Baker–Campbell–Hausdorff formula, we obtain[13]

$$\rho(t) = \rho_{00}|0\rangle\langle 0| + \rho_{11}|1\rangle\langle 1|$$
$$+\ \rho_{01}e^{i\omega_{\mathrm{q}}t}\Big(\int e^{i\Phi_{\mathbf{x}}(t)}\,p(\mathbf{x})d\mathbf{x}\Big)\,|0\rangle\langle 1| + \rho_{10}e^{-i\omega_{\mathrm{q}}t}\Big(\int e^{-i\Phi_{\mathbf{x}}(t)}\,p(\mathbf{x})d\mathbf{x}\Big)\,|1\rangle\langle 0|$$
$$= \begin{bmatrix} \rho_{00} & \rho_{01}e^{i\omega_{\mathrm{q}}t}\left\langle e^{i\Phi(t)}\right\rangle \\ \rho_{10}e^{-i\omega_{\mathrm{q}}t}\left\langle e^{-i\Phi(t)}\right\rangle & \rho_{11} \end{bmatrix}. \tag{8.84}$$

As usual, the brackets $\langle \cdot \rangle = \int \cdot\, p(\mathbf{x})d\mathbf{x}$ indicate an average over many experiments. Equation (8.84) indicates that random fluctuations of the qubit frequency leave the diagonal elements of the density matrix unchanged, while the off-diagonal elements vary in time. Since the random variable $\delta\omega_{\mathrm{q}}(t)$ has a Gaussian distribution, $\Phi(t)$ has a Gaussian distribution too. As shown in Exercise 8.8, for a generic random variable Z with Gaussian distribution it holds that $\left\langle e^{iZ}\right\rangle = e^{-\left\langle Z^2\right\rangle/2}$. Therefore, the off-diagonal element $\rho_{01}(t)$ is given by

$$\boxed{\rho_{01}(t) = \rho_{01}e^{i\omega_{\mathrm{q}}t}e^{-\left\langle \Phi_{\mathbf{x}}^2(t)\right\rangle/2} = \rho_{01}e^{i\omega_{\mathrm{q}}t}e^{-\gamma(t)},}$$

where in the last step we defined the **dephasing function**,

$$\boxed{\gamma_\phi(t) = \left\langle \frac{1}{2}\Phi^2(t)\right\rangle.}$$

This function can be written in a more explicit form as

$$\gamma_\phi(t) = \frac{1}{2}\langle\Phi^2(t)\rangle = \frac{1}{2}\sum_{i=1}^{n}\sum_{j=1}^{n}\langle\delta\omega_{\mathrm{q}}(t_i)\delta\omega_{\mathrm{q}}(t_j)\rangle\Delta t\Delta t$$
$$= \frac{1}{2}\left(\frac{\partial\omega_{\mathrm{q}}}{\partial\lambda}\right)^2\sum_{i=1}^{n}\sum_{j=1}^{n}\langle\delta\lambda(t_i)\delta\lambda(t_j)\rangle\Delta t\Delta t$$
$$= \frac{1}{2}\left(\frac{\partial\omega_{\mathrm{q}}}{\partial\lambda}\right)^2\sum_{i=1}^{n}\sum_{j=1}^{n}G_{\lambda\lambda}(t_i - t_j)\Delta t\Delta t, \tag{8.85}$$

where we used the definition of the auto-correlation function (8.83). If we now assume that the duration of each step $\Delta t = T/n$ is infinitesimal (i.e. $n \to \infty$), we can convert the sums in (8.85) into integrals,

[13]Here, we are using the identities $e^{\alpha_{\mathbf{x}}\hat{\sigma}_z}|0\rangle\langle 1|e^{-\alpha_{\mathbf{x}}\hat{\sigma}_z} = |0\rangle\langle 1|e^{2\alpha_{\mathbf{x}}}$ and $e^{\alpha_{\mathbf{x}}\hat{\sigma}_z}|1\rangle\langle 0|e^{-\alpha_{\mathbf{x}}\hat{\sigma}_z} = |1\rangle\langle 0|e^{-2\alpha_{\mathbf{x}}}$.

$$\gamma_\phi(t) = \frac{1}{2}\langle\Phi^2(t)\rangle = \frac{1}{2}\left(\frac{\partial\omega_{\rm q}}{\partial\lambda}\right)^2 \int_0^t\int_0^t G_{\lambda\lambda}(t'-t'')dt'dt''.$$

Recall that the auto-correlation function is the Fourier transform of the power spectral density $S_{\lambda\lambda}(\omega)$,

$$\begin{aligned}\gamma_\phi(t) &= \frac{1}{2}\left(\frac{\partial\omega_{\rm q}}{\partial\lambda}\right)^2 \int_0^t\int_0^t \left(\frac{1}{2\pi}\int_{-\infty}^{\infty} e^{i\omega(t'-t'')}S_{\lambda\lambda}(\omega)d\omega\right) dt'dt'' \\ &= \frac{1}{4\pi}\left(\frac{\partial\omega_{\rm q}}{\partial\lambda}\right)^2 \int_{-\infty}^{\infty} S_{\lambda\lambda}(\omega)\left[\int_0^t e^{i\omega t'}dt' \int_0^t e^{-i\omega t''}dt''\right]d\omega.\end{aligned}$$

Computing the integrals in the square brackets, the dephasing function becomes

$$\boxed{\gamma_\phi(t) = \frac{t^2}{4\pi}\left(\frac{\partial\omega_{\rm q}}{\partial\lambda}\right)^2 \int_{-\infty}^{\infty} S_{\lambda\lambda}(\omega)\,\mathrm{sinc}^2\left(\frac{\omega t}{2}\right)d\omega,} \tag{8.86}$$

where $\mathrm{sinc}(x) = \sin x/x$. The dephasing function is a weighted integral of the power spectral density. Equation (8.86) shows that under free evolution the weight function, also called the **filter function**, is given by

$$g_0(\omega, t) = \mathrm{sinc}^2(\omega t/2).$$

This function filters the power spectral density around the origin such that only the value of the power spectral density around $\omega \approx 0$ determines the dephasing rate (see Fig. 8.11). As we shall see in the following sections, specific pulse sequences applied to the qubit during the free evolution might change the filter function.

To summarize, under free evolution, random fluctuations of the qubit frequency transform the initial state of the qubit

$$\begin{bmatrix}\rho_{00} & \rho_{01}\\ \rho_{10} & \rho_{11}\end{bmatrix} \mapsto \begin{bmatrix}\rho_{00} & \rho_{01}e^{i\omega_{\rm q}t}e^{-\gamma_\phi(t)}\\ \rho_{10}e^{-i\omega_{\rm q}t}e^{-\gamma_\phi(t)} & \rho_{11}\end{bmatrix},$$

where $\gamma_\phi(t)$ is given by (8.86). Clearly, in absence of fluctuations $\gamma_\phi(t) = 0$, and we obtain the trivial dynamics $\rho_{01} \mapsto \rho_{01}e^{i\omega_{\rm q}t}$.

Example 8.7 Let us suppose that the qubit frequency is coupled to a noise source λ,

$$\hat{H} = -\frac{\hbar\omega_{\rm q}}{2}\hat{\sigma}_z + \beta\lambda\hat{\sigma}_z,$$

where β is a coupling constant. If the fluctuations of λ are characterized by white noise, the power spectral density is frequency-independent, $S_{\lambda\lambda}(\omega) = A^2$. Let us assume that the qubit is initially prepared in the superposition state $|+\rangle$. Pure dephasing will transform this state into a classical state after a sufficient amount of time. This can be shown by calculating the explicit form of the dephasing function (8.86),

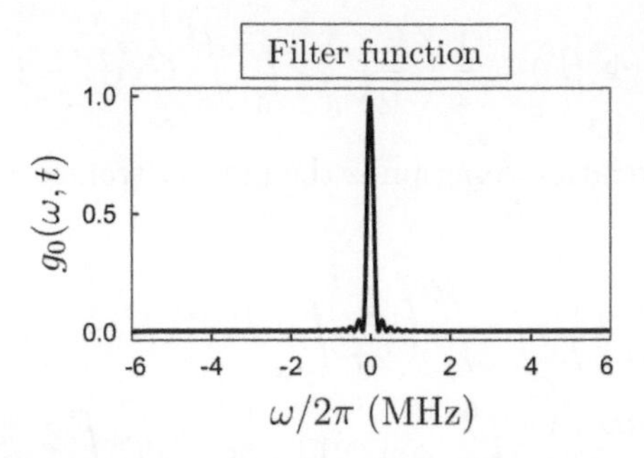

Fig. 8.11 Filter function. Plot of the filter function $g_0(\omega, t) = \text{sinc}^2(\omega t/2)$ for a Ramsey experiment of duration $t = 5\ \mu s$. The filter function has a pronounced peak in the origin and samples the power spectral density at a low frequency. Thus, a Ramsey experiment is most sensitive to low-frequency noise.

$$\gamma_\phi(t) = \frac{t^2}{4\pi\hbar^2}\Big(\frac{\partial H_{\rm q}}{\partial\lambda}\Big)^2 \int_{-\infty}^{\infty} A^2\, \text{sinc}^2\Big(\frac{\omega t}{2}\Big) d\omega = \frac{t^2}{4\pi\hbar^2}\Big(\frac{\partial H_{\rm q}}{\partial\lambda}\Big)^2 \frac{2\pi A^2}{t} = \frac{t}{T_\phi},$$

where we introduced the dephasing time

$$T_\phi = \frac{2\hbar^2}{A^2} \frac{1}{\big(\frac{\partial H_{\rm q}}{\partial\lambda}\big)^2} = \frac{2\hbar^2}{A^2\beta^2}.$$

The initial qubit state is mapped into

$$\rho(0) = |+\rangle\langle+| = \frac{1}{2}\begin{bmatrix} 1 & 1 \\ 1 & 1 \end{bmatrix} \qquad \mapsto \qquad \tilde{\rho}(t) = \frac{1}{2}\begin{bmatrix} 1 & e^{-t/T_\phi} \\ e^{-t/T_\phi} & 1 \end{bmatrix}.$$

In the presence of white noise the coherences decay exponentially. The dephasing time T_ϕ can be measured with a Ramsey experiment (see Fig. 8.12a). Once the dephasing time has been determined, the PSD can be calculated using the expression $S_{FF}(\omega) = A^2 = 2\hbar^2/\beta^2 T_\phi$.

Example 8.8 Suppose that the fluctuations of the noise source are characterized by $1/f$ noise, $S_{\lambda\lambda}(\omega) = 2\pi A^2/|\omega|$. In this case, the decay of the coherences will no longer be exponential. To see this, let us calculate the explicit form of the dephasing function (8.86),

$$\begin{aligned} \gamma_\phi(t) &= \frac{t^2}{4\pi\hbar^2}\Big(\frac{\partial H}{\partial\lambda}\Big)^2 \int_{-\infty}^{\infty} \frac{2\pi A^2}{|\omega|} \text{sinc}^2\Big(\frac{\omega t}{2}\Big) d\omega \\ &= \frac{t^2}{4\pi\hbar^2}\Big(\frac{\partial H}{\partial\lambda}\Big)^2 \int_{0}^{\infty} \frac{4\pi A^2}{\omega} \text{sinc}^2\Big(\frac{\omega t}{2}\Big) d\omega, \end{aligned} \tag{8.87}$$

where we used the fact that the integrand is symmetric with respect to the y axis. Unfortunately, the integral diverges around the origin. However, in a real experiment,

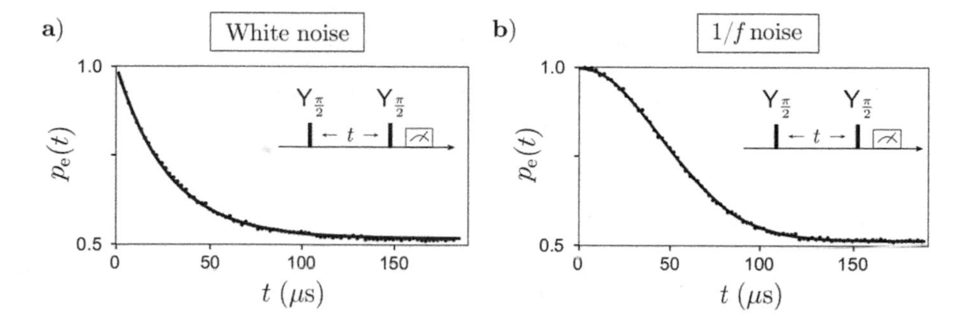

Fig. 8.12 Effect of white and $1/f$ noise on a Ramsey experiment. a) Decay of the population of the excited state in a Ramsey experiment. In the presence of white noise, the coherences decay exponentially. The fitting function is of the form $\frac{1}{2} + \frac{1}{2}e^{-t/T_\phi}e^{-t/2T_1}$. **b)** In the presence of $1/f$ noise, the decay of the coherences is Gaussian. The dephasing time can be extracted from the fit $\frac{1}{2} + \frac{1}{2}e^{-t^2/T_\phi^2}e^{-t/2T_1}$. In both plots, we assumed $T_1 \gg T_\phi$.

the bounds are finite,

$$\gamma_\phi(t) = \frac{t^2}{\hbar^2}\left(\frac{\partial H}{\partial \lambda}\right)^2 \int_{\omega_{\text{low}}}^{\omega_{\text{high}}} \frac{A^2}{\omega}\text{sinc}^2\left(\frac{\omega t}{2}\right) d\omega,$$

where $\omega_{\text{low}} = 2\pi/t_{\text{acq}}$ and $\omega_{\text{high}} = 2\pi/T$. Here, T is the duration of the free evolution. In the field of superconducting devices, the duration of the free evolution in a Ramsey experiment is of the order of $T \approx 100$ μs. The parameter t_{acq} is the total acquisition time. A typical Ramsey experiment has at least $n_{\text{pts}} = 10$ points, each data point is the average of $m = 1000$ shots and each run takes 1 ms. Therefore, the total acquisition time is $t_{\text{acq}} \approx n_{\text{pts}} \cdot m \cdot 1 \text{ ms} = 10$ s. Using these approximations for ω_{low} and ω_{high}, the integral reduces to

$$\gamma_\phi(t) \approx \frac{t^2}{\hbar^2}\left(\frac{\partial H}{\partial \lambda}\right)^2 A^2[-\gamma_{\text{E}} - \ln(\omega_{\text{low}} t)],$$

where γ_{E} is Euler's constant. The term in the square brackets gives a slowly-varying contribution that can be approximated with a constant $c = -\gamma_{\text{E}} - \ln(2\pi\omega_{\text{low}}/\omega_{\text{high}})$. Thus, we have

$$\gamma_\phi(t) \approx \frac{cA^2}{\hbar^2}\left(\frac{\partial H}{\partial \lambda}\right)^2 t^2 = \frac{t^2}{T_\phi^2},$$

where we introduced the dephasing time

$$\boxed{T_\phi = \frac{\hbar}{\sqrt{c}A}\frac{1}{\left|\frac{\partial H}{\partial \lambda}\right|}.} \tag{8.88}$$

This means that in the presence of $1/f$ noise the coherences decay with a Gaussian profile

$$\boxed{\rho_{01}(0) \to \rho_{01}(t)e^{-t^2/T_\phi^2}.}$$

Figure 8.12b shows the decay of the excited state population in a Ramsey experiment when the fluctuations of the qubit frequency are subject to $1/f$ noise.

8.8 Hahn-echo experiment

The decay of the coherences not only depends on the analytical form of the power spectral density but also on the control pulse sequence applied while the qubit is subject to the noise process. Noise spectroscopy (the experimental procedure required to determine the power spectral density of a noise source) is based on **echo experiments**. These types of experiments have been studied for decades in different fields, from NMR [234] to superconducting devices [235, 236].

In this section, we study one particular pulse sequence known as the **Hahn-echo experiment**. As reported in Fig. 8.13a, this experiment comprises two $\mathsf{Y}_{\frac{\pi}{2}}$ pulses separated by a time interval t with an additional X_π halfway through the experiment. For simplicity, we assume that the pulses are instantaneous. Between the pulses, the qubit frequency fluctuates in time[14]

$$H_{\mathrm{q}}(t') = -\frac{\hbar}{2}\delta\omega(t')\sigma_z.$$

The time evolution operator for a Hahn-echo experiment is given by

$$U_{\mathrm{H}} = \mathsf{Y}_{\frac{\pi}{2}} U_{\mathbf{x}}(t, t/2)\mathsf{X}_\pi U_{\mathbf{x}}(t/2, 0)\mathsf{Y}_{\frac{\pi}{2}}. \tag{8.89}$$

The time evolution operator in the time interval $[0, t/2]$ reads

$$U_{\mathbf{x}}(t/2, 0) = e^{\frac{i}{2}\sigma_z \int_0^{t/2} \delta\omega_{\mathbf{x}}(t')dt'} = e^{\frac{i}{2}\sigma_z \Phi_{\mathbf{x}}(0,t/2)},$$

where $\Phi_{\mathbf{x}}(0, t/2) = \int_0^{t/2} \delta\omega_{\mathbf{x}}(t')dt'$ is the phase accumulated due to the fluctuations of the qubit frequency. Thus, eqn (8.89) becomes

$$U_{\mathrm{H}} = \mathsf{Y}_{\frac{\pi}{2}} e^{\frac{i}{2}\sigma_z \Phi_{\mathbf{x}}(t/2,t)} \mathsf{X}_\pi e^{\frac{i}{2}\sigma_z \Phi_{\mathbf{x}}(0,t/2)} \mathsf{Y}_{\frac{\pi}{2}}.$$

With a simple manipulation[15], the operator X_π can be moved to the left of the exponential to obtain

$$U_{\mathrm{H}} = \mathsf{Y}_{\frac{\pi}{2}} \mathsf{X}_\pi e^{\frac{i}{2}\sigma_z[\Phi_{\mathbf{x}}(0,t/2) - \Phi_{\mathbf{x}}(t/2,t)]} \mathsf{Y}_{\frac{\pi}{2}}. \tag{8.90}$$

The effect of the refocusing pulse X_π is to flip the sign of the random phase for the second part of the free evolution with respect to the one in the first part. If the qubit frequency varies on a time scale much longer than t, the phases $\Phi_{\mathbf{x}}(0, t/2)$ and $\Phi_{\mathbf{x}}(t/2, t)$ are identical. This means that the exponential in eqn (8.90) vanishes and the dephasing mechanism disappears too. On the contrary, if the fluctuations

[14] Here, we are working in a rotating frame.

[15] This step is clarified in Exercise 8.11.

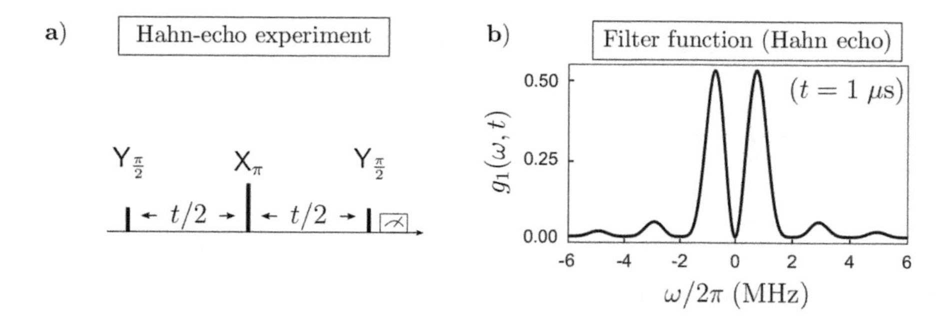

Fig. 8.13 Hahn-echo experiment. a) This experiment comprises two $\mathsf{Y}_{\frac{\pi}{2}}$ pulses with an additional X_π in the middle. **b)** Filter function for the Hahn-echo experiment. In this plot, we set the duration of the experiment to $t = 1\ \mu$s. The filter shows a pronounced peak around ± 1 MHz.

are characterized by high-frequency noise, the two cumulative phases are significantly different and the qubit state is subject to dephasing. Hence, the refocusing pulse can only filter out the low-frequency component of the fluctuations.

Let us focus our attention on the exponential in eqn (8.90),

$$V_{\mathbf{x}}(t) \equiv e^{\frac{i}{2}\sigma_z[\Phi_{\mathbf{x}}(0,t/2)-\Phi_{\mathbf{x}}(t/2,t)]}.$$

This operator describes the free evolution of the qubit state in a Hahn-echo experiment. It transforms a qubit state $\rho(0)$ into

$$\rho(0) \mapsto V_{\mathbf{x}}(t)\rho(0)V_{\mathbf{x}}^\dagger(t) = \begin{bmatrix} \rho_{00} & \rho_{01}e^{i[\Phi_{\mathbf{x}}(0,t/2)-\Phi_{\mathbf{x}}(t/2,t)]} \\ \rho_{10}e^{i[\Phi_{\mathbf{x}}(0,t/2)-\Phi_{\mathbf{x}}(t/2,t)]} & \rho_{11} \end{bmatrix}.$$

The Hahn-echo experiment is usually repeated several times to measure the decay of the coherences. By performing an ensemble average, the quantum state of the qubit becomes[16]

$$\rho(t) = \begin{bmatrix} \rho_{00} & \rho_{01}e^{-\gamma_\phi(t)} \\ \rho_{10}e^{-\gamma_\phi(t)} & \rho_{11} \end{bmatrix},$$

where we introduced the dephasing function $\gamma_\phi(t) = \frac{1}{2}\langle[\Phi(0,t/2) - \Phi(t/2,t)]^2\rangle$. Let us calculate this function explicitly. We start from

$$\begin{aligned}\Phi(0,t/2) - \Phi(t/2,t) &= \int_0^{t/2} \delta\omega_{\mathbf{x}}(t')dt' - \int_{t/2}^{t} \delta\omega_{\mathbf{x}}(t')dt' \\ &= \frac{\partial\omega_{\mathrm{q}}}{\partial\lambda}\Big[\int_0^{t/2} \delta\lambda_{\mathbf{x}}(t')dt' - \int_{t/2}^{t} \delta\lambda_{\mathbf{x}}(t')dt'\Big].\end{aligned}$$

The square of this quantity is given by

[16]Here, we are using the identity $\langle e^{iZ}\rangle = e^{-\langle Z^2\rangle/2}$ valid for any random variable Z with Gaussian distribution (see Exercise 8.8).

$$[\Phi(0,t/2)-\Phi(t/2,t)]^2=\left(\frac{\partial\omega_{\mathrm{q}}}{\partial\lambda}\right)^2\Big[\int_0^{t/2}dt'\int_0^{t/2}dt''\delta\lambda_{\mathbf{x}}(t')\delta\lambda_{\mathbf{x}}(t'')$$
$$-\int_0^{t/2}dt'\int_{t/2}^{t}dt''\delta\lambda_{\mathbf{x}}(t')\delta\lambda_{\mathbf{x}}(t'')-\int_{t/2}^{t}dt'\int_0^{t/2}dt''\delta\lambda_{\mathbf{x}}(t')\delta\lambda_{\mathbf{x}}(t'')$$
$$+\int_{t/2}^{t}dt'\int_{t/2}^{t}dt''\delta\lambda_{\mathbf{x}}(t')\delta\lambda_{\mathbf{x}}(t'')\Big].$$

Using eqns (8.83, 8.58), the dephasing function becomes

$$\gamma_\phi(t)=\frac{1}{2}\langle[\Phi(0,t/2)-\Phi(t/2,t)]^2\rangle$$
$$=\frac{1}{2}\left(\frac{\partial\omega_{\mathrm{q}}}{\partial\lambda}\right)^2\int_{-\infty}^{\infty}d\omega\frac{1}{2\pi}S_{\lambda\lambda}(\omega)\Big[\int_0^{t/2}dt'\int_0^{t/2}dt'e^{i\omega(t'-t'')}$$
$$-\int_0^{t/2}dt'\int_{t/2}^{t}dt''e^{i\omega(t'-t'')}-\int_{t/2}^{t}dt'\int_0^{t/2}dt''e^{i\omega(t'-t'')}$$
$$+\int_{t/2}^{t}dt'\int_{t/2}^{t}dt''e^{i\omega(t'-t'')}\Big].$$

With a bit of patience, one can verify that these time integrals lead to

$$\boxed{\gamma_\phi(t)=\frac{t^2}{4\pi}\left(\frac{\partial\omega_{\mathrm{q}}}{\partial\lambda}\right)^2\int_{-\infty}^{\infty}d\omega S_{\lambda\lambda}(\omega)\frac{\sin^4(\omega t/4)}{(\omega t/4)^2}.}\tag{8.91}$$

This shows that for a Hahn-echo experiment the filter function is of the form $g_1(\omega,t)=\sin^4(\omega t/4)/\ (\omega t/4)^2$ (the subscript 1 indicates that only one refocusing pulse has been applied between the two $\pi/2$ pulses). The filter function is plotted in Fig. 8.13b. As anticipated, the filter function peaks at higher frequencies with respect to a Ramsey experiment.

As an example, suppose that the fluctuations of the qubit frequency are subject to $1/f$ noise, $S_{\lambda\lambda}(\omega)=2\pi A^2/|\omega|$. Then, the dephasing function becomes

$$\gamma_\phi(t)=\frac{t^2}{4\pi}\left(\frac{\partial\omega_{\mathrm{q}}}{\partial\lambda}\right)^2\int_{-\infty}^{\infty}d\omega\frac{2\pi A^2}{|\omega|}\frac{\sin^4(\omega t/4)}{(\omega t/4)^2}$$
$$=\left(\frac{\partial\omega_{\mathrm{q}}}{\partial\lambda}\right)^2A^2\ln(2)\,t^2=t^2/T_{\phi\mathrm{E}}^2,$$

where we defined the dephasing time of the Hahn-echo experiment as

$$\boxed{T_{\phi\mathrm{E}}=\frac{1}{\left|\frac{\partial\omega_{\mathrm{q}}}{\partial\lambda}\right|A\sqrt{\ln(2)}}.}\tag{8.92}$$

If the fluctuations of the qubit frequency are subject to $1/f$ noise only, the dephasing

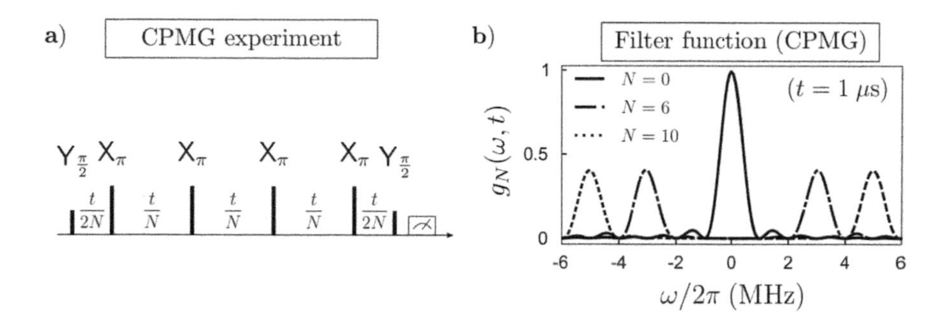

Fig. 8.14 CPMG experiment. a) A CPMG experiment comprises N X_π pulses applied at times $t_j = t/2N + jt/N$ where $j \in \{0, \ldots, N-1\}$. The pulse sequence starts and ends with a $\mathsf{Y}_{\frac{\pi}{2}}$ rotation. **b)** Plot of the filter function $g_N(\omega, t)$ of eqn (8.94) for some values of N. In this graph, we assumed that the duration of the entire experiment is $t = 1$ μs. For $N = 0$ and $N = 1$, one recovers the filter function of the Ramsey experiment and the Hahn-echo experiment respectively.

time can be extracted from a Hahn-echo experiment by fitting the decay with $\frac{1}{2} + \frac{1}{2}e^{-t^2/T_{\phi\mathrm{E}}^2}e^{-t/2T_1}$.

8.9 CPMG sequence

A **Carr–Purcell–Meiboom–Gill (CPMG) sequence** is a generalization of the pulse sequences explained in the previous sections and allows us to analyze the intensity of the noise at higher frequencies. As shown in Fig. 8.14a, this sequence consists of two $\mathsf{Y}_{\frac{\pi}{2}}$ pulses with a sequence of N equidistant X_π pulses in between. It can be shown that in a CPMG experiment, the dephasing function is given by [237, 235]

$$\boxed{\gamma_\phi(t) = \frac{t^2}{4\pi}\left(\frac{\partial \omega_{\mathrm{q}}}{\partial \lambda}\right)^2 \int_{-\infty}^{\infty} d\omega S_{\lambda\lambda}(\omega) g_N(\omega, t)} \tag{8.93}$$

where the filter function $g_N(\omega, t)$ is

$$g_N(\omega, t) = \frac{1}{\omega^2 t^2}\left|1 + (-1)^{1+N}e^{i\omega t} + 2\sum_{j=1}^{N}(-1)^j e^{i\omega\delta_j t}\cos(\omega t_{\mathrm{p}}/2)\right|^2. \tag{8.94}$$

Here, t_{p} is the duration of a single pulse (in the previous sections, we assumed $t_{\mathrm{p}} = 0$ for simplicity). The parameter δ_j indicates the normalized location of the center of the j-th X_π pulse,

$$\delta_j = \frac{1}{2N} + \frac{j-1}{N}.$$

The filter function of the CPMG experiment is shown in Fig. 8.14b. As N increases, the filter function shifts to higher frequencies. The transverse relaxation time of a CPMG experiment is usually indicated with the symbol $T_2^{(N)}$, where N indicates the

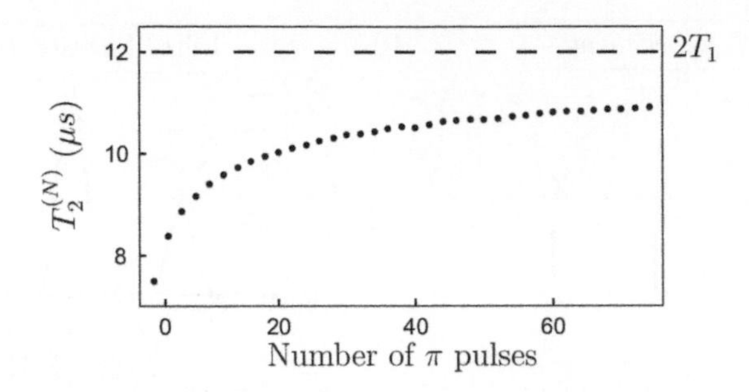

Fig. 8.15 CPMG experiment. Plot of $T_2 = (1/2T_1 + 1/T_\phi)^{-1}$ as a function of the number of refocusing pulses in a CPMG experiment. In this plot, we set $T_1 = 6\ \mu$s and we determined $T_\phi = \sqrt{t^2/\gamma_\phi(t)}$ using eqns (8.93, 8.94) for $1/f$ noise.

number of refocusing pulses. If the fluctuations of the qubit frequency are subject to $1/f$ noise, the value of $T_2^{(N_1)}$ is higher than $T_2^{(N_2)}$ for $N_1 > N_2$. This behavior is shown in Fig. 8.15.

8.10 Decoherence under driven evolution

Until now, we have only considered the decoherence of a qubit state under free evolution. How do the coherences evolve when the qubit is subject to a resonant pulse? For simplicity, let us start our discussion assuming that the qubit is not affected by decoherence. The Hamiltonian of a qubit interacting with a resonant drive is

$$H = \underbrace{-\frac{\hbar\omega_{\mathrm{q}}}{2}\sigma_z}_{\text{qubit } H_{\mathrm{q}}} + \underbrace{\hbar\Omega\sin(\omega_{\mathrm{d}}t + \alpha)\sigma_y}_{\text{drive } H_{\mathrm{d}}},$$

where ω_{d} is the drive frequency, α is the phase of the drive and Ω [rad/s] is called the **Rabi frequency**. This parameter describes the strength of the drive and is proportional to the amplitude of the signal. This Hamiltonian does not include any dissipative coupling to external systems and leads to unitary dynamics. We will assume that the drive is in resonance with the qubit, $\omega_{\mathrm{d}} = \omega_{\mathrm{q}}$. It is convenient to study the time evolution in the interaction picture. With the change of basis $R = e^{-\frac{1}{i\hbar}H_{\mathrm{q}}t}$, the Hamiltonian becomes

$$\tilde{H} = RHR^{-1} - i\hbar R\frac{d}{dt}R^{-1} = \hbar\Omega\sin(\omega_{\mathrm{q}}t + \alpha)\tilde{\sigma}_y,$$

where $\tilde{\sigma}_y = R\sigma_y R^{-1} = \sigma_y\cos\omega_{\mathrm{q}}t + \sigma_x\sin\omega_{\mathrm{q}}t$, see eqn (3.64). The last term of this Hamiltonian can be expanded as

$$\begin{aligned}\sin(\omega_{\mathrm{q}}t + \alpha)\tilde{\sigma}_y &= \sin(\omega_{\mathrm{q}}t + \alpha)(\sigma_y\cos\omega_{\mathrm{q}}t + \sigma_x\sin\omega_{\mathrm{q}}t)\\ &= \frac{\sigma_x}{2}(-\cos(2\omega_{\mathrm{q}}t + \alpha) + \cos\alpha) + \frac{\sigma_y}{2}(\sin(2\omega_{\mathrm{q}}t + \alpha) + \sin\alpha).\end{aligned}$$

The terms proportional to $\cos(2\omega_q t)$ and $\sin(2\omega_q t)$ oscillate very rapidly and give a small contribution to the dynamics. Hence, we will drop them (this is called the rotating wave approximation, see Appendix A.1). The phase of the drive can be set to any value. For convenience, we choose $\alpha = -\pi$ such that $\cos\alpha = -1$ and $\sin\alpha = 0$. The Hamiltonian in the interaction picture reduces to $\tilde{H} = -\frac{\hbar\Omega}{2}\sigma_x$ and the equation of motion becomes

$$\boxed{\frac{d\tilde{\rho}}{dt} = \frac{1}{i\hbar}[\tilde{H}, \tilde{\rho}] = i\frac{\Omega}{2}[\sigma_x, \tilde{\rho}].}$$

Since $\tilde{\rho} = \frac{1}{2}(\mathbb{1} + r_x\sigma_x + r_y\sigma_y + r_z\sigma_z)$, the time derivative of the density operator can be expressed as

$$\begin{aligned}\frac{d\tilde{\rho}}{dt} = \frac{1}{2}\Big(\frac{dr_x}{dt}\sigma_x + \frac{dr_y}{dt}\sigma_y + \frac{dr_z}{dt}\sigma_z\Big) &= i\frac{\Omega}{2}[\sigma_x, \tilde{\rho}] = \frac{i\Omega}{4}[\sigma_x, r_y\sigma_y + r_z\sigma_z] \\ &= \frac{1}{2}(\Omega r_z\sigma_y - \Omega r_y\sigma_z)\end{aligned}$$

from which we obtain the Bloch equations in the rotating frame

$$\frac{dr_x}{dt} = 0, \qquad \frac{dr_y}{dt} = \Omega r_z, \qquad \frac{dr_z}{dt} = -\Omega r_y.$$

The solution of these differential equations leads to the well-known Rabi oscillations, i.e. continuous rotations of the qubit state about the x axis of the Bloch sphere. Assuming that the qubit is initially in the ground state, the probability of measuring the qubit in the excited state, $p_e(t) = (1 - r_z(t))/2$, is given by

$$\boxed{p_e(t) = \frac{1}{2} - \frac{1}{2}\cos(\Omega t).} \tag{8.95}$$

Now suppose the time evolution is affected by amplitude damping and dephasing simultaneously. The equation of motion in the interaction picture is given by

$$\frac{d\tilde{\rho}}{dt} = i\frac{\Omega}{2}[\sigma_x, \tilde{\rho}] + \gamma\mathcal{D}[\sigma^-](\tilde{\rho}) + \frac{\gamma_\phi}{2}\mathcal{D}[\sigma_z](\tilde{\rho}),$$

and the Bloch equations take the form

$$\frac{dr_x}{dt} = \Big(-\frac{\gamma}{2} - \gamma_\phi\Big)r_x, \tag{8.96}$$

$$\frac{dr_y}{dt} = \Omega r_z + \Big(-\frac{\gamma}{2} - \gamma_\phi\Big)r_y, \tag{8.97}$$

$$\frac{dr_z}{dt} = -\Omega r_y - \gamma r_z + \gamma, \tag{8.98}$$

where we used (8.79, 8.81). Suppose that the qubit is initially in the ground state and $\Omega \gg \{\gamma, \gamma_\phi\}$. Then, the probability of measuring the qubit in the excited state

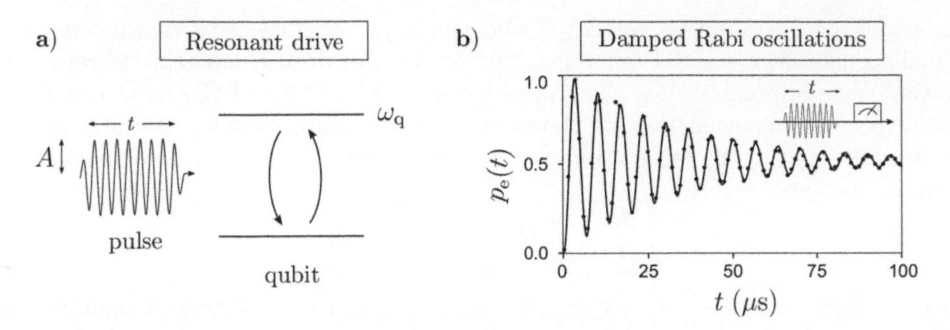

Fig. 8.16 Damped Rabi oscillations. a) An electromagnetic wave interacts with a qubit. When the drive frequency ω_d matches the qubit frequency, the qubit state rotates about an axis lying on the equator of the Bloch sphere. The period of the oscillations is $2\pi/\Omega$ where Ω is proportional to the pulse amplitude A. **b)** Plot of the Rabi oscillations in the presence of decoherence. As shown in the inset, a time Rabi experiment comprises a resonant pulse of duration t immediately followed by a measurement. The experiment is repeated for different pulse lengths. Assuming $\Omega \gg \{\gamma = 1/T_1, \gamma_\phi = 1/T_\phi\}$, decoherence attenuates the oscillations and the qubit state converges toward the center of the Bloch sphere. In this plot, we fitted the decay using (8.99) with $\Omega/2\pi = 150$ kHz, $T_1 = 1/\gamma = 40$ μs, and $T_\phi = 1/\gamma_\phi = 50$ μs.

immediately after a resonant pulse of duration t is[17]

$$\boxed{p_e(t) = \frac{1}{2} - \frac{1}{2}e^{-\frac{1}{4}(3\gamma+2\gamma_\phi)t}\cos(\mu t),} \tag{8.99}$$

where $\mu = \frac{1}{4}\sqrt{16\Omega^2 - (2\gamma_\phi - \gamma)^2}$. The plot of this equation is shown in Fig. 8.16b and its mathematical derivation can be found in Exercise 8.12. From (8.99), we see that the Rabi oscillations decay with a characteristic time [235]

$$\boxed{T_{2\rho} = \frac{4}{3\gamma + 2\gamma_\phi} = \frac{4}{\frac{3}{T_1} + \frac{2}{T_\phi}}.}$$

This is called the **Rabi time**. This parameter indicates the transverse relaxation time of the qubit state under driven evolution.

Further reading

For a general introduction to quantum noise, we suggest reading the comprehensive review by Clerk, Devoret, Girvin, Marquardt, Schoelkopf [230]. Some studies of a qubit coupled to a quantum noise source can be found in Refs [238, 231, 239]. For more information about pulse sequences, we suggest reading the papers by Biercuk et al. [237] and Bylander et al. [235].

[17]Note that this expression reduces to eqn (8.95) for $\gamma_\phi = \gamma = 0$.

Summary

Markovian master equation

$\mathcal{E}_t(\rho) = \mathrm{Tr}_{\mathrm{B}}[U(t)(\rho \otimes \rho_{\mathrm{B}})U^{\dagger}(t)]$	Quantum dynamical map
$\mathcal{E}_{t_2}(\mathcal{E}_{t_1}(\rho)) = \mathcal{E}_{t_1+t_2}(\rho)$	Semigroup property
$\frac{d\rho(t)}{dt} = \mathcal{L}(\rho(t))$	Markov master equation
A_k	Lindblad operators
$\gamma_k > 0$	Decay rates
$\frac{d\rho(t)}{dt} = \frac{1}{i\hbar}[H, \rho] + \sum_k \gamma_k \mathcal{D}[A_k](\rho)$	Markov master eqn in Lindblad form
$\mathcal{D}[A_k](\rho) = A_k \rho A_k^{\dagger} - \frac{1}{2}\{A_k^{\dagger} A_k, \rho\}$	Dissipator

Redfield equation:
$\frac{d\tilde{\rho}(t)}{dt} = \frac{1}{i\hbar}\mathrm{Tr}_{\mathrm{B}}([\tilde{V}(t), \rho(0) \otimes \rho_{\mathrm{B}}]) - \frac{1}{\hbar^2}\int_0^t \mathrm{Tr}_{\mathrm{B}}([\tilde{V}(t), [\tilde{V}(t'), \tilde{\rho}(t) \otimes \rho_{\mathrm{B}}]])dt'$

Amplitude damping

$\frac{d\rho(t)}{dt} - i\frac{\omega_{\mathrm{q}}'}{2}[\sigma_z, \rho] + \gamma_{\downarrow}\mathcal{D}[\sigma^-](\rho) + \gamma_{\uparrow}\mathcal{D}[\sigma^+](\rho)$	Master equation
$\gamma_{\uparrow} = \gamma \bar{n}_{\mathrm{q}}$	Rate from $\|0\rangle$ to $\|1\rangle$
$\gamma_{\downarrow} = \gamma(\bar{n}_{\mathrm{q}} + 1)$	Rate from $\|1\rangle$ to $\|0\rangle$
$T_1 = \frac{1}{\gamma_{\downarrow} + \gamma_{\uparrow}}$	Relaxation time
$\tilde{\rho}_{11}(t) = \tilde{\rho}_{11}(0)e^{-\gamma t}$	Assuming $k_{\mathrm{B}}T \ll \hbar\omega_{\mathrm{q}}$
$\tilde{\rho}_{01}(t) = \tilde{\rho}_{01}(0)e^{-\gamma t/2}$	Assuming $k_{\mathrm{B}}T \ll \hbar\omega_{\mathrm{q}}$

Noise source

$H = H_{\mathrm{A}} + H_{\mathrm{B}} + A \otimes F$	Noise source F coupled to the system
$G_{FF}(t', t'') = \mathrm{Tr}_{\mathrm{B}}[F(t')\rho_{\mathrm{B}}F(t'')]$	Auto-correlation function
$G_{FF}(t', t'') = G_{FF}(t' - t'')$	Stationary process
$\gamma_{i \to f} \approx \frac{1}{\hbar^2}\|\langle f\|A\|i\rangle\|^2 S_{FF}(\omega_i - \omega_f)$	Rate from state $\|i\rangle$ to state $\|f\rangle$
$S_{FF}(\omega) = \int_{-\infty}^{\infty} d\tau e^{-i\omega\tau} G_{FF}(\tau)$	Power spectral density
$S_{FF}(-\omega_{\mathrm{q}})$	Absorption process
$S_{FF}(+\omega_{\mathrm{q}})$	Emission process
$S_{FF}(-\omega) = S_{FF}(+\omega)e^{-\hbar\omega/k_{\mathrm{B}}T}$	
$S_{FF}(\omega) = A$	White noise
$S_{FF}(\omega) = 2\pi A^2/\|\omega\|$	$1/f$ noise (or pink noise)

Dephasing

$\frac{d\rho}{dt} = i\frac{\omega_{\mathrm{q}}}{2}[\sigma_z, \rho] + \frac{\gamma_{\phi}}{2}(\sigma_z \rho \sigma_z - \rho)$	Master equation for white noise
$\tilde{\rho}_{01}(t) = \tilde{\rho}_{01}(0)e^{-\gamma_{\phi} t}$	
$\tilde{\rho}_{11}(t) = \tilde{\rho}_{11}(0)$	
$T_{\phi} = 1/\gamma_{\phi}$	Dephasing time

Decoherence under free evolution

$\frac{d\tilde{\rho}}{dt} = \frac{1}{T_1}\mathcal{D}[\sigma^-](\tilde{\rho}) + \frac{1}{2T_\phi}\mathcal{D}[\sigma_z](\tilde{\rho})$	Master equation ($k_BT \ll \hbar\omega_q$)
$\tilde{\rho}_{11}(t) = \tilde{\rho}_{11}(0)e^{-t/T_1}$	Evolution of the population
$\tilde{\rho}_{01}(t) = \tilde{\rho}_{01}(0)e^{-\left(\frac{1}{T_\phi}+\frac{1}{2T_1}\right)t}$	Evolution of the coherence
$T_2 = \left(\frac{1}{T_\phi} + \frac{1}{2T_1}\right)^{-1}$	Transverse relaxation time
$\dot{r}_x = -r_x/T_2$ $\dot{r}_y = -r_y/T_2$ $\dot{r}_z = -r_z/T_1 + 1/T_1$	Bloch equations (assuming $k_BT \ll \hbar\omega_q$)

T_ϕ and the power spectral density

$\hat{H}_q(t) = -\frac{\hbar(\omega_q+\delta\omega_q(t))}{2}\hat{\sigma}_z$	Hamiltonian of a qubit subject to noise
λ	Noise source
$\delta\omega_q(t) = \frac{\partial\omega_q}{\partial\lambda}\delta\lambda(t)$	Fluctuations of the qubit frequency
$\rho_{11}(t) = \rho_{11}$	The populations remain constant
$\rho_{01}(t) = \rho_{01}e^{i\omega_q t}e^{-\gamma_\phi(t)}$	Evolution of the coherence
$\gamma_\phi(t) = \frac{t^2}{4\pi}\left(\frac{\partial\omega_q}{\partial\lambda}\right)^2\int_{-\infty}^{\infty} S_{\lambda\lambda}(\omega)\mathrm{sinc}^2\left(\frac{\omega t}{2}\right)$	Dephasing function for a Ramsey exp.
$\gamma_\phi(t) = t/T_\phi$	White noise
$\gamma_\phi(t) = t^2/T_\phi^2$	$1/f$ noise
$\gamma_\phi(t) = \frac{t^2}{4\pi}\left(\frac{\partial\omega_q}{\partial\lambda}\right)^2\int_{-\infty}^{\infty} S_{\lambda\lambda}(\omega)\frac{\sin^4(\omega t/4)}{(\omega t/4)^2}$	Dephasing function for a Hahn-echo exp.
$g_1(\omega,t) = \sin^4(\omega t/4)/(\omega t/4)^2$	Filter function for a Hahn-echo exp.

Hahn-echo experiment

$\gamma_\phi(t) = \frac{t^2}{4\pi}\left(\frac{\partial\omega_q}{\partial\lambda}\right)^2\int_{-\infty}^{\infty} S_{\lambda\lambda}(\omega)\frac{\sin^4(\omega t/4)}{(\omega t/4)^2}$	Dephasing function for a Hahn-echo exp.
$\gamma_\phi(t) = t^2/T_\phi^2$	Dephasing function for $1/f$ noise
$T_\phi = \left[\left\vert\frac{\partial\omega_q}{\partial\lambda}\right\vert A\sqrt{\ln 2}\right]^{-1}$ ($1/f$ noise)	Dephasing time for a Hahn-echo exp.

CPMG sequence

$\gamma_\phi(t) = \frac{t^2}{4\pi}\left(\frac{\partial\omega_q}{\partial\lambda}\right)^2\int_{-\infty}^{\infty} S_{\lambda\lambda}(\omega)g_N(\omega,t)$	Dephasing function for a CPMG sequence
$g_N(\omega,t) = \frac{1}{\omega^2t^2}\left\vert 1 + (-1)^{1+N}e^{i\omega t} + 2\sum_j(-1)^j e^{i\omega\delta_j t}\cos(\omega t_p/2)\right\vert^2$	Filter function
$\delta_j = \frac{1}{2N} + \frac{j-1}{N}$	Center of the j-th π pulse
t_p	Duration of a single π pulse

Driven-evolution decoherence

$\frac{d\tilde{\rho}}{dt} = i\frac{\Omega}{2}[\sigma_x,\tilde{\rho}] + \gamma\mathcal{D}[\sigma^-](\tilde{\rho}) + \frac{\gamma_\phi}{2}\mathcal{D}[\sigma_z](\tilde{\rho})$	Master equation under driven evolution
$T_{2\rho} = \frac{4}{\frac{3}{T_1}+\frac{2}{T_\phi}}$	Decay in a time Rabi experiment

Exercises

Exercise 8.1 Show that the dynamical map $\mathcal{E}_t(\rho)$ defined in eqn (8.1) is a quantum map. **Hint:** Show that it can be expressed in Kraus form.

Exercise 8.2 Consider a function $\mathcal{E}_t$ that transforms the initial state of a qubit ρ into

$$\rho(t) = \mathcal{E}_t(\rho) = f(t)\rho + (1 - f(t))\frac{\mathbb{1}}{2}.$$

Note the similarity between this function and the depolarizing channel presented in Section 7.2. Find the analytical expression of $f(t)$ such that $\mathcal{E}_t$ satisfies the semigroup property $\mathcal{E}_t(\mathcal{E}_s(\rho)) = \mathcal{E}_{t+s}(\rho)$.

Exercise 8.3 In Exercise 8.2, we showed that the quantum map $\mathcal{E}_t(\rho) = e^{-\gamma t}\rho + (1 - e^{-\gamma t})\frac{\mathbb{1}}{2}$ satisfies the semigroup property. Using the identity $\mathbb{1}/2 = \frac{1}{4}(\sigma_x\rho\sigma_x + \sigma_y\rho\sigma_y + \sigma_z\rho\sigma_z + \rho)$, the quantum map $\mathcal{E}_t$ can be written as

$$\mathcal{E}_t(\rho) = \frac{3e^{-\gamma t} + 1}{4}\rho + \frac{1 - e^{-\gamma t}}{4}(\sigma_x\rho\sigma_x + \sigma_y\rho\sigma_y + \sigma_z\rho\sigma_z).$$

Derive the Markovian master equation associated with $\mathcal{E}_t$. **Hint:** Use eqn (8.8) and write $\mathcal{E}_t$ in the form $\mathcal{E}_t(\rho) = \sum_{i=1}^{4}\sum_{j=1}^{4} c_{ij}F_i\rho F_j^\dagger$ where F_i are the elements of the Pauli basis.

Exercise 8.4 Consider a system coupled to a bath of harmonic oscillators

$$H = H_{\mathrm{A}} + H_{\mathrm{B}} + V$$

where $H = \sum_j \hbar\omega_j b^\dagger b$, and $V = \sum_j A_j \otimes B_j$. The bath is in a thermal state ρ_{B}. By readjusting some terms in the Hamiltonian, show that it is always possible to set the first term of the Redfield equation (8.16) to zero.

Exercise 8.5 Show that the matrix c_{ij} defined in eqn (8.6) is positive and Hermitian.

Exercise 8.6 The Magnus expansion [240] states that $\mathcal{T}e^{\frac{1}{i\hbar}\int_0^t dt' H(t')} = e^{\sum_{n=1}^{\infty} S_n}$ where

$$S_1(t) = \frac{1}{i\hbar}\int_0^t dt_1 H(t),$$

$$S_2(t) = \frac{1}{2!}\frac{1}{(i\hbar)^2}\int_0^t dt_1\int_0^t dt_2 [H(t_1), H(t_2)],$$

$$S_3(t) = \frac{1}{3!}\frac{1}{(i\hbar)^3}\int_0^t dt_1\int_0^{t_1} dt_2\int_0^{t_2} dt_3 [H(t_1), [H(t_2), H(t_3)]] + [H(t_3), [H(t_2), H(t_1)]],$$

and so on. Using the Magnus expansion, show that $U(t)$ of eqn (8.67) can be expressed as $U(t) = e^{\frac{1}{i\hbar}\int_0^t H(t_1)dt_1}e^{i\phi(t)}$ where $\phi(t)$ is a phase. **Hint:** Start from the commutator $[H(t_1), H(t_2)]$.

Exercise 8.7 In Section 8.5, we presented a model for pure dephasing. Starting from (8.68), show that the diagonal elements of the density matrix remain unchanged. **Hint:** Use the properties $D^\dagger(\alpha) = D(-\alpha)$ and $D(\beta)D(\alpha) = D(\alpha + \beta)\exp(-\frac{1}{2}(\alpha\beta^* - \alpha^*\beta))$.

Exercise 8.8 Suppose that the random variable Z has a Gaussian distribution with average $\langle Z\rangle = 0$ and variance $\langle Z^2\rangle = \sigma^2$. Show that the expectation value $\langle e^{iZ}\rangle$ is equal to $e^{i\langle Z^2\rangle/2}$.

Exercise 8.9 In Section 8.3, we derived the dynamics of a qubit coupled to a bath of oscillators at thermal equilibrium. Now consider a harmonic oscillator coupled to a bath of harmonic oscillators

$$H = \hbar\omega_{\mathrm{r}} a^\dagger a + \sum_{k=1}^{\infty} \hbar\omega_k b_k^\dagger b_k + \sum_{k=1}^{\infty} \hbar g_k (ab^\dagger + a^\dagger b).$$

Assuming that the oscillators are in a thermal state, show that the master equation for the harmonic oscillator is given by

$$\frac{d\rho(t)}{dt} = -i\omega_{\mathrm{r}}'[a^\dagger a, \rho] + \kappa(\bar{n}_{\mathrm{r}} + 1)\mathcal{D}[\hat{a}](\rho) + \kappa\bar{n}_{\mathrm{r}}\mathcal{D}[a^\dagger]\rho,$$

where ω_{r}' is the renormalized oscillator frequency, $\bar{n}_{\mathrm{r}} = 1/(e^{\hbar\omega_{\mathrm{r}}/k_{\mathrm{B}}T} - 1)$ is the number of thermal photons in the oscillators with frequency ω_{r}, and κ is a real parameter.

Exercise 8.10 In eqn (8.72), we derived the decay of the coherences of a qubit assuming that the bath was characterized by a constant spectral density. Now suppose that the spectral density of the bath is $J(\omega) = A\omega e^{-\omega/\Omega}$ where Ω is a cutoff frequency. Starting from (8.70) show that the decay of the coherences is not exponential.

Exercise 8.11 Show that

$$e^{\frac{i}{2}\alpha\sigma_z}\mathsf{X}_\pi e^{\frac{i}{2}\beta\sigma_z} = \mathsf{X}_\pi e^{-\frac{i}{2}\alpha\sigma_z} e^{\frac{i}{2}\beta\sigma_z}.$$

Hint: Use the relation $e^{iUAU^\dagger} = Ue^{iA}U^\dagger$ valid for any unitary operator U.

Exercise 8.12 Calculate the solution of the Bloch equations (8.96,8.98) assuming that at $t = 0$ the qubit is in the ground state and the drive amplitude is much greater than the decay rates, $\Omega \gg \{\gamma, \gamma_\phi\}$.

Solutions

Solution 8.1 To prove that $\mathcal{E}_t(\rho)$ can be written in Kraus form, we will follow the same steps presented in Section 7.8.1 when we demonstrated Stinespring's dilation theorem. We decompose the initial state of the environment as $\rho_{\mathrm{B}} = \sum_k \lambda_k |k\rangle\langle k|$ where $\{|k\rangle\}_{k=1}^{d_{\mathrm{B}}}$ is an orthonormal basis for the Hilbert space of the environment, while λ_k are non-negative numbers summing up to one. Substituting ρ_{B} into eqn (8.1), we obtain

$$\begin{aligned}
\mathcal{E}_t(\rho) &= \sum_{k=1}^{d_{\mathrm{B}}} \lambda_k \mathrm{Tr}_{\mathrm{B}}[U(t)(\rho \otimes |k\rangle\langle k|)U^\dagger(t)] = \sum_{k,l=1}^{d_{\mathrm{B}}} \lambda_k \langle l|U(t)(\rho \otimes |k\rangle\langle k|)U^\dagger(t)|l\rangle \\
&= \sum_{k,l=1}^{d_{\mathrm{B}}} \underbrace{\sqrt{\lambda_k}\langle l|U(t)|k\rangle}_{E_{kl}(t)} \rho \langle k|U^\dagger(t)|l\rangle\sqrt{\lambda_k} = \sum_{k,l=1}^{d_{\mathrm{B}}} E_{kl}(t)\rho E_{kl}^\dagger(t),
\end{aligned} \tag{8.100}$$

where we defined the Kraus operators $E_{kl}(t) = \sqrt{\lambda_k}\langle l|U(t)|k\rangle$. The function $\mathcal{E}_t$ is a quantum map only if $\sum_{kl} E_{kl}^\dagger(t)E_{kl}(t) = \mathbb{1}$. This is indeed the case since

$$\sum_{k,l} E_{kl}^\dagger(t)E_{kl}(t) = \sum_{k,l} \lambda_k \langle k|U^\dagger(t)|l\rangle\langle l|U(t)|k\rangle = \sum_k \lambda_k \langle k|U^\dagger(t)U(t)|k\rangle = \mathbb{1}.$$

In conclusion, the function $\mathcal{E}_t$ of eqn (8.1) is a quantum map.

Solution 8.2 Let us calculate the combination $\mathcal{E}_t \circ \mathcal{E}_s$,

$$\begin{aligned}\mathcal{E}_t(\mathcal{E}_s(\rho)) = \mathcal{E}_t\Big[f(s)\rho + (1 - f(s))\frac{\mathbb{1}}{2}\Big] &= f(t)\Big[f(s)\rho + [1 - f(s)]\frac{\mathbb{1}}{2}\Big] + [1 - f(t)]\frac{\mathbb{1}}{2} \\ &= f(t)f(s)\rho + [1 - f(t)f(s)]\frac{\mathbb{1}}{2}.\end{aligned} \tag{8.101}$$

The function $\mathcal{E}_t$ satisfies the semigroup property only if $\mathcal{E}_t(\mathcal{E}_s(\rho))$ is equal to

$$\mathcal{E}_{t+s}(\rho) = f(t+s)\rho + [1 - f(t+s)]\frac{\mathbb{1}}{2}. \tag{8.102}$$

By comparing (8.101) with (8.102), we see that $f(t+s) = f(t)f(s)$. Thus, $f(t) = e^{-\gamma t}$ where $\gamma > 0$. The solution $f(t) = e^{\gamma t}$ with $\gamma > 0$ does not have a physical meaning.

Solution 8.3 The quantum map $\mathcal{E}_t$ can be written as

$$\mathcal{E}_t = \sum_{i=1}^{4}\sum_{j=1}^{4} c_{ij} F_i \rho F_j^\dagger$$

where the operators F_i belong to the orthonormal basis $(\mathbb{1}/\sqrt{2}, \sigma_x/\sqrt{2}, \sigma_y/\sqrt{2}, \sigma_z/\sqrt{2})$ and c_{ij} is given by

$$c_{ij} = \begin{bmatrix} \frac{3e^{-\gamma t}+1}{2} & 0 & 0 & 0 \\ 0 & \frac{1-e^{-\gamma t}}{2} & 0 & 0 \\ 0 & 0 & \frac{1-e^{-\gamma t}}{2} & 0 \\ 0 & 0 & 0 & \frac{1-e^{-\gamma t}}{2} \end{bmatrix}.$$

From eqn (8.8), we know that

$$\frac{d\rho}{dt} = \alpha\rho + \sum_{i\geq 2}^{4}\Big(\frac{\alpha_i}{\sqrt{2}}F_i\rho + \frac{\alpha_i^*}{\sqrt{2}}\rho F_i^\dagger\Big) + \sum_{i,j\geq 2}^{4} \beta_{ij} F_i \rho F_j^\dagger \tag{8.103}$$

where

$$\begin{aligned}\alpha &= \lim_{\varepsilon\to 0}\frac{c_{11}(\varepsilon) - 2}{\varepsilon 2} = \lim_{\varepsilon\to 0}\frac{\frac{3e^{-\gamma\varepsilon}+1}{2} - 2}{\varepsilon 2} = -\frac{3\gamma}{4}, \\ \alpha_i &= \lim_{\varepsilon\to 0}\frac{c_{i1}(\varepsilon)}{\varepsilon} = 0, \\ \beta_{ij} &= \lim_{\varepsilon\to 0}\frac{c_{ij}(\varepsilon)}{\varepsilon} = \delta_{ij}\lim_{\varepsilon\to 0}\frac{c_{ii}(\varepsilon)}{\varepsilon} = \delta_{ij}\lim_{\varepsilon\to 0}\frac{\frac{1-e^{-\gamma\varepsilon}}{2}}{\varepsilon} = \delta_{ij}\frac{\gamma}{2}.\end{aligned}$$

Thus, the master equation (8.103) becomes

$$\frac{d\rho}{dt} = -\frac{3\gamma}{4}\rho + \sum_{i=2}^{4}\sum_{j=2}^{4}\delta_{ij}\frac{\gamma}{2}F_i\rho F_j^\dagger = -\frac{3\gamma}{4}\rho + \sum_{j=1}^{3}\frac{\gamma}{4}\sigma_j\rho\sigma_j.$$

Solution 8.4 Let us calculate the term $\mathrm{Tr}_\mathrm{B}[\tilde{V}(t), \rho \otimes \rho_\mathrm{B}]$ of the Redfield equation (8.16), where $\tilde{V}(t)$ is the potential energy in the interaction picture,

$$\tilde{V} = e^{-\frac{1}{i\hbar}(H_\mathrm{A}+H_\mathrm{B})t}Ve^{\frac{1}{i\hbar}(H_\mathrm{A}+H_\mathrm{B})t} = e^{-\frac{1}{i\hbar}(H_\mathrm{A}+H_\mathrm{B})t}\Big(\sum_j A_j \otimes B_j\Big)e^{\frac{1}{i\hbar}(H_\mathrm{A}+H_\mathrm{B})t}$$

$$= \sum_j e^{-\frac{1}{i\hbar}H_\mathrm{A}t}A_j e^{\frac{1}{i\hbar}H_\mathrm{A}t} \otimes e^{-\frac{1}{i\hbar}H_\mathrm{B}t}B_j e^{\frac{1}{i\hbar}H_\mathrm{B}t} = \sum_j \tilde{A}_j(t) \otimes \tilde{B}_j(t).$$

We have

$$\begin{aligned}
\mathrm{Tr}_\mathrm{B}[\tilde{V}(t), \rho \otimes \rho_\mathrm{B}] &= \mathrm{Tr}_\mathrm{B}[\sum_j \tilde{A}_j(t) \otimes \tilde{B}_j(t), \rho \otimes \rho_\mathrm{B}] \\
&= \sum_j [\tilde{A}_j(t), \rho]\mathrm{Tr}_\mathrm{B}[e^{-\frac{1}{i\hbar}H_\mathrm{B}t}B_j e^{\frac{1}{i\hbar}H_\mathrm{B}t}\rho_\mathrm{B}] \\
&= \sum_j [\tilde{A}_j(t), \rho]\mathrm{Tr}_\mathrm{B}[B_j e^{\frac{1}{i\hbar}H_\mathrm{B}t}\rho_\mathrm{B} e^{-\frac{1}{i\hbar}H_\mathrm{B}t}] \\
&= \sum_j [\tilde{A}_j(t), \rho]\underbrace{\mathrm{Tr}_\mathrm{B}[B_j\rho_\mathrm{B}]}_{b_j} = \frac{1}{i\hbar}\sum_j [\tilde{A}_j(t), \rho]b_j,
\end{aligned}$$

where we defined some complex numbers $b_j(t) = \mathrm{Tr}_\mathrm{B}[\tilde{B}_j(t), \rho_\mathrm{B}]$ and we used the fact that thermal states are stationary states, $e^{\frac{1}{i\hbar}b^\dagger bt}\rho_\mathrm{B}e^{-\frac{1}{i\hbar}b^\dagger bt} = \rho_\mathrm{B}$. To set this trace to zero, we can add and subtract from the Hamiltonian the quantity $S = \sum_j b_j A_j \otimes \mathbb{1}$,

$$\begin{aligned}
H &= H_\mathrm{A} + S + H_\mathrm{B} + V - S \\
&= \underbrace{H_\mathrm{A} + \sum_j b_j A_j \otimes \mathbb{1}}_{H'_\mathrm{A}} + H_\mathrm{B} + \underbrace{\sum_j A_j \otimes (B_j - b_j\mathbb{1})}_{V'} = H'_\mathrm{A} + H_\mathrm{B} + V'.
\end{aligned}$$

In the interaction picture, the Hamiltonian becomes

$$\begin{aligned}
\tilde{V}' &= e^{-\frac{1}{i\hbar}(H'_\mathrm{A}+H_\mathrm{B})t}V'e^{\frac{1}{i\hbar}(H'_\mathrm{A}+H_\mathrm{B})t} \\
&= \sum_j e^{-\frac{1}{i\hbar}H'_\mathrm{A}t}A_j e^{\frac{1}{i\hbar}H'_\mathrm{A}t} \otimes e^{-\frac{1}{i\hbar}H_\mathrm{B}t}(B_j - b_j\mathbb{1})e^{\frac{1}{i\hbar}H_\mathrm{B}t} = \sum_j \tilde{A}'_j \otimes (\tilde{B}_j - b_j\mathbb{1}),
\end{aligned}$$

and therefore

$$\begin{aligned}
\mathrm{Tr}_\mathrm{B}[\tilde{V}'(t), \rho \otimes \rho_\mathrm{B}] &= \mathrm{Tr}_\mathrm{B}[\sum_j \tilde{A}'_j \otimes (\tilde{B}_j - b_j\mathbb{1}), \rho \otimes \rho_\mathrm{B}] \\
&= \sum_j [\tilde{A}'_j, \rho](\mathrm{Tr}_\mathrm{B}[\tilde{B}_j\rho_\mathrm{B}] - b_j\mathrm{Tr}[\rho_\mathrm{B}]) = 0.
\end{aligned}$$

Solution 8.5 A positive matrix must satisfy $\langle v|c|v\rangle \geq 0$ for any vector $|v\rangle$. It is straightforward to see that the matrix c satisfies this property,

$$\begin{aligned}\langle v|c|v\rangle &= \sum_{i,j} c_{ij}\langle v|i\rangle\langle j|v\rangle = \sum_{i,j} c_{ij}v_i^* v_j = \sum_{ij}\sum_{k,l}\langle F_i|E_{kl}\rangle\langle F_j|E_{kl}\rangle^* v_i^* v_j \\ &= \sum_{k,l}\Big(\sum_i \langle F_i|E_{kl}\rangle v_i^*\Big)\Big(\sum_j \langle F_j|E_{kl}\rangle^* v_j\Big) \\ &= \sum_{k,l}\Big|\sum_i \langle F_i|E_{kl}\rangle v_i^*\Big|^2 \geq 0.\end{aligned}$$

This shows that c is a positive matrix. All positive matrices are Hermitian.

Solution 8.6 S_1 is simply the time integral of the Hamiltonian. To derive S_2, we need to calculate the commutator of the interaction Hamiltonian at two different times,

$$\begin{aligned}[\tilde{H}(t_1), \tilde{H}(t_2)] &= \sum_{k,j=1}^{\infty} \hbar^2 g_k[\sigma_z \otimes (\tilde{b}_k(t_1) + \tilde{b}_k^\dagger(t_1)), \sigma_z \otimes (\tilde{b}_j(t_2) + \tilde{b}_j^\dagger(t_2))] \\ &= \sum_{k,j=1}^{\infty} \hbar^2 g_k g_j \Big([\tilde{b}_k(t_1), \tilde{b}_j^\dagger(t_2)] - [\tilde{b}_j(t_2), \tilde{b}_k^\dagger(t_1)]\Big) \\ &= \sum_{k,j=1}^{\infty} \hbar^2 g_k g_j \Big(\delta_{kj} e^{-i\omega_k t_1} e^{i\omega_j t_2} - \delta_{kj} e^{-i\omega_j t_2} e^{i\omega_k t_1}\Big) \\ &= i\underbrace{\sum_{j=1}^{\infty} 2\hbar^2 g_j^2 \sin(\omega_j(t_2 - t_1))}_{f(t_2-t_1)} = if(t_2 - t_1).\end{aligned}$$

Substituting this commutator into the expression of $S_2(t)$, we obtain

$$S_2(t) = \frac{1}{2!}\frac{1}{(i\hbar)^2}\int_0^t dt_1 \int_0^t dt_2 [H(t_1), H(t_2)] = i\underbrace{\Big(-\frac{1}{2\hbar^2}\int_0^t dt_1 \int_0^t dt_2 f(t_2 - t_1)\Big)}_{\phi(t)} = i\phi(t).$$

Since the commutator $[\tilde{H}(t_1), \tilde{H}(t_2)]$ is proportional to the identity, the terms S_k with $k \geq 3$ are all zero. Therefore,

$$U(t) = Te^{\frac{1}{i\hbar}\int_0^t dt' H(t')} = e^{S_1 + S_2} = e^{S_1}e^{S_2} = e^{\frac{1}{i\hbar}\int_0^t dt_1 H(t_1)} e^{i\phi(t)}.$$

Solution 8.7 The probability of measuring the qubit in the excited state stays constant

$$\begin{aligned}\tilde{\rho}_{11}(t) &= \rho_{11}(0)\mathrm{Tr}_\mathrm{B}[D_1(\vec{\alpha})|0\rangle\langle 0|D_1^\dagger(\vec{\alpha})] \\ &= \rho_{11}(0)\mathrm{Tr}_\mathrm{B}[D_1^\dagger(\vec{\alpha})D_1(\vec{\alpha})|0\rangle\langle 0|] \\ &= \rho_{11}(0)\mathrm{Tr}_\mathrm{B}[D_1(-\vec{\alpha})D_1(\vec{\alpha})|0\rangle\langle 0|] = \rho_{11}(0)\mathrm{Tr}_\mathrm{B}[|0\rangle\langle 0|] = \rho_{11}(0).\end{aligned}$$

With similar calculations, one can verify that $\tilde{\rho}_{00}(t) = \rho_{00}(0)$.

Solution 8.8 Let us compute the expectation value $\langle e^{iZ} \rangle$,

$$\langle e^{iZ} \rangle = \frac{1}{\sqrt{2\pi\sigma^2}} \int_{-\infty}^{\infty} e^{-\frac{z^2}{2\sigma^2}} e^{iz} dz = \frac{1}{\sqrt{2\pi\sigma^2}} \int_{-\infty}^{\infty} e^{-\frac{1}{2\sigma^2}(z^2 - 2i\sigma^2 z)} dz$$
$$= e^{-\frac{\sigma^2}{2}} \frac{1}{\sqrt{2\pi\sigma^2}} \int_{-\infty}^{\infty} e^{-\frac{1}{2\sigma^2}(z - i\sigma^2)^2} dz.$$

Now perform the change of variable $\{y = z - i\sigma^2\}$ and obtain

$$\langle e^{iZ} \rangle = e^{-\frac{\sigma^2}{2}} \frac{1}{\sqrt{2\pi\sigma^2}} \int_{-\infty}^{\infty} e^{-\frac{1}{2\sigma^2}y^2} dy = e^{-\frac{\sigma^2}{2}} = e^{-\frac{\langle Z^2 \rangle}{2}}.$$

Solution 8.9 The Hamiltonian of an oscillator coupled to a bath of harmonic oscillators is structurally identical to eqn (8.17) with the substitutions

$$-\frac{\omega_{\mathrm{q}}\sigma_z}{2} \mapsto \omega_{\mathrm{r}} a^\dagger a, \qquad \sigma^+ \mapsto a^\dagger, \qquad \sigma_- \mapsto a.$$

Applying these substitutions to the master equation (8.46), one obtains

$$\frac{d\rho(t)}{dt} = -i\omega_{\mathrm{r}}'[a^\dagger a, \rho] + \kappa(\overline{n}_{\mathrm{r}} + 1)\mathcal{D}[a](\rho) + \kappa\overline{n}_{\mathrm{r}}\mathcal{D}[a^\dagger](\rho). \tag{8.104}$$

For more information about the derivation of this master equation, see for example Chapter 1 of Ref. [228].

Solution 8.10 The integral (8.70) gives

$$\int_0^{\infty} A\omega e^{-\omega/\Omega} \frac{1 - \cos(\omega t)}{\omega^2} d\omega = \frac{1}{2} A \ln(1 + \Omega^2 t^2).$$

Therefore, the coherence decays as

$$\rho_{01}(t) = \rho_{01}(0) e^{-\frac{1}{2} A \ln(1 + \Omega^2 t^2)}.$$

Solution 8.11 Since $\sigma_x \sigma_x = \mathbb{1}$ and $\mathsf{X}_\pi = -i\sigma_x$, we have

$$e^{\frac{i}{2}\alpha\sigma_z} \mathsf{X}_\pi e^{\frac{i}{2}\beta\sigma_z} = \sigma_x \sigma_x e^{\frac{i}{2}\alpha\sigma_z} \mathsf{X}_\pi e^{\frac{i}{2}\beta\sigma_z} = -i\sigma_x \sigma_x e^{\frac{i}{2}\alpha\sigma_z} \sigma_x e^{\frac{i}{2}\beta\sigma_z}.$$

Now recall that for any unitary operator U, $Ue^{iV}U^\dagger = e^{iUVU^\dagger}$ (see Exercise 11.2). Therefore,

$$e^{\frac{i}{2}\alpha\sigma_z} \mathsf{X}_\pi e^{\frac{i}{2}\beta\sigma_z} = -i\sigma_x e^{\frac{i}{2}\alpha\sigma_x\sigma_z\sigma_x} e^{\frac{i}{2}\beta\sigma_z} = -i\sigma_x e^{-\frac{i}{2}\alpha\sigma_z} e^{\frac{i}{2}\beta\sigma_z} = \mathsf{X}_\pi e^{-\frac{i}{2}\alpha\sigma_z} e^{\frac{i}{2}\beta\sigma_z},$$

where we used $\sigma_x \sigma_z \sigma_x = -\sigma_z$.

Solution 8.12 The Bloch equations can be written in a more compact form as

$$\frac{d\vec{r}}{dt} = M\vec{r} + \vec{b} \tag{8.105}$$

where

$$M = \begin{bmatrix} -\gamma/2 - \gamma_\phi & 0 & 0 \\ 0 & -\gamma/2 - \gamma_\phi & \Omega \\ 0 & -\Omega & -\gamma \end{bmatrix}, \qquad \vec{b} = \begin{bmatrix} 0 \\ 0 \\ \gamma \end{bmatrix}.$$

The Bloch vector $\vec{r}_\infty = \vec{r}(t \to \infty)$ describes the qubit state at $t \to \infty$. This vector can be obtained by setting the time derivative to zero,

$$0 = M\vec{r}_\infty + \vec{b} \qquad \to \qquad \vec{r}_\infty = -M^{-1}\vec{b} = \begin{bmatrix} 0 \\ \frac{2\gamma\Omega}{\gamma^2 + 2\gamma\gamma_\phi + 2\Omega^2} \\ \frac{\gamma^2 + 2\gamma\gamma_\phi}{\gamma^2 + 2\gamma\gamma_\phi + 2\Omega^2} \end{bmatrix}.$$

We will focus on the case in which the strength of the drive is much higher than the decay rates, $\Omega \gg \{\gamma, \gamma_\phi\}$. The qubit decays to the center of the Bloch sphere after sufficiently long times, $\vec{r}_\infty = (0, 0, 0)$. By defining $\vec{r}\,' = \vec{r} - \vec{r}_\infty$, eqn (8.105) can be written as $d\vec{r}\,'/dt = M\vec{r}\,'$ whose solution is

$$\vec{r}\,'(t) = e^{Mt}\vec{r}\,'(0). \tag{8.106}$$

We will assume that the qubit starts in the ground state $\vec{r}(0) = (0, 0, 1)$, i.e. $\vec{r}\,'(0) = \vec{r}(0) - r_\infty = (0, 0, 1)$. The analytical expression of $\vec{r}\,'(t)$ can be obtained by diagonalizing the matrix M. The eigenvalues and normalized eigenvectors of M are

$$\begin{aligned} \lambda_1 &= -\frac{\gamma}{2} - \gamma_\phi, & \vec{m}_1 &= (1, 0, 0), \\ \lambda_2 &= -\frac{1}{4}(3\gamma + 2\gamma_\phi) - i\mu, & \vec{m}_2 &= \frac{1}{\sqrt{2}}\Big(0, \frac{2\gamma_\phi - \gamma}{4\Omega} + i\frac{\mu}{\Omega}, 1\Big), \\ \lambda_3 &= -\frac{1}{4}(3\gamma + 2\gamma_\phi) + i\mu, & \vec{m}_3 &= \frac{1}{\sqrt{2}}\Big(0, \frac{2\gamma_\phi - \gamma}{4\Omega} - i\frac{\mu}{\Omega}, 1\Big), \end{aligned}$$

where $\mu = \frac{1}{4}\sqrt{16\Omega^2 - (2\gamma_\phi - \gamma)^2}$. Since $\vec{r}\,'(0) = (0, 0, 1)$, the decomposition of $\vec{r}\,'(0)$ in the eigenbasis $\{\vec{m}_1, \vec{m}_2, \vec{m}_3\}$ is given by $\vec{r}\,'(0) = \alpha_1\vec{m}_1 + \alpha_2\vec{m}_2 + \alpha_3\vec{m}_3$ where

$$\alpha_1 = \vec{m}_1 \cdot \vec{r}\,'(0) = 0, \qquad \alpha_2 = \vec{m}_2 \cdot \vec{r}\,'(0) = 1/\sqrt{2}, \qquad \alpha_3 = \vec{m}_3 \cdot \vec{r}\,'(0) = 1/\sqrt{2}.$$

Thus, eqn (8.106) can be written as

$$\begin{aligned} \vec{r}\,'(t) = e^{Mt}\vec{r}\,'(0) &= e^{Mt}(\alpha_1\vec{m}_1 + \alpha_2\vec{m}_2 + \alpha_3\vec{m}_3) \\ &= \alpha_1 e^{\lambda_1 t}\vec{m}_1 + \alpha_2 e^{\lambda_2 t}\vec{m}_2 + \alpha_3 e^{\lambda_3 t}\vec{m}_3 \\ &= \frac{1}{\sqrt{2}} e^{-\frac{1}{4}(3\gamma + 2\gamma_\phi)t}\big(e^{-i\mu t}\vec{m}_2 + e^{i\mu t}\vec{m}_3\big). \end{aligned}$$

Since $\vec{r}(t) = \vec{r}\,'(t) + \vec{r}_\infty$, we have

$$\vec{r}(t) = \frac{1}{\sqrt{2}} e^{-\frac{1}{4}(3\gamma+2\gamma_\phi)t} \left(e^{-i\mu t}\vec{m}_2 + e^{i\mu t}\vec{m}_3\right) + \vec{r}_\infty.$$

Let us derive the r_z component (this will allow us to calculate the probability of measuring the qubit in the excited state as a function of time),

$$r_z(t) = \frac{1}{2} e^{-\frac{1}{4}(3\gamma+2\gamma_\phi)t} \left(e^{-i\mu t} + e^{i\mu t}\right) = e^{-\frac{1}{4}(3\gamma+2\gamma_\phi)t} \cos(\mu t)$$

Finally,

$$p_\mathrm{e}(t) = \rho_{11}(t) = \frac{1}{2} - \frac{1}{2} r_z(t) = \frac{1}{2} - \frac{1}{2} e^{-\frac{1}{4}(3\gamma+2\gamma_\phi)t} \cos(\mu t).$$

PART III
APPLICATIONS

9 Entanglement

Not all entangled states violate
Bell's inequality.

Entanglement played a key role in the development of quantum physics. The notion of entanglement dates back to 1935 when Schrödinger investigated measurement operations on the individual subsystems of a composite system [241]. It received particular attention when Einstein, Podolski, and Rosen published an influential paper on the incongruence between entanglement and physical theories based on locality and realism [242]. At that time, entanglement was considered a mysterious phenomenon of quantum physics and a quantitative analysis was still missing. In 1964, this gap was partially filled by the work of John Bell [243]. Bell's inequalities helped to transition the discussion about entanglement from metaphysics to quantitative predictions of the correlations between measurement outcomes. The first experimental tests on entangled states [244, 245] culminated with the work of Aspect et al. who showed a first convincing violation of Bell's inequalities in favor of quantum mechanical predictions [246, 247]. Technological progress over the last decades has made it possible to prepare, control, and measure entangled states with high precision [248, 249, 250]. Entanglement is now a resource actively used in quantum information processing to perform tasks that are impossible or less efficient with classical means.

As the amount of published work about entanglement is considerable [251, 252, 253, 254, 111], we do not aspire to outline an exhaustive review of this topic and we will limit our discussion to some central concepts. Our treatment starts with a definition of entanglement for pure states of bipartite systems. In Section 9.2, we present the Schmidt decomposition, a mathematical tool that can be used to understand whether a pure state is entangled or not. Section 9.3 explains how to use the von Neumann entropy to quantify the entanglement of a pure state. This part is followed by the discussion of some quantum protocols that use entanglement as a resource. Section 9.7 includes a presentation of locality and realism and a derivation of the CHSH inequality, an extended version of Bell's inequalities. The separability problem, namely understanding whether a density operator is entangled or not, is addressed in Section 9.8. Some techniques to solve the separability problem are discussed, including the PPT criterion, the reduction criterion, and entanglement witnesses. Section 9.9 explains how to quantify the entanglement of a mixed state.

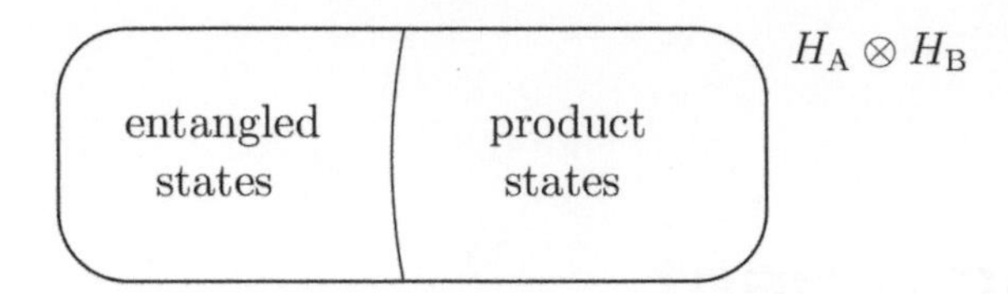

Fig. 9.1 Entangled states vs. product states. The Hilbert space of pure bipartite states, $H_A \otimes H_B$, is separated into two disjoint sets, one containing entangled states and one containing product states.

9.1 Definition and examples

Entanglement is a property of composite systems that can be quantified with suitable techniques. Consider a bipartite system with Hilbert space $H_A \otimes H_B$, where H_A and H_B are the Hilbert spaces of the individual subsystems. A pure state $|\psi\rangle \in H_A \otimes H_B$ is a **product state** if and only if it can be written as

$$|\psi\rangle = |\alpha\rangle_A \otimes |\beta\rangle_B, \tag{9.1}$$

where $|\alpha\rangle_A \in H_A$ and $|\beta\rangle_B \in H_B$. If a pure bipartite state cannot be expressed as a tensor product like in (9.1) in any basis, it is said to be[1] **entangled**. Product states and entangled states divide the Hilbert space $H_A \otimes H_B$ into two disjoint sets as shown in Fig. 9.1.

If the state of a bipartite system is a product state, it cannot be converted into an entangled state employing **local operations** (LO), i.e. transformations that act on the two subsystems separately. In mathematical terms, if $\mathcal{E}_A$ and $\mathcal{E}_B$ are two quantum maps acting on the individual subsystems and $|\psi\rangle = |\alpha\rangle \otimes |\beta\rangle$ is a product state, then $(\mathcal{E}_A \otimes \mathcal{E}_B)|\psi\rangle\langle\psi|$ is still a product state.

Important two-qubit entangled states are the **Bell states**:

$$|\beta_{00}\rangle = \frac{|00\rangle + |11\rangle}{\sqrt{2}}, \qquad |\beta_{01}\rangle = \frac{|01\rangle + |10\rangle}{\sqrt{2}}, \tag{9.2}$$

$$|\beta_{10}\rangle = \frac{|00\rangle - |11\rangle}{\sqrt{2}}, \qquad |\beta_{11}\rangle = \frac{|01\rangle - |10\rangle}{\sqrt{2}}. \tag{9.3}$$

They form a basis of the Hilbert space $\mathbb{C}^2 \otimes \mathbb{C}^2$ usually called the **Bell basis**. These states have an interesting feature: a projective measurement on one of the two qubits in *any* orthonormal basis produces two results with equal probability indicating a complete lack of information about the individual subsystems.

The Bell states can be generated using the quantum circuit shown in Fig. 9.2a. Any Bell state can be obtained from another by applying a unitary transformation to *just one* of the two qubits. Indeed, $|\beta_{ij}\rangle = \hat{U}_{ij} \otimes \mathbb{1}|\beta_{00}\rangle$ where $\hat{U}_{01} = \hat{X}$, $\hat{U}_{10} = \hat{Z}$ and $\hat{U}_{11} = \hat{Z}\hat{X}$ as illustrated in Fig. 9.2b. A **Bell measurement** is a projective

[1] In this chapter, we will mainly focus on entangled states of distinguishable particles, i.e. particles that are very far away from each other. The definition of entanglement for indistinguishable particles is much more subtle [255, 256].

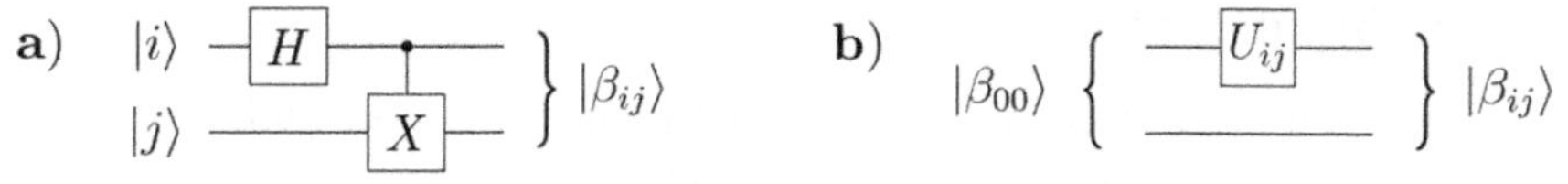

Fig. 9.2 Bell states. **a)** A quantum circuit for generating the Bell states $|\beta_{ij}\rangle$. **b)** A quantum circuit for the transformation of the Bell state $|\beta_{00}\rangle$ into any other Bell state $|\beta_{ij}\rangle$. Observe that the unitary U_{ij} operates on one qubit only.

measurement in the Bell basis $\{|\beta_{00}\rangle, |\beta_{01}\rangle, |\beta_{10}\rangle, |\beta_{11}\rangle\}$. To perform a measurement in such a basis, it is necessary to apply a CNOT and a Hadamard gate to the two qubits (i.e. the quantum circuit in Fig. 9.2a in reversed order) before executing a standard measurement in the computational basis.

In an experimental setting, entangled photons can be generated with multiple techniques [257]. A pioneering method to create polarization-entangled photons is based on a cascaded two-photon emission from a single atom, such as calcium [244, 245, 247]. In this scheme, the atom is excited from the ground state $|\mathrm{g}\rangle = 4\mathrm{s}^2\,{}^1\mathrm{S}_0$ to the excited state $|\mathrm{e}\rangle = 4\mathrm{p}^2\,{}^1\mathrm{S}_0$, which contains two electrons in the 4p orbital with zero total angular momentum. The excited state decays to the ground state through the $4\mathrm{p}^2\,{}^1S_0 \to 4\mathrm{s}4\mathrm{p}\,{}^1P_1 \to 4\mathrm{s}^2\,{}^1S_0$ cascade reaction, emitting two photons in opposite directions with wavelengths $\lambda_1 = 551.3$ nm and $\lambda_2 = 422.7$ nm. The polarization of the emitted photons is $\frac{1}{\sqrt{2}}(|\mathrm{HH}\rangle - |\mathrm{VV}\rangle)$, where H and V indicate the horizontal and vertical polarization. By defining $|\mathrm{H}\rangle = |0\rangle$ and $|\mathrm{V}\rangle = |1\rangle$, this entangled state is equivalent to the Bell state $|\beta_{10}\rangle$. Once the entangled state has been created, the two particles remain entangled until they eventually interact with other systems. It is worth pointing out that entanglement is a property not related to the distance between the particles.

The generation of entangled photons from atomic sources has some disadvantages, including the fact the photon pairs are emitted randomly in all directions and it is quite difficult to collect the emitted photons. A more practical method to generate polarization-entangled photons is based on parametric down-conversion (PDC), a nonlinear optical process in which two photons are produced from a parent photon. The interested reader can find more information about PDC in Refs [257, 258, 259, 260, 261]. An alternative way to prepare an entangled state is to run the quantum circuit shown in Fig. 9.2a on a quantum computer.

Example 9.1 Consider the two-qubit state $\frac{1}{\sqrt{2}}(|0\rangle|1\rangle + |1\rangle|1\rangle)$. This state is not entangled because it can be expressed as a product state $|\alpha\rangle|\beta\rangle$, where $|\alpha\rangle = \frac{1}{\sqrt{2}}(|0\rangle + |1\rangle)$ and $|\beta\rangle = |1\rangle$. Also the two-qubit state $\frac{1}{2}(|00\rangle + |01\rangle + |10\rangle + |11\rangle)$ is not entangled, since it can be expressed as a product state $\frac{1}{\sqrt{2}}(|0\rangle + |1\rangle) \otimes \frac{1}{\sqrt{2}}(|0\rangle + |1\rangle)$.

Example 9.2 Consider the two-qubit state $\frac{1}{\sqrt{2}}(|0\rangle|0\rangle - i|1\rangle|1\rangle)$. This state is entangled because it cannot be expressed in the form $|\alpha\rangle|\beta\rangle$ for any choice of $|\alpha\rangle$ and $|\beta\rangle$.

9.2 The Schmidt decomposition

Is there a standard procedure to understand whether a pure state of a bipartite system is entangled or not? We now introduce a mathematical tool, called the **Schmidt decomposition**, and we show how to use it to answer this question. Recall that a generic quantum state in $H_{\rm A} \otimes H_{\rm B}$ can be written as

$$|\psi\rangle = \sum_{i=1}^{d_{\rm A}} \sum_{j=1}^{d_{\rm B}} c_{ij} |i\rangle_{\rm A} |j\rangle_{\rm B}, \tag{9.4}$$

where $\{|i\rangle_{\rm A}\}_{i=1}^{d_{\rm A}}$ and $\{|j\rangle_{\rm B}\}_{j=1}^{d_{\rm B}}$ are orthonormal bases. The key idea of the Schmidt decomposition is that the entries c_{ij} define a $d_{\rm A} \times d_{\rm B}$ matrix c that can be expressed as $c = uav$, where u and v are unitary matrices and a is a rectangular diagonal matrix (this is nothing but the singular value decomposition presented in Section 3.2.13). The state $|\psi\rangle$ is entangled if and only if the matrix a has at least two non-zero entries on the diagonal. Let us present the Schmidt decomposition of a pure bipartite state.

Theorem 9.1 Consider the Hilbert space $H_{\rm A} \otimes H_{\rm B}$ and suppose $\dim(H_{\rm A}) = d_{\rm A}$ and $\dim(H_{\rm B}) = d_{\rm B}$ where $d_{\rm A} \leq d_{\rm B}$ without loss of generality. Then, any pure state $|\psi\rangle \in H_{\rm A} \otimes H_{\rm B}$ can be written in the Schmidt form [262]

$$\boxed{|\psi\rangle = \sum_{i=1}^{d_{\rm A}} a_i |\phi_i\rangle_{\rm A} |\varphi_i\rangle_{\rm B},} \tag{9.5}$$

where a_i are real numbers and the sets $\{|\phi_i\rangle_{\rm A}\}_i$ and $\{|\varphi_i\rangle_{\rm B}\}_i$ contain orthonormal vectors. This is called the Schmidt decomposition of $|\psi\rangle$. The number of non-zero terms in this sum is called the **Schmidt number** S and clearly $1 \leq S \leq d_{\rm A}$.

Proof For simplicity, let us assume that $\dim(H_{\rm A}) = \dim(H_{\rm B}) = d$. Consider a generic pure state $|\psi\rangle \in H_{\rm A} \otimes H_{\rm B}$ with representation $|\psi\rangle = \sum_{ij} c_{ij} |i\rangle|j\rangle$. The elements c_{ij} define a matrix c with $d \times d$ entries. Using the singular value decomposition, c can be written as $c = uav$ where a is a diagonal matrix with non-negative real elements and u and v are unitary matrices. By substituting $c_{ij} = \sum_k u_{ik} a_k v_{kj}$ into (9.4), we obtain

$$\begin{aligned} |\psi\rangle &= \sum_{i=1}^{d} \sum_{j=1}^{d} c_{ij} |i\rangle_{\rm A} |j\rangle_{\rm B} = \sum_{i=1}^{d} \sum_{j=1}^{d} \sum_{k=1}^{d} u_{ik} a_k v_{kj} |i\rangle_{\rm A} |j\rangle_{\rm B} \\ &= \sum_{k=1}^{d} a_k \Big(\sum_{i=1}^{d} u_{ik} |i\rangle_{\rm A} \Big) \Big(\sum_{j=1}^{d} v_{kj} |j\rangle_{\rm B} \Big) = \sum_{k=1}^{d} a_k |\phi_k\rangle_{\rm A} |\varphi_k\rangle_{\rm B}, \end{aligned}$$

where in the last step we defined the states $|\phi_k\rangle_{\rm A} = \sum_i u_{ik} |i\rangle_{\rm A}$ and $|\varphi_k\rangle_{\rm B} = \sum_i v_{kj} |j\rangle_{\rm B}$. The number of non-zero elements on the diagonal of a is the Schmidt number of the state $|\psi\rangle$. The general case $\dim H_{\rm A} \neq \dim H_{\rm B}$ is left as an exercise to the reader. □

The Schmidt decomposition provides an equivalent way to define entangled states: a pure bipartite state is entangled if and only if the Schmidt number S is greater or equal to two. The Schmidt number is invariant under local unitary transformations acting on subsystems A and B alone[2]. This is consistent with the fact that a product state cannot be converted into an entangled state by applying local transformations to the subsystems. Even if the Schmidt decomposition provides an easy way to understand whether a pure bipartite state is entangled or classically correlated, determining if a mixed state is entangled is much more complicated, as we shall see in Section 9.8.

The Schmidt decomposition is useful to calculate the partial traces of a pure bipartite state. Consider a pure bipartite state $|\psi\rangle \in H_{\mathrm{A}} \otimes H_{\mathrm{B}}$ and the associated density operator[3]

$$\rho = |\psi\rangle\langle\psi| = \sum_{k=1}^{d}\sum_{l=1}^{d} a_k a_l |kk\rangle\langle ll|,$$

where $|\psi\rangle = \sum_k a_k |k\rangle|k\rangle$ is the Schmidt decomposition of $|\psi\rangle$ and $d = \min\{d_{\mathrm{A}}, d_{\mathrm{B}}\}$. The reduced density operator ρ_{B} is given by

$$\begin{aligned}
\rho_{\mathrm{B}} = \mathrm{Tr}_{\mathrm{A}}[\rho] &= \mathrm{Tr}_{\mathrm{A}}\left[\sum_{k,l=1}^{d} a_k a_l \; |k\rangle\langle l|_{\mathrm{A}} \otimes |k\rangle\langle l|_{\mathrm{B}}\right] \\
&= \sum_{k,l=1}^{d}\sum_{i=1}^{d} a_k a_l \; \langle i|k\rangle\langle l|i\rangle \; |k\rangle\langle l|_{\mathrm{B}} \\
&= \sum_{k,l=1}^{d}\sum_{i=1}^{d} a_k a_l \; \delta_{ik}\delta_{li} \; |k\rangle\langle l|_{\mathrm{B}} \; = \sum_{k=1}^{d} a_k^2 \; |k\rangle\langle k|_{\mathrm{B}}.
\end{aligned} \tag{9.6}$$

The quantities a_k^2 are the eigenvalues of the reduced density operator ρ_{B}. With similar calculations, one can verify that the reduced density operator $\rho_{\mathrm{A}} = \mathrm{Tr}_{\mathrm{B}}[\rho]$ is $\rho_{\mathrm{A}} = \sum_k a_k^2 |k\rangle\langle k|_{\mathrm{A}}$. This indicates that ρ_{A} and ρ_{B} have the same eigenvalues and the same von Neumann entropy,

$$\boxed{S(\rho_{\mathrm{A}}) = -\sum_{k=1}^{d} a_k^2 \log_2\left(a_k^2\right) = S(\rho_{\mathrm{B}}).} \tag{9.7}$$

This result will be helpful in Section 9.3 when we introduce an entanglement measure for pure bipartite states.

Example 9.3 Consider the Bell state $|\beta_{00}\rangle = \frac{1}{\sqrt{2}}(|00\rangle + |11\rangle)$. The Schmidt decomposition of this state is $\sum_{i=0}^{1} a_i |i\rangle|i\rangle$ where $a_0 = a_1 = 1/\sqrt{2}$. This sum has two non-zero terms. Thus, $S = 2$ and $|\beta_{00}\rangle$ is entangled.

[2]Indeed, the states $\sum_k a_k |\phi_k\rangle|\varphi_k\rangle$ and $\sum_k a_k (\hat{U}_{\mathrm{A}}|\phi_k\rangle) \otimes (\hat{U}_{\mathrm{B}}|\varphi_k\rangle)$ have the same Schmidt number.

[3]To simplify the notation, we will often write the Schmidt decomposition as $\sum_k a_k |k\rangle|k\rangle$ instead of $\sum_k a_k |\phi_k\rangle|\varphi_k\rangle$.

Example 9.4 Consider the two-qubit state $|\psi\rangle = \frac{1}{\sqrt{2}}(|01\rangle + |11\rangle)$. As seen in Example 9.1, this is not an entangled state. The decomposition of this state in the canonical basis is $|\psi\rangle = \sum_{ij} c_{ij}|i\rangle|j\rangle$ where

$$c = \begin{bmatrix} c_{00} & c_{01} \\ c_{10} & c_{11} \end{bmatrix} = \begin{bmatrix} 0 & 1/\sqrt{2} \\ 0 & 1/\sqrt{2} \end{bmatrix}.$$

The singular value decomposition of this matrix is given by

$$c = uav = \begin{bmatrix} 1/\sqrt{2} & -1/\sqrt{2} \\ 1/\sqrt{2} & 1/\sqrt{2} \end{bmatrix} \begin{bmatrix} 1 & 0 \\ 0 & 0 \end{bmatrix} \begin{bmatrix} 0 & 1 \\ 1 & 0 \end{bmatrix}.$$

Since a has only one non-zero entry on the diagonal, $|\psi\rangle$ is not entangled.

9.3 Quantifying entanglement for pure states

Some states are "more entangled" than others. This concept can be formalized by introducing an **entanglement measure**, i.e. a function $E(\rho)$ that operates on density operators of bipartite systems and outputs a real number. If $E(\rho_1) > E(\rho_2)$, then ρ_1 is more entangled than ρ_2.

Let us focus our attention on pure states. It is reasonable to require that $E(\rho) = 0$ for product states because these states are not entangled. Since product states are invariant under local operations, it is also reasonable to require that $E(\rho) = E((\hat{U} \otimes \hat{V})\rho(\hat{U}^\dagger \otimes \hat{V}^\dagger))$ where $\hat{U}$ and $\hat{V}$ are two unitaries acting on the two subsystems separately. Lastly, only product states lead to reduced density operators that are pure. Therefore, determining the "mixedness" of the reduced density operator is a good way of quantifying entanglement. The von Neumann entropy of the reduced density operator $S(\rho_\mathrm{A})$ meets these requirements and is a common choice for quantifying the entanglement of pure states[4],

$$\boxed{E(\rho) = S(\mathrm{Tr}_\mathrm{B}[\rho]) = S(\rho_\mathrm{A}) = -\mathrm{Tr}_\mathrm{A}[\rho_\mathrm{A} \log_2 \rho_\mathrm{A}].} \tag{9.8}$$

If we substitute the spectral decomposition $\rho_\mathrm{A} = \sum_i \lambda_i |i\rangle\langle i|$ into (9.8), we arrive at

$$E(\rho) = -\sum_{i=1}^{d} \lambda_i \log_2 \lambda_i.$$

One can verify that this function is equal to zero for product states and reaches its maximum value $\log_2 d$ when $\rho_\mathrm{A} = \mathbb{1}/d$. When $E(\rho) = \log_2 d$ (or equivalently $\rho_\mathrm{A} = \mathbb{1}/d$), the quantum state ρ is said to be **maximally entangled**.

It is not immediately clear why the function $E(\rho)$ defined in (9.8) includes the reduced density operator ρ_A and not ρ_B. As seen in eqn (9.7), the von Neumann

[4]The von Neumann entropy was introduced in Section 6.1.2. It is defined as $S(\rho) = -\mathrm{Tr}[\rho \log_2 \rho]$. It is useful to remember that for a normal operator $\hat{A}$, $S(\hat{A}) = -\sum_i a_i \log_2(a_i)$, where a_i are the eigenvalues of $\hat{A}$.

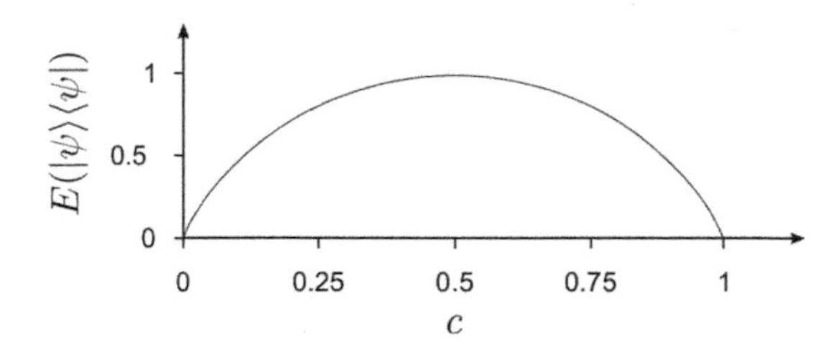

Fig. 9.3 Entanglement of a pure bipartite state. Plot of the entanglement of $|\psi\rangle = \sqrt{c}|00\rangle + \sqrt{1-c}|11\rangle$ as a function of c.

entropies $S(\rho_{\rm A})$ and $S(\rho_{\rm B})$ are identical. Thus, the entanglement of a bipartite state does not depend on the reduced density operator chosen to calculate the von Neumann entropy. We warn the reader that the function $S(\rho_{\rm A})$ cannot be used to quantify the entanglement of *mixed* states and other entanglement measures must be adopted. We will come back to this point in Section 9.9.

Example 9.5 The Bell state $|\beta_{00}\rangle$ is an example of a maximally entangled state. To see this, compute the reduced density operator

$$\rho_{\rm A} = {\rm Tr}_{\rm B}\left[|\beta_{00}\rangle\langle\beta_{00}|\right] = \frac{1}{2}|0\rangle\langle 0|_{\rm A} + \frac{1}{2}|1\rangle\langle 1|_{\rm A}.$$

Since $\rho_{\rm A}$ is proportional to the identity, $|\beta_{00}\rangle$ is maximally entangled. Indeed, the entanglement of $|\beta_{00}\rangle$ is

$$E(|\beta_{00}\rangle\langle\beta_{00}|) = S\left(\rho_{\rm A}\right) = -{\rm Tr}_{\rm A}\left[\rho_{\rm A}\log_2\rho_{\rm A}\right] = -\frac{1}{2}\log_2\left(\frac{1}{2}\right) - \frac{1}{2}\log_2\left(\frac{1}{2}\right) = 1.$$

With similar calculations, it is possible to show that the Bell states $|\beta_{01}\rangle$, $|\beta_{10}\rangle$, and $|\beta_{11}\rangle$ are maximally entangled too.

Example 9.6 Let us calculate the entanglement of the two-qubit state $|\psi\rangle = \sqrt{c}|00\rangle + \sqrt{1-c}|11\rangle$ where $0 \leq c \leq 1$ is a real parameter. The reduced density operator of the state $|\psi\rangle\langle\psi|$ is given by

$$\begin{aligned}\rho_{\rm A} &= {\rm Tr}_{\rm B}\left[\left(\sqrt{c}|00\rangle + \sqrt{1-c}|11\rangle\right)\left(\sqrt{c}\langle 00| + \sqrt{1-c}\langle 11|\right)\right] \\ &= c|0\rangle\langle 0|_{\rm A} + (1-c)|1\rangle\langle 1|_{\rm A}.\end{aligned}$$

Therefore,

$$E(|\psi\rangle\langle\psi|) = -c\log_2 c - (1-c)\log_2(1-c).$$

This function is plotted in Fig. 9.3. It is evident that $|\psi\rangle$ is not entangled for $c = 0$ and $c = 1$. When $c = 0.5$, $|\psi\rangle$ is maximally entangled.

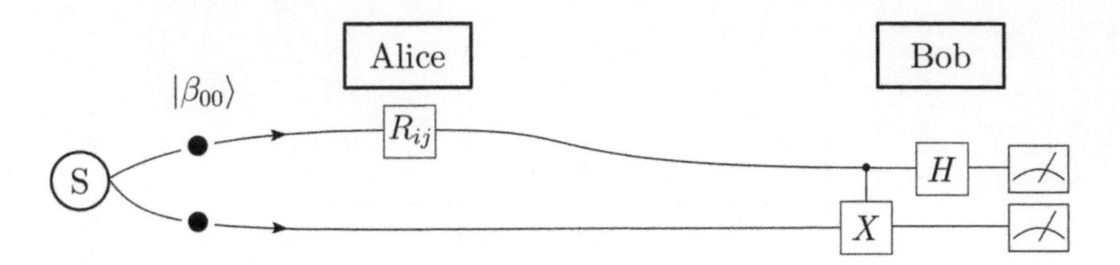

Fig. 9.4 Quantum dense coding. A source S prepares two particles in an entangled state. One particle moves toward Alice, the other toward Bob. After performing a unitary transformation, Alice sends her particle to Bob. He applies some gates to the two particles followed by a projective measurement. The measurement outcomes indicate the message Alice intended to transmit: 00, 01, 10, or 11.

Example 9.7 Consider the quantum state $\rho = \mathbb{1}/2 \otimes \mathbb{1}/2$. This is a classical state in which two parties prepare their qubit 50% of the time in $|0\rangle$ and 50% of the time in $|1\rangle$. The state ρ can be prepared by two parties with local operations on the two subsystems and thus it is not entangled. Nevertheless, $S(\mathrm{Tr}_{\mathrm{B}}[\rho]) = S(\mathbb{1}/2) = \log_2 2 = 1$. This shows that von Neumann entropy cannot be used as an entanglement measure for mixed states.

9.4 Quantum dense coding

Quantum dense coding is the simplest communication protocol that exploits quantum entanglement as a valuable resource [263]. This protocol allows Alice to send Bob **two bits** of classical information by sending him only **one qubit**. This is quite remarkable since it is generally believed that a qubit can only transmit one bit of information. The dense coding protocol is illustrated in Fig. 9.4. A source prepares two entangled qubits, such as two photons, in the Bell state $|\beta_{00}\rangle$. One of the two qubits is sent to Alice, the other one to Bob. Alice can choose to transmit one of these four messages: 00, 01, 10, and 11. To communicate this information, she operates a transformation $\hat{R}_{ij}$ on her qubit that depends on which message she would like to transmit:

Alice's message is		Alice applies		The state becomes
00	$\to$	$\hat{R}_{00} = \mathbb{1}$	$\to$	$\mathbb{1} \otimes \mathbb{1}|\beta_{00}\rangle = |\beta_{00}\rangle$
01	$\to$	$\hat{R}_{01} = \hat{X}$	$\to$	$\hat{X} \otimes \mathbb{1}|\beta_{00}\rangle = |\beta_{01}\rangle$
10	$\to$	$\hat{R}_{10} = \hat{Z}$	$\to$	$\hat{Z} \otimes \mathbb{1}|\beta_{00}\rangle = |\beta_{10}\rangle$
11	$\to$	$\hat{R}_{11} = \hat{Z}\hat{X}$	$\to$	$\hat{Z}\hat{X} \otimes \mathbb{1}|\beta_{00}\rangle = |\beta_{11}\rangle$

Alice sends her qubit to Bob, who will receive two qubits (one from her and one from the source) in an entangled state. As shown in the table above, this entangled state is one of the four Bell states. Bob now performs a Bell measurement: he applies a CNOT gate and a Hadamard gate to his qubits before measuring them in the computational basis as illustrated in Fig. 9.4. To summarize:

Bob receives		Bob applies		Bob obtains
$\|\beta_{00}\rangle$	$\to$	$(\hat{H}\otimes\mathbb{1})\mathsf{CNOT}$	$\to$	$\|00\rangle$
$\|\beta_{01}\rangle$	$\to$	$(\hat{H}\otimes\mathbb{1})\mathsf{CNOT}$	$\to$	$\|01\rangle$
$\|\beta_{10}\rangle$	$\to$	$(\hat{H}\otimes\mathbb{1})\mathsf{CNOT}$	$\to$	$\|10\rangle$
$\|\beta_{11}\rangle$	$\to$	$(\hat{H}\otimes\mathbb{1})\mathsf{CNOT}$	$\to$	$\|11\rangle$

Bob's measurement will project the two qubits in one of the four states $|00\rangle$, $|01\rangle$, $|10\rangle$, $|11\rangle$. These states correspond to the two classical bits that Alice wanted to communicate. It is worth mentioning that this protocol works perfectly without any classical communication between Alice and Bob. More information about the experimental implementation of quantum dense coding with trapped ions, optical fiber links, and nuclear magnetic resonance can be found in Refs [264, 265, 266].

9.5 Quantum teleportation of a qubit state

One of the most interesting communication schemes that exploits entangled states is the **quantum teleportation** protocol [267]. The words "quantum teleportation" might induce the reader to think that this protocol makes it possible to teleport an object from point A to point B instantly. Let us be clear about this point: this is not what quantum teleportation is about. This procedure instead allows for the transmission of quantum information between two parties even though they can only share classical information. Quantum teleportation has been experimentally implemented with different platforms, including entangled photons [268, 269, 270, 271, 250], and superconducting qubits on a silicon chip [249].

Let us begin our discussion by considering the following scenario: Alice has a qubit in an **unknown** quantum state $|\psi\rangle = \alpha|0\rangle + \beta|1\rangle$. She can communicate with Bob solely through a classical communication, such as a **telephone**. Alice would like to tell Bob all the information contained in the qubit state. However, she does not know the coefficients α and β and she has only **one copy** of $|\psi\rangle$. How can Alice faithfully communicate the qubit state to Bob relying only on a classical communication channel? In principle, Alice can measure her state in the computational basis and inform Bob of the result. However, this will not provide Bob with enough information to reconstruct the original state. This is because the description of $|\psi\rangle$ might require an infinite number of bits if α and β have an infinite decimal expansion and one measurement provides only one bit of information. Hence, the perfect communication of an unknown qubit state faces some challenges when Alice has only one copy of it, and the two parties can only communicate over the phone.

The situation changes completely if Alice and Bob share two qubits in an entangled state, such as two photons in the Bell state $|\beta_{00}\rangle$ as illustrated in Fig. 9.5. These two particles are generated by a source S located between them. Therefore, Alice owns two qubits (the qubit that she would like to transmit and half of the entangled pair) and Bob owns just one qubit (the other half of the entangled pair). As shown in Fig. 9.6, the initial state of the three qubits is

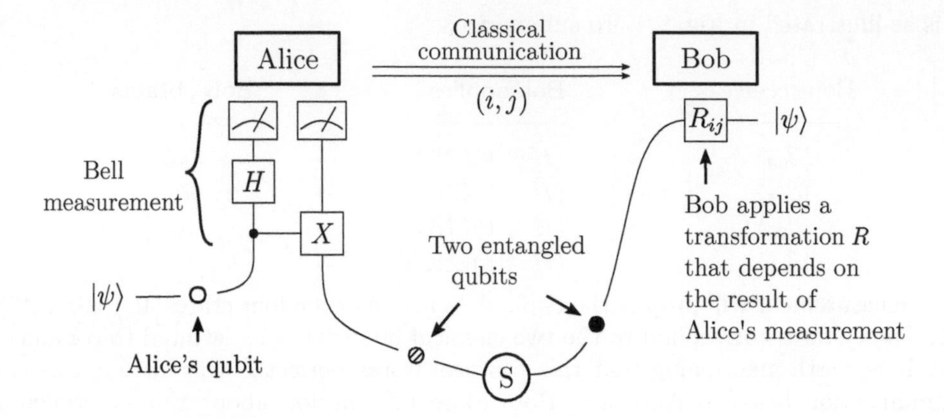

Fig. 9.5 Quantum teleportation. Alice wishes to transmit a single-qubit state $|\psi\rangle$ to Bob. The two parties share an entangled pair generated by a source S and can only communicate over the phone. For a faithful transmission, Alice entangles her particle with half of the entangled pair. She then performs a Bell measurement on her two particles and tells Bob the outcome of her measurement. Depending on this result, Bob performs a unitary transformation $\hat{R}_{ij}$ onto his particle. This operation leaves Bob's particle in the original quantum state Alice intended to transfer.

$$
\begin{aligned}
|\psi\rangle|\beta_{00}\rangle &= (\alpha|0\rangle + \beta|1\rangle)\,\frac{1}{\sqrt{2}}(|00\rangle + |11\rangle) \\
&= \frac{\alpha}{\sqrt{2}}|0\rangle\,(|00\rangle + |11\rangle) + \frac{\beta}{\sqrt{2}}|1\rangle(|00\rangle + |11\rangle).
\end{aligned} \tag{9.9}
$$

Alice receives the incoming particle and makes it interact with her qubit. This interaction consists of a CNOT followed by a Hadamard. The first CNOT transforms (9.9) into

$$
\frac{\alpha}{\sqrt{2}}|0\rangle\,(|00\rangle + |11\rangle) + \frac{\beta}{\sqrt{2}}|1\rangle(|10\rangle + |01\rangle),
$$

and the additional Hadamard yields

$$
\begin{aligned}
&\frac{\alpha}{\sqrt{2}}|+\rangle\,(|00\rangle + |11\rangle) + \frac{\beta}{\sqrt{2}}|-\rangle(|10\rangle + |01\rangle) = \\
&\quad = \frac{1}{2}\Big[\,|00\rangle\,(\alpha|0\rangle + \beta|1\rangle) + |01\rangle\,(\alpha|1\rangle + \beta|0\rangle) \\
&\qquad + |10\rangle\,(\alpha|0\rangle - \beta|1\rangle) + |11\rangle\,(\alpha|1\rangle - \beta|0\rangle)\Big].
\end{aligned} \tag{9.10}
$$

Alice now measures her two qubits in the computational basis. It is evident from (9.10) that this measurement will produce one of the states $\{|00\rangle, |01\rangle, |10\rangle, |11\rangle\}$ with equal probability. If the outcome is $|00\rangle$, the state of Bob's photon becomes $\alpha|0\rangle + \beta|1\rangle$, which is exactly the state Alice wished to transmit. If instead the outcome of Alice's measurement is $|01\rangle$, $|10\rangle$, or $|11\rangle$, Bob's particle will be projected onto $\alpha|1\rangle + \beta|0\rangle$,

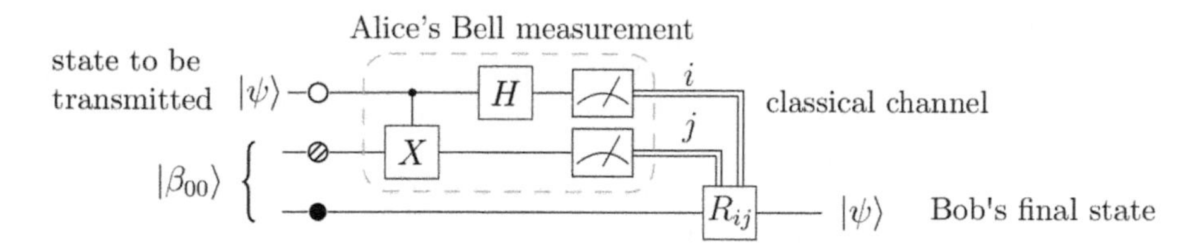

Fig. 9.6 The quantum teleportation circuit. This diagram represents the quantum circuit of the quantum teleportation protocol. A single line indicates a qubit; a double line represents a classical bit.

$\alpha|0\rangle - \beta|1\rangle$, and $\alpha|1\rangle - \beta|0\rangle$ respectively. These states do not correspond to the state Alice intended to transfer.

The crucial point is that Alice's measurement provides two bits of information that can easily be communicated to Bob over the phone. Depending on the bits received, Bob will apply a unitary transformation $\hat{R}_{ij}$ to his qubit in order to reconstruct the original state $|\psi\rangle$. The following table summarizes this process:

Bob receives over the phone		Bob applies		Bob obtains
00	$\rightarrow$	$\hat{R}_{00} = \mathbb{1}$	$\rightarrow$	$\mathbb{1}\,(\alpha|0\rangle + \beta|1\rangle) = |\psi\rangle$
01	$\rightarrow$	$\hat{R}_{01} = \hat{X}$	$\rightarrow$	$\hat{X}\,(\alpha|1\rangle + \beta|0\rangle) = |\psi\rangle$
10	$\rightarrow$	$\hat{R}_{10} = \hat{Z}$	$\rightarrow$	$\hat{Z}\,(\alpha|0\rangle - \beta|1\rangle) = |\psi\rangle$
11	$\rightarrow$	$\hat{R}_{11} = \hat{Z}\hat{X}$	$\rightarrow$	$\hat{Z}\hat{X}\,(\alpha|1\rangle - \beta|0\rangle) = |\psi\rangle$.

In a real experiment random noise will affect the manipulation and measurement of the qubits. Hence, Bob's final state ρ^{Bob} might be different from the original state $|\psi\rangle$ and the Jozsa fidelity of the overall protocol $F = \langle\psi|\rho^{\text{Bob}}|\psi\rangle$ might be less than one. In principle, the quantum teleportation protocol should perform equally well for all states $|\psi\rangle$ transmitted by Alice.

Some additional comments are in order. First, the quantum teleportation protocol does **not** allow the two parties to communicate any information faster than light. This is because Alice must call Bob over the phone and tell him the outcomes of her measurement. If this does not happen, Bob will not be able to reconstruct $|\psi\rangle$. Another interesting observation is that this protocol does not violate the no-cloning theorem presented in Section 4.1.1. Indeed, at the end of the protocol Alice's qubit is left either in $|0\rangle$ or $|1\rangle$ and the unknown quantum state $|\psi\rangle$ is transferred to Bob and not copied. Lastly, it is important to observe that Alice and Bob could store the entangled pair for unlimited time and use it only when necessary. However, this is challenging in an experimental situation since entangled pairs tend to have a short lifetime.

One can construct a much simpler protocol to transfer quantum information between two locations. Alice could send the qubit directly to Bob using a quantum communication channel. However, these types of channels are usually noisy and degrade the transmitted signal. The advantage of using the quantum teleportation protocol is that only a classical communication channel is required when the two parties share an entangled state.

It is interesting to compute the maximum fidelity achievable in a teleportation protocol when the two parties do **not** share an entangled state but can still communicate over the phone. The optimal strategy is for Alice to measure the unknown state $|\psi\rangle = \cos\frac{\theta}{2}|0\rangle + e^{i\phi}\sin\frac{\theta}{2}|1\rangle$ in the computational basis and guess the original qubit state based on whether it was found in $|0\rangle$ or $|1\rangle$. In practice, Alice will find her qubit in $|0\rangle$ with probability $p_0 = \cos^2\frac{\theta}{2}$. In this case, Bob will prepare his qubit in $|0\rangle$, and the Jozsa fidelity will be $F_0 = |\langle 0|\psi\rangle|^2 = \cos^2\frac{\theta}{2}$. Similarly, if Alice finds her qubit in $|1\rangle$ (which occurs with probability $p_1 = \sin^2\frac{\theta}{2}$), Bob will prepare his qubit in $|1\rangle$, and the Jozsa fidelity will be $F_1 = |\langle 1|\psi\rangle|^2 = \sin^2\frac{\theta}{2}$. The fidelity of the protocol is given by the weighted average of these two fidelities:

$$\boxed{F_{\text{classical}} = \frac{p_0 F_0 + p_1 F_1}{p_0 + p_1} = \cos^4\frac{\theta}{2} + \sin^4\frac{\theta}{2}.} \tag{9.11}$$

The fidelity is equal to 1 only when $\theta = 0$ or $\theta = \pi$, i.e. when the qubit state to be transferred is a classical bit. The average fidelity of this teleportation protocol "without entanglement" is given by the mean value of the fidelity over all possible initial states[5],

$$\overline{F}_{\text{classical}} = \frac{1}{4\pi}\int_0^{2\pi} d\phi \int_0^{\pi} d\theta \sin\theta \Big(\cos^4\frac{\theta}{2} + \sin^4\frac{\theta}{2}\Big) = 2/3. \tag{9.12}$$

In conclusion, if a teleportation protocol that exploits an entangled state does not reach an average fidelity greater than 2/3, then there is no actual advantage in using an entangled state.

Example 9.8 Suppose that the source generating the entangled state is faulty and generates the completely depolarized state $\frac{\mathbb{1}}{2} \otimes \frac{\mathbb{1}}{2}$ instead of an entangled state. The measurement performed by Alice does not affect Bob's subsystem and his qubit remains in the state $\frac{\mathbb{1}}{2}$. In addition, the transformation $\hat{R}_{ij}$ will leave Bob's qubit unchanged because $\hat{R}_{ij}\frac{\mathbb{1}}{2}\hat{R}_{ij}^\dagger = \frac{\mathbb{1}}{2}$. Thus, when the two parties share a completely depolarized state, the fidelity of the quantum teleportation protocol becomes $F = \langle\psi|\frac{\mathbb{1}}{2}|\psi\rangle = 1/2$ and the average fidelity is $\overline{F} = 1/2$.

Example 9.9 Suppose Alice does not communicate her measurement results to Bob. Hence, Bob will not apply any unitary transformation to his qubit. From eqns (6.52, 9.10), the final state of Bob's qubit is given by

$$\begin{aligned}\rho^{\text{Bob}} = \frac{1}{4}\text{Tr}_{\text{AB}}\Big[&|00\rangle\langle 00| \otimes |\psi\rangle\langle\psi| + |01\rangle\langle 01| \otimes \hat{X}|\psi\rangle\langle\psi|\hat{X} \\ &+|10\rangle\langle 10| \otimes \hat{Z}|\psi\rangle\langle\psi|\hat{Z} + |11\rangle\langle 11| \otimes \hat{X}\hat{Z}|\psi\rangle\langle\psi|\hat{Z}\hat{X}\Big] = \frac{\mathbb{1}}{2},\end{aligned}$$

[5] It can be shown that if Alice can measure N copies of her qubit state, then the maximum fidelity of this classical communication protocol is $\overline{F}_{\text{classical}} = N + 1/(N+2)$ [272]. In our case, $N = 1$ since she has only one copy of $|\psi\rangle$.

where in the last step we used $\sum_{k=0}^{3} \hat{\sigma}_k |\psi\rangle\langle\psi| \hat{\sigma}_k = 2\mathbb{1}$. In conclusion, Bob's qubit is left in a completely depolarized state and there is no information transfer.

9.5.1 Quantum teleportation for mixed states

It is natural to extend the quantum teleportation protocol to mixed states. Suppose that Alice would like to teleport a quantum state ρ and suppose that the qubit moving toward Bob is affected by some noise described by the quantum map $\mathcal{E}$ as shown in Fig. 9.7. The effect of this noise is to degrade the fidelity of the overall protocol. In the following, A and B will denote the two particles owned by Alice, while the letter C will indicate the qubit moving toward Bob. The initial state of the three particles is given by

$$\underbrace{\rho}_{\text{Alice's qubit}} \otimes \underbrace{[(\mathbb{1}_\mathrm{B} \otimes \overset{\text{Noise}}{\overset{\downarrow}{\mathcal{E}}}) \, |\beta_{00}\rangle\langle\beta_{00}|]}_{\text{Shared entangled state}}.$$

Alice performs a Bell measurement on her two qubits. This measurement is described by the projectors $|\beta_{ij}\rangle\langle\beta_{ij}|$. Using the notation presented in Appendix B.1, the projectors $|\beta_{ij}\rangle\langle\beta_{ij}|$ can be expressed as $\frac{1}{2}|\sigma_k\rangle\langle\sigma_k|$ where $k \in \{0, 1, 2, 3\}$. Alice's measurement yields one of four possible outcomes: 0, 1, 2, or 3. After this measurement, the state of Bob's qubit is the conditional state[6]

$$\rho_{\mathrm{C},k} = \frac{1}{p_k} \mathrm{Tr}_{\mathrm{AB}} \Big[\underbrace{\Big(\rho \otimes (\mathbb{1}_\mathrm{B} \otimes \mathcal{E}) |\beta_{00}\rangle\langle\beta_{00}| \Big)}_{\text{State of the three qubits}} \underbrace{\Big(\frac{1}{2} |\sigma_k\rangle\langle\sigma_k| \otimes \mathbb{1}_\mathrm{C} \Big)}_{\text{Bell measurement}} \Big], \tag{9.13}$$

where p_k is a probability distribution. By calculating this trace, it can be shown that the state of Bob's qubit after Alice's measurement is given by $\rho_{\mathrm{C},k} = \mathcal{E}(\hat{\sigma}_k \rho \hat{\sigma}_k)$ (see Exercise 9.7 for a detailed derivation). To summarize, we have the following scenarios:

Alice measures		Bob's qubit becomes	with probability
$\vert\beta_{00}\rangle$	$\rightarrow$	$\rho_{\mathrm{C},0} = \mathcal{E}(\rho)$	$p_0 = 1/4$
$\vert\beta_{01}\rangle$	$\rightarrow$	$\rho_{\mathrm{C},1} = \mathcal{E}(\hat{\sigma}_x \rho \hat{\sigma}_x)$	$p_1 = 1/4$
$\vert\beta_{11}\rangle$	$\rightarrow$	$\rho_{\mathrm{C},2} = \mathcal{E}(\hat{\sigma}_y \rho \hat{\sigma}_y)$	$p_2 = 1/4$
$\vert\beta_{10}\rangle$	$\rightarrow$	$\rho_{\mathrm{C},3} = \mathcal{E}(\hat{\sigma}_z \rho \hat{\sigma}_z)$	$p_3 = 1/4$

We note that when $\rho = |\psi\rangle\langle\psi|$ and $\mathcal{E} = \mathbb{1}$, the density operators $\rho_{\mathrm{C},k}$ reduce to the states $\alpha|0\rangle + \beta|1\rangle$, $\alpha|1\rangle + \beta|0\rangle$, $\alpha|1\rangle - \beta|0\rangle$ and $\alpha|0\rangle - \beta|1\rangle$ encountered in eqn (9.10).

Alice tells Bob the outcome of her measurement over the phone. Based on this information, Bob applies a transformation $\hat{\sigma}_k$ to recover Alice's original state. This operation will transform Bob's qubit into $\hat{\sigma}_k \mathcal{E}(\hat{\sigma}_k \rho \hat{\sigma}_k) \hat{\sigma}_k$. Thus, the final state of his qubit over many experiments is given by the weighted average

[6] The conditional state was presented in eqn (6.52).

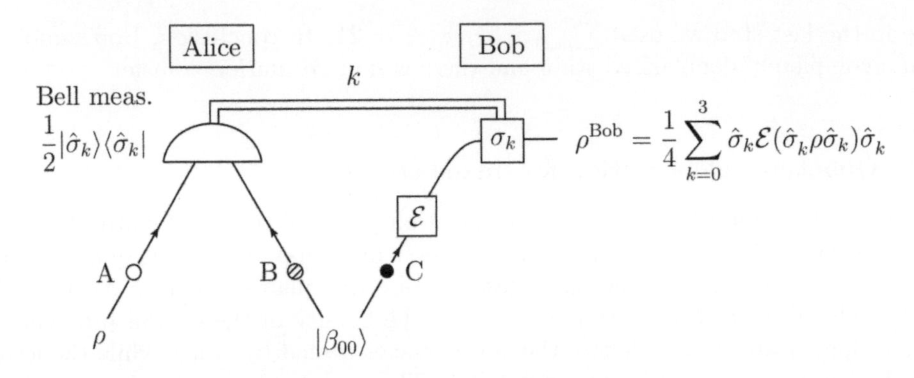

Fig. 9.7 Quantum teleportation protocol for mixed states. In this generalized scheme, Alice intends to transmit a density operator ρ to Bob. The channel connecting the source S to Bob is affected by some noise described by the quantum map $\mathcal{E}$. The qubits A, B, and C are indicated with black dots.

$$\boxed{\rho^{\text{Bob}} = \frac{1}{4}\sum_{k=0}^{3}\hat{\sigma}_k\mathcal{E}(\hat{\sigma}_k\rho\hat{\sigma}_k)\hat{\sigma}_k.} \tag{9.14}$$

Assuming that the state Alice wishes to teleport is pure $\rho = |\psi\rangle\langle\psi|$, the Jozsa fidelity of the overall protocol can be quantified as $F = \langle\psi|\rho^{\text{Bob}}|\psi\rangle$ and the average fidelity is given by $\overline{F} = \frac{1}{4\pi}\int_0^{2\pi} d\phi \int_0^{\pi} d\theta \sin\theta\, \langle\psi|\rho^{\text{Bob}}|\psi\rangle$.

Example 9.10 Suppose Alice wishes to teleport a pure state $\rho = |\psi\rangle\langle\psi|$ to Bob, and they share an entangled state $|\beta_{00}\rangle$. Let us assume that Bob's qubit is affected by the depolarizing channel $\mathcal{E}_{\text{depo}}(\rho) = p\rho + (1-p)\mathbb{1}/2$. From eqn (9.14), the teleported state will be

$$\begin{aligned}\rho^{\text{Bob}} &= \frac{1}{4}\Big[\mathcal{E}_{\text{depo}}(\rho) + \hat{\sigma}_x\mathcal{E}_{\text{depo}}(\hat{\sigma}_x\rho\hat{\sigma}_x)\hat{\sigma}_x + \hat{\sigma}_y\mathcal{E}_{\text{depo}}(\hat{\sigma}_y\rho\hat{\sigma}_y)\hat{\sigma}_y + \hat{\sigma}_z\mathcal{E}_{\text{depo}}(\hat{\sigma}_z\rho\hat{\sigma}_z)\hat{\sigma}_z\Big] \\ &= \mathcal{E}_{\text{depo}}(\rho).\end{aligned} \tag{9.15}$$

The fidelity of the protocol is $F = \langle\psi|\mathcal{E}_{\text{depo}}(|\psi\rangle\langle\psi|)|\psi\rangle = (1+p)/2$ and the average fidelity is $\overline{F} = \int F d\psi = (1+p)/2$. When $p < 1/3$, the noise is so severe that the fidelity of the quantum teleportation protocol is lower than the classical limit $\overline{F}_{\text{classical}} = 2/3$.

Example 9.11 The quantum teleportation protocol provides a beautiful interpretation of the Choi–Jamiolkowski isomorphism discussed in Section 7.9. To appreciate this physical interpretation, let us go through the main steps of the quantum teleportation protocol. Alice aims to transmit an unknown qubit state $\rho \in \text{D}(H_\text{A})$. The two parties are in contact over a classical communication channel and they share a pair of qubits in the state $\hat{R} \in \text{D}(H_\text{B} \otimes H_\text{C})$ (ideally, $\hat{R}$ is a maximally entangled state such as $|\beta_{00}\rangle\langle\beta_{00}|$). Alice performs a Bell measurement on her two qubits. Let us assume that her measurement produces $k = 0$, i.e. the two qubits are projected onto the Bell state $|\beta_{00}\rangle$. The state of Bob's qubit becomes

$$\rho_{\mathrm{C},0} = 2\mathrm{Tr}_\mathrm{C}[(\rho^\mathrm{T} \otimes \mathbb{1}_\mathrm{C})\hat{R}]. \tag{9.16}$$

Here, we used (9.45) with $k = 0$ and $\hat{R} = (\mathbb{1} \otimes \mathcal{E})|\beta_{00}\rangle\langle\beta_{00}|$. Equation (9.16) is the Choi–Jamiolkowski isomorphism $\mathrm{J}^{-1}[\hat{R}]$ presented in Section 7.9.

9.6 Entanglement swapping

Is it possible to entangle two particles even without them interacting directly? Yes, this is possible. **Entanglement swapping** is a technique to transfer the entanglement between two particles to two other particles that have never interacted in the past [273, 267, 274, 275]. Consider a source S_1 that emits *two particles* in the entangled state $|\beta_{00}\rangle$ and consider another source S_2 that emits *two particles* in the same state (see Fig. 9.8). One of the particles emitted by S_1 is sent to Alice and one particle generated by S_2 is sent to Bob. The other two particles instead are sent to Charlie, a third party. The initial state of the four particles is given by

$$\begin{aligned} |\beta_{00}\rangle|\beta_{00}\rangle &= \frac{1}{\sqrt{2}}\big(|00\rangle + |11\rangle\big)\frac{1}{\sqrt{2}}\big(|00\rangle + |11\rangle\big) \\ &= \frac{1}{2}[|0\rangle_\mathrm{A}\,|00\rangle_\mathrm{C}\,|0\rangle_\mathrm{B} + |0\rangle_\mathrm{A}\,|01\rangle_\mathrm{C}\,|1\rangle_\mathrm{B} + |1\rangle_\mathrm{A}\,|10\rangle_\mathrm{C}\,|0\rangle_\mathrm{B} + |1\rangle_\mathrm{A}\,|11\rangle_\mathrm{C}\,|1\rangle_\mathrm{B}]. \end{aligned} \tag{9.17}$$

We added the subscripts A, B, and C for clarity. Charlie performs a measurement in the Bell basis. This procedure consists in applying a CNOT gate and a Hadamard gate to his two qubits followed by a measurement in the computational basis. The CNOT and Hadamard gates transform (9.17) into

$$\begin{aligned} &\frac{1}{2}\left[|0\rangle\,\frac{|00\rangle + |10\rangle}{\sqrt{2}}\,|0\rangle + |0\rangle\,\frac{|01\rangle + |11\rangle}{\sqrt{2}}\,|1\rangle + |1\rangle\,\frac{|01\rangle - |11\rangle}{\sqrt{2}}\,|0\rangle + |1\rangle\,\frac{|00\rangle - |10\rangle}{\sqrt{2}}\,|1\rangle\right] = \\ &\quad= \frac{1}{2\sqrt{2}}\,[\,|0\rangle\,|00\rangle\,|0\rangle + |1\rangle\,|00\rangle\,|1\rangle + |0\rangle\,|01\rangle\,|1\rangle + |1\rangle\,|01\rangle\,|0\rangle \\ &\qquad +|0\rangle\,|10\rangle\,|0\rangle - |1\rangle\,|10\rangle\,|1\rangle + |0\rangle\,|11\rangle\,|1\rangle - |1\rangle\,|11\rangle\,|0\rangle\,]. \end{aligned}$$

This equation indicates that the measurement performed by Charlie will project the qubits owned by Alice and Bob onto one of the four Bell states:

Charlie measures		The state of particles A and B becomes
$\|00\rangle$	$\rightarrow$	$\frac{1}{\sqrt{2}}[\|00\rangle + \|11\rangle] = \|\beta_{00}\rangle$
$\|01\rangle$	$\rightarrow$	$\frac{1}{\sqrt{2}}[\|01\rangle + \|10\rangle] = \|\beta_{01}\rangle$
$\|10\rangle$	$\rightarrow$	$\frac{1}{\sqrt{2}}[\|00\rangle - \|11\rangle] = \|\beta_{10}\rangle$
$\|11\rangle$	$\rightarrow$	$\frac{1}{\sqrt{2}}[\|01\rangle - \|10\rangle] = \|\beta_{11}\rangle$

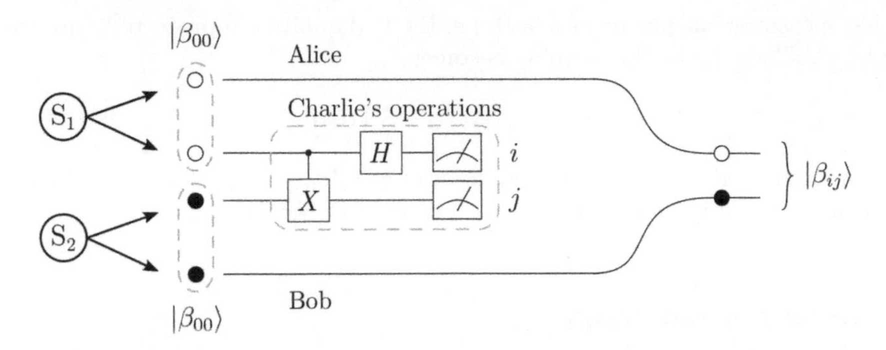

Fig. 9.8 Entanglement swapping. Two sources S_1 and S_2 generate two entangled pairs for a total of four particles. Two particles are sent to Charlie; the other two are shared between Alice and Bob. Charlie performs a Bell measurement on his two qubits. This operation leaves Alice and Bob's particles in an entangled state.

This protocol creates two entangled qubits (one owned by Alice and one owned by Bob) without them interacting. If Charlie tells Bob the outcome of his measurement, Bob can perform a local operation on his qubit and transform the entangled state into any other Bell state. This procedure can be repeated several times, enabling the distribution of entanglement between locations arbitrarily distant from one another. More details about the experimental implementation of entanglement swapping with an all-photonic scheme, a hybrid approach with photons and spins, and with quantum dots can be found in Refs [276, 277, 278].

9.7 Realism and locality

One of the most intriguing aspects of quantum entanglement is that it defies realism and locality, two seemingly natural assumptions of our rational thought.

Realism It is possible to associate definite values with all properties of a physical system. Measurement operations reveal these values without altering them.

According to this principle, measurement outcomes are determined by properties of the system prior to, and independent of, the measurement process. This worldview is aligned with our daily experience. For instance, suppose that a dark veil covers a chair. If we want to find out the color of the chair, we can simply remove the veil and have a look at the chair. If this observation tells us that the chair is red, we can confidently conclude that the chair must have been red *even before* our observation.

Locality is another concept inherited from our everyday experience.

Locality The outcome of a measurement in a region A is not affected by a spacelike separated event[7] happening in a region B.

[7]Consider two events with spacetime coordinates $(x_\mathrm{A}, t_\mathrm{A})$ and $(x_\mathrm{B}, t_\mathrm{B})$. These two events are said to be **spacelike separated** if $|x_\mathrm{A} - x_\mathrm{B}| > c|t_\mathrm{A} - t_\mathrm{B}|$, where c is the speed of light. In other words,

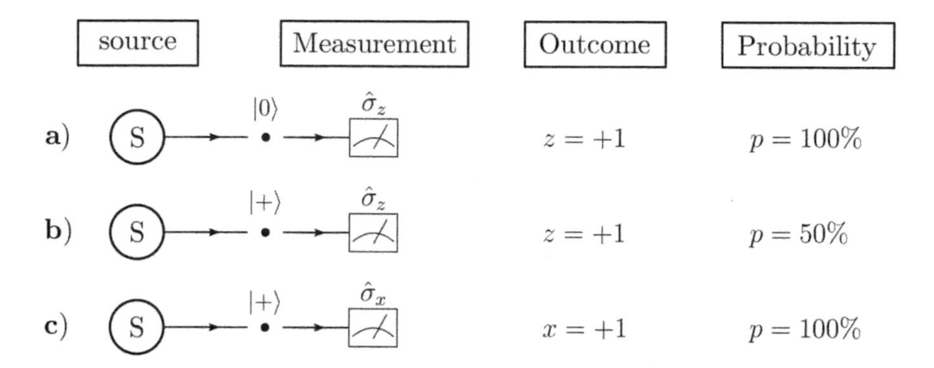

Fig. 9.9 Measurement of a qubit a) A source prepares a qubit in the state $|0\rangle$. A measurement of the observable $\hat{\sigma}_z$ will give $+1$ with 100% probability. **b)** The measurement of the superposition state $|+\rangle = (|0\rangle + |1\rangle)/\sqrt{2}$ in the computational basis will produce the state $|0\rangle$ with 50% probability. **c)** A source prepares a qubit in the state $|+\rangle$. Since this is an eigenvector of $\hat{\sigma}_x$ with eigenvalue $+1$, the measurement of this observable will give $+1$ with 100% probability.

The concept of locality can be better grasped with a simple example. Consider two chairs both covered with a dark veil. One chair is located next to Alice, the other chair is located next to Bob. Alice and Bob are very far away from one another. Suppose they remove the two veils at the same time. Alice discovers that her chair is red and Bob notices that his chair is blue. The measurement performed by Alice has not affected Bob's observation. This experiment is in agreement with the principle of locality.

In the remainder of this section, we will present some experiments in quantum mechanics that challenge the principles of locality and realism. Let us first review how measurement operations work in quantum mechanics. Suppose that a source S prepares a qubit in state $|0\rangle$ as shown in Fig. 9.9a. If this qubit is measured in the $\hat{\sigma}_z$ basis, the measurement outcome $z = +1$ will be observed with 100% probability. On the other hand, if the source prepares a superposition state $|+\rangle = \frac{1}{\sqrt{2}}[|0\rangle + |1\rangle]$, a measurement in the computational basis will collapse the state into $|0\rangle$ or $|1\rangle$ with equal probability (see Fig. 9.9b). This is because the particle polarization along the z direction is completely undetermined before the measurement and it is impossible to predict the outcome of each observation. Only the probability distribution over several observations acquires a precise meaning. However, if the measurement of the state $|+\rangle$ is performed in the $\hat{\sigma}_x$ basis, the outcome $x = +1$ will be observed with 100% probability (see Fig. 9.9c). This holds in general: given a pure qubit state, there always exists a projective measurement that gives a specific outcome with certainty[8].

two events are spacelike separated if there exists a reference frame in which the two events occur simultaneously but in two different places. It is worth mentioning that if $|x_A - x_B| \leq c|t_A - t_B|$, the two events are **timelike separated**, meaning that one event can influence the other.

[8]If the qubit state is $|\psi\rangle = \cos\frac{\theta}{2}|0\rangle + e^{i\phi}\sin\frac{\theta}{2}|1\rangle$, then the measurement of $\hat{\sigma}_{\hat{n}} = \hat{n}\cdot\vec{\sigma}$ with $\hat{n} = (\sin\theta\cos\phi, \sin\theta\sin\phi, \cos\theta)$ will give $+1$ with 100% probability.

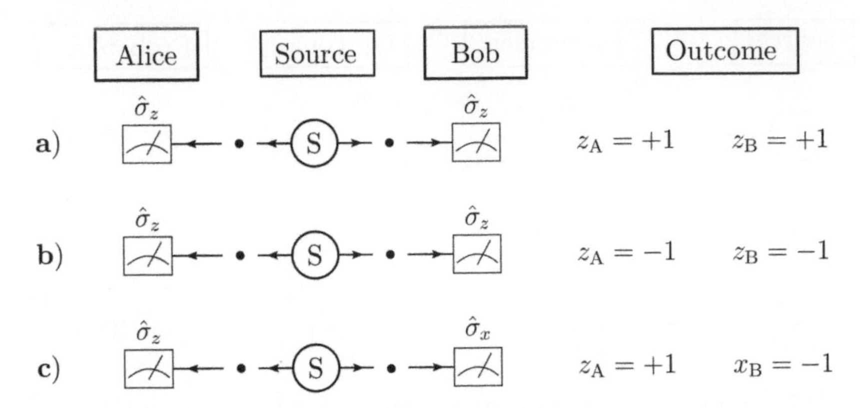

Fig. 9.10 Measurements of an entangled state. A source S prepares two particles in the entangled state $|\beta_{00}\rangle$. If Alice and Bob decide to measure $\hat{\sigma}_z$, they will both obtain +1 (panel **a**) or −1 (panel **b**). However, if Alice decides to measure $\hat{\sigma}_z$ and Bob measures $\hat{\sigma}_x$, their measurement outcomes will be completely uncorrelated (panel **c**).

Now consider a more interesting source S that prepares two photons in an entangled state of the polarization

$$|\psi\rangle = \frac{1}{\sqrt{2}}(|00\rangle + |11\rangle). \tag{9.18}$$

The two photons, depicted in Fig. 9.10, move in opposite directions toward Alice and Bob, who are located far apart. Both parties can measure any observable of the form $\hat{\sigma}_{\hat{n}}$, where $\hat{n}$ represents an arbitrary direction. In other words, they can align their detector along any axis. Suppose they both measure $\hat{\sigma}_z$ at the same time. According to eqn (9.18), Alice will obtain $z_A = +1$ and $z_A = -1$ with equal probability. If Alice's measurement yields $z_A = +1$, the global state collapses into $|00\rangle$ and she can predict with certainty that Bob's outcome will be $z_B = +1$. Similarly, if Alice's measurement gives $z_A = -1$, the global state collapses into $|11\rangle$ and Alice can predict with certainty that Bob's outcome will be $z_B = -1$. Therefore, their measurement outcomes are perfectly correlated when they both measure $\hat{\sigma}_z$. This is not surprising: also in classical physics many experiments show perfectly correlated results. For instance, suppose that there is a bag containing two balls, one red and one blue, placed in front of Alice and Bob. They both draw a ball from the bag. If Alice observes that her ball is blue, she can immediately deduce that Bob's ball is red. Their results are perfectly correlated.

What is really surprising is that the state (9.18) can also be expressed as[9]

$$|\psi\rangle = \frac{1}{\sqrt{2}}\left(|++\rangle + |--\rangle\right). \tag{9.19}$$

Therefore, if Alice decides to measure $\hat{\sigma}_x$ and she obtains $x_A = +1$, Bob will obtain $x_B = +1$ with certainty. Alice can decide which observable to measure while the two photons are still flying and well separated: if Alice measures $\hat{\sigma}_z$, Bob's particle

[9]The reader can check this by substituting $|+\rangle = (|0\rangle+|1\rangle)/\sqrt{2}$ and $|-\rangle = (|0\rangle-|1\rangle)/\sqrt{2}$ into (9.19).

collapses in an eigenstate of $\hat{\sigma}_z$ and it acquires a precise polarization along the z axis. If instead Alice measures $\hat{\sigma}_x$, Bob's particle collapses in an eigenstate of $\hat{\sigma}_x$ and acquires a precise polarization along the x axis. The operators $\hat{\sigma}_x$ and $\hat{\sigma}_z$ do not commute, meaning that if one of these observables is well defined, the other one is completely undetermined. This is quite remarkable: the polarization of Bob's particle is influenced by the measurement Alice performs on her subsystem! These arguments are clearly against the concept of locality since an event in region A instantly affects the outcome of a measurement performed in a spacelike separated region B. It is interesting to note that if the two observations happen at the same time, then there exists an inertial frame of reference with respect to which Alice's measurement happens before Bob's observation and one in which Bob's measurement precedes Alice's. This means that observers in different inertial frames will disagree on which measurement collapsed the entangled state first.

It is important to highlight that the Bell state (9.18) can be expressed as

$$\boxed{|\psi\rangle = \frac{1}{\sqrt{2}}(|0_{\hat{n}}0_{\hat{n}}\rangle + |1_{\hat{n}}1_{\hat{n}}\rangle),} \tag{9.20}$$

where $|0_{\hat{n}}\rangle$ and $|1_{\hat{n}}\rangle$ are the eigenstates of $\hat{\sigma}_{\hat{n}} = \hat{n} \cdot \vec{\sigma}$. Equation (9.20) indicates that there does not exist *any* direction of Alice's detector with respect to which a given polarization will be observed with certainty. In other words, Alice's particle does not have a definite polarization in *any* direction before the measurement. This is in contrast with the principle of realism.

These conclusions are quite unexpected and many scientists, including Einstein, did not find them convincing [242]. The principles of realism and locality are so well ingrained in our everyday life that it is hard to accept a physical theory that completely rejects them. There are only two possible explanations: either quantum theory has something missing or the way we look at our world must completely change.

9.7.1 CHSH inequality

At first, these arguments might seem a bit metaphysical and not easily verifiable. Does there exist an experiment that can tell us whether or not nature obeys the principles of locality and realism? In 1964 John Bell advanced the hypothesis that some correlation measurements on entangled states might not satisfy an inequality (now known as the **Bell's inequality**) which he derived from the plausible assumptions of realism and locality [243]. In this section, we present the derivation of the **CHSH inequality**, an extended version of Bell's inequality, obtained by Clauser, Horne, Shimony, and Holt [279].

To understand the derivation of the CHSH inequality, it is important that the reader completely forgets about quantum mechanics for a moment and pictures a source that emits two classical particles, such as two billiard balls, in opposite directions. One particle moves toward Alice, the other one toward Bob. As usual, Alice and Bob are far from one another. Alice can measure two observables denoted as A_1 and A_2. Similarly, Bob can measure two observables that will be indicated as B_1 and B_2. Their measurements will produce one of two possible outcomes, namely $+1$ or -1.

In more concrete terms, the observables A_1, A_2, B_1, and B_2 can be considered to be the sign of the angular momentum along an arbitrary direction. Alice can decide to measure either A_1 or A_2 during one run of the experiment. Similarly, Bob is free to measure either B_1 or B_2.

Let us assume that Alice and Bob measure one of the two observables at the same time. If the source prepares the two particles in a state with physical properties $A_1 = -1$ and $A_2 = B_1 = B_2 = 1$, the possible results are

A_1	A_2	B_1	B_2
-1		1	
-1			1
	1	1	
	1		1

Here, we are assuming that Alice's measurement does not affect Bob's one, i.e. the quantities B_1 and B_2 do not depend on which observable Alice decides to measure and vice versa (this is the principle of locality). In addition, we assume that the physical quantities A_1, A_2, B_1, and B_2 are well defined before the measurement and the observation of the two particles merely reveals these pre-existing properties (this is the concept of realism). Repeating the experiment under the same conditions will produce the same results. Lastly, if Alice and Bob decide to measure one of the two observables with uniform probability, the outcomes $\{A_1 = -1, B_1 = 1\}$, $\{A_1 = -1, B_2 = 1\}$, $\{A_2 = 1, B_1 = 1\}$, and $\{A_2 = 1, B_2 = 1\}$ will be observed with the same probability.

From the table above one can calculate the products A_1B_1, A_1B_2, A_2B_1, and A_2B_2 and define a **correlation function** C of the form

$$C = A_1B_1 + A_1B_2 + A_2B_1 - A_2B_2 = -2.$$

Now suppose that the source emits the two particles in a different state with physical properties $A_1 = A_2 = B_1 = 1$ and $B_2 = -1$. The possible results are

A_1	A_2	B_1	B_2
1		1	
1			-1
	1	1	
	1		-1

From this table, we can calculate C once again

$$C = A_1B_1 + A_1B_2 + A_2B_1 - A_2B_2 = 2.$$

After this experiment, the source prepares the two particles in a different state and Alice and Bob perform their measurements again. This procedure is repeated a large number of times N. We can define $p(a_1, a_2, b_1, b_2)$ as the probability that the source prepares the two particles in a state with physical properties $A_1 = a_1$, $A_2 = a_2$, $B_1 = b_1$, and $B_2 = b_2$. It is straightforward to show that the average of the random variable C is smaller than 2. Indeed,

$$\begin{aligned}\langle C\rangle &= \sum_{a_1a_2b_1b_2} p(a_1,a_2,b_1,b_2)(a_1b_1+a_1b_2+a_2b_1-a_2b_2)\\ &\le 2\sum_{a_1a_2b_1b_2} p(a_1,a_2,b_1,b_2) = 2,\end{aligned} \tag{9.21}$$

where we used the fact that[10] $a_1b_1+a_1b_2+a_2b_1-a_2b_2 \le 2$. This classical average can be expressed as

$$\begin{aligned}\langle C\rangle &= \sum_{a_1a_2b_1b_2} p(a_1,a_2,b_1,b_2)a_1b_1 + \sum_{a_1a_2b_1b_2} p(a_1,a_2,b_1,b_2)a_1b_2\\ &\quad+\sum_{a_1a_2b_1b_2} p(a_1,a_2,b_1,b_2)a_2b_1 - \sum_{a_1a_2b_1b_2} p(a_1,a_2,b_1,b_2)a_2b_2\\ &= \langle A_1B_1\rangle + \langle A_1B_2\rangle + \langle A_2B_1\rangle - \langle A_2B_2\rangle.\end{aligned} \tag{9.22}$$

By comparing (9.21) with (9.22), we arrive at

$$\boxed{\langle C\rangle = \langle A_1B_1\rangle + \langle A_1B_2\rangle + \langle A_2B_1\rangle - \langle A_2B_2\rangle \le 2.} \tag{9.23}$$

This is called the **CHSH inequality**. If Alice and Bob want to verify that their results are in agreement with (9.23), they can compare their N measurements, estimate the sample averages of the random variables A_iB_j,

$$\langle A_iB_j\rangle = \frac{A_i^{(1)}B_j^{(1)} + \ldots + A_i^{(N)}B_j^{(N)}}{N},$$

calculate the sample average of $\langle C\rangle$, and check whether the CHSH inequality (9.23) is valid within statistical uncertainty. If their results demonstrate that $\langle C\rangle > 2$, it would mean that a classical model relying on the principles of realism and locality is inadequate in accurately predicting the correlations. It is worth noting that the CHSH inequality does not stem from quantum mechanics principles but rather from a physical model that is based on the concepts of realism and locality [109].

Let us now go back to quantum mechanics and see whether the predictions of quantum theory are aligned with the classical ones. Suppose that the source prepares the two particles in the entangled state $|\psi\rangle = \frac{1}{\sqrt{2}}(|00\rangle + |11\rangle)$. As usual, the two particles move in opposite directions toward Alice and Bob, and their measurements occur simultaneously. They decide to measure the observables

$$\hat{A}_1 = \hat{\sigma}_z, \qquad \hat{A}_2 = \hat{\sigma}_x, \qquad \hat{B}_1 = \frac{\hat{\sigma}_x + \hat{\sigma}_z}{\sqrt{2}}, \qquad \hat{B}_2 = \frac{\hat{\sigma}_z - \hat{\sigma}_x}{\sqrt{2}}. \tag{9.24}$$

[10]It is easy to show that the quantity C is either $+2$ or -2. This follows from the fact that C can be written as $C = A_1(B_1+B_2) + A_2(B_1-B_2)$. Since A_1, A_2, B_1, and B_2 can either be $+1$ or -1, this means that

$$\text{if } B_1+B_2 = 0, \text{ then } C = A_2(B_1-B_2) = \pm 2,$$
$$\text{if } B_1-B_2 = 0, \text{ then } C = A_1(B_1+B_2) = \pm 2.$$

The expectation value of $\hat{A}_1 \otimes \hat{B}_1$ is

$$\langle \hat{A}_1 \otimes \hat{B}_1 \rangle_\psi = \frac{1}{2\sqrt{2}} \left(\langle 00| + \langle 11|\right) \hat{\sigma}_z \otimes (\hat{\sigma}_x + \hat{\sigma}_z) \left(|00\rangle + |11\rangle\right)$$
$$= \frac{1}{2\sqrt{2}} [(\langle 00| + \langle 11|) (|01\rangle - |10\rangle) + (\langle 00| + \langle 11|) (|00\rangle + |11\rangle)] = \frac{1}{\sqrt{2}}.$$

With similar calculations, one can verify that

$$\langle \hat{A}_1 \otimes \hat{B}_2 \rangle_\psi = \frac{1}{\sqrt{2}}, \qquad \langle \hat{A}_2 \otimes \hat{B}_1 \rangle_\psi = \frac{1}{\sqrt{2}}, \qquad \langle \hat{A}_2 \otimes \hat{B}_2 \rangle_\psi = -\frac{1}{\sqrt{2}}.$$

Therefore, the expectation value of the operator

$$\boxed{\hat{C} = \hat{A}_1 \otimes \hat{B}_1 + \hat{A}_1 \otimes \hat{B}_2 + \hat{A}_2 \otimes \hat{B}_1 - \hat{A}_2 \otimes \hat{B}_2}$$

is given by

$$\boxed{\langle \hat{C} \rangle_\psi = 2\sqrt{2}.}$$

According to quantum mechanics, the average value of $\hat{C}$ can be greater than 2, in contrast with the classical bound (9.23). Which prediction is correct? To answer this question, it is necessary to perform an experiment and test the models. In the last decades, multiple experiments have repeatedly shown that $\langle \hat{C} \rangle$ can be greater than 2 when the two particles are prepared in an entangled state and suitable dichotomic observables are measured[11]. Bell test experiments indicate that our world is not local *and* realistic [280, 281, 282]. This leads to three scenarios: 1) the principle of locality is incorrect, 2) the principle of realism is incorrect, or 3) both are incorrect. Abandoning the principle of locality goes against the fundamentals of special relativity (as a signal cannot travel faster than the speed of light). Since there is no experimental evidence supporting faster-than-light signal transmission, it is more reasonable to reject the assumption of realism.

9.7.2 The Tsirelson bound

What is the maximum value of $\langle \hat{C} \rangle_\psi$ predicted by quantum mechanics? Let us show that quantum theory predicts that $\langle \hat{C} \rangle_\psi$ cannot be greater than $2\sqrt{2}$. This is called the **Tsirelson bound** [283]. The observables measured by Alice and Bob can be expressed as

$$\hat{A}_1 = \vec{\sigma} \cdot \mathbf{a}_1, \qquad \hat{A}_2 = \vec{\sigma} \cdot \mathbf{a}_2, \qquad \hat{B}_1 = \vec{\sigma} \cdot \mathbf{b}_1, \qquad \hat{B}_2 = \vec{\sigma} \cdot \mathbf{b}_2,$$

where $\mathbf{a}_i$ and $\mathbf{b}_i$ are three-dimensional unit vectors. Since $\hat{A}_1^2 = \hat{A}_2^2 = \hat{B}_1^2 = \hat{B}_2^2 = \mathbb{1}$, the operator $\hat{C}^2$ is given by

[11]A dichotomic observable is an observable that can take only two values, such as +1 and −1. A Pauli operator is an example of a dichotomic observable. We note that if the source prepares the entangled state $|\beta_{00}\rangle$ and the two parties measure $\hat{A}_1 = \hat{A}_2 = \hat{B}_1 = \hat{B}_2 = \hat{\sigma}_z$, then $|\langle \hat{C} \rangle_\psi| \leq 2$ and the CHSH inequality is not violated. This shows that a pure entangled state might not violate the CHSH inequality if the dichotomic observables $\hat{A}_1$, $\hat{A}_2$, $\hat{B}_1$ and $\hat{B}_2$ are not chosen in a suitable way.

$$\hat{C}^2 = (\hat{A}_1 \otimes \hat{B}_1 + \hat{A}_2 \otimes \hat{B}_1 + \hat{A}_1 \otimes \hat{B}_2 - \hat{A}_2 \otimes \hat{B}_2)^2$$
$$= 4(\mathbb{1} \otimes \mathbb{1}) - [\hat{A}_1, \hat{A}_2] \otimes [\hat{B}_1, \hat{B}_2]. \tag{9.25}$$

The operator norm of this Hermitian operator is bounded by[12]

$$\begin{aligned}\|\hat{C}^2\|_o &\leq \|4(\mathbb{1} \otimes \mathbb{1})\|_o + \|[\hat{A}_1, \hat{A}_2] \otimes [\hat{B}_1, \hat{B}_2]\|_o \\ &= 4 + \|[\hat{A}_1, \hat{A}_2]\|_o \cdot \|[\hat{B}_1, \hat{B}_2]\|_o \\ &\leq 4 + \left(\|\hat{A}_1\hat{A}_2\|_o + \|\hat{A}_2\hat{A}_1\|_o\right) \cdot \left(\|\hat{B}_1\hat{B}_2\|_o + \|\hat{B}_2\hat{B}_1\|_o\right) \\ &\leq 4 + 2\|\hat{A}_1\|_o \cdot \|\hat{A}_2\|_o \cdot 2\|\hat{B}_1\|_o \cdot \|\hat{B}_2\|_o \leq 8,\end{aligned} \tag{9.26}$$

where in the last step we used the fact that the operator norms of $\hat{A}_i$ and $\hat{B}_i$ are smaller than one. Equation (9.26) implies that $\|\hat{C}\|_o \leq 2\sqrt{2}$. Since $\|\hat{C}\|_o$ is the largest eigenvalue of $\hat{C}$, the expectation value $\langle\hat{C}\rangle_\psi$ satisfies

$$\boxed{\langle\hat{C}\rangle_\psi \leq |\langle\psi|\hat{C}|\psi\rangle| \leq \|\hat{C}\|_o \leq 2\sqrt{2}.}$$

When either $\hat{A}_1, \hat{A}_2$ or $\hat{B}_1, \hat{B}_2$ commute, eqn (9.25) reduces to $\hat{C} = 2\,(\mathbb{1} \otimes \mathbb{1})$ and we recover the classical bound $\langle\hat{C}\rangle_\psi \leq 2$.

A natural question arises at this point: have Bell test experiments ever shown correlations greater than $2\sqrt{2}$? All the investigations conducted so far have always produced correlations not greater than $2\sqrt{2}$ in agreement with quantum mechanical predictions [280, 281, 282]. In conclusion, not only does quantum theory predict correlations greater than the classical ones, but it also correctly predicts their maximum value.

9.7.3 Gisin theorem

Does an entangled state always violate a CHSH inequality? Yes, if the entangled state is pure. This is known as the **Gisin theorem.**

Theorem 9.2 (Gisin) Consider an entangled state $|\psi\rangle \in H_\mathrm{A} \otimes H_\mathrm{B}$. Then, there exist four dichotomic observables $\hat{A}_1$, $\hat{A}_2$, $\hat{B}_1$, $\hat{B}_2$ such that the operator

$$\hat{C} = \hat{A}_1 \otimes \hat{B}_1 + \hat{A}_1 \otimes \hat{B}_2 + \hat{A}_2 \otimes \hat{B}_1 - \hat{A}_2 \otimes \hat{B}_2 \tag{9.27}$$

has expectation value $\langle\psi|\hat{C}|\psi\rangle > 2$.

---

[12]In these calculations, we used the triangle inequality $\|\hat{A} + \hat{B}\| \leq \|\hat{A}\| + \|\hat{B}\|$. We also used the properties $\|\hat{A} \otimes \hat{B}\| = \|\hat{A}\| \cdot \|\hat{B}\|$ and $\|\hat{A} \cdot \hat{B}\| \leq \|\hat{A}\| \cdot \|\hat{B}\|$ which are valid for several norms, including the operator norm and the Hilbert–Schmidt norm presented in Section 3.2.2 and 3.2.14.

Proof We prove this theorem for the simplified case $H_A = H_B = \mathbb{C}^2$. For a more general treatment see Refs [284, 285]. By using the Schmidt decomposition, any two-qubit state can be expressed in the form $|\psi\rangle = c_0|00\rangle + c_1|11\rangle$, where c_i are non-negative numbers that satisfy the normalization condition $c_0^2 + c_1^2 = 1$. Since $|\psi\rangle$ is entangled, both c_0 and c_1 are different from zero. The proof of the theorem relies on the identity

$$\begin{aligned}\langle\psi|\mathbf{a}\cdot\vec{\sigma}\otimes\mathbf{b}\cdot\vec{\sigma}|\psi\rangle &= \langle\psi|\,(a_x\hat{\sigma}_x + a_y\hat{\sigma}_y + a_z\hat{\sigma}_z)\otimes(b_x\hat{\sigma}_x + b_y\hat{\sigma}_y + b_z\hat{\sigma}_z)\,|\psi\rangle \\ &= a_z b_z + 2c_0c_1(a_x b_x - a_y b_y), \end{aligned} \tag{9.28}$$

that the reader can check with a direct calculation. Now consider the observables

$$\begin{aligned} \hat{A}_1 &= \hat{\sigma}_z, & \hat{A}_2 &= \hat{\sigma}_x, \\ \hat{B}_1 &= \sin(\beta_1)\,\hat{\sigma}_x + \cos(\beta_1)\,\hat{\sigma}_z, & \hat{B}_2 &= \sin(\beta_2)\,\hat{\sigma}_x + \cos(\beta_2)\,\hat{\sigma}_z. \end{aligned}$$

By using (9.28), we have

$$\begin{aligned}\langle\hat{C}\rangle_\psi &= \langle\hat{A}_1\otimes\hat{B}_1\rangle_\psi + \langle\hat{A}_1\otimes\hat{B}_2\rangle_\psi + \langle\hat{A}_2\otimes\hat{B}_1\rangle_\psi - \langle\hat{A}_2\otimes\hat{B}_2\rangle_\psi \\ &= \cos\beta_1 + \cos\beta_2 + 2c_0c_1\sin\beta_1 - 2c_0c_1\sin\beta_2. \end{aligned}$$

We can set the angles $\beta_1 = \arctan(2c_0c_1)$ and $\beta_2 = -\arctan(2c_0c_1)$. Since $\cos(\arctan x) = 1/\sqrt{1+x^2}$ and $\sin(\arctan x) = x/\sqrt{1+x^2}$, we finally have $\langle\hat{C}\rangle_\psi = 2\sqrt{1+(2c_0c_1)^2}$. This expectation value is greater than 2 when c_0 and c_1 are both non-zero, i.e. when $|\psi\rangle$ is entangled. □

The Gisin theorem does not specify the dimensionality of H_A and H_B. This suggests that particles with arbitrarily large spin can violate a CHSH inequality. This corroborates the interpretation that classical properties do not simply emerge from quantum systems with "large" quantum numbers [284].

9.8 Entanglement for mixed states

So far, we have only considered pure states. In this section, we generalize the notion of entanglement to mixed states using the definition given by Werner in 1989 [286].

Definition 9.1 A quantum state of a bipartite system is separable if it can be prepared with local operations on the subsystems that might be coordinated via a classical communication channel (these types of operations are usually indicated with the acronym LOCC, see Fig. 9.11). In mathematical terms, a quantum state $\rho \in \mathrm{D}\{H_A \otimes H_B\}$ is called **separable** if it can be written as[13]

$$\boxed{\rho = \sum_i p_i \rho_i^{\mathrm{A}} \otimes \rho_i^{\mathrm{B}},} \tag{9.29}$$

[13]To simplify the calculations, it is often convenient to express eqn (9.29) as $\rho = \sum_i p_i|a_i\rangle\langle a_i|\otimes|b_i\rangle\langle b_i|$, where $|a_i\rangle$ and $|b_i\rangle$ are purifications of ρ_i^{A} and ρ_i^{B}.

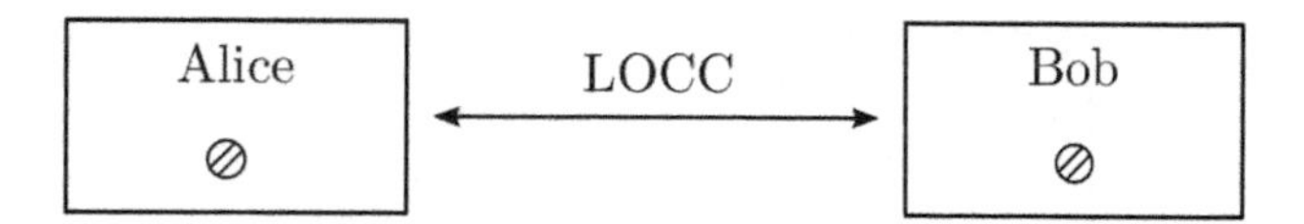

Fig. 9.11 Separable states. Alice and Bob are in two different labs and each owns a quantum system. They can perform any quantum operation on their local subsystem coordinated via a classical communication channel, such as a phone. These LOCC operations are not sufficient to create an entangled state.

where $\{p_i\}_i$ is a probability distribution and ρ_i^{A} and ρ_i^{B} are generic density operators. The set of separable states will be denoted as $\mathrm{S}(H \otimes H)$.

Quantum states that cannot be decomposed as in eqn (9.29) are called **entangled states**, and determining whether a given density operator is entangled or separable is known as the **separability problem**. This is an NP hard problem. We will discuss various techniques to address the separability problem in the next sections.

Interestingly, a linear combination of entangled states might not be entangled. Let us explain this point with a simple example. Consider a source S_1 that emits two qubits in the entangles states $|\beta_{00}\rangle$ and $|\beta_{10}\rangle$ with the same probability. The density operator describing this ensemble is

$$\rho_1 = \frac{1}{2}|\beta_{00}\rangle\langle\beta_{00}| + \frac{1}{2}|\beta_{10}\rangle\langle\beta_{10}| = \frac{1}{2}|00\rangle\langle 00| + \frac{1}{2}|11\rangle\langle 11| = \rho_2.$$

The state ρ_2 is not entangled: Alice and Bob can create this state by agreeing over the phone to prepare their qubits in $|00\rangle$ or in $|11\rangle$. Since the convex sum of two entangled states does not always produce an entangled state, the set of entangled states is not convex.

On the other hand, the set of separable states is convex. In other words, if ρ_1 and ρ_2 are separable and $0 \leq \lambda \leq 1$ is a real number, then the state $\rho = \lambda\rho_1 + (1-\lambda)\rho_2$ is still separable. This is because

$$\begin{aligned}
\rho = \lambda\rho_1 + (1-\lambda)\rho_2 &= \sum_{k=1}^{N_1} \lambda p_k |a_k\rangle\langle a_k| \otimes |b_k\rangle\langle b_k| + \sum_{j=1}^{N_2} (1-\lambda) q_j |c_j\rangle\langle c_j| \otimes |d_j\rangle\langle d_j| \\
&= \sum_{i=1}^{N_1+N_2} s_i |e_i\rangle\langle e_i| \otimes |f_i\rangle\langle f_i|,
\end{aligned}$$

where $\{s_i\}_i$ is a probability distribution, $\{|e_i\rangle = |a_i\rangle, |f_i\rangle = |b_i\rangle\}$ for $i \in (1, \ldots, N_1)$, and $\{|e_i\rangle = |c_i\rangle, |f_i\rangle = |d_i\rangle\}$ for $i \in (N_1+1, \ldots, N_1+N_2)$. Entangled states and separable states divide the set of density operators $\mathrm{D}(H \otimes H)$ into two disjoint sets as illustrated in Fig. 9.12.

Example 9.12 Suppose that Alice and Bob have one qubit each. Alice prepares her qubit 50% of the time in $|0\rangle$ and 50% of the time in $|1\rangle$. Bob does the same thing with

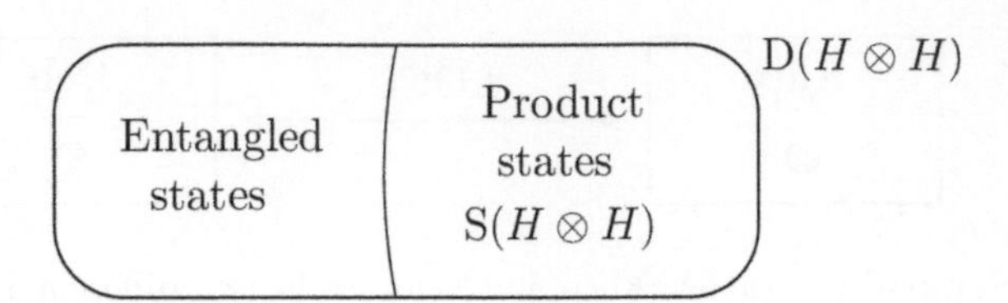

Fig. 9.12 Entangled states vs. product states. The set of density operators $\mathrm{D}(H \otimes H)$ is divided into two disjoint subsets: the set of separable states and entangled states. The former is convex, the latter is not.

---

his qubit. The state of the global system is

$$\rho = \frac{1}{2}(|0\rangle\langle 0|_{\mathrm{A}} + |1\rangle\langle 1|_{\mathrm{A}}) \otimes \frac{1}{2}(|0\rangle\langle 0|_{\mathrm{B}} + |1\rangle\langle 1|_{\mathrm{B}}) = \mathbb{1}_{\mathrm{A}}/2 \otimes \mathbb{1}_{\mathrm{B}}/2.$$

This state is separable since the two parties can create it by performing local operations on the two separate subsystems. Furthermore, it is expressed as in eqn (9.29).

Example 9.13 Consider the two-qubit state $\frac{1}{2}(|00\rangle\langle 00| + |11\rangle\langle 11|)$. This state is separable because it is written as in (9.29). It is interesting to note that two parties cannot prepare this state without sharing some classical information. Whenever Alice prepares her qubit in $|0\rangle$ ($|1\rangle$), Bob must prepare his qubit in $|0\rangle$ ($|1\rangle$) as well. The preparation of the states $|00\rangle$ and $|11\rangle$ is possible only if the actions of Alice and Bob are coordinated. This is an example of a classical state that can only be created if the two parties are in contact over a classical communication channel.

9.8.1 The PPT criterion

In Section 9.2, we presented the Schmidt decomposition, a mathematical tool that can be used to understand whether a pure state is entangled or not. The separability problem, namely deciding whether a mixed state is entangled or not, is a much more difficult problem. Several criteria have been developed but none of them provides a necessary and sufficient condition for the general case [252].

The **positive partial transpose (PPT) criterion** is a method that provides a sufficient condition for a mixed state to be entangled. This method was introduced by Peres in 1996 [287, 207] and it boils down to applying the identity to one subsystem and the transposition to the other subsystem. The PPT criterion states that[14]

$$\boxed{\text{if } (\mathbb{1}_{\mathrm{A}} \otimes \mathrm{T})\,\rho < 0, \text{ then } \rho \text{ is entangled,}} \tag{9.30}$$

or equivalently, if ρ is separable, then $(\mathbb{1}_{\mathrm{A}} \otimes \mathrm{T})\,\rho \geq 0$. Let us explain how the PPT criterion works. The key idea is that the transposition is a positive map but not

[14] See Section 7.5 for a discussion of the transposition map. Note that if $\mathbb{1}_{\mathrm{A}} \otimes \mathrm{T}$ is replaced with $\mathrm{T} \otimes \mathbb{1}_{\mathrm{B}}$, the PPT criterion does not change.

completely positive. The application of $\mathbb{1}_A \otimes T$ onto a separable state ρ_{sep} always produces a positive operator,

$$(\mathbb{1}_A \otimes T)\, \rho_{sep} = (\mathbb{1}_A \otimes T) \sum_i p_i |a_i\rangle\langle a_i| \otimes |b_i\rangle\langle b_i|$$
$$= \sum_i p_i \underbrace{|a_i\rangle\langle a_i|}_{\text{positive}} \otimes \underbrace{T(|b_i\rangle\langle b_i|)}_{\text{positive}} \geq 0.$$

The PPT criterion is simple: one just needs to compute the partial transpose of the density matrix, $(\mathbb{1} \otimes T)\rho$, and check its eigenvalues. If at least one is negative, then the state is entangled. Unfortunately, the PPT criterion sometimes fails to detect an entangled state. For instance, the quantum state $\rho \in D(\mathbb{C}^3 \otimes \mathbb{C}^3)$ with density matrix

$$\rho = \frac{1}{10} \left[\begin{array}{ccc|ccc|ccc} 1 & 0 & 0 & 0 & 1 & 0 & 0 & 0 & 1 \\ 0 & 1 & 0 & 0 & 0 & 0 & 0 & 0 & 0 \\ 0 & 0 & 1 & 0 & 0 & 0 & 0 & 0 & 0 \\ \hline 0 & 0 & 0 & 1 & 0 & 0 & 0 & 0 & 0 \\ 1 & 0 & 0 & 0 & 1 & 0 & 0 & 0 & 1 \\ 0 & 0 & 0 & 0 & 0 & 1 & 0 & 0 & 0 \\ \hline 0 & 0 & 0 & 0 & 0 & 0 & \frac{3}{2} & 0 & \frac{\sqrt{3}}{2} \\ 0 & 0 & 0 & 0 & 0 & 0 & 0 & 1 & 0 \\ 1 & 0 & 0 & 0 & 1 & 0 & \frac{\sqrt{3}}{2} & 0 & \frac{3}{2} \end{array}\right]$$

is entangled and yet all of the eigenvalues of $(\mathbb{1}_A \otimes T)\, \rho$ are greater or equal to zero. Other examples of entangled states not detected by the PPT criterion can be found in Ref. [288]. Since the partial transpose technique fails to signal some entangled states, it is sometimes necessary to use other methods to solve the separability problem.

The good news is that the PPT criterion is a necessary and sufficient condition for mixed states defined on the Hilbert space $\mathbb{C}^2 \otimes \mathbb{C}^2$ or on the Hilbert space $\mathbb{C}^2 \otimes \mathbb{C}^3$. Thus, the separability problem for two-qubit states can easily be solved by computing the partial transpose of the density matrix and checking its eigenvalues. If one of them is negative, the two-qubit state is entangled.

For our later discussion, it is useful to note that if a density operator $\rho \in D(H_A \otimes H_B)$ is expressed with square sub-matrices,

$$\rho = \begin{bmatrix} A_{11} & \cdots & A_{1d_A} \\ \vdots & \ddots & \vdots \\ A_{d_A 1} & \cdots & A_{d_A d_A} \end{bmatrix},$$

where the matrices A_{ij} have dimension $d_B \times d_B$, then the partial transpose of ρ is

$$(\mathbb{1}_A \otimes T)\rho = \begin{bmatrix} A_{11}^T & \cdots & A_{1d_A}^T \\ \vdots & \ddots & \vdots \\ A_{d_A 1}^T & \cdots & A_{d_A d_A}^T \end{bmatrix}. \tag{9.31}$$

This trick will be used several times in the remainder of this chapter.

Example 9.14 Consider the Bell state $|\beta_{00}\rangle$. As seen in Section 7.5, the operator $(\mathbb{1}_\mathrm{A} \otimes \mathrm{T})|\beta_{00}\rangle\langle\beta_{00}|$ has a negative eigenvalue. Thus, $|\beta_{00}\rangle$ is entangled.

Example 9.15 Consider the **Werner state** defined as

$$\boxed{\rho_\mathrm{W} = p|\beta_{00}\rangle\langle\beta_{00}| + (1-p)\frac{\mathbb{1}}{2} \otimes \frac{\mathbb{1}}{2},}$$

where p is a real parameter in the interval $[0, 1]$. Clearly, this state is not entangled for $p = 0$. Let us prove that the Werner state is entangled for all $p > 1/3$. To show this, compute the associated density matrix,

$$\begin{aligned}\rho_\mathrm{W} &= \frac{p}{2}(|00\rangle\langle 00| + |11\rangle\langle 00| + |11\rangle\langle 00| + |11\rangle\langle 11|) + \frac{1-p}{4}\mathbb{1} \otimes \mathbb{1} \\ &= \frac{p}{2}\begin{bmatrix} 1 & 0 & 0 & 1 \\ 0 & 0 & 0 & 0 \\ 0 & 0 & 0 & 0 \\ 1 & 0 & 0 & 1 \end{bmatrix} + \frac{1-p}{4}\begin{bmatrix} 1 & 0 & 0 & 0 \\ 0 & 1 & 0 & 0 \\ 0 & 0 & 1 & 0 \\ 0 & 0 & 0 & 1 \end{bmatrix} = \begin{bmatrix} \frac{1+p}{4} & 0 & 0 & \frac{p}{2} \\ 0 & \frac{1-p}{4} & 0 & 0 \\ 0 & 0 & \frac{1-p}{4} & 0 \\ \frac{p}{2} & 0 & 0 & \frac{1+p}{4} \end{bmatrix}.\end{aligned}$$

Let us compute the partial transpose using eqn (9.31). This is done by calculating the transpose of the 2×2 sub-matrices,

$$(\mathbb{1}_\mathrm{A} \otimes \mathrm{T})\rho_\mathrm{W} = \begin{bmatrix} \frac{1+p}{4} & 0 & 0 & 0 \\ 0 & \frac{1-p}{4} & \frac{p}{2} & 0 \\ 0 & \frac{p}{2} & \frac{1-p}{4} & 0 \\ 0 & 0 & 0 & \frac{1+p}{4} \end{bmatrix}.$$

This matrix has two eigenvalues. One eigenvalue is $\lambda_1 = (1+p)/4$, which is positive for all p. The other eigenvalue is $\lambda_2 = (1-3p)/4$, which is negative for $p > 1/3$. Since the PPT criterion is a necessary and sufficient condition for two-qubit states, the Werner state is entangled if and only if $p > 1/3$.

Example 9.16 Consider the Bell state $|\beta_{00}\rangle$ and suppose that the dephasing channel $\mathcal{E}_\mathrm{deph}(\hat{A}) = (1-p)\hat{A} + p\hat{\sigma}_z\hat{A}\hat{\sigma}_z$ operates on one of the two qubits. This leads to

$$\begin{aligned}\rho = (\mathbb{1} \otimes \mathcal{E}_\mathrm{deph})\left(|\beta_{00}\rangle\langle\beta_{00}|\right) &= \frac{1}{2}\big(|0\rangle\langle 0| \otimes \mathcal{E}_\mathrm{deph}(|0\rangle\langle 0|) + |0\rangle\langle 1| \otimes \mathcal{E}_\mathrm{deph}(|0\rangle\langle 1|) \\ &\quad + |1\rangle\langle 0| \otimes \mathcal{E}_\mathrm{deph}(|1\rangle\langle 0|) + |1\rangle\langle 1| \otimes \mathcal{E}_\mathrm{deph}(|1\rangle\langle 1|)\big) \\ &= (1-p)\,|\beta_{00}\rangle\langle\beta_{00}| + p|\beta_{10}\rangle\langle\beta_{10}|.\end{aligned}$$

It is evident that this state is entangled for $p = 0$ and $p = 1$. For $p = 1/2$, ρ reduces to $\frac{1}{2}(|00\rangle\langle 00| + |11\rangle\langle 11|)$ which is not entangled.

Let us study the action of the partial transpose on this state. First of all, let us express ρ in matrix form,

$$\rho = (\mathbb{1} \otimes \mathcal{E}_{\text{deph}})\,|\beta_{00}\rangle\langle\beta_{00}| = \begin{bmatrix} \frac{1}{2} & 0 & 0 & \frac{1-2p}{2} \\ 0 & 0 & 0 & 0 \\ 0 & 0 & 0 & 0 \\ \frac{1-2p}{2} & 0 & 0 & \frac{1}{2} \end{bmatrix}. \tag{9.32}$$

From eqn (9.31), the partial transpose of ρ is given by

$$(\mathbb{1}_{\text{A}} \otimes \text{T})\rho = \begin{bmatrix} \frac{1}{2} & 0 & 0 & 0 \\ 0 & 0 & \frac{1-2p}{2} & 0 \\ 0 & \frac{1-2p}{2} & 0 & 0 \\ 0 & 0 & 0 & \frac{1}{2} \end{bmatrix}. \tag{9.33}$$

This matrix has three eigenvalues: $\lambda_1 = 1/2$, $\lambda_2 = 1/2 - p$, and $\lambda_3 = p - 1/2$. When λ_2 is positive, λ_3 is negative, and vice versa. Thus, ρ is always entangled except when $p = 1/2$, because only in this case are the eigenvalues all non-negative.

9.8.2 The reduction criterion

The **reduction criterion** is another interesting method to solve the separability problem. This criterion is based on the reduction function

$$\boxed{M(\rho) = \mathbb{1}\text{Tr}[\rho] - \rho.}$$

This function is positive since the trace of a positive operator is never smaller than one of its eigenvalues. The reduction criterion states that

$$\boxed{\text{if } (\mathbb{1}_{\text{A}} \otimes \text{M})\,\rho < 0, \text{ then } \rho \text{ is entangled.}} \tag{9.34}$$

Let us study in more detail this criterion. The action of the operator $\mathbb{1}_{\text{A}} \otimes \text{M}$ on a bipartite state can be expressed as

$$\begin{aligned} (\mathbb{1} \otimes M)\rho &= \sum_i p_i |a_i\rangle\langle a_i| \otimes M(|b_i\rangle\langle b_i|) \\ &= \sum_i p_i |a_i\rangle\langle a_i| \otimes \mathbb{1}\text{Tr}[|b_i\rangle\langle b_i|] - \sum_i p_i |a_i\rangle\langle a_i| \otimes |b_i\rangle\langle b_i| \\ &= \sum_i p_i |a_i\rangle\langle a_i| \otimes \mathbb{1} - \rho \;=\; \text{Tr}_{\text{B}}[\rho] \otimes \mathbb{1} - \rho. \end{aligned}$$

The reduction criterion boils down to calculating the eigenvalues of the matrix $\text{Tr}_{\text{B}}[\rho] \otimes \mathbb{1} - \rho$. If one of them is negative, ρ is entangled. For two-qubit states, the reduction criterion, like the PPT criterion, provides a separability condition that is both necessary and sufficient [252].

The PPT criterion and the reduction criterion have something important in common: they are both based on functions that are positive (P) but not completely positive (CP). In general, if a function $\mathcal{E}$ is P but not CP and $(\mathbb{1} \otimes \mathcal{E})\rho < 0$, then ρ is entangled. This follows from the observation that if ρ_{sep} is separable,

$$(\mathbb{1}\otimes\mathcal{E})\rho_{\text{sep}} = (\mathbb{1}\otimes\mathcal{E})\sum_i p_i|a_i\rangle\langle a_i|\otimes|b_i\rangle\langle b_i|$$
$$= \sum_i p_i\underbrace{|a_i\rangle\langle a_i|}_{\text{positive}}\otimes\underbrace{\mathcal{E}(|b_i\rangle\langle b_i|)}_{\text{positive}} \geq 0.$$

The last step is justified by the fact that the tensor product of two positive operators is positive.

Positive maps can be divided into two categories: the ones that are decomposable and the ones that are not. A **decomposable map** is a map that is related in a simple way to the transposition map. More formally, a map $\mathcal{E}_{\text{dec}}$ is decomposable if it can be expressed as a sum of a CP map $\mathcal{E}_1$ and another CP map $\mathcal{E}_2$ combined with the transposition map T,

$$\boxed{\mathcal{E}_{\text{dec}}(\rho) = \mathcal{E}_1(\rho) + \mathcal{E}_2(\text{T}(\rho)).} \tag{9.35}$$

It is evident that the transposition map is decomposable: it can be obtained by setting $\mathcal{E}_1 = 0$ and $\mathcal{E}_2 = \mathbb{1}$. Interestingly, all positive maps acting on the space $\mathbb{C}^2\otimes\mathbb{C}^2$ or $\mathbb{C}^2\otimes\mathbb{C}^3$ are decomposable [193, 289]. This explains why the PPT criterion is a necessary and sufficient condition for density operators defined on these Hilbert spaces. In 1996, Horodecki et al. showed that the transposition map is the most efficient decomposable map for the detection of entangled states: if the partial transpose does not detect an entangled state, all other decomposable maps will fail in this task too [290].

Example 9.17 The reduction map $M(\rho)$ on single-qubit states is decomposable because

$$M(\rho) = \mathbb{1}\text{Tr}[\rho] - \rho = \hat{\sigma}_y\text{T}(\rho)\hat{\sigma}_y \tag{9.36}$$

and this expression has the same form of eqn (9.35), where $\mathcal{E}_1 = 0$ and $\mathcal{E}_2(\rho) = \hat{\sigma}_y\rho\hat{\sigma}_y$. Since $M(\rho)$ and $T(\rho)$ differ by a unitary transformation, the reduction criterion and the PPT criterion are equivalent for two-qubit gates.

9.8.3 Entanglement witnesses

The methods explained in the previous sections can solve the separability problem in many interesting cases. Still, they all have a crucial disadvantage: they are based on positive maps that are not completely positive. Unfortunately, these types of transformations cannot be easily implemented in a real experiment. An **entanglement witness** $\hat{W}$ is an observable that can detect an entangled state in an experimental setting. This can be done by measuring $\hat{W}$ multiple times and estimating the expectation value $\langle\hat{W}\rangle = \text{Tr}[\rho\hat{W}]$: if this quantity is negative, then ρ is entangled. If instead $\langle\hat{W}\rangle \geq 0$, then there is no guarantee that ρ is separable and other entanglement witnesses must be considered.

Entanglement witnesses $\hat{W}$ are Hermitian operators satisfying two conditions:

1. $\hat{W}$ must have at least one negative eigenvalue.
2. For all product states $|i\rangle\otimes|j\rangle$, the expectation value $\langle ij|\hat{W}|ij\rangle \geq 0$.

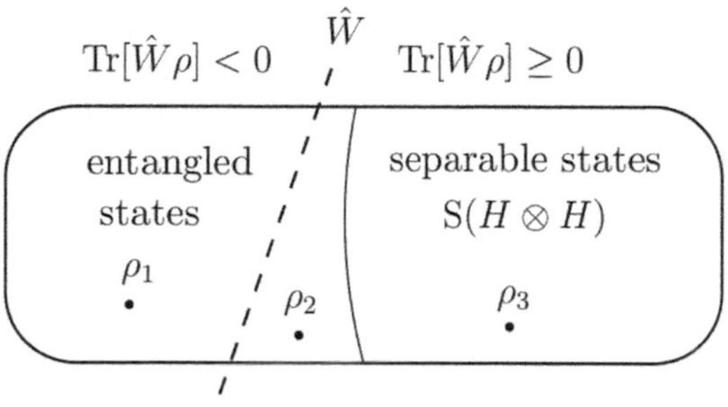

Fig. 9.13 Entanglement witnesses. The entanglement witness $\hat{W}$ divides the set of density operators $\mathrm{D}(H \otimes H)$ in two regions, one with positive and one with negative eigenvalues. In this diagram, the quantum state ρ_1 is entangled and $\hat{W}$ successfully detects it. This is not the case for ρ_2: this is an entangled state and yet $\mathrm{Tr}[\hat{W}\rho_2] \geq 0$. This entangled state can be detected with a different entanglement witness that cuts the set of density operators along a different hyperplane. As regards ρ_3, this state belongs to the set of separable states $\mathrm{S}(H \otimes H)$ and by definition $\mathrm{Tr}[\hat{W}\rho_3] \geq 0$. Ideally, one would replace the hyperplane defined by $\hat{W}$ with a family of hyperplanes tangent to the set of separable states.

The second requirement implies that

$$\boxed{\text{if } \rho \text{ is separable, then } \mathrm{Tr}[\hat{W}\rho] \geq 0,} \tag{9.37}$$

or equivalently if $\mathrm{Tr}[\hat{W}\rho] < 0$, ρ is entangled. If ρ is entangled and $\mathrm{Tr}[\hat{W}\rho] < 0$, we say that the entangled state ρ is **detected** by the witness $\hat{W}$.

An entanglement witness $\hat{W}$ can be associated with a hyperplane that cuts the set of bipartite states into two regions, one with positive and one with negative eigenvalues. Given an entangled state ρ_1, it is always possible to construct an entanglement witness that detects it. We will not demonstrate this statement in detail—we will simply give a geometric interpretation[15]. As illustrated in Fig. 9.13, the set of separable states $\mathrm{S}(H \otimes H)$ is convex and compact. The entangled state ρ_1 lies outside this set. Intuitively, we can always draw a hyperplane, associated with a Hermitian operator $\hat{W}$, that separates the vector ρ_1 from $\mathrm{S}(H \otimes H)$. It might happen that some quantum states (such as ρ_2 in the figure) are not detected by the witness $\hat{W}$.

A simple example of an entanglement witness for two-qubit states is the SWAP gate,

$$\mathsf{SWAP} = |00\rangle\langle 00| + |01\rangle\langle 10| + |10\rangle\langle 01| + |11\rangle\langle 11| = \begin{bmatrix} 1 & 0 & 0 & 0 \\ 0 & 0 & 1 & 0 \\ 0 & 1 & 0 & 0 \\ 0 & 0 & 0 & 1 \end{bmatrix}.$$

[15]A formal proof follows from the fact that the set of separable states $\mathrm{S}(H_\mathrm{A} \otimes H_\mathrm{B})$ is convex and compact and is contained in $\mathrm{Her}(H_\mathrm{A} \otimes H_\mathrm{B})$, the set of Hermitian operators. The set $\mathrm{Her}(H_\mathrm{A} \otimes H_\mathrm{B})$, along with the Hilbert–Schmidt inner product, forms a real inner product space. Thus, we can apply the hyperplane separation theorem. This theorem states that there always exists a hyperplane separating a point $\rho \notin \mathrm{S}(H_\mathrm{A} \otimes H_\mathrm{B})$ from the convex set $\mathrm{S}(H_\mathrm{A} \otimes H_\mathrm{B})$.

This Hermitian operator satisfies both requirements: it has a negative eigenvalue and $\langle ij|\mathsf{SWAP}|ij\rangle \geq 0$ for all product states $|i\rangle|j\rangle \in \mathbb{C}^2 \otimes \mathbb{C}^2$. It is simple to show that this witness can detect some entangled states. As an example, consider the Bell state $|\beta_{11}\rangle = \frac{1}{\sqrt{2}}[|01\rangle - |10\rangle]$. One can verify that $\mathrm{Tr}[|\beta_{11}\rangle\langle\beta_{11}|\mathsf{SWAP}] < 0$. Thus, $|\beta_{11}\rangle$ is entangled.

The Choi–Jamiolkowki isomorphism, presented in Section 7.9, provides a natural way to construct entanglement witnesses from positive but not completely positive maps: if the function $\mathcal{E}$ is P but not CP, then the operator $\hat{W}_{\mathcal{E}} = (\mathbb{1} \otimes \mathcal{E})|\Omega\rangle\langle\Omega|$ is an entanglement witness[16]. To see this, we show that the entanglement witness associated with the transposition map is nothing but the SWAP gate,

$$\begin{aligned}\hat{W}_{\mathrm{T}} = (\mathbb{1} \otimes \mathrm{T})|\beta_{00}\rangle\langle\beta_{00}| &= |0\rangle\langle 0| \otimes \mathrm{T}(|0\rangle\langle 0|) + |0\rangle\langle 1| \otimes \mathrm{T}(|0\rangle\langle 1|)\\ &\quad + |1\rangle\langle 0| \otimes \mathrm{T}(|1\rangle\langle 0|) + |1\rangle\langle 1| \otimes \mathrm{T}(|1\rangle\langle 1|)\\ &= \frac{1}{2}(|00\rangle\langle 00| + |01\rangle\langle 10| + |10\rangle\langle 01| + |11\rangle\langle 11|) = \frac{1}{2}\mathsf{SWAP}.\end{aligned} \tag{9.38}$$

The SWAP gate is called a decomposable entanglement witness because the associated positive map T is decomposable.

9.8.4 Violation of the CHSH inequality for two-qubit states

Do all entangled states violate a CHSH inequality? In Section 9.7.3, we showed that this is the case for pure bipartite states. This is not always true for mixed states. The goal of this section is to show that the Werner state (a mixed state of two qubits) is entangled but does not violate the CHSH inequality. First of all, recall that any two-qubit state can be decomposed in the Pauli basis as

$$\begin{aligned}\rho &= \sum_{i=0}^{4}\sum_{j=0}^{4} c_{ij}\hat{\sigma}_i \otimes \hat{\sigma}_j\\ &= \frac{1}{4}[\mathbb{1} \otimes \mathbb{1} + (\mathbf{r}\cdot\vec{\sigma}) \otimes \mathbb{1} + \mathbb{1} \otimes (\mathbf{s}\cdot\vec{\sigma}) + \sum_{i=1}^{3}\sum_{j=1}^{3} t_{ij}\hat{\sigma}_i \otimes \hat{\sigma}_j].\end{aligned} \tag{9.39}$$

The coefficients t_{ij} define a 3×3 matrix that will be denoted as T,

$$T = \begin{bmatrix} t_{11} & t_{12} & t_{13} \\ t_{21} & t_{22} & t_{23} \\ t_{31} & t_{32} & t_{33} \end{bmatrix}.$$

Since ρ is Hermitian, the parameters t_{ij} are real. Let us now introduce some useful quantities. The transpose of the matrix T will be indicated as T^{T}. The matrix $A = T^{\mathrm{T}}T$ is symmetric and can be diagonalized. The eigenvalues of A are a_1, a_2, and a_3 and we will assume that $a_1 \leq a_2 \leq a_3$.

[16] Here, $|\Omega\rangle$ is the maximally entangled state $|\Omega\rangle = \frac{1}{\sqrt{d}}\sum_i |ii\rangle$.

To understand whether a two-qubit state ρ violates a CHSH inequality, we need to determine the observables $\hat{A}_1$, $\hat{A}_2$, $\hat{B}_1$, and $\hat{B}_2$ that maximize the quantity[17] $\langle \hat{C} \rangle = \mathrm{Tr}[\hat{C}\rho]$ and check whether $\max_{\hat{C}} \langle \hat{C} \rangle > 2$. In 1995, Horodecki showed that the maximum of $\langle \hat{C} \rangle$ is given by [291]

$$\boxed{\max_{\hat{C}} \langle \hat{C} \rangle = 2\sqrt{a_2 + a_3}.} \tag{9.40}$$

Thus, a two-qubit state violates the CHSH inequality if and only if $\sqrt{a_2 + a_3} > 1$.

This result can be used to show that some entangled states do *not* violate the CHSH inequality. Consider the Werner state defined as

$$\begin{aligned} \rho_{\mathrm{W}} &= p|\beta_{00}\rangle\langle\beta_{00}| + \frac{1-p}{4}\mathbb{1} \otimes \mathbb{1} \\ &= \frac{1}{4}(\mathbb{1} \otimes \mathbb{1} + p\,\hat{\sigma}_x \otimes \hat{\sigma}_x - p\,\hat{\sigma}_y \otimes \hat{\sigma}_y + p\,\hat{\sigma}_z \otimes \hat{\sigma}_z). \end{aligned}$$

The matrix T for the Werner state is given by

$$T_{\mathrm{W}} = \begin{bmatrix} p & 0 & 0 \\ 0 & -p & 0 \\ 0 & 0 & p \end{bmatrix} = T_{\mathrm{W}}^{\mathrm{T}}.$$

The eigenvalues of $A = T_{\mathrm{W}}^{\mathrm{T}} T_{\mathrm{W}}$ are $a_1 = a_2 = a_3 = p^2$ and from eqn (9.40) we have $\max \langle \hat{C} \rangle = 2\sqrt{2}p$. Hence, the Werner state violates the CHSH inequality only when $p > 1/\sqrt{2}$. In Example 9.15, we showed that the Werner state is entangled if and only if $p > 1/3$. Thus, all Werner states with $p \in (1/3, 1/\sqrt{2})$ are entangled but do not violate the CHSH inequality.

9.9 Quantifying entanglement of mixed states

In Section 9.3, we learned that the von Neumann entropy of the reduced density operator is a good entanglement measure for pure states. Unfortunately, this function is not a valid entanglement measure for mixed states (we discussed this point in Example 9.7). To extend our treatment to mixed states, we introduce the entanglement measure $E(\rho)$, a function that operates on density operators of bipartite systems and outputs a non-negative real number. This function must satisfy two conditions:

E1. $E(\rho) = 0$ if and only if ρ is separable.

E2. The entanglement of a quantum state cannot increase under local operations coordinated through classical communication,

$$E(\Lambda(\rho)) \leq E(\rho), \tag{9.41}$$

where Λ is a LOCC operation.

[17] As usual $\hat{C} = \hat{A}_1 \otimes (\hat{B}_1 + \hat{B}_2) + \hat{A}_2 \otimes (\hat{B}_1 - \hat{B}_2)$.

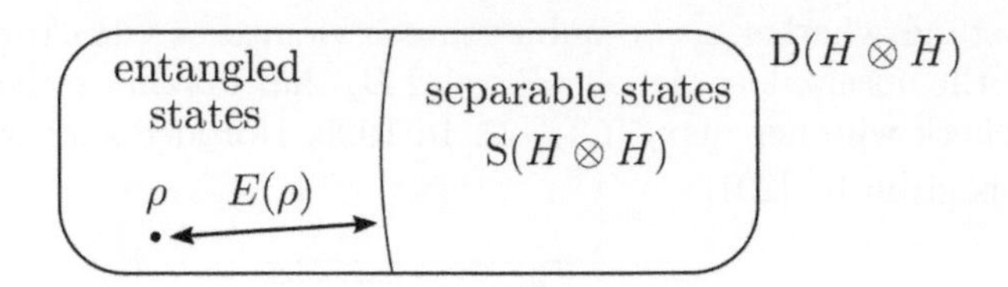

Fig. 9.14 Quantifying the entanglement of a mixed state. The entanglement of a mixed state ρ can be quantified by calculating the minimum distance from the set of separable states $\mathrm{S}(H \otimes H)$.

The analytical form of the quantum map Λ (describing a LOCC operation, i.e. local operations on the subsystems supported by classical communication) is rather complicated to derive: Alice and Bob might communicate over the phone before or after a local operation [254, 292]. Luckily, a class of operations called **separable operations** includes the class of LOCC operations and is much easier to handle. A separable operation is defined as a quantum map of the form[18]

$$\boxed{\Lambda_{\text{sep}}(\rho) = \sum_i (\hat{A}_i \otimes \hat{B}_i)\rho(\hat{A}_i^\dagger \otimes \hat{B}_i^\dagger),} \tag{9.42}$$

where $\hat{A}_i$ and $\hat{B}_i$ are Kraus operators satisfying $\sum_i \hat{A}_i^\dagger \hat{A}_i = \mathbb{1}_\mathrm{A}$ and $\sum_i \hat{B}_i^\dagger \hat{B}_i = \mathbb{1}_\mathrm{B}$. All LOCC operations can be expressed as in eqn (9.42). The converse is not true: there exist some separable operations that cannot be performed with local transformations even with the help of classical communication between the two parties [293].

In the remaining part of this section, we present an interesting entanglement measure based on the intuition that the closer the state is to the set of separable states, the less entangled it is (see Fig. 9.14). In mathematical terms,

$$\boxed{E(\rho) = \min_{\sigma \in \mathrm{S}} \mathrm{D}(\rho, \sigma),} \tag{9.43}$$

where D is a distance. The function (9.43) is a valid entanglement measure only if D is contractive, i.e. $\mathrm{D}(\Lambda(\rho), \Lambda(\sigma)) \leq \mathrm{D}(\rho, \sigma)$ for any quantum map Λ.

Theorem 9.3 Consider two quantum states $\rho, \sigma \in \mathrm{D}(\mathrm{H_A} \otimes \mathrm{H_B})$, a quantum map Λ and a distance D satisfying $\mathrm{D}(\Lambda(\rho), \Lambda(\sigma)) \leq \mathrm{D}(\rho, \sigma)$. Then, the function $E(\rho) = \min_{\sigma \in S} \mathrm{D}(\rho, \sigma)$ is an entanglement measure.

Proof To prove this theorem, we need to show that $E(\rho) = \min_{\sigma \in S} D(\rho, \sigma)$ satisfies properties **E1** and **E2**. Let us start from **E1**. Suppose ρ is a separable state. Then, $E(\rho) = D(\rho, \rho) = 0$. On the other hand, if $E(\rho) = 0$, there exists a separable state

[18] Suppose that Alice and Bob perform some local operations on their subsystems without sharing any classical information. This experiment is described in mathematical terms by the operation $\Lambda(\rho) = (\mathcal{E}_\mathrm{A} \otimes \mathcal{E}_\mathrm{B})\rho$ where $\mathcal{E}_\mathrm{A}$ and $\mathcal{E}_\mathrm{B}$ are two quantum maps acting on the individual subsystems. This is the simplest example of LOCC operation.

σ such that $D(\rho, \sigma) = 0$. Since D is a distance, this implies that $\rho = \sigma$, i.e. ρ is separable. Let us prove property **E2**. Let $\tilde{\sigma}$ be the closest separable state to ρ. Thus, $E(\rho) = D(\rho, \tilde{\sigma})$. Suppose that Λ is a quantum map associated with a LOCC operation. Then,

$$E(\rho) = D(\rho, \tilde{\sigma}) \geq D(\Lambda(\rho), \Lambda(\tilde{\sigma})).$$

Since separable states are invariant under LOCC operations, the state $\Lambda(\tilde{\sigma}) = \tau$ is still separable. This observation leads to

$$E(\rho) \geq D(\Lambda(\rho), \tau) \geq \min_{\tau \in \mathrm{S}} D(\Lambda(\rho), \tau) = E(\Lambda(\rho)). \qquad \square$$

It is rather difficult to calculate $\min_{\sigma \in \mathrm{S}} D(\rho, \sigma)$ because this requires to find the separable state closest to ρ. An analytical approach is possible only in the presence of some special symmetries [252].

Further reading

The theory of entanglement is presented extensively in Refs [251, 252, 253, 254, 111]. In this chapter, we focused on systems of distinguishable particles for pedagogical reasons; the zero and finite temperature properties of entanglement in interacting fermions and bosons are discussed in the review by Amico et al. [294]. An important result of entanglement theory is the Lieb–Robinson bound. This bound quantifies the speed at which information can propagate in non-relativistic quantum systems [295]. Another important result is the area law [296, 297, 298, 299], showing that the entanglement between two subsystems grows at most proportionally with the boundary between the two subsystems for certain wavefunctions.

The theory of entanglement has profound implications for computational physics, chemistry, and computer science. Some wavefunctions have weak or structured entanglement, and therefore they can be represented efficiently by tensor networks. A tensor network is constructed with tensors that encode local information. They are connected to create a large network describing a complex quantum state. The way in which the tensors are connected defines the entanglement structure of the system. The formalism of tensor networks ushered a variety of algorithms for the approximation of ground, thermal, and excited states of many-particle systems, see Refs [300, 301, 302, 303, 304, 305, 306] for more information.

Summary

Entanglement for pure states

$\vert\psi\rangle = \vert\alpha\rangle_A\vert\beta\rangle_A$	Product state
$\vert\beta_{ij}\rangle = c_1\hat{X}_0\hat{H}_1\vert ij\rangle$	Bell states
$\vert\psi\rangle = \sum_k a_k\vert k\rangle_A \otimes \vert k\rangle_B$	Schmidt decomposition
$\rho_A = \mathrm{Tr}_B(\vert\psi\rangle\langle\psi\vert) = \sum_k a_k^2\vert k\rangle\langle k\vert$	Reduced density operator
$E(\rho) = S(\rho_A)$	Entanglement of a pure bipartite state
$E(\rho) = \log_2 d$	ρ is maximally entangled

Bell's inequality

Realism: It is possible to associate definite values to all properties of a physical system. Measurement operations reveal these values without altering them.

Locality: The outcome of a measurement in region A is not affected by a spacelike separated event happening in a region B.

CHSH inequality: $\langle C\rangle = \langle A_1B_1\rangle + \langle A_1B_2\rangle + \langle A_2B_1\rangle - \langle A_2B_2\rangle \leq 2$

Tsirelson bound: $\langle\psi|C|\psi\rangle \leq 2\sqrt{2}$ for any state $|\psi\rangle$.

Gisin theorem: for any entangled state $|\psi\rangle \in H_A \otimes H_B$, there exist four dichotomic observables $\hat{A}_1$, $\hat{A}_2$, $\hat{B}_1$, $\hat{B}_2$ such that $\langle\psi|\hat{C}|\psi\rangle \geq 2$, where $C = \hat{A}_1 \otimes \hat{B}_1 + \hat{A}_1 \otimes \hat{B}_2 + \hat{A}_2 \otimes \hat{B}_1 - \hat{A}_2 \otimes \hat{B}_2$.

Entanglement for mixed states

$\rho = \sum_k p_k\rho_k^A \otimes \rho_k^B$	Separable state for a bipartite system

A convex sum of entangled states might not be entangled.

A convex sum of separable states is always separable.

PPT criterion: if $(\mathbb{1}_A \otimes T)\rho < 0$, then ρ is entangled.

The PPT is a necessary and sufficient condition for quantum states $\rho \in \mathrm{D}(\mathbb{C}^2 \otimes \mathbb{C}^2)$.

$M(\rho) = \mathrm{Tr}[\rho]\mathbb{1} - \rho$ is the reduction map.

The reduction criterion: if $(\mathbb{1} \otimes M)\rho < 0$, then ρ is entangled.

$\mathcal{E}$ is decomposable if $\mathcal{E}(\rho) = \mathcal{E}_1(\rho) + \mathcal{E}_2(T(\rho))$, where $\mathcal{E}_1$, $\mathcal{E}_2$ are completely positive.

$\hat{W} \in B(H \otimes H)$ is Hermitian, has at least one negative eigenvalue
and $\langle ij|\hat{W}|ij\rangle \geq 0$ for all product states $|ij\rangle$ $\Rightarrow \hat{W}$ is an entanglement witness.

If $\text{Tr}[\rho\hat{W}] < 0$, then ρ is entangled.

$\rho_{\text{W}} = p|\beta_{00}\rangle\langle\beta_{00}| + \frac{1-p}{4}\mathbb{1} \otimes \mathbb{1}$ Werner state.
ρ_{W} with $1/3 < p < 1/\sqrt{2}$ is entangled but does not violate the CHSH inequality.

$E(\rho) = 0$ if and only if ρ is separable.
$E(\Lambda(\rho)) \leq E(\rho)$, where Λ is a LOCC operation.

$\Lambda(\rho) = \sum_i (\hat{A}_i \otimes \hat{B}_i)\rho(\hat{A}_i^\dagger \otimes \hat{B}_i^\dagger)$ Λ is a separable operation.

The set of LOCC operations is contained in the set of separable operations.

Theorem: $E(\rho) = \min_{\sigma \in S} D(\rho, \sigma)$ is an entanglement measure if D is contractive.

Exercises

Exercise 9.1 Show that the reduction function $M(\rho) = \mathbb{1}\text{Tr}[\rho] - \rho$ acting on single-qubit states maps a generic Bloch vector (r_x, r_y, r_z) into $(-r_x, -r_y, -r_z)$.

Exercise 9.2 Consider a function Λ that is positive but not completely positive. In Section 9.8.3, we explained that the operator $\hat{W}_\Lambda = (\mathbb{1} \otimes \Lambda)|\Omega\rangle\langle\Omega|$ is an entanglement witness. It is tempting to conclude that the two criteria:

$$\text{if } (\mathbb{1} \otimes \Lambda)\rho < 0, \text{ then } \rho \text{ is entangled,}$$
$$\text{if } \text{Tr}[\hat{W}_\Lambda \rho] < 0, \text{ then } \rho \text{ is entangled,}$$

are identical. This is not the case: the second criterion is weaker. Show this by proving that $(\mathbb{1} \otimes \text{T})|\beta_{00}\rangle\langle\beta_{00}| < 0$, but $\text{Tr}[\hat{W}_\text{T}|\beta_{00}\rangle\langle\beta_{00}|] > 0$ where T is the transposition map.

Exercise 9.3 Consider a two-qubit state $|\psi\rangle = \sum_{ij} c_{ij}|ij\rangle$. The entries c_{ij} define a 2×2 matrix c. Show that $|\psi\rangle$ is entangled if and only if $\det(c) \neq 0$ (be careful, this trick only works for two-qubit states).

Exercise 9.4 Show that the maximum of the entanglement $E(\rho) = S(\text{Tr}_\text{B}[\rho])$ for for pure bipartite states $\rho = |\psi\rangle\langle\psi| \in \text{D}(H \otimes H)$ is $\log_2 d$. **Hint:** Use the method of Lagrange multipliers.

Exercise 9.5 In the quantum teleportation protocol, Alice wishes to transmit a pure state $\rho = |\psi\rangle\langle\psi|$. Suppose that the entangled pair shared by Alice and Bob is affected by some noise $(\mathbb{1} \otimes \mathcal{E}_{\text{deph}})|\beta_{00}\rangle\langle\beta_{00}|$ where $\mathcal{E}_{\text{deph}}$ is the dephasing channel. Compute the average fidelity $\bar{F}$ of the teleportation protocol. **Hint:** Start from eqn (9.14).

Exercise 9.6 The quantum circuit shown below

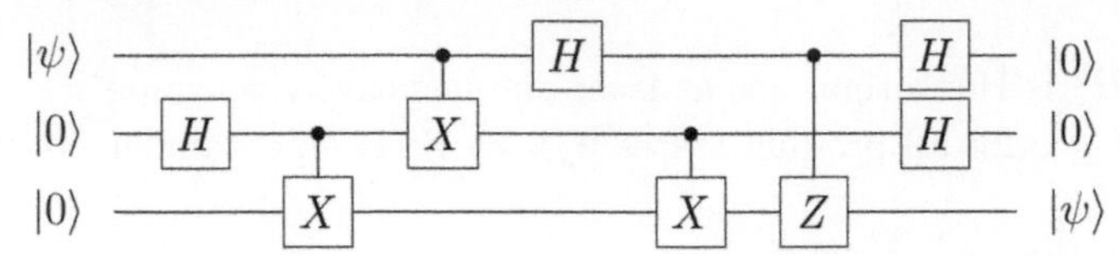

can be used to teleport an unknown state $|\psi\rangle$ from one qubit to another within the same circuit. This quantum circuit is sometimes called the **quantum intraportation**. Check that the initial state $|\psi\rangle|0\rangle|0\rangle$ is transformed into the final state $|0\rangle|0\rangle|\psi\rangle$ for any qubit state $|\psi\rangle = \alpha|0\rangle + \beta|1\rangle$.

Exercise 9.7 Calculate the partial traces in eqn (9.13) and show that $\rho_{\mathrm{C},k} = \mathcal{E}(\hat{\sigma}_k \rho \hat{\sigma}_k)$. **Hint**: Use the notation and results presented in Appendix B.1.

Solutions

Solution 9.1 Apply the function M to a qubit state $\rho = (\mathbb{1} + r_x\hat{\sigma}_x + r_y\hat{\sigma}_y + r_z\hat{\sigma}_z)/2$,

$$M(\rho) = \mathbb{1}\mathrm{Tr}[\rho] - \rho = \mathbb{1} - \frac{\mathbb{1} + r_x\hat{\sigma}_x + r_y\hat{\sigma}_y + r_z\hat{\sigma}_z}{2} = \frac{\mathbb{1} - r_x\hat{\sigma}_x - r_y\hat{\sigma}_y - r_z\hat{\sigma}_z}{2} = \rho',$$

where ρ' is described by the Bloch vector $(-r_x, -r_y, -r_z)$.

Solution 9.2 We already proved in Section 7.5 that $(\mathbb{1} \otimes \mathrm{T})|\beta_{00}\rangle\langle\beta_{00}| < 0$. Let us show that $\mathrm{Tr}[\hat{W}_\mathrm{T}|\beta_{00}\rangle\langle\beta_{00}|] > 0$. First, express the state $|\beta_{00}\rangle = (|00\rangle + |11\rangle)/\sqrt{2}$ in matrix form,

$$|\beta_{00}\rangle\langle\beta_{00}| = \frac{1}{2}[|00\rangle\langle 00| + |00\rangle\langle 11| + |11\rangle\langle 00| + |11\rangle\langle 11|] = \frac{1}{2}\begin{bmatrix} 1 & 0 & 0 & 1 \\ 0 & 0 & 0 & 0 \\ 0 & 0 & 0 & 0 \\ 1 & 0 & 0 & 1 \end{bmatrix}.$$

From eqn (9.38), we know that $\hat{W}_\mathrm{T} = \frac{1}{2}\mathrm{SWAP}$. Thus,

$$\mathrm{Tr}[\hat{W}_\mathrm{T}|\beta_{00}\rangle\langle\beta_{00}|] = \frac{1}{2}\mathrm{Tr}[\mathrm{SWAP}|\beta_{00}\rangle\langle\beta_{00}|] = \frac{1}{4}\mathrm{Tr}\begin{bmatrix} 1 & 0 & 0 & 0 \\ 0 & 0 & 1 & 0 \\ 0 & 1 & 0 & 0 \\ 0 & 0 & 0 & 1 \end{bmatrix}\begin{bmatrix} 1 & 0 & 0 & 1 \\ 0 & 0 & 0 & 0 \\ 0 & 0 & 0 & 0 \\ 1 & 0 & 0 & 1 \end{bmatrix} = \frac{1}{2}.$$

Solution 9.3 The entries c_{ij} define a 2×2 matrix c which can be expressed as $c = uav$ where u and v are unitary matrices and a is a 2×2 diagonal matrix with non-negative entries (this is the singular value decomposition). According to Theorem 9.1, $|\psi\rangle$ is entangled if at least two entries on the diagonal of a are non-zero. For a 2×2 matrix, this requirement is equivalent to $\det a \neq 0$. The determinant of c is proportional to the determinant of a because $\det(c) = \det(u)\det(a)\det(v)$ where $\det(u)$ and $\det(v)$ are two complex numbers with modulus 1 since u and v are unitary matrices. This means that $\det(a) \neq 0$ if and only if $\det(c) \neq 0$.

Solution 9.4 The Schmidt decomposition of a pure bipartite state $\rho = |\psi\rangle\langle\psi| \in \mathrm{D}(H \otimes H)$ is given by

$$\rho = \sum_{k=1}^{d}\sum_{l=1}^{d} a_k a_l |kk\rangle\langle ll|.$$

As seen in eqn (9.6), the reduced density operator takes the form $\rho_{\rm A} = {\rm Tr}_{\rm B}[\rho] = \sum_{k=1}^{d} a_k^2 |k\rangle\langle k|$, where $\sum_k a_k^2 = 1$. The von Neumann entropy of this state is

$$E(\rho) = S(\rho_{\rm A}) = -\sum_{k=1}^{d} a_k^2 \log_2(a_k^2) \equiv f(\vec{a}).$$

Our task is to maximize the function $f(\vec{a}) = -\sum_k a_k^2 \log_2(a_k^2)$ with the constraint $\sum_k a_k^2 = 1$. This can be done with the method of Lagrange multipliers. First, consider the Lagrangian $\mathcal{L} = f(\vec{a}) - \lambda g(\vec{a})$, where $g(\vec{a}) = \sum_k a_k^2$. The extreme points of the function $f(\vec{a})$ satisfy

$$\frac{\partial}{\partial a_k} f(\vec{a}) = \lambda \frac{\partial}{\partial a_k} g(\vec{a}),$$

where λ is a Lagrangian multiplier. This relation can be written as

$$-\frac{2}{\ln 2} a_k(\log_2(a_k^2) + 1) = 2\lambda a_k,$$

which is solved by $a_k = 2^{-\lambda/2}/\sqrt{e}$. This shows that the entanglement of a pure bipartite state is maximized when all of the Schmidt coefficients a_k are identical, i.e. $a_k = 1/\sqrt{d}$. In conclusion,

$$E(\rho) = -\sum_{k=1}^{d} a_k^2 \log_2(a_k^2) \le -\sum_{k=1}^{d} \frac{1}{d} \log_2 \frac{1}{d} = -\log_2 \frac{1}{d} = \log_2 d.$$

Solution 9.5 As explained in eqn (9.14), at the end of the protocol the state of Bob's qubit is

$$\begin{aligned}\rho^{\rm Bob} &= \frac{1}{4}\Big[\mathcal{E}_{\rm deph}(\rho) + \hat{\sigma}_x \mathcal{E}_{\rm deph}(\hat{\sigma}_x \rho \hat{\sigma}_x)\hat{\sigma}_x + \hat{\sigma}_y \mathcal{E}_{\rm deph}(\hat{\sigma}_y \rho \hat{\sigma}_y)\hat{\sigma}_y + \hat{\sigma}_z \mathcal{E}_{\rm deph}(\hat{\sigma}_z \rho \hat{\sigma}_z)\hat{\sigma}_z\Big] \\ &= \mathcal{E}_{\rm deph}(\rho). \end{aligned} \tag{9.44}$$

The fidelity of the protocol is $F = \langle\psi|\mathcal{E}_{\rm deph}(|\psi\rangle\langle\psi|)|\psi\rangle = 1 - p + p\cos\frac{\theta}{2}$, and the average fidelity is $\overline{F} = 1 - p + \frac{p}{4\pi}\int_0^{2\pi} d\phi \int_0^{\pi} d\theta \sin\theta \cos^2\frac{\theta}{2} = 1 - 2p/3$. When $p > 1/2$, the noise is so intense that the teleportation protocol underperforms the classical protocol.

Solution 9.6 First, decompose the state $|\psi\rangle$ in the computational basis, $|\psi\rangle = \sum_{j=0}^{1} c_j |j\rangle$. Therefore

$$|\psi_{\rm A}\rangle = |\psi, 0, 0\rangle = \sum_{j=0}^{1} c_j |j, 0, 0\rangle.$$

The quantum circuit transforms this state into

$$|\psi_{\rm B}\rangle = \sum_{j,k=0}^{1} \frac{c_j}{\sqrt{2}} |j,k,0\rangle \qquad \text{(Hadamard gate)},$$

$$|\psi_{\rm C}\rangle = \sum_{j,k=0}^{1} \frac{c_j}{\sqrt{2}} |j,(j+k)\%2,k\rangle \qquad \text{(CNOT gates)},$$

$$|\psi_{\rm D}\rangle = \sum_{j,k,\ell=0}^{1} \frac{c_j}{2} (-1)^{j\ell} |\ell,(j+k)\%2,k\rangle \qquad \text{(Hadamard gate)},$$

$$|\psi_{\rm E}\rangle = \sum_{j,k,\ell=0}^{1} \frac{c_j}{2} |\ell,(j+k)\%2,j\rangle \qquad \text{(CNOT and CZ gates)},$$

$$|\psi_{\rm F}\rangle = \sum_{j,k,\ell,m,n=0}^{1} \frac{c_j}{4} (-1)^{\ell n} (-1)^{jm+km} |n,d,j\rangle \quad \text{(Hadamard gates)}.$$

To conclude the exercise, we need to compute the sums $\sum_{k=0}^{1}(-1)^{kn} = 2\delta_{n0}$ and $\sum_{k=0}^{1}(-1)^{km} = 2\delta_{m0}$, leading to $|\psi_{\rm F}\rangle = \sum_{j=0}^{1} c_j|0,0,j\rangle = |0,0,\psi\rangle$.

Solution 9.7 To simplify the calculations, we will adopt the matrix formalism for bipartite states presented in Appendix B.1. Let us start from the trace over subsystem A:

$$\begin{aligned}
\rho_{{\rm C},k} &= \frac{1}{2p_k} {\rm Tr}_{\rm AB}\Big[\big(\mathbb{1}_{\rm A} \otimes (\mathbb{1}_{\rm B} \otimes \mathcal{E})|\beta_{00}\rangle\langle\beta_{00}|\big)\ \big(\rho \otimes \mathbb{1}_{\rm B} \otimes \mathbb{1}_{\rm C}\big)\ \big(|\sigma_k\rangle\langle\sigma_k| \otimes \mathbb{1}_{\rm C}\big)\Big] \\
&= \frac{1}{2p_k} {\rm Tr}_{\rm AB}\Big[\big(\mathbb{1}_{\rm A} \otimes (\mathbb{1}_{\rm B} \otimes \mathcal{E})|\beta_{00}\rangle\langle\beta_{00}|\big)\ \big(|\rho\sigma_k\rangle\langle\sigma_k| \otimes \mathbb{1}_{\rm C}\big)\Big] \\
&= \frac{1}{2p_k} {\rm Tr}_{\rm B}\Big[(\mathbb{1}_{\rm B} \otimes \mathcal{E})|\beta_{00}\rangle\langle\beta_{00}|\ \big({\rm Tr}_{\rm A}[|\rho\sigma_k\rangle\langle\sigma_k|] \otimes \mathbb{1}_{\rm C}\big)\Big] \\
&= \frac{1}{2p_k} {\rm Tr}_{\rm B}\Big[\big(\hat\sigma_k^{\rm T}\rho^{\rm T}\hat\sigma_k^* \otimes \mathbb{1}_{\rm C}\big)\ (\mathbb{1}_{\rm B} \otimes \mathcal{E})|\beta_{00}\rangle\langle\beta_{00}|\Big],
\end{aligned} \tag{9.45}$$

where in the first and last step we used properties **C1.1** and **C1.4** explained in Appendix B.1. Let us proceed with the trace over subsystem B,

$$\begin{aligned}
\rho_{{\rm C},k} &= \frac{1}{4p_k} \sum_{j=0}^{1}\sum_{l=0}^{1} {\rm Tr}_{\rm B}\Big[\big(\hat\sigma_k^{\rm T}\rho^{\rm T}\hat\sigma_k^* \otimes \mathbb{1}_{\rm C}\big)\ \big(|j\rangle\langle l|_{\rm B} \otimes \mathcal{E}(|j\rangle\langle l|_{\rm C})\big)\Big] \\
&= \frac{1}{4p_k} \sum_{j=0}^{1}\sum_{l=0}^{1} {\rm Tr}_{\rm B}\Big[\hat\sigma_k^{\rm T}\rho^{\rm T}\hat\sigma_k^*|j\rangle\langle l|_{\rm B} \otimes \mathcal{E}(|j\rangle\langle l|_{\rm C})\Big] \\
&= \frac{1}{4p_k} \sum_{j=0}^{1}\sum_{l=0}^{1} \langle l|\hat\sigma_k^{\rm T}\rho^{\rm T}\hat\sigma_k^*|j\rangle\mathcal{E}(|j\rangle\langle l|_{\rm C}) \\
&= \frac{1}{4p_k} \sum_{j=0}^{1}\sum_{l=0}^{1} \langle j|\hat\sigma_k^{\dagger}\rho\hat\sigma_k|l\rangle\mathcal{E}(|j\rangle\langle l|_{\rm C}) = \frac{1}{4p_k}\mathcal{E}(\hat\sigma_k\rho\hat\sigma_k).
\end{aligned} \tag{9.46}$$

Since ${\rm Tr}[\rho_{{\rm C},k}] = 1$, we must have $p_k = 1/4$ for a correct normalization.

10
Early quantum algorithms

Early quantum algorithms showed
the potential advantage of quantum computers.

Quantum computation is an active field of research investigating the computational power of physical machines exploiting quantum principles. It combines two important discoveries of the last century: quantum mechanics and computer science. One of the objectives of the field is to develop new quantum algorithms that can solve useful computational problems with fewer resources than classical algorithms.

The possibility for a quantum computer to outperform a classical machine was advanced at the beginning of the '80s by Benioff [10, 307], Manin [308], and Feynman [11, 12]. The work of Benioff focused on a quantum mechanical description of the functioning of a standard Turing machine. Feynman correctly predicted that the evolution of a quantum system can be efficiently simulated on a quantum computer. At the same time, Manin noted that a quantum automaton would be able to represent the state of a molecule with far fewer resources (see appendix of Ref. [309]).

A formal definition of a universal quantum computer was introduced in 1985 by Deutsch [13]. In his seminal work, Deutsch described the first quantum algorithm for a black-box problem with a clear speed-up over classical approaches. The following decade saw the development of other quantum algorithms for black-box problems, including the ones by Simon [17], Bernstein–Vazirani [16], and Grover [19]. The interest in quantum computation significantly increased in 1994 when Peter Shor discovered an efficient quantum algorithm for the prime factorization of large numbers [18]. The difficulty of factoring large semiprimes is at the core of modern public-key cryptography. Shor's breakthrough showed that when a universal quantum computer becomes available, any messages that had been encrypted using the RSA protocol could be opened by anyone who had recorded the encrypted message and the public key. Recent studies indicate that a noisy quantum computer with $10^5 - 10^6$ qubits would be able to break modern public-key cryptography [61, 62]. However, it is reasonable to expect that by the time a fully-functional quantum computer becomes available, modern cryptography will transition to new quantum-resistant cryptographic protocols [310]. The National Institute of Standards and Technology has already started evaluating proposals for a post-quantum cryptography standard. The selection of the new standard is expected to be completed by 2030 [311].

The chapter opens with an explanation of how to execute Boolean functions on

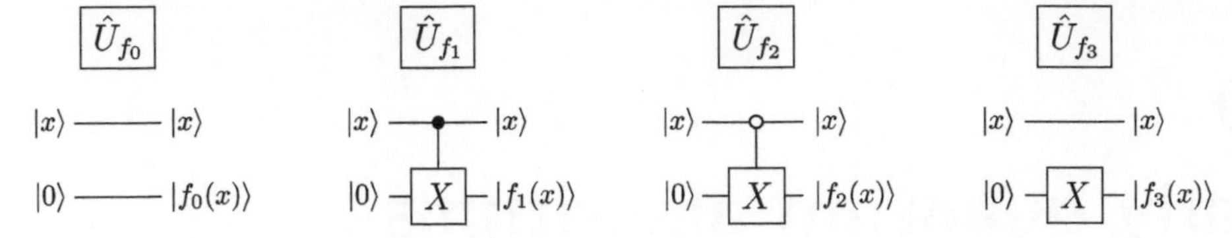

$\hat{U}_{f_0}$ $\hat{U}_{f_1}$ $\hat{U}_{f_2}$ $\hat{U}_{f_3}$

$|x\rangle$ — $|x\rangle$ $|x\rangle$ — $|x\rangle$ $|x\rangle$ — $|x\rangle$ $|x\rangle$ — $|x\rangle$

$|0\rangle$ — $|f_0(x)\rangle$ $|0\rangle$ — X — $|f_1(x)\rangle$ $|0\rangle$ — X — $|f_2(x)\rangle$ $|0\rangle$ — X — $|f_3(x)\rangle$

Fig. 10.1 Quantum circuits for single-bit Boolean functions. Each of these circuits implements a single-bit Boolean function $f_i : \mathbb{B} \to \mathbb{B}$. The input qubit and the ancilla are initialized in $|x\rangle$ and $|0\rangle$, respectively. The black circle indicates that the $\hat{X}$ gate is applied to the target qubit only when the control qubit is in $|1\rangle$. The white circle indicates that the $\hat{X}$ gate is applied to the target qubit only when the control qubit is in $|0\rangle$. At the end of the circuit, the input qubit is left unchanged, while the ancilla qubit stores $f_i(x)$.

a quantum computer. This section is a natural extension of the concepts presented in Section 2.3 where we learned how to implement Boolean functions using classical circuits. In Section 10.3, we present the Deutsch–Jozsa algorithm, one of the first quantum algorithms that showed a quantum advantage for a simple black-box problem. In Section 10.4, we introduce the Bernstein–Vazirani algorithm, a more interesting quantum algorithm that solves a black-box problem with a significant speed-up. In Section 10.5, we explain Grover's algorithm, a quantum algorithm that finds an item in an unstructured database with a quadratic speed-up over classical algorithms. In Section 10.7, we briefly recall the definition of discrete Fourier transform and present its quantum counterpart. The remaining part of the chapter will focus on two applications of the quantum Fourier transform: the period finding algorithm and Shor's algorithm for prime factorization.

10.1 Function evaluation with a quantum computer

In Section 2.3, we learned how to implement Boolean functions using the operations AND, OR, NOT, and COPY on a classical computer. Let us show how to implement these functions on a quantum computer. For simplicity, we first consider Boolean functions operating on a single bit

x		$f_0(x)$	$f_1(x)$	$f_2(x)$	$f_3(x)$
0	$\mapsto$	0	0	1	1
1		0	1	0	1

The functions f_0 and f_3 project the input bit onto 0 and 1 respectively and thus they are not invertible. Quantum computers can only implement unitary gates, intrinsically invertible operations. A simple strategy to implement these functions using unitary gates is to add an extra qubit to the circuit. This auxiliary qubit is usually called an **ancilla qubit**. A quantum computer implements a Boolean function $f : \mathbb{B} \to \mathbb{B}$ by reading the qubit state $|x\rangle$ and storing the output $f(x)$ in the state of the ancilla,

$$\boxed{\hat{U}_f|x\rangle|y\rangle = |x\rangle|y \oplus f(x)\rangle.}$$

Here, the ancilla qubit is initialized in $|y\rangle$ and its final state is $|y \oplus f(x)\rangle$ where $\oplus$ indicates the usual sum modulo 2. If the initial state of the ancilla is $|0\rangle$, the final state of the two qubits is $|x\rangle|f(x)\rangle$. The transformation $\hat{U}_f$ is invertible because

$$\hat{U}_f\hat{U}_f|x\rangle|y\rangle = \hat{U}_f|x\rangle|y \oplus f(x)\rangle = |x\rangle|y \oplus f(x) \oplus f(x)\rangle = |x\rangle|y\rangle.$$

The quantum circuits implementing the single-bit Boolean functions $f_i : \mathbb{B} \to \mathbb{B}$ are illustrated in Fig. 10.1. In these circuits, the black and white dots should be interpreted as "apply an X gate to the ancilla only if the qubits with the black dots are in 1 and qubits with the white dots are in 0".

Let us take this a step further and consider two-bit Boolean operations $f_i : \mathbb{B}^2 \to \mathbb{B}$. These operations transform two bits into a bit,

x_1	x_0		f_0	f_1	f_2	f_3	f_4	f_5	f_6	f_7	f_8	f_9	f_{10}	f_{11}	f_{12}	f_{13}	f_{14}	f_{15}
0	0		0	0	0	0	0	0	0	0	1	1	1	1	1	1	1	1
0	1	$\mapsto$	0	0	0	0	1	1	1	1	0	0	0	0	1	1	1	1
1	0		0	0	1	1	0	0	1	1	0	0	1	1	0	0	1	1
1	1		0	1	0	1	0	1	0	1	0	1	0	1	0	1	0	1

These functions are not invertible. A quantum computer can evaluate them by using an extra ancilla initialized in $|0\rangle$,

$$\hat{U}_f|x_1x_0\rangle|0\rangle = |x_1x_0\rangle|f(x_1, x_0)\rangle.$$

Figure 10.2 illustrates the quantum circuits implementing some of these functions. We only show how to construct eight of them because the quantum circuit associated with $\hat{U}_{f_i}$ is the same as that of $\hat{U}_{f_{15-i}}$ with an extra $\hat{X}$ gate on the ancilla.

Let us generalize these concepts to arbitrary functions $f : \mathbb{B}^n \to \mathbb{B}^m$. As shown in Section 1.1.7, the output string $f(\mathbf{x}) = (f_{m-1}(\mathbf{x}), \ldots, f_0(\mathbf{x}))$ can be expressed with minterms

$$f_i(\mathbf{x}) = m_{\mathbf{c}_1}(\mathbf{x}) \vee \ldots \vee m_{\mathbf{c}_k}(\mathbf{x}),$$

where k is the number of strings $\mathbf{x}$ for which $f_i(\mathbf{x}) = 1$. Each minterm can be implemented on a quantum computer with controlled unitaries. As an example, consider the three-bit function $f : \mathbb{B}^3 \to \mathbb{B}^2$,

$\mathbf{x}$				$f(\mathbf{x})$	
x_2	x_1	x_0		$f_1(\mathbf{x})$	$f_0(\mathbf{x})$
0	0	0		0	0
0	0	1		1	0
0	1	0		0	0
0	1	1	$\mapsto$	0	0
1	0	0		0	1
1	0	1		0	1
1	1	0		0	1
1	1	1		0	0

This function can be expressed with minterms as $f(\mathbf{x}) = (f_1(\mathbf{x}), f_0(\mathbf{x}))$ where

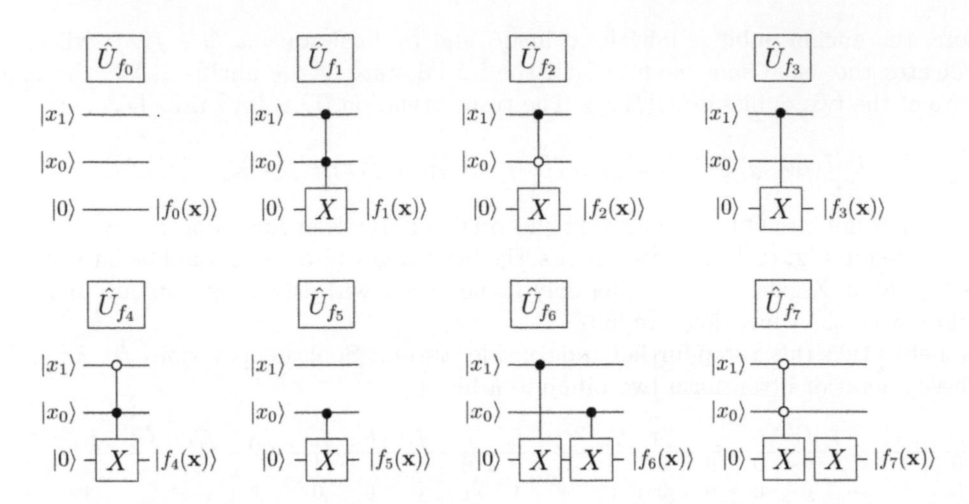

Fig. 10.2 Quantum circuits for two-bit Boolean functions. These circuits implement the two-bit Boolean functions $f : \mathbb{B}^2 \to \mathbb{B}$. The input qubits are initialized in $|x_1\rangle$ and $|x_0\rangle$, the ancilla qubit is initialized in $|0\rangle$. At the end of the circuit, the two input qubits are left unchanged and the ancilla qubit stores $f(x_1, x_0)$. Note that $\hat{U}_{f_1}$ is nothing but the Toffoli gate.

$$
\begin{aligned}
f_0(\mathbf{x}) &= m_{100}(\mathbf{x}) \vee m_{101}(\mathbf{x}) \vee m_{110}(\mathbf{x}), \\
f_1(\mathbf{x}) &= m_{001}(\mathbf{x}).
\end{aligned}
$$

One can verify that the quantum circuit associated with $f(\mathbf{x})$ is given by

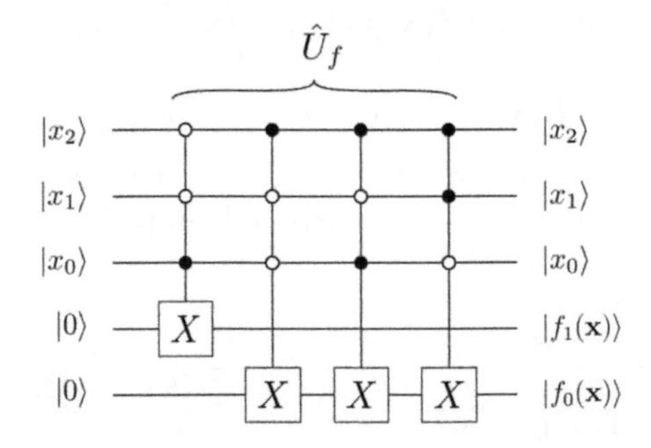

It is worth mentioning that the number of controlled operations in this circuit can be reduced by using the relation

Therefore,

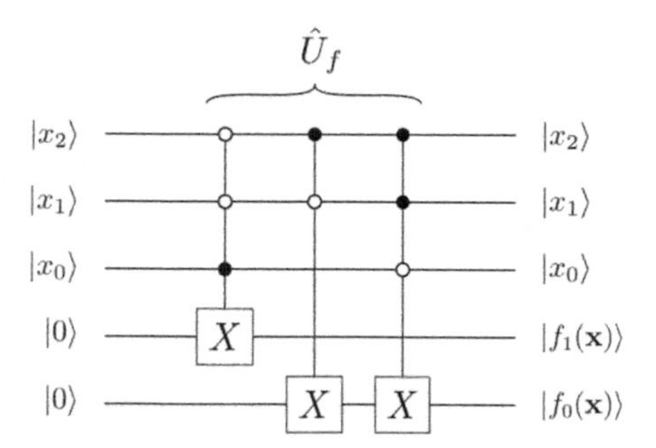

The reader should now be convinced that any function $f : \mathbb{B}^n \to \mathbb{B}^m$ can be implemented on a quantum computer as long as enough qubits are provided. Since a quantum computer can transform a binary string into any other binary string, quantum machines can execute any classical algorithm.

10.1.1 Quantum parallelism

In this section, we illustrate how to construct a quantum circuit for the evaluation of the function $f(x) = 2x$ assuming that the input string has only two bits. We will make use of this example to introduce quantum parallelism, an important concept in quantum computation. The function f takes an input string $\mathbf{x} = (x_1, x_0)$ and it converts into a three-bit string,

$$\begin{array}{ccccccccccc} & & x_1 & x_0 & & x_2' & x_1' & x_0' & & \\ \hline 0 & = & 0 & 0 & & 0 & 0 & 0 & = & 0 \\ 1 & = & 0 & 1 & \longmapsto & 0 & 1 & 0 & = & 2 \\ 2 & = & 1 & 0 & & 1 & 0 & 0 & = & 4 \\ 3 & = & 1 & 1 & & 1 & 1 & 0 & = & 6 \end{array}$$

The quantum circuit associated with $\hat{U}_{2x}$ is illustrated in Fig. 10.3a. The two input qubits are left unchanged, while the output of the computation is stored in three ancilla qubits. Four circuit runs are necessary to evaluate $f(x)$ for all of the input strings.

Can we evaluate $f(x)$ for all of the input strings with a *single* circuit run? To answer this question, let us investigate what happens when two Hadamard gates are applied to the input qubits at the beginning of the circuit (see Fig. 10.3b). These two operations transform the initial state $|\psi_{\text{A}}\rangle = |00000\rangle$ into

$$\begin{aligned} |\psi_{\text{B}}\rangle = \hat{H}_4\hat{H}_3|00000\rangle &= \frac{|0\rangle + |1\rangle}{\sqrt{2}}\frac{|0\rangle + |1\rangle}{\sqrt{2}}|000\rangle \\ &= \frac{1}{2}\big(|00\rangle|000\rangle + |01\rangle|000\rangle + |10\rangle|000\rangle + |11\rangle|000\rangle\big). \end{aligned}$$

The unitary $\hat{U}_{2x}$ is applied to the state $|\psi_{\text{B}}\rangle$ leading to

$$\begin{aligned} |\psi_{\text{C}}\rangle = \hat{U}_{2x}|\psi_{\text{B}}\rangle &= \frac{1}{2}\big(|00\rangle|000\rangle + |01\rangle|010\rangle + |10\rangle|100\rangle + |11\rangle|110\rangle\big) \\ &= \frac{1}{2}\big(|0\rangle|0\rangle + |1\rangle|2\rangle + |2\rangle|4\rangle + |3\rangle|6\rangle\big). \end{aligned}$$

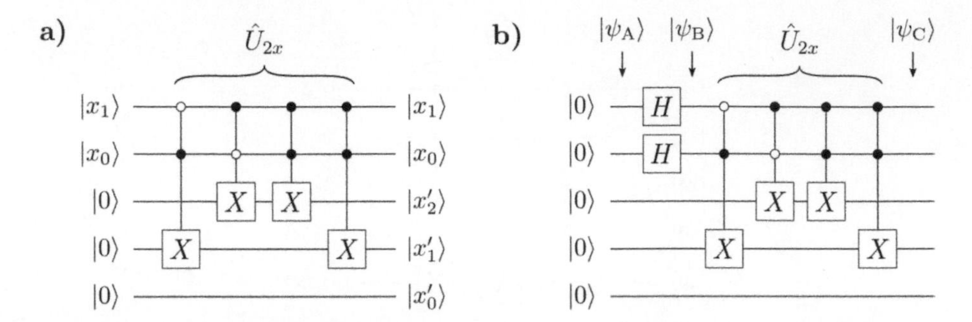

Fig. 10.3 Quantum parallelism. a) Quantum circuit implementing the unitary $\hat{U}_{2x}$. The circuit has been built following the procedure presented at the end of Section 10.1. The circuit operates on five qubits, two for the input and three for the output. At the end of the computation, the three output qubits store $2x$ in binary representation. Note that the number of controlled operations can be reduced, see Exercise 10.1. **b)** By preparing the two input qubits in a superposition state, the final state $|\psi_C\rangle$ is a superposition of the four possible solutions.

This is a remarkable result: we have applied $\hat{U}_{f(x)}$ only once and we have obtained a superposition state encoding $f(x)$ for four different inputs. This phenomenon is known as **quantum parallelism**. It is tempting to conclude that a quantum computer can evaluate a function for an arbitrary number of inputs in only one step. There is a big caveat though: if we want to extract some useful information from the final state $|\psi_C\rangle$, we must measure it. This operation will collapse $|\psi_C\rangle$ into one of the states $\{|0\rangle|0\rangle, |1\rangle|2\rangle, |2\rangle|4\rangle, |3\rangle|6\rangle\}$ with equal probability. The final measurement completely wipes away any apparent speed-up arising from quantum parallelism.

Fortunately, this is not the end of the story: in the following sections, we will see that some mathematical problems can be solved with fewer elementary gates than on a classical computer by taking advantage of quantum parallelism. The trick is to create constructive interference between the qubit states such that the bit string encoding the solution to the problem has a very high chance of being measured at the end of the computation.

10.2 Black-box problems

The following sections will present some quantum algorithms for **black-box problems** [69]. In these types of problems, we are given a black box, i.e. a physical machine that executes a function $f : \mathbb{B}^n \to \mathbb{B}^m$. In the classical case, the black box is simply a logic circuit. In the quantum case, the black box is a quantum circuit that implements the unitary $\hat{U}_f|\mathbf{x}\rangle|\mathbf{y}\rangle = |\mathbf{x}\rangle|\mathbf{y} \oplus f(\mathbf{x})\rangle$ where $|\mathbf{x}\rangle$ is the initial state of n qubits and $|\mathbf{y}\rangle$ is the initial state of m ancilla qubits (see Fig. 10.4). The black box is sometimes called the **oracle**, because when an input string is fed into the circuit, it produces an output string reliably. Our goal is to find some properties of f using the black box the minimum number of times (the most natural way to proceed is to feed some input strings into the machine and analyze the output strings). The number of times an algorithm

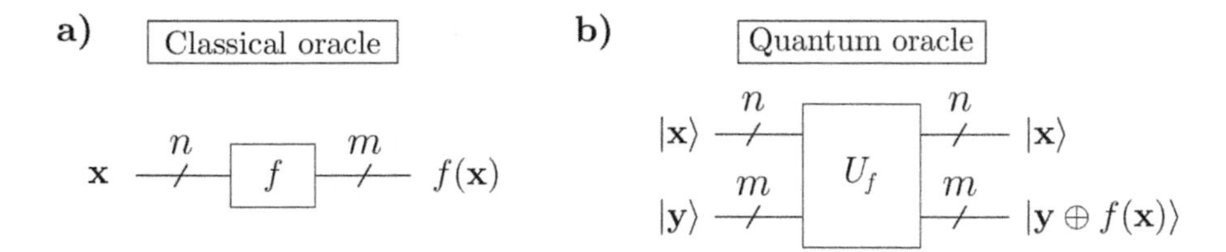

Fig. 10.4 Classical and quantum oracles. a) In the classical case, the oracle is a circuit that transforms an n-bit string $\mathbf{x}$ into an m-bit string $f(\mathbf{x})$. **b)** In the quantum case, the oracle is a unitary operation $\hat{U}_f$ that transforms an $n+m$ qubit state $|\mathbf{x}\rangle|\mathbf{y}\rangle$ into an $n+m$ qubit state $|\mathbf{x}\rangle|\mathbf{y} \oplus f(\mathbf{x})\rangle$.

uses the black box is called the **query complexity**. The Deutsch–Jozsa, Bernstein–Vazirani, Simon, and Grover algorithms are examples of quantum algorithms that solve a black-box problem with a lower query complexity than classical algorithms.

All quantum algorithms for black-box problems presented in this chapter start with a set of Hadamard gates applied to a set of qubits. For this reason, we consider it necessary to first present some mathematical results common to all of these algorithms in this section. First of all, the action of a Hadamard gate on a qubit state $|x\rangle$ where $x \in \mathbb{B}$ can be concisely expressed with a sum,

$$\boxed{\hat{H}|x\rangle = \begin{cases} \frac{|0\rangle+|1\rangle}{\sqrt{2}} & x=0 \\ \frac{|0\rangle-|1\rangle}{\sqrt{2}} & x=1 \end{cases} = \frac{|0\rangle + (-1)^x|1\rangle}{\sqrt{2}} = \sum_{a=0}^{1} \frac{(-1)^{ax}}{\sqrt{2}}|a\rangle,}$$

where ax is the usual product between integers. This equation can be generalized to a set of n qubits,

$$\begin{aligned} \hat{H}^{\otimes n}|\mathbf{x}\rangle &= \frac{|0\rangle + (-1)^{x_{n-1}}|1\rangle}{\sqrt{2}} \otimes \ldots \otimes \frac{|0\rangle + (-1)^{x_0}|1\rangle}{\sqrt{2}} \\ &= \frac{1}{2^{n/2}}\Big(\sum_{a_{n-1}=0}^{1} (-1)^{a_{n-1}x_{n-1}}|a_{n-1}\rangle \Big) \otimes \ldots \otimes \Big(\sum_{a_0=0}^{1} (-1)^{a_0 x_0}|a_0\rangle \Big) \\ &= \frac{1}{2^{n/2}} \sum_{\mathbf{a}\in\mathbb{B}^n} (-1)^{\mathbf{a}\cdot\mathbf{x}}|\mathbf{a}\rangle, \end{aligned} \tag{10.1}$$

where in the last step we introduced the **scalar product** between two binary strings,

$$\boxed{\mathbf{a}\cdot\mathbf{x} = a_{n-1}x_{n-1} \oplus \ldots \oplus a_0 x_0.} \tag{10.2}$$

This operation returns either 0 or 1. Recall that the Hadamard gate satisfies $\hat{H}^2 = \mathbb{1}$. This means that the operation $\hat{H}^{\otimes n}\hat{H}^{\otimes n}$ acts like the identity,

$$\hat{H}^{\otimes n}\hat{H}^{\otimes n}|\mathbf{x}\rangle = \hat{H}^{\otimes n}\Big(\frac{1}{2^{n/2}}\sum_{\mathbf{a}\in\mathbb{B}^n}(-1)^{\mathbf{x}\cdot\mathbf{a}}|\mathbf{a}\rangle\Big)$$
$$= \frac{1}{2^n}\sum_{\mathbf{k}\in\mathbb{B}^n}\sum_{\mathbf{a}\in\mathbb{B}^n}(-1)^{\mathbf{k}\cdot\mathbf{a}}(-1)^{\mathbf{x}\cdot\mathbf{a}}|\mathbf{k}\rangle \;=\; |\mathbf{x}\rangle. \qquad (10.3)$$

This relation will be useful when we discuss the Deutsch–Jozsa and Bernstein–Vazirani algorithms.

As we shall see in the next sections, a common technique in quantum computation is the **phase kickback method** [312]. This technique consists of applying a unitary transformation $\hat{U}_f$ to the state $|\mathbf{x}\rangle|-\rangle$. This is a state of $n+1$ qubits where n qubits are prepared in $|x_{n-1}\rangle\dots|x_0\rangle$ and one ancilla qubit is in the superposition state $|-\rangle$. The unitary $\hat{U}_f$ transforms this state into

$$\hat{U}_f|\mathbf{x}\rangle\otimes|-\rangle = \hat{U}_f|\mathbf{x}\rangle\otimes\frac{|0\rangle-|1\rangle}{\sqrt{2}} = |\mathbf{x}\rangle\otimes\frac{|f(\mathbf{x})\rangle-|1\oplus f(\mathbf{x})\rangle}{\sqrt{2}}$$
$$= \begin{cases}|\mathbf{x}\rangle\frac{|0\rangle-|1\rangle}{\sqrt{2}} & \text{if } f(\mathbf{x})=0\\ |\mathbf{x}\rangle\frac{|1\rangle-|0\rangle}{\sqrt{2}} & \text{if } f(\mathbf{x})=1\end{cases} = (-1)^{f(\mathbf{x})}|\mathbf{x}\rangle\otimes|-\rangle. \qquad (10.4)$$

The important point here is that when $\hat{U}_f$ operates on a superposition state $|\psi\rangle|-\rangle$, the information about $f(\mathbf{x})$ is stored in the relative phases of the output state. We are ready to present some quantum algorithms for black-box problems.

10.3 Deutsch's algorithm

Deutsch's algorithm is one of the simplest quantum algorithms that achieves a quantum advantage for a black-box problem [13]. This algorithm does not have any practical applications. However, it is interesting from a pedagogical point of view since more sophisticated quantum algorithms are based on similar techniques. **Deutsch's problem** is formulated as follows:

Given: A black box that implements a Boolean function $f : \mathbb{B} \to \mathbb{B}$.

Goal: Determine whether f is constant or not.

The goal is to understand whether the black box is implementing a constant function such as f_0 and f_3 in Fig. 10.5 or a balanced function like f_1 and f_2 represented in the same figure.

A classical computer can solve this problem by evaluating $f(0)$ and storing the result in memory. Subsequently, it evaluates $f(1)$ and compares this quantity with $f(0)$. If $f(0) = f(1)$, f is constant. This simple procedure requires two evaluations of the function. Thus, the query complexity of the algorithm is 2. A quantum computer can solve this problem with only one evaluation of the function, bringing the query complexity down to 1. Let us understand how this is possible.

Unlike in the classical case, in the quantum case the black box is a quantum device that executes one of the four unitary operations shown in Fig. 10.5 and it is our task to

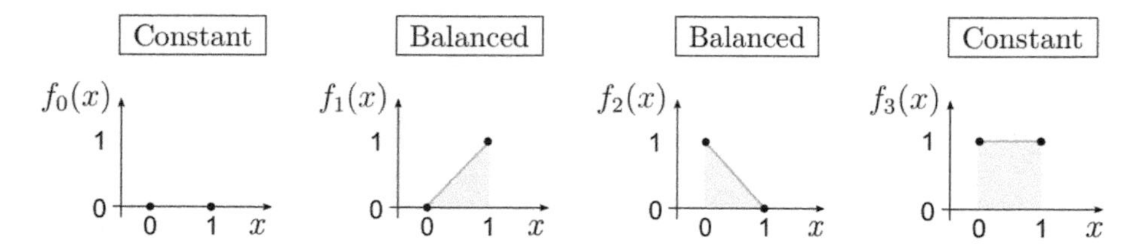

Fig. 10.5 Single-bit Boolean functions. Graphical representation of the single-bit Boolean functions f_0, f_1, f_2, and f_3. A constant function assigns the same output to all of the strings in the domain. A balanced function assigns 0 to half of the strings in the domain and 1 to the remaining strings.

---

determine whether the operation is either constant or balanced. Note that the goal of the algorithm is not to identify which of the four functions is implemented. Figure 10.6 shows the quantum circuit for **Deutsch's algorithm**. The application of an $\hat{X}$ gate to the ancilla qubit transforms the initial state $|\psi_{\rm A}\rangle = |0\rangle|0\rangle$ into $|\psi_{\rm B}\rangle = |0\rangle|1\rangle$. The algorithm proceeds with two Hadamard gates applied to the register,

$$|\psi_{\rm C}\rangle = \hat{H}_1\hat{H}_0|0\rangle|1\rangle = \frac{|0\rangle + |1\rangle}{\sqrt{2}} \otimes \frac{|0\rangle - |1\rangle}{\sqrt{2}} = \frac{1}{\sqrt{2}}(|0\rangle \otimes |-\rangle + |1\rangle \otimes |-\rangle). \quad (10.5)$$

The unitary $\hat{U}_f$ (encoding the unknown function f) is now applied to the two qubits,

$$\begin{aligned}|\psi_{\rm D}\rangle = \hat{U}_f|\psi_{\rm C}\rangle &= \frac{1}{\sqrt{2}}\left[(-1)^{f(0)}\,|0\rangle \otimes |-\rangle + (-1)^{f(1)}\,|1\rangle \otimes |-\rangle\right] \\ &= \frac{1}{\sqrt{2}}\left[(-1)^{f(0)}\,|0\rangle + (-1)^{f(1)}\,|1\rangle\right] \otimes |-\rangle.\end{aligned}$$

In these steps, we used the phase kickback method. The state of the ancilla qubit will now be ignored since it will not affect the rest of the computation. The application of a Hadamard gate produces the final state

$$\begin{aligned}|\psi_{\rm E}\rangle = \hat{H}|\psi_{\rm D}\rangle &= \frac{1}{\sqrt{2}}\left[(-1)^{f(0)}\,\frac{|0\rangle + |1\rangle}{\sqrt{2}} + (-1)^{f(1)}\,\frac{|0\rangle - |1\rangle}{\sqrt{2}}\right] \\ &= \frac{1}{2}\left\{\left[(-1)^{f(0)} + (-1)^{f(1)}\right]|0\rangle + \left[(-1)^{f(0)} - (-1)^{f(1)}\right]|1\rangle\right\}.\end{aligned}$$

This state can be written as

$$|\psi_{\rm E}\rangle = \begin{cases}|0\rangle & \text{if } f(0) = f(1), \\ |1\rangle & \text{if } f(0) \neq f(1).\end{cases}$$

A final measurement of the top qubit will produce $|0\rangle$ if f is constant and $|1\rangle$ if f is not constant. Deutsch's algorithm finds out whether f is constant or not with only one oracle query.

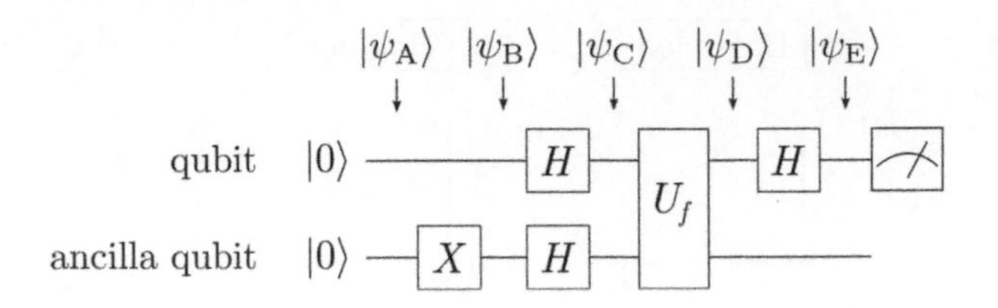

Fig. 10.6 Deutsch's algorithm.

10.3.1 Deutsch–Jozsa algorithm

The Deutsch–Jozsa algorithm is a generalization of Deutsch's algorithm [15]. It does not have any practical application, but it is still interesting from a pedagogical point of view. The problem is formulated as:

Given: A black box that implements an unknown function $f : \mathbb{B}^n \to \mathbb{B}$.

Assumptions: The function is either constant (i.e. $f(\mathbf{x}) = c$ for all $\mathbf{x}$) or balanced[1] ($f(\mathbf{x}) = 0$ for half of the strings in the domain and $f(\mathbf{x}) = 1$ for the remaining strings).

Goal: Determine whether f is constant or balanced.

A classical computer solves this problem by evaluating the quantity $f(0, \dots, 0)$ and storing it in memory. It then evaluates $f(\mathbf{x})$ for other strings in the domain. If there is one string $\tilde{\mathbf{x}}$ such that $f(\tilde{\mathbf{x}}) \neq f(0, \dots, 0)$, then f is not constant. In the worst case, the algorithm must evaluate f for half of the strings in the domain plus one, meaning that $2^n/2 + 1$ oracle queries are required to solve the problem.

Can a classical probabilistic machine solve this problem with fewer queries? To answer this question, suppose that a classical machine evaluates the function k times where $k \geq 2$. If the function is balanced, the probability that the evaluations give the same result is $2/2^k$. This probability drops exponentially with the number of evaluations. If 10 evaluations give the same output, we conclude that the function is constant with $1 - 2/2^{10} \approx 99.8\,\%$ probability of being right. We can arbitrarily increase the success probability by increasing the number of evaluations. Thus, a probabilistic machine can solve this problem with a number of queries determined by the desired success probability.

A quantum computer can solve this problem with *only one application* of the oracle. The quantum circuit for the Deutsch–Jozsa algorithm is shown in Fig. 10.7. First, an $\hat{X}$ gate on the ancilla qubit maps the initial state of the register into $|\psi_B\rangle = |0 \dots 0\rangle|1\rangle$. Similarly to the Deutsch algorithm, a Hadamard gate is applied to each qubit (including the ancilla) yielding

[1]The two-bit Boolean functions f_0 and f_{15} presented in Section 10.1 are constant, whereas the functions f_3, f_5, f_6, f_9, f_{10}, and f_{12} are balanced.

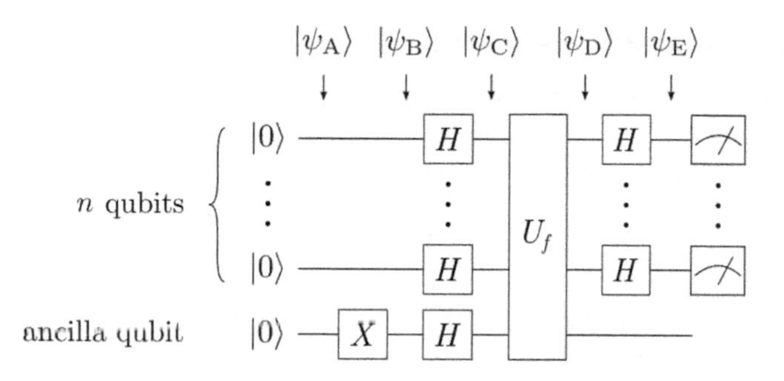

Fig. 10.7 Deutsch–Jozsa algorithm. Quantum circuit of the Deutsch–Jozsa algorithm. Interestingly, this quantum circuit also implements the Bernstein–Vazirani algorithm.

$$|\psi_C\rangle = (|+\rangle \otimes \ldots \otimes |+\rangle) \otimes |-\rangle = \Big(\frac{1}{2^{n/2}} \sum_{\mathbf{x}\in\mathbb{B}^n} |\mathbf{x}\rangle\Big) \otimes |-\rangle, \qquad (10.6)$$

where in the last step we used eqn (10.1). The oracle $\hat{U}_f$ transforms the state of the quantum register into

$$|\psi_D\rangle = \hat{U}_f|\psi_C\rangle = \Big(\frac{1}{2^{n/2}} \sum_{\mathbf{x}\in\mathbb{B}^n} (-1)^{f(\mathbf{x})}\,|\mathbf{x}\rangle\Big) \otimes |-\rangle.$$

The state of the ancilla will now be ignored since it will not influence the rest of the computation. The algorithm ends with a Hadamard gate applied to each qubit of the top register,

$$|\psi_E\rangle = \hat{H} \otimes \ldots \otimes \hat{H}|\psi_D\rangle = \frac{1}{2^n} \sum_{\mathbf{k}\in\mathbb{B}^n} \sum_{\mathbf{x}\in\mathbb{B}^n} (-1)^{\mathbf{x}\cdot\mathbf{k}}\,(-1)^{f(\mathbf{x})}\,|\mathbf{k}\rangle. \qquad (10.7)$$

If f is constant, the state $|\psi_E\rangle$ can be written as[2]

$$|\psi_E\rangle = \frac{1}{2^n} \sum_{\mathbf{k}\in\mathbb{B}^n} \sum_{\mathbf{x}\in\mathbb{B}^n} (-1)^{\mathbf{x}\cdot\mathbf{k}}|\mathbf{k}\rangle = |0\ldots0\rangle,$$

where we used eqn (10.3). If the final measurement shows that the state of the register is $|0\ldots0\rangle$, f is constant. On the other hand, if the register ends up in a state other than $|0\ldots0\rangle$, then f is balanced. The Deutsch–Jozsa algorithm can identify a specific property of a Boolean function with just one oracle query.

[2]To be precise, if $f(\mathbf{x}) = 0$ for all $\mathbf{x}$, the state of the top register becomes $|0\ldots0\rangle$. If instead $f(\mathbf{x}) = 1$ for all $\mathbf{x}$, the state of the top register becomes $-|0\ldots0\rangle$; the two states are identical up to an irrelevant global phase.

10.4 Bernstein–Vazirani algorithm

Similarly to the Deutsch–Jozsa algorithm, the Bernstein–Vazirani algorithm aims to determine a global property of a Boolean function with fewer queries than a classical algorithm [16]. The problem is formulated as:

Given: A black box that implements a Boolean function $f_{\mathbf{a}} : \mathbb{B}^n \to \mathbb{B}$.

Assumption: The function returns the scalar product $f_{\mathbf{a}}(\mathbf{x}) = \mathbf{a}\cdot\mathbf{x}$ as in eqn (10.2).

Goal: Find $\mathbf{a}$.

A classical computer solves this problem by evaluating $f_{\mathbf{a}}(\mathbf{x})$ for n different input strings:

$$\begin{aligned} f_{\mathbf{a}}(0,\ldots,0,1) &= a_0, \\ f_{\mathbf{a}}(0,\ldots,1,0) &= a_1, \\ &\vdots \\ f_{\mathbf{a}}(1,\ldots,0,0) &= a_{n-1}. \end{aligned}$$

Each evaluation provides a different entry of $\mathbf{a}$ meaning that n evaluations are necessary to solve the problem. A quantum computer can determine $\mathbf{a}$ with *only one oracle query.* The circuit for the Bernstein–Vazirani algorithm is the same as that of the Deutsch–Jozsa algorithm shown in Fig. 10.7. From eqn (10.7), the final state of the top register is

$$|\psi_{\mathrm{E}}\rangle = \frac{1}{2^n} \sum_{\mathbf{k}\in\mathbb{B}^n} \sum_{\mathbf{x}\in\mathbb{B}^n} (-1)^{\mathbf{k}\cdot\mathbf{x}} (-1)^{f_{\mathbf{a}}(\mathbf{x})} |\mathbf{k}\rangle. \tag{10.8}$$

The only difference between the Deutsch–Jozsa algorithm and the Bernstein–Vazirani algorithm is that in the latter the unknown function is of the form $f_{\mathbf{a}}(\mathbf{x}) = \mathbf{a}\cdot\mathbf{x}$. By substituting this expression into (10.8), we obtain

$$|\psi_{\mathrm{E}}\rangle = \frac{1}{2^n} \sum_{\mathbf{k}\in\mathbb{B}^n} \sum_{\mathbf{x}\in\mathbb{B}^n} (-1)^{\mathbf{k}\cdot\mathbf{x}} (-1)^{\mathbf{a}\cdot\mathbf{x}} |\mathbf{k}\rangle = |\mathbf{a}\rangle.$$

where in the last step we used (10.3). A final measurement will give the bit string $\mathbf{a}$ with unit probability. Thus, in the quantum case only one evaluation of $f_{\mathbf{a}}$ is necessary, bringing the query complexity from n down to 1.

10.5 Grover's algorithm

Grover's algorithm is a quantum search algorithm first devised by Lov Grover in 1996 [19]. This algorithm allows us to solve an unstructured search problem with a quadratic speed-up over a classical approach. Before diving into the technical details, it is necessary to clarify the concept of a search problem. A search problem is a computational problem in which a specific element must be identified within a large set.

Search problems can be divided into structured and unstructured. A simple example of a **structured search problem** is searching for a word in an alphabetically ordered dictionary. This problem can be easily solved by a classical machine [32] and a quantum computer would not offer a significant speed-up in this case. In an **unstructured search problem**, the set in which the search is performed is not ordered. A classical computer can solve these types of problems with a brute force approach: the items are checked one after another until the right element is found. On average, if the set contains N elements, the correct element will be identified after $N/2$ steps. Grover's algorithm requires only $O(\sqrt{N})$ steps to find the item.

In mathematical terms, the problem is formulated as:

Given: A black box that implements a Boolean function $f_{\mathbf{a}} : \mathbb{B}^n \to \mathbb{B}$.

Assumption: $f_{\mathbf{a}}$ assigns 1 only to string $\mathbf{a}$,

$$f_{\mathbf{a}}(\mathbf{x}) = \delta_{\mathbf{xa}} = \begin{cases} 1 & \text{for } \mathbf{x} = \mathbf{a} \\ 0 & \text{for } \mathbf{x} \neq \mathbf{a}. \end{cases}$$

Goal: Find $\mathbf{a}$.

The function of Grover's problem is a Kronecker delta that assigns 1 to only one string of the domain and 0 to all others. This string can be found with a classical computer by calculating $f_{\mathbf{a}}$ for some strings in the domain (there is a total of $N = 2^n$ of them). In the worst case, it is necessary to evaluate the function $N - 1$ times. On average, the string will be found after $N/2$ queries. A quantum computer can determine $\mathbf{a}$ with high probability with only $O(\sqrt{N})$ oracle queries providing a quadratic speed-up over classical algorithms.

For simplicity, let us first explain Grover's algorithm when the unknown function is the two-bit operation

x_1	x_0		$f_{\mathbf{a}}(\mathbf{x})$
0	0		0
0	1	$\mapsto$	0
1	0		1
1	1		0

In this example, the string that we are looking for is $\mathbf{a} = 10$. The quantum circuit implementing Grover's algorithm is shown in Fig. 10.8. An $\hat{X}$ gate on the ancilla qubit followed by a set of Hadamard gates transforms the initial state $|\psi_{\mathrm{A}}\rangle = |000\rangle$ into

$$|\psi_{\mathrm{B}}\rangle = \hat{H}_2\hat{H}_1\hat{H}_0\hat{X}|000\rangle = \frac{1}{2}(|00\rangle + |01\rangle + |10\rangle + |11\rangle) \otimes |-\rangle.$$

The transformation $\hat{U}_{f_{\mathbf{a}}}$ with $\mathbf{a} = 10$ is applied to the register[3],

[3]This unitary operator corresponds to the controlled gate $\hat{U}_{f_2}$ shown in Fig. 10.2.

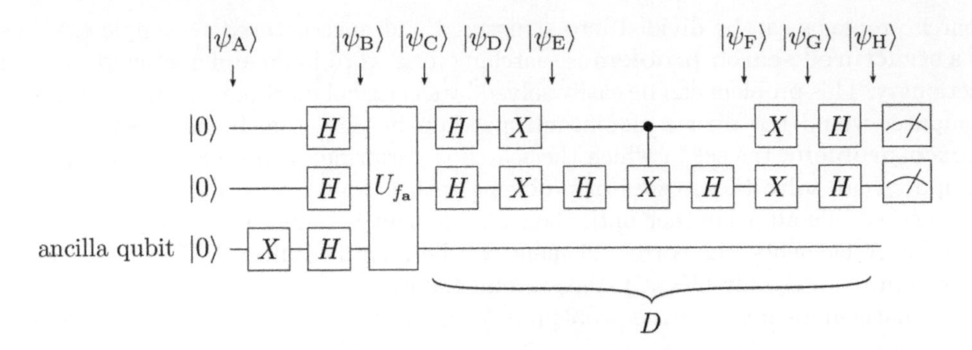

Fig. 10.8 Grover's algorithm for two-bit Boolean functions. Quantum circuit implementing Grover's algorithm for Boolean functions $f_{\mathbf{a}} : \mathbb{B}^2 \to \mathbb{B}$. In this register, qubit 0 is an auxiliary qubit. At the end of the computation, qubit 1 and qubit 2 store the final result.

$$|\psi_C\rangle = \hat{U}_{f_{\mathbf{a}}}|\psi_B\rangle = \frac{1}{2}(|00\rangle + |01\rangle - |10\rangle + |11\rangle) \otimes |-\rangle.$$

The vector $|10\rangle$ has a relative phase which is different from the other three vectors. The solution to the problem is encoded in this relative phase. The remaining part of the algorithm converts this phase into an amplitude, i.e. transforms the superposition state $|\psi_C\rangle$ into

$$|\psi_H\rangle = \Big[c_{00}|00\rangle + c_{01}|01\rangle + c_{10}|10\rangle + c_{11}|11\rangle\Big] \otimes |-\rangle,$$

where $|c_{10}| \gg |c_{00}|, |c_{01}|, |c_{11}|$. This procedure is known as **quantum amplitude amplification** [313]. A measurement of $|\psi_H\rangle$ will give $|10\rangle$ (the solution to our problem) with high probability.

Let us explain step by step how Grover's algorithm transforms $|\psi_C\rangle$ into $|\psi_H\rangle$. Hereafter, the state of the ancilla will be ignored since it does not affect the final result. We will make use of the following operator (see Fig. 10.8)

$$\hat{D} = (\hat{H}_2\hat{H}_1)(\hat{X}_2\hat{X}_1)\hat{H}_1(\mathsf{c}_2\hat{X}_1)\hat{H}_1(\hat{X}_2\hat{X}_1)(\hat{H}_2\hat{H}_1).$$

The first two Hadamard gates transform ψ_C into

$$|\psi_D\rangle = \hat{H}_2\hat{H}_1|\psi_D\rangle = \frac{1}{2}\big[|10\rangle + |11\rangle + |00\rangle - |01\rangle\big].$$

An $\hat{X}$ gate on each qubit gives

$$|\psi_E\rangle = \hat{X}_2\hat{X}_1|\psi_E\rangle = \frac{1}{2}\left[|01\rangle + |00\rangle + |11\rangle - |10\rangle\right].$$

The operators $\hat{H}_1(\mathsf{c}_2\hat{X}_1)\hat{H}_1$ transform this state into

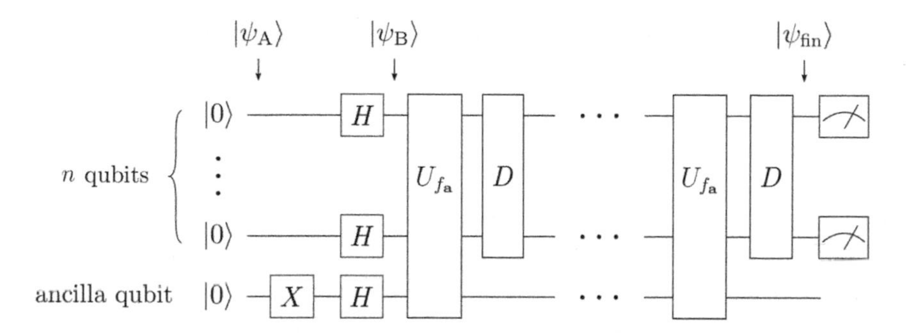

Fig. 10.9 Grover's algorithm. Quantum circuit of Grover's algorithm for the general case. The operator $\hat{D}\hat{U}_{f_{\mathbf{a}}}$ must be applied to the register $k \approx \pi\sqrt{2^n}/4$ times to obtain a final state $|\psi_{\text{fin}}\rangle$ that is similar to state $|\mathbf{a}\rangle$, the solution of the problem.

$$|\psi_F\rangle = \hat{H}_1(\mathsf{c}_2\hat{X}_1)\hat{H}_1|\psi_E\rangle = \frac{1}{2}[|00\rangle + |01\rangle - |10\rangle - |11\rangle].$$

Two $\hat{X}$ gates on this state produce

$$|\psi_G\rangle = \hat{X}_2\hat{X}_1|\psi_F\rangle = \frac{1}{2}[|11\rangle + |10\rangle - |01\rangle - |00\rangle],$$

and, finally, two more Hadamard gates yield $|\psi_H\rangle = |10\rangle$. We have obtained the solution to our problem with only one query of the function $f_{\mathbf{a}}$.

10.5.1 Grover's algorithm: the general case

Let us present the quantum circuit that solves the search problem for the general case $f_{\mathbf{a}} : \mathbb{B}^n \to \mathbb{B}$. The unknown function is a Kronecker delta

$$f_{\mathbf{a}}(\mathbf{x}) = \delta_{\mathbf{xa}} = \begin{cases} 1 & \text{for } \mathbf{x} = \mathbf{a} \\ 0 & \text{for } \mathbf{x} \neq \mathbf{a} \end{cases}$$

and we do not have any information about $\mathbf{a}$. The quantum circuit shown in Fig. 10.9 allows us to determine $\mathbf{a}$ with high probability. The algorithm starts with all qubits in the ground state $|\psi_A\rangle = |0\ldots0\rangle|0\rangle$. An $\hat{X}$ gate on the ancilla qubit followed by a set of $n+1$ Hadamard gates yield

$$|\psi_B\rangle = \hat{H}_n \ldots \hat{H}_0\hat{X}_0|\psi_A\rangle = \underbrace{\left(\frac{1}{2^{n/2}}\sum_{\mathbf{x}\in\mathbb{B}^n}|\mathbf{x}\rangle\right)}_{|s\rangle} \otimes |-\rangle = |s\rangle \otimes |-\rangle, \tag{10.9}$$

where in the last step we introduced $|s\rangle = \hat{H}_n \ldots \hat{H}_0|0\ldots0\rangle$, the balanced superposition state of n qubits. It is not immediately clear why we have prepared all the qubits into a superposition state. The rationale behind this choice is that we do not have any information about the string we are looking for. Hence, it is reasonable to initialize the register into a superposition state with equal weights to not penalize some bit strings over others. The next step consists in applying the black box to the register,

$$\hat{U}_{f_{\mathbf{a}}}|\psi_{\mathrm{B}}\rangle = \frac{1}{2^{n/2}} \sum_{\mathbf{x}\in\mathbb{B}^n} (-1)^{f_{\mathbf{a}}(\mathbf{x})}|\mathbf{x}\rangle \otimes |-\rangle.$$

Unlike the simple example discussed in the previous section, it is not possible to obtain **a** with a single query of the oracle, and several applications of $\hat{U}_{f_{\mathbf{a}}}$ are necessary as shown in Fig. 10.9. In this circuit, $\hat{D}$ is defined as

$$\boxed{\hat{D} = \hat{H}^{\otimes n}\,(2|0\ldots0\rangle\langle0\ldots0| - \mathbb{1})\,\hat{H}^{\otimes n} = 2|s\rangle\langle s| - \mathbb{1}.}$$

Grover's algorithm involves the iterative application of the operator $\hat{D}\hat{U}_{f_{\mathbf{a}}}$ until the final state $|\psi_{\mathrm{fin}}\rangle$ is very similar to the solution, $|\langle\mathbf{a}|\psi_{\mathrm{fin}}\rangle| \approx 1$. In the next section, we prove that the operator $\hat{D}\hat{U}_{f_{\mathbf{a}}}$ must be applied approximately $\sqrt{N}$ times in order to obtain a final state similar to $|\mathbf{a}\rangle$. It is important to mention that $\hat{D}$ can be decomposed into a polynomial number of elementary gates [103].

10.5.2 Graphical visualization of Grover's algorithm

In Grover's algorithm, the top register starts from a superposition of all basis vectors, $|s\rangle = N^{-1/2}\sum_{\mathbf{x}}|\mathbf{x}\rangle$. The circuit transforms $|s\rangle$ into a state $|s_k\rangle \approx |\mathbf{a}\rangle$ by means of a sequence of k rotations. To show this, let us start with two important observations:

1) From eqn (10.4), the action of $\hat{U}_{f_{\mathbf{a}}}$ on $|\mathbf{x}\rangle|-\rangle$ produces

$$\hat{U}_{f_{\mathbf{a}}}|\mathbf{x}\rangle|-\rangle = (-1)^{f_{\mathbf{a}}(x)}|\mathbf{x}\rangle|-\rangle = \begin{cases} -|\mathbf{a}\rangle|-\rangle & \text{for } \mathbf{x} = \mathbf{a} \\ |\mathbf{x}\rangle|-\rangle & \text{for } \mathbf{x} \neq \mathbf{a}. \end{cases}$$

The action of $\hat{U}_{f_{\mathbf{a}}}$ on the $n+1$ qubit state $|\mathbf{x}\rangle|-\rangle$ is equivalent to the action of the operator $\hat{O}_{\mathbf{a}} \equiv \mathbb{1} - 2|\mathbf{a}\rangle\langle\mathbf{a}|$ on the n-qubit state $|\mathbf{x}\rangle$,

$$\hat{O}_{\mathbf{a}}|\mathbf{x}\rangle = (\mathbb{1} - 2|\mathbf{a}\rangle\langle\mathbf{a}|)\,|\mathbf{x}\rangle = \begin{cases} -|\mathbf{a}\rangle & \text{for } \mathbf{x} = \mathbf{a} \\ |\mathbf{x}\rangle & \text{for } \mathbf{x} \neq \mathbf{a}. \end{cases} \tag{10.10}$$

Since $\hat{U}_{f_{\mathbf{a}}}$ and $\hat{O}_{\mathbf{a}}$ operate in the same way on the vectors $|\mathbf{x}\rangle$, hereafter we will only deal with the operator $\hat{O}_{\mathbf{a}}$ and will neglect the ancilla qubit since it will not influence the rest of the computation. Figure 10.10 illustrates the quantum circuit of Grover's algorithm without the ancilla qubit and $\hat{U}_{f_{\mathbf{a}}}$ replaced by $\hat{O}_{\mathbf{a}}$.

2) The second observation is that the operators $\hat{O}_{\mathbf{a}}$ and $\hat{D}$ map the vectors $|\mathbf{a}\rangle$ and $|s\rangle$ into linear combinations of the same vectors:

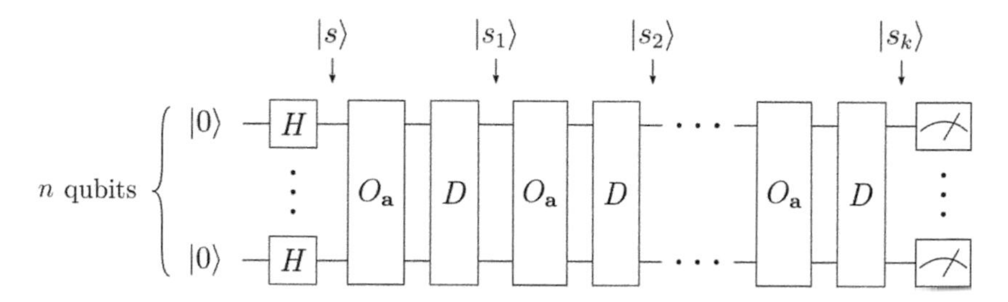

Fig. 10.10 Amplitude amplification. This quantum circuit is equivalent to that shown in Fig. 10.9. Here, however, the ancilla qubit is omitted and the oracle $\hat{U}_{f_{\mathbf{a}}}$ is substituted with the operator $\hat{O}_{\mathbf{a}} = \mathbb{1} - 2|\mathbf{a}\rangle\langle\mathbf{a}|$.

$$\begin{aligned}
\hat{O}_{\mathbf{a}}|s\rangle &= (\mathbb{1} - 2|\mathbf{a}\rangle\langle\mathbf{a}|)\,|s\rangle = |s\rangle - 2\langle\mathbf{a}|s\rangle\,|\mathbf{a}\rangle, \\
\hat{O}_{\mathbf{a}}|\mathbf{a}\rangle &= (\mathbb{1} - 2|\mathbf{a}\rangle\langle\mathbf{a}|)\,|\mathbf{a}\rangle = -\,|\mathbf{a}\rangle, \\
\hat{D}|s\rangle &= (2|s\rangle\langle s| - \mathbb{1})\,|s\rangle = |s\rangle, \\
\hat{D}|\mathbf{a}\rangle &= (2|s\rangle\langle s| - \mathbb{1})\,|\mathbf{a}\rangle = 2\langle s|\mathbf{a}\rangle\,|s\rangle - |\mathbf{a}\rangle.
\end{aligned}$$

In other other words, the operators $\hat{O}_{\mathbf{a}}$ and $\hat{D}$ rotate the vectors $|s\rangle$ and $|\mathbf{a}\rangle$ in a two-dimensional space.

We are ready to discuss the iterative action of the operator $\hat{D}\hat{O}_{\mathbf{a}}$ on the quantum register. As shown in Fig. 10.10, the register is initially prepared in the superposition state

$$\begin{aligned}
|s\rangle &= \frac{1}{\sqrt{N}}\sum_{\mathbf{x}}|\mathbf{x}\rangle = \frac{1}{\sqrt{N}}\Big(|\mathbf{a}\rangle + \sum_{\mathbf{x}\neq\mathbf{a}}|\mathbf{x}\rangle\Big) = \frac{1}{\sqrt{N}}|\mathbf{a}\rangle + \frac{\sqrt{N-1}}{\sqrt{N}}\underbrace{\frac{1}{\sqrt{N-1}}\sum_{\mathbf{x}\neq\mathbf{a}}|\mathbf{x}\rangle}_{|\mathbf{a}_\perp\rangle} \\
&= \frac{1}{\sqrt{N}}|\mathbf{a}\rangle + \frac{\sqrt{N-1}}{\sqrt{N}}|\mathbf{a}_\perp\rangle,
\end{aligned}$$

where $|\mathbf{a}_\perp\rangle$ is proportional to the sum of all basis vectors orthogonal[4] to $|\mathbf{a}\rangle$. The state $|s\rangle$ can be expressed in a more compact form as

$$|s\rangle = \sin\theta|\mathbf{a}\rangle + \cos\theta|\mathbf{a}_\perp\rangle, \tag{10.11}$$

where

$$\sin\theta = \langle\mathbf{a}|s\rangle = 1/\sqrt{N}, \qquad \cos\theta = \langle\mathbf{a}_\perp|s\rangle = \sqrt{N-1}/\sqrt{N}. \tag{10.12}$$

[4]For example, $|s\rangle = \frac{1}{\sqrt{4}}(|00\rangle + |01\rangle + |10\rangle + |11\rangle)$ for two qubits. If $|\mathbf{a}\rangle = |01\rangle$, then $|\mathbf{a}_\perp\rangle = \frac{1}{\sqrt{3}}(|00\rangle + |10\rangle + |11\rangle)$. Therefore, $|s\rangle = \sin\theta|\mathbf{a}\rangle + \cos\theta|\mathbf{a}_\perp\rangle$ where $\sin\theta = 1/\sqrt{4}$ and $\cos\theta = \sqrt{3/4}$.

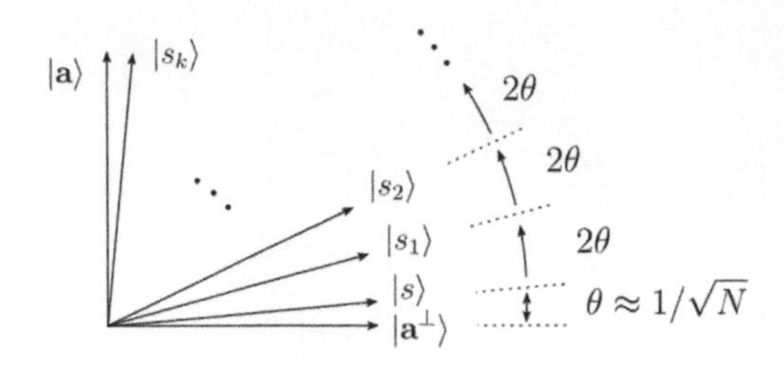

Fig. 10.11 Graphical visualization of Grover's algorithm. The operator $\hat{D}\hat{O}_{\mathbf{a}}$ rotates the initial state $|s_i\rangle$ anti-clockwise by an angle 2θ where $\sin\theta = \langle s|\mathbf{a}\rangle$.

Therefore, $|s\rangle$ points in a direction very similar to $|\mathbf{a}_{\perp}\rangle$ for $N \gg 1$ (see Fig. 10.11). One application of the operator $\hat{O}_{\mathbf{a}}$ reflects $|s\rangle$ about the vector $|\mathbf{a}_{\perp}\rangle$,

$$\hat{O}_{\mathbf{a}}|s\rangle = (\mathbb{1} - 2|\mathbf{a}\rangle\langle\mathbf{a}|)\,(\sin\theta|\mathbf{a}\rangle + \cos\theta|\mathbf{a}_{\perp}\rangle) = -\sin\theta|\mathbf{a}\rangle + \cos\theta|\mathbf{a}_{\perp}\rangle.$$

The operator $\hat{D}$ transforms this state into

$$\begin{aligned}
|s_1\rangle \equiv \hat{D}\hat{O}_{\mathbf{a}}|s\rangle &= (2|s\rangle\langle s| - \mathbb{1})(-\sin\theta|\mathbf{a}\rangle + \cos\theta|\mathbf{a}_{\perp}\rangle) \\
&= -2\sin\theta\langle s|\mathbf{a}\rangle\,|s\rangle + 2\cos\theta\langle s|\mathbf{a}_{\perp}\rangle\,|s\rangle + \sin\theta|\mathbf{a}\rangle - \cos\theta|\mathbf{a}_{\perp}\rangle \\
&= -2\sin^2\theta|s\rangle + 2\cos^2\theta|s\rangle + \sin\theta|\mathbf{a}\rangle - \cos\theta|\mathbf{a}_{\perp}\rangle \\
&= \sin(3\theta)|\mathbf{a}\rangle + \cos(3\theta)|\mathbf{a}_{\perp}\rangle,
\end{aligned}$$

where we used eqns (10.11, 10.12). In short, the transformation $\hat{D}\hat{O}_{\mathbf{a}}$ rotates the state $|s\rangle$ counterclockwise by an angle 2θ. After k applications, we have

$$|s_k\rangle \;\equiv\; (\hat{D}\hat{O}_{\mathbf{a}})^k|s\rangle \;=\; \sin[(2k+1)\theta]|\mathbf{a}\rangle + \cos[(2k+1)\theta]|\mathbf{a}_{\perp}\rangle.$$

The vector $|s_k\rangle$ is similar to $|\mathbf{a}\rangle$ when $\sin[(2k+1)\theta] \approx 1$, i.e. when $(2k+1)\theta \approx \pi/2$. In conclusion, the minimum number of oracle queries is[5]

$$\boxed{k = \text{Round}\Big(\frac{\pi}{4\theta} - \frac{1}{2}\Big) \approx \text{Round}\Big(\frac{\pi}{4}\sqrt{N} - \frac{1}{2}\Big),} \tag{10.13}$$

where Round(x) indicates the nearest integer to the real number x. The computation ends with a projective measurement of $|s_k\rangle$ which produces a bit string $\mathbf{b}$. This solution can be verified by computing $f_{\mathbf{a}}(\mathbf{b})$. If $f_{\mathbf{a}}(\mathbf{b}) = 1$, then the desired bit string has been found; otherwise, the quantum circuit must be run again. In contrast to the simple case presented in Section 10.5, the success probability of the circuit is close to but not exactly one. However, a minor modification to Grover's algorithm has been proposed to ensure a deterministic result [314]. This unstructured search problem can be extended to scenarios where the unknown function f assigns 1 to multiple strings in the domain. For further information, refer to Ref. [312].

[5]Here, we are using the relation $1/\sqrt{N} = \sin\theta \approx \theta$ which is only valid for $N \gg 1$.

Some mathematical problems are challenging to solve on classical machines when the input of the problem is significantly large. In particular, *some* NP complete problems do not have any structure and the most efficient way to solve them is a brute force approach in which all possible solutions are tried one after the other. Grover's algorithm can offer a quadratic speed-up for such problems.

It is important to mention that a quantum computer cannot solve an NP complete problem in polynomial time by trying all possible solutions. This is because Grover's algorithm is the optimal quantum algorithm for unstructured search problems (this point will be discussed in detail in the next section). It is widely believed that quantum computers will never solve NP complete problems with polynomial resources, see Refs [315, 316] for more information.

10.5.3 Grover's algorithm is optimal

Can a quantum algorithm solve an unstructured search problem with fewer oracle queries than Grover's algorithm? No, Grover's algorithm is the most efficient quantum algorithm for unstructured search problems, i.e. it is not possible to determine $\mathbf{a}$ with high probability with less than $O(\sqrt{N})$ oracle queries [95, 317]. We will now dwell on this point.

In an unstructured search problem, we are given a quantum circuit that implements an oracle $\hat{O}_{\mathbf{a}} = \mathbb{1} - 2|\mathbf{a}\rangle\langle\mathbf{a}|$ and our goal is to find $\mathbf{a}$ using the minimum number of queries. Since we do not have any information about $\mathbf{a}$, it is reasonable to initialize the quantum computer in a superposition state of all strings $|s\rangle = N^{-1/2}\sum_{\mathbf{x}\in\mathbb{B}^n}|\mathbf{x}\rangle$ where as usual $N = 2^n$. If we measure $|s\rangle$ in the computational basis, the probability that the measurement outcome yields the correct string is

$$|\langle\mathbf{a}|s\rangle|^2 = \left|\frac{1}{\sqrt{N}}\sum_{\mathbf{x}\in\mathbb{B}^n}\langle\mathbf{a}|\mathbf{x}\rangle\right|^2 = \frac{1}{N}.$$

To increase the success probability, we must apply some transformations to $|s\rangle$.

Naively, we could transform $|s\rangle$ with a sequence of transformations $\hat{U}_k \ldots \hat{U}_1$ that *do not contain the oracle* and hope to obtain a state $|\phi_k\rangle = \hat{U}_k \ldots \hat{U}_1|s\rangle$ which is very similar to $|\mathbf{a}\rangle$. The probability of observing the correct string $\mathbf{a}$ is

$$\boxed{q_{\mathbf{a}} = |\langle\mathbf{a}|\phi_k\rangle|^2 = \left|\frac{1}{\sqrt{N}}\sum_{\mathbf{x}\in\mathbb{B}^n}\langle\mathbf{a}|\hat{U}_k \ldots \hat{U}_1|\mathbf{x}\rangle\right|^2.}$$

If we are lucky, $|\phi_k\rangle = |\mathbf{a}\rangle$ and $q_{\mathbf{a}} = 1$. If we are not, $|\phi_k\rangle$ will be significantly different from $|\mathbf{a}\rangle$, and the success probability $q_{\mathbf{a}}$ will be low. Ideally, the algorithm should work well for all strings $|\mathbf{a}\rangle$ and not just one in particular. Hence, it is more appropriate to consider the average success probability for a randomly chosen $\mathbf{a}$,

$$\overline{q} = \frac{1}{N} \sum_{\mathbf{a} \in \mathbb{B}^n} |\langle \mathbf{a} | \phi_k \rangle|^2 = \frac{1}{N} \sum_{\mathbf{a} \in \mathbb{B}^n} \left| \frac{1}{\sqrt{N}} \sum_{\mathbf{x} \in \mathbb{B}^n} \langle \mathbf{a} | \hat{U}_k \ldots \hat{U}_1 | \mathbf{x} \rangle \right|^2$$
$$= \frac{1}{N^2} \sum_{\mathbf{a},\mathbf{x},\mathbf{y} \in \mathbb{B}^n} \langle \mathbf{y} | \hat{U}_1^\dagger \ldots \hat{U}_k^\dagger | \mathbf{a} \rangle \langle \mathbf{a} | \hat{U}_k \ldots \hat{U}_1 | \mathbf{x} \rangle$$
$$= \frac{1}{N^2} \sum_{\mathbf{x} \in \mathbb{B}^n} \sum_{\mathbf{y} \in \mathbb{B}^n} \langle \mathbf{y} | \mathbf{x} \rangle = \frac{1}{N^2} \sum_{\mathbf{x} \in \mathbb{B}^n} 1 = \frac{1}{N}. \tag{10.14}$$

If we do not query the oracle, the average success probability approaches zero for $N \gg 1$. The bottom line is that we must use the oracle to increase the average success probability.

The most generic quantum circuit that employs the oracle $\hat{O}_{\mathbf{a}}$ has the form

$$|\psi_k^{\mathbf{a}}\rangle = \hat{U}_k \hat{O}_{\mathbf{a}} \hat{U}_{k-1} \hat{O}_{\mathbf{a}} \ldots \hat{U}_1 \hat{O}_{\mathbf{a}} |s\rangle,$$

where $\hat{U}_j$ are generic unitary operations and the parameter k indexes the number of oracle queries. It is evident that the average success probability

$$\overline{p} = \frac{1}{N} \sum_{\mathbf{a} \in \mathbb{B}^n} |\langle \mathbf{a} | \psi_k^{\mathbf{a}} \rangle|^2$$

is large only if the state $|\psi_k^{\mathbf{a}}\rangle$ deviates considerably from $|\phi_k\rangle$ (we already know from eqn (10.14) that the measurement of $|\phi_k\rangle$ leads to a small average success probability). Hence, the distance $\| \, |\psi_k^{\mathbf{a}}\rangle - |\phi_k\rangle \, \|^2$ must increase as k increases and this should hold for any string $\mathbf{a}$. The quantity that should be maximized is the average distance

$$\boxed{Z_k \equiv \sum_{\mathbf{a} \in \mathbb{B}^n} \| \, |\psi_k^{\mathbf{a}}\rangle - |\phi_k\rangle \, \|^2,} \tag{10.15}$$

under the constraint $0 < \overline{p} \leq 1$. The sum in (10.15) ensures that the algorithm performs equally well for all strings $\mathbf{a}$. As explained in Exercises (10.5, 10.6), a method based on Lagrange multipliers shows that Z_k is bounded by two inequalities [317],

$$\boxed{4k^2 \; \geq \; Z_k \; \geq \; 2N - 2\sqrt{N\overline{p}} - 2N\sqrt{1 - \overline{p}}.} \tag{10.16}$$

To achieve an average success probability of $\overline{p} = 0.5$, the number of oracle queries k must satisfy $4k^2 \geq 2N - \sqrt{2N} - \sqrt{2}N$. This formula can be rearranged as

$$k \; \geq \; \frac{1}{2}\sqrt{\left(2 - \sqrt{2}\right) N - \sqrt{2N}} \; \approx \; 0.38\sqrt{N},$$

where in the last step we assumed $N \gg 1$. In conclusion, $O(\sqrt{N})$ oracle queries are necessary to achieve an average success probability of 50 %. This is the same number of queries used by Grover's algorithm (up to a constant factor). This proves that Grover's algorithm is optimal as regards the query complexity.

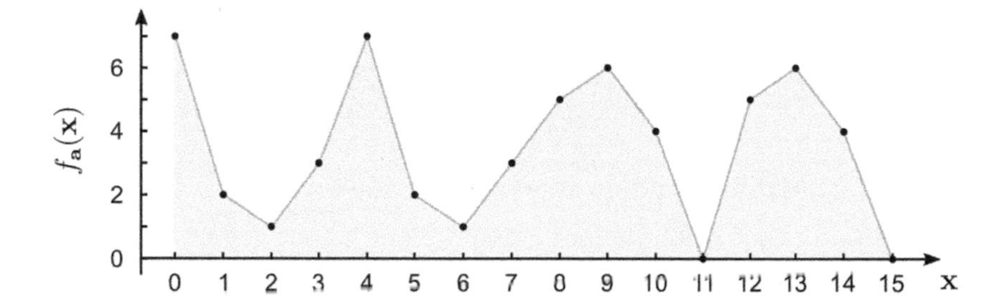

Fig. 10.12 Plot of a function $f_\mathbf{a} : \mathbb{B}^4 \to \mathbb{B}^3$ for Simon's problem. The function assigns the same value to two strings $\mathbf{x}$ and $\mathbf{x}'$ satisfying $\mathbf{x} \oplus \mathbf{a} = \mathbf{x}'$. If two strings $\mathbf{x}_1$ and $\mathbf{x}_2$ do not satisfy $\mathbf{x}_1 = \mathbf{x}_2 \oplus \mathbf{a}$, then their images are different, $f(\mathbf{x}_1) \neq f(\mathbf{x}_2)$. In this example, we set $\mathbf{a} = 0100_2 = 4$. For example, $f_\mathbf{a}(9) = f_\mathbf{a}(13)$ because $9 \oplus 4 = 1001_2 \oplus 0100_2 = 1101_2 = 13$.

10.6 Simon's algorithm

Simon's problem is a computational problem that a quantum computer can solve with an exponential speed-up with respect to classical approaches [17]. Even if Simon's algorithm does not have a practical application, this algorithm laid the foundations for the discovery of an efficient quantum algorithm for the period finding problem, an important subroutine of Shor's algorithm [318].

Simon's problem can be formulated as follows:

Given: A black box that implements a function $f_\mathbf{a} : \mathbb{B}^n \to \mathbb{B}^{n-1}$.

Assumption: The function satisfies

$$f_\mathbf{a}(\mathbf{x}) = f_\mathbf{a}(\mathbf{x}') \qquad \Leftrightarrow \qquad \mathbf{x}' = \mathbf{x} \oplus \mathbf{a}, \tag{10.17}$$

where $\mathbf{a} \neq \mathbf{0}$ and the operation $\oplus$ is defined as $\mathbf{x} \oplus \mathbf{a} = (x_{n-1} \oplus a_{n-1}, \ldots, x_0 \oplus a_0)$. The relation (10.17) indicates that the domain contains pairs of strings $\mathbf{x}$ and $\mathbf{x}' = \mathbf{x} \oplus \mathbf{a}$ satisfying $f_\mathbf{a}(\mathbf{x}) = f_\mathbf{a}(\mathbf{x}')$. Pairs in the domain, such as $(\mathbf{x}_1, \mathbf{x}_1 \oplus \mathbf{a})$ and $(\mathbf{x}_2, \mathbf{x}_2 \oplus \mathbf{a})$, have different images.

Goal: Find the bit string $\mathbf{a}$.

Figure 10.12 shows an example of function $f_\mathbf{a}(\mathbf{x})$ for Simon's problem. A classical computer can solve Simon's problem by calculating $f_\mathbf{a}(\mathbf{x})$ multiple times. After some evaluations, we might find that $f(\mathbf{x}_1)$ and $f(\mathbf{x}_2)$ are identical, which implies that the solution of the problem is $\mathbf{a} = \mathbf{x}_1 \oplus \mathbf{x}_2$. In the worst case, $2^n/2+1$ evaluations are required to find two identical outputs. We now show that a quantum computer can solve this problem with a polynomial number of oracle queries.

As shown in Fig. 10.13, **Simon's algorithm** starts with the application of a set of Hadamard gates and the oracle $\hat{U}_f$ to the initial state $|\psi_\mathrm{A}\rangle = |\mathbf{0}\rangle|\mathbf{0}\rangle$. This leads to

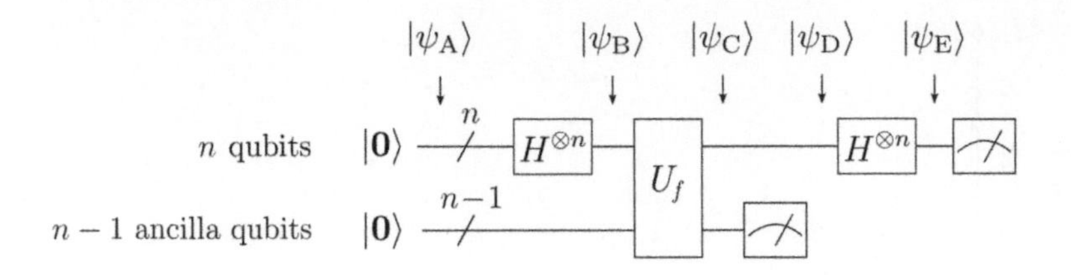

Fig. 10.13 Quantum circuit for Simon's algorithm.

$$|\psi_C\rangle = \frac{1}{\sqrt{2^n}} \sum_{\mathbf{x}\in\mathbb{B}^n} |\mathbf{x}\rangle|f(\mathbf{x})\rangle. \tag{10.18}$$

A measurement of the bottom register collapses the state of the ancilla qubits into a particular state $|\mathbf{z}\rangle$,

$$|\psi_D\rangle = \frac{1}{\sqrt{2}}\Big(|\mathbf{x}\rangle + |\mathbf{x}\oplus\mathbf{a}\rangle\Big)|\mathbf{z}\rangle,$$

where $\mathbf{z} = f(\mathbf{x}) = f(\mathbf{x}\oplus\mathbf{a})$. We will now ignore the ancilla qubits since they will not affect the remaining calculations. The application of a Hadamard gate to each qubit of the top register produces

$$\begin{aligned}
|\psi_E\rangle &= \frac{1}{\sqrt{2}}\Big(\frac{1}{\sqrt{2^n}}\sum_{\mathbf{k}\in\mathbb{B}^n}(-1)^{\mathbf{k}\cdot\mathbf{x}}|\mathbf{k}\rangle + (-1)^{\mathbf{k}\cdot(\mathbf{x}\oplus\mathbf{a})}|\mathbf{k}\rangle\Big) \\
&= \frac{1}{\sqrt{2^{n+1}}}\sum_{\mathbf{k}\in\mathbb{B}^n}\big[(-1)^{\mathbf{k}\cdot\mathbf{x}} + (-1)^{\mathbf{k}\cdot(\mathbf{x}\oplus\mathbf{a})}\big]|\mathbf{k}\rangle \\
&= \frac{1}{\sqrt{2^{n+1}}}\sum_{\mathbf{k}\in\mathbb{B}^n}(-1)^{\mathbf{k}\cdot\mathbf{x}}\cdot\big[1+(-1)^{\mathbf{k}\cdot\mathbf{a}}\big]|\mathbf{k}\rangle,
\end{aligned}$$

where in the last step we used $\mathbf{k}\cdot(\mathbf{x}\oplus\mathbf{a}) = (\mathbf{k}\cdot\mathbf{x})\oplus(\mathbf{k}\cdot\mathbf{a})$. Since the scalar product $\mathbf{k}\cdot\mathbf{a}$ is either 0 or 1, the final state can be expressed as

$$|\psi_E\rangle = \frac{1}{\sqrt{2^{n-1}}}\sum_{\mathbf{k}\in\mathbb{B}^n} c_{\mathbf{k}}|\mathbf{k}\rangle, \quad \text{where } c_{\mathbf{k}} = \begin{cases}(-1)^{\mathbf{k}\cdot\mathbf{x}} & \text{if } \mathbf{k}\cdot\mathbf{a}=0\\ 0 & \text{if } \mathbf{k}\cdot\mathbf{a}=1.\end{cases}$$

A measurement of the top register will project the state of the qubits onto a random state $|\mathbf{w}\rangle$ satisfying $\mathbf{w}\cdot\mathbf{a} = 0$. If we run the quantum circuit $O(n)$ times[6], we will eventually obtain n different bit strings $\{\mathbf{w}_1,\ldots,\mathbf{w}_n\}$. We then solve the set of equations

$$\mathbf{w}_1\cdot\mathbf{a}=0, \qquad \mathbf{w}_2\cdot\mathbf{a}=0, \qquad \ldots \qquad \mathbf{w}_n\cdot\mathbf{a}=0, \tag{10.19}$$

with a classical computer and obtain the string $\mathbf{a}$. To summarize, with high probability,

[6] After repeating the computation $2n$ times, we will obtain n linearly independent strings with 50 % probability.

the string **a** can be determined with $O(n)$ oracle queries, followed by $O(n^2)$ classical operations to solve the set of equations (10.19). This is an exponential speed-up with respect to classical approaches.

Simon's problem is strictly related to the period finding problem, an important computational problem with many applications in signal processing and number theory. A function f is said to be periodic if there exists a number $a > 0$ such that

$$f(x + a) = f(x) \tag{10.20}$$

for all x in the domain. The **period finding problem** consists of determining the smallest a satisfying the relation (10.20). The next sections present a quantum algorithm that solves the period finding problem with polynomial resources. To this end, we first need to introduce the concepts of the discrete Fourier transform and quantum Fourier transform. At that point, we will show that the quantum algorithm for the period finding problem is very similar to Simon's algorithm: the only difference is that some Hadamard gates are replaced with a quantum Fourier transform.

10.7 The discrete Fourier transform

The **discrete Fourier transform** has numerous applications in signal processing and number theory. This transformation maps a complex vector $\mathbf{x} = (x_{N-1}, \ldots, x_0)$ into a complex vector $\mathbf{y} = (y_{N-1}, \ldots, y_0)$. The components of $\mathbf{y}$ are given by

$$y_k = \frac{1}{\sqrt{N}} \sum_{j=0}^{N-1} e^{\frac{2\pi i}{N} jk} x_j, \tag{10.21}$$

where jk is the usual product between integers. This relation can be written in matrix form as

$$\begin{bmatrix} y_0 \\ \vdots \\ y_{N-1} \end{bmatrix} = \frac{1}{\sqrt{N}} \begin{bmatrix} e^{\frac{2\pi i}{N} 0\cdot 0} & \ldots & e^{\frac{2\pi i}{N} 0\cdot(N-1)} \\ \vdots & \ddots & \vdots \\ e^{\frac{2\pi i}{N}(N-1)\cdot 0} & \ldots & e^{\frac{2\pi i}{N}(N-1)\cdot(N-1)} \end{bmatrix} \begin{bmatrix} x_0 \\ \vdots \\ x_{N-1} \end{bmatrix}. \tag{10.22}$$

Hereafter, we will assume that the length of the vector $\mathbf{x}$ is $N = 2^n$. If this is not the case, it is always possible to add some extra zeros to make its length an integer power of 2. This technique is usually called zero padding.

If $\mathbf{x}$ has all entries equal to zero apart from one, the entries of the output vector $\mathbf{y}$ all have the same magnitude. This concept is illustrated in Fig. 10.14a. In addition, if the input vector $\mathbf{x}$ is periodic with an integer period $r = 2^n/s$, eqn (10.21) indicates that the non-zero components of $\mathbf{y}$ are equally spaced by s. This concept is illustrated in Fig. 10.14b and will be important when we discuss the period finding algorithm later in this chapter.

Example 10.1 The discrete Fourier transform of the vector $\mathbf{x} = (x_1, x_0)$ is the vector $\mathbf{y} = (y_1, y_0)$, where

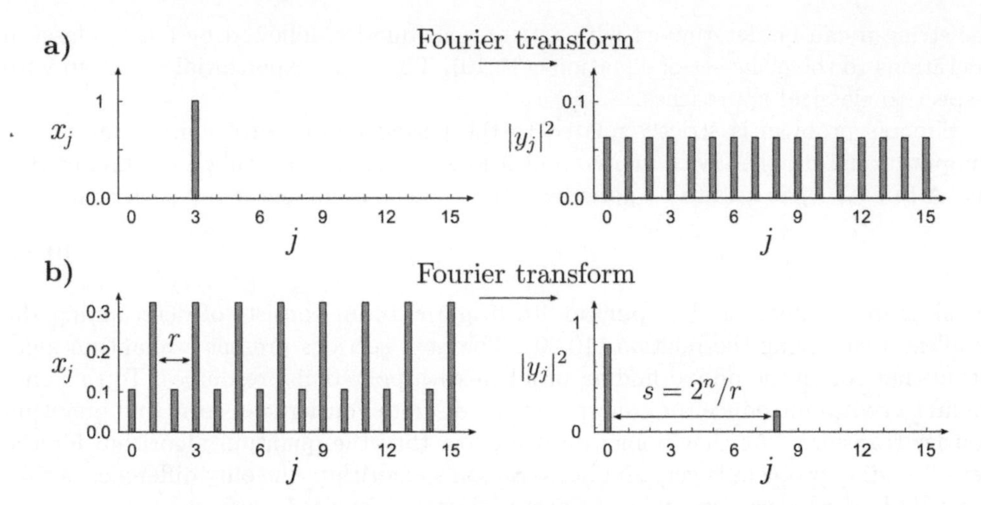

Fig. 10.14 The discrete Fourier transform. a) Graphical representation of the discrete Fourier transform of a real vector **x** with only one non-zero component. The entries of **y** have all the same magnitude. **b)** Graphical representation of the discrete Fourier transform of a real vector **x** whose entries are periodic with period $r = 2$. Since the vector length $2^n = 16$ is a multiple of the period, the non-zero components of **y** are spaced by $2^n/r = 8$.

$$y_0 = \frac{1}{\sqrt{2}}\left(e^{\frac{2\pi i}{2}0\cdot 0}x_0 + e^{\frac{2\pi i}{2}0\cdot 1}x_1\right) = \frac{x_0 + x_1}{\sqrt{2}}, \tag{10.23}$$

$$y_1 = \frac{1}{\sqrt{2}}\left(e^{\frac{2\pi i}{2}1\cdot 0}x_0 + e^{\frac{2\pi i}{2}1\cdot 1}x_1\right) = \frac{x_0 - x_1}{\sqrt{2}}. \tag{10.24}$$

The next section will show how to calculate the components y_j with a quantum computer.

10.7.1 The quantum Fourier transform

The **quantum Fourier transform** (QFT) is the quantum version of the discrete Fourier transform [319]. The quantum Fourier transform $\hat{F}$ is a unitary transformation that maps a basis vector $|j\rangle$ into[7]

$$\boxed{\hat{F}|j\rangle = \frac{1}{2^{n/2}}\sum_{k=0}^{2^n-1} e^{\frac{2\pi i}{2^n}jk}|k\rangle.} \tag{10.25}$$

Using (10.25), we can calculate the QFT of a generic n-qubit state $|\psi\rangle = \sum_j x_j|j\rangle$ as

$$\hat{F}|\psi\rangle = \sum_{j=0}^{2^n-1} x_j\,\hat{F}|j\rangle = \frac{1}{2^{n/2}}\sum_{j=0}^{2^n-1}\sum_{k=0}^{2^n-1} e^{\frac{2\pi i}{2^n}jk}x_j|k\rangle.$$

[7] Here, we are using the notation $|0\rangle \equiv |0\ldots000\rangle$, $|1\rangle \equiv |0\ldots001\rangle$, ..., $|2^n-1\rangle \equiv |1\ldots111\rangle$.

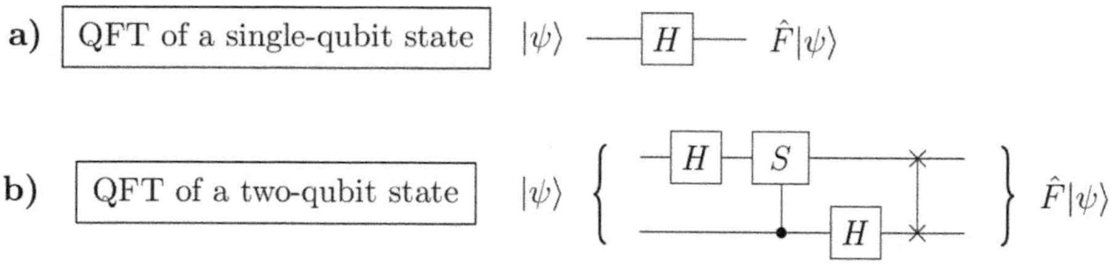

a) QFT of a single-qubit state $|\psi\rangle$ —[H]— $\hat{F}|\psi\rangle$

b) QFT of a two-qubit state $|\psi\rangle$ { H, S, H, SWAP } $\hat{F}|\psi\rangle$

Fig. 10.15 The quantum Fourier transform. a) The QFT of a single-qubit state can be performed with a Hadamard gate. **b)** The QFT of a two-qubit state can be performed with Hadamard gates, a controlled phase gate, and a SWAP gate.

The inverse of the QFT is defined in a similar way,

$$\boxed{\hat{F}^{-1}|j\rangle = \frac{1}{2^{n/2}} \sum_{k=0}^{2^n-1} e^{-\frac{2\pi i}{2^n}jk}|k\rangle.} \tag{10.26}$$

The reader can verify that $\hat{F}\hat{F}^{-1} = \hat{F}^{-1}\hat{F} = \mathbb{1}$. The quantum Fourier transform can be implemented with a quantum circuit comprising $O(n^2)$ elementary gates. This point is explained in Appendix C.

Example 10.2 The QFT on a single qubit can be implemented with a Hadamard gate as shown in Fig. 10.15a. This is because

$$\begin{aligned}
\hat{F}|0\rangle &= \frac{1}{\sqrt{2}}\Big(e^{\frac{2\pi i}{2}0\cdot 0}|0\rangle + e^{\frac{2\pi i}{2}0\cdot 1}|1\rangle\Big) = \frac{1}{\sqrt{2}}(|0\rangle + |1\rangle) = \hat{H}|0\rangle, \\
\hat{F}|1\rangle &= \frac{1}{\sqrt{2}}\Big(e^{\frac{2\pi i}{2}1\cdot 0}|0\rangle + e^{\frac{2\pi i}{2}1\cdot 1}|1\rangle\Big) = \frac{1}{\sqrt{2}}(|0\rangle - |1\rangle) = \hat{H}|1\rangle.
\end{aligned}$$

Therefore, the QFT of a generic qubit state $|\psi\rangle = c_0|0\rangle + c_1|1\rangle$ is

$$\hat{F}(c_0|0\rangle + c_1|1\rangle) = \frac{1}{\sqrt{2}}[(c_0 + c_1)|0\rangle + (c_0 - c_1)|1\rangle].$$

This result is consistent with the classical case eqns (10.23, 10.24).

Example 10.3 Let us calculate the QFT of the two-qubit state $|01\rangle$. We have

$$\begin{aligned}
\hat{F}|01\rangle &= \frac{1}{2}\Big(e^{\frac{2\pi i}{4}1\cdot 0}|00\rangle + e^{\frac{2\pi i}{4}1\cdot 1}|01\rangle + e^{\frac{2\pi i}{4}1\cdot 2}|10\rangle + e^{\frac{2\pi i}{4}1\cdot 3}|11\rangle\Big) \\
&= \frac{1}{2}(|00\rangle + i|01\rangle - |10\rangle - i|11\rangle).
\end{aligned}$$

The reader can compute the QFT of the basis vectors $|00\rangle$, $|01\rangle$, $|10\rangle$, $|11\rangle$ and check that the quantum circuit shown in Fig. 10.15b outputs the expected state. This example indicates that the QFT transforms a basis vector into a balanced superposition of all basis vectors. This is consistent with the classical case in which a vector with only one non-zero component is mapped into a vector whose entries are all non-zero (see Fig. 10.14a).

10.8 Period finding

An important application of the quantum Fourier transform is finding the period of a function. This problem is formulated as:

Given: A black box that implements a function $f : \mathbb{B}^n \to \mathbb{B}^m$.

Assumptions: The function f is periodic with period $r \in \{2, \ldots, 2^n - 1\}$. The function assigns different outputs to elements of the domain within a period.

Goal: Estimate the period of f, i.e. the smallest integer r satisfying $f(x+r) = f(x)$.

Determining the period of a function using a classical computer is not simple. We could in principle compute $f(0)$, $f(1)$, $f(2)$, and so on until we find an integer x_0 such that $f(x_0) = f(0)$. From this result, we conclude that $r = x_0$. Unfortunately, this brute force approach is not efficient since r could be very large. It is widely accepted that a classical computer cannot find the period of a function with less than $O(r)$ oracle queries. We now show that a quantum computer can determine the periodicity of f with high probability with a constant number of queries, bringing the query complexity from $O(r)$ to $O(1)$.

Figure 10.16 shows the quantum circuit for the period finding problem. This circuit is very similar to the circuit of Simon's algorithm presented in Section 10.6, the main difference being the substitution of a set of Hadamard gates with a quantum Fourier transform. The quantum circuit comprises two registers, one with n qubits and one with m ancilla qubits[8]. A set of Hadamard gates and the oracle $\hat{U}_f$ transform the initial state $|\psi_{\text{A}}\rangle = |0\rangle|0\rangle$ into

$$|\psi_{\text{C}}\rangle = \frac{1}{\sqrt{2^n}} \sum_{x=0}^{2^n-1} |x\rangle|f(x)\rangle. \tag{10.27}$$

A measurement of the bottom register projects the ancilla qubits into a random state $|z\rangle$,

$$|\psi_{\text{D}}\rangle = \frac{1}{\sqrt{s}} \sum_{x:f(x)=z} |x\rangle|z\rangle = \left(\frac{1}{\sqrt{s}} \sum_{j=0}^{s-1} |x_0 + jr\rangle\right)|z\rangle, \tag{10.28}$$

where x_0 is the smallest element in the domain satisfying $f(x_0) = z$ and s is the number of strings x in the domain satisfying $f(x) = z$. What is the value of s? Well, it depends on the measurement outcome z. The value of s can be either

[8]We will assume that the number of qubits n satisfies $2^n \gg r$.

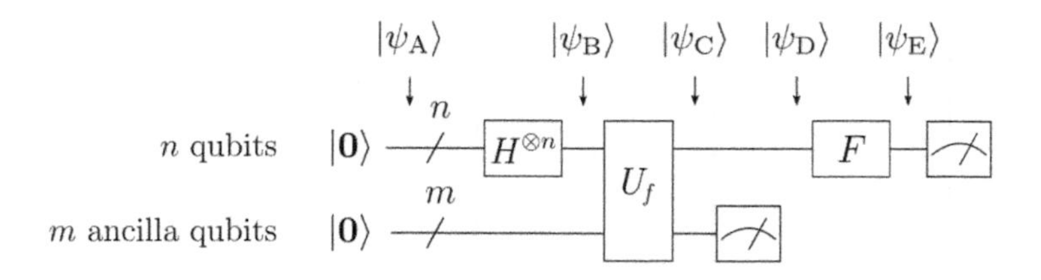

Fig. 10.16 Period finding. Quantum circuit for the period finding problem. Note the similarity with Simon's algorithm, Fig. 10.13.

$$s = \left\lfloor \frac{2^n}{r} \right\rfloor \qquad \text{or} \qquad s = \left\lfloor \frac{2^n}{r} \right\rfloor + 1.$$

We can now neglect the ancilla qubits since they will not influence the following calculations. The rest of the computation boils down to determining the period r from the quantum state (10.28), a superposition state of equally spaced strings. This can be done by applying the QFT to the state $|\psi_D\rangle$,

$$\begin{aligned} |\psi_E\rangle = \frac{1}{\sqrt{s}} \sum_{j=0}^{s-1} \hat{F}|x_0 + jr\rangle &= \frac{1}{\sqrt{s}} \sum_{j=0}^{s-1} \frac{1}{\sqrt{2^n}} \sum_{k=0}^{2^n-1} e^{\frac{2\pi i}{2^n}(x_0+jr)k} |k\rangle \\ &= \frac{1}{\sqrt{2^n s}} \sum_{k=0}^{2^n-1} e^{\frac{2\pi i}{2^n}x_0 k} \sum_{j=0}^{s-1} e^{\frac{2\pi i}{2^n}jrk} |k\rangle. \end{aligned}$$

A measurement of $|\psi_E\rangle$ produces a random integer $x \in [0, 2^n - 1]$ with probability

$$\begin{aligned} p_x = |\langle x|\psi_E\rangle|^2 &= \left| \frac{1}{\sqrt{2^n s}} \sum_{k=0}^{2^n-1} e^{\frac{2\pi i}{2^n}x_0 k} \sum_{j=0}^{s-1} e^{\frac{2\pi i}{2^n}jrk}\, \delta_{xk} \right|^2 \\ &= \left| \frac{1}{\sqrt{2^n s}} e^{\frac{2\pi i}{2^n}x_0 x} \sum_{j=0}^{s-1} e^{\frac{2\pi i}{2^n}jrx} \right|^2 = \frac{1}{2^n s} \left| \sum_{j=0}^{s-1} \left(e^{\frac{2\pi i}{2^n}rx}\right)^j \right|^2. \end{aligned} \tag{10.29}$$

The crucial point is that the probability distribution p_x has pronounced maxima close to integer multiples of $2^n/r$ (see Fig. 10.17). As explained in detail in Exercise 10.3, the cumulative probability of measuring an integer close to a multiple of $2^n/r$ is greater than 40%.

The measurement of the top register produces with high probability an integer[9] $x = \text{Round}(k2^n/r)$. How do we extract the period r from x? First, the difference between the real number $k \cdot 2^n/r$ and its integer approximation x is smaller than half,

$$\left| x - \frac{k \cdot 2^n}{r} \right| \leq \frac{1}{2},$$

[9]Here, the notation Round(a) indicates the closest integer to the real number a. By definition, $|a - \text{Round}(a)| \leq 1/2$.

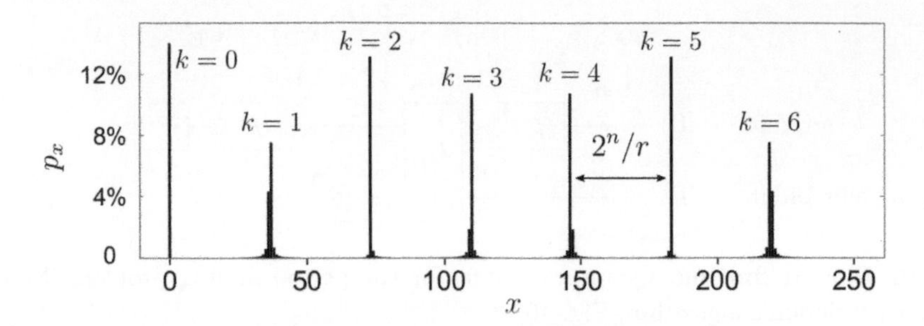

Fig. 10.17 Period finding. Histogram of the probability distribution p_x of eqn (10.29). In this example, we set $2^n = 256$ and $r = 7$. The histogram shows r peaks close to multiples of $2^n/r$. It is most likely that the measurement returns $x = \text{Round}(k2^n/r)$ where the parameter $k \in \{0, \ldots, r-1\}$ indexes the peaks. The sum of the labeled peaks is always greater than 40 %.

which implies that

$$\left| \frac{x}{2^n} - \frac{k}{r} \right| \leq \frac{1}{2 \cdot 2^n}. \tag{10.30}$$

Here, x and n are known, while k and r are not. We now assume that the number of qubits in the top register is high enough such that $2^n > r^2$. Therefore, the inequality (10.30) becomes

$$\left| \frac{x}{2^n} - \frac{k}{r} \right| < \frac{1}{2r^2}. \tag{10.31}$$

Theorem 184 of Ref. [320] shows that the condition (10.31) implies that one of the convergents of the continued fraction expansion of $x/2^n$ is equal to k/r. Let us explain this point step by step. The **continued fraction expansion** of $x/2^n$ is given by[10]

$$\frac{x}{2^n} = a_0 + \frac{1}{a_1 + \frac{1}{a_2 + \frac{1}{a_3 + \ldots}}}.$$

This formula allows us to calculate the **convergents** p_j/q_j of the fraction $x/2^n$,

$$\frac{p_1}{q_1} = a_0 + \frac{1}{a_1}, \qquad \frac{p_2}{q_2} = a_0 + \frac{1}{a_1 + \frac{1}{a_2}}, \qquad \frac{p_3}{q_3} = a_0 + \frac{1}{a_1 + \frac{1}{a_2 + \frac{1}{a_3}}}, \qquad \ldots$$

The convergents p_j/q_j are fractions that approximate the rational number $x/2^n$ with an accuracy that increases as j increases. According to Theorem 184 of Ref. [320], the condition (10.31) ensures that the convergent p/q with the largest q satisfying the inequalities

[10] The continued fraction expansion of a real number x can be efficiently calculated with a classical algorithm: $a_0 = \lfloor x \rfloor$, $a_1 = \lfloor x_1 \rfloor$, $a_2 = \lfloor x_2 \rfloor$, $a_3 = \lfloor x_3 \rfloor$ ... where $x_1 = 1/(x - a_0)$, $x_2 = 1/(x_1 - a_1)$, $x_3 = 1/(x_2 - a_2)$, and so on.

$$\left|\frac{x}{2^n} - \frac{p}{q}\right| < \frac{1}{2q^2} \qquad \text{and} \qquad 1 < q^2 < 2^n$$

is equal to k/r. If k and r are coprime[11], the identity $k/r = p/q$ implies that $k = p$ and $r = q$. The last step is to compute $f(0)$ and $f(q)$ and check that they are identical. If they are not, the algorithm must be run again.

The period finding algorithm can fail for two reasons:

1) The measurement produces an integer $x \neq \text{Round}(k2^n/r)$.

2) The measurement produces an integer $x = \text{Round}(k2^n/r)$, but the integer k is not coprime with r.

This means that the success probability for a single run of the circuit is given by

$$p_{\text{success}} = \sum_{k=0}^{r-1} p_{x=\text{Round}(k2^n/r)} \cdot p(k \text{ and } r \text{ are coprime}). \tag{10.32}$$

Since $k < r$, the probability that k is coprime with r is given by the **totient function** $\varphi(r)$ (which returns the number of integers smaller than r that are coprime with r) divided by the number of integers smaller than r,

$$p(k \text{ and } r \text{ are coprime}) = \frac{\varphi(r)}{r}.$$

Substituting this expression into (10.32), we obtain

$$p_{\text{success}} = \frac{\varphi(r)}{r} \sum_{k=0}^{r-1} p_{x=\text{Round}(k2^n/r)} > 0.4 \frac{\varphi(r)}{r}, \tag{10.33}$$

where in the last step we used $\sum_k p_{x=\text{Round}(k2^n/r)} > 0.4$ (see Exercise 10.3). As demonstrated in Theorem 328 of Ref. [320], the totient function has an interesting asymptotic behavior,

$$\liminf_{r\to\infty} \frac{\varphi(r)}{r} \ln(\ln r) = e^{-\gamma},$$

where $\gamma \approx 0.577$ is the Euler–Mascheroni constant (see Fig. 10.18 for a graphical visualization of this limit). Using this limit, eqn (10.33) becomes

$$p_{\text{success}} > 0.4 \frac{\varphi(r)}{r} > 0.4 \frac{e^{-\gamma}}{\log(\log r)}. \tag{10.34}$$

This expression indicates that the algorithm must be run $O(\log \log r)$ times to achieve a high success probability. In conclusion, the period finding problem can be solved

[11]Two integers are coprime when their greatest common divisor is 1. If k and r are not coprime, the identity $k/r = p/q$ does not imply that $k = r$ and $q = r$, but only that q is a divisor of r.

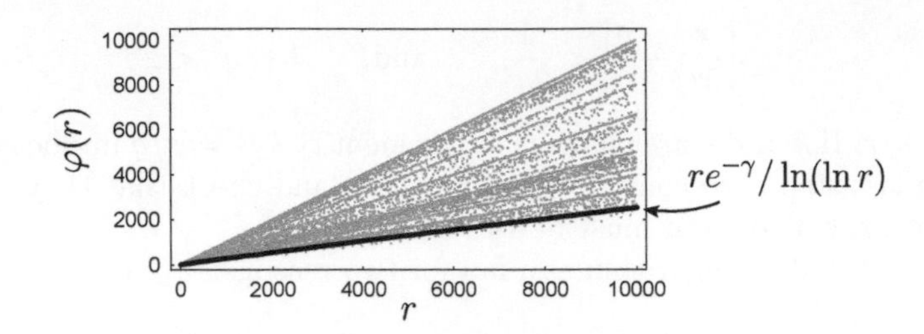

Fig. 10.18 The totient function. Plot of the totient function $\varphi(r)$ (gray dots). This function returns the number of integers smaller than r that do not share common factors with r. As shown in Theorem 328 of Ref. [320], $\liminf_{r\to\infty}\varphi(r)\frac{\ln(\ln r)}{r} = e^{-\gamma}$ where $\gamma \approx 0.577$ is the Euler–Mascheroni constant. The function $re^{-\gamma}/\ln(\ln r)$ is plotted with a solid curve.

---

with $O(\log\log r)$ oracle queries.

A query complexity that scales logarithmically is already a great achievement, but it can be further improved. Indeed, it can be decreased to $O(1)$, meaning that the quantum circuit of the period finding algorithm must be run only a constant number of times to achieve a high success probability. Let us explain this point in detail. Suppose that the algorithm is run twice. After two trials, we obtain two fractions p_1/q_1 and p_2/q_2 satisfying

$$\frac{p_1}{q_1} = \frac{k_1}{r}, \qquad \frac{p_2}{q_2} = \frac{k_2}{r}.$$

Here, the integers p_1, q_1, p_2, and q_2 are known, while k_1, k_2, and r are not. If both q_1 and q_2 are not the period of the function, it is likely that the their least common multiple[12] $\mathrm{lcm}(q_1, q_2)$ is the period of the function. To be precise, if k_1 and k_2 are coprime, which happens with a probability greater than 35 % (see Exercise 10.8), then $\mathrm{lcm}(q_1, q_2)$ is definitely the period of the function (see Exercise 10.7). In conclusion, we just need to repeat the quantum circuit a constant number of times to find the function's period.

How many gates are required to implement the period finding algorithm? We need $O(n)$ Hadamard gates plus $O(n^2)$ elementary gates for the quantum Fourier transform (see Appendix C). Provided the oracle can be decomposed into $O(t_n)$ elementary gates, the entire quantum circuit requires $O(n^2)+O(t_n)$ gates. The continued fraction algorithm requires $O(n^3)$ classical operations [312]. Since the period can be determined by running the algorithm a constant number of times, the total cost is $O(n^3)+O(t_n)$.

Example 10.4 As an example, consider the function $f\colon \mathbb{B}^4 \to \mathbb{B}^2$ shown in Fig. 10.19a. This transformation is periodic with period $r = 3$. The state $|\psi_C\rangle$ of eqn (10.27) is given by

[12]The least common multiple between two positive integers a and b is the smallest integer that is divisible by both a and b, and is related in a simple way to the greatest common divisor: $\mathrm{lcm}(a,b) = ab/\mathrm{gcd}(a,b)$.

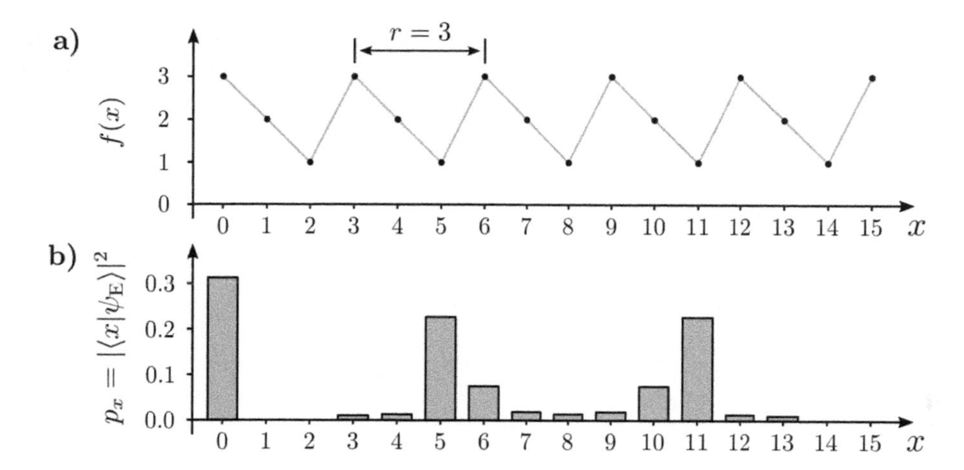

Fig. 10.19 Example of a period finding problem. **a)** Plot of a periodic function $f : \mathbb{B}^4 \to \mathbb{B}^2$ with period $r = 3$. The size of the domain $2^n = 16$ satisfies $2^n > r^2$. **b)** Plot of the probability distribution $p_x = |\langle x|\psi_E\rangle|^2$ when the ancilla qubits are measured in $y_0 = 2$. The outcomes 0, 5 and 11 are the most likely to be measured. If the measurement outcome is 5 or 11, we can extract the period of the function from the fractions $5/2^4$ and $11/2^4$.

$$|\psi_C\rangle = \frac{1}{\sqrt{2^4}}\left[|0\rangle|3\rangle + |1\rangle|2\rangle + |2\rangle|1\rangle + |3\rangle|3\rangle + |4\rangle|2\rangle + \ldots + |15\rangle|3\rangle\right].$$

Suppose that a measurement of the ancilla qubits returns $z = 2$. This measurement collapses the top register into

$$|\psi_D\rangle = \frac{1}{\sqrt{s}}\sum_{j=0}^{s-1}|x_0 + jr\rangle = \frac{1}{\sqrt{5}}[|1\rangle + |4\rangle + |7\rangle + |10\rangle + |13\rangle].$$

A QFT applied to this state produces

$$\begin{aligned}|\psi_E\rangle = &\frac{1}{\sqrt{2^4\cdot 5}}\left[|0\rangle + e^{\frac{\pi i}{8}}|1\rangle + e^{\frac{\pi i}{8}2}|2\rangle + \ldots + e^{\frac{\pi i}{8}15}|15\rangle\right]\\ &+ \frac{1}{\sqrt{2^4\cdot 5}}\left[|0\rangle + e^{\frac{\pi i}{8}4}|1\rangle + e^{\frac{\pi i}{8}8}|2\rangle + \ldots + e^{\frac{\pi i}{8}60}|15\rangle\right]\\ &+ \frac{1}{\sqrt{2^4\cdot 5}}\left[|0\rangle + e^{\frac{\pi i}{8}7}|1\rangle + e^{\frac{\pi i}{8}14}|2\rangle + \ldots + e^{\frac{\pi i}{8}105}|15\rangle\right]\\ &+ \frac{1}{\sqrt{2^4\cdot 5}}\left[|0\rangle + e^{\frac{\pi i}{8}10}|1\rangle + e^{\frac{\pi i}{8}20}|2\rangle + \ldots + e^{\frac{\pi i}{8}150}|15\rangle\right]\\ &+ \frac{1}{\sqrt{2^4\cdot 5}}\left[|0\rangle + e^{\frac{\pi i}{8}13}|1\rangle + e^{\frac{\pi i}{8}26}|2\rangle + \ldots + e^{\frac{\pi i}{8}195}|15\rangle\right].\end{aligned}$$

A measurement in the computational basis projects $|\psi_E\rangle$ onto one of the basis states $|x\rangle$ with probability $p_x = |\langle x|\psi_E\rangle|^2$ plotted in Fig. 10.19b. From the histogram, it is evident that the states $|0\rangle$, $|5\rangle$, $|11\rangle$ are the most likely to be observed[13]. Suppose that the measurement outcome is $x = 5$. The continued fraction expansion of $x/2^n = 5/16$ is

$$\frac{5}{16} = a_0 + \frac{1}{a_1 + \frac{1}{a_2}}, \qquad \text{where } a_0 = 0,\ a_1 = 3,\ a_2 = 5.$$

Now calculate the convergents p_j/q_j,

$$\frac{p_1}{q_1} = a_0 + \frac{1}{a_1} = \frac{1}{3}, \qquad \frac{p_2}{q_2} = a_0 + \frac{1}{a_1 + \frac{1}{a_2}} = \frac{5}{16}.$$

The fraction p_1/q_1 is the only one satisfying the inequalities

$$\left| \frac{x}{2^n} - \frac{p_1}{q_1} \right| \leq \frac{1}{2q_1^2}, \qquad \text{and} \qquad 1 < q_1^2 < 16.$$

Therefore, the candidate for the period is $q_1 = 3$. Calculate the quantities $f(0)$ and $f(3)$ and check that they are identical. Since $f(0) = f(3)$, the period is $r = 3$.

10.9 Shor's algorithm

The prime factorization of a composite number N is one of the most fascinating problems in number theory. To date, no classical algorithm is known that can factor arbitrarily large numbers in an efficient way. The number field sieve, one of the best performing classical algorithms, solves prime factorization with a running time[14]

$$\exp\left[c(\log N)^{1/3}(\log\log N)^{2/3}\right],$$

where $c \approx 2$. The security of some modern cryptographic protocols, such as the RSA protocol, is based on the conjecture that efficient *classical* algorithms for prime factorization do not exist [56].

In 1994, Peter Shor proposed a quantum algorithm that can solve prime factorization with polynomial resources [18]. This quantum algorithm is based on the observation that prime factorization can be reduced to the period finding problem discussed in Section 10.8. Since a quantum computer can easily solve the period finding problem, a quantum computer can efficiently solve prime factorization. The first experimental implementation of Shor's algorithm was performed with an NMR quantum computer in 2001 [322].

Before presenting Shor's algorithm, we will make a reasonable assumption. We will assume that N (the number we are trying to factorize) is the product of two large prime numbers of similar length since this is the most challenging case to deal with and has direct applications in cryptography. The prime factorization problem can be

[13]The probability distribution p_x does not significantly change even when the measurement of the ancilla qubits returns $z = 1$ or $z = 3$.

[14]For more information about the historical development of fast classical algorithms for prime factorization, read the beautiful review by Pomerance [321].

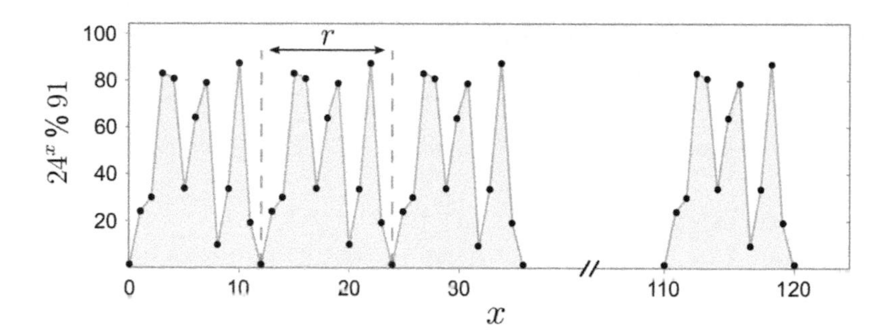

Fig. 10.20 Modular exponentiation. Plot of the function $f(x) = a^x \,\%\, N$ with $a = 24$ and $N = 91$. The function is periodic with period $r = 12$.

---

formulated as:

Given: A large composite number $N = pq$ where $p \neq q$ are two prime numbers of similar length.

Goal: Find the factors p and q.

Shor's algorithm can be used to find the factors of N:

Step 1 Choose a random integer a from the set $\{2, \ldots, N-1\}$.

Step 2 Calculate the greatest common divisor[15] between a and N.

- If $\gcd(a, N) = 1$, carry on to Step 3.
- If $\gcd(a, N) \neq 1$, the algorithm stops immediately. The factors of N are $p = \gcd(a, N)$ and $q = N/p$.

Step 3 Since a and N are coprime, the sequence of numbers[16]

$$a^0 \,\%\, N, \qquad a^1 \,\%\, N, \qquad a^2 \,\%\, N, \qquad a^3 \,\%\, N, \qquad \ldots$$

is periodic. In other words, the function

$$f(x) = a^x \,\%\, N \tag{10.35}$$

is periodic[17] with period r (see Fig. 10.20 for a graphical representation). Calculate the period r, i.e. the smallest integer satisfying $a^r \,\%\, N = 1$.

[15] The greatest common divisor between two integers a and b is the largest divisor shared by a and b. If $\gcd(a, b) = 1$, a and b are said to be **coprime**.

[16] Here, the function $a \,\%\, b$ returns the remainder of the division of a by b.

[17] The fact that $f(x)$ is periodic is a direct consequence of Euler's theorem. See Exercise 10.4 for more details.

- If r is odd, start over from Step 1.
- If r is even, carry on to Step 4.

Step 4 Since r is even, $a^{r/2}$ is a positive integer. Therefore, we can write

$$\begin{aligned} a^r \,\%\, N = 1 \qquad &\Rightarrow \qquad (a^{r/2})^2 \,\%\, N = 1, \\ &\Rightarrow \qquad (a^{r/2})^2 - 1 = k \cdot N, \\ &\Rightarrow \qquad (a^{r/2} - 1)(a^{r/2} + 1) = k \cdot N, \end{aligned} \tag{10.36}$$

for some positive integer k. Equation (10.36) indicates that the product between $a^{r/2} - 1$ and $a^{r/2} + 1$ is a multiple of N. On one hand, the integer $a^{r/2} - 1$ is not a multiple of N[18]. On the other hand, $a^{r/2} + 1$ might be a multiple of N.

- If $a^{r/2} + 1$ is a multiple of N, start the algorithm again from Step 1.
- If $a^{r/2} + 1$ is not a multiple of N, carry on to Step 5.

Step 5 In the last step of the algorithm, we calculate the factors p and q satisfying $p \cdot q = N$. Since N is neither a factor of $a^{r/2} - 1$ nor $a^{r/2} + 1$, but their product is a multiple of N, both these numbers must share a non-trivial factor with N. In conclusion,

$$\boxed{p = \gcd(a^{r/2} - 1, N),} \qquad \boxed{q = \gcd(a^{r/2} + 1, N).}$$

All of the steps of the algorithm can be efficiently performed by a classical computer apart from Step 3: no known classical algorithm can find the period of the function $a^x \,\%\, N$ in a reasonable time when $N \gg 1$. This step is executed with the aid of a quantum computer. See Ref. [323] for a detailed explanation on how to implement the oracle $f(x) = a^x \,\%\, N$ on a quantum computer.

Example 10.5 Let us study the factorization of $N = 21$.

1. Suppose we choose $a = 4$. This number satisfies $\gcd(4, 21) = 1$. The period of the function $a^x \,\%\, N$ is 3; indeed

$$4^0 \,\%\, 21 = 1, \qquad 4^1 \,\%\, 21 = 4, \qquad 4^2 \,\%\, 21 = 16, \qquad 4^3 \,\%\, 21 = 1.$$

Unfortunately, r is odd, and the algorithm must be run again.

2. Suppose we choose $a = 5$. This number satisfies $\gcd(5, 21) = 1$ and the period of $a^x \,\%\, N$ is $r = 6$, an even number. Therefore,

[18] If $a^{r/2} - 1$ is a multiple of N, then

$$a^{r/2} - 1 = kN \qquad \Rightarrow \qquad (a^{r/2} - 1) \,\%\, N = 0 \qquad \Rightarrow \qquad a^{r/2} \,\%\, N = 1.$$

This result is in contrast with the assumption that r is the smallest integer satisfying $a^r \,\%\, N = 1$.

$$a^{r/2} - 1 = 5^3 - 1 = 124,$$
$$a^{r/2} + 1 = 5^3 + 1 = 126.$$

Unfortunately, $a^{r/2} + 1 = 126$ is a multiple of 21. Hence, the algorithm must be run again.

3. Suppose we choose $a = 8$. This number satisfies $\gcd(8, 21) = 1$ and the period of $a^x \% N$ is $r = 6$, an even number. Therefore,

$$a^{r/2} - 1 = 8^3 - 1 = 1330,$$
$$a^{r/2} + 1 = 8^3 + 1 = 1332.$$

The number 1332 is not a multiple of N. The integers $p = \gcd(1330, 21) = 7$ and $p = \gcd(1332, 21) = 3$ are the factors of N.

Example 10.6 How many qubits are required to factorize a composite number N with a quantum computer? Factoring N boils down to finding the period of the function $a^x \% N$ using the quantum algorithm presented in Section 10.8. Since the period r is smaller than N and the period finding algorithm has a high success probability when the number of qubits n in the top register satisfies $2^n > N^2 > r^2$, we need at least $n = \lceil 2 \log_2 N \rceil$ qubits. Furthermore, the algorithm requires m ancilla qubits to store the quantity $a^x \% N$. Since this quantity is always smaller than N, we need $m = \lceil \log_2 N \rceil$ ancilla qubits. In short, the factorization of a semi-prime N requires

$$n + m = \lceil 2 \log_2 N \rceil + \lceil \log_2 N \rceil \text{ qubits.}$$

Modern RSA protocols usually adopt semiprimes N with 1024 binary digits. This means that $N < 2^{1024}$ which implies that $\log_2 N < 1024$. Therefore, to break RSA-1024 we need at least

$$n + m = 2 \cdot 1024 + 1024 = 3072 \text{ qubits.}$$

Note that a quantum computer is prone to errors and the actual quantum computation might require more qubits than this estimate. Recent studies suggest that a noisy quantum computer with $10^5 - 10^6$ qubits can break modern RSA cryptography [61, 62]. The number of qubits strongly depends on the performance of single and two-qubit gates and the connectivity of the device.

Further reading

Many excellent textbooks explain quantum algorithms at various levels and from different viewpoints. For an introductory presentation, readers can refer to the textbooks by Rieffel [324] and Lipton [325]. Another excellent resource is the textbook by Benenti, Casati, and Strini [209]. The textbook by Yanofsky [326] presents quantum algorithms from a computer science perspective.

The review by Mosca [327] provides a summary of early quantum algorithms and a presentation of the abelian hidden subgroup problem (a generalization of Shor's algorithm), quantum searching, and amplitude amplification. The review by Montanaro [328] surveys quantum algorithms with an emphasis on their applications rather than their technical details. The review by Childs [329] describes in detail quantum algorithms for algebraic problems, focusing on algorithms with super-polynomial speed-up over known classical approaches. The review by Santha [330] presents the formalism of quantum algorithms based on quantum walks.

Summary

Useful formulas

$\hat{U}_f\vert\mathbf{x}\rangle\vert\mathbf{y}\rangle = \vert\mathbf{x}\rangle\vert\mathbf{y}\oplus f(\mathbf{x})\rangle$	Evaluation of a function on a quantum comp.
$\hat{H}\vert x\rangle = \sum_{a=0}^{1}\frac{(-1)^{ax}}{\sqrt{2}}\vert a\rangle$	Hadamard gate on a basis state
$\hat{H}^{\otimes n}\vert\mathbf{x}\rangle = \frac{1}{2^{n/2}}\sum_{\mathbf{a}\in\mathbb{B}^n}(-1)^{\mathbf{a}\cdot\mathbf{x}}\vert\mathbf{a}\rangle$	n Hadamard gates on n qubits
$\mathbf{a}\cdot\mathbf{x} = a_{n-1}x_{n-1}\oplus\ldots\oplus a_0x_0$	Scalar product
$\hat{U}_f\vert\mathbf{x}\rangle\vert-\rangle = (-1)^{f(\mathbf{x})}\vert\mathbf{x}\rangle\vert-\rangle$	Phase kickback method

Fourier transform

$F(\mathbf{x}) = \mathbf{y}$	Discrete Fourier transform
$y_k = \sum_{j=0}^{N-1} e^{\frac{2\pi i}{N}jk}x_j/\sqrt{N}$	Components of the output vector
$\hat{F}\vert j\rangle = \frac{1}{2^{n/2}}\sum_{k=0}^{2^n-1} e^{\frac{2\pi i}{2^n}jk}\vert k\rangle$	Quantum Fourier transform

Query complexity: classical versus quantum

	Determ. machine (worst case)	Quantum computer
Deutsch	2	1
Deutsch–Jozsa	$2^n/2+1$	1
Bernstein–Vazirani	n	1
Grover	2^n-1	$O(\sqrt{2^n})$
Simon	$2^n/2+1$	$O(n)$
Period finding	$O(r)$	$O(1)$

Exercises

Exercise 10.1 Using the relation

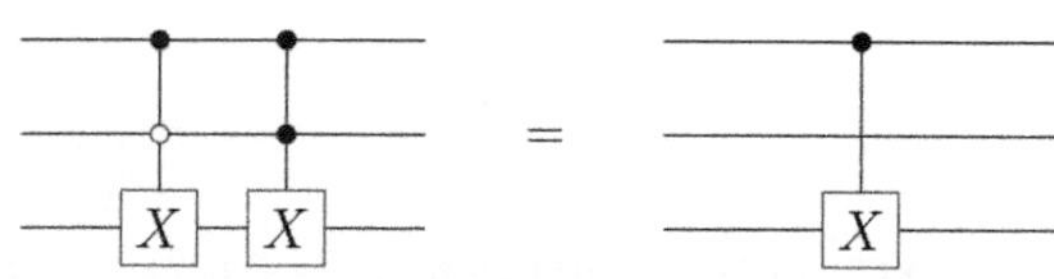

show that the circuit in Fig. 10.3a can be implemented with two controlled operations instead of four.

Exercise 10.2 With a high probability the period finding algorithm returns a number $x = \text{Round}(k\cdot 2^n/r)$ where r is the unknown period, k is an unknown integer in the set $\{0,\ldots,r-1\}$ and $2^n > r^2$ is the size of the domain. Write a Mathematica script that calculates the convergents p_j/q_j of the fraction $x/2^n$ and finds a candidate for the period.

Exercise 10.3 As seen in Section 10.8, the final measurement of the period finding algorithm produces an integer x with probability

$$p_x = \frac{1}{2^n s}\frac{\sin^2(\frac{\pi}{2^n} srx)}{\sin^2(\frac{\pi}{2^n} rx)}.$$

Show that the probability of measuring an integer x that is close to a multiple of $2^n/r$, i.e. $p_{x=\text{Round}(k\cdot 2^n/r)}$, is greater than $4/\pi^2 r$.

Exercise 10.4 The totient function $\varphi(N)$ returns the number of positive integers smaller than N that are coprime with N (for example, $\varphi(5) = 4$). Euler's theorem states that for two coprime numbers a and $N > a$, it holds that $a^{\varphi(N)} \,\%\, N = 1$. Using Euler's theorem, show that $f(x) = a^x \,\%\, N$ is periodic. **Hint:** Using the relation $(a \cdot b) \,\%\, N = [(a \,\%\, N) \cdot (b \,\%\, N)] \,\%\, N$, show that $a^{k\varphi(N)} \,\%\, N = 1$.

Exercise 10.5 Show that

$$\boxed{Z_k = \sum_{\mathbf{a}\in\mathbb{B}^n} \| \, |\psi_k^{\mathbf{a}}\rangle - |\phi_k\rangle \, \|^2 \leq 4k^2,}$$

where

$$|\psi_k^{\mathbf{a}}\rangle = \hat{U}_k \hat{O}_{\mathbf{a}} \hat{U}_{k-1} \hat{O}_{\mathbf{a}} \ldots \hat{U}_1 \hat{O}_{\mathbf{a}} |s\rangle, \qquad |\phi_k\rangle = \hat{U}_k \ldots \hat{U}_1 |s\rangle, \qquad |s\rangle = \frac{1}{\sqrt{N}} \sum_{\mathbf{x}\in\mathbb{B}^n} |\mathbf{x}\rangle,$$

and $N = 2^n$. Here, the operators $\hat{U}_j$ are generic unitary transformations, while $\hat{O}_{\mathbf{a}} = \mathbb{1} - 2|\mathbf{a}\rangle\langle\mathbf{a}|$, see eqn (10.10). **Hint:** Prove that $Z_k \leq 4k^2$ for $k = 0$ and then proceed by induction.

Exercise 10.6 Show that

$$\boxed{Z_k = \sum_{\mathbf{a}\in\mathbb{B}^n} \| \, |\psi_k^{\mathbf{a}}\rangle - |\phi_k\rangle \, \|^2 \geq 2N - 2\sqrt{N\overline{p}} - 2N\sqrt{(1-\overline{p})},}$$

where $|\phi_k\rangle = \hat{U}_k \ldots \hat{U}_1 |s\rangle$, $|s\rangle = \frac{1}{\sqrt{N}} \sum_{\mathbf{x}\in\mathbb{B}^n} |\mathbf{x}\rangle$, $|\psi_k^{\mathbf{a}}\rangle = \hat{U}_k \hat{O}_{\mathbf{a}} \hat{U}_{k-1} \hat{O}_{\mathbf{a}} \ldots \hat{U}_1 \hat{O}_{\mathbf{a}} |s\rangle$ and the average success probability is defined as

$$\overline{p} = \frac{1}{N} \sum_{\mathbf{a}\in\mathbb{B}^n} |\langle\mathbf{a}|\psi_k\rangle|^2.$$

Hint: Find the stationary points of the Lagrangian $L_1 = \| \, |\psi_k^{\mathbf{a}}\rangle - |\phi_k\rangle \, \|^2 - \lambda_1 \| \, |\psi_k^{\mathbf{a}}\rangle \, \|^2 - \lambda_2 |\langle\mathbf{a}|\psi_k^{\mathbf{a}}\rangle|^2$ where λ_1 and λ_2 are two real parameters.

Exercise 10.7 The period finding algorithm returns a fraction $p_1/q_1 = k_1/r$. If k_1 and r are coprime, $p_1 = k_1$ and $q_1 = r$. If k_1 and r are not coprime, then q_1 is a divisor of r and the algorithm must be run again. Suppose the second run returns p_2/q_2 and once again q_2 is not the period of the function, i.e. q_2 is a divisor of r. Then, we have

$$\frac{p_1}{q_1} = \frac{k_1}{r}, \qquad \frac{p_2}{q_2} = \frac{k_2}{r}, \tag{10.37}$$

with $q_1 \neq r$ and $q_2 \neq r$. Show that if k_1 and k_2 are coprime, then $r = \text{lcm}(q_1, q_2)$. **Hint:** Since q_1 and q_2 are divisors of r, $\text{lcm}(q_1, q_2)$ is a divisor of r.

Exercise 10.8 Show that the probability that two positive integers $k_1, k_2 < r$ are coprime is greater than 35 %.

Exercise 10.9 Consider the integer

$$\begin{aligned} N = \; & 1634587289413122283582999525224832493882479520134329818905911930346530 \\ & 6186145144874702707989499682830590987205641646215356703963669791190345 \\ & 5066128232237375306568631856212225231818458743527593500949226205812461 \\ & 1165464062568716355577908938387369696561119570555060097087194724422825 \\ & 1950292465930038155996192542 7. \end{aligned}$$

This number has 309 decimal digits and can be expressed with 1024 binary digits. Run Shor's algorithm on a quantum computer and find the primes p and q such that $p \cdot q = N$.

Solutions

Solution 10.1 The circuit can be implemented with two controlled operations because

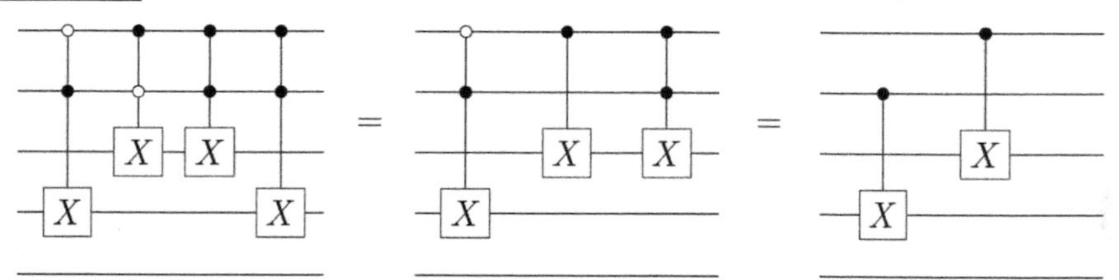

Solution 10.2 Suppose that the number of qubits in the register is $n = 10$ and the measurement outcome is $x = 118$. Then, the candidate for the period can be calculated with a Mathematica script:

```
n=10; x=118; q={};
convergent=Convergents[ContinuedFraction[x/2^n]];
For[i=1, i<Length[convergent], i++,
   If[Abs[x/2^n-convergent[[i]]]<1/(2*Denominator[convergent[[i]]]^2)
      &&1<Denominator[convergent[[i]]]^2<2^n,
   AppendTo[q,Denominator[convergent[[i]]]]]]
candidate=Max[q]
```

This script returns `candidate` = 26. The last step is to calculate $f(0)$ and $f(26)$ and check that they are identical. If they are not, the quantum algorithm must be run again.

Solution 10.3 An integer $x = \text{Round}(k \cdot 2^n/r)$ can be expressed as $x = k \cdot 2^n/r + \varepsilon$ where $|\varepsilon| \leq 1/2$. The probability of measuring x is given by

$$\begin{aligned} p_{x=\text{Round}(k \cdot 2^n/r)} &= \frac{1}{2^n s} \frac{\sin^2(\frac{\pi}{2^n} sr(k \cdot 2^n/r + \varepsilon))}{\sin^2(\frac{\pi}{2^n} r(k \cdot 2^n/r + \varepsilon))} \\ &= \frac{1}{2^n s} \frac{\sin^2(\pi sk + \frac{\pi}{2^n} sr\varepsilon)}{\sin^2(\pi k + \frac{\pi}{2^n} r\varepsilon)} = \frac{1}{2^n s} \frac{\sin^2(\frac{\pi}{2^n} sr\varepsilon)}{\sin^2(\frac{\pi}{2^n} r\varepsilon)}. \end{aligned}$$

Since $\sin^2(x) \le x^2$ for any x, we can upper bound the denominator,

$$p_{x=\text{Round}(k\cdot 2^n/r)} \ge \frac{1}{2^n s}\frac{\sin^2(\frac{\pi}{2^n}sr\varepsilon)}{(\frac{\pi}{2^n}r\varepsilon)^2}.$$

The argument of the sine in the numerator is $\frac{\pi}{2^n}sr\varepsilon$. Since $\frac{\pi}{2^n}sr\varepsilon \le \pi/2$, we can use the relation $\sin^2(x) \ge 4x^2/\pi^2$ to lower bound the numerator,

$$p_{x=\text{Round}(k\cdot 2^n/r)} \ge \frac{1}{2^n s}\frac{\frac{4}{\pi^2}(\frac{\pi}{2^n}sr\varepsilon)^2}{(\frac{\pi}{2^n}r\varepsilon)^2} = \frac{s}{2^n}\frac{4}{\pi^2} = \frac{4}{\pi^2 r}.$$

Recall that the probability distribution p_x has r peaks. Therefore, the probability of measuring one of the multiples of $2^n/r$ is

$$\sum_{k=0}^{r-1} p_{x=\text{Round}(k\cdot 2^n/r)} \ge r\cdot\frac{4}{\pi^2 r} > 40\,\%.$$

Solution 10.4 Using Euler's theorem, for any positive integer k, we have

$$\begin{aligned} a^{k\varphi(N)}\,\%\,N = (a^{\varphi(N)})^k\,\%\,N &= (a^{\varphi(N)}\cdot\ldots\cdot a^{\varphi(N)})\,\%\,N \\ &= [(a^{\varphi(N)}\,\%\,N)\cdot\ldots\cdot(a^{\varphi(N)}\,\%\,N)]\,\%\,N \\ &= (1\cdot\ldots\cdot 1)\,\%\,N = 1, \end{aligned}$$

where we used $(a\cdot b)\,\%\,N = [(a\,\%\,N)\cdot(b\,\%\,N)]\,\%\,N$. This shows that the function $a^x\,\%\,N$ is periodic. The period is either $\varphi(N)$ or a factor of $\varphi(N)$. For example, the period of $2^x\,\%\,13$ is $\varphi(13) = 12$. The period of $2^x\,\%\,17$ is 8, a factor of $\varphi(17) = 16$.

Solution 10.5 Let us prove this inequality by induction starting from $k = 0$. If $k = 0$, then $|\psi_0^{\mathbf{a}}\rangle = |\phi_0\rangle = |s\rangle$ and therefore,

$$Z_0 = \sum_{\mathbf{a}\in\mathbb{B}^n} \|\,|\psi_0^{\mathbf{a}}\rangle - |\phi_0\rangle\,\|^2 = \sum_{\mathbf{a}\in\mathbb{B}^n}\|\,|s\rangle - |s\rangle\,\|^2 = 0.$$

This shows that the statement $Z_k \le 4k^2$ is true for $k = 0$. We now make the hypothesis that the statement $Z_k \le 4k^2$ is valid for all $k \ge 0$ and prove step by step that this leads to the inequality $Z_{k+1} \le 4(k+1)^2$. To this end, we need to express Z_{k+1} in terms of Z_k. Let us start with the expression

$$\begin{aligned} Z_{k+1} &= \sum_{\mathbf{a}\in\mathbb{B}^n}\|\,\hat{U}_{k+1}(\hat{O}_{\mathbf{a}}|\psi_k^{\mathbf{a}}\rangle - |\phi_k\rangle)\,\|^2 = \sum_{\mathbf{a}\in\mathbb{B}^n}\|\,\hat{O}_{\mathbf{a}}|\psi_k^{\mathbf{a}}\rangle - |\phi_k\rangle\,\|^2 \\ &= \sum_{\mathbf{a}\in\mathbb{B}^n}\|\,\underbrace{\hat{O}_{\mathbf{a}}(|\psi_k^{\mathbf{a}}\rangle - |\phi_k\rangle)}_{|b\rangle} + \underbrace{(\hat{O}_{\mathbf{a}} - \mathbb{1})|\phi_k\rangle}_{|c\rangle}\,\|^2 \\ &\le \sum_{\mathbf{a}\in\mathbb{B}^n}\|\,|b\rangle\,\|^2 + \|\,|c\rangle\,\|^2 + 2\||b\rangle\|\cdot\||c\rangle\|, \end{aligned} \tag{10.38}$$

where in the last step we used the triangle inequality $\||b\rangle+|c\rangle\|^2 \leq \||b\rangle\|^2+\||c\rangle\|^2+2\||b\rangle\|\cdot\||c\rangle\|$. Using the relation

$$|c\rangle = (\hat{O}_{\mathbf{a}} - \mathbb{1})|\phi_k\rangle = (\mathbb{1} - 2|\mathbf{a}\rangle\langle\mathbf{a}| - \mathbb{1})|\phi_k\rangle = -2\langle\mathbf{a}|\phi_k\rangle\,|\mathbf{a}\rangle,$$

eqn (10.38) becomes

$$\begin{aligned}
Z_{k+1} &\leq \sum_{\mathbf{a}\in\mathbb{B}^n}\Big[\|\,\hat{O}_{\mathbf{a}}(|\psi_k^{\mathbf{a}}\rangle - |\phi_k\rangle)\,\|^2 + 4\|\,\langle\mathbf{a}|\phi_k\rangle\,|\mathbf{a}\rangle\,\|^2 + 4\|\,\hat{O}_{\mathbf{a}}(|\psi_k^{\mathbf{a}}\rangle - |\phi_k\rangle)\,\|\cdot\|\,\langle\mathbf{a}|\phi_k\rangle\,|\mathbf{a}\rangle\,\|\Big] \\
&= \sum_{\mathbf{a}\in\mathbb{B}^n}\|\,|\psi_k^{\mathbf{a}}\rangle - |\phi_k\rangle\,\|^2 + 4\sum_{\mathbf{a}\in\mathbb{B}^n}|\langle\mathbf{a}|\phi_k\rangle|^2 + 4\sum_{\mathbf{a}\in\mathbb{B}^n}\|\,|\psi_k^{\mathbf{a}}\rangle - |\phi_k\rangle\,\|\cdot|\langle\mathbf{a}|\phi_k\rangle| \\
&= Z_k + 4 + 4\sum_{\mathbf{a}\in\mathbb{B}^n}\underbrace{\|\,|\psi_k^{\mathbf{a}}\rangle - |\phi_k\rangle\,\|}_{z_{\mathbf{a}}}\cdot\underbrace{|\langle\mathbf{a}|\phi_k\rangle|}_{w_{\mathbf{a}}}, \qquad (10.39)
\end{aligned}$$

where we used $\|\hat{O}_{\mathbf{a}}|\mathbf{v}\rangle\| = \||\mathbf{v}\rangle\|$ valid for all vectors $|\mathbf{v}\rangle$ and in the last step we used $\sum_{\mathbf{a}}|\langle\mathbf{a}|\phi_k\rangle|^2 = 1$. Recall that for two non-negative real numbers $z_{\mathbf{a}}$ and $w_{\mathbf{a}}$, it holds that $\sum_{\mathbf{a}} z_{\mathbf{a}}w_{\mathbf{a}} \leq \sqrt{\sum_{\mathbf{a}} z_{\mathbf{a}}^2}\sqrt{\sum_{\mathbf{a}} w_{\mathbf{a}}^2}$. As a consequence, eqn (10.39) simplifies to

$$\begin{aligned}
Z_{k+1} &\leq Z_k + 4 + 4\Big(\sum_{\mathbf{a}\in\mathbb{B}^n}\|\,|\psi_k^{\mathbf{a}}\rangle - |\phi_k\rangle\,\|^2\Big)^{1/2}\Big(\sum_{\mathbf{a}\in\mathbb{B}^n}|\langle\mathbf{a}|\phi_k\rangle|^2\Big)^{1/2} \\
&= Z_k + 4 + 4\Big(\sum_{\mathbf{a}\in\mathbb{B}^n}\|\,|\psi_k^{\mathbf{a}}\rangle - |\phi_k\rangle\,\|^2\Big)^{1/2} = Z_k + 4 + 4\sqrt{Z_k}. \qquad (10.40)
\end{aligned}$$

Now that we have expressed Z_{k+1} in terms of Z_k, we make use of the hypothesis that $Z_k \leq 4k^2$. By substituting $Z_k \leq 4k^2$ into eqn (10.40), we obtain

$$Z_{k+1} \leq 4k^2 + 4 + 8k = 4(k+1)^2.$$

This demonstrates that $Z_k \leq 4k^2$ for all $k \geq 0$.

Solution 10.6 Let us decompose the vectors $|\psi_k^{\mathbf{a}}\rangle$ and $|\phi_k\rangle$ in the computational basis, $|\psi_k^{\mathbf{a}}\rangle = \sum_{\mathbf{x}} b_{\mathbf{x}}|\mathbf{x}\rangle$ and $|\phi_k\rangle = \sum_{\mathbf{x}} c_{\mathbf{x}}|\mathbf{x}\rangle$. The first step is to minimize the quantity $\||\psi_k^{\mathbf{a}}\rangle - |\phi_k\rangle\|^2$ appearing in the definition of Z_k. This minimization must be performed under two constraints,

$$\|\,|\psi_k^{\mathbf{a}}\rangle\,\|^2 = \sum_{\mathbf{x}}|b_{\mathbf{x}}|^2 = 1, \qquad \text{and} \qquad p_{\mathbf{a}} \equiv |\langle\mathbf{a}|\psi_k^{\mathbf{a}}\rangle|^2 = |b_{\mathbf{a}}|^2 = \text{const}.$$

The second constraint ensures that $\mathbf{a}$ is observed with a finite probability. The next step is to calculate the stationary points of the Lagrangian

$$\begin{aligned}
L_1 &= \|\,|\psi_k^{\mathbf{a}}\rangle - |\phi_k\rangle\,\|^2 - \lambda_1\|\,|\psi_k^{\mathbf{a}}\rangle\,\|^2 - \lambda_2|\langle\mathbf{a}|\psi_k^{\mathbf{a}}\rangle|^2 \\
&= \sum_{\mathbf{x}\in\mathbb{B}^n}|b_{\mathbf{x}} - c_{\mathbf{x}}| - \sum_{\mathbf{x}\in\mathbb{B}^n}\lambda_1|b_{\mathbf{x}}|^2 - \lambda_2|b_{\mathbf{a}}|^2 \\
&= \sum_{\mathbf{x}\in\mathbb{B}^n}(b_{\mathbf{x}}b_{\mathbf{x}}^* - b_{\mathbf{x}}c_{\mathbf{x}}^* - c_{\mathbf{x}}b_{\mathbf{x}}^* - c_{\mathbf{x}}c_{\mathbf{x}}^*) - \sum_{\mathbf{x}\in\mathbb{B}^n}(\lambda_1 b_{\mathbf{x}}^* b_{\mathbf{x}}) - \lambda_2 b_{\mathbf{a}}^* b_{\mathbf{a}},
\end{aligned}$$

where λ_1 and λ_2 are real parameters. The stationary points can be found by setting $\partial L_1/\partial b_1^* = 0$, $\partial L_1/\partial b_2^* = 0 \ldots$ This produces $N = 2^n$ equations:

$$\frac{\partial L_1}{\partial b_{\mathbf{x}}^*} = b_{\mathbf{x}} - c_{\mathbf{x}} - \lambda_1 b_{\mathbf{x}} = 0, \qquad \text{for } \mathbf{x} \neq \mathbf{a}$$
$$\frac{\partial L_1}{\partial b_{\mathbf{a}}^*} = b_{\mathbf{a}} - c_{\mathbf{a}} - \lambda_1 b_{\mathbf{a}} - \lambda_2 b_{\mathbf{a}} = 0.$$

The stationary points of the Lagrangian satisfy:

$$c_{\mathbf{x}} = (1-\lambda_1) b_{\mathbf{x}}, \qquad \text{for x} \neq \mathbf{a} \tag{10.41}$$
$$c_{\mathbf{a}} = (1-\lambda_1-\lambda_2) b_{\mathbf{a}}. \tag{10.42}$$

These two expressions can be used to determine the values of the Lagrange multipliers. Taking the sum of the moduli of eqn (10.41), we have

$$\sum_{\mathbf{x}\neq\mathbf{a}} |c_{\mathbf{x}}|^2 = (1-\lambda_1)^2 \sum_{\mathbf{x}\neq\mathbf{a}} |b_{\mathbf{x}}|^2 \qquad \Rightarrow$$
$$\Rightarrow \qquad (1-\lambda_1)^2 = \frac{\sum_{\mathbf{x}\neq\mathbf{a}} |c_{\mathbf{x}}|^2}{\sum_{\mathbf{x}\neq\mathbf{a}} |b_{\mathbf{x}}|^2} = \frac{1-|c_{\mathbf{a}}|^2}{1-b_{\mathbf{a}}^2} = \frac{1-|c_{\mathbf{a}}|^2}{1-p_{\mathbf{a}}}. \tag{10.43}$$

On the other hand, eqn (10.42) can be rearranged as

$$(1-\lambda_1-\lambda_2)^2 = \frac{|c_{\mathbf{a}}|^2}{b_{\mathbf{a}}^2} = \frac{|c_{\mathbf{a}}|^2}{p_{\mathbf{a}}}. \tag{10.44}$$

Using eqns (10.43, 10.44), we can finally calculate the minimum of $\|\,|\psi_k^{\mathbf{a}}\rangle - |\phi_k\rangle\,\|^2$,

$$\|\,|\psi_k^{\mathbf{a}}\rangle - |\phi_k\rangle\,\|^2 = \underbrace{\sum_{\mathbf{x}\in\mathbb{B}^n} |b_{\mathbf{x}}|^2}_{1} - \sum_{\mathbf{x}\in\mathbb{B}^n} b_{\mathbf{x}} c_{\mathbf{x}}^* - \sum_{\mathbf{x}\in\mathbb{B}^n} c_{\mathbf{x}} b_{\mathbf{x}}^* - \underbrace{\sum_{\mathbf{x}\in\mathbb{B}^n} |c_{\mathbf{x}}|^2}_{1} \tag{10.45}$$
$$= 2 - b_{\mathbf{a}} c_{\mathbf{a}}^* - c_{\mathbf{a}} b_{\mathbf{a}}^* - \sum_{\mathbf{x}\neq\mathbf{a}} b_{\mathbf{x}} c_{\mathbf{x}}^* - \sum_{\mathbf{x}\neq\mathbf{a}} c_{\mathbf{x}} b_{\mathbf{x}}^*$$
$$= 2 - 2(1-\lambda_1-\lambda_2)|b_{\mathbf{a}}|^2 - 2(1-\lambda_1)\sum_{\mathbf{x}\neq\mathbf{a}} |b_{\mathbf{x}}|^2$$
$$= 2 - 2(1-\lambda_1-\lambda_2)p_{\mathbf{a}} - 2(1-\lambda_1)(1-p_{\mathbf{a}})$$
$$= 2 \pm 2|c_{\mathbf{a}}|\sqrt{p_{\mathbf{a}}} \pm 2\sqrt{1-|c_{\mathbf{a}}|^2}\sqrt{1-p_{\mathbf{a}}}$$
$$= 2 \pm 2\sqrt{q_{\mathbf{a}}}\sqrt{p_{\mathbf{a}}} \pm 2\sqrt{1-q_{\mathbf{a}}}\sqrt{1-p_{\mathbf{a}}}, \tag{10.46}$$

where in the last step we defined $q_{\mathbf{a}} = |\langle \mathbf{a}|\phi_k\rangle|^2 = |c_{\mathbf{a}}|^2$. Since $0 \leq q_{\mathbf{a}}, p_{\mathbf{a}} \leq 1$, the minimum is obtained when the signs in eqn (10.46) are both negative,

$$\min_{b_{\mathbf{x}}, c_{\mathbf{x}}} \|\,|\psi_k^{\mathbf{a}}\rangle - |\phi_k\rangle\,\|^2 = 2 - 2\sqrt{q_{\mathbf{a}}}\sqrt{p_{\mathbf{a}}} - 2\sqrt{1-q_{\mathbf{a}}}\sqrt{1-p_{\mathbf{a}}}.$$

By performing a sum over $\mathbf{a}$, we arrive at

$$\boxed{Z_k = \sum_{\mathbf{a}\in\mathbb{B}^n} \| \, |\psi_k^{\mathbf{a}}\rangle - |\phi_k\rangle \, \|^2 \geq 2N - 2\sum_{\mathbf{a}\in\mathbb{B}^n} \sqrt{q_{\mathbf{a}}}\sqrt{p_{\mathbf{a}}} - 2\sum_{\mathbf{a}\in\mathbb{B}^n} \sqrt{1-q_{\mathbf{a}}}\sqrt{1-p_{\mathbf{a}}}.} \tag{10.47}$$

The next step is to minimize Z_k using the technique of Lagrange multipliers once again. This time the constraints are

$$\sum_{\mathbf{a}\in\mathbb{B}^n} q_{\mathbf{a}} = 1, \qquad \text{and} \qquad \frac{1}{N}\sum_{\mathbf{a}\in\mathbb{B}^n} p_{\mathbf{a}} = \overline{p}.$$

To perform this minimization, we can introduce the Lagrangian

$$L_2 = 2N - 2\sum_{\mathbf{a}\in\mathbb{B}^n} \sqrt{q_{\mathbf{a}}}\sqrt{p_{\mathbf{a}}} - 2\sum_{\mathbf{a}\in\mathbb{B}^n} \sqrt{1-q_{\mathbf{a}}}\sqrt{1-p_{\mathbf{a}}} - \mu_1 \sum_{\mathbf{a}\in\mathbb{B}^n} q_{\mathbf{a}} - \frac{\mu_2}{N}\sum_{\mathbf{a}\in\mathbb{B}^n} p_{\mathbf{a}}.$$

We now impose the constraint that the Lagrangian is stationary with respect to the variables $q_{\mathbf{a}}$ and $p_{\mathbf{a}}$:

$$\frac{\partial L_2}{\partial q_{\mathbf{a}}} = -\frac{\sqrt{p_{\mathbf{a}}}}{\sqrt{q_{\mathbf{a}}}} + \frac{\sqrt{1-p_{\mathbf{a}}}}{\sqrt{1-q_{\mathbf{a}}}} - \mu_1 = 0, \tag{10.48}$$

$$\frac{\partial L_2}{\partial p_{\mathbf{a}}} = -\frac{\sqrt{q_{\mathbf{a}}}}{\sqrt{p_{\mathbf{a}}}} + \frac{\sqrt{1-q_{\mathbf{a}}}}{\sqrt{1-p_{\mathbf{a}}}} - \frac{\mu_2}{N} = 0. \tag{10.49}$$

The Lagrange multipliers μ_1 and μ_2 must satisfy:

$$\frac{\sqrt{1-p_{\mathbf{a}}}}{\sqrt{1-q_{\mathbf{a}}}} = \mu_1 + \frac{\sqrt{p_{\mathbf{a}}}}{\sqrt{q_{\mathbf{a}}}}, \qquad \text{and} \qquad \frac{\sqrt{1-q_{\mathbf{a}}}}{\sqrt{1-p_{\mathbf{a}}}} = \frac{\mu_2}{N} + \frac{\sqrt{q_{\mathbf{a}}}}{\sqrt{p_{\mathbf{a}}}}.$$

Multiplying these equations together, we obtain

$$1 = \frac{\mu_1\mu_2}{N} + \mu_1\frac{\sqrt{q_{\mathbf{a}}}}{\sqrt{p_{\mathbf{a}}}} + \frac{\mu_2}{N}\frac{\sqrt{p_{\mathbf{a}}}}{\sqrt{q_{\mathbf{a}}}} + 1,$$

whose solution is

$$\frac{\sqrt{q_{\mathbf{a}}}}{\sqrt{p_{\mathbf{a}}}} = \frac{-\mu_1\mu_2 + \sqrt{\mu_1^2\mu_2^2 - 4\mu_1\mu_2 N}}{2\mu_1 N} \equiv \sqrt{\ell}.$$

The crucial point is that the ratio $q_{\mathbf{a}}/p_{\mathbf{a}} = \ell$ is a constant that does not depend on $\mathbf{a}$. Substituting $q_{\mathbf{a}} = \ell p_{\mathbf{a}}$ into eqn (10.48), we have

$$\frac{\sqrt{1-p_{\mathbf{a}}}}{\sqrt{1-\ell p_{\mathbf{a}}}} = \mu_1 + \frac{1}{\sqrt{\ell}}.$$

This expression shows that $p_{\mathbf{a}}$ is a constant independent of $\mathbf{a}$. The only constant that is compatible with the constraint $\frac{1}{N}\sum_{\mathbf{a}} p_{\mathbf{a}} = \overline{p}$ is $p_{\mathbf{a}} = \overline{p}$. In a similar way, substituting $p_{\mathbf{a}} = q_{\mathbf{a}}/\ell$ into eqn (10.48), we have

$$\frac{\sqrt{1-q_{\mathbf{a}}/\ell}}{\sqrt{1-q_{\mathbf{a}}}} = \mu_1 + \frac{1}{\sqrt{\ell}}.$$

The variable $q_{\mathbf{a}}$ is a constant independent of $\mathbf{a}$. The only constant that is compatible with the constraint $\sum_{\mathbf{a}} q_{\mathbf{a}} = 1$ is $q_{\mathbf{a}} = 1/N$. By substituting $p_{\mathbf{a}} = \overline{p}$ and $q_{\mathbf{a}} = 1/N$ into (10.47), we arrive at

$$Z_k \geq 2N - 2\sum_{\mathrm{a}\in\mathbb{B}^n}\sqrt{\frac{\overline{p}}{N}} - 2\sum_{\mathrm{a}\in\mathbb{B}^n}\sqrt{1-\frac{1}{N}}\sqrt{1-\overline{p}} = 2N - 2\sqrt{N\overline{p}} - \sqrt{N-1}\sqrt{N}\sqrt{1-\overline{p}}.$$

Since $N > \sqrt{N-1}\sqrt{N}$, this inequality becomes

$$\boxed{Z_k \geq 2N - 2\sqrt{N\overline{p}} - N\sqrt{1-\overline{p}}.}$$

Solution 10.7 Since q_1 and q_2 are divisors of r, $\mathrm{lcm}(q_1, q_2)$ is a divisor of r. Therefore, there exists a positive integer ℓ satisfying $r = \ell\,\mathrm{lcm}(q_1, q_2)$. Let us substitute this expression into (10.37),

$$k_1 = r\frac{p_1}{q_1} = \frac{p_1}{q_1}\ell\,\mathrm{lcm}(q_1, q_2), \qquad\qquad k_2 = r\frac{p_2}{q_2} = \frac{p_2}{q_2}\ell\,\mathrm{lcm}(q_1, q_2).$$

The greatest common divisor between k_1 and k_2 is

$$\gcd(k_1, k_2) = \gcd\Big[\frac{p_1}{q_1}\ell\,\mathrm{lcm}(q_1, q_2),\ \frac{p_2}{q_2}\ell\,\mathrm{lcm}(q_1, q_2)\Big].$$

Using the property $\gcd(ax, bx) = x\gcd(a, b)$ valid for any integers a, b, x, we obtain

$$\gcd(k_1, k_2) = \ell\ \underbrace{\gcd\Big[\frac{p_1}{q_1}\mathrm{lcm}(q_1, q_2), \frac{p_2}{q_2}\mathrm{lcm}(q_1, q_2)\Big]}_{w} = \ell \cdot w. \tag{10.50}$$

If k_1 and k_2 are coprime, i.e. $\gcd(k_1, k_2) = 1$, eqn (10.50) indicates that the product between the positive integers ℓ and w is 1. This means that $\ell = 1$ and $w = 1$. In conclusion, if k_1 and k_2 are coprime, then $\ell = 1$ and $r = \ell\,\mathrm{lcm}(q_1, q_2) = \mathrm{lcm}(q_1, q_2)$.

Solution 10.8 Two integers k_1 and k_2 randomly chosen in the interval $[0, \dots, r-1]$ are not coprime if they share a common factor, i.e. there exists a prime number m smaller than k_1 and k_2 that is a divisor of both of them. This happens with probability

$$p(\gcd(k_1, k_2) \neq 1) = \sum_{k_1=0}^{r-1} p(k_1) \sum_{k_2=0}^{r-1} p(k_1) \sum_{\substack{m\in\text{primes},\\ 2\leq m\leq\min(k_1,k_2)}} p(m|k_1)p(m|k_2). \tag{10.51}$$

Here, $p(k_1) = p(k_2) = 1/r$, since these two integers are generated with uniform probability and the notation $m|k_1$ indicates that m is a divisor of k_1. What is the probability that a prime number m smaller than k_1 is a divisor of k_1? It is given by the ratio between t (defined as the number of multiples of m smaller than r) and the number of integers smaller than r,

$$p(m|k_1) = \frac{t}{r} < \frac{r/m}{r} = \frac{1}{m}.$$

Using this inequality, eqn (10.51) becomes

$$p(\gcd(k_1, k_2) \neq 1) \leq \sum_{k_1=0}^{r-1} \frac{1}{r} \sum_{k_2=0}^{r-1} \frac{1}{r} \sum_{\substack{m \in \text{primes}, \\ 2 \leq m \leq \min(k_1,k_2)}} \frac{1}{m^2}$$

$$= \sum_{\substack{m \in \text{primes}, \\ 2 \leq m \leq \min(k_1,k_2)}} \frac{1}{m^2} < \sum_{m=2}^{\infty} \frac{1}{m^2} = \frac{\pi^2}{6} - 1.$$

In conclusion, the probability that k_1 and k_2 are coprime is bounded by

$$p(\gcd(k_1, k_2) = 1) = 1 - p(\gcd(k_1, k_2) \neq 1) > 2 - \frac{\pi^2}{6} > 35\,\%.$$

This bound can be heuristically checked with a Mathematica script:

```
r = 10^7; trials = 10^6; j = 0;
For[i = 0, i < trials, i++,
  k1 = RandomInteger[r]; k2 = RandomInteger[r];
  If[GCD[k1, k2] == 1, j = j + 1.0;]];
Print["probability=", j/trials]
```

This script returns `probability` $\approx$ 0.607, which is larger than $35\,\%$.

Solution 10.9 If you can provide the solution to this problem, it means you have access to a quantum computer with very low error rates and at least 1000 qubits.

11 Quantum simulation of Hamiltonian dynamics

The simulation of the dynamics of a quantum system is one of the most important applications of quantum computers.

An important application of quantum computers is the simulation of the dynamics of quantum systems [11]. A simulation is a computational process in which one physical system mimics the behavior of another. Let us consider the simulation of the time evolution of a system with N classical particles subject to external mechanical forces. The dynamics are described by Newton's law

$$\mathbf{F}_i = m_i \frac{d}{dt}\mathbf{v}_i.$$

The velocity of the particles and the external forces are known at the initial time and the goal is to determine the velocity of the particles at a future time t. A classical computer can numerically solve this problem by discretizing the system in space and time so that the configuration of the particles can be encoded into a binary string. The time interval $[0, t]$ is divided into m steps of duration $\Delta t = t/m$. The velocity of the particles is iteratively calculated at each step using the approximation $\mathbf{v}_i(t_j + \Delta t) \approx \mathbf{v}_i(t_j) + \mathbf{F}_i \Delta t/m_i$. The error of this approximation scales polynomially with the number of steps m under certain conditions.

In quantum physics, the dynamics of N particles are described by the Schrödinger equation

$$\boxed{i\frac{d}{dt}|\phi(t)\rangle = \hat{H}_{\mathrm{s}}|\phi(t)\rangle,}$$

where we set $\hbar = 1$ (we will keep this convention for the rest of the chapter). Given an initial state $|\phi(0)\rangle$, our goal is to derive

$$\boxed{|\phi(t)\rangle = e^{-i\hat{H}_{\mathrm{s}}t}|\phi(0)\rangle}$$

and calculate the expectation value of a physical observable $\hat{O}$ at time t. This problem can be numerically solved following the same procedure as the classical case: the time

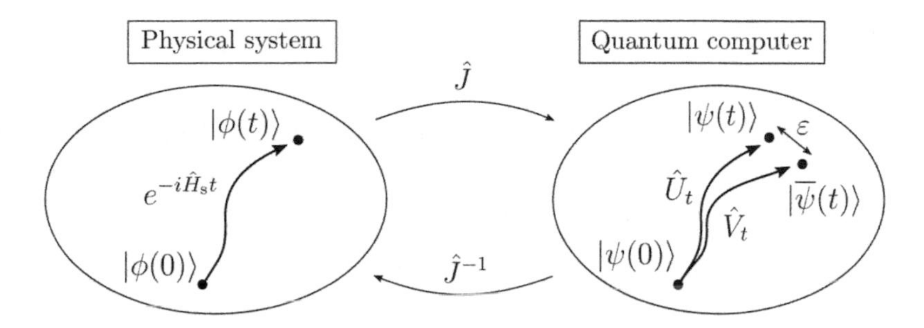

Fig. 11.1 Digital quantum simulation. A quantum system is in a quantum state $|\phi(0)\rangle$ and is subject to a Hamiltonian $\hat{H}_s$. The state $|\phi(0)\rangle$ is mapped into the quantum state of a quantum computer $|\psi(0)\rangle$ using a bijective function $\hat{J}$. The dynamics of the qubits are governed by the time evolution operator $\hat{U}_t = e^{-i\hat{H}t}$ where $\hat{H} = \hat{J}\hat{H}_s\hat{J}^{-1}$. The actual simulation is performed by a quantum circuit $\hat{V}_t$ that approximates the time evolution operator $\hat{U}_t$. The final state of the quantum computer is $|\overline{\psi}(t)\rangle$ which differs from the ideal state $|\psi(t)\rangle$ by no more than $\varepsilon = \|\hat{U}_t - \hat{V}_t\|_o$.

interval $[0, t]$ is divided into m intervals and the computer calculates $|\phi(t_j + \Delta t)\rangle$ at each step with suitable approximations. For instance, the time evolution operator can be approximated as $e^{-i\hat{H}_s\Delta t} \approx \mathbb{1} - i\hat{H}_s\Delta t$ so that

$$|\phi(t_j + \Delta t)\rangle \approx \alpha(\mathbb{1} - i\hat{H}_s\Delta t)|\phi(t_j)\rangle,$$

where α is a normalization constant. Once the final state has been determined, one can evaluate the expectation value $\langle\phi(t)|\hat{O}|\phi(t)\rangle$.

Simulating the time evolution of a quantum system with a classical computer without approximations is not practical. Consider a system with N spin-1/2 particles described by the wavefunction $|\phi(0)\rangle = \sum_i c_i|i\rangle$. This quantum state can be represented by a vector with 2^N components where each component corresponds to a probability amplitude c_i. If the system under investigation has $N = 100$ particles, the vector has 2^{100} entries. Assuming that each probability amplitude can be expressed with 32 bits, the classical memory required to encode the wavefunction is $32 \cdot 2^{100}$ bit $= 5 \cdot 10^{18}$ TB. This quantity of information exceeds the processing capabilities of any classical computer. On the other hand, 100 qubits are sufficient to represent this vector. A quantum computer often allows us to dramatically reduce the volume of the physical support required to store the wavefunction.

In this chapter, we will focus on the simulation of the time evolution of an N-particle system. This procedure is divided into three parts:

Step 1 The particles are initially in a known state $|\phi(0)\rangle$ and are subject to a Hamiltonian $\hat{H}_s$. The quantum state of the particles is encoded into a vector $|\psi(0)\rangle$ of a quantum computer with n qubits using an invertible map $\hat{J}$ as illustrated in Fig. 11.1,

$$\underset{\text{Physical system}}{|\phi(0)\rangle} \quad \overset{\hat{J}}{\longmapsto} \quad \underset{\text{Quantum computer}}{|\psi(0)\rangle}.$$

In the worst case, the preparation of $|\psi(0)\rangle$ requires a number of elementary operations that scale exponentially with the system size. However, in many practical cases, it is possible to prepare $|\psi(0)\rangle$ with a polynomial number of gates[1].

Step 2 The function $\hat{J}$ can be used to derive the Hamiltonian acting on the qubits,

$$\underset{\text{Physical system}}{\hat{H}_s} \quad \overset{\hat{J}}{\longmapsto} \quad \underset{\text{Quantum computer}}{\hat{H} = \hat{J}\hat{H}_s\hat{J}^{-1}}.$$

For now, we will assume that $\hat{H}$ is k-**local**, i.e. it can be decomposed as[2]

$$\hat{H} = \sum_{j=1}^{L} \hat{H}_j, \tag{11.1}$$

where the Hermitian operators $\hat{H}_j$ act on no more than k qubits and k is a constant. This means that the total number of terms in the sum scales polynomially with the number of qubits, $L = \text{poly}(n)$. Let us suppose that $\hat{H}$ is time-independent and each pair of terms in (11.1) commutes, $[\hat{H}_i, \hat{H}_j] = 0$. Then, the time evolution operation is simply given by (see Exercise 11.1)

$$\hat{U}_t = e^{-i\hat{H}t} = e^{-i(\hat{H}_1+\ldots+\hat{H}_L)t} = e^{-i\hat{H}_1 t} \ldots e^{-i\hat{H}_L t}. \tag{11.2}$$

The exponentials on the right act on at most k qubits. This makes it possible to decompose them into a small number of single and two-qubit gates. The goal of an optimal algorithm for quantum simulation is to find a quantum circuit $\hat{V}_t$ that not only implements $\hat{U}_t$ with a small error, but also minimizes the number of elementary gates required.

Step 3 The circuit $\hat{V}_t$ is applied to the initial state $|\psi(0)\rangle$ and the final state $|\psi(t)\rangle$ is measured in the computational basis. This procedure must be repeated multiple times to determine the probability amplitudes with high precision. In general, $O(2^n)$ projective measurements are required to reconstruct all of the information in the wavefunction $|\psi(t)\rangle$. However, an exponential number of measurements is not necessary if we are only interested in the expectation value of some specific physical observables [332, 333].

It is important to define which kind of Hamiltonians can be efficiently simulated on a quantum computer. The dynamics generated by a Hamiltonian $\hat{H}$ acting on n qubits

[1] An example of a quantum state that can be efficiently encoded on a quantum computer is the Hartree–Fock state of a molecule, see Section 12.9 and Ref. [331].

[2] As seen in Section 2.8.1, the Ising and Heisenberg Hamiltonian are 2-local.

can be simulated by a quantum computer if for all $\varepsilon \in [0, 2]$ and $t > 0$ there exists a quantum circuit $\hat{V}_t$ that simulates the time evolution operator with a small error. In other words, the final state produced by the quantum computer $|\overline{\psi}(t)\rangle = \hat{V}_t|\psi(0)\rangle$ should be similar[3] to the ideal one $|\psi(t)\rangle = \hat{U}_t|\psi(0)\rangle$,

$$\boxed{\| |\overline{\psi}(t)\rangle - |\psi(t)\rangle \| = \| (\hat{V}_t - \hat{U}_t)|\psi_0\rangle \| \leq \| \hat{V}_t - \hat{U}_t \|_\text{o} < \varepsilon \in [0, 2].}$$

The simulation is said to be **efficient** if the number of elementary gates in the circuit scales at most polynomially with respect to the number of qubits n, the desired precision ε^{-1}, and the evolution time t,

$$n_{\text{gates}} = \text{poly}(n, \varepsilon^{-1}, t).$$

Let us list some useful properties about quantum simulations [334]:

H1. If $\hat{H}$ can be simulated efficiently, $c\hat{H}$ with $c \in \mathbb{R}$ can be simulated efficiently.
H2. If $\hat{H}_1$ and $\hat{H}_2$ can be simulated efficiently, $\hat{H}_1 + \hat{H}_2$ can be simulated efficiently.
H3. If $\hat{H}$ is diagonal, $\hat{H}$ can be simulated efficiently.
H4. If $\hat{H}$ can be simulated efficiently and $\hat{U}$ is a unitary operator that can be implemented efficiently on a quantum computer, then $\hat{H}' = \hat{U}\hat{H}\hat{U}^{-1}$ can be simulated efficiently.

These properties are proven in the exercises at the end of the chapter. Property **H1** indicates that the dynamics generated by two Hamiltonians differing by a multiplicative constant can be simulated with the same resources. Hence, it is often convenient to rescale the Hamiltonian such that all of its eigenvalues lie in the range $[-1, 1]$, i.e.

$$\hat{H} \qquad \mapsto \qquad \hat{H}' = \frac{\hat{H}}{\|\hat{H}\|_\text{o}},$$

so that the time evolution operator to be simulated becomes

$$\hat{U} = e^{-i\hat{H}t} = e^{-i\hat{H}'\tau},$$

where $\tau = t\|\hat{H}\|_\text{o}$ is a dimensionless parameter.

Example 11.1 Consider three particles with spin-1/2 prepared in a state $|\uparrow\downarrow\downarrow\rangle$ and suppose that the Hamiltonian of the system is given by

$$\hat{H}_\text{s} = \omega(\hat{\sigma}_x \otimes \hat{\sigma}_y \otimes \hat{\sigma}_z), \tag{11.3}$$

where ω is a real number. The dynamics of the particles can be simulated with a quantum computer with three qubits. Since the qubits themselves can be considered

[3] See Section 3.2.2 for more information about the operator norm $\|\cdot\|_\text{o}$. For a normal operator, the operator norm is nothing but the highest-magnitude eigenvalue, which is $\|\hat{U}\|_\text{o} = 1$ for unitary operators $\hat{U}$. Due to the triangle inequality, the distance between two unitary operators $\hat{U}$ and $\hat{V}$ satisfies $0 \leq \|\hat{U} - \hat{V}\|_\text{o} \leq \|\hat{U}\|_\text{o} + \|\hat{V}\|_\text{o} = 2$.

spin-1/2 particles, the map $\hat{J}$ is nothing but the identity: $|100\rangle = \hat{J}|\uparrow\downarrow\downarrow\rangle$ and $\hat{H} = \hat{H}_s$. As shown in Fig. 11.2, the simulation starts with an $\hat{X}$ gate on the third qubit so that the state of the quantum computer becomes $|100\rangle$. Let us introduce three operators: $\hat{A}_x = \mathsf{H}$, $\hat{A}_y = \hat{S}\mathsf{H}$, $\hat{A}_z = \mathbb{1}$ where H is the Hadamard gate and $\hat{S}$ is the phase gate. With simple matrix multiplications, one can verify that

$$\hat{\sigma}_x = \hat{A}_x\hat{\sigma}_z A_x^\dagger, \qquad \hat{\sigma}_y = \hat{A}_y\hat{\sigma}_z A_y^\dagger, \qquad \hat{\sigma}_z = \hat{A}_z\hat{\sigma}_z\hat{A}_z^\dagger.$$

Substituting these expressions into (11.3), we arrive at

$$\begin{aligned}\hat{H} &= \omega(\hat{A}_x \otimes \hat{A}_y \otimes \hat{A}_z)(\hat{\sigma}_z \otimes \hat{\sigma}_z \otimes \hat{\sigma}_z)(\hat{A}_x^\dagger \otimes \hat{A}_y^\dagger \otimes \hat{A}_z^\dagger) \\ &= \omega\hat{A}(\hat{\sigma}_z \otimes \hat{\sigma}_z \otimes \hat{\sigma}_z)\hat{A}^\dagger,\end{aligned}$$

where in the last step we defined $\hat{A} = \hat{A}_x \otimes \hat{A}_y \otimes \hat{A}_z$. We note that the operator in parentheses can be expressed as a $\hat{\sigma}_z$ gate preceded and followed by two CNOT gates,

$$\hat{\sigma}_z \otimes \hat{\sigma}_z \otimes \hat{\sigma}_z = \underbrace{\mathsf{c}_2\hat{X}_1\,\mathsf{c}_1\hat{X}_0}_{\hat{B}}\,(\mathbb{1} \otimes \mathbb{1} \otimes \hat{\sigma}_z)\,\underbrace{\mathsf{c}_1\hat{X}_0\,\mathsf{c}_2\hat{X}_1}_{\hat{B}^\dagger} = \hat{B}(\mathbb{1} \otimes \mathbb{1} \otimes \hat{\sigma}_z)\hat{B}^\dagger.$$

In conclusion, $\hat{H} = \omega\hat{A}\hat{B}(\mathbb{1} \otimes \mathbb{1} \otimes \hat{\sigma}_z)\hat{B}^\dagger\hat{A}^\dagger$ and the time evolution operator becomes

$$\hat{U}_t = e^{-iHt} = e^{-i\omega\hat{A}\hat{B}(\mathbb{1}\otimes\mathbb{1}\otimes\hat{\sigma}_z)\hat{B}^\dagger\hat{A}^\dagger t} = \hat{A}\hat{B}e^{-i\omega t(\mathbb{1}\otimes\mathbb{1}\otimes\hat{\sigma}_z)}\hat{B}^\dagger\hat{A}^\dagger,$$

where in the last step we used a property explained in Exercise 11.2. The unitary $\hat{U}_t$ can be implemented on a quantum computer using the circuit

$$\hat{V}_t = \hat{A}\hat{B}(\mathbb{1} \otimes \mathbb{1} \otimes \mathsf{Z}_{2\omega t})\hat{B}^\dagger\hat{A}^\dagger.$$

This circuit simulates exactly the dynamics for any time t and can be implemented with 11 elementary gates as illustrated in Fig. 11.2. The final state $|\psi(t)\rangle = \hat{V}_t|100\rangle$ must be measured multiple times to acquire some information about the probability amplitudes. Once $|\psi(t)\rangle$ has been determined with the desired accuracy, the state of the particles can be derived by applying the inverse map, $|\phi(t)\rangle = \hat{J}^{-1}|\psi(t)\rangle = |\psi(t)\rangle$ since $\hat{J} = \mathbb{1}$.

11.0.1 Comparison between quantum simulation methods

In most cases, the time evolution operator $e^{-i\hat{H}t}$ is a mathematical operator that cannot be calculated *exactly* with a classical computer using polynomial resources. Over the last three decades, scientists have developed a variety of quantum simulation algorithms that can *approximate* the time evolution operator efficiently on a quantum computer. The number of gates used by these algorithms has a precise scaling as a function of the evolution time t and the accuracy ε. Before diving into the technical details, it is worth summarizing the properties of these algorithms.

The quantum algorithms for the simulation of the dynamics discovered in the '90s are known as the "product formulas" (or Trotterization methods, or Lie–Trotter–Suzuki formulas) because the exponential $e^{-i\hat{H}t}$ is approximated with a product of

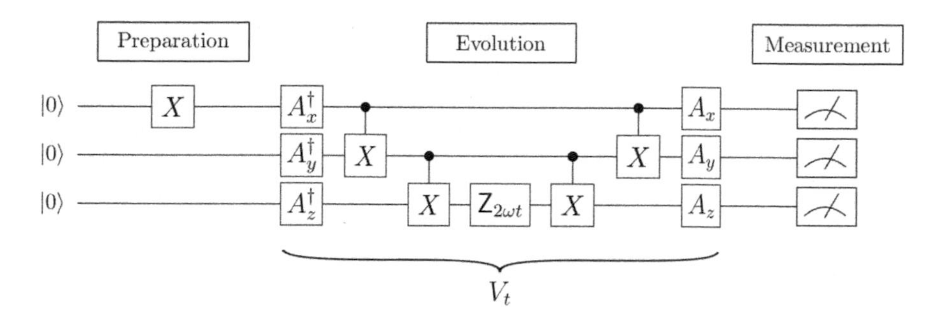

Fig. 11.2 Example of a quantum simulation. Simulation of the dynamics generated by the Hamiltonian $\hat{H} = \omega(\hat{\sigma}_x \otimes \hat{\sigma}_y \otimes \hat{\sigma}_z)$.

exponentials. They were originally introduced by Lie and later generalized by Trotter and Suzuki. In Section 11.1, we will study in detail the primitive Trotter formula and show that its computational cost scales as $O(t^2\varepsilon^{-1})$. A more refined product formula, based on a second-order expansion, is derived in Section 11.2. This analysis will show that the time evolution operator can be approximated with $O(t^{1.5}\varepsilon^{-0.5})$ gates. In general, a product formula to the $2k$-th order can approximate the time evolution operator with $O(5^{k-1}t^{1+\frac{1}{2k}}\varepsilon^{-\frac{1}{2k}})$ gates. Recent studies have investigated the scaling of product formulas and showed that the number of gates required to approximate $e^{-i\hat{H}t}$ for particular Hamiltonians (for example, k-local Hamiltonians and Hamiltonians with a low-rank decomposition) can be further reduced [335]. Product formulas are appealing because they are simple to implement on quantum hardware and do not require extra ancilla qubits. This is particularly important for noisy intermediate-scale quantum devices, which do not have a substantial number of qubits and are still prone to errors.

A well-known theoretical result in the field of quantum simulation, known as the **no-fast-forwarding theorem** [336, 337], states that the computational cost cannot be sub-linear in t. A subsequent study showed that the optimal scaling for the simulation of the dynamics is $O(t + \log\varepsilon^{-1})$ [28]. A disadvantage of the product formulas is that the number of gates required for the decomposition is not linear in t. Only when $k \to \infty$, the computational cost becomes linear in t, albeit with a divergent prefactor 5^{k-1}.

After discussing product formulas, we explain the Taylor series and qubitization, two simulation methods based on the linear combination of unitaries (LCU) lemma. These quantum algorithms simulate the dynamics with more qubits than Trotterization methods but with fewer gates. In particular, the Taylor series method uses $O(t\log\varepsilon^{-1})$ gates, whereas the qubitization reaches the optimal value $O(t + \log\varepsilon^{-1})$. Even if the qubitization method has an optimal scaling, it is still unclear how to generalize this method to time-dependent Hamiltonians. For this reason, product formulas and the Taylor series are preferred for simulating the dynamics generated by time-dependent Hamiltonians. For an overview of quantum simulation methods, we suggest the review by Martyn et al. [338].

All of the simulation methods presented in this chapter have a polynomial scaling in terms of evolution time and accuracy. This observation will naturally lead to a fundamental concept of quantum information: the problem of simulating the time evolution of a quantum system is BQP complete (in other words, any quantum circuit can be mapped into a quantum circuit simulating the dynamics generated by a Hamiltonian). In this sense, Hamiltonian simulation is the hardest problem that a quantum computer can solve in polynomial time with a bounded error probability.

11.1 Trotterization

In many practical cases, the operators $\hat{H}_j$ in the local Hamiltonian (11.1) do not commute, $[\hat{H}_i, \hat{H}_j] \neq 0$. This means that the identity (11.2) does not hold and other techniques must be adopted to efficiently decompose $\hat{U}_t$ into elementary operations. The first-order Lie–Suzuki–Trotter formula [339, 25] is an approximation that expresses the time evolution operator as a product of separate exponentials that act on a limited number of qubits. This method is also known as the **Trotterization**. The idea behind this procedure is to divide the time interval $[0,t]$ into m intervals of duration $\Delta t = t/m$ (see Fig. 11.3). Assuming that the terms in the Hamiltonian do not commute, the Trotter formula states that the time evolution operator for a time interval Δt can be approximated with a product of exponentials

$$\boxed{\hat{U}_{\Delta t} = e^{-i(\hat{H}_1 + \ldots + \hat{H}_L)\Delta t} \approx e^{-i\hat{H}_1 \Delta t} \ldots e^{-i\hat{H}_L \Delta t} \equiv \hat{V}_{\Delta t}.} \tag{11.4}$$

Each exponential $e^{-i\hat{H}_j \Delta t}$ acts on no more than k qubits since we assumed $\hat{H}$ to be k-local.

Let us quantify the error introduced by this approximation. The exact time evolution operator for a time interval Δt is

$$\hat{U}_{\Delta t} = e^{-i(\hat{H}_1 + \ldots + \hat{H}_L)\Delta t}.$$

This operator can be expanded with a Taylor series to second order,

$$\hat{U}_{\Delta t} \approx \mathbb{1} + (-i\Delta t)\sum_{j=1}^{L} \hat{H}_j + \frac{(-i\Delta t)^2}{2}\Big[\sum_{j=1}^{L} \hat{H}_j^2 + \sum_{i<j}^{L} \hat{H}_j \hat{H}_i + \sum_{i<j}^{L} \hat{H}_i \hat{H}_j\Big]. \tag{11.5}$$

Let us now consider the actual quantum circuit implemented by the Trotterization,

$$\boxed{\hat{V}_{\Delta t} = e^{-i\hat{H}_1 \Delta t} \ \ldots \ e^{-i\hat{H}_L \Delta t}.}$$

Again, we can perform a Taylor expansion up to the second order and obtain

$$\begin{aligned} \hat{V}_{\Delta t} &\approx \Big[\mathbb{1} + (-i\Delta t)\hat{H}_1 + \frac{(-i\Delta t)^2}{2}\hat{H}_1^2\Big] \ \ldots \ \Big[\mathbb{1} + (-i\Delta t)\hat{H}_L + \frac{(-i\Delta t)^2}{2}\hat{H}_L^2\Big] \\ &= \mathbb{1} + (-i\Delta t)\sum_{j=1}^{L} \hat{H}_j + \frac{(-i\Delta t)^2}{2}\Big[\sum_{j=1}^{L} \hat{H}_j^2 + 2\sum_{i<j}^{L} \hat{H}_i \hat{H}_j\Big]. \end{aligned} \tag{11.6}$$

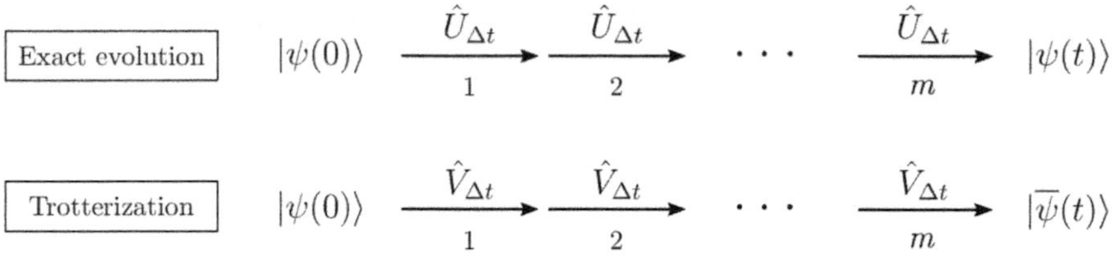

$$\boxed{\text{Exact evolution}} \quad |\psi(0)\rangle \xrightarrow[1]{\hat{U}_{\Delta t}} \xrightarrow[2]{\hat{U}_{\Delta t}} \cdots \xrightarrow[m]{\hat{U}_{\Delta t}} |\psi(t)\rangle$$

$$\boxed{\text{Trotterization}} \quad |\psi(0)\rangle \xrightarrow[1]{\hat{V}_{\Delta t}} \xrightarrow[2]{\hat{V}_{\Delta t}} \cdots \xrightarrow[m]{\hat{V}_{\Delta t}} |\overline{\psi}(t)\rangle$$

Fig. 11.3 Trotterization. The time interval $[0, t]$ is divided into m intervals of duration Δt. At each step, the circuit $\hat{V}_{\Delta t} = e^{-i\hat{H}_1\Delta t} \ldots e^{-i\hat{H}_L\Delta t}$ is applied to the register of qubits. This operator is a first-order approximation of $\hat{U}_{\Delta t} = e^{-i(\hat{H}_1 + \ldots + \hat{H}_L)\Delta t}$.

The operator norm $\|\hat{U}_{\Delta t} - \hat{V}_{\Delta t}\|_{\text{o}}$ provides the error of the approximation at each step. By comparing (11.6) with (11.5), we see that the error of the approximation is[4]

$$\begin{aligned} \|\hat{U}_{\Delta t} - \hat{V}_{\Delta t}\|_{\text{o}} &= \left\| \frac{(-i\Delta t)^2}{2} \Big(\sum_{i<j}^{L} \hat{H}_j \hat{H}_i - \sum_{i<j}^{L} \hat{H}_i \hat{H}_j \Big) + O(\Delta t^3) \right\|_{\text{o}} \leq \\ &\leq \frac{1}{2} \sum_{i<j}^{L} \left\| [\hat{H}_j, \hat{H}_i] \right\|_{\text{o}} \Delta t^2 + \|O(\Delta t^3)\|_{\text{o}} = c\Delta t^2 + \|O(\Delta t^3)\|_{\text{o}}, \end{aligned} \tag{11.7}$$

where $c = \frac{1}{2} \sum_{i<j} \|[\hat{H}_j, \hat{H}_i]\|_{\text{o}}$ is a constant. In conclusion, the error at each step is not greater than $c\Delta t^2$ (the contribution $O(\Delta t^3)$ is small and can be ignored).

What is the error of the entire simulation? From eqn (11.7) we know that the error at each step is not greater than $c\Delta t^2$. It is reasonable to conclude that the error of the entire simulation ε scales as $\varepsilon \leq mc\Delta t^2$ where m is the number of steps. Let us rigorously prove this. On one hand, the ideal unitary $\hat{U}_{\Delta t}$ transforms $|\psi_j\rangle$ into $|\psi_{j+1}\rangle = \hat{U}_{\Delta t}|\psi_j\rangle$ where j indicates the step number. On the other hand, the circuit $\hat{V}_{\Delta t}$ transforms $|\psi_j\rangle$ into

$$\begin{aligned} \hat{V}_{\Delta t}|\psi_j\rangle &= (\hat{V}_{\Delta t} - \hat{U}_{\Delta t} + \hat{U}_{\Delta t})|\psi_j\rangle \\ &= \underbrace{(\hat{V}_{\Delta t} - \hat{U}_{\Delta t})|\psi_j\rangle}_{|E_{j+1}\rangle} + \underbrace{\hat{U}_{\Delta t}|\psi_j\rangle}_{|\psi_{j+1}\rangle} = |\psi_{j+1}\rangle + |E_{j+1}\rangle. \end{aligned}$$

The state of the quantum computer at each step will be

$$\begin{aligned} |\overline{\psi}_1\rangle &= \hat{V}_{\Delta t}|\psi_0\rangle = |E_1\rangle + |\psi_1\rangle, \\ |\overline{\psi}_2\rangle &= \hat{V}_{\Delta t}|\overline{\psi}_1\rangle = |\psi_2\rangle + |E_2\rangle + \hat{V}_{\Delta t}|E_1\rangle, \\ &\vdots \\ |\overline{\psi}_m\rangle &= \hat{V}_{\Delta t}|\overline{\psi}_{m-1}\rangle = |\psi_m\rangle + |E_m\rangle + \hat{V}_{\Delta t}|E_{m-1}\rangle + \ldots + \hat{V}_{\Delta t}^{m-1}|E_1\rangle. \end{aligned} \tag{11.8}$$

[4]Here, we used the triangle inequality $\|\hat{A} + \hat{B}\|_{\text{o}} \leq \|\hat{A}\|_{\text{o}} + \|\hat{B}\|_{\text{o}}$ and the relation $\|\sum_i \hat{A}_i\|_{\text{o}} \leq \sum_i \|\hat{A}_i\|_{\text{o}}$.

Rearranging eqn (11.8), one can calculate an upper bound of the error of the entire simulation,

$$\begin{aligned}\varepsilon = \||\overline{\psi}_m\rangle - |\psi_m\rangle\| &= \||E_m\rangle + \hat{V}_{\Delta t}|E_{m-1}\rangle + \ldots + \hat{V}_{\Delta t}^{m-1}|E_1\rangle\| \\ &\leq \||E_m\rangle\| + \|\hat{V}_{\Delta t}|E_{m-1}\rangle\| + \ldots + \|\hat{V}_{\Delta t}^{m-1}|E_1\rangle\| \\ &\leq \||E_m\rangle\| + \||E_{m-1}\rangle\| + \ldots + \||E_1\rangle\|, \end{aligned} \tag{11.9}$$

where we used the triangle inequality and in the last step we used the fact that $\|\hat{V}|\psi\rangle\| = \||\psi\rangle\|$ for any unitary $\hat{V}$. The length of the vectors $|E_j\rangle$ is bounded by

$$\||E_j\rangle\| = \|(\hat{V}_{\Delta t} - \hat{U}_{\Delta t})|\psi_{j-1}\rangle\| \leq \sup_{\psi} \|(\hat{V}_{\Delta t} - \hat{U}_{\Delta t})|\psi\rangle\| = \|\hat{V}_{\Delta t} - \hat{U}_{\Delta t}\|_{\text{o}}, \tag{11.10}$$

where in the last step we used the definition of operator norm. Since the error at each step satisfies $\|\hat{V}_{\Delta t} - \hat{U}_{\Delta t}\|_{\text{o}} \leq c\Delta t^2$, eqn (11.9) indicates that the cumulative error is not greater than $\varepsilon \leq mc\Delta t^2 = ct^2/m$. In summary, if our goal is to simulate $\hat{U}_t$ with an error lower than ε using a Trotterization, the time interval $[0, t]$ must be divided into at least $m = ct^2/\varepsilon$ steps.

How many elementary gates are required to simulate the dynamics with an error smaller than ε? First, recall that the exponentials in the Trotter formula act on no more than k qubits, i.e. they operate on 2^k dimensional subspaces. Each exponential can be implemented with $O(4^k)$ single and two-qubit gates[5] and thus each simulation step requires $O(4^k L)$ elementary gates. Since there are m steps, the overall simulation can be executed with $O(4^k Lm)$ gates. Recall that $m = ct^2/\varepsilon$ steps are necessary to reach a precision ε which implies that the number of gates scales as $n_{\text{gate}} = O(4^k Lct^2/\varepsilon)$. The last step is to explicitly calculate an upper bound of the constant c,

$$\begin{aligned} c &= \frac{1}{2}\sum_{i<j}^{L} \|\hat{H}_j\hat{H}_i - \hat{H}_i\hat{H}_j\|_{\text{o}} \leq \frac{1}{2}\sum_{i<j}^{L} \|\hat{H}_j\|_{\text{o}} \cdot \|\hat{H}_i\|_{\text{o}} + \|\hat{H}_i\|_{\text{o}} \cdot \|\hat{H}_j\|_{\text{o}} \\ &= \frac{1}{2}\sum_{i<j}^{L} 2\|\hat{H}_j\|_{\text{o}} \cdot \|\hat{H}_i\|_{\text{o}} \leq \sum_{i<j}^{L} \left[\max_i \|\hat{H}_i\|_{\text{o}}\right]^2 = \sum_{i<j}^{L} \nu^2 = \frac{L(L-1)}{2}\nu^2 < L^2\nu, \end{aligned}$$

where $\nu = \max_i \|\hat{H}_i\|_{\text{o}}$ is the eigenvalue with maximum magnitude among all terms of the Hamiltonian. Therefore, a first-order Trotterization requires a number of gates that scales as

$$\boxed{n_{\text{gate}} = O(4^k L^3 \nu t^2/\varepsilon).} \qquad \textbf{Trotter (first order)}$$

The number of gates has a quadratic dependence on time and an inverse dependence with respect to the accuracy ε. We will see in the next sections that more advanced methods, such as the Taylor series and qubitization, require asymptotically fewer gates at the cost of adding some extra qubits to the circuit. In particular, the qubitization

[5] See the discussion after the universality theorem in Section 5.5.

reaches the optimal scaling with respect to time and accuracy, $n_{\text{gate}}^{\text{qubitization}} = O(t + \log(1/\varepsilon))$.

Example 11.2 Consider a two-site Ising Hamiltonian,

$$\hat{H} = \omega_1 \hat{\sigma}_x \otimes \mathbb{1} + \omega_2 \mathbb{1} \otimes \hat{\sigma}_x + g \hat{\sigma}_z \otimes \hat{\sigma}_z,$$

where ω_1, ω_2, and g are real numbers. Suppose that the initial state is $|\phi(0)\rangle = |\uparrow\downarrow\rangle$. Let us simulate the time evolution using a Trotterization. The simulation can be performed on a quantum computer with two qubits. The quantum computer is prepared in the state $|\psi(0)\rangle = |10\rangle$ and the time interval $[0, t]$ is divided into m intervals of duration Δt. The time evolution operator $\hat{U}_{\Delta t} = e^{-i\hat{H}\Delta t}$ can be approximated with the quantum circuit

$$\hat{V}_{\Delta t} = e^{-ig(\hat{\sigma}_z \otimes \hat{\sigma}_z)\Delta t} e^{-i\omega_1(\hat{\sigma}_x \otimes \mathbb{1})\Delta t} e^{-i\omega_2(\mathbb{1} \otimes \hat{\sigma}_x)\Delta t}. \tag{11.11}$$

The two operators on the right can be implemented with single-qubit rotations as illustrated in Fig. 11.4,

$$e^{-i\omega_1(\hat{\sigma}_x \otimes \mathbb{1})\Delta t} = \mathsf{X}_{2\omega_1 \Delta t} \otimes \mathbb{1},$$
$$e^{-i\omega_2(\mathbb{1} \otimes \hat{\sigma}_x)\Delta t} = \mathbb{1} \otimes \mathsf{X}_{2\omega_2 \Delta t}.$$

In eqn (11.11), the operator $e^{-ig(\hat{\sigma}_z \otimes \hat{\sigma}_z)\Delta t}$ can be implemented with two CNOT gates and a single-qubit rotation about the z axis,

$$e^{-ig(\hat{\sigma}_z \otimes \hat{\sigma}_z)\Delta t} = \begin{bmatrix} e^{-ig\Delta t} & 0 & 0 & 0 \\ 0 & e^{ig\Delta t} & 0 & 0 \\ 0 & 0 & e^{ig\Delta t} & 0 \\ 0 & 0 & 0 & e^{-ig\Delta t} \end{bmatrix} = \mathsf{c}_1 X_0 (\mathbb{1} \otimes \mathsf{Z}_{2g\Delta t}) \mathsf{c}_1 X_0.$$

The quantum circuit $\hat{V}_{\Delta t}$ is applied m times to the initial state. The final state $|\overline{\psi}(t)\rangle = \hat{V}_{\Delta t}^m |10\rangle = c_{00}|00\rangle + c_{01}|01\rangle + c_{10}|10\rangle + c_{11}|11\rangle$ is measured multiple times until the probabilities $|c_{ij}|^2$ are estimated with the desired accuracy.

11.2 Trotterization at higher orders

There are several ways to improve the standard Trotterization method. The simplest one is based on the symmetrized decomposition [340, 341]. The **symmetrized decomposition** approximates the dynamics using fewer elementary gates for a fixed accuracy. For simplicity, let us consider a Hamiltonian composed of only two non-commuting terms,

$$\boxed{\hat{H} = \hat{H}_1 + \hat{H}_2.}$$

As usual, the time interval $[0, t]$ is divided into m_s steps of duration Δt. The exact time evolution operator for a time interval Δt is

$$\hat{U}_{\Delta t} = e^{-i(\hat{H}_1 + \hat{H}_2)\Delta t}.$$

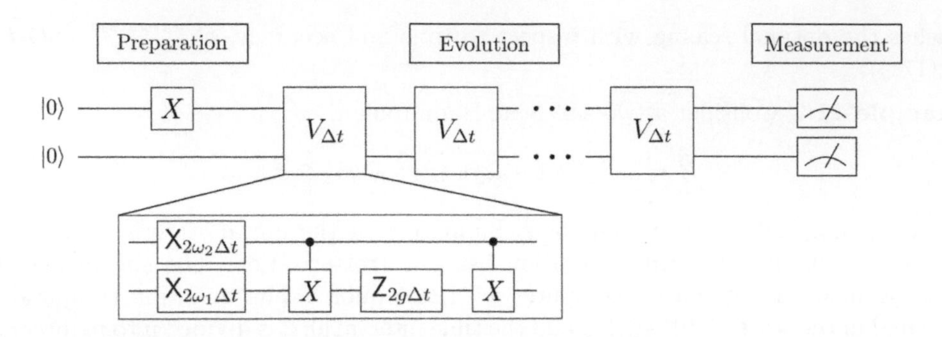

Fig. 11.4 Example of a Trotterization. The quantum circuit $\hat{V}_{\Delta t}$ is applied to the quantum computer m times. Since $\hat{V}_{\Delta t}$ requires five elementary gates, the entire simulation requires $5m$ gates where $m = ct^2/\varepsilon$.

We can implement this operator on a quantum computer using a circuit $\hat{V}_{\Delta t}^{\text{Sym}}$ based on the symmetrized decomposition, which is nothing but a Trotter formula that approximates the time evolution operator at higher orders,

$$\boxed{\hat{V}_{\Delta t}^{\text{Sym}} = e^{-i\hat{H}_1\frac{\Delta t}{2}}e^{-i\hat{H}_2\Delta t}e^{-i\hat{H}_1\frac{\Delta t}{2}}.} \tag{11.12}$$

Let us quantify the error of this approximation. The Taylor expansion of $\hat{V}_t$ to third order can be calculated by multiplying together three terms,

$$\begin{aligned}\hat{V}_{\Delta t}^{\text{Sym}} \approx &\left[\mathbb{1} + \hat{H}_1\frac{\Delta t/2}{i} + \frac{1}{2}\hat{H}_1^2\left(\frac{\Delta t/2}{i}\right)^2 + \frac{1}{6}\hat{H}_1^3\left(\frac{\Delta t/2}{i}\right)^3\right]\\ &\left[\mathbb{1} + \hat{H}_2\frac{\Delta t}{i} + \frac{1}{2}\hat{H}_2^2\left(\frac{\Delta t}{i}\right)^2 + \frac{1}{6}\hat{H}_2^3\left(\frac{\Delta t}{i}\right)^3\right]\\ &\left[\mathbb{1} + \hat{H}_1\frac{\Delta t/2}{i} + \frac{1}{2}\hat{H}_1^2\left(\frac{\Delta t/2}{i}\right)^2 + \frac{1}{6}\hat{H}_1^3\left(\frac{\Delta t/2}{i}\right)^3\right].\end{aligned}$$

The product of these terms leads to

$$\begin{aligned}\hat{V}_{\Delta t}^{\text{Sym}} \approx\ &\mathbb{1} + \frac{\Delta t}{i}(\hat{H}_1 + \hat{H}_2) + \frac{1}{2}\left(\frac{\Delta t}{i}\right)^2(\hat{H}_1 + \hat{H}_2)^2\\ &+ \frac{1}{6}\left(\frac{\Delta t}{i}\right)^3\left[\hat{H}_1^3 + \frac{3}{2}\left(\hat{H}_1\hat{H}_2^2 + \hat{H}_2^2\hat{H}_1 + \hat{H}_1\hat{H}_2\hat{H}_1\right) + \frac{3}{4}\left(\hat{H}_2\hat{H}_1^2 + \hat{H}_1^2\hat{H}_2\right) + \hat{H}_2^3\right].\end{aligned} \tag{11.13}$$

The Taylor expansion of the exact time evolution operator is

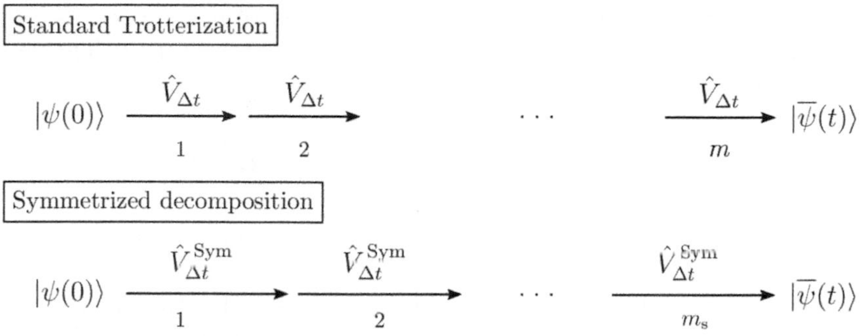

Standard Trotterization

$$|\psi(0)\rangle \xrightarrow[1]{\hat{V}_{\Delta t}} \xrightarrow[2]{\hat{V}_{\Delta t}} \cdots \xrightarrow[m]{\hat{V}_{\Delta t}} |\overline{\psi}(t)\rangle$$

Symmetrized decomposition

$$|\psi(0)\rangle \xrightarrow[1]{\hat{V}^{\text{Sym}}_{\Delta t}} \xrightarrow[2]{\hat{V}^{\text{Sym}}_{\Delta t}} \cdots \xrightarrow[m_{\text{s}}]{\hat{V}^{\text{Sym}}_{\Delta t}} |\overline{\psi}(t)\rangle$$

Fig. 11.5 Symmetrized decomposition. The standard Trotterization approximates the time evolution operator $\hat{U}_{\Delta t} = e^{-i(\hat{H}_1+\ldots+\hat{H}_L)\Delta t}$ with the quantum circuit $\hat{V}_{\Delta t} = e^{-i\hat{H}_1\Delta t}\ldots e^{-i\hat{H}_L\Delta t}$. The method based on the symmetrized decomposition uses instead $\hat{V}^{\text{Sym}}_{\Delta t} = (e^{-i\hat{H}_1\frac{\Delta t}{2}}\ldots e^{-i\hat{H}_L\frac{\Delta t}{2}})(e^{-i\hat{H}_L\frac{\Delta t}{2}}\ldots e^{-i\hat{H}_1\frac{\Delta t}{2}})$. The number of steps and the total number of gates required to simulate the dynamics for a fixed accuracy ε and simulation time t is lower for the symmetrized decomposition.

$$\hat{U}_{\Delta t} \approx \mathbb{1} + \frac{\Delta t}{i}(\hat{H}_1 + \hat{H}_2) + \frac{1}{2}\left(\frac{\Delta t}{i}\right)^2 (\hat{H}_1 + \hat{H}_2)^2 + \frac{1}{6}\left(\frac{\Delta t}{i}\right)^3 (\hat{H}_1 + \hat{H}_2)^3. \quad (11.14)$$

Comparing (11.14) with (11.13), we see that the error at each step is given by

$$\|\hat{U}_{\Delta t} - \hat{V}^{\text{Sym}}_{\Delta t}\|_{\text{o}} \leq \underbrace{\frac{1}{24}\left\|\hat{H}_2^2\hat{H}_1 + \hat{H}_1\hat{H}_2^2 + 2\hat{H}_1\hat{H}_2\hat{H}_1 - 4\hat{H}_2\hat{H}_1\hat{H}_2\right\|_{\text{o}}}_{c_{\text{s}}} \Delta t^3 = c_{\text{s}}\Delta t^3$$

where c_{s} is a constant[6]. The error of the entire simulation is bounded by

$$\boxed{\varepsilon \leq m_{\text{s}} c_{\text{s}} \Delta t^3 = c_{\text{s}}\frac{t^3}{m_{\text{s}}^2},}$$

where we used $\Delta t = t/m_{\text{s}}$. In summary, if our goal is to simulate $\hat{U}_t$ with an error not greater than ε using a symmetrized decomposition, the time interval $[0, t]$ must be divided into at least $m_{\text{s}} = \sqrt{c_{\text{s}}}t^{1.5}/\varepsilon^{0.5}$ steps. The number of gates required to implement the operator $\hat{V}^{\text{Sym}}_{\Delta t}$ is $2n_1 + n_2$, where n_1 (n_2) is the number of elementary gates needed to decompose $e^{-i\hat{H}_1\Delta t/2}$ ($e^{-i\hat{H}_2\Delta t}$). Thus, the number of gates required to simulate the entire time evolution is

$$\boxed{n^{\text{Sym}}_{\text{gates}} = (2n_1 + n_2)m_{\text{s}} = (2n_1 + n_2)\sqrt{c_{\text{s}}}\frac{t^{1.5}}{\varepsilon^{0.5}} = O(t^{1.5}/\varepsilon^{0.5}).} \qquad \textbf{Sym. decomp.} \quad (11.15)$$

[6]Here, we are ignoring higher contributions. Proving that these contributions are negligible is quite complicated. For a more rigorous analysis, see Refs [336, 342].

As a comparison, let us compute the number of gates needed to simulate the dynamics with the standard Trotterization method presented in Section 11.1. The quantum circuit $e^{-i\hat{H}_1\Delta t}e^{-i\hat{H}_2\Delta t}$ is implemented with $n_1 + n_2$ gates and therefore

$$n_{\text{gates}}^{\text{Trott}} = (n_1 + n_2)m = (n_1 + n_2)\frac{ct^2}{\varepsilon} = O(t^2/\varepsilon).$$

This shows that the symmetrized decomposition requires fewer gates, $n_{\text{gates}}^{\text{Sym}}/n_{\text{gates}}^{\text{Trott}} = O(\varepsilon^{0.5}/t^{0.5})$. This concept is illustrated schematically in Fig. 11.5.

In this section, we only considered Hamiltonians expressed as the sum of two terms. The generalization to Hamiltonians with L terms is straightforward. The quantum circuit that approximates the time evolution operator $\hat{U}_{\Delta t} = e^{-i(\hat{H}_1+\ldots+\hat{H}_L)\Delta t}$ is given by

$$\hat{V}_{\Delta t}^{\text{Sym}} = \left(e^{-i\hat{H}_1\frac{\Delta t}{2}}\ldots e^{-i\hat{H}_L\frac{\Delta t}{2}}\right)\left(e^{-i\hat{H}_L\frac{\Delta t}{2}}\ldots e^{-i\hat{H}_1\frac{\Delta t}{2}}\right). \tag{11.16}$$

The number of gates scales more favorably than the standard Trotterization as a function of the accuracy and simulation time.

11.3 Trotterization of sparse Hamiltonians

In many practical cases, the Hamiltonian of the system is not provided as a sum of a polynomial number of terms as in eqn (11.1). Instead, we only have access to the **matrix** associated with the Hamiltonian. A straightforward method to simulate the dynamics consists of decomposing the matrix associated with the Hamiltonian into a sum of Hermitian matrices and then proceeding with a standard Trotterization [343, 336]. In this section, we will focus on Hamiltonians whose associated matrix is a d-sparse matrix with real components[7]. A d-sparse matrix is a matrix in which each row has at most d non-zero elements. First, let us write the Hamiltonian in the computational basis,

$$\hat{H} = \sum_{i=0}^{2^n-1}\sum_{j=0}^{2^n-1} h_{i,j}|i\rangle\langle j|,$$

where $h_{i,j} = h_{j,i}$ because we have assumed $\hat{H}$ to be Hermitian with real entries. We will assume that the matrix elements $h_{i,j}$ can be approximated with binary strings of length p. The Hamiltonian can be written as the sum of two terms: the diagonal part and the off-diagonal part,

$$\hat{H} = \underbrace{\sum_{i=0}^{2^n-1} h_{i,i}|i\rangle\langle i|}_{\hat{H}_{\text{d}}} + \underbrace{\sum_{i<j}^{2^n-1} h_{i,j}|i\rangle\langle j| + h_{j,i}|j\rangle\langle i|}_{\hat{H}_{\text{o}}} = \hat{H}_{\text{d}} + \hat{H}_{\text{o}}.$$

The diagonal part can be simulated in a simple way and for this reason it will be ignored (see Exercise 11.5). To better understand how to simulate the off-diagonal

[7]We deal with these types of matrices because in quantum chemistry the matrix associated with the Hamiltonian is often a d-sparse matrix with real components [344, 345].

part $\hat{H}_{\text{o}}$, we will consider a concrete example. Suppose that $\hat{H}_{\text{o}}$ acts on $n = 3$ qubits and the associated matrix is

$$H_{\text{o}} = \begin{bmatrix} 0 & 0 & 0 & h_{0,3} & 0 & 0 & 0 & 0 \\ 0 & 0 & 0 & h_{1,3} & h_{1,4} & h_{1,5} & 0 & 0 \\ 0 & 0 & 0 & 0 & h_{2,4} & h_{2,5} & 0 & 0 \\ h_{3,0} & h_{3,1} & 0 & 0 & 0 & 0 & h_{3,6} & 0 \\ 0 & h_{4,1} & h_{4,2} & 0 & 0 & 0 & 0 & h_{4,7} \\ 0 & h_{5,1} & h_{5,2} & 0 & 0 & 0 & 0 & h_{5,7} \\ 0 & 0 & 0 & h_{6,3} & 0 & 0 & 0 & 0 \\ 0 & 0 & 0 & 0 & h_{7,4} & h_{7,5} & 0 & 0 \end{bmatrix}. \tag{11.17}$$

This Hamiltonian is 3-sparse. It is tempting to decompose $\hat{H}_{\text{o}}$ into the Pauli basis

$$\hat{H}_{\text{o}} = \sum_{i=1}^{4^n} c_i \hat{P}_i,$$

where $c_i = \text{Tr}[\hat{P}_i \hat{H}]$, and then proceed with a Trotterization. The problem is that the Pauli decomposition of a generic d-sparse matrix contains an exponential number of terms[8]. This does not allow us to efficiently simulate the dynamics using the Trotter formula $e^{-i\Delta t \sum_i c_i \hat{P}_i \Delta t} \approx e^{-i\Delta t c_1 \hat{P}_1} \dots e^{-i\Delta t c_{4^n} \hat{P}_{4^n}}$ because the decomposition has too many terms. We first need to find a decomposition of $\hat{H}_{\text{o}}$ with a polynomial number of terms.

An efficient decomposition of a d-sparse Hamiltonian can be obtained by noticing that a graph $G = (V, E)$ can be associated with the matrix H_{o}. The number of nodes in the graph is $|V| = 2^n$, i.e. the number of rows in H_{o}. Two nodes of the graph a and b are connected by an edge $e = \{a, b\}$ if and only if $h_{a,b} \neq 0$. Since H_{o} is symmetric and has at most d non-zero elements per row, the number of edges connected to each node is not greater than d. The graph associated with H_{o} of eqn (11.17) is shown in Fig. 11.6. Since each edge corresponds to a matrix element, one can express the Hamiltonian as

$$\hat{H}_{\text{o}} = \sum_{(a,b)\in E} h_{a,b}|a\rangle\langle b| + h_{b,a}|b\rangle\langle a|.$$

Let us color the edges of the graph with a set of colors c such that edges connected to the same node have different colors[9]. In our example, $c(a, b) \in \{\text{r}, \text{b}, \text{g}\}$ where "r", "b", and "g" stand for red, blue, and green. Using this notation, we have

[8]Consider for example the n-qubit operator $|\mathbf{x}\rangle\langle\mathbf{y}| = \otimes_{i=1}^{n} |x_i\rangle\langle y_i|$. Each dyad $|x_i\rangle\langle y_i|$ is a linear combination of two Pauli operators: $|0\rangle\langle 0| = (\mathbb{1} + \hat{\sigma}_z)/2$, $|1\rangle\langle 1| = (\mathbb{1} - \hat{\sigma}_z)/2$, $|0\rangle\langle 1| = (\hat{\sigma}_x + i\hat{\sigma}_y)/2$ and $|1\rangle\langle 0| = (\hat{\sigma}_x - i\hat{\sigma}_y)/2$). Therefore, $|\mathbf{x}\rangle\langle\mathbf{y}|$ is a linear combination of 2^n Pauli operators.

[9]Vizing's theorem states that if G is a graph whose nodes have maximum degree d, then $d + 1$ colors are sufficient to color the graph. There are multiple algorithms to *efficiently* color the edges of a graph, see for example [346].

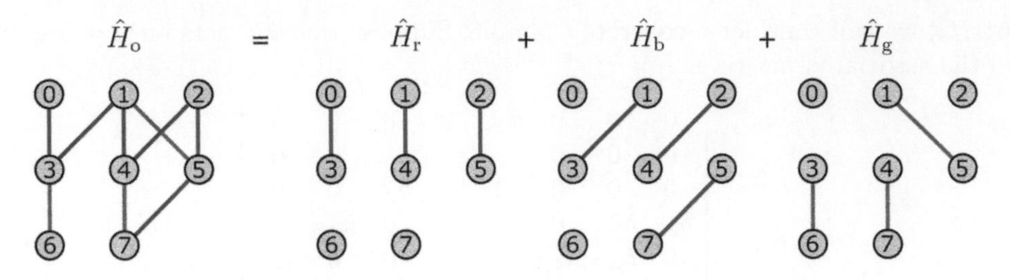

Fig. 11.6 Example of a graph associated with a sparse Hamiltonian. Graph associated with the Hamiltonian (11.17). The graph has eight nodes, one for each matrix row. The matrix elements in the upper-right triangle $h_{0,3}$, $h_{1,3}$, $h_{1,4}$, $h_{1,5}$, $h_{2,4}$, $h_{2,5}$, $h_{3,6}$, $h_{4,7}$, and $h_{5,7}$, are non-zero. Hence, the graph contains nine edges. The segments are colored such that each node is connected to edges with different colors. According to our notation, $c(1,4) = \text{r}$, $v_\text{b}(3) = 1$, and $h_{4,v_\text{b}(4)} = h_{4,2}$.

$$\hat{H}_\text{o} = \sum_{c\in\{\text{r,b,g}\}} \underbrace{\sum_{c(a,b)=c} h_{a,b}|a\rangle\langle b| + h_{b,a}|b\rangle\langle a|}_{\hat{H}_c} = \sum_{c\in\{\text{r,b,g}\}} \hat{H}_c.$$

We have decomposed the Hamiltonian as the sum of three Hamiltonians

$$\boxed{\hat{H}_\text{o} = \hat{H}_\text{r} + \hat{H}_\text{b} + \hat{H}_\text{g},} \tag{11.18}$$

where:

$$\begin{aligned}
\hat{H}_\text{r} &= h_{0,3}|0\rangle\langle 3| + h_{1,4}|1\rangle\langle 4| + h_{2,5}|2\rangle\langle 5| + \text{h.c.},\\
\hat{H}_\text{b} &= h_{1,3}|1\rangle\langle 3| + h_{2,4}|2\rangle\langle 4| + h_{5,7}|5\rangle\langle 7| + \text{h.c.},\\
\hat{H}_\text{g} &= h_{1,5}|1\rangle\langle 5| + h_{3,6}|3\rangle\langle 6| + h_{4,7}|4\rangle\langle 7| + \text{h.c.}
\end{aligned}$$

We note that the Pauli decomposition of $\hat{H}_\text{o}$ would have required 26 terms and not just three.

Now that we have an efficient decomposition of $\hat{H}_\text{o}$, we need to demonstrate that the time evolution governed by the Hamiltonian $\hat{H}_\text{o} = \hat{H}_\text{r} + \hat{H}_\text{b} + \hat{H}_\text{g}$ can be simulated with polynomial resources. We will use the notation $v_c(a)$ to indicate a node connected to a by an edge of color c (for example, $v_\text{g}(3) = 6$). To simulate the time evolution operator $e^{-i\hat{H}_\text{o}t}$ we will use a register of n qubits, one register with n ancilla qubits, and another register with p ancilla qubits (where p is the number of bits needed to approximate the elements $h_{a,b}$). Thus, the basis of the Hilbert space is given by $\{|i,j,k\rangle\}_{i,j,k}$, where $i \in \mathbb{B}^n$, $j \in \mathbb{B}^n$, $k \in \mathbb{B}^p$. Let us now introduce three unitary operators $\hat{V}_c$, one per color. These transformations operate on the basis vectors in the following way,

$$\boxed{\hat{V}_c|a\rangle\,|b\rangle\,|z\rangle = |a\rangle\,|b \oplus v_c(a)\rangle\,|z \oplus h_{a,v_c(a)}\rangle,} \tag{11.19}$$

where $\oplus$ denotes bitwise addition modulo 2 and the letters a and b label the nodes. Note that $\hat{V}_c^\dagger = \hat{V}_c$. Let us define the Hermitian operator $\hat{S}$ as

$$\boxed{\hat{S}|a\rangle\,|b\rangle\,|z\rangle = z|b\rangle\,|a\rangle\,|z\rangle.}$$

This operator is simply a sequence of SWAP gates and a diagonal transformation on the p ancilla qubits. These definitions allow us to introduce a new Hamiltonian

$$\boxed{\hat{\mathbf{H}}_\mathrm{o} = \hat{V}_\mathrm{r}^\dagger \hat{S} \hat{V}_\mathrm{r} + \hat{V}_\mathrm{b}^\dagger \hat{S} \hat{V}_\mathrm{b} + \hat{V}_\mathrm{g}^\dagger \hat{S} \hat{V}_\mathrm{g}.}$$

The main difference between the original Hamiltonian $\hat{H}_\mathrm{o}$ and $\hat{\mathbf{H}}_\mathrm{o}$ is that the former operates on n qubits, while the latter acts on $n+n+p$ qubits. These two Hamiltonians act in the same way on the top register. To see this, let us show that $\hat{\mathbf{H}}_\mathrm{o}|a,0,0\rangle = \hat{H}_\mathrm{o} \otimes \mathbb{1} \otimes \mathbb{1}|a,0,0\rangle$. Indeed, we have

$$\begin{aligned}
\hat{\mathbf{H}}_\mathrm{o}|a,0,0\rangle &= \sum_{c\in\{\mathrm{r,g,b}\}} \hat{V}_c^\dagger \hat{S} \hat{V}_c|a,0,0\rangle = \sum_{c\in\{\mathrm{r,g,b}\}} \hat{V}_c^\dagger \hat{S}|a, v_c(a), h_{a,v_c(a)}\rangle \\
&= \sum_{c\in\{\mathrm{r,g,b}\}} h_{a,v_c(a)} \hat{V}_c^\dagger |v_c(a), a, h_{a,v_c(a)}\rangle \\
&= \sum_{c\in\{\mathrm{r,g,b}\}} h_{a,v_c(a)} |v_c(a), a \oplus v_c(v_c(a)), h_{a,v_c(a)} \oplus h_{a,v_c(a)}\rangle \\
&= \sum_{c\in\{\mathrm{r,g,b}\}} h_{a,v_c(a)} |v_c(a), 0, 0\rangle = (\hat{H}_\mathrm{o} \otimes \mathbb{1} \otimes \mathbb{1})|a,0,0\rangle,
\end{aligned}$$

where we used $v_c(v_c(a)) = a$ and in the last step we used the relation $h_{a,v_c(a)}|v_c(a)\rangle = \hat{H}_c|a\rangle$. The time evolution operator $e^{-i\hat{\mathbf{H}}_\mathrm{o}\Delta t}$ can be approximated with a Trotterization

$$\begin{aligned}
e^{-i\hat{\mathbf{H}}_\mathrm{o}\Delta t} &= e^{-i(\hat{V}_\mathrm{r}^\dagger \hat{S}\hat{V}_\mathrm{r} + \hat{V}_\mathrm{g}^\dagger \hat{S}\hat{V}_\mathrm{g} + \hat{V}_\mathrm{b}^\dagger \hat{S}\hat{V}_\mathrm{b})\Delta t} \\
&\approx e^{-i\hat{V}_\mathrm{r}^\dagger \hat{S}\hat{V}_\mathrm{r}\Delta t}\, e^{-i\hat{V}_\mathrm{b}^\dagger \hat{S}\hat{V}_\mathrm{b}\Delta t}\, e^{-i\hat{V}_\mathrm{g}^\dagger \hat{S}\hat{V}_\mathrm{g}\Delta t} \\
&= \hat{V}_\mathrm{r}^\dagger e^{-i\hat{S}\Delta t}\hat{V}_\mathrm{r}\; \hat{V}_\mathrm{b}^\dagger e^{-i\hat{S}\Delta t}\hat{V}_\mathrm{b}\; \hat{V}_\mathrm{g}^\dagger e^{-i\hat{S}\Delta t}\hat{V}_\mathrm{g},
\end{aligned} \tag{11.20}$$

where in the last step we used the identity $e^{i\hat{U}\hat{A}\hat{U}^\dagger} = \hat{U}e^{i\hat{A}}\hat{U}^\dagger$ valid for any unitary $\hat{U}$. Equation (11.20) can be expressed in a more concise form as

$$\boxed{e^{-i\hat{\mathbf{H}}_\mathrm{o}\Delta t} \approx \hat{V}_{\Delta t} = \prod_\mathrm{c} \hat{V}_\mathrm{c}^\dagger e^{-i\hat{S}\Delta t}\hat{V}_\mathrm{c}.} \tag{11.21}$$

The unitary operators in (11.21) can be decomposed with a polynomial number of elementary gates [347]. After m applications of the quantum circuit $\hat{V}_{\Delta t}$, the final state $|\psi(t)\rangle$ will be encoded in the third register (the line in the upper part of Fig. 11.7). The method presented in this section shows that a d-sparse matrix with real components can be simulated efficiently. The generalization to d-sparse matrices with complex

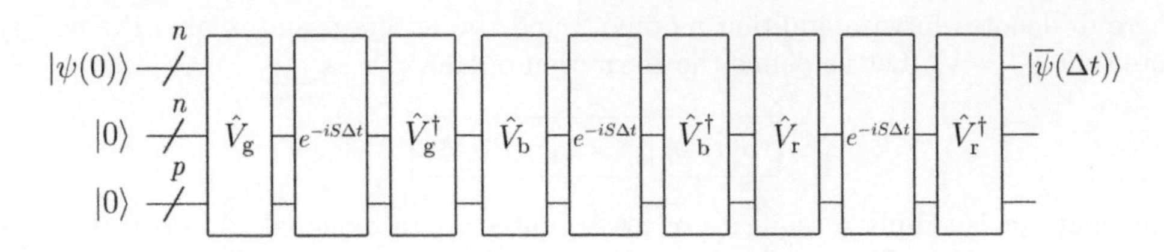

Fig. 11.7 Simulation of a sparse Hamiltonian. Simulation of the time evolution operator $e^{-i\hat{H}_\text{o}\Delta t}$ where $\hat{H}_\text{o} = \hat{H}_\text{r} + \hat{H}_\text{b} + \hat{H}_\text{g}$. The n-qubit register in the upper part of the circuit starts in $|\psi(0)\rangle$ and ends up in $|\overline{\psi}(\Delta t)\rangle = e^{-i\hat{H}_\text{r}\Delta t}e^{-i\hat{H}_\text{b}\Delta t}e^{-i\hat{H}_\text{g}\Delta t}|\psi(0)\rangle$.

---

entries is explained in Ref. [347].

Example 11.3 Consider a Hamiltonian acting on two qubits and its associated graph

$$H_\text{o} = \begin{bmatrix} 0 & 1 & 1 & 0 \\ 1 & 0 & 0 & 1 \\ 1 & 0 & 0 & 1 \\ 0 & 1 & 1 & 0 \end{bmatrix}$$

The Hamiltonian can be decomposed as $\hat{H}_\text{o} = \hat{H}_\text{r} + \hat{H}_\text{b}$ where $\hat{H}_\text{r} = |0\rangle\langle 2| + |2\rangle\langle 0| + |1\rangle\langle 3| + |3\rangle\langle 1|$, $\hat{H}_\text{b} = |0\rangle\langle 1| + |1\rangle\langle 0| + |2\rangle\langle 3| + |3\rangle\langle 2|$. These expressions can be written in binary representation as

$$\hat{H}_\text{r} = |00\rangle\langle 10| + |10\rangle\langle 00| + |01\rangle\langle 11| + |11\rangle\langle 01|,$$
$$\hat{H}_\text{b} = |00\rangle\langle 01| + |01\rangle\langle 00| + |10\rangle\langle 11| + |11\rangle\langle 10|.$$

Let us simulate the dynamics using three registers: one with $n = 2$ qubits, one with $n = 2$ ancilla qubits and one with $p = 1$ ancilla (the third register has only one ancilla qubit because the entries of H_o are either 0 or 1). The quantum circuit $\hat{V}_{\Delta t}$ that simulates the evolution for a time interval Δt is shown in Fig. 11.8. In this figure, the two n-qubit registers are interleaved, and the ancilla register is represented with a line at the bottom of the circuit. Using eqn (11.21), the quantum circuit $\hat{V}_{\Delta t}$ is given by

$$\hat{V}_{\Delta t} = \hat{V}_\text{b}^\dagger e^{-i\hat{S}\Delta t}\hat{V}_\text{b}\hat{V}_\text{r}^\dagger e^{-i\hat{S}\Delta t}\hat{V}_\text{r}, \tag{11.22}$$

where the operators $\hat{V}_\text{c}$ are defined as in eqn (11.19),

$$\hat{V}_\text{c}|a\rangle|b\rangle|z\rangle = |a\rangle|b \oplus v_\text{c}(a)\rangle|z \oplus h_{a,v_\text{c}(a)}\rangle.$$

Here, a and b are bit strings of length two. As shown in Fig. 11.8, the operators $\hat{V}_\text{r}$ and $\hat{V}_\text{b}$ can be implemented with some CNOT and NOT gates (revisit Section 10.1 to understand how to decompose $\hat{V}_\text{r}$ and $\hat{V}_\text{b}$ into elementary gates). The transformation $\hat{S}$ swaps the content of the n-qubit registers and leaves the ancilla qubit unchanged,

$$\hat{S} = \mathsf{SWAP} \otimes \mathsf{SWAP} \otimes \mathbb{1}.$$

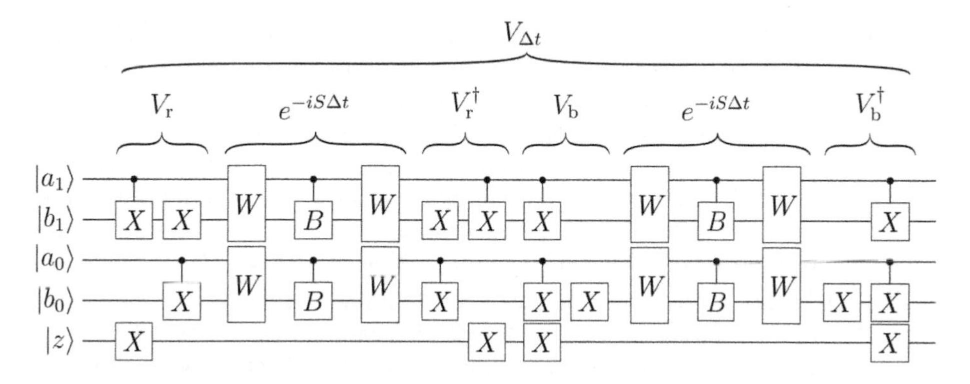

Fig. 11.8 Example of a Trotterization. This quantum circuit implements $\hat{V}_{\Delta t}$ of eqn (11.22). The two $n = 2$ qubit registers are interleaved so it is easier to represent the transformation $\hat{W}$. The line at the bottom of the circuit represents the ancilla qubit.

To implement $e^{-i\hat{S}\Delta t}$, we first need to diagonalize the SWAP gate with a change of basis, $\mathsf{SWAP} = \hat{W}\hat{D}\hat{W}^\dagger$ where

$$W = \begin{bmatrix} 1 & 0 & 0 & 0 \\ 0 & 1/\sqrt{2} & 1/\sqrt{2} & 0 \\ 0 & 1/\sqrt{2} & -1/\sqrt{2} & 0 \\ 0 & 0 & 0 & 1 \end{bmatrix}, \qquad D = \begin{bmatrix} 1 & 0 & 0 & 0 \\ 0 & 1 & 0 & 0 \\ 0 & 0 & -1 & 0 \\ 0 & 0 & 0 & 1 \end{bmatrix}.$$

Therefore, the exponential $e^{-i\hat{S}\Delta t}$ in (11.22) can be implemented as

$$\begin{aligned} e^{-i\hat{S}\Delta t} &= e^{-i(\mathsf{SWAP}\otimes\mathsf{SWAP}\otimes\mathbb{1})\Delta t} = e^{-i(\hat{W}\hat{D}\hat{W}^\dagger\otimes\hat{W}\hat{D}\hat{W}^\dagger\otimes\mathbb{1})\Delta t} \\ &= \hat{W}e^{-i\hat{D}\Delta t}\hat{W}^\dagger \otimes \hat{W}e^{-i\hat{D}\Delta t}\hat{W}^\dagger \otimes \mathbb{1}. \end{aligned}$$

Lastly, since D is diagonal, the operator $e^{-i\hat{D}\Delta t}$ can be implemented with a controlled operation $\mathsf{c}B$ where

$$B = \begin{bmatrix} e^{i2\Delta t} & 0 \\ 0 & 1 \end{bmatrix}.$$

As usual, the circuit $\hat{V}_{\Delta t}$ must be applied m times before a measurement in the computational basis. Note that this Hamiltonian can be simulated in a much more efficient way: one can decompose $\hat{H}_{\rm o}$ in the Pauli basis, $\hat{H}_{\rm o} = \mathbb{1} \otimes \hat{\sigma}_x + \hat{\sigma}_x \otimes \mathbb{1}$, and simulate the dynamics by applying two single-qubit rotations to a register of two qubits, $V_t = e^{-i(\mathbb{1}\otimes\hat{\sigma}_x+\hat{\sigma}_x\otimes\mathbb{1})t} = \mathsf{X}_{2t} \otimes \mathsf{X}_{2t}$.

11.4 LCU: linear combination of unitaries

In the previous sections, we learned how to simulate the dynamics generated by local Hamiltonians and sparse Hamiltonians using the Trotterization method. This process requires a number of gates that scales as $O(t^2/\varepsilon)$ for the standard Trotterization method and as $O(t^{1.5}/\varepsilon^{0.5})$ for the symmetrized decomposition. Is it possible to simulate the dynamics with fewer gates as a function of t and ε? In the last two decades, new simulation techniques with better asymptotic scaling than the Trotterization have been discovered, including the qubitization [28] and algorithms based on the Taylor series [27]. Both these methods use a radically different approach from Trotterization. In a Trotterization, the time evolution operator $\hat{U}_t$ is approximated by another unitary operator $\hat{V}_t$. Instead, the Taylor series and the qubitization method simulate the time evolution operator with a linear combination of unitaries $\tilde{U} = \sum_j c_j \hat{V}_j$. The substantial difference is that, in general, $\tilde{U}$ is not unitary. How can one implement a non-unitary operator on a quantum computer? This can be done by exploiting a block encoding of $\tilde{U}$ [348, 349], i.e. by finding a unitary operator $\hat{G}$ acting on a larger vector space having $\tilde{U}$ as a sub-block. Let us pause for a second and explain these concepts in more detail. We will then carry on with our presentation about simulation methods.

Block encoding We first assume that $\tilde{U}$ is a generic operator. We will then make the explicit assumption that $\tilde{U}$ is a linear combination of unitaries. Given a generic operator $\tilde{U}$ acting on n qubits and a real number $s > 0$, a **block encoding** of $\tilde{U}$ is a unitary operator $\hat{G}$ acting on $n + p$ qubits satisfying the relation [350]

$$\boxed{\langle a|\tilde{U}|b\rangle = s\langle \mathbf{0}, a|\hat{G}|\mathbf{0}, b\rangle}$$

where $|\mathbf{0}\rangle$ is the state of p ancilla qubits and $|a\rangle$, $|b\rangle$ are two generic n-qubit states. The block encoding $\hat{G}$ can be used to apply $\tilde{U}$ to a state $|\psi\rangle$ as shown in the figure below:

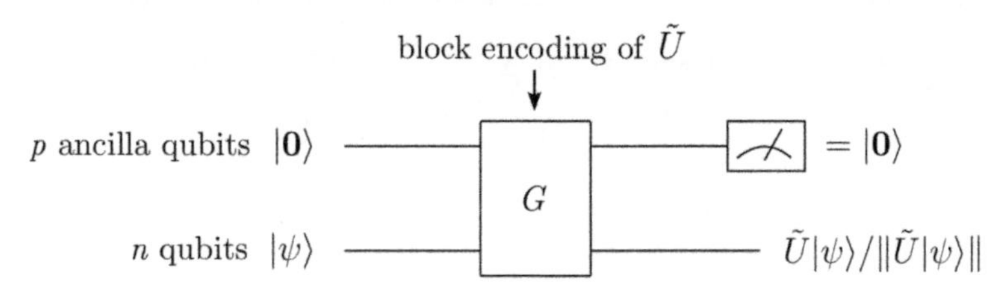

At the end of the circuit, the state of the n qubits collapses into a state proportional to $\tilde{U}|\psi\rangle$ only if the ancilla qubits are measured in $|\mathbf{0}\rangle$. Let us show that this is indeed the case. The ancilla qubits are projected onto the state $|\mathbf{0}\rangle$ with probability

$$\begin{aligned}
p(\mathbf{0}) &= \mathrm{Tr}\left[(|\mathbf{0}\rangle\langle\mathbf{0}| \otimes \mathbb{1})(\hat{G}|\mathbf{0}\psi\rangle\langle\mathbf{0}\psi|\hat{G}^\dagger)\right] = \sum_{j=1}^{2^n} \mathrm{Tr}\left[(|\mathbf{0}j\rangle\langle\mathbf{0}j|)(\hat{G}|\mathbf{0}\psi\rangle\langle\mathbf{0}\psi|\hat{G}^{\dagger)}\right] \\
&= \sum_{j=1}^{2^n} \langle\mathbf{0}j|\hat{G}|\mathbf{0}\psi\rangle\langle\mathbf{0}\psi|\hat{G}^\dagger|\mathbf{0}j\rangle = \sum_{j=1}^{2^n} |\langle\mathbf{0}j|\hat{G}|\mathbf{0}\psi\rangle|^2 \\
&= \sum_{j=1}^{2^n} \frac{1}{s^2}|\langle j|\tilde{U}|\psi\rangle|^2 = \frac{\|\tilde{U}|\psi\rangle\|^2}{s^2}.
\end{aligned}$$

If the ancilla qubits are measured in the state $|\mathbf{0}\rangle$, the n-qubit register collapses into

$$\begin{aligned}\rho_{\mathbf{0}} &= \frac{1}{p(\mathbf{0})}\mathrm{Tr}_{\mathrm{ancilla}}\big[(|\mathbf{0}\rangle\langle\mathbf{0}|\otimes\mathbb{1})(\hat{G}|\mathbf{0}\psi\rangle\langle\mathbf{0}\psi|\hat{G}^\dagger)\big]\\ &= \frac{1}{p(\mathbf{0})}\sum_{k=1}^{2^p}\langle k|\big[(|\mathbf{0}\rangle\langle\mathbf{0}|\otimes\mathbb{1})(\hat{G}|\mathbf{0}\psi\rangle\langle\mathbf{0}\psi|\hat{G}^\dagger)\big]|k\rangle\\ &= \frac{1}{p(\mathbf{0})}\langle\mathbf{0}|\big[(|\mathbf{0}\rangle\langle\mathbf{0}|\otimes\mathbb{1})(\hat{G}|\mathbf{0}\psi\rangle\langle\mathbf{0}\psi|\hat{G}^\dagger)\big]|\mathbf{0}\rangle\\ &= \frac{1}{p(\mathbf{0})}\langle\mathbf{0}|\hat{G}|\mathbf{0}\psi\rangle\langle\mathbf{0}\psi|\hat{G}^\dagger|\mathbf{0}\rangle = \frac{1}{p(\mathbf{0})}\frac{\tilde{U}|\psi\rangle\langle\psi|\tilde{U}^\dagger}{s^2} = \frac{\tilde{U}|\psi\rangle\langle\psi|\tilde{U}^\dagger}{\|\tilde{U}|\psi\rangle\|^2},\end{aligned}$$

which is proportional to $\tilde{U}|\psi\rangle$ as desired.

LCU lemma Now we make the assumption that $\tilde{U}$ is a **linear combination of unitaries** acting on n qubits,

$$\boxed{\tilde{U} = \sum_{j=0}^{M-1} c_j\hat{V}_j}$$

where the operators $\hat{V}_j$ are M unitary operators, the coefficients c_j are positive and their sum is $s = \sum_j c_j$. Even if the operators $\hat{V}_j$ are unitary, the linear combination $\tilde{U}$ might not be. The **LCU lemma** is a quantum circuit designed to apply a linear combination of unitaries on a n-qubit state $|\psi\rangle$. This is done by defining a block encoding $\hat{G} = (\hat{A}^\dagger\otimes\mathbb{1})\hat{B}(\hat{A}\otimes\mathbb{1})$ as shown in the circuit of Fig. 11.9. The quantum circuit contains n qubits and $p = \lceil\log_2 M\rceil$ ancilla qubits initialized in $|\mathbf{0}\rangle = |0\ldots0\rangle$. In this circuit, the operator $\hat{A}$ prepares the ancilla qubits in a superposition of M states,

$$\boxed{\hat{A}|\mathbf{0}\rangle = \frac{1}{\sqrt{s}}\sum_{j=0}^{M-1}\sqrt{c_j}|j\rangle.} \tag{11.23}$$

The transformation $\hat{B}$ is a controlled operation: it applies the operator $\hat{V}_j$ onto the n-qubit register when the state of the ancilla qubits is $|j\rangle$,

$$\boxed{\hat{B} = \sum_{j=0}^{M-1}|j\rangle\langle j|\otimes\hat{V}_j.} \tag{11.24}$$

The action of this operator on the basis vectors reads

$$\hat{B}|i\rangle\otimes|k\rangle = \sum_{j=0}^{M-1}|j\rangle\langle j|i\rangle\otimes\hat{V}_j|k\rangle = |i\rangle\otimes\hat{V}_i|k\rangle. \tag{11.25}$$

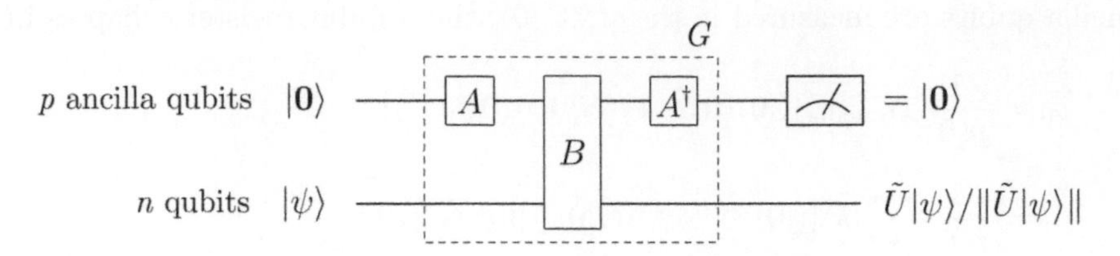

Fig. 11.9 Linear combination of unitaries. Quantum circuit for the probabilistic implementation of a linear combination of unitaries $\tilde{U} = \sum_j c_j \hat{V}_j$ where $c_j > 0$. If the state of the ancillas collapses into $|\mathbf{0}\rangle = |0\ldots0\rangle$, the state of the n-qubit register becomes $\tilde{U}|\psi\rangle/\|\tilde{U}|\psi\rangle\|$. If the state of the ancillas collapses into a different state, the protocol must be repeated. This quantum circuit has success probability $p(\mathbf{0}) = \|\tilde{U}|\psi\rangle\|^2/s^2$ where $s = \sum_j c_j$.

It is straightforward to verify that $\hat{G} = (\hat{A}^\dagger \otimes \mathbb{1})\hat{B}(\hat{A} \otimes \mathbb{1})$ is a block encoding of $\tilde{U}$,

$$s\langle\mathbf{0}k|\hat{G}|\mathbf{0}\ell\rangle = s\langle\mathbf{0}k|(\hat{A}^\dagger \otimes \mathbb{1})\hat{B}(\hat{A} \otimes I)|\mathbf{0}\ell\rangle = \sum_{m=0}^{M-1}\sum_{j=0}^{M-1}\sqrt{c_m c_j}\langle mk|\hat{B}|j\ell\rangle$$

$$= \sum_{m=0}^{M-1}\sum_{j=0}^{M-1}\sqrt{c_m c_j}\langle mk|\mathbb{1} \otimes \hat{V}_j|j\ell\rangle \sum_{j=0}^{M-1} c_j\langle k|\hat{V}_j|\ell\rangle = \langle k|\tilde{U}|\ell\rangle.$$

The quantum circuit ends with a measurement of the ancilla qubits. If their state collapses into $|\mathbf{0}\rangle = |0\ldots0\rangle$, the state of the n qubits becomes $\tilde{U}|\psi\rangle/\|\tilde{U}|\psi\rangle\|$ as desired (we proved this in the previous paragraph). If the ancilla qubits collapse into a different state, the entire circuit must be run again until they are measured all in $|0\rangle$. This protocol requires n qubits plus $p = \log_2 M$ ancilla qubits and the success probability is $p(\mathbf{0}) = \|\tilde{U}|\psi\rangle\|^2/s^2$.

How many gates are required to implement an LCU? An LCU boils down to the application of three operators: $\hat{A}$ and $\hat{A}^\dagger$ that operate on p qubits and $\hat{B}$ that operates on $n+p$ qubits. In the worst case, $\hat{A}$ requires $n_A = O(2^p)$ elementary gates [170]. Since $p = \log_2 M$, we have $n_A = O(M)$. The operator $\hat{B}$ comprises M unitary operators $\hat{V}_j$ acting on n qubits controlled by p ancillae. Assuming that the unitary operators $\hat{V}_j$ are **Pauli operators**, each controlled operation can be implemented with $O(np)$ elementary gates [103] (note that a more efficient implementation based on a binary tree construction is known [351]). Therefore, $n_B = O(Mnp) = O(Mn\log_2 M)$. The number of gates to implement an LCU scales as

$$\boxed{n_{\text{gates}} = 2n_A + n_B = O(M) + O(nM\log_2 M) = O(nM\log_2 M).} \qquad \textbf{LCU}$$

In summary, an LCU can be decomposed into a number of elementary gates that scales as $O(nM\log_2 M)$ where M is the number of terms in the linear combination. The circuit has to be repeated multiple times until all ancilla qubits have been measured in $|0\rangle$.

Example 11.4 Consider the LCU $\tilde{U} = c_1\hat{Y} + c_2\hat{Z}$ where $c_1 = \cos^2\theta$, $c_2 = \sin^2\theta$ and

$s = c_1 + c_2 = 1$. Here, θ is a parameter in the interval $[0, 2\pi)$. The transformation $\tilde{U}$ is the sum of two unitary operations and can be implemented with the following circuit,

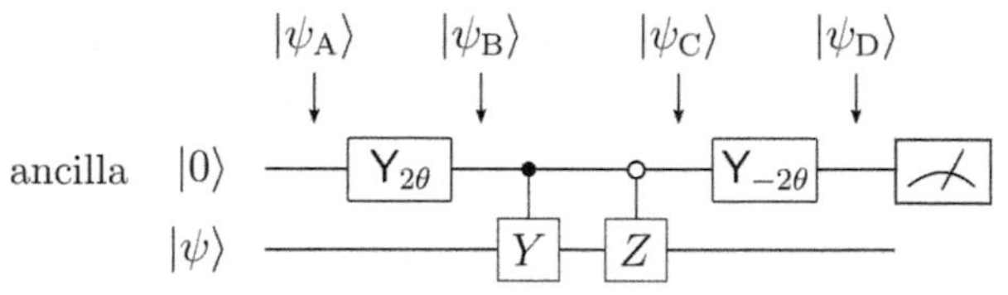

The ancilla and the qubit are initialized in $|0\rangle$ and $|\psi\rangle$ respectively. The final state of the circuit $|\psi_{\rm D}\rangle$ is given by

$$\begin{aligned}|\psi_{\rm D}\rangle = &|0\rangle \otimes (\cos^2\theta \hat{Y}|\psi\rangle + \sin^2\theta \hat{Z}|\psi\rangle) \\ &+ |1\rangle \otimes (-\cos\theta\sin\theta \hat{Y}|\psi\rangle + \cos\theta\sin\theta \hat{Z}|\psi\rangle).\end{aligned}$$

If the measurement reveals the ancilla in $|0\rangle$, the qubit state becomes $\tilde{U}|\psi\rangle/\|\tilde{U}|\psi\rangle\|$. The probability of this event is

$$\begin{aligned}p(0) &= {\rm Tr}_{\rm AB}[(|0\rangle\langle 0| \otimes \mathbb{1})|\psi_{\rm D}\rangle\langle\psi_{\rm D}|] \\ &= {\rm Tr}_{\rm B}[(\cos^2\theta \hat{Y}|\psi\rangle + \sin^2\theta \hat{Z}|\psi\rangle)(\cos^2\theta\langle\psi|\hat{Y} + \sin^2\theta\langle\psi|\hat{Z})] = \cos^4\theta + \sin^4\theta.\end{aligned}$$

If the state of the ancilla collapses onto $|1\rangle$, the circuit must be run again.

11.5 Taylor series

Let us continue our discussion about simulation methods. In this section, we will study a method based on the Taylor series [27]. This algorithm simulates Hamiltonian dynamics with lower scaling in the number of gates than Trotterization, at the cost of adding some extra qubits to the circuit. As we shall see later, this algorithm boils down to implementing an LCU.

Suppose that the Hamiltonian is decomposed into a basis of unitary operators $\hat{U}_\ell$,

$$\boxed{\hat{H} = \sum_{\ell=1}^{L} b_\ell \hat{U}_\ell,} \tag{11.26}$$

where the coefficients b_ℓ are positive and their sum is $\gamma \equiv \sum_\ell b_\ell$. Here, the unitary operations $\hat{U}_\ell$ are transformations that can be easily implemented on a quantum computer, such as the Pauli operators. As usual, the interval $[0, t]$ is divided into m steps of duration Δt. The exact time evolution operator $\hat{U}_{\Delta t} = e^{-i\hat{H}\Delta t}$ can be expanded with a Taylor series. Truncating the Taylor series at the order $r - 1$, we obtain an approximation of the unitary operator that we call $\hat{V}_{\Delta t}$,

$$\hat{U}_{\Delta t} = e^{-i\hat{H}\Delta t} = \sum_{k=0}^{\infty} \frac{(-i\Delta t)^k}{k!}\hat{H}^k \approx \sum_{k=0}^{r-1} \frac{(-i\Delta t)^k}{k!}\hat{H}^k = \hat{V}_{\Delta t}. \tag{11.27}$$

Let us explicitly calculate $\hat{V}_{\Delta t}$. Substituting (11.26) into (11.27), we have

$$\hat{V}_{\Delta t} = \underbrace{\mathbb{1}}_{1\text{ operator}} + \underbrace{\Delta t \sum_{\ell=1}^{L} b_\ell(-i\hat{U}_\ell)}_{L\text{ operators}} + \underbrace{\frac{(\Delta t)^2}{2!}\sum_{\ell=1}^{L}\sum_{m=1}^{L} b_\ell b_m(-i\hat{U}_\ell)(-i\hat{U}_m)}_{L^2\text{ operators}} + \ldots \quad (11.28)$$

The number of operators in this expression depends on the order of the Taylor expansion,

$$M_r = 1 + L + L^2 + \ldots + L^{r-1} = \frac{L^r - 1}{L - 1}.$$

The crucial point is that the operator $\hat{V}_{\Delta t}$ in eqn (11.28) is written as a linear combination of unitaries,

$$\boxed{\hat{V}_{\Delta t} = \sum_{j=0}^{M_r-1} c_j \hat{V}_j,} \quad (11.29)$$

where the coefficients c_j are positive and their sum is $s = \sum_j c_j$. The unitary operators $\hat{V}_j$ are nothing but products of operators of the form $-i\hat{U}_\ell$. For clarity, we report the first coefficients and operators

$$\begin{aligned} &c_0 = 1, && c_1 = b_1\Delta t, && c_2 = b_2\Delta t, && \ldots \\ &\hat{V}_0 = \mathbb{1}, && \hat{V}_1 = -i\hat{U}_1, && \hat{V}_2 = -i\hat{U}_2, && \ldots \end{aligned}$$

The LCU $\hat{V}_{\Delta t}$ can be implemented with the technique presented in the previous section. The simulation of the dynamics of a quantum system using the Taylor series is shown in Fig. 11.10. The simulation is successful only if the ancilla qubits are measured in the state $|\mathbf{0}\rangle$ at each step.

What must be the value of r such that the error of the entire simulation $\|\hat{U}_t - \hat{V}_t\|_{\text{o}}$ is smaller than $\varepsilon \in [0, 2]$? To answer this question, let us start from (11.27) and calculate the operator norm

$$\begin{aligned} \|\hat{U}_{\Delta t} - \hat{V}_{\Delta t}\|_{\text{o}} &= \Big\| \sum_{k=r}^{\infty} \frac{(-i\Delta t)^k}{k!} \hat{H}^k \Big\|_{\text{o}} \leq \sum_{k=r}^{\infty} \frac{\Delta t^k}{k!} \Big\| \Big(\sum_{\ell=1}^{L} b_\ell \hat{U}_\ell\Big)^k \Big\|_{\text{o}} \\ &\leq \sum_{k=r}^{\infty} \frac{\Delta t^k}{k!} \Big(\sum_{\ell=1}^{L} b_\ell \|\hat{U}_\ell\|_{\text{o}}\Big)^k \\ &= \sum_{k=r}^{\infty} \frac{\Delta t^k}{k!} \Big(\sum_{\ell=1}^{L} b_\ell\Big)^k = \sum_{k=r}^{\infty} \frac{\Delta t^k}{k!}\gamma^k = \sum_{k=r}^{\infty} \frac{x^k}{k!}, \end{aligned}$$

where in the last step we defined the parameter $x = \gamma\Delta t$. Let us perform a change of variables so that the sum starts from zero,

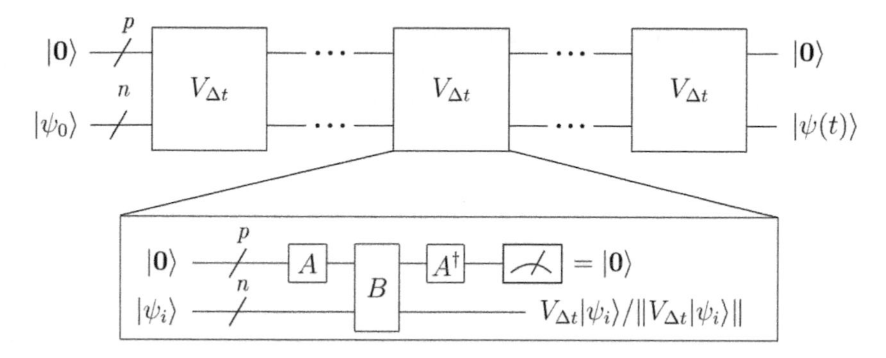

Fig. 11.10 Quantum simulation based on the Taylor series. The transformation $\hat{V}_{\Delta t}$ is a linear combination of M_r unitaries. This LCU can be implemented on a quantum computer with n qubits plus $\lceil \log_2 M_r \rceil$ ancilla qubits initially prepared in $|\mathbf{0}\rangle$. Only if the ancilla qubits are measured in $|\mathbf{0}\rangle$ at each step of the simulation does the state of the n-qubit register end up in the desired state $|\psi(t)\rangle$.

$$\begin{aligned}\|\hat{U}_{\Delta t} - \hat{V}_{\Delta t}\|_{\text{o}} &\leq \sum_{k=r}^{\infty} \frac{x^k}{k!} = \sum_{k'=0}^{\infty} \frac{x^{k'+r}}{(k'+r)!} \\ &\leq \sum_{k'=0}^{\infty} \frac{x^{k'+r}}{r!} = \frac{x^r}{r!} \sum_{k'=0}^{\infty} x^{k'} = \frac{x^r}{r!} \frac{1}{1-x}.\end{aligned} \tag{11.30}$$

The last step is valid only if $x < 1$, which implies $\gamma\Delta t = \gamma t/m < 1$. In other words, the sum does not diverge as long as the number of steps m satisfies

$$\boxed{m > \gamma t.} \tag{11.31}$$

If $r \gg 1$, the factorial in (11.30) can be approximated as $r! \approx r^r e^{-r}$. Therefore,

$$\|\hat{U}_{\Delta t} - \hat{V}_{\Delta t}\|_{\text{o}} \leq \frac{x^r}{r^r e^{-r}} \frac{1}{1-x}.$$

To ensure that the error of the entire simulation is smaller than ε, the error at each step $\|\hat{U}_{\Delta t} - \hat{V}_{\Delta t}\|_{\text{o}}$ must be smaller than ε/m,

$$\|\hat{U}_{\Delta t} - \hat{V}_{\Delta t}\|_{\text{o}} \leq \frac{x^r}{r^r e^{-r}} \frac{1}{1-x} \leq \frac{\varepsilon}{m} \qquad \Rightarrow \qquad \frac{x^r}{r^r e^{-r}} \leq \frac{\varepsilon}{m}(1-x). \tag{11.32}$$

In Exercise 11.8, we show that the solution to this inequality is given by

$$\boxed{r \in O\left(\frac{\log(m/\varepsilon)}{\log(\log(m/\varepsilon))}\right).} \tag{11.33}$$

In summary, the order of the Taylor expansion has a logarithmic dependence on the desired accuracy ε.

It is important to remember that the linear combination of unitaries $\hat{V}_{\Delta t}$ must be applied successfully m times in a row. As seen in the previous section, the probability that an LCU is successfully applied in a single run is given by

$$\begin{aligned} p(\mathbf{0}) = \frac{\|\hat{V}_{\Delta t}|\psi\rangle\|^2}{s^2} &= \frac{\left\|(\hat{V}_{\Delta t} - \hat{U}_{\Delta t} + \hat{U}_{\Delta t})|\psi\rangle\right\|^2}{s^2} \\ &\leq \frac{\left\|(\hat{V}_{\Delta t} - \hat{U}_{\Delta t})|\psi\rangle\right\|^2 + \|\hat{U}_{\Delta t}|\psi\rangle\|^2}{s^2} \\ &\leq \frac{\left\|\hat{V}_{\Delta t} - \hat{U}_{\Delta t}\right\|_{\mathrm{o}}^2 + \|\hat{U}_{\Delta t}\|_{\mathrm{o}}^2}{s^2} \leq \frac{\varepsilon^2/m^2 + 1}{s^2}, \end{aligned} \tag{11.34}$$

where in the last step we used (11.32). Provided that $r \gg 1$ and the number of steps $m = \gamma t/\log(2)$, then the constant s is given by

$$\begin{aligned} s &= \sum_{j=0}^{M_r-1} c_j = 1 + \Delta t(b_1 + \ldots + b_L) + \frac{(\Delta t)^2}{2}(b_1\gamma + \ldots + b_L\gamma) \\ &= \sum_{j=0}^{r-1} \frac{(\gamma\Delta t)^k}{k!} = \sum_{j=0}^{r-1} \frac{(\gamma t/m)^k}{k!} = \sum_{j=0}^{r-1} \frac{(\log 2)^k}{k!} \approx \sum_{j=0}^{\infty} \frac{(\log 2)^k}{k!} = 2. \end{aligned}$$

Using this result, we see that the success probability of the entire simulation p_{success} is bounded by

$$p_{\text{success}} = [p(\mathbf{0})]^m \leq \left(\frac{\varepsilon^2/m^2 + 1}{s^2}\right)^m \approx 4^{-m}(\varepsilon^2/m^2 + 1)^m.$$

This probability decreases exponentially with m, meaning that an exponential number of repetitions of the algorithm is necessary to achieve a reasonable success probability. Such an exponential scaling is prohibitive. This problem is fixed by introducing the the oblivious amplitude amplification (OAA) method, presented in Appendix E. This technique boosts the success probability of the quantum simulation up to $p_{\text{success}} = O(1)$ with a polynomial cost. In conclusion, only if the Taylor series method is combined with the OAA technique, we obtain a scalable algorithm to simulate Hamiltonian dynamics.

How many gates are required to implement a simulation based on the Taylor series? First, recall from Section 11.4 that the LCU $\hat{V}_{\Delta t}$ can be implemented with $O(nM_r \log_2 M_r)$ gates. The quantum simulation requires m applications of this transformation. Therefore, $n_{\text{gates}} = O(mnM_r \log_2 M_r)$. Since $M_r = (L^r - 1)/(L - 1)$, we have $n_{\text{gates}} = O(mnL^{r-1}(r-1)\log_2 L)$ where L is the number of terms in the Hamiltonian. Lastly, we know from eqn (11.31) that the number of steps must be at least $m > \gamma t$, which implies that

$$n_{\text{gates}} = O(\gamma t n L^{r-1}(r-1)\log_2 L) = O\Big(\gamma t n L^{O\left(\frac{\log(\gamma t/\varepsilon)}{\log(\log(\gamma t/\varepsilon))}\right)} O\Big(\frac{\log(\gamma t/\varepsilon)}{\log(\log(\gamma t/\varepsilon))}\Big)\log_2 L\Big),$$

where in the last step we used (11.33). The takeaway is that fewer gates are required with respect to the Trotterization in the asymptotic limit. This comes at the cost of adding $\log M_r$ ancilla qubits to the circuit.

The number of gates can be decreased by adding more qubits to the circuit. As explained in Ref. [27], if the circuit contains a register with n qubits plus a register with $(r-1)+(r-1)\log_2 L$ ancilla qubits, the number of gates goes down to

$$\boxed{n_{\text{gates}} \in O\left(\gamma t \frac{L(n+\log L)\log(\gamma t/\varepsilon)}{\log\log(\gamma t/\varepsilon)}\right).} \qquad \textbf{Taylor series} \qquad (11.35)$$

This improved version of the algorithm uses a clever encoding of the ancilla states based on the unary representation.

11.6 Qubitization

Qubitization is an algorithm aimed at simulating the dynamics of a quantum system with optimal scaling with respect to accuracy and time [28, 350]. It is based on a block encoding of the Hamiltonian, a very different paradigm from the Trotterization methods. Let us start our discussion by considering a time-independent Hamiltonian decomposed into the Pauli basis

$$\boxed{\hat{H} = \sum_{j=0}^{L-1} c_j \hat{P}_j,}$$

where $c_j > 0$ and[10] $\sum_j c_j = 1$. The Hamiltonian acts on n qubits and has 2^n eigenvalues and eigenvectors defined as

$$\underset{\text{eigenvalues}}{\{\lambda_1, \ldots, \lambda_{2^n}\}}, \qquad \underset{\text{eigenvectors}}{\{|\lambda_1\rangle, \ldots, |\lambda_{2^n}\rangle\}}.$$

Both the eigenvalues and the eigenvectors are unknown. However, since $\|\hat{H}\|_{\text{o}} \leq 1$, then $-1 \leq \lambda_j \leq 1$. The first step of the qubitization method is to write the time evolution operator with a Jacobi–Anger expansion,

$$\boxed{e^{-i\hat{H}t} = \sum_{k=0}^{\infty} d_k(t) T_k(\hat{H}),} \qquad (11.36)$$

[10]Recall that it is always possible to rescale the Hamiltonian by a multiplicative factor without changing the resources required to simulate the dynamics: indeed, simulating the dynamics of $e^{-i\hat{H}t}$ for a generic Hamiltonian $\hat{H}$ is equivalent to simulating the dynamics of $e^{-i\hat{H}'\tau}$ where $\hat{H}' = \hat{H}/\|\hat{H}\|_{\text{o}}$ and $\tau = t\|\hat{H}\|_{\text{o}}$. Hence, the assumption $\sum_j c_j = 1$ is general enough. In addition, taking the coefficients to be positive is not a problem, since if $c_j < 0$ we can absorb the negative sign into the definition of $\hat{P}_j$.

where $T_k(x)$ are the Chebyshev polynomials of the first kind[11], while the complex coefficients $d_k(t)$ are given by

$$d_k(t) = \begin{cases} J_0(-t) & k = 0 \\ 2i^k J_k(-t) & k \geq 1. \end{cases}$$

Here, $J_k(x)$ are the Bessel functions of the first kind. The goal of the qubitization method is to construct a block encoding of the operator

$$\boxed{\hat{V}_t = \sum_{k=0}^{r} d_k(t) T_k(\hat{H})} \qquad \text{where } r > 0.$$

This operator approximates the time evolution operator (11.36) to the r-th order. Once a block encoding has been found, one can follow the procedure explained in Section 11.4 to simulate the dynamics.

For pedagogical reasons, we first show how to construct the block encoding of $\hat{V}_t$ when $r = 1$,

$$\hat{V}_t = J_0(-t) + 2iJ_1(-t)\hat{H}. \tag{11.37}$$

The generalization to any order r will be explained in Section 11.6.2. Figure 11.11 shows the circuit that implements $\hat{V}_t$ to first order using the qubitization method. It is important to note that this method does not require dividing the time interval $[0, t]$ into small steps; the circuit is built so that the entire simulation can be performed in a single step. The circuit has three registers: the first has n qubits, the second comprises $p = \lfloor \log L \rfloor + 1$ ancilla qubits and the third has one ancilla. The circuit contains two controlled gates preceded by the unitary operator $\mathsf{X}_{-2\theta} \otimes \hat{A} \otimes \mathbb{1}_n$ and followed by the unitary operator $\mathsf{H} \otimes \hat{A}^\dagger \otimes \mathbb{1}_n$. The operation $\hat{A}$ transforms the state of the ancilla qubits from $|0 \ldots 0\rangle_p$ into a superposition of L states[12] defined as

$$\boxed{|g\rangle \equiv \hat{A}|\mathbf{0}\rangle_p = \sum_{j=0}^{L-1} \sqrt{c_j}|j\rangle.}$$

The gates Z_ϕ are **multi-qubit rotations** acting on the p ancilla qubits,

$$\boxed{\mathsf{Z}_\phi \equiv e^{i\phi}|g\rangle\langle g| \otimes \mathbb{1}_n + e^{-i\phi}\left(\mathbb{1}_p - |g\rangle\langle g|\right) \otimes \mathbb{1}_n\,,} \tag{11.38}$$

while the operator $\hat{W}$ (also known as the **qubiterate**) is defined as

$$\boxed{\hat{W} \equiv \hat{R}_g \hat{B}}$$

[11]The first three Chebyshev polynomials of the first kind are $T_0(\hat{H}) = \mathbb{1}$, $T_1(\hat{H}) = \hat{H}$, $T_2(x) = 2\hat{H}^2 - 1$.

[12]Note the similarity between this expression and eqn (11.23).

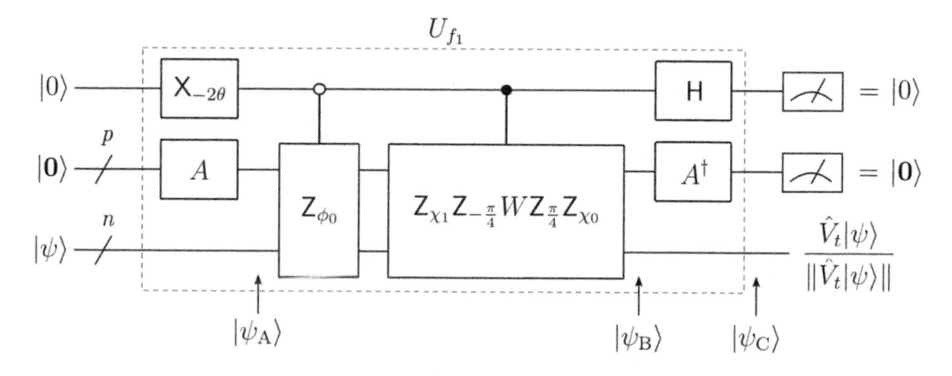

Fig. 11.11 Qubitization to first order. Simulation of the time evolution $|\psi\rangle \mapsto \hat{V}_t|\psi\rangle$ with $\hat{V}_t = J_0(-t) + 2iJ_1(-t)\hat{H}$ using the qubitization method. The circuit comprises three registers with a total of $n + p + 1$ qubits where $p = \log L$. The operator $\hat{A}^\dagger \hat{W} \hat{A}$ is a block encoding of the Hamiltonian, while the unitary operator $\hat{U}_{f_1}$ is a block encoding of $\hat{V}_t$. If at the end of the circuit the ancilla qubits are measured in $|0\rangle|\mathbf{0}\rangle$, the state of the n-qubit register collapses into $\hat{V}_t|\psi\rangle / \|\hat{V}_t|\psi\rangle\|$.

---

where

$$\hat{R}_g = 2|g\rangle\langle g| \otimes \mathbb{1}_n - \mathbb{1}_p \otimes \mathbb{1}_n, \qquad \hat{B} = \sum_{j=0}^{L-1} |j\rangle\langle j| \otimes \hat{P}_j.$$

The operator $\hat{R}_g$ is nothing but a **reflection** about the vector $|g\rangle$. The unitary operator $\hat{B}$ applies the Pauli operator $\hat{P}_j$ to the n qubits only when the p ancilla qubits are in the state $|j\rangle$. Before showing that the circuit in Fig. 11.11 correctly simulates the time evolution to first order, let us pause for a second and explore the properties of the qubiterate $\hat{W}$.

Properties of W Let us show that the operator $\hat{A}^\dagger \hat{W} \hat{A}$ is a block encoding of the Hamiltonian, i.e. it acts on a larger vector space, but its projection onto a particular subspace is equal to the Hamiltonian,

$$\begin{aligned}
\langle \mathbf{0}i|\hat{A}^\dagger \hat{W} \hat{A}|\mathbf{0}j\rangle &= \langle gi|\hat{W}|gj\rangle = \langle gi|\hat{R}_g \hat{B}|gj\rangle = \langle gi|\hat{B}|gj\rangle \\
&= \sum_{m=0}^{L-1} \sum_{k=0}^{L-1} \sqrt{c_m}\sqrt{c_k}\langle mi|\hat{B}|kj\rangle \\
&= \sum_{m=0}^{L-1} \sum_{k=0}^{L-1} \sum_{q=0}^{L-1} \sqrt{c_m}\sqrt{c_k}\langle mi|\big(|q\rangle\langle q| \otimes \hat{P}_q\big)|kj\rangle \\
&= \sum_{q=0}^{L-1} c_q \langle i|\hat{P}_q|j\rangle = \langle i|\hat{H}|j\rangle.
\end{aligned} \tag{11.39}$$

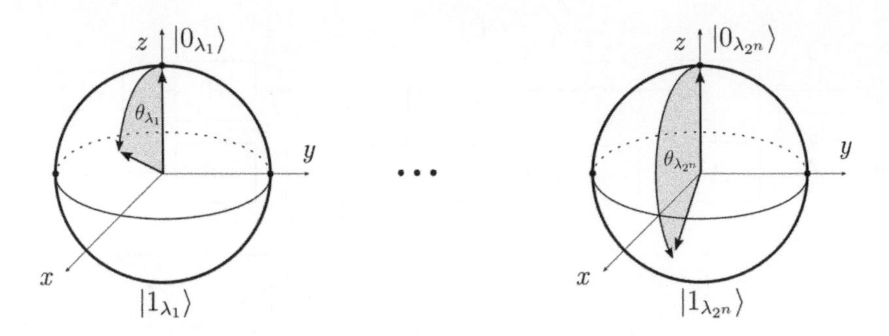

Fig. 11.12 The operator W. The unitary operator $\hat{W}$ rotates the states $\alpha|0_\lambda\rangle + \beta|1_\lambda\rangle$ about the y axis by an angle $\theta_\lambda = -2\arccos\lambda$.

The operator $\hat{W}$ has another important property: it can be expressed as the direct sum of rotations acting on two-dimensional subspaces. To prove this, let us introduce 2^n pairs of $n+p$ qubit states:

$$\{|0_{\lambda_1}\rangle, |1_{\lambda_1}\rangle\}, \quad \{|0_{\lambda_2}\rangle, |1_{\lambda_2}\rangle\}, \quad \ldots \quad \{|0_{\lambda_{2^n}}\rangle, |1_{\lambda_{2^n}}\rangle\},$$

where the indices λ_i correspond to the eigenvalues of the Hamiltonian. These states are defined as

$$|0_\lambda\rangle = |g\rangle_p|\lambda\rangle_n, \qquad\qquad |1_\lambda\rangle = \frac{\lambda|g\rangle_p|\lambda\rangle_n - \hat{B}|g\rangle_p|\lambda\rangle_n}{\sqrt{1-\lambda^2}},$$

where $|\lambda\rangle$ are the unknown eigenvectors of the Hamiltonian. In Exercise 11.9, we show that these $n+p$ qubit states are orthonormal and thus each pair $\{|0_\lambda\rangle, |1_\lambda\rangle\}$ can be considered as a qubit. In the same exercise, we prove that the action of $\hat{W}$ onto these states produces

$$\hat{W}|0_\lambda\rangle = \lambda|0_\lambda\rangle + \sqrt{1-\lambda^2}|1_\lambda\rangle, \qquad\qquad \hat{W}|1_\lambda\rangle = -\sqrt{1-\lambda^2}|0_\lambda\rangle + \lambda|1_\lambda\rangle. \quad (11.40)$$

This relation can be written in matrix form as

$$\hat{W}(\alpha|0_\lambda\rangle + \beta|1_\lambda\rangle) = \begin{bmatrix} \lambda & \sqrt{1-\lambda^2} \\ -\sqrt{1-\lambda^2} & \lambda \end{bmatrix} \begin{bmatrix} \alpha \\ \beta \end{bmatrix}.$$

Therefore, the action of $\hat{W}$ onto a state $\alpha|0_\lambda\rangle + \beta|1_\lambda\rangle$ is equivalent to a rotation about the y axis of the Bloch sphere by an angle $\theta_\lambda = -2\arccos\lambda$ (see Fig. 11.12 for a graphical visualization). The action of $\hat{W}$ on a linear combination of states of the form $\alpha|0_\lambda\rangle + \beta|1_\lambda\rangle$ can be expressed with a direct sum[13]

[13] $\hat{W}$ is defined on a vector space of dimension 2^{n+p}, while eqn (11.41) defines an operator acting on a subspace of dimension $2 \cdot 2^n$. Hereafter, the operator $\hat{W}$ will only be applied to states of the form $|g\rangle \sum_j c_j|\lambda_j\rangle$ and hence it is not necessary to define its action on states outside this subspace.

$$\boxed{\hat{W} = \bigoplus_{\lambda=1}^{2^n} \begin{bmatrix} \lambda & \sqrt{1-\lambda^2} \\ -\sqrt{1-\lambda^2} & \lambda \end{bmatrix} = \bigoplus_{\lambda=1}^{2^n} e^{-\frac{i}{2}\theta_\lambda \hat{\sigma}_y}.} \tag{11.41}$$

In summary, the qubiterate $\hat{W}$ has two important properties: $\hat{A}^\dagger \hat{W} \hat{A}$ is a block encoding of the Hamiltonian, and $\hat{W}$ acts separately on 2^n two-dimensional subspaces, each labeled by an eigenvalue λ.

We can now revisit the quantum circuit depicted in Fig. 11.11 and demonstrate that the operator $\hat{U}_{f_1}$ is a block encoding of $\hat{V}_t$. If this is the case, then for what we have seen in Section 11.4, the final state of the n-qubit register will be proportional to $\hat{V}_t|\psi\rangle$ when all ancillary qubits are measured in the $|0\rangle$ state. Let us prove that for any pair of n-qubit states $|\phi\rangle$ and $|\psi\rangle$ and for some constant $s > 0$, the operator $\hat{U}_f$ is a block encoding of $\hat{V}_t$,

$$\boxed{\langle 0\mathbf{0}\phi|\hat{U}_{f_1}|0\mathbf{0}\psi\rangle = s\langle\phi|\hat{V}_t|\psi\rangle.}$$

First, we decompose the states $|\phi\rangle$ and $|\psi\rangle$ into the eigenbasis of the Hamiltonian, $|\psi\rangle = \sum_\lambda \alpha_\lambda|\lambda\rangle$ and $|\phi\rangle = \sum_{\lambda'} \beta_{\lambda'}|\lambda'\rangle$. The gate $\mathsf{X}_{-2\theta} \otimes \hat{A} \otimes \mathbb{1}_n$ transforms the initial state $|0\mathbf{0}\psi\rangle$ into

$$\begin{aligned}
|\psi_{\mathrm{A}}\rangle = \mathsf{X}_{-2\theta}|0\rangle\hat{A}|\mathbf{0}\rangle|\psi\rangle &= \Big(\cos\theta|0\rangle + i\sin\theta|1\rangle\Big)|g\rangle|\psi\rangle \\
&= \sum_\lambda \alpha_\lambda \Big(\cos\theta|0\rangle + i\sin\theta|1\rangle\Big)|g\rangle|\lambda\rangle \\
&= \sum_\lambda \alpha_\lambda \Big(\cos(\theta)|0\rangle|0_\lambda\rangle + i\sin(\theta)|1\rangle|0_\lambda\rangle\Big).
\end{aligned}$$

The controlled gates yield the state

$$|\psi_{\mathrm{B}}\rangle = \sum_\lambda \alpha_\lambda \Big[\cos(\theta)|0\rangle \mathsf{Z}_{\phi_0}|0_\lambda\rangle + i\sin(\theta)|1\rangle\Big(\mathsf{Z}_{\chi_1-\frac{\pi}{4}}\hat{W}\mathsf{Z}_{\chi_0+\frac{\pi}{4}}|0_\lambda\rangle\Big)\Big]. \tag{11.42}$$

To calculate the state $\mathsf{Z}_{\chi_1-\frac{\pi}{4}}\hat{W}\mathsf{Z}_{\chi_0+\frac{\pi}{4}}|0_\lambda\rangle$ explicitly, it is important to note that the multi-qubit rotation Z_ϕ defined in eqn (11.38) can be written as a direct sum. Indeed, since

$$\mathsf{Z}_\phi|0_\lambda\rangle = e^{i\phi}|0_\lambda\rangle, \qquad \mathsf{Z}_\phi|1_\lambda\rangle = e^{-i\phi}|1_\lambda\rangle,$$

we have

$$\mathsf{Z}_\phi = \bigoplus_{\lambda=1}^{2^n} \begin{bmatrix} e^{i\phi} & 0 \\ 0 & e^{-i\phi} \end{bmatrix}.$$

Therefore, the product between Z_ϕ and the qubiterate $\hat{W}$ can be computed by simply multiplying 2×2 matrices. For example,

$$\mathsf{Z}_{\chi_1-\frac{\pi}{4}}\hat{W}\mathsf{Z}_{\chi_0+\frac{\pi}{4}} = \bigoplus_{\lambda=1}^{2^n} \begin{bmatrix} e^{i(\chi_0+\chi_1)}\lambda & ie^{-i(\chi_0-\chi_1)}\sqrt{1-\lambda^2} \\ ie^{i(\chi_0-\chi_1)}\sqrt{1-\lambda^2} & e^{-i(\chi_0+\chi_1)}\lambda \end{bmatrix},$$

which implies

$$\mathsf{Z}_{\chi_1-\frac{\pi}{4}}\hat{W}\mathsf{Z}_{\chi_0+\frac{\pi}{4}}|0_\lambda\rangle = e^{i(\chi_0+\chi_1)}\lambda|0_\lambda\rangle + ie^{i(\chi_0-\chi_1)}\sqrt{1-\lambda^2}|1_\lambda\rangle.$$

Using this result, the state (11.42) becomes

$$|\psi_{\mathrm{B}}\rangle = \sum_\lambda \alpha_\lambda\big[\cos(\theta)|0\rangle e^{i\phi_0}|0_\lambda\rangle + i\sin(\theta)|1\rangle\big(e^{i(\chi_0+\chi_1)}\lambda|0_\lambda\rangle + ie^{i(\chi_0-\chi_1)}\sqrt{1-\lambda^2}|1_\lambda\rangle\big)\big].$$

We can finally check that U_{f_1} is a block encoding of $\hat{V}_t$,

$$\begin{aligned}
\langle 00\phi|\,\hat{U}_{f_1}\,|00\psi\rangle &= \langle 00\phi|\psi_{\mathrm{C}}\rangle = \sum_{\lambda'}\beta^*_{\lambda'}\langle 00\lambda'|(\mathsf{H}\otimes\hat{A}^\dagger\otimes\mathbb{1}_n)|\psi_{\mathrm{B}}\rangle \\
&= \sum_{\lambda\lambda'}\beta^*_{\lambda'}\alpha_\lambda\langle +|\langle 0_{\lambda'}|\Big[\cos(\theta)|0\rangle e^{i\phi_0}|0_\lambda\rangle\Big] \\
&\quad + \sum_{\lambda\lambda'}\beta^*_{\lambda'}\alpha_\lambda\langle +|\langle 0_{\lambda'}|\Big[i\sin(\theta)|1\rangle\left(e^{i(\chi_0+\chi_1)}\lambda|0_\lambda\rangle + ie^{i(\chi_0-\chi_1)}\sqrt{1-\lambda^2}|1_\lambda\rangle\right)\Big] \\
&= \sum_\lambda \beta^*_\lambda\alpha_\lambda\left(\frac{\cos(\theta)}{\sqrt{2}}e^{i\phi_0} + i\sin(\theta)e^{i(\chi_0+\chi_1)}\lambda\right).
\end{aligned}$$

Since $\sum_\lambda \beta^*_\lambda\alpha_\lambda = \langle\phi|\psi\rangle$ and $\sum_\lambda \beta^*_\lambda\alpha_\lambda\lambda = \langle\phi|\hat{H}|\psi\rangle$, we arrive at

$$\boxed{\langle 00\phi|\hat{U}_{f_1}|00\psi\rangle = \langle\phi|\frac{\cos(\theta)}{\sqrt{2}}e^{i\phi_0}\mathbb{1} + i\frac{\sin(\theta)}{\sqrt{2}}e^{i(\chi_0+\chi_1)}\hat{H}|\psi\rangle.} \tag{11.43}$$

The unitary $\hat{U}_{f_1}$ is a block encoding of a linear combination $a\mathbb{1} + b\hat{H}$. This is the desired operator $\hat{V}_t = J_0(-t) + 2iJ_1(-t)\hat{H}$ of eqn (11.37) when

$$\boxed{\frac{\cos(\theta)}{\sqrt{2}}e^{i\phi_0} = s\,J_0(-t), \qquad i\frac{\sin(\theta)}{\sqrt{2}}e^{i(\chi_0+\chi_1)} = s\,2iJ_1(-t),} \tag{11.44}$$

where $1/(2s^2) = J_0(-t)^2 + 4J_1(-t)^2$. In conclusion, the circuit shown in Fig. 11.11 simulates the time evolution operator $\hat{U}_t$ to first order when the angles θ, ϕ_0, χ_0, and χ_1 satisfy eqn (11.44).

Example 11.5 Consider a single-qubit Hamiltonian

$$\hat{H} = \frac{1}{2}\hat{\sigma}_x + \frac{1}{2}\hat{\sigma}_z. \tag{11.45}$$

The coefficients in this decomposition are positive and add up to one. The circuit in Fig. 11.13 shows how to simulate the dynamics using the first-order qubitization method. The state $|g\rangle$ is prepared with a Hadamard gate

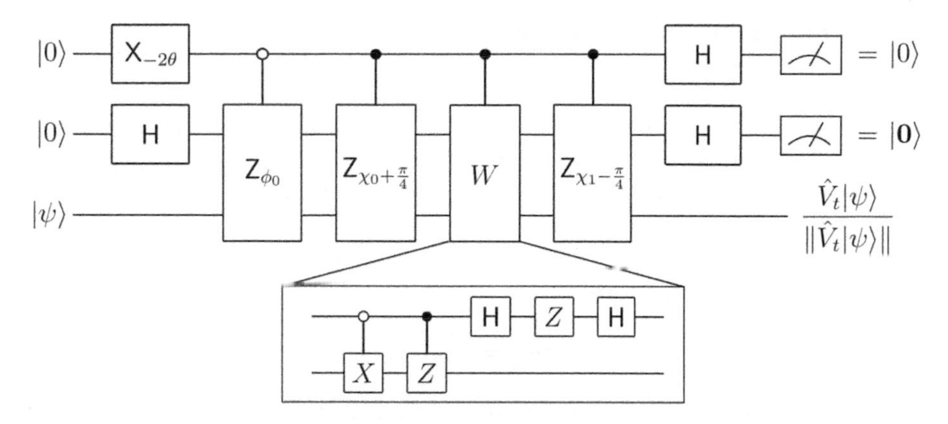

Fig. 11.13 Example of a first-order qubitization. Quantum circuit for the simulation of the dynamics generated by the Hamiltonian $\hat{H} = \hat{\sigma}_x/2 + \hat{\sigma}_z/2$ using the qubitization method at first order. The register has three qubits initialized in $|00\psi\rangle$. The operator $\hat{W}$ is implemented with a sequence of single and two-qubit gates.

$$|g\rangle = \mathsf{H}|0\rangle = \frac{1}{\sqrt{2}}(|0\rangle + |1\rangle). \tag{11.46}$$

The operator $\hat{W}$ is given by the product $\hat{W} = \hat{R}_g\hat{B}$ where

$$\hat{R}_g = 2|g\rangle\langle g| \otimes \mathbb{1} - \mathbb{1} \otimes \mathbb{1} = (2\mathsf{H}|0\rangle\langle 0|\mathsf{H} - \mathbb{1}) \otimes \mathbb{1} = \mathsf{H}\hat{Z}\mathsf{H} \otimes \mathbb{1}$$

and $\hat{B} = \mathsf{c}_{\bar{1}}X\mathsf{c}_1Z$.

Example 11.6 The qubiterate has two interesting properties: it is a block encoding of the Hamiltonian and acts separately on 2^n two-dimensional subspaces, each labeled by an eigenvalue λ. This property allows us to use the qubiterate to simulate the **eigenstates** of the Hamiltonian with the phase estimation algorithm as explained in Exercise 11.10. In Exercise 11.11, we also show that

$$\hat{W}^k = \bigoplus_{\lambda=1}^{2^n} (\mathsf{Y}_{\theta_\lambda})^k = \begin{bmatrix} T_k(\lambda) & -\sqrt{1-\lambda^2}U_{k-1}(\lambda) \\ \sqrt{1-\lambda^2}U_{k-1}(\lambda) & T_k(\lambda) \end{bmatrix},$$

where $U_k(x)$ are the Chebyshev polynomials of the second kind. This equation indicates that the powers of the qubiterate allow us to construct a block encoding for the Chebyshev polynomials $T_k(\hat{H})$ of the Hamiltonian. This is quite remarkable because the Chebyshev polynomials are a complete orthogonal system. Thus, linear combinations of these polynomials can be used to approximate a large class of functions $f(\hat{H})$ where f is an analytical function.

11.6.1 Quantum signal processing

In the previous section, we used the qubiterate $\hat{W}$ to simulate the dynamics to first order. We now generalize this construction using the formalism of **quantum signal processing** [352, 353, 28, 354, 355, 356]. Given a Hamiltonian $\hat{H}$ acting on a n-qubit system and the corresponding qubiterate operator $\hat{W}$ acting on $n+p$ qubits, the quantum signal processing technique constructs a block encoding $\hat{U}_f$ of a function $f(\hat{H})$, i.e. a unitary operator satisfying

$$\langle \mathbf{0}a|\hat{U}_f|\mathbf{0}b\rangle = s\langle a|f(\hat{H})|b\rangle,$$

where $s > 0$ is a real number and a and b are two generic n-qubit states. As we saw in Section 11.4, once $\hat{U}_f$ has been found, one can implement the transformation

$$|\psi\rangle \quad \mapsto \quad \frac{f(\hat{H})|\psi\rangle}{\|f(\hat{H})|\psi\rangle\|}.$$

In the context of Hamiltonian dynamics, we are interested in the particular case $f(\hat{H}) = e^{-it\hat{H}}$. However, quantum signal processing formalism can handle a wide variety of functions.

Let us start our discussion of quantum signal processing assuming, for simplicity, that the function f can be approximated by a real-valued polynomial

$$\boxed{f(x) \approx p_r(x) = a_0 + a_1 x + \cdots + a_r x^r.}$$

This polynomial must satisfy two requirements: its norm must be

$$\|p_r\|_\infty = \max_{\lambda\in[-1,1]} |p_r(\lambda)| \leq 1, \tag{11.47}$$

and it is also required that

$$\text{if } r \text{ is even:} \quad p_r(x) = p_r(-x), \tag{11.48}$$
$$\text{if } r \text{ is odd:} \quad p_r(x) = -p_r(-x). \tag{11.49}$$

If we are able to construct a block encoding $\hat{U}_{p_r}$ of the polynomial $p_r(\hat{H})$, then $\hat{U}_{p_r}$ is an approximation of $\hat{U}_f$,

$$\langle \mathbf{0}a|\hat{U}_{p_r}|\mathbf{0}b\rangle = s\langle a|p_r(\hat{H})|b\rangle \approx s\langle a|f(\hat{H})|b\rangle = \langle \mathbf{0}a|\hat{U}_f|\mathbf{0}b\rangle.$$

How can we build $\hat{U}_{p_r}$? One option is to define a family of unitary transformations containing some free parameters $\Phi = (\phi_0, \phi_1, \ldots, \phi_r)$,

$$\boxed{\hat{U}_\Phi = (\hat{A}^\dagger \otimes \mathbb{1}_n)\left(\prod_{i=1}^{r} \mathsf{Z}_{\phi_i}\mathsf{Z}_{-\frac{\pi}{4}}\hat{W}\mathsf{Z}_{\frac{\pi}{4}}\right)\mathsf{Z}_{\phi_0}(\hat{A}\otimes\mathbb{1}_n),}$$

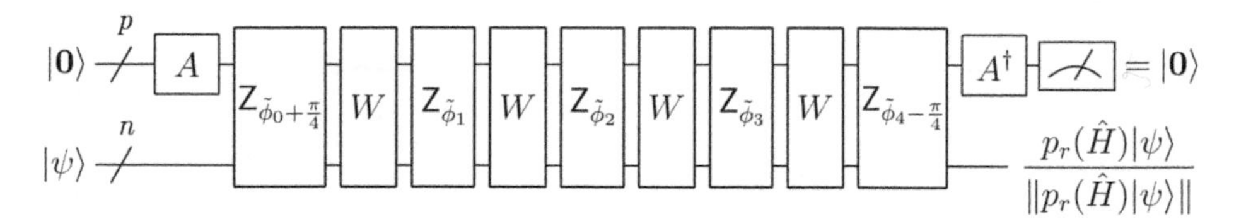

Fig. 11.14 Quantum signal processing. Quantum circuit implementing the transformation $|\psi\rangle \mapsto p_r(\hat{H})|\psi\rangle$ for a polynomial p_r of degree $r = 4$ using the quantum signal processing method.

where $\phi_j \in [-\pi, \pi)$. As explained in Appendix E.1, for a particular vector $\tilde{\Phi} = (\tilde{\phi}_0, \ldots, \tilde{\phi}_r)$, the unitary operator $\hat{U}_{\tilde{\Phi}}$ reduces to $\hat{U}_{p_r}$:

$$\hat{U}_{\tilde{\Phi}} = \hat{U}_{p_r} \approx \hat{U}_f.$$

The quantum circuit implementing $\hat{U}_{\tilde{\Phi}}$ for $r = 4$ is illustrated in Fig. 11.14.

11.6.2 Qubitization: the general case

The qubitization method uses the quantum signal processing technique presented in the previous section to construct a block-encoding for $f(\hat{H}) = e^{-it\hat{H}}$. The qubitization method relies on an important observation: the time evolution operator can be approximated with a Jacobi–Anger expansion (a series that converges much faster than the Taylor series),

$$e^{-i\hat{H}t} \approx \sum_{k=0}^{r} d_k(t) T_k(\hat{H}) = \hat{V}_t \,. \tag{11.50}$$

As always, the objective is to simulate the dynamics generated by $e^{-i\hat{H}t}$ reasonably well. What is the value of r such that $\|\hat{V}_t - e^{-it\hat{H}}\|_{\text{o}} < \varepsilon \in [0, 2]$? To answer this question, we need to calculate the distance

$$\begin{aligned}
\|\hat{V}_t - e^{-it\hat{H}}\|_{\text{o}} &= \|\sum_{k=0}^{r} d_k(t) T_k(\hat{H}) - \sum_{k=0}^{\infty} d_k(t) T_k(\hat{H})\|_{\text{o}} \\
&= \|\sum_{k=r+1}^{\infty} d_k(t) T_k(\hat{H})\|_{\text{o}} \leq \sum_{k=r+1}^{\infty} |d_k(t)| \, \|T_k(\hat{H})\|_{\text{o}} \\
&\leq \sum_{k=r+1}^{\infty} |d_k(t)| \leq \sum_{k=r}^{\infty} |d_k(t)| \leq \sum_{k=r}^{\infty} 2|J_k(t)|.
\end{aligned}$$

Recall that the Bessel functions satisfy $|J_k(t)| \leq (t/2)^k / k!$. *Assuming* $t < r$, then

$$\sum_{k=r}^{\infty} 2|J_k(t)| \leq \sum_{k=r}^{\infty} 2 \frac{(t/2)^k}{k!} \leq \frac{4(t/2)^r}{r!}.$$

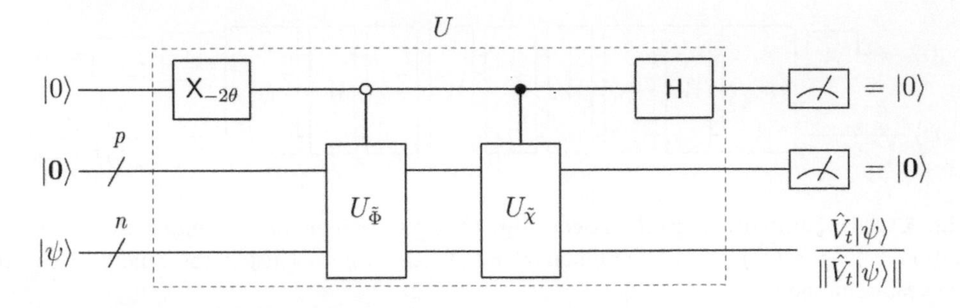

Fig. 11.15 The qubitization method. The gates $\hat{U}_{\tilde{\Phi}}$ and $\hat{U}_{\tilde{\chi}}$ are block encodings of the polynomials $p(\hat{H})$ and $q(\hat{H})$ respectively, which are related to $\hat{V}_t$ through $\hat{V}_t = \nu\cos(\theta)p(\hat{H}) + i\nu\sin(\theta)q(\hat{H})$. The unitary operator $\hat{U}$ enclosed in the dashed rectangle is a block encoding of $\hat{V}_t$.

If r is sufficiently large, one can write $r! \approx r^r e^{-r}$ and obtain

$$\|\hat{V}_t - e^{-it\hat{H}}\|_{\text{o}} \leq \frac{4(t/2)^r}{r!} \approx \frac{4(t/2)^r}{r^r e^{-r}} < \varepsilon.$$

This equation is solved by

$$r(\varepsilon, t) = \frac{\log(4/\varepsilon)}{W[\frac{2}{et}\log(4/\varepsilon)]} = \frac{et}{2} + \log(4/\varepsilon) + o(1), \tag{11.51}$$

where W is the Lambert function. Note that this is a valid solution because $r(\varepsilon, t) > t$.

Let us construct a block encoding of $\hat{V}_t$ using the quantum signal processing technique. First, we write

$$\begin{aligned} V_t(\lambda) &= \sum_{k=0}^{r} d_k(t)T_k(\lambda) = \sum_{k=0}^{\lfloor r/2\rfloor} d_{2k}(t)T_{2k}(\lambda) + \sum_{k=0}^{\lfloor (r-1)/2\rfloor} d_{2k+1}(t)T_{2k+1}(\lambda) \\ &= \underbrace{\sum_{k=0}^{\lfloor r/2\rfloor} d_{2k}(t)T_{2k}(\lambda)}_{A(\lambda)} + i\underbrace{\sum_{k=0}^{\lfloor (r-1)/2\rfloor} 2i^{2k}(t)J_{2k+1}(-t)T_{2k+1}(\lambda)}_{B(\lambda)} = A(\lambda) + iB(\lambda). \end{aligned}$$

Multiplying and dividing by the norms $\|A\|_\infty = \nu\cos\theta$ and $\|B\|_\infty = \nu\sin\theta$, we obtain

$$\begin{aligned} V_t(\lambda) &= \|A\|_\infty \frac{A(\lambda)}{\|A\|_\infty} + i\|B\|_\infty \frac{B(\lambda)}{\|B\|_\infty} \\ &= \nu\left[\cos\theta\underbrace{\frac{A(\lambda)}{\|A\|_\infty}}_{p(\lambda)} + i\sin\theta\underbrace{\frac{B(\lambda)}{\|B\|_\infty}}_{q(\lambda)}\right] = \nu\left[\cos\theta p(\lambda) + i\sin\theta q(\lambda)\right]. \end{aligned}$$

The functions $p(\lambda)$ and $q(\lambda)$ are real-valued polynomials with degrees $\deg(p) = r -$

$(r\%2)$ and $\deg(q) = r - 1 + (r\%2)$, and norm 1. These polynomials satisfy the conditions (11.47–11.49). Therefore, as discussed in the previous section, there exist some angles $\tilde{\Phi} = (\tilde{\phi}_0, \ldots, \tilde{\phi}_{\deg(p)})$ and $\tilde{\chi} = (\tilde{\chi}_0, \ldots, \tilde{\chi}_{\deg(q)})$, so that the unitary operators $\hat{U}_{\tilde{\Phi}}$ and $\hat{U}_{\tilde{\chi}}$ implement a block encoding of the polynomials $p(\hat{H})$ and $q(\hat{H})$ respectively. These block encodings allow us to build a block encoding for $\hat{V}_t$ (see Fig. 11.15). Indeed, for any pair of n-qubit states ϕ and ψ,

$$\begin{aligned}
\langle \mathbf{0}0\phi|\hat{U}|\mathbf{0}0\psi\rangle &= \langle +\mathbf{0}\phi| \left[\cos\theta|0\rangle\hat{U}_{\tilde{\Phi}}|\mathbf{0}\psi\rangle + i\sin\theta|1\rangle\hat{U}_{\tilde{\chi}}|\mathbf{0}\psi\rangle\right] \\
&= \frac{\cos\theta}{\sqrt{2}}\langle \mathbf{0}\phi|\hat{U}_{\tilde{\Phi}}|\mathbf{0}\psi\rangle + i\frac{\sin\theta}{\sqrt{2}}\langle \mathbf{0}\phi|\hat{U}_{\tilde{\chi}}|\mathbf{0}\psi\rangle \\
&= \frac{\cos\theta}{\sqrt{2}}\langle \phi|p(\hat{H})|\psi\rangle + i\frac{\sin\theta}{\sqrt{2}}\langle \phi|q(\hat{H})|\psi\rangle \\
&= \frac{1}{\sqrt{2}}\langle \phi|\cos\theta p(\hat{H}) + i\sin\theta q(\hat{H})|\psi\rangle = \frac{1}{\sqrt{2}\nu}\langle \phi|\hat{V}_t|\psi\rangle.
\end{aligned}$$

How many gates are required to implement the quantum circuit in Fig. 11.15? The circuits $\hat{U}_{\tilde{\Phi}}$ and $\hat{U}_{\tilde{\chi}}$ contain $\deg(p) + 1 + \deg(q) + 1$ controlled $\hat{W}$ and Z_ϕ gates. Since $\deg(p) + 1 + \deg(q) + 1 = 2r + 1$, the number of gates scales as

$$\boxed{n_{\text{gates}} = O(r) = O(t + \log(1/\varepsilon)),} \qquad \textbf{qubitization}$$

where we used eqn (11.51). This is the optimal scaling for the simulation of the dynamics.

11.7 Hamiltonian simulation is BQP complete

The HAMILTONIAN problem is the problem of simulating the time evolution of a quantum system. In this chapter, we have studied different quantum algorithms that solve this problem with polynomial resources, and therefore HAMILTONIAN is in BQP. In a seminal lecture in 1985, Richard Feynman suggested that any quantum computation can be efficiently mapped into the simulation of the dynamics of a quantum system, meaning that HAMILTONIAN is BQP complete [12]. In this sense, Hamiltonian simulation is the hardest problem that a quantum computer can solve in polynomial time with a bounded error probability.

We conclude this chapter with a detailed proof that HAMILTONIAN is BQP complete following the original work by Feynman [12]. To this end, we need to show that any quantum circuit with a polynomial number of gates, $\hat{U}_1 \ldots \hat{U}_n$, is equivalent to a time evolution operator $e^{-i\hat{H}t}$ acting on a set of qubits. For pedagogical reasons, we consider a circuit comprising only $T = 3$ unitary operators $\hat{U}_1$, $\hat{U}_2$, $\hat{U}_3$ acting on a register of n qubits,

$$|\mathbf{0}\rangle \xrightarrow{\;/^{n}\;} \boxed{U_1} - \boxed{U_2} - \boxed{U_3} - |\phi\rangle$$

These operators are single and two-qubit gates that transform the initial state $|\mathbf{0}\rangle =$

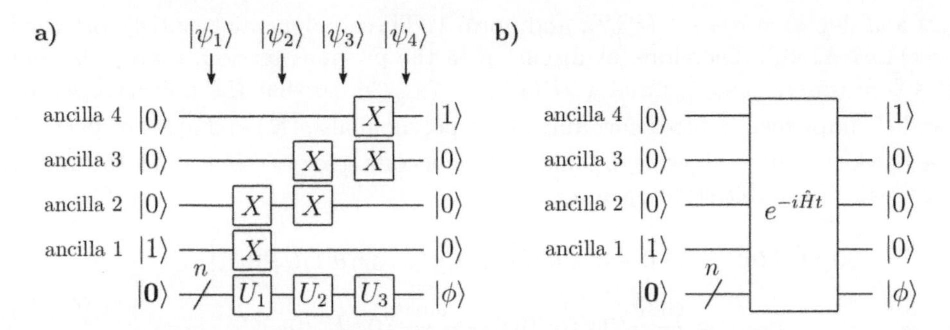

Fig. 11.16 Hamiltonian simulation is BQP complete. a) A quantum circuit comprising three unitary operators acting on n qubits plus four ancilla qubits initialized in $|0001\rangle$. The ancilla qubits work as a counter. The unitaries are single or two-qubit gates. **b)** A quantum circuit simulating the computation $\hat{U}_3\hat{U}_2\hat{U}_1$. The circuit contains n qubits plus four ancilla qubits initialized in $|0001\rangle$. The time evolution of the register is determined by the Hamiltonian $\hat{H}$ defined in eqn (11.52). The two circuits are equivalent when $t = \pi$.

$|0\ldots0\rangle$ into $|\phi\rangle \equiv \hat{U}_3\hat{U}_2\hat{U}_1|\mathbf{0}\rangle$. Let us show that performing this quantum computation is equivalent to simulating the dynamics generated by a Hamiltonian $\hat{H}$. First, we add $T+1$ ancilla qubits to the circuit as shown in Fig. 11.16a and we define $T+1$ states

$$|\psi_1\rangle = |0001\rangle|\mathbf{0}\rangle, \qquad |\psi_2\rangle = |0010\rangle\hat{U}_1|\mathbf{0}\rangle,$$
$$|\psi_3\rangle = |0100\rangle\hat{U}_2\hat{U}_1|\mathbf{0}\rangle, \qquad |\psi_4\rangle = |1000\rangle\hat{U}_3\hat{U}_2\hat{U}_1|\mathbf{0}\rangle = |1000\rangle|\phi\rangle.$$

Second, we define a Hamiltonian acting on n qubits plus $T+1=4$ ancilla qubits,

$$\begin{aligned}\hat{H} = \frac{\sqrt{3}}{2}&(\hat{\sigma}_2^+\hat{\sigma}_1^- \otimes \hat{U}_1 + \hat{\sigma}_2^-\hat{\sigma}_1^+ \otimes \hat{U}_1^\dagger)\\ &+ (\hat{\sigma}_3^+\hat{\sigma}_2^- \otimes \hat{U}_2 + \hat{\sigma}_3^-\hat{\sigma}_2^+ \otimes \hat{U}_2^\dagger)\\ &+ \frac{\sqrt{3}}{2}(\hat{\sigma}_4^+\hat{\sigma}_3^- \otimes \hat{U}_3 + \hat{\sigma}_4^-\hat{\sigma}_3^+ \otimes \hat{U}_3^\dagger).\end{aligned} \tag{11.52}$$

The operator $\hat{\sigma}_j^+ = |1\rangle\langle 0|$ ($\hat{\sigma}_j^- = |0\rangle\langle 1|$) is a lowering (raising) operator acting on the j-th ancilla qubit. $\hat{H}$ is Hermitian because it is the sum of unitary operators and their Hermitian conjugates. Furthermore, $\hat{H}$ is 4-local since each term acts on no more than four qubits[14]. Our goal is to demonstrate that the two circuits shown in Fig. 11.16 are equivalent, i.e. $e^{-i\hat{H}t}|\psi_1\rangle = |\psi_4\rangle$ for a specific evolution time t. The action of $\hat{H}$ on one of the states $|\psi_i\rangle$ produces a linear combination of $|\psi_1\rangle$, $|\psi_2\rangle$, $|\psi_3\rangle$, and $|\psi_4\rangle$. For example, $\hat{H}$ transforms the state $|\psi_2\rangle$ into

[14]The operator $\sigma_j^+\sigma_{j-1}^-$ acts on two qubits, while $\hat{U}_j$ is a single or two-qubit gate.

$$\hat{H}|\psi_2\rangle = \hat{H}|0010\rangle\hat{U}_1|\mathbf{0}\rangle = \Big(\frac{\sqrt{3}}{2}\hat{\sigma}_2^-\hat{\sigma}_1^+ \otimes \hat{U}_1^\dagger + \hat{\sigma}_3^+\hat{\sigma}_2^- \otimes \hat{U}_2\Big)|0010\rangle\hat{U}_1|\mathbf{0}\rangle$$
$$= \frac{\sqrt{3}}{2}|0001\rangle|\mathbf{0}\rangle + |0100\rangle\hat{U}_2\hat{U}_1|\mathbf{0}\rangle \;=\; \frac{\sqrt{3}}{2}|\psi_1\rangle + |\psi_3\rangle.$$

Thus, the subspace generated by the vectors $\{|\psi_1\rangle, |\psi_2\rangle, |\psi_3\rangle, |\psi_4\rangle\}$ is closed under the action of $\hat{H}$. In the subspace $B = \{|\psi_1\rangle, |\psi_2\rangle, |\psi_3\rangle, |\psi_4\rangle\}$, the matrix associated with $\hat{H}$ is a $(T+1) \times (T+1)$ matrix of the form,

$$H_B = \begin{bmatrix} 0 & \sqrt{3}/2 & 0 & 0 \\ \sqrt{3}/2 & 0 & 1 & 0 \\ 0 & 1 & 0 & \sqrt{3}/2 \\ 0 & 0 & \sqrt{3}/2 & 0 \end{bmatrix}.$$

Note that the eigenvalues of H_B are the same as the eigenvalues of a spin operator for a particle with spin $s = T/2$. Indeed, the eigenvalues of H_B are

$$\lambda_1 = -3/2, \qquad \lambda_2 = -1/2, \qquad \lambda_3 = 1/2, \qquad \lambda_4 = 3/2,$$

and the corresponding eigenvectors are

$$|\lambda_1\rangle = \begin{bmatrix} -1 \\ \sqrt{3} \\ -\sqrt{3} \\ 1 \end{bmatrix}, \quad |\lambda_2\rangle = \begin{bmatrix} 1 \\ -1/\sqrt{3} \\ -1/\sqrt{3} \\ 1 \end{bmatrix}, \quad |\lambda_3\rangle = \begin{bmatrix} -1 \\ -1/\sqrt{3} \\ 1/\sqrt{3} \\ 1 \end{bmatrix}, \quad |\lambda_4\rangle = \begin{bmatrix} 1 \\ \sqrt{3} \\ \sqrt{3} \\ 1 \end{bmatrix}.$$

It is convenient to introduce the reflection matrix R to simplify future calculations. In the subspace generated by $\{|\psi_1\rangle, |\psi_2\rangle, |\psi_3\rangle, |\psi_4\rangle\}$, the matrix associated with R is defined as

$$R = |\psi_1\rangle\langle\psi_4| + |\psi_2\rangle\langle\psi_3| + |\psi_3\rangle\langle\psi_2| + |\psi_4\rangle\langle\psi_1| = \begin{bmatrix} 0 & 0 & 0 & 1 \\ 0 & 0 & 1 & 0 \\ 0 & 1 & 0 & 0 \\ 1 & 0 & 0 & 0 \end{bmatrix}.$$

With a simple matrix multiplication, one can verify that $\hat{R}$ leaves the eigenvectors $|\lambda_1\rangle$ and $|\lambda_3\rangle$ unchanged, while it multiplies the eigenvectors $|\lambda_2\rangle$ and $|\lambda_4\rangle$ by a factor -1,

$$\hat{R}|\lambda_1\rangle = |\lambda_1\rangle, \qquad \hat{R}|\lambda_3\rangle = |\lambda_3\rangle, \qquad \hat{R}|\lambda_2\rangle = -|\lambda_2\rangle, \qquad \hat{R}|\lambda_4\rangle = -|\lambda_4\rangle. \tag{11.53}$$

Using these relations, we can finally show that $e^{-i\hat{H}t}|\psi_1\rangle = |\psi_4\rangle$ when $t = \pi$. Let us start from the decomposition of $|\psi_1\rangle$ in the eigenbasis $\{|\lambda_i\rangle\}_i$,

$$|\psi_1\rangle = c_1|\lambda_1\rangle + c_3|\lambda_3\rangle + c_2|\lambda_2\rangle + c_4|\lambda_4\rangle.$$

The action of $e^{-i\hat{H}t}$ transforms this state into

$$\begin{aligned} e^{-i\hat{H}t}|\psi_1\rangle &= c_{\lambda_1}e^{-i\lambda_1 t}|\lambda_1\rangle + c_{\lambda_3}e^{-i\lambda_3 t}|\lambda_3\rangle + c_{\lambda_2}e^{-i\lambda_2 t}|\lambda_2\rangle + c_{\lambda_4}e^{-i\lambda_4 t}|\lambda_4\rangle \\ &= c_{\lambda_1}e^{-i\lambda_1 t}|\lambda_1\rangle + c_{\lambda_3}e^{-i(\lambda_1+2)t}|\lambda_3\rangle + c_{\lambda_2}e^{-i\lambda_2 t}|\lambda_2\rangle + c_{\lambda_4}e^{-i(\lambda_2+2)t}|\lambda_4\rangle, \end{aligned}$$

where in the last step we used the fact that $\lambda_3 = \lambda_1 + 2$ and $\lambda_4 = \lambda_2 + 2$. By setting the simulation time to $t = \pi$, we obtain

$$e^{-i\hat{H}\pi}|\psi_1\rangle = e^{-i\lambda_1\pi}[c_{\lambda_1}|\lambda_1\rangle + c_{\lambda_3}|\lambda_3\rangle] + e^{-i\lambda_2\pi}[c_{\lambda_2}|\lambda_2\rangle + c_{\lambda_4}|\lambda_4\rangle]. \tag{11.54}$$

Recalling that $\lambda_2 = \lambda_1 + 1$ and using (11.53), eqn (11.54) becomes

$$\begin{aligned} e^{-i\hat{H}\pi}|\psi_1\rangle &= e^{-i\lambda_1\pi}\big[c_{\lambda_1}|\lambda_1\rangle + c_{\lambda_3}|\lambda_3\rangle\big] + e^{-i(\lambda_1+1)\pi}\big[c_{\lambda_2}|\lambda_2\rangle + c_{\lambda_4}|\lambda_4\rangle\big] \\ &= e^{-i\lambda_1\pi}\big[c_{\lambda_1}|\lambda_1\rangle + c_{\lambda_3}|\lambda_3\rangle - c_{\lambda_2}|\lambda_2\rangle - c_{\lambda_4}|\lambda_4\rangle\big] \\ &= e^{-i\lambda_1\pi}\hat{R}\big[c_{\lambda_1}|\lambda_1\rangle + c_{\lambda_3}|\lambda_3\rangle + c_{\lambda_2}|\lambda_2\rangle + c_{\lambda_4}|\lambda_4\rangle\big] \\ &= e^{-i\lambda_1\pi}\hat{R}|\psi_1\rangle \;=\; e^{-i\lambda_1\pi}|\psi_4\rangle. \end{aligned}$$

The global phase $e^{-i\lambda_1\pi}$ is irrelevant and can be ignored. This proves that the simulation of the circuit $\hat{U}_3\hat{U}_2\hat{U}_1$ is equivalent to the simulation of the dynamics generated by $\hat{H}$ for a finite time. Since $\hat{H}$ is 4-local, its dynamics can be simulated efficiently on a quantum computer using a Trotterization or any other method presented in this chapter. The generalization to circuits with a polynomial number of single and two-qubit gates is presented in Exercise 11.12. In conclusion, since any quantum circuit with a polynomial number of elementary gates can be efficiently mapped into the Hamiltonian simulation problem, HAMILTONIAN is BQP complete.

Further reading

Quantum simulation of Hamiltonian dynamics has been extensively covered in excellent articles and reviews [357, 358, 359, 351].

The first algorithm for the simulation of the dynamics was published in 1996 [25] when Lloyd proposed to simulate the unitary evolution of a k-local Hamiltonian using a Trotter formula. Around the same year, Wiesner and Zalka [360, 361] investigated the dynamics of many-body systems using a quantum computer. Dodd et al. and Nielsen et al. [362, 363] extended Lloyd's algorithm to the simulation of Hamiltonians expressed as a linear combination of Pauli operators. Independently, Aharonov and Ta-Shma applied the Trotter formula to sparse Hamiltonians [343]. Further research on the Trotterization led to formulas at higher orders [336, 364] and alternative approaches [365, 366, 367, 342]. Recently, Childs and Su [335] revisited the algorithms based on the Trotter formula and demonstrated that their performance is much better than what was indicated by simple error bounds. We should also mention the existence of alternatives to high-order Lie–Trotter–Suzuki formulas [368], where approximations to the exponential of an operator are produced by solving a set of coupled algebraic equations rather than by a recursive construction. The simulation of the single-particle Schrödinger equation was investigated by Benenti and Strini [369] and Somma [370].

Circuit compilation techniques aimed at reducing the number of gates required in a quantum simulation are receiving particular attention because NISQ devices have limited resources and are still error prone [371, 372, 341, 373].

An interesting approach to quantum simulation is based on the equivalence between quantum walks in discrete steps and in continuous time [374], which improves the resources required for the simulation of sparse Hamiltonians. The simulation of the time evolution with a truncated Taylor series and the LCU lemma ware introduced by Berry et al. in Refs [346, 27]. The Jacobi–Anger expansion was originally used for an LCU approach to quantum simulation using a linear combination of powers of the each walk step [375]. Low and Chuang developed approaches based on the quantum signal processing [352, 353] and the qubitization [28]. Low and Wiebe extended these results to the simulation of the dynamics in the interaction picture [376].

Summary

Quantum simulation of the dynamics

$\lvert\phi(0)\rangle \mapsto \lvert\psi(0)\rangle = \hat{J}\lvert\phi(0)\rangle$	Mapping
$\hat{H}_{\mathrm{s}} \mapsto \hat{H} = \hat{J}\hat{H}_{\mathrm{s}}\hat{J}^{-1}$	Mapping
$\hat{H} = \sum_j \hat{H}_j$	k-local Hamiltonian
$\hat{U}_t = e^{-i\hat{H}t}$	Exact time evolution operator
$e^{-i(\hat{H}_1+\hat{H}_2)t} = e^{-i\hat{H}_1 t}e^{-i\hat{H}_2 t}$	When $[\hat{H}_1, \hat{H}_2] = 0$
$\lVert\hat{V}_t - \hat{U}_t\rVert_{\mathrm{o}} \leq \varepsilon$	The circuit $\hat{V}_t$ simulates $\hat{U}_t = e^{-i\hat{H}t}$
$n_{\mathrm{gates}} = \mathrm{poly}(n, \varepsilon^{-1}, t)$	$\hat{V}_t$ simulates $\hat{U}_t$ efficiently

If $\hat{H}$ can be simulated efficiently, $c\hat{H}$ with $c \in \mathbb{R}$ can be simulated efficiently.
If $\hat{H}_1$ and $\hat{H}_2$ can be simulated efficiently, $\hat{H}_1 + \hat{H}_2$ can be simulated efficiently.
If $\hat{H}$ is diagonal, $\hat{H}$ can be simulated efficiently.
If $\hat{H}$ is diagonal and $\hat{U}$ can be easily implemented on a quantum computer, then $\hat{U}\hat{H}\hat{U}^{-1}$ can be simulated efficiently.

Trotterization

$\Delta t = t/m$	Duration of each step
$\hat{U}_{\Delta t} = e^{-i(\hat{H}_1+\ldots+\hat{H}_L)\Delta t}$	Exact time evolution operator
$\hat{V}_{\Delta t} = e^{-i\hat{H}_1\Delta t}\ldots e^{-i\hat{H}_L\Delta t}$	Trotter approximation
$\lVert\hat{U}_{\Delta t} - \hat{V}_{\Delta t}\rVert_{\mathrm{o}} \leq c\Delta t^2$	Error at each step
$n_{\mathrm{gates}} = O(t^2/\varepsilon)$	Number of gates

Trotterization with symmetrized decomposition

$\Delta t = t/m_{\mathrm{s}}$	Duration of each step
$\hat{U}_{\Delta t} = e^{-i(\hat{H}_1+\ldots+\hat{H}_L)\Delta t}$	Exact time evolution operator
$\hat{V}_{\Delta t}^{(\mathrm{Sym})} = \left(e^{-i\hat{H}_1\frac{\Delta t}{2}}\ldots e^{-i\hat{H}_L\frac{\Delta t}{2}}\right)\left(e^{-i\hat{H}_L\frac{\Delta t}{2}}\ldots e^{-i\hat{H}_1\frac{\Delta t}{2}}\right)$	Symm. decomposition
$\lVert\hat{U}_{\Delta t} - \hat{V}_{\Delta t}\rVert_{\mathrm{o}} \leq c_{\mathrm{s}}\Delta t^3$	Error at each step
$n_{\mathrm{gates}} = O(t^{3/2}/\sqrt{\epsilon})$	Number of gates

Trotterization of sparse Hamiltonians

$\hat{H}_{\mathrm{o}}$ is d sparse with real entries	
$\hat{H}_{\mathrm{o}} \leftrightarrow G$	Mapping into a connected graph G
$\hat{H}_{\mathrm{o}} = \sum_c \hat{H}_c$	Efficient decomposition
$\hat{V}_c\lvert a\rangle\lvert b\rangle\lvert z\rangle = \lvert a\rangle\lvert b \oplus v_c(a)\rangle\lvert z \oplus h_{a,v_c(a)}\rangle$	Definition of $\hat{V}_c$
$\hat{S}\lvert a\rangle\lvert b\rangle\lvert z\rangle = z\lvert b\rangle\lvert a\rangle\lvert z\rangle$	Definition of $\hat{S}$
$\mathbf{H}_{\mathrm{o}} = \sum_c \hat{V}_c^\dagger\hat{S}\hat{V}_c$	Embedding of the Hamiltonian
$e^{-i\mathbf{H}_{\mathrm{o}}\Delta t} \approx \hat{V}_{\Delta t} = \prod_c \hat{V}_c^\dagger e^{-i\hat{S}\Delta t}\hat{V}_c$	Approx. of the time-evolution operator

Linear combination of unitaries

$\tilde{U} = \sum_{j=0}^{M-1} c_j \hat{V}_j$	LCU
$s = \sum_j c_j$	Sum of the coefficients
$\hat{A}\lvert\mathbf{0}\rangle = \frac{1}{\sqrt{s}} \sum_{j=0}^{M-1} \sqrt{c_j}\lvert j\rangle$	Prepare operator
$\hat{B} = \sum_{j=0}^{M-1} \lvert j\rangle\langle j\rvert \otimes \hat{V}_j$	Select operator
$\lvert\psi_{\text{fin}}\rangle = \sqrt{p_{\text{s}}}\lvert\mathbf{0}\rangle \otimes \frac{\tilde{U}\lvert\psi\rangle}{\lVert\tilde{U}\lvert\psi\rangle\rVert} + \sqrt{1-p_{\text{s}}}\lvert\Phi\rangle$	Final state
$p_{\text{s}} = \langle\psi\rvert\tilde{U}^\dagger\tilde{U}\lvert\psi\rangle/s^2$	Success probability

Taylor series

$\hat{H} = \sum_{j=1}^{L} b_j \hat{U}_j$	$\hat{U}_j$ can be implemented efficiently
$\hat{U}_{\Delta t} = \sum_{k=0}^{\infty} \frac{(-i\Delta t)^k}{k!} \hat{H}^k$	Exact time evolution operator
$\hat{V}_{\Delta t} = \sum_{k=0}^{r-1} \frac{(-i\Delta t)^k}{k!} \hat{H}^k$	Truncated Taylor series
$\hat{V}_{\Delta t} = \sum_{k=0}^{M_r-1} c_j \hat{V}_j$	LCU
$M_r = (L^r - 1)/(L-1)$	Number of terms in the LCU
$n_{\text{gates}} = O\Big(t \frac{\log^2(t/\varepsilon)}{\log\log(t/\varepsilon)}\Big)$	Number of gates

Qubitization

$\hat{H} = \sum_{j=1}^{L} c_j \hat{P}_j$	Decomposition of the Hamiltonian
$\lvert g\rangle = \hat{A}\lvert\mathbf{0}\rangle = \sum_{j=0}^{L-1} \sqrt{c_j}\lvert j\rangle$	A is the prepare operator
$\hat{R}_g = (2\lvert g\rangle\langle g\rvert \otimes \mathbb{1}_n - \mathbb{1}_p \otimes \mathbb{1}_n)$	Reflection operator
$\hat{B} = \sum_{j=0}^{L-1} \lvert j\rangle\langle j\rvert \otimes \hat{P}_j$	Select operator
$\langle gi\rvert\hat{B}\lvert gj\rangle = \langle i\rvert\hat{H}\lvert j\rangle$	$\hat{B}$ is an encoding of the Hamiltonian
$\hat{W} = \hat{R}_g\hat{B}$	Definition of $\hat{W}$
$\lvert 0_\lambda\rangle = \lvert g\rangle\lvert\lambda\rangle$	Ground state of the qubit
$\lvert 1_\lambda\rangle = (\lambda\lvert g\rangle\lvert\lambda\rangle - \hat{B}\lvert g\rangle\lvert\lambda\rangle)/\sqrt{1-\lambda^2}$	Excited state of the qubit
$\hat{W} = \bigoplus_\lambda e^{-\frac{i}{2}\theta_\lambda\hat{\sigma}_y}$ with $\theta_\lambda = -2\arccos\lambda$	Decomp. of $\hat{W}$ into single-qubit rotations

HAMILTONIAN is BQP complete

$\hat{C}$	Circuit with T gates $\hat{U}_1, \ldots, \hat{U}_T$
$\hat{H} = \frac{2}{T+1}\sum_{j=1}^{T}\sqrt{j(T+1-j)}\left(\hat{\sigma}_{j+1}^{+}\hat{\sigma}_{j}^{-}\otimes\hat{U}_j + \hat{\sigma}_{j+1}^{-}\hat{\sigma}_{j}^{+}\otimes\hat{U}_j^{\dagger}\right)$	Hamiltonian
$\lvert\psi_j\rangle = \lvert 0_{T+1}\ldots 1_j \ldots 0_1\rangle\hat{U}_j\ldots\hat{U}_1\lvert 0\ldots 0\rangle$	Intermediate states
$B = \{\lvert\psi_1\rangle, \ldots, \lvert\psi_{T+1}\rangle\}$	Basis of a subspace
$(\hat{H})_{ab} = \frac{1}{2}(\delta_{a,b+1} + \delta_{a+1,b})\sqrt{(T/2+1)(a+b-1) - ab}$	Hamilt. in the basis B
$\hat{R} = \sum_j \lvert\psi_j\rangle\langle\psi_{T+1-j}\rvert$	Reflection operator
$e^{-i\hat{H}t}\lvert\psi_1\rangle = \lvert\psi_{T+1}\rangle$	Final state

Exercises

Exercise 11.1 Derive eqn (11.2). **Hint:** Start from a Hamiltonian with two terms, $\hat{H} = \hat{H}_1 + \hat{H}_2$.

Exercise 11.2 Consider a linear operator $\hat{A}$ and a unitary operator $\hat{U}$. Show that $e^{\hat{U}\hat{A}\hat{U}^{\dagger}} = \hat{U}e^{\hat{A}}\hat{U}^{\dagger}$. **Hint:** Expand $e^{\hat{U}\hat{A}\hat{U}^{\dagger}}$ with a Taylor series.

Exercise 11.3 Consider a real number $c \neq 0$. Show that if $\hat{H}$ can be simulated efficiently, $c\hat{H}$ can be simulated efficiently.

Exercise 11.4 Consider a unitary operator $\hat{U}$ that can be implemented on a quantum computer with a polynomial number of elementary gates. Show that if $\hat{H}$ can be simulated efficiently, then $\hat{H}' = \hat{U}\hat{H}\hat{U}^{\dagger}$ can be simulated efficiently. **Hint:** Use the formula $e^{\hat{U}\hat{H}\hat{U}^{\dagger}} = \hat{U}e^{\hat{H}}\hat{U}^{\dagger}$.

Exercise 11.5 Consider a diagonal Hamiltonian $\hat{H} = \text{diag}(h_0, \ldots, h_{2^n-1})$, where each entry $h_{\mathbf{x}}$ is a non-negative real number that can be approximated with p bits,

$$h_{\mathbf{x}} = c_{\mathbf{x}_{p-1}}2^{p-1} + \ldots + c_{\mathbf{x}_0}2^0,$$

where $c_{\mathbf{x}_j} \in \{0, 1\}$. Build a quantum circuit that simulates $\hat{U}(t) = e^{-i\hat{H}t}$ efficiently.

Exercise 11.6 Suppose $\hat{H}$ can be efficiently diagonalized on a classical computer, i.e. we can easily calculate its eigenvalues. Show that $\hat{H}$ can be simulated efficiently. **Hint:** Combine the results of Exercises 11.5 and 11.4.

Exercise 11.7 Consider the linear combination of unitaries: $\tilde{U} = \cos^2\theta\cos^2\phi\hat{Z}\otimes\mathbb{1} + \cos^2\theta\sin^2\phi\mathbb{1}\otimes\hat{Z} + \sin^2\theta\hat{X}\otimes\hat{X}$. Build a quantum circuit with two qubits and two ancilla qubits that implements this LCU. **Hint:** Use eqns (11.23, 11.24).

Exercise 11.8 Show that the solution of $(ex/r)^r < (1-x)\varepsilon/m$ is $r \in O\left(\frac{\ln(m/\varepsilon)}{\ln(\ln(m/\varepsilon))}\right)$.

Exercise 11.9 Show that the states

$$|0_\lambda\rangle = |g\rangle|\lambda\rangle, \qquad |1_\lambda\rangle = \frac{\lambda|g\rangle|\lambda\rangle - B|g\rangle|\lambda\rangle}{\sqrt{1-\lambda^2}}$$

are orthonormal. Show that the action of the operator $\hat{W} = \hat{R}_g\hat{B}$ onto these states produces

$$\hat{W}|0_\lambda\rangle = \lambda|0_\lambda\rangle + \sqrt{1-\lambda^2}|1_\lambda\rangle, \qquad \hat{W}|1_\lambda\rangle = -\sqrt{1-\lambda^2}|0_\lambda\rangle + \lambda|1_\lambda\rangle.$$

Hint: Recall from eqn (11.39) that $\langle gi|B|gj\rangle = \langle i|\hat{H}|j\rangle$.

Exercise 11.10 Show that the qubiterate operator $\hat{W}$ can be used to simulate the eigenstates of the Hamiltonian using the quantum phase estimation presented in Section 12.1.

Exercise 11.11 Consider the operator

$$\hat{W} = \bigoplus_{\lambda=1}^{2^n} \mathsf{Y}_{\theta_\lambda} = \bigoplus_{\lambda=1}^{2^n} e^{-i\frac{\theta_\lambda}{2}\hat{\sigma}_y} = \bigoplus_{\lambda=1}^{2^n} \left[\cos\left(\frac{\theta_\lambda}{2}\right)\mathbb{1} - i\sin\left(\frac{\theta_\lambda}{2}\right)\hat{\sigma}_y\right],$$

where $\theta_\lambda = -2\arccos\lambda$. Show that

$$\hat{W}^k = \bigoplus_{\lambda=1}^{2^n} (\mathsf{Y}_{\theta_\lambda})^k = \bigoplus_{\lambda=1}^{2^n} \begin{bmatrix} T_k(\lambda) & -\sqrt{1-\lambda^2}U_{k-1}(\lambda) \\ \sqrt{1-\lambda^2}U_{k-1}(\lambda) & T_k(\lambda) \end{bmatrix},$$

where $T_k(x)$ and $U_k(x)$ are the Chebyshev polynomials of the first and second kind.

Exercise 11.12 Consider a family of quantum circuits $\{\hat{C}_i\}_i$. The circuits in the family contain a polynomial number of single and two-qubit gates. Show that each circuit C_i can be mapped into the simulation of the dynamics generated by a Hamiltonian.

Solutions

Solution 11.1 Consider the Hamiltonian $\hat{H} = \hat{H}_1 + \hat{H}_2$ and suppose that $[\hat{H}_1, \hat{H}_2] = 0$. The exponentials e^{-iH_1t} and e^{-iH_2t} can be expanded with a Taylor series

$$e^{-i\hat{H}_1t} = \sum_{n=0}^{\infty} \frac{(-i\hat{H}_1t)^n}{n!}, \qquad e^{-i\hat{H}_2t} = \sum_{n=0}^{\infty} \frac{(-i\hat{H}_2t)^n}{n!}.$$

Since $\hat{H}_1$ and $\hat{H}_2$ commute, we have

$$e^{-i\hat{H}_1t}e^{-i\hat{H}_2t} = \sum_{n,m=0}^{\infty} \frac{(-i\hat{H}_1t)^n}{n!}\frac{(-i\hat{H}_2t)^m}{m!} = \sum_{n=0}^{\infty} \frac{(-i(\hat{H}_1+\hat{H}_2)t)^n}{n!} = e^{-i(\hat{H}_1+\hat{H}_2)t}.$$

This result is valid for any pair of commuting operators. Thus, $e^{-i(\hat{H}_1+\ldots+\hat{H}_L)t} = \prod_{j=1}^{L} e^{-i\hat{H}_jt}$.

Solution 11.2 By expanding $e^{\hat{U}\hat{A}\hat{U}^\dagger}$ with a Taylor series, we have

$$\begin{aligned} e^{\hat{U}\hat{A}\hat{U}^\dagger} &= \mathbb{1} + \hat{U}\hat{A}\hat{U}^\dagger + \frac{1}{2!}\hat{U}\hat{A}\hat{U}^\dagger\hat{U}\hat{A}\hat{U}^\dagger + \frac{1}{3!}\hat{U}\hat{A}\hat{U}^\dagger\hat{U}\hat{A}\hat{U}^\dagger\hat{U}\hat{A}\hat{U}^\dagger + O(\hat{A}^4) \\ &= \mathbb{1} + \hat{U}\hat{A}\hat{U}^\dagger + \frac{1}{2!}\hat{U}A^2\hat{U}^\dagger + \frac{1}{3!}\hat{U}\hat{A}^3\hat{U}^\dagger + O(\hat{A}^4) \\ &= \hat{U}\Big[\mathbb{1} + \hat{A} + \frac{1}{2!}\hat{A}^2 + \frac{1}{3!}\hat{A}^3 + O(\hat{A}^4)\Big]\hat{U}^\dagger = \hat{U}e^{\hat{A}}\hat{U}^\dagger. \end{aligned}$$

Solution 11.3 One can rescale both the Hamiltonian and the simulation time by the same factor

$$\hat{H}' = c\hat{H}, \qquad t' = t/c.$$

The simulation of H' for a time t' is equivalent to the simulation of $\hat{H}$ for a time t. Since $\hat{H}$ can be simulated efficiently for any time t, H' can also be simulated efficiently for any time t'.

Solution 11.4 Suppose that the quantum circuit $\hat{V}_t$ simulates the time evolution $e^{-i\hat{H}t}$ efficiently. Since $e^{-i\hat{H}'t} = e^{-i\hat{U}\hat{H}\hat{U}^\dagger} = \hat{U}e^{-i\hat{H}}\hat{U}^\dagger$, the quantum circuit $\hat{U}\hat{V}_tU^\dagger$ simulates the dynamics generated by the Hamiltonian $\hat{H}'$ efficiently.

Solution 11.5 The goal is to build a quantum circuit that implements the transformation $e^{-i\hat{H}t}$ with a polynomial number of gates. To this end, we will use a register of n qubits and p ancilla qubits prepared in $|0\dots0\rangle$. The circuit shown below simulates the dynamics exactly with a polynomial number of gates

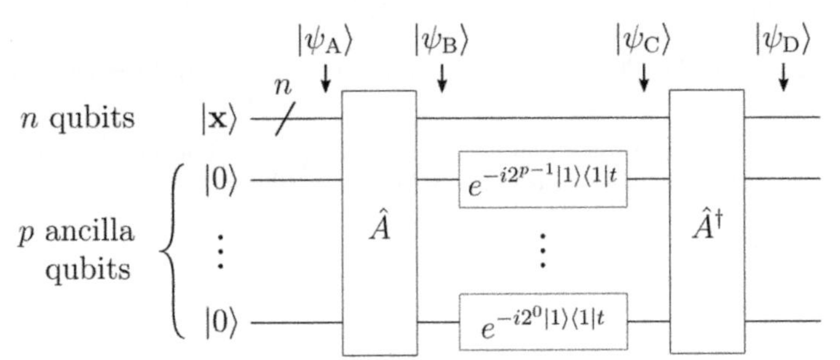

The operator $\hat{A}$ is defined as $\hat{A}|\mathbf{x}\rangle|\mathbf{y}\rangle = |\mathbf{x}\rangle|\mathbf{y} \oplus h_{\mathbf{x}}\rangle$. The initial state $|\mathbf{x}\rangle|\mathbf{0}\rangle$ is transformed into

$$|\psi_B\rangle = \hat{A}|\mathbf{x}\rangle|\mathbf{0}\rangle = |\mathbf{x}\rangle|h_{\mathbf{x}}\rangle = |x_{n-1}\dots x_0\rangle|c_{\mathbf{x}_{p-1}}\dots c_{\mathbf{x}_0}\rangle.$$

The states of the ancilla qubits are now rotated with single-qubit operations,

$$\begin{aligned} |\psi_C\rangle &= |\mathbf{x}\rangle\big(e^{-it2^{p-1}|1\rangle\langle 1|} \otimes \dots \otimes e^{-it2^0|1\rangle\langle 1|}|c_{\mathbf{x}_{p-1}}\dots c_{\mathbf{x}_0}\rangle\big) \\ &= |\mathbf{x}\rangle e^{-it2^{p-1}c_{\mathbf{x}_{p-1}}}\dots e^{-it2^0 c_{\mathbf{x}_0}}|c_{\mathbf{x}_{p-1}}\dots c_{\mathbf{x}_0}\rangle \\ &= |\mathbf{x}\rangle e^{-ih_{\mathbf{x}}t}|c_{\mathbf{x}_{p-1}}\dots c_{\mathbf{x}_0}\rangle = e^{-ih_{\mathbf{x}}t}|\mathbf{x}\rangle|h_{\mathbf{x}}\rangle. \end{aligned}$$

The application of $\hat{A}^\dagger$ produces $|\psi_D\rangle = e^{-ih_{\mathbf{x}}t}|\mathbf{x}\rangle|0\dots0\rangle$, because $A^\dagger|\mathbf{x}\rangle|h_{\mathbf{x}}\rangle = |\mathbf{x}\rangle|h_{\mathbf{x}} \oplus h_{\mathbf{x}}\rangle = |\mathbf{x}\rangle|\mathbf{0}\rangle$. At the end of the simulation, the ancilla qubits are all in 0 while the state of the n qubits is $e^{-ih_{\mathbf{x}}t}|\mathbf{x}\rangle = e^{-i\hat{H}t}|\mathbf{x}\rangle$. The circuit uses p ancilla qubits plus two controlled unitaries.

This simulation is exact for all times t.

Solution 11.6 $\hat{H}$ can be easily diagonalized, i.e. we can easily construct a unitary operation $\hat{U}$ such that $\hat{H}' = \hat{U}\hat{H}\hat{U}^{-1}$ is diagonal. From Exercise 11.5, we know that the diagonal matrix $\hat{H}'$ can be simulated efficiently. Since $\hat{H} = \hat{U}^{-1}\hat{H}'\hat{U}$, then $\hat{H}$ can be simulated efficiently (see Exercise 11.4).

Solution 11.7 The operator $\tilde{U}$ is an LCU with $M = 3$ coefficients and unitary operators,

$$c_0 = \cos^2\theta\cos^2\phi, \qquad c_1 = \cos^2\theta\sin^2\phi, \qquad c_2 = \sin^2\theta,$$
$$\hat{V}_0 = \hat{Z}\otimes\mathbb{1}, \qquad \hat{V}_1 = \mathbb{1}\otimes\hat{Z}, \qquad \hat{V}_2 = \hat{X}\otimes\hat{X},$$

with $s = \sum_i c_i = 1$. To implement this LCU, we need to initialize $\lceil\log_2 M\rceil = 2$ ancilla qubits in the state (see eqn (11.23))

$$\hat{A}|00\rangle = \cos(\theta)\cos(\phi)|00\rangle + \cos(\theta)\sin(\phi)|01\rangle + \sin(\theta)|10\rangle,$$

and then apply the unitary operation (see eqn (11.24))

$$\hat{B}|00\rangle = |00\rangle\langle 00|\otimes\hat{V}_0 + |01\rangle\langle 01|\otimes\hat{V}_1 + |10\rangle\langle 10|\otimes\hat{V}_2.$$

This can be done with the quantum circuit shown below

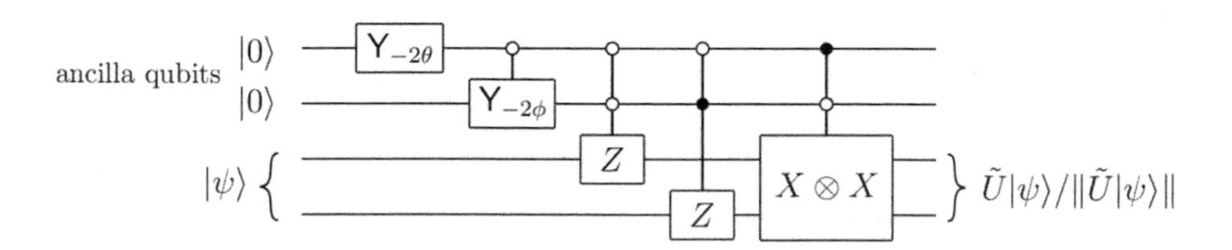

Solution 11.8 The inequality $(ex/r)^r$ can be simplified by removing the factor $1 - x$: $(ex/r)^r < \varepsilon/m$. The solution to this inequality is

$$r > \frac{\ln(m/\varepsilon)}{W(\frac{1}{ex}\ln(m/\varepsilon))},$$

where $W(x)$ is the Lambert function. Assuming $\varepsilon \ll 1$, we can perform the approximation

$$W\Big(\frac{1}{ex}\ln(m/\varepsilon)\Big) \approx \ln\Big(\frac{1}{ex}\ln(m/\varepsilon)\Big) - \ln\Big(\ln\Big(\frac{1}{ex}\ln(m/\varepsilon)\Big)\Big) + c,$$

where c is a small constant. Thus, we have

$$r \in O\left(\frac{\ln(m/\varepsilon)}{\ln(\ln(m/\varepsilon))}\right).$$

Solution 11.9 Let us show that $|0_\lambda\rangle$ and $|1_\lambda\rangle$ are orthonormal,

$$\langle 0_\lambda | 1_\lambda \rangle = \langle 0_\lambda | \frac{\lambda |0_\lambda\rangle - B|0_\lambda\rangle}{\sqrt{1-\lambda^2}} = \frac{1}{\sqrt{1-\lambda^2}}(\lambda - \langle 0_\lambda | B | 0_\lambda \rangle)$$
$$= \frac{1}{\sqrt{1-\lambda^2}}(\lambda - \langle g\lambda | B | g\lambda \rangle) = \frac{1}{\sqrt{1-\lambda^2}}(\lambda - \langle \lambda | H | \lambda \rangle) = 0.$$

Now we need to calculate the action of $\hat{W}$ onto $|0_\lambda\rangle = |g\lambda\rangle$. Using eqn (11.39), we have

$$\hat{W}|0_\lambda\rangle = (2|g\rangle\langle g| \otimes \mathbb{1}_n - \mathbb{1}_p \otimes \mathbb{1}_n)\hat{B}|g\lambda\rangle = (2|g\rangle\langle g| \otimes \mathbb{1}_n)\hat{B}|g\lambda\rangle - \hat{B}|g\lambda\rangle$$
$$= \sum_k 2|gk\rangle\langle gk|\hat{B}|g\lambda\rangle - \hat{B}|g\lambda\rangle = \sum_k 2|gk\rangle\langle k|\hat{H}|\lambda\rangle - \hat{B}|g\lambda\rangle$$
$$= \sum_k 2\lambda|gk\rangle\langle k|\lambda\rangle - \hat{B}|g\lambda\rangle = 2\lambda|g\lambda\rangle - \hat{B}|g\lambda\rangle = 2\lambda|0_\lambda\rangle - \hat{B}|0_\lambda\rangle.$$

Let us write this equation in the form

$$\hat{W}|0_\lambda\rangle = 2\lambda|0_\lambda\rangle - \hat{B}|0_\lambda\rangle = \lambda|0_\lambda\rangle + \underbrace{\lambda|0_\lambda\rangle - \hat{B}|0_\lambda\rangle}_{\sqrt{1-\lambda^2}|1_\lambda\rangle} = \lambda|0_\lambda\rangle + \sqrt{1-\lambda^2}|1_\lambda\rangle. \tag{11.55}$$

Since the coefficients of this state sum up to one ($\langle 0_\lambda|\hat{W}|0_\lambda\rangle^2 + \langle 1_\lambda|\hat{W}|0_\lambda\rangle^2 = 1$) there cannot be other states in the decomposition of $\hat{W}|0_\lambda\rangle$. We are left to demonstrate that the application of $\hat{W}$ onto the state $|1_\lambda\rangle$ produces the linear combination

$$\hat{W}|1_\lambda\rangle = -\sqrt{1-\lambda^2}|0\rangle + \lambda|1_\lambda\rangle.$$

To this end, let us calculate $\langle 0_\lambda|\hat{W}|1_\lambda\rangle$ and $\langle 1_\lambda|\hat{W}|1_\lambda\rangle$ separately. We have

$$\langle 0_\lambda|\hat{W}|1_\lambda\rangle = \underbrace{\langle 0_\lambda|(2|g\rangle\langle g| \otimes \mathbb{1}_n - \mathbb{1}_p \otimes \mathbb{1}_n)}_{\langle 0_\lambda|}\hat{B}|1_\lambda\rangle = \langle 0_\lambda|\hat{B}|1_\lambda\rangle = \langle 0_\lambda|\frac{\lambda\hat{B}|0_\lambda\rangle - \hat{B}^2|0_\lambda\rangle}{\sqrt{1-\lambda^2}}$$
$$= \frac{1}{\sqrt{1-\lambda^2}}[\lambda\langle 0_\lambda|\hat{B}|0_\lambda\rangle - \langle 0_\lambda|\underbrace{\hat{B}^2}_{\mathbb{1}}|0_\lambda\rangle]$$
$$= \frac{1}{\sqrt{1-\lambda^2}}[\lambda^2 - 1] = -\sqrt{1-\lambda^2},$$

where we used the fact that $\hat{B}^2 = \mathbb{1}_p \otimes \mathbb{1}_n$,

$$\hat{B}^2 = \sum_j \sum_k (|j\rangle\langle j| \otimes \hat{P}_j)(|k\rangle\langle k| \otimes \hat{P}_k)$$
$$= \sum_j \sum_k |j\rangle\langle j|k\rangle\langle k| \otimes \hat{P}_j\hat{P}_k = \sum_j |j\rangle\langle j| \otimes \hat{P}_j^2 = \sum_j |j\rangle\langle j| \otimes \mathbb{1}_n = \mathbb{1}_p \otimes \mathbb{1}_n.$$

The last step is to compute $\langle 1_\lambda|\hat{W}|1_\lambda\rangle$,

$$\langle 1_\lambda|\hat{W}|1_\lambda\rangle = \frac{1}{1-\lambda^2}((\langle 0_\lambda|\lambda - \langle 0_\lambda|\hat{B})\hat{W}(\lambda|0_\lambda\rangle - \hat{B}|0_\lambda\rangle) =$$

$$= \frac{1}{1-\lambda^2}\big(\lambda^2 \underbrace{\langle 0_\lambda|\hat{W}|0_\lambda\rangle}_{a} - \lambda \underbrace{\langle 0_\lambda|\hat{W}\hat{B}|0_\lambda\rangle}_{b} - \lambda \underbrace{\langle 0_\lambda|\hat{B}\hat{W}|0_\lambda\rangle}_{c} + \underbrace{\langle 0_\lambda|\hat{B}\hat{W}\hat{B}|0_\lambda\rangle}_{d}\big).$$

Let us calculate the terms a, b, c, and d one by one. First, from eqn (11.55) we know that $a = \langle 0_\lambda|\hat{W}|0_\lambda\rangle = \lambda$. The term b is given by

$$b = \langle 0_\lambda|\hat{W}\hat{B}|0_\lambda\rangle = \underbrace{\langle 0_\lambda|(2|g\rangle\langle g| \otimes \mathbb{1} - \mathbb{1} \otimes \mathbb{1})}_{\langle 0_\lambda|}\hat{B}\hat{B}|0_\lambda\rangle$$
$$= \langle 0_\lambda|\hat{B}\hat{B}|0_\lambda\rangle = \langle 0_\lambda|0_\lambda\rangle = 1.$$

The term c can be derived as follows

$$c = \langle 0_\lambda|\hat{B}\hat{W}|0_\lambda\rangle = \langle 0_\lambda|\hat{B}(2\lambda|0_\lambda\rangle - \hat{B}|0_\lambda\rangle)$$
$$= 2\lambda\langle 0_\lambda|\hat{B}|0_\lambda\rangle - \langle 0_\lambda|\hat{B}^2|0_\lambda\rangle = 2\lambda^2 - 1.$$

Lastly,

$$d = \langle 0_\lambda|\hat{B}\hat{W}\hat{B}|0_\lambda\rangle = \langle 0_\lambda|B(2|g\rangle\langle g| \otimes \mathbb{1} - \mathbb{1} \otimes \mathbb{1})\hat{B}\hat{B}|0_\lambda\rangle$$
$$= \langle 0_\lambda|\hat{B}\underbrace{(2|g\rangle\langle g| \otimes \mathbb{1} - \mathbb{1} \otimes \mathbb{1})|0_\lambda\rangle}_{|0_\lambda\rangle} = \langle 0_\lambda|\hat{B}|0_\lambda\rangle = \lambda.$$

Putting everything together, we finally obtain

$$\langle 1_\lambda|\hat{W}|1_\lambda\rangle = \frac{1}{1-\lambda^2}(\lambda^3 - \lambda - \lambda(2\lambda^2 - 1) + \lambda) = \lambda.$$

Since $\langle 0_\lambda|\hat{W}|1_\lambda\rangle^2 + \langle 1_\lambda|\hat{W}|1_\lambda\rangle^2 = 1$, there cannot be other states in the decomposition of $\hat{W}|1_\lambda\rangle$.

Solution 11.10 It is convenient to introduce the eigenstates and eigenvalues of $\hat{W}$. Since $\hat{W}$ acts as a rotation about the y axis, its eigenvectors are equal to those of $\hat{\sigma}_y$:

$$\hat{W}|1_{y,\lambda}\rangle = e^{-\frac{i}{2}\theta_\lambda}|1_{y,\lambda}\rangle, \qquad \text{where} \quad |1_{y,\lambda}\rangle = \frac{1}{\sqrt{2}}(|0_\lambda\rangle + i|1_\lambda\rangle),$$
$$\hat{W}|-1_{y,\lambda}\rangle = e^{\frac{i}{2}\theta_\lambda}|-1_{y,\lambda}\rangle, \qquad \text{where} \quad |-1_{y,\lambda}\rangle = \frac{1}{\sqrt{2}}(|0_\lambda\rangle - i|1_\lambda\rangle),$$

where $\lambda \in (-1, 1)$ is an eigenvalue of the Hamiltonian. The eigenvalues of $\hat{W}$ can be written as

$$e^{\mp\frac{i}{2}\theta_\lambda} = e^{\pm i\arccos(\lambda)} = e^{i2\pi u_\lambda^\pm} \tag{11.56}$$

where

$$u_\lambda^+ = \frac{\arccos(\lambda)}{2\pi} \in (0, 1/2)\ ,\ u_\lambda^- = 1 - \frac{\arccos(\lambda)}{2\pi} \in (1/2, 1)\ . \tag{11.57}$$

The quantum phase estimation algorithm, executed with $|0_\lambda\rangle$ as initial state and $\hat{W}$ as unitary operator, will return the phases $u_\lambda^\pm$ with equal probabilities. If the measured phase is a number $u \in (0, 1/2)$, then $u = u_\lambda^+$ and the eigenvalue of the Hamiltonian is $\lambda = \cos(2\pi u)$. If the measured phase is a number $u \in (1/2, 1)$, then $u = u_\lambda^-$ and the eigenvalue of the

Hamiltonian is $\lambda = \cos(2\pi(1-u))$.

Solution 11.11 Recall that

$$T_k(\cos x) = \cos(kx),$$
$$\sin x U_{k-1}(\cos x) = \sin(kx),$$

where $T_k(x)$ and $U_k(x)$ are the Chebyshev polynomials of the first and second kind. Then,

$$\begin{aligned}\hat{W}^k &= \bigoplus_{\lambda=1}^{2^n}\left[\cos\left(k\frac{\theta_\lambda}{2}\right)\mathbb{1} - i\sin\left(k\frac{\theta_\lambda}{2}\right)\hat{\sigma}_y\right] \\ &= \bigoplus_{\lambda=1}^{2^n}\left[T_k(\cos\frac{\theta_\lambda}{2})\mathbb{1} - i\sin\frac{\theta_\lambda}{2}U_{k-1}(\cos\frac{\theta_\lambda}{2})\hat{\sigma}_y\right] \\ &= \bigoplus_{\lambda=1}^{2^n}\left[T_k(\lambda)\mathbb{1} - i\sqrt{1-\lambda^2}U_{k-1}(\lambda)\hat{\sigma}_y\right].\end{aligned}$$

Solution 11.12 Suppose that $\hat{C}_i$ contains T single and two-qubit gates $\hat{U}_1, \ldots, \hat{U}_T$. The Hamiltonian that simulates $\hat{C}_i$ acts on n qubits plus $T+1$ ancilla qubits,

$$\hat{H} = \frac{2}{T+1}\sum_{j=1}^{T}\sqrt{j(T+1-j)}\left(\hat{\sigma}_{j+1}^{+}\hat{\sigma}_j^{-}\otimes\hat{U}_j + \hat{\sigma}_{j+1}^{-}\hat{\sigma}_j^{+}\otimes\hat{U}_j^{\dagger}\right).$$

The states $|\psi_i\rangle$ are defined as

$$\begin{aligned}|\psi_1\rangle &= |0\ldots 01\rangle|0\ldots 0\rangle, \\ |\psi_2\rangle &= |0\ldots 10\rangle\hat{U}_1|0\ldots 0\rangle, \\ &\vdots \\ |\psi_{T+1}\rangle &= |1\ldots 00\rangle\hat{U}_T\ldots\hat{U}_1|0\ldots 0\rangle.\end{aligned}$$

The matrix associated to $\hat{H}$ in the basis $B = \{|\psi_1\rangle, \ldots, |\psi_{T+1}\rangle\}$ is a $(T+1)\times(T+1)$ matrix of the form

$$(H_B)_{ab} = \frac{1}{2}(\delta_{a,b+1} + \delta_{a+1,b})\sqrt{(T/2+1)(a+b-1)-ab}.$$

This is nothing but the spin matrix S_x for a particle with spin $s = T/2$. The eigenvalues $\lambda_1, \ldots, \lambda_{T+1}$ are equally spaced ($\lambda_{j+1} - \lambda_j = 1$) and the corresponding eigenvectors are defined as $|\lambda_j\rangle$. Let us introduce the reflection matrix R in the basis B,

$$R = \begin{bmatrix} & & & 1 \\ & & 1 & \\ & \cdot^{\cdot^{\cdot}} & & \\ 1 & & & \end{bmatrix}.$$

This operator leaves the eigenvectors $|\lambda_{\text{odd}}\rangle$ unchanged and multiplies the eigenvectors $|\lambda_{\text{even}}\rangle$ by -1,

$$R|\lambda_{\text{odd}}\rangle = |\lambda_{\text{odd}}\rangle, \qquad R|\lambda_{\text{even}}\rangle = -|\lambda_{\text{even}}\rangle.$$

The decomposition of the state $|\psi_1\rangle$ in the eigenbasis $\{|\lambda_j\rangle\}_j$ is given by

$$|\psi_1\rangle = \sum_{j\in\{1,3,\ldots\}} c_j|\lambda_j\rangle + \sum_{j\in\{2,4,\ldots\}} c_j|\lambda_j\rangle.$$

The application of the operator $e^{-i\hat{H}t}$ leads to

$$\begin{aligned}
e^{-i\hat{H}t}|\psi_1\rangle &= \sum_{j\in\{1,3,\ldots\}} c_j e^{-i\lambda_j t}|\lambda_j\rangle + \sum_{j\in\{2,4,\ldots\}} c_j e^{-i\lambda_j t}|\lambda_j\rangle \\
&= \sum_{j\in\{1,3,\ldots\}} c_j e^{-i(\lambda_1+(j-1))t}|\lambda_j\rangle + \sum_{j\in\{2,4,\ldots\}} c_j e^{-i(\lambda_2+(j-2))t}|\lambda_j\rangle \\
&= \sum_{j\in\{1,3,\ldots\}} c_j e^{-i(\lambda_1+(j-1))t}|\lambda_j\rangle + \sum_{j\in\{2,4,\ldots\}} c_j e^{-i(\lambda_1+1+(j-2))t}|\lambda_j\rangle.
\end{aligned}$$

If we now set $t = \pi$, we obtain

$$\begin{aligned}
e^{-i\hat{H}t}|\psi_1\rangle &= e^{-i\lambda_1\pi}\Big[\sum_{j\in\{1,3,\ldots\}} c_j e^{-i(j-1)\pi}|\lambda_j\rangle + \sum_{j\in\{2,4,\ldots\}} c_j e^{-i(1+(j-2))\pi}|\lambda_j\rangle\Big] \\
&= e^{-i\lambda_1\pi}\Big[\sum_{j\in\{1,3,\ldots\}} c_j|\lambda_j\rangle - \sum_{j\in\{2,4,\ldots\}} c_j|\lambda_j\rangle\Big] \\
&= e^{-i\lambda_1\pi}R\Big[\sum_{j\in\{1,3,\ldots\}} c_j|\lambda_j\rangle + \sum_{j\in\{2,4,\ldots\}} c_j|\lambda_j\rangle\Big] = e^{-i\lambda_1\pi}R|\psi_1\rangle = |\psi_{T+1}\rangle.
\end{aligned}$$

12 Quantum simulation of Hamiltonian eigenstates

If a simulation method is accurate enough,
a theoretical model can be tested
without running an actual experiment.

In the previous chapter we studied the simulation of the time evolution of a quantum system. In this chapter we will turn our attention to exploring the simulation of the **eigenstates and eigenvalues of a Hamiltonian**, an essential step in determining the energy levels of a quantum system, such as a molecule. This is particularly important in the field of quantum chemistry since it enables the calculation of the rate of a chemical reaction or the spectroscopic properties of the molecule. Despite the existence of accurate classical methods for studying the properties of a molecule, no general numerically exact method exists that can handle systems with a large number of electrons with polynomial resources. The known methods either have systematic errors that cannot be easily quantified or their computational cost scales very rapidly with the system size (see Table 12.1).

In the first part of the chapter, we introduce several quantum algorithms that allow us to simulate the eigenvalues of a Hamiltonian, starting with quantum phase estimation (QPE) and iterative phase estimation (IPE), followed by adiabatic state preparation (ASP), quantum approximate optimization algorithm (QAOA), and the variational quantum eigensolver (VQE). At the end of the chapter, we will apply the IPE algorithm to calculate the ground state energy of a molecule. However, before doing so, we need to introduce some technical notions:

1) We need to choose a finite basis of electronic orbitals. For pedagogical reasons, we present the minimal STO-nG basis sets in Section 12.7.5. While minimal bases do not produce accurate results, they are simple to understand and contain the minimum number of elements.

2) Quantum algorithms for simulating the eigenstates of a Hamiltonian often start from an approximation produced by a classical algorithm, such as Hartree–Fock. For this reason, we will explain the Hartree–Fock algorithm in Section 12.7. This algorithm

Method	Scaling	Maximum K
HF	K^4	$\approx 10^4$
MP2	$N^2K^3 + N^2K^2$	$\approx 10^3$
CCSD	$N^2K^4 + N^3K^3$	$\approx 10^3$
CCSD(T)	N^3K^4	$\approx 10^3$
DMRG	$D^3K^3 + D^2K^4$	$\approx 10^2$
FCI	$\frac{K!}{N!(K-N)!}$	≈ 20

Table 12.1 Scaling of some classical methods for calculating the ground state of a Hamiltonian describing a quantum system with N electrons and K spin-orbitals. The acronyms stand for: Hartree–Fock (HF), Møller–Plesset second-order perturbation theory (MP2) [377], coupled-cluster singles and doubles (CCSD) [378], coupled-cluster singles, doubles, and perturbative triples (CCSD(T)), density matrix renormalization group (DMRG) [379] and full configuration interaction (FCI). The column on the right reports the maximum number of orbitals that can be simulated with a conventional computer with 128 GB of memory [380].

is the starting point for more sophisticated classical methods such as those listed in Table 12.1.

3) To apply the IPE algorithm to a molecule, we need to express the electronic Hamiltonian in terms of Pauli operators. To this end, we will introduce the formalism of second quantization in Section 12.8 and the Jordan–Wigner transformation in Section 12.8.2.

The chapter will conclude with a quantum simulation of the ground state energy of the hydrogen molecule. It is worth noting that the properties of this molecule can be easily simulated on a classical computer. The real challenges lie in simulating more complex electronic systems with many electrons and strong electronic correlation. Quantum computers will be able to support the study of these systems once hardware devices have many qubits and low error rates.

Lastly, while simulating the time evolution generated by a broad class of Hamiltonians is a BQP problem (see Section 11.7), calculating the eigenvalues for a variety of Hamiltonians (such as k-local Hamiltonians) is a QMA-complete problem. It is conjectured that quantum computers will never decide QMA-complete problems with polynomial resources.

12.1 Quantum phase estimation

Quantum phase estimation is a subroutine that allows us to find the eigenvalues u of a unitary operator $\hat{U}$ provided that we can prepare[1] a set of qubits in the corresponding eigenstates $|u\rangle$ and implement $\hat{U}$ as a controlled gate [381]. When the unitary

[1]In many practical situations, the eigenvectors $|u\rangle$ of a unitary operator are not known, and one can only provide an approximation $\sum_i c_i|u_i\rangle$ as input of the QPE algorithm. We will address this issue in more detail towards the end of this section.

operator is the time evolution operator, $\hat{U} = e^{\frac{1}{i\hbar}\hat{H}t}$, QPE returns an approximation of the energy levels of the corresponding Hamiltonian $\hat{H}$.

The quantum phase estimation algorithm has the following structure:

Given: A set of n qubits prepared in an eigenvector $|u\rangle$ of a unitary operator $\hat{U}$.

Assumption: The controlled gates $\mathsf{c}U^{2^j}$ can be implemented efficiently.

Goal: Find the eigenvalue u associated with $|u\rangle$.

We know from Section 3.2.11 that an eigenvalue of a unitary matrix can be expressed as $u = e^{i2\pi\lambda}$ for some real number $\lambda \in [0, 1)$. Therefore, the QPE algorithm is aptly named phase estimation, as determining u is equivalent to finding the corresponding phase λ.

Figure 12.1 shows the quantum circuit implementing the QPE algorithm. The circuit contains two registers. The bottom register, which contains n qubits, is necessary to represent the eigenstate $|u\rangle$. The top register contains m qubits: the integer m determines the accuracy of the estimate of λ. Let us go through each step of the circuit. The initial state $|\psi_{\mathrm{A}}\rangle = |0\ldots 0\rangle|u\rangle$ is transformed into

$$|\psi_{\mathrm{B}}\rangle = \frac{1}{\sqrt{2}}(|0\rangle + |1\rangle)\ldots\frac{1}{\sqrt{2}}(|0\rangle + |1\rangle)\frac{1}{\sqrt{2}}(|0\rangle + |1\rangle)|u\rangle$$

by a set of Hadamard gates. The controlled gate $\mathsf{c}U$ is now applied to the registers. Since $\hat{U}|u\rangle = e^{i2\pi\lambda}|u\rangle$, we have

$$|\psi_{\mathrm{C}}\rangle = \frac{1}{\sqrt{2}}(|0\rangle + |1\rangle)\ldots\frac{1}{\sqrt{2}}(|0\rangle + |1\rangle)\frac{1}{\sqrt{2}}(|0\rangle + e^{i2\pi\lambda}|1\rangle)|u\rangle.$$

A sequence of controlled gates $\mathsf{c}U^{2^j}$ produces

$$\begin{aligned}|\psi_{\mathrm{D}}\rangle &= \frac{1}{\sqrt{2}}(|0\rangle + e^{i2\pi 2^{m-1}\lambda}|1\rangle)\ldots\frac{1}{\sqrt{2}}(|0\rangle + e^{i2\pi 2\lambda}|1\rangle)\frac{1}{\sqrt{2}}(|0\rangle + e^{i2\pi\lambda}|1\rangle)|u\rangle \\ &= \frac{1}{2^{m/2}}\sum_{j=0}^{2^m-1} e^{i2\pi j\lambda}|j\rangle|u\rangle.\end{aligned}$$

From now on, we can ignore the state of the bottom register. The last step consists in applying the inverse of the quantum Fourier transform to the top register[2],

$$|\psi_{\mathrm{E}}\rangle = \frac{1}{2^{m/2}}\sum_{j=0}^{2^m-1} e^{i2\pi j\lambda}\hat{F}^{-1}|j\rangle = \frac{1}{2^m}\sum_{j=0}^{2^m-1}\sum_{k=0}^{2^m-1} e^{i2\pi j(\lambda-\frac{k}{2^m})}|k\rangle, \tag{12.1}$$

[2]See Section 10.7.1 for more information about the quantum Fourier transform.

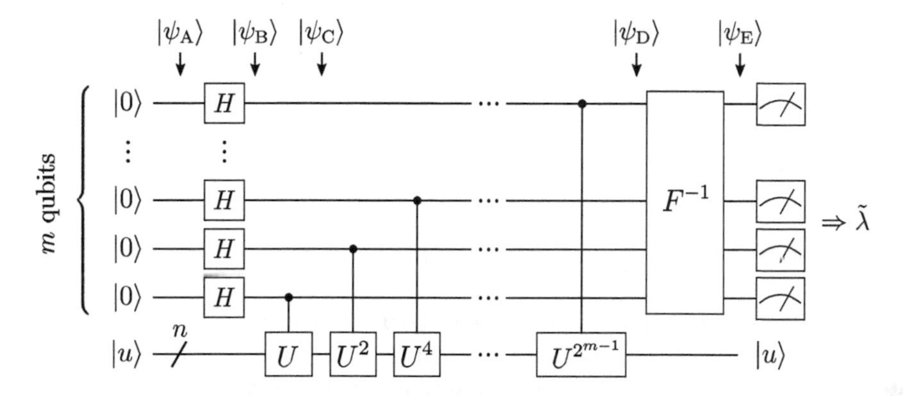

Fig. 12.1 Quantum phase estimation algorithm. The measurement of the top register projects the state of the m qubits onto the state $|\tilde{\lambda}\rangle$ with high probability. The best approximation of the real number λ is calculated with the formula $\tilde{\lambda}/2^m$.

where we used eqn (10.26). The circuit ends with a measurement in the computational basis, which collapses the state of the top register onto a basis vector $|\ell\rangle$ with probability

$$p(\ell) = |\langle \ell|\psi_E\rangle|^2 = \left|\frac{1}{2^m}\sum_{j=0}^{2^m-1} e^{i2\pi j(\lambda - \frac{\ell}{2^m})}\right|^2. \tag{12.2}$$

It is convenient to express the real number λ in a different way. Let $\tilde{\lambda}$ be the integer in the interval $[0, 2^m - 1]$ such that the rational number $\tilde{\lambda}/2^m$ is the best approximation[3] of λ. Therefore,

$$\lambda = \frac{\tilde{\lambda}}{2^m} + \delta, \tag{12.3}$$

where δ is a real number in the interval $[-1/2^{m+1}, 1/2^{m+1}]$ representing the error of the estimate due to the finite number of qubits in the top register. Substituting eqn (12.3) into (12.2), we have

$$p(\ell) = \left|\frac{1}{2^m}\sum_{j=0}^{2^m-1} e^{i2\pi j(\tilde{\lambda}-\ell)/2^m} e^{i2\pi j\delta}\right|^2. \tag{12.4}$$

The important point is that the probability distribution $p(\ell)$ has a well-defined peak around $\tilde{\lambda}$ as shown in Fig. 12.2. Using eqn (12.4), the probability that the measurement returns $\tilde{\lambda}$ is given by

[3]The integer $\tilde{\lambda}$ can be defined as $\tilde{\lambda} = \text{Round}(\lambda 2^m)$. For example, if $\lambda = 0.249$ and $m = 5$, then $\tilde{\lambda} = \text{Round}(0.249 \cdot 2^5) = 8$. The difference between λ and the approximation $\tilde{\lambda}/2^m$ is $\delta = \lambda - \tilde{\lambda}/2^m = -0.001$.

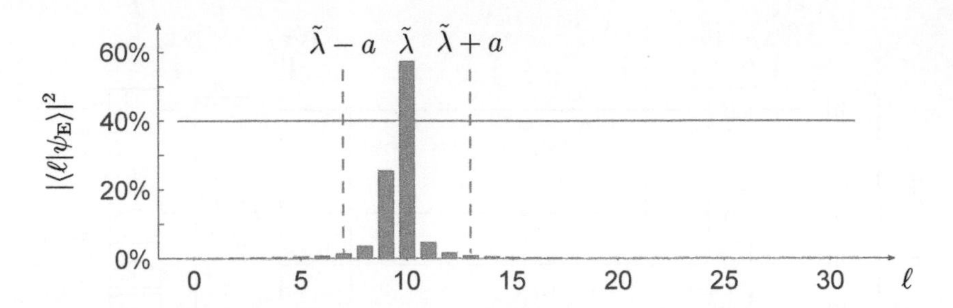

Fig. 12.2 Probability distribution of the QPE algorithm. Probability distribution $p(\ell) = |\langle \ell|\psi_{\rm E}\rangle|^2$ of eqn (12.2) with $\lambda = 0.3$ and $m = 5$. The peak of the distribution is located at $\tilde{\lambda} = \text{Round}(\lambda 2^m) = 10$ and the probability $p(\tilde{\lambda})$ is greater than 40%. Assuming a is a positive integer, the probability of measuring a number that lies within the interval $[\tilde{\lambda} - a, \tilde{\lambda} + a]$ is always greater than $1 - 1/(2a-2)$ (see Exercise 12.1).

$$p(\tilde{\lambda}) = \left|\frac{1}{2^m}\sum_{j=0}^{2^m-1} e^{i2\pi j\delta}\right|^2 = \frac{1}{2^{2m}}\left|\sum_{j=0}^{2^m-1}\left(e^{i2\pi\delta}\right)^j\right|^2 = \frac{1}{2^{2m}}\left|\frac{1-e^{i2\pi\delta 2^m}}{1-e^{i2\pi\delta}}\right|^2$$
$$= \frac{1}{2^{2m}}\left[\frac{\sin^2(\pi\delta 2^m)}{\sin^2(\pi\delta)}\right].$$

The probability $p(\tilde{\lambda})$ is always greater than 40%. This is because the numerator and the denominator in the square bracket satisfy

$$\sin^2(\pi\delta 2^m) \geq (2\delta 2^m)^2, \qquad \sin^2(\pi\delta) \leq (\pi\delta)^2, \tag{12.5}$$

where we used the fact that $-1/2^{m+1} \leq \delta \leq 1/2^{m+1}$. These relations lead to

$$\boxed{p(\tilde{\lambda}) \geq \frac{4}{\pi^2} > 40\%.}$$

In conclusion, the measurement projects the state of the top register onto $|\tilde{\lambda}\rangle$ with high probability. The best approximation of the phase λ is obtained with the formula $\tilde{\lambda}/2^m$. Once an estimate of λ has been determined, one can calculate the eigenvalue $u = e^{i2\pi\lambda}$.

Increasing the number of qubits in the top register improves the accuracy of the estimate. Let us explain this point in more detail. The measurement produces an integer ℓ from which one extracts the best approximation of λ using $\ell/2^m$. Our goal is to obtain an accuracy $|\lambda - \ell/2^m| \leq \varepsilon$ with a probability greater than $1-q$, where q is a fixed constant and ε is a desired accuracy. As shown in Exercise 12.1, the probability of measuring an integer in the interval $[\tilde{\lambda} - a, \tilde{\lambda} + a]$ is

$$p(\ell \in [\tilde{\lambda} - a, \tilde{\lambda} + a]) > 1 - \frac{1}{2a-2}, \tag{12.6}$$

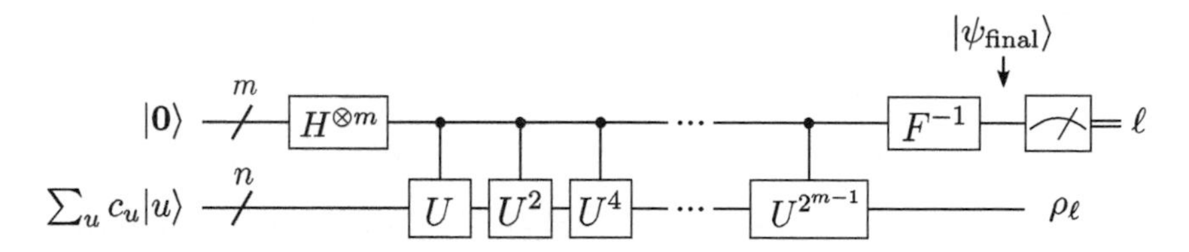

Fig. 12.3 QPE algorithm. The bottom register is prepared in a quantum state $|\psi\rangle = \sum_u c_u|u\rangle$ which approximates the eigenstate $|\tilde{u}\rangle$ of a Hamiltonian $\hat{H}$. When $\hat{H}$ is the molecular Hamiltonian, the state $|\psi\rangle$ is usually approximated with classical numerical methods, such as the Hartree–Fock method. With high probability, the QPE algorithm projects the top register onto $|\tilde{\lambda}\rangle$, leaving the bottom register in a state $\rho_{\tilde{\lambda}}$ which is very similar to the true eigenstate of the Hamiltonian.

---

where a is a positive integer. In the same exercise, we show that if the number of ancilla qubits m satisfies [382]

$$\boxed{m > \log\left[\frac{1}{\varepsilon}\left(\frac{1}{2q}+2\right)\right],} \tag{12.7}$$

the measurement returns an approximation of λ with accuracy ε and probability greater than $1-q$, i.e.

$$p(|\lambda - \ell/2^m| \leq \varepsilon) > 1-q. \tag{12.8}$$

As anticipated, increasing the number of qubits in the top register allows one to obtain an approximation of λ with high probability and arbitrary accuracy.

In many practical cases, we do not know the exact eigenstates of $\hat{U}$ and we can only prepare the bottom register in a linear superposition $|\psi\rangle = \sum_{u=1}^{2^n} c_u|u\rangle$, where $|u\rangle$ are the eigenstates of the unitary operator with eigenvalues $u = e^{i2\pi\lambda_u}$. If the number of ancilla qubits is sufficiently large, the final measurement produces a good approximation of λ_u with probability greater than $|c_u|^2(1-q)$ where q is a constant. Let us prove this step by step. As shown in Fig. 12.3, the QPE algorithm transforms the initial state $|\mathbf{0}\rangle\sum_u c_u|u\rangle$ into an entangled state $|\psi_{\text{final}}\rangle = \sum_{u=1}^{2^n} c_u|\psi_{\text{E},u}\rangle|u\rangle$. The measurement of the ancilla qubits outputs an integer ℓ with probability

$$p(\ell) = \text{Tr}[(|\ell\rangle\langle\ell| \otimes \mathbb{1}_n)|\psi_{\text{final}}\rangle\langle\psi_{\text{final}}|] = \sum_{u=1}^{2^n} \underbrace{|c_u|^2}_{p(u)} \underbrace{|\langle\ell|\psi_{\text{E},u}\rangle|^2}_{p(\ell|u)} = \sum_{u=1}^{2^n} p(u)p(\ell \mid u),$$

while the state of the n qubits is projected onto (see eqn (6.52))

$$\rho_\ell = \frac{1}{p(\ell)}\text{Tr}_{\text{ancilla}}[|\psi_{\text{final}}\rangle\langle\psi_{\text{final}}|(|\ell\rangle\langle\ell| \otimes \mathbb{1}_n)], \tag{12.9}$$

which might not be an eigenstate of $\hat{U}$. Let us show that the probability distribution $p(\ell)$ is peaked around the integers $\tilde{\lambda}_u$, where $\tilde{\lambda}_u/2^m$ are the best approximations of the real numbers λ_u. Assuming that the number of ancilla qubits is sufficiently large, we can use eqns (12.7, 12.8) to obtain

$$p(|\ell/2^m - \lambda_u| \leq \varepsilon) \;=\; p(u)p(|\ell/2^m - \lambda_u| \leq \varepsilon \mid u) \;\geq\; p(u) \cdot (1-q) \;=\; |c_u|^2(1-q).$$

This expression indicates that the probability that an integer ℓ gives an approximation of λ_u with an accuracy smaller than ε is greater than $|c_u|^2(1-q)$. In conclusion, the distribution $p(\ell)$ has well-defined peaks centered around λ_u [383].

An interesting case study is when the operator $\hat{U}$ is the time evolution operator $\hat{U} = e^{\frac{1}{i\hbar}\hat{H}t}$ where $\hat{H}$ is the Hamiltonian of a physical system. If the initial state of the quantum computer $|\psi\rangle = \sum_u c_u|u\rangle$ is sufficiently similar to one eigenstate $|u\rangle$ (which is both an eigenstate of $\hat{U}$ and $\hat{H}$), QPE returns a good approximation of the corresponding eigenvalue[4] $e^{\frac{1}{i\hbar}E_u t}$ from which one can extract the energy E_u (see Fig. 12.3). This has interesting applications in quantum chemistry. In a typical experiment, the n qubits are prepared in a quantum state that approximates one of the eigenstates of the Hamiltonian. Provided that the input state has a *significant* overlap with $|u\rangle$, the final measurement produces with high probability an approximation of the eigenpair $\{u, |u\rangle\}$. This approximation improves exponentially with the number of ancilla qubits.

Determining whether the QPE algorithm provides an exponential advantage over classical methods for computing the ground-state energy of a quantum system is a delicate matter. To approximate the ground-state energy, QPE requires that the initial state $|\psi\rangle$ has a significant overlap with the ground state. However, the states $|\psi\rangle$ are not always easy to prepare, and the implications of this limitation on the potential for quantum speedup are still under investigation. For more details, refer to Ref. [384].

12.1.1 Iterative phase estimation

The quantum phase estimation algorithm presented in the previous section requires a substantial number of ancilla qubits to obtain an accurate estimate of the eigenvalue. At the time of writing, quantum computers still have a modest number of physical qubits, which are prone to errors. One strategy to reduce the qubit count is to run the **iterative phase estimation algorithm** (IPE). This algorithm is designed to estimate u with a single ancilla qubit at the expense of running the circuit multiple times (the accuracy of the estimate is related to the number of iterations of the circuit rather than the number of ancilla qubits). Following Ref. [385], we write the eigenvalue as $u = e^{i2\pi\phi}$ where the phase $\phi \in [0, 1]$ can be expressed with exactly t binary digits,

$$\boxed{\phi = \sum_{j=1}^{t} \frac{\phi_j}{2^j}, \qquad \text{where } \phi_1, \ldots, \phi_t \in \{0, 1\}.}$$

We assume it is possible to prepare the bottom register in an eigenstate $|u\rangle$. As usual, the goal is to find the phase ϕ, i.e. the bits ϕ_j.

[4] The time t must be chosen such that the product $E_u t/\hbar \in [0, 2\pi)$.

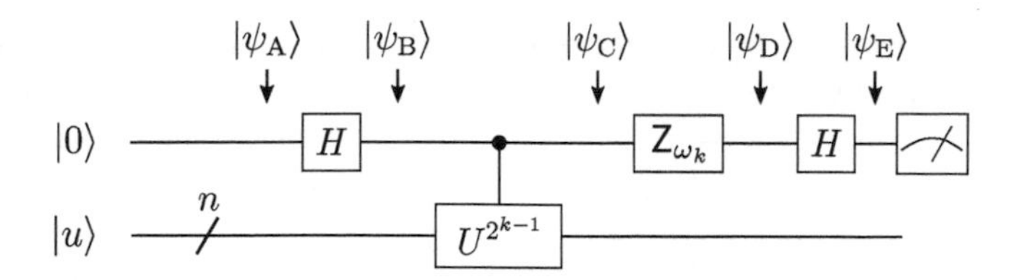

Fig. 12.4 Iterative phase estimation algorithm. The bottom register is prepared in an eigenstate $|u\rangle$ of the unitary operator $\hat{U}$. One circuit iteration produces one bit of the phase ϕ.

The quantum circuit implementing the iterative phase estimation algorithm is shown in Fig. 12.4. The top register contains one ancilla qubit, while the bottom register has n qubits. One iteration of the circuit produces one bit ϕ_j at a time in a deterministic way. Let us go through the algorithm step by step. A Hadamard gate applied to the initial state $|\psi_A\rangle = |0\rangle|u\rangle$ leads to $|\psi_B\rangle = \frac{1}{\sqrt{2}}[|0\rangle|u\rangle + |1\rangle|u\rangle]$. A controlled operation $cU^{2^{k-1}}$ produces

$$|\psi_C\rangle = \frac{1}{\sqrt{2}}[|0\rangle|u\rangle + e^{i2\pi\phi 2^{k-1}}|1\rangle|u\rangle.$$

The state of the top qubit is now rotated about the z axis by an angle ω_k,

$$|\psi_D\rangle = \frac{1}{\sqrt{2}}[|0\rangle + e^{i(\omega_k + 2\pi\phi 2^{k-1})}|1\rangle]|u\rangle = \frac{1}{\sqrt{2}}[|0\rangle + e^{i2\gamma}|1\rangle]|u\rangle,$$

where we defined $\gamma = \omega_k/2 + \pi\phi 2^{k-1}$. A Hadamard gate generates the final state

$$|\psi_E\rangle = \left[\frac{1 + e^{i2\gamma}}{2}|0\rangle + \frac{1 - e^{i2\gamma}}{2}|1\rangle\right]|u\rangle.$$

The measurement of the top qubit collapses the state into $|0\rangle|u\rangle$ or $|1\rangle|u\rangle$ with probabilities:

$$p(0) = \left|\frac{1 + e^{i2\gamma}}{2}\right|^2 = \cos^2\gamma, \qquad p(1) = \left|\frac{1 - e^{i2\gamma}}{2}\right|^2 = \sin^2\gamma.$$

Several iterations of the circuit are executed from $k = t$ down to $k = 1$.

Iteration $k = t$ In the first iteration, we set $\omega_k = \omega_t = 0$. This allows us to determine the first bit ϕ_t. Indeed, the parameter γ is given by

$$\gamma = \frac{\omega_k}{2} + \pi\phi 2^{k-1} = \pi\phi 2^{t-1} = \pi\sum_{j=1}^{t}\frac{\phi_j}{2^j}2^{t-1} = \pi\underbrace{\left(\sum_{j=1}^{t-1}\frac{\phi_j}{2^j}2^{t-1}\right)}_{c_t} + \pi\frac{\phi_t}{2} = \pi c_t + \pi\frac{\phi_t}{2}.$$

Note that the quantity πc_t is a multiple of π and the function $\cos^2 x$ is periodic with period π. Therefore, $p(0) = \cos^2\gamma = \cos^2(\pi\phi_t/2)$. There are only two possibilities: if $\phi_t = 0$, $p(0) = 1$, while if $\phi_t = 1$, $p(1) = 1$. In conclusion, the measurement of the ancilla qubit reveals the value of ϕ_t in a deterministic way.

Iteration $k < t$ We use the values of the bits $\phi_t, \phi_{t-1}, \ldots, \phi_{k+1}$ found in the previous iterations to set the angle

$$\omega_k = -2\pi \sum_{j=k+1}^{t} \frac{\phi_j}{2^j} 2^{k-1}.$$

The parameter γ becomes

$$\gamma = \frac{\omega_k}{2} + \pi\phi 2^{k-1} = \pi\left(-\sum_{j=k+1}^{t} \frac{\phi_j}{2^j} 2^{k-1} + \sum_{j=1}^{t} \frac{\phi_j}{2^j} 2^{k-1} \right)$$

$$= \pi \sum_{j=1}^{k} \frac{\phi_j}{2^j} 2^{k-1} = \pi \underbrace{\left(\sum_{j=1}^{k-1} \frac{\phi_j}{2^j} 2^{k-1} \right)}_{c_k} + \pi\frac{\phi_k}{2} = \pi c_k + \pi\frac{\phi_k}{2}.$$

The product πc_k is a multiple of π and therefore

$$p(0) = \cos^2\gamma = \cos^2(\pi\phi_k/2) = \begin{cases} 0\,\% & \phi_k = 1 \\ 100\,\% & \phi_k = 0. \end{cases}$$

If the final state of the ancilla qubit is $|0\rangle$, then $\phi_k = 0$. If instead the state of the ancilla qubit is $|1\rangle$, $\phi_k = 1$.

Iteration $k = 1$ The last iteration produces the bit ϕ_1. Once all of the bits have been determined, one can reconstruct $\phi = \sum_j \phi_j/2^j$ and the eigenvalue $u = e^{i2\pi\phi}$. In principle, the phase ϕ might have an infinite decimal expansion and this slightly complicates the analysis. The interested reader can find more information about the general case in Exercise 12.2.

When the unitary operator $\hat{U}$ is the time evolution operator of a quantum system, $\hat{U} = e^{\frac{1}{i\hbar}\hat{H}t}$, and the bottom register is prepared in a state $|\psi\rangle$ sufficiently close to an eigenstate $|u\rangle$ of the Hamiltonian, IPE returns an estimate of the corresponding eigenvalue E_u with high probability. However, the state of the bottom register does not collapse into a state similar to $|u\rangle$. Hence, the iterative phase estimation algorithm (unlike QPE) returns only an estimate of the eigenvalue but not of the corresponding eigenvector.

12.2 Adiabatic state preparation

An alternative method to approximate the ground state of a quantum system is **adiabatic state preparation** (ASP). This method is based on the **adiabatic theorem** of quantum mechanics: if a system is prepared in the ground state of a Hamiltonian (and the Hamiltonian changes slowly), the system remains in the instantaneous ground state under certain conditions.

Adiabatic state preparation has the following structure:

Given: We are given a Hamiltonian $\hat{H} = \hat{H}_0 + \hat{V}$ where $\hat{H}_0$ has known eigenstates, while the eigenstates of $\hat{H}$ are not known.

Assumptions: It is possible to prepare a quantum computer in the ground state of the Hamiltonian $\hat{H}_0$ efficiently.

Goal: Find the ground state of $\hat{H}$.

The first step of the adiabatic state preparation is to define a time-dependent Hamiltonian of the form

$$\boxed{\hat{H}(t) = \hat{H}_0 + \frac{t}{T}\hat{V},} \tag{12.10}$$

with eigenvalues and eigenvectors defined by

$$\hat{H}(t)|E_k(t)\rangle = E_k(t)|E_k(t)\rangle.$$

We then prepare a quantum computer in the state $|\psi(0)\rangle = |E_0(0)\rangle$ and simulate the time evolution of this state from 0 to T. If the evolution time is sufficiently long and the energy gap $E_1(t) - E_0(t)$ between the ground state and first excited state is always positive and not too small, the final state $|\psi(T)\rangle$ accurately approximates the state $|E_0(T)\rangle$, i.e. the ground state of the Hamiltonian $\hat{H}(T) = \hat{H}$.

Let us explain how to simulate the dynamics from 0 to T. In this section, we show how to do it with a Trotterization, but other methods are also viable. We divide the interval $[0, T]$ in n steps of length Δt and define a sequence of Hamiltonians,

$$\hat{H}_k = \hat{H}_0 + \frac{k}{n}\hat{V} = \hat{H}_0 + s_k\hat{V},$$

where $s_k = k/n$ and k is an integer in the range $[0, n-1]$. We then write the time evolution operator as

$$\begin{aligned}\hat{U}(T,0) = \mathcal{T}\exp\left(\int_0^T \frac{dt}{i\hbar}\hat{H}(t)\right) &\approx e^{\frac{\Delta t}{i\hbar}\hat{H}_{n-1}} \dots e^{\frac{\Delta t}{i\hbar}\hat{H}_0} \\ &= e^{\frac{\Delta t}{i\hbar}(\hat{H}_0 + s_{n-1}\hat{V})} \dots e^{\frac{\Delta t}{i\hbar}(\hat{H}_0 + s_0\hat{V})}.\end{aligned}$$

One can approximate $\hat{U}(T, 0)$ with a Trotterization,

$$\hat{U}(T,0) \approx e^{\frac{\Delta t}{i\hbar}\hat{H}_0} e^{\frac{\Delta t}{i\hbar}s_{n-1}\hat{V}} \dots e^{\frac{\Delta t}{i\hbar}\hat{H}_0} e^{\frac{\Delta t}{i\hbar}s_0\hat{V}} \tag{12.11}$$

and apply this gate to the initial state $|\psi(0)\rangle$ to obtain

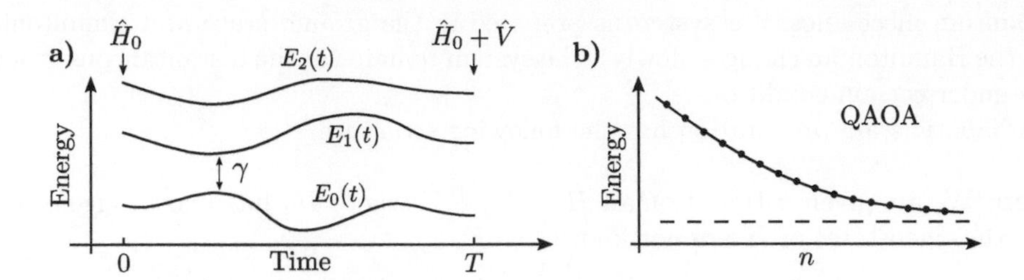

Fig. 12.5 Adiabatic state preparation and QAOA. a) Plot of the first three energy levels of the Hamiltonian $\hat{H}(t)$. When $t = 0$ (or $t = T$), the Hamiltonian is $\hat{H} = \hat{H}_0$ (or $\hat{H} = \hat{H}_0 + \hat{V}$). If the initial state of a quantum computer is the ground-state of $\hat{H}_0$ and the Hamiltonian changes sufficiently slowly, the system remains in the instantaneous ground state. **b)** For a fixed evolution time T, the expectation value of the energy produced by QAOA monotonically decreases as a function of the Trotter steps n. The dashed line represents the actual ground-state energy of the quantum system.

$$\boxed{|\tilde{\psi}(T)\rangle = e^{\frac{\Delta t}{i\hbar}\hat{H}_0} e^{\frac{\Delta t}{i\hbar}s_{n-1}\hat{V}} \dots e^{\frac{\Delta t}{i\hbar}\hat{H}_0} e^{\frac{\Delta t}{i\hbar}s_0\hat{V}} |\psi(0)\rangle.} \tag{12.12}$$

This state approximates the state $|\psi(T)\rangle$. The algorithm uses polynomial resources if the operators $e^{\frac{1}{i\hbar}\hat{H}_0\Delta t}$ and $e^{\frac{1}{i\hbar}\hat{V}\Delta t}$ can be decomposed into single and two-qubit gates efficiently. Clearly, if the number of Trotter steps n increases, the distance $\||\tilde{\psi}(T)\rangle - |\psi(T)\rangle\|$ decreases. To summarize, the approximation of the ground state of the Hamiltonian $\hat{H} = \hat{H}_0 + \hat{V}$ can be reduced to studying the dynamics generated by the Hamiltonian (12.10).

A delicate question is how slowly the Hamiltonian should be evolved to prevent undesired excitations. In other words, what value of T ensures that the system remains in its ground state? The adiabatic theorem provides some insight into this question, suggesting that T is inversely proportional to the energy gap between the ground state and the first excited state along the adiabatic path (see Fig. 12.5a). To be more quantitative, let us define the energy gap as $\gamma(t) = |E_1(t) - E_0(t)|$. If the energy gap is always greater than zero, then $\gamma \equiv \min_t \gamma(t) > 0$ and the distance between the vectors $|\psi(T)\rangle$ and $|E_0(T)\rangle$ is bounded by [386]

$$\||\psi(T)\rangle - |E_0(T)\rangle\| \leq \frac{1}{T}\left[\frac{f_1(T)}{\gamma^2(T)} + \frac{f_1(0)}{\gamma^2(0)} + \int_0^1 \frac{5f_1^2(s)}{\gamma^3(s)} + \frac{f_2(s)}{\gamma^2(s)}ds\right],$$

where $s = t/T$ and $f_k(s) = \|d^k H(s)/ds^k\|$. In the particular case $\hat{H} = \hat{H}_0 + t/T\hat{V}$, this equation reduces to

$$\||\psi(T)\rangle - |E_0(T)\rangle\| \leq \frac{1}{T}\left[\frac{\|\hat{V}\|}{\gamma^2(T)} + \frac{\|\hat{V}\|}{\gamma^2(0)} + \int_0^1 \frac{5\|\hat{V}\|^2}{\gamma^3(s)}ds\right].$$

This inequality gives some information about the scaling of the time evolution. Readers can find more information about the adiabatic theorem in Refs [387, 388, 389, 390, 391].

The ASP has been proposed as a viable method to solve some combinatorial problems [22, 392] and has interesting applications in molecular chemistry [393, 394]. It has been demonstrated that the adiabatic state preparation gives rise to an actual computational model for universal quantum computation that is equivalent to the circuit model [104, 395, 396]. The computational cost of the ASP is strictly related to the energy gap γ. If the gap stays constant or decreases polynomially as a function of the number of qubits, the quantum circuit is efficient. However, if it decays exponentially, $\gamma = O(1/\exp n)$, the algorithm is not efficient [397, 398].

12.3 Quantum approximate optimization algorithm

In the previous section, we learned that the ASP allows us to calculate the ground state of the Hamiltonian by studying the time evolution of a quantum state. However, adiabatic state preparation requires a substantial number of elementary gates. One strategy to obtain better results than the ASP with the same circuit depth is the **quantum approximate optimization algorithm** (QAOA) [22, 392]. This quantum algorithm is very similar to the ASP, the main difference being the substitution of the quantities $\Delta t/\hbar$ and $\Delta t s_k/\hbar$ in eqn (12.12) with some free parameters β and γ. With this substitution, the final state becomes

$$\boxed{|\psi(\boldsymbol{\beta},\boldsymbol{\gamma})\rangle = e^{-i\beta_{n-1}\hat{H}_0}e^{-i\gamma_{n-1}\hat{V}}\ldots e^{-i\beta_0\hat{H}_0}e^{-i\gamma_0\hat{V}}|\psi(0)\rangle.}$$

At the end of the circuit, $|\psi(\boldsymbol{\beta},\boldsymbol{\gamma})\rangle$ is measured in a suitable basis to determine the expectation value of the energy $E_n(\boldsymbol{\beta},\boldsymbol{\gamma}) = \langle\psi(\boldsymbol{\beta},\boldsymbol{\gamma})|\hat{H}|\psi(\boldsymbol{\beta},\boldsymbol{\gamma})\rangle$ for different vectors $\boldsymbol{\beta}$ and $\boldsymbol{\gamma}$. An optimization routine executed on a classical computer optimizes $\boldsymbol{\beta}$ and $\boldsymbol{\gamma}$ until the expectation value of the energy is minimized. In this sense, QAOA is a hybrid quantum-classical algorithm: the expectation value of the energy is calculated by a quantum computer, while a classical machine performs the optimization routine.

QAOA has some interesting properties. Suppose that for a fixed number of Trotter steps n, the minimum of the energy with respect to the parameters $\boldsymbol{\beta}$ and $\boldsymbol{\gamma}$ is

$$M_n = \min_{\boldsymbol{\beta},\boldsymbol{\gamma}} E_n(\boldsymbol{\beta},\boldsymbol{\gamma}).$$

One can demonstrate that increasing the number of Trotter steps decreases the expectation value of the energy, $M_{n+1} \leq M_n$. For $T \to \infty$ and $n \to \infty$, the quantity M_n is equal to the ground-state energy of the system. This concept is illustrated in Fig. 12.5b.

Example 12.1 Consider a two-qubit Hamiltonian of the form

$$\hat{H} = \underbrace{-\hat{\sigma}_{z0} - \hat{\sigma}_{z1}}_{\hat{H}_0} + \underbrace{g\hat{\sigma}_{x0} \otimes \hat{\sigma}_{x1}}_{\hat{V}} = \begin{bmatrix} -2 & 0 & 0 & g \\ 0 & 0 & g & 0 \\ 0 & g & 0 & 0 \\ g & 0 & 0 & 2 \end{bmatrix}.$$

The ground state of the Hamiltonian $\hat{H}_0$ is $|\psi(0)\rangle = |00\rangle$. By fixing the number of Trotter steps to $n = 1$, QAOA returns

$$\begin{aligned}|\psi(\beta_0,\gamma_0)\rangle &= e^{-i\beta_0\hat{H}_0}e^{-i\gamma_0\hat{V}}|00\rangle = e^{-i\beta_0\sigma_{z0}}e^{-i\beta_0\sigma_{z1}}e^{-i\gamma_0 g\sigma_{x0}\otimes\sigma_{x1}}|00\rangle \\ &= e^{-i\beta_0\sigma_{z0}}e^{-i\beta_0\sigma_{z1}}[\cos(c\gamma_0)\mathbb{1} - i\sin(c\gamma_0)\sigma_{x0}\otimes\sigma_{x1}]|00\rangle \\ &= e^{-i(\beta_0+\beta_1)}\cos(c\gamma_0)|00\rangle - i\sin(c\gamma_0)e^{i(\beta_0+\beta_1)}|11\rangle \\ &= \cos(c\gamma_0)|00\rangle - i\sin(c\gamma_0)e^{i2(\beta_0+\beta_1)}|11\rangle,\end{aligned}$$

where in the last step we dropped a global phase. With some matrix multiplications, one can verify that the expectation value of the energy is

$$E(\beta_0,\gamma_0) = \langle\psi(\beta_0,\gamma_0)|\hat{H}|\psi(\beta_0,\gamma_0)\rangle = -2\cos(2g\gamma_0) + g\sin[2(\beta_0+\beta_1)]\sin(2g\gamma_0).$$

The minimum of this function can be found with a classical computer by calculating $\partial E(\beta,\gamma)/\partial(\beta_0+\beta_1) = 0$ and $\partial E(\beta,\gamma)/\partial\gamma_0 = 0$. The minimum is located at $\beta_0+\beta_1 = \pi/4$ and $\gamma_0 = [2\pi - \arctan(g/2)]/2g$.

12.4 Variational quantum eigensolver

The **variational quantum eigensolver** (VQE) is an extension of QAOA, in which the variational principle is applied to a generic parametrized circuit. The initial state $|\psi_0\rangle$ (approximating the ground state of the physical system) is typically generated with a classical method. The final state of the circuit is given by

$$|\psi(\boldsymbol{\theta})\rangle = \hat{U}(\boldsymbol{\theta})|\psi_0\rangle.$$

In the literature, this quantum state is called the **Ansatz**. The quantum circuit $\hat{U}$ can often be expressed as

$$\hat{U}(\boldsymbol{\theta}) = e^{i\theta_L\hat{h}_L}\ldots e^{i\theta_1\hat{h}_1},$$

where each $\hat{h}_j$ is an arbitrary operator and the exponentials $e^{i\theta_j\hat{h}_j}$ can be efficiently decomposed into single and two-qubit gates. As in QAOA, the expectation value of the energy $\langle\psi(\boldsymbol{\theta})|\hat{H}|\psi(\boldsymbol{\theta})\rangle$ is optimized with a classical routine. Some important remarks are in order:

1. The user must choose the Ansatz $|\psi(\boldsymbol{\theta})\rangle$ in order to find a balance between computational cost, accuracy, and ease of optimization.
2. VQE, like QAOA, is a heuristic algorithm and therefore it has some systematic errors related to the chosen Ansatz that cannot be easily quantified.

This concludes the first part of the chapter on quantum algorithms for simulating the ground state of a Hamiltonian. The next step is to apply these methods to a real molecule. However, before doing so, it is necessary to compute an approximation of the ground state of the Hamiltonian with a classical method, such as the Hartree-Fock method. Hence, the following sections will cover the basics of quantum chemistry and the Hartree–Fock method. Additionally, we need to represent the Hamiltonian in terms of Pauli operators, a topic that will be covered in Section 12.8.2. After that, we will be ready to apply the IPE algorithm to the hydrogen molecule.

12.5 Introduction to quantum chemistry

Quantum chemistry focuses on applying quantum theory to chemical systems, particularly the electronic contributions to the chemical properties of molecules and solids. The calculations performed in quantum chemistry often involve some approximations intended to lower the computational cost while still capturing the main properties of the system. An important branch of quantum chemistry deals with the the ground state and excited states of many-electron systems (such as individual atoms and molecules) and the prediction of chemical reaction rates. These studies often start from the Born–Oppenheimer approximation, in which the nuclei are considered fixed points. The classical methods for approximating the electronic wavefunction include Hartree–Fock, quantum Monte Carlo, coupled-cluster methods, and perturbation theory [345, 399].

A molecule consists of M positively charged nuclei and N negatively charged electrons (see Fig. 12.6 for a schematic representation). The **molecular Hamiltonian** is given by

$$\boxed{\hat{H}_{\mathrm{mol}} = \hat{T}_{\mathrm{n}} + \hat{T}_{\mathrm{e}} + \hat{V}_{\mathrm{ne}} + \hat{V}_{\mathrm{ee}} + \hat{V}_{\mathrm{nn}},} \tag{12.13}$$

where $\hat{T}_{\mathrm{n}}$ and $\hat{T}_{\mathrm{e}}$ are the kinetic energies of the nuclei and electrons, while the terms $\hat{V}_{\mathrm{ne}}$, $\hat{V}_{\mathrm{ee}}$, and $\hat{V}_{\mathrm{nn}}$ describe the electrostatic attraction and repulsion between the nuclei and electrons. The molecular Hamiltonian can be expressed in a more explicit form as

$$\begin{aligned}\hat{H}_{\mathrm{mol}} = &\sum_{A=1}^{M} \frac{\mathbf{P}_A^2}{2M_A} + \sum_{i=1}^{N} \frac{\mathbf{p}_i^2}{2m_{\mathrm{e}}} - \sum_{i=1}^{N}\sum_{A=1}^{M} \frac{e^2}{4\pi\epsilon_0}\frac{Z_A}{|\mathbf{r}_i - \mathbf{R}_A|} + \sum_{i=1}^{N}\sum_{j=i+1}^{N} \frac{e^2}{4\pi\epsilon_0}\frac{1}{|\mathbf{r}_i - \mathbf{r}_j|} \\ &+ \sum_{A=1}^{M}\sum_{B=A+1}^{M} \frac{e^2}{4\pi\epsilon_0}\frac{Z_A Z_B}{|\mathbf{R}_A - \mathbf{R}_B|},\end{aligned} \tag{12.14}$$

where $\mathbf{p}_i$, $\mathbf{r}_i$, and m_{e} are the momentum, position, and mass of the i-th electron, whereas $\mathbf{P}_A$, $\mathbf{R}_A$, M_A, and Z_A are the momentum, position, mass, and number of protons of the A-th nucleus (see Fig. 12.6). Before studying this Hamiltonian in detail, it is useful to summarize the typical molecular scales. The typical distance between nuclei and electrons is on the order of the **Bohr radius** defined as

$$a_0 = 4\pi\epsilon_0 \frac{\hbar^2}{m_{\mathrm{e}} e^2} = 0.529 \cdot 10^{-10}\ \mathrm{m}.$$

The energy required to remove one electron from an isolated atom is of the order of the **Hartree energy**,

$$E_{\mathrm{h}} = \frac{m_{\mathrm{e}} e^4}{(4\pi\epsilon_0)^2 \hbar^2} = 4.3597 \times 10^{-18}\ \mathrm{J}\ =\ 27.211\ \mathrm{eV}.$$

Recall that the visible spectrum is in the range $1.6 - 3$ eV and an energy of 27.211 eV corresponds to the energy of an ultraviolet photon. The typical speed of an electron

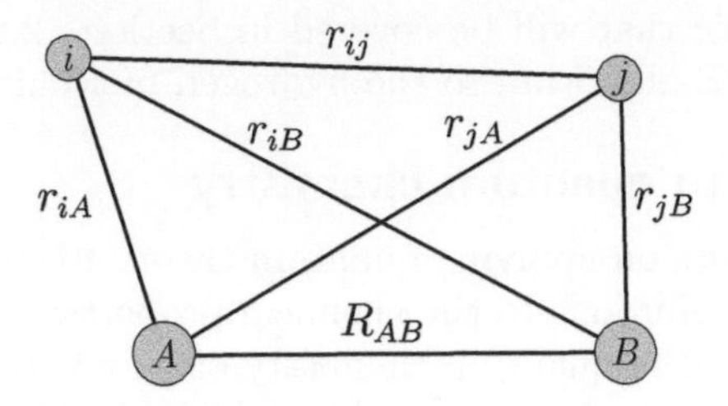

Fig. 12.6 A molecular geometry. Schematic representation of two nuclei (A and B) and two electrons (i and j) in a molecule.

in a molecule can be calculated from the relation

$$v_{\mathrm{e}} \approx \sqrt{\frac{2E_{\mathrm{h}}}{m_{\mathrm{e}}}} = 3 \cdot 10^6 \text{ m/s},$$

which is approximately 1% of the speed of light. The slow speed of electrons means that relativistic effects can often be neglected when studying molecular properties. Since nuclei move much slower than electrons due to their larger mass, they are typically approximated as fixed points. Finally, the spin-orbit energy

$$E_{\text{S-O}} \approx Z^4 \alpha^2 E_{\mathrm{h}}, \qquad (\alpha = 137.035)$$

is often neglected. This approximation is particularly suited for light nuclei. For example, when $Z \approx 1$, the spin-orbit interaction is several orders of magnitude smaller than typical molecular energies, $E_{\text{S-O}}/E_{\mathrm{h}} \approx 10^{-5}$.

In quantum chemistry, it is common to work with atomic units because it significantly simplifies analytical and numerical calculations. In this chapter, we will use the "Hartree type" of atomic units summarized in Table 12.2. The molecular Hamiltonian (12.14) in atomic units is given by[5]

$$\boxed{\hat{H}_{\mathrm{mol}} = -\sum_A \frac{\nabla_A^2}{2M_A} - \sum_j \frac{\nabla_j^2}{2} - \sum_{i,A} \frac{Z_A}{r_{iA}} + \sum_{i<j} \frac{1}{r_{ij}} + \sum_{A<B} \frac{Z_A Z_B}{R_{AB}}.} \qquad (12.15)$$

Here, we introduced the notations $r_{ij} = |\mathbf{r}_i - \mathbf{r}_j|$, $r_{iA} = |\mathbf{r}_i - \mathbf{R}_A|$, $R_{AB} = |\mathbf{R}_A - \mathbf{R}_B|$ and we expressed the momentum operators in terms of the gradient, $\mathbf{p}_j \to -i\nabla_j$ and $\mathbf{P}_A \to -i\nabla_A$. In eqn (12.15), all constants disappeared and the structure of the Hamiltonian is much easier to read. Hereafter, all quantities will be in atomic units.

Until now, we have not taken into account the spin of the electrons. This is justified by the fact that the interactions between the electron spins are much smaller than the typical molecular bonds. Since the Hamiltonian does not contain spin operators, $\hat{H}_{\mathrm{mol}}$ commutes with $\hat{S}_z$ and $\hat{S}^2$. This means that the eigenfunctions of $\hat{H}_{\mathrm{mol}}$ are eigenfunctions of $\hat{S}_z$ and $\hat{S}^2$ as well.

[5]For convenience, we introduced the short hand notation $\sum_{i<j} = \sum_{i=1}^{N} \sum_{j=i+1}^{N}$.

Dimension	Symbol	SI units	Atomic units
Mass	m_e	9.1095×10^{-31} kg	$1\ m_e$
Charge	e	1.6022×10^{-19} C	$1\ e$
Action	$\hbar$	1.0546×10^{-34} J s	$1\ \hbar$
Energy	E_h	4.3598×10^{-18} J	$1\ E_\text{h}$
Length	a_0	5.2918×10^{-11} m	$1\ a_0$
Permittivity^{-1}	$k_\text{e} = 1/4\pi\epsilon_0$	8.987×10^9 J m/C^2	$1\ k_\text{e}$

Table 12.2 Conversion between atomic units and SI units.

12.5.1 The Born–Oppenheimer approximation

An important problem in quantum chemistry is the calculation of the energy levels of molecules, namely the eigenvalues of the Hamiltonian (12.15). A simple analysis shows that this problem is impossible to solve *analytically* when the molecule contains more than one electron and *numerical* approaches are necessary.

Since calculating the eigenvalues of the molecular Hamiltonian is significantly difficult, it is reasonable to consider some approximations. The first approximation is based on the observation that the nuclei move much slower than the electrons because they are more massive. This often allows us to approximate the nuclei as fixed points. This approximation, known as **the Born–Oppenheimer approximation**, consists in removing the nuclear kinetic energy and considering the nuclear repulsion constant,

$$\hat{T}_\text{n} = -\sum_A \frac{\nabla^2}{2M_A} = 0 \text{ because the nuclei are not moving,}$$

$$V_{\text{nn},\mathbf{R}} = \sum_{A<B} \frac{Z_A Z_B}{R_{AB}} = \text{const because } R_{AB} = |\mathbf{R}_A - \mathbf{R}_B| \text{ is constant.}$$

Thus, the Hamiltonian (12.15) becomes

$$\boxed{\hat{H}_\mathbf{R}^{(\text{elec})} = -\sum_i \frac{\nabla_i^2}{2} - \sum_{i,A} \frac{Z_A}{r_{iA}} + \sum_{i<j} \frac{1}{r_{ij}}.} \tag{12.16}$$

This is called the **electronic Hamiltonian**. The eigenfunctions and eigenvalues of the electronic Hamiltonian are defined by the equation,

$$\hat{H}_\mathbf{R}^{(\text{elec})} |\psi_\mathbf{R}^{(\text{elec})}\rangle = E_\mathbf{R}^{(\text{elec})} |\psi_\mathbf{R}^{(\text{elec})}\rangle.$$

The subscript $\mathbf{R}$ reminds us that the electronic Hamiltonian has been derived for some specific positions $\mathbf{R} = \{\mathbf{R}_1, \ldots, \mathbf{R}_M\}$ of the nuclei.

Let us explain the steps required to derive the molecular energy as a function of the nuclear geometry. The typical procedure consists of the following steps:

1) We first fix the position of the nuclei $\mathbf{R} = \{\mathbf{R}_1, \ldots, \mathbf{R}_M\}$ and we explicitly write

1) Fix the positions of the nuclei $\mathbf{R} = \{\mathbf{R}_1, \mathbf{R}_2\}$

2) Calculate the minimum energy $E^{(\text{elec})}_{\mathbf{R}}$ of the Hamiltonian $H^{(\text{elec})}_{\mathbf{R}}$

3) Calculate the real number $V_{\text{nn},\mathbf{R}} + E^{(\text{elec})}_{\mathbf{R}} = E^{(\text{tot})}_{\mathbf{R}}$

4) Go back to the first step, change the positions of the nuclei and cycle

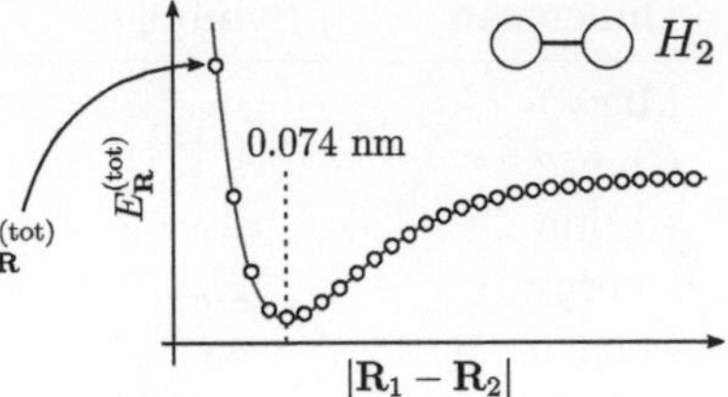

Fig. 12.7 Potential energy curve of the hydrogen molecule. To obtain the total energy of the hydrogen molecule, we first calculate the lowest eigenvalue of the electronic Hamiltonian $E^{(\text{elec})}_{\mathbf{R}}$ and use it compute $E^{(\text{tot})}_{\mathbf{R}} = V_{nn,\mathbf{R}} + E^{\text{elec}}_{\mathbf{R}}$. This procedure is repeated for different nuclear configurations. The plot on the right illustrates the total energy $E^{(\text{tot})}_{\mathbf{R}}$ for the hydrogen molecule as a function of the interatomic distance. At equilibrium, the distance between the two nuclei is $|\mathbf{R}_1 - \mathbf{R}_2| = 0.074$ nm [400].

---

the electronic Hamiltonian,

$$\hat{H}^{(\text{elec})}_{\mathbf{R}} = \hat{T}_{\text{e}} + \hat{V}_{\text{ee}} + \hat{V}_{\text{ne},\mathbf{R}}.$$

With a numerical method, one can calculate the ground state $\psi^{(\text{elec})}_{\mathbf{R}}(\mathbf{r})$ and the corresponding eigenenergy $E^{(\text{elec})}_{\mathbf{R}}$. The total energy of the molecule is given by

$$E^{(\text{tot})}_{\mathbf{R}} = V_{\text{nn},\mathbf{R}} + E^{(\text{elec})}_{\mathbf{R}}.$$

As usual, the subscript $\mathbf{R}$ reminds us that the positions of the nuclei are fixed.

2) We go back to step 1 and change the position of the nuclei, $\{\mathbf{R}_1, \ldots, \mathbf{R}_M\} \to \{\mathbf{R}'_1, \ldots, \mathbf{R}'_M\}$. With a numerical method, we calculate once again the total energy $E^{(\text{tot})}_{\mathbf{R}'}$. This procedure is repeated multiple times until we have enough points to determine the potential perceived by the nuclei as a function of their positions (see Fig. 12.7).

3) Once the potential perceived by the nuclei has been determined, one can compute equilibrium geometries and energy differences (such as the binding energy) by studying the nuclear Hamiltonian $\hat{H}^{(\text{nucl})}$. Since the speed of the electrons is much greater than the one of the nuclei, one can substitute the electronic Hamiltonian $\hat{H}^{(\text{elec})}_{\hat{\mathbf{R}}}$ with its expectation value,

$$\begin{aligned}
\hat{H}^{(\text{nucl})} &= \hat{T}_{\text{n}} + \hat{V}_{\text{nn},\hat{\mathbf{R}}} + \underbrace{\langle \psi^{(\text{elec})}_{\hat{\mathbf{R}}}(\mathbf{r}) | \hat{H}^{(\text{elec})}_{\hat{\mathbf{R}}} | \psi^{(\text{elec})}_{\hat{\mathbf{R}}}(\mathbf{r}) \rangle}_{E^{(\text{elec})}_{\hat{\mathbf{R}}}} \\
&= \hat{T}_{\text{n}} + \underbrace{\hat{V}_{\text{nn},\hat{\mathbf{R}}} + E^{(\text{elec})}_{\hat{\mathbf{R}}}}_{\hat{V}^{(\text{nucl})}_{\mathbf{R}}} = \hat{T}_{\text{n}} + \hat{V}^{(\text{nucl})}_{\hat{\mathbf{R}}}.
\end{aligned}$$

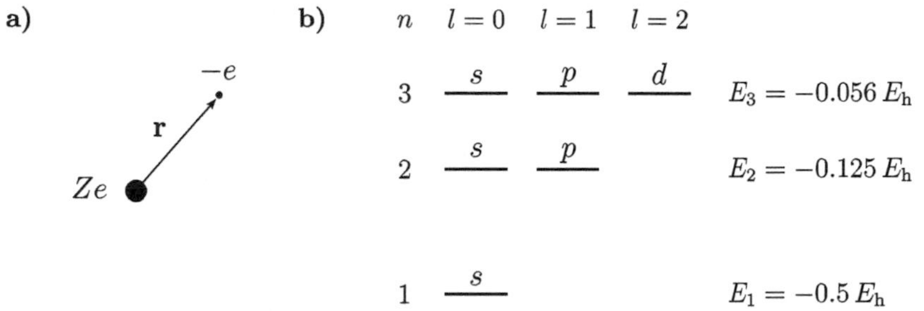

a) b)

n	$l=0$	$l=1$	$l=2$	
3	s	p	d	$E_3 = -0.056\,E_\text{h}$
2	s	p		$E_2 = -0.125\,E_\text{h}$
1	s			$E_1 = -0.5\,E_\text{h}$

$-e$, $\mathbf{r}$, Ze

Fig. 12.8 The hydrogen-like atom. a) Hydrogen-like atoms consist of a single electron orbiting around the nucleus. **b)** The energy levels depend on the integer n as indicated by eqn (12.20). Neglecting the spin-orbit interaction, the states with different l but the same n have the same energy. The energies E_k reported on the right assume $Z = 1$.

The operator $\hat{V}_{\mathbf{R}}^{(\text{nucl})}$ is the potential energy that governs the dynamics of the nuclei. Lastly, one can calculate the vibrational and rotational modes of the molecule by solving the eigenvalue problem $\hat{H}^{(\text{nucl})}|\psi^{\text{nucl}}\rangle = E^{\text{nucl}}|\psi^{\text{nucl}}\rangle$. These steps will be explained in more detail in Setion 12.7.6 when we study the Hartree–Fock method for the hydrogen molecule.

In the remaining part of this chapter, we will present classical approaches and quantum algorithms to calculate the ground state of the electronic Hamiltonian (12.16). We will omit the superscript "elec" and the subscript "**R**" to simplify the notation. We will start by revisiting one of the simplest systems in quantum physics, the hydrogen atom. This analysis will allow us to introduce fundamental concepts, such as orbitals and spin functions.

12.5.2 The hydrogen atom

As shown in Fig. 12.8a, the hydrogen atom consists of a proton and an electron. The hydrogen-like Hamiltonian in atomic units is given by

$$\boxed{\hat{H} = -\frac{\nabla^2}{2} - \frac{Z}{r}.} \tag{12.17}$$

In this equation, we kept the parameter Z indicating the number of protons in the nucleus (for the hydrogen atom $Z = 1$). We did it because the results obtained for the hydrogen atom can be extended to other ions, such as He^{+} and Li^{2+}, with the substitution $Z = 2$, $Z = 3$, and so on. The eigenvalues of the Hamiltonian (12.17) are the energy levels of the atom, whereas the squared absolute value of the wavefunctions, $|\psi(\mathbf{r})|^2$, is the probability density of measuring an electron at a given point in space. The eigenvalues and eigenfunctions can be derived by solving the time-independent Schrödinger equation

$$\left(-\frac{\nabla^2}{2} - \frac{Z}{r}\right)\psi(\mathbf{r}) = E\psi(\mathbf{r}). \tag{12.18}$$

Since the potential energy has a spherical symmetry (it only depends on the distance between the nucleus and the electron), it is obviously advantageous to solve this problem in spherical coordinates (r, θ, ϕ). The Laplacian in spherical coordinates is

$$\nabla^2 = \frac{1}{r^2}\frac{\partial}{\partial r}\left(r^2\frac{\partial}{\partial r}\right) + \frac{1}{r^2 \sin\theta}\frac{\partial}{\partial \theta}\left(\sin\theta\frac{\partial}{\partial \theta}\right) + \frac{1}{r^2 \sin^2\theta}\frac{\partial^2}{\partial \phi^2}. \tag{12.19}$$

The solution of the differential equation (12.18) is presented in many books of quantum mechanics (see for example Refs [401, 402]). Therefore, we will limit ourselves to writing the solution without an explicit calculation. The eigenvalues are

$$\boxed{E_n = -\frac{1}{2}\frac{Z^2}{n^2}\,E_{\text{h}},} \tag{12.20}$$

where n is a positive integer called the principal quantum number. The energy levels of the hydrogen atom are shown schematically in Fig. 12.8b. Setting $n = 1$ and $Z = 1$, we see that the lowest-energy level is $E_1 = -0.5\ E_{\text{h}}$. This means that $0.5\ E_{\text{h}}$ is the energy required to remove an electron from the hydrogen atom when the system is in the ground state.

The analysis in spherical coordinates allows us to derive the eigenfunctions with the method of separation of variables. It can be shown that the eigenfunctions are the product between a radial function and an angular function,

$$\boxed{\psi_{nlm}(r, \theta, \phi) = R_{nl}(r)Y_{lm}(\theta, \phi).} \tag{12.21}$$

The eigenfunctions ψ_{nlm} are called the **hydrogen-like orbitals**. In atomic units, the radial part is given by

$$R_{nl}(r) = \sqrt{\frac{4Z^3}{n^4}\frac{(n-l-1)!}{(n+l)!}}\left(\frac{2Zr}{n}\right)^l e^{-Zr/n}\,L_{n-l-1}^{2l+1}\left(\frac{2Zr}{n}\right) \tag{12.22}$$

where $L_b^a(x)$ are the generalized Laguerre polynomials[6]. The angular part is given by

$$Y_{lm}(\theta, \phi) = \sqrt{\frac{2l+1}{4\pi}}\sqrt{\frac{(l-m)!}{(l+m)!}}P_l^m(\cos\theta)e^{im\phi}, \tag{12.23}$$

where $P_b^a(x)$ are the associated Legendre polynomials. The functions $Y_{lm}(\theta, \phi)$ are called **spherical harmonics**.

The integers n, l, and m appearing in equations (12.22) and (12.23) satisfy certain inequalities. In particular, n can assume any non-negative value, whereas l ranges over $[0, n-1]$ for a fixed n. The integer m takes on integer values within the interval $[-l, l]$. It is common to write an eigenstate of the hydrogen atom with the notation $|nlm\rangle$, and therefore

[6] The generalized Laguerre polynomials are defined as $L_a^b(x) = \sum_{k=0}^{a}\binom{a+b}{a-k}\frac{(-x)^k}{k!}$.

n	l	m	Eigenfunctions in atomic units	Orbital
1	0	0	$\psi_{100} = \frac{1}{\sqrt{\pi}} Z^{3/2} e^{-Zr}$	1s
2	0	0	$\psi_{200} = \frac{1}{4\sqrt{2\pi}} Z^{3/2}(2 - Zr)e^{-Zr/2}$	2s
2	1	-1	$\psi_{21-1} = \frac{1}{8\sqrt{\pi}} Z^{5/2} r e^{-Zr/2} \sin\theta e^{-i\phi}$	2p
2	1	0	$\psi_{210} = \frac{1}{4\sqrt{2\pi}} Z^{5/2} r e^{-Zr/2} \cos\theta$	2p
2	1	1	$\psi_{211} = -\frac{1}{8\sqrt{\pi}} Z^{5/2} r e^{-Zr/2} \sin\theta e^{i\phi}$	2p

Table 12.3 Eigenfunctions (12.21) of the hydrogen atom in atomic units for the first energy levels.

$$\langle r\theta\phi|nlm\rangle = \psi_{nlm}(r, \theta, \phi).$$

The states with $l = 0, 1, 2$ are denoted with the letters s, p, and d. The eigenfunctions for the first few energy levels are listed in Table 12.3.

The hydrogen-like eigenfunctions depend on three spatial variables, making it difficult to plot them. Typically, the radial part and the angular part are illustrated separately. When the atom is prepared in an eigenstate ψ_{nlm}, the probability of measuring the electron within a small volume $d\tau$ around the point (r, θ, ϕ) is given by

$$|\psi_{nlm}(r, \theta, \phi)|^2 d\tau = |R_{nl}(r)|^2 \, |Y_{lm}(\theta, \phi)|^2 r^2 \sin\theta dr d\theta d\phi,$$

and the probability of observing the electron in a small volume enclosed by two spheres with radii r and $r + dr$ is $|R_{nl}(r)|^2 r^2 dr$. The probability density $|R_{nl}(r)|^2 r^2$ is shown in Fig. 12.9 for different wavefunctions. As n increases, it is more likely to find the electron further from the nucleus.

Figure 12.10 shows the angular part for some orbitals. The spherical harmonic associated with the quantum number $l = 0$ is simple to visualize since it has a spherical symmetry (this is because the function $Y_{00} = 1/\sqrt{4\pi}$ does not depend on θ nor ϕ). The spherical harmonics with $l = 1$ are slightly more difficult to illustrate since Y_{1-1} and Y_{11} are complex functions:

$$Y_{1-1} = \sqrt{\frac{3}{8\pi}} \sin\theta e^{i\phi}, \qquad Y_{10} = \sqrt{\frac{3}{4\pi}} \cos\theta, \qquad Y_{11} = \sqrt{\frac{3}{8\pi}} \sin\theta e^{-i\phi}.$$

The probability densities $|Y_{1-1}|^2$ and $|Y_{11}|^2$ are equal even if the two wavefunctions have a different relative phase. Hence, it is common to perform a change of basis,

$$p_x = \frac{1}{\sqrt{2}}(Y_{11} + Y_{1-1}) = \sqrt{\frac{3}{4\pi}} \sin\theta \cos\phi,$$

$$p_y = \frac{1}{\sqrt{2}i}(Y_{11} - Y_{1-1}) = \sqrt{\frac{3}{4\pi}} \sin\theta \sin\phi,$$

$$p_z = Y_{10} = \sqrt{\frac{3}{4\pi}} \cos\theta.$$

The vectors of the basis $\{p_x, p_y, p_z\}$ are nothing but a linear combination of the vectors

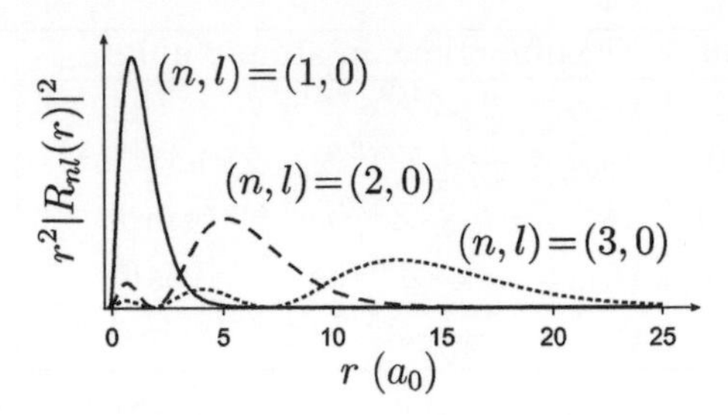

Fig. 12.9 Radial part of the eigenfunctions. As the integer n increases, the center of the probability density function $|R_{nl}(r)|^2 r^2$ moves away from the nucleus. In this plot, we set $Z = 1$.

$\{Y_{1-1}, Y_{10}, Y_{11}\}$. The absolute square of the real functions p_i is illustrated in Fig.12.10. It is worth mentioning that these figures are not faithful representations of the orbitals because they do not contain any radial dependence.

To obtain a complete description of the quantum state of the hydrogen atom, the wavefunction must contain some information about the electron spin. In our non-relativistic treatment, a quantum state of the spin is introduced as a normalized vector $|\sigma\rangle = c_0|0\rangle + c_1|1\rangle$ with complex coefficients, where

$$\textbf{spin up:}\quad |0\rangle = \begin{bmatrix} 1 \\ 0 \end{bmatrix}, \qquad \textbf{spin down:}\quad |1\rangle = \begin{bmatrix} 0 \\ 1 \end{bmatrix}.$$

The states $|0\rangle$ and $|1\rangle$ correspond to the spin pointing up and down along the z axis, respectively. The projection of a generic spin state $|\sigma\rangle = c_0|0\rangle + c_1|1\rangle$ along a vector $|\omega\rangle$ where $\omega \in \{0, 1\}$ is given by

$$\sigma(\omega) = \langle\omega|\sigma\rangle = c_0\underbrace{\langle\omega|0\rangle}_{\alpha(\omega)} + c_1\underbrace{\langle\omega|1\rangle}_{\beta(\omega)} = c_0\alpha(\omega) + c_1\beta(\omega).$$

Here, we have introduced a different notation for the Kronecker delta, which is widely used in quantum chemistry,

$$\textbf{spin up:}\quad \delta_{\omega 0} \equiv \alpha(\omega), \qquad \textbf{spin down:}\quad \delta_{\omega 1} \equiv \beta(\omega).$$

If the quantum state of the spin is $|\sigma\rangle = |0\rangle$ or $|\sigma\rangle = |1\rangle$, then the projection $\sigma(\omega)$ reduces to $\alpha(\omega)$ or $\beta(\omega)$.

The electron wavefunction is the tensor product of a spatial wavefunction and a spin wavefunction,

$$\boxed{|\chi\rangle = \underset{\text{spatial}}{|\psi\rangle} \otimes \underset{\text{spin}}{|\sigma\rangle}.} \tag{12.24}$$

The global wavefunction χ is called a **spin-orbital**. The projection of χ in the basis $\langle r\theta\phi\omega|$ is written as

$$\chi(\mathbf{r}, \omega) = \langle r\theta\phi\omega|\chi\rangle = \psi(\mathbf{r})\sigma(\omega).$$

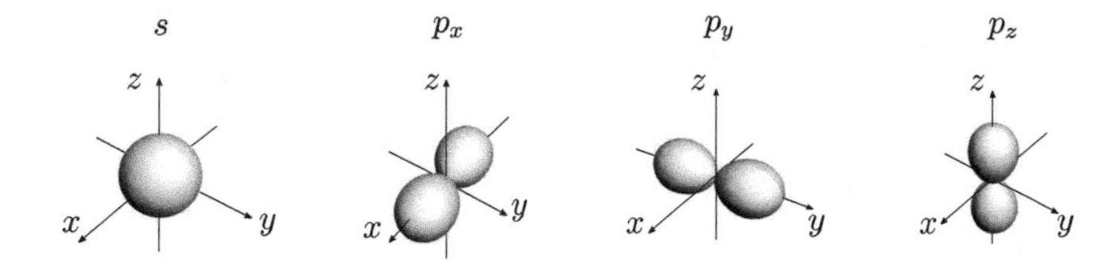

Fig. 12.10 Angular part of the eigenfunctions. Representation of the angular part for the s orbital and the p orbitals.

To simplify the notation, the spatial and spin variables will be denoted with a single variable $x = \{\mathbf{r}, \omega\}$ so that the spin-orbital becomes $\chi(x) = \chi(\mathbf{r}, \omega)$. The integral over the entire 3D space and the spin variables will be indicated as $\int dx \equiv \sum_{\omega=0}^{1} \int d\tau$. Thus, the expectation value of an observable $\hat{O}$ when the system is in a quantum state $|\chi\rangle$ is given by

$$\langle \chi | \hat{O} | \chi \rangle \equiv \int dx \chi^*(x) \hat{O} \chi(x) = \sum_{\omega=0}^{1} \int d\tau \chi^*(\mathbf{r}, \omega) \hat{O} \chi(\mathbf{r}, \omega).$$

Example 12.2 Let us consider a hydrogen atom in a quantum state $|nlm\rangle$ with the electron spin pointing up. The spin-orbital is given by

$$|\chi\rangle = |nlm\rangle|0\rangle, \quad \text{or equivalently} \quad \chi(x) = \psi_{nlm}(\mathbf{r})\alpha(\omega).$$

The probability of measuring an electron within a small volume $d\tau$ around the point $\mathbf{r}$ with the spin pointing in the direction ω is given by $|\psi_{nlm}(\mathbf{r})|^2 \alpha(\omega) d\tau$. If, instead, the spin is in a superposition state of up and down, we have

$$|\chi\rangle = |nlm\rangle \frac{|0\rangle + |1\rangle}{\sqrt{2}}, \quad \text{or equivalently} \quad \chi(x) = \psi_{nlm}(\mathbf{r}) \frac{\alpha(\omega) + \beta(\omega)}{\sqrt{2}}.$$

Example 12.3 In quantum chemistry, the **Slater orbitals** are defined as

$$\boxed{S_{nlm}(\mathbf{r}, \zeta) = A\, r^{n-1} e^{-\zeta r}\, Y_{lm}(\theta, \phi)} \tag{12.25}$$

where $A = (2\zeta)^{n+1/2}/\sqrt{(2n)!}$ is a normalization constant and ζ is a fixed parameter. The only difference between the hydrogen-like orbitals (12.21) and the Slater orbitals is the radial part (the angular part is described by the spherical harmonics). As usual, the integers n, l, m must satisfy the relations: $n \in \{0, 1, 2, 3, \ldots\}$, $l \in \{0, 1, 2, \ldots, n-1\}$, and $m \in \{-l, -l+1, \ldots, l-1, l\}$. Note that $\psi_{100}(\mathbf{r}) = S_{100}(\mathbf{r}, \zeta = 1)$.

12.5.3 Helium atom

Let us consider a system with more than one electron. It is natural to start with the helium atom, which consists of two electrons orbiting around a nucleus. The electronic Hamiltonian of the helium atom is given by

$$\boxed{\hat{H} = -\frac{\nabla_1^2}{2} - \frac{2}{r_1} - \frac{\nabla_2^2}{2} - \frac{2}{r_2} + \frac{1}{r_{12}}.} \tag{12.26}$$

Even if the Hamiltonian only contains five terms, it is impossible to derive the eigenfunctions analytically and some approximations must be considered.

A simple approximation consists in ignoring the Coulomb repulsion between the electrons, i.e. the term $1/r_{12}$. This is equivalent to assuming that the electrons move independently in a Coulomb potential generated by the nucleus. Therefore, the Hamiltonian becomes

$$\hat{H}_0 = -\frac{\nabla_1^2}{2} - \frac{2}{r_1} - \frac{\nabla_2^2}{2} - \frac{2}{r_2}.$$

The lowest-energy wavefunction Ψ_0 is nothing but the product of two 1s orbitals, $\Psi_0(\mathbf{r}_1, \mathbf{r}_2) = \psi_{100}(\mathbf{r}_1)\psi_{100}(\mathbf{r}_2)$. The energy associated with this wavefunction is

$$\langle \Psi_0 | \hat{H}_0 | \Psi_0 \rangle = E_{n_1} + E_{n_2} = -0.5\frac{4}{n_1^2} - 0.5\frac{4}{n_2^2} = -4\ E_{\mathrm{h}}, \tag{12.27}$$

where we used eqn (12.20) with $n_1 = n_2 = 1$. This energy is significantly different from the experimental value of the ground-state energy, $-2.9033\ E_{\mathrm{h}}$. This is because we completely ignored the Coulomb repulsion between the electrons which adds a positive term. We will consider this contribution later on in this section.

Before presenting a perturbative technique to obtain a more accurate estimate of the ground state of the helium atom, it is necessary to introduce an important concept. In classical physics, each particle can be labeled and its motion can be tracked in time. Unlike classical mechanics, in quantum mechanics it is not possible to assign a label to each particle, especially when their wavefunctions overlap significantly. We see that there is a profound difference in the description of classical systems and quantum systems. An effective description of a quantum state must be based on a mathematical theory that takes into account the **indistinguishability of particles**. This mathematical framework must satisfy a precise condition: a measurement outcome cannot depend on a specific label assigned to each particle.

To make things more concrete, let us consider a quantum system composed of two electrons. Suppose that electron 1 occupies a hydrogen-like orbital described by the quantum numbers n_1, l_1, m_1, m_{s1}, while the state of electron 2 is described by the quantum numbers n_2, l_2, m_2, m_{s2}. We can write these states with the notation

$$\chi_a(x_1) = \langle r\theta\phi\omega | n_1 l_1 m_1 m_{s1} \rangle, \qquad \chi_b(x_2) = \langle r\theta\phi\omega | n_2 l_2 m_2 m_{s2} \rangle,$$

where the letters $a = \{n_1, l_1, m_1, m_{s1}\}$, $b = \{n_2, l_2, m_2, m_{s2}\}$ concisely express four quantum numbers. Let us assume that the global wavefunction is given by the product

$$\Psi(x_1, x_2) = \chi_a(x_1)\chi_b(x_2), \tag{12.28}$$

and check whether a measurement of Ψ produces results that depend on the label assigned to each electron. The simplest physical observable to consider is the probability density. The probability density associated with the wavefunction (12.28) is given by

$$|\Psi(x_1, x_2)|^2 = |\chi_a(x_1)|^2|\chi_b(x_2)|^2.$$

Since the electrons are indistinguishable, we should be able to exchange their labels without changing the probability density. However, an exchange of the labels leads to

$$|\Psi(x_1, x_2)|^2 = |\chi_a(x_1)|^2|\chi_b(x_2)|^2 \quad \underset{1\leftrightarrow 2}{\longrightarrow} \quad |\chi_a(x_2)|^2|\chi_b(x_1)|^2. \tag{12.29}$$

The quantity on the left-hand side differs from the quantity on the right-hand side: an exchange of labels has changed the probability density. We thus conclude that the global wavefunction $\Psi(x_1, x_2)$ is not an admissible wavefunction for the description of a system with two indistinguishable electrons.

It is possible to write a wavefunction that leaves the probability density unchanged when the labels of the electrons are swapped. For example, we can consider the **antisymmetric wavefunction**

$$\Psi(x_1, x_2) = \frac{1}{\sqrt{2}}[\chi_a(x_1)\chi_b(x_2) - \chi_b(x_1)\chi_a(x_2)]. \tag{12.30}$$

The factor $1/\sqrt{2}$ ensures that Ψ is correctly normalized when χ_a and χ_b are orthonormal. Let us show that the probability density does not change when the particles are relabeled. Indeed, we have

$$\begin{aligned}\Psi(x_1, x_2) = \frac{1}{\sqrt{2}}[\chi_a(x_1)\chi_b(x_2) - \chi_b(x_1)\chi_a(x_2)] \quad &\underset{1\leftrightarrow 2}{\longrightarrow} \\ \frac{1}{\sqrt{2}}[\chi_a(x_2)\chi_b(x_1) - \chi_b(x_2)\chi_a(x_1)] &= -\Psi(x_1, x_2).\end{aligned}$$

A relabeling of the particles just introduces a minus sign (this is why it is called antisymmetric) and the probability density remains unchanged,

$$|\Psi(x_1, x_2)|^2 = \Psi^*(x_1, x_2)\Psi(x_1, x_2) \quad \underset{1\leftrightarrow 2}{\longrightarrow} \quad (-\Psi^*(x_1, x_2))(-\Psi(x_1, x_2)) = |\Psi(x_1, x_2)|^2.$$

One can show that the expectation value of any physical observable evaluated for an antisymmetric wavefunction does not change when the particles are relabeled. In more general terms, a system of N electrons is always described by an antisymmetric wavefunction Ψ satisfying the relation

$$\boxed{\Psi(x_1, \ldots, x_i, \ldots, x_j, \ldots, x_N) = -\Psi(x_1, \ldots, x_j, \ldots, x_i, \ldots, x_N).}$$

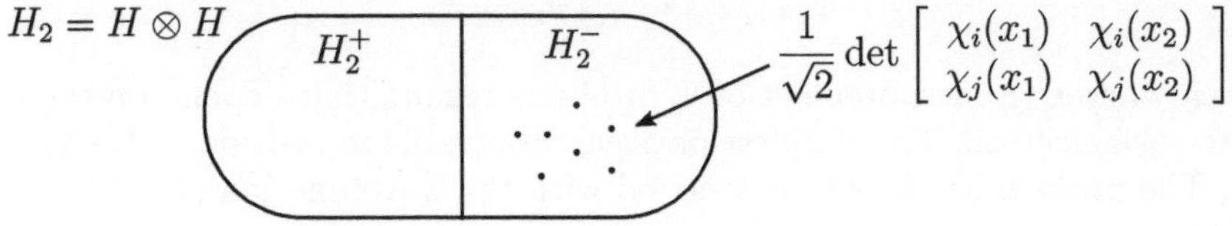

$H_2 = H \otimes H$ H_2^+ H_2^- $\frac{1}{\sqrt{2}} \det \begin{bmatrix} \chi_i(x_1) & \chi_i(x_2) \\ \chi_j(x_1) & \chi_j(x_2) \end{bmatrix}$

Fig. 12.11 Antisymmetric wavefunctions. The Hilbert space $H_2 = H \otimes H$ is generated by the basis $\{|\chi_i\rangle|\chi_j\rangle\}_{i,j=1}^K$. The space H_2 contains the Hilbert space of antisymmetric wavefunctions H_2^- and symmetric wavefunctions H_2^+. All wavefunctions in H_2^- can be written as a linear combination of Slater determinants (black dots in the diagram).

We highlight that the antisymmetry requirement applies to an exchange of both spatial and spin coordinates.

Given two normalized wavefunctions $\chi_a(x_1)$, $\chi_b(x_2)$ is straightforward to create a two-electron wavefunction with the correct antisymmetry using the **Slater determinant**,

$$\frac{1}{\sqrt{2!}} \det \begin{bmatrix} \chi_a(x_1) & \chi_a(x_2) \\ \chi_b(x_1) & \chi_b(x_2) \end{bmatrix} = \frac{1}{\sqrt{2}} [\chi_a(x_1)\chi_b(x_2) - \chi_b(x_1)\chi_a(x_2)].$$

If the quantum numbers a and b are identical, the determinant is zero. This is consistent with the Pauli exclusion principle which asserts that the state of two particles cannot be described by the same quantum numbers. It is important to highlight that any antisymmetric wavefunction can be expressed as a linear combination of Slater determinants (see Fig. 12.11 for a graphical illustration of this concept). To show this, let us consider a basis of spin-orbitals $\{\chi_j(x)\}_{j=1}^K$ with K elements and an antisymmetric wavefunction for two electrons,

$$\Psi(x_1, x_2) = \sum_{i=1}^{K} \sum_{j=1}^{K} b_{ij} \chi_i(x_1) \chi_j(x_2).$$

Since Ψ is antisymmetric, the elements on the diagonal of the matrix b are zero, $b_{jj} = 0$, and the off-diagonal elements have opposite sign, $b_{ij} = -b_{ji}$. Thus,

$$\begin{aligned} \Psi(x_1, x_2) &= \sum_{i=1}^{K} \sum_{j=i+1}^{K} b_{ij} \Big[\chi_i(x_1)\chi_j(x_2) - \chi_j(x_1)\chi_i(x_2) \Big] \\ &= \sqrt{2} \sum_{i=1}^{K} \sum_{j=i+1}^{K} b_{ij} \frac{1}{\sqrt{2}} \det \begin{bmatrix} \chi_i(x_1) & \chi_i(x_2) \\ \chi_j(x_1) & \chi_j(x_2) \end{bmatrix}. \end{aligned}$$

This shows that any two-electron wavefunction can be expressed as a linear combination of Slater determinants. This argument can be generalized to a system with N electrons: any antisymmetric wavefunction of N electrons can be written as a linear combination of Slater determinants generated by a complete set of spin-orbitals. We will study the properties of Slater determinants in Section 12.7.2.

Let us go back to the helium atom. To calculate the ground-state energy more accurately, one can take into account the Coulomb repulsion and perform a perturbative calculation up to the first order. Let us divide the electronic Hamiltonian of the helium atom into two parts, $\hat{H} = \hat{H}_0 + \hat{V}_{\mathrm{ee}}$ where

$$\hat{H}_0 = -\frac{\nabla_1^2}{2} - \frac{2}{r_1} - \frac{\nabla_2^2}{2} - \frac{2}{r_2}, \qquad \hat{V}_{\mathrm{ee}} = \frac{1}{r_{12}}.$$

We will treat the term $\hat{V}_{\mathrm{ee}}$ as a small perturbation. The ground state of the Hamiltonian $\hat{H}_0$ is an antisymmetric state in which both electrons occupy the 1s orbital, i.e.

$$\begin{aligned}\Psi(x_1, x_2) &= \psi_{100}(r_1)\,\psi_{100}(r_2)\,\sigma(\omega_1, \omega_2)\\ &= \frac{2^{3/2}}{\sqrt{\pi}} e^{-2r_1}\,\frac{2^{3/2}}{\sqrt{\pi}} e^{-2r_2}\,\frac{\alpha(\omega_1)\beta(\omega_2) - \beta(\omega_1)\alpha(\omega_2)}{\sqrt{2}}.\end{aligned} \tag{12.31}$$

Here, $\sigma(\omega_1, \omega_2)$ is an antisymmetric spin function so that the global wavefunction is antisymmetric. The ground-state energy can be approximated with first-order perturbation theory[7],

$$E_{\min} \approx \underbrace{\langle\Psi|H_0|\Psi\rangle}_{-0.5\cdot 4 - 0.5\cdot 4} + \langle\Psi|\hat{V}_{\mathrm{ee}}|\Psi\rangle = -4\,E_{\mathrm{h}} + \langle\Psi|\hat{V}_{\mathrm{ee}}|\Psi\rangle,$$

where we used eqn (12.20). We now need to calculate the quantity (see Exercise 12.4),

$$\langle\Psi|\hat{V}_{\mathrm{ee}}|\Psi\rangle = \int d\tau_1 \int d\tau_2 |\psi_{100}(\mathbf{r}_1)|^2 |\psi_{100}(\mathbf{r}_2)|^2 \frac{1}{r_{12}} = \frac{5}{4}\,E_{\mathrm{h}}.$$

Therefore, first-order perturbation theory predicts that the ground-state energy of the helium atom is

$$E_{\min} \approx -4\,E_{\mathrm{h}} + \langle\Psi|\hat{V}_{\mathrm{ee}}|\Psi\rangle = -2.75\,E_{\mathrm{h}}, \tag{12.32}$$

which is 0.1533 E_{h} above the experimental value. We could, in principle, improve this estimate by calculating higher-order contributions. In 1963, Scherr and Knight calculated the ground-state energy of the helium atom up to the 13th perturbation order, obtaining $-2.9037\,E_{\mathrm{h}}$ [403]. In the next section, we will introduce alternative ways to approximate the ground state of a quantum system based on the variational method.

With first-order perturbation theory, we found that the approximation of the ground-state energy is 0.1533 E_{h} above the experimental value. Is this an acceptable approximation? The answer is absolutely not. In quantum chemistry, an acceptable approximation for energy differences is within 0.0016 E_{h} from the experimental value. Let us explain this point in more detail. In a typical chemical reaction (see Fig. 12.12), two reactants transition from an initial state A to a transition state B with a reaction rate

$$\boxed{R = e^{-\Delta_{\mathrm{act}}/k_{\mathrm{B}}T},}$$

[7] See Ref. [128] for an introduction to perturbation theory.

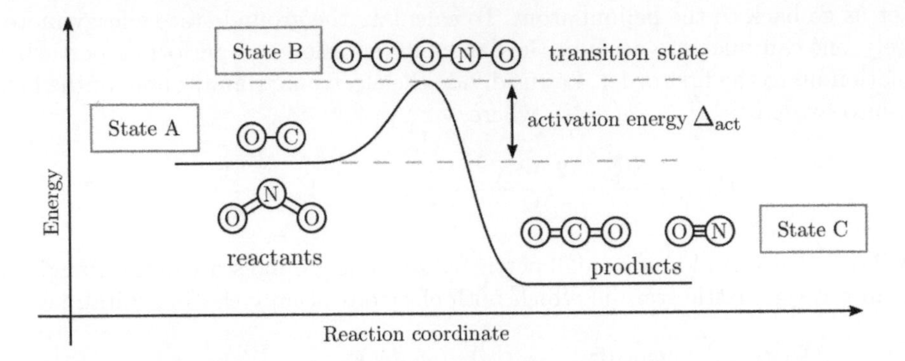

Fig. 12.12 Chemical reaction. Plot of the energy of a chemical reaction as a function of a reaction coordinate (such as a geometric distance).

where T is the temperature of the environment and Δ_{act} is the activation energy. Let us suppose that the reaction takes place at 500 K and therefore $k_{\text{B}}T = 0.0016\ E_{\text{h}}$. If the difference between the energy levels calculated with a numerical method, Δ_{theory}, differs from the experimental value of the activation energy by an amount $\Delta_{\text{theory}} = \Delta_{\text{act}} \pm 0.0016\ E_{\text{h}}$, the reaction rate predicted by the theory is approximately three times the experimental one,

$$R_{\text{theory}} = e^{-\Delta_{\text{theory}}/k_{\text{B}}T} = e^{-0.0016/k_{\text{B}}T} e^{-\Delta_{\text{exp}}/k_{\text{B}}T} \approx 3R.$$

In quantum chemistry, an acceptable approximation of energy differences must be within 0.0016 E_{h} of the experimental value [404]. Hereafter, the quantity $0.0016E_{\text{h}}$ will be called the **chemical accuracy**.

Example 12.4 Let us consider a two-electron *symmetric* wavefunction,

$$\psi(\mathbf{r}_1, \mathbf{r}_2) = \frac{1}{\sqrt{2}}(\psi_a(\mathbf{r}_1)\psi_b(\mathbf{r}_2) + \psi_b(\mathbf{r}_1)\psi_a(\mathbf{r}_2))$$

and an antisymmetric spin function, $\sigma(\omega_1, \omega_2) = (\alpha\beta - \beta\alpha)/\sqrt{2}$. The total wavefunction $\Psi = \psi(\mathbf{r}_1, \mathbf{r}_2)\sigma(\omega_1, \omega_2)$ is antisymmetric because it is the product of a symmetric function and an antisymmetric function. Therefore, Ψ is a valid wavefunction for the description of a system of two electrons. One can show that Ψ can be written as the sum of two Slater determinants,

$$\begin{aligned}\Psi(x_1, x_2) &= \frac{1}{2}\left[\psi_a(\mathbf{r}_1)\psi_b(\mathbf{r}_2)\alpha\beta - \psi_b(\mathbf{r}_1)\psi_a(\mathbf{r}_2)\beta\alpha + \psi_a(\mathbf{r}_1)\psi_b(\mathbf{r}_2)\beta\alpha - \psi_b(\mathbf{r}_1)\psi_a(r_2)\alpha\beta\right] \\ &= \frac{\sqrt{2}}{2}\left(\frac{1}{\sqrt{2}}\det\begin{bmatrix}\psi_a(\mathbf{r}_1)\alpha & \psi_a(\mathbf{r}_2)\alpha \\ \psi_b(\mathbf{r}_1)\beta & \psi_b(\mathbf{r}_2)\beta\end{bmatrix} + \frac{1}{\sqrt{2}}\det\begin{bmatrix}\psi_a(\mathbf{r}_1)\beta & \psi_a(\mathbf{r}_2)\beta \\ \psi_b(\mathbf{r}_1)\alpha & \psi_b(\mathbf{r}_2)\alpha\end{bmatrix}\right),\end{aligned}$$

where we used the definition of determinant, eqn (3.46).

12.6 The variational method

The variational method is a mathematical procedure for the approximation of the **ground state** of a quantum system (as we shall see later, the Hartree-Fock method is itself a variational method). To begin, we consider a Hamiltonian $\hat{H}$ with an unknown eigenvalues and eigenvectors. We could try to guess the lowest-energy wavefunction by considering a trial function $|\phi\rangle$ (which is not necessarily normalized) and evaluating the expectation value of the Hamiltonian,

$$E_\phi = \frac{\langle\phi|\hat{H}|\phi\rangle}{\langle\phi|\phi\rangle}.$$

It is straightforward to show that the energy E_ϕ is always greater than the ground-state energy:

Theorem 12.1 Consider a trial state $|\phi\rangle$ and a Hamiltonian $\hat{H}$ with eigenvalues and eigenvectors defined by $\hat{H}|k\rangle = E_k|k\rangle$ (where $|0\rangle$ is the ground state with energy E_0 while the energy levels $E_k > E_0$ for $k \geq 1$). Then,

$$E_\phi = \frac{\langle\phi|\hat{H}|\phi\rangle}{\langle\phi|\phi\rangle} \geq E_0.$$

Proof Let us decompose the quantum state $|\phi\rangle$ in the eigenbasis of the Hamiltonian,

$$|\phi\rangle = \sum_k |k\rangle\underbrace{\langle k|\phi\rangle}_{c_k} = \sum_k c_k|k\rangle.$$

We have

$$\begin{aligned} E_\phi = \frac{\langle\phi|\hat{H}|\phi\rangle}{\langle\phi|\phi\rangle} &= \frac{\sum_{kj}\langle\phi|k\rangle\langle k|\hat{H}|j\rangle\langle j|\phi\rangle}{\sum_k\langle\phi|k\rangle\langle k|\phi\rangle} = \frac{\sum_{kj} E_j\delta_{kj}c_k^*c_j}{\sum_k |c_k|^2} = \frac{\sum_k E_k|c_k|^2}{\sum_k |c_k|^2} \\ &= \frac{\sum_k |c_k|^2(E_k - E_0 + E_0)}{\sum_k |c_k|^2} = \frac{\sum_k |c_k|^2(E_k - E_0)}{\sum_k |c_k|^2} + E_0 \geq E_0, \end{aligned}$$

where we used the completeness relation and the fact that $E_k - E_0 \geq 0$. Only when $|\phi\rangle$ coincides with the ground state $|0\rangle$, then $E_\phi = E_0$. □

The variational method does not indicate which wavefunction we should try to find a good approximation of the ground-state energy. In practice, we can parametrize the trial wavefunction $\phi = \phi(\lambda_1 \ldots \lambda_n)$ with some real parameters λ_j. We then calculate analytically or numerically the stationary points of the functional $E_\phi = \langle\phi|\hat{H}|\phi\rangle/\langle\phi|\phi\rangle$, by setting the partial derivatives equal to zero,

$$\frac{\partial E_\phi}{\partial\lambda_1} = 0, \qquad \ldots \qquad \frac{\partial E_\phi}{\partial\lambda_n} = 0.$$

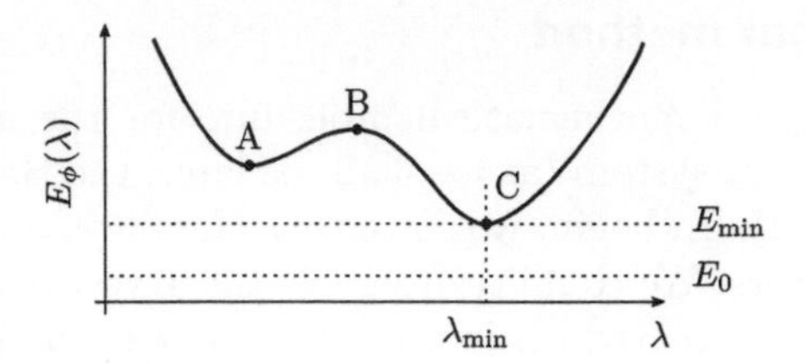

Fig. 12.13 Variational method. Plot of the expectation value $E_\phi = \langle\phi|E|\phi\rangle/\langle\phi|\phi\rangle$ as a function of a variational parameter λ. This graph shows three stationary points where $\partial E/\partial\lambda = 0$. Points A and C are local minima, while point B is a local maximum. The global minimum is point C. The approximation of the ground-state energy is $E_{\rm min}$ which is greater than the real ground-state energy E_0 as stated by Theorem 12.1.

Not all stationary points correspond to true energy minima (some of them may be local maxima or saddle points, see Fig. 12.13). Finally, we identify the global minimum $\lambda_{\rm min}$. Let us study two concrete examples.

Example 12.5 The first example deals with the hydrogen atom. Let us suppose that we do not know the eigenfunction with minimum energy. Our goal is to find the ground state of the Hamiltonian $\hat{H} = -\nabla^2/2 - 1/r$. We can guess the trial wavefunction that approximates the ground state and consider a Gaussian function centered around the origin,

$$\phi(r) = \left(\frac{2\lambda}{\pi}\right)^{3/4} e^{-\lambda r^2}, \tag{12.33}$$

where $\lambda > 0$ is a variational parameter. Note that this wavefunction is normalized, $\langle\phi|\phi\rangle = \int |\phi(r)|^2 d\tau = 1$. The expectation value of the energy is given by (see Exercise 12.5),

$$E_\phi = \langle\phi|\hat{H}|\phi\rangle = \int d\tau \phi^*(r)\hat{H}\phi(r) = \frac{3}{2}\lambda - 2\sqrt{\frac{2}{\pi}}\lambda^{1/2}. \tag{12.34}$$

Let us minimize this quantity with respect to the parameter λ. By imposing $\partial E_\phi/\partial\lambda = 0$, we find

$$\frac{dE_\phi}{d\lambda} = \frac{3}{2} - \sqrt{\frac{2}{\pi\lambda}} = 0 \qquad \Rightarrow \qquad \lambda_{\rm min} = \frac{8}{9\pi}.$$

By substituting $\lambda_{\rm min}$ into eqn (12.34), we see that the minimum energy associated with $\phi(r)$ is $E_\phi^{(\rm min)} = -0.424\ E_{\rm h}$. This value is greater than the experimental value by $0.076\,E_{\rm h}$. This approximation is not sufficiently accurate since it exceeds the chemical accuracy by two orders of magnitude. In Fig. 12.14, we show a comparison between the normalized wavefunction $\phi(r) = (2\lambda_{\rm min}/\pi)^{3/4}\,e^{-\lambda_{\rm min}r^2}$ and the exact ground state $\psi_{100}(r) = e^{-r}/\sqrt{\pi}$.

Example 12.6 This example deals with the helium atom. Let us start from the electronic Hamiltonian of the helium atom,

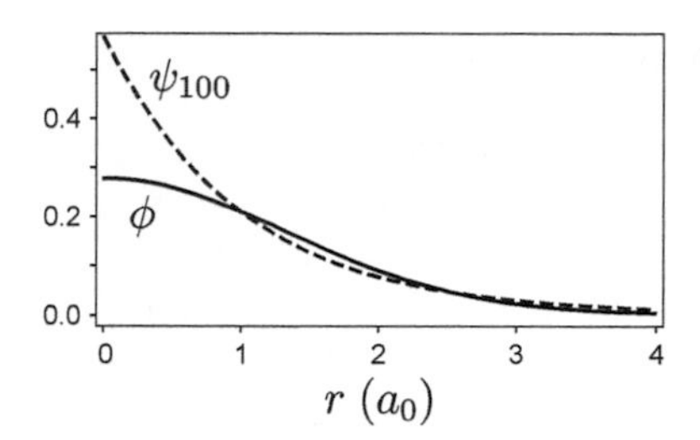

Fig. 12.14 Variational method. Comparison between the exact ground-state wavefunction for the hydrogen atom $\psi_{100} = 1/\sqrt{\pi}e^{-r}$ (dashed curve) and the trial function ϕ of eqn (12.33) with $\lambda = 8/9\pi$ (solid curve).

$$\hat{H} = -\frac{\nabla_1^2}{2} - \frac{\nabla_2^2}{2} - \frac{2}{r_1} - \frac{2}{r_2} + \frac{1}{r_{12}}.$$

Our goal is to find an approximation of the ground state. We can try as a trial state the antisymmetric wavefunction

$$\begin{aligned}\Psi(x_1, x_2) &= \psi_{100}(r_1)\psi_{100}(r_2)\,\sigma(\omega_1, \omega_2) \\ &= \frac{Z^{3/2}}{\sqrt{\pi}}e^{-Zr_1}\frac{Z^{3/2}}{\sqrt{\pi}}e^{-Zr_2}\,\frac{\alpha(\omega_1)\beta(\omega_2) - \beta(\omega_1)\alpha(\omega_2)}{\sqrt{2}}. \qquad (12.35)\end{aligned}$$

The parameter Z will be treated as a variational parameter. The expectation value of the energy for the trial wavefunction is given by (see Exercise 12.6),

$$E_\Psi = \frac{\langle\Psi|\hat{H}|\Psi\rangle}{\langle\Psi|\Psi\rangle} = \left(Z^2 - \frac{27}{8}Z\right) E_\mathrm{h}. \qquad (12.36)$$

This quantity is minimized when

$$\frac{dE_\Psi}{dZ} = 0, \qquad \Rightarrow \qquad Z_\mathrm{min} = \frac{27}{16}.$$

Substituting Z_min into eqn (12.36), we find that the minimum energy associated with the wavefunction $\Psi(x_1, x_2)$ is $E_\Psi^{(\mathrm{min})} = -2.8477\ E_\mathrm{h}$. This energy is $0.0556\ E_\mathrm{h}$ above the experimental value, $E_\mathrm{exp} = -2.9033\ E_\mathrm{h}$. This approximation is not sufficiently accurate since it exceeds the chemical accuracy by one order of magnitude. We could, in principle, improve our approximation by considering a wavefunction with a greater number of variational parameters. It is important to note that although the helium nucleus contains two protons, the variational method predicts that the value of Z that minimizes the energy is smaller than two. This phenomenon is consistent with the fact that the electrons in the atom partially shield the nucleus, resulting in an effective nuclear charge that is lower than the actual number of protons.

12.6.1 Linear variational method

The variational method presented in the previous section produces an expectation value that might depend on many variational parameters. In general, $E_\phi(\lambda)$ is a complicated function and it might be difficult to calculate its global minimum. However, if the trial wavefunction is a linear combination of functions, $|\phi\rangle = \sum_i c_i|e_i\rangle$, and the variational parameters are the complex coefficients c_i, the minimization of the functional E_ϕ reduces to the diagonalization of a square matrix. This variational approach is known as the **Ritz method** [405].

Let us show a systematic procedure to find an approximation of the minimum energy of a quantum system using the Ritz method. We write the wavefunction as a linear combination of m wavefunctions,

$$\boxed{|\phi\rangle = \sum_{p=1}^{m} c_p|e_p\rangle,}$$

where c_p are complex variational parameters, while the quantum states $|e_p\rangle$ are not necessarily orthonormal. The expectation value of the energy $E_\phi = \langle\phi|\hat{H}|\phi\rangle/\langle\phi|\phi\rangle$ has two main contributions:

$$\langle\phi|\hat{H}|\phi\rangle = \sum_p\sum_q c_p^* c_q h_{pq}, \qquad\qquad \langle\phi|\phi\rangle = \sum_p\sum_q c_p^* c_q S_{pq},$$

where we defined $h_{pq} = \langle e_p|\hat{H}|e_q\rangle$ and the overlap matrix $S_{pq} = \langle e_p|e_q\rangle$. Therefore,

$$\boxed{E_\phi = \frac{\sum_{p,q} c_p^* c_q h_{pq}}{\sum_{p,q} c_p^* c_q S_{pq}}.}$$

To avoid quotients in the differentiation we multiply both sides by $\sum_{pq} c_p^* c_q S_{pq}$,

$$\sum_{p,q} c_p^* c_q h_{pq} = E_\phi \sum_{p,q} c_p^* c_q S_{pq}. \tag{12.37}$$

To find the minimum of the functional E_ϕ, we can either differentiate with respect to c_i^* or[8] c_i. For example, by performing the partial differentiation with respect to c_i^*, the left-hand side and the right-hand side of eqn (12.37) become:

$$\frac{\partial}{\partial c_i^*}\sum_{p,q} c_p^* c_q h_{pq} = \sum_{p,q} \delta_{pi} c_q h_{pq} = \sum_q h_{iq} c_q,$$

$$\frac{\partial E_\phi}{\partial c_i^*} + E_\phi \frac{\partial}{\partial c_i^*}\sum_{p,q} c_p^* c_q S_{pq} = \frac{\partial E_\phi}{\partial c_i^*} + E_\phi \sum_{p,q} \delta_{pi} c_q S_{pq} = \frac{\partial E_\phi}{\partial c_i^*} + E_\phi \sum_q S_{iq} c_q.$$

[8]It does not matter which coefficient we choose, since they are not independent: $c_i = (c_i^*)^*$.

By imposing $\partial E_\phi / \partial c_i^* = 0$, eqn (12.37) leads to a set of m linear equations. The i-th equation is given by

$$\sum_q h_{iq} c_q = E_\phi \sum_q S_{iq} c_q \qquad (i\text{-th equation}).$$

Note that this system of equations can be written in a more compact form as

$$\boxed{(\mathsf{h} - E_\phi \mathsf{S})\,\mathbf{c} = \mathbf{0},}$$

where h corresponds to the Hamiltonian matrix h_{pq}, S corresponds to the overlap matrix S_{pq}, and $\mathbf{c}$ is a column vector. This system of equations has a non-trivial solution for the vector $\mathbf{c}$ if and only if $\det(\mathsf{h} - E_\phi \mathsf{S}) = 0$. The solution of the equation $\det(\mathsf{h} - E_\phi \mathsf{S}) = 0$ provides m values $E_{\phi,i}$. The smallest value $E_{\phi,0}$ is an approximation of the ground-state energy, while the other $E_{\phi,i}$'s are upper bounds of the excited-state energies.

Example 12.7 In this example, we calculate an approximation of the ground-state energy of the hydrogen atom using the trial wavefunction $\phi(r) = c_1 e_1(r) + c_2 e_2(r)$ where

$$e_1(r) = e^{-\lambda r^2}, \qquad e_2(r) = e^{-2\lambda r^2}, \qquad \lambda = 8/9\pi.$$

The terms h_{ij} and S_{ij} are given by (see Exercise 12.7):

$$h_{11} = -5.55165, \quad h_{12} = -3.37295, \quad S_{11} = 13.0808, \quad S_{12} = 7.12027, \qquad (12.38)$$

$$h_{21} = -3.37295, \quad h_{22} = -1.62604, \quad S_{21} = 7.12027, \quad S_{22} = 4.62475. \qquad (12.39)$$

The solution to the equation

$$\det \begin{bmatrix} h_{11} - E_\phi S_{11} & h_{12} - E_\phi S_{12} \\ h_{21} - E_\phi S_{21} & h_{22} - E_\phi S_{22} \end{bmatrix} = 0,$$

is given by $E_{\phi,0} = -0.4373\ E_\text{h}$ and $E_{\phi,1} = 0.5484\ E_\text{h}$. The lowest eigenvalue $E_{\phi,0}$ is an approximation of the true ground-state energy $E_0 = -0.5\ E_\text{h}$.

12.7 The Hartree–Fock method

The Hartree–Fock method is a computational method used in quantum chemistry to obtain an approximate solution of the ground state of a many-body quantum system. It is based on the variational principle and is a self-consistent field (SCF) method, meaning that it involves an iterative process where the electronic wavefunction is updated until self-consistency is achieved. For pedagogical reasons, we first study the Hartree–Fock method for the helium atom and then consider a molecule with N electrons.

12.7.1 The Hartree–Fock method for the helium atom

Let us apply the Hartree–Fock method to the helium atom. Similarly to Example 12.6, the first step is writing the wavefunction that approximates the ground state of the helium atom as a *single* Slater determinant,

$$|\Psi\rangle = \frac{1}{\sqrt{2}}\begin{bmatrix}|\chi_1\rangle & |\chi_1\rangle \\ |\chi_2\rangle & |\chi_2\rangle\end{bmatrix} = \frac{1}{\sqrt{2}}\left[|\chi_1\rangle|\chi_2\rangle - |\chi_2\rangle|\chi_1\rangle\right]. \tag{12.40}$$

For simplicity, we choose two spin-orbitals $|\chi_1\rangle$ and $|\chi_2\rangle$ with the same spatial part, $|\chi_1\rangle = |\psi\rangle|0\rangle$, $|\chi_2\rangle = |\psi\rangle|1\rangle$, so that the global wavefunction becomes,

$$|\Psi\rangle = |\psi\rangle|\psi\rangle \otimes \frac{|0\rangle|1\rangle - |1\rangle|0\rangle}{\sqrt{2}}.$$

The goal is to find a parametrization of the spatial orbital $|\psi\rangle$ that best approximates the ground-state energy.

Following the usual variational procedure, we first write the spatial orbital as a linear combination of m functions, $|\psi\rangle = \sum_{p=1}^{m} c_p|e_p\rangle$. We will assume that the coefficients c_p are real and the functions $\langle \mathbf{r}|e_p\rangle = e_p(\mathbf{r})$ are real as well. The next step is to calculate the coefficients c_p that minimize the expectation value of the energy $E_\Psi = \langle\Psi|\hat{H}|\Psi\rangle/\langle\Psi|\Psi\rangle$, where

$$\hat{H} = -\frac{\nabla_1^2}{2} - \frac{2}{r_1} - \frac{\nabla_2^2}{2} - \frac{2}{r_2} + \frac{1}{r_{12}} \tag{12.41}$$

is the Hamiltonian of the helium atom in atomic units. Using eqns (12.40, 12.41), we have

$$E_\Psi = \frac{\langle\psi\psi|\hat{H}|\psi\psi\rangle_{12}}{\langle\psi\psi|\psi\psi\rangle_{12}} = 2\frac{\langle\psi|\left(-\frac{\nabla_1^2}{2} - \frac{2}{r_1}\right)|\psi\rangle_1}{\langle\psi|\psi\rangle_1} + \frac{\langle\psi\psi|\frac{1}{r_{12}}|\psi\psi\rangle_{12}}{\langle\psi|\psi\rangle_1\langle\psi|\psi\rangle_2}, \tag{12.42}$$

where we used the notation $\langle\psi|\hat{O}|\psi\rangle_j = \int \psi(\mathbf{r}_j)\hat{O}\psi(\mathbf{r}_j)d\tau_j$. Substituting $|\psi\rangle = \sum_p c_p|e_p\rangle$ into (12.42), the expectation value becomes

$$E_\Psi = 2\frac{\sum_{pq} c_p c_q \langle e_p|\left(-\frac{\nabla_1^2}{2} - \frac{2}{r_1}\right)|e_q\rangle_1}{\sum_{pq} c_p c_q \langle e_p|e_q\rangle_1} + \frac{\sum_{pq} c_p c_q \sum_{rs} c_r c_s \langle e_p e_r|\frac{1}{r_{12}}|e_s e_q\rangle_{12}}{\sum_{pq} c_p c_q \langle e_p|e_q\rangle_1 \sum_{rs} c_r c_s \langle e_r|e_s\rangle_2}.$$

To further simplify the notation, we introduce the matrices:

$$h_{pq} = \langle e_p|\left(-\frac{\nabla_1^2}{2} - \frac{2}{r_1}\right)|e_q\rangle_1, \quad S_{pq} = \langle e_p|e_q\rangle_1, \quad g_{pq}(\mathbf{c}) = \frac{\sum_{rs} c_r c_s \langle e_p e_r|\frac{1}{r_{12}}|e_s e_q\rangle_{12}}{\sum_{rs} c_r c_s \langle e_r|e_s\rangle_2}. \tag{12.43}$$

Therefore, E_Ψ becomes

$$\boxed{E_\Psi = \frac{\sum_{pq} c_p c_q \left(2h_{pq} + g_{pq}(\mathbf{c})\right)}{\sum_{pq} c_p c_q S_{pq}}.} \tag{12.44}$$

Let us calculate the minimum of this quantity. To avoid quotients in the differentiation we multiply both sides by $\sum_{pq} c_p c_q S_{pq}$ obtaining,

$$E_\Psi\Big(\sum_{pq} c_p c_q S_{pq}\Big) = 2\sum_{pq} c_p c_q h_{pq} + \sum_{pq} c_p c_q g_{pq}(\mathbf{c}). \tag{12.45}$$

To determine the stationary points, we differentiate with respect to c_i,

$$\frac{\partial E_\Psi}{\partial c_i}\sum_{pq} c_p c_q S_{pq} + E\frac{\partial}{\partial c_i}\sum_{pq} c_p c_q S_{pq} = 2\frac{\partial}{\partial c_i}\sum_{pq} c_p c_q h_{pq} + \frac{\partial}{\partial c_i}\sum_{pq} c_p c_q g_{pq}(\mathbf{c}). \tag{12.46}$$

By imposing the condition $\partial E_\Psi/\partial c_i = 0$, we obtain a system of m linear equations. The i-th equation is given by (see Exercise 12.8)

$$\boxed{\sum_{q=1}^{m} \left[h_{iq} + g_{iq}(\mathbf{c}) - \gamma(\mathbf{c}) S_{iq}\right] c_q = 0} \qquad (i\text{-th equation}) \tag{12.47}$$

where $\gamma(\mathbf{c})$ is defined as

$$\gamma(\mathbf{c}) = \frac{1}{2}\left(E_\Psi + \frac{\sum_{pq} c_p c_q g_{pq}(\mathbf{c})}{\sum_{pq} c_p c_q S_{pq}}\right).$$

In the literature, the system of equations (12.47) is often expressed in a more concise form. By defining the **Fock matrix** as

$$F_{pq}(\mathbf{c}) = h_{pq} + g_{pq}(\mathbf{c}), \tag{12.48}$$

eqn (12.47) becomes $\sum_q \left[F_{iq}(\mathbf{c}) - \gamma(\mathbf{c}) S_{iq}\right] c_q = 0$, or in matrix form

$$\boxed{[\mathsf{F}(\mathbf{c}) - \gamma(\mathbf{c})\mathsf{S}]\mathbf{c} = \mathbf{0},} \tag{12.49}$$

where $\mathbf{c}$ is a column vector and S is the overlap matrix. The solution of this system of equations gives a stationary point $\mathbf{c}$. This vector can be used to calculate an approximation of the ground-state energy using eqn (12.44).

Some comments are in order. The system of equations (12.49) cannot be solved in a straightforward way, because the quantities $\mathsf{F}(\mathbf{c})$ and $\gamma(\mathbf{c})$ depend on the coefficients $\mathbf{c}$ themselves! Hence, the solution must be calculated iteratively until the coefficients c_i have converged[9]. This iterative approach is called a **self-consistent field** (SCF) method and is represented in a schematic way in Fig. 12.15. Furthermore, the condition $\partial E_\Psi/\partial c_i = 0$ does not guarantee that the stationary point found corresponds to a local minimum because it could be a saddle point. This should be checked by calculating the Hessian matrix $M_{ij} = \partial E/\partial c_i \partial c_j$: if all of the eigenvalues of M are positive, the point found is a local minimum. Lastly, there is no guarantee that the minimum found is a global minimum: it is necessary to repeat the procedure with different initial values for the coefficients c_i and find additional local minima. By comparing the expectation

[9]In any iterative scheme, convergence and stability of the solutions are a delicate topic. We will not cover this problem in detail in this textbook.

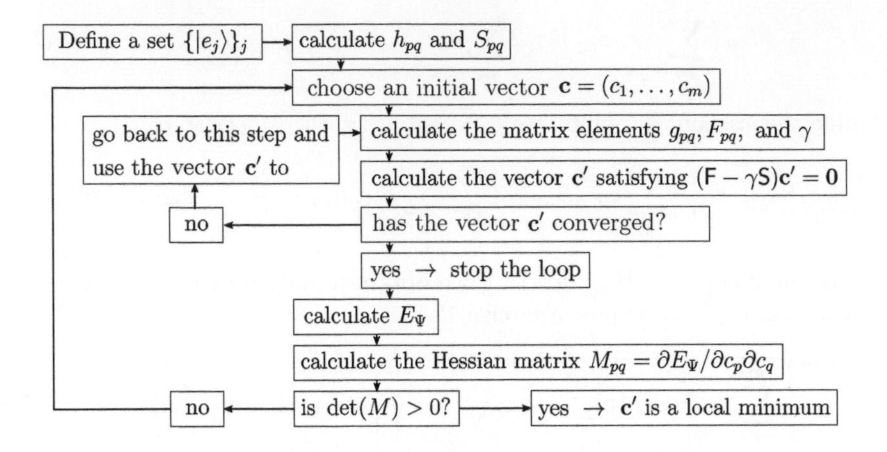

Fig. 12.15 Hartree–Fock method for the helium atom.

value of the energy, one can determine which of the local minima returns the lowest value. We encourage readers not familiar with the Hartree–Fock method to study Exercise 12.11, which includes some explicit calculations.

A simple way to improve the accuracy of the ground-state energy predicted by the HF method is to consider a larger basis $\{e_p\}_p$. Larger and larger basis sets will keep lowering the Hartree–Fock energy until a limit is reached, known as the **Hartree–Fock limit**. In practice, a finite basis set will always produce a ground-state energy higher than the asymptotic limit.

12.7.2 Slater determinants

Before presenting the general case of the Hartree–Fock method, it is important to dwell on some properties of Slater determinants. Consider a system of N electrons and N orthonormal spin-orbitals $\{\chi_1, \ldots, \chi_N\}$. A Slater determinant is an antisymmetric wavefunction defined as

$$\Psi(x_1, \ldots, x_N) = \frac{1}{\sqrt{N!}} \det \begin{bmatrix} \chi_1(x_1) & \chi_1(x_2) & \cdots & \chi_1(x_N) \\ \chi_2(x_1) & \chi_2(x_2) & \cdots & \chi_2(x_N) \\ \vdots & \vdots & & \vdots \\ \chi_N(x_1) & \chi_N(x_2) & \cdots & \chi_N(x_N) \end{bmatrix}.$$

In the ket notation, a Slater determinant can be expressed as

$$\boxed{|\Psi\rangle = \frac{1}{\sqrt{N!}} \sum_{\sigma \in S_N} (-1)^\sigma |\chi_{\sigma(1)} \cdots \chi_{\sigma(N)}\rangle,}$$

where S_N is a multiset containing all permutations σ of the sequence $(1, 2, \ldots, N)$. Examples of permutations are $\sigma_1 = (1, 2, 3, 4, \ldots, N)$ and $\sigma_2 = (2, 1, 3, 4, \ldots, N)$. The notation $\sigma(j)$ denotes the j-th element of σ, while $(-1)^\sigma$ is the signature of the permutation, a quantity that is $+1$ when the permutation σ can be obtained from $(1,\ldots,N)$ with an even number of transpositions of adjacent numbers and -1 whenever it can be achieved with an odd number of such interchanges. In the literature, Slater determinants are often written with a more compact notation

$$\boxed{|\Psi\rangle = |\chi_1 \ldots \chi_N\rangle_-.}$$

Here, the subscript reminds us that the wavefunction is antisymmetric. Since the determinant of a matrix changes sign when two rows are transposed, Slater determinants satisfy the relation

$$|\chi_1 \ldots \chi_i \chi_{i+1} \ldots \chi_N\rangle = (-1)|\chi_1 \ldots \chi_{i+1}\chi_i \ldots \chi_N\rangle.$$

This property is useful to reorder spin-orbitals in ascending order.

A generic product state can be transformed into an antisymmetric wavefunction by applying the **antisymmetrizer** $\hat{A}$. The action of the antisymmetrizer on a product state $|\chi_1 \ldots \chi_N\rangle$ returns the corresponding Slater determinant,

$$\hat{A}|\chi_1 \ldots \chi_N\rangle = |\chi_1 \ldots \chi_N\rangle_-.$$

The antisymmetrizer commutes with any observable $\hat{O}$, including the electronic Hamiltonian.

Example 12.8 Consider the Slater determinant $|\chi_2\chi_3\chi_1\rangle_-$. This Slater determinant can be expressed in ascending order by moving χ_1 to the first position,

$$|\chi_2\chi_3\chi_1\rangle_- = (-1)|\chi_2\chi_1\chi_3\rangle_- = (-1)^2|\chi_1\chi_2\chi_3\rangle_- = |\chi_1\chi_2\chi_3\rangle_-.$$

Since the number of adjacent transpositions is even, the overall phase is one.

12.7.3 Energy associated with a single Slater determinant

To lay the foundation for the derivation of the Hartree-Fock equations, we first need to derive the energy associated with a Slater determinant for a molecule containing N electrons. Consider a Slater determinant $|\Psi\rangle = |\chi_1 \ldots \chi_N\rangle_-$ where $\{\chi_j\}_{j=1}^N$ are N orthonormal spin-orbitals. The expectation value of the energy is given by $\langle\Psi|\hat{H}|\Psi\rangle$, where $\hat{H}$ is the electronic Hamiltonian

$$\boxed{\hat{H} = \sum_{j=1}^{N}\left(-\frac{\nabla_j^2}{2} - \sum_{A=1}^{M}\frac{Z}{|\mathbf{r}_j - \mathbf{R}_A|}\right) + \sum_{j=1}^{N}\sum_{i=j+1}^{N}\frac{1}{|\mathbf{r}_i - \mathbf{r}_j|} = \hat{O}_1 + \hat{O}_2.} \quad (12.50)$$

Here, we have written the electronic Hamiltonian as the sum of two parts: $\hat{O}_1$ contains operators acting on individual electrons, while $\hat{O}_2$ contains operators acting on pairs of electrons. For this reason, they are referred to **one-body** and **two-body operators**, respectively. The expectation value $\langle\Psi|\hat{H}|\Psi\rangle = \langle\Psi|\hat{O}_1|\Psi\rangle + \langle\Psi|\hat{O}_2|\Psi\rangle$ has two contributions. After some lengthy calculations, one can verify that the first term is given by (see Exercise 12.9),

$$\boxed{\langle\Psi|\hat{O}_1|\Psi\rangle = \sum_{j=1}^{N}\langle\chi_j|\hat{h}|\chi_j\rangle,} \tag{12.51}$$

where

$$\hat{h} = -\frac{\nabla^2}{2} - \sum_{A=1}^{M}\frac{Z}{|\mathbf{r}-\mathbf{R}_A|}, \qquad \langle\chi_j|\hat{h}|\chi_j\rangle = \int \chi_j^*(x)\hat{h}\chi_j(x)dx.$$

Similarly, the second term is given by (see Exercise 12.10),

$$\boxed{\langle\Psi|\hat{O}_2|\Psi\rangle = \frac{1}{2}\sum_{i=1}^{N}\sum_{j=1}^{N}\Big[\langle\chi_i\chi_j|\frac{1}{r_{12}}|\chi_i\chi_j\rangle - \langle\chi_j\chi_i|\frac{1}{r_{12}}|\chi_i\chi_j\rangle\Big],} \tag{12.52}$$

where[10]

$$\langle\chi_i\chi_j|\frac{1}{r_{12}}|\chi_k\chi_l\rangle = \iint \chi_i^*(x_1)\chi_j^*(x_2)\frac{1}{|\mathbf{r}_1-\mathbf{r}_2|}\chi_k(x_1)\chi_l(x_2)dx_1dx_2.$$

The integrals (12.51, 12.52) are usually calculated with numerical methods.

Up to this point, we have not made any assumptions about the spin-orbitals χ_j. It is common to study two separate cases:

- **Unrestricted Hartree–Fock**: the molecule contains N electrons. The electrons with the spin function α occupy the spatial orbitals ψ_j, while the electrons described by the spin function β occupy the spatial orbitals ϕ_j. Some of the spatial orbitals ψ_j and ϕ_j might be identical.
- **Closed-shell restricted Hartree–Fock**: the system contains an even number of electrons N occupying $N/2$ spatial orbitals ψ_j. Half of the electrons have a spin function α, and half have a spin function β.

The unrestricted method offers more flexibility in the choice of spatial orbitals. This often allows us to obtain better approximations of the ground state with respect to the restricted method. However, the unrestricted method does not guarantee that the variational wavefunction is an eigenstate of the total spin[11]. We will focus on the closed-shell restricted Hartree–Fock method (RHF). For more information about the unrestricted method, see for example Ref. [345].

In the closed-shell restricted HF, N electrons occupy $N/2$ spatial orbitals ψ_j,

$$\chi_1 = \psi_1\alpha, \qquad \chi_2 = \psi_1\beta, \qquad \ldots, \qquad \chi_{N-1} = \psi_{N/2}\alpha, \qquad \chi_N = \psi_{N/2}\beta. \tag{12.53}$$

Now that we have written the explicit form of the spin-orbitals, we can further simplify

[10]Note that $\langle\chi_i\chi_j|1/r_{12}|\chi_k\chi_l\rangle = \langle\chi_j\chi_i|1/r_{12}|\chi_l\chi_k\rangle$. This property simplifies some calculations.

[11]We expect the ground state to be an eigenstate of the total spin since the electronic Hamiltonian under consideration does not contain spin operators.

the terms $\langle\Psi|\hat{O}_1|\Psi\rangle$ and $\langle\Psi|\hat{O}_2|\Psi\rangle$. Substituting (12.53) into eqn (12.51) and summing over the spin components, we obtain a set of integrals over the spatial components only

$$\begin{aligned}\langle\Psi|\hat{O}_1|\Psi\rangle &= \sum_{j=\text{odd}}\langle\chi_j|\hat{h}|\chi_j\rangle + \sum_{j=\text{even}}\langle\chi_j|\hat{h}|\chi_j\rangle = \sum_{j=1}^{N/2}\langle\psi_j\alpha|\hat{h}|\psi_j\alpha\rangle + \sum_{j=1}^{N/2}\langle\psi_j\beta|\hat{h}|\psi_j\beta\rangle \\ &= \sum_{j=1}^{N/2}\langle\alpha|\alpha\rangle\langle\psi_j|\hat{h}|\psi_j\rangle + \sum_{j=1}^{N/2}\langle\beta|\beta\rangle\langle\psi_j|\hat{h}|\psi_j\rangle = 2\sum_{j=1}^{N/2}\langle\psi_j|\hat{h}|\psi_j\rangle. \end{aligned} \tag{12.54}$$

In the literature, this step is sometimes called the **spin integration**. In a similar way, eqn (12.52) becomes (see Exercise 12.12),

$$\langle\Psi|\hat{O}_2|\Psi\rangle = \sum_{i=1}^{N/2}\sum_{j=1}^{N/2}\Big[2\langle\psi_i\psi_j|\frac{1}{r_{12}}|\psi_i\psi_j\rangle - \langle\psi_i\psi_j|\frac{1}{r_{12}}|\psi_j\psi_i\rangle\Big]. \tag{12.55}$$

The factors of 2 in eqns (12.54, 12.55) occur because we are investigating a system with an even number of electrons N occupying $N/2$ spatial orbitals. Putting everything together, we finally obtain the energy associated with a single Slater determinant in the closed-shell restricted HF method,

$$\boxed{E_\Psi = \langle\Psi|\hat{O}_1|\Psi\rangle + \langle\Psi|\hat{O}_2|\Psi\rangle = 2\sum_{j=1}^{N/2} h_j + \sum_{i,j=1}^{N/2}\left[2J_{ij} - K_{ij}\right],} \tag{12.56}$$

where we introduced the quantities,

$$\boxed{h_j = \langle\psi_j|\hat{h}|\psi_j\rangle, \qquad J_{ij} = \langle\psi_i\psi_j|\frac{1}{r_{12}}|\psi_i\psi_j\rangle, \qquad K_{ij} = \langle\psi_i\psi_j|\frac{1}{r_{12}}|\psi_j\psi_i\rangle.} \tag{12.57}$$

Equation (12.56) approximates the ground-state energy of a molecule with a single Slater determinant. The integrals J_{ij} are called the **Coulomb integrals**, while the K_{ij} integrals are called the **exchange integrals**[12].

Example 12.9 The beryllium atom contains four electrons. An approximation of the ground state of the beryllium atom is given by the Slater determinant,

$$\Psi(x_1,x_2,x_3,x_4) = \frac{1}{\sqrt{4!}}\det\begin{bmatrix}\psi_{100}(\mathbf{r}_1)\alpha & \psi_{100}(\mathbf{r}_2)\alpha & \psi_{100}(\mathbf{r}_3)\alpha & \psi_{100}(\mathbf{r}_4)\alpha \\ \psi_{100}(\mathbf{r}_1)\beta & \psi_{100}(\mathbf{r}_2)\beta & \psi_{100}(\mathbf{r}_3)\beta & \psi_{100}(\mathbf{r}_4)\beta \\ \psi_{200}(\mathbf{r}_1)\alpha & \psi_{200}(\mathbf{r}_2)\alpha & \psi_{200}(\mathbf{r}_3)\alpha & \psi_{200}(\mathbf{r}_4)\alpha \\ \psi_{200}(\mathbf{r}_1)\beta & \psi_{200}(\mathbf{r}_2)\beta & \psi_{200}(\mathbf{r}_3)\beta & \psi_{200}(\mathbf{r}_4)\beta\end{bmatrix}.$$

Here, we are assuming that the electrons occupy the 1s and 2s orbitals, two with spin up and two with spin down. Let us calculate the energy associated with this Slater

[12]Not all of these integrals are independent. One can verify that: $J_{ij} = J_{ji}$, $K_{ij} = K_{ji}$, and $K_{jj} = J_{jj}$.

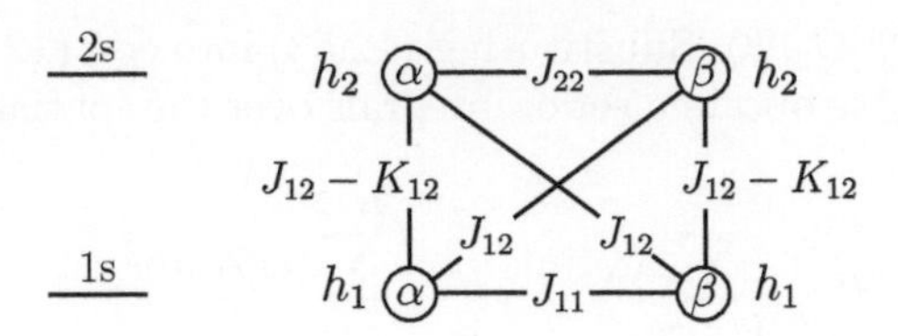

Fig. 12.16 Closed-shell restricted Hartree–Fock. Schematic representation of the electrons in a beryllium atom. Two electrons have spin function α and two have spin function β. Each electron is associated with an h_j integral. A segment connecting two spin-orbitals with opposite spins is associated with a "J_{ij} integral". A segment connecting two spin-orbitals with the same spin is associated with a "$J_{ij} - K_{ij}$ integral".

---

determinant using (12.56),

$$E_\Psi = 2h_1 + 2h_2 + J_{11} + 4J_{12} + J_{22} - 2K_{12}.$$

The diagram in Fig. 12.16 helps keep track of all the integrals involved. The value of E_Ψ can be calculated with a Mathematica script,

```
Z = 4; Inf = Infinity;
fun = {Z^(3/2) Exp[-Z r]/Sqrt[Pi],
       Z^(3/2) (2 - Z r) Exp[-Z r/2]/(4 Sqrt[2 Pi])};
H[x_]:=-(1/2) Laplacian[x, {r,\[Theta],\[Phi]}, "Spherical"]-Z x/r;
h = Table[Integrate[4 Pi r^2 fun[[i]] H[fun[[i]]],{r,0,Inf}],{i,1,2}];
J = Table[Integrate[4 Pi r1^2 4 Pi r2^2 (fun[[i]]/.r->r1)
      (fun[[j]]/.r->r2) 1/Max[r1,r2](fun[[i]]/.r->r1)
      (fun[[j]]/.r->r2), {r1,0,Inf}, {r2,0,Inf}], {i,1,2}, {j,1,2}];
K = Table[Integrate[4 Pi r1^2 4 Pi r2^2 (fun[[j]]/.r->r1)
      (fun[[i]]/.r->r2) 1/Max[r1,r2](fun[[i]]/.r->r1)
      (fun[[j]]/.r->r2), {r1,0,Inf}, {r2,0,Inf}], {i,1,2}, {j,1,2}];
Epsi=2 h[[1]]+2 h[[2]]+J[[1,1]]+4 J[[1,2]]+J[[2,2]]-2 K[[1,2]]
```

This script returns $E_\Psi = -13.716\,E_{\rm h}$, while the experimental value is $-14.66737\,E_{\rm h}$.

12.7.4 Derivation of the closed-shell Hartree–Fock equations

In this section, we discuss the derivation of the Hartree-Fock equations for a closed-shell system consisting of N electrons. The aim is to minimize the expectation value of the ground-state energy E_Ψ (eqn 12.56) by varying the spatial orbitals ψ_j.

We begin by expressing the spatial orbitals as a linear combination of a finite number of vectors $\{|e_p\rangle\}_{p=1}^K$,

$$|\psi_j\rangle = \sum_{p=1}^{K} c_{pj}|e_p\rangle,$$

where $K > N/2$ and c_{pj} is a $K \times N/2$ matrix. Although any basis can be chosen, the basis vectors $e_p(\mathbf{r}) = \langle\mathbf{r}|e_p\rangle$ are typically atomic wavefunctions centered around the nuclei[13]. The next step is to calculate the expectation value E_Ψ explicitly. The

[13]Note that the vectors $|e_p\rangle$ do not need to be orthonormal, meaning that the $K \times K$ matrix $S_{pr} \equiv \langle e_p|e_r\rangle$ does not necessarily need to be the identity matrix.

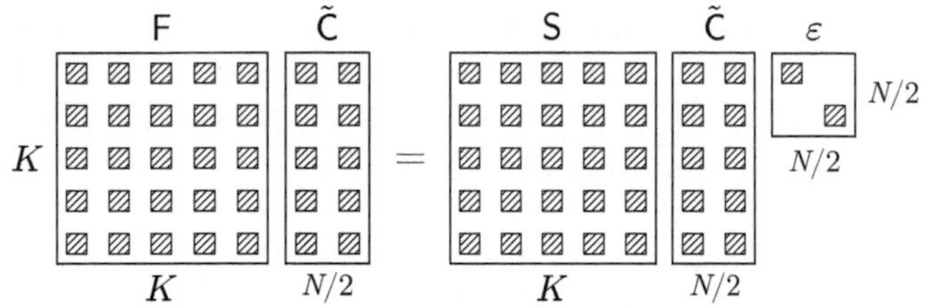

Fig. 12.17 Roothaan equations. Matrix form of the Roothaan equations. F and S are $K \times K$ matrices, $\tilde{\mathsf{C}}$ is a $K \times N/2$ and ε is a diagonal $N/2 \times N/2$ matrix.

quantities in eqn (12.56) are given by:

$$h_j = \langle \psi_j | \hat{h} | \psi_j \rangle = \sum_{p,q=1}^{K} c_{pj}^* c_{qj} \langle e_p | \hat{h} | e_q \rangle, \tag{12.58}$$

$$J_{ij} = \langle \psi_i \psi_j | \frac{1}{r_{12}} | \psi_i \psi_j \rangle = \sum_{p,q,r,s=1}^{K} c_{pi}^* c_{ri} c_{qj}^* c_{sj} \langle e_p e_q | \frac{1}{r_{12}} | e_r e_s \rangle, \tag{12.59}$$

$$K_{ij} = \langle \psi_i \psi_j | \frac{1}{r_{12}} | \psi_j \psi_i \rangle = \sum_{p,q,r,s=1}^{K} c_{pi}^* c_{si} c_{qj}^* c_{rj} \langle e_p e_q | \frac{1}{r_{12}} | e_r e_s \rangle. \tag{12.60}$$

Therefore,

$$\begin{aligned} E_\Psi = {} & 2 \sum_{p,q=1}^{K} \sum_{j=1}^{N/2} c_{pj}^* c_{qj} \langle e_p | \hat{h} | e_q \rangle \\ & + \sum_{p,q,r,s=1}^{K} \sum_{i,j=1}^{N/2} [2 c_{pi}^* c_{ri} c_{qj}^* c_{sj} - c_{pi}^* c_{si} c_{qj}^* c_{rj}] \langle e_p e_q | \frac{1}{r_{12}} | e_r e_s \rangle. \end{aligned} \tag{12.61}$$

To simplify the notation, we introduce a $K \times K$ matrix $\rho_{pq} = \sum_{j=1}^{N/2} c_{pj}^* c_{qj}$, so that the expectation value of the ground-state energy can be written as

$$E_\Psi = 2 \sum_{p,q=1}^{K} \rho_{pq} \langle e_p | \hat{h} | e_q \rangle + \sum_{p,q,r,s=1}^{K} [2 \rho_{pr} \rho_{qs} - \rho_{ps} \rho_{qr}] \langle e_p e_q | \frac{1}{r_{12}} | e_r e_s \rangle.$$

Let us minimize E_Ψ under the constraint that the spatial orbitals remain orthonormal, $\langle \psi_i | \psi_j \rangle = \delta_{ij}$. This can be done with the method of Lagrange multipliers. The Lagrangian is written as the sum of the energy and a set of orthonormality constraints,

$$\boxed{L = E_\Psi - 2 \sum_{i=1}^{N/2} \sum_{j=1}^{N/2} \lambda_{ji} \langle \psi_i | \psi_j \rangle.}$$

The factor of 2 is not relevant and is introduced for later convenience. The $N/2 \times N/2$ matrix λ_{ji} must be Hermitian because the Lagrangian is a real quantity. Hence, it can be diagonalized, $\lambda_{ji} = \sum_{k=1}^{N/2} U_{jk}\varepsilon_k U_{ik}^*$ where ε is a $N/2 \times N/2$ diagonal matrix. This allows us to remove one sum from the second term,

$$\sum_{i,j=1}^{N/2} \lambda_{ji}\langle\psi_i|\psi_j\rangle = \sum_{k=1}^{N/2}\sum_{i,j=1}^{N/2} U_{jk}\varepsilon_k U_{ik}^*\langle\psi_i|\psi_j\rangle = \sum_{k=1}^{N/2}\sum_{p,q=1}^{K}\underbrace{\sum_{j=1}^{N/2} c_{qj}U_{jk}}_{\tilde{c}_{qk}}\varepsilon_k\underbrace{\sum_{i=1}^{N/2} c_{pi}^* U_{ik}^*}_{\tilde{c}_{pk}^*}\langle e_p|e_q\rangle$$

$$= \sum_{k=1}^{N/2}\varepsilon_k \underbrace{\sum_{p=1}^{K}\langle e_p|\tilde{c}_{pk}^*}_{\langle\tilde{\psi}_k|}\underbrace{\sum_{q=1}^{K}\tilde{c}_{qk}|e_q\rangle}_{|\tilde{\psi}_k\rangle} = \sum_{k=1}^{N/2}\varepsilon_k\langle\tilde{\psi}_k|\tilde{\psi}_k\rangle.$$

The Lagrangian simplifies to[14]

$$L = E_\Psi - 2\sum_{k=1}^{N/2}\varepsilon_k\langle\tilde{\psi}_k|\tilde{\psi}_k\rangle. \tag{12.62}$$

By imposing the condition $\partial L/\partial\tilde{c}_{tj} = 0$, one obtains the restricted closed-shell Hartree–Fock equations also known as the **Roothaan equations** (see Exercise 12.13),

$$\boxed{\sum_{r=1}^{K} F_{pr}\tilde{c}_{rj} = \sum_{r=1}^{K} S_{pr}\tilde{c}_{rj}\varepsilon_j,} \tag{12.63}$$

where the $K \times K$ **Fock matrix** F is defined as

$$F_{pr} = \Big[\langle e_p|\hat{h}|e_r\rangle + 2\sum_{q,s}\rho_{qs}\langle e_pe_q|\frac{1}{r_{12}}|e_re_s\rangle - \sum_{q,s}\rho_{qs}\langle e_pe_q|\frac{1}{r_{12}}|e_se_r\rangle\Big].$$

These equations are usually expressed in a more compact form as

$$\boxed{\mathsf{F}\tilde{\mathsf{C}} = \mathsf{S}\tilde{\mathsf{C}}\varepsilon.} \tag{12.64}$$

Figure 12.17 shows the dimensionality of the matrices involved in the Roothaan equations.

The solution of the Roothaan equations corresponds to the coefficients $\tilde{c}_{rj}$ that minimize the energy. Since both the matrix F and the matrix ε depend on the coefficients c_{rj} themselves, the equations (12.64) must be solved iteratively until the coefficients c_{pq} have converged to a stable value. We will solve the Roothaan equations for the hydrogen molecule in Section 12.7.6.

[14]Note that $\rho_{pq}(\mathbf{c}) = \rho_{pq}(\tilde{\mathbf{c}})$ and therefore the expectation value of the energy does not change under the transformation $\mathbf{c} \to \tilde{\mathbf{c}}$.

Atom	ζ_{1s}	ζ_{2s}	ζ_{2p}
H	1.24		
He	1.69		
Li	2.69	0.80	0.80
Be	3.68	1.15	1.15
B	4.68	1.50	1.50
C	5.67	1.72	1.72
N	6.67	1.95	1.95
O	7.66	2.25	2.25

Table 12.4 Exponents of the Slater-type orbitals in molecular calculations.

12.7.5 STO-nG basis

Before applying the Hartree–Fock method to the hydrogen molecule, it is necessary to highlight some important points about the basis functions $\{e_p\}_p$. A minimal basis for an atom is a basis that contains all orbitals up to the valence shell (the valence shell is defined as the set of orbitals involved in forming chemical bonds and depends on the element's location in the periodic table). Examples of minimal basis sets are

$$\begin{aligned} \text{H, He:} &\quad 1s \\ \text{Li to Ne:} &\quad 1s, 2s, 2p_x, 2p_y, 2p_z \\ \text{Na to Ar:} &\quad 1s, 2s, 2p_x, 2p_y, 2p_z, 3s, 3p_x, 3p_y, 3p_z. \end{aligned}$$

Each orbital in a basis set can be approximated with a Slater-type orbital

$$S_{nlm}(\mathbf{r}, \zeta) = A\, r^{n-1} e^{-\zeta r}\, Y_{lm}(\theta, \phi). \tag{12.65}$$

Here, $A = (2\zeta)^{n+1/2}/\sqrt{(2n)!}$ is a normalization factor, and the exponent ζ is ideally a variational parameter used to best approximate each orbital. However, optimizing the exponents can be computationally inefficient. Empirically, it has been observed that a fixed set of exponents is sufficient to approximate some important quantities in quantum chemistry, such as the atomization energy (the energy required to break the molecular bonds to obtain its constituent atoms in the gas phase). Choosing a fixed set of exponents significantly reduces the computational cost [406]. Table 12.4 shows typical exponents for Slater-type orbitals used in molecular calculations.

Finding the solution to the Hartree–Fock equations involves the explicit calculation of the integrals J_{ij} and K_{ij} in eqns 12.59–12.60. These integrals contain terms of the form

$$\langle e_p e_q | \frac{1}{r_{12}} | e_r e_s \rangle = \int e_p^*(\mathbf{r}_1) e_q^*(\mathbf{r}_2) \frac{1}{r_{12}} e_r(\mathbf{r}_1) e_s(\mathbf{r}_2) d\tau_1 d\tau_2, \tag{12.66}$$

where $e_p(\mathbf{r})$ are the elements of a finite basis. When the functions $e_p(\mathbf{r})$ are Slater-type orbitals centered around the nuclei, the integral (12.66) is very complicated and inefficient to compute. In 1950, Boys observed that this integral can be calculated *analytically* and efficiently when the basis functions are Gaussian functions [407]. Hence,

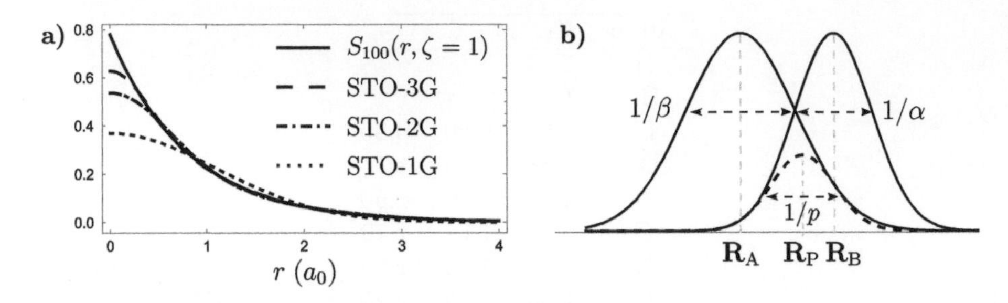

Fig. 12.18 STO-nG. a) Approximation of the Slater-type orbital $S_{100}(\mathbf{r}, 1) = e^{-r}/\sqrt{\pi}$ with a linear combination of Gaussian functions. **b)** The product between two Gaussian functions is still a Gaussian function centered between them with width $1/p = 1/(\alpha + \beta)$. This property can be leveraged to simplify many calculations.

in quantum chemistry calculations it is often convenient to approximate Slater-type orbitals with a finite number of Gaussian functions (this is analogous to the method employed in Example 12.5, where we approximated the ground state of the hydrogen atom with a Gaussian function).

Let us investigate Gaussian functions in more detail. A **Gaussian function** is defined as

$$f_\alpha(\mathbf{r}) = \left(\frac{2\alpha}{\pi}\right)^{3/4} e^{-\alpha \mathbf{r}^2}.$$

The prefactor $(2\alpha/\pi)^{3/4}$ guarantees the correct normalization, $\int |f_\alpha(\mathbf{r})|^2 d\tau = 1$. Let us approximate the radial part of a Slater-type orbital with a linear combination of Gaussian functions. For example, the Slater-type orbital $S_{100}(\mathbf{r}, \zeta = 1) = e^{-|\mathbf{r}|}/\sqrt{\pi}$ can be approximated with 1, 2, or 3 Gaussian functions (see Fig. 12.18a),

$$\begin{aligned}
&S_{100}(\mathbf{r}, 1) \approx f_{0.27095}(\mathbf{r}) && \text{(STO-1G)},\\
&S_{100}(\mathbf{r}, 1) \approx 0.6789 f_{0.1516}(\mathbf{r}) + 0.4301 f_{0.8518}(\mathbf{r}) && \text{(STO-2G)},\\
&S_{100}(\mathbf{r}, 1) \approx 0.4446 f_{0.1098}(\mathbf{r}) + 0.5353 f_{0.4057}(\mathbf{r}) + 0.1543 f_{2.2276}(\mathbf{r}) && \text{(STO-3G)}.
\end{aligned}$$

The approximation improves as the number of Gaussian functions increases. These approximations are called **STO-nG**, because a single Slater-type orbital is approximated with N Gaussian functions. It is worth mentioning that if the radial part of a Slater-type orbital $S_\alpha(\mathbf{r}, 1)$ is approximated by $\sum_j c_j f_{\alpha_j}(\mathbf{r})$, then the radial part of $S_\alpha(\mathbf{r}, \zeta)$ is approximated by

$$\sum_j c_j f_{\zeta^2 \alpha_j}(\mathbf{r}). \tag{12.67}$$

In the remaining part of this section, we show how to calculate the integral $\langle e_p e_q | \frac{1}{r_{12}} | e_r e_s \rangle$ **analytically** when $e_p(\mathbf{r})$ are Gaussian functions in the STO-1G approximation. Following Ref. [345], we define four *non-normalized* Gaussian functions centered around the nuclei $\mathbf{R}_\mathrm{A}$, $\mathbf{R}_\mathrm{B}$, $\mathbf{R}_\mathrm{C}$, and $\mathbf{R}_\mathrm{D}$,

$$g_\alpha(\mathbf{r}-\mathbf{R}_\mathrm{A}) = e^{-\alpha|\mathbf{r}-\mathbf{R}_\mathrm{A}|^2}, \qquad g_\beta(\mathbf{r}-\mathbf{R}_\mathrm{B}) = e^{-\beta|\mathbf{r}-\mathbf{R}_\mathrm{B}|^2},$$
$$g_\gamma(\mathbf{r}-\mathbf{R}_\mathrm{C}) = e^{-\gamma|\mathbf{r}-\mathbf{R}_\mathrm{C}|^2}, \qquad g_\delta(\mathbf{r}-\mathbf{R}_\mathrm{D}) = e^{-\delta|\mathbf{r}-\mathbf{R}_\mathrm{D}|^2},$$

where the positive real numbers α, β, γ, δ determine the width of the Gaussian functions. The integral $\langle g_\alpha g_\gamma|\frac{1}{r_{12}}|g_\beta g_\delta\rangle$ becomes

$$\langle g_\alpha g_\gamma|\frac{1}{r_{12}}|g_\beta g_\delta\rangle = \int g_\alpha(\mathbf{r}_1-\mathbf{R}_\mathrm{A})g_\gamma(\mathbf{r}_2-\mathbf{R}_\mathrm{C})\frac{1}{r_{12}}g_\beta(\mathbf{r}_1-\mathbf{R}_\mathrm{B})g_\delta(\mathbf{r}_2-\mathbf{R}_\mathrm{D})d\mathbf{r}_1 d\mathbf{r}_2.$$

Let us calculate the products $g_\alpha g_\beta$ and $g_\gamma g_\delta$. The product of two Gaussian functions is still a Gaussian function (see Fig. 12.18b),

$$g_\alpha(\mathbf{r}_1-\mathbf{R}_\mathrm{A})g_\beta(\mathbf{r}_1-\mathbf{R}_\mathrm{B}) = K_{\alpha\beta}\, g_p(\mathbf{r}_1-\mathbf{R}_\mathrm{P}),$$
$$g_\gamma(\mathbf{r}_2-\mathbf{R}_\mathrm{C})g_\delta(\mathbf{r}_2-\mathbf{R}_\mathrm{D}) = K_{\gamma\delta}\, g_q(\mathbf{r}_2-\mathbf{R}_\mathrm{Q}),$$

where

$$K_{\alpha\beta} = e^{-\frac{\alpha\beta}{\alpha+\beta}|\mathbf{R}_\mathrm{A}-\mathbf{R}_\mathrm{B}|^2}, \qquad p=\alpha+\beta, \qquad \mathbf{R}_\mathrm{P} = \frac{\alpha\mathbf{R}_\mathrm{A}+\beta\mathbf{R}_\mathrm{B}}{\alpha+\beta},$$
$$K_{\gamma\delta} = e^{-\frac{\gamma\delta}{\gamma+\delta}|\mathbf{R}_\mathrm{C}-\mathbf{R}_\mathrm{D}|^2}, \qquad q=\gamma+\delta, \qquad \mathbf{R}_\mathrm{Q} = \frac{\gamma\mathbf{R}_\mathrm{C}+\delta\mathbf{R}_\mathrm{D}}{\gamma+\delta}.$$

Using these expressions, we have

$$\langle g_\alpha g_\gamma|\frac{1}{r_{12}}|g_\beta g_\delta\rangle = K_{\alpha\beta}K_{\gamma\delta}\int g_p(\mathbf{r}_1-\mathbf{R}_\mathrm{P})\frac{1}{r_{12}}g_q(\mathbf{r}_2-\mathbf{R}_\mathrm{Q})d\mathbf{r}_1 d\mathbf{r}_2. \tag{12.68}$$

This integral can be calculated analytically (see Exercise 12.14),

$$\boxed{\langle g_\alpha g_\gamma|\frac{1}{r_{12}}|g_\beta g_\delta\rangle = \frac{K_{\alpha\beta}K_{\gamma\delta}}{(2\pi)^3}\left(\frac{\pi^2}{pq}\right)^{3/2}\frac{8\pi^3}{|\mathbf{R}_\mathrm{P}-\mathbf{R}_\mathrm{Q}|}\mathrm{Erf}\Big[|\mathbf{R}_\mathrm{P}-\mathbf{R}_\mathrm{Q}|\sqrt{\frac{pq}{p+q}}\Big].} \tag{12.69}$$

The bottom line is that integrals of the form $\langle e_p e_q\,|\frac{1}{r_{12}}|\,e_r e_s\rangle$ can be calculated analytically with a clever choice of basis sets. We will use this result in the next section when we discuss the restricted Hartree–Fock method of the hydrogen molecule.

12.7.6 Hartree–Fock for the hydrogen molecule

The hydrogen molecule is the simplest molecule in nature. It is composed of two protons and two electrons. The ground state of the hydrogen molecule can be approximated with excellent accuracy by writing a few lines of code with advanced quantum chemistry software, such as PySCF. In this section, we present the Hartree–Fock calculations for the hydrogen molecule to offer the reader a pedagogical explanation and show a typical numerical analysis that modern computational packages operate in the background.

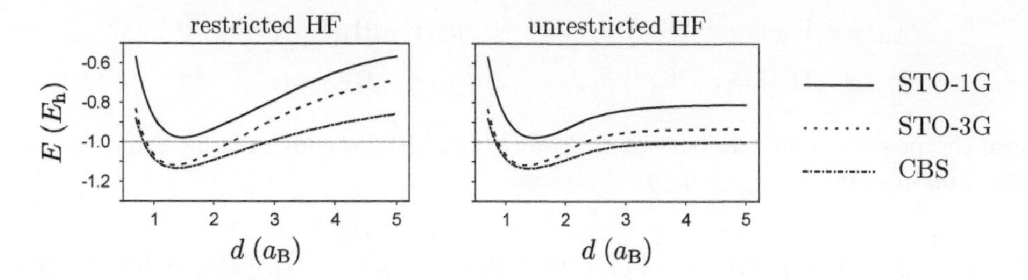

Fig. 12.19 Hartree–Fock method. Restricted and unrestricted Hartree–Fock methods for the hydrogen molecule with the STO-1G approximation (solid curves), with the STO-3G approximation (dashed curves), and with a complete basis set (dot-dashed curves). These simulations were executed with PySCF.

Let us calculate the ground state of the hydrogen molecule with the restricted Hartree–Fock method and the STO-1G approximation. First, we fix the position of the nuclei so that their distance is d:

$$\mathbf{R}_{\mathrm{A}} = (0,0,0), \qquad \mathbf{R}_{\mathrm{B}} = (d,0,0).$$

We write the electron wavefunction as a single Slater determinant,

$$\Psi(x_1,x_2) = \frac{1}{\sqrt{2}} \begin{bmatrix} \psi(\mathbf{r}_1)\alpha & \psi(\mathbf{r}_2)\alpha \\ \psi(\mathbf{r}_1)\beta & \psi(\mathbf{r}_2)\beta \end{bmatrix} = \psi(\mathbf{r}_1)\psi(\mathbf{r}_2)\frac{\alpha\beta - \beta\alpha}{\sqrt{2}}.$$

Let us decompose the spatial orbital in a minimal basis $B = \{e_1(\mathbf{r}), e_2(\mathbf{r})\}$ with $K = 2$ elements,

$$\psi(\mathbf{r}) = c_1 e_1(\mathbf{r}) + c_2 e_2(\mathbf{r})$$

where[15]

$$e_1(\mathbf{r}) = \frac{1}{\sqrt{\pi}} e^{-\alpha|\mathbf{r}-\mathbf{R}_{\mathrm{A}}|^2}, \qquad e_2(\mathbf{r}) = \frac{1}{\sqrt{\pi}} e^{-\alpha|\mathbf{r}-\mathbf{R}_{\mathrm{B}}|^2}, \qquad \alpha = 0.416613.$$

The functions $e_1(\mathbf{r})$ and $e_2(\mathbf{r})$ are two non-normalized Gaussian functions centered around nuclei A and B, respectively. Once the basis has been fixed, it is necessary to calculate the integrals $h_{pq} = \langle e_p|\hat{h}|e_q\rangle$, $S_{pq} = \langle e_p|e_q\rangle$ and $v_{pqrs} = \langle e_p e_q|\frac{1}{r_{12}}|e_r e_s\rangle$. The calculations of the matrix elements v_{pqrs} are particularly simple because the basis functions are Gaussian. This allows us to use the analytical result (12.69).

Since the molecule contains $N = 2$ electrons and the basis has $K = 2$ elements, the Hartree–Fock equations (12.64) become,

$$\begin{bmatrix} F_{11} & F_{12} \\ F_{21} & F_{22} \end{bmatrix} \begin{bmatrix} c_1 \\ c_2 \end{bmatrix} = \begin{bmatrix} S_{11} & S_{12} \\ S_{21} & S_{22} \end{bmatrix} \begin{bmatrix} c_1 \\ c_2 \end{bmatrix} \varepsilon. \tag{12.70}$$

[15] In principle, the parameter α should be varied as well, but this would further complicate the calculations. We choose $\alpha = 0.416613$ because a Slater-type orbital with $\zeta = 1.24$ is well approximated by a Gaussian with $\alpha = 1.24 \cdot 0.27095$ (see Table 12.4 and eqn (12.67)).

As usual, the matrix elements F_{ij} depend on the coefficients c_1 and c_2. Hence, eqn (12.70) must be solved iteratively. We set the initial values of the coefficients c_1 and c_2 to $1/\sqrt{2}$ and calculate the matrix elements $F_{ij}(c_1, c_2)$. We then calculate the minimum value of ε satisfying the equation $\det[\mathsf{F}(c_1, c_2) - \varepsilon \mathsf{S}] = 0$ and solve the system of equations

$$
\begin{aligned}
[F_{11}(c_1, c_2) - \varepsilon_0 S_{11}]\, c_1' + [F_{12}(c_1, c_2) - \varepsilon_0 S_{12}]\, c_2' &= 0, \\
\langle \psi | \psi \rangle = c_1'^2 S_{11} + 2c_1' c_2' S_{12} + c_2'^2 S_{22} &= 1,
\end{aligned}
$$

in order to determine the new coefficients c_1' and c_2'. This procedure is repeated until the new coefficients are similar to the ones calculated in the previous step. The final step is to calculate the electronic energy using eqn (12.61),

$$
E_\Psi = 2 \sum_{p,q=1}^{2} h_{pq} c_p c_q + \sum_{p,q=1}^{2} (2c_p c_r c_q c_s - c_p c_s c_q c_r) v_{pqrs},
$$

and the molecular energy

$$
E_{\text{mol}} = E_\Psi + V_{nn} = E_\Psi + \frac{1}{|\mathbf{R}_\text{A} - \mathbf{R}_\text{B}|} = E_\Psi + \frac{1}{d},
$$

where V_{nn} is the Coulomb repulsion between the nuclei. This procedure is executed with a Mathematica script in Exercise 12.15. The molecular energy is calculated for different distances between the nuclei to obtain the potential illustrated with a solid curve in Fig. 12.19a. It is important to mention that from the point-group symmetries of the hydrogen molecule, it can be shown that the Hartree–Fock state is given by the Slater determinant $\Psi(x_1, x_2) = \frac{1}{\sqrt{2}}(\chi_0(x_1)\chi_1(x_2) - \chi_1(x_1)\chi_0(x_2))$ where the spin-orbitals χ_j are

$$
\chi_0(r) = \frac{e_1(r) + e_2(r)}{\sqrt{2 + 2\langle e_1 | e_2 \rangle}} \alpha, \qquad \chi_1(r) = \frac{e_1(r) + e_2(r)}{\sqrt{2 + 2\langle e_1 | e_2 \rangle}} \beta,
$$

and e_1 and e_2 are Gaussian functions centered around the nuclei A and B, respectively.

This numerical method can be improved by approximating the 1s orbital with a larger number of Gaussian functions. For example, the STO-3G approximation uses the basis $\{e_1, e_2\}$, where

$$
\begin{aligned}
e_1(\mathbf{r}) &= c_1 e^{-\alpha_1 |\mathbf{r} - \mathbf{R}_\text{A}|^2} + c_2 e^{-\alpha_2 |\mathbf{r} - \mathbf{R}_\text{A}|^2} + c_3 e^{-\alpha_3 |\mathbf{r} - \mathbf{R}_\text{A}|^2}, \\
e_2(\mathbf{r}) &= c_1 e^{-\alpha_1 |\mathbf{r} - \mathbf{R}_\text{B}|^2} + c_2 e^{-\alpha_2 |\mathbf{r} - \mathbf{R}_\text{B}|^2} + c_3 e^{-\alpha_3 |\mathbf{r} - \mathbf{R}_\text{B}|^2}.
\end{aligned}
$$

In Figure 12.19a, it is evident that the STO-3G approximation is superior to the STO-1G approximation. Moreover, one can note that the RHF method diverges as the internuclear distance tends to infinity, indicating that this technique is inadequate in describing the dissociation of the molecule into two free atoms for large separations between the nuclei. This problem is overcome by the unrestricted HF method, as shown in Figure 12.19b. It is important to mention that calculations performed with minimal basis sets produce highly inaccurate or even unrealistic results. To improve the simulation of the hydrogen molecule, a **complete basis set** (CBS) consisting of an infinite number of orbitals can be taken into account.

This chapter has provided a comprehensive explanation of the HF method, which serves as a fundamental classical approach for calculating the ground-state energy of a molecule. While the HF method is a useful starting point, its results are typically less accurate than those obtained with more sophisticated techniques, such as the coupled-cluster method. Due to the limitations of space and scope, this chapter does not cover advanced quantum chemistry techniques. We recommend interested readers refer to Refs [345, 399] for further study.

12.8 Second quantization

In the remaining part of this chapter, we explain how to simulate the ground state of a molecule with a quantum computer. The first problem is how to represent the electronic wavefunction on a set of qubits. The occupation number representation, also known as **second quantization**, is a powerful mathematical formalism for describing quantum many-body systems (it has been used to investigate various physical phenomena, from superfluidity to superconductivity). This formalism allows us to express Slater determinants as binary strings and represent the electronic wavefunction on a register of qubits in a straightforward way.

Let us introduce the second quantization by considering a single electron. The position and spin of an electron are indicated by the variable $x = \{\mathbf{r}, \omega\}$. The Hilbert space of an electron H_1 is generated by an orthonormal basis with K spin-orbitals,

$$\boxed{B = \{\chi_0(x), \ldots, \chi_{K-1}(x)\}.} \tag{12.71}$$

In second quantization, when an electron occupies a spin-orbital $\chi_p(x)$, the state of the system is described by a binary string

$$\underset{\text{Spin-orbital}}{\chi_p(x)} \quad \leftrightarrow \quad \underset{\text{Second quantization}}{|0\ldots01_p0\ldots0\rangle}.$$

If the quantum state of the electron is a superposition of spin-orbitals, in second quantization we have

$$\sum_{j=0}^{K-1} a_j \chi_j(x) \quad \leftrightarrow \quad a_0|00\ldots01\rangle + \ldots + a_{K-1}|10\ldots00\rangle.$$

Let us now consider the quantum state of a pair of electrons. In quantum chemistry, the electronic wavefunctions must be antisymmetric. If two electrons occupy the orbitals χ_{p_1} and χ_{p_2}, the antisymmetric wavefunction is given by the Slater determinant

$$\Psi_{p_1,p_2}(x_1, x_2) = \frac{1}{\sqrt{2}}\left[\chi_{p_1}(x_1)\chi_{p_2}(x_2) - \chi_{p_1}(x_1)\chi_{p_2}(x_2)\right].$$

In the occupation number representation, this quantum state is expressed as

$$\Psi_{p_1,p_2}(x_1,x_2) \text{ with } p_1 < p_2 \qquad \leftrightarrow \qquad |0\dots01_{p_2}0\dots01_{p_1}0\dots0\rangle,$$

where the length of the binary string is K. In second quantization the wavefunction Ψ_{p_1,p_2} is represented by a string where each integer indicates the number of electrons occupying each spin-orbital. Since a Slater determinant is zero when two electrons occupy the same spin-orbital, the occupational numbers must be either zero or one.

One can generalize this analysis to a system with N indistinguishable particles. The Hilbert space H_N^- (containing all of the antisymmetric wavefunctions with N electrons) is a subspace of $H_1 \otimes \dots \otimes H_1$ (see Fig. 12.20a). Assuming that $N \leq K$ where $K = \dim(H_1)$, the Slater determinant for N electrons becomes

$$\Psi_{p_1\dots p_N}(x_1,\dots,x_N) = \frac{1}{\sqrt{N!}}\det\begin{bmatrix} \chi_{p_1}(x_1) & \chi_{p_1}(x_2) & \cdots & \chi_{p_1}(x_N) \\ \chi_{p_2}(x_1) & \chi_{p_2}(x_2) & \cdots & \chi_{p_2}(x_N) \\ \vdots & \vdots & & \vdots \\ \chi_{p_N}(x_1) & \chi_{p_N}(x_2) & \cdots & \chi_{p_N}(x_N) \end{bmatrix}.$$

Here, $\chi_{p_1}\dots\chi_{p_N}$ are N spin-orbitals taken from the basis B of eqn (12.71). The maximum number of independent Slater determinants is given by the binomial coefficient $\binom{K}{N}$. In second quantization, a Slater determinant is expressed with a binary string

$$\boxed{\Psi_{p_1\dots p_N}(x_1,\dots,x_N) \text{ with } p_1 < \dots < p_N \qquad \leftrightarrow \qquad |\mathbf{n}\rangle = |n_{K-1}\dots n_0\rangle,}$$

where $n_j = 1$ if the spin-orbital χ_j is present in the Slater determinant, otherwise $n_j = 0$. The correspondence between Slater determinants and binary strings is very convenient for mapping an electronic wavefunction onto a quantum computer: one just needs to apply a sequence of $\hat{X}$ gates to a register of qubits initialized in $|0\dots0\rangle$ to prepare a Slater determinant $|\mathbf{n}\rangle = |n_{k-1}\dots n_0\rangle$ in second quantization.

12.8.1 Fock spaces and electronic Hamiltonian in second quantization

The binary strings $|\mathbf{n}\rangle$ belong to the Fock space H_F. The basis of the Fock space is given by

$$\boxed{B_F = \{|\mathbf{n}\rangle\}_{\mathbf{n}\in\mathbb{B}^K} = \{|0\dots00\rangle, |0\dots01\rangle, \dots, |1\dots11\rangle\}.}$$

Therefore, the dimension of the Fock space is 2^K. The vector $|\mathbf{0}\rangle = |0\dots0\rangle$ is the vacuum state, i.e. the quantum state with no particles. In the Fock space, one can define an inner product between two binary strings as

$$\langle\mathbf{m}|\mathbf{n}\rangle = \delta_{\mathbf{n},\mathbf{m}} = \begin{cases} 1 & \mathbf{n} = \mathbf{m}, \\ 0 & \mathbf{n} \neq \mathbf{m}. \end{cases}$$

The inner product between two generic vectors $|\Psi\rangle = \sum_{\mathbf{n}} a_{\mathbf{n}}|\mathbf{n}\rangle$ and $|\Phi\rangle = \sum_{\mathbf{n}} b_{\mathbf{n}}|\mathbf{n}\rangle$ is given by

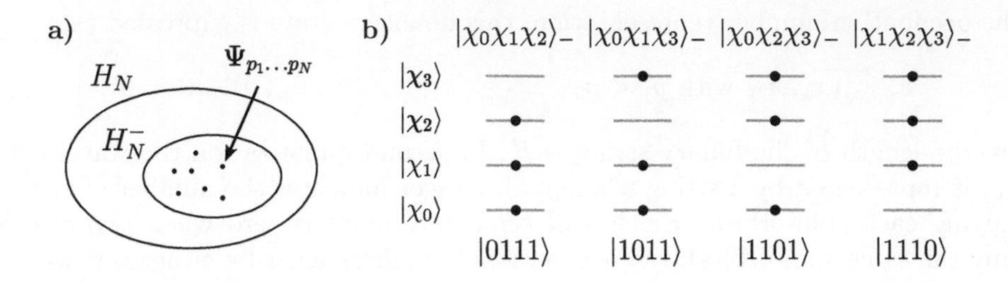

Fig. 12.20 Fock space. a) The Hilbert space H_N^-, containing all antisymmetric wavefunctions of N electrons, is a subspace of $H_N = H_1 \otimes \ldots \otimes H_1$. The vector space H_N^- is generated by $\binom{K}{N}$ Slater determinants $\Psi_{p_1 \ldots p_N}$ (black dots in the diagram). **b)** Three electrons in four spin-orbitals $\{\chi_0, \chi_1, \chi_2, \chi_3\}$ can be in $\binom{4}{3} = 4$ different configurations. Each configuration corresponds to a Slater determinant $|n_3 n_2 n_1 n_0\rangle$.

$$\langle \Psi | \Phi \rangle = \sum_{\mathbf{n} \in \mathbb{B}^K} \sum_{\mathbf{m} \in \mathbb{B}^K} a_{\mathbf{n}}^* b_{\mathbf{m}} \langle \mathbf{n} | \mathbf{m} \rangle.$$

As the reader might expect, the completeness relation is written in the usual way $\mathbb{1} = \sum_{\mathbf{n} \in \mathbb{B}^k} |\mathbf{n}\rangle\langle\mathbf{n}|$. The strings containing only one non-zero integer are indicated as

$$\boxed{|\mathbf{e}_j\rangle = |0 \ldots 010 \ldots 0\rangle}$$

where the one is at the position j. The sum between two binary strings is defined with the usual sum modulo 2,

$$|\mathbf{n} \oplus \mathbf{m}\rangle = |(n_1 + m_1)\%2, \ldots, (n_K + m_K)\%2\rangle.$$

This property will be used in the exercises at the end of this chapter to simplify some calculations.

In second quantization, the operation of adding a particle to a quantum system is described by the **creation operator** $a_p^\dagger$. This operator adds an electron to the spin-orbital χ_p if it is not already occupied,

$$a_p^\dagger|\mathbf{n}\rangle = a_p^\dagger|n_{K-1} \ldots 0_p \ldots n_0\rangle = (-1)^{[\mathbf{n}]_p}|n_{K-1} \ldots 1_p \ldots n_0\rangle \tag{12.72}$$

$$a_p^\dagger|\mathbf{n}\rangle = a_p^\dagger|n_{K-1} \ldots 1_p \ldots n_0\rangle = 0. \tag{12.73}$$

Here, the term

$$\boxed{(-1)^{[\mathbf{n}]_p} \equiv (-1)^{\sum_{i=0}^{p-1} n_i}}$$

is equal to $+1$ if there is an even number of electrons occupying the spin-orbitals $i < p$ and it's equal to -1 if there's an odd number of electrons occupying the orbitals χ_i where $i < p$. Equations (12.72, 12.72) can be written in a more compact form as

$$\boxed{a_p^\dagger|\mathbf{n}\rangle = \delta_{n_p,0}(-1)^{[\mathbf{n}]_p}|n_{K-1} \ldots 1_p \ldots n_0\rangle;} \tag{12.74}$$

an alternative form of this equality is

$$a_p^\dagger|\mathbf{n}\rangle = \delta_{n_p,0}(-1)^{[\mathbf{n}]_p}|\mathbf{n} \oplus \mathbf{e}_p\rangle. \tag{12.75}$$

By applying the creation operator to the vacuum state, one can create an arbitrary binary string,

$$(a_1^\dagger)^{n_{K-1}} \dots (a_K^\dagger)^{n_0}|\mathbf{0}\rangle = \prod_{i=0}^{K-1} (a_i^\dagger)^{n_i}|\mathbf{0}\rangle = |\mathbf{n}\rangle.$$

It is important to mention that the application of two identical creation operators produces the null vector,

$$a_p^\dagger a_p^\dagger|\mathbf{n}\rangle = \delta_{n_p,0}(-1)^{[\mathbf{n}]_p} a_p^\dagger|n_{K-1}\dots 1_p \dots n_0\rangle = 0. \tag{12.76}$$

This property is consistent with the fact that two electrons cannot occupy the same spin-orbital. Since the relation (12.76) is valid for any basis vector, we conclude that $a_p^\dagger a_p^\dagger = 0$.

We now proceed to study the properties of the **destruction operator** a_p. This operator removes one particle from the spin-orbital χ_p and its action on a Slater determinant $|\mathbf{n}\rangle$ is defined as

$$\boxed{a_p|\mathbf{n}\rangle = \delta_{n_p,1}(-1)^{[\mathbf{n}]_p}|n_{K-1}\dots 0_p \dots n_0\rangle} \tag{12.77}$$

or equivalently

$$a_p|\mathbf{n}\rangle = \delta_{n_p,1}(-1)^{[\mathbf{n}]_p}|\mathbf{n} \oplus \mathbf{e}_p\rangle. \tag{12.78}$$

The application of two destruction operators produces the null vector, $a_p a_p = 0$. One can show that the creation and destruction operators satisfy the anti-commutation relations [402],

$$\boxed{\{a_p, a_q^\dagger\} = \delta_{pq}\mathbb{1}} \qquad \boxed{\{a_p^\dagger, a_q^\dagger\} = 0} \qquad \boxed{\{a_p, a_q\} = 0.} \tag{12.79}$$

The operators a_p and $a_p^\dagger$ are called the **Fermionic operators**.

Let us use these notions to express the electronic Hamiltonian in second quantization. Assuming that the basis set is $B = \{\chi_j\}_{j=0}^{K-1}$, it is possible to demonstrate that the projection of the electronic Hamiltonian (12.50) onto the Fock space defined by the basis B is given by (see Appendix D.1)

$$\boxed{\hat{H} = \sum_{p,q=1}^{K} h_{pq} a_p^\dagger a_q + \frac{1}{2} \sum_{p,q,r,s=1}^{K} h_{pqrs} a_p^\dagger a_q^\dagger a_r a_s,} \tag{12.80}$$

where

$$h_{pq} = \langle \chi_p | \hat{h} | \chi_q \rangle = \int dx \chi_p^*(x) \left(\frac{\nabla^2}{2} - \sum_{A=1}^{M} \frac{Z_A}{|\mathbf{R}_A - \mathbf{r}|} \right) \chi_q(x),$$

$$h_{pqrs} = \langle \chi_p \chi_q | \frac{1}{r_{12}} | \chi_s \chi_r \rangle = \iint dx_1 dx_2 \frac{\chi_p^*(x_1)\chi_q^*(x_2)\chi_s(x_1)\chi_r(x_2)}{|\mathbf{r}_1 - \mathbf{r}_2|}.$$

In the next section, we show how to represent a Slater determinant and the electronic Hamiltonian on a quantum computer using the formalism of second quantization.

Example 12.10 Consider a basis of spin-orbitals with $K = 4$ elements,

$$B = \{\chi_0 = \psi_{100}\alpha,\ \chi_1 = \psi_{100}\beta,\ \chi_2 = \psi_{200}\alpha,\ \chi_3 = \psi_{200}\beta\}.$$

If the system contains $N = 3$ electrons, there is a total of $\binom{K}{N} = \binom{4}{3} = 4$ independent Slater determinants which are

$$\begin{aligned} |\chi_0\chi_1\chi_2\rangle_- &\quad\leftrightarrow\quad |0111\rangle, \\ |\chi_0\chi_1\chi_3\rangle_- &\quad\leftrightarrow\quad |1011\rangle, \\ |\chi_0\chi_2\chi_3\rangle_- &\quad\leftrightarrow\quad |1101\rangle, \\ |\chi_1\chi_2\chi_3\rangle_- &\quad\leftrightarrow\quad |1110\rangle. \end{aligned}$$

A generic wavefunction Φ with three electrons can be decomposed as

$$|\Phi\rangle = c_0|0111\rangle + c_1|1011\rangle + c_2|1101\rangle + c_3|1110\rangle,$$

where the complex coefficients c_j must satisfy $|c_0|^2 + |c_1|^2 + |c_2|^2 + |c_3|^2 = 1$. When the destruction operator a_3 and the creation operator $a_1^\dagger$ operate on these states, they produce

$$\begin{aligned} a_3|0111\rangle &= 0, & a_1^\dagger|0111\rangle &= 0, \\ a_3|1011\rangle &= (-1)^2|0011\rangle, & a_1^\dagger|1011\rangle &= 0, \\ a_3|1101\rangle &= (-1)^2|0101\rangle, & a_1^\dagger|1101\rangle &= (-1)^1|1111\rangle, \\ a_3|1110\rangle &= (-1)^2|0110\rangle, & a_1^\dagger|1110\rangle &= 0, \end{aligned}$$

where we used eqns (12.74, 12.77).

12.8.2 Jordan–Wigner transformation

There exist different methods to map the electronic Hamiltonian onto a quantum computer. Some examples are the Jordan–Wigner transformation, the Bravyi-Kitaev transformation, and the parity encoding. In this section, we will present the Jordan–Wigner transformation since it is the most intuitive. We refer the interested reader to Refs [408, 409, 410] for more information about other encodings.

The Jordan–Wigner transformation [411] stores each occupation number in the state of a qubit,

$$\underbrace{|\mathbf{n}\rangle = |n_{K-1}\dots n_0\rangle}_{\text{Slater determinant}} \quad \overset{\text{JW}}{\longleftrightarrow} \quad \underbrace{|\psi\rangle = |n_{K-1}\dots n_0\rangle}_{\text{state of } K \text{ qubits}}.$$

A quantum computer can be prepared in the state $|\psi\rangle$ by applying some $\hat{X}$ gates to a register initialized in $|0\dots0\rangle$. The Jordan–Wigner transformation maps the Fermionic operators a_p and $a_p^\dagger$ into tensor products of Pauli operators [371],

$$a_p^\dagger \quad \overset{\text{JW}}{\longleftrightarrow} \quad \underbrace{\mathbb{1}\otimes\dots\otimes\mathbb{1}}_{K-p}\otimes\sigma_p^+\otimes\underbrace{Z\otimes\dots\otimes Z}_{p-1},$$

$$a_p \quad \overset{\text{JW}}{\longleftrightarrow} \quad \underbrace{\mathbb{1}\otimes\dots\otimes\mathbb{1}}_{K-p}\otimes\sigma_p^-\otimes\underbrace{Z\otimes\dots\otimes Z}_{p-1},$$

where $\sigma^- = |0\rangle\langle 1|$ and $\sigma^+ = |1\rangle\langle 0|$. The Z operators are necessary to ensure that the commutation relations (12.79) still hold. To simplify the notation, we will write

$$a_p^\dagger \quad \overset{\text{JW}}{\longleftrightarrow} \quad \sigma_p^+ Z_{p-1}\dots Z_0,$$

$$a_p \quad \overset{\text{JW}}{\longleftrightarrow} \quad \sigma_p^- Z_{p-1}\dots Z_0,$$

where it is assumed that the identity operator acts on the remaining qubits. These relations allow us to express the second-quantized electronic Hamiltonian (12.80) in terms of Pauli operators. For example, the operator $a_p^\dagger a_q$ is mapped into (see Exercise 12.16):

$$a_p^\dagger a_q \quad \overset{\text{JW}}{\longleftrightarrow} \quad \sigma_q^-\sigma_p^+ Z_{q+1}\dots Z_{p-1} \qquad (q<p), \tag{12.81}$$

$$a_p^\dagger a_q \quad \overset{\text{JW}}{\longleftrightarrow} \quad \sigma_p^+\sigma_q^- Z_{p+1}\dots Z_{q-1} \qquad (q>p), \tag{12.82}$$

$$a_p^\dagger a_q \quad \overset{\text{JW}}{\longleftrightarrow} \quad \sigma_p^+\sigma_p = \frac{\mathbb{1}-Z_p}{2} \qquad (q=p). \tag{12.83}$$

12.9 Simulating the hydrogen molecule on a quantum computer

We conclude this chapter with an instructive example: the quantum simulation of the ground-state energy of the hydrogen molecule on a quantum computer. Even if a classical computer can easily solve this problem, it is still an interesting example that shows the typical steps required to calculate the energy levels of a molecule on quantum hardware.

We will follow these steps:

1. We write the second-quantized electronic Hamiltonian of the hydrogen molecule in terms of Pauli operators using the Jordan–Wigner transformation.
2. We prepare a register of qubits in a quantum state $|\psi\rangle$ that represents the HF state of the hydrogen molecule calculated with classical means.
3. We then use the iterative phase estimation algorithm to approximate the ground-state energy.

Step 1 We fix the nuclei at specific positions: $\mathbf{R}_\mathrm{A} = (0,0,0)$ and $\mathbf{R}_\mathrm{B} = (d,0,0)$ where $d = 1.401\ a_0$ is the equilibrium distance for the hydrogen molecule. We define two normalized Gaussian functions centered around each nucleus,

$$g_1 = \left(\frac{2\zeta}{\pi}\right)^{3/4} e^{-\zeta|\mathbf{r}-\mathbf{R}_A|^2}, \qquad g_2 = \left(\frac{2\zeta}{\pi}\right)^{3/4} e^{-\zeta|\mathbf{r}-\mathbf{R}_B|^2}, \qquad \zeta = 0.4166123,$$

and construct an orthonormal basis of spin-orbitals $B = \{\chi_0, \chi_1, \chi_2, \chi_3\}$ where

$$\chi_0 = \frac{g_1 + g_2}{\sqrt{2 + 2\langle g_1|g_2\rangle}}\alpha, \qquad \chi_1 = \frac{g_1 + g_2}{\sqrt{2 + 2\langle g_1|g_2\rangle}}\beta,$$
$$\chi_2 = \frac{g_1 - g_2}{\sqrt{2 - 2\langle g_1|g_2\rangle}}\alpha, \qquad \chi_3 = \frac{g_1 - g_2}{\sqrt{2 - 2\langle g_1|g_2\rangle}}\beta.$$

Here, $\langle g_1|g_2\rangle$ is the overlap between the Gaussian functions, $\langle g_1|g_2\rangle = \int d\tau_1 d\tau_2 g_1 g_2 = 0.664404$. The Hamiltonian of the hydrogen molecule in second quantization is given by (see eqn (12.80))

$$\boxed{\hat{H} = \sum_{p,q=0}^{3} h_{pq} a_p^\dagger a_q + \frac{1}{2} \sum_{p,q,r,s=0}^{3} h_{pqrs} a_p^\dagger a_q^\dagger a_r a_s.} \tag{12.84}$$

The first sum produces 16 terms, while the second one generates 256 terms. However, many of the matrix elements h_{pq} and h_{pqrs} are zero because the spin functions α and β are orthogonal. Hence, the Hamiltonian reduces to (see Exercise 12.17)

$$\begin{aligned} H =& b_0 \left[a_0^\dagger a_0 + a_1^\dagger a_1\right] + b_1 \left[a_2^\dagger a_2 + a_3^\dagger a_3\right] + b_2\, a_0^\dagger a_0 a_1^\dagger a_1 \\ &+ b_3\, a_2^\dagger a_2 a_3^\dagger a_3 + b_4 \left[a_0^\dagger a_0 a_2^\dagger a_2 + a_0^\dagger a_0 a_3^\dagger a_3 + a_1^\dagger a_1 a_2^\dagger a_2 + a_1^\dagger a_1 a_3^\dagger a_3\right] \\ &+ b_5 \left[-a_0^\dagger a_1^\dagger a_2 a_3 - a_0^\dagger a_0 a_2^\dagger a_2 + a_0^\dagger a_1 a_2 a_3^\dagger + a_0 a_1^\dagger a_2^\dagger a_3 - a_1^\dagger a_1 a_3^\dagger a_3 - a_0 a_1 a_2^\dagger a_3^\dagger\right], \end{aligned} \tag{12.85}$$

where the real numbers b_j are given by $b_0 = -1.18328$, $b_1 = -0.499382$, $b_2 = 0.677391$, $b_3 = 0.578317$, $b_4 = 0.586907$, $b_5 = 0.141412$. Using the Jordan–Wigner transformation, one can express the electronic Hamiltonian in terms of Pauli operators. The operators $a_p^\dagger a_q$ transform as in eqns (12.81–12.83) and the two-body operators transform similarly. For example, $a_3 a_2 a_1^\dagger a_0^\dagger$ is mapped into

$$\begin{aligned} a_3 a_2 a_1^\dagger a_0^\dagger \quad &\overset{\text{JW}}{\longleftrightarrow} \quad (\sigma_3^- Z_2 Z_1 Z_0)\,(\mathbb{1}\sigma_2^- Z_1 Z_0)\,(\mathbb{1}\mathbb{1}\sigma_1^+ Z_0)\,(\mathbb{1}\mathbb{1}\mathbb{1}\sigma_0^+) \\ &\qquad = \sigma_3^- Z_2 \sigma_2^- \sigma_1^+ Z_0 \sigma_0^+ = -\sigma_3^- \sigma_2^- \sigma_1^+ \sigma_0^+. \end{aligned}$$

where we used the relations $Z\sigma^+ = -\sigma^+$, $Z\sigma^- = \sigma^-$ and $ZZ = \mathbb{1}$. With similar substitutions, one can verify that

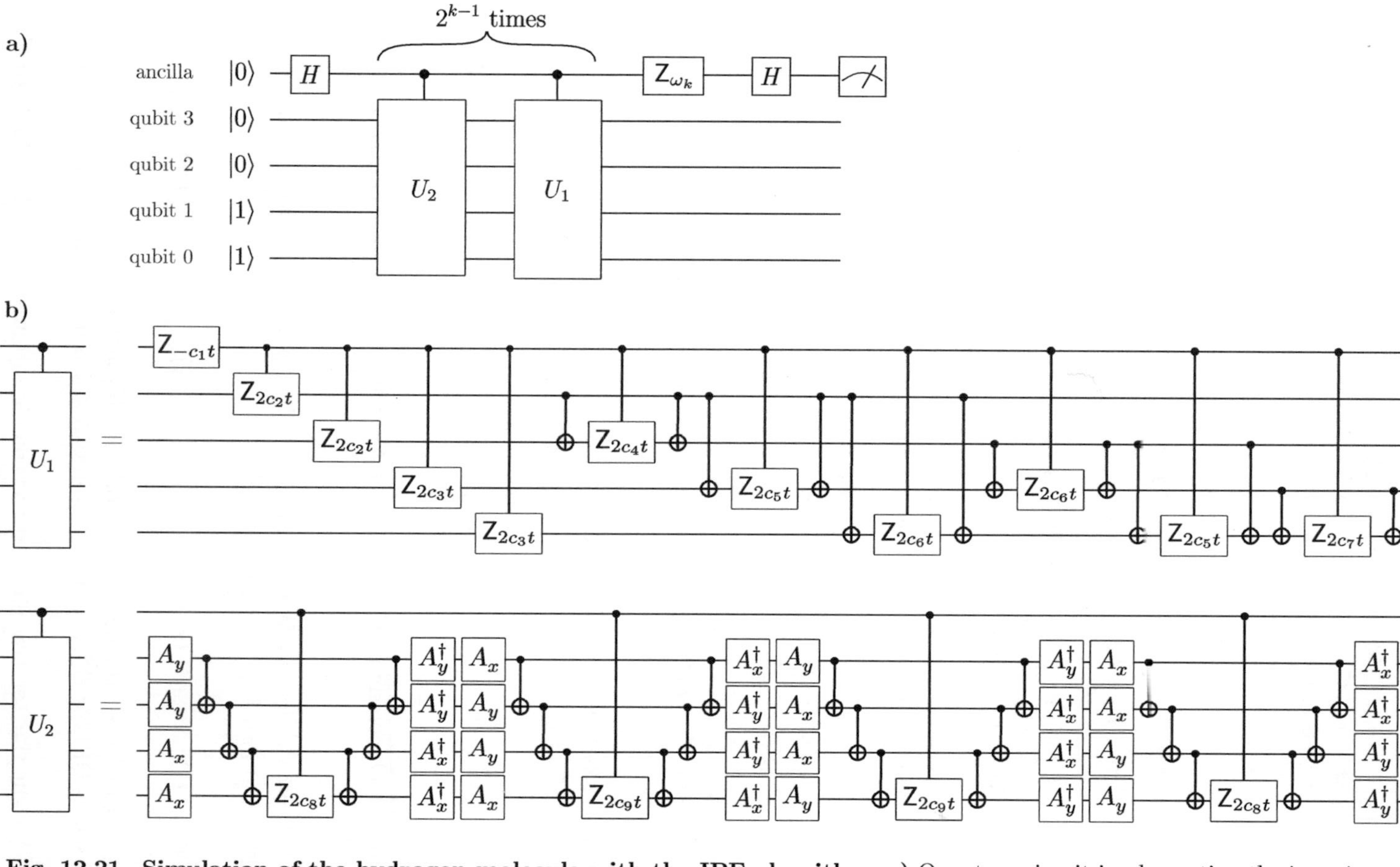

Fig. 12.21 Simulation of the hydrogen molecule with the IPE algorithm. a) Quantum circuit implementing the iterative phase estimation algorithm for the simulation of the hydrogen molecule using a primitive Trotter approximation. **b)** Quantum circuit for the controlled gates cU_1 and cU_2 shown in panel **a**. The coefficients c_j are defined in Table 12.5.

$c_1 = -0.138754$	$c_2 = -0.152989$	$c_3 = 0.164190$
$c_4 = 0.144579$	$c_5 = 0.111373$	$c_6 = 0.146726$
$c_7 = 0.169348$	$c_8 = -0.035353$	$c_9 = 0.035353$

Table 12.5 Coefficients of the Hamiltonian (12.86) in the STO-1G approximation.

$$\begin{aligned}\hat{H} &= b_0 \left[\sigma_0^+\sigma_0 + \sigma_1^+\sigma_1\right] + b_1 \left[\sigma_2^+\sigma_2 + \sigma_3^+\sigma_3\right] + b_2 \left[\sigma_1^+\sigma_1\sigma_0^+\sigma_0\right] \\ &+ b_3\sigma_3^+\sigma_3\sigma_2^+\sigma_2 + b_4\left[\sigma_2^+\sigma_2\sigma_0^+\sigma_0 + \sigma_3^+\sigma_3\sigma_0^+\sigma_0 + \sigma_2^+\sigma_2\sigma_1^+\sigma_1 + \sigma_3^+\sigma_3\sigma_1^+\sigma_1\right] \\ &+ b_5\left[\sigma_3^-\sigma_2^-\sigma_1^+\sigma_0^+ - \sigma_2^+\sigma_2\sigma_0^+\sigma_0 - \sigma_3^+\sigma_2^-\sigma_1^-\sigma_0^+ - \sigma_3^-\sigma_2^+\sigma_1^+\sigma_0^- \right. \\ &\left. - \sigma_3^+\sigma_3\sigma_1^+\sigma_1 + \sigma_3^+\sigma_2^+\sigma_1^-\sigma_0^-\right].\end{aligned}$$

The last step consists of replacing σ^+ and σ^- with $(X - iY)/2$ and $(X + iY)/2$ so that the Hamiltonian only contains Pauli operators,

$$\begin{aligned}\hat{H} &= c_1\mathbb{1} + c_2Z_2 + c_2Z_3 + c_3Z_0 + c_3Z_1 \\ &+c_4Z_3Z_2 + c_5Z_3Z_1 + c_5Z_2Z_0 + c_6Z_2Z_1 + c_6Z_3Z_0 + c_7Z_1Z_0 \\ &+c_8Y_3Y_2X_1X_0 + c_8X_3X_2Y_1Y_0 + c_9X_3Y_2Y_1X_0 + c_9Y_3X_2X_1Y_0. \end{aligned} \tag{12.86}$$

This is the Hamiltonian of the hydrogen molecule in a minimal basis expressed in terms of Pauli operators (the coefficients c_j are reported in Table 12.5). This Hamiltonian can be written as the sum of two terms, $\hat{H} = \hat{O}_1 + \hat{O}_2$, where

$$\begin{aligned}\hat{O}_1 &= c_1\mathbb{1} + c_2Z_2 + c_2Z_3 + c_3Z_0 + c_3Z_1 \\ &+ c_4Z_3Z_2 + c_5Z_3Z_1 + c_5Z_2Z_0 + c_6Z_2Z_1 + c_6Z_3Z_0 + c_7Z_1Z_0 \quad \textbf{(group 1)} \\ \hat{O}_2 &= c_8Y_3Y_2X_1X_0 + c_8X_3X_2Y_1Y_0 + c_9X_3Y_2Y_1X_0 + c_9Y_3X_2X_1Y_0. \quad \textbf{(group 2)}\end{aligned} \tag{12.87}$$

The operators in $\hat{O}_1$ commute with each other. Similarly, the operators in $\hat{O}_2$ commute with each other.

Step 2 We use the Jordan–Wigner transformation to map the Hartree–Fock wavefunction $\Psi(x_1, x_2)$ into a quantum computer,

$$\Psi(x_1, x_2) = \frac{1}{\sqrt{2}}[\chi_0(x_1)\chi_1(x_2) - \chi_0(x_2)\chi_1(x_1)] \qquad \rightarrow \qquad |0011\rangle.$$

Now one can run one of the quantum algorithms presented at the beginning of this chapter (such as QPE or IPE) to find an approximation of the ground-state energy.

Step 3 We run the iterative phase estimation algorithm shown in Fig. 12.21 to obtain an approximation of the ground-state energy. The circuit contains the controlled-operation $\mathsf{c}\hat{U}$ where $\hat{U}$ is the time evolution operator[16],

$$\hat{U} = e^{-i\hat{H}t} = e^{-i(\hat{O}_1+\hat{O}_2)t} \approx e^{-i\hat{O}_1 t}e^{-i\hat{O}_2 t}.$$

Since the operators in $\hat{O}_1$ commute with each other and the same holds for the operators in $\hat{O}_2$, we have

$$\begin{aligned}\hat{U} \approx e^{-i\hat{O}_1 t}e^{-i\hat{O}_2 t} &= \exp\Big(it \sum_{\hat{h}_j \in \text{group 1}} \hat{h}_j \Big) \exp\Big(-it \sum_{\hat{h}_j \in \text{group 2}} \hat{h}_j \Big) \\ &= \prod_{\hat{h}_j \in \text{group 1}} e^{-i\hat{h}_j t} \prod_{\hat{h}_j \in \text{group 2}} e^{-i\hat{h}_j t}.\end{aligned}$$

Each of these exponentials can be implemented exactly with single and two-qubit gates as explained in Example 11.1. For instance, let us consider the exponential $e^{-itc_4 Z_3 Z_2}$. This operator can be expressed as

$$e^{-itc_4 Z_3 Z_2} = \mathsf{c}_2 X_3 e^{-itc_4 Z_3} \mathsf{c}_2 X_3 = \mathsf{c}_2 X_3 \, \mathsf{Z}_{2tc_4} \, \mathsf{c}_2 X_3.$$

Therefore, it can be implemented with a sequence of CNOT and controlled rotations, as shown in Fig. 12.21b. Similarly, the exponential $e^{-itc_{14} Y_3 X_2 X_1 Y_0}$ can be written as

$$e^{-itc_{14} Y_3 X_2 X_1 Y_0} = A_{y3}^\dagger A_{x2}^\dagger A_{x1}^\dagger A_{y0}^\dagger e^{-itc_{14} Z_3 Z_2 Z_1 Z_0} A_{y3} A_{x2} A_{x1} A_{y0},$$

where $A_x = \mathsf{H}$, $A_y = S\mathsf{H}$ (H and S are the Hadamard and phase gates, respectively). The circuit is run multiple times to measure the bits of the phase ϕ that approximates $-E_0 t$ (the fixed parameter t must be chosen such that the product $E_0 t$ is in the interval $[0, 2\pi)$). The algorithm can be repeated for the different internuclear distances to obtain the potential energy curve illustrated in Fig. 12.7. For more information about the iterative phase estimation algorithm for the simulation of the hydrogen molecule, see Ref. [412].

In this chapter, we have explored the simulation of H_2 in a minimal basis set, which serves as an excellent pedagogical example in electronic structure. However, it is crucial to note that this molecule is the simplest system to study, and it can be easily solved on a classical computer. The real challenges lie in simulating more complex electronic systems with many electrons and strong electronic correlation. In such scenarios, the electronic wavefunction cannot be adequately described by a single Slater determinant, as in the case of radicals and transition metal atoms. In quantum chemistry, electronic systems with interesting applications include the chromium dimer [413], the active site of Mo-dependent nitrogenase, the iron-molybdenum cofactor (FeMoCo) [414, 415], and the cytochrome P450 enzymes (CYPs) [416]. Simulating these systems using up to 34, 76, and 58 orbitals, respectively, requires approximately 10^6 qubits and 10^{11} fault-tolerant logical gates. Achieving such computational capabilities remains a long-term goal in the field of quantum computation.

[16] Here, we set $\hbar = 1$ because we are working in atomic units.

Example 12.11 Let us show how to simulate the ground state of the hydrogen molecule using the QAOA method with $n = 1$. The goal is to calculate the ground state of the Hamiltonian (12.86) using the Ansatz $|\psi(\beta,\gamma)\rangle = e^{-i\beta\hat{H}_0}e^{-i\beta\hat{V}}|0011\rangle$ where $\hat{H}_0 = \hat{O}_1$ and $\hat{V} = \hat{O}_2$ are defined in eqn (12.87).

First, we write the QAOA Ansatz as

$$|\psi(\beta,\gamma)\rangle = e^{-i\beta\hat{H}_0}e^{-i\gamma\hat{V}}|0011\rangle, \tag{12.88}$$

where $\hat{H}_0 = \hat{O}_1$ and $\hat{V} = \hat{O}_2$ as in eqn (12.87). Let us calculate the QAOA wavefunction explicitly

$$\begin{aligned}|\psi(\beta,\gamma)\rangle = {} & e^{-i(c_1+2c_2-2c_3+c_4-2(c_5+c_6)+c_7)\beta}\cos(2(c_8-c_9)\gamma)|0011\rangle \\ & +ie^{-i(c_1-2c_2+2c_3+c_4-2(c_5+c_6)+c_7)\beta}\sin(2(c_8-c_9)\gamma)|1100\rangle.\end{aligned}$$

The expectation value of the energy is given by

$$\begin{aligned}E(\beta,\gamma) &= \langle\psi(\beta,\gamma)|\hat{H}|\psi(\beta,\gamma)\rangle \\ &= c_1+c_4-2(c_5+c_6)+c_7+2(c_2-c_3)\cos[4(c_8-c_9)\gamma] \\ &\quad +2(c_8-c_9)\sin[4(c_2-c_3)\beta]\sin[4(c_8-c_9)\gamma].\end{aligned}$$

At the equilibrium geometry $d = 1.401\ a_0$, using the coefficients listed in Table 12.5, the minimum of $E(\beta,\gamma)$ is given by

$$E(\beta_{\text{min}},\gamma_{\text{min}}) = -0.990954\ E_{\text{h}} \qquad \text{with } (\beta_{\text{min}},\gamma_{\text{min}}) = (1.2381, 0.7755). \tag{12.89}$$

Further reading

The research field of quantum simulation of Hamiltonian eigenstates is evolving very rapidly. Some reviews of the field can be found in Refs [344, 417, 358, 418, 359, 419, 410, 420]. For more information about the algorithms, we list some useful resources:

- The quantum phase estimation (QPE) algorithm and the iterative phase estimation (IPE) algorithm were introduced in Refs [381, 421] and described in great detail in [382, 385]. Aspuru-Guzik et al. originally proposed to employ QPE in the context of quantum simulation of molecular eigenstates [383]. Some applications of this approach are described in Refs [422, 412, 423, 424, 425]. More recent works focused on the optimization of QPE [426, 423, 427, 424, 428, 429].
- The adiabatic theorem was enunciated by several authors [387, 388, 389]. Modern presentations of this theorem can be found in Refs [430, 391, 386]. Fahri proposed to use the quantum adiabatic algorithm to tackle combinatorial optimization problems [22, 392]. This work was later generalized to chemistry problems [431, 393, 394]. See Ref. [397, 398] for a detailed analysis of the computational cost of this algorithm.
- QAOA was introduced in Ref. [23]. In recent years, we witnessed significant progress on both the experimental and theoretical fronts [432, 433, 434, 435, 436,

437, 438, 439, 440, 441, 442, 443]. Theoretical studies have explored the potential advantage of QAOA over classical optimization algorithms [444], but also its limitations [445, 446].

- VQE, introduced in [24], is currently a widely used strategy to approximate Hamiltonian eigenstates. The accuracy and computational cost of VQE calculations are determined by the underlying Ansatz, which is typically problem-dependent. In the context of electronic systems, Ansätze are usually generated with the unitary coupled-cluster method [447, 24, 448, 449, 450, 412, 451, 452, 453, 454, 451].
- For an introduction to theoretical and computational quantum chemistry, we recommend the books by Szabo and Ostlund [345], and Helgaker [399]. For a deeper understanding of many-body physics and second quantization, we refer our readers to the textbook by Fetter and Walecka [455] and Martin [456]. Classical methods for the electronic structure problem are extensively reviewed in Refs [457, 378, 458, 459, 460, 379].

Summary

Quantum phase estimation

$u = e^{i2\pi\lambda}$	Eigenvalue of $\hat{U}$
$\lambda = \tilde{\lambda}/2^m + \delta$	$\tilde{\lambda}/2^m$ is the best approximation of λ
$p(\tilde{\lambda}) = \frac{1}{2^{2m}} \sin^2(\pi\delta 2^m)/\sin^2(\pi\delta) > 40\%$	Prob. of measuring $\tilde{\lambda}$
$m > \log[(1/2q + 2)\varepsilon]$	m is the number of ancilla qubits so that $p(\|\ell/2^m - \lambda\| \leq \varepsilon) > 1 - q$

Iterative phase estimation

$u = e^{i2\pi\phi}$	Eigenvalue of $\hat{U}$
$\phi = \sum_{j=1}^{t} \phi_j/2^j$	Binary expansion of ϕ
$\omega_k = 0$	Rotation angle of the Z gate ($k = t$)
$\omega_k = -2\pi \sum_{j=k+1}^{t} 2^{k-1}\phi_j/2^j$	Rotation angle of the Z gate ($k < t$)

Adiabatic state preparation

$\hat{H} = \hat{H}_0 + \hat{V}$	$\hat{H}_0$ ($\hat{H}_0 + \hat{V}$) has known (unknown) eigenvalues
$\hat{H}(t) = \hat{H}_0 + \frac{t}{T}\hat{V}$	Time-dependent Hamiltonian
$\hat{H}(t)\|E_k(t)\rangle = E_k(t)\|E_k(t)\rangle$	Eigenvalues and eigenvectors of $\hat{H}(t)$
$\|\psi(0)\rangle = \|E_0(0)\rangle$	Initial state of ASP
$\|\psi(T)\rangle \approx \|E_0(T)\rangle$	$\|\psi(T)\rangle$ approximates the ground state of $\hat{H}_0 + \hat{V}$

The evolution from $|\psi(0)\rangle$ to $|\psi(T)\rangle$ can be approximated with a Trotterization:
$|\tilde{\psi}(T)\rangle = e^{\frac{\Delta t}{i\hbar}\hat{H}_0} e^{\frac{\Delta t}{i\hbar}s_{n-1}\hat{V}} \dots e^{\frac{\Delta t}{i\hbar}\hat{H}_0} e^{\frac{\Delta t}{i\hbar}s_0\hat{V}} |\psi(0)\rangle$ where $s_k = k/n$
and $k \in \{0, \dots, n-1\}$.

Adiabatic theorem:
$\| |\psi(T)\rangle - |E_0(T)\rangle \| \leq \frac{1}{T}\left(\frac{f_1(T)}{\gamma^2(T)} + \frac{f_1(0)}{\gamma^2(0)} + \int_0^1 \frac{5f_1^2(s)}{\gamma^3(s)} + \frac{f_2(s)}{\gamma^2(s)} ds\right)$
where $s = t/T, f_k(s) = \|d^k H(s)/ds^k\|$.

QAOA

$|\psi(\boldsymbol{\beta},\boldsymbol{\gamma})\rangle = e^{-i\frac{\Delta t}{i\hbar}\hat{H}_0} e^{\frac{\Delta t}{i\hbar}s_{n-1}\hat{V}} \dots e^{\frac{\Delta t}{i\hbar}\hat{H}_0} e^{\frac{\Delta t}{i\hbar}s_0\hat{V}} |\psi(0)\rangle$

$M_n = \min_{\boldsymbol{\beta},\boldsymbol{\gamma}} E_n(\boldsymbol{\beta},\boldsymbol{\gamma})$	Minimum of the energy for a fixed n
$M_{n+1} \leq M_n$	The approximation improves as n increases

Introduction to quantum chemistry

$\hat{H}_{\text{mol}} = \hat{T}_{\text{n}} + \hat{T}_{\text{e}} + \hat{T}_{\text{ne}} + \hat{T}_{\text{ee}} + \hat{V}_{\text{nn}}$ Molecular Hamiltonian

$\hat{H}_{\text{mol}} = -\sum_A \frac{\nabla_A^2}{2M_A} - \sum_j \frac{\nabla_j^2}{2} - \sum_{i,A} \frac{Z_A}{r_{iA}} + \sum_{i<j} \frac{1}{r_{ij}} + \sum_{A<B} \frac{Z_A Z_B}{R_{AB}}$

$\hat{T}_n = 0, \hat{V}_{\text{nn}} = 0 = \text{const}$ Born–Oppenheimer approximation

$\hat{H}_{\mathbf{R}}^{(\text{elec})} = -\sum_i \frac{\nabla_i^2}{2} - \sum_{i,A} \frac{Z_A}{r_{iA}} + \sum_{i<j} \frac{1}{r_{ij}}$ Electronic Hamiltonian

$\hat{H} = -\nabla^2/2 - Z/r$ Hydrogen-like Hamiltonian

$E_n = -0.5Z/n^2 \ (E_{\text{h}})$ Energy levels of the hydrogen atom

$\psi_{nlm}(r,\theta,\phi) = R_{nl}(r) Y_{lm}(\theta,\phi)$ Hydrogen-like orbitals

$\sigma(\omega) = c_0\alpha(\omega) + c_1\beta(\omega)$ Spin function

$\alpha(\omega) = \delta_{\omega 0}, \ \beta(\omega) = \delta_{\omega 1}$ Spin up and spin down

$|\chi\rangle = |\psi\rangle \otimes |\sigma\rangle$ or $\chi(x) = \psi(\mathbf{r})\sigma(\omega)$ Electron spin-orbital

$S_{nlm}(\mathbf{r},\zeta) = Ar^{n-1}e^{-\zeta r} Y_{lm}(\theta,\phi)$ Slater-type orbital

$0.0016 \ E_{\text{h}}$ Chemical accuracy

Variational method

$|\phi\rangle$ Trial wavefunction

$E_\phi = \langle\phi|\hat{H}|\phi\rangle / \langle\phi|\phi\rangle$ Expectation value of the energy

$E_\phi \geq E_0$ E_0 is the ground-state energy

Linear variational method

$|\phi\rangle = \sum_{p=1}^m c_p |e_p\rangle$ Trial wavefunction

$h_{pq} = \langle e_p|\hat{H}|e_q\rangle, \ S_{pq} = \langle e_p|e_q\rangle$ Matrix elements

$E_\phi = \sum_{p,q} c_p^* c_q h_{pq} / \sum_{p,q} c_p^* c_q S_{pq}$ Expectation value of the energy

$\det(\mathsf{h} - E_\phi \mathsf{S}) = 0$ The minimum E_ϕ satisfying this equation is an approximation of the ground-state energy

---

Slater determinants

$|\Psi\rangle = \frac{1}{\sqrt{N!}} \sum_{\sigma \in S_N} (-1)^\sigma |\chi_{\sigma(1)} \cdots \chi_{\sigma(N)}\rangle$ Slater determinant

$|\Psi\rangle = |\chi_1 \cdots \chi_N\rangle_-$ Shorthand notation

$|\chi_1 \cdots \chi_i \chi_{i+1} \cdots \chi_N\rangle_- = (-1)|\chi_1 \cdots \chi_{i+1}\chi_i \cdots \chi_N\rangle_-$ Transposition

$\hat{O}_1 = \sum_{j=1}^N \hat{h}_j, \quad \hat{h}_j = -\nabla_j^2/2 - \sum_{A=1}^M Z/|\mathbf{r}_j - \mathbf{R}_A|$ One-body operator

$\hat{O}_2 = \sum_{j=1}^N \sum_{i=j+1}^N 1/|\mathbf{r}_i - \mathbf{r}_j|$ Two-body operator

$E_\Psi = \langle\Psi|\hat{H}|\Psi\rangle = \langle\Psi|\hat{O}_1|\Psi\rangle + \langle\Psi|\hat{O}_2|\Psi\rangle$ Expectation value of the energy

$\langle\Psi|\hat{O}_1|\Psi\rangle = \sum_{j=1}^N \langle\chi_j|\hat{h}|\chi_j\rangle$ One-body contrib.

$\langle\Psi|\hat{O}_2|\Psi\rangle = \sum_{i,j=1}^N [\langle\chi_i\chi_j|1/r_{12}|\chi_i\chi_j\rangle - \langle\chi_j\chi_{ji}|1/r_{12}|\chi_i\chi_j\rangle]$ Two-body contrib.

Closed-shell Hartree–Fock

$\chi_1 = \psi_1\alpha, \dots, \chi_{N-1} = \psi_{N/2}\alpha, \chi_N = \psi_{N/2}\beta$	Basis of spin-orbitals
$\langle\Psi\|\hat{O}_1\|\Psi\rangle = 2\sum_{j=1}^{N/2}\langle\psi_j\|\hat{h}\|\psi_j\rangle$	One-body contribution
$\langle\Psi\|\hat{O}_2\|\Psi\rangle = \sum_{i,j=1}^{N}[2\langle\psi_i\psi_j\|1/r_{12}\|\psi_i\psi_j\rangle - \langle\psi_i\psi_j\|1/r_{12}\|\psi_j\psi_i\rangle]$	Two-body contrib.
$E_\Psi = 2\sum_{j=1}^{N/2} h_j + \sum_{i,j=1}^{N/2}[2J_{ij} - K_{ij}]$	Expectation value of the energy
$h_j = \langle\psi_j\|\hat{h}\|\psi_j\rangle$	Single electron contrib.
$J_{ij} = \langle\psi_i\psi_j\|1/r_{12}\|\psi_i\psi_j\rangle$	Coulomb integral
$K_{ij} = \langle\psi_i\psi_j\|1/r_{12}\|\psi_j\psi_i\rangle$	Exchange integral
$\mathsf{F}\tilde{\mathsf{C}} = \mathsf{S}\tilde{\mathsf{C}}\varepsilon$	Closed-shell Hartree–Fock equations

Second quantization

$B = \{\chi_0(x), \dots, \chi_{K-1}(x)\}$	Basis of spin-orbitals
$\chi_p(x) \leftrightarrow \|0\dots01_p0\dots0\rangle$	Mapping
$\Psi_{p_1,p_2}(x_1,\dots,x_N)$ with $p_1 < \dots < p_N \leftrightarrow \|n_{K-1}\dots n_0\rangle$	Mapping
$B_F = \{\|\mathbf{n}\rangle\}_{\mathbf{n}\in\mathbb{B}^K}$	Basis of the Fock space
$\mathbb{1} = \sum_{\mathbf{n}=1}^{K}\|\mathbf{n}\rangle\langle\mathbf{n}\|$	Completeness relation
$\|\mathbf{e}_p\rangle = \|0\dots01_p0\dots0\rangle$	State with only one occupation
$\|\mathbf{n}\oplus\mathbf{m}\rangle = \|(n_1+m_1)\%2,\dots,(n_K+m_K)\%2\rangle$	Sum between two binary strings
$a_p^\dagger\|\mathbf{n}\rangle = \delta_{n_p,0}(-1)^{[\mathbf{n}]_p}\|\mathbf{n}\oplus\mathbf{e}_p\rangle$	Creation operator
$a_p\|\mathbf{n}\rangle = \delta_{n_p,1}(-1)^{[\mathbf{n}]_p}\|\mathbf{n}\oplus\mathbf{e}_p\rangle$	Destruction operator
$\{a_p, a_q^\dagger\} = \delta_{pq}\mathbb{1}$, $\{a_p^\dagger, a_q^\dagger\} = 0$, $\{a_p, a_q\} = 0$	Anti-commutation relations
$\hat{H} = \sum_{p,q=1}^{K} h_{pq}a^\dagger a_q + \frac{1}{2}\sum_{p,q,r,s}^{K} h_{pqrs}a_p^\dagger a_q^\dagger a_r a_s$	Electronic Hamiltonian
$h_{pq} = \langle\chi_p\|\hat{h}\|\chi_q\rangle$, $h_{pqrs} = \langle\chi_p\chi_q\|1/r_{12}\|\chi_s\chi_r\rangle$	Tensors

Jordan–Wigner transformation

$\|\mathbf{n}\rangle = \|n_{K-1}\dots n_0\rangle \leftrightarrow \|\psi\rangle = \|x_{K-1}\dots x_0\rangle$	Mapping
$a_p^\dagger \leftrightarrow \sigma_p^+ Z_{p-1}\dots Z_0$	Creation operator
$a_p \leftrightarrow \sigma_p^- Z_{p-1}\dots Z_0$	Destruction operator
$a_p^\dagger a_q \leftrightarrow \sigma_q^-\sigma_p^+ Z_{q+1}\dots Z_{p-1}\quad (q < p)$	Mapping
$a_p^\dagger a_q \leftrightarrow \sigma_p^+\sigma_q^- Z_{p+1}\dots Z_{q-1}\quad (q > p)$	Mapping
$a_p^\dagger a_q \leftrightarrow \sigma_p^+\sigma_p = (\mathbb{1} - Z_p)/2\quad (q = p)$	Mapping

Exercises

Exercise 12.1 At the end of the QPE algorithm, the probability of measuring an integer $\ell \in [0, 2^m - 1]$ is given by $p(\ell) = |\langle\ell|\psi_{\mathrm{E}}\rangle|^2$ (see eqn (12.2)). Show that the probability of measuring an integer in the interval $[\tilde{\lambda} - a, \tilde{\lambda} + a]$ satisfies $p(\ell \in [\tilde{\lambda} - a, \tilde{\lambda} + a]) > 1 - 1/(2a - 2)$ where a is a positive integer.

Exercise 12.2 When we presented the iterative phase estimation algorithm, we assumed that the phase ϕ had a finite binary expansion. Let us consider a more general case. Suppose that ϕ can be expressed as

$$\phi = \sum_{j=1}^{t} \frac{\phi_j}{2^j} + \delta 2^{-t},$$

where $\delta \in [0, 1)$ is an error related to the finite number of iterations of the algorithm. Show that the IPE algorithm returns the correct bits ϕ_j with a probability greater than 45%.

Exercise 12.3 Evaluate the hydrogen-like orbitals $\psi_{nlm}(r, \theta, \phi)$ of eqn (12.21) with Mathematica and check that they are correctly normalized,

$$\int_0^{\pi} d\theta \int_0^{2\pi} d\phi \int_0^{\infty} dr |\psi(r, \theta, \phi)|^2 r^2 \sin\theta = 1.$$

Exercise 12.4 Using Mathematica, show that

$$E = \int d\mathbf{r}_1 \int d\mathbf{r}_2 |\psi_{100}(\mathbf{r}_1)|^2 \frac{1}{r_{12}} |\psi_{100}(\mathbf{r}_2)|^2 = \frac{5Z}{8} E_{\mathrm{h}}, \tag{12.90}$$

where $\psi_{100}(\mathbf{r}_1)$ and $\psi_{100}(\mathbf{r}_2)$ are the 1s orbitals,

$$\psi_{100}(\mathbf{r}_1) = \frac{Z^{3/2}}{\sqrt{\pi}} e^{-Zr_1}, \qquad \psi_{100}(\mathbf{r}_2) = \frac{Z^{3/2}}{\sqrt{\pi}} e^{-Zr_2}.$$

Hint: Use the law of cosines to express $r_{12} = |\mathbf{r}_1 - \mathbf{r}_2|$ as $r_{12} = \sqrt{r_1^2 + r_2^2 - 2r_1 r_2 \cos\alpha}$ where α is the angle between $\mathbf{r}_1$ and $\mathbf{r}_2$.

Exercise 12.5 Write a Mathematica script to calculate the integral $E_\phi = \langle \phi | \hat{H} | \phi \rangle$ of eqn (12.34). **Hint:** Write the Laplacian in spherical coordinates.

Exercise 12.6 Consider the electronic Hamiltonian of the helium atom in atomic units

$$\hat{H} = -\frac{\nabla_1^2}{2} - \frac{\nabla_2^2}{2} - \frac{2}{r_1} - \frac{2}{r_2} + \frac{1}{r_{12}},$$

and the antisymmetric wavefunction

$$\Psi(x_1, x_2) = \psi_{100}(\mathbf{r}_1)\psi_{100}(\mathbf{r}_1) \otimes \sigma(\omega_1, \omega_2),$$

where $\psi_{100}(\mathbf{r}) = Z^{3/2} e^{-Zr}/\sqrt{\pi}$, while σ is an antisymmetric spin wavefunction. Using Mathematica, show that $E_\phi = \langle \Psi | \hat{H} | \Psi \rangle = \left(Z^2 - \frac{27}{8} Z\right) E_{\mathrm{h}}$.

Exercise 12.7 Using a Mathematica script, calculate the terms

$$H_{ij} = \langle e_i | \hat{H} | e_j \rangle = \int e_i^*(r) \hat{H} e_j(r) d\tau, \qquad S_{ij} = \langle e_i | e_j \rangle = \int e_i^*(r) e_j(r) d\tau$$

where $e_1(r) = e^{-\lambda r^2}$, $e_2(r) = e^{-2\lambda r^2}$, $\lambda = 8/9\pi$ and $\hat{H} = -\nabla^2/2 - 1/r$ is the Hamiltonian of the hydrogen atom.

Exercise 12.8 Starting from eqn (12.46), derive eqn (12.47),

$$\sum_{q=1}^{m} \left[h_{iq} + g_{iq}(\mathbf{c}) - \gamma(\mathbf{c})S_{iq}\right] c_q = 0.$$

Hint: Use the fact that $h_{pq} = h_{qp}$, $S_{pq} = S_{qp}$, and $g_{pq}(\mathbf{c}) = g_{qp}(\mathbf{c})$.

Exercise 12.9 Consider a Slater determinant $|\Psi\rangle = \frac{1}{\sqrt{N!}} \sum_{\sigma \in S_N} (-1)^\sigma |\chi_{\sigma(1)} \dots \chi_{\sigma(N)}\rangle$ and the one-body operator $\hat{O}_1 = \sum_j \hat{h}(j)$ defined in eqn (12.50). Show that

$$\langle\Psi|\hat{O}_1|\Psi\rangle = \sum_{j=1}^{N} \langle\chi_j|\hat{h}|\chi_j\rangle.$$

Exercise 12.10 Consider a Slater determinant $|\Psi\rangle = \frac{1}{\sqrt{N!}} \sum (-1)^\sigma |\chi_{\sigma(1)} \dots \chi_{\sigma(N)}\rangle$ and the two-body operator

$$\hat{O}_2 = \sum_{j=1}^{N} \sum_{i=j+1}^{N} \frac{1}{|\mathbf{r}_i - \mathbf{r}_j|} = \sum_{i<j}^{N} \hat{v}(i,j),$$

where $\hat{v}(i,j) = \frac{1}{|\mathbf{r}_i - \mathbf{r}_j|}$. Show that

$$\langle\Psi|\hat{O}_2|\Psi\rangle = \frac{1}{2} \sum_{i=1}^{N} \sum_{j=1}^{N} \left[\langle\chi_i\chi_j|\hat{v}|\chi_i\chi_j\rangle - \langle\chi_j\chi_i|\hat{v}|\chi_i\chi_j\rangle\right].$$

Exercise 12.11 In this exercise, we study the Hartree–Fock method for the helium atom. Let us approximate the ground state of the helium atom with a single Slater determinant

$$\Psi(x_1, x_2) = \psi(\mathbf{r}_1)\psi(\mathbf{r}_2)\sigma$$

where the spatial orbital ψ is a linear combination of two functions $\psi(\mathbf{r}) = c_1 e_1(\mathbf{r}) + c_2 e_2(\mathbf{r})$. Here, e_1 and e_2 are Slater-type orbitals,

$$e_1(\mathbf{r}) = \sqrt{\frac{\zeta_1^3}{\pi}} e^{-\zeta_1 r}, \qquad e_2(\mathbf{r}) = \sqrt{\frac{\zeta_2^3}{\pi}} e^{-\zeta_2 r}, \tag{12.91}$$

and $\zeta_1 = 1.45$ and $\zeta_2 = 2.90$ are fixed parameters. Write a Mathematica script to perform the Hartree–Fock method of Fig. 12.15.

Exercise 12.12 Starting from $\langle\Psi|\hat{O}_2|\Psi\rangle = \frac{1}{2}\sum_{i,j} \left[\langle\chi_i\chi_j|\frac{1}{r_{12}}|\chi_i\chi_j\rangle - \langle\chi_j\chi_i|\frac{1}{r_{12}}|\chi_i\chi_j\rangle\right]$, show that

$$\langle\Psi|\hat{O}_2|\Psi\rangle = \sum_{i=1}^{N/2} \sum_{j=1}^{N/2} \left[2\langle\psi_i\psi_j|\frac{1}{r_{12}}|\psi_i\psi_j\rangle - \langle\psi_i\psi_j|\frac{1}{r_{12}}|\psi_j\psi_i\rangle\right], \tag{12.92}$$

assuming that the odd (even) spin-orbitals are $\psi_i\alpha$ ($\psi_i\beta$).

Exercise 12.13 Starting from eqn (12.62), derive the Hartree–Fock equations (12.63).

Exercise 12.14 Starting from

$$\langle g_\alpha g_\gamma | \frac{1}{r_{12}} | g_\beta g_\delta \rangle = K_{\alpha\beta} K_{\gamma\delta} \int g_p(\mathbf{r}_1 - \mathbf{R}_\mathrm{P}) \frac{1}{r_{12}} g_q(\mathbf{r}_2 - \mathbf{R}_\mathrm{Q}) d\mathbf{r}_1 d\mathbf{r}_2,$$

show that

$$\langle g_\alpha g_\gamma | \frac{1}{r_{12}} | g_\beta g_\delta \rangle = \frac{K_{\alpha\beta} K_{\gamma\delta}}{(2\pi)^3} \left(\frac{\pi^2}{pq} \right)^{3/2} \frac{8\pi^3}{|\mathbf{R}_\mathrm{P} - \mathbf{R}_\mathrm{Q}|} \mathrm{Erf}\left[|\mathbf{R}_\mathrm{P} - \mathbf{R}_\mathrm{Q}| \sqrt{\frac{pq}{p+q}} \right].$$

Hint: Calculate these integrals in the Fourier space.

Exercise 12.15 Calculate the ground-state energy of the hydrogen molecule with the restricted Hartree–Fock method using Mathematica. For this exercise, use a minimal basis $B = \{e_1(\mathbf{r}), e_2(\mathbf{r})\}$ where e_1 and e_2 are Gaussian functions centered around the nuclei A and B, respectively. **Hint:** Use the results from Exercise 12.14 to calculate the matrix elements $v_{pqrs} = \langle e_p e_q | \frac{1}{r_{12}} | e_r e_s \rangle$.

Exercise 12.16 Using the Jordan–Wigner transformation,

$$a_p^\dagger \overset{\mathrm{JW}}{\longleftrightarrow} \sigma_p^+ Z_{p-1} \dots Z_0, \qquad a_p \overset{\mathrm{JW}}{\longleftrightarrow} \sigma_p^- Z_{p-1} \dots Z_0,$$

show that

$$\begin{aligned} a_p^\dagger a_q &\overset{\mathrm{JW}}{\longleftrightarrow} \sigma_q^- \sigma_p^+ Z_{q+1} \dots Z_{p-1} && (q < p), \\ a_p^\dagger a_q &\overset{\mathrm{JW}}{\longleftrightarrow} \sigma_p^+ \sigma_q^- Z_{p+1} \dots Z_{q-1} && (q > p), \\ a_p^\dagger a_q &\overset{\mathrm{JW}}{\longleftrightarrow} \sigma_p^+ \sigma_p = \frac{\mathbb{1} - Z_p}{2} && (q = p). \end{aligned}$$

Exercise 12.17 Starting from eqn (12.84), derive eqn (12.85).

Exercise 12.18 Show that the ground state of the Hamiltonian (12.86) (describing the energy of a hydrogen molecule) has the form

$$|\psi\rangle = a|0011\rangle + b|1100\rangle, \tag{12.93}$$

where a, b are real numbers. **Hint:** Show that the subspace generated by $\{|0011\rangle, |1100\rangle\}$ is closed under the action of $\hat{H}$.

Exercise 12.19 Calculate the ground state of the Hamiltonian (12.86) with the VQE method using the Ansatz

$$|\psi(\theta)\rangle = e^{-i\theta X_3 X_2 X_1 Y_0} |0011\rangle.$$

Solutions

Solution 12.1 We first define the integer $\tilde{\lambda}$ as $\tilde{\lambda} = \lfloor 2^m \lambda \rfloor$. In other words, $\tilde{\lambda}/2^m$ is the best approximation of λ that is smaller than λ. The parameter δ, defined as $\delta = \lambda - \tilde{\lambda}/2^m$, is a non-negative real number in the interval $[0, 2^{-m}]$. Next, we need to calculate the probability of measuring an integer ℓ that lies outside the interval $[\tilde{\lambda}-a, \tilde{\lambda}+a]$, where a is a small positive number. The probability of measuring an integer ℓ that lies *outside* the interval $[\tilde{\lambda}-a, \tilde{\lambda}+a]$ is given by

$$p(\ell \notin [\tilde{\lambda}-a, \tilde{\lambda}+a]) = \sum_{\ell=0}^{\tilde{\lambda}-a-1} p(\ell) + \sum_{\ell=\tilde{\lambda}+a+1}^{2^m-1} p(\ell).$$

When a and $p(\ell \notin [\tilde{\lambda}-a, \tilde{\lambda}+a])$ are both small, the QPE algorithm returns an integer close to $\tilde{\lambda}$. To estimate an upper limit of the probability $p(\ell \notin [\tilde{\lambda}-a, \tilde{\lambda}+a])$, we can shift the probability distribution so that $\tilde{\lambda}$ is centered at zero, as shown in the figure below

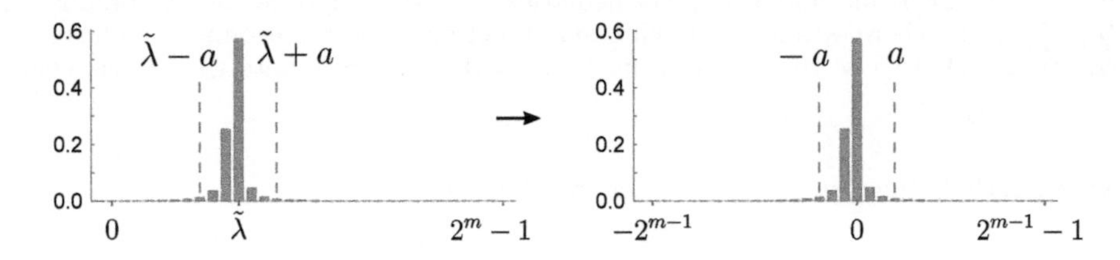

Therefore,

$$p(\ell \notin [\tilde{\lambda}-a, \tilde{\lambda}+a]) = \sum_{\ell=-2^{m-1}}^{-a-1} p[(\ell+\tilde{\lambda})\%2^m] + \sum_{\ell=a+1}^{2^{m-1}-1} p[(\ell+\tilde{\lambda})\%2^m], \tag{12.94}$$

where (see eqn (12.2))

$$p[(\ell+\tilde{\lambda})\%2^m] = \left| \frac{1}{2^m} \sum_{j=0}^{2^m-1} e^{i2\pi j(\lambda - \frac{\tilde{\lambda}}{2^m} - \frac{\ell}{2^m})} \right|^2 = \frac{1}{2^{2m}} \left| \sum_{j=0}^{2^m-1} e^{i2\pi j(\delta - \frac{\ell}{2^m})} \right|^2$$

$$= \frac{1}{2^{2m}} \left| \sum_{j=0}^{2^m-1} [e^{i2\pi(\delta - \frac{\ell}{2^m})}]^j \right|^2 = \frac{1}{2^{2m}} \frac{|1 - e^{i2\pi(2^m\delta - \ell)}|^2}{|1 - e^{i2\pi(\delta - \frac{\ell}{2^m})}|^2}.$$

Here, $\lambda - \tilde{\lambda}/2^m = \delta$ and the definition of geometric series. Since $|1 - e^{ix}| \leq 2$ for all real numbers x, we have

$$p[(\ell+\tilde{\lambda})\%2^m] \leq \frac{1}{2^{2m}} \frac{4}{|1 - e^{i2\pi(\delta - \frac{\ell}{2^m})}|^2}.$$

Now we need to deal with the denominator. Recall that $|1 - e^{ix}| \geq 2|x|/\pi$ when $-\pi \leq x \leq \pi$. Since the integer ℓ varies in the interval $[-2^{m-1}, 2^{m-1}-1]$, the condition $-\pi \leq 2\pi(\delta - \ell/2^m) \leq \pi$ is satisfied. This allows us to bound the denominator $|1 - e^{i2\pi(\delta - \frac{\ell}{2^m})}| \geq 2|2\pi(\delta - \frac{\ell}{2^m})|/\pi$ and write

$$p[(\ell+\tilde{\lambda})\%2^m] \leq \frac{1}{2^{2m}} \frac{1}{4(\delta - \ell/2^m)^2}.$$

Substituting this inequality into eqn (12.94), we obtain

$$p(\ell \notin [\tilde{\lambda} - a, \tilde{\lambda} + a]) \le \sum_{\ell=-2^{m-1}}^{-a-1} \frac{1}{2^{2m}} \frac{1}{4(\delta - \ell/2^m)^2} + \sum_{\ell=a+1}^{2^{m-1}-1} \frac{1}{2^{2m}} \frac{1}{4(\delta - \ell/2^m)^2}$$
$$= \frac{1}{4} \left[\sum_{\ell=-2^{m-1}}^{-a-1} \frac{1}{(2^m\delta - \ell)^2} + \sum_{\ell=a+1}^{2^{m-1}-1} \frac{1}{(2^m\delta - \ell)^2} \right].$$

Since $0 \le \delta \le 1/2^m$, we arrive at

$$p(\ell \notin [\tilde{\lambda} - a, \tilde{\lambda} + a]) \le \frac{1}{4} \left[\sum_{\ell=-2^{m-1}}^{-a-1} \frac{1}{\ell^2} + \sum_{\ell=a+1}^{2^{m-1}-1} \frac{1}{(\ell-1)^2} \right] = \frac{1}{4} \left[\sum_{\ell=a+1}^{2^{m-1}} \frac{1}{\ell^2} + \sum_{\ell=a+1}^{2^{m-1}-1} \frac{1}{(\ell-1)^2} \right]$$
$$= \frac{1}{4} \left[\sum_{\ell=a+1}^{2^{m-1}} \frac{1}{\ell^2} + \sum_{\ell'=a}^{2^{m-1}-2} \frac{1}{\ell'^2} \right] < \frac{1}{4} \left[\sum_{\ell=a}^{2^{m-1}} \frac{1}{\ell^2} + \sum_{\ell'=a}^{2^{m-1}} \frac{1}{\ell'^2} \right] = \frac{1}{2} \sum_{\ell=a}^{2^{m-1}} \frac{1}{\ell^2}$$
$$< \frac{1}{2} \sum_{\ell=a}^{\infty} \frac{1}{\ell^2} < \frac{1}{2} \int_{a-1}^{\infty} \frac{1}{\ell^2} d\ell = \frac{1}{2(a-1)}.$$

This means that the probability of measuring an integer ℓ in the interval $[\tilde{\lambda} - a, \tilde{\lambda} + a]$ is bounded by

$$\boxed{p(\ell \in [\tilde{\lambda} - a, \tilde{\lambda} + a]) = 1 - p(\ell \notin [\tilde{\lambda} - a, \tilde{\lambda} + a]) > 1 - \frac{1}{2(a-1)}.} \tag{12.95}$$

Let us calculate the number of ancilla qubits required to achieve a target accuracy. The algorithm outputs an integer ℓ, from which we estimate the best approximation of λ with the formula $\ell/2^m$. Suppose we want to estimate λ to an accuracy ε, i.e. $|\lambda - \ell/2^m| \le \varepsilon$. This expression can be written as

$$\left|\lambda - \frac{\ell}{2^m}\right| \le \left|\lambda - \frac{\tilde{\lambda}}{2^m}\right| + \left|\frac{\tilde{\lambda}}{2^m} - \frac{\ell}{2^m}\right| = \delta + \left|\frac{\tilde{\lambda}}{2^m} - \frac{\ell}{2^m}\right| \le \frac{1}{2^m} + \left|\frac{\tilde{\lambda}}{2^m} - \frac{\ell}{2^m}\right| \le \varepsilon$$

and therefore $|\tilde{\lambda} - \ell| \le 2^m\varepsilon - 1$. Thus, to achieve an accuracy ε, the measurement should output an integer ℓ that falls within the interval $[\tilde{\lambda} - 2^m\varepsilon + 1, \tilde{\lambda} + 2^m\varepsilon - 1]$ with high probability. Suppose we aim for a success probability of $1-q$ where q is a constant. Then from eqn (12.95), we have

$$p(\ell \in [\tilde{\lambda} - 2^m\varepsilon + 1, \tilde{\lambda} + 2^m\varepsilon - 1]) > 1 - \frac{1}{2(2^m\varepsilon - 2)} > 1 - q.$$

Rearranging this expression, we can obtain a condition on the number of ancilla qubits required for the QPE algorithm, $m > \log\left[\frac{1}{\varepsilon}\left(\frac{1}{2q} + 2\right)\right]$.

Solution 12.2 In the first iteration, we set $k = t$ and $\omega_k = 0$. Therefore, the parameter γ becomes

$$\gamma = \frac{\omega_k}{2} + \pi\phi 2^{k-1} = \pi\phi 2^{t-1} = \pi\sum_{j=1}^{t}\frac{\phi_j}{2^j}2^{t-1} + \pi\delta 2^{-1}$$
$$= \underbrace{\pi\sum_{j=1}^{t-1}\frac{\phi_j}{2^j}2^{t-1}}_{c_t} + \pi\phi_t 2^{-1} + \pi\delta 2^{-1} = \pi c_t + \pi\frac{\phi_t}{2} + \pi\frac{\delta}{2}.$$

If $\phi_t = 0$, then

$$p(0) = \cos^2(\gamma) = \cos^2\left(\pi c_t + \pi\frac{\delta}{2}\right) = \cos^2\left(\pi\frac{\delta}{2}\right).$$

Similarly, if $\phi_t = 1$, then

$$p(1) = \sin^2(\gamma) = \sin^2\left(\pi c_t + \frac{\pi}{2} + \pi\frac{\delta}{2}\right) = \sin^2\left(\frac{\pi}{2} + \pi\frac{\delta}{2}\right) = \cos^2\left(\pi\frac{\delta}{2}\right).$$

The correct bit is measured with probability $\cos^2(\pi\delta/2)$. For the iteration $k < t$, we set $\omega_k = -2\pi\sum_{j=k+1}^{t}\phi_j 2^{k-1}/2^j$, so that

$$\gamma = \pi c_k + \pi\frac{\phi_k}{2} + \pi\delta 2^{-t}2^{k-1}.$$

If $\phi_k = 0$, then

$$p(0) = \cos^2(\gamma) = \cos^2\left(\pi c_k + \pi\delta 2^{k-1-t}\right) = \cos^2\left(\pi\delta 2^{k-1-t}\right).$$

If $\phi_k = 1$, then

$$p(1) = \sin^2(\gamma) = \sin^2\left(\pi c_k + \frac{\pi}{2} + \pi\delta 2^{k-1-t}\right) = \cos^2\left(\pi\delta 2^{k-1-t}\right).$$

In brief, the probability of measuring the correct bit ϕ_k is $\cos^2(\pi\delta 2^{k-1-t})$.

The probability that each iteration returns the correct bit is given by the product

$$p_{\text{success}}(\delta, t) = \cos^2(\pi\delta 2^{-1})\cos^2(\pi\delta 2^{-2})\dots\cos^2(\pi\delta 2^{-t})$$
$$= \prod_{j=1}^{t}\cos^2(\pi\delta 2^{j-1-t}) = \frac{\sin^2\pi\delta}{4^t\sin^2(\pi\delta 2^{-t})}.$$

If the binary expansion is sufficiently large, we can perform the approximation $p_{\text{success}}(\delta) = \lim_{t\to\infty} p_{\text{success}}(\delta, t) = \text{sinc}^2(\pi\delta)$. If the error δ is uniformly distributed in the range $[0, 1)$, then

$$p_{\text{success}} = \int_0^1 p_{\text{success}}(\delta)d\delta = \int_0^1 \text{sinc}^2(\pi\delta)d\delta \approx 0.45.$$

Solution 12.3 In Mathematica, the function **LaguerreL[a, b, x]** returns the generalized Laguerre polynomials, while the spherical harmonics are given by **SphericalHarmonicY[l, m, θ, ϕ]**. The following code can be used to build the hydrogen-like orbitals,

```
fun[n_,l_,Z_] := Sqrt[4 Z^3 (n-l-1)!/(n^4 * (n+l)!)];
R[n_,l_,Z_,r_]:= fun[n, l, Z] (2 Z r/n)^l Exp[-Z r/n] LaguerreL[n-l-1,
```

```
                  2 l+1, 2 Z r/n];
Psi[n_, l_, m_, Z_, r_, \[Theta]_, \[Phi]_] :=
R[n, l, Z, r] * SphericalHarmonicY[l, m, \[Theta], \[Phi]]
```

where we used eqns (12.22, 12.23). The first eigenfunctions of the hydrogen atom can be calculated by printing the output of Psi[n, l, m, Z, r, θ, ϕ] for different values of n, l, and m. To check that the eigenfunctions are correctly normalized, one can execute the following integral

```
Integrate[r^2 Sin[\[Theta]] Abs[Psi[n,m,l,Z,r, \[Theta],\[Phi]]]^2,
        {\[Phi], 0, 2 Pi},{\[Theta], 0, Pi}, {r, 0, Infinity}]
```

for different values of n, l, m, Z and check that the output is one.

Solution 12.4 Without loss of generality, we assume that the vector $\mathbf{r}_1 = (r_1, \theta_1, \phi_1)$ points along the z direction as shown in the figure below,

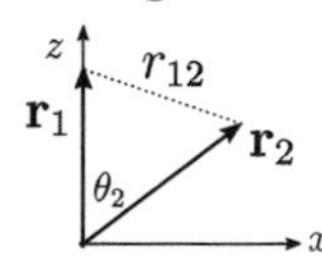

The distance between the vectors $\mathbf{r}_1$ and $\mathbf{r}_2$ can be expressed as $r_{12} = |\mathbf{r}_1 - \mathbf{r}_2| = \sqrt{r_1^2 + r_2^2 - 2r_1 r_2 \cos\theta_2}$, where we used the law of cosines. Since $\int d\theta_1 \sin\theta_1^2 \int d\phi_1 = 4\pi$, the integral (12.90) becomes

$$E = 4\pi \int_0^\infty dr_1 r_1^2 |\psi_{100}(\mathbf{r}_1)|^2 \int_0^\infty dr_2 r_2^2 |\psi_{100}(\mathbf{r}_2)|^2 \int_0^{2\pi} d\phi_2 \int_0^\pi d\theta_2 \frac{\sin\theta_2}{\sqrt{r_1^2 + r_2^2 - 2r_1 r_2 \cos\theta_2}}. \tag{12.96}$$

Substituting $\cos\theta_2 = x$, the integral over θ_2 returns

$$\int_0^\pi d\theta_2 \frac{\sin\theta_2}{\sqrt{r_1^2 + r_2^2 - 2r_1 r_2 \cos\theta_2}} = \int_1^{-1} dx \frac{-1}{\sqrt{r_1^2 + r_2^2 - 2r_1 r_2 x}} = \frac{r_1 + r_2 - |r_1 - r_2|}{r_1 r_2}$$
$$= \frac{2}{\max(r_1, r_2)}.$$

Replacing this expression into (12.96) and noting that $\int d\phi_2 = 2\pi$, one obtains

$$E = 4\pi \int_0^\infty dr_1 r_1^2 |\psi_{100}(\mathbf{r}_1)|^2 4\pi \int_0^\infty dr_2 r_2^2 |\psi_{100}(\mathbf{r}_2)|^2 \frac{1}{\max(r_1, r_2)}.$$

This integral can be calculated with Mathematica,

```
psi[r_] := Z^(3/2)/Sqrt[Pi] Exp[-Z r];
Integrate[4 Pi r1^2 4 Pi r2^2 (psi[r1] * psi[r2])^2/Max[r1, r2],
        {r1, 0, Infinity}, {r2, 0, Infinity}, Assumptions -> Z > 0].
```

The solution is $E = 5Z/8\, E_\mathrm{h}$.

Solution 12.5 Here is a Mathematica script to calculate the integral $\langle\phi|\hat{H}|\phi\rangle$:

```
phi[r_] := ((2 \[Lambda])/Pi)^(3/4) Exp[-\[Lambda]*r^2];
H[x_]:=-Laplacian[x,{r,\[Theta],\[Phi]},"Spherical"]/2-x/r;
Simplify[4 Pi Integrate[phi[r]*H[phi[r]]*r^2, {r, 0, Infinity},
        Assumptions -> \[Lambda] > 0]]
```

Solution 12.6 Let us divide the Hamiltonian into three parts, $\hat{H} = \hat{H}_1 + \hat{H}_2 + \hat{V}_{\text{ee}}$ where $\hat{V}_{\text{ee}} = 1/r_{12}$ while

$$\hat{H}_1 = -\frac{\nabla_1^2}{2} - \frac{2}{r_1}, \qquad \hat{H}_2 = -\frac{\nabla_2^2}{2} - \frac{2}{r_2}.$$

Therefore,

$$E_\Psi = \langle\Psi|\hat{H}_1|\Psi\rangle + \langle\Psi|\hat{H}_2|\Psi\rangle + \langle\Psi|\hat{V}_{\text{ee}}|\Psi\rangle.$$

The first two terms are identical and can be calculated with these few lines of code:

```
psi[r_] := Z^(3/2)/Sqrt[Pi] Exp[-Z r];
H[x_] := -Laplacian[x, {r,\[Theta],\[Phi]},"Spherical"]/2 - 2 x/r;
4 Pi Integrate[psi[r] H[psi[r]] r^2, {r,0,Infinity}, Assumptions -> Z>0]
```

This script returns $\langle\Psi|H_1|\Psi\rangle = \langle\Psi|H_2|\Psi\rangle = Z(Z-4)/2$. The term $\langle\Psi|V_{\text{ee}}|\Psi\rangle$ has already been calculated in Exercise 12.4. Putting everything together

$$E_\phi = \frac{Z(Z-4)}{2} + \frac{Z(Z-4)}{2} + \frac{5Z}{8} = \left(Z^2 - \frac{27}{8}Z\right) E_{\text{h}}.$$

Solution 12.7 Here is a Mathematica script to solve this problem:

```
a = 8.0/(9 Pi); fun = Table[Exp[-j a r^2], {j, 1, 2}];
Ham[x_] := -Laplacian[x, {r, \[Theta], \[Phi]}, "Spherical"]/2 - x/r;
H = Table[4 Pi Integrate[fun[[i]]*Ham[fun[[j]]] r^2, {r, 0, Infinity}],
        {i,1,2}, {j,1,2}];
S = Table[4 Pi Integrate[fun[[i]]*fun[[j]] r^2, {r, 0, Infinity}],
        {i,1,2}, {j,1,2}];
```

Solution 12.8 By imposing $\partial E_\Psi/\partial c_t = 0$, eqn (12.46) becomes

$$\underbrace{E_\Psi \frac{\partial}{\partial c_t}\sum_{pq} c_p c_q S_{pq}}_{A} = \underbrace{2\frac{\partial}{\partial c_t}\sum_{pq} c_p c_q h_{pq}}_{B} + \underbrace{\frac{\partial}{\partial c_t}\sum_{pq} c_p c_q g_{pq}(\mathbf{c})}_{C}. \tag{12.97}$$

Let us calculate these terms one by one:

$$\begin{aligned}
A &= E_\Psi \sum_{pq} \frac{\partial c_p}{\partial c_t} c_q S_{pq} + E_\Psi \sum_{pq} c_p \frac{\partial c_q}{\partial c_t} S_{pq} = 2E_\Psi \sum_q c_q S_{tq}, \\
B &= 2\sum_{pq} \frac{\partial c_p}{\partial c_t} c_q h_{pq} + 2\sum_{pq} c_p \frac{\partial c_q}{\partial c_t} h_{pq} = 4\sum_q c_q h_{tq}, \\
C &= \sum_{pq} \frac{\partial c_p}{\partial c_t} c_q g_{pq}(\mathbf{c}) + \sum_{pq} c_p c_q \frac{\partial g_{pq}(\mathbf{c})}{\partial c_t} = 2\sum_q c_q g_{tq}(\mathbf{c}) + \sum_{pq} c_p c_q \frac{\partial g_{pq}(\mathbf{c})}{\partial c_t},
\end{aligned} \tag{12.98}$$

where we used the fact that $S_{pq} = S_{qp}$, $h_{pq} = h_{qp}$, and $g_{pq}(\mathbf{c}) = g_{qp}(\mathbf{c})$. We now need to calculate $\partial g_{pq}(\mathbf{c})/\partial c_t$ explicitly,

$$\frac{\partial g_{pq}(\mathbf{c})}{\partial c_t} = \frac{\frac{\partial}{\partial c_t}\left(\sum_{rs} c_r c_s \langle e_p e_r|\frac{1}{r_{12}}|e_s e_q\rangle\right)}{\sum_{rs} c_r c_s S_{rs}} - \frac{\sum_{rs} c_r c_s \langle e_p e_r|\frac{1}{r_{12}}|e_s e_q\rangle}{\left(\sum_{rs} c_r c_s S_{rs}\right)^2}\frac{\partial}{\partial c_t}\sum_{pq} c_p c_q S_{pq}$$

$$= \frac{\left(\sum_s c_s \langle e_p e_t|\frac{1}{r_{12}}|e_s e_q\rangle + \sum_r c_r \langle e_p e_r|\frac{1}{r_{12}}|e_t e_q\rangle\right)}{\sum_{rs} c_r c_s S_{rs}} - \frac{g_{pq}(\mathbf{c})}{\sum_{rs} c_r c_s S_{rs}} 2\sum_q c_q S_{tq}.$$

Therefore, the last term in (12.98) becomes

$$\sum_{pq} c_p c_q \frac{\partial g_{pq}(\mathbf{c})}{\partial c_t} = 2\sum_s c_s g_{ts}(\mathbf{c}) - \frac{\left(\sum_{pq} c_p c_q g_{pq}(\mathbf{c})\right)\left(2\sum_q c_q S_{tq}\right)}{\sum_{rs} c_r c_s S_{rs}}.$$

Putting everything together, eqn (12.97) reduces to

$$2E_\Psi \sum_q c_q S_{tq} = 4\sum_q c_q h_{tq} + 4\sum_q c_q g_{tq}(\mathbf{c}) - \frac{\left(\sum_{pq} c_p c_q g_{pq}(\mathbf{c})\right)\left(2\sum_q c_q S_{tq}\right)}{\sum_{rs} c_r c_s S_{rs}}.$$

This expression can be rearranged in the form

$$\sum_q c_q \left(h_{tq} + g_{tq}(\mathbf{c}) - \frac{1}{2}E_\Psi S_{tq} - \frac{1}{2}\frac{\sum_{pq} c_p c_q g_{pq}(\mathbf{c}))}{\sum_{rs} c_r c_s S_{rs}} S_{tq}\right) = 0,$$

which is equivalent to (12.47) after renaming some dummy indexes.

Solution 12.9 Let us start from $\langle\Psi|\hat{O}_1|\Psi\rangle = \langle\Psi|\sum_j \hat{h}(j)|\Psi\rangle = \sum_j \langle\Psi|\hat{h}(j)|\Psi\rangle$. The term $\hat{h}(j)|\Psi\rangle$ can be written as

$$\hat{h}(j)|\Psi\rangle = \frac{1}{\sqrt{N!}}\sum_{\sigma\in S_N}(-1)^\sigma |\chi_{\sigma(1)}\rangle \ldots \hat{h}|\chi_{\sigma(j)}\rangle \ldots |\chi_{\sigma(N)}\rangle.$$

Therefore,

$$\langle\Psi|\hat{O}_1|\Psi\rangle = \sum_j \sum_{\sigma,\tau\in S_N} \frac{(-1)^\sigma(-1)^\tau}{N!}\left(\langle\chi_{\tau(1)}\ldots\chi_{\tau(N)}|\right)\left(|\chi_{\sigma(1)}\rangle\ldots\hat{h}|\chi_{\sigma(j)}\rangle\ldots|\chi_{\sigma(N)}\rangle\right)$$

$$= \sum_j \sum_{\sigma,\tau\in S_N} \frac{(-1)^\sigma(-1)^\tau}{N!}\langle\chi_{\tau(1)}|\chi_{\sigma(1)}\rangle\ldots\langle\chi_{\tau(j-1)}|\chi_{\sigma(j-1)}\rangle\langle\chi_{\tau(j)}|\hat{h}|\chi_{\sigma(j)}\rangle\ldots\langle\chi_{\tau(N)}|\chi_{\sigma(N)}\rangle$$

$$= \sum_j \sum_{\sigma,\tau\in S_N} \frac{(-1)^\sigma(-1)^\tau}{N!}\langle\chi_{\tau(j)}|\hat{h}|\chi_{\sigma(j)}\rangle \prod_{\ell\neq j}\langle\chi_{\tau(\ell)}|\chi_{\sigma(\ell)}\rangle.$$

The term $\prod_{\ell\neq j}\langle\chi_{\tau(\ell)}|\chi_{\sigma(\ell)}\rangle$ is different from zero if and only if $\sigma=\tau$. Since the spin-orbitals are orthonormal $\prod_{\ell\neq j}\langle\chi_{\tau(\ell)}|\chi_{\sigma(\ell)}\rangle = \delta_{\tau(\ell),\sigma(\ell)}$, we have

$$\langle\Psi|\hat{O}_1|\Psi\rangle = \frac{1}{N!}\sum_{\sigma\in S_N}\sum_{j=1}^N \langle\chi_{\sigma(j)}|\hat{h}|\chi_{\sigma(j)}\rangle =$$

$$= \frac{1}{N!} \sum_{\sigma \in S_N} \sum_{j=1}^{N} \langle \chi_j | \hat{h} | \chi_j \rangle = \frac{1}{N!} N! \sum_{j=1}^{N} \langle \chi_j | \hat{h} | \chi_j \rangle = \sum_{j=1}^{N} \langle \chi_j | \hat{h} | \chi_j \rangle .$$

Solution 12.10 Let us start from

$$\langle \Psi | \hat{O}_2 | \Psi \rangle = \frac{1}{2} \sum_{i \neq j} \sum_{\tau, \sigma \in S_n} \frac{(-1)^{\tau}(-1)^{\sigma}}{N!} \langle \chi_{\tau(1)} \dots \chi_{\tau(N)} | v(i,j) | \chi_{\sigma(1)} \dots \chi_{\sigma(N)} \rangle . \tag{12.99}$$

The bra-ket can be simplified by noting that for three particles it holds that

$$\langle \chi_4 \chi_5 \chi_9 | v(1,2) | \chi_1 \chi_3 \chi_7 \rangle = \langle \chi_4 \chi_5 | v(1,2) | \chi_1 \chi_3 \rangle \langle \chi_9 | \chi_7 \rangle .$$

In light of this observation, eqn (12.99) becomes

$$\langle \Psi | \hat{O}_2 | \Psi \rangle = \frac{1}{2} \sum_{i \neq j} \sum_{\tau, \sigma \in S_n} \frac{(-1)^{\tau}(-1)^{\sigma}}{N!} \langle \chi_{\tau(i)} \chi_{\tau(j)} | v | \chi_{\sigma(i)} \chi_{\sigma(j)} \rangle \prod_{k \neq i,j} \langle \chi_{\tau(k)} | \chi_{\sigma(k)} \rangle .$$

Since $\langle \chi_{\tau(k)} | \chi_{\sigma(k)} \rangle = \delta_{\tau(k),\sigma(k)}$, the only non-zero terms are

$$\begin{aligned} \langle \Psi | \hat{O}_2 | \Psi \rangle &= \frac{1}{2} \sum_{i \neq j} \sum_{\sigma \in S_n} \frac{(-1)^{\sigma}}{N!} \left[(-1)^{\sigma} \langle \chi_{\sigma(i)} \chi_{\sigma(j)} | v | \chi_{\sigma(i)} \chi_{\sigma(j)} \rangle - (-1)^{\sigma} \langle \chi_{\sigma(j)} \chi_{\sigma(i)} | v | \chi_{\sigma(i)} \chi_{\sigma(j)} \rangle \right] \\ &= \frac{1}{2} \sum_{i \neq j} \sum_{\sigma \in S_n} \frac{1}{N!} \left[\langle \chi_{\sigma(i)} \chi_{\sigma(j)} | v | \chi_{\sigma(i)} \chi_{\sigma(j)} \rangle - \langle \chi_{\sigma(j)} \chi_{\sigma(i)} | v | \chi_{\sigma(i)} \chi_{\sigma(j)} \rangle \right] . \end{aligned}$$

The sum runs over $i \neq j$. However, when $i = j$, the argument of the sum is zero. This allows us to sum over all values of i and j,

$$\langle \Psi | \hat{O}_2 | \Psi \rangle = \frac{1}{2} \sum_{i,j} \sum_{\sigma \in S_n} \frac{1}{N!} \left[\langle \chi_{\sigma(i)} \chi_{\sigma(j)} | v | \chi_{\sigma(i)} \chi_{\sigma(j)} \rangle - \langle \chi_{\sigma(j)} \chi_{\sigma(i)} | v | \chi_{\sigma(i)} \chi_{\sigma(j)} \rangle \right] .$$

The dummy indexes i and j can be replaced with $\sigma^{-1}(k)$ and $\sigma^{-1}(\ell)$ respectively. In conclusion,

$$\begin{aligned} \langle \Psi | \hat{O}_2 | \Psi \rangle &= \frac{1}{2} \sum_{k,\ell} \sum_{\sigma \in S_n} \frac{1}{N!} \left[\langle \chi_{\sigma(\sigma^{-1}(k))} \chi_{\sigma(\sigma^{-1}(\ell))} | v | \chi_{\sigma(\sigma^{-1}(k))} \chi_{\sigma(\sigma^{-1}(\ell))} \rangle \right. \\ &\qquad \left. - \langle \chi_{\sigma(\sigma^{-1}(\ell))} \chi_{\sigma(\sigma^{-1}(k))} | v | \chi_{\sigma(\sigma^{-1}(k))} \chi_{\sigma(\sigma^{-1}(\ell))} \rangle \right] \\ &= \frac{1}{2} \sum_{k,\ell} \frac{1}{N!} \left[\langle \chi_k \chi_\ell | v | \chi_k \chi_\ell \rangle - \langle \chi_\ell \chi_k | v | \chi_k \chi_\ell \rangle \right] \sum_{\sigma \in S_n} 1 \\ &= \frac{1}{2} \sum_{k,\ell} \left[\langle \chi_k \chi_\ell | v | \chi_k \chi_\ell \rangle - \langle \chi_\ell \chi_k | v | \chi_k \chi_\ell \rangle \right] . \end{aligned}$$

Solution 12.11 The matrix elements h_{pq} and S_{pq} of eqn (12.43) can be calculated with a Mathematica script:

```
fun[r_]  := {Sqrt[1.45^3/Pi] Exp[-1.45 r], Sqrt[2.90^3/Pi] Exp[-2.90 r]};
H[x_]  := -Laplacian[x, {r, \[Theta], \[Phi]}, "Spherical"]/2 - 2 x/r;
```

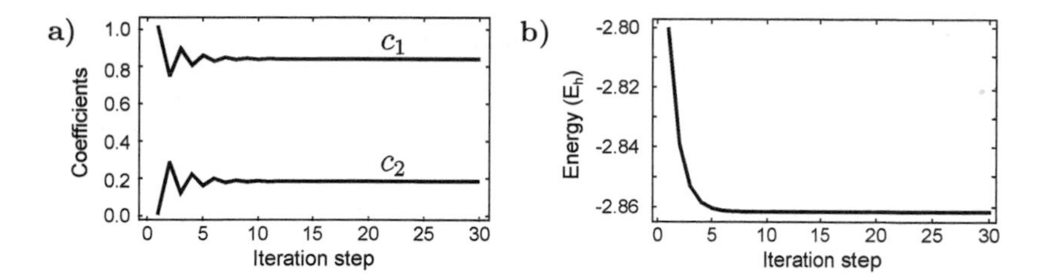

Fig. 12.22 Hartree–Fock method for the helium atom. a) Plot of the coefficients c_1 and c_2 as a function of the iteration step. The coefficients converge to $c_1 = 0.840852$ and $c_2 = 0.183882$. **b)** Plot of the ground-state energy E_Ψ as a function of the iteration step. The energy converges to $-2.86167\, E_\text{h}$.

```
S = Table[Integrate[4 Pi r^2 fun[r][[i]]*fun[r][[j]], {r, 0, Infinity}],
        {i,1,2}, {j,1,2}];
h = Table[Integrate[4 Pi r^2 fun[r][[i]]*H[fun[r][[j]]],{r,0,Infinity}],
        {i,1,2}, {j,1,2}];
```

The next step is to calculate the matrix elements $v_{ijkl} \equiv \langle e_i e_j | \frac{1}{r_{12}} | e_k e_l \rangle$ where $i, j, k, l \in \{1, 2\}$. This can be done with the command

```
v=Table[Integrate[(4 Pi r1^2 4 Pi r2^2) fun[r1][[i]] fun[r2][[j]]
      1/Max[r1, r2] fun[r1][[k]] fun[r2][[l]], {r1,0,Infinity},
      {r2,0,Infinity}], {i,1,2}, {j,1,2}, {k,1,2}, {l,1,2}]
```

Let us define the functions $g_{pq}(\mathbf{c})$ and $F_{pq}(\mathbf{c})$ as in eqns (12.43, 12.48):

$$g_{pq}(\mathbf{c}) = \frac{c_1^2 v_{p11q} + c_1 c_2 v_{p12q} + c_2 c_1 v_{p21q} + c_2^2 v_{p22q}}{c_1^2 S_{11} + 2c_1 c_2 S_{12} + c_2^2 S_{22}}, \qquad F_{pq}(\mathbf{c}) = h_{pq} + g_{pq}(\mathbf{c}).$$

In Mathematica, this would be

```
den[c1_,c2_]:=c1 c1 S[[1,1]]+2 c1 c2 S[[1,2]]+c2 c2 S[[2,2]];
g11[c1_,c2_]:=(c1^2 v[[1,1,1,1]]+c1 c2 v[[1,1,2,1]]+c2 c1 v[[1,2,1,1]]+
              c2^2 v[[1,2,2,1]])/den[c1,c2];
g12[c1_,c2_]:=(c1^2 v[[1,1,1,2]]+c1 c2 v[[1,1,2,2]]+c2 c1 v[[1,2,1,2]]+
              c2^2 v[[1,2,2,2]])/den[c1,c2];
g22[c1_,c2_]:=(c1^2 v[[2,1,1,2]]+c1 c2 v[[2,1,2,2]]+c2 c1 v[[2,2,1,2]]+
              c2^2 v[[2,2,2,2]])/den[c1,c2];
F11[c1_,c2_]:=h[[1,1]]+g11[c1,c2]; F12[c1_,c2_]:=h[[1,2]]+g12[c1,c2];
F21[c1_,c2_]:=F12[c1,c2]; F22[c1_,c2_]:=h[[2, 2]]+g22[c1,c2];
```

where we used the fact that $g_{pq} = g_{qp}$ and $F_{pq} = F_{qp}$. The vector $\mathbf{c} = (c_1, c_2)$ satisfying $(\mathsf{F} - \gamma\mathsf{S})\mathbf{c} = \mathbf{0}$ can be derived by calculating the smallest value of γ satisfying the expression $\det(\mathsf{F} - \gamma\mathsf{S}) = 0$, and then using γ to compute the coefficients c_1, c_2 from the system of equations:

$$(F_{11} - \gamma S_{11})c_1 + (F_{12} - \gamma S_{12})c_2 = 0,$$
$$c_1^2 + 2c_1 c_2 S_{12} + c_2^2 = \langle \psi | \psi \rangle = 1.$$

Once the new coefficients have been determined, the ground-state energy of the helium atom can be calculated from (12.44). This procedure is repeated until the coefficients have converged. In Mathematica, these steps can be executed with a for loop

```
c1 = 0.5; c2 = 0.5; Iterations = 30; Results = {};
For[i = 0, i < Iterations, i++,
sol1 = NSolve[Det[{{F11[c1, c2] - x S[[1, 1]],
         F12[c1, c2] - S[[1, 2]] x}, {F21[c1, c2] - S[[2, 1]] x,
         F22[c1, c2] - x S[[2, 2]]}}] == 0, x];
gamma = x /. sol1[[1]];
sol2 = NSolve[{(F11[c1, c2] - gamma S[[1,1]]) x1 + (F12[c1, c2] -
    gamma S[[1,2]]) x2==0,x1^2+2 x1 x2 S[[1,2]] + x2^2==1}, {x1,x2}];
c1 = Abs[ x1 /. sol2[[1]]]; c2 = Abs[x2 /. sol2[[1]]];
Energy = (c1^2(2 h[[1,1]]+g11[c1,c2])+2 c1 c2 (2h[[1,2]]+g12[c1,c2]) +
     c2^2 (2 h[[2,2]]+g22[c1,c2]))/den[c1,c2];
AppendTo[Results, {c1, c2, gamma, Energy}];]
```

Here, we chose the initial parameters $c_1 = 0.5$, $c_2 = 0.5$, but other values would have worked as well. Figure 12.22 shows the coefficients c_1 and c_2 as well as the ground-state energy E_Ψ as a function of the iteration step. The ground-state energy converges to $-2.8616\ E_\mathrm{h}$ which is $0.0417\ E_\mathrm{h}$ above the experimental value.

Solution 12.12 Let us separate the sum into odd and even sums,

$$\begin{aligned}\langle\Psi|\hat{O}_2|\Psi\rangle = &\frac{1}{2}\sum_{i=\mathrm{odd}}\sum_{j=\mathrm{odd}}[\langle\chi_i\chi_j|\hat{v}|\chi_i\chi_j\rangle - \langle\chi_j\chi_i|\hat{v}|\chi_i\chi_j\rangle]\\ &+\frac{1}{2}\sum_{i=\mathrm{even}}\sum_{j=\mathrm{odd}}[\langle\chi_i\chi_j|v|\chi_i\chi_j\rangle - \langle\chi_j\chi_i|v|\chi_i\chi_j\rangle]\\ &+\frac{1}{2}\sum_{i=\mathrm{odd}}\sum_{j=\mathrm{even}}[\langle\chi_i\chi_j|v|\chi_i\chi_j\rangle - \langle\chi_j\chi_i|v|\chi_i\chi_j\rangle]\\ &+\frac{1}{2}\sum_{i=\mathrm{even}}\sum_{j=\mathrm{even}}[\langle\chi_i\chi_j|v|\chi_i\chi_j\rangle - \langle\chi_j\chi_i|v|\chi_i\chi_j\rangle]\,.\end{aligned}$$

Since the spin-orbitals are $\psi_i\alpha$ ($\psi_i\beta$) when i is odd (even),

$$\begin{aligned}\langle\Psi|\hat{O}_2|\Psi\rangle = &\frac{1}{2}\sum_{i=1}^{N/2}\sum_{j=1}^{N/2}[\langle\psi_i\alpha\psi_j\alpha|\hat{v}|\psi_i\alpha\psi_j\alpha\rangle - \langle\psi_j\alpha\psi_i\alpha|\hat{v}|\psi_i\alpha\psi_j\alpha\rangle]\\ &+\frac{1}{2}\sum_{i=1}^{N/2}\sum_{j=1}^{N/2}\left[\langle\psi_i\beta\psi_j\alpha|\hat{v}|\psi_i\beta\psi_j\alpha\rangle - \cancel{\langle\psi_j\alpha\psi_i\beta|\hat{v}|\psi_i\beta\psi_j\alpha\rangle}\right]\\ &+\frac{1}{2}\sum_{i=1}^{N/2}\sum_{j=1}^{N/2}\left[\langle\psi_i\alpha\psi_j\beta|\hat{v}|\psi_i\alpha\psi_j\beta\rangle - \cancel{\langle\psi_j\beta\psi_i\alpha|\hat{v}|\psi_i\alpha\psi_j\beta\rangle}\right]\\ &+\frac{1}{2}\sum_{i=1}^{N/2}\sum_{j=1}^{N/2}[\langle\psi_i\beta\psi_j\beta|\hat{v}|\psi_i\beta\psi_j\beta\rangle - \langle\psi_j\beta\psi_i\beta|\hat{v}|\psi_i\beta\psi_j\beta\rangle]\,,\end{aligned}$$

where we used $\langle\alpha|\beta\rangle = \langle\beta|\alpha\rangle = 0$. In conclusion,

$$\langle\Psi|\hat{O}_2|\Psi\rangle = \sum_{i=1}^{N/2}\sum_{j=1}^{N/2}[2\langle\psi_i\psi_j|\hat{v}|\psi_i\psi_j\rangle - \langle\psi_j\psi_i|\hat{v}|\psi_i\psi_j\rangle]\,.$$

Solution 12.13 Let us impose the condition

$$\frac{\partial L}{\partial \tilde{c}^*_{tj}} = \frac{\partial E_\Psi}{\partial \tilde{c}^*_{tj}} - \sum_{k=1}^{N/2} \varepsilon_k \frac{\partial \langle \tilde{\psi}_k | \tilde{\psi}_k \rangle}{\partial \tilde{c}^*_{tj}} = 0. \tag{12.100}$$

The derivative of the first term is

$$\frac{\partial E_\Psi}{\partial \tilde{c}^*_{tj}} = \sum_{u,v=1}^{M} \frac{\partial E_\Psi}{\partial \rho_{uv}} \frac{\partial \rho_{uv}}{\partial \tilde{c}^*_{tj}},$$

where we used the chain rule. Since

$$\tilde{\rho}_{uv} = \sum_{j=1}^{N/2} \tilde{c}^*_{uj} \tilde{c}_{vj} = \sum_{j=1}^{N/2} \sum_{i=1}^{N/2} c^*_{ui} U^*_{ij} \sum_{k=1}^{N/2} c_{vk} U_{kj} = \sum_{i=1}^{N/2} c^*_{ui} \sum_{k=1}^{N/2} c_{vk} \underbrace{\sum_{j=1}^{N/2} U^*_{ij} U_{kj}}_{\delta_{ik}} = \sum_{i=1}^{N/2} c^*_{ui} c_{vi} = \rho_{uv}$$

we have

$$\begin{aligned}
\frac{\partial E_\Psi}{\partial \tilde{c}^*_{tj}} &= \sum_{u,v=1}^{M} \frac{\partial E_\Psi}{\partial \rho_{uv}} \frac{\partial \tilde{\rho}_{uv}}{\partial \tilde{c}^*_{tj}} = \sum_{u,v=1}^{M} \frac{\partial E_\Psi}{\partial \rho_{uv}} \sum_{k=1}^{N/2} \frac{\partial \tilde{c}^*_{uk}}{\partial \tilde{c}^*_{tj}} \tilde{c}_{vk} \\
&= \sum_{u,v=1}^{M} \frac{\partial E_\Psi}{\partial \rho_{uv}} \sum_{k=1}^{N/2} \delta_{ut} \delta_{kj} \tilde{c}_{vk} = \sum_{v=1}^{M} \frac{\partial E_\Psi}{\partial \rho_{tv}} \tilde{c}_{vj}.
\end{aligned}$$

Let us calculate the derivative $\partial E_\Psi / \partial \rho_{tv}$,

$$\begin{aligned}
\frac{\partial E_\Psi}{\partial \rho_{tv}} &= \frac{\partial}{\partial \rho_{tv}} \left[2 \sum_{p,q} \rho_{pq} \langle e_p | \hat{h} | e_q \rangle + \sum_{p,q,r,s} [2\rho_{pr}\rho_{qs} - \rho_{ps}\rho_{qr}] \langle e_p e_q | \frac{1}{r_{12}} | e_r e_s \rangle \right] \\
&= 2 \sum_{p,q} \delta_{pt} \delta_{qv} \langle e_p | \hat{h} | e_q \rangle \\
&\quad + \sum_{p,q,r,s} [2\delta_{pt}\delta_{rv}\rho_{qs} + 2\rho_{pr}\delta_{qt}\delta_{sv} - \delta_{pt}\delta_{sv}\rho_{qr} - \rho_{ps}\delta_{qt}\delta_{rv}] \langle e_p e_q | \frac{1}{r_{12}} | e_r e_s \rangle \\
&= 2 \Big[\langle e_t | \hat{h} | e_v \rangle + 2 \sum_{q,s} \rho_{qs} \langle e_t e_q | \frac{1}{r_{12}} | e_v e_s \rangle - \sum_{q,r} \rho_{qr} \langle e_t e_q | \frac{1}{r_{12}} | e_r e_v \rangle \Big] = 2F_{tv},
\end{aligned}$$

where we used $\langle e_t e_q | 1/r_{12} | e_v e_s \rangle = \langle e_q e_t | 1/r_{12} | e_s e_v \rangle$ and in the last step we introduced the Fock operator F_{tv}. This leads to

$$\frac{\partial E_\Psi}{\partial \tilde{c}^*_{tj}} = \sum_{v=1}^{M} \frac{\partial E_\Psi}{\partial \rho_{tv}} \tilde{c}_{vj} = \sum_{v=1}^{M} 2F_{tv} \tilde{c}_{vj}.$$

The second term of eqn (12.100) is given by

$f(\mathbf{r})$	$F(\mathbf{k})$
$1/r$	$4\pi/k^2$
$e^{-\alpha r^2}$	$\left(\frac{\pi}{\alpha}\right)^{3/2} e^{-k^2/4\alpha}$
$\delta(\mathbf{r})$	1
1	$(2\pi)^3\delta(\mathbf{k})$

Table 12.6 Relations between distributions in the Fourier space. The Fourier transform is defined as $f(\mathbf{r}) = \frac{1}{(2\pi)^3}\int d\mathbf{k}F(\mathbf{k})e^{i\mathbf{k}\cdot\mathbf{r}}$ while its inverse is given by $F(\mathbf{k}) = \int d\mathbf{r}f(\mathbf{r})e^{-i\mathbf{k}\cdot\mathbf{r}}$.

$$\sum_{k=1}^{N/2} \varepsilon_k \frac{\partial\langle\tilde{\psi}_k|\tilde{\psi}_k\rangle}{\partial\tilde{c}_{tj}^*} = \sum_{k=1}^{N/2} \varepsilon_k \frac{\partial}{\partial\tilde{c}_{tj}^*} \sum_{p,q=1}^{M} \tilde{c}_{pk}^*\tilde{c}_{qk}\langle e_p|e_q\rangle = \sum_{k=1}^{N/2} \varepsilon_k \sum_{p,q=1}^{M} \delta_{tp}\delta_{jk}\tilde{c}_{qk}\langle e_p|e_q\rangle$$

$$= \varepsilon_j \sum_{q=1}^{M} \langle e_t|e_q\rangle\tilde{c}_{qj} = \varepsilon_j \sum_{q=1}^{M} S_{tq}\tilde{c}_{qj}.$$

Putting everything together, we obtain the closed-shell Hartree–Fock equations,

$$\boxed{\sum_{v=1}^{M} F_{tv}\tilde{c}_{vj} = \varepsilon_j \sum_{q=1}^{M} S_{tq}\tilde{c}_{qj}.}$$

Solution 12.14 The integral $\langle g_\alpha g_\gamma|\frac{1}{r_{12}}|g_\beta g_\delta\rangle$ can be calculated analytically by performing a Fourier transform. Using the relations shown in Table 12.6, we have

$$\langle g_\alpha g_\gamma|\frac{1}{r_{12}}|g_\beta g_\delta\rangle = \frac{K_{\alpha\beta}K_{\gamma\delta}}{(2\pi)^9}\left(\frac{\pi^2}{pq}\right)^{3/2} \int d\mathbf{r}_1 d\mathbf{r}_2 d\mathbf{k}_1 d\mathbf{k}_2 d\mathbf{k}_3 e^{-k_1^2/4p} e^{i\mathbf{k}_1\cdot(\mathbf{r}_1-\mathbf{R}_P)}$$
$$\frac{4\pi}{k_3} e^{i\mathbf{k}_3\cdot(\mathbf{r}_1-\mathbf{r}_2)} e^{-k_2^2/4q} e^{i\mathbf{k}_2\cdot(\mathbf{r}_2-\mathbf{R}_Q)}.$$

The relation $\int d\mathbf{r}e^{-i\mathbf{r}\cdot(\mathbf{k}-\mathbf{k}')} = (2\pi)^3\delta(\mathbf{k}-\mathbf{k}')$ allows us to calculate the integrals over $d\mathbf{r}_1$ and $d\mathbf{r}_2$ in a straightforward way,

$$\langle g_\alpha g_\gamma|\frac{1}{r_{12}}|g_\beta g_\delta\rangle = \frac{K_{\alpha\beta}K_{\gamma\delta}}{(2\pi)^9}\left(\frac{\pi^2}{pq}\right)^{3/2} \int d\mathbf{k}_1 d\mathbf{k}_2 d\mathbf{k}_3 \overbrace{\int d\mathbf{r}_1 e^{i\mathbf{r}_1\cdot(\mathbf{k}_1+\mathbf{k}_3)}}^{(2\pi)^3\delta(\mathbf{k}_1+\mathbf{k}_3)} \overbrace{\int d\mathbf{r}_2 e^{i\mathbf{r}_2\cdot(\mathbf{k}_2-\mathbf{k}_3)}}^{(2\pi)^3\delta(\mathbf{k}_2-\mathbf{k}_3)}$$
$$e^{-k_1^2/4p} e^{-i\mathbf{k}_1\cdot\mathbf{R}_P} \frac{4\pi}{k_3^2} e^{-k_2^2/4q} e^{-i\mathbf{k}_2\cdot\mathbf{R}_Q}$$
$$= \frac{K_{\alpha\beta}K_{\gamma\delta}}{(2\pi)^3}\left(\frac{\pi^2}{pq}\right)^{3/2} \int d\mathbf{k}_3 e^{i\mathbf{k}_3\cdot(\mathbf{R}_P-\mathbf{R}_Q)} \frac{4\pi}{k_3^2} e^{-k_3^2(p+q)/4pq}.$$

One can now substitute $d\mathbf{k}_3 = k_3^2 \sin\theta d\theta d\phi dk_3$ and perform the integration over θ and ϕ,

$$\langle g_\alpha g_\gamma|\frac{1}{r_{12}}|g_\beta g_\delta\rangle = \frac{K_{\alpha\beta}K_{\gamma\delta}}{(2\pi)^3}\left(\frac{\pi^2}{pq}\right)^{3/2}\int dk_3 \sin\theta d\theta d\phi e^{ik_3|\mathbf{R}_P-\mathbf{R}_Q|\cos\theta} 4\pi e^{-k_3^2(p+q)/4pq}$$
$$= \frac{K_{\alpha\beta}K_{\gamma\delta}}{(2\pi)^3}\left(\frac{\pi^2}{pq}\right)^{3/2}\int dk_3 \frac{(4\pi)^2\sin(k_3|\mathbf{R}_P-\mathbf{R}_Q|)}{k_3|\mathbf{R}_P-\mathbf{R}_Q|}e^{-k_3^2(p+q)/4pq}.$$

This integral produces

$$\boxed{\langle g_\alpha g_\gamma|\frac{1}{r_{12}}|g_\beta g_\delta\rangle = \frac{K_{\alpha\beta}K_{\gamma\delta}}{(2\pi)^3}\left(\frac{\pi^2}{pq}\right)^{3/2}\frac{8\pi^3}{|\mathbf{R}_P-\mathbf{R}_Q|}\mathrm{Erf}\left[|\mathbf{R}_P-\mathbf{R}_Q|\sqrt{\frac{pq}{p+q}}\right].}$$

This equality is helpful for the calculation of the ground-state energy of the hydrogen molecule with the Hartree–Fock method and an STO-1G basis (see Exercise 12.15).

Solution 12.15 This Mathematica script simulates the ground-state energy of the hydrogen molecule as a function of the internuclear distance. The parameters a1, a2, a3, and a4 correspond to α, β, γ, and δ of the main text.

```
Z = 1; Inf = Infinity; EnergyMol = {};
a1 = 0.4166123; a2 = a3 = a4 = a1; p = a1 + a2; q = a3 + a4;
For[k = 0, k < 35, k++,
 d = 0.3 + 0.2*k; Vnn = 1/d;
 RA = {0, 0, 0}; RB = {0, 0, d}; pos = {RA, RB};
 F0[x_] := Piecewise[{{16 Pi^(5/2) Sqrt[(p q)/(p + q)], x == 0}},
         8 Pi^3 Erf[x Sqrt[(p q)/(p + q)]]*1/x];
 v[RA_,RB_,RC_,RD_]:=Exp[-((a1 a2)/p)*(RA-RB).(RA-RB)
         -(a3 a4)/q*(RC-RD).(RC-RD)]*(1/(p*q))^(3/2)*1/(8 Pi^2)*
         F0[Norm[((a1 RA + a2 RB)/p - (a3 RC + a4 RD)/q)]];
 fun[x_,y_,z_]:={1/Sqrt[Pi] Exp[-a1 ({x,y,z}-RA).({x,y,z}-RA)],
                 1/Sqrt[Pi] Exp[-a2 ({x,y,z}-RB).({x,y,z}-RB)]};
 H[X_] := -Laplacian[X, {x, y, z}]/2 -
        Z X/Sqrt[({x, y, z} - RA).({x, y, z} - RA)] -
        Z X/Sqrt[({x, y, z} - RB).({x, y, z} - RB)];
 rho[c1_, c2_] := {{c1^2, c1*c2}, {c2*c1, c2^2}};
 F[c1_,c2_]:= Table[h[[p,r]]+2 Sum[rho[c1,c2][[q,s]]*Ve[[p,r,q,s]],
            {q,1,2}, {s,1,2}]-Sum[rho[c1,c2][[q,s]]*Ve[[p,s,q,r]],
            {q, 1, 2}, {s, 1, 2}], {p,1, 2}, {r, 1, 2}];
 h =Table[NIntegrate[fun[x, y, z][[i]]*H[fun[x, y, z][[j]]],
        {x, -Inf, Inf}, {y, -Inf, Inf}, {z, -Inf, Inf}],
        {i, 1, 2}, {j, 1, 2}] //Quiet;
 S =Table[NIntegrate[fun[x, y, z][[i]]*fun[x, y, z][[j]],
        {x, -Inf, Inf}, {y, -Inf, Inf}, {z, -Inf, Inf}],
        {i, 1, 2}, {j, 1, 2}] // Quiet;
 Ve=Table[v[pos[[i]], pos[[j]], pos[[k]], pos[[l]]],
        {i, 1, 2}, {j, 1, 2}, {k, 1, 2}, {l, 1, 2}];
 c1=1/Sqrt[2]; c2=1/Sqrt[2]; Iterations=5; Results={};
 For[i = 0, i < Iterations, i++,
   sol1 = NSolve[Det[{{F[c1, c2][[1, 1]] - x S[[1, 1]],
          F[c1, c2][[1, 2]] - S[[1, 2]] x}, {F[c1, c2][[2, 1]] -
          S[[2, 1]] x, F[c1, c2][[2, 2]] - x S[[2, 2]]}}] == 0, x];
   eps = x /. sol1[[1]];
   sol2 = NSolve[{(F[c1, c2][[1, 1]] - eps S[[1, 1]]) x1 +
          (F[c1, c2][[1, 2]] - eps S[[1, 2]]) x2 == 0,
          S[[1,1]] x1^2 + 2 x1 x2 S[[1,2]]+S[[2,2]] x2^2==1},{x1,x2}];
   c1 = x1 /. sol2[[1]]; c2 = (x2 /. sol2[[1]])/Sign[c1];
   c1 = c1/Sign[c1]; Cvec = {c1, c2};
```

```
    Energy=Vnn + 2 Sum[h[[p,q]]*Cvec[[p]]*Cvec[[q]], {p,1,2}, {q,1,2}] +
            Sum[(2 Cvec[[p]] *Cvec[[r]]*Cvec[[q]]*Cvec[[s]] -
            Cvec[[p]]* Cvec[[s]]*Cvec[[q]]*Cvec[[r]]) Ve[[p,q,r,s]],
            {p,1,2}, {q,1,2}, {r,1,2}, {s,1,2}];
    AppendTo[Results, {c1, c2, eps, Energy}];];
    Print["distance = ", d, "(a0), Molecular energy = ",
         Energy, "(Eh), c1 = ", c1, ", c2 = ", c2];
    AppendTo[EnergyMol, {d, Energy}]]
ListPlot[EnergyMol, Joined -> True,
       PlotRange -> {{0.0, 8}, {-1, 0.3}}, Frame -> True,
       FrameLabel -> {Style["d (a0)", 14], Style["E (Eh)", 14]}]
```

Solution 12.16 If $q < p$, we have

$$\begin{aligned} a_p^\dagger a_q \quad \longleftrightarrow \quad & Z_1 \dots Z_{p-1}\sigma_p^+ Z_1 \dots Z_{q-1}\sigma_q^- \\ &= Z_1 \dots Z_{q-1} Z_q Z_{q+1} \dots Z_{p-1}\sigma_p^+ Z_1 \dots Z_{q-1}\sigma_q^- \\ &= Z_1 Z_1 \dots Z_{q-1} Z_{q-1} Z_q Z_{q+1} \dots Z_{p-1}\sigma_p^+ \sigma_q^- \\ &= Z_q Z_{q+1} \dots Z_{p-1}\sigma_p^+ \sigma_q^- \\ &= Z_q \sigma_q^- \sigma_p^+ Z_{q+1} \dots Z_{p-1} = \sigma_q^- \sigma_p^+ Z_{q+1} \dots Z_{p-1}, \end{aligned}$$

where in the last step we used $Z_q\sigma_q^- = \sigma_q^-$. If $p > q$, we obtain

$$a_p^\dagger a_q = (a_q^\dagger a_p)^\dagger = (\sigma_p^- \sigma_q^+ Z_{p+1} \dots Z_{q-1})^\dagger = \sigma_p^+ \sigma_q^- Z_{p+1} \dots Z_{q-1}.$$

If $p = q$, then

$$a_p^\dagger a_q = Z_1 \dots Z_{p-1}\sigma_p^+ Z_1 \dots Z_{p-1}\sigma_p^- = \sigma_p^+ \sigma_p^- = \frac{\mathbb{1} - Z_p}{2}.$$

Solution 12.17 The indices p, q, r, s can be separated in spatial part and spin part:

$$p = p'\sigma, \qquad q = q'\tau, \qquad r = r'\nu, \qquad s = s'\rho.$$

For example, for the hydrogen molecule in the minimal basis, we have

old notation			
p	spin-orbital	orbital	spin
1	$\|\chi_1\rangle = a_1^\dagger\|\mathbf{0}\rangle$	ψ_1	α
2	$\|\chi_2\rangle = a_2^\dagger\|\mathbf{0}\rangle$	ψ_1	β
3	$\|\chi_3\rangle = a_3^\dagger\|\mathbf{0}\rangle$	ψ_2	α
4	$\|\chi_4\rangle = a_4^\dagger\|\mathbf{0}\rangle$	ψ_2	β

new notation				
p'	σ	spin-orbital	orbital	spin
1	α	$\|\chi_1\rangle = a_{1\alpha}^\dagger\|\mathbf{0}\rangle$	ψ_1	α
1	β	$\|\chi_2\rangle = a_{1\beta}^\dagger\|\mathbf{0}\rangle$	ψ_1	β
2	α	$\|\chi_3\rangle = a_{2\alpha}^\dagger\|\mathbf{0}\rangle$	ψ_2	α
2	β	$\|\chi_4\rangle = a_{2\beta}^\dagger\|\mathbf{0}\rangle$	ψ_2	β

With the substitutions $p = p'\sigma, q = q'\tau, r = r'\nu, s = s'\rho$, the Hamiltonian (12.84) becomes

$$\hat{H} = \sum_{p',q'=1}^{2} \sum_{\sigma,\tau=\alpha}^{\beta} h_{p'\sigma,q'\tau} a_{p'\sigma}^\dagger a_{q'\tau} + \frac{1}{2} \sum_{p',q',r',s'=1}^{2} \sum_{\sigma,\tau,\rho,\nu=\alpha}^{\beta} h_{p'\sigma,q'\tau,r'\rho,s'\nu} a_{p'\sigma}^\dagger a_{q'\tau}^\dagger a_{r'\rho} a_{s'\nu}. \tag{12.101}$$

The tensors $h_{p'\sigma,q'\tau}$ and $h_{p'\sigma,q'\tau,r'\rho,s'\nu}$ can be expressed as

$$h_{p'\sigma,q'\tau} = h_{pq} = \langle \chi_p|\hat{h}|\chi_q\rangle = \langle \psi_{p'}\sigma|\hat{h}|\psi_{q'}\tau\rangle = \underbrace{\langle \psi_{p'}|\hat{h}|\psi_{q'}\rangle}_{\tilde{h}_{pq}}\delta_{\sigma\tau} = \tilde{h}_{pq}\delta_{\sigma\tau}, \tag{12.102}$$

$$h_{p'\sigma,q'\tau,r'\rho,s'\nu} = h_{pqrs} = \langle \chi_p\chi_q|\frac{1}{r_{12}}|\chi_s\chi_r\rangle = \underbrace{\langle \psi_{p'}\psi_{q'}|\frac{1}{r_{12}}|\psi_{s'}\psi_{r'}\rangle}_{\tilde{h}_{p'q'r's'}}\delta_{\sigma\nu}\delta_{\tau\rho} = \tilde{h}_{p'q'r's'}\delta_{\sigma\nu}\delta_{\tau\rho}. \tag{12.103}$$

The symbol ˜ indicates that the quantities $\tilde{h}_{pq}$ and $\tilde{h}_{pqrs}$ are matrix elements calculated with an integration of the spatial part only. Therefore, the Hamiltonian (12.101) reduces to

$$\begin{aligned} H &= \sum_{p',q'=1}^{2}\sum_{\sigma,\tau=\alpha}^{\beta} \tilde{h}_{p'q'}\delta_{\sigma\tau}a^\dagger_{p',\sigma}a_{q',\tau} + \frac{1}{2}\sum_{p',q',r',s'=1}^{2}\sum_{\sigma,\tau,\rho,\nu=\alpha}^{\beta} \tilde{h}_{p'q'r's'}\delta_{\sigma\nu}\delta_{\tau\rho}a^\dagger_{p'\sigma}a^\dagger_{q'\tau}a_{r'\rho}a_{s'\nu} \\ &= \sum_{p',q'=1}^{2}\sum_{\sigma=\alpha}^{\beta} \tilde{h}_{p'q'}a^\dagger_{p',\sigma}a_{q',\sigma} + \frac{1}{2}\sum_{p',q',r',s'=1}^{2}\sum_{\sigma,\tau=\alpha}^{\beta} \tilde{h}_{p'q'r's'}a^\dagger_{p'\sigma}a^\dagger_{q'\tau}a_{r'\tau}a_{s'\sigma}. \end{aligned}$$

The important thing to note is that the tensors $\tilde{h}_{pq}$, $\tilde{h}_{pqrs}$ are sparse. With a Mathematica script, one can calculate the quantities $\tilde{h}_{pq}$:

```
Z=1; Inf=Infinity; a1=0.4166123; d=1.401; RA={0,0,0}; RB={d,0,0};
g1[x_,y_,z_]=(2 a1/Pi)^(3/4) Exp[-a1 ({x, y, z} - RA).({x, y, z} - RA)];
g2[x_,y_,z_]=(2 a1/Pi)^(3/4) Exp[-a1 ({x, y, z} - RB).({x, y, z} - RB)];
g1g2 = Chop[Integrate[g1[x,y,z]*g2[x,y,z], {x,-Inf,Inf}, {y,-Inf,Inf},
          {z,-Inf,Inf}], 10^-7];
chi1[x_, y_, z_] = (g1[x, y, z] + g2[x, y, z])/(Sqrt[2 + 2 g1g2]);
chi2[x_, y_, z_] = (g1[x, y, z] - g2[x, y, z])/(Sqrt[2 - 2 g1g2]);
fun[x_, y_, z_] := {chi1[x, y, z], chi2[x, y, z]};
H[X_] := -Laplacian[X, {x, y, z}]/2 -
         Z X/Sqrt[({x, y, z} - RA).({x, y, z} - RA)] -
         Z X/Sqrt[({x, y, z} - RB).({x, y, z} - RB)];
h = Table[Chop[NIntegrate[fun[x, y, z][[i]]*H[fun[x, y, z][[j]]],
         {x, -Inf, Inf}, {y, -Inf,Inf}, {z, -Inf, Inf}], 10^-7],
         {i, 1, 2}, {j, 1, 2}] // Quiet
```

and verify that

$$\tilde{h}_{p'q'} = \begin{bmatrix} \tilde{h}_{11} & \tilde{h}_{12} \\ \tilde{h}_{21} & \tilde{h}_{22} \end{bmatrix} = \begin{bmatrix} -1.18328 & 0 \\ 0 & -0.499382 \end{bmatrix}.$$

Similarly, the components $\tilde{h}_{pqrs}$ can be computed analytically with Mathematica. For example, the element $\tilde{h}_{1122} = \langle \psi_1\psi_1|\frac{1}{r_{12}}|\psi_2\psi_2\rangle$ is given by

```
RA = {0,0,0}; RB = {d,0,0}; Inf = Infinity;
g1[x_,y_,z_] = (2 a1/Pi)^(3/4) Exp[-a1 ({x,y,z} - RA).({x,y,z} - RA)];
g2[x_,y_,z_] = (2 a1/Pi)^(3/4) Exp[-a1 ({x,y,z} - RB).({x,y,z} - RB)];
g1g2 = Chop[Integrate[g1[x, y, z]*g2[x, y, z], {x, -Inf, Inf},
       {y,-Inf,Inf}, {z,-Inf,Inf}] /. {a1->0.416613,d->1.401},10^-7]
chi1[x_,y_,z_] = (g1[x, y, z] + g2[x, y, z])/(Sqrt[2] Sqrt[1 + a2]);
chi2[x_,y_,z_] = (g1[x, y, z] - g2[x, y, z])/(Sqrt[2] Sqrt[1 - a2]);
fun[x_, y_, z_]:= {chi1[x, y, z], chi2[x, y, z]};
J[p_,q_,r_,s_] := fun[x1, y1, z1][[p]]*fun[x2, y2, z2][[q]]*
                  fun[x1, y1, z1][[s]]*fun[x2, y2, z2][[r]]*
                  1/Sqrt[(x1-x2)^2+(y1-y2)^2+(z1-z2)^2]//FullSimplify
A1 = Assuming[a1 > 0, Integrate[J[1,1,2,2]/. {x2->x1+dx, y2->y1+dy,
     z2->z1+dz}, {x1,-Inf,Inf}, {y1,-Inf,Inf}, {z1,-Inf,Inf}]];
```

```
A2 = Assuming[r > 0, A1 /. {dx -> r Sin[\[Theta]] Cos[\[Phi]],
     dy->r Sin[\[Theta]] Sin[\[Phi]],dz->r Cos[\[Theta]]}//FullSimplify];
A3 = Assuming[a1>0 && d>0,Integrate[A2 r^2 Sin[\[Theta]], {r,0,Inf},
     {\[Theta], 0, \[Pi]}, {\[Phi], 0, 2 \[Pi]}]];
A3 /. {d -> 1.401, a1 -> 0.416613, a2 -> g1g2}
```

The elements $\tilde{h}_{pqrs}$ can be reshaped into a 4×4 matrix,

$$\tilde{h}_{p'q'r's'} = \begin{bmatrix} \tilde{h}_{1111} & \tilde{h}_{1112} & \tilde{h}_{1121} & \tilde{h}_{1122} \\ \tilde{h}_{1211} & \tilde{h}_{1212} & \tilde{h}_{1221} & \tilde{h}_{1222} \\ \tilde{h}_{2111} & \tilde{h}_{2112} & \tilde{h}_{2121} & \tilde{h}_{2122} \\ \tilde{h}_{2211} & \tilde{h}_{2112} & \tilde{h}_{2221} & \tilde{h}_{2222} \end{bmatrix} = \begin{bmatrix} 0.677391 & 0 & 0 & 0.141412 \\ 0 & 0.141412 & 0.586907 & 0 \\ 0 & 0.586907 & 0.141412 & 0 \\ 0.141412 & 0 & 0 & 0.578317 \end{bmatrix}.$$

Since most of the elements $\tilde{h}_{p'q'}$ and $\tilde{h}_{p'q'r's'}$ are zero, the Hamiltonian (12.103) can be expressed as

$$\begin{aligned} \hat{H} = \sum_{\sigma=\alpha}^{\beta} \left[\tilde{h}_{11} a_{1\sigma}^\dagger a_{1\sigma} + \tilde{h}_{22} a_{2\sigma}^\dagger a_{2\sigma}\right] + \frac{1}{2} \sum_{\sigma,\tau=\alpha}^{\beta} \Big[\tilde{h}_{1111} a_{1\sigma}^\dagger a_{1\tau}^\dagger a_{1\tau} a_{1\sigma} \\ + \tilde{h}_{2222} a_{2\sigma}^\dagger a_{2\tau}^\dagger a_{2\tau} a_{2\sigma} + \tilde{h}_{1221} a_{1\sigma}^\dagger a_{2\tau}^\dagger a_{2\tau} a_{1\sigma} \\ + \tilde{h}_{2112} a_{2\sigma}^\dagger a_{1\tau}^\dagger a_{1\tau} a_{2\sigma} + \tilde{h}_{1122} a_{1\sigma}^\dagger a_{1\tau}^\dagger a_{2\tau} a_{2\sigma} \\ + \tilde{h}_{1212} a_{1\sigma}^\dagger a_{2\tau}^\dagger a_{1\tau} a_{2\sigma} + \tilde{h}_{2121} a_{2\sigma}^\dagger a_{1\tau}^\dagger a_{2\tau} a_{1\sigma} + \tilde{h}_{2211} a_{2\sigma}^\dagger a_{2\tau}^\dagger a_{1\tau} a_{1\sigma}\Big]. \end{aligned}$$

After some lengthy calculations, one can verify that

$$\begin{aligned} \hat{H} =& \tilde{h}_{11} \left[a_1^\dagger a_1 + a_2^\dagger a_2\right] + \tilde{h}_{22} \left[a_3^\dagger a_3 + a_4^\dagger a_4\right] + \tilde{h}_{1111} \left[a_1^\dagger a_1 a_2^\dagger a_2\right] \\ &+ \tilde{h}_{2222} \left[a_3^\dagger a_3 a_4^\dagger a_4\right] + \tilde{h}_{1221} \left[a_1^\dagger a_1 a_3^\dagger a_3 + a_1^\dagger a_1 a_4^\dagger a_4 + a_2^\dagger a_2 a_3^\dagger a_3 + a_2^\dagger a_2 a_4^\dagger a_4\right] \\ &+ \tilde{h}_{1122} \left[-a_1^\dagger a_2^\dagger a_3 a_4 - a_1^\dagger a_1 a_3^\dagger a_3 + a_1^\dagger a_2 a_3 a_4^\dagger + a_1 a_2^\dagger a_3^\dagger a_4 - a_2^\dagger a_2 a_4^\dagger a_4 - a_1 a_2 a_3^\dagger a_4^\dagger\right], \end{aligned}$$

where we used the relations $\tilde{h}_{1122} = \tilde{h}_{1212} = \tilde{h}_{2121} = \tilde{h}_{2211}$, $\tilde{h}_{1221} = \tilde{h}_{2112}$, the anti-commutation relations (12.79) and we expressed the subscripts 1α, 1β, 2α, and 2β in the old notation: $1\alpha \leftrightarrow 1$, $1\beta \leftrightarrow 2$, $2\alpha \leftrightarrow 3$, and $2\beta \leftrightarrow 4$.

Solution 12.18 The Hamiltonian transforms the state $|0011\rangle$ into

$$\hat{H}|0011\rangle = (c_1 + 2c_2 - 2c_3 + c_4 - 2c_5 - 2c_6 + c_7)|0011\rangle + (2c_9 - 2c_8)|1100\rangle, \tag{12.104}$$

where $|1100\rangle$ is associated with the Slater determinant

$$|1100\rangle \quad \longleftrightarrow \quad \frac{1}{\sqrt{2!}} \det \begin{bmatrix} \chi_2(x_1) & \chi_2(x_2) \\ \chi_3(x_1) & \chi_3(x_2) \end{bmatrix}. \tag{12.105}$$

Similarly, $\hat{H}$ transforms $|1100\rangle$ into

$$\hat{H}|1100\rangle = (c_1 - 2c_2 + 2c_3 + c_4 - 2c_5 - 2c_6 + c_7)|1100\rangle + (2c_9 - 2c_8)|0011\rangle\,. \tag{12.106}$$

Therefore, the subspace span($|0011\rangle, |1100\rangle$) is closed under the action of $\hat{H}$. Within this subspace, the Hamiltonian is described by the matrix

$$H = \begin{pmatrix} E_0 & G \\ G & E_1 \end{pmatrix} \tag{12.107}$$

where

$$\begin{aligned} E_0 &= c_1 + 2c_2 - 2c_3 + c_4 - 2c_5 - 2c_6 + c_7, \\ G &= 2c_9 - 2c_8, \\ E_1 &= c_1 - 2c_2 + 2c_3 + c_4 - 2c_5 - 2c_6 + c_7. \end{aligned}$$

The eigenstate with minimum energy is a quantum state $|\psi\rangle = a|0011\rangle + b|1100\rangle$ where a and b are real coefficients. At the equilibrium geometry $d = 1.401\ a_0$, using the coefficients listed in Table 12.5, we obtain $a = -0.993993$ and $b = 0.109448$ and the ground-state energy is given by $E_{\rm g} = -0.990954\ E_{\rm h}$.

Solution 12.19 Let us calculate the expectation value of the energy $E(\theta) = \langle\psi(\theta)|\hat{H}|\psi(\theta)\rangle$ and minimize it as a function of the parameter θ. The VQE wavefunction is given by

$$|\psi(\theta)\rangle = \cos\theta|0011\rangle - \sin\theta|1100\rangle.$$

The expectation value of the energy is

$$E(\theta) = c_1 + c_4 - 2(c_5 + c_6) + c_7 + 2(c_2 - c_3)\cos(2\theta) + 2(c_8 - c_9)\sin(2\theta).$$

At the equilibrium geometry, using the coefficients listed in Table 12.5, we find that the minimum of $E(\theta)$ is

$$E(\theta_{\rm min}) = -0.990954\ E_{\rm h} \qquad \text{with } \theta_{\rm min} = (0.109668), \tag{12.108}$$

which is precisely the ground-state energy calculated in Exercise 12.18.

PART IV
QUANTUM ENGINEERING

13 Microwave resonators for superconducting devices

A microwave resonator in the quantum regime can be described by a quantum harmonic oscillator.

Solid-state quantum computers based on superconducting devices are composed of two main elements: superconducting qubits and microwave resonators. The qubits perform the actual computation, while the resonators are used to measure the state of the qubits [138]. We will first explain microwave resonators since their working principles are relatively simple to understand. Superconducting qubits and their interactions will be discussed in the next chapter.

13.1 Planar resonators

Microwave resonators are millimeter-scale electrical devices that confine electromagnetic energy in the microwave region of the spectrum. In the field of superconducting devices, their resonance frequency is typically in the range 4–10 GHz. The equivalent circuit of a microwave resonator comprises two fundamental components: an inductance and a capacitance. The first stores the magnetic energy of the circuit, whereas the second stores the electrical energy. When the resonator is excited by an electrical pulse, the energy of the system oscillates between electrical and magnetic following a periodic pattern. This resembles the behavior of an oscillator in which the kinetic energy is cyclically converted into potential energy as the mass swings back and forth.

Microwave resonators can be constructed with different geometries and shapes and can be grouped into several categories, as shown in Fig. 13.1. The two main categories are planar resonators and 3D cavities. **Planar resonators** consist of a thin superconducting layer patterned on a substrate (such as silicon or sapphire) with standard photolithographic techniques. Instead, **3D cavities** are metallic enclosures built with micromachining tools and tend to have higher quality factors with the disadvantage of being bulkier [461]. Compact versions of 3D cavities built within stacks of bonded silicon wafers have recently shown high performance at low temperatures and are promising options for multilayer microwave integration [462, 463, 464]. In

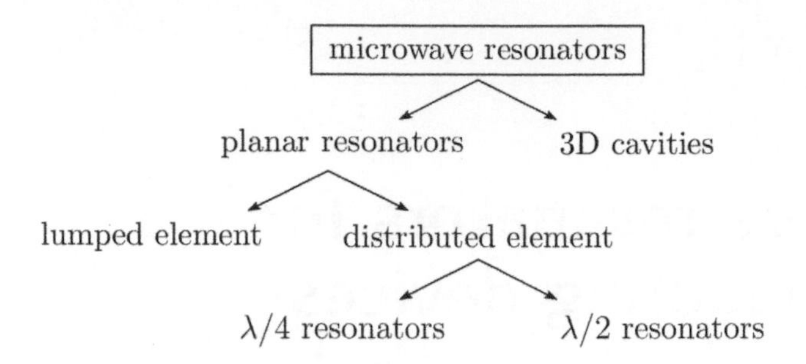

Fig. 13.1 Main categories of electromagnetic resonators for superconducting devices. Note that this diagram is not exhaustive.

this chapter, we will focus on planar resonators and refer the interested reader to Refs [465, 466, 467] for more information about 3D cavities.

Planar microwave resonators can be divided into lumped element and distributed element resonators. Some examples of **lumped element resonators** are shown in Fig. 13.2a–c. In these representations, the inductor is depicted in red, and the capacitor is in blue. These types of resonators are smaller than the electrical wavelength of the applied signal[1]. **Distributed element resonators** are microwave resonators whose size is comparable to the wavelength of the electrical signal. Figures 13.2d–e show an illustration of a $\lambda/4$ coplanar waveguide resonator (CPWR) [468], a prime example of a distributed element resonator. In these types of resonators, the electrical and magnetic energy are not concentrated in one specific part of the circuit, but rather distributed over the device. Lumped and distributed element resonators have similar performance and the choice of one or the other depends on the architecture and the purpose of the device.

As we shall see later in this chapter, the transmission coefficient of a microwave resonator can be measured by probing it with a microwave drive supplied by a signal generator. When the frequency of the signal is significantly different from the resonator frequency, most of the energy is either reflected or transmitted through the waveguide. On the other hand, when the drive frequency is close to resonance, part of the energy is absorbed by the resonator. Hence, the transmission coefficient as a function of frequency shows a pronounced peak at a specific frequency f_{r} called the **resonance frequency**. The width and depth of the resonance give valuable information about the performance of the device.

When a microwave resonator is well isolated from the environment and cooled to low temperatures such that $hf_{\mathrm{r}} \gg k_{\mathrm{B}}T$, the resonator is in its quantum ground state. This is called the **quantum regime**. Assuming that the frequency of the resonator is in the range $f_{\mathrm{r}} = 4 - 10$ GHz, the quantum regime can be reached by cooling the device at temperatures of the order of 10 mK. Nowadays, these temperatures can be routinely reached inside modern dilution refrigerators. As we shall explain at the end

[1]The wavelength of a microwave signal is $\lambda = v/f$ where f is the drive frequency and v is the phase velocity on the medium. For coplanar waveguides built on silicon $v \approx 1.2 \cdot 10^8$ m/s at low temperatures. Thus, a 6 GHz microwave signal has a wavelength $\lambda \approx 2$ cm.

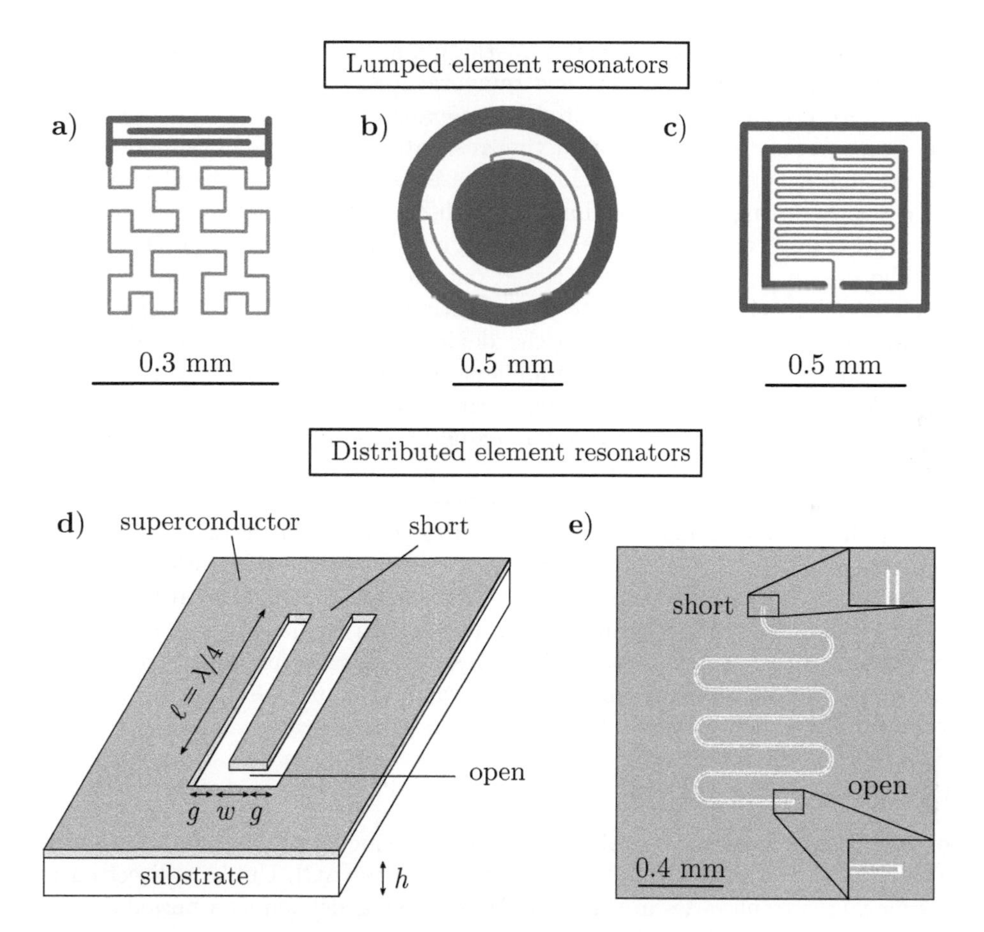

Fig. 13.2 Lumped and distributed element resonators. a) Top view of a lumped element resonator with an interdigitated capacitor (blue) connected to a thin meander-type inductor (red). This type of resonator is usually used in microwave kinetic inductance detectors for particle detection [469]. **b)** Top view of a lumped element resonator with concentric geometry. This resonator was used in Ref. [470] to read out the state of a superconducting qubit in a double-sided coaxial architecture. **c)** Top view of a lumped element resonator with a thin meander line surrounded by a capacitor. This resonator was used in Ref. [471] as a readout component in a multi-qubit superconducting device. **d)** A representation of a short-circuited $\lambda/4$ CPWR consisting of a thin superconducting layer (gray) on top of a substrate (white). The resonator is open on one end and shorted on the opposite end. This representation is not to scale: the substrate is usually $h \approx 0.5$ mm thick and the thickness of the superconducting layer varies between $100 - 300$ nm. The width of the center conductor is around $w \approx 10$ μm and the gap g is of the same order. The resonance frequency f_r is given by $f_r = v/4\ell$ where v is the phase velocity of the waveguide. **e)** A scale representation of a short-circuited $\lambda/4$ CPWR. The resonator length ℓ is the length of the winding path from the short to the open. A $\lambda/4$ CPWR with a resonance frequency of $f_r = 7$ GHz has a length of $\ell = v/4f_r = 4.4$ mm where we assumed $v = 1.2 \cdot 10^8$m/s for silicon.

of this chapter, a microwave resonator in the quantum regime can be described as a quantum harmonic oscillator. At low temperatures, the resonator remains in the ground state if not intentionally excited by an external drive.

The measurement of a superconducting resonator at millikelvin temperatures follows a systematic procedure [472]. In most cases, the device is affixed to a printed circuit board featuring some sub-miniature push-on connectors on the perimeter. The on-chip waveguides are wire-bonded to pads on the circuit board with aluminum wires. The circuit board is finally screwed onto a sample holder, which is mechanically anchored to the bottom plate of a dilution refrigerator. The room temperature instrumentation is connected to the device using coaxial cables with suitable microwave filters, attenuators, and amplifiers as illustrated in Fig. 13.3. The fridge usually takes 24–36 hours to reach a temperature of ≈ 10 mK. For more information about cryogenic measurements of superconducting devices, we refer the interested reader to Refs [473, 474, 475, 476, 477].

Planar resonators for superconducting devices are usually fabricated with thin layers of aluminum or niobium, which have a critical temperature of 1.2 K and 9 K, respectively. At 0.01 K, these metals are superconducting and dissipations are significantly reduced. One might wonder why these devices are cooled down at temperatures much lower than the critical temperature. The reason is that if the temperature of the environment is too hot, $T = 0.5\text{ K} > hf_{\text{r}}/k_{\text{B}}$, thermal photons would excite a resonator with an operating frequency in the range $f_{\text{r}} = 6$–7 GHz. This would make it more difficult to initialize the resonator into its ground state.

13.1.1 Alternating currents

Superconducting devices are probed with radio-frequency signals and the currents circulating in these circuits are **alternating currents** (AC). Unlike a direct current, an alternating current varies in time. An AC current is defined as a periodic current, $I(t) = I(t+T)$, whose time integral over a period is equal to zero,

$$\frac{1}{T}\int_t^{t+T} I(t')dt' = 0,$$

where T is the period and $f = \omega/2\pi = 1/T$ is the frequency of the signal. A sinusoidal current $I(t) = I_0\cos(\omega t + \phi)$ is an example of an alternating current. To simplify the calculations, it is often convenient to express a sinusoidal current in complex form,

$$I_0\cos(\omega t + \phi) \qquad \leftrightarrow \qquad Ie^{i\omega t},$$

where $I = I_0 e^{i\phi}$ is a complex amplitude. By taking the real part of the expression on the right-hand side, we obtain the expression on the left-hand side.

The basic elements of an electrical circuit are resistors, capacitors, and inductors. When a microwave source applies a sinusoidal current $I(t) = Ie^{i\omega t}$ to a **resistor** R, a voltage develops across this element according to Ohm's law,

$$V_R(t) = RI(t) = \underbrace{RI}_{V} e^{i\omega t} = Ve^{i\omega t}. \tag{13.1}$$

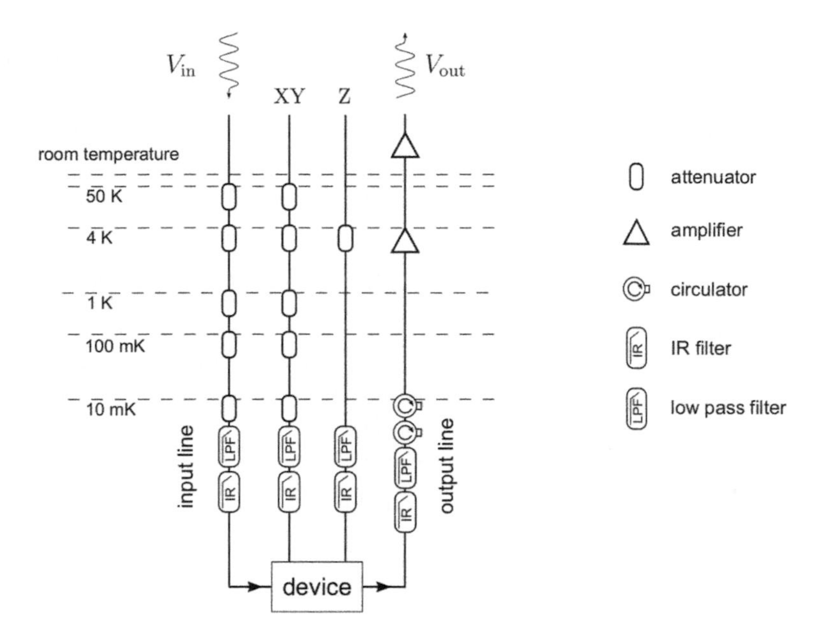

Fig. 13.3 Schematic of a dilution refrigerator. The measurement of the transmission coefficient of a microwave resonator at cryogenic temperatures requires a sophisticated experimental apparatus. The signal V_{in} produced by a room temperature signal generator is attenuated and filtered before reaching the microwave resonator installed at the bottom of the dilution refrigerator. The signal leaving the device proceeds along the output line. After a sequence of filters and amplifiers, the signal reaches the room temperature instrumentation. The transmission coefficient $V_{\text{out}}/V_{\text{in}}$ is finally acquired.

The voltage varies in phase with the current. The impedance of a resistor is defined as the ratio

$$\boxed{Z_R = \frac{V}{I} = R.} \tag{13.2}$$

Thus, the impedance of a resistor is a real quantity and, most importantly, does not depend on frequency.

An inductor is an electrical component that stores energy in the form of a magnetic field. The **inductance** L quantifies the ratio between the magnetic flux threading a closed circuit and the current flowing through that circuit[2], $L = \Phi/I$. If the current through the inductor varies in a sinusoidal way $I(t) = Ie^{i\omega t}$, the voltage across this element is

$$V_L(t) = L\frac{dI}{dt} = \underbrace{i\omega LI}_{V}\, e^{i\omega t} = Ve^{i\omega t}.$$

[2]The magnetic flux is defined as $\Phi = \int_\Sigma \vec{B} \cdot d\vec{\Sigma}$ where Σ is any surface enclosed by a closed loop and $\vec{B}$ is the magnetic field threading that loop.

	Applied current: $I_0\cos(\omega t+\phi)=\mathrm{Re}[Ie^{i\omega t}]$	Impedance
R	$V_R(t)=RI_0\cos(\omega t+\phi)=\mathrm{Re}[RIe^{i\omega t}]$	$Z_R=R$
L	$V_L(t)=\omega LI_0\cos(\omega t+\phi+\frac{\pi}{2})=\mathrm{Re}[i\omega LIe^{i\omega t}]$	$Z_L=i\omega L$
C	$V_C(t)=\dfrac{I_0}{\omega C}\cos(\omega t+\phi-\frac{\pi}{2})=\mathrm{Re}[Ie^{i\omega t}/i\omega C]$	$Z_C=\dfrac{1}{i\omega C}$

Fig. 13.4 Summary of the relations between currents and voltages for resistances, inductances, and capacitances. The column on the right lists the impedances of these electrical components. The quantity I is defined as $I=I_0e^{i\phi}$.

The impedance is given by the ratio between the complex amplitudes associated with the voltage and the current,

$$\boxed{Z_L=\frac{V}{I}=i\omega L.} \tag{13.3}$$

The impedance of an ideal inductor is purely imaginary and increases linearly with frequency.

A capacitor is an electrical component that stores electrical energy in the form of an electric field. The **capacitance** C is the ratio between the charge accumulated on one capacitor plate and the voltage across it, $C=Q/V$. When a sinusoidal current $I(t)=Ie^{i\omega t}$ is applied to a capacitor, the voltage across this element is given by

$$V_C(t)=\frac{1}{C}\int I(t)dt=\underbrace{\frac{I}{i\omega C}}_{V}e^{i\omega t}=Ve^{i\omega t}.$$

Therefore, the impedance of a capacitor is

$$\boxed{Z_C=\frac{V}{I}=\frac{1}{i\omega C}.} \tag{13.4}$$

Unlike a resistor, the impedance of a capacitor is purely imaginary and is inversely proportional to the frequency of the applied signal. The relations between currents and voltages for the electric elements discussed in this section are summarized in Fig. 13.4.

The impedance of n electrical components connected in series is given by the sum of the impedances of each component, $Z_{\mathrm{tot}}=Z_1+\ldots+Z_n$. In contrast, the impedance of n electrical components connected in parallel is given by $Z_{\mathrm{tot}}=(1/Z_1+\ldots+1/Z_n)^{-1}$. We now have all the ingredients to present RLC resonators.

13.2 Parallel resonant circuit

The frequency response of a microwave resonator close to resonance can be modeled with an **RLC circuit** comprising a resistance, a capacitance and an inductance. In a parallel RLC resonator, these electrical components are connected as shown in Fig. 13.5a. Suppose a microwave generator (not shown in the figure) applies an oscillating voltage to the circuit. The voltage across each component is the same and has the sinusoidal behavior $V_0 \cos(\omega t + \phi)$. The average power dissipated by the resistor is

$$\bar{P}_R = \frac{1}{2}\frac{V_0^2}{R},$$

while the average power dissipated by the capacitor and inductor are both zero (see Exercise 13.1). The average electrical and magnetic energy, $\bar{E}_\mathrm{C}$ and $\bar{E}_\mathrm{L}$, stored in the capacitance and inductance are (see Exercise 13.2)

$$\bar{E}_\mathrm{C} = \frac{1}{4}CV_0^2, \qquad \bar{E}_\mathrm{L} = \frac{1}{4}\frac{V_0^2}{\omega^2 L}.$$

The **resonance frequency** f_r is defined as the frequency at which the electric energy $\bar{E}_\mathrm{C}$ is the same as the magnetic energy $\bar{E}_\mathrm{L}$. Therefore,

$$\boxed{\omega_\mathrm{r} = \frac{1}{\sqrt{LC}} = 2\pi f_\mathrm{r},} \tag{13.5}$$

where ω_r has units of rad/s and f_r is in Hz. The **internal quality factor** Q_i is a dimensionless quantity defined as the ratio between the average energy stored in the resonator and the power loss per cycle

$$\boxed{Q_\mathrm{i} \equiv \omega_\mathrm{r}\frac{\bar{E}_\mathrm{C} + \bar{E}_\mathrm{L}}{\bar{P}_R} = \omega_\mathrm{r} RC = \frac{R}{\omega_\mathrm{r} L}.} \tag{13.6}$$

The higher the internal quality factor, the higher the performance of the resonator. It should not be surprising that the internal quality factor of a parallel RLC circuit is proportional to the resistance; indeed, the power loss is significantly reduced for $R \gg 1$ since most of the current flows through the capacitance and inductance.

The internal quality factor is a dimensionless quantity that describes the intrinsic performance of the resonator. In the field of superconducting devices, Q_i is usually in the range 10^5–10^6 for planar resonators and is dictated by multiple factors that the experimenter cannot precisely control. The most relevant ones are the parameters of the fabrication process [478, 479, 480, 481, 482], the participation ratio of the electric field [483, 484], the presence of two-level systems in the interfaces [485, 486, 487], and the constant bombardment of atmospheric muons [488]. In the quantum regime, the internal quality factor of planar resonators also depends on the power of the applied signal, see Section 13.5.

Let us derive the impedance of a parallel RLC resonator. From eqns (13.2, 13.3, 13.4), the impedance of a parallel resonant circuit as a function of frequency can be

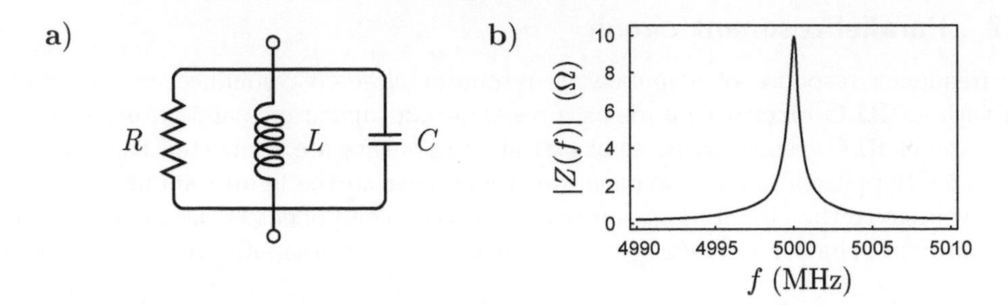

Fig. 13.5 Parallel RLC resonator. a) Electrical circuit of a parallel RLC resonator. **b)** Magnitude of the impedance of a parallel RLC circuit close to resonance. The solid line is a plot of the absolute value of eqn (13.8) with $R = 10\ \Omega$, $Q_\mathrm{i} = 10000$ and $f_\mathrm{r} = 5$ GHz. At resonance, $\Delta\omega = 0$ and $Z(f) = R$.

expressed as

$$Z_\mathrm{r}(\omega) = \left[\frac{1}{R} + \frac{1}{i\omega L} + i\omega C\right]^{-1}. \tag{13.7}$$

We are interested in the frequency response close to resonance, i.e. when ω is similar to ω_r. It is convenient to introduce the **detuning parameter** $\Delta\omega$ defined as $\omega = \omega_\mathrm{r} + \Delta\omega$. Substituting this expression into (13.7), we obtain

$$\begin{aligned} Z_\mathrm{r}(\omega) &= \left[\frac{1}{R} + \frac{1}{i\omega_\mathrm{r} L}\left(1 + \frac{\Delta\omega}{\omega_\mathrm{r}}\right)^{-1} + i\omega_\mathrm{r} C + i\Delta\omega C\right]^{-1} \\ &\approx \left[\frac{1}{R} + \frac{1}{i\omega_\mathrm{r} L}\left(1 - \frac{\Delta\omega}{\omega_\mathrm{r}}\right) + i\omega_\mathrm{r} C + i\Delta\omega C\right]^{-1}, \end{aligned}$$

where in the last step we used the approximation $\Delta\omega/\omega_\mathrm{r} \ll 1$. Since $\omega_\mathrm{r} = 1/\sqrt{LC}$, we can write the impedance of the resonator as

$$\begin{aligned} Z_\mathrm{r}(\omega) &\approx \left[\frac{1}{R} + i\frac{\Delta\omega}{\omega_\mathrm{r}^2 L} + i\Delta\omega C\right]^{-1} = \left[\frac{1}{R} + i2\Delta\omega C\right]^{-1} \\ &= \frac{R}{1 + i2\Delta\omega RC} = \frac{R}{1 + i2Q_\mathrm{i}\frac{\Delta\omega}{\omega_\mathrm{r}}}, \end{aligned} \tag{13.8}$$

where in the last step we used (13.6). The magnitude of the impedance close to resonance is shown in Fig. 13.5b. Equation (13.8) will be important later on in this chapter when we introduce short-circuited $\lambda/4$ resonators.

We devote the following section to the discussion of transmission lines. In particular, we will focus on coplanar waveguides since these are the most common transmission lines employed in superconducting devices. This treatment will lead to the presentation of coplanar waveguide resonators.

13.3 Transmission lines

The most common transmission lines encountered in a scientific laboratory are **coaxial cables**. These cables comprise an inner cylindrical conductor surrounded by a concentric metallic shield with a dielectric layer in between, as shown in Fig. 13.6a–b. The outer conductor is often covered with flexible plastic and its outer diameter ranges between 0.2–5 mm. Coaxial cables can be used to transmit microwave signals up to 20 GHz between two points of an electrical network.

Another interesting example of a transmission line is a **coplanar waveguide**. This transmission line can be imagined as the longitudinal section of a coaxial cable patterned on a substrate, as illustrated in Fig. 13.6c–d. The width of an on-chip coplanar waveguide is usually $w + 2g \approx 20$ μm. These types of waveguides are often employed in superconducting devices to deliver electrical pulses to the qubits and the resonators. The losses of coaxial cables and coplanar waveguides slightly increase as a function of frequency due to a phenomenon known as the **skin effect**. This loss mechanism is strictly related to the metallic conductors of the transmission line.

A schematic model of an ideal **transmission line** is typically represented with two parallel metallic wires that extend to infinity. Unlike lumped element components, the voltage and current might vary significantly along the line. Nevertheless, transmission lines can still be described with an equivalent lumped element model. The idea behind this model is to divide the line into small sections of length Δx as shown in Fig. 13.7. Each section comprises a model of the waveguide characterized by a resistance R, an inductance L, a conductance G, and a capacitance C per unit length. The resistance R represents the Ohmic losses of the two conductors in Ω/m (for a superconductor, the resistance per unit length is zero). The inductance L describes the self-inductance of the two wires in H/m. The shunt capacitance C takes into account that the two conductors are usually close to one another and this leads to a finite capacitance per unit length in F/m. Lastly, the conductance G models the resistive losses of the dielectric between the two conductors in S/m. We will assume that R, L, G, and C do not depend on x; this is often a good approximation for on-chip waveguides.

In a section Δx of the transmission line much smaller than the electrical wavelength, the voltage and current at the coordinates x and $x + \Delta x$ satisfy the relations[3]:

$$V(x + \Delta x, t) = V(x, t) - R\Delta x I(x, t) - L\Delta x \frac{\partial I(x, t)}{\partial t},$$
$$I(x + \Delta x, t) = I(x, t) - G\Delta x V(x + \Delta x, t) - C\Delta x \frac{\partial V(x + \Delta x, t)}{\partial t}.$$

If we rearrange these terms, divide by Δx and take the limit $\Delta x \to 0^+$, we obtain:

$$\frac{\partial V(x, t)}{\partial x} = -RI(x, t) - L\frac{\partial I(x, t)}{\partial t}, \tag{13.9}$$
$$\frac{\partial I(x, t)}{\partial x} = -GV(x, t) - C\frac{\partial V(x, t)}{\partial t}. \tag{13.10}$$

[3]These relations can be derived from Kirchhoff's circuit laws.

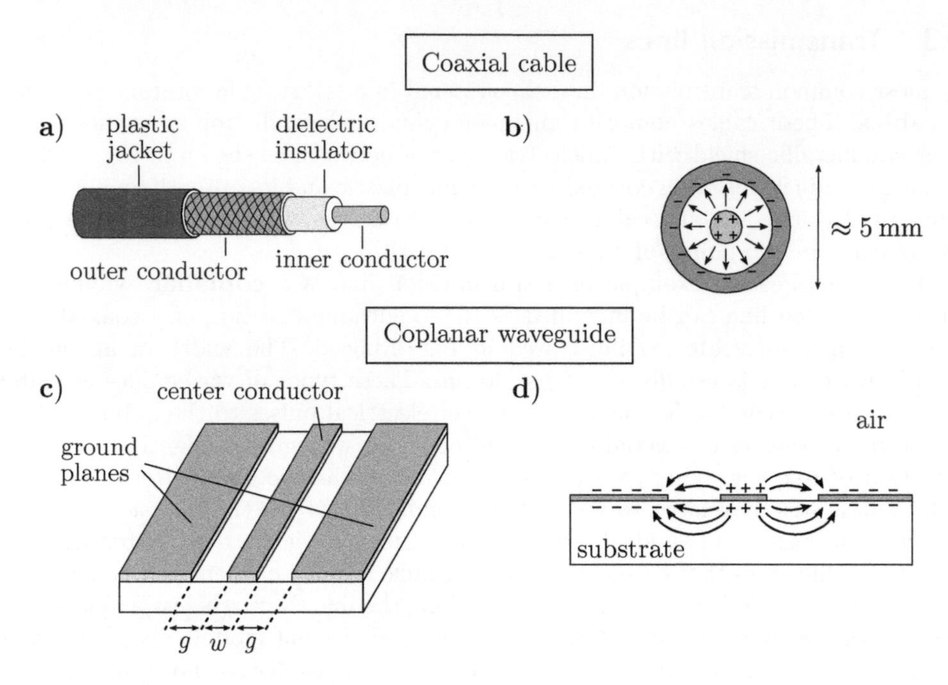

Fig. 13.6 Transmission lines. a) A representation of a coaxial cable. **b)** The cross-section of a coaxial cable and the distribution of the oscillating electric field at a specific moment in time. **c)** A section of a coplanar waveguide. The width of the center conductor is denoted as w, and the gap between the center conductor and the ground planes is indicated as g. The typical width of on-chip waveguides is $w + 2g = 20\ \mu m$. **d)** Cross section of a coplanar waveguide and the electric field distribution at a specific moment in time. The electric field is concentrated in the gap between the center conductor and the ground planes.

These are called the **telegrapher's equations**. We now assume that the currents and voltages vary sinusoidally with time, i.e. $I(x,t) = I(x)e^{i\omega t}$, and $V(x,t) = V(x)e^{i\omega t}$. Substituting these expressions into (13.9, 13.10) leads to:

$$\frac{\partial V(x)}{\partial x} = -(R + i\omega L)I(x), \tag{13.11}$$

$$\frac{\partial I(x)}{\partial x} = -(G + i\omega C)V(x). \tag{13.12}$$

The solutions of this system of differential equations are **traveling waves** of the form:

$$V(x) = V^+ e^{-\gamma x} + V^- e^{\gamma x}, \tag{13.13}$$

$$I(x) = I^+ e^{-\gamma x} + I^- e^{\gamma x}. \tag{13.14}$$

The terms proportional to $e^{-\gamma x}$ ($e^{\gamma x}$) correspond to waves propagating in the $+x$ ($-x$) direction. The complex amplitudes $V^{\pm}$ and $I^{\pm}$ are determined by the boundary

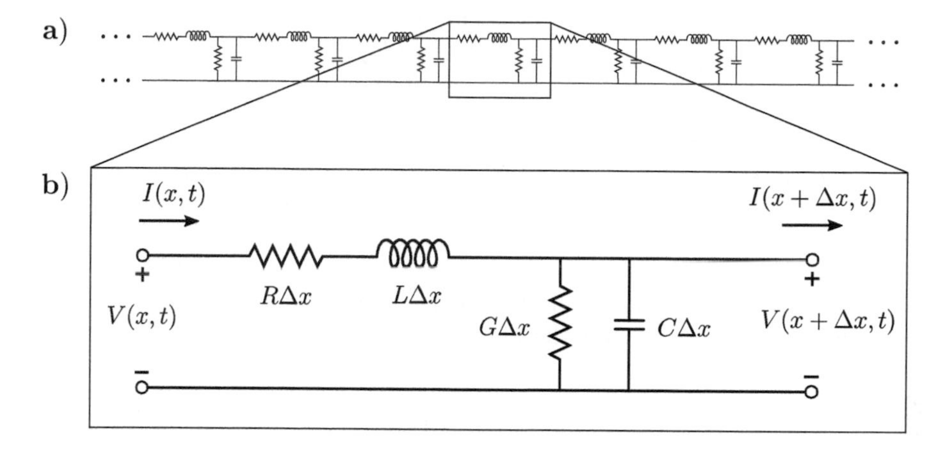

Fig. 13.7 Lumped element model of a transmission line. a) Model of a transmission line, such as a coaxial cable or a coplanar waveguide, in terms of lumped element components. **b)** A close-up of a small section Δx of a transmission line.

conditions on the two ends of the line. The parameter γ is the **propagation constant**, defined as

$$\boxed{\gamma = \sqrt{(R + i\omega L)(G + i\omega C)} = \alpha + ik.}$$

The real part of the propagation constant α describes the **losses** of the line. The imaginary part is called the **wavenumber** and is inversely proportional to the wavelength of the signal, $k = 2\pi/\lambda$. For a lossless transmission line $R = G = 0$ and the propagation constant is purely imaginary.

A waveguide is characterized by two main parameters: the propagation constant and the characteristic impedance. The **characteristic impedance** is given by the ratio between the amplitudes of the voltage and the current along the line. To derive the analytical expression of this parameter, first compute the spatial derivative of (13.13) and combine the result with (13.11),

$$\frac{\partial V(x)}{\partial x} = -\gamma V^+ e^{-\gamma x} + \gamma V^- e^{\gamma x} = -(R + i\omega L)I(x).$$

Rearranging this equation, we obtain the current along the transmission line in terms of the voltage,

$$I(x) = \underbrace{\frac{\gamma V^+}{(R + i\omega L)}}_{I^+} e^{-\gamma x} + \underbrace{\frac{-\gamma V^-}{(R + i\omega L)}}_{I^-} e^{\gamma x}.$$

The **characteristic impedance of the line** Z_0 is defined as the ratio between the amplitudes of the voltage and the current[4]

[4]Note that Z_0 is a constant because we assumed that R, G, L, and C do not depend on the position.

$$\boxed{Z_0 = \frac{V^+}{I^+} = \frac{R + i\omega L}{\gamma} = \sqrt{\frac{R + i\omega L}{G + i\omega C}}.}$$

In general, this is a complex quantity. However, when the losses are negligible, $R = G = 0$ and therefore $Z_0 = \sqrt{L/C}$. Most commercial coaxial cables have a characteristic impedance of 50 Ω. The reasoning behind this choice is that the attenuation of a coaxial line filled with air is minimized when $Z_0 = 77$ Ω, but its power capacity is maximized when $Z_0 = 30$ Ω [465]. A characteristic impedance of 50 Ω is a good compromise between these contrasting optimal points. Since by definition $I^\pm = V^\pm/Z_0$, the equations for the voltage and the current along the line (13.13, 13.14) can be written as

$$V(x) = V^+ e^{-\gamma x} + V^- e^{\gamma x}, \tag{13.15}$$

$$I(x) = \frac{V^+}{Z_0} e^{-\gamma x} - \frac{V^-}{Z_0} e^{\gamma x}. \tag{13.16}$$

These two equations will be the starting point for our discussion of transmission line resonators.

Example 13.1 As shown in Fig. 13.6c, the main geometric parameters of a coplanar waveguide are the width w of the center conductor and the gap g between the center conductor and the ground planes. Assuming that w and g are much smaller than the thickness of the substrate, it can be shown that the **characteristic impedance of a coplanar waveguide** is given by [489]

$$Z_0 = \frac{30\pi}{\sqrt{(1 + \varepsilon_r)/2}} \frac{K(b)}{K(a)},$$

where ε_r is the relative permittivity of the substrate, $K(x)$ is the complete elliptic integral of the first kind and

$$a = \frac{w}{w + 2g}, \qquad b = \sqrt{1 - a^2}.$$

The characteristic impedance of a superconducting coplanar waveguide on silicon with relative permittivity $\varepsilon_r = 11.5$, width $w = 10$ μm, and gap $g = 5$ μm is approximately 50 Ω. This impedance is commonly chosen to minimize any mismatch with other 50 Ω waveguides in the network. A coplanar waveguide with these dimensions can be patterned with standard photolithographic techniques. The impedance of a coplanar waveguide can be simulated with commercial software, such as TX-LINE from Cadence.

13.3.1 Terminated lossy line

In this section, we study the frequency response of a transmission line terminated with an arbitrary load, such as a resistor. We will discover that when a microwave signal travels along the line and reaches the load, it might get reflected back depending on the impedance of the load.

A microwave generator produces a continuous wave propagating in the $+x$ direction and an electrical component ("the load") with impedance Z_L is positioned at $x = 0$ as shown in Fig. 13.8a. This boundary condition requires that the impedance of the line must be Z_L at the origin

$$\frac{V(0)}{I(0)} = \frac{V^+ + V^-}{V^+ - V^-} Z_0 = Z_\mathrm{L},$$

where we used eqns (13.15–13.16). Rearranging this expression, we can introduce the **reflection coefficient** Γ as the ratio between the amplitude of the reflected and incident wave[5]

$$\boxed{\Gamma = \frac{V^-}{V^+} = \frac{Z_\mathrm{L} - Z_0}{Z_\mathrm{L} + Z_0}.} \tag{13.17}$$

Intuitively, the absolute value of the reflection coefficient quantifies how much of the incoming wave is reflected. When $Z_\mathrm{L} \neq Z_0$, the reflection coefficient is not zero and a reflected wave propagates in the $-x$ direction. Substituting (13.17) into (13.15, 13.16), we find the voltage and current along a terminated line:

$$V(x) = V^+(e^{-\gamma x} + \Gamma e^{\gamma x}), \tag{13.18}$$

$$I(x) = \frac{V^+}{Z_0}(e^{-\gamma x} - \Gamma e^{\gamma x}). \tag{13.19}$$

Thus, the voltage and current are a linear superposition of two waves propagating in opposite directions. Three interesting cases can be distinguished:

1. When the waveguide is connected to a **short**, $Z_\mathrm{L} = 0$ and the reflection coefficient is $\Gamma = -1$. The reflected wave has an opposite phase to the incident wave.
2. When the load is **matched** with the line, $Z_\mathrm{L} = Z_0$, the reflection coefficient is zero (there is no reflected wave). The power of the signal is entirely transferred to the load. This happens, for example, when a commercial coaxial cable is connected to a 50 Ω resistor.
3. When the end of the line is left open, $Z_\mathrm{L} = +\infty$ and $\Gamma = +1$. The reflected wave has the same phase as the incident wave and no power is absorbed by the load.

When a transmission line is connected to a short (or an open) and a continuous wave propagates along the line, standing waves are formed by the superposition of two waves propagating in opposite directions.

[5] In optics, a light beam propagating in a medium with refractive index n_1 is partly reflected when it impinges against a medium with refractive index $n_2 \neq n_1$. For normal incidence, the reflectivity is given by $R = |n_1 - n_2|^2/|n_1 + n_2|^2$. Note the similarity between R and the reflection coefficient Γ.

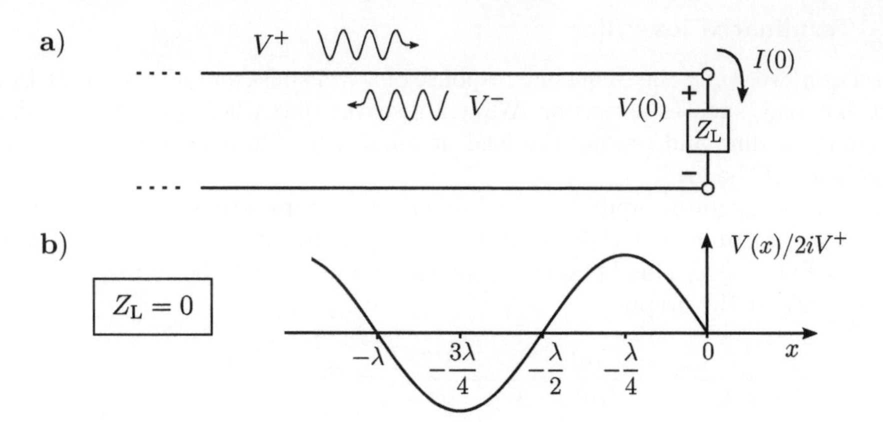

Fig. 13.8 Terminated transmission line. a) A portion of a transmission line terminated with an arbitrary load Z_L. The incident wave is generated by a source on the far left (not shown in the figure). **b)** Plot of the voltage of a transmission line terminated with a short at $x = 0$. The superposition of the incident and reflected waves generates standing waves on the line. The voltage is equal to zero at $x = -n\lambda/2$ and reaches its extrema points at $x = -(2n+1)\lambda/4$ where $n \in \mathbb{Z}_{\geq 0}$. These points are called nodes and antinodes, respectively.

It is instructive to calculate the impedance at each point of a terminated transmission line

$$Z(x) = \frac{V(x)}{I(x)} = Z_0 \frac{V_0^+(e^{-\gamma x} + \Gamma e^{\gamma x})}{V_0^+(e^{-\gamma x} - \Gamma e^{\gamma x})} = \frac{1 + \Gamma e^{2\gamma x}}{1 - \Gamma e^{2\gamma x}} Z_0. \tag{13.20}$$

Recall that the reflection coefficient is $\Gamma = (Z_L - Z_0)/(Z_L + Z_0)$. Substituting this formula into (13.20), we arrive at

$$\begin{aligned} Z(x) &= \frac{(Z_L + Z_0) + (Z_L - Z_0)e^{2\gamma x}}{(Z_L + Z_0) - (Z_L - Z_0)e^{2\gamma x}} Z_0 \\ &= \frac{Z_L(1 + e^{2\gamma x}) - Z_0(1 - e^{2\gamma x})}{Z_0(1 + e^{2\gamma x}) - Z_L(1 - e^{2\gamma x})} Z_0 \\ &= \frac{Z_L - Z_0 \tanh(\gamma x)}{Z_0 - Z_L \tanh(\gamma x)} Z_0. \end{aligned}$$

We are interested in the impedance of the line at a distance ℓ from the load,

$$\boxed{Z(-\ell) = \frac{Z_L + Z_0 \tanh(\gamma \ell)}{Z_0 + Z_L \tanh(\gamma \ell)} Z_0.} \tag{13.21}$$

This is the impedance of a lossy line terminated with an arbitrary load as a function of the distance from the load. This expression will be very important for deriving the impedance of transmission line resonators in the next section.

Example 13.2 Let us calculate the voltage and current along a lossless transmission line terminated with a short. This corresponds to connecting the center pin of a coaxial cable to ground or short-circuiting the end of a coplanar waveguide. For a short-circuited transmission line, $Z_\mathrm{L} = 0$, $\Gamma = -1$ and eqns (13.18, 13.19) reduce to

$$V(x) = V^+(e^{-\gamma x} - e^{\gamma x}) = -2iV^+ \sin(kx),$$
$$I(x) = \frac{V^+}{Z_0}(e^{-\gamma x} + e^{\gamma x}) = \frac{2V^+}{Z_0}\cos(kx),$$

where in the last step we assumed that the line is lossless, i.e. the propagation constant is purely imaginary $\gamma = ik$ and $Z_0 = \sqrt{L/C}$. As expected, the voltage at the short is zero and the current is maximum. The plot of the voltage is illustrated in Fig. 13.8b.

13.3.2 Transmission line resonators

Transmission line resonators consist of a finite section of a transmission line terminated with a short or an open. In superconducting devices, two types of transmission line resonators are often employed: short-circuited quarter-wave CPWRs and open-circuited half-wave CPWRs. The former has similar performance to the latter, but it has the advantage of being much more compact. Hence, we will focus our attention on quarter-wave resonators.

A **short-circuited quarter-wave resonator** (or $\lambda/4$ resonator) is a transmission line connected to a short whose length is a quarter of the electrical wavelength. A prime example is the short-circuited $\lambda/4$ CPWR illustrated in Fig. 13.2d–e and in Fig. 13.9a. Let us show that these distributed element resonators support multiple resonant modes that can be modeled with a parallel RLC circuit close to resonance [465]. From (13.21) the impedance of a short-circuited line at a distance ℓ from the load is given by

$$Z(-\ell) = \frac{Z_\mathrm{L} + Z_0 \tanh(\gamma\ell)}{Z_0 + Z_\mathrm{L}\tanh(\gamma\ell)} Z_0 = Z_0 \tanh(\gamma\ell), \tag{13.22}$$

where in the last step we set $Z_\mathrm{L} = 0$, since the end of the line is shorted to ground. If we now substitute the propagation constant $\gamma = \alpha + ik$ into (13.22), we obtain[6]

$$\begin{aligned} Z(-\ell) &= Z_0 \tanh(\alpha\ell + ik\ell) \\ &= Z_0 \frac{\tanh(\alpha\ell)\cos(k\ell) + i\sin(k\ell)}{\cos(k\ell) + i\tanh(\alpha\ell)\sin(k\ell)} = Z_0 \frac{1 - i\tanh(\alpha\ell)\cot(k\ell)}{\tanh(\alpha\ell) - i\cot(k\ell)}, \end{aligned}$$

where in the last step we divided both the numerator and the denominator by $i \sin k\ell$. In practical cases, the losses of a superconducting transmission line are very small, $\alpha \ll 1$. Hence, $\tanh(\alpha\ell) \approx \alpha\ell$ and we can make the approximation

$$Z(-\ell) = Z_0 \frac{1 - i\alpha\ell\cot(k\ell)}{\alpha\ell - i\cot(k\ell)}.$$

[6] Recall that $\tanh(x+iy) = (\tanh x \cos y + i \sin y)/(\cos y + i \tanh x \sin y)$ and $\cot x = \cos x/\sin x$.

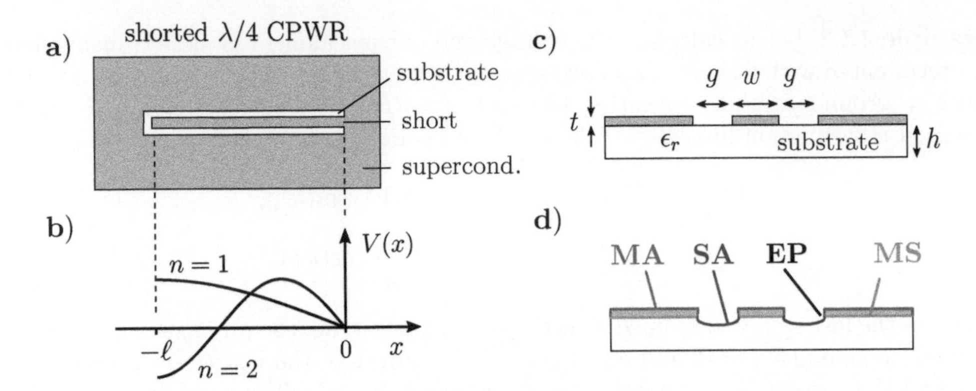

Fig. 13.9 Short-circuited $\lambda/4$ coplanar waveguide resonator. a) Top view representation of a short-circuited $\lambda/4$ CPWR. This schematic is not to scale. **b)** Voltage distribution of the first and second modes of a $\lambda/4$ CPWR. **c)** Cross section of a CPWR resonator. In superconducting devices, $h = 0.3 - 0.7$ mm, and $t = 50 - 300$ nm. The parameters g and w are in the range 5–50 μm. The substrate is usually silicon or sapphire. **d)** A representation of the main interfaces of a coplanar waveguide: the metal-air (MA), the substrate-air (SA), and the metal-substrate (MS) interface. Engineering these interfaces and the etching profile (EP) is crucial to increase the internal quality factor of CPWRs.

Recall that the wavenumber is inversely proportional to the wavelength of the applied signal, $k = 2\pi/\lambda$, and that the length of the line is a fixed quantity ℓ which defines a characteristic wavelength $\lambda_\mathrm{r} \equiv 4\ell$. We are interested in the impedance close to resonance, i.e. when $\omega = \omega_\mathrm{r} + \Delta\omega$ where $\Delta\omega$ is a small detuning parameter. Using these relations, the product $k\ell$ is given by[7]

$$k\ell = \frac{\omega}{v}\ell = (\omega_\mathrm{r} + \Delta\omega)\frac{\ell}{v} = (\omega_\mathrm{r} + \Delta\omega)\frac{\lambda_\mathrm{r}}{4v} = \frac{\pi}{2} + \frac{\pi\Delta\omega}{2\omega_\mathrm{r}},$$

and therefore[8]

$$\begin{aligned} Z(-\ell) &= Z_0 \frac{1 - i\alpha\ell \cot\left(\frac{\pi}{2} + \frac{\pi\Delta\omega}{2\omega_\mathrm{r}}\right)}{\alpha\ell - i\cot\left(\frac{\pi}{2} + \frac{\pi\Delta\omega}{2\omega_\mathrm{r}}\right)} \\ &\approx Z_0 \frac{1 + i\alpha\ell\, \pi\Delta\omega/2\omega_\mathrm{r}}{\alpha\ell + i\,\pi\Delta\omega/2\omega_\mathrm{r}} \approx \frac{Z_0}{\alpha\ell + i\pi\frac{\Delta\omega}{2\omega_\mathrm{r}}}, \end{aligned} \tag{13.23}$$

where in the last step we dropped a term in the numerator proportional to $\alpha\ell\Delta\omega/\omega_\mathrm{r}$ since we are assuming $\Delta\omega/\omega_\mathrm{r} \ll 1$ and $\alpha\ell \ll 1$.

The crucial point is that eqn (13.23) has the same analytical form as the impedance of a parallel RLC resonator. Indeed, if we introduce the parameters

[7] Here, we are using the identities $k = \omega/v$, $\omega = \omega_\mathrm{r} + \Delta\omega$, $\ell = \lambda_\mathrm{r}/4$ and $\omega_\mathrm{r} = k_\mathrm{r} v$.

[8] Here, we used the relation $\cot\left(\frac{\pi}{2} + \frac{\pi\Delta\omega}{2\omega_\mathrm{r}}\right) = -\tan\left(\frac{\pi\Delta\omega}{2\omega_\mathrm{r}}\right) \approx -\frac{\pi\Delta\omega}{2\omega_\mathrm{r}}$ for $\Delta\omega \ll 1$.

$$R = \frac{Z_0}{\alpha \ell}, \qquad C = \frac{\pi}{4\omega_r Z_0}, \qquad L = \frac{1}{\omega_r^2 C}, \qquad Q_i = \omega_r RC = \frac{\pi}{4\alpha\ell}, \tag{13.24}$$

eqn (13.23) becomes

$$\boxed{Z(-\ell) = \frac{R}{1 + i2Q_i \frac{\Delta\omega}{\omega_r}}.} \tag{13.25}$$

This expression is identical to the impedance of a parallel RLC resonator, eqn (13.8). The frequency response of a $\lambda/4$ resonator is the same as a lumped element resonator when probed by an external drive.

It is important to note that the boundary condition $\{V(0) = \text{node}, V(-\ell) = \text{antinode}\}$ is also satisfied for $\ell = n\lambda/4$ where $n = 1, 3, 5$, and so on. This means that a $\lambda/4$ resonator resonates at $\omega_n = (2n+1)\omega_r$ where n is a non-negative integer. The voltage profile for the first two modes is shown in Fig. 13.9b. In superconducting devices, only the fundamental mode ω_r is relevant. However, the higher modes play an important role in the field of multimode circuit quantum electrodynamics [490].

The most popular type of microwave resonator in the field of superconducting devices is the short-circuited $\lambda/4$ CPWR shown in Fig. 13.2d-e and Fig. 13.9a. This is because this resonator is much smaller than other transmission line resonators, such as open-circuited $\lambda/2$ CPWRs, and its performance is not significantly affected by its reduced size[9]. In addition, its resonant frequency is less sensitive to fabrication variations as compared to lumped element resonators [473].

The design of a CPWR starts from its resonance frequency. In superconducting devices, the resonance frequency is chosen to be in the range 4–10 GHz mainly for two reasons: 1) commercial electronics work efficiently at these frequencies and 2) in quantum devices comprising superconducting qubits, the resonator frequency is usually set at least one gigahertz higher than the qubit frequency so that the resonator acts as a band-pass filter protecting the qubit from noise. For a $\lambda/4$ CPWR, the resonant condition is $f_r = v/4\ell$ where $v = c/\sqrt{\varepsilon_{\text{eff}}}$ is the phase velocity and ε_{eff} is the **effective relative permittivity**. The parameter ε_{eff} has two main contributions, one from the substrate and one from the air on top of the device. As shown in Fig. 13.6d, approximately half of the electrical field is distributed in the substrate and half in air. Hence, $\varepsilon_{\text{eff}} = 1/2 + \varepsilon_r/2$, where ε_r is the relative permittivity of the substrate. To achieve a resonance frequency $f_r = 7$ GHz on a silicon substrate with $\varepsilon_r = 11.5$ at low temperature, the resonator length must be[10]

$$\boxed{\ell = \frac{c}{\sqrt{1/2 + \varepsilon_r/2}} \frac{1}{4f_r} = 4.4 \text{ mm.}}$$

To decrease the space occupied on the chip, the CPWR is usually designed to follow a winding path, where each meander has a radius of curvature greater or equal to $4(w + 2g)$ to minimize spurious capacitances between different parts of the resonator.

[9] One advantage of $\lambda/2$ CPWRs is that they can be used as intrinsic Purcell filters [491].

[10] An open-circuited $\lambda/2$ CPWR would be double this length for the same resonance frequency.

The internal quality factor Q_i of a CPWR quantifies the ratio between the amount of energy stored inside the resonator and the power dissipated per cycle. When a CPWR is employed as a measurement tool of the state of a superconducting qubit, Q_i should be as high as possible so as not to compromise the readout fidelity. High internal quality factors can be obtained with two complementary strategies: a robust fabrication process and a well-thought design [492]. It is well known that the losses of a CPWR mainly derive from defects and impurities in the metal-air, substrate-air, and metal-substrate **interfaces** illustrated in Fig. 13.9d [486, 485]. The impurities in these interfaces mainly come from surface oxides and organic contaminants. To reduce defects and impurities in the metal-substrate interface, etching processes should be favored with respect to lift-off recipes since etching procedures make it possible to clean the substrate with aggressive acids before the deposition of the superconducting film. The metal-air interface is usually treated with a buffered oxide etch at the end of the photolithography procedure to remove any organic residue. Recent studies indicate that the thickness of the oxide layer on top of the metal strongly affects the internal quality factor of the resonator [493, 494, 495, 496]. Furthermore, the etching profile of CPWRs plays an important role: trenching techniques into silicon have been developed to reduce the amount of electric field penetrating lossy interfaces [479, 497]. Lastly, the shielding of the device from cosmic rays and environmental radioactivity is essential for the frequency stability of the resonator [498, 499].

Another methodology to improve the performance of a microwave resonator is to design it so that most of the electrical energy is stored in the air and the substrate and a minimal amount penetrates the interfaces, which are considerably more lossy. The ratio between the electrical energy stored in a lossy interface and the total electrical energy is called the **participation ratio** [500],

$$\boxed{p_i = \frac{\frac{1}{2}\varepsilon_0 \int \varepsilon_{\mathrm{r}i} |\mathbf{E}_i|^2 dV_i}{U_{\mathrm{tot}}},}$$

where $i =$ MA, SA and MS are the interfaces, V_i is the volume of each interface and U_{tot} is the electrical energy in the entire space (dominated by energy in the substrate and air). An optimal design should minimize the participation ratio for each lossy interface[11]. Improving the performance of planar resonators at the single photon level is a crucial step toward the fabrication of highly coherent superconducting qubits, since these circuit components are built with similar shapes and fabrication techniques.

13.4 Microwave resonators coupled to a feedline

In this section, we present an experimental procedure to measure planar resonators. Figure 13.10a shows a CPWR capacitively coupled to a waveguide with a characteristic impedance of 50 Ω. A wave generated by an external voltage source (not shown in the figure) with complex amplitude V_1^+ travels along the feedline. Some of the incident power is absorbed by the resonator, some is reflected, and some is transmitted.

[11] 3D cavities tend to have higher quality factors with respect to planar resonators because a larger fraction of the electric field is stored in the vacuum inside the cavity rather than in the lossy interfaces [467].

The amplitude of the reflected wave is $V_1^- = S_{11}V_1^+$ where $S_{11}(\omega)$ is the **reflection coefficient**. The amplitude of the transmitted signal is $V_2^+ = S_{21}V_1^+$ where $S_{21}(\omega)$ is the **transmission coefficient**. A measurement of a microwave resonator involves acquiring either the reflection coefficient or the transmission coefficient. This is usually done by connecting the input and output of the feedline to a vector network analyzer (VNA) or any other transceiver. The goal of this section is to derive the analytical expression of the reflection and transmission coefficients, $S_{11}(\omega)$ and $S_{21}(\omega)$, for a parallel resonator capacitively coupled to a feedline. As we shall see later, these network parameters show a pronounced peak close to resonance.

Figure 13.10b shows the equivalent electrical circuit modeling the coupling between a microwave resonator and a waveguide. This circuit comprises a resonator with impedance Z_r connected to a feedline through a coupling capacitor C. The energy stored inside the resonator is not only lost due to some intrinsic losses (captured by the internal quality factor) but also leaks out through the coupling capacitor. This phenomenon is quantified by the **external quality factor** Q_e, a dimensionless parameter that takes into account this additional loss mechanism. It can be shown that for a resonator capacitively coupled to a feedline [473]

$$\boxed{Q_\mathrm{e} = \frac{\pi}{2(\omega_\mathrm{r} C Z_0)^2}.}$$

Unlike the internal quality factor, this parameter can be accurately engineered by the experimenter by adjusting the shape of the coupling capacitor.

The total rate of energy loss from the resonator is given by the sum of the decay rates: $\kappa = \kappa_\mathrm{i} + \kappa_\mathrm{e}$, where κ_i captures internal losses and κ_e takes into account the losses through the coupling capacitor. Since the decay rate is inversely proportional to the quality factor, the **total quality factor** Q is given by

$$\boxed{\kappa = \kappa_i + \kappa_e \quad \rightarrow \quad \frac{1}{Q} = \frac{1}{Q_\mathrm{i}} + \frac{1}{Q_\mathrm{e}}.} \tag{13.26}$$

This equation indicates that Q is mostly determined by the lowest value between Q_i and Q_e. When a resonator is employed as a readout component of a superconducting qubit, Q_i should be as high as possible (typically 10^5–10^6) while Q_e is usually set around 10^4. In this way, the energy inside the resonator can leak out from the device in approximately $Q_\mathrm{e}/\omega_\mathrm{r} \approx 1\ \mu$s. This sets the timescale of the readout process when the resonator is used as a readout component in a quantum processor.

Let us compute the reflection and transmission coefficients for a parallel resonator capacitively coupled to a feedline. Using the ABCD matrix formalism, it can be shown that[12]

[12] Equation (13.27) has the expected behavior: when $Z \to +\infty$, $S_{21} = 1$ and the power of the incoming signal is completely transmitted through the waveguide. For $Z = 0$, the incoming wave impinges against a short. In this case, $S_{21} = 0$ and the wave is entirely reflected. See Chapter 4 of Ref. [465] for more information.

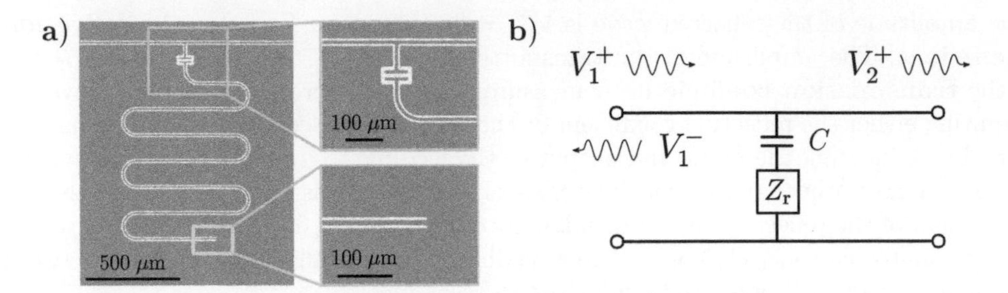

Fig. 13.10 A CPWR capacitively coupled to a feedline. a) A representation of a short-circuited $\lambda/4$ CPWR capacitively coupled to a feedline. The superconducting metal is shown in gray, and the silicon substrate underneath in white. This figure is to scale. The resonator is 4.8 mm long, the width of the center conductor is $w = 10$ μm, and the width of the gap is $g = 5$ μm. The two boxes show a close-up of the capacitive coupling between the resonator and the coplanar waveguide (top) and the short at the end of the CPWR (bottom). **b)** The equivalent electrical circuit of a resonator with impedance Z_r capacitively coupled to a feedline. An external voltage source (not shown in the figure) generates a wave propagating along the feedline with amplitude V_1^+. The incident power is partly absorbed by the resonator, reflected, and transmitted along the waveguide. The amplitude of the reflected and transmitted waves are indicated as V_1^- and V_2^+, respectively.

$$S_{21}(\omega) = \frac{2}{2 + Z_0/Z(\omega)}, \qquad S_{11}(\omega) = \frac{-Z_0/Z}{2 + Z_0/Z(\omega)}. \tag{13.27}$$

Here, $Z(\omega)$ is the sum of the resonator and coupling capacitor impedances,

$$Z(\omega) = Z_\mathrm{r} + Z_\mathrm{C} = \frac{R}{1 + i2Q_\mathrm{i}\frac{\Delta\omega_\mathrm{r}}{\omega_\mathrm{r}}} + \frac{1}{i\omega C}, \tag{13.28}$$

where we used (13.4, 13.25) and $\Delta\omega_\mathrm{r} = \omega - \omega_\mathrm{r}$. With some calculations and approximations, it can be shown that (see Exercise 13.3):

$$Z(\omega) = Z_0\sqrt{\frac{2Q_\mathrm{e}}{\pi}}\,\frac{-i + 2Q_\mathrm{i}\frac{\Delta\omega_0}{\omega_0}}{1 + 2iQ_\mathrm{i}\frac{\Delta\omega_0}{\omega_0} - 2iQ_\mathrm{i}\sqrt{\frac{2}{\pi Q_\mathrm{e}}}}. \tag{13.29}$$

Here, $\Delta\omega_0 = \omega - \omega_0$ and ω_0 is the renormalized frequency of the resonator. The frequency ω_0 is slightly different from the bare resonance frequency ω_r due to the coupling capacitor,

$$\frac{\Delta\omega_0}{\omega_0} = \frac{\Delta\omega_\mathrm{r}}{\omega_\mathrm{r}} + \sqrt{\frac{2}{\pi Q_\mathrm{e}}}. \tag{13.30}$$

Substituting (13.29) into (13.27), we obtain the transmission coefficient

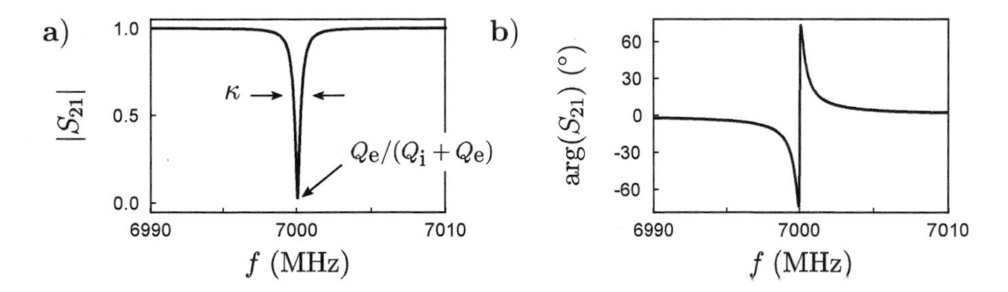

Fig. 13.11 The transmission coefficient S_{21}. a) Magnitude of the complex transmission coefficient $S_{21}(f)$ for a parallel resonator capacitively coupled to a feedline, eqn (13.31). At resonance $|S_{21}| = Q_e/(Q_i + Q_e)$. The linewidth of the resonance is the total decay rate κ that is related to the total quality factor by the formula $Q = 2\pi f_r/\kappa$. **b)** Phase of the complex transmission coefficient $S_{21}(f)$ for a parallel RLC resonator coupled to a feedline, eqn (13.31). In both plots, $f_0 = \omega_0/2\pi = 7$ GHz, $Q_e = 7000$ and $Q_i = 300000$.

$$S_{21}(\omega) = \frac{V_2^+}{V_1^+} = \frac{2}{2 + Z_0/Z(\omega)} = \frac{2}{2 + \frac{1+2iQ_i\frac{\Delta\omega_0}{\omega_0} - 2iQ_i\sqrt{\frac{2}{\pi Q_e}}}{\sqrt{\frac{2Q_e}{\pi}}\left(-i+2Q_i\frac{\Delta\omega_0}{\omega_0}\right)}}.$$

After some lengthy calculations and dropping some small contributions, we can write the transmission coefficient in a simpler form,

$$\boxed{S_{21}(\omega) = \frac{\frac{Q_e}{Q_i+Q_e} + 2iQ\frac{\Delta\omega_0}{\omega_0}}{1 + 2iQ\frac{\Delta\omega_0}{\omega_0}},} \tag{13.31}$$

where $Q = (1/Q_i + 1/Q_e)^{-1}$ is the total quality factor introduced in (13.26). Figure 13.11 shows the magnitude and phase of the complex transmission coefficient for a parallel resonator capacitively coupled to a feedline. At resonance, $\Delta\omega_0 = 0$ and the transmission coefficient reduces to $S_{21}(\omega_0) = Q_e/(Q_i + Q_e)$. This means that if two resonators have the same external quality factor, the one with the deepest resonance has the highest internal quality factor. With similar calculations, one can verify that the reflection coefficient at resonance takes the form

$$\boxed{S_{11}(\omega_0) = \frac{-Z_0/Z(\omega_0)}{2 + Z_0/Z(\omega_0)} = \frac{-Q_i}{Q_i + Q_e}.} \tag{13.32}$$

For an ideal microwave resonator ($Q_i \to +\infty$) the incoming signal is entirely reflected ($|S_{11}(\omega = \omega_0)| = 1$). In a real experiment, the resonator might be coupled to a mismatched transmission line, leading to an asymmetric response. In this case, the diameter correction method provides a more accurate fitting function for the transmission and reflection coefficients [501].

Let us derive the power required to populate a resonator with an average of a single photon. Portions of the incident power P_1^+ get reflected, transmitted, and absorbed by the resonator. The **reflected power** will be denoted as P_1^- and the **transmitted power** as P_2^+. At resonance, the reflected and transmitted powers are given by

$$P_1^- = P_1^+|S_{11}|^2 = P_1^+\frac{Q_\mathrm{i}^2}{(Q_\mathrm{i}+Q_\mathrm{e})^2} \qquad \text{(reflected)} \tag{13.33}$$

$$P_2^+ = P_1^+|S_{21}|^2 = P_1^+\frac{Q_\mathrm{e}^2}{(Q_\mathrm{i}+Q_\mathrm{e})^2} \qquad \text{(transmitted)} \tag{13.34}$$

where we used (13.31, 13.32). The **power absorbed by the resonator** is the incident power minus the reflected and transmitted powers,

$$P_\mathrm{abs} = P_1^+ - P_1^- - P_2^+ = P_1^+\frac{2Q_\mathrm{i}Q_\mathrm{e}}{(Q_\mathrm{i}+Q_\mathrm{e})^2}, \qquad \text{(absorbed)} \tag{13.35}$$

where we used (13.33, 13.34). The quantity P_abs indicates the amount of power dissipated by the resonator due to intrinsic losses (this power does not go back into the transmission line). The absorbed power is also equal to the energy inside the resonator E multiplied by the internal decay rate,

$$P_\mathrm{abs} = E\cdot\kappa_\mathrm{i} = \hbar\omega_0\overline{n}\cdot\frac{\omega_0}{Q_\mathrm{i}} = \overline{n}\frac{\hbar\omega_0^2}{Q_\mathrm{i}}. \tag{13.36}$$

Here, the quantity $E=\hbar\omega_0$ is the energy of a single photon, and $\overline{n}$ is the average number of photons inside the resonator. Equating (13.35) with (13.36), we can calculate the **number of photons** inside a superconducting resonator as a function of incident power

$$\boxed{\overline{n} = P_1^+\frac{Q_\mathrm{i}^2Q_\mathrm{e}}{(Q_\mathrm{i}+Q_\mathrm{e})^2}\frac{2}{\hbar\omega_0^2}.}$$

Thus, the power required to populate a superconducting resonator with a single photon is

$$P_1^+(\overline{n}=1) = \frac{(Q_\mathrm{i}+Q_\mathrm{e})^2}{2Q_\mathrm{i}^2Q_\mathrm{e}}\hbar\omega_0^2. \tag{13.37}$$

As we shall see in Example 13.4, for typical microwave resonators $P_1^+(\overline{n}=1)$ is of the order of 10^{-17} W.

It is common practice to plot the magnitude of the transmission coefficient $S_{21}(\omega)$ in **decibels (dB)**. This allows us to compare transmission coefficients varying by several orders of magnitude on the same plot. The ratio between the power of the transmitted and incoming waves can be expressed in decibels as $10\log_{10}P_2^+/P_1^+$ dB. Since these powers are proportional to $|V_1^+|^2/Z_0$ and $|V_2^+|^2/Z_0$, we have

$$10\log_{10}\frac{P_2^+}{P_1^+} = 20\log_{10}\frac{|V_2^+|}{|V_1^+|} = 20\log_{10}|S_{21}|.$$

For a passive element $0\leq|V_2^+|/|V_1^+|\leq 1$ (the transmitted power cannot be greater than the incident power), and for this reason the quantity $20\log_{10}|S_{21}|$ is never greater than zero for microwave resonators. It is simple to add up attenuation values in dB. For example, if an electrical component attenuates the incident power by 10 dB and another component by 5 dB, the attenuation of these two components in series is 15 dB.

While the decibel is a dimensionless unit that quantifies the logarithmic ratio between two powers, the **dBm (decibel-milliwatt)** is a unit of measure of the power. A power in watt can be converted in dBm with the formula $x = 10 \log_{10} \frac{P}{1\text{ mW}}$. Thus, 1 mW and 0 dBm are the same power.

Example 13.3 A microwave resonator has the parameters $f_r = 5$ GHz, $Q_e = 5000$ and $Q_i = 100000$. Suppose a microwave signal with power $P_1^+ = 10^{-15}$ W is applied to the resonator. From eqns (13.33, 13.35), the reflected power is $P_1^- = 90.7\% \, P_1^+$, the transmitted power is $P_2^+ = 0.23\% \, P_1^+$ and the absorbed power is $P_{\text{abs}} = 9.07\% \, P_1^+$.

Example 13.4 Let us calculate the power required to populate a microwave resonator with an average of one photon. Suppose $f_r = 5$ GHz, $Q_e = 5000$ and $Q_i = 100000$. Then, from eqn (13.37) the incident power must be $P_1^+ \approx 6 \cdot 10^{-18}$ $W = -142$ dBm. If the microwave generator is connected to the device with a coaxial line with an overall attenuation of -70 dB at *resonance*, the power of the signal generator at room temperature should be set to -72 dBm. This way, the power that reaches the device inside the dilution refrigerator is -142 dBm.

13.4.1 A microwave resonator inductively coupled to a feedline

Another common way to measure a CPWR is to inductively couple the resonator to a feedline, as shown in Fig. 13.12a. The part of the resonator short-circuited to ground must be placed close to the feedline. This electrical system can be studied with two different methods: 1) one method consists in sketching a lumped element equivalent circuit and following a similar analytical derivation to the one presented in the previous section, 2) another strategy is to simulate the device with a finite element method (FEM) software, such as HFSS by Ansys [473, 474]. Let us focus on the latter method.

A FEM simulation provides both the frequency of the resonator and the external quality factor. These parameters can be easily adjusted at the design stage. As shown in Fig. 13.12b, only the region close to the feedline has to be simulated to extract Q_e. The electrical circuit can be modeled as a 3-port network. Port 1 and port 2 are the input and output ports. Port 3 is at the base of the CPWR. The signal traveling along the waveguide penetrates the resonator and bounces back. Half of the energy will be transferred to port 2 and a half to port 3. The power leaking out from the resonator in both directions is given by [474]

$$P_e = 2E(|S_{13}|^2 + |S_{23}|^2) f_0, \tag{13.38}$$

where S_{13} (S_{23}) is the scattering matrix element denoting the voltage transmission from port 3 to 1 (from port 3 to port 2). The factor of two in (13.38) takes into account the fact that the wave encounters the inductive coupler twice per cycle in a

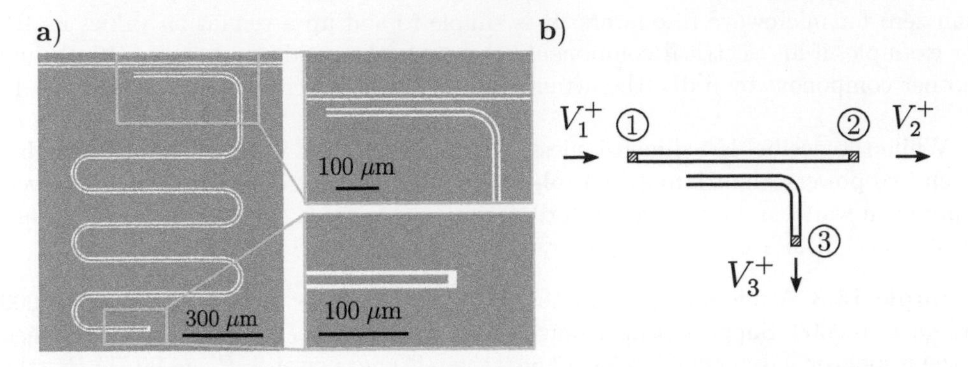

Fig. 13.12 Inductive coupling to a feedline. a) A short-circuited $\lambda/4$ CPWR inductively coupled to a waveguide. The inductive coupler, shown in the top panel, is approximately 300 μm long and only 10 μm away from the waveguide. As usual the length of the resonator ℓ determines the resonance frequency $f_r = v/4\ell$, where the phase velocity $v \approx 1.21 \cdot 10^8$ m/s for silicon at low temperatures. In this figure, $\ell = 4.4$ mm. **b)** Electrical network modeling the inductive coupling between the resonator and the feedline. The microwave generator is connected to port 1. Part of the incoming wave gets transmitted to the resonator (port 3). Part of the signal is transmitted down the line (port 2). The transmission coefficient is given by the ratio $S_{21} = V_2^+/V_1^+$ where V_1^+ is the amplitude of the applied signal and V_2^+ is the amplitude of the transmitted wave.

$\lambda/4$ resonator. Since the circuit is almost symmetric $|S_{13}| \approx |S_{23}|$, the external quality factor is

$$Q_e = \omega_0 \frac{E}{P_e} = 2\pi f_0 \frac{E}{2E(|S_{13}|^2 + |S_{23}|^2) f_0} = \frac{\pi}{2|S_{13}|^2}. \tag{13.39}$$

A FEM software can easily simulate the S_{13} parameter at resonance. Once $S_{13}(\omega_r)$ has been extracted from the simulation, the external quality factor can be calculated from eqn (13.39). It is intuitive that the external quality factor decreases when the length of the inductive coupler increases and decreases when the gap between the coupler and the feedline decreases.

Multiple resonators can be connected to a common feedline and measured with a single signal generator. This allows for a reduction in the number of coaxial cables needed to connect the room-temperature instrumentation to the device located inside the dilution refrigerator. This **multiplexing** strategy is widespread in the field of superconducting devices [502, 503, 504, 497]. Figure 13.13a shows a sequence of CPWR inductively coupled to a common feedline. The length of the resonators can be set such that their frequencies are spaced by 50 MHz as shown in Fig. 13.13b. The resonator with the longest (shortest) length will show a resonance at the lowest (highest) frequency. Assuming that the external quality factor is the same for all the resonators, the deepest dip will correspond to the resonator with the highest internal quality factor. This follows from the fact that at resonance $S_{21}(f_0) = Q_e/(Q_i + Q_e)$.

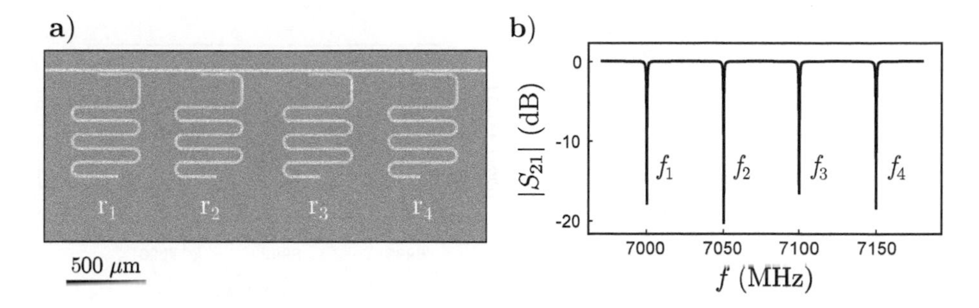

Fig. 13.13 Multiplexing. a) Four short-circuited $\lambda/4$ CPWRs with different lengths are inductively coupled to a common feedline. This multiplexing strategy permits reducing the number of coaxial lines required to measure each resonator. The resonators are in descending order of length: r_1 is the longest, r_4 is the shortest. **b)** Magnitude of the transmission coefficient for four microwave resonators connected to a common feedline. Each resonance corresponds to a different resonator. In this graph, we are plotting the quantity $20\log_{10}|S_{21}|$. The resonances f_1, f_2, f_3, f_4 correspond to the resonators r_1, r_2, r_3, r_4 respectively.

13.5 Temperature and power dependence

The measurement of a superconducting coplanar waveguide resonator in the frequency domain does not show any resonance at room temperature. This is simply because the losses of the device at ambient conditions are too high. When the temperature of the cryostat is lower than the critical temperature of the superconductor, the conductor's losses dramatically decrease and the measurement of the transmission coefficient shows a pronounced peak. As the temperature of the dilution refrigerator decreases further, the resonance frequency shifts to a higher frequency. To understand this phenomenon, let us start from the frequency of the resonator (13.5),

$$f_{\mathrm{r}}(T) = \frac{1}{2\pi\sqrt{L(T)C}} = \frac{1}{2\pi\sqrt{(L_{\mathrm{geom}} + L_{\mathrm{kin}}(T))C}}, \tag{13.40}$$

where T is the temperature of the device and in the last step we expressed the inductance of the resonator as the sum of two contributions: the geometric inductance and the kinetic inductance. The **geometric inductance** is related to the shape of the resonator. For a coplanar waveguide resonator, it is given by [489, 505]

$$L_{\mathrm{geom}} = \ell\frac{\mu_0}{4}\frac{K(a)}{K(b)},$$

where ℓ is the length of the resonator, $a = w/(w+2g)$, $b = \sqrt{1-a^2}$ and $K(x)$ is the complete elliptic integral of the first kind. On the other hand, the **kinetic inductance** is related to the motion of the superconducting electrons under an alternating electric field. As shown in Fig. 13.14a, the electric field accelerates the superconducting electrons in one direction. When the field is reversed, the electrons slow down until they finally change their direction of motion. Therefore, the velocity of the electrons,

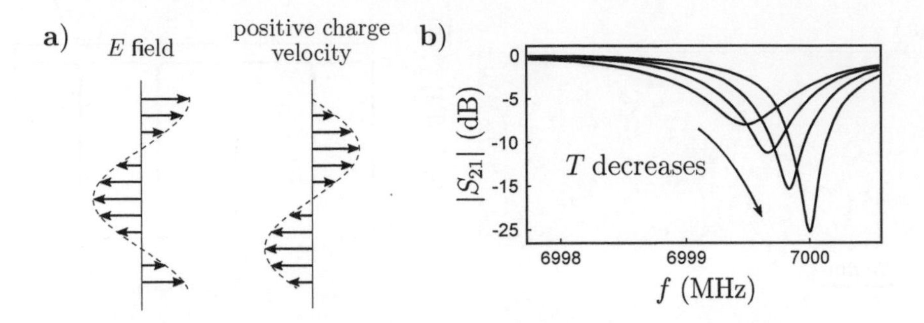

Fig. 13.14 Temperature dependence. a) When an electric field is applied to a superconductor, the Cooper pairs accelerate in one direction. When the electric field is reversed, the Cooper pairs slow down until they finally change their direction of motion. A 90° phase difference between the electric field and the displacement of the charge carriers is the typical behavior of an inductive component. **b)** As the cryostat temperature decreases from T_C to $T_C/10$, the kinetic inductance decreases and the frequency of the resonator shifts to higher frequencies. The resonance becomes narrower as the losses due to quasiparticles decrease.

i.e. the electrical current, lags the electric field by 90 degrees (see Fig. 13.14a). This is the typical behavior of an inductive component and is what gives rise to the kinetic inductance. Unlike the geometric inductance, the kinetic inductance is temperature dependent and is proportional to $\lambda^2(T)$ where $\lambda(T)$ is the temperature-dependent London penetration depth[13] [505, 506, 507]. Figure 13.14b shows the typical shift of the resonator frequency from T_C down to $T_C/10$. At 10 mK, the kinetic inductance of common superconductors, such as aluminum and niobium, is two orders of magnitude lower than the geometric inductance and can be neglected [468].

All superconducting devices have oxide layers covering the substrate surface and the metallic parts. In most cases, the oxide layers have an amorphous structure containing multiple defects (see Fig. 13.15a). Each defect can be modeled as a **two-level system** (TLS), i.e. a quantum system with two energy levels [508]. Even if the microscopic origin of TLSs is still a matter of debate [509], it has been shown in multiple works that the internal quality factor of a microwave resonator can be significantly impacted by spurious interactions between the resonator and nearby TLSs.

One experimental evidence of the presence of TLSs is related to the measurement of the internal quality factor as a function of the power of the applied signal. As shown in Fig. 13.15b, the internal quality factor typically shows an S-shape dependence as a function of the applied power [510, 485, 478, 511, 492]. This behavior is consistent with a theoretical model in which the resonator interacts with a bath of TLSs. When the signal power is low, the electrical pulse populates the resonator with a few microwave photons. This energy can in principle be transferred to nearby TLSs with a transition frequency similar to that of the resonator, thereby reducing the quality factor. However, when the power is high, the microwave drive brings most of the TLSs to an

[13]The London penetration depth is the characteristic distance to which a magnetic field penetrates a superconductor. Typical values range from 50 to 500 nm.

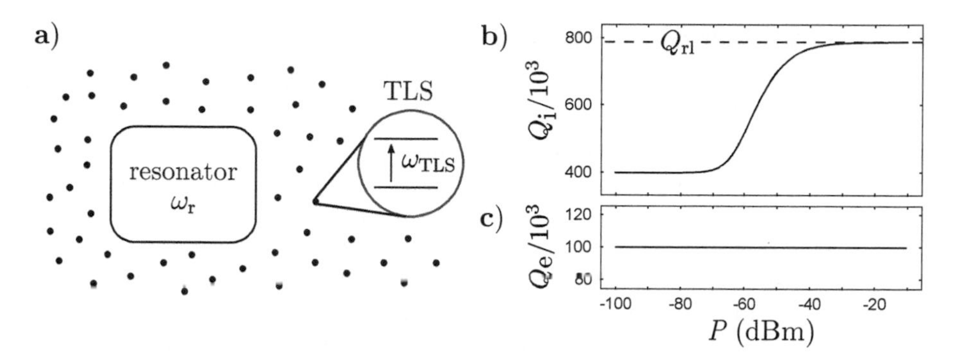

Fig. 13.15 Power dependence. a) A schematic of a resonator surrounded by two-level systems. If the frequency of a TLS is similar to that of the resonator, the two systems exchange energy, leading to a shorter decay time of the resonator. **b)** Plot of the internal quality of a coplanar waveguide resonator as a function of the drive power. The data points follow an S-shape curve that can be fitted to eqn (13.41). Unfortunately, in circuit quantum electrodynamics the resonator is often operated at low powers where the internal quality factor is the lowest. **c)** Plot of the external quality factor as a function of applied drive power. This parameter does not show any power dependence.

excited state, preventing them from absorbing additional energy from the resonator. This leads to a higher internal quality factor. The model predicts that the internal quality factor as a function of power is given by [510, 485, 492]

$$Q_{\mathrm{i}}(P) = \left(\frac{1}{Q_{\mathrm{rl}}} + \frac{1}{Q_{\mathrm{TLS}}(P)} \right)^{-1}, \tag{13.41}$$

where Q_{rl} takes into account remaining losses, while the losses due to the TLSs are captured by the parameter

$$\frac{1}{Q_{\mathrm{TLS}}(P)} = F\delta^0_{\mathrm{TLS}} \left(1 + \frac{P}{P_{\mathrm{c}}}\right)^{\gamma} \tanh\left(\frac{\hbar\omega_{\mathrm{r}}}{2k_{\mathrm{B}}T}\right). \tag{13.42}$$

Here, P is the power of the applied signal, P_{c} is a characteristic power at which the saturation of the TLS takes place, δ^0_{TLS} is the intrinsic TLS loss, and the exponent $\gamma \approx -1/2$ slightly depends on the location of the TLS [484]. The parameter F, called the filling factor, is the fraction of the electrical energy stored in the interface hosting the TLSs. The term $\tanh(\hbar\omega_{\mathrm{r}}/2k_{\mathrm{B}}T)$ reflects the thermal population difference between the ground and excited state of the TLS. Unlike the internal quality factor, the external quality factor Q_{e} is not affected by the presence of TLSs and is not power dependent (see Fig. 13.15c). This section ends the classical description of microwave resonators. In the next section, we will study the quantum nature of microwave resonators when operated at millikelvin temperatures.

13.6 Quantization of an LC resonator

When a microwave resonator is well isolated from the environment and cooled to low temperatures such that $k_B T \ll hf_r$, it can approach its quantum ground state where it may be used for fundamental applications in quantum information. This is called the quantum regime. Reaching the quantum regime not only requires low temperatures but also that the loss rate κ is much smaller compared to the resonance frequency, $\kappa \ll \omega_r$. In 2004, it was experimentally demonstrated that a microwave resonator in the quantum regime can exchange energy with a superconducting qubit in a coherent way. This architecture is known as **circuit quantum electrodynamics** (QED). It is routinely employed in quantum information processing for the dispersive read out of the qubits and for coupling superconducting qubits to each other using the resonator as a bus coupler [139, 138, 512, 513].

Our discussion of circuit quantum electrodynamics will start with the quantization of an LC resonator. We will discover that this system can be described by a simple quantum harmonic oscillator when losses do not play an important role. In the next chapter, we will study the coupling between a superconducting qubit and a microwave resonator. This will lead to the Jaynes–Cummings Hamiltonian, the mathematical model at the heart of circuit-QED.

Figure 13.16 shows the equivalent circuit of a parallel LC resonator. This circuit has two nodes, indicated with the letters a and b, connected by two distinct branches. One branch comprises a capacitor and the other one an inductor. The voltage across the capacitor and the current flowing through the inductor will be denoted as V and I, respectively. In general, these quantities are time-dependent. It is convenient to introduce a new variable related to the voltage: the **generalized flux** Φ is defined as the time integral of the voltage across the capacitor,

$$\boxed{\Phi(t) = \int_{-\infty}^{t} V(t')dt'.} \tag{13.43}$$

From this definition, one can write the energy stored in the capacitive and inductive components as[14]:

$$E_C = \frac{1}{2}CV^2 = \frac{1}{2}C\dot{\Phi}^2, \tag{13.44}$$

$$E_L = \frac{1}{2}LI^2 = \frac{\Phi^2}{2L}, \tag{13.45}$$

where we used $V = d\Phi/dt$ and $I = \Phi/L$. The Lagrangian of an LC resonator is given by the difference between the kinetic and potential energies,

[14]The reader might have noticed a similarity between E_C (E_L) and the kinetic (potential) energy of a harmonic oscillator. Here, the flux Φ and its derivative $\dot{\Phi}$ play the roles of position and speed. Similarly, the parameters C and $1/L$ can be considered to be the mass and spring constant of the oscillator.

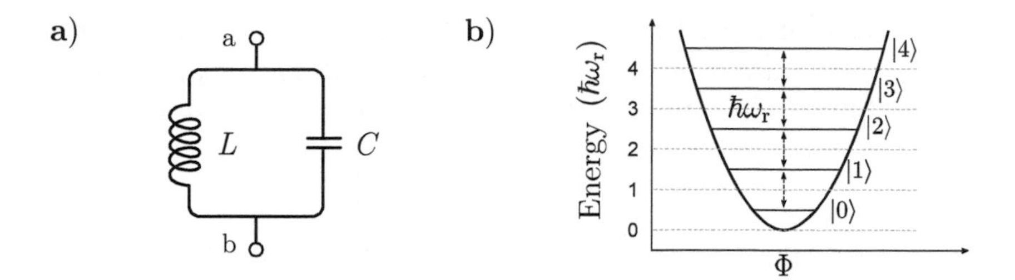

Fig. 13.16 Quantization of an LC resonator. a) Electrical circuit of a parallel LC resonator. **b)** Representation of the quadratic potential of a harmonic oscillator and the eigenstates. **c)** Phase space representation of the vacuum state $|0\rangle$.

$$L_{\rm r} = \frac{1}{2}C\dot{\Phi}^2 - \frac{\Phi^2}{2L}. \tag{13.46}$$

The Hamiltonian of an LC resonator can be derived from the Lagrangian in a straightforward way. The first step is to calculate the **canonical conjugate variable** of the flux, $Q = \partial L_{\rm r}/\partial\dot{\Phi} = C\dot{\Phi}$ where Q is the charge. The second step is to derive the Hamiltonian from the Legendre transform $H_{\rm r} = Q\dot{\Phi} - L_{\rm r}$. This procedure yields

$$H_{\rm r} = \frac{Q^2}{2C} + \frac{\Phi^2}{2L} = \frac{Q^2}{2C} + \frac{1}{2}C\omega_{\rm r}^2\Phi^2, \tag{13.47}$$

where $\omega_{\rm r} = 1/\sqrt{LC}$ is the resonance frequency. The Hamiltonian (13.47) is equivalent to the Hamiltonian of a classical oscillator where Φ and Q play the roles of position and momentum.

The **quantization of an LC circuit** involves replacing the classical variables Φ and Q with their corresponding Hermitian operators $\hat{\Phi}$ and $\hat{Q}$. Since the flux and charge are conjugate variables (i.e. their Poisson bracket satisfies $\{\Phi, Q\} = 1$), Dirac's quantization procedure establishes that the commutator of the corresponding operators satisfies $[\hat{\Phi}, \hat{Q}] = i\hbar\mathbb{1}$ [125]. This is the same commutation relation of the position and momentum in ordinary quantum mechanics. The quantum version of (13.47) is

$$\boxed{\hat{H}_{\rm r} = \frac{\hat{Q}^2}{2C} + \frac{1}{2}C\omega_{\rm r}^2\hat{\Phi}^2.} \tag{13.48}$$

This is the Hamiltonian of a quantum harmonic oscillator where $\hat{\Phi}$ plays the role of the position and $\hat{Q}$ of the momentum. As explained in Section 4.5, to find the eigenvectors and eigenvalues of $\hat{H}_{\rm r}$, it is often convenient to introduce the raising and lowering operators $\hat{a}$ and $\hat{a}^\dagger$, two non-Hermitian operators defined as:

$$\hat{\Phi} = \Phi_{\rm zpf}(\hat{a} + \hat{a}^\dagger), \tag{13.49}$$

$$\hat{Q} = -iQ_{\rm zpf}(\hat{a} - \hat{a}^\dagger), \tag{13.50}$$

where $\Phi_{\text{zpf}} = \sqrt{\hbar/2C\omega_{\text{r}}}$ and $Q_{\text{zpf}} = \sqrt{C\hbar\omega_{\text{r}}/2}$ are the **zero-point fluctuations.** Since $[\hat{Q}, \hat{P}] = i\hbar\mathbb{1}$, then $[\hat{a}, \hat{a}^\dagger] = 1$. By substituting (13.49, 13.50) into (13.48), we finally obtain the Hamiltonian of a microwave resonator in the quantum regime

$$\boxed{\hat{H}_{\text{r}} = \hbar\omega_{\text{r}}(\hat{a}^\dagger\hat{a} + 1/2).} \tag{13.51}$$

The eigenvectors of the Hamiltonian are the number states $\{|n\rangle\}_{n=0}^{\infty}$. The corresponding eigenvalues are $E_n = \hbar\omega_{\text{r}}(n + 1/2)$. The energy levels of a microwave resonator are equally spaced, as shown in Fig. 13.16b. When the system is in the state $|n\rangle$, the resonator is populated with n excitations at the frequency ω_{r}. For electromagnetic resonators, these excitations are called **photons.**

It is sometimes convenient to express the Hamiltonian (13.48) in the form

$$\hat{H}_{\text{r}} = 4E_{\text{C}}\hat{n}^2 + \frac{1}{2}E_L\hat{\phi}^2, \tag{13.52}$$

where the operators $\hat{n}$ and $\hat{\phi}$ are dimensionless versions of the charge and flux operators: $\hat{n} = \hat{Q}/2e$, and $\hat{\phi} = 2\pi\hat{\Phi}/\Phi_0$. Here, Φ_0 is defined as $\Phi_0 = h/2e$ and is called the flux quantum. The quantity $E_{\text{C}} = e^2/2C$ is called the **charging energy**, whereas $E_{\text{L}} = (\Phi_0/2\pi)^2/L$ is the **inductive energy**. Equation (13.52) will be useful in the next chapter when we present the Cooper pair box.

Since a microwave resonator has multiple energy levels, one might argue that the first two energy states, $|0\rangle$ and $|1\rangle$, can be used as a qubit. Unfortunately, the constant separation between adjacent energy levels makes it impossible to excite the system from $|0\rangle$ to $|1\rangle$ without also exciting higher levels. Hence, two eigenstates of a quantum harmonic oscillator do not form a controllable qubit. However, if the potential is not perfectly parabolic, the energy levels are not equally spaced and a subspace spanned by two states can be used as a qubit. As we shall see in Section 14.4, the potential energy of the transmon, a widespread superconducting qubit, slightly deviates from a harmonic behavior due to a quartic perturbation. This perturbation is large enough to separate adjacent levels so that an electrical pulse can individually address them.

When the temperature of the environment satisfies $k_{\text{B}}T \approx \hbar\omega_{\text{r}}$, the resonator is in a thermal state with an average photon number[15] $\bar{n} = 1/(e^{\hbar\omega_{\text{r}}/k_{\text{B}}T} - 1)$. By lowering the temperature of the environment such that $k_{\text{B}}T \ll \hbar\omega_{\text{r}}$, the resonator approaches the vacuum state. In the next section, we show that the state of a resonator can be transformed from a vacuum state into a coherent state by applying a short electrical pulse. We suggest our readers not familiar with coherent states read Section 4.6 and come back to this part of the chapter once they have grasped the main technical aspects.

13.7 Driving a microwave resonator

In this section, we study the excitation of an LC resonator using a microwave generator. We show that when a resonant microwave pulse is applied to a resonator initially

[15] See eqn (6.41) for a derivation of this result.

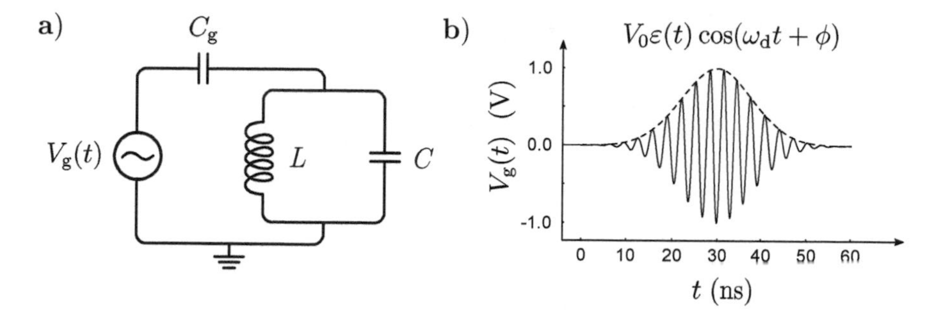

Fig. 13.17 Driving an ideal LC resonator. a) Equivalent circuit of a parallel LC resonator capacitively connected to an AC voltage source. **b)** An electrical pulse with a Gaussian profile. The solid line indicates the voltage delivered by the microwave generator as a function of time. The dashed line is the modulation of the pulse, $\varepsilon_g(t) = e^{-\frac{1}{2}(t-t_0)^2/\sigma^2}$. In this example, $V_0 = 1$ V, the center of the pulse is $t_0 = 30$ ns and the width is $\sigma = 8$ ns.

prepared in the vacuum state, the state of the resonator becomes a coherent state. The generator is coupled to the resonator via a gate capacitance C_g as shown in Fig. 13.17a. The charge on the resonator's capacitor is determined by the applied voltage $V_g(t)$. The Hamiltonian (13.48) becomes

$$\hat{H}(t) = \frac{(\hat{Q} + C_g V_g(t))^2}{2C_\Sigma} + \frac{\hat{\Phi}^2}{2L}, \tag{13.53}$$

where $C_\Sigma = C_g + C$. By expanding the parentheses and dropping some terms that do not depend on $\hat{Q}$ and $\hat{\Phi}$, we arrive at

$$\boxed{\hat{H}(t) = \frac{\hat{Q}^2}{2C_\Sigma} + \frac{\hat{\Phi}^2}{2L} + \frac{C_g}{C_\Sigma} V_g(t)\hat{Q}.} \tag{13.54}$$

The electric pulse can be written in the form $V_g(t) = V_0\varepsilon(t)\cos(\omega_d t + \phi)$ where V_0 is an amplitude in volts, ω_d is the drive frequency, ϕ is the phase and $\varepsilon(t)$ is a dimensionless function that defines the profile of the pulse. Two typical modulation profiles are the square pulse and the Gaussian pulse:

$$\text{square pulse:} \quad \varepsilon_{sq}(t) = \begin{cases} 1 & 0 \le t \le t_g \\ 0 & \text{else} \end{cases}$$

$$\text{Gaussian pulse:} \quad \varepsilon_g(t) = e^{-\frac{1}{2\sigma^2}(t-t_0)^2}.$$

An electrical pulse with a Gaussian profile is plotted in Fig. 13.17a. Clearly, the drive frequency ω_d must be close to the resonance frequency ω_r in order to excite the resonator.

By substituting the expression of $V_{\rm g}(t)$ and eqns (13.49, 13.50) into (13.54), we obtain[16]

$$\hat{H}(t) = \hbar\omega_{\rm r}\hat{a}^\dagger\hat{a} + \frac{C_{\rm g}}{C_\Sigma}V_0\varepsilon(t)\cos(\omega_{\rm d}t+\phi)\left[-iQ_{\rm zpf}(\hat{a}-\hat{a}^\dagger)\right]$$
$$= \hbar\omega_{\rm r}\hat{a}^\dagger\hat{a} - i\hbar\lambda(t)\left(e^{i(\omega_{\rm d}t+\phi)} + e^{-i(\omega_{\rm d}t+\phi)}\right)(\hat{a}-\hat{a}^\dagger),$$

where we defined

$$\lambda(t) = \frac{C_{\rm g}V_0Q_{\rm zpf}}{2\hbar C_\Sigma}\,\varepsilon(t).$$

It is convenient to study the dynamics of the system in a rotating frame. The Hamiltonian in the new frame is given by (see eqn (4.37))

$$\hat{H}_{\rm I}(t) = \hat{R}(t)\hat{H}(t)\hat{R}^{-1}(t) - i\hbar\hat{R}(t)\frac{d}{dt}\hat{R}^{-1}(t),$$

where $\hat{R}(t) = e^{i\omega_{\rm r}\hat{a}^\dagger\hat{a}t}$ is a time-dependent change of basis. Thus[17]

$$\hat{H}_{\rm I}(t) = -i\hbar\lambda(t)\left(e^{i(\omega_{\rm d}t+\phi)} + e^{-i(\omega_{\rm d}t+\phi)}\right)\left(\hat{a}e^{-i\omega_{\rm r}t} - \hat{a}^\dagger e^{i\omega_{\rm r}t}\right)$$
$$= -i\hbar\lambda(t)\left(\hat{a}e^{i\phi} - \hat{a}^\dagger e^{-i\phi} + \hat{a}e^{i(2\omega_{\rm r}t+\phi)} - \hat{a}^\dagger e^{-i(2\omega_{\rm r}t+\phi)}\right), \tag{13.55}$$

where in the last step we assumed that the drive frequency $\omega_{\rm d}$ is equal to the resonator frequency $\omega_{\rm r}$. The last two terms in (13.55) oscillate very fast and give a small contribution to the dynamics of the system. Hence, they can be ignored (this approximation is usually called the **rotating wave approximation**, see Appendix A.1). This approximation yields

$$\hat{H}_{\rm I}(t) = -i\hbar\lambda(t)(\hat{a}e^{i\phi} - \hat{a}^\dagger e^{-i\phi}).$$

The phase ϕ of the electrical pulse can be set by the experimenter. If the resonator is initially in the vacuum state $|0\rangle$, the state of the resonator at $t > 0$ is given by

$$|\psi_{\rm I}(t)\rangle = e^{\frac{1}{i\hbar}\int_0^t \hat{H}_{\rm I}(t')dt'}|0\rangle = e^{(\hat{a}^\dagger e^{-i\phi} - \hat{a}e^{i\phi})\int_0^t \lambda(t')dt'}|0\rangle$$
$$= e^{\alpha(t)\hat{a}^\dagger - \alpha^*(t)\hat{a}}|0\rangle = \hat{D}(\alpha(t))|0\rangle = |\alpha(t)\rangle,$$

where we defined $\alpha(t) = e^{i\phi}\int_0^t \lambda(t')dt'$ and we used the definition of the displacement operator $\hat{D}(\alpha)$, eqn (4.79). The state of the resonator evolves from the vacuum state into a coherent state (see Fig. 13.18). In this plot, we are representing the state of the resonator in the interaction picture. Note that in the Schrödinger picture, $|\psi(t)\rangle = \hat{R}^{-1}|\psi_{\rm I}(t)\rangle = e^{-i\omega_{\rm r}t}|\alpha(t)\rangle$, i.e. the state rotates in the phase space clockwise about the origin with frequency $\omega_{\rm r}$.

[16] Here, we ignored the vacuum energy $\hbar\omega_{\rm r}/2$.

[17] To derive $\hat{H}_{\rm I}(t)$, use the Baker–Campbell–Hausdorff formula (3.59): $\hat{R}(t)\hat{a}\hat{R}^\dagger(t) = \hat{a}e^{-i\omega_{\rm r}t}$ and $\hat{R}(t)\hat{a}^\dagger\hat{R}^\dagger(t) = \hat{a}^\dagger e^{i\omega_{\rm r}t}$.

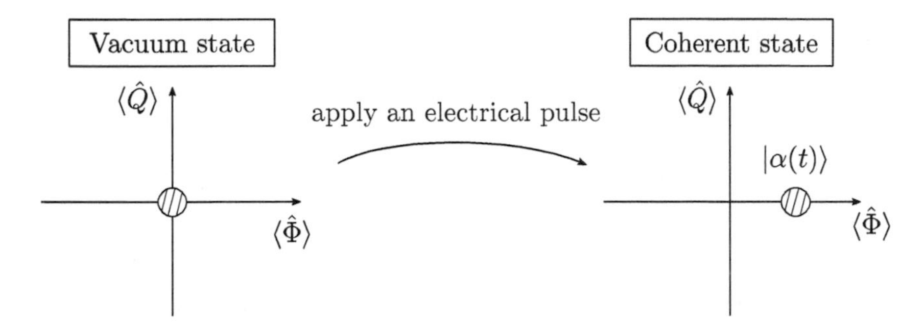

Fig. 13.18 Creating a coherent state. At a temperature of 10 mK, a microwave resonator with a working frequency of 7 GHz is in the ground state. This state is depicted in the phase space as a circle centered at the origin. When an electrical pulse with frequency $\omega_{\rm d} = \omega_{\rm r}$ is applied to the resonator, the state of the resonator becomes a coherent state $|e^{i\phi}\alpha\rangle$, where ϕ is the phase of the pulse and α is determined by the amplitude and duration of the pulse.

In summary, when a short microwave pulse is applied to a microwave resonator in the quantum regime, the action of the pulse on the state of the resonator is described by a displacement operator $\hat{D}(\alpha(t)) = e^{\alpha(t)\hat{a}^\dagger - \alpha^*(t)\hat{a}}$. If the resonator is initially in the vacuum state, the state of the resonator *at the end* of the pulse is $|\alpha\rangle$ where $\alpha = e^{i\phi}\int_0^\infty \lambda(t')dt'$ depends on the duration, amplitude, phase, and modulation of the pulse. By adjusting these parameters, the resonator can in principle be prepared in an arbitrary coherent state. Photon losses tend to bring the state of the resonator back to the vacuum state.

Example 13.5 Suppose a Gaussian pulse with width $\sigma = 100$ ns and phase $\phi = 0$ is applied to a microwave resonator. Assume that the parameters of the circuit are $C = 450$ fF, $C_{\rm g} = 0.5$ fF and $\omega_{\rm r}/2\pi = 7$ GHz. Let us calculate the amplitude V_0 required to create the coherent state $|\alpha = 2\rangle$,

$$2 = \alpha = e^{i\phi}\int_{-\infty}^{\infty}\lambda(t)dt = \frac{C_{\rm g}V_0 Q_{\rm zpf}}{2\hbar(C_{\rm g}+C)}\int_{-\infty}^{\infty}\varepsilon_{\rm g}(t)dt = \frac{C_{\rm g}V_0 Q_{\rm zpf}}{2\hbar(C_{\rm g}+C)}\sqrt{2\pi}\sigma.$$

Since $Q_{\rm zpf} = \sqrt{C\hbar\omega_{\rm r}/2} = 1.02 \cdot 10^{-18}$ C, the amplitude of the electrical pulse should be $V_0 = 1.48\ \mu$V. The resonator is usually connected to the room temperature instrumentation with a highly attenuated coaxial line. If the overall attenuation of the line is -80 dBm at the resonance frequency, the generator at room temperature should deliver a pulse with amplitude $\tilde{V}_0 = V_0\, 10^{80\ {\rm dB}/20\ {\rm dB}} = 14.8$ mV. Note that if the phase of the pulse is set to $\phi = -\pi$, the final state of the resonator is $|\alpha = -2\rangle$.

13.8 Photon losses

The energy stored in the resonator can leak into the environment due to intrinsic losses and the coupling to the transmission line. This process can be described by a master equation of the form[18]

$$\frac{d\rho}{dt} = \kappa \hat{a}\rho\hat{a}^\dagger - \frac{1}{2}\kappa\hat{a}^\dagger\hat{a}\rho - \frac{1}{2}\kappa\rho\hat{a}^\dagger\hat{a}. \tag{13.56}$$

Here, we assumed that the temperature of the environment is much lower than the transition frequency of the resonator, $k_B T \ll \hbar\omega_r$. The parameter κ is the loss rate of the microwave resonator, which is usually in the range $\kappa/2\pi = 0.5 - 10$ MHz.

Let us show that a resonator initially prepared in a coherent state $|\alpha_0\rangle$ decays to the vacuum state after a sufficiently long time. By substituting $\rho = |\alpha\rangle\langle\alpha|$ into (13.56), we obtain

$$\begin{aligned} \frac{d|\alpha\rangle\langle\alpha|}{dt} &= \kappa\hat{a}|\alpha\rangle\langle\alpha|\hat{a}^\dagger - \frac{1}{2}\kappa\hat{a}^\dagger\hat{a}|\alpha\rangle\langle\alpha| - \frac{1}{2}\kappa|\alpha\rangle\langle\alpha|\hat{a}^\dagger\hat{a} \\ &= \kappa|\alpha|^2|\alpha\rangle\langle\alpha| - \frac{1}{2}\kappa\alpha\hat{a}^\dagger|\alpha\rangle\langle\alpha| - \frac{1}{2}\kappa\alpha^*|\alpha\rangle\langle\alpha|\hat{a}. \end{aligned} \tag{13.57}$$

As shown in Exercise 13.4, the time derivative of a coherent state is given by

$$\frac{d}{dt}|\alpha\rangle\langle\alpha| = \dot{\alpha}a^\dagger|\alpha\rangle\langle\alpha| + |\alpha\rangle\langle\alpha|a\dot{\alpha}^* - 2\Big(\frac{d}{dt}|\alpha|^2\Big)|\alpha\rangle\langle\alpha|. \tag{13.58}$$

Comparing (13.57) with (13.58), we see that $\dot{\alpha}(t) = -k\alpha(t)/2$. Using the initial condition $\alpha(0) = \alpha_0$, the state of the resonator at time t is

$$\boxed{\alpha(t) = \alpha_0 e^{-\kappa_e t/2}.}$$

In conclusion, the state of the resonator approaches the vacuum state with a decay time $\tau = 2/\kappa$. If $\kappa/2\pi = 1$ MHz, the microwave resonator will return to the vacuum state in ≈ 1 μs.

Until now, we have only talked about coherent states, classical states of a quantum harmonic oscillator. In the next section, we present a protocol to prepare a microwave resonator in a non-classical state.

13.9 Preparing a cat state

We conclude this chapter with a discussion of cat states, highly non-classical states of a quantum harmonic oscillator[19]. Cat states play an essential role in quantum computing architectures based on 3D cavities [134, 135]. It is fascinating to investigate how to prepare a microwave resonator in a cat state exploiting a dispersive interaction with a superconducting qubit. Our treatment will follow Ref. [134]. Figure 13.19 shows the quantum circuit for the preparation of a cat state. This circuit consists of a $\pi/2$ rotation, three displacement operators, and two controlled unitary operations, $\mathsf{c_e}U$ and $\mathsf{c}_m\mathsf{X}_\pi$, acting on a superconducting qubit and a resonator. Let us investigate these operators in more detail.

[18]See Exercise 8.9 for a microscopic derivation of this equation.

[19]See Section 4.6 for an introduction to cat states.

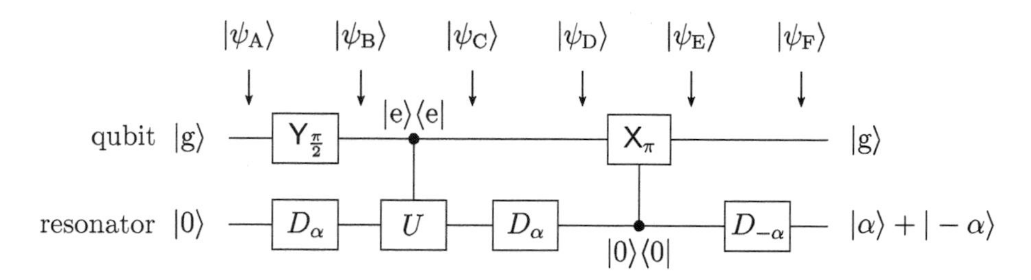

Fig. 13.19 Preparing a cat state. Quantum circuit for the creation of a cat state in a microwave resonator. Experimentally, the displacement operators are implemented by sending short microwave pulses to the resonator (see Section 13.7). In contrast, the $\mathsf{Y}_{\frac{\pi}{2}}$ rotation is implemented by applying an electrical pulse to the qubit (see Section 14.6). The entangling gate $\mathsf{c_e}U$ originates from the free evolution of the system under the Jaynes–Cummings Hamiltonian in the dispersive regime. The conditional rotation $\mathsf{c_0X_\pi}$ is made possible by the dependence of the qubit frequency on the number of photons inside the resonator.

The transformation $\mathsf{c_e}U$ applies the operator $e^{i\pi\hat{a}^\dagger\hat{a}}$ to the state of the resonator only when the qubit is excited,

$$\mathsf{c_e}U = |\mathrm{g}\rangle\langle\mathrm{g}| \otimes \mathbb{1} + |\mathrm{e}\rangle\langle\mathrm{e}| \otimes e^{i\pi\hat{a}^\dagger\hat{a}}. \tag{13.59}$$

This controlled operation originates from the free evolution of the Jaynes–Cummings Hamiltonian in the dispersive regime and is derived in Exercise 13.6.

When a superconducting qubit is dispersively coupled to a microwave resonator, the qubit frequency depends linearly on the number of photons inside the resonator (see eqn 4.95). If the dispersive shift χ is much larger than the qubit and the resonator linewidths, it is possible to apply a single-qubit rotation X_π conditional on the number of photons inside the cavity [134, 514],

$$\mathsf{c}_m\mathsf{X}_\pi = \mathsf{R}_{\hat{n}}(\alpha) \otimes |m\rangle\langle m| + \mathbb{1} \otimes \sum_{n\neq m}^{\infty} |n\rangle\langle n|. \tag{13.60}$$

The effect of this gate is to rotate the qubit state about the x axis by an angle π only when the cavity is populated with m photons. Experimentally, this is done by adjusting the frequency of the electrical pulse ω_d such that $\omega_\mathrm{d} = \omega_\mathrm{q} + 2\chi n$.

We have all the ingredients to study the quantum circuit of Fig. 13.19. The circuit starts with the qubit and the resonator in the ground state. A $\pi/2$ rotation and a displacement operator bring the system into $|\psi_\mathrm{B}\rangle = \frac{1}{\sqrt{2}}(|\mathrm{g}\rangle + |\mathrm{e}\rangle) \otimes |\alpha\rangle$. The controlled unitary $\mathsf{c_e}U$ transforms this state into[20]

[20]Here, we used $e^{i\phi\hat{a}^\dagger\hat{a}}|\alpha\rangle = e^{-\frac{1}{2}|\alpha|^2}\sum_n \frac{\alpha^n}{\sqrt{n!}} e^{i\phi\hat{n}}|n\rangle = e^{-\frac{1}{2}|\alpha|^2}\sum_n \frac{(\alpha e^{i\phi})^n}{\sqrt{n!}}|n\rangle = |e^{i\phi}\alpha\rangle$.

$$|\psi_{\rm C}\rangle = \mathsf{c_e}U|\psi_{\rm B}\rangle = \left(|{\rm g}\rangle\langle{\rm g}|\otimes\mathbb{1} + |{\rm e}\rangle\langle{\rm e}|\otimes e^{i\pi\hat{a}^\dagger\hat{a}}\right)\frac{|{\rm g}\rangle|\alpha\rangle + |{\rm e}\rangle|\alpha\rangle}{\sqrt{2}}$$
$$= \frac{|{\rm g}\rangle|\alpha\rangle + |{\rm e}\rangle e^{i\pi\hat{a}^\dagger\hat{a}}|\alpha\rangle}{\sqrt{2}} = \frac{|{\rm g}\rangle|\alpha\rangle + |{\rm e}\rangle|-\alpha\rangle}{\sqrt{2}}.$$

The second displacement operator produces

$$|\psi_{\rm D}\rangle = \frac{1}{\sqrt{2}}(|{\rm g}\rangle|2\alpha\rangle + |{\rm e}\rangle|0\rangle).$$

This entangled state between the qubit and the resonator is called a **Schrödinger cat state** [515, 516]. The next step is to disentangle the state of the qubit from that of the resonator. This can be done by applying a π rotation to the qubit conditioned on having no photons in the cavity, $\mathsf{c_0X_\pi}$ [517, 518]. This transformation maps $|{\rm e}\rangle|0\rangle$ into $|{\rm g}\rangle|0\rangle$, whereas the other half of the entangled state in $|\psi_{\rm D}\rangle$ becomes

$$\mathsf{c_0X_\pi}|{\rm g}\rangle|2\alpha\rangle = e^{-\frac{1}{2}|2\alpha|^2}\sum_{n=0}^{\infty}\frac{(2\alpha)^n}{\sqrt{n!}}\mathsf{c_0X_\pi}|{\rm g}\rangle|n\rangle$$
$$= e^{-\frac{1}{2}|2\alpha|^2}\left[\mathsf{X_\pi}|{\rm g}\rangle|0\rangle + \frac{(2\alpha)^1}{\sqrt{1}}|{\rm g}\rangle|1\rangle + \frac{(2\alpha)^2}{\sqrt{2}}|{\rm g}\rangle|2\rangle + \ldots\right]$$
$$= e^{-\frac{1}{2}|2\alpha|^2}\left[|{\rm e}\rangle|0\rangle + |{\rm g}\rangle\sum_{n=1}^{\infty}\frac{(2\alpha)^n}{\sqrt{n!}}|n\rangle\right]$$
$$= e^{-\frac{1}{2}|2\alpha|^2}|{\rm e}\rangle|0\rangle + |{\rm g}\rangle|2\alpha\rangle - e^{-\frac{1}{2}|2\alpha|^2}|{\rm g}\rangle|0\rangle \approx |{\rm g}\rangle|2\alpha\rangle.$$

In the last step, we assumed that the modulus of α is sufficiently large such that $e^{-\frac{1}{2}|2\alpha|^2} \approx 0$. Thus, the state of the qubit and the resonator turns into

$$|\psi_{\rm E}\rangle = \frac{1}{\sqrt{2}}(|{\rm g}\rangle|2\alpha\rangle + |{\rm g}\rangle|0\rangle).$$

Lastly, a displacement operator $\hat{D}_{-\alpha}$ applied to the cavity produces

$$|\psi_{\rm F}\rangle = |{\rm g}\rangle\otimes\frac{1}{\sqrt{2}}(|\alpha\rangle + |-\alpha\rangle).$$

At the end of the circuit, the resonator is in a cat state, while the qubit is left in the ground state.

Further reading

A comprehensive introduction to microwave resonators is the book by Pozar [465]. The most relevant sections for the field of superconducting devices are Chapters 2, 4, and 6. Other good references on planar resonators are the PhD theses by Mazin [473], Barends [474], and Göppl [475].

The impact of two-level systems on the performance of microwave resonators at low temperatures was investigated in Refs [486, 485, 487]. A detailed presentation of two-level systems and how they interact with bulk phonons can be found in the work of Phillips [508]. A modern reference is the review by Muller, Cole, and Lisenfeld [509]. The behavior of CPWRs under an applied magnetic field was studied by different groups [519, 520, 521]. These works investigate the effect of flux vortices on resonator performance as well. References [483, 484] include some simulations of the electric field distribution for coplanar waveguide resonators. Some details on the fabrication procedure to increase the internal quality factors of CPWRs can be found in Refs [478, 479, 480, 481, 494, 495, 493, 496]. Finally, the engineering of phonon traps to mitigate the detrimental impact of quasiparticles on the frequency stability can be found in Ref. [488]. This study suggests that the interaction between atmospheric muons and microwave resonators might affect their spectral noise.

A pedagogical introduction to the quantization of an electrical circuit can be found in the Les Houches lectures by Devoret [522]. Other useful resources on this topic are the lectures by Girvin on circuit-QED [523] and the more recent quantum engineer's guide to superconducting qubits [477]. The first investigation of the coherent interaction between a microwave resonator and a superconducting qubit was performed by the Yale group in 2004 [138].

Summary

Parallel RLC resonator

$\bar{E}_{\mathrm{C}} = C\lvert V\rvert^2/4$	Average elec. energy stored in the resonator
$\bar{E}_{\mathrm{L}} = V_0^2/4\omega^2 L$	Average magn. energy stored in the resonator
$\bar{P} = \lvert V\rvert^2/2R$	Power loss per cycle
$\omega_{\mathrm{r}} = 1/\sqrt{LC}$	Resonance frequency
$Q_{\mathrm{i}} = \omega_{\mathrm{r}} RC$	Internal quality factor
$Z_{\mathrm{r}} = R/(1 + i2Q_{\mathrm{i}}\Delta\omega/\omega_{\mathrm{r}})$	Impedance close to resonance
$\hbar\omega_{\mathrm{r}} \ll k_{\mathrm{B}}T$ and $\omega_{\mathrm{r}} \gg \kappa$	Quantum regime

---

Transmission lines

$\partial V/\partial x = -(R + i\omega L)I$	First telegrapher equation
$\partial I/\partial x = -(G + i\omega C)V$	Second telegrapher equation
$V = V^+e^{-\gamma x} + V^-e^{\gamma x}$	Voltage along a transmission line
$I = (V^+e^{-\gamma x} - V^-e^{\gamma x})/Z_0$	Current along a transmission line
$\gamma = \alpha + ik$	Propagation constant
$\gamma = \sqrt{(R + i\omega L)(G + i\omega C)}$	Propagation constant
$Z_0 = \sqrt{(R + i\omega L)/(G + i\omega C)}$	Characteristic impedance

---

Terminated lossy line

$\Gamma = (Z_{\mathrm{L}} - Z_0)\,/\,(Z_{\mathrm{L}} + Z_0)$	Reflection coefficient
$Z(-\ell) = \frac{Z_{\mathrm{L}} + Z_0\tanh(\gamma\ell)}{Z_0 + Z_{\mathrm{L}}\tanh(\gamma\ell)}$	Impedance along the line

---

Short-circuited $\lambda/4$ resonator

$Z(-\ell) \approx Z_0/(\alpha\ell + i\pi\frac{\Delta\omega}{2\omega_{\mathrm{r}}})$	Impedance close to resonance
$\ell \approx \frac{c}{\sqrt{1/2 + \varepsilon_{\mathrm{r}}/2}}\frac{1}{4f_{\mathrm{r}}}$	Length of a $\lambda/4$ CPWR
$\varepsilon_{\mathrm{r}} \approx 11.45$ for Si at $T = 10$ mK	Permittivity of the substrate

---

Parallel RLC resonator capacitive coupled to a feedline

$Q_{\mathrm{e}} = \pi/2(\omega C Z_0)^2$	External quality factor
$1/Q = 1/Q_{\mathrm{i}} + 1/Q_{\mathrm{e}} = \kappa/\omega_{\mathrm{r}}$	Total quality factor
$S_{21} = 2/(2 + Z_0/Z)$	Transmission coefficient
$S_{21} = \frac{\frac{Q_{\mathrm{e}}}{Q_{\mathrm{i}} + Q_{\mathrm{e}}} + 2iQ\frac{\Delta\omega_0}{\omega_0}}{1 + 2iQ\frac{\Delta\omega_0}{\omega_0}}$	Transmission coefficient
$\overline{n} = P_1^+\frac{2Q_{\mathrm{i}}^2 Q_{\mathrm{e}}}{(Q_{\mathrm{i}} + Q_{\mathrm{e}})^2}\frac{1}{\hbar\omega_0^2}$	Photons inside the resonator vs. incident power P_1^+

Quantization of an LC resonator

$\Phi(t) = \int_{-\infty}^{t} V(t')dt'$	Flux across an electrical component
$E_C = C\dot{\Phi}^2/2$	Electrical energy
$E_L = \Phi^2/2L$	Magnetic energy
$L_r = C\dot{\Phi}^2/2 - \Phi^2/2L$	Lagrangian of an LC resonator
$Q = \partial L_r/\partial\dot{\Phi} = C\dot{\Phi}$	Charge (canonical conjugate variable)
$H_r = Q^2/2C + C\omega_r^2\Phi^2/2$	Hamiltonian of an LC resonator
$\hat{\Phi} = \sqrt{\hbar/2C\omega_r}(\hat{a} + \hat{a}^\dagger)$	Flux operator
$\hat{Q} = -i\sqrt{C\hbar\omega_r/2}(\hat{a} - \hat{a}^\dagger)$	Charge operator
$\hat{H}_r = \hbar\omega_r(\hat{a}^\dagger\hat{a} + 1/2)$	Hamiltonian of an LC resonator

Driving an ideal microwave resonator

$V_g(t) = V_0\varepsilon(t)\cos(\omega_d t + \phi)$	Applied voltage
$\varepsilon(t)$	Envelope of the pulse (normalized to one)
ω_d	Pulse frequency
V_0	Pulse amplitude
$\hat{H} = \hat{Q}^2/2C_\Sigma + \hat{\Phi}^2/2L + \hat{Q}C_g V_g(t)/C_\Sigma$	Hamiltonian of the system
$\hat{H}_D = -i\hbar\lambda(t)(\hat{a}e^{i\phi} - \hat{a}^\dagger e^{-i\phi})$	Hamiltonian in the interaction picture
$\lambda(t) = C_g Q_{zpf} V_0 \varepsilon(t)/2\hbar C_\Sigma$	Time-dependent parameter
$\lvert\alpha\rangle$, $\alpha(t) = e^{i\phi}\int_0^\infty \lambda(t')dt'$	Coherent state at the end of the pulse

Exercises

Exercise 13.1 A voltage source applies a sinusoidal signal $V(t) = V_0\cos(\omega t + \phi)$ to a resistor. Calculate the time average of the instantaneous power $P(t) = V(t)I(t)$ and show that the average power dissipated by a resistor is $\bar{P}_R = I_0^2 R/2$. Show that the average power dissipated by an inductor and a capacitor is zero. **Hint:** Calculate the time average of $V(t)I(t)$ over one period.

Exercise 13.2 A microwave generator applies a sinusoidal voltage to a capacitor. Calculate the average energy stored in the capacitor. Repeat the same calculations for an inductor.

Exercise 13.3 Derive eqn (13.29) from eqn (13.28).

Exercise 13.4 Prove eqn (13.58).

Exercise 13.5 Derive the voltage along a lossless transmission line terminated with an open circuit ($Z_L = +\infty$). **Hint**: Start from eqn (13.18).

Exercise 13.6 The Jaynes–Cummings Hamiltonian in the dispersive regime eqn (4.95) can be expressed in the form

$$\hat{H} = \underbrace{\hbar\omega_{\mathrm{r}}\hat{a}^{\dagger}\hat{a} + \hbar\tilde{\omega}_{\mathrm{q}}|\mathrm{e}\rangle\langle\mathrm{e}| - \hbar\chi|\mathrm{g}\rangle\langle\mathrm{g}| \otimes \hat{a}^{\dagger}\hat{a}}_{\hat{H}_0} + \underbrace{\hbar\chi|\mathrm{e}\rangle\langle\mathrm{e}| \otimes \hat{a}^{\dagger}\hat{a}}_{\hat{V}} = \hat{H}_0 + \hat{V},$$

where $\tilde{\omega}_{\mathrm{q}} = \omega_{\mathrm{q}} + \chi$ and we used $\hat{\sigma}_z = |\mathrm{g}\rangle\langle\mathrm{g}| - |\mathrm{e}\rangle\langle\mathrm{e}|$. Show that in the interaction picture, the time evolution operator $e^{\frac{1}{i\hbar}\hat{V}t}$ with $t = \pi/\chi$ is equivalent to the controlled unitary

$$\mathrm{c}U = |\mathrm{g}\rangle\langle\mathrm{g}| \otimes \mathbb{1} + |\mathrm{e}\rangle\langle\mathrm{e}| \otimes e^{i\chi t\hat{a}^{\dagger}\hat{a}}.$$

The controlled unitary cU can be interpreted as a quantum gate $e^{i\chi t\hat{a}^{\dagger}\hat{a}}$ which is applied to the resonator only when the qubit is excited. **Hint**: Expand $e^{\frac{1}{i\hbar}\hat{V}t}$ with a Taylor series.

Solutions

Solution 13.1 From eqn (13.1), we know that for a resistor the current oscillates in phase with the voltage: $I_R(t) = I_0\cos(\omega t + \phi)$ where $I_0 = V_0/R$. The average power dissipated by a resistor can be calculated with a time integral over one period,

$$\begin{aligned}
\bar{P}_R &= \frac{1}{T}\int_0^T V(t)I_R(t)dt = \frac{1}{T}\int_0^T V_0\cos(\omega t + \phi)I_0\cos(\omega t + \phi)dt \\
&= \frac{V_0 I_0}{T}\int_0^T \cos^2(\omega t + \phi)dt \\
&= \frac{V_0 I_0}{T}\int_0^T \cos^2(2\pi t/T + \phi)dt = \frac{V_0 I_0}{2} = \frac{V_0^2}{2R},
\end{aligned}$$

where we used $\omega = 2\pi/T$. For a capacitor, the current oscillates as $I_C(t) = CdV(t)/dt = -V_0 C\omega\sin(\omega t + \phi)$. Therefore,

$$\begin{aligned}
\bar{P}_C = \frac{1}{T}\int_0^T V(t)I_C(t)dt &= -\frac{1}{T}\int_0^T V_0\cos(\omega t + \phi)V_0 C\omega\sin(\omega t + \phi)dt \\
&= -\frac{V_0^2}{T}C\omega\int_0^T \cos(2\pi t/T + \phi)\sin(2\pi t/T + \phi)dt = 0.
\end{aligned}$$

This indicates that an ideal capacitor is a dissipationless component. With similar calculations, one can verify that an ideal inductor is dissipationless too ($\bar{P}_L = 0$).

Solution 13.2 The energy stored in a capacitor biased with a constant voltage V_0 is $E_C = \frac{1}{2}CV_0^2$. In contrast, when the signal oscillates in a sinusoidal way $V(t) = V_0\cos(\omega t + \phi)$, the average energy stored in the capacitor is given by

$$\bar{E}_C = \frac{1}{2}C\frac{1}{T}\int_0^T V_0^2\cos^2(\omega t + \phi)dt = \frac{CV_0^2}{2}\frac{1}{T}\int_0^T \cos^2(2\pi t/T + \phi)dt = \frac{CV_0^2}{4},$$

where T is the period of the oscillations. Similarly, when an oscillating current is applied to an inductor, the current flowing through this electrical element is $I(t) = \frac{1}{L}\int V(t)dt = V_0\sin(\omega t + \phi)/L\omega$ and the average energy stored is

$$\bar{E}_L = \frac{1}{2}L\frac{1}{T}\int_0^T I^2(t)dt = \frac{1}{2}L\frac{1}{T}\int_0^T \frac{V_0^2}{L^2\omega^2}\sin(\omega t+\phi)dt$$
$$= \frac{1}{2}\frac{V_0^2}{L\omega^2}\frac{1}{T}\int_0^T \sin^2(2\pi t/T+\phi)dt = \frac{1}{4}\frac{V_0^2}{\omega^2 L}.$$

Solution 13.3 Let us express (13.28) in a more convenient form. From (13.24), we know that $R = 4Z_0Q_i/\pi$ and therefore

$$\frac{Z}{Z_0} = \frac{\frac{4Q_\mathrm{i}}{\pi}}{1+i2Q_\mathrm{i}\frac{\Delta\omega_\mathrm{r}}{\omega_\mathrm{r}}} + \frac{-i}{\omega CZ_0} = \frac{\frac{4Q_\mathrm{i}}{\pi} - i\frac{8Q_\mathrm{i}^2}{\pi}\frac{\Delta\omega_\mathrm{r}}{\omega_\mathrm{r}}}{1+4Q_\mathrm{i}^2\frac{\Delta\omega_\mathrm{r}}{\omega_\mathrm{r}}} + \frac{-i}{\omega CZ_0}. \tag{13.61}$$

We can write these two terms as a single fraction,

$$\frac{Z}{Z_0} = \frac{\frac{4Q_\mathrm{i}}{\pi} - i\frac{8Q_\mathrm{i}^2}{\pi}\frac{\Delta\omega_\mathrm{r}}{\omega_\mathrm{r}} - \frac{i}{\omega CZ_0}\left(1+4Q_\mathrm{i}^2\frac{\Delta\omega_\mathrm{r}}{\omega_\mathrm{r}}\right)}{1+4Q_\mathrm{i}^2\frac{\Delta\omega_\mathrm{r}}{\omega_\mathrm{r}}}. \tag{13.62}$$

Substituting $\omega CZ_0 = \sqrt{\pi/2Q_\mathrm{e}}$ into (13.62) yields

$$\frac{Z}{Z_0} = \frac{\frac{4Q_\mathrm{i}}{\pi} - i\frac{8Q_\mathrm{i}^2}{\pi}\frac{\Delta\omega_\mathrm{r}}{\omega_\mathrm{r}} - i\sqrt{\frac{2Q_\mathrm{e}}{\pi}}\left(1+4Q_\mathrm{i}^2\frac{\Delta\omega_\mathrm{r}}{\omega_\mathrm{r}}\right)}{1+4Q_\mathrm{i}^2\frac{\Delta\omega_\mathrm{r}}{\omega_\mathrm{r}}}.$$

The resonance occurs when the imaginary part of the impedance is equal to zero. For $Q_\mathrm{i} \gg 1$, this happens when $\Delta\omega_\mathrm{r}/\omega_\mathrm{r} = -\sqrt{2/\pi Q_\mathrm{e}}$. This means that the coupling capacitor slightly shifts the resonance to a lower frequency ω_0, where

$$\frac{\Delta\omega_0}{\omega_0} = \frac{\Delta\omega_\mathrm{r}}{\omega_\mathrm{r}} + \sqrt{\frac{2}{\pi Q_\mathrm{e}}}. \tag{13.63}$$

From eqn (13.61), the impedance Z can be expressed as

$$\frac{Z}{Z_0} = \frac{\frac{4Q_\mathrm{i}}{\pi} - i\sqrt{\frac{2Q_\mathrm{e}}{\pi}}\left(1+2iQ_\mathrm{i}\frac{\Delta\omega_\mathrm{r}}{\omega_\mathrm{r}}\right)}{1+i2Q_\mathrm{i}\frac{\Delta\omega_\mathrm{r}}{\omega_\mathrm{r}}} = \sqrt{\frac{2Q_\mathrm{e}}{\pi}}\frac{-i+2Q_\mathrm{i}\frac{\Delta\omega_\mathrm{r}}{\omega_\mathrm{r}}+2Q_\mathrm{i}\sqrt{\frac{2}{\pi Q_\mathrm{e}}}}{1+i2Q_\mathrm{i}\frac{\Delta\omega_\mathrm{r}}{\omega_\mathrm{r}}}$$
$$= \sqrt{\frac{2Q_\mathrm{e}}{\pi}}\frac{-i+2Q_\mathrm{i}\frac{\Delta\omega_0}{\omega_0}}{1+i2Q_\mathrm{i}\frac{\Delta\omega_0}{\omega_0} - 2iQ_\mathrm{i}\sqrt{\frac{2}{\pi Q_\mathrm{e}}}},$$

where in the last step we used eqn (13.63).

Solution 13.4 The time derivative of a coherent state is given by

$$\frac{d}{dt}|\alpha\rangle\langle\alpha| = \frac{d}{dt}\left[D(\alpha)|0\rangle\langle 0|D^\dagger(\alpha)\right]$$
$$= \left(\frac{d}{dt}D(\alpha)\right)|0\rangle\langle 0|D^\dagger(\alpha) + D(\alpha)|0\rangle\langle 0|\left(\frac{d}{dt}D^\dagger(\alpha)\right). \tag{13.64}$$

To simplify this equation, recall that

$$\frac{d}{dt}D(\alpha) = \frac{d}{dt}e^{\alpha a - \alpha^* a^\dagger} = \underbrace{\left[\dot{\alpha}(a^\dagger - \alpha^*) - \dot{\alpha}^*(a - \alpha) - \frac{1}{2}\frac{d}{dt}|\alpha|^2\right]}_{m} D(\alpha) = mD(\alpha). \tag{13.65}$$

Now substitute (13.65) into (13.64) and obtain

$$\begin{aligned}\frac{d}{dt}|\alpha\rangle\langle\alpha| &= mD(\alpha)|0\rangle\langle 0|D^\dagger(\alpha) + D(\alpha)|0\rangle\langle 0|D^\dagger(\alpha)m^\dagger \\ &= m|\alpha\rangle\langle\alpha| + |\alpha\rangle\langle\alpha|m^\dagger \\ &= \dot{\alpha}a^\dagger|\alpha\rangle\langle\alpha| + |\alpha\rangle\langle\alpha|a\dot{\alpha}^* - 2\left(\frac{d}{dt}|\alpha|^2\right)|\alpha\rangle\langle\alpha|.\end{aligned}$$

Solution 13.5 From eqn (13.18), we have $V(x) = V^+(e^{-\gamma x} + \Gamma e^{\gamma x})$ where

$$\Gamma = \frac{Z_L - Z_0}{Z_L + Z_0} = 1.$$

Since the transmission line is lossless, $\gamma = ik$ and therefore

$$V(x) = V^+(e^{-\gamma x} + e^{\gamma x}) = V^+(e^{-ikx} + e^{ikx}) = 2V^+\cos(kx).$$

Solution 13.6 Expanding $e^{\frac{1}{i\hbar}\hat{V}t}$ with a Taylor series, one obtains

$$\begin{aligned}e^{\frac{1}{i\hbar}\hat{V}t} = e^{-i\chi t|\mathrm{e}\rangle\langle\mathrm{e}|\otimes\hat{a}^\dagger\hat{a}} &= \mathbb{1}\otimes\mathbb{1} + \sum_{k=1}^{\infty}\frac{(-i\chi t)^k}{k!}|\mathrm{e}\rangle\langle\mathrm{e}|\otimes(\hat{a}^\dagger\hat{a})^k \\ &= (|\mathrm{g}\rangle\langle\mathrm{g}| + |\mathrm{e}\rangle\langle\mathrm{e}|)\otimes\mathbb{1} + \sum_{k=1}^{\infty}\frac{(-i\chi t)^k}{k!}|\mathrm{e}\rangle\langle\mathrm{e}|\otimes(\hat{a}^\dagger\hat{a})^k \\ &= |\mathrm{g}\rangle\langle\mathrm{g}|\otimes\mathbb{1} + |\mathrm{e}\rangle\langle\mathrm{e}|\otimes\mathbb{1} + \sum_{k=1}^{\infty}\frac{(-i\chi t)^k}{k!}|\mathrm{e}\rangle\langle\mathrm{e}|\otimes(\hat{a}^\dagger\hat{a})^k \\ &= |\mathrm{g}\rangle\langle\mathrm{g}|\otimes\mathbb{1} + \sum_{k=0}^{\infty}\frac{(-i\chi t)^k}{k!}|\mathrm{e}\rangle\langle\mathrm{e}|\otimes(\hat{a}^\dagger\hat{a})^k \\ &= |\mathrm{g}\rangle\langle\mathrm{g}|\otimes\mathbb{1} + |\mathrm{e}\rangle\langle\mathrm{e}|\otimes e^{i\chi t\hat{a}^\dagger\hat{a}}.\end{aligned}$$

When $t = \pi/\chi$, this transformation reduces to $|\mathrm{g}\rangle\langle\mathrm{g}|\otimes\mathbb{1} + |\mathrm{e}\rangle\langle\mathrm{e}|\otimes e^{i\pi\hat{a}^\dagger\hat{a}} = \mathrm{c}U$.

14
Superconducting qubits

Superconducting devices are
a promising platform
for large-scale quantum computers.

The theory of complexity was revolutionized in the mid-'90s by the discovery of quantum algorithms that can solve some computational problems with an exponential speed-up with respect to their classical counterparts. This theoretical result motivated many scientists to build a scalable quantum computer, a machine capable of running quantum algorithms in a reliable way. Despite significant progress, constructing a quantum processor is still a challenging engineering problem. While qubits, the building blocks of quantum devices, must be strongly coupled with control lines and readout components to perform fast gates and measurements, at the same time they must be well isolated from the environment to achieve high coherence times. These challenges are absent for classical computers where the functioning of the processor is governed by the classical laws of physics.

Many architectures for quantum computers, including trapped ions [524, 525, 526, 527], ultracold atoms [528, 529, 530, 531], spins in silicon [532, 533, 534, 535, 536, 537], nitrogen-vacancies in diamonds [538, 539], quantum dots [540, 541, 542, 543], and polarized photons [544, 545, 546, 547], store and manipulate quantum information on a microscopic scale. Superconducting devices take an entirely different approach. These devices are based on millimeter-scale circuits built with lithographic techniques borrowed from conventional integrated circuits. The fabrication involves a multi-step additive and subtractive patterning process of thin metallic layers that allows high flexibility of the circuit design. Another advantage of superconducting circuits is that they operate in the microwave regime and most of our current technology has been developed at these frequencies. A downside is that they must be operated at millikelvin temperatures inside expensive dilution refrigerators. The limited number of microwave lines that can be routed inside a refrigerator might pose some challenges to scale superconducting processors beyond ≈ 10000 qubits [476]. To overcome this limitation, new strategies must be adopted, including moving part of the control electronics inside the refrigerator itself [548, 549, 550, 551, 552, 553], using photonic links to connect the room-temperature instrumentation to the quantum device [554] and/or connecting multiple dilution refrigerators with cryogenic microwave channels [555].

The main purpose of this chapter is to present the principles of superconducting

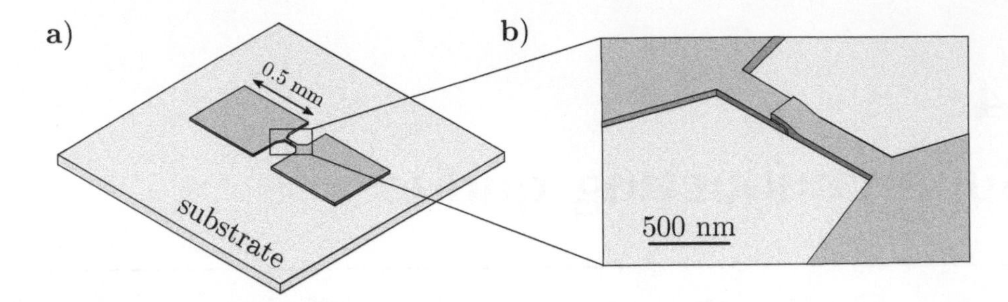

Fig. 14.1 The transmon. a) A transmon qubit with two rectangular pads connected by a Josephson junction. The rectangular pads are typically 100 nm thick and the size of the qubit is around 1 mm^2. The superconducting pads are usually made of Al, Nb, NbTi, NbTiN, or Ta (dark gray), whereas the substrate is either silicon or sapphire (light gray). **b)** A zoom in of the Josephson junction connecting the two superconducting pads.

qubits. We will start our discussion with a presentation of the Josephson effect. We will then introduce the Cooper pair box (CPB) and the transmon, the most widespread superconducting qubit. We will study how to manipulate and read out the state of a transmon. The last part of the chapter will be devoted to the presentation of some two-qubit gate schemes.

We do not aspire to give a historical review of the entire field of superconducting qubits. We have intentionally narrowed our focus on the transmon, because it is currently the most used superconducting qubit for gate-based quantum computation and its working principles are relatively simple to understand. We refer our readers to Refs [556, 557, 146, 558, 559, 560, 523, 561, 562, 563, 564, 565, 477, 566] for more details about superconducting devices. We will assume that the readers are familiar with the basic concepts of superconductivity, electromagnetism, electrical engineering, and quantum mechanics.

14.1 Introduction

A **transmon** consists of two superconducting pads connected by a Josephson junction (see Fig. 14.1a). The size of the metallic pads is on the millimeter scale and their thickness is on the order of 100 nm. The **Josephson junction** is a thin oxide layer sandwiched between the two superconductors as illustrated in Fig. 14.1b. The area of the junction can range from 100×100 nm^2 to 2×2 μm^2.

From an electrical point of view, a transmon can be considered a planar capacitor connected by a nonlinear inductor. The transmon capacitance is usually in the range 70–150 fF and its geometry varies depending on the architecture of the quantum computer. Figure 14.2 shows some common transmon geometries. The rectangular transmon shown in Fig. 14.2a is relatively simple to model: its geometry depends only on a few parameters and it has been employed by several research groups [567, 568, 569, 570, 571, 572, 573]. The transmon represented in Fig. 14.2b is formed by an interdigitated capacitor. This shape was investigated in the early days of circuit quantum electrodynamics [513, 144, 574, 575], but was soon abandoned because it is

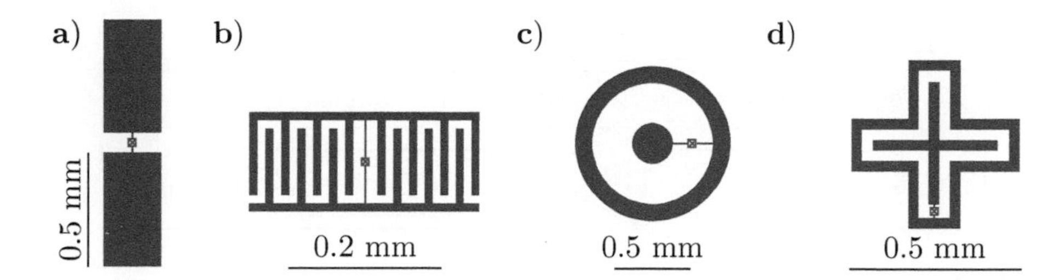

Fig. 14.2 Top view of some transmons. a) A rectangular transmon. **b)** An interdigitated transmon. **c)** A transmon with a concentric shape. **d)** A transmon with the shape of a cross. In all these representations, the capacitor is illustrated in blue and the junction in red.

not optimal for the qubit coherence [576, 500, 577]. The circular geometry in Fig. 14.2c was used in Refs [578, 470] in a coaxial architecture and for gradiometric SQUIDs [579]. Lastly, the superconducting qubit depicted in Fig. 14.2d is sometimes called the **X-mon** due to its shape. This transmon was first designed by the UCSB group [580, 581] and is now widely employed [582, 583, 584].

Superconducting qubits comprise at least one Josephson junction. We present the working principles of this electrical component and show that it behaves like a nonlinear inductor. We then discuss the Cooper pair box, a superconducting qubit formed by a Josephson junction shunted by a capacitor. The next step will be to show that a transmon is nothing but a Cooper pair box operating in a specific regime.

14.2 The Josephson effect

The Josephson junction is the only nonlinear lossless electrical component at low temperatures and is the most important element of superconducting circuits. A **Josephson junction** is formed by two superconducting pads separated by a thin insulating barrier. In the field of superconducting devices, the metal is usually aluminum and the barrier is a layer of aluminum oxide with a thickness of 1 nm.

The behavior of a Josephson junction is described by two equations known as the Josephson equations after Brian Josephson, who derived them in 1962 [585]. He correctly predicted that a small direct current flowing through a junction does not produce any voltage drop. He also predicted that when a small DC voltage is applied across the junction, an alternating current flows through it. We present the Josephson equations following Feynman's derivation [586]. Despite its simplicity, this model is accurate enough to capture the physics of the system. For a more elegant derivation based on the BCS theory, see for example Ref. [587].

When the temperature is much smaller than the critical temperature[1], all the Cooper pairs in a superconducting material are described by the same **macroscopic wavefunction** $\psi(\mathbf{r}) = \sqrt{\rho(\mathbf{r})}e^{i\theta(\mathbf{r})}$ where θ is the phase common to all the particles. The density of Cooper pairs in the superconductor $\rho(\mathbf{r})$ is constant in the bulk of the

[1]The critical temperature of a thin layer of aluminum (niobium) is $T_{\mathrm{c}}^{\mathrm{Al}} = 1.2$ K ($T_{\mathrm{c}}^{\mathrm{Nb}} = 9$ K). The critical temperature slightly depends on the film thickness.

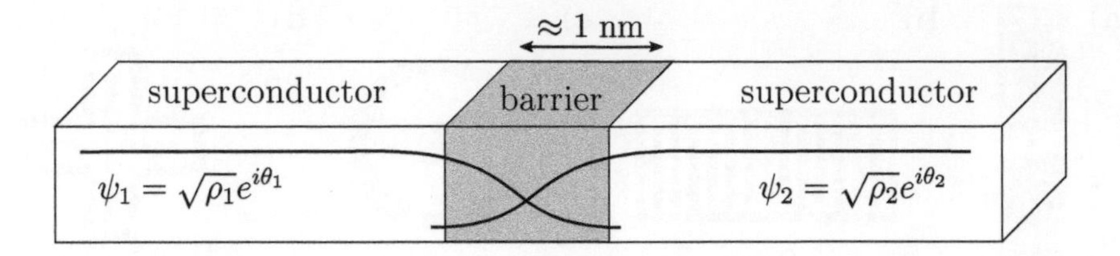

Fig. 14.3 A representation of a Josephson junction. When the temperature of the environment is lower than the critical temperature of the superconductor and external magnetic fields are absent, the electrons in the two electrodes are all paired up in Cooper pairs. The electrons on the two sides of the junction can be described by two macroscopic wavefunctions, which slightly overlap inside the barrier.

---

material and for aluminum $\rho \approx 4 \cdot 10^{24}$ m^{-3} [568, 499]. The wavefunctions on the two sides of the junction will be indicated with the notation

$$\psi_1(\mathbf{r}, t) = \sqrt{\rho_1(\mathbf{r}, t)}e^{i\theta_1(\mathbf{r},t)}, \qquad \psi_2(\mathbf{r}, t) = \sqrt{\rho_2(\mathbf{r}, t)}e^{i\theta_2(\mathbf{r},t)}. \tag{14.1}$$

These wavefunctions are null outside the superconductor.

The two wavefunctions slightly overlap inside the junction, as shown in Fig. 14.3. The Hamiltonian of a Josephson junction is given by $\hat{H} = \hat{H}_1 + \hat{H}_2 + \hat{H}_K$ where $\hat{H}_1$ and $\hat{H}_2$ are the Hamiltonians for the two electrodes and $\hat{H}_K$ is the coupling term. In the basis $\{|\psi_1\rangle, |\psi_2\rangle\}$, the Hamiltonian can be expressed as

$$\hat{H} = E_1|\psi_1\rangle\langle\psi_1| + E_2|\psi_2\rangle\langle\psi_2| + K\left[|\psi_1\rangle\langle\psi_2| + |\psi_2\rangle\langle\psi_1|\right] = \begin{bmatrix} E_1 & K \\ K & E_2 \end{bmatrix},$$

where E_1 and E_2 are the potential energies of a Cooper pair on the left and right side of the junction, respectively. The real parameter K takes into account the weak coupling between the two electrodes. The Schrödinger equation for the wavefunction $\Psi = (\psi_1, \psi_2)$ is $i\hbar\frac{\partial\Psi}{\partial t} = \hat{H}\Psi$, which can be written in a more explicit form as:

$$i\hbar\frac{d\psi_1}{dt} = E_1\psi_1 + K\psi_2, \tag{14.2}$$

$$i\hbar\frac{d\psi_2}{dt} = E_2\psi_2 + K\psi_1. \tag{14.3}$$

By substituting $\psi_i = \sqrt{\rho_i}e^{i\theta_i}$ into (14.2, 14.3), we obtain:

$$-\hbar\frac{d\theta_1}{dt} + i\frac{\hbar}{2\rho_1}\frac{d\rho_1}{dt} = E_1 + K\sqrt{\frac{\rho_2}{\rho_1}}e^{i(\theta_2-\theta_1)}, \tag{14.4}$$

$$-\hbar\frac{d\theta_2}{dt} + i\frac{\hbar}{2\rho_2}\frac{d\rho_2}{dt} = E_2 + K\sqrt{\frac{\rho_1}{\rho_2}}e^{i(\theta_1-\theta_2)}. \tag{14.5}$$

The real and imaginary parts of these two expressions give rise to the first and second Josephson equation. Let us first consider the imaginary parts:

$$\frac{d\rho_1}{dt} = \frac{2K}{\hbar}\sqrt{\rho_1\rho_2}\sin(\theta_2 - \theta_1),$$
$$\frac{d\rho_2}{dt} = -\frac{2K}{\hbar}\sqrt{\rho_1\rho_2}\sin(\theta_2 - \theta_1). \tag{14.6}$$

These expressions obey the charge conservation relation, $d\rho_1/dt = -d\rho_2/dt$ (if a Cooper pair moves from the first electrode to the second, ρ_1 must decrease and ρ_2 must increase at the same rate). Since the time derivative of the charge density $d\rho_1/dt$ is proportional to the current, eqn (14.6) leads to the **first Josephson equation**

$$\boxed{I = I_c \sin\phi,} \tag{14.7}$$

where we introduced the **superconducting phase difference** $\phi = \theta_2 - \theta_1$, and the **critical current** I_c, a parameter that depends on the material, thickness, and area of the junction. Let us now focus on the real part of eqns (14.4, 14.5)

$$-\frac{d\theta_1}{dt} = \frac{E_1}{\hbar} + \frac{K}{\hbar}\sqrt{\frac{\rho_2}{\rho_1}}\cos(\theta_2 - \theta_1), \tag{14.8}$$

$$\frac{d\theta_2}{dt} = -\frac{E_2}{\hbar} - \frac{K}{\hbar}\sqrt{\frac{\rho_1}{\rho_2}}\cos(\theta_2 - \theta_1). \tag{14.9}$$

Suppose that the energy difference between the two superconductors is supplied by an external voltage source such that $E_1 - E_2 = 2eV(t)$, where $V(t)$ is the voltage across the junction. The sum of eqns (14.8, 14.9) yields

$$\frac{d\phi}{dt} = \frac{2e}{\hbar}V(t) + \frac{K}{\hbar}\left(\sqrt{\frac{\rho_2}{\rho_1}} - \sqrt{\frac{\rho_1}{\rho_2}}\right)\cos\phi, \tag{14.10}$$

where as usual $\phi = \theta_2 - \theta_1$ is the phase difference. If the two electrodes have the same density of Cooper pairs $\rho_1 = \rho_2$, eqn (14.10) reduces to

$$\boxed{\frac{d\phi}{dt} = \frac{2e}{\hbar}V(t).} \tag{14.11}$$

This is the **second Josephson equation**. We note that the condition $\rho_1 = \rho_2$ is not compatible with the conservation of charge, $d\rho_1/dt = -d\rho_2/dt$. For this reason (as admitted by Feynman himself), this model is incomplete [586]. Technically, the Hamiltonian should also include a term related to the external voltage source. This point was thoroughly investigated by Ohta [588, 589].

In summary, the two Josephson equations characterizing the electrical behavior of a weak link connecting two superconductors at low temperature are

$$I(t) = I_c \sin\phi(t), \qquad \textbf{1st Josephson equation} \tag{14.12}$$

$$\frac{d\phi(t)}{dt} = \frac{2\pi}{\Phi_0}V(t), \qquad \textbf{2nd Josephson equation} \tag{14.13}$$

where we introduced the **superconducting flux quantum**

$$\boxed{\Phi_0 = h/2e = 2.07 \cdot 10^{-15}\ \mathrm{Tm}^2.} \tag{14.14}$$

This parameter has the units of the magnetic flux. Let us consider two interesting cases at a temperature $T \ll T_\mathrm{c}$:

1) When the voltage across the junction is zero, eqn (14.13) indicates that the phase difference is constant, $\phi = \phi_0$. From (14.12), we conclude that the current is constant too, $I = I_\mathrm{c} \sin \phi_0$, and can take any value in the interval $[-I_\mathrm{c}, I_\mathrm{c}]$. Thus, we have a DC current flowing through the junction with no voltage drop. This phenomenon is called **the DC Josephson effect.**

If the junction is biased with a current $I \gg I_\mathrm{c}$, this electrical component shows a resistive behavior. According to Ohm's law, the voltage across the junction will be $V = R_n I$, where R_n is called the **normal state resistance**. The Ambegaokar–Baratoff formula establishes a simple relation between the critical current and the normal state resistance [590]

$$I_\mathrm{c} = \frac{\pi \Delta(T)}{2eR_n} \tanh\left(\frac{\Delta(T)}{2k_\mathrm{B}T}\right),$$

where $\Delta(T)$ is the temperature dependent **superconducting gap**. In a typical experiment, the temperature of the superconductor is much lower than the critical temperature and the Ambegaokar-Baratoff formula simplifies to $I_\mathrm{c} \approx \pi\Delta(0)/2eR_n$. According to BCS theory, the superconducting energy gap is related to the critical temperature of a superconductor by $\Delta(0) = 1.76 k_\mathrm{B} T_\mathrm{c}$. Therefore, at low temperatures we have

$$\boxed{I_\mathrm{c} \approx \frac{\pi}{2eR_n} 1.76 k_\mathrm{B} T_\mathrm{c}.} \tag{14.15}$$

It is worth mentioning that the critical temperature T_c slightly depends on the film thickness.

2) Let us now consider the case in which a DC voltage source is connected to the junction such that the potential difference across this element is lower than the superconducting gap, $e|V| < 2\Delta$. The integration of (14.13) leads to $\phi = \phi_0 + 2\pi V t/\Phi_0$. Substituting this expression into (14.12), we find that the current through the junction has a sinusoidal behavior in time, $I = I_\mathrm{c} \sin(\phi_0 + 2\pi V t/\Phi_0)$. This is the **AC Josephson effect**: a constant potential difference gives rise to an oscillating current with frequency V/Φ_0. For instance, a voltage difference of only 10 μV generates a current that oscillates at 5 GHz. The high frequency of these small oscillations makes it difficult to observe this phenomenon experimentally [591].

14.2.1 The Josephson junction as a nonlinear inductor

Figure 14.4a shows the electrical circuit of a Josephson junction. The cross represents an ideal electrical element governed by the two Josephson equations. The capacitance

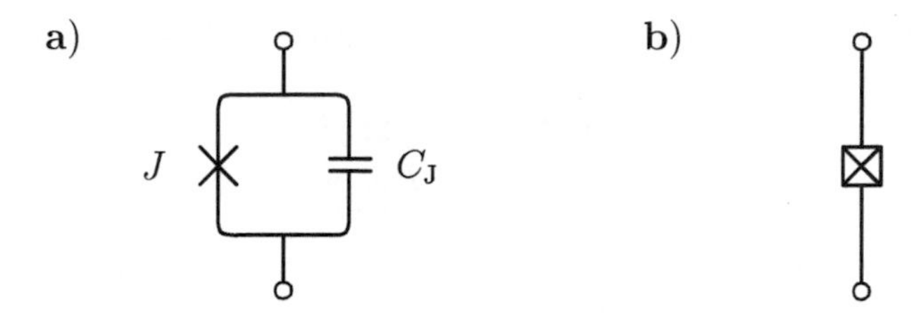

Fig. 14.4 The equivalent circuit of a Josephson junction. **a)** The electrical circuit of a Josephson junction includes an ideal element (the cross on the left) described by the two Josephson equations (14.12, 14.13) shunted by a capacitance C_J. **b)** The ideal junction J and the intrinsic capacitance C_J are usually combined in one symbol, a boxed-X.

C_J models the finite capacitance between the two metallic electrodes,

$$C_J = \epsilon A/d,$$

where A is the area of the junction, d is the thickness of the junction, and ϵ is the permittivity of the oxide[2]. Since any Josephson junction has a small capacitance, the symbols of the junction and the intrinsic capacitance are usually combined in a single electrical symbol as shown in Fig. 14.4b.

A Josephson junction behaves like a nonlinear inductor. This can be better understood by calculating the derivative of the first Josephson equation with respect to time

$$\frac{dI}{dt} = I_c \cos\phi \frac{d\phi}{dt} = \frac{2\pi I_c \cos\phi}{\Phi_0} V.$$

In this formula, the time derivative of the current is proportional to the voltage. This is the typical behavior of an inductor. By defining the **Josephson inductance** L_J according to the usual definition $dI/dt = V/L_J$, we have

$$L_J = \frac{\Phi_0}{2\pi I_c \cos\phi} = \frac{L_{J0}}{\sqrt{1-\sin^2\phi}} = \frac{L_{J0}}{\sqrt{1-I^2/I_c^2}},$$

where we introduced the constant $L_{J0} = \Phi_0/2\pi I_c$. The Josephson inductance is **nonlinear** in the phase difference. This nonlinear behavior makes the Josephson junction a unique component in electrical engineering. The electrical energy stored by the Josephson inductance can be calculated with a time integral of the product IV as follows,

$$\begin{aligned} U = \int_0^t IV dt &= \frac{\Phi_0 I_c}{2\pi} \int_0^t \sin(\phi) \frac{d\phi}{dt} dt \\ &= \frac{\Phi_0 I_c}{2\pi} \int_0^\phi \sin(\phi')\, d\phi' = -E_J \cos\phi + \text{const}, \end{aligned} \qquad (14.16)$$

[2]The typical value of the capacitance of a Josephson junction with an area of $200 \times 200\ \text{nm}^2$ and a thickness of 1 nm is $C_J \approx 10$ fF.

where we used (14.12, 14.13) and in the last step we introduced the **Josephson energy**

$$\boxed{E_{\rm J} = \frac{\Phi_0 I_{\rm c}}{2\pi} = \frac{(\Phi_0/2\pi)^2}{L_{\rm J0}}.} \tag{14.17}$$

This parameter is determined by the room-temperature resistance of the junction, R_n. Indeed, combining eqns (14.15, 14.17), we see that [472]

$$\boxed{E_{\rm J} = \frac{\Phi_0}{2\pi}\frac{\pi}{2eR_n}1.76k_{\rm B}T_{\rm c}.} \tag{14.18}$$

If the junction is made of aluminum with a critical temperature $T_{\rm c} = 1.2$ K, this formula reduces to $E_{\rm J} = h\alpha/R_n$ where $\alpha = 142000$ GHz/Ω. The room-temperature resistance can be adjusted by varying the area and thickness of the oxide barrier. Now that we have covered the basic concepts of the Josephson junction, we are ready to present the working principles of the Cooper pair box.

14.3 The Cooper pair box

The Cooper pair box (CPB) is a superconducting qubit that was investigated by Buttiker in 1987 [592] and experimentally implemented by the Saclay group in 1998 [593]. The coherent control of a CPB in the time domain was achieved for the first time by Nakamura and coworkers in 1999 [594]. In this section, we will describe some properties of a Cooper pair box including its energy levels and main drawbacks. We will then show that a transmon is nothing but a Cooper pair box operating in a specific regime.

A **Cooper pair box** consists of a small superconducting island connected to a reservoir of superconducting electrons via a Josephson junction [595]. Figure 14.5a shows the electrical circuit of a CPB. The number of Cooper pairs on the island (reservoir) will be denoted as $n_{\rm I}$ ($n_{\rm R}$). The tunneling of one Cooper pair from the island to the reservoir changes the state of the system from $|n_{\rm I}, n_{\rm R}\rangle$ to $|n_{\rm I}-1, n_{\rm R}+1\rangle$. To keep the notation concise, we can identify the state of a CPB with the number of Cooper pairs in excess on the island

$$|n\rangle = |n_{\rm I} + n, n_{\rm R} - n\rangle.$$

We note that n can be negative since a Cooper pair can also tunnel from the island back to the reservoir.

The Hamiltonian of a Cooper pair box can be derived by drawing an analogy with an LC resonator. As explained in Section 13.6, the Hamiltonian of an LC resonator is the sum of a capacitive and inductive term

$$\hat{H}_{\rm r} = 4E_{\rm C}\hat{n}^2 + \frac{1}{2}E_{\rm L}\hat{\phi}^2,$$

where $E_{\rm C}$ is the charging energy and $E_{\rm L}$ is the inductive energy. The circuit of a CPB resembles that of a parallel LC resonator. The only difference is that the inductor has

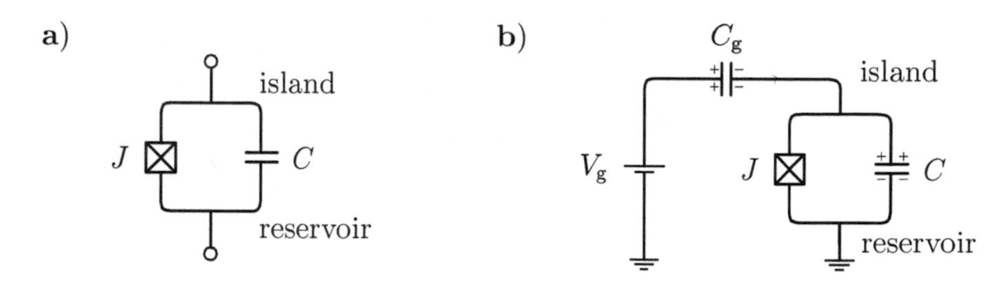

Fig. 14.5 The electrical circuit of a Cooper pair box. a) The circuit of a CPB consists of a Josephson junction shunted by a capacitor C. **b)** Electrical circuit of a CPB capacitively coupled to a DC voltage source by a gate capacitor C_g.

been replaced by a Josephson junction. Therefore, **the Hamiltonian of a CPB** is given by

$$\boxed{\hat{H} = 4E_C\hat{n}^2 - E_J \cos\hat{\phi},} \tag{14.19}$$

where we used eqn (14.16). The two conjugate variables are the number of Cooper pairs $\hat{n}$ and the superconducting phase difference $\hat{\phi}$. The energy scale of the Hamiltonian is determined by the charging energy $E_C = e^2/2C$ and the Josephson energy $E_J = \Phi_0 I_c/2\pi$. The former is set by the geometry of the qubit, the latter by the room-temperature resistance of the junction. The ratio E_J/E_C significantly affects the energy levels of the Hamiltonian as we shall see later.

To control the number of Cooper pairs on the island, one can connect a DC voltage source to the system via a gate capacitor C_g as shown in Fig. 14.5b. When $V_g = 0$, both the gate capacitor and the qubit capacitor are discharged. When the applied DC voltage is slightly greater than zero, some positive charge $Q_g = C_g V_g$ will accumulate on the left plate of the gate capacitor. For the island to stay neutral, some negative charge $-Q_g$ will accumulate on the right plate of C_g and the same amount of positive charge will accumulate on the top plate of the qubit capacitor. When V_g is large enough such that $Q_g \approx 2e$, a Cooper pair will tunnel from the reservoir to the island and the qubit capacitor will be discharged again (it is energetically more favorable for a Cooper pair to tunnel through the junction than for the qubit capacitor to accumulate more charge). The tunneling of Cooper pairs from and to the island takes place whenever $Q_g = 2en$ where $n \in \mathbb{Z}$. This quantized behavior is illustrated in Fig. 14.6a.

The remaining part of this section will focus on the derivation of the energy levels of a CPB. The external voltage source is an additional control knob on the number of Cooper pairs on the island. Hence, the Hamiltonian of a Cooper pair box connected to a DC voltage source takes the form

$$\boxed{\hat{H} = 4E_C(\hat{n} - n_g)^2 - E_J \cos\hat{\phi},}$$

where $n_g = C_g V_g/2e$ is the **gate charge**, i.e. the normalized charge that accumulates on the gate capacitor due to the applied DC voltage. To calculate the eigenvalues of the CPB Hamiltonian, it is useful to express the number operator in the phase

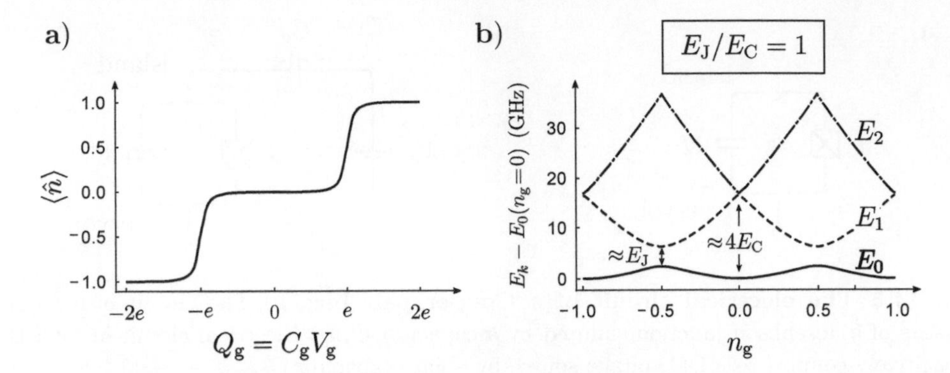

Fig. 14.6 A Cooper pair box connected to a DC voltage source. a) Plot of the average number of Cooper pairs on the island, $\langle\hat{n}\rangle$, as a function of the normalized charge on the gate capacitor, $n_g = C_g V_g/2e$. The analytical dependence between $\langle\hat{n}\rangle$ and n_g can be found in Ref. [593]. **b)** Plot of the first three energy levels of a CPB as a function of the gate charge. The energy levels are periodic in n_g. In this example, $E_J/h = E_C/h = 4$ GHz. The qubit frequency is $f_{01} \approx 16$ GHz at $n_g = 0$ and $f_q \approx 4$ GHz at the sweet spots $n_g = \pm 0.5$.

basis[3]: $\langle\phi|\hat{n}|\psi\rangle = -i\frac{\partial}{\partial\phi}\psi(\phi)$. In this basis, the time-independent Schrödinger equation $\hat{H}\psi_k(\phi) = E_k\psi_k(\phi)$ becomes

$$4E_C\Big(-i\frac{\partial}{\partial\phi} - n_g\Big)^2\psi_k(\phi) - E_J\cos(\phi)\psi_k(\phi) = E_k\psi_k(\phi).$$

The solution of this differential equation provides the eigenenergies and eigenfunctions. The eigenenergies E_k can be expressed in terms of the Mathieu characteristic function $\mathcal{M}_A$. This is an analytical function that depends on two variables [593, 596]:

$$\boxed{E_k = E_C\,\mathcal{M}_A(x_k,\, y),}$$

where $y = -E_J/2E_C$ and

$$x_k = \lfloor 2n_g\rfloor - 2n_g - (-1)^{k+\lfloor 2n_g\rfloor}k + (k + \lfloor 2n_g\rfloor) \bmod 2.$$

In Fig. 14.6b, we plot the first three energy levels[4] as a function of the gate charge n_g.

[3]This is a direct consequence of the commutation relation $[\hat{\phi},\hat{n}] = i\mathbb{1}$. For example, the commutation relation between the position and the momentum is $[\hat{x},\hat{p}] = i\hbar\mathbb{1}$ from which follows that $\langle x|\hat{p}|\psi\rangle = -i\hbar\frac{\partial}{\partial x}\psi(x)$ (see for example Ref. [128], Section 1.7).

[4]In Mathematica, these eigenenergies can be expressed as

```
Energy[k_, ng_] := Ec MathieuCharacteristicA[x[k, ng], y],
```

where y=−EJ/(2EC) and

```
x[k_, ng_] := Floor[2ng] − 2ng − (−1)^(k+Floor[2ng]) k + Mod[k + Floor[2ng], 2].
```

Unlike a quantum harmonic oscillator, the energy levels are not equally spaced and the subspace formed by the first two states can be used as a qubit. The qubit frequency is $f_{01}(n_g) = (E_1 - E_0)/h$ and can be tuned by varying the applied DC voltage.

The coherence of a qubit is characterized by two parameters, T_1 and T_ϕ. The **relaxation time** T_1 is the timescale in which the qubit relaxes from the excited state to the ground state. The **dephasing time** T_ϕ is the typical timescale in which the information about the relative phase of the qubit is lost. As explained in Section 8.7, the dephasing time is affected by the fluctuations of the qubit frequency. For this reason, a CPB is usually operated at the **sweet spot** $n_g = 0.5$ because at this working point the qubit frequency is less sensitive to fluctuations of the charge on the gate capacitor. In superconducting devices, the charge fluctuations are characterized by $1/f$ noise with power spectral density $S_{n_g n_g}(\omega) = 2\pi A_Q^2/|\omega|$ [597, 598]. In eqn (8.88), we showed that the dephasing time is inversely proportional to the derivative of the qubit frequency with respect to the noise source, i.e. $T_\phi \propto 1/|dE_{01}/dn_g|$. However, this formula cannot be used to calculate the dephasing time of a Cooper pair box at the sweet spot because at this operating point $dE_{01}/dn_g = 0$. The second derivative must be taken into account and it can be shown that [599, 600, 601]

$$T_\phi \approx \frac{\hbar}{A_Q^2 \pi^2} \frac{1}{\left|\frac{\partial^2 E_{01}}{\partial n_g^2}\right|}. \tag{14.20}$$

Since the first transition energy of a CPB close to the sweet spot can be approximated by the function

$$E_{01}(n_g) = \sqrt{[4E_C(2n_g - 1)]^2 + E_J^2}\,,$$

the second derivative at $n_g = 0.5$ reduces to $d^2E_{01}/dn_g^2 = 8E_C^2/E_J$. Thus, the upper limit of T_ϕ due to charge noise is

$$\boxed{T_\phi \approx \frac{\hbar}{A^2\pi^2}\frac{E_J}{(8E_C)^2}.}$$

Using realistic values for the qubit parameters $E_J/h = E_C/h = 6$ GHz [595, 594], and for the charge noise $A_Q = 2 \times 10^{-4}\ e/\sqrt{\text{Hz}}$ [597, 598], we obtain an upper limit of T_ϕ on the order of 1 μs. Even if at the sweet spot the first derivative with respect to the gate charge is zero, the contribution from the second derivative is significant and severely limits the dephasing time of a CPB. In the next section, we present a strategy to decrease the sensitivity to charge noise.

14.3.1 From the Cooper pair box to the transmon

One method to improve the dephasing time of a Cooper pair box is to flatten the energy levels, so they do not strongly depend on the gate charge. This can be done by increasing the ratio between the Josephson energy E_J and the charging energy E_C. A Cooper pair box with a ratio $E_J/E_C \gg 1$ is called a **transmon** [599]. A simple way to increase this ratio is to increase the size of the qubit capacitor so that E_C decreases. More sophisticated approaches have also been explored [602].

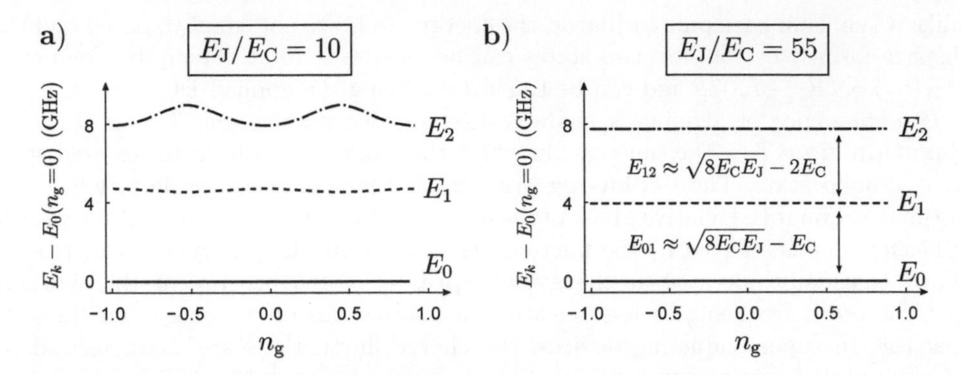

Fig. 14.7 Energy levels of a transmon. a) Plot of the energy levels of a transmon as a function of the gate charge (here, $E_C/h = 0.6$ GHz and $E_J/h = 6$ GHz). The energy levels show a sinusoidal behavior that can be approximated with eqn 14.21. This approximation is particularly accurate when $E_J/E_C > 20$. **b)** Plot of the energy levels with $E_C/h = 0.2$ GHz and $E_J/h = 11$ GHz. The energy levels are almost flat.

The first three energy levels of a Cooper pair box with $E_J/E_C = 10$ and $E_J/E_C = 55$ are shown in Fig. 14.7. From this figure, it is evident that the energy levels do not show a significant dependence on the gate charge for large values of E_J/E_C. When $E_J/E_C > 20$, the energy levels can be approximated by a cosine [599],

$$\boxed{E_k(n_g) \approx E_k(n_g{=}1/4) - \frac{\epsilon_k}{2}\cos(2\pi n_g),} \tag{14.21}$$

where the parameter ϵ_k is the amplitude of the cosine

$$\epsilon_k \approx (-1)^k E_C \frac{2^{4k+5}}{k!}\sqrt{\frac{2}{\pi}}\left(\frac{E_J}{2E_C}\right)^{\frac{k}{2}+\frac{3}{4}} e^{-\sqrt{8E_J/E_C}}. \tag{14.22}$$

The subspace formed by the first two energy levels can be used as a qubit. The transition energy from the ground state to the first excited state is given by

$$\begin{aligned} E_{01}(n_g) &= E_1(n_g) - E_0(n_g) \\ &\approx E_1(n_g{=}1/4) - E_0(n_g{=}1/4) - \frac{\epsilon_1}{2}\cos(2\pi n_g), \end{aligned} \tag{14.23}$$

where we used (14.21) and $\epsilon_0 \approx 0$. Using eqns (8.88, 14.23), one can calculate the upper limit on T_ϕ due to charge noise

$$T_\phi(n_g) \approx \frac{\hbar}{\sqrt{c}A_Q}\frac{1}{|\frac{\partial E_{01}}{\partial n_g}|} = \frac{\hbar}{\sqrt{c}A_Q}\frac{1}{|\pi\epsilon_1\sin(2\pi n_g)|}, \tag{14.24}$$

where $c \approx 10$ and $A_Q = 2\times10^{-4}e/\sqrt{\text{Hz}}$ [597]. The lowest value of the dephasing time is obtained when $\sin(2\pi n_g) = \pm1$. Therefore, the upper limit of the dephasing time of a transmon due to charge noise is

$$\boxed{T_\phi = \frac{\hbar}{\sqrt{c}A_Q}\frac{1}{\pi|\epsilon_1|}.} \tag{14.25}$$

Using realistic values $E_J/h = 11$ GHz, and $E_C/h = 0.2$ GHz, we obtain $T_\phi \approx 20$ ms. This shows that in the transmon limit the qubit becomes less sensitive to charge noise. Hence, there is no need to connect a DC voltage source to bias the transmon at a specific sweet spot since any value of the gate charge n_g leads to a good dephasing time.

It is important to emphasize that the dephasing time reported in eqn (14.25) is just an upper limit of the total dephasing time. The total dephasing time T_ϕ^{tot} is affected by multiple contributions that add up together,

$$\frac{1}{T_\phi^{\text{tot}}} = \frac{1}{T_\phi^{(1)}} + \frac{1}{T_\phi^{(2)}} + \ldots + \frac{1}{T_\phi^{(n)}},$$

where each term $1/T_\phi^{(j)}$ is caused by a different noise source, such as charge noise, flux noise, critical current noise, quasi particle tunneling and thermal fluctuations [599]. The lowest $T_\phi^{(j)}$ has the largest impact on T_ϕ^{tot}.

14.4 The transmon

Since a Cooper pair box and a transmon have the same electrical circuit, they are described by the same Hamiltonian[5],

$$\boxed{\hat{H} = 4E_C\hat{n}^2 - E_J\cos\hat{\phi}.} \tag{14.26}$$

The first term accounts for the energy stored in the capacitor, while the second arises from the inductive nature of the Josephson junction. It is interesting to observe that the Hamiltonian of a transmon is equivalent to that of a classical **pendulum** [599]. Consider a rod of length ℓ connected to a mass m. The mass is free to swing back and forth. The system is subject to a gravitational field pointing down, $\mathbf{g} = -g\mathbf{e}_y$. The potential energy due to the gravitation field is given by $U = -mg\ell\cos\phi$ where ϕ is the angle of the oscillations with respect to a vertical line. The kinetic energy of the mass can be expressed in terms of the angular momentum $L_z = m\ell\dot{\phi}$ as $T = L_z^2/2m\ell^2$. The Hamiltonian of a pendulum is given by

$$H = T + U = \frac{L_z^2}{2m\ell^2} - mg\ell\cos\phi. \tag{14.27}$$

By comparing (14.27) with (14.26), one can outline the following correspondence between a transmon and a pendulum: $E_C \leftrightarrow \hbar^2/8m\ell^2$ and $E_J \leftrightarrow mg\ell$. It is interesting to

[5] Here, we are assuming that the transmon is not connected to a DC voltage source, and therefore $n_g = 0$.

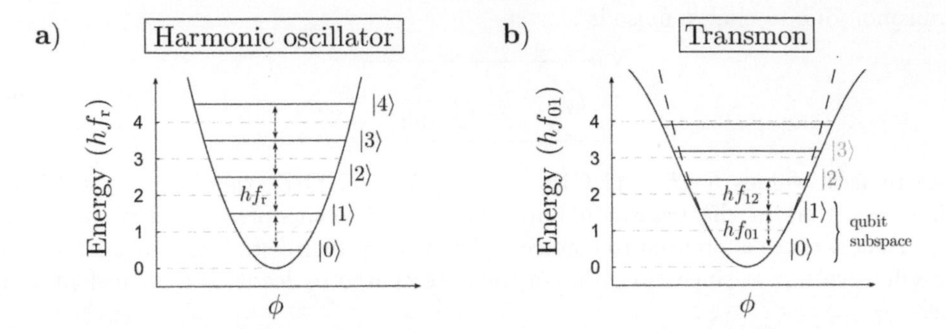

Fig. 14.8 Energy potentials. a) The potential energy of a microwave resonator, $U(\phi) = \frac{1}{2}E_L\phi^2$. The energy levels are evenly spaced. **b)** The potential energy of a transmon $U(\phi) = -E_J \cos\phi$ (solid line) compared to a harmonic potential (dashed line). The transition energies are not equidistant. This allows us to selectively address the subspace formed by energy levels. Typically, the states $|0\rangle$ and $|1\rangle$ are used as a qubit.

see how the Josephson energy E_J acts like the "gravitational force" of the pendulum. Since for a transmon $E_J/E_C \gg 1$, the strong gravitational field restricts the system to perform small oscillations around the equilibrium point[6]. This constraint allows us to expand the cosine term in (14.26) with a Taylor series and neglect higher order terms,

$$\boxed{\hat{H} \approx 4E_C\hat{n}^2 + \frac{1}{2}E_J\hat{\phi}^2 - \frac{E_J}{4!}\hat{\phi}^4.} \tag{14.28}$$

The first two terms represent the Hamiltonian of a quantum harmonic oscillator. The term proportional to $\hat{\phi}^4$ is a small contribution that makes the potential of a transmon slightly anharmonic. The potential energy of a harmonic oscillator and a transmon are shown side by side in Fig. 14.8.

Let us calculate the energy levels E_k of the transmon Hamiltonian (14.28) by solving the eigenvalue problem $\hat{H}|k\rangle = E_k|k\rangle$. Since the operators $\hat{\phi}$ and $\hat{n}$ obey the canonical commutation relation $[\hat{\phi}, \hat{n}] = i\mathbb{1}$, we can introduce the lowering and raising operators, $\hat{b}$ and $\hat{b}^\dagger$, as

$$\hat{\phi} = \sqrt{\xi}(\hat{b} + \hat{b}^\dagger), \tag{14.29}$$

$$\hat{n} = -\frac{i}{2\sqrt{\xi}}(\hat{b} - \hat{b}^\dagger), \tag{14.30}$$

where

$$\xi = \sqrt{2E_C/E_J}$$

[6]This is not true for a Cooper pair box because in this case $E_J \approx E_C$. Hence, the Taylor series expansion (14.28) cannot be performed for a CPB.

is a constant parameter[7]. Since $[\hat{\phi},\hat{n}] = i\mathbb{1}$, then $[\hat{b},\hat{b}^\dagger] = \mathbb{1}$. Substituting eqn (14.29, 14.30) into (14.28), the transmon Hamiltonian becomes

$$\boxed{\hat{H} = \sqrt{8E_{\mathrm{J}}E_{\mathrm{C}}}\,\hat{b}^\dagger\hat{b} - \frac{E_{\mathrm{C}}}{12}(\hat{b}+\hat{b}^\dagger)^4.} \tag{14.31}$$

The first term is the Hamiltonian of a quantum harmonic oscillator with eigenvectors $|k\rangle$ and eigenvalues $\sqrt{8E_{\mathrm{J}}E_{\mathrm{C}}}k$. Since $E_{\mathrm{C}} \ll E_{\mathrm{J}}$, the second term $\hat{V} = -E_{\mathrm{C}}(\hat{b}+\hat{b}^\dagger)^4/12$ is a small contribution that can be treated with perturbation theory. The first order correction to the energy levels arises from the diagonal elements of the perturbation $\langle k|\hat{V}|k\rangle$ (this is a well-known result of perturbation theory, see for example Ref. [128]). One can verify that[8] $\langle k|\hat{V}|k\rangle = -E_{\mathrm{C}}(6k^2+6k+3)/12$. Thus, the eigenenergies of the transmon Hamiltonian are

$$\boxed{E_k \approx \sqrt{8E_{\mathrm{J}}E_{\mathrm{C}}}\,k - \frac{E_{\mathrm{C}}}{2}(k^2+k),} \tag{14.32}$$

where we dropped some constant terms of no physical significance. As illustrated in Fig. 14.8b, the energy levels of a transmon are not equally spaced and the subspace formed by the two lowest states can be used as a qubit. The qubit frequency is defined as $f_{01} = (E_1 - E_0)/h$ where h is the Planck constant. From eqn (14.32), we have

$$\boxed{f_{01} \approx (\sqrt{8E_{\mathrm{J}}E_{\mathrm{C}}} - E_{\mathrm{C}})/h.} \tag{14.33}$$

A more accurate analytical expression of the qubit frequency that takes into account higher order corrections can be found in Refs [603, 604].

A large $E_{\mathrm{J}}/E_{\mathrm{C}}$ ratio makes the transmon less sensitive to charge noise. However, this comes at the cost of a much smaller anharmonicity. We define the **anharmonicity** as the difference between the second transition energy and the first transition energy,

$$\boxed{\eta = \frac{(E_2 - E_1) - (E_1 - E_0)}{\hbar} = \omega_{12} - \omega_{01}.}$$

For a transmon, η is negative. Ideally, the absolute value of the anharmonicity should be as large as possible so that external microwave pulses can selectively drive the first excited state without unintentionally exciting higher energy levels. Unfortunately, the anharmonicity of a transmon is relatively small: using eqn (14.32), one can verify that $\eta = -E_{\mathrm{C}}/\hbar$, meaning that $\eta/2\pi \approx 300$ MHz for a typical transmon. For this reason, special techniques must be adopted to prevent a microwave signal from exciting the transmon to higher energy levels [605]. We will come back to this point in Section 14.6.3 when we present the DRAG technique.

[7]For a typical transmon, $E_{\mathrm{J}}/E_{\mathrm{C}} \approx 50$ and therefore $\xi \approx 0.2$.

[8]To show this, expand $\hat{V} = -E_{\mathrm{C}}(\hat{a}+\hat{a}^\dagger)^4/12$ and remember that $[\hat{a},\hat{a}^\dagger] = \mathbb{1}$, and $\langle k|\hat{a}|k\rangle = \langle k|\hat{a}^\dagger|k\rangle = 0$.

Example 14.1 Let us go through the steps required to design a transmon with a frequency $f_{01} = 4$ GHz. Combining eqn (14.33) with the constraint $E_J/E_C = 50$, we see that $E_C/h = 0.21$ GHz and $E_J/h = 10.52$ GHz. Since $E_C = e^2/2C$, this means that the transmon capacitance must be $C = 92$ fF. The capacitor can be designed with CAD software, while the capacitance value can be simulated with FEM software. The design can be adjusted until the desired capacitance has been obtained. The next step is calculating the room-temperature resistance that sets the right value of E_J. Since $E_J = h\alpha/R$, where $\alpha \approx 142000$ GHz/Ω for aluminum junctions, the room-temperature resistance of the junction must be $R = 13.5$ kΩ. After fabrication, the junction's resistance increases over time; this phenomenon is usually called "junction aging". Different methods to mitigate junction aging have been explored, see for example Refs [606, 607].

14.5 The tunable transmon

The transition frequency of a transmon (14.33) is determined by the charging energy and the Josephson energy. These two parameters are fixed by the geometry of the capacitor and the room-temperature resistance of the junction, respectively. Hence, the qubit frequency cannot be easily changed after fabrication[9].

To implement some two-qubit gate schemes, such as swap interactions, it is necessary to tune the qubit frequency on short time scales. A common technique to create a **tunable transmon** is to add an extra junction to the qubit as shown in Fig. 14.9. The two junctions in parallel form a loop known as a **superconducting quantum interference device** (SQUID). The qubit frequency can be varied by changing the intensity of the magnetic field through the SQUID. To see this, let us start from the Hamiltonian of a tunable transmon

$$\boxed{\hat{H} = 4E_C\hat{n}^2 - E_{J1}\cos\hat{\phi}_1 - E_{J2}\cos\hat{\phi}_2,} \tag{14.34}$$

where E_{J1} and E_{J2} are the Josephson energies of the two junctions, and the operators $\hat{\phi}_1$ and $\hat{\phi}_2$ are the phase differences across the junctions. The quantities $\hat{\phi}_1$ and $\hat{\phi}_2$ are not independent due to the quantization of the magnetic flux through the SQUID. As discussed in Exercise 14.3, the difference between $\hat{\phi}_1$ and $\hat{\phi}_2$ is proportional to the applied magnetic flux Φ_{ext},

$$\hat{\phi}_1 - \hat{\phi}_2 = \frac{2\pi}{\Phi_0}\Phi_{\text{ext}}\mathbb{1} \pmod{2\pi}. \tag{14.35}$$

The magnetic flux is defined as the surface integral of the magnetic field over the SQUID area,

$$\Phi_{\text{ext}} = \int_\Sigma \mathbf{B} \cdot d\mathbf{\Sigma}.$$

[9] A sophisticated technique to change the qubit frequency after fabrication is based on laser annealing, see Ref. [607].

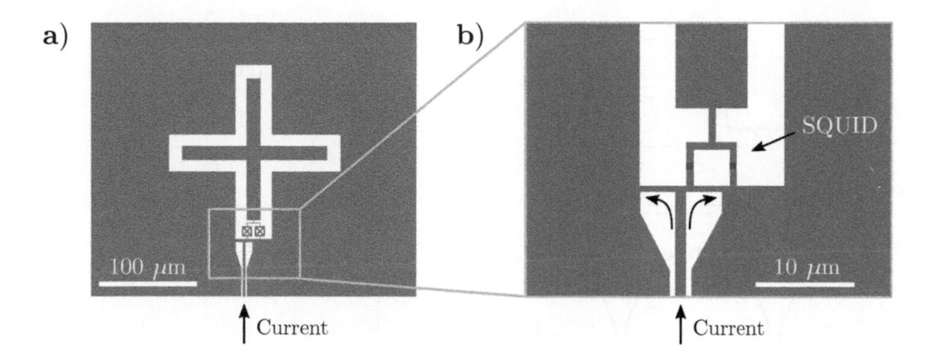

Fig. 14.9 Tunable transmons and flux bias lines. a) A tunable transmon is composed of two superconducting pads connected by a SQUID. A SQUID is a small superconducting loop interrupted by two Josephson junctions. **b)** The qubit frequency can be tuned by delivering a DC current down a flux bias line, creating a small magnetic field through the SQUID. In these images, the silicon is represented in white, the superconducting metal in blue, and the junctions in red.

Defining the effective phase difference as $\hat{\varphi} = (\hat{\phi}_1 + \hat{\phi}_2)/2$ and the constant operator $\phi_{\text{ext}} = \mathbb{1}2\pi\Phi_{\text{ext}}/\Phi_0 = \hat{\phi}_1 - \hat{\phi}_2$, the Hamiltonian becomes

$$\hat{H} = 4E_\text{C}\hat{n}^2 - E_{\text{J}1}\cos(\hat{\varphi} + \phi_{\text{ext}}/2) - E_{\text{J}2}\cos(\hat{\varphi} - \phi_{\text{ext}}/2)\,. \tag{14.36}$$

In Exercise 14.2, we show how to simplify this expression with some trigonometric identities to obtain

$$\boxed{\hat{H} = 4E_\text{C}\hat{n}^2 - E_\text{J}(\Phi_{\text{ext}})\cos\hat{\varphi}\,.} \tag{14.37}$$

This equation is very similar to the Hamiltonian of a transmon with a single junction, (eqn 14.26). The only difference is that the Josephson energy is now flux dependent

$$E_\text{J}(\Phi_{\text{ext}}) = (E_{\text{J}1} + E_{\text{J}2})\left|\cos\left(\pi\frac{\Phi_{\text{ext}}}{\Phi_0}\right)\right|\sqrt{1 + d^2\tan^2\left(\pi\frac{\Phi_{\text{ext}}}{\Phi_0}\right)},$$

where $d = (E_{\text{J}1} - E_{\text{J}2})/(E_{\text{J}1} + E_{\text{J}2})$ is the relative junction asymmetry. The qubit frequency is flux dependent too

$$\boxed{f_{01} = \frac{1}{h}(\sqrt{8E_\text{C}E_\text{J}(\Phi_{\text{ext}})} - E_\text{C}).} \tag{14.38}$$

Figure 14.10b shows the frequency of a tunable transmon with symmetric junctions ($E_{\text{J}1} = E_{\text{J}2}$) as a function of the applied magnetic flux. The qubit frequency shows a clear periodicity. The magnetic field is usually generated by a current flowing through an on-chip bias line close to the SQUID. This line is usually called a **flux bias line** or Z line. Assuming that the mutual inductance between the flux bias line and the SQUID is $M = 700$ fH, the current required to move the qubit to its lowest frequency

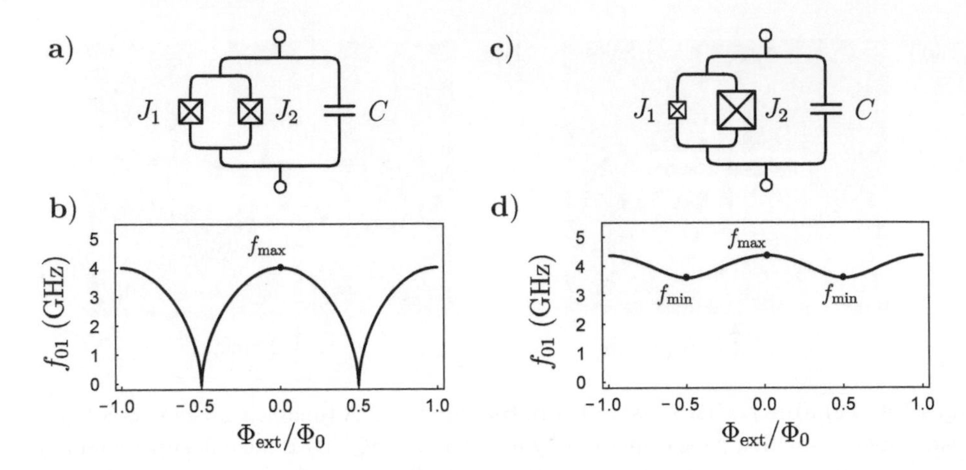

Fig. 14.10 Frequency of a tunable transmon. a) The electrical circuit of a transmon with symmetric junctions, $E_{\mathrm{J1}} = E_{\mathrm{J2}}$. **b)** Plot of f_{01} of eqn (14.38) versus the applied magnetic flux for $E_{\mathrm{J1}}/h = E_{\mathrm{J2}}/h = 5.5$ GHz and $E_{\mathrm{C}}/h = 200$ MHz. **c)** The electrical circuit of a transmon with asymmetric junctions. **d)** Plot of f_{01} of eqn (14.38) versus the applied magnetic flux for $E_{\mathrm{J1}}/h = 11$ GHz, $E_{\mathrm{J2}}/h = 1.2$ GHz and $E_{\mathrm{C}}/h = 200$ MHz. The tunability is smaller compared to the symmetric case.

is $I = \Phi_0/2M = 1.5$ mA. Ideally, the amount of current needed to tune the qubit frequency should be small so that the power dissipated on the coolest plate is low.

To improve the performance of some two-qubit gate schemes, it is often necessary to have a reduced tunability of the qubit frequency. This can be accomplished by building the junctions with different resistances such that $E_{\mathrm{J1}} \neq E_{\mathrm{J2}}$. As shown in Fig. 14.10c–d, the frequency of a tunable transmon with asymmetric junctions has a smaller tunability compared to the symmetric case and varies from a maximum value $f_{\max}$ to a minimum value $f_{\min} > 0$. The qubit is usually operated either at $f_{\max}$ or $f_{\min}$ because at these working points the qubit frequency is less sensitive to fluctuations of the magnetic flux leading to an improved dephasing time [608].

14.5.1 Flux noise

While tunable transmons enable frequency tunability, they are susceptible to an additional dephasing mechanism due to random fluctuations of the magnetic field through the SQUID. This type of noise is called the **flux noise** [609, 235]. The origin of this noise has been attributed to magnetic two-level systems localized in oxide layers surrounding the SQUID [610, 611]. The theoretical model assumes that flux noise is caused by unpaired electrons that move from one defect center to another by thermal activation. The spin of the electron points in a specific direction when it occupies a given site; this direction varies randomly from site to site leading to a $1/f$ power spectral density

$$S_{\Phi\Phi}(\omega) = \frac{2\pi A_\Phi^2}{|\omega|},$$

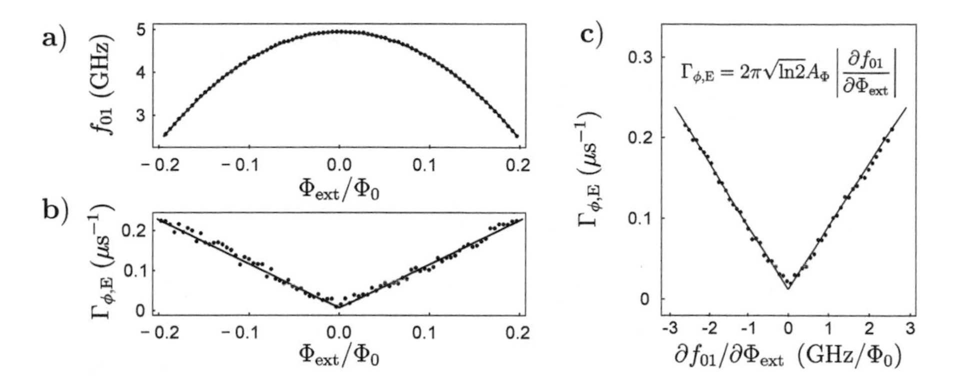

Fig. 14.11 Flux noise. a) Frequency of a transmon as a function of the applied magnetic flux close to the DC sweet spot. The solid line is a fit to eqn (14.38). **b)** Dephasing rate $\Gamma_{\phi,\mathrm{E}}$ measured with Hahn-echo experiments as a function of the magnetic flux. **c)** Dephasing rate versus the derivative of the qubit frequency. The solid line is a fit to eqn (14.39). The slope of the line gives a direct estimate of the noise amplitude A_Φ.

where $A_\phi \approx 1\ \mu\Phi_0/\sqrt{\mathrm{Hz}}$ [612, 613]. Since random fluctuations of the magnetic field threading the SQUID cause fluctuations of the qubit frequency, flux noise introduces an extra dephasing channel. Proper shielding of the device from spurious magnetic fields is crucial to reduce the flux noise.

The impact of flux noise on the qubit coherence can be measured by performing a Hahn-echo experiment as a function of the qubit frequency (see Fig. 14.11a–b). A Hahn-echo experiment comprises two $\mathsf{Y}_{\pi/2}$ pulses separated by a time interval t with an additional X_π halfway through the experiment. Each Hahn-echo experiment can be fitted to

$$Ae^{-\Gamma_{\mathrm{exp}}t-\Gamma_{\phi,\mathrm{E}}^2t^2}+B,$$

where the fitting parameter Γ_{exp} is the decay rate related to the intrinsic relaxation time T_1 and residual white noise, while the Gaussian decay rate $\Gamma_{\phi,\mathrm{E}}$ is a distinctive feature of $1/f$ noise. In Fig. 14.11b, we plot the fitting parameter $\Gamma_{\phi,\mathrm{E}}$ as a function of flux. At $\Phi_{\mathrm{ext}}=0$, the qubit is first order insensitive to fluctuations of the flux and the decoherence rate is the lowest. The qubit becomes more sensitive to flux noise away from this optimal point, and the decoherence rate increases. As explained in eqn (8.92), the dephasing rate measured with a Hahn-echo experiment in the presence of $1/f$ noise is proportional to the derivative of the qubit frequency with respect to flux,

$$\boxed{\Gamma_{\phi\mathrm{E}}=\sqrt{\ln 2}A_\Phi\left|\frac{\partial\omega_{01}}{\partial\Phi_{\mathrm{ext}}}\right|.} \tag{14.39}$$

The derivative can be calculated from (14.38),

$$\frac{d\omega_{01}}{d\Phi_{\mathrm{ext}}}=-\frac{\pi}{\hbar}\sqrt{2E_{\mathrm{C}}(E_{\mathrm{J1}}+E_{\mathrm{J2}})}\tan\left(\pi\frac{\Phi_{\mathrm{ext}}}{\Phi_0}\right),$$

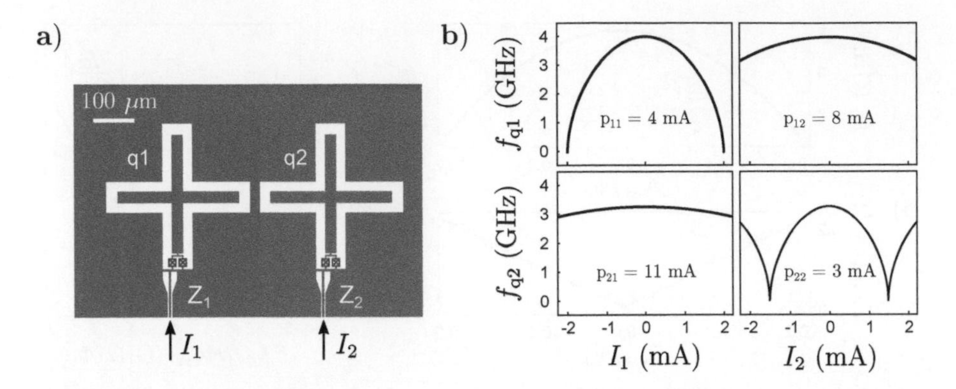

Fig. 14.12 Flux crosstalk. a) Two transmons connected to two separate flux bias lines, Z_1 and Z_2. **b)** Plot of the qubit frequency as a function of the applied current. The top right and bottom left plots would be flat without flux crosstalk. The parameters p_{ij} indicate the periodicity of the curves. The flux crosstalk is given by the ratio p_{12}/p_{22} and p_{21}/p_{11}.

where we assumed that the SQUID has symmetric junctions, $E_{J1} = E_{J2}$. The experimental value of the noise amplitude A_Φ can be extracted from a fit of the dephasing rate as a function of the slope of the qubit frequency as illustrated in Fig. 14.11c. The data points are fitted to eqn (14.39).

It has been shown that SQUID loops with a small perimeter are characterized by a smaller noise amplitude A_Φ [614]. Furthermore, a large asymmetry between the two junctions ($E_{J1} \gg E_{J2}$) produces a smaller tunability and a decreased sensitivity to flux noise [608]. This is advantageous for some two-qubit gate schemes, such as the cross-resonance gate, where a large tunability of the qubit frequency is not strictly required.

14.5.2 DC flux crosstalk

In a quantum computer based on tunable qubits, each qubit is connected to a flux bias line. Ideally, the current flowing through a specific flux bias line produces a localized magnetic field that changes the frequency of a single superconducting qubit. However, when the current returns to ground, it spreads along the ground planes of the device and can inadvertently flow close to the SQUID of another superconducting qubit, thereby changing its frequency. This effect, known as the **DC flux crosstalk**, generates undesired effects between distant qubits. If this effect is too severe, it is challenging to run simultaneous high-fidelity two-qubit gates using DC flux pulses.

A simple method to quantify the DC flux crosstalk is to measure the qubit frequency as a function of the DC flux applied to other elements on the device [615]. For instance, suppose that the device comprises two tunable qubits as shown in Fig. 14.12a. The frequency of each qubit is measured as a function of the applied flux (see top left and bottom right panels in Fig. 14.12b). These measurements provide the flux periodicities p_{11} and p_{22}; in our example, $p_{11} = 4$ mA and $p_{22} = 3$ mA. These measurements are repeated by applying the current through the other Z line. This provides the periodicities $p_{12} = 8$ mA and $p_{21} = 11$ mA shown in the top-right and bottom-left panels.

The periodicities can be used to construct the DC crosstalk matrix,

$$C = \begin{bmatrix} p_{11}/p_{11} & p_{22}/p_{12} \\ p_{11}/p_{21} & p_{22}/p_{22} \end{bmatrix} = \begin{bmatrix} 1 & 37.5\% \\ 36.3\% & 1 \end{bmatrix}.$$

Typical crosstalk values for state-of-the-art superconducting quantum processors are in the range $0.1 - 5\%$ [503, 616, 617]. Once the DC crosstalk matrix has been determined, it can be used to calibrate the applied current on each line to compensate for any undesired flux crosstalk.

Multiple works have reported that the DC crosstalk matrix is not symmetric [503, 617]. In addition, it has been observed that the crosstalk might be more pronounced between two remote qubits rather than between two neighboring qubits [617]. The lack of a clear pattern indicates that the geometry of the device significantly affects the flow of the supercurrent. Experimentally, flux crosstalk can be mitigated by optimizing the path of the current back to ground.

The method described in this section to measure the DC flux crosstalk is relatively slow when the device contains more than 10 qubits. A faster procedure relies on the measurement of the periodicity of the resonators coupled to the qubits [617]. A systematic way to measure the crosstalk between fast flux pulses is explained in the supplementary material of Ref. [616].

14.6 Driving a transmon

In this section, we explain how to implement single-qubit gates. First of all, at millikelvin temperatures a transmon with frequency $f_{01} > 3$ GHz is in its quantum ground state[10]. The qubit can be brought into an arbitrary superposition state by applying an electrical pulse with a specific amplitude, duration and phase. The pulse is generated by an AC voltage source placed outside the dilution refrigerator. The signal travels along a coaxial cable and reaches an on-chip waveguide capacitively coupled to the qubit pads as shown in Fig. 14.13a–b. This line is usually called a **control line** or an XY line. The pulse reaching the device has the analytical form

$$V_{\mathrm{d}}(t) = A\varepsilon(t)\sin(\omega_{\mathrm{d}}t + \alpha),$$

where A is the pulse amplitude in volts, ω_{d} is the drive frequency in rad/s, α is the phase of the pulse and $\varepsilon(t)$ is the modulation of the pulse (the maximum of this function is fixed to one). The envelope $\varepsilon(t)$ of the square pulse illustrated in Fig. 14.14a is

$$\textbf{square pulse}: \qquad \varepsilon_{\mathrm{s}}(t) = \begin{cases} 1 & 0 \le t \le t_{\mathrm{s}} \\ 0 & \text{else} \end{cases}$$

[10]Recall from Section 6.5.2 that at thermal equilibrium the probability of measuring the qubit in the excited state is $p_1 = \frac{1}{Z}e^{-hf_{01}/2k_{\mathrm{B}}T}$ where $Z = 2\cosh(\frac{hf_{01}}{2k_{\mathrm{B}}T})$. If $f_{01} > 3$ GHz and $T \approx 10$ mk, then $p_1 \approx 0$.

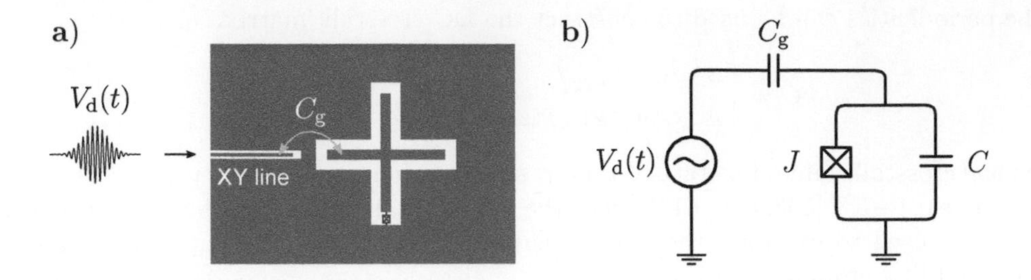

Fig. 14.13 Driving a qubit. a) Transmon capacitively coupled to a transmission line for microwave control. The capacitance between the XY line and the qubit is denoted as C_g. Typical values are $C_g = 0.2 - 1$ fF. **b)** Equivalent circuit of a transmon connected to a voltage source via a gate capacitor C_g. The waveform generator is usually located outside the fridge and is connected to the device with a highly attenuated coaxial line (not shown in the figure).

where t_s is the pulse duration. Another interesting example of electrical pulse is the Gaussian pulse illustrated in Fig. 14.14b,

$$\textbf{Gaussian pulse:} \qquad \varepsilon_G(t) = e^{-\frac{1}{2\sigma^2}(t-t_0)^2}.$$

The parameter σ controls the width of the Gaussian and t_0 indicates the center of the pulse. Gaussian pulses are usually preferred to square pulses due to their smaller frequency bandwidth which minimizes the excitation of higher energy levels.

From the equivalent circuit shown in Fig. 14.13b, one can derive the Hamiltonian of a transmon capacitively coupled to a control line

$$\hat{H} = 4E_C\Big(\hat{n} + \frac{C_g V_d(t)}{2e}\Big)^2 - E_J \cos\hat{\phi},$$

where $E_C = e^2/2C_\Sigma$ and $C_\Sigma = C_g + C$. The value of the gate capacitance C_g is usually in the range $0.1 - 1$ fF and is much smaller than the qubit capacitance, $C \approx 100$ fF. By expanding the parenthesis and dropping a constant, we obtain

$$\hat{H} = \underbrace{4E_C\hat{n}^2 - E_J\cos\hat{\phi}}_{\hat{H}_q} + \underbrace{2e\frac{C_g}{C_\Sigma}V_d(t)\hat{n}}_{\hat{H}_d}, \tag{14.40}$$

where $\hat{H}_d$ is the drive Hamiltonian. Since $\hat{n} = -i(\hat{b} - \hat{b}^\dagger)/2\sqrt{\xi}$ (see eqn 14.30), the drive Hamiltonian can be expressed as

$$\boxed{\hat{H}_d = -i\frac{e}{\sqrt{\xi}}\frac{C_g}{C_\Sigma}V_d(t)(\hat{b} - \hat{b}^\dagger).}$$

For simplicity, we restrict our analysis to the first two energy states of the transmon. In

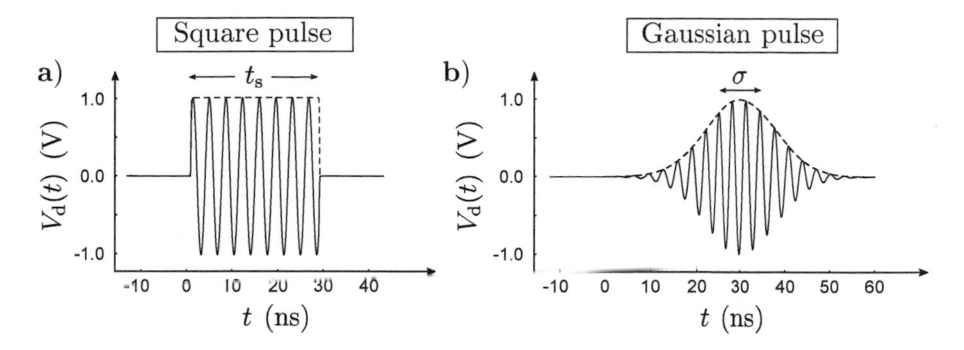

Fig. 14.14 Electrical pulses. a) Plot of $V_d(t)$ for a square pulse (solid line). The modulation function $\varepsilon_s(t)$ is illustrated with a dashed line. In this example, $A = 1$ V, $\omega_d = 2.44 \cdot 10^9$ rad/s and $t_s = 30$ ns. **b)** Plot of $V_d(t)$ for a Gaussian pulse (solid line). The modulation function $\varepsilon_G(t)$ is illustrated with a dashed line. In this example, $A = 1$ V, $\sigma = 8$ ns, $t_G = 60$ ns and $\omega_d = 2 \cdot 10^9$ rad/s.

this subspace, the qubit Hamiltonian is given by $\hat{H}_q = -\frac{1}{2}\hbar\omega_q\hat{\sigma}_z$, while the operators in $\hat{H}_d$ can be expressed in terms of a Pauli operator, $\hat{b} - \hat{b}^\dagger = |0\rangle\langle 1| - |1\rangle\langle 0| = i\hat{\sigma}_y$. Thus, eqn (14.40) becomes

$$\begin{aligned}\hat{H} = \hat{H}_q + \hat{H}_d &= -\frac{\hbar\omega_q}{2}\hat{\sigma}_z + \frac{e}{\sqrt{\xi}}\frac{C_g}{C_\Sigma}A\varepsilon(t)\sin(\omega_d t + \alpha)\hat{\sigma}_y \\ &= -\frac{\hbar\omega_q}{2}\hat{\sigma}_z + \hbar\Omega\,\varepsilon(t)\sin(\omega_d t + \alpha)\hat{\sigma}_y,\end{aligned}$$

where in the last step we introduced the **Rabi frequency**

$$\boxed{\Omega = \frac{e}{\hbar\sqrt{\xi}}\frac{C_g}{C_\Sigma}A,} \qquad (14.41)$$

with units of rad/s. This parameter quantifies the coupling strength between the control line and the qubit. We note that the Rabi frequency is proportional to the amplitude of the applied pulse.

Let us derive the evolution of the qubit state when the applied pulse is in resonance with the qubit frequency. To simplify the calculations, it is convenient to study the dynamics in a frame rotating about the z axis of the Bloch sphere at a frequency ω_d. The qubit state in the **rotating frame** is given by $|\psi'\rangle = \hat{R}|\psi\rangle$ where $\hat{R} = e^{-i\omega_d t\hat{\sigma}_z/2}$ is a time-dependent change of basis. The Hamiltonian becomes[11]

[11] Use (3.64) to derive this equation.

$$\hat{H}' = \hat{R}(\hat{H}_{\text{q}} + \hat{H}_{\text{d}})\hat{R}^{-1} - i\hbar\hat{R}\Big(\frac{d}{dt}\hat{R}^{-1}\Big)$$
$$= -\frac{\hbar(\omega_{\text{q}} - \omega_{\text{d}})}{2}\hat{\sigma}_z + \hbar\Omega\,\varepsilon(t)\sin(\omega_{\text{d}}t + \alpha)\big(\hat{\sigma}_x \sin\omega_{\text{d}}t + \hat{\sigma}_y \cos\omega_{\text{d}}t\big)\,.$$

The second term can be expanded as

$$\sin(\omega_{\text{d}}t + \alpha)(\hat{\sigma}_x \sin\omega_{\text{d}}t + \hat{\sigma}_y \cos\omega_{\text{d}}t) =$$
$$= \frac{\hat{\sigma}_x}{2}\big(-\cos(2\omega_{\text{d}} + \alpha) + \cos\alpha\big) + \frac{\hat{\sigma}_y}{2}\big(\sin(2\omega_{\text{d}} + \alpha) + \sin\alpha\big).$$

Assuming that the frequency of the applied pulse is similar to the qubit frequency $\omega_{\text{d}} \approx \omega_{\text{q}}$, the terms $\cos(2\omega_{\text{d}}t + \alpha)$ and $\sin(2\omega_{\text{d}}t + \alpha)$ oscillate rapidly and their contribution to the dynamics can be neglected. This approximation is called the **rotating wave approximation** (RWA) (see Appendix A.1 for more information). This approximation leads to

$$\boxed{\hat{H}'(t) = -\frac{\hbar(\omega_{\text{q}} - \omega_{\text{d}})}{2}\hat{\sigma}_z + \frac{\hbar\Omega}{2}\varepsilon(t)\big(\hat{\sigma}_x \cos\alpha + \hat{\sigma}_y \sin\alpha\big).} \tag{14.42}$$

In literature, the quantities $I \equiv \cos\alpha$ and $Q \equiv \sin\alpha$ are called the **quadratures**. By calibrating the parameters of the applied pulse (in particular, the pulse amplitude A, the drive frequency ω_{d}, the pulse duration and phase), the qubit state can be rotated about the x and y axis of the Bloch sphere by an arbitrary angle. This point will be explained in detail in the next section.

14.6.1 Single-qubit rotations

Suppose that an electrical pulse $V_{\text{d}}(t) = A\varepsilon(t)\sin(\omega_{\text{d}}t + \alpha)$ with frequency $\omega_{\text{d}} = \omega_{\text{q}}$ and phase $\alpha = \pi/2$ is applied to a transmon initially prepared in the ground state. Under these assumptions, the Hamiltonian (14.42) becomes $\hat{H}'(t) = \frac{1}{2}\hbar\Omega\varepsilon(t)\hat{\sigma}_y$ and the qubit state at the end of the pulse is

$$|\psi\rangle = e^{\frac{1}{i\hbar}\int_0^\infty \hat{H}'(t')dt'}|0\rangle = e^{-i\frac{\Omega}{2}\hat{\sigma}_y\int_0^\infty \varepsilon(t')dt'}|0\rangle = e^{-i\frac{\theta}{2}\hat{\sigma}_y}|0\rangle, \tag{14.43}$$

where we defined the rotation angle as

$$\boxed{\theta = \Omega\int_0^\infty \varepsilon(t')dt'.} \tag{14.44}$$

The operator $e^{-i\frac{\theta}{2}\hat{\sigma}_y}$ in eqn (14.43) is a rotation[12] about the y axis of the Bloch sphere by an angle θ. The rotation angle θ is the integral of the envelope multiplied by a constant. If the parameters of the pulse are calibrated such that $\theta = \pi$, the pulse implements a rotation of 180° about the y axis; this is called a π **pulse**. Similarly, when the parameters of the pulse are calibrated such that $\theta = \pi/2$, the pulse rotates the qubit state by 90° about the y axis; this is called a $\pi/2$ **pulse**. It is important to

[12]See Section 5.3 for more information about single-qubit rotations.

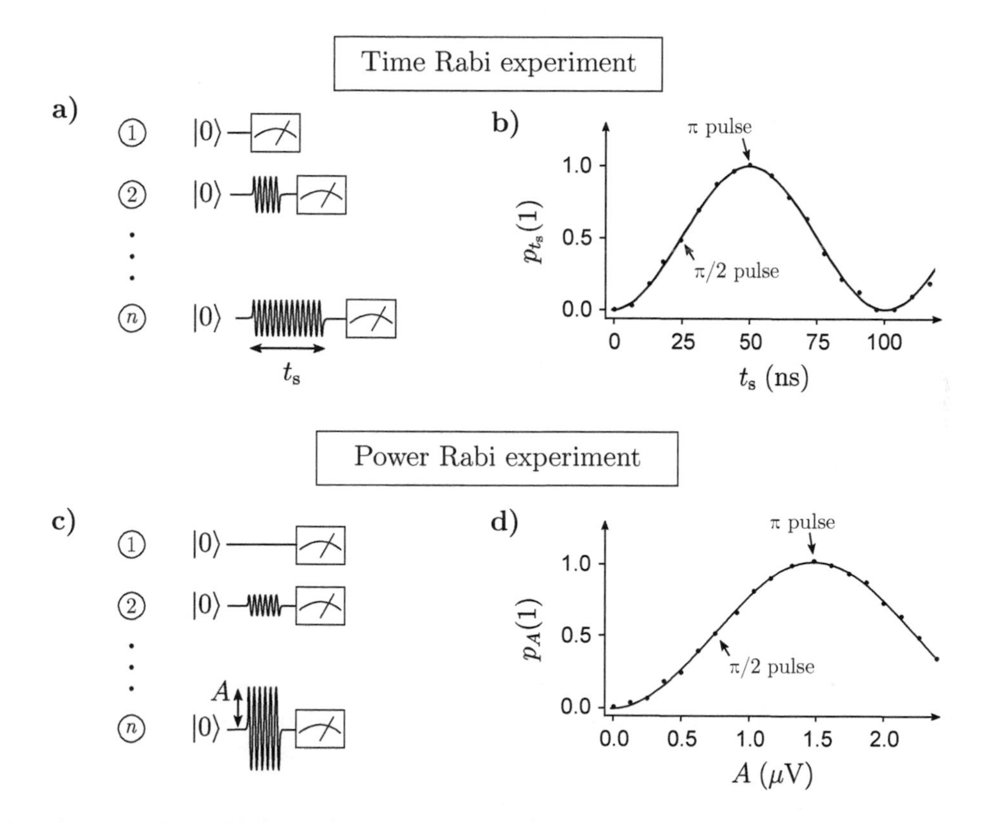

Fig. 14.15 Time Rabi experiment. a) A square pulse with fixed amplitude $A = 1\ \mu V$, frequency $\omega_\text{d} = \omega_\text{q}$, phase $\phi = \pi/2$, and duration t_s is applied to a qubit initialized in the ground state. A total of n experiments are performed with different pulse durations t_s. Each experiment is repeated N times to acquire some statistics. The qubit will be found N_0 times in the ground state and N_1 times in the excited state. **b)** Plot of the probability of finding the qubit in the excited state $p_{t_\text{s}}(1) = N_1/N$ as a function of the pulse duration. When $t_\text{s} \approx 50$ ns, the electrical pulse brings the qubit from the ground state to the excited state. Thus, the electrical pulse with parameters $(A = 1\ \mu\text{V}, t_\text{s} = 50\ \text{ns}, \phi = \pi/2, \omega_\text{d} = \omega_\text{q})$ implements a Y_π gate. Similarly, an electrical pulse with parameters $(A = 1\ \mu\text{V}, t_\text{s} = 25\ \text{ns}, \phi = \pi/2, \omega_\text{d} = \omega_\text{q})$ implements a $\mathsf{Y}_{\pi/2}$ gate. **Power Rabi experiment. c)** In this experiment, the pulse duration is kept fixed and the pulse amplitude varies. **d)** Plot of the probability of finding the qubit in the excited state $p_A(1) = N_1/N$ versus pulse amplitude. When $A = 0.75\ \mu\text{V}$ $(A = 1.5\ \mu\text{V})$, the pulse implements a $\mathsf{Y}_{\pi/2}$ (Y_π) gate. In both experiments, the fitting function is $\sin^2(\Omega t_\text{s}/2)$. The pulse is usually generated by an instrument located outside the fridge. In order to obtain at the chip level a pulse amplitude of $x\ \mu\text{V}$, the instrument should output an amplitude of $x \cdot 10^{G/20\ \text{dB}}\ \mu\text{V}$ where G is the attenuation of the line (typically, $G \approx 70$ dB at the qubit frequency).

mention that if the pulse frequency is slightly different from the qubit frequency, the Hamiltonian (14.42) contains a term proportional to $\hat{\sigma}_z$. This term causes an additional rotation about the z axis.

If the phase of the pulse is set to $\alpha = 0$ instead of $\pi/2$, the microwave pulse implements a rotation about the x axis. In general, a microwave pulse $V_{\text{d}} = A\varepsilon(t)\sin(\omega_{\text{d}}t+\alpha)$ implements a single-qubit rotation $\mathsf{R}_{\hat{n}(\alpha)}(\theta)$ along an axis $\hat{n}$ lying on the equator of the Bloch sphere,

$$\mathsf{R}_{\hat{n}(\alpha)}(\theta) = e^{-\frac{i}{2}\hat{n}(\alpha)\cdot\vec{\sigma}\theta} = e^{-\frac{i}{2}(\hat{\sigma}_x \cos\alpha + \hat{\sigma}_y \sin\alpha)\theta},$$

where the unit vector $\hat{n}(\alpha) = (\cos\alpha, \sin\alpha, 0)$. A train of k pulses $V_{\text{d}1}(t)$, ..., $V_{\text{d}k}(t)$ activates a sequence of single-qubit rotations,

$$\mathsf{R}_{\hat{n}(\alpha_k)}(\theta_k)\ldots\mathsf{R}_{\hat{n}(\alpha_2)}(\theta_2)\mathsf{R}_{\hat{n}(\alpha_1)}(\theta_1) = \prod_{j=0}^{k} e^{-\frac{i}{2}(\hat{\sigma}_x \cos\alpha_j + \hat{\sigma}_y \sin\alpha_j)\theta_j}.$$

Any single-qubit gate can be implemented by combining single-qubit rotations about the x and y axes.

The calibration of the amplitude and duration of the electrical pulse required to implement a specific single-qubit rotation is performed with a **time Rabi experiment** as shown in Fig. 14.15a. In this experiment, an electrical pulse of varying duration is applied to a qubit followed by a measurement. The plot in Fig. 14.15b shows the probability of finding the qubit in the excited state as a function of the pulse duration. The parameter Ω can be extracted by fitting these oscillations with

$$p_{t_s}(1) = |\langle 1|\psi\rangle|^2 = |\langle 1|e^{-i\frac{\Omega t_s}{2}\hat{\sigma}_y}|0\rangle|^2 = \sin^2(\Omega t_s/2), \tag{14.45}$$

where we assumed that the applied pulse is a square pulse. In an actual experiment, it is often preferred to calibrate microwave pulses associated with single-qubit rotations by fixing the pulse duration and varying the pulse amplitude. This type of experiment is known as **power Rabi experiment** and is illustrated in Fig. 14.15c–d. The qubit state at the end of the pulse is still described by eqn (14.43).

Example 14.2 Let us calculate the amplitude required to implement a π rotation with a 50 ns square pulse. We will assume that: the gate capacitance is $C_{\text{g}} = 0.3$ fF, the qubit capacitance is $C = 100$ fF, the ratio $E_{\text{J}}/E_{\text{C}} = 50$, and therefore $\xi = 0.2$. To implement a π pulse, we must have $\Omega t_{\text{s}} = \pi$. Using eqn (14.41), we obtain the expression

$$\theta = \Omega t_{\text{s}} = \frac{e}{\hbar\sqrt{0.2}}\frac{0.3}{0.3+100}A \times 50 = \pi.$$

This relation is satisfied when the pulse amplitude is set to $A = 6.2\ \mu$V. Let us assume that the microwave generator is connected to the device inside the dilution refrigerator via a coaxial cable with an overall attenuation of -70 dB at the qubit frequency. Therefore, the instrument at room temperature should generate a pulse with amplitude $6.2\ \mu\text{V} \times 10^{70\text{ dB}/20\text{ dB}} = 19$ mV.

14.6.2 Control line effect on T_1

The capacitive coupling between the qubit and the control line shown in Fig. 14.13a might limit the relaxation time of the qubit. This is because the energy stored in the qubit can in principle leak out into the control line. To investigate this phenomenon, we will start from the drive Hamiltonian (14.40),

$$\hat{H}_{\mathrm{d}} = 2e\frac{C_{\mathrm{g}}}{C_{\Sigma}}V_{\mathrm{d}}\hat{n}. \tag{14.46}$$

Suppose that the voltage along the waveguide fluctuates in time with power spectral density $S_{V_{\mathrm{d}}V_{\mathrm{d}}}(\omega)$. The qubit decay rate can be calculated using eqn (8.57),

$$\boxed{\gamma = \frac{1}{T_1} \approx \frac{1}{\hbar^2}|\langle 0|\hat{A}|1\rangle|^2 S_{V_{\mathrm{d}}V_{\mathrm{d}}}(\omega_{01}),} \tag{14.47}$$

where $\hat{A} = \partial\hat{H}/\partial V_{\mathrm{d}} = 2eC_{\mathrm{g}}\hat{n}/C_{\Sigma}$. Approximating the transmon as a two-level system, the matrix element $|\langle 0|\hat{A}|1\rangle|^2$ can be written as

$$|\langle 0|\hat{A}|1\rangle|^2 = \frac{(2eC_{\mathrm{g}})^2}{C_{\Sigma}^2}\left|\langle 0|\frac{i}{2\sqrt{\xi}}(\hat{b}^{\dagger} - \hat{b})|1\rangle\right|^2 = \frac{(2eC_{\mathrm{g}})^2}{C_{\Sigma}^2}\frac{1}{4\xi}.$$

Substituting this expression into (14.47), we obtain

$$\frac{1}{T_1} \approx \frac{1}{\hbar^2}\frac{(2eC_{\mathrm{g}})^2}{C_{\Sigma}^2}\frac{1}{4\xi}S_{V_{\mathrm{d}}V_{\mathrm{d}}}(\omega_{01}). \tag{14.48}$$

At low temperature, the voltage fluctuations across a resistor are characterized by the power spectral density

$$S_{V_{\mathrm{d}}V_{\mathrm{d}}}[\omega_{01}] = 2\hbar\omega_{01}Z_0, \tag{14.49}$$

where Z_0 is the impedance of the load connected to the line. Thus,

$$\frac{1}{T_1} \approx \frac{1}{\hbar^2}\frac{(2eC_{\mathrm{g}})^2}{C_{\Sigma}^2}\frac{1}{4\xi}2\hbar\omega_{01}Z_0. \tag{14.50}$$

This formula can be simplified recalling that $\xi = \sqrt{2E_{\mathrm{C}}/E_{\mathrm{J}}}$, $E_{\mathrm{C}} = e^2/2C_{\Sigma}$ and $\omega_{01} \approx \hbar\sqrt{8E_{\mathrm{J}}E_{\mathrm{C}}}$. Therefore, the upper limit on the qubit relaxation time due to the coupling with the control line is given by

$$\boxed{T_1 \approx \frac{C_{\Sigma}}{\omega_{01}^2 C_{\mathrm{g}}^2 Z_0}.} \tag{14.51}$$

Using realistic values for the qubit frequency $\omega_{01}/2\pi = 5$ GHz, its capacitance $C_{\Sigma} = C + C_{\mathrm{g}} = 100$ fF, the capacitance between the qubit and the XY line $C_{\mathrm{g}} = 0.1$ fF, and the impedance of the load $Z_0 = 50\ \Omega$, we obtain an upper limit on T_1 of the order of 200 μs. This value can be increased by decreasing the gate capacitance at the

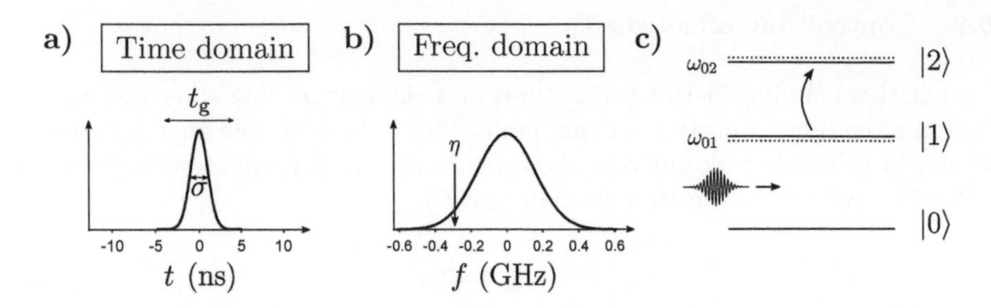

Fig. 14.16 Leakage to the second level and phase errors. a) Envelope of a Gaussian pulse with width $\sigma = 1$ ns. **b)** Fourier transform of the Gaussian pulse shown in panel **a**. The bandwidth of the pulse overlaps with the anharmonicity of the qubit η. **c)** When a qubit is in the state $|1\rangle$, a Gaussian pulse with a short gate time can inadvertently excite the qubit to state $|2\rangle$. In addition, an electrical pulse at the qubit frequency induces a repulsion between levels $|1\rangle$ and $|2\rangle$ (dotted lines). This repulsion leads to phase errors.

cost of using pulses with larger amplitude to implement single-qubit gates (for a fixed attenuation of the control line).

It is important to mention that the measured value of the relaxation time is affected by multiple contributions that add up together,

$$\frac{1}{T_1^{\text{tot}}} = \frac{1}{T_1^{(1)}} + \frac{1}{T_1^{(2)}} + \ldots + \frac{1}{T_1^{(n)}},$$

where each term $1/T_1^{(j)}$ is caused by a different noise source, such as the coupling to the XY line and the Z line, the interaction with two-level systems, quasiparticle tunneling, and the coupling with other circuits on the chip. The lowest $T_1^{(j)}$ has the most significant impact on the measured relaxation time.

14.6.3 The DRAG technique

The manipulation of the qubit state requires the application of an electrical pulse. To implement the maximum number of gates within the qubit coherence time, the gate time should be as short as possible. However, the pulse duration cannot be arbitrarily short. As illustrated in Fig. 14.16a–b, the Fourier transform of a Gaussian pulse $\Omega(t) = \Omega e^{-t^2/2\sigma^2}$ with width $\sigma = 1$ ns has a frequency bandwidth larger than the transmon anharmonicity. This leads to some undesired effects:

- A qubit initially prepared in state $|1\rangle$ might be excited to state $|2\rangle$ by a short π pulse. Furthermore, a qubit in the ground state $|0\rangle$ might be directly excited to state $|2\rangle$[13]. This is usually referred to as leakage to the second level.

[13] This phenomenon is similar to the excitation of a microwave resonator by a resonant drive (see Section 13.7). Since the energy levels of a microwave resonator are equally spaced, a pulse with frequency $\omega_d = \omega_r$ brings it into a superposition of multiple states. This is what would happen to a transmon with a very small anharmonicity.

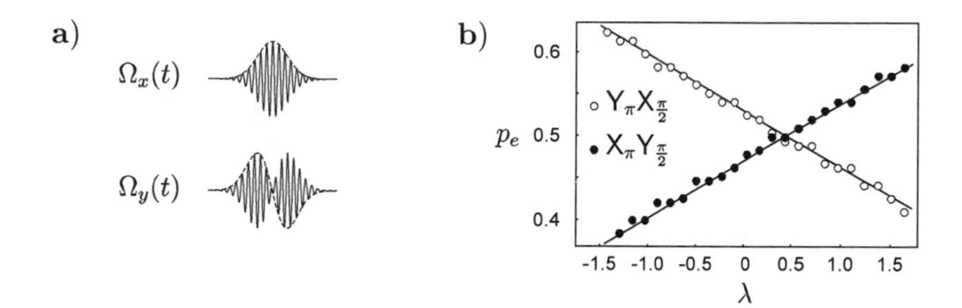

Fig. 14.17 DRAG pulse. a) In the DRAG technique, a rotation about the x axis of the Bloch sphere can be implemented by applying a Gaussian pulse along the x axis and a pulse whose envelope is the derivative of a Gaussian along the y axis. This mitigates leakage to the second level and phase errors. **b)** Experimentally, the value of λ that minimizes phase errors can be calibrated by measuring the probability of finding the qubit in the excited state with two different pulse sequences: $\mathsf{Y}_\pi \mathsf{X}_{\frac{\pi}{2}}$ and $\mathsf{X}_\pi \mathsf{Y}_{\frac{\pi}{2}}$. These two experiments generate two lines that intersect at the optimal value [623].

- A microwave pulse at the qubit frequency generates an AC Stark shift of the transition $1 \leftrightarrow 2$ leading to phase errors [618, 619].

One method to solve these issues is the **DRAG technique** (Derivative Reduction by Adiabatic Gate) [605, 620]. This technique implements a single-qubit rotation by applying a pulse with envelope $\Omega(t)$ on one quadrature and a second pulse with envelope $d\Omega(t)/dt$ on the opposite quadrature. For example, a rotation about the x axis can be implemented with the drive Hamiltonian

$$\hat{H}_{\mathrm{d}}(t) = \hbar\Omega_x(t)\sin(\omega_{\mathrm{d}}t)\hat{\sigma}_x + \hbar\Omega_y(t)\sin(\omega_{\mathrm{d}}t + \pi/2)\hat{\sigma}_y,$$

where

$$\Omega_x(t) = \Omega e^{-t^2/2\sigma^2}, \qquad \Omega_y(t) = -\frac{\lambda}{\eta}\frac{d}{dt}\Omega_x(t).$$

The parameter λ is a scaling factor and η is the qubit anharmonicity. Theory predicts that the optimal value of λ to minimize leakage to the second level is $\lambda = 1$, while the optimal value to compensate phase errors is $\lambda = 0.5$ [605, 620, 621]. In an experimental setting, the optimal value might deviate from these predictions due to the distortion of the pulse and the presence of the readout resonator [622]. A simple method to calibrate this parameter is to perform two separate experiments. The first experiment consists in applying two DRAG pulses $\mathsf{Y}_\pi \mathsf{X}_{\frac{\pi}{2}}$ parametrized by the coefficient λ. The second experiment is equal to the first one with the only difference that the pulse sequence is $\mathsf{X}_\pi \mathsf{Y}_{\frac{\pi}{2}}$. The value of λ that minimizes phase errors is the one that gives the same probability of measuring the qubit in the excited state as shown in Fig. 14.17b [623].

Now that we have understood the general idea behind the DRAG technique, let us be more quantitative. Suppose that a simple Gaussian pulse is applied to a qubit *without* DRAG correction,

$$V_{\mathrm{d}}(t) = A_x \varepsilon_x(t) \sin(\omega_{\mathrm{d}} t + \phi_0)$$

where A_x is the pulse amplitude and $\varepsilon_x(t) = e^{-\frac{t^2}{2\sigma^2}}$ is the envelope. The phase of the pulse ϕ_0 can be set to any value without loss of generality. Using eqns (14.40, 14.30), the Hamiltonian of a transmon excited by a microwave pulse is given by

$$\begin{aligned}\hat{H} &= 4E_{\mathrm{C}}\hat{n} - E_{\mathrm{J}}\cos\hat{\phi} + 2e\beta V_{\mathrm{d}}(t)\hat{n} \\ &= \sum_{j=0}^{\infty} \hbar\omega_j |j\rangle\langle j| + \frac{e\beta}{\sqrt{\xi}} V_{\mathrm{d}}(t) \sum_{j=0}^{\infty} \sqrt{j+1}\Big(-i|j\rangle\langle j+1| + i|j+1\rangle\langle j|\Big),\end{aligned}$$

where we used $\hat{b} = \sum_j \sqrt{j+1}|j\rangle\langle j+1|$. Let us truncate the Hamiltonian to the first three energy levels,

$$\hat{H} = \hbar\omega_{01}|1\rangle\langle 1| + \hbar(2\omega_{01} + \eta)|2\rangle\langle 2| + \hbar\Omega_x(t)\sin(\omega_{\mathrm{d}} t + \phi_0)(\hat{\sigma} + \hat{\sigma}^\dagger), \tag{14.52}$$

where we introduced the operator $\hat{\sigma} = -i(|0\rangle\langle 1| + \sqrt{2}|1\rangle\langle 2|)$ and we defined $\Omega_x = e\beta A_x \varepsilon_x(t)/\hbar\sqrt{\xi}$. It is convenient to study the dynamics in a rotating frame by performing the change of basis $\hat{R} = e^{i(\omega_{\mathrm{d}}|1\rangle\langle 1| + 2\omega_{\mathrm{d}}|2\rangle\langle 2|)t}$. The Hamiltonian in the new frame becomes (see eqn (4.37))

$$\begin{aligned}\hat{H}' &= \hat{R}\hat{H}\hat{R}^{-1} - i\hbar\hat{R}\frac{d}{dt}\hat{R}^{-1} \\ &= \hbar\Delta|1\rangle\langle 1| + \hbar(2\Delta + \eta)|2\rangle\langle 2| - \hbar\Omega_x \sin(\omega_{\mathrm{d}} t)(\hat{\sigma}e^{-i\omega_{\mathrm{d}} t} + \hat{\sigma}^\dagger e^{i\omega_{\mathrm{d}} t}),\end{aligned}$$

where we set $\phi_0 = \pi$ for convenience and we introduced the detuning parameter $\Delta = \omega_{01} - \omega_{\mathrm{d}}$. Assuming that $\Omega_x \ll \{\omega_{\mathrm{d}}, \omega_{01}\}$, we can take advantage of the RWA approximation and neglect the terms proportional to $e^{\pm i2\omega_{\mathrm{d}} t}$. Under this approximation, the Hamiltonian can be expressed as [624, 620, 619]

$$\hat{H}'/\hbar = \Delta|1\rangle\langle 1| + (2\Delta + \eta)|2\rangle\langle 2| + \frac{\Omega_x}{2}(|0\rangle\langle 1| + |1\rangle\langle 0| + \sqrt{2}|1\rangle\langle 2| + \sqrt{2}|2\rangle\langle 1|)$$

Let us assume that the drive frequency is in resonance with the qubit frequency, $\Delta = 0$,

$$\hat{H}'/\hbar = \eta|2\rangle\langle 2| + \frac{\Omega_x}{2}(\sqrt{2}|1\rangle\langle 2| + \sqrt{2}|2\rangle\langle 1|) = \begin{bmatrix} 0 & 0 & 0 \\ 0 & 0 & \frac{\sqrt{2}\Omega_x}{2} \\ 0 & \frac{\sqrt{2}\Omega_x}{2} & \eta \end{bmatrix}. \tag{14.53}$$

We can now show that an electrical pulse at the qubit frequency induces an AC Stark shift. This can be seen by calculating the eigenvalues of the matrix $\hat{H}'$ (see Exercise 14.1),

$$\omega_0 = 0, \qquad \omega_1 \approx -\frac{\Omega^2}{2\eta}, \qquad \omega_2 \approx \eta + \frac{\Omega^2}{2\eta},$$

where we assumed $|\Omega_x| \ll |\eta|$ since in a real experiment $\Omega_x/2\pi \approx 10$ MHz, while $\eta/2\pi \approx -250$ MHz. Therefore, the presence of the higher energy levels introduces

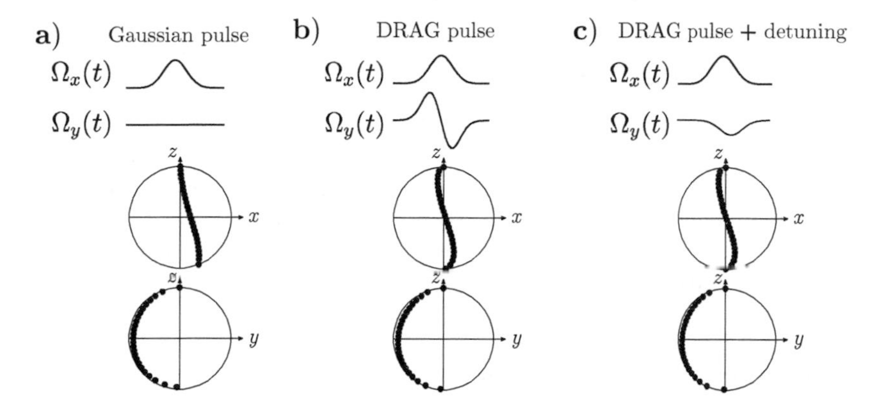

Fig. 14.18 Dynamics in the Bloch sphere. a) Time evolution of the qubit state in the Bloch sphere subject to a resonant π pulse with a Gaussian profile. The width of the pulse is $\sigma = 12$ ns. Phase errors and leakage to the second level distort the trajectory. **b)** DRAG pulse with $\lambda = 0.5$ and $\Delta = 0$. At the end of the pulse, the qubit state is closer to the desired state. **c)** The gate error can be reduced by slightly detuning the pulse from the qubit frequency, $\Delta = 1$ MHz. All of these simulations have been performed numerically with a Trotterization method setting $\omega_{01}/2\pi = 4$ GHz and $\eta/2\pi = -0.23$ GHz.

a shift of the qubit frequency with magnitude $|\Omega^2/2\eta|$. This means that when the experimenter attempts to perform a rotation about the x axis with a short Gaussian pulse without DRAG correction, the actual rotation takes place about an axis tilted toward the z axis leading to an undesired Z rotation. This is the origin of phase errors.

Example 14.3 Consider a qubit with frequency $f_{01} = 4$ GHz and anharmonicity $\eta/2\pi = -0.23$ GHz. Suppose that the qubit starts from the ground state $|0\rangle$ and a π pulse with a Gaussian profile and no DRAG correction

$$\Omega_x(t) = \pi \frac{e^{-t^2/2\sigma^2}}{\sqrt{2\pi\sigma^2}}$$

is applied to the qubit. The qubit state at time t is given by $|\psi(t)\rangle = \mathcal{T} e^{\frac{1}{i\hbar}\int_0^t \hat{H}(t')dt'}|0\rangle$ where $\hat{H}(t)$ is reported in eqn (14.53). The time evolution can be simulated with a Trotterization: the time interval $[0, t]$ is divided into short time intervals $\Delta t = t/N$ and at each time step the qubit state is approximated as $|\psi(t+dt)\rangle = e^{\frac{1}{i\hbar}\hat{H}(t)\Delta t}|\psi(t)\rangle$. As shown in Fig. 14.18a, if the pulse width is $\sigma = 12$ ns and the detuning is $\Delta = 0$, leakage to the second level significantly distorts the trajectory of the rotation. This undesired effect can be compensated using the DRAG technique. An additional pulse with an envelope

$$\Omega_y(t) = \frac{\lambda}{\eta} \frac{d\Omega_x}{dt}$$

is applied simultaneously on the opposite quadrature. The drive Hamiltonian in the rotating frame becomes

$$\hat{H}'_{\mathrm{d}}/\hbar = \begin{bmatrix} 0 & \frac{\Omega_x(t)+i\Omega_y(t)}{2} & 0 \\ \frac{\Omega_x(t)-i\Omega_y(t)}{2} & \Delta & \frac{\Omega_x(t)+i\Omega_y(t)}{2/\sqrt{2}} \\ 0 & \frac{\Omega_x(t)-i\Omega_y(t)}{2/\sqrt{2}} & 2\Delta+\eta \end{bmatrix} \tag{14.54}$$

Figure 14.18b shows the trajectory of the qubit state in the Bloch sphere using a DRAG pulse with $\lambda = 0.5$ and $\Delta = 0$. The qubit state ends up much closer to the south pole. The gate error can be further reduced by slightly detuning the pulse frequency from the qubit frequency, as shown in Fig. 14.18c.

14.6.4 Rotations about the z axis

Rotations about the z axis, also known as **Z gates**, correspond to a change in the relative phase between the $|0\rangle$ and $|1\rangle$ states[14],

$$\mathsf{Z}_\alpha|\psi\rangle = e^{-i\frac{\alpha}{2}\hat{\sigma}_z}(c_0|0\rangle + c_1|1\rangle) = e^{i\frac{\alpha}{2}}c_0|0\rangle + e^{-i\frac{\alpha}{2}}c_1|1\rangle = c_0|0\rangle + e^{-i\alpha}c_1|1\rangle.$$

This gate can be implemented in multiple ways.

1) A simple way to implement a Z gate is to combine rotations about the x and y axis

$$\mathsf{Z}_\alpha = \mathsf{X}_{\frac{\pi}{2}}\mathsf{Y}_\alpha\mathsf{X}_{-\frac{\pi}{2}}. \tag{14.55}$$

The downside of this method is that the gate time is relatively long: if a $\mathsf{X}_{\pi/2}$ rotation takes t_{g} nanoseconds, a Z_α rotation takes at least $2t_{\mathrm{g}}$ nanoseconds.

2) If the superconducting qubit under investigation is flux tunable, a Z gate can be implemented by varying the qubit frequency on a short timescale with a flux pulse [625, 626]. A square DC flux pulse has the form

$$\Phi_{\mathrm{ext}}(t) = \begin{cases} \Phi_{\mathrm{dc}} & 0 \le t \le t_{\mathrm{g}}, \\ 0 & \text{else.} \end{cases}$$

The flux pulse shifts the qubit frequency from ω_{01} to $\omega_{01}(\Phi_{\mathrm{dc}})$ and then back to the initial value (see Fig. 14.19a). The relative phase accumulated during the excursion depends on the pulse amplitude and duration,

$$\alpha = \big[\omega_{01} - \omega_{01}(\Phi_{\mathrm{dc}})\big]t_g. \tag{14.56}$$

The pulse duration required to implement a rotation by a desired angle α can be calibrated with a Ramsey-type experiment as shown in Fig. 14.19b. The DC flux pulse is applied to the qubit between two $\pi/2$ pulses [626]. The experiment is repeated as a function of the pulse duration. The probability of measuring the qubit in the excited state is fitted to a sinusoidal curve. One can extract the pulse duration required to

[14]In the last step we dropped a global phase of no physical significance.

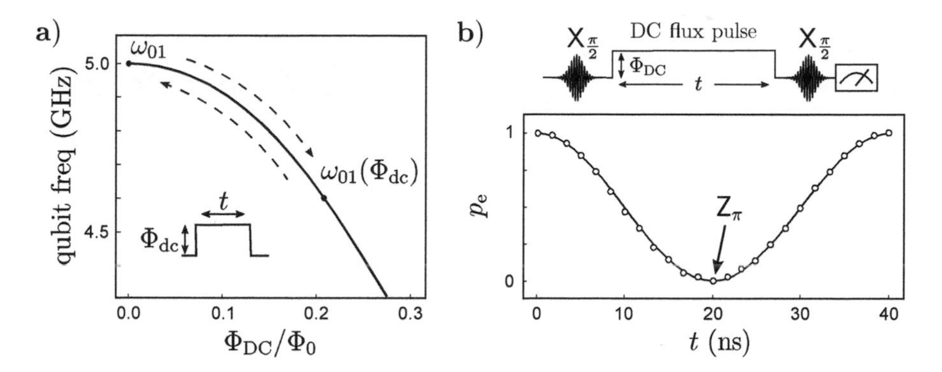

Fig. 14.19 Z rotations. a) A flux pulse with amplitude Φ_{dc} and duration t_g is applied to a tunable transmon. The pulse shifts the qubit frequency from ω_{01} to $\omega_{01}(\Phi_{dc})$. The qubit frequency moves back to the initial value at the end of the flux pulse. **b)** The relative phase accumulated during the excursion can be measured with a Ramsey-type experiment. This experiment consists of two $\pi/2$ pulses separated by a DC flux pulse. The plot shows the probability of measuring the qubit in the excited state as a function of the flux pulse duration. The $\pi/2$ pulses are applied through the XY line, while the DC flux pulse is delivered through the Z line.

implement a desired Z_α gate from the fit.

3) If the superconducting qubit is not tunable, a rotation about the z axis can be executed by shifting the qubit frequency with a detuned microwave tone. For pulse amplitudes sufficiently small compared to the detuning $\Omega \ll |\omega_d - \omega_{01}|$, the electrical pulse shifts the qubit frequency without inducing direct excitations, a phenomenon known as the **AC Stark shift** [512]. A detuned microwave pulse with a small amplitude shifts the qubit frequency to (see Exercise 14.7)

$$\omega_{01} \quad \rightarrow \quad \omega_{01} + \delta_{\text{Stark}} \approx \omega_{01} + \frac{\eta\Omega^2}{2\Delta(\Delta+\eta)}, \tag{14.57}$$

where $\Delta = \omega_{01} - \omega_d$. If the drive frequency is greater than the qubit frequency, the qubit shifts to lower frequencies as shown in Fig. 14.20a. This produces a rotation about the z axis by an angle $\alpha \approx -\delta_{\text{Stark}} t_g$ where t_g is the microwave pulse duration. The rotation angle can be calibrated with a Ramsey-type experiment, see Fig. 14.20b.

4) As explained in Section 14.6.1, rotations about the x axis and the y axis are implemented by changing the phase of the microwave pulse by 90°. This change of phase is reminiscent of the action of a Z_π rotation: this gate simply rotates the Bloch sphere by 90°. This analogy can be exploited to implement **virtual Z gates** [627, 628, 629, 630, 477]. For example, let us consider a simple circuit consisting of a Z_α rotation followed by a X_β rotation:

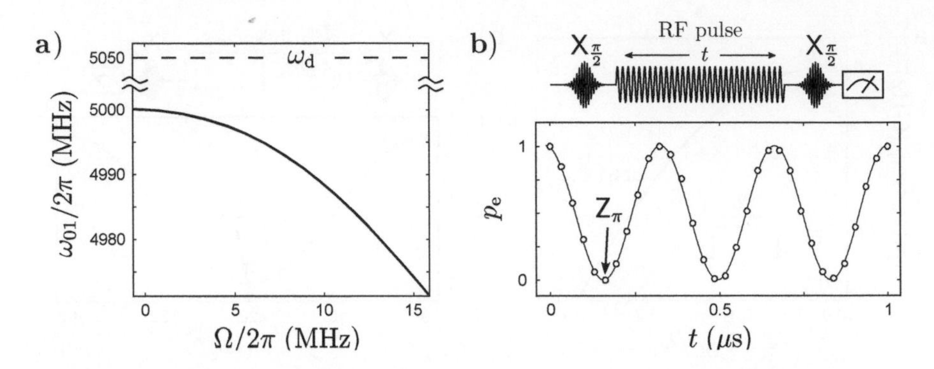

Fig. 14.20 AC Stark shift. a) An electrical pulse at $\omega_d/2\pi = 5050$ is applied to a qubit with frequency $\omega_{01}/2\pi = 5000$ MHz. The microwave pulse shifts the qubit frequency to $\omega_{01} \to \omega_{01}+\delta_{\text{Stark}}$. The solid line is a plot of eqn (14.57). The frequency shift has a quadratic dependence on the pulse amplitude. **b)** The calibration of the rotation angle is executed with a Ramsey-type experiment. The detuned microwave pulse is applied between the $\pi/2$ pulses (all of the pulses are delivered through the XY line). The probability of measuring the qubit in the excited state as a function of the pulse duration shows periodic oscillations with frequency δ_{Stark}. The rotation angle can be adjusted by tuning the amplitude and duration of the RF pulse.

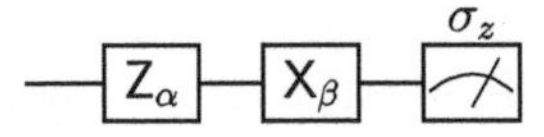

In the field of superconducting devices, the qubits can only be measured in the σ_z basis. This allows us to add a $Z_{-\alpha}$ rotation right before the measurement without changing the measurement result,

$$—[Z_\alpha]—[X_\beta]—[\sigma_z] \quad = \quad —[Z_\alpha]—[X_\beta]—[Z_{-\alpha}]—[\sigma_z]$$

The operator associated with this sequence of rotations is $Z_{-\alpha}X_\beta Z_\alpha$. With some matrix multiplications, one can verify that (see Exercise 14.16),

$$Z_{-\alpha}X_\beta Z_\alpha = e^{i\frac{\alpha}{2}\sigma_z}e^{-i\frac{\beta}{2}\sigma_x}e^{-i\frac{\alpha}{2}\sigma_z} = e^{-i\frac{\beta}{2}(\cos(-\alpha)\sigma_x+\sin(-\alpha)\sigma_y)} = R_{\hat{n}(-\alpha)}(\beta),$$

where $R_{\hat{n}(-\alpha)}(\beta)$ is a rotation by an angle β about the axis $\hat{n} = (\cos(-\alpha), \sin(-\alpha), 0)$ lying on the equator. In Section 14.6.1, we showed that the rotation $R_{\hat{n}(-\alpha)}(\beta)$ can be activated with a single microwave pulse!

Let us consider a slightly more complicated circuit,

$$—[Z_\alpha]—[X_\beta]—[Z_\gamma]—[Y_\delta]—[\sigma_z]$$

Now, let us add and subtract a Z_α rotation after the X_β gate,

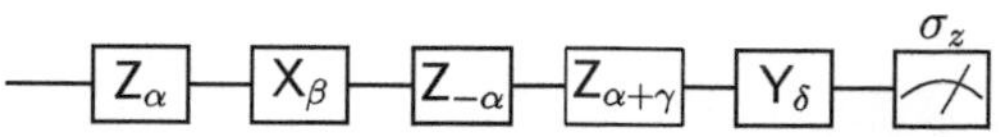

$$-\boxed{\mathsf{Z}_\alpha}-\boxed{\mathsf{X}_\beta}-\boxed{\mathsf{Z}_{-\alpha}}-\boxed{\mathsf{Z}_{\alpha+\gamma}}-\boxed{\mathsf{Y}_\delta}-\boxed{\sigma_z\text{ meas.}}$$

Lastly, a $\mathsf{Z}_{-\alpha-\gamma}$ rotation can be added right before the measurement without affecting the measurement outcome,

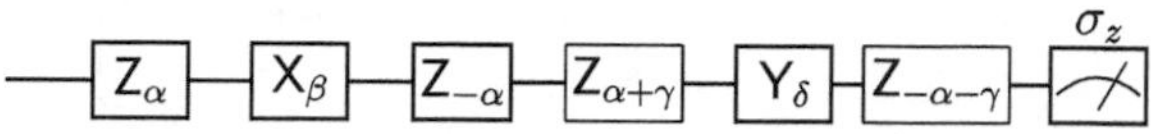

$$-\boxed{\mathsf{Z}_\alpha}-\boxed{\mathsf{X}_\beta}-\boxed{\mathsf{Z}_{-\alpha}}-\boxed{\mathsf{Z}_{\alpha+\gamma}}-\boxed{\mathsf{Y}_\delta}-\boxed{\mathsf{Z}_{-\alpha-\gamma}}-\boxed{\sigma_z\text{ meas.}}$$

This sequence of gates is equivalent to

$$\mathsf{Z}_{-\alpha-\gamma}\mathsf{Y}_\delta\mathsf{Z}_{\alpha+\gamma}\mathsf{Z}_{-\alpha}\mathsf{X}_\beta\mathsf{Z}_\alpha = \mathsf{R}_{\hat{n}(-\alpha-\gamma+\pi/2)}(\delta)\mathsf{R}_{\hat{n}(-\alpha)}(\beta).$$

This shows that Z gates can be applied in a virtual way by changing the phase of subsequent pulses. This strategy reduces the number of gates in the circuit. In addition, virtual Z gates do not introduce any error because they are directly implemented in software.

As discussed in Section 14.6.3, driving a transmon might produce phase errors and leakage outside the computational subspace. These problems are usually addressed with the DRAG technique. The DRAG pulse is usually calibrated to mitigate phase errors and not leakage (which requires measuring the state $|2\rangle$ [618]). Reference [630] shows that virtual Z gates can be used to correct phase errors in single-qubit gates, so that DRAG pulses can be calibrated to minimize leakage only.

Example 14.4 In this example, we show that any unitary operator U acting on a qubit can be decomposed (up to a global phase) with Z_α rotations combined with $\mathsf{X}_{\frac{\pi}{2}}$ and $\mathsf{X}_{-\frac{\pi}{2}}$ rotations. From Theorem 5.1, we know that any unitary operator acting on a qubit can be decomposed as $U = e^{i\delta}\mathsf{Z}_\alpha\mathsf{Y}_\theta\mathsf{Z}_\beta$. The reader can verify that $\mathsf{Y}_\theta = \mathsf{X}_{-\frac{\pi}{2}}\mathsf{Z}_\theta\mathsf{X}_{\frac{\pi}{2}}$ with some matrix multiplications. Therefore, any single-qubit gate can be expressed as $U = e^{i\delta}\mathsf{Z}_\alpha\mathsf{X}_{-\frac{\pi}{2}}\mathsf{Z}_\theta\mathsf{X}_{\frac{\pi}{2}}\mathsf{Z}_\beta$ where $e^{i\delta}$ is a global phase. This indicates that on a physical machine, only Z rotations and $\pi/2$ rotations about the x axis must be calibrated to implement single-qubit gates. This simplifies the calibration procedure because only a single pulse amplitude must be accurately calibrated if the Z gates are implemented virtually.

14.7 Circuit quantum electrodynamics

Now that we learned how to operate single-qubit gates on a quantum computer based on superconducting qubits, we will discuss how to measure the qubit state. The measurement process must satisfy several criteria [631]:

1) The readout fidelity should be on the order of 99% for fault-tolerant quantum computing.

2) The measurement time should be much shorter than the qubit relaxation time so that the qubit state does not vary during the measurement.

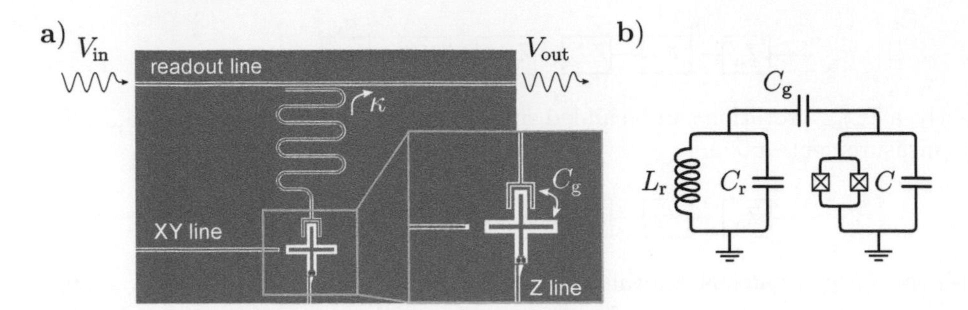

Fig. 14.21 Circuit QED. a) A tunable transmon capacitively coupled to a $\lambda/4$ CPWR. The resonator is inductively coupled to a readout line. The transmon is connected to an XY line (for the manipulation of the qubit state) and a Z line (for the control of the qubit frequency). **b)** Equivalent circuit of a tunable transmon capacitively coupled to a microwave resonator.

3) In a scalable architecture, the size of the measurement device should be comparable to or smaller than the qubit size.

4) The coupling with the measurement device should not significantly affect the qubit relaxation time.

5) Two consecutive measurements of the qubit state (with no delay between them) should give the same result.

In **circuit quantum electrodynamics**, the qubit state is measured via the dispersive interaction between the qubit and a far-detuned microwave resonator [139, 138, 137]. A change in the qubit state from $|0\rangle$ to $|1\rangle$ causes a frequency shift of the resonator from $\omega_{\text{r},|0\rangle}$ to $\omega_{\text{r},|1\rangle}$. A microwave signal applied to the resonator at a frequency between $\omega_{\text{r},|0\rangle}$ and $\omega_{\text{r},|1\rangle}$ acquires a phase shift dependent on the qubit state. By accurately measuring this phase shift, one can infer the qubit state. To decrease the impact on the qubit dephasing time, the signal power is usually set so that the microwave resonator is populated with a small number of photons. The signal is amplified by cryogenic amplifiers and room-temperature amplifiers before being acquired with standard heterodyne mixing at room temperature. The first demonstration of this measurement procedure was performed in 2004 and it is now widely employed in superconducting devices [138, 632, 137].

Figure 14.21a–b shows a superconducting qubit capacitively coupled to a $\lambda/4$ CPWR. From the electrical circuit, one can write the Hamiltonian of the system as

$$\begin{aligned}\hat{H} &= \hbar\omega_{\text{r}}\hat{a}^{\dagger}\hat{a} + 4E_{\text{C}}\Big(\hat{n} + \frac{C_{\text{g}}\hat{V}}{2e}\Big)^2 - E_{\text{J}}\cos\hat{\phi} \\ &= \hbar\omega_{\text{r}}\hat{a}^{\dagger}\hat{a} + 4E_{\text{C}}\hat{n}^2 - E_{\text{J}}\cos\hat{\phi} + \frac{4E_{\text{C}}}{e}\hat{n}C_{\text{g}}\hat{V},\end{aligned} \tag{14.58}$$

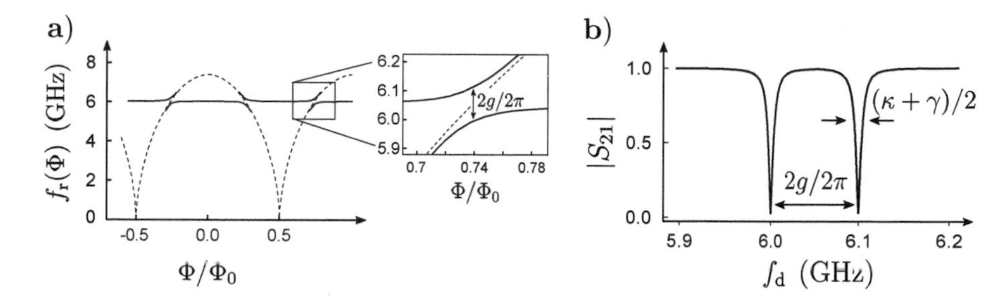

Fig. 14.22 Measuring the coupling strength. a) Plot of the resonator frequency (solid line) as a function of the magnetic field applied to the tunable qubit. When the qubit frequency (dashed line) crosses the resonator frequency, an anti-crossing appears. The magnitude of the splitting is proportional to the coupling strength between the qubit and the resonator. **b)** Transmission coefficient of a resonator in resonance with a superconducting qubit, $\omega_r = \omega_q$. From the separation of the resonances, we extract $g/2\pi = 50$ MHz. Since the system's excitation is a balanced superposition of the single photon state and the excited qubit state, the decay rate of the dressed state is an average of the two, $(\kappa+\gamma)/2$.

---

where ω_r is the resonator frequency and $\hat{V} = \hat{Q}/C_r$ is the voltage across the resonator capacitor. Using eqns (13.50, 14.30, 14.31), the Hamiltonian becomes

$$\hat{H} = \hbar\omega_r \hat{a}^\dagger \hat{a} + \sqrt{8E_J E_C}\hat{b}^\dagger \hat{b} - \frac{E_C}{12}(\hat{b} + b^\dagger)^4 + \hbar g(\hat{b}^\dagger - \hat{b})(\hat{a} - \hat{a}^\dagger),$$

where we introduced the **coupling strength**

$$\boxed{g = \frac{2E_C}{\hbar e}\frac{C_g}{C_r}\frac{Q_{zpf}}{\sqrt{\xi}} = \frac{E_C}{\hbar e}\left(\frac{E_J}{2E_C}\right)^{1/4}\frac{C_g}{C_r}\sqrt{2\hbar\omega_r C_r}.}$$

This parameter quantifies the strength of the interaction between the qubit and the resonator and can be adjusted by varying the capacitance between the two elements (typically, $g/2\pi = 50 - 200$ MHz). If $g \ll \{\omega_q, \omega_r\}$, one can perform the rotating wave approximation and write the Hamiltonian in the form

$$\hat{H} = \hbar\omega_r \hat{a}^\dagger \hat{a} + \sqrt{8E_J E_C}\hat{b}^\dagger \hat{b} - \frac{E_C}{12}(\hat{b} + \hat{b}^\dagger)^4 + \hbar g(\hat{b}^\dagger \hat{a} + \hat{b}\hat{a}^\dagger).$$

For simplicity, we restrict our attention to the first two energy levels of the transmon. This leads to the Jaynes–Cummings Hamiltonian presented in Section 4.7,

$$\hat{H} = \hbar\omega_r \hat{a}^\dagger \hat{a} - \frac{\hbar\omega_{01}}{2}\hat{\sigma}_z + \hbar g(\hat{\sigma}^+ \hat{a} + \hat{\sigma}^- \hat{a}^\dagger).$$

When the coupling strength g is much smaller than the detuning between the qubit and the resonator, $\Delta = \omega_q - \omega_r$, the composite system operates in the **dispersive regime**.

The Jaynes–Cummings Hamiltonian in the dispersive regime can be approximated as (see Section 4.7 for a full derivation)

$$\boxed{\hat{H}_{\text{disp}} = \hbar(\omega_{\text{r}} - \chi\hat{\sigma}_z)\hat{a}^\dagger\hat{a} - \frac{\hbar}{2}(\omega_{01} + \chi)\hat{\sigma}_z,} \tag{14.59}$$

where we introduced the **dispersive shift**

$$\boxed{\chi = \frac{g^2}{\Delta}}. \tag{14.60}$$

Note that the dispersive shift is negative for $\omega_{\text{r}} > \omega_{\text{q}}$. Equation (14.59) shows that the resonator frequency is $\omega_{\text{r},|0\rangle} = \omega_{\text{r}} - \chi$ $(\omega_{\text{r},|1\rangle} = \omega_{\text{r}} + \chi)$ when the qubit is in the ground state (excited state). The separation between the two frequencies is $|\omega_{\text{r},|0\rangle} - \omega_{\text{r},|1\rangle}| = |2\chi|$. Thus, the qubit state can be inferred from a precise measurement of the resonator response. This point will be explained in detail in Section 14.7.1.

The dispersive shift (14.60) was derived assuming that the qubit can be approximated as a two-level system. Taking into account the higher energy levels of the transmon, a more accurate expression for the dispersive shift is given by

$$\boxed{\chi = \frac{g^2}{\Delta(1 + \Delta/\eta)},} \tag{14.61}$$

where η is the qubit anharmonicity. The derivation of this expression can be found in Exercise 14.5.

Experimentally, the coupling strength g can be measured in two ways:

1) If the superconducting qubit is tunable and its frequency crosses the resonator frequency, g can be determined by measuring the resonator's transmission coefficient as a function of flux. When the qubit is in resonance with the resonator, the eigenstates of the composite system are entangled states of the form[15],

$$|\psi_{1-}\rangle = \tfrac{1}{\sqrt{2}}|\text{g}, 1\rangle - \tfrac{1}{\sqrt{2}}|\text{e}, 0\rangle, \qquad E_{1-} = \hbar\omega_{\text{r}} - \hbar g,$$

$$|\psi_{1+}\rangle = \tfrac{1}{\sqrt{2}}|\text{g}, 1\rangle + \tfrac{1}{\sqrt{2}}|\text{e}, 0\rangle, \qquad E_{1+} = \hbar\omega_{\text{r}} + \hbar g.$$

Spectroscopic measurements of the resonator show an anti-crossing of the energy levels as illustrated in Fig. 14.22a–b. The splitting of the energy levels is $2g$. Close to the degeneracy point, the frequency of the two resonances can be fitted to

$$\omega_{\text{r}} - \frac{1}{2}\Big[(\omega_{\text{r}} - \omega_{01}(\Phi_{\text{ext}})) \pm \sqrt{(\omega_{\text{r}} - \omega_{01}(\Phi_{\text{ext}}))^2 + 4g^2}\Big], \tag{14.62}$$

where we used eqns (4.88, 4.89). Here, ω_{r} is the bare resonator frequency, while $\omega_{01}(\Phi_{\text{ext}})$ is the qubit frequency as a function of flux, eqn (14.38).

[15]This eigenstates were derived in Section 4.7.

2) If the qubit is not tunable, the value of g can be estimated from a measurement of the dispersive shift χ. First, the resonator is measured with the qubit in the ground state: this provides the frequency $\omega_{\mathrm{r},|0\rangle}$. Second, the resonator is measured with the qubit in the excited state (a π pulse is applied to the qubit before a microwave drive is sent to the resonator). The transmission coefficient will show a peak at $\omega_{\mathrm{r},|1\rangle}$ from which one can extract the dispersive shift, $\chi = \frac{1}{2}(\omega_{\mathrm{r},|1\rangle} - \omega_{\mathrm{r},|0\rangle})$. Finally, the coupling g is derived by inverting eqn (14.61).

14.7.1 Transmission measurements

In the previous section, we learned that the qubit state induces a frequency shift of the resonator. Now we just need to show how to measure this frequency shift. As shown in Fig. 14.21a and Fig. 14.23, the resonator is coupled to a transmission line, usually called the readout line. The presence of the resonator locally changes the impedance of the line, thus creating a scattering site. The incoming electrical pulse with frequency ω and amplitude V_{in} is partly reflected and partly transmitted. The voltage of the reflected signal is $S_{11}(\omega)V_{\mathrm{in}}$, while the voltage of the transmitted wave is $S_{21}(\omega)V_{\mathrm{in}}$. The amplitude and phase of the transmitted signal reveal whether the resonator frequency has shifted or not, i.e. whether the qubit is in the ground or excited state.

In Section 13.4, we showed that the transmission coefficient of a resonator coupled to a transmission line is given by

$$S_{21}(\omega) = \frac{\frac{Q_{\mathrm{e}}}{Q_{\mathrm{e}}+Q_{\mathrm{i}}} + 2iQ\frac{(\omega-\omega_{\mathrm{r}})}{\omega_{\mathrm{r}}}}{1 + 2iQ\frac{(\omega-\omega_{\mathrm{r}})}{\omega_{\mathrm{r}}}}, \tag{14.63}$$

where Q_{i} and Q_{e} are the internal and external quality factors, while $1/Q = 1/Q_{\mathrm{i}} + 1/Q_{\mathrm{e}}$ is the loaded quality factor. Figure 14.24a–b shows a plot of the magnitude and phase of the transmission coefficient as a function of frequency when the qubit is in the ground and excited state. The magnitude of the transmission coefficient shows a sharp dip and its minimum value is $S_{\mathrm{min}} = Q_{\mathrm{e}}/(Q_{\mathrm{e}} + Q_{\mathrm{i}})$. The full width at half maximum, i.e. the distance between the points A and C, is $\kappa = \omega_{\mathrm{r}}/Q$. In the complex plane, the frequency response has a circular shape with a diameter $1 - S_{\mathrm{min}}$ (see Fig. 14.24c).

If the signal probes the resonator between the two frequencies, $(\omega_{\mathrm{r},|0\rangle} + \omega_{\mathrm{r},|1\rangle})/2$, the transmission coefficient will be $S_{21}(\omega_{\mathrm{r}} - \chi)$ if the qubit is in the ground state and $S_{21}(\omega_{\mathrm{r}} + \chi)$ if the qubit is in the excited state. An accurate measurement of the change in the transmission coefficient is critical to infer the state of the qubit. To increase the visibility of the signal, the resonator parameters must be set so that the points $S_{21}(\omega_{\mathrm{r}} - \chi)$ and $S_{21}(\omega_{\mathrm{r}} + \chi)$ are diametrically opposed on the circle illustrated in Fig. 14.24c. Points A and C, at the top and bottom of the circle, are the opposite points that require the smallest frequency shift. Since these points occur at $\omega_{\mathrm{r}} \pm \kappa/2$, the condition of maximum visibility is obtained when

$$\boxed{2\chi = \kappa.} \tag{14.64}$$

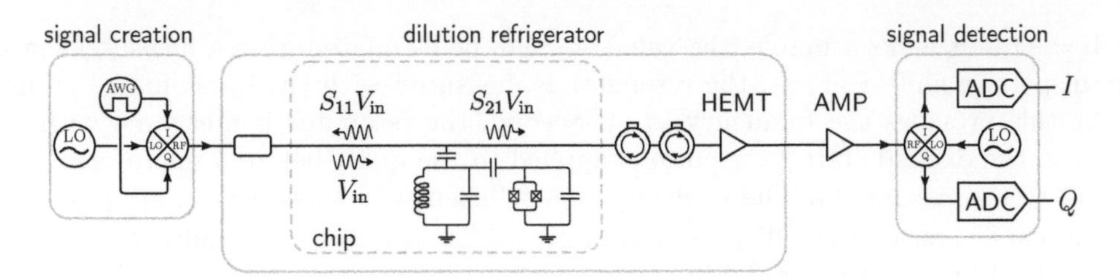

Fig. 14.23 Transmission measurement of a resonator. A microwave source generates a high-frequency sinusoidal signal (ω_{LO}), while an arbitrary waveform generator (AWG) generates the pulse envelope $\varepsilon(t)$. The two signals are combined by an IQ-mixer which produces a shaped electrical pulse $V_{\text{in}}(t)$ with frequency $\omega_{\text{d}} \approx \omega_{\text{r}}$. The signal is carried with coaxial cables into the dilution refrigerator, where it is filtered and attenuated before reaching the actual device. The transmitted signal $S_{21}V_{\text{in}}$ passes through two circulators and a cryogenic low-noise amplifier (HEMT). The circulators prevent noise generated by the HEMT from going toward the device. The signal is amplified with a room-temperature amplifier before it is downconverted using heterodyne mixing and finally sampled in a digitizer. The in-phase and quadrature components of the output signal, $I = \text{Re}(S_{21}V_{\text{in}})$ and $Q = \text{Im}(S_{21}V_{\text{in}})$, provide information about the resonator frequency and consequently about the qubit state.

At the design stage, the experimenter must adjust the capacitive coupling between the resonator and the qubit as well as the coupling between the resonator and the feedline so that the condition (14.64) is satisfied.

The signal from the readout line is amplified and acquired with standard heterodyne mixing. The final result is a single point in the IQ plane with coordinates $I = \text{Re}(S_{21})$ and $Q = \text{Im}(S_{21})$. In the ideal case, the measured point should correspond to either point A or C′ shown in Fig. 14.24c. If the distance between the two points is greater than zero, one should be able to distinguish the two states with no errors. In practice, however, random noise adds some statistical fluctuations. If the measurement of the qubit state is repeated multiple times, instead of observing single points perfectly aligned with points A and C′, one obtains two clouds as illustrated in Fig. 14.25. The projection of these IQ clouds onto a line connecting their centers produces two Gaussians centered around x_0 and x_1 with standard deviations σ_0 and σ_1. For a high-fidelity readout, the ratio between the distance of the Gaussians and their width should be as large as possible. This ratio is called the **signal-to-noise ratio**,

$$\boxed{\text{SNR} = \frac{(x_0 - x_1)^2}{2\sigma_1\sigma_2}.} \tag{14.65}$$

When SNR $\gg 1$, the two distributions are well separated and one can unambiguously distinguish the qubit states. The **separation error** ε_{sep} originates from the finite overlap between the two distributions. Assuming that the two Gaussians have the same standard deviation σ, the separation error can be defined as half of the overlap

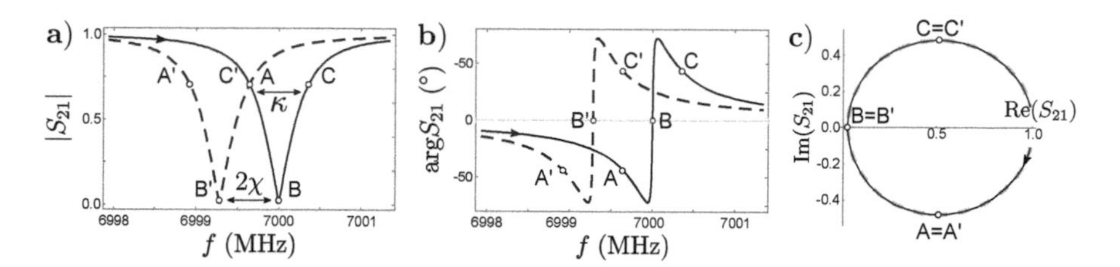

Fig. 14.24 Transmission coefficient. a) Magnitude of the transmission coefficient $S_{21}(f)$ of eqn (14.63) when the qubit is in the ground state (solid curve) and in the excited state (dashed curve). The linewidth is $\kappa = \omega_r/Q$, while the separation between the two resonances is $|2\chi|$. In this plot, we set $|2\chi| = \kappa$ so that points A and C$'$ coincide. The points B and B$'$ indicate the resonances $\omega_{r,|0\rangle}$ and $\omega_{r,|1\rangle}$, respectively. **b)** Phase of the transmission coefficient as a function of frequency. Note the big separation between the points A and C$'$. **c)** Plot of the transmission coefficient S_{21} in the complex plane when the qubit is in the ground state (solid circle) and the excited state (dashed gray circle). The two circles overlap. If the resonator is probed at $(\omega_{r,|0\rangle} + \omega_{r,|1\rangle})/2$, the quadratures of the transmitted signal, $I = \mathrm{Re}(S_{21})$ and $Q = \mathrm{Re}(S_{21})$, will be aligned with point A when the qubit is in the ground state and with point C$'$ when the qubit is the excited state.

between the two Gaussians[16]

$$\varepsilon_{\mathrm{sep}} = \frac{1}{\sqrt{2\pi\sigma^2}} \int_{\frac{x_0+x_1}{2}}^{\infty} e^{\frac{-(x-x_1)^2}{2\sigma^2}} dx = \frac{1}{2}\left(1 + \mathrm{Erf}\left[\frac{|x_0 - x_1|}{2\sqrt{2}\sigma}\right]\right).$$

Using the definition of SNR, the separation error can be expressed in a more compact form as

$$\boxed{\varepsilon_{\mathrm{sep}} = \frac{1}{2}\left(1 + \mathrm{Erf}\left[\frac{\mathrm{SNR}}{2}\right]\right).}$$

One method to decrease the separation error is to amplify the output signal using a Josephson parametric amplifier [633, 634, 635, 636].

The dispersive coupling between the qubit and the resonator unlocks many interesting experiments, such as qubit spectroscopy. **Qubit spectroscopy** is a pulse sequence designed to measure the energy levels of a qubit. As shown in Fig. 14.26, this measurement entails sending a long pulse[17] with drive frequency f_d through the XY line followed by a measurement pulse applied through the readout line[18]. The transition frequency of the qubit is measured by detecting a change in the cavity

[16]The Erf function is defined as $\mathrm{Erf}(x) = \frac{2}{\sqrt{\pi}} \int_0^x e^{-t^2} dt$.

[17]In the frequency domain, the bandwidth of the pulse is around $\approx 1/T$ where T is the pulse duration. Qubit spectroscopy can only resolve features that are wider than $1/T$ Hz.

[18]This technique is known as pulsed spectroscopy. An alternative method to execute qubit spectroscopy is based on continuous waves. In this approach, the excitation pulse and the measurement pulse are long continuous drives that *overlap* with each other. This strategy leads to an improved SNR. However, the qubit frequency might get Stark shifted by a few MHz since the resonator is populated with a finite number of photons. This phenomenon is captured by eqn (14.70).

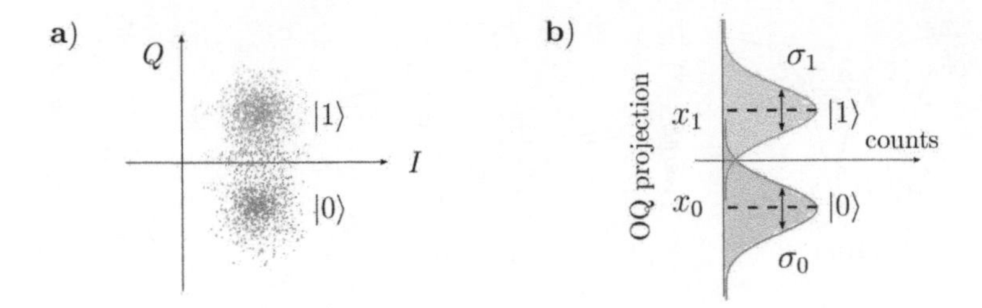

Fig. 14.25 IQ clouds. a) Distribution of the transmitted signal in the IQ plane when the qubit is in the ground state and excited state in the presence of noise. The x axis represents the in-phase component of the output signal $I = \mathrm{Re}(S_{21})$, while the y axis represents the quadrature component of the output signal $Q = \mathrm{Im}(S_{21})$. The clouds usually follow a two-dimensional Gaussian distribution. **b)** Projection of the clouds onto the optimal quadrature, i.e. the line passing through the center of the two distributions. The separation error is related to the overlap between the distributions. The SNR is the ratio between the distance and the width of the two Gaussian distributions.

transmission coefficient at resonance, $S_{21}(\omega_\mathrm{r})$, as the drive frequency is swept across a defined frequency range. When f_d is detuned from the qubit frequency, the value of the transmission coefficient remains unchanged. When f_d is close to the qubit frequency, the resonator peak shifts by 2χ, producing a change in the transmission coefficient. Assuming that the drive amplitude is small, the probability of measuring the qubit in the excited state as a function of frequency has a characteristic Lorentzian shape [637] (see Exercise 14.6),

$$\boxed{p_\mathrm{e}(\omega_\mathrm{d}) = \frac{T_1 T_2 \Omega^2}{(\omega_\mathrm{d} - \omega_\mathrm{q})^2 + \frac{1}{T_2^2}(1 + T_1 T_2 \Omega^2)},} \tag{14.66}$$

where the Rabi frequency Ω is proportional to the drive amplitude. The Lorentzian peak is centered around ω_q with a FWHM $= \frac{2}{T_2}\sqrt{1 + T_1 T_2 \Omega^2}$. Note that for small drive amplitudes $\Omega \ll 1/\sqrt{T_1 T_2}$, the linewidth of the peak becomes $2/T_2$. Thus, the inverse of the FWHM gives a rough estimate of the qubit transverse relaxation time T_2. This parameter can be measured more accurately with a Ramsey experiment.

14.7.2 Purcell effect

The relaxation time of a superconducting qubit coupled to a microwave resonator depends on the detuning between the qubit frequency and the resonator frequency. This effect is known as the **Purcell effect**, named after E. M. Purcell [638]. To study this decay mechanism, we will consider a qubit coupled to a resonator which is in turn coupled to the environment,

$$\hat{H} = \hat{H}_\mathrm{q} + \hat{H}_\mathrm{r} + \hat{H}_\mathrm{e} + \hat{H}_{\mathrm{r,q}} + \hat{H}_{\mathrm{r,e}},$$

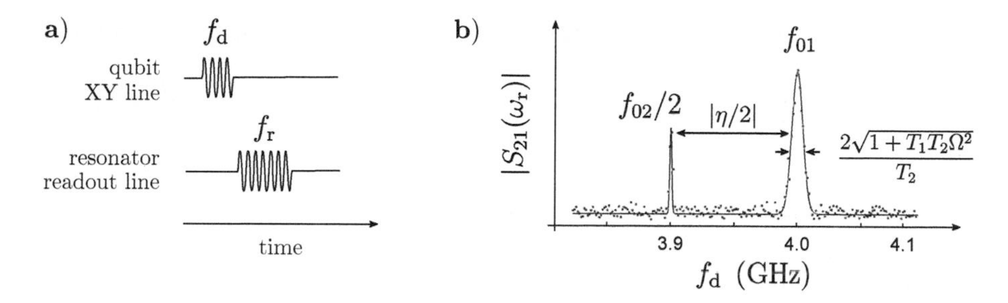

Fig. 14.26 Qubit spectroscopy. a) In pulsed spectroscopy, an electrical pulse with frequency $f_{\rm d}$ and duration ≈ 20 μs is applied to the qubit followed by a measurement pulse at the resonator frequency, $f_{\rm r}$. The duration of the latter is typically in the range $0.1 - 5$ μs. **b)** Plot of the magnitude of the transmission coefficient as a function of the drive frequency $f_{\rm d}$. The transmission coefficient shows a Lorentzian peak close to the qubit frequency with a FWHM $= 2\sqrt{1+T_1T_2\Omega^2}/T_2$. If the drive amplitude is sufficiently large, the $|0\rangle \to |2\rangle$ transition can be excited with a two-photon process producing a Lorentzian peak at $f_{02}/2$. The frequency difference $f_{02}/2 - f_{01}$ is half of the anharmonicity.

---

where $\hat{H}_{\rm q} = -\hbar\omega_{\rm q}\hat{\sigma}_z/2$ is the qubit Hamiltonian, $\hat{H}_{\rm r} = \hbar\omega_{\rm r}\hat{a}^\dagger\hat{a}$ is the Hamiltonian of the resonator, and $\hat{H}_{\rm r,q} = \hbar g(\hat{\sigma}^+\hat{a} + \hat{\sigma}^-\hat{a}^\dagger)$ describes the interaction between the qubit and the resonator. As usual, the environment is modeled with an infinite set of harmonic oscillators at thermal equilibrium, $\hat{H}_{\rm e} = \sum_j \hbar\omega_j\hat{c}_j^\dagger\hat{c}_j$. The Hamiltonian describing the interaction between the resonator and the environment is given by

$$\hat{H}_{\rm r,e} = \hbar\sum_{j=1}^{\infty}\lambda_j(\hat{a}\hat{c}_j^\dagger + \hat{a}^\dagger\hat{c}_j), \tag{14.67}$$

where $\hat{c}_j$ and $\hat{c}_j^\dagger$ are the destruction and creation operators of the j-th harmonic mode, and λ_j expresses the coupling strength between the resonator and the j-th mode. Let us assume that the system operates in the dispersive regime and the initial state of the global system $|i\rangle$ starts with the qubit in the excited state $|{\rm e}\rangle$ (with a small photonic component) and the thermal oscillators in the vacuum state $|\underline{0}\rangle = |00\ldots\rangle$,

$$|i\rangle = |{\rm e}, 0, \underline{0}\rangle + \frac{g}{\Delta}|{\rm g}, 1, \underline{0}\rangle, \tag{14.68}$$

where we used eqn (4.91) and ignored the normalization factor. This state has energy E_i. A transfer of energy brings an oscillator in the environment into the excited state leaving the qubit and resonator in the ground state. The energy of the final state $|f\rangle = |{\rm g}, 0, 1_k\rangle$ is $E_f = E_i$. If the modes in the environment are very dense, we can perform a continuum limit and derive the decay rate using Fermi's golden rule (8.60),

$$\frac{1}{T_1} = \frac{2\pi}{\hbar^2}p(\omega_k)|\langle f|\hat{H}_{\rm r,e}|i\rangle|^2,$$

where $p(\omega_k)$ indicates the density of states in the bath with frequency ω_k. Since

$$
\begin{aligned}
|\langle f|\hat{H}_{\mathrm{r,e}}|i\rangle|^2 &= \Big|\hbar\sum_j \lambda_j \langle \mathrm{g}, 0, 1_k| \big(\mathbb{1}\otimes\hat{a}\otimes\hat{c}_j^\dagger + \mathbb{1}\otimes\hat{a}^\dagger\otimes\hat{c}_j\big)\big(|\mathrm{e},0,\underline{0}\rangle - \frac{g}{\Delta}|\mathrm{g},1,\underline{0}\rangle\big)\Big|^2 \\
&= \Big|\hbar\sum_j \lambda_j \frac{g}{\Delta}\delta_{jk}\Big|^2 = \hbar^2\lambda_k^2\frac{g^2}{\Delta^2},
\end{aligned}
$$

we arrive at [639]

$$
\boxed{T_1 = \frac{\Delta^2}{\kappa g^2},} \tag{14.69}
$$

where we introduced the decay rate of the resonator $\kappa = 2\pi p(\omega_k)\lambda_k^2$. This is called the **Purcell limit** on T_1. In conclusion, the upper limit on T_1 is large when the detuning between the qubit and the resonator is large and when the resonator linewidth κ is small.

Equation (14.69) shows that there are two competing effects at play. On the one hand, a small decay rate κ leads to a larger upper limit on T_1. On the other hand, the energy should leak out from the resonator as fast as possible to enable fast measurement cycles, meaning that $\kappa \gg 1$ would be desirable. This issue can be solved by introducing a microwave filter, called the **Purcell filter**, between the resonator and the readout line. This microwave component allows for fast readout while protecting the qubit from releasing energy into the readout line [623, 640, 491].

14.7.3 Photon number fluctuations due to thermal effects

The fluctuations of the number of photons in the resonator can degrade the dephasing time of the qubit. To better understand this point, we start once again from the Jaynes–Cummings Hamiltonian in the dispersive regime (14.59). By rearranging some terms, we obtain

$$
\boxed{\hat{H} = \hbar\omega_{\mathrm{r}}\hat{a}^\dagger\hat{a} - \frac{\hbar}{2}(\omega_{01} + \chi + 2\chi\hat{a}^\dagger\hat{a})\hat{\sigma}_z} \tag{14.70}
$$

This expression shows that the coupling between the qubit and the resonator shifts the qubit frequency by a fixed amount χ, known as the **Lamb shift**, and by an amount that depends on the number of photons inside the resonator: the addition of one photon to the cavity shifts the qubit frequency by 2χ. This is known as the **AC Stark shift** [139]. Fluctuations of the photon number in the resonator cause random shifts of the qubit frequency and thus contribute to qubit dephasing [637, 145, 641]. Let us investigate this phenomenon in more detail.

In theory, a 6 GHz microwave resonator at 10 mK should be in its quantum ground state. In practice, the device might not be adequately thermalized and thus the resonator might be populated with some residual thermal photons [642, 643, 644]. This has a detrimental effect on the qubit dephasing time. To quantify this effect, we will model the environment as an infinite set of oscillators at thermal equilibrium. The state of the composite system "qubit + resonator" ρ_{tot} varies in time according to the master equation

$$\frac{d\rho_{\text{tot}}}{dt} = \frac{1}{i\hbar}[\hat{H}, \rho_{\text{tot}}] + \kappa(1 + n_{\text{th}})\mathcal{D}[\hat{a}](\rho_{\text{tot}}) + \kappa n_{\text{th}}\mathcal{D}[\hat{a}^\dagger]\rho_{\text{tot}}, \tag{14.71}$$

where $\hat{H}$ is the Jaynes–Cummings Hamiltonian in the dispersive regime (14.70), $n_{\text{th}} = 1/(e^{\hbar\omega_{\text{r}}/k_{\text{B}}T} - 1)$, and κ is the coupling strength between the resonator and the environment. The last two terms in (14.71) describe the energy exchange between the resonator and the environment. Assuming that the temperature of the device is low $k_{\text{B}}T \ll \hbar\omega_{\text{r}}$, a lengthy mathematical calculation shows that the off-diagonal element of the qubit density matrix evolves as $\rho_{01}(t) = \rho_{01}(0)e^{-\gamma_\phi t}$ where the dephasing rate is given by [567, 645]

$$\boxed{\gamma_\phi = \frac{\kappa}{2} - \frac{1}{2}\text{Re}\Big[\sqrt{(\kappa - 2i\chi)^2 + i8\chi\kappa n_{\text{th}}}\Big].} \tag{14.72}$$

The derivation of this expression can be found in Exercise 14.8. Using realistic values for the dispersive shift $\chi/2\pi = 1$ MHz, the resonator linewidth $\kappa/2\pi = 2$ MHz, the resonator frequency $\omega_{\text{r}}/2\pi = 6$ GHz and the effective temperature of the resonator $T = 40$ mK, eqn (14.72) predicts that the maximum value of the qubit dephasing time is $T_\phi = 1/\gamma_\phi \approx 210$ μs.

With some manipulations, eqn (14.72) can be expressed in a simpler form. By introducing the parameter $r = 2\chi/\kappa$, we have

$$\gamma_\phi = \frac{\kappa}{2} - \frac{\kappa}{2}\text{Re}\big[\sqrt{Z}\big]$$

where $Z = 1 - r^2 + i2r(2n_{\text{th}} - 1)$ is a complex number. Let us calculate the quantity[19],

$$\text{Re}\big[\sqrt{Z}\big] = \sqrt{\frac{1 - r^2 + \sqrt{(1 - r^2)^2 + 4r^2(1 + 4n_{\text{th}}^2 + 4n_{\text{th}})}}{2}}.$$

If the number of thermal photons in the resonator is small $n_{\text{th}}^2 \ll n_{\text{th}}$, we can perform the approximation $\text{Re}\big[\sqrt{Z}\big] \approx 1 + 2r^2 n_{\text{th}}/(1 + r^2)$. This leads to

$$\boxed{\gamma_\phi = \frac{\bar{n}_{\text{th}}\kappa 4\chi^2}{\kappa^2 + 4\chi^2}.}$$

At low temperature, this expression is an excellent approximation of eqn (14.72).

14.8 Two-qubit gates

Two-qubit gates are logical operations that permit entangling the state of two qubits. In the field of superconducting devices, these types of gates are implemented by applying RF pulses and/or flux pulses to the qubits. A two-qubit gate scheme should satisfy

[19] Recall that the square root of a complex number $Z = c + id$ is given by $\sqrt{Z} = a + ib$ where $a = \sqrt{\frac{1}{2}(c + \sqrt{c^2 + d^2})}$.

multiple criteria. First, the interaction between the qubits should be small when the qubits are idling. This improves the performance of simultaneous single-qubit gates. Second, the duration of the two-qubit gate should be much shorter than the coherence time (as a rule of thumb, the gate time should be at least 1000 times faster than the coherence time to reach two-qubit gate fidelities of 99.9%). Lastly, the number of input lines required to implement the two-qubit gates should be as small as possible to minimize the complexity of the fridge build-out. In the last decade, several two-qubit gate schemes have been explored. They can be grouped into three categories:

Flux tunable This two-qubit gate scheme requires the application of DC or RF flux pulses through on-chip flux bias lines. The flux pulse changes the qubit frequency and brings it into resonance with a neighboring qubit enacting a swap interaction or a controlled-phase interaction [646, 513, 647, 648, 649, 650, 651, 652, 653, 654]. The advantage of this scheme is that the gate time is relatively short compared to other two-qubit gate implementations, $t_{\text{gate}} = 30 - 150$ ns. However, this comes at the cost of having a flux bias line per qubit as well as exposing the qubits to flux noise. The iSWAP gate presented in Section 14.8.2 is an example of a two-qubit gate activated with DC fast flux pulses.

All-microwave control This type of two-qubit gate can be implemented without the need for flux bias lines [655, 656, 574, 657, 658, 659]. This significantly decreases the number of fridge lines in a large-scale quantum computer. In addition, since the qubits do not need to be tunable, they are not exposed to flux noise and their dephasing time tends to be longer. A downside is that the gate time is relatively slow compared to other two-qubit gate implementations, $t_{\text{gate}} = 150 - 300$ ns. Furthermore, the qubit frequencies must be separated by no more than 300 MHz, otherwise the gate time would be too long[20], limiting the number of gates that can be executed within the coherence time. Thus, all-microwave two-qubit gates rely on a very accurate fabrication of the Josephson junctions so that the qubit frequency ends up in the desired frequency band[21]. The relatively small qubit-qubit detunings can cause an additional issue: due to microwave crosstalk, an electrical pulse delivered to a qubit might excite a neighboring qubit, a problem known as the **frequency crowding problem**. The cross-resonance gate presented in Section 14.8.3 is an example of an all-microwave gate.

Tunable couplers One strategy to decrease the duration of a two-qubit gate is to increase the capacitive coupling between the qubits. However, this comes at the cost of intensifying an always-on $\hat{\sigma}_z \otimes \hat{\sigma}_z$ coupling that significantly degrades simultaneous single-qubit operations (see Section 14.8.1). To overcome this problem, the qubits can be connected by a tunable coupler, i.e. a tunable element, such as a superconducting qubit or a tunable resonator, that mediates the interaction between two qubits [660, 661, 662, 663, 664, 665, 666]. The tunable coupler frequency directly affects

[20] In all-microwave control gates, the gate time is inversely proportional to the detuning.

[21] Recall that the qubit frequency is determined by E_{C} and E_{J}. The latter parameter is related to the room-temperature resistance of the junction (see eqn 14.18), which is determined by the thickness and area of the Josephson junction.

the coupling strength between the two qubits, providing several advantages over fixed coupling designs. First, tunable couplers help mitigate the frequency crowding problem because the interaction between neighboring qubits can be switched off by biasing the coupler to a specific frequency. Since the coupling can be turned off, one can execute high-fidelity single-qubit gates while keeping the qubits at their flux-insensitive point. Second, since the on/off coupling ratio does not depend on the maximum detuning between the qubits, one can increase the coupling strength enabling faster gates with reduced decoherence error [581, 663, 20, 667]. A downside is that each edge of the quantum processor requires an additional flux bias line, thereby increasing the complexity of the fridge build-out for large-scale devices. Section 14.8.4 explains how to implement an iSWAP gate using a tunable coupler.

The table below summarizes the number of fridge lines necessary to operate a grid of $d = n \times n$ superconducting qubits for different two-qubit gate schemes:

	# of XY lines	# of Z lines	# of readout lines	Total
Flux tunable	d	d	$2d/10$	$2.2d$
All-microwave	d	0	$2d/10$	$1.2d$
Tunable couplers	d	$3d - 2\sqrt{d}$	$2d/10$	$4.2d - 2\sqrt{d}$

These estimates are based on the assumption that the qubits are arranged in a square grid and groups of 10 resonators are measured with a single readout line. This table does not take into account the lines needed to pump the Josephson parametric amplifiers for the amplification of the signal exiting the readout line [633, 634, 635, 636].

14.8.1 Hamiltonian of two coupled transmons

To implement a two-qubit gate between a pair of superconducting qubits, the qubits are usually capacitively coupled together, as shown in Fig. 14.27a–b. Let us study the interaction Hamiltonian between two transmons. As explained in Exercise 14.9, the Hamiltonian of the coupled system is given by

$$\begin{aligned}\hat{H} &= \hat{H}_1 + \hat{H}_2 + \hat{H}_{12} \\ &= 4E_{C1}\hat{n}_1^2 - E_{J1}\cos\hat{\phi}_1 + 4E_{C2}\hat{n}_2^2 - E_{J2}\cos\hat{\phi}_2 + 4e^2\frac{C_{12}}{C_1C_2}\hat{n}_1\hat{n}_2.\end{aligned}$$

This expression comprises three terms. The first two terms, $\hat{H}_1$ and $\hat{H}_2$, are nothing but the Hamiltonians of the individual transmons, while the interaction term $\hat{H}_{12}$ is proportional to the product between the charge operators $\hat{n}_1$ and $\hat{n}_2$. Using $\hat{n}_j = -i(\hat{b}_j - \hat{b}_j^\dagger)/2\sqrt{\xi_j}$, the Hamiltonian becomes

$$\boxed{\hat{H} = \hat{H}_1 + \hat{H}_2 - \hbar g(\hat{b}_1 - \hat{b}_1^\dagger)(\hat{b}_2 - \hat{b}_2^\dagger),} \qquad (14.73)$$

where we introduced the **coupling strength**

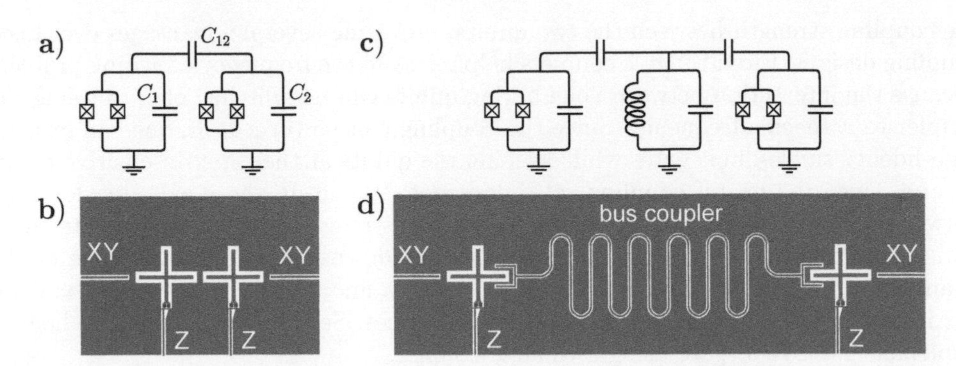

Fig. 14.27 Capacitive coupling between two superconducting qubits. a) Equivalent circuit of a tunable transmon capacitively coupled to another tunable transmon. **b)** Representation of two transmons capacitively coupled together. The distance between the qubit pads determines the value of the capacitance C_{12} and therefore the coupling strength g. The Z lines are used to bring the qubits into resonance, while the XY lines are used to manipulate the qubit state. **c)** Equivalent circuit of two transmons capacitively coupled to a common LC resonator. **d)** Representation of two transmons capacitively coupled to a $\lambda/2$ CPWR. The resonator, also known as the bus coupler, mediates the interaction between the qubits. The capacitance between the qubits and the resonator determines the coupling strength between the qubits.

$$g = \frac{C_{12}}{C_1 C_2} \frac{e^2}{\hbar\sqrt{\xi_1}\sqrt{\xi_2}}.$$

This parameter is proportional to the rate at which the two qubits exchange energy when they are in resonance. The coupling strength can be adjusted by changing the capacitance between the qubits (its value is typically in the range $g/2\pi = 1-20$ MHz). Suppose the anharmonicity of the transmons is sufficiently large and leakage to the second level can be ignored. In that case, it is possible to truncate the Hamiltonian to the first two energy levels. Using the substitutions $\hat{b} = \hat{\sigma}^-$ and $\hat{b}^\dagger = \hat{\sigma}^+$, we obtain

$$\hat{H} = -\frac{\hbar\omega_1}{2}\hat{\sigma}_{z1} - \frac{\hbar\omega_2}{2}\hat{\sigma}_{z2} - \hbar g(\hat{\sigma}^- - \hat{\sigma}^+)(\hat{\sigma}^- - \hat{\sigma}^+), \tag{14.74}$$

where ω_1 and ω_2 are the qubit frequencies. If the qubit frequencies are much greater than the coupling strength, one can perform the rotating wave approximation and ignore the terms $\hat{\sigma}^-\hat{\sigma}^-$ and $\hat{\sigma}_1^+\hat{\sigma}_2^+$. This leads to

$$\boxed{\hat{H} = -\frac{\hbar\omega_1}{2}\hat{\sigma}_{z1} - \frac{\hbar\omega_2}{2}\hat{\sigma}_{z2} + \hbar g(\hat{\sigma}^-\hat{\sigma}^+ + \hat{\sigma}^+\hat{\sigma}^-).} \tag{14.75}$$

This Hamiltonian will be the starting point for studying swap interactions in Section 14.8.2. Note the similarity between eqn (14.75) and the Jaynes–Cummings Hamiltonian (14.7).

In some architectures, the coupling between the qubits is mediated by a microwave resonator, typically a $\lambda/2$ coplanar waveguide resonator as shown in Fig. 14.27c–d. Since coplanar waveguide resonators are relatively long, they make it possible to couple qubits separated by more than 1 cm. The coupling Hamiltonian in the interaction picture is still given by eqn (14.73), with the only difference that the parameter g is replaced by [139, 512, 513, 648]

$$J = -\frac{1}{2} g_1 g_2 \frac{\delta_1 + \delta_2}{\delta_1 \delta_2}, \tag{14.76}$$

where g_j is the coupling strength between qubit j and the resonator, while $\delta_j = \omega_{\mathrm{r}} - \omega_j$ is the detuning between qubit j and the resonator. A detailed derivation of eqn (14.76) can be found in Exercise 14.12.

When the two qubits are sufficiently detuned $|\omega_1 - \omega_2| \gg g$, the system is said to be in the **dispersive regime**. In this regime, there is no energy exchange between the qubits. However, an interesting phenomenon occurs: the qubit frequency shifts when the neighbor is excited. This can be shown by diagonalizing the Hamiltonian (14.73) with a Schrieffer–Wolff transformation. The truncated Hamiltonian in the dispersive regime is given by

$$\boxed{\hat{H}/\hbar = -\frac{\tilde{\omega}_1}{2}\hat{\sigma}_{z1} - \frac{\tilde{\omega}_2}{2}\hat{\sigma}_{z2} + \frac{\zeta}{4}\hat{\sigma}_z \otimes \hat{\sigma}_z,} \tag{14.77}$$

where $\tilde{\omega}_1$ and $\tilde{\omega}_2$ are the renormalized qubit frequencies and the **ZZ coupling** ζ is given by [648] (see Exercise 14.10-14.11)

$$\boxed{\zeta \approx 2g^2\Big(\frac{1}{\Delta - \eta_2} - \frac{1}{\Delta + \eta_1}\Big),} \qquad \text{(first order approx.)} \tag{14.78}$$

where $\Delta = \omega_1 - \omega_2$. The ZZ coupling can be either positive or negative depending on the qubit frequencies and anharmonicities. The absolute value of the ZZ coupling reaches a maximum when the detuning between the qubits is either $\Delta = \eta_2$ or $\Delta = -\eta_1$. To better understand why the ZZ coupling induces a frequency shift of a qubit when the neighbor is excited, let us express eqn (14.77) in a different form

$$\hat{H} = -\frac{\hbar\tilde{\omega}_1}{2}\hat{\sigma}_{z1} - \frac{\hbar}{2}\Big(\tilde{\omega}_2 + \frac{\zeta}{2}\hat{\sigma}_{z1}\Big)\hat{\sigma}_{z2} = -\frac{\hbar}{2}\hat{\sigma}_{z1}\Big(\tilde{\omega}_1 + \frac{\zeta}{2}\hat{\sigma}_{z2}\Big) - \frac{\hbar\tilde{\omega}_2}{2}\hat{\sigma}_{z2}.$$

This expression shows that the frequency of qubit j is $\tilde{\omega}_j + \zeta/2$ ($\tilde{\omega}_j - \zeta/2$) when the neighboring qubit is in the ground state (excited state). If the ZZ coupling is large, this state-dependent frequency shift significantly degrades the performance of simultaneous single-qubit gates [668, 669].

Experimentally, the coupling strength g between two qubits can be measured in three ways:

1) If the qubits are tunable, they can be brought into resonance by adjusting the magnetic flux threading the SQUID of one of the two qubits. Qubit spectroscopy will

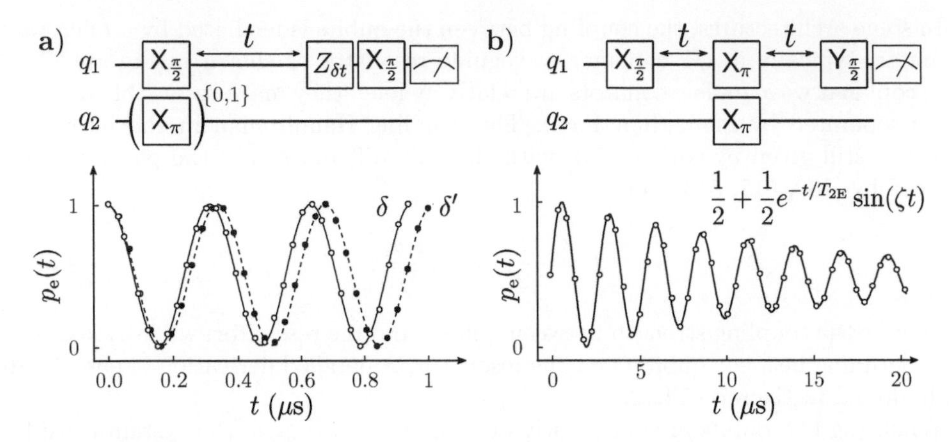

Fig. 14.28 Measuring the ZZ coupling. a) The ZZ coupling ζ can be measured with two Ramsey experiments. The first Ramsey experiment determines the frequency of qubit 1 with high accuracy (solid curve). From the fit of the oscillations, one can extract δ, the artificial detuning introduced by the z rotation. The second Ramsey experiment is executed with a π pulse applied to qubit 2 at the beginning of the experiment. This will produce oscillations with frequency δ' (dashed curve). The ZZ coupling is extracted from the difference $\zeta = \delta' - \delta$. **b)** An alternative method to measure the ZZ coupling is based on the JAZZ protocol [570, 670, 671]. This measurement consists of a Hahn-echo experiment on qubit 1 with a π pulse applied to qubit 2 in the middle of the experiment. The probability of measuring qubit 1 as a function of the delay shows periodic oscillations with frequency ζ. If $T_{2\text{E}}$ is much shorter than $1/\zeta$, one can introduce a virtual Z rotation before the second $\pi/2$ pulse to speed up the oscillations and obtain an accurate fit from a faster sequence.

show a splitting of the energy levels similar to what we have seen for a qubit coupled to a resonator in Fig. 14.22a. The magnitude of the splitting is $2g$ and gives a direct estimate of the coupling strength.

2) If the qubits are not tunable and the detuning is much greater than the coupling strength, the parameter g can be determined from a measurement of the ZZ coupling ζ. The parameter ζ can be measured with two Ramsey experiments, which involve probing the frequency of qubit 1 with qubit 2 either in the ground or excited state [624, 205]. As shown in Fig. 14.28a, the two measurements will show periodic oscillations with frequencies δ_1 and δ_2, and the difference between them is nothing but ζ. Once the ZZ coupling has been determined, the coupling strength g can be backed out from eqn (14.78).

3) Another method to measure the ZZ coupling is shown in Fig. 14.28b. This protocol is usually called a **joint amplification of ZZ (JAZZ)** measurement [570, 670, 671]. This method is based on a Hahn-echo experiment on qubit 1 with a π pulse applied to qubit 2 in the middle of the experiment. The probability of measuring qubit 1 in the excited state shows periodic oscillations with frequency ζ.

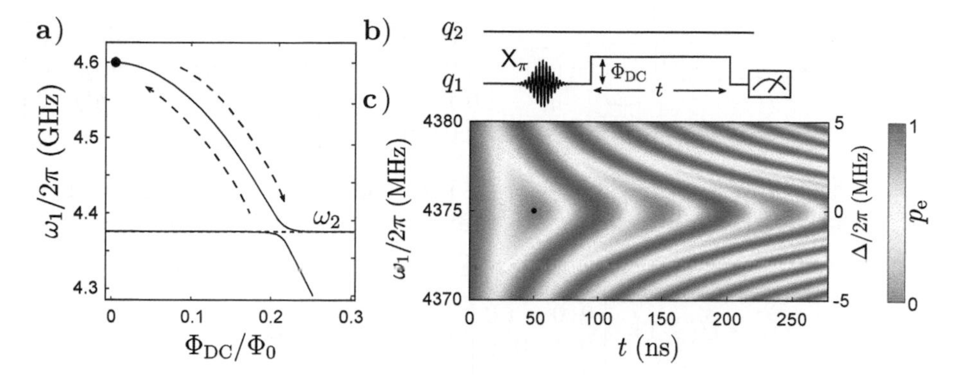

Fig. 14.29 SWAP spectroscopy. a) A superconducting qubit q_1 is parked at its maximum frequency $\omega_1/2\pi = 4.8$ GHz. A DC flux pulse brings it into resonance with qubit q_2 with frequency $\omega_2/2\pi = 4.375$ GHz. At the end of the flux pulse, q_1 returns to its initial frequency. **b)** Pulse sequence to calibrate the parameters of an iSWAP gate. Qubit 1 is excited with a π pulse. Subsequently, a flux pulse with varying amplitude and duration is applied to the qubit before a projective measurement. **c)** Probability of measuring qubit 1 in the excited state as a function of the flux pulse duration and detuning. In this example, an iSWAP gate is activated in 50 ns.

14.8.2 The iSWAP and $\sqrt{\text{iSWAP}}$ gates

A simple strategy to execute an entangling gate between two tunable qubits consists in bringing them into resonance for a specific amount of time [672, 513, 673, 649, 652]. Suppose two qubits (with frequencies ω_1 and ω_2) are capacitively coupled. As we have seen in the previous section, the Hamiltonian of the coupled system is given by

$$\hat{H} = -\frac{\hbar\omega_1}{2}\hat{\sigma}_{z1} - \frac{\hbar\omega_2}{2}\hat{\sigma}_{z2} + \hbar g(\sigma^+\sigma^- + \hat{\sigma}^-\hat{\sigma}^+).$$

It is easier to analyze the dynamics in the interaction picture. By performing the change of basis $\hat{R} = e^{-\frac{1}{i\hbar}\hat{H}_0 t}$, where $\hat{H}_0 = -\hbar\omega_1\hat{\sigma}_{z1}/2 - \hbar\omega_2\hat{\sigma}_{z2}/2$, we obtain

$$\hat{H}' = \hat{R}\hat{H}\hat{R}^{-1} - i\hbar\hat{R}\frac{d}{dt}\hat{R}^{-1} = \hbar g(\hat{\sigma}^-\hat{\sigma}^+ e^{-i\Delta t} + \hat{\sigma}^+\hat{\sigma}^- e^{i\Delta t}), \tag{14.79}$$

where $\Delta = \omega_1 - \omega_2$. If the two qubits are brought into resonance with a DC flux pulse, $\Delta = 0$, and the Hamiltonian reduces to

$$\hat{H} = \hbar g(\hat{\sigma}^+\hat{\sigma}^- + \hat{\sigma}^-\hat{\sigma}^+) = \hbar g \begin{bmatrix} 0 & 0 & 0 & 0 \\ 0 & 0 & 1 & 0 \\ 0 & 1 & 0 & 0 \\ 0 & 0 & 0 & 0 \end{bmatrix}.$$

This Hamiltonian generates an iSWAP gate if the qubits are left to interact for the right amount of time. Indeed, the time evolution operator is given by

$$\hat{U}(t) = e^{-ig(\sigma^+\sigma^- + \sigma^-\sigma^+)t} = \sum_{n=0}^{\infty} \frac{(-igt)^n}{n!} \begin{bmatrix} 0\,0\,0\,0 \\ 0\,0\,1\,0 \\ 0\,1\,0\,0 \\ 0\,0\,0\,0 \end{bmatrix}^n = \begin{bmatrix} 1 & 0 & 0 & 0 \\ 0 & \cos(gt) & -i\sin(gt) & 0 \\ 0 & -i\sin(gt) & \cos(gt) & 0 \\ 0 & 0 & 0 & 1 \end{bmatrix},$$

If the DC flux pulse bringing qubit 1 into resonance with qubit 2 has a duration $t = \pi/2g$, the effective interaction is locally equivalent to a iSWAP gate,

$$\hat{U}(\tfrac{\pi}{2g}) = \begin{bmatrix} 1 & 0 & 0 & 0 \\ 0 & 0 & -i & 0 \\ 0 & -i & 0 & 0 \\ 0 & 0 & 0 & 1 \end{bmatrix} = \mathsf{iSWAP}(\hat{Z} \otimes \hat{Z}).$$

If the coupling between the two qubits is $g/2\pi = 10$ MHz, the duration of the two-qubit gate is $t = \pi/2g = 25$ ns. If the qubits are left to interact for half of the duration, the resulting gate is locally equivalent to a $\sqrt{\mathsf{iSWAP}}$ gate,

$$\hat{U}(\tfrac{\pi}{4g}) = \begin{bmatrix} 1 & 0 & 0 & 0 \\ 0 & 1/\sqrt{2} & -i/\sqrt{2} & 0 \\ 0 & -i/\sqrt{2} & 1/\sqrt{2} & 0 \\ 0 & 0 & 0 & 1 \end{bmatrix} = (\hat{Z} \otimes \mathbb{1})\sqrt{\mathsf{iSWAP}}(\hat{Z} \otimes \mathbb{1}).$$

The crucial point is that both the iSWAP and $\sqrt{\mathsf{iSWAP}}$ gates are universal for quantum computation when combined with single-qubit rotations [674] (see Section 5.5). The activation of iSWAP and $\sqrt{\mathsf{iSWAP}}$ gates requires a fast tuning of the qubit frequencies. As seen in Section 14.6.4, this induces a single-qubit rotation about the z axis, meaning that each qubit acquires a local phase. These single-qubit phases can be calibrated with a variety of methods, such as quantum process tomography, unitary tomography [667], cross-entropy benchmarking [20], and Floquet calibration [675, 676]. The maximum fidelity of an iSWAP gate is limited by the coherence times of the two qubits [677, 678],

$$F = 1 - \frac{2t_{\text{gate}}}{5} \sum_{k=1}^{2} \left(\frac{1}{T_{1,k}} + \frac{1}{T_{\phi,k}} \right),$$

where $T_{1,k}$ and $T_{\phi,k}$ are the relaxation and dephasing times of qubit k. Hence, the gate time should be as fast as possible.

Experimentally, an iSWAP gate between tunable transmons is activated by applying a DC flux pulse to a superconducting qubit so that its frequency matches that of the neighboring qubit (see Fig. 14.29a). The parameters of the flux pulse can be calibrated by performing swap spectroscopy as illustrated in Fig. 14.29b. This circuit begins with a π pulse on qubit 1 followed by a DC flux pulse which shifts the qubit frequency for a time t. If the flux pulse brings qubit 1 into resonance with qubit 2, the excitation moves back and forth between the two qubits. Figure 14.29c shows the probability of measuring qubit 1 in the excited state as a function of the flux pulse amplitude and duration. This contour plot is usually called a **chevron** due to its characteristic shape. The chevron pattern can be fitted to (see Exercise 14.14)

$$p_{\mathrm{e}}(t,\Delta) = \frac{\Delta^2}{\Delta^2+4g^2} + \frac{4g^2}{\Delta^2+4g^2}\cos^2\left(\frac{\sqrt{\Delta^2+4g^2}}{2}t\right), \tag{14.80}$$

where $\Delta = \omega_1 - \omega_2$. The black dot in the figure indicates the point where the excitation is completely transferred from qubit 1 to qubit 2. A chevron measurement provides an alternative method to estimate the coupling strength between two qubits. Indeed, the fit of the oscillations at resonance returns the gate time t_{gate} for an iSWAP gate from which one can extract the coupling strength using the formula $g = \pi/2t_{\text{gate}}$.

14.8.3 The cross-resonance gate

The cross-resonance gate is an all-microwave two-qubit gate that relies on directly driving one of the qubits at the frequency of the other [679, 680, 656, 574, 659]. This type of two-qubit gate does not require flux tunable qubits, thereby reducing the number of fridge lines needed to operate the device and avoiding the problem of flux noise.

Experimentally, the cross-resonance gate is implemented by applying an electrical pulse to qubit 1 at the frequency of qubit 2. This technique is remarkably simple as activating an entangling gate requires the same type of electronics used for single-qubit gates. However, the presence of the drive introduces some undesired effects. First, the electrical pulse will cause an off-resonant drive of qubit 1. Second, qubit 2 will be excited by the drive due to the presence of higher energy levels (this effect is quantified by a quantum crosstalk factor ν). Third, as we have seen in Section 14.8.1, there is an always-on $\sigma_z \otimes \sigma_z$ interaction between the qubits when they operate in the dispersive regime. This virtual interaction degrades the performance of simultaneous single-qubit operations if the ZZ coupling is too large [681, 578].

The Hamiltonian of the system is the sum of four terms: the Hamiltonian of the two qubits, the interaction term, and the drive applied to qubit 1,

$$H/\hbar = \sum_{j\geq 0}\omega_j|j\rangle\langle j|\otimes \mathbb{1} + \sum_{k\geq 0}\omega_k \mathbb{1}\otimes|k\rangle\langle k| + g(b_1b_2^\dagger + b_1^\dagger b_2) + \Omega(t)(b_1+b_1^\dagger),$$

where $b = \sum_j \sqrt{j+1}|j\rangle\langle j+1|$ and $\Omega(t) = \Omega\sin(\omega_{\mathrm{d}}t+\phi)$ is the drive. Here, Ω is proportional to the amplitude of the drive, see eqn (14.41). For simplicity, let us focus our analysis on the first three energy levels of the transmon,

$$\begin{aligned} H/\hbar = {} & \omega_1|1\rangle\langle 1|\otimes\mathbb{1} + (2\omega_1+\eta_1)|2\rangle\langle 2|\otimes\mathbb{1} + \omega_2\mathbb{1}\otimes|1\rangle\langle 1| + (2\omega_2+\eta_2)\mathbb{1}\otimes|2\rangle\langle 2| \\ & + \Omega(t)(c_1+c_1^\dagger) + g(c_1\otimes c_2^\dagger + c_1^\dagger\otimes c_2), \end{aligned}$$

where we introduced the operator $c = |0\rangle\langle 1| + \sqrt{2}|1\rangle\langle 2|$. If the system operates in the dispersive regime $g \gg |\Delta|$ where $\Delta = \omega_1 - \omega_2$, one can diagonalize the Hamiltonian to second order with a Schrieffer–Wolff transformation and obtain (see Exercise 14.15)

$$\boxed{H'/\hbar \approx -\frac{\tilde{\omega}_1}{2}\sigma_{z1} - \frac{\tilde{\omega}_2}{2}\sigma_{z2} + \frac{\zeta}{4}\sigma_z\otimes\sigma_z + \Omega(t)\Big[\sigma_x\otimes\mathbb{1} + \nu\mathbb{1}\otimes\sigma_x + \mu\sigma_z\otimes\sigma_x\Big].} \tag{14.81}$$

The parameter ζ is the familiar ZZ coupling of eqn (14.78), while the **quantum crosstalk factor** ν and the **cross-resonance factor** μ are given by [681, 682, 683]

$$\boxed{\nu \approx -\frac{g}{\Delta + \eta_1},} \qquad \boxed{\mu \approx -\frac{g\eta_1}{\Delta(\Delta + \eta_1)}.}$$

The term proportional to $\sigma_z \otimes \sigma_x$ in eqn (14.81) is called the **cross-resonance term**. This term is the one that actually activates an entangling interaction between the two qubits. Let us assume that the drive applied to qubit 1 is in resonance with qubit 2,

$$\Omega(t) = \Omega \sin(\tilde{\omega}_2 t + \phi).$$

The square bracket in eqn (14.81) indicates that the electrical pulse causes an off-resonant drive of qubit 1 and a direct drive of qubit 2 with a Rabi rate dependent on the state of qubit 1. It is important to mention that there is always a finite microwave crosstalk between neighboring qubits. This results in a *classical* crosstalk factor ε that adds up to the quantum crosstalk factor producing a term $\nu' \mathbb{1} \otimes \sigma_x$ where $\nu' = \nu + \varepsilon$ [574, 578].

The unwanted term $\sigma_x \otimes \mathbb{1}$ in eqn (14.81) can be neglected if the drive amplitude is much smaller than the detuning between the qubits, $\Omega \ll \Delta$. Furthermore, the undesired terms $\mathbb{1} \otimes \sigma_x$ and $\sigma_z \otimes \sigma_z$ can be removed with an echo scheme [659, 684, 578]. Under these assumptions, the Hamiltonian can be approximated as

$$\boxed{H' \approx -\frac{\hbar\tilde{\omega}_1}{2}\sigma_{z1} - \frac{\hbar\tilde{\omega}_2}{2}\sigma_{z2} + \hbar\Omega(t)\mu(\sigma_z \otimes \sigma_x).}$$

It is convenient to study the dynamics in the interaction picture. By performing the change of basis $R = e^{-it(\tilde{\omega}_1\sigma_{z1} + \tilde{\omega}_2\sigma_{z2})/2}$, the Hamiltonian in the new frame is given by

$$H' = RHR^{-1} - i\hbar R\frac{d}{dt}R^{-1} = \hbar\mu\Omega\sin(\tilde{\omega}_2 t + \phi)\Big[\cos(\tilde{\omega}_2 t)\sigma_z \otimes \sigma_x - \sin(\tilde{\omega}_2 t)\sigma_z \otimes \sigma_y\Big].$$

The RWA approximation allows us to neglect rapidly oscillating terms,

$$\sin(\tilde{\omega}_2 t + \phi)\cos(\tilde{\omega}_2 t) = \frac{1}{2}[\sin(2\tilde{\omega}_2 t + \phi) + \sin(\phi)] \;\rightarrow\; \frac{\sin\phi}{2},$$
$$\sin(\tilde{\omega}_2 t + \phi)\sin(\tilde{\omega}_2 t) = \frac{1}{2}[\cos(\phi) - \cos(2\tilde{\omega} t + \phi)] \;\rightarrow\; \frac{\cos\phi}{2}.$$

This approximation leads to

$$H' = \frac{\hbar}{2}\mu\Omega\Big[\sin(\phi)\sigma_z \otimes \sigma_x - \cos(\phi)\sigma_z \otimes \sigma_y\Big].$$

The phase of the microwave pulse can be set to any value by the experimenter. For convenience, we choose $\phi = \pi/2$ so that the Hamiltonian becomes

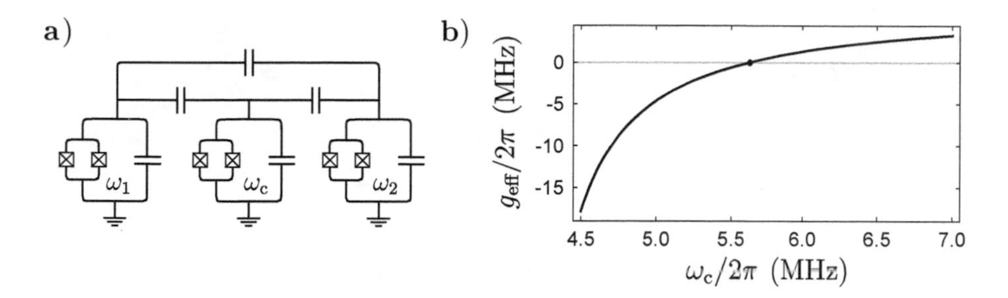

Fig. 14.30 Tunable couplers. a) Electrical circuit of two qubits coupled through a tunable coupler. **b)** Plot of the effective coupling $g_{\rm eff}$ of eqn (14.83) as a function of the tunable coupler frequency. When the qubits are idling, the coupler is parked at a specific frequency so that the effective coupling between the qubits is zero (black dot). In this example, we set $\omega_1/2\pi = 4$ GHz, $\omega_2/2\pi = 4.1$ GHz, $g_{12} = 7$ MHz, and $g_1 = g_2 = 105$ MHz.

$$H' = \frac{\hbar\mu\Omega}{2}(\sigma_z \otimes \sigma_x).$$

If the duration and amplitude of the microwave pulse are correctly tuned, an entangling operation takes place. To see this, let us calculate the time evolution operator

$$U(t) = e^{-i\frac{\mu\Omega t}{2}(\sigma_z\otimes\sigma_x)} = \begin{bmatrix} \cos\frac{\theta}{2} & -i\sin\frac{\theta}{2} & 0 & 0 \\ -i\sin\frac{\theta}{2} & \cos\frac{\theta}{2} & 0 & 0 \\ 0 & 0 & \cos\frac{\theta}{2} & i\sin\frac{\theta}{2} \\ 0 & 0 & i\sin\frac{\theta}{2} & \cos\frac{\theta}{2} \end{bmatrix} \equiv \mathsf{ZX}_\theta,$$

where $\theta = \mu\Omega t$. If the pulse amplitude and duration satisfy the relation $\mu\Omega t = -\pi/2$, we obtain the two-qubit gate $\mathsf{ZX}_{-\pi/2}$ which is equivalent to a CNOT gate up to two single-qubit rotations. This is evident from the quantum circuit shown below:

q_1 — $\mathsf{Z}_{\frac{\pi}{2}}$ — $\mathsf{ZX}_{-\frac{\pi}{2}}$ — $=$ — control
q_2 — $\mathsf{X}_{\frac{\pi}{2}}$ — $\mathsf{ZX}_{-\frac{\pi}{2}}$ — $=$ — X

It is tempting to conclude that the gate time $t = -\pi/2\mu\Omega$ can be made arbitrarily small by increasing the pulse amplitude Ω. However, a large pulse amplitude $\Omega \gg \Delta$, would induce an AC Stark shift on qubit 1 leading to phase errors.

14.8.4 Tunable couplers

In large-scale quantum processors, undesired couplings and frequency crowding can limit the performance of entangling operations. Architectures with tunable couplers have been proposed as a viable option to tune the qubit-qubit coupling to zero when the qubits are idling [660, 661, 662, 663, 665, 666]. By changing the frequency of the tunable coupler with a fast flux pulse, the coupling strength between neighboring

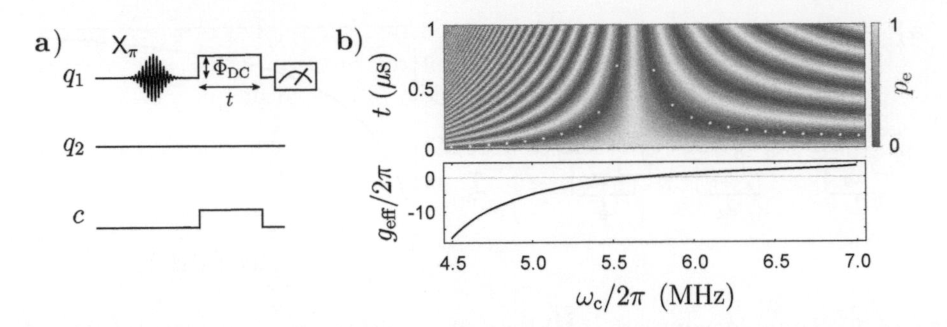

Fig. 14.31 Energy exchange vs. coupler frequency. a) Pulse sequence to calibrate the amplitude and duration of the flux pulses to implement an **iSWAP** gate. **a)** Top panel: probability of measuring qubit 1 in the excited state as a function of the coupler frequency, $p_e(t)$, eqn (14.84). The white dots indicate the point where the iSWAP gate takes place. Bottom panel: plot of the effective coupling strength as a function of the coupler frequency, eqn (14.83).

qubits can be increased, making it possible to execute two-qubit gates with fidelities above 99% [581, 663, 20].

Figure 14.30 shows the electrical circuit of two qubits (ω_1 and ω_2) connected by a tunable coupler (ω_c), i.e. a tunable transmon. The direct coupling is denoted as g_{12} while the coupling between the qubit j and the coupler is denoted g_j. The Hamiltonian of the system is given by

$$\begin{aligned} H/\hbar = &-\frac{\omega_1}{2}\sigma_{z1} - \frac{\omega_1}{2}\sigma_{z2} - \frac{\omega_c}{2}\sigma_{zc} + g_{12}(\sigma_1^-\sigma_2^+ + \sigma_1^+\sigma_2^-) \\ &+ g_1(\sigma_1^+\sigma_c^- + \sigma_1^-\sigma_c^+) + g_2(\sigma_2^+\sigma_c^- + \sigma_2^-\sigma_c^+). \end{aligned} \tag{14.82}$$

Typical values of the coupling strengths are $g_1 = g_2 = 80 - 140$ MHz, and $g_{12} = 5 - 15$ MHz. If the system operates in the dispersive regime $g_j \ll |\omega_c - \omega_j|$, one can diagonalize the Hamiltonian up to second order in $g_j/(\omega_j - \omega_c)$ with a Schrieffer–Wolff transformation resulting in (see Exercise 14.13)

$$H'/\hbar \approx -\frac{\tilde{\omega}_1}{2}\hat{\sigma}_{z1} - \frac{\tilde{\omega}_2}{2}\hat{\sigma}_{z2} - \frac{\tilde{\omega}_c}{2}\hat{\sigma}_{zc} + \hbar g_{\text{eff}}(\omega_c)(\sigma_1^+\sigma_2^- + \sigma_1^-\sigma_2^+),$$

where the **effective coupling** g_{eff} depends on the coupler frequency,

$$g_{\text{eff}}(\omega_c) \approx g_{12} + \frac{g_1 g_2}{2}\Big(\frac{1}{\omega_1 - \omega_c} + \frac{1}{\omega_2 - \omega_c}\Big). \tag{14.83}$$

The effective coupling is the sum of two terms: the first is always positive while the second can be negative for some values of ω_c. If the tunable coupler frequency is adjusted so that the second term has the same magnitude as the first but of opposite sign, the coupling between the qubits is zero. Figure 14.30b shows a plot of the effective coupling strength as a function of the coupler frequency.

An iSWAP gate can be implemented by bringing the qubits into resonance while applying an additional flux pulse to the coupler so that the coupling strength reaches the desired value (see Fig. 14.31a). The probability of measuring qubit 1 in the excited state oscillates periodically,

$$p_{\mathrm{e}}(t) = \cos^2(g_{\mathrm{eff}}(\omega_{\mathrm{c}})t). \tag{14.84}$$

Figure 14.31b shows the energy exchange between the qubits as a function of the tunable coupler frequency. The flux pulse is calibrated so that the excitation moves from qubit 1 to qubit 2 (white dots in the figure).

Further reading

Excellent reviews about superconducting devices can be found in Refs [556, 557, 558, 559, 523, 561, 562, 563, 564, 565, 477, 566].

In this chapter, we focused on the transmon. However, other types of superconducting qubits have been explored, including phase qubits [685], flux qubits [686, 687, 688, 689, 235], C-shunt flux qubits [236], fluxoniums [690, 691, 692], and $0-\pi$ qubits [693, 694, 695].

Summary

Josephson junction

$\psi_k(\mathbf{r},t) = \sqrt{\rho_k(\mathbf{r},t)}e^{i\theta_k(\mathbf{r},t)}$	Macroscopic wavefunction of the electrode k
$\phi = \theta_2 - \theta_1$	Phase difference
$I = I_c \sin\phi$	First Josephson equation
$\frac{d\phi}{dt} = \frac{2\pi}{\Phi_0}V$	Second Josephson equation
$\Phi_0 = h/2e = 2.07 \cdot 10^{-15}\ \mathrm{Tm}^2$	Flux quantum
$I_c \approx \pi 1.76 k_B T_c/2eR_n$	Junction critical current for a BCS superconductor
T_c	Critical temperature
R_n	Normal state resistance
$L_J = \Phi_0/2\pi I_c \cos\phi$	Josephson inductance
$E_J = \Phi_0 I_c/2\pi$	Josephson energy
$E_J \approx \Phi_0 \pi 1.76 k_B T_c/4\pi e R_n$	Josephson energy for BCS superconductors

Cooper pair box

n_I	Number of Cooper pairs on the island
n_R	Number of Cooper pairs in the reservoir
n	Number of Cooper pairs in excess on the island
$\hat{H} = 4E_C\hat{n}^2 - E_J\cos\hat{\phi}$	Hamiltonian of a Cooper pair box
$\hat{H} = 4E_C(\hat{n} - n_g)^2 - E_J\cos\hat{\phi}$	Hamiltonian of a CPB with a DC source
$n_g = C_g V_g/2e$	Gate charge
C_g	Gate capacitance
$E_k = E_C\mathcal{M}_A(x_k, y)$	Energy levels of a CPB

where $y = -E_J/2E_C$ and $x_k = \lfloor 2n_g \rfloor - 2n_g - (-1)^{k+\lfloor 2n_g \rfloor}k + (k + \lfloor 2n_g \rfloor) \bmod 2$.

$T_\phi \approx \hbar E_J/A_Q^2\pi^2(8E_C)^2$	Dephasing time due to charge noise

Transmon

A transmon is a CPB with $E_J/E_C \gg 1$.

$T_\phi \approx \hbar/\sqrt{c}A_Q\pi\lvert\epsilon_1\rvert$	Dephasing time due to charge noise
$\hat{H} = 4E_C\hat{n}^2 - E_J\cos\hat{\phi}$	Hamiltonian of a transmon
$\hat{\phi} = \sqrt{\xi}(\hat{b} + \hat{b}^\dagger)$	Phase operator
$\hat{n} = -i(\hat{b} - \hat{b}^\dagger)/2\sqrt{\xi}$	Charge operator
$\xi = \sqrt{2E_C/E_J}$	Dimensionless parameter
$\hat{H} \approx \sqrt{8E_J E_C}\hat{b}^\dagger\hat{b} - E_C(\hat{b} + \hat{b}^\dagger)^4/12$	Hamiltonian of a transmon
$E_k \approx \sqrt{8E_J E_C}k - E_C(k^2 + k)/2$	Energy levels of a transmon
$f_{01} \approx (\sqrt{8E_J E_C} - E_C)/h$	First transition frequency
$\eta = \omega_{12} - \omega_{01} \approx -E_C/\hbar$	Anharmonicity
$\omega_j - \omega_{j-1} = \omega_{01} + \eta j$	Transition frequencies

Tunable transmon

$\hat{H} = 4E_{\rm C}\hat{n}^2 - E_{\rm J1}\cos\hat{\phi}_1 - E_{\rm J2}\cos\hat{\phi}_1$	Hamiltonian of a tunable transmon
Σ	Area of the squid ($\approx$ 20 $\times$ 20 μm^2)
$\Phi_{\rm ext} = \int_\Sigma \mathbf{B}\cdot d\mathbf{\Sigma}$	Applied magnetic flux
$\hat{H} = 4E_{\rm C}\hat{n}^2 - E_{\rm J}(\Phi_{\rm ext})\cos\hat{\varphi}$	Hamiltonian of a tunable transmon

where $E_{\rm J}(\Phi_{\rm ext}) = (E_{\rm J1} + E_{\rm J2})|\cos(\pi\Phi_{\rm ext}/\Phi_0)|\sqrt{1 + d^2\tan^2(\pi\Phi_{\rm ext}/\Phi_0)}$,
$\hat{\varphi} = (\hat{\phi}_1 + \hat{\phi}_2)/2$, and $d = I_{\rm c1} - I_{\rm c2}/(I_{\rm c1} + I_{\rm c2})$.

$f_{01} \approx [\sqrt{8E_{\rm C}E_{\rm J}(\Phi_{\rm ext})} - E_{\rm C}]/h$	First transition frequency
$S_{\Phi\Phi}(\omega) = 2\pi A_\Phi^2/\lvert\omega\rvert$	PSD of flux noise
$A_\Phi \approx 1\ \mu\Phi_0/\sqrt{\rm Hz}$	Typical noise amplitude

---

Driving a transmon

$V_{\rm d}(t) = A\varepsilon(t)\sin(\omega_{\rm d}t + \alpha)$	Applied electrical pulse
$\varepsilon(t)$	Envelope
$\hat{H}_{\rm d} = 2eC_{\rm g}V_{\rm d}(t)\hat{n}/C_\Sigma$	Drive Hamiltonian
$\hat{H}_{\rm d} = -ieC_{\rm g}V_{\rm d}(t)(\hat{b} - \hat{b}^\dagger)/\sqrt{\xi}C_\Sigma$	Drive Hamiltonian
$\Omega = eC_{\rm g}A/\hbar\sqrt{\xi}C_\Sigma$	Rabi rate
$\hat{H}'(t) = -\frac{\hbar}{2}(\omega_{\rm q} - \omega_{\rm d})\hat{\sigma}_z + \hbar\Omega\varepsilon(t)(\hat{\sigma}_x\cos\alpha + \hat{\sigma}_y\sin\alpha)$	$\hat{H}_{\rm q} + \hat{H}_{\rm d}$ in the rotating frame
$\theta = \Omega\int_0^\infty \varepsilon(t')dt'$	Rotation angle
$p(1) = \sin^2(\Omega t_{\rm s}/2)$	Rabi oscillations
$T_1 \approx C_\Sigma/\omega_{01}^2 C_g^2 Z_0$	Upper limit on T_1 due to the control line
$\Omega_x(t)$, $\Omega_y(t) \propto \dot{\Omega}_x(t)$	DRAG technique

---

Z gates (rotations about the Z axis)

$\mathsf{Z}_\alpha = e^{-i\alpha\hat{\sigma}_z/2}$	Definition of Z gate
$\mathsf{Z}_\alpha = \mathsf{X}_{\frac{\pi}{2}}\mathsf{Y}_\alpha\mathsf{X}_{-\frac{\pi}{2}}$	Decomposition of a Z gate
$\alpha = \int_0^{t_{\rm g}} \omega_{01} - \omega_{01}(\Phi(t))dt$	Single-qubit phase accumulation during a flux pulse
$\delta_{\rm Stark} \approx \eta\Omega^2/2\Delta(\Delta + \eta)$	AC Stark shift

Circuit QED

Hamiltonian of a transmon capacitively coupled to a microwave resonator:

$$\hat{H} = \hbar\omega_{\rm r}\hat{a}^\dagger\hat{a} + \sqrt{8E_{\rm J}E_{\rm C}}\hat{b}^\dagger\hat{b} - E_{\rm C}(\hat{b}+\hat{b}^\dagger)^4/12 + \hbar g(\hat{b}^\dagger - \hat{b})(\hat{a}^\dagger - \hat{a})$$

$\hat{H} = \hbar\omega_{\rm r}\hat{a}^\dagger\hat{a} - \hbar\omega_{01}\hat{\sigma}_z/2 + \hbar g(\hat{\sigma}^+\hat{a} + \hat{\sigma}^-\hat{a}^\dagger)$	Jaynes–Cummings Hamiltonian
$\hat{H} = \hbar(\omega_{\rm r} - \chi\hat{\sigma}_z)\hat{a}^\dagger\hat{a} - \hbar(\omega_{01}+\chi)\hat{\sigma}_z/2$	JC Hamiltonian in the dispers. regime
$\chi \approx g^2/[\Delta(1+\Delta/\eta)]$	Dispersive shift
$S_{21}(\omega)$	Transmission coefficient
$\kappa = \omega_{\rm r}/Q$	Resonator linewidth
$2\chi = \kappa$	Optimal readout parameters
$\mathrm{SNR} = (x_0 - x_1)^2/2\sigma_1\sigma_2$	Signal-to-noise ratio
$p_{\rm e}(\omega_{\rm d}) = A\frac{T_1T_2\Omega^2}{(\omega_{\rm d}-\omega_{\rm q})^2+(1+T_1T_2\Omega^2)/T_2^2}$	Lorentzian peak in qubit spec.
$T_1 = \Delta^2/\kappa g^2$	Purcell limit on T_1
$1/T_\phi = \bar{n}_{\rm th}\kappa 4\chi^2/(\kappa^2+4\chi^2)$	Dephasing rate due to thermal noise

---

Coupled qubits

$\hat{H}/\hbar = -\omega_1\sigma_{z1}/2 - \omega_2\sigma_{z2}/2 + g(\sigma^-\sigma^+ + \sigma^+\sigma^-)$	Coupled qubits
$g = C_{12}e^2/\hbar\sqrt{\xi_1\xi_2}C_1C_2$	Coupling strength
$\hat{H}/\hbar = -\omega_1\sigma_{z1}/2 - \omega_2\sigma_{z2}/2 + J(\sigma^-\sigma^+ + \sigma^+\sigma^-)$	Coupling via a bus coupler
$J = -g_1g_2(\delta_1+\delta_2)/2\delta_1\delta_2$	Effective coupling
$\hat{H}/\hbar = -\omega_1\sigma_{z1}/2 - \omega_2\sigma_{z2}/2 + \zeta(\sigma_z\otimes\sigma_z)/4$	Coupled qubits in the disp. regime
$\zeta \approx 2g^2[1/(\Delta-\eta_2) - 1/(\Delta+\eta_1)]$	ZZ coupling
$\hat{U}(t) = e^{-ig(\sigma_1^-\sigma_2^+ + \sigma_1^+\sigma_2^-)t}$ when $\omega_1 = \omega_2$	Time evolution operator
$\hat{U}(\pi/2g)$	iSWAP gate
$\hat{U}(\pi/4g)$	$\sqrt{\mathsf{iSWAP}}$ gate
$p_{\rm e}(t,\Delta) = \frac{\Delta^2}{\Delta^2+4g^2} + \frac{4g^2}{\Delta^2+4g^2}\cos^2\left(\frac{\sqrt{\Delta^2+4g^2}}{2}t\right)$	Chevron pattern

---

Cross-resonance gate

$\Omega\sin(\omega_{\rm d}t+\phi)$	Drive on qubit 1
$\hbar\Omega(t)\mu(\sigma_z\otimes\sigma_x)$	Cross-resonance term
$\mu \approx -g\eta_1/\Delta(\Delta+\eta_1)$	Cross-resonance factor
$U(t) = e^{-i\frac{\theta}{2}(\sigma_z\otimes\sigma_x)} \equiv \mathsf{ZX}_\theta$	ZX_θ gate
$\mathsf{CNOT} = \mathsf{ZX}_{-\pi/2}(\mathsf{Z}_{\pi/2}\otimes\mathsf{X}_{\pi/2})$	CNOT gates in terms of the $\mathsf{ZX}_{-\pi/2}$ gate

Exercises

Exercise 14.1 Find the eigenvalues of the Hamiltonian (14.53) to second order in Ω_x.

Exercise 14.2 Derive eqn (14.37) from eqn (14.36).

Exercise 14.3 From superconductivity theory, we know that the Cooper pairs in a superconductor are described by a macroscopic wavefunction of the form $\psi(\mathbf{r}) = \sqrt{\rho(\mathbf{r})}e^{i\theta(\mathbf{r})}$. The phase of the wavefunction varies according to the relation [73]

$$\nabla\theta = \frac{2\pi}{\Phi_0}(\Lambda\mathbf{J} + \mathbf{A}), \tag{14.85}$$

where Λ is a constant, $\mathbf{J}$ is the current density, and $\mathbf{A}$ is the vector potential associated with an external magnetic field, $\nabla \times \mathbf{A} = \mathbf{B}$. We also know that the phase difference across a Josephson junction is $\phi = \theta_{\rm B} - \theta_{\rm A}$ where $\theta_{\rm A}$ and $\theta_{\rm B}$ are the phases of the wavefunctions on the two sides of the junction. In the presence of an external magnetic field, this relation becomes [73]

$$\phi = \theta_{\rm B} - \theta_{\rm A} - \frac{2\pi}{\Phi_0}\int_{\rm A}^{\rm B} \mathbf{A}\cdot d\mathbf{l}. \tag{14.86}$$

By calculating the line integral $\oint \nabla\theta \cdot d\mathbf{l}$ on the perimeter of the SQUID, show that

$$2\pi n = \phi_1 - \phi_2 + \frac{2\pi}{\Phi_0}\Phi_{\rm ext},$$

where $n \in \mathbb{Z}$, ϕ_1 and ϕ_2 are the phase differences across the two junctions, and $\Phi_{\rm ext}$ is the magnetic flux through the SQUID.

Exercise 14.4 A square pulse is applied to a superconducting qubit initially prepared in the ground state. Suppose that the frequency of the pulse $\omega_{\rm d}$ is slightly detuned from the qubit frequency, $\Delta = \omega_{\rm q} - \omega_{\rm d} \neq 0$ (this might happen when the control signal is not properly calibrated). From eqn (14.42), the Hamiltonian in the rotating frame becomes $\hat{H}' = -\frac{1}{2}\hbar\Delta\hat{\sigma}_z + \frac{1}{2}\hbar\Omega\varepsilon_{\rm s}(t)\hat{\sigma}_y$ where we assumed that the phase of the pulse is $\alpha = \pi/2$. The time evolution operator is given by

$$\hat{U} = e^{\frac{1}{i\hbar}\int_0^t \hat{H}'(t')dt'} = e^{-\frac{i}{2}(-\Delta t_{\rm s}\hat{\sigma}_z + \Omega t_{\rm s}\hat{\sigma}_y)} = e^{-\frac{i}{2}(\alpha_z\hat{\sigma}_z + \alpha_y\hat{\sigma}_y)},$$

where we introduced $\alpha_z = -\Delta t_{\rm s}$ and $\alpha_y = \Omega t_{\rm s}$. Show that the probability of measuring the qubit in the excited state as a function of the pulse length is

$$p(1) = |\langle 1|\hat{U}|0\rangle|^2 = n_y^2 \sin^2\frac{\theta}{2} = \frac{\Omega^2}{\Omega^2+\Delta^2}\sin^2\left(\sqrt{\Omega^2+\Delta^2}\frac{t_{\rm s}}{2}\right).$$

Exercise 14.5 Consider the Hamiltonian of a transmon coupled to a microwave resonator under the RWA approximation,

$$\hat{H} = \hbar\omega_{\rm r}\hat{a}^\dagger\hat{a} + \sum_{j=0}\hbar\omega_j|j\rangle\langle j| + \hbar g(\hat{a}\hat{b}^\dagger + \hat{a}^\dagger\hat{b}),$$

where ω_r is the resonator frequency, $\hbar\omega_j = \sqrt{8E_\mathrm{J}E_\mathrm{C}}j - \frac{E_\mathrm{c}}{2}(j^2 + j)$ are the transmon energy levels and $|j\rangle$ are the eigenstates. Writing the creation and destruction operators of the transmon as $\hat{b} = \sum_j \sqrt{j+1}|j\rangle\langle j+1|$ and $\hat{b}^\dagger = \sum_j \sqrt{j+1}|j+1\rangle\langle j|$, the Hamiltonian becomes

$$\hat{H} = \underbrace{\hbar\omega_\mathrm{r}\hat{a}^\dagger\hat{a} + \sum_{j=0}\hbar\omega_j|j\rangle\langle j|}_{\hat{H}_0} + \underbrace{\hbar g\sum_{j=0}(\hat{a}\sqrt{j+1}|j+1\rangle\langle j| + \hat{a}^\dagger\sqrt{j+1}|j\rangle\langle j+1|)}_{\hat{H}_\mathrm{g}}.$$

Using the Schrieffer–Wolff transformation $\hat{U} = e^{\hat{\alpha}}$ where

$$\hat{\alpha} = g\sum_{j=0}\frac{\sqrt{j+1}}{\omega_\mathrm{r} - (\omega_{01} + \eta)}(a^\dagger|j\rangle\langle j+1| - \hat{a}|j+1\rangle\langle j|),$$

calculate $\hat{H}' = \hat{U}\hat{H}\hat{U}^{-1}$ to second order in g and show that the dispersive shift is given by

$$\chi = \frac{g^2}{\Delta(1+\Delta/\eta)},$$

where $\Delta = \omega_{01} - \omega_\mathrm{r}$, $\omega_{01} = \sqrt{8E_\mathrm{J}E_\mathrm{C}} - E_\mathrm{C}$, and $\eta \approx -E_\mathrm{C}/\hbar$ is the anharmonicity of the qubit. **Hint:** Start from $\hat{H}' = e^{\hat{\alpha}}\hat{H}e^{-\hat{\alpha}} = \hat{H} + [\hat{\alpha},\hat{H}] + \frac{1}{2}[\hat{\alpha},[\hat{\alpha},\hat{H}]] + \dots$ and recall that the energy levels of a transmon satisfy $\omega_j - \omega_{j-1} = \omega_{01} + \eta j$.

Exercise 14.6 Consider the Hamiltonian of a qubit excited by a continuous drive,

$$\hat{H} = -\frac{\hbar\omega_\mathrm{q}}{2}\hat{\sigma}_z + \hbar\Omega\sin(\omega_\mathrm{d}t + \alpha)\hat{\sigma}_y,$$

where the Rabi rate Ω is proportional to the drive amplitude. In a frame rotating at the drive frequency, $\hat{R} = e^{-i\omega_\mathrm{d}t\hat{\sigma}_z/2}$, the Hamiltonian becomes (see eqn 14.42),

$$\hat{H} = -\frac{\hbar(\omega_\mathrm{q} - \omega_\mathrm{d})}{2}\hat{\sigma}_z + \frac{\hbar\Omega}{2}(\hat{\sigma}_x\cos\alpha + \hat{\sigma}_y\sin\alpha) = -\frac{\hbar\Delta}{2}\hat{\sigma}_z - \frac{\hbar\Omega}{2}\hat{\sigma}_x$$

where we introduced the detuning $\Delta = \omega_\mathrm{q} - \omega_\mathrm{d}$ and we assumed that the phase of the drive is $\alpha = -\pi/2$. Using the von Neumann equation $\frac{d\rho}{dt} = \frac{1}{i\hbar}[\hat{H},\rho]$, one can derive the equation of motion for the components of the Bloch vector,

$$\frac{d}{dt}(\mathbb{1} + r_x\hat{\sigma}_x + r_y\hat{\sigma}_y + r_z\hat{\sigma}_z) = \Delta(-r_x\hat{\sigma}_y + r_y\hat{\sigma}_x) + \Omega(-r_y\hat{\sigma}_z + r_z\hat{\sigma}_y),$$

where we used $\rho = \frac{1}{2}(\mathbb{1} + \vec{r}\cdot\vec{\sigma})$ and the commutation relations of the Pauli operators. These equations of motion can be written in matrix form as

$$\begin{bmatrix}\dot{r}_x\\ \dot{r}_y\\ \dot{r}_z\end{bmatrix} = \begin{bmatrix}0 & \Delta & 0\\ -\Delta & 0 & \Omega\\ 0 & -\Omega & 0\end{bmatrix}\begin{bmatrix}r_x\\ r_y\\ r_z\end{bmatrix}.$$

In the presence of decoherence, the Bloch equations at low temperature become

$$\begin{bmatrix} \dot{r}_x \\ \dot{r}_y \\ \dot{r}_z \end{bmatrix} = \begin{bmatrix} -\frac{\gamma}{2} - \gamma_\phi & \Delta & 0 \\ -\Delta & -\frac{\gamma}{2} - \gamma_\phi & \Omega \\ 0 & -\Omega & -\gamma \end{bmatrix} \begin{bmatrix} r_x \\ r_y \\ r_z \end{bmatrix} + \begin{bmatrix} 0 \\ 0 \\ \gamma \end{bmatrix}. \tag{14.87}$$

where $\gamma = 1/T_1$ and $1/\gamma_\phi = T_\phi$. Find the stationary solution $\vec{r}(t \to \infty)$ of this system of differential equations and show that the probability of measuring the qubit in the excited state is given by

$$p_{\mathrm{e}}(\omega_{\mathrm{d}}) = \frac{1 - r_z(t \to \infty)}{2} = \frac{T_1 T_2 \Omega^2}{(\omega_{\mathrm{d}} - \omega_{\mathrm{q}})^2 + \frac{1 + T_1 T_2 \Omega^2}{T_2^2}}.$$

Exercise 14.7 A continuous drive is applied to a superconducting qubit. From eqn (14.52), the Hamiltonian of the system is given by

$$\hat{H} = \hbar\omega_{\mathrm{q}}|1\rangle\langle 1| + \hbar(2\omega_{\mathrm{q}} + \eta)|2\rangle\langle 2| + \frac{\hbar\Omega}{2}\sin(\omega_{\mathrm{d}}t + \phi_0)(\hat{\sigma} + \hat{\sigma}^\dagger),$$

where ω_{01} is the qubit frequency, η is the anharmonicity, Ω is the Rabi rate, ω_{d} is the drive frequency, $\hat{\sigma} = -i(|0\rangle\langle 1| + \sqrt{2}|1\rangle\langle 2|)$ and we truncated the Hamiltonian to the first three energy levels for simplicity. Assuming that the drive amplitude is small $\Omega \ll |\omega_{\mathrm{q}} - \omega_{\mathrm{d}}|$, show that the microwave drive induces a shift of the qubit frequency.

Exercise 14.8 Starting from the master equation (14.71), derive the dephasing rate of the qubit, eqn (14.72), when the number of thermal photons in the resonator is small, $\langle \hat{n} \rangle \ll 1$.

Exercise 14.9 The circuit below represents two transmon qubits capacitively coupled together:

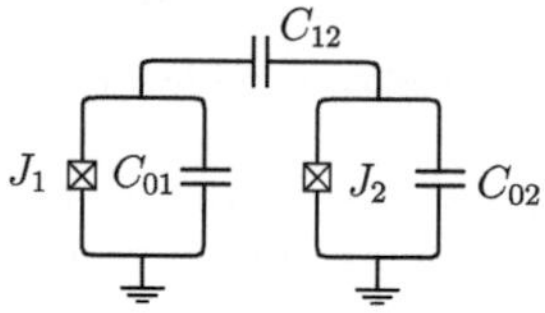

C_{12}

J_1 C_{01} J_2 C_{02}

Following Refs [522, 523], derive the Hamiltonian of the coupled system and show that

$$\boxed{\hat{H} \approx 4E_{\mathrm{C1}}\hat{n}_1^2 - E_{\mathrm{J1}}\cos\hat{\phi}_1 + 4E_{\mathrm{C2}}\hat{n}_2^2 - E_{\mathrm{J2}}\cos\hat{\phi}_2 + 4e^2\frac{C_{12}}{C_{01}C_{02}}\hat{n}_1\hat{n}_2.}$$

Exercise 14.10 Consider the Hamiltonian of a transmon coupled to another transmon under the RWA approximation

$$\hat{H} = \sum_{j=0}^{\infty} \hbar\omega_j |j\rangle\langle j| \otimes \mathbb{1} + \sum_{k=0}^{\infty} \hbar\omega_k \mathbb{1} \otimes |k\rangle\langle k| + \hbar g(b_1^\dagger b_2 + b_1 b_2^\dagger).$$

If we restrict our attention to the first three energy levels, the Hamiltonian reduces to

$$\hat{H} = \hat{H}_0 + \hat{H}_g, \tag{14.88}$$

where

$$\hat{H}_0 = \hbar\omega_1|1\rangle\langle 1| \otimes \mathbb{1} + \hbar(2\omega_1 + \eta_1)|2\rangle\langle 2| \otimes \mathbb{1} + \hbar\omega_2 \mathbb{1} \otimes |1\rangle\langle 1| + \hbar(2\omega_2 + \eta_2)\mathbb{1} \otimes |2\rangle\langle 2|,$$
$$\hat{H}_g = \hbar g(\hat{c} \otimes \hat{c}^\dagger + \hat{c}^\dagger \otimes \hat{c}).$$

Here, the operator $\hat{c}$ is defined as $\hat{c} = |0\rangle\langle 1| + \sqrt{2}|1\rangle\langle 2|$. Suppose that the detuning is much larger than the coupling strength, $|\omega_2 - \omega_1| \gg g$, i.e. the system operates in the dispersive regime. Diagonalize the Hamiltonian to second order using a Schrieffer–Wolff $e^{\hat{\alpha}}\hat{H}e^{-\hat{\alpha}}$ transformation and show that

$$e^{\hat{\alpha}}\hat{H}e^{-\hat{\alpha}} \approx -\hbar\Big(\tilde{\omega}_1 + \frac{\zeta}{2}\Big)\frac{\hat{\sigma}_{z1}}{2} - \hbar\Big(\tilde{\omega}_2 + \frac{\zeta}{2}\Big)\frac{\hat{\sigma}_{z2}}{2} + \frac{\hbar\zeta}{4}\hat{\sigma}_z \otimes \hat{\sigma}_z,$$

where $\tilde{\omega}_1 = \omega_1 + g^2/(\omega_1 - \omega_2)$, $\tilde{\omega}_2 = \omega_2 - g^2/(\omega_1 - \omega_2)$ and the ZZ coupling is given by

$$\zeta \approx 2g^2\Big(\frac{1}{\omega_1 - \omega_2 - \eta_2} - \frac{1}{\omega_1 - \omega_2 + \eta_1}\Big). \tag{14.89}$$

Hint: Set $\hat{\alpha} = \alpha_0(|10\rangle\langle 01| - |01\rangle\langle 10|) + \alpha_1(|11\rangle\langle 20| - |20\rangle\langle 11|) + \alpha_2(|11\rangle\langle 02| - |02\rangle\langle 11|) + \alpha_3(|21\rangle\langle 12| - |12\rangle\langle 21|)$.

Exercise 14.11 Derive the analytical expression of the ZZ coupling ζ eqn (14.89) with a Mathematica script.

Exercise 14.12 Consider the Hamiltonian of two qubits coupled to a common microwave resonator under the RWA approximation, $\hat{H} = \hat{H}_0 + \hat{H}_g$ where

$$\hat{H}_0 = \hbar\omega_r\hat{a}^\dagger\hat{a} - \frac{\hbar\omega_1}{2}\hat{\sigma}_{z1} - \frac{\hbar\omega_1}{2}\hat{\sigma}_{z2},$$
$$\hat{H}_g = \hbar g_1(\hat{\sigma}_1^+\hat{a} + \hat{\sigma}_1^-\hat{a}^\dagger) + \hbar g_2(\hat{\sigma}_2^+\hat{a} + \hat{\sigma}_2^-\hat{a}^\dagger).$$

Assuming $g_k \ll |\omega_k - \omega_r|$, diagonalize the Hamiltonian using a Schrieffer–Wolff transformation $e^{\hat{\alpha}}\hat{H}e^{-\hat{\alpha}}$ and show that the interaction term in the dispersive regime reduces to $\hat{H}'_g = \hbar J(\hat{\sigma}_1^-\hat{\sigma}_2^+ + \hat{\sigma}_1^+\hat{\sigma}_2^-)$, where

$$J = -\frac{1}{2}g_1 g_2\frac{\delta_1 + \delta_2}{\delta_1\delta_2}$$

and $\delta_k = \omega_r - \omega_k$. **Hint**: Set $\hat{\alpha} = \frac{g_1}{\delta_1}(\hat{\sigma}_1^-\hat{a}^\dagger - \hat{\sigma}_1^+\hat{a}) + \frac{g_2}{\delta_2}(\hat{\sigma}_2^-\hat{a}^\dagger - \hat{\sigma}_2^+\hat{a})$.

Exercise 14.13 The Hamiltonian of two qubits capacitively coupled to a tunable coupler is given by $\hat{H} = \hat{H}_0 + \hat{H}_g$ where:

$$\hat{H}_0 = -\frac{\omega_1}{2}\hat{\sigma}_{z1} - \frac{\omega_2}{2}\hat{\sigma}_{z2} - \frac{\omega_c}{2}\hat{\sigma}_{zc},$$
$$\hat{H}_g = g_1(\hat{\sigma}_1^+\hat{\sigma}_c^- + \hat{\sigma}_1^-\hat{\sigma}_c^+) + g_2(\hat{\sigma}_2^+\hat{\sigma}_c^- + \hat{\sigma}_2^-\hat{\sigma}_c^+) + g_{12}(\hat{\sigma}_1^-\hat{\sigma}_2^+ + \hat{\sigma}_1^+\hat{\sigma}_2^-),$$

and $\hat{\sigma}^-$ and $\hat{\sigma}^+$ are the lowering and raising operators. Using a Schrieffer–Wolff transformation $e^{\hat{\alpha}}\hat{H}e^{-\hat{\alpha}}$, show that in the dispersive regime $g_j \ll |\Delta_j|$ the Hamiltonian can be approximated as

$$\hat{H}'/\hbar = \frac{1}{\hbar}e^{\hat{\alpha}}\hat{H}e^{-\hat{\alpha}} \approx -\frac{\tilde{\omega}_1}{2}\hat{\sigma}_{z1} - \frac{\tilde{\omega}_2}{2}\hat{\sigma}_{z2} - \frac{\tilde{\omega}_c}{2}\hat{\sigma}_{zc} + g_{\text{eff}}(\hat{\sigma}_1^+\hat{\sigma}_2^- + \hat{\sigma}_1^-\hat{\sigma}_2^+),$$

where

$$g_{\text{eff}} = g_{12} + \frac{g_1 g_2}{2}\left(\frac{1}{\omega_1 - \omega_c} + \frac{1}{\omega_2 - \omega_c}\right), \tag{14.90}$$

$$\tilde{\omega}_k = \omega_k + \frac{g_j^2}{\omega_k - \omega_c}, \tag{14.91}$$

$$\tilde{\omega}_c = \omega_c - \frac{g_1^2}{\omega_1 - \omega_c} - \frac{g_2^2}{\omega_2 - \omega_c}. \tag{14.92}$$

Exercise 14.14 Consider two qubits coupled together,

$$\hat{H} = -\frac{\hbar\omega_1}{2}\hat{\sigma}_{z1} - \frac{\hbar\omega_2}{2}\hat{\sigma}_{z2} + \hbar g(\hat{\sigma}^+\hat{\sigma}^- + \hat{\sigma}^-\hat{\sigma}^+). \tag{14.93}$$

As seen in eqn (14.79), the Hamiltonian in the interaction picture becomes

$$\hat{H}' = \hbar g(\hat{\sigma}^-\hat{\sigma}^+ e^{-i\Delta t} + \hat{\sigma}^+\hat{\sigma}^- e^{i\Delta t}), \tag{14.94}$$

where $\Delta = \omega_1 - \omega_2$. Suppose that qubit 1 starts in the excited state and qubit 2 in the ground state, $|\psi(0)\rangle = |10\rangle$. Derive the wavefunction $|\psi(t)\rangle$ and show that the probability of measuring qubit 1 in the excited state is given by

$$p_{\text{e}}(t) = \frac{\Delta^2}{\Delta^2 + 4g^2} + \frac{4g^2}{\Delta^2 + 4g^2}\cos^2\left(\frac{\sqrt{\Delta^2 + 4g^2}}{2}t\right).$$

Exercise 14.15 Consider the Hamiltonian of two coupled transmons with a drive applied to qubit 1, $\hat{H} = \hat{H}_0 + \hat{H}_g + \hat{H}_{\text{d}}$ where:

$$\hat{H}_0 = \hbar\omega_1|10\rangle\langle 10| + \hbar(2\omega_1 + \eta_1)|20\rangle\langle 20| + \hbar\omega_2|01\rangle\langle 01| + \hbar(2\omega_2 + \eta_2)|02\rangle\langle 02|,$$
$$\hat{H}_g = \hbar g(\hat{c}_1\hat{c}_2^\dagger + \hat{c}_1^\dagger c_2), \quad \hat{H}_{\text{d}} = \hbar\Omega(t)(\hat{c}_1 + \hat{c}_1^\dagger),$$

and $\hat{c} = |0\rangle\langle 1| + \sqrt{2}|1\rangle\langle 2|$. Using the Schrieffer–Wolff transformation of Exercise 14.10, diagonalize the Hamiltonian to second order and show that

$$\hat{H}'/\hbar \approx -\frac{\tilde{\omega}_1}{2}\hat{\sigma}_{z1} - \frac{\tilde{\omega}_1}{2}\hat{\sigma}_{z2} + \frac{\zeta}{4}\hat{\sigma}_z \otimes \hat{\sigma}_z + \Omega(t)\big(\hat{\sigma}_x \otimes \mathbb{1} + \nu\mathbb{1} \otimes \hat{\sigma}_x + \mu\hat{\sigma}_z \otimes \hat{\sigma}_x\big),$$

where ζ is the ZZ coupling of eqn (14.89) and

$$\nu = -\frac{g}{\Delta + \eta_1}, \qquad \mu = -\frac{g\eta_1}{\Delta(\Delta + \eta_1)}, \qquad \Delta = \omega_1 - \omega_2.$$

Exercise 14.16 Using eqns (5.15, 5.16, 5.17, 5.18), show with a Mathematica script that

$$\mathsf{Z}_{-\alpha}\mathsf{X}_{\beta}\mathsf{Z}_{\alpha} = e^{i\frac{\alpha}{2}\sigma_z}e^{-i\frac{\beta}{2}\sigma_x}e^{-i\frac{\alpha}{2}\sigma_z} = e^{-i\frac{\beta}{2}(\cos(-\alpha)\sigma_x+\sin(-\alpha)\sigma_y)} = R_{\hat{n}(-\alpha)}(\beta),$$

where $\hat{n}(\alpha) = (\cos(-\alpha), \sin(-\alpha), 0)$.

Solutions

Solution 14.1 The eigenvalues can be calculated with a Mathematica script:

```
H = {{0, 0, 0}, {0, 0, Sqrt[2] w/2}, {0, Sqrt[2] w/2, n}};}
eigenvalues = Eigenvalues[H];
a0 = Normal[FullSimplify[Series[eigenvalues[[1]], w, 0, 3], n < 0]]
a1 = Normal[FullSimplify[Series[eigenvalues[[3]], w, 0, 3], n < 0]]
a2 = Normal[FullSimplify[Series[eigenvalues[[2]], w, 0, 3], n < 0]]
```

Solution 14.2 The last two terms in eqn (14.36) can be written as

$$\begin{aligned} A =& - E_{\mathrm{J1}}\cos(\hat{\varphi} + \phi_{\mathrm{ext}}/2) - E_{\mathrm{J2}}\cos(\hat{\varphi} - \phi_{\mathrm{ext}}/2) \\ =& - E_{\mathrm{J1}}\Big(\cos\hat{\varphi}\cos\frac{\phi_{\mathrm{ext}}}{2} - \sin\hat{\varphi}\sin\frac{\phi_{\mathrm{ext}}}{2}\Big) - E_{\mathrm{J2}}\Big(\cos\hat{\varphi}\cos\frac{\phi_{\mathrm{ext}}}{2} + \sin\hat{\varphi}\sin\frac{\phi_{\mathrm{ext}}}{2}\Big). \end{aligned}$$

Introducing the parameter $d = (E_{\mathrm{J2}} - E_{\mathrm{J1}})/(E_{\mathrm{J1}} + E_{\mathrm{J2}})$, the quantity A becomes

$$\begin{aligned} A &= -(E_{\mathrm{J1}} + E_{\mathrm{J2}})\cos\frac{\phi_{\mathrm{ext}}}{2}\Big(\cos\hat{\varphi} + d\tan\frac{\phi_{\mathrm{ext}}}{2}\sin\hat{\varphi}\Big) \\ &= -(E_{\mathrm{J1}} + E_{\mathrm{J2}})\cos\frac{\phi_{\mathrm{ext}}}{2}\sqrt{1 + d^2\tan^2\frac{\phi_{\mathrm{ext}}}{2}}\left(\frac{\cos\hat{\varphi}}{\sqrt{1 + d^2\tan^2\frac{\phi_{\mathrm{ext}}}{2}}} + \frac{d\tan\frac{\phi_{\mathrm{ext}}}{2}\sin\hat{\varphi}}{\sqrt{1 + d^2\tan^2\frac{\phi_{\mathrm{ext}}}{2}}}\right) \\ &= -(E_{\mathrm{J1}} + E_{\mathrm{J2}})\cos\frac{\phi_{\mathrm{ext}}}{2}\sqrt{1 + x^2}\Big(\frac{1}{\sqrt{1 + x^2}}\cos\hat{\varphi} + \frac{x}{\sqrt{1 + x^2}}\sin\hat{\varphi}\Big), \end{aligned} \tag{14.95}$$

where $x = d\tan(\phi_{\mathrm{ext}}/2)$. Using the identities $\frac{1}{\sqrt{1+x^2}} = \cos(\arctan x)$ and $\frac{x}{\sqrt{1+x^2}} = \sin(\arctan x)$, we obtain

$$\begin{aligned} A &= (E_{\mathrm{J1}} + E_{\mathrm{J2}})\cos\frac{\phi_{\mathrm{ext}}}{2}\sqrt{1 + x^2}\big(\cos\hat{\varphi}\cos(\arctan x) + \sin\hat{\varphi}\sin(\arctan x)\big) \\ &= (E_{\mathrm{J1}} + E_{\mathrm{J2}})\cos\frac{\phi_{\mathrm{ext}}}{2}\sqrt{1 + x^2}\big(\cos\hat{\varphi}\cos\varphi_0 + \sin\hat{\varphi}\sin\varphi_0\big) \\ &= \underbrace{(E_{\mathrm{J1}} + E_{\mathrm{J2}})\cos\Big(\pi\frac{\Phi_{\mathrm{ext}}}{\Phi_0}\Big)\sqrt{1 + d^2\tan^2(\pi\Phi_{\mathrm{ext}}/\Phi_0)}}_{E_{\mathrm{J}}(\Phi_{\mathrm{ext}})}\cos(\hat{\varphi} - \varphi_0) \\ &= E_{\mathrm{J}}(\Phi_{\mathrm{ext}})\cos(\hat{\varphi} - \varphi_0), \end{aligned}$$

where we introduced the parameter $\varphi_0 = \arctan x$ and we used $\phi_{\mathrm{ext}} = 2\pi\Phi_{\mathrm{ext}}/\Phi_0$. Thus, the Hamiltonian of a tunable transmon is

$$\hat{H} = 4E_{\mathrm{C}}\hat{n}^2 - E_{\mathrm{J}}(\Phi_{\mathrm{ext}})\cos(\hat{\varphi} - \varphi_0).$$

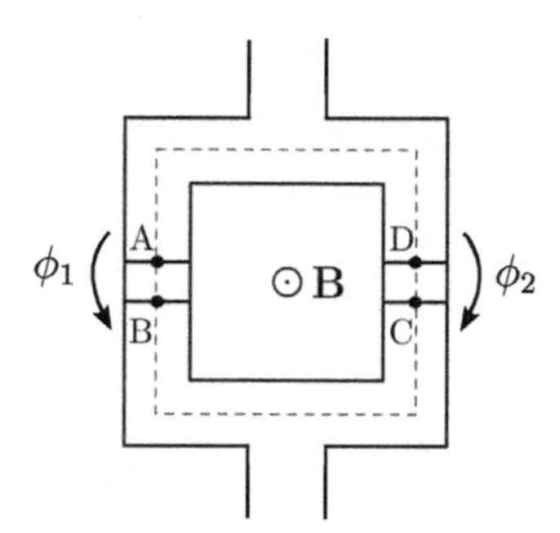

Fig. 14.32 Top view of a SQUID. The line integral is performed along the dashed line. The contour is taken well inside the superconductor, where the current density is zero.

The last step is to perform a change of basis $\hat{U} = e^{i\varphi_0 \hat{n}}$ to absorb the offset φ_0. The Hamiltonian in the new basis reads

$$\hat{U}\hat{H}\hat{U}^\dagger = 4E_C\hat{n}^2 - E_J(\Phi_{\text{ext}})\cos\hat{\varphi},$$

where we used $e^{i\varphi_0\hat{n}}\hat{\varphi}e^{-i\varphi_0\hat{n}} = \hat{\varphi}+\varphi_0$ which follows from the commutation relation $[\hat{\varphi}, \hat{n}] = i\mathbb{1}$ and the Baker–Campbell–Hausdorf formula.

Solution 14.3 Figure 14.32 shows a schematic of a SQUID. Points A, B, C, and D are on the edge of the junctions. Since the macroscopic wavefunction $\psi(\mathbf{r}) = \sqrt{\rho(\mathbf{r})}e^{i\theta(\mathbf{r})}$ is single-valued, the line integral $\oint \nabla\theta \cdot d\mathbf{l}$ must be a multiple of 2π. Therefore,

$$\oint \nabla\theta \cdot d\mathbf{l} = (\theta_B - \theta_A) + (\theta_C - \theta_B) + (\theta_D - \theta_C) + (\theta_A - \theta_D) = 2\pi n. \tag{14.96}$$

Using eqn (14.85), it is straightforward to calculate the differences $\theta_C - \theta_B$ and $\theta_A - \theta_D$:

$$\theta_C - \theta_B = \int_B^C \frac{2\pi}{\Phi_0}(\Lambda\mathbf{J} + \mathbf{A}) \cdot d\mathbf{l},$$
$$\theta_A - \theta_D = \int_D^A \frac{2\pi}{\Phi_0}(\Lambda\mathbf{J} + \mathbf{A}) \cdot d\mathbf{l}.$$

The differences $\theta_B - \theta_A$ and $\theta_D - \theta_C$ can be derived from (14.86):

$$\theta_B - \theta_A = \phi_1 + \frac{2\pi}{\Phi_0}\int_A^B \mathbf{A} \cdot d\mathbf{l},$$
$$\theta_D - \theta_C = -\phi_2 + \frac{2\pi}{\Phi_0}\int_C^D \mathbf{A} \cdot d\mathbf{l}.$$

Combining these results together, eqn (14.96) becomes

$$\phi_1 - \phi_2 + \frac{2\pi}{\Phi_0}\oint \mathbf{A} \cdot d\mathbf{l} + \int_B^C \frac{2\pi}{\Phi_0}\Lambda\mathbf{J} \cdot d\mathbf{l} + \int_D^A \frac{2\pi}{\Phi_0}\Lambda\mathbf{J} \cdot d\mathbf{l} = 2\pi n. \tag{14.97}$$

If the SQUID consists of a superconducting material with a thickness much larger than the London penetration depth $\lambda \approx 100$ nm, the integration path can be taken well inside the superconductor where the current density $\mathbf{J}$ is negligible. Setting $\mathbf{J} = \underline{\mathbf{0}}$, we finally obtain

$$\phi_1 - \phi_2 + \frac{2\pi}{\Phi_0} \oint (\nabla \times \mathbf{B}) \cdot d\mathbf{l} = \phi_1 - \phi_2 + \frac{2\pi}{\Phi_0} \int_\Sigma \mathbf{B} \cdot d\mathbf{\Sigma}$$
$$= \phi_1 - \phi_2 + \frac{2\pi}{\Phi_0} \Phi_{\text{ext}} = 2\pi n,$$

where we introduced the magnetic flux $\Phi_{\text{ext}} = \int_\Sigma \mathbf{B} \cdot d\mathbf{\Sigma}$.

Solution 14.4 The time evolution operator can be expressed in a more compact form by introducing the variable $\theta = \sqrt{\alpha_y^2 + \alpha_z^2}$ and the unit vector $\vec{n} = (0, \alpha_y/\theta, \alpha_z/\theta)$. From these definitions, we have

$$\hat{U} = e^{-i\frac{\theta}{2}\vec{n}\cdot\vec{\sigma}} = \cos\left(\frac{\theta}{2}\right) \mathbb{1} - i \sin\left(\frac{\theta}{2}\right) (n_y \hat{\sigma}_y + n_z \hat{\sigma}_z),$$

where we used eqn (5.18). The probability of finding the qubit in the excited state is

$$p(1) = |\langle 1|\hat{U}|0\rangle|^2 = n_y^2 \sin^2 \frac{\theta}{2} = \frac{\Omega^2}{\Omega^2 + \Delta^2} \sin^2 \left(\sqrt{\Omega^2 + \Delta^2} \frac{t_s}{2}\right).$$

This expression reduces to eqn (14.45) when $\Delta = 0$.

Solution 14.5 The Schrieffer–Wolff transformation is a transformation that allows us to diagonalize the Jaynes–Cummings Hamiltonian to any desired order. Using the BKH formula, we have

$$\hat{H}' = e^{\hat{\alpha}} H e^{-\hat{\alpha}} \approx \hat{H}_0 + \hat{H}_{\text{g}} + [\hat{\alpha}, \hat{H}_0] + [\hat{\alpha}, \hat{H}_{\text{g}}] + \frac{1}{2}[\hat{\alpha}, [\hat{\alpha}, \hat{H}_0]]. \tag{14.98}$$

Let us compute the commutator $[\hat{\alpha}, \hat{H}_0] = [\hat{\alpha}, \hbar\omega_{\text{r}} \hat{a}^\dagger \hat{a}] + [\hat{\alpha}, \sum_j \hbar\omega_j |j\rangle\langle j|]$ and show that it is equal to $-\hat{H}_{\text{g}}$. The first term is given by

$$[\hat{\alpha}, \hbar\omega_{\text{r}} \hat{a}^\dagger \hat{a}] = -\hbar g \sum_{j=0}^{\infty} \frac{\sqrt{j+1}\omega_{\text{r}}}{\omega_{\text{r}} - (\omega_{01} + \eta j)} (\hat{a}^\dagger |j\rangle\langle j+1| + \hat{a}|j+1\rangle\langle j|).$$

The second term is

$$[\hat{\alpha}, \sum_j \hbar\omega_j |j\rangle\langle j|] = \hbar g \sum_{j=0}^{\infty} \frac{\sqrt{j+1}(\omega_{j+1} - \omega_j)}{\omega_{\text{r}} - (\omega_{01} + \eta j)} (\hat{a}^\dagger |j\rangle\langle j+1| + \hat{a}|j+1\rangle\langle j|).$$

In addition, we know that for a transmon $\omega_{j+1} - \omega_j = \omega_{01} + \eta j$ (see eqn 14.32), and therefore $[\hat{\alpha}, \hat{H}_0] = -\hat{H}_{\text{g}}$ as anticipated. This means that eqn (14.98) simplifies to $\hat{H}' = \hat{H}_0 + \frac{1}{2}[\hat{\alpha}, \hat{H}_{\text{g}}]$. Now it remains to derive the term $\frac{1}{2}[\hat{\alpha}, \hat{H}_{\text{g}}]$. With some calculations, one can verify that

$$\begin{aligned}\frac{1}{2}[\hat{\alpha},\hat{H}_\mathrm{g}] =&\hbar g^2\sum_j \frac{j+1}{\omega_\mathrm{r}-(\omega_{j+1}-\omega_j)}(\hat{a}^\dagger\hat{a}|j\rangle\langle j| - \hat{a}^\dagger\hat{a}|j+1\rangle\langle j+1| - |j+1\rangle\langle j+1|)\\ &+\frac{\hbar g^2}{2}\sum_j\Big[\frac{\sqrt{j+1}\sqrt{j+2}}{\omega_\mathrm{r}-(\omega_{j+1}-\omega_j)} - \frac{\sqrt{j+1}\sqrt{j+2}}{\omega_\mathrm{r}-(\omega_{j+2}-\omega_{j+1})}\Big]\hat{a}^\dagger\hat{a}^\dagger|j\rangle\langle j+2|\\ &+\frac{\hbar g^2}{2}\sum_j\Big[\frac{\sqrt{j+1}\sqrt{j+2}}{\omega_\mathrm{r}-(\omega_{j+1}-\omega_j)} - \frac{\sqrt{j+1}\sqrt{j+2}}{\omega_\mathrm{r}-(\omega_{j+2}-\omega_{j+1})}\Big]\hat{a}\hat{a}|j+2\rangle\langle j|.\end{aligned}$$

The terms in square brackets are very small and can be neglected (the denominators are almost equal to each other). Therefore,

$$\frac{1}{2}[\hat{\alpha},\hat{H}_\mathrm{g}] = \hbar g^2\sum_j(\mu_{j+1}-\mu_j)|j\rangle\langle j|\hat{a}^\dagger\hat{a} - \hbar g^2\sum_j \mu_j|j\rangle\langle j|, \tag{14.99}$$

where $\mu_j = \frac{j}{\omega_\mathrm{r}-(\omega_j-\omega_{j-1})}$. The first term in eqn (14.99) is the Stark shift; the second term is the Lamb shift. Using these results, we arrive at

$$\boxed{\hat{H}' = \hat{H}_0 + \frac{1}{2}[\hat{\alpha},\hat{H}_\mathrm{g}] = \hbar\Big(\omega_\mathrm{r} + \sum_j \chi_j|j\rangle\langle j|\Big)\hat{a}^\dagger\hat{a} + \hbar\sum_j\tilde{\omega}_j|j\rangle\langle j|,} \tag{14.100}$$

where

$$\chi_j = g^2(\mu_{j+1}-\mu_j) = g^2\frac{\omega_\mathrm{r}-\omega_{01}+\eta}{[\omega_\mathrm{r}-\omega_{01}-\eta j][\omega_\mathrm{r}-\omega_{01}-\eta(j-1)]} \qquad \text{and} \qquad \tilde{\omega}_j = \omega_j - \mu_j g^2.$$

Equation (14.100) shows that the frequency of the resonator is $\omega_{\mathrm{r},|j\rangle} = \omega_\mathrm{r} + \chi_j$ when the qubit is in the state $|j\rangle$. The shift of the resonator frequency due to the excitation of the qubit from $|0\rangle$ to $|1\rangle$ is

$$2\chi \equiv \omega_{\mathrm{r},|1\rangle} - \omega_{\mathrm{r},|0\rangle} = \chi_1 - \chi_0 = \frac{2g^2}{(\omega_\mathrm{r}-\omega_{01})(\omega_\mathrm{r}-\omega_{01}-\eta)} = \frac{2g^2}{\Delta(1+\Delta/\eta)},$$

where in the last step we introduced the detuning $\Delta = \omega_{01} - \omega_\mathrm{r}$.

Solution 14.6 The stationary state $\vec{r}(t\to\infty)$ can be calculated by setting the time derivative of the Bloch vector to zero, $(\dot{r}_x,\dot{r}_y,\dot{r}_z) = (0,0,0)$. From eqn (14.87) we have

$$\begin{aligned}\begin{bmatrix} r_x(t\to\infty)\\ r_y(t\to\infty)\\ r_z(t\to\infty)\end{bmatrix} &= \begin{bmatrix} -\frac{\gamma}{2}-\gamma_\phi & \Delta & 0\\ -\Delta & -\frac{\gamma}{2}-\gamma_\phi & \Omega\\ 0 & -\Omega & -\gamma\end{bmatrix}^{-1}\begin{bmatrix}0\\0\\-\gamma\end{bmatrix}\\ &= \frac{1}{\gamma((\gamma+2\gamma_\phi)^2+4\Delta^2)+2(\gamma+2\gamma_\phi)\Omega^2}\begin{bmatrix}4\gamma\Omega\Delta\\ 2\gamma(\gamma+2\gamma_\phi)\Omega\\ \gamma((\gamma+2\gamma_\phi)^2+4\Delta^2)\end{bmatrix}.\end{aligned}$$

Let us write the r_z component in a different way,

$$r_z(t \to \infty) = \frac{\frac{1}{T_1}(\frac{4}{T_2^2} + 4\Delta^2)}{\frac{1}{T_1}(\frac{4}{T_2^2} + 4\Delta^2) + \frac{4}{T_2}\Omega^2}$$

where we used $\gamma = 1/T_1$ and $\gamma + 2\gamma_\phi = 2/T_2$. The probability of measuring the qubit in the excited state as a function of the drive frequency is

$$p_{\rm e}(\omega_{\rm d}) = \frac{1 - r_z(t \to \infty)}{2} = \frac{T_1 T_2 \Omega^2}{(\omega_{\rm d} - \omega_{\rm q})^2 + \frac{1 + T_1 T_2 \Omega^2}{T_2^2}}.$$

This is a Lorentzian function centered around $\omega_{\rm q}$ with a FWHM $= 2\sqrt{1 + T_1 T_2 \Omega^2}/T_2$. Note that for small drive amplitudes Ω, the qubit linewidth reduces to $2/T_2$.

Solution 14.7 It is convenient to analyze the Hamiltonian in the rotating frame by performing the change of basis

$$\hat{R} = e^{i(\omega_{\rm d}|1\rangle\langle 1| + 2\omega_{\rm d}|2\rangle\langle 2|)t}.$$

Using the Baker–Campbell–Hausdorff formula (3.59), one can verify that the Hamiltonian in the new frame becomes

$$\begin{aligned}\hat{H}' &= \hat{R}\hat{H}\hat{R}^{-1} - i\hbar\hat{R}\frac{d}{dt}\hat{R}^{-1} \\ &= \hbar\Delta|1\rangle\langle 1| + \hbar(2\Delta + \eta)|2\rangle\langle 2| - \hbar\Omega\sin(\omega_{\rm d}t)(\hat{\sigma}e^{-i\omega_{\rm d}t} + \hat{\sigma}^\dagger e^{i\omega_{\rm d}t}),\end{aligned}$$

where we set $\phi_0 = \pi$ for convenience and we introduced the detuning parameter $\Delta = \omega_{01} - \omega_{\rm d}$. Since $\Omega \ll \{\omega_{\rm d}, \omega_{01}\}$, one can perform the RWA approximation and neglect the terms that are proportional to $e^{\pm i2\omega_{\rm d}t}$. Under this approximation, the Hamiltonian in the rotating frame simplifies to

$$\begin{aligned}\tilde{H} &= \hbar\Delta|1\rangle\langle 1| + \hbar(2\Delta + \eta)|2\rangle\langle 2| + \frac{\hbar\Omega}{2}(|0\rangle\langle 1| + |1\rangle\langle 0| + \sqrt{2}|1\rangle\langle 2| + \sqrt{2}|2\rangle\langle 1|) \\ &= \begin{bmatrix} 0 & \Omega/2 & 0 \\ \Omega/2 & \Delta & \Omega/\sqrt{2} \\ 0 & \Omega/\sqrt{2} & 2\Delta + \eta \end{bmatrix},\end{aligned}$$

where we introduced the detuning parameter $\Delta = \omega_{\rm q} - \omega_{\rm d}$. The eigenvalues of this matrix are

$$\tilde{\omega}_0 \approx -\frac{\Omega^2}{4\Delta}, \qquad \tilde{\omega}_1 \approx \Delta - \frac{\Delta - \eta}{4\Delta(\Delta + \eta)}\Omega^2, \qquad \tilde{\omega}_2 \approx 2\Delta + \eta + \frac{1}{2(\Delta + \eta)}\Omega^2.$$

Thus, the qubit frequency becomes

$$\tilde{\omega}_{01} = \tilde{\omega}_1 - \tilde{\omega}_0 = \Delta + \frac{\eta\Omega^2}{2\Delta(\Delta + \eta)}.$$

In conclusion, the qubit frequency shifts by an amount $\delta_{\rm Stark} = \eta\Omega^2/[2\Delta(\Delta + \eta)]$ [512, 137, 696].

Solution 14.8 Our goal is to derive the time evolution of the qubit coherence, $\rho_{01}(t)$. To this end, let us calculate the time derivative of $\rho_{01}(t)$ by tracing out the resonator,

$$\frac{d\rho_{01}}{dt} = \frac{d}{dt}\langle 0|\mathrm{Tr}_{\mathrm{A}}[\rho_{\mathrm{tot}}]|1\rangle = \sum_{m=0}^{\infty} \frac{d}{dt}\underbrace{\langle 0m|\rho_{\mathrm{tot}}|m1\rangle}_{b_m} = \sum_{m=0}^{\infty} \frac{db_m}{dt}.$$

Using the master equation (14.71), we have

$$\begin{aligned}\frac{db_m}{dt} = \langle 0m|\frac{d\rho_{\mathrm{tot}}}{dt}|m1\rangle &= \frac{1}{i\hbar}\langle m0|\hat{H}\rho_{\mathrm{tot}} - \rho_{\mathrm{tot}}\hat{H}|m1\rangle \\ &+ \kappa(1+n_{\mathrm{th}})\langle m0|\mathcal{D}_{\hat{a}}(\rho_{\mathrm{tot}})|m1\rangle \\ &+ \kappa n_{\mathrm{th}}\langle m0|\mathcal{D}_{\hat{a}^\dagger}(\rho_{\mathrm{tot}})|m1\rangle. \end{aligned} \tag{14.101}$$

The first term is given by

$$\frac{1}{i\hbar}\langle m0|\hat{H}\rho_{\mathrm{tot}} - \rho_{\mathrm{tot}}\hat{H}|m1\rangle = \frac{1}{i\hbar}\Big(E_{m0} - E_{m1}\Big)b_m,$$

where we used (14.70). Here, $E_{m0} = \hbar\omega_{\mathrm{r}}m - \frac{\hbar}{2}(\tilde{\omega}_{01} + 2\chi m)$, $E_{m1} = \hbar\omega_{\mathrm{r}}m + \frac{\hbar}{2}(\tilde{\omega}_{01} + 2\chi m)$, and $\tilde{\omega}_{01} = \omega_{01} + \chi$. The second term is given by

$$\begin{aligned}\langle m0|\mathcal{D}_{\hat{a}}(\rho_{\mathrm{tot}})|m1\rangle &= \langle m0|\hat{a}\rho_{\mathrm{tot}}\hat{a}^\dagger|m1\rangle - \langle m0|\frac{\hat{a}^\dagger\hat{a}}{2}\rho_{\mathrm{tot}}|m1\rangle - \langle m0|\rho_{\mathrm{tot}}\frac{\hat{a}^\dagger\hat{a}}{2}|m1\rangle \\ &= (m+1)\langle m+1,0|\rho_{\mathrm{tot}}|m+1,1\rangle - m\langle m0|\rho_{\mathrm{tot}}|m1\rangle \\ &= (m+1)b_{m+1} - mb_m.\end{aligned}$$

The third term is

$$\begin{aligned}\langle m0|\mathcal{D}_{\hat{a}^\dagger}(\rho_{\mathrm{tot}})|m1\rangle &= \langle m0|\hat{a}^\dagger\rho_{\mathrm{tot}}\hat{a}|m1\rangle - \langle m0|\frac{\hat{a}\hat{a}^\dagger}{2}\rho_{\mathrm{tot}}|m1\rangle - \langle m0|\rho_{\mathrm{tot}}\frac{\hat{a}\hat{a}^\dagger}{2}|m1\rangle \\ &= m\langle m-1,0|\rho_{\mathrm{tot}}|m-1,1\rangle - \langle m0|\frac{\hat{a}^\dagger\hat{a}+1}{2}\rho_{\mathrm{tot}}|m1\rangle - \langle m0|\rho_{\mathrm{tot}}\frac{\hat{a}^\dagger\hat{a}+1}{2}|m1\rangle \\ &= mb_{m-1} - (m+1)b_m,\end{aligned}$$

where we used $\hat{a}\hat{a}^\dagger = \hat{a}^\dagger\hat{a} + 1$. Putting everything together, eqn (14.101) becomes

$$\frac{db_m}{dt} = i(\tilde{\omega}_{01} + 2\chi m)b_m + \kappa(1+n_{\mathrm{th}})\Big[(m+1)b_{m+1} - mb_m\Big] + \kappa n_{\mathrm{th}}\Big[mb_{m-1} - (m+1)b_m\Big]. \tag{14.102}$$

For future calculations, it is convenient to define the parameter $N_{01} = \sum_m mb_m$.
We now use eqn (14.102) to derive two coupled differential equations:

1) The first differential equation can be derived by summing both members of (14.102) over m,

$$\frac{d\rho_{01}}{dt} = \sum_{m=0}^{\infty} \frac{db_m}{dt} = \sum_{m=0}^{\infty} i(\tilde{\omega}_{01} + 2\chi m)a_m + \sum_{m=0}^{\infty} \kappa(1+n_{\text{th}})\Big[(m+1)a_{m+1} - ma_m\Big]$$
$$+ \sum_{m=0}^{\infty} \kappa n_{\text{th}}\Big[ma_{m-1} - (m+1)a_m\Big]$$
$$= i(\tilde{\omega}_{01}\rho_{01} + 2\chi N_{01}) + \kappa(1+n_{\text{th}})\Big[\sum_{\ell=1}^{\infty} \ell a_\ell - \sum_{m=0}^{\infty} ma_m\Big]$$
$$+ \kappa n_{\text{th}}\Big[\sum_{\ell=0}^{\infty}(\ell+1)a_\ell - \sum_{m=0}^{\infty}(m+1)a_m\Big] = i\tilde{\omega}_{01}\rho_{01} + i2\chi N_{01}.$$

2) The second differential equation can be derived by multiplying both members of (14.102) by m and then summing over m,

$$\frac{dN_{01}}{dt} = \frac{d}{dt}\sum_{m=0}^{\infty} ma_m = \sum_{m=0}^{\infty} i(\tilde{\omega}_{01} + 2\chi m)ma_m + \kappa(1+n_{\text{th}})\Big[\sum_{m=0}^{\infty}(m+1)ma_{m+1} - \sum_{m=0}^{\infty} m^2 a_m\Big]$$
$$+ \kappa n_{\text{th}}\Big[\sum_{m=0}^{\infty} m^2 a_{m-1} - \sum_{m=0}^{\infty}(m+1)ma_m\Big]$$
$$= i\tilde{\omega}_{01}N_{01} + i2\chi\Big[\sum_{m=0}^{\infty} m^2 a_m\Big] + \kappa(1+n_{\text{th}})\Big[\sum_{\ell=0}^{\infty}(\ell^2 a_\ell - \ell a_\ell) - \sum_{m=0}^{\infty} m^2 a_m\Big]$$
$$+ \kappa n_{\text{th}}\Big[\sum_{\ell=0}^{\infty}(\ell^2 + 2\ell + 1)a_\ell - \sum_{m=0}^{\infty}(m^2+m)a_m\Big]$$
$$= i\tilde{\omega}_{01}N_{01} + i2\chi\Big[\sum_{m=0}^{\infty} m^2 a_m\Big] + \kappa(1+n_{\text{th}})\Big[\sum_{\ell=0}^{\infty} -\ell a_\ell\Big] + \kappa n\Big[\sum_{\ell=0}^{\infty}(\ell+1)a_\ell\Big]$$
$$= i\tilde{\omega}_{01}N_{01} + i2\chi\Big[\sum_{m=0}^{\infty} m^2 a_m\Big] - \kappa N_{01} + \kappa n_{\text{th}}\rho_{01}. \tag{14.103}$$

Since for a thermal state at low temperature $\langle\hat{n}^2\rangle \approx \langle\hat{n}\rangle$ (see Exercise 6.7), the sum in the last line of (14.103) can be approximated with N_{01}. In conclusion, we have obtained two coupled differential equations

$$\frac{dN_{01}}{dt} = (i\tilde{\omega}_{01} + i2\chi - \kappa)N_{01} + \kappa n_{\text{th}}\rho_{01}, \tag{14.104}$$
$$\frac{d\rho_{01}}{dt} = i\tilde{\omega}_{01}\rho_{01} + i2\chi N_{01}. \tag{14.105}$$

The time derivative of (14.105) gives

$$i2\chi\frac{dN_{01}}{dt} = \frac{d^2\rho_{01}}{dt^2} - i\tilde{\omega}_{01}\frac{d\rho_{01}}{dt}. \tag{14.106}$$

By multiplying eqn (14.104) by $i2\chi$, we obtain

$$i2\chi\frac{dN_{01}}{dt} = (i\tilde{\omega}_{01} + 2i\chi - \kappa)\Big(\frac{d\rho_{01}}{dt} - i\tilde{\omega}_{01}\rho_{01}\Big) + i2\chi\kappa n_{\text{th}}\rho_{01}. \tag{14.107}$$

Combining eqns (14.106, 14.107), we finally arrive at

$$\frac{d^2\rho_{01}}{dt^2} + (\kappa - i2\tilde{\omega}_{01} - 2i\chi)\frac{d\rho_{01}}{dt} + (-\tilde{\omega}_{01}^2 - 2\chi\tilde{\omega}_{01} - i\kappa\tilde{\omega}_{01} - i2\chi\kappa n_{\text{th}})\rho_{01} = 0.$$

This differential equation is of the form $\ddot{\rho}(t) + \alpha\dot{\rho}(t) + \beta\rho(t) = 0$. The solution is given by

$$\rho_{01}(t) = c_0 e^{-\Gamma_1 t} + c_1 e^{-\Gamma_2 t},$$

where $\Gamma_{1,2} = \frac{\alpha \pm \sqrt{\alpha^2 - 4\beta}}{2}$, and the coefficients c_0, c_1 are determined by the initial conditions. The analytical form of Γ_1 and Γ_2 is

$$\Gamma_{1,2} = \frac{1}{2}\left[(\kappa - i2\tilde{\omega}_{01} - 2i\chi) \pm \sqrt{(\kappa - 2i\chi)^2 + i8\chi\kappa n_{\text{th}}}\right].$$

The real part of these complex coefficients gives the qubit dephasing rate. If we substitute some realistic values for $\kappa/2\pi = 2$ MHz, $\chi/2\pi = 1$ MHz and $n_{\text{th}} = 0.001$, we see that $\text{Re}[\Gamma_2] \ll \text{Re}[\Gamma_1]$. This means that Γ_1 can be ignored since it gives a negligible contribution to the decay. The dephasing rate of the qubit is

$$\gamma_\phi = \text{Re}[\Gamma_2] = \frac{\kappa}{2} - \frac{1}{2}\text{Re}\left[\sqrt{(\kappa - 2i\chi)^2 + i8\chi\kappa n_{\text{th}}}\right].$$

Solution 14.9 The derivation of the Hamiltonian of the coupled system starts from the Lagrangian $L = T - U$ written in terms of the node fluxes,

$$T = \frac{1}{2}C_{01}\dot{\Phi}_1^2 + \frac{1}{2}C_{02}\dot{\Phi}_2^2 + \frac{1}{2}C_{12}(\dot{\Phi}_2 - \dot{\Phi}_1)^2,$$
$$U = -E_{\text{J}1}\cos(2\pi\Phi_1/\Phi_0) - E_{\text{J}2}\cos(2\pi\Phi_2/\Phi_0).$$

The Lagrangian can be expressed in a more compact form as $L = \frac{1}{2}\dot{\mathbf{\Phi}}^{\text{T}}C\dot{\mathbf{\Phi}} - U$, where

$$\dot{\mathbf{\Phi}} = \begin{bmatrix} \dot{\Phi}_1 \\ \dot{\Phi}_2 \end{bmatrix}, \qquad C = \begin{bmatrix} C_{01} + C_{12} & -C_{12} \\ -C_{12} & C_{02} + C_{12} \end{bmatrix}.$$

The Hamiltonian is calculated from the Legendre transform $\hat{H} = \frac{1}{2}\mathbf{Q}^{\text{T}}C^{-1}\mathbf{Q} + U$ where $\mathbf{Q} = C\dot{\mathbf{\Phi}}$, and

$$C^{-1} = \frac{1}{A}\begin{bmatrix} C_{02} + C_{12} & C_{12} \\ C_{12} & C_{01} + C_{12} \end{bmatrix}.$$

Here, A is defined as $A = C_{01}C_{02} + C_{01}C_{12} + C_{02}C_{12}$. Thus, the Hamiltonian becomes

$$\hat{H} = \frac{1}{2A}(C_{02} + C_{12})\hat{Q}_1^2 + \frac{1}{2A}(C_{01} + C_{12})\hat{Q}_2^2 + \frac{1}{A}C_{12}\hat{Q}_1\hat{Q}_2 - E_{\text{J}1}\cos\hat{\phi}_1 - E_{\text{J}2}\cos\hat{\phi}_2,$$

where we introduced the operators $\hat{\phi}_1 = 2\pi\hat{\Phi}_1/\Phi_0$ and $\hat{\phi}_2 = 2\pi\hat{\Phi}_2/\Phi_0$. Using realistic values, $C_{01} = 100$ fF, $C_{02} = 110$ fF and $C_{12} = 5$ fF, we see that $C_{12} \ll C_{01}, C_{02}$. Therefore, we can approximate the Hamiltonian as

$$\hat{H} \approx \frac{1}{2C_{01}}\hat{Q}_1^2 + \frac{1}{2C_{02}}\hat{Q}_2^2 + \frac{C_{12}}{C_{01}C_{02}}\hat{Q}_1\hat{Q}_2 - E_{J1}\cos\hat{\phi}_1 - E_{J2}\cos\hat{\phi}_2.$$

It is convenient to express the charge operators in terms of dimensionless quantities, $\hat{Q}_i = 2e\hat{n}_i$. Using this relation, we finally obtain

$$\hat{H} \approx 4E_{C1}\hat{n}_1^2 - E_{J1}\cos\hat{\phi}_1 + 4E_{C2}\hat{n}_2^2 - E_{J2}\cos\hat{\phi}_2 + 4e^2\frac{C_{12}}{C_{01}C_{02}}\hat{n}_1\hat{n}_2,$$

where $E_{C1} = e^2/2C_{01}$, and $E_{C2} = e^2/2C_{02}$ are the charging energies of the two qubits. In conclusion, the Hamiltonian of two transmons capacitively coupled together is $\hat{H} = \hat{H}_1 + \hat{H}_2 + \hat{H}_{12}$, where $\hat{H}_1$ and $\hat{H}_2$ are the Hamiltonians of the individual transmons, whereas the interaction term is given by $\hat{H}_{12} = 4e^2 C_{12}\hat{n}_1\hat{n}_2/C_{01}C_{02}$.

Solution 14.10 The Schrieffer–Wolff transformation $e^{\hat{\alpha}}\hat{H}e^{-\hat{\alpha}}$ is a transformation that diagonalizes the Hamiltonian of a composite system to any order. The transformed Hamiltonian takes the form

$$\hat{H} = e^{\hat{\alpha}}\hat{H}e^{-\hat{\alpha}} \approx \hat{H}_0 + \hat{H}_g + [\hat{\alpha}, \hat{H}_0] + [\hat{\alpha}, \hat{H}_g] + \frac{1}{2}[\hat{\alpha}, [\hat{\alpha}, \hat{H}_g]], \tag{14.108}$$

where we truncated the expansion to the second order (this approximation is valid only if $g \ll |\omega_2 - \omega_1|$). To diagonalize the Hamiltonian, we must impose that $[\hat{\alpha}, \hat{H}_0] = -\hat{H}_g$. What is the value of $\hat{\alpha}$ that satisfies this relation? A good Ansatz is

$$\begin{aligned}\hat{\alpha} = \alpha_0(|10\rangle\langle 01| - |01\rangle\langle 10|) + \alpha_1(|11\rangle\langle 20| - |20\rangle\langle 11|)\\ + \alpha_2(|11\rangle\langle 02| - |02\rangle\langle 11|) + \alpha_3(|21\rangle\langle 12| - |12\rangle\langle 21|),\end{aligned}$$

where $\alpha_j \in \mathbb{R}$. Let us calculate the real parameters α_j satisfying $[\hat{\alpha}, \hat{H}_0] = -\hat{H}_g$. With some calculations, one can verify that

$$\begin{aligned}[\hat{\alpha}, \hat{H}_0] = (|10\rangle\langle 01| + |01\rangle\langle 10|)\alpha_0(\omega_2 - \omega_1)\\ + (|11\rangle\langle 20| + |20\rangle\langle 11|)\alpha_1(\omega_1 + \eta_1 - \omega_2)\\ + (|11\rangle\langle 02| + |02\rangle\langle 11|)\alpha_2(\omega_2 + \eta_2 - \omega_1)\\ + (|21\rangle\langle 12| + |12\rangle\langle 21|)\alpha_3(\omega_2 - \omega_1 - \eta_1 + \eta_2).\end{aligned}$$

Since

$$\begin{aligned}-\hat{H}_g = -(|10\rangle\langle 01| + |01\rangle\langle 10|)g \; - \; (|11\rangle\langle 20| + |20\rangle\langle 11|)\sqrt{2}g\\ - (|11\rangle\langle 02| + |02\rangle\langle 11|)\sqrt{2}g \; - \; (|21\rangle\langle 12| + |12\rangle\langle 21|)2g,\end{aligned}$$

then the relation $[\hat{\alpha}, \hat{H}_0] = -\hat{H}_g$ suggests that

$$\alpha_0 = \frac{g}{\Delta}, \qquad \alpha_1 = -\frac{\sqrt{2}g}{\Delta + \eta_1}, \qquad \alpha_2 = \frac{\sqrt{2}g}{\Delta - \eta_2}, \qquad \alpha_3 = \frac{2g}{\Delta + \eta_1 - \eta_2},$$

where $\Delta = \omega_1 - \omega_2$. Substituting $[\hat{\alpha}, \hat{H}_0] = -\hat{H}_g$ into (14.108), the transformed Hamiltonian becomes

$$\hat{H}' = \hat{H}_0 + \frac{1}{2}[\hat{\alpha}, \hat{H}_g].$$

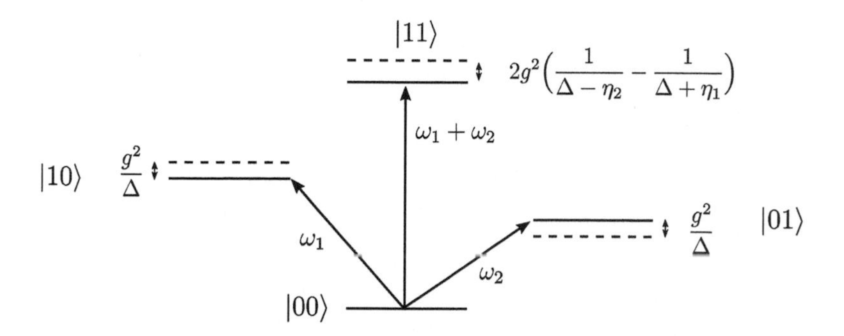

Fig. 14.33 Energy levels. Energy levels of two coupled transmons in the dispersive regime (not to scale). The solid lines indicate the bare frequency of the levels $|00\rangle$, $|01\rangle$, $|10\rangle$, and $|11\rangle$, while the dashed lines indicate the frequencies of the dressed states.

Now we just need to derive the term $\frac{1}{2}[\hat{\alpha}, \hat{H}_g]$. With some calculations, one can verify that

$$\begin{aligned}\frac{1}{2}[\hat{\alpha}, \hat{H}_g] &= \alpha_0 g(|10\rangle\langle 10| - |01\rangle\langle 01|) - \alpha_1\sqrt{2}g|20\rangle\langle 20| - \alpha_2\sqrt{2}g|02\rangle\langle 02| \\ &+ (\alpha_1\sqrt{2}g + \alpha_2\sqrt{2}g)|11\rangle\langle 11| - \alpha_3 2g|12\rangle\langle 12| + \alpha_3 2g|21\rangle\langle 21| \\ &- \frac{\sqrt{2}}{2}g(\alpha_1 + \alpha_2)(|02\rangle\langle 20| + |20\rangle\langle 02|).\end{aligned}$$

Therefore,

$$\begin{aligned}\hat{H}' = \hat{H}_0 + \frac{1}{2}[\hat{\alpha}, \hat{H}_g] &= (\omega_1 + \alpha_0 g)|10\rangle\langle 10| + (\omega_2 - \alpha_0 g)|01\rangle\langle 01| \\ &+ (2\omega_1 + \eta_1 - \alpha_1\sqrt{2}g)|20\rangle\langle 20| + (2\omega_2 + \eta_2 - \alpha_2\sqrt{2}g)|02\rangle\langle 02| \\ &+ (2\omega_2 + \eta_2 - \alpha_3 2g)|12\rangle\langle 12| + (2\omega_1 + \eta_1 + \alpha_3 2g)|21\rangle\langle 21| \\ &+ (\omega_1 + \omega_2 + g\sqrt{2}(\alpha_1 + \alpha_2))|11\rangle\langle 11| - \frac{\sqrt{2}}{2}g(\alpha_1 + \alpha_2)(|02\rangle\langle 20| + |20\rangle\langle 02|).\end{aligned}$$

The energy levels of the qubits in the dispersive regime are shown in Fig. 14.33. By truncating the Hamiltonian to the qubit subspaces, we obtain

$$\hat{H}' = (\omega_1 + \alpha_0 g)|10\rangle\langle 10| + (\omega_2 - \alpha_0 g)|01\rangle\langle 01| + (\omega_1 + \omega_2 + \sqrt{2}g(\alpha_1 + \alpha_2))|11\rangle\langle 11|.$$

This equation can be recast in a more familiar form using the relation $|ab\rangle\langle cd| = |a\rangle\langle c| \otimes |b\rangle\langle d|$ and introducing the Pauli operators $|0\rangle\langle 0| = (\mathbb{1} + \hat{\sigma}_z)/2$, $|1\rangle\langle 1| = (\mathbb{1} - \hat{\sigma}_z)/2$. Making these substitutions, we finally arrive at

$$\hat{H}' = -\Big(\omega_1 + \lambda + \frac{\zeta}{2}\Big)\frac{\hat{\sigma}_{z1}}{2} - \Big(\omega_2 - \lambda + \frac{\zeta}{2}\Big)\frac{\hat{\sigma}_{z2}}{2} + \frac{\zeta}{4}\hat{\sigma}_{z1} \otimes \hat{\sigma}_{z2}, \tag{14.109}$$

where $\lambda = g^2/\Delta$ and

$$\zeta = 2g^2\Big(\frac{1}{\Delta - \eta_2} - \frac{1}{\Delta + \eta_1}\Big).$$

Solution 14.11 A simple Mathematica script can be used to diagonalize the Hamiltonian (14.88) and derive the ZZ coupling:

```
(v_List) ⊗ (w_List) := KroneckerProduct[v,w]; Id3 = IdentityMatrix[3];
zero = {{1},{0},{0}}; one = {{0},{1},{0}}; two = {{0},{0},{1}};
sigma = zero ⊗ Transpose[one] + Sqrt[2] one ⊗ Transpose[two];
sigmaDagger = Conjugate[Transpose[sigma]];
OneId = (one⊗Transpose[one]) ⊗ Id3;
IdOne = Id3 ⊗ (one⊗Transpose[one]);
TwoId = (two⊗Transpose[two]) ⊗ Id3;
IdTwo = Id3 ⊗ (two⊗Transpose[two]);
Hamiltonian = f1 OneId + f2 IdOne +
              (2f1+η1) TwoId + (2f2+η2) IdTwo +
              g (sigma⊗sigmaDagger + sigmaDagger⊗sigma);
eigenvalues = Eigenvalues[Hamiltonian];
f01 = Normal[FullSimplify[Series[eigenvalues[[2]], {g,0,2}],{f1>f2}]];
f10 = Normal[FullSimplify[Series[eigenvalues[[3]], {g,0,2}],{f1>f2}]];
f11 = Normal[FullSimplify[Series[eigenvalues[[7]], {g,0,2}]]];
ZZcoupling = FullSimplify[f11 - f10 - f01]
```

Solution 14.12 The transformed Hamiltonian takes the form

$$\hat{H}' = e^{\hat{\alpha}}\hat{H}e^{-\hat{\alpha}} \approx \hat{H}_0 + \hat{H}_\mathrm{g} + [\hat{\alpha}, \hat{H}_0] + [\hat{\alpha}, \hat{H}_\mathrm{g}] + \frac{1}{2}[\hat{\alpha}, [\hat{\alpha}, H_\mathrm{g}]].$$

By setting $\hat{\alpha} = \frac{g_1}{\delta_1}(\hat{\sigma}_1^-\hat{a}^\dagger - \hat{\sigma}_1^+\hat{a}) + \frac{g_2}{\delta_2}(\hat{\sigma}_2^-\hat{a}^\dagger - \hat{\sigma}_2^+\hat{a})$, one can verify that $[\hat{\alpha}, \hat{H}_0] = -\hat{H}_\mathrm{g}$. Hence, the Hamiltonian simplifies to $\hat{H}' = \hat{H}_0 + \frac{1}{2}[\hat{\alpha}, \hat{H}_\mathrm{g}]$. We now need to calculate the commutator

$$\begin{aligned}\frac{1}{2}[\hat{\alpha}, \hat{H}_\mathrm{g}] &= \frac{\hbar}{2}\sum_{j=1}^{2}\sum_{k=1}^{2}\frac{g_j g_k}{\delta_j}[\hat{\sigma}_j^-\hat{a}^\dagger - \hat{\sigma}_j^+\hat{a}, (\hat{\sigma}_k^+\hat{a} + \hat{\sigma}_k^-\hat{a}^\dagger)] \\ &= \sum_{j=1}^{2}\hbar\frac{g_j^2}{\delta_j}\left[(\hat{\sigma}_j^-\hat{\sigma}_j^+ - \hat{\sigma}_j^+\hat{\sigma}_j^-)\hat{a}^\dagger\hat{a} - \hat{\sigma}_j^+\hat{\sigma}_j^-\right] - \frac{\hbar}{2}g_1g_2\frac{\delta_1+\delta_2}{\delta_1\delta_2}(\hat{\sigma}_1^+\hat{\sigma}_1^- + \hat{\sigma}_1^-\hat{\sigma}_2^+).\end{aligned}$$

Thus, the Hamiltonian becomes

$$\hat{H}' = \hbar\sum_{j=1}^{2}\left[\omega_\mathrm{r} + \chi_j(\hat{\sigma}_j^-\hat{\sigma}_j^+ + \hat{\sigma}_j^+\hat{\sigma}_j^-)\right]\hat{a}^\dagger\hat{a} + (\omega_j - \chi_j)\hat{\sigma}_j^+\hat{\sigma}_j^- - \frac{\hbar}{2}g_1g_2\frac{\delta_1+\delta_2}{\delta_1\delta_2}(\hat{\sigma}_1^+\hat{\sigma}_1^- + \hat{\sigma}_1^-\hat{\sigma}_2^+),$$

where $\chi_j = g_j^2/\delta_j$ is the dispersive shift. In conclusion, the coupling Hamiltonian in the dispersive regime is given by $\hat{H}_{12} = \hbar J(\hat{\sigma}_1^+\hat{\sigma}_1^- + \hat{\sigma}_1^-\hat{\sigma}_2^+)$ where $J = -\frac{1}{2}g_1g_2(\delta_1+\delta_2)/\delta_1\delta_2$.

Solution 14.13 The Hamiltonian can be diagonalized to second order using the transformation

$$\hat{H}' = e^{\hat{\alpha}}\hat{H}e^{-\hat{\alpha}} \approx \hat{H}_0 + \hat{H}_\mathrm{g} + [\hat{\alpha}, \hat{H}_0] + [\hat{\alpha}, \hat{H}_g] + \frac{1}{2}[\hat{\alpha}, [\hat{\alpha}, \hat{H}_0]]$$

where the operator $\hat{\alpha}$ must satisfy $[\hat{\alpha}, \hat{H}_0] = -\hat{H}_g$ so that the transformed Hamiltonian reduces to $\hat{H}' \approx \hat{H}_0 + \frac{1}{2}[\hat{\alpha}, \hat{H}_\mathrm{g}]$. The transformation α that does the job is given by

$$\hat{\alpha} = \frac{g_1}{\Delta_1}(\hat{\sigma}_1^+\hat{\sigma}_\mathrm{c}^- - \hat{\sigma}_1^-\hat{\sigma}_\mathrm{c}^+) + \frac{g_2}{\Delta_2}(\hat{\sigma}_2^+\hat{\sigma}_\mathrm{c}^- - \hat{\sigma}_2^-\hat{\sigma}_\mathrm{c}^+),$$

where $\Delta_1 = \omega_1 - \omega_\mathrm{c}$ and $\Delta_2 = \omega_2 - \omega_\mathrm{c}$. With some calculations, one can verify that

$$\hat{H}' \approx \hat{H}_0 + \frac{1}{2}[\hat{\alpha}, \hat{H}_\mathrm{g}] = \hat{H}_0 + \frac{\hbar g_1^2}{2\Delta_1}(\hat{\sigma}_{zc} - \hat{\sigma}_{z1}) + \frac{\hbar g_2^2}{2\Delta_2}(\hat{\sigma}_{zc} - \hat{\sigma}_{z2})$$
$$+ \frac{\hbar g_1 g_2}{2}\left(\frac{1}{\Delta_1} + \frac{1}{\Delta_2}\right)(\hat{\sigma}_1^+\hat{\sigma}_2^- + \hat{\sigma}_1^-\hat{\sigma}_2^+)\hat{\sigma}_{zc}. \quad (14.110)$$

Assuming that the tunable coupler is always in the ground state, the last operator in (14.110) can be replaced with its expectation value $\sigma_{zc} \to +1$. This leads to

$$\hat{H}'/\hbar \approx -\frac{\tilde{\omega}_1}{2}\hat{\sigma}_{z1} - \frac{\tilde{\omega}_2}{2}\hat{\sigma}_{z2} - \frac{\tilde{\omega}_\mathrm{c}}{2}\hat{\sigma}_{zc} + g_\mathrm{eff}(\hat{\sigma}_1^+\hat{\sigma}_2^- + \hat{\sigma}_1^-\hat{\sigma}_2^+)$$

where g_eff, $\tilde{\omega}_1$, $\tilde{\omega}_2$, and $\tilde{\omega}_\mathrm{c}$ are given by eqns (14.90, 14.91, 14.92).

Solution 14.14 This problem can be solved in two different ways:

Method 1 One way consists in integrating the Schrödinger equation $i\hbar\frac{d}{dt}|\psi(t)\rangle = \hat{H}'|\psi(t)\rangle$, where $|\psi(t)\rangle = c_{00}(t)|00\rangle + c_{01}(t)|01\rangle + c_{10}(t)|10\rangle + c_{11}(t)|11\rangle$. It is convenient to restrict our analysis to the subspace $|01\rangle, |10\rangle$ (the dynamics of the states $|00\rangle$, $|11\rangle$ is trivial). Using eqn (14.94), we have

$$i\hbar\Big(\dot{c}_{01}(t)|01\rangle + \dot{c}_{10}(t)|10\rangle\Big) = \hbar g(\hat{\sigma}^-\hat{\sigma}^+e^{-i\Delta t} + \hat{\sigma}^+\hat{\sigma}^-e^{i\Delta t})\Big(c_{01}(t)|01\rangle + c_{10}(t)|10\rangle\Big)$$
$$= \hbar g c_{01}(t)e^{i\Delta t}|10\rangle + \hbar g c_{10}(t)e^{-i\Delta t}|01\rangle.$$

This expression produces two coupled differential equations,

$$i\dot{c}_{01}(t) = g c_{10}(t)e^{-i\Delta t},$$
$$i\dot{c}_{10}(t) = g c_{01}(t)e^{i\Delta t}.$$

Assuming that the initial conditions are $c_{10}(1) = 0$ and $c_{01}(0) = 0$, the time evolution of the coefficient $c_{01}(t)$ is

$$c_{10}(t) = \cos\left(\frac{1}{2}\sqrt{\Delta^2 + 4g^2}t\right) - i\frac{\Delta}{\sqrt{\Delta^2 + 4g^2}}\sin\left(\frac{1}{2}\sqrt{\Delta^2 + 4g^2}t\right). \quad (14.111)$$

Therefore, the probability of measuring qubit 1 in the excited state is given by

$$p_\mathrm{e}(t) = |c_{10}(t)|^2 = \frac{\Delta^2}{\Delta^2 + 4g^2} + \frac{4g^2}{\Delta^2 + 4g^2}\cos^2\left(\frac{\sqrt{\Delta^2 + 4g^2}}{2}t\right).$$

Method 2 Using eqn 14.93, we note that:

$$\hat{H}|01\rangle = -\frac{\hbar\Delta}{2}|01\rangle + \hbar g|10\rangle,$$
$$\hat{H}|10\rangle = \frac{\hbar\Delta}{2}|10\rangle + \hbar g|01\rangle.$$

The Hamiltonian maps the vectors in the subspace $\{|01\rangle, |10\rangle\}$ into vectors of the same subspace. In the subspace $\{|01\rangle, |10\rangle\}$, the Hamiltonian can be expressed with a 2×2 matrix

$$\hat{H}_{\text{subspace}} = \frac{\hbar}{2}\begin{bmatrix} -\Delta & 2g \\ 2g & \Delta \end{bmatrix} = -\frac{\hbar\Delta}{2}\hat{Z} + \hbar g\hat{X},$$

where $\hat{Z}$ and $\hat{X}$ are the Pauli operators and $\Delta = \omega_1 - \omega_2$. The time evolution operator is given by

$$\hat{U}(t) = e^{\frac{1}{i\hbar}\hat{H}t} = e^{-i\left(-\frac{\Delta\hat{Z}}{2}+g\hat{X}\right)t} = \sum_{n=0}^{\infty}\frac{(-it)^n}{n!}\left(-\frac{\Delta\hat{Z}}{2}+g\hat{X}\right)^n = \sum_{n=0}^{\infty}\frac{(-it)^n}{n!}\hat{A}^n,$$

where we defined $\hat{A} = -\Delta\hat{Z}/2 + g\hat{X}$. Since $\hat{A}^0 = \mathbb{1}$, $\hat{A}^1 = \hat{A}$, $\hat{A}^2 = \left(g^2 + \frac{\Delta^2}{4}\right)\mathbb{1}$, $\hat{A}^3 = \left(g^2 + \frac{\Delta^2}{4}\right)\hat{A}$ and so on, we have

$$\hat{U}(t) = \Big[1 - \frac{t^2}{2!}(g^2 + \tfrac{\Delta^2}{4}) + O(t^2)\Big]\mathbb{1} - i\Big[\frac{t}{1!}(g^2 + \tfrac{\Delta^2}{4})^{1/2} - \frac{t^3}{3!}(g^2 + \tfrac{\Delta^2}{4})^{3/2} + O(t^5)\Big]\frac{\hat{A}}{\sqrt{g^2 + \frac{\Delta^2}{4}}}$$
$$= \cos\left(\sqrt{g^2 + \tfrac{\Delta^2}{4}}\,t\right)\mathbb{1} - \frac{i}{\sqrt{g^2 + \frac{\Delta^2}{4}}}\sin\left(\sqrt{g^2 + \tfrac{\Delta^2}{4}}\,t\right)\hat{A}.$$

This means that if the initial state is $|\psi(0)\rangle = |10\rangle$, the final state will be

$$\hat{U}(t)|10\rangle = \left[\cos\left(\sqrt{g^2 + \tfrac{\Delta^2}{4}}\,t\right) - \frac{i\Delta/2}{\sqrt{g^2 + \frac{\Delta^2}{4}}}\sin\left(\sqrt{g^2 + \tfrac{\Delta^2}{4}}\,t\right)\right]|10\rangle$$
$$- \frac{i}{\sqrt{g^2 + \frac{\Delta^2}{4}}}\sin\left(\sqrt{g^2 + \tfrac{\Delta^2}{4}}\,t\right)g|01\rangle.$$

The coefficient $c_{10}(t)$ is consistent with eqn (14.111).

Solution 14.15 The transformed Hamiltonian is given by

$$\hat{H}' = e^{\hat{\alpha}}\hat{H}e^{-\hat{\alpha}} \approx \hat{H}_0 + \hat{H}_g + [\hat{\alpha}, \hat{H}_0] + [\hat{\alpha}, \hat{H}_g] + \frac{1}{2}[\hat{\alpha}, [\hat{\alpha}, \hat{H}_g]] + \hat{H}_d + [\hat{\alpha}, \hat{H}_d] \tag{14.112}$$

where

$$\hat{\alpha} = \alpha_0(|10\rangle\langle 01| - |01\rangle\langle 10|) + \alpha_1(|11\rangle\langle 20| - |20\rangle\langle 11|)$$
$$+ \alpha_2(|11\rangle\langle 02| - |02\rangle\langle 11|) + \alpha_3(|21\rangle\langle 12| - |12\rangle\langle 21|)$$

and

$$\alpha_0 = \frac{g}{\Delta}, \qquad \alpha_1 = -\frac{\sqrt{2}g}{\Delta + \eta_1}, \qquad \alpha_2 = \frac{\sqrt{2}g}{\Delta - \eta_2}, \qquad \alpha_3 = \frac{2g}{\Delta + \eta_1 - \eta_2}, \qquad \Delta = \omega_1 - \omega_2.$$

In Exercise 14.10, we have already derived the commutators in eqn (14.112) apart from the last one. With some calculations, one can verify that

$$[\hat{\alpha}, \hat{H}_{\rm d}] = \hbar\Omega(t)\Big\{\alpha_0\Big[|10\rangle\langle 11| - \sqrt{2}|01\rangle\langle 20| - |00\rangle\langle 01| - |01\rangle\langle 00| - \sqrt{2}|20\rangle\langle 01| + |11\rangle\langle 10|\Big]$$
$$+ \alpha_1\Big[-\sqrt{2}|20\rangle\langle 21| - |01\rangle\langle 20| + \sqrt{2}|10\rangle\langle 11| + \sqrt{2}|11\rangle\langle 10| - |20\rangle\langle 01| - \sqrt{2}|21\rangle\langle 10|\Big]$$
$$+ \alpha_2\Big[|11\rangle\langle 12| - \sqrt{2}|02\rangle\langle 21| - |01\rangle\langle 02| - |02\rangle\langle 01| - \sqrt{2}|21\rangle\langle 02| + |12\rangle\langle 11|\Big]$$
$$+ \alpha_3\Big[\sqrt{2}|21\rangle\langle 22| + |02\rangle\langle 21| - \sqrt{2}|11\rangle\langle 12| + |21\rangle\langle 02| - \sqrt{2}|12\rangle\langle 11| + \sqrt{2}|22\rangle\langle 21|\Big]\Big\}.$$

Let us focus on the subspace spanned by the states $\{|00\rangle, |01\rangle, |10\rangle, |11\rangle\}$,

$$[\hat{\alpha}, \hat{H}_{\rm d}] = \hbar\Omega(t)\Big\{-\alpha_0\Big[|00\rangle\langle 01| + |01\rangle\langle 00|\Big] + (\alpha_0 + \sqrt{2}\alpha_1)\Big[|10\rangle\langle 11| + |11\rangle\langle 10|\Big]\Big\}.$$

Using the relation $|ab\rangle\langle cd| = |a\rangle\langle c| \otimes |b\rangle\langle d|$ and the identities $\hat{\sigma}_x = |0\rangle\langle 1| + |1\rangle\langle 0|$ and $\hat{\sigma}_z = |0\rangle\langle 0| - |1\rangle\langle 1|$, we obtain

$$[\hat{\alpha}, \hat{H}_{\rm d}] = \hbar\Omega(t)\Big[\nu\mathbb{1} \otimes \hat{\sigma}_x + \mu\hat{\sigma}_z \otimes \hat{\sigma}_x\Big] \tag{14.113}$$

where $\nu = -g/(\Delta + \eta_1)$ and $\mu = -g\eta_1/\Delta(\Delta + \eta_1)$. From eqns (14.112, 14.113, 14.109), the transformed Hamiltonian becomes

$$\hat{H}'/\hbar \approx -\frac{\tilde{\omega}_1}{2}\hat{\sigma}_{z1} - \frac{\tilde{\omega}_2}{2}\hat{\sigma}_{z2} + \frac{\zeta}{4}\hat{\sigma}_z \otimes \hat{\sigma}_z + \Omega(t)\big(\hat{\sigma}_x \otimes \mathbb{1} + \nu\mathbb{1} \otimes \hat{\sigma}_x + \mu\hat{\sigma}_z \otimes \hat{\sigma}_x\big).$$

Solution 14.16

```
X[a_] := {{Cos[a/2], -I Sin[a/2]}, {-I Sin[a/2], Cos[a/2]}}
Y[a_] := {{Cos[a/2], -Sin[a/2]}, {Sin[a/2], Cos[a/2]}}
Z[a_] := {{Exp[-I a/2], 0}, {0, Exp[I a/2]}};
Id = {{1, 0}, {0, 1}}; x = {{0, 1}, {1, 0}};
y = {{0, -I}, {I, 0}}; z = {{1, 0}, {0, -1}};
R[a_, nx_, ny_, nz_] := Cos[a/2] Id - I Sin[a/2]*(nx x + ny y + nz z);
MatrixForm[Simplify[Z[-a].X[b].Z[a] - R[b, Cos[-a], Sin[-a], 0]]]
```

Appendix A
The rotating wave approximation

A.1 The Rabi model

The Hamiltonian of the Rabi model is given by

$$H = H_{\mathrm{q}} + H_{\mathrm{r}} + H_{\mathrm{g}},$$

where

$$H_{\mathrm{q}} = -\frac{\omega_{\mathrm{q}}}{2}\sigma_z, \qquad H_{\mathrm{r}} = \omega_{\mathrm{r}} a^\dagger a, \qquad H_{\mathrm{g}} = g(\sigma^+ a^\dagger + \sigma^- a + \sigma^+ a + \sigma^- a^\dagger).$$

In many practical cases, the qubit frequency and the oscillator frequency are much greater than the coupling strength g. This allows us to neglect the terms $\sigma^+ a^\dagger$ and $\sigma^- a$ because they give a small contribution to the dynamics. This is called the rotating wave approximation. To clarify this approximation, it is convenient to study the dynamics in the interaction picture. Using the transformation $\tilde{H} = RHR^{-1}$ where $R = e^{i(H_{\mathrm{q}}+H_{\mathrm{r}})t}$, the Hamiltonian in the interaction picture becomes

$$\begin{aligned}\tilde{H} &= R^{-1} H_{\mathrm{g}} R^{-1} \\ &= g\left[\sigma^+ a^\dagger e^{i(\omega_{\mathrm{r}}+\omega_{\mathrm{q}})t} + \sigma^- a e^{-i(\omega_{\mathrm{r}}+\omega_{\mathrm{q}})t} + \sigma^+ a e^{-i(\omega_{\mathrm{r}}-\omega_{\mathrm{q}})t} + \sigma^- a^\dagger e^{i(\omega_{\mathrm{r}}-\omega_{\mathrm{q}})t}\right],\end{aligned}$$

where we used the relations

$$\begin{aligned} RaR^{-1} &= ae^{-i\omega_{\mathrm{r}}t}, \qquad & Ra^\dagger R^{-1} &= a^\dagger e^{i\omega_{\mathrm{r}}t}, \\ R\sigma^- R^{-1} &= \sigma^- e^{-i\omega_{\mathrm{q}}t}, \qquad & R\sigma^+ R^{-1} &= \sigma^+ e^{i\omega_{\mathrm{q}}t}, \end{aligned}$$

which can be derived with the Baker-Campbell-Hausdorff formula.The time evolution operator is given by the Dyson series

$$U(t) = \mathcal{T} e^{-i\int_0^t \tilde{H}(t')dt'} = \mathbb{1} - i\int_0^t \tilde{H}(t')dt' + O(t^2).$$

Let us calculate the first order contribution

$$\int_0^t \tilde{H}(t')dt' = g\left[\int_0^t \sigma^+ a^\dagger e^{i(\omega_r+\omega_q)t'}dt' + \int_0^t \sigma^- a e^{-i(\omega_r+\omega_q)t'}dt'\right.$$
$$\left. + \int_0^t \sigma^+ a e^{-i(\omega_r-\omega_q)t'}dt' + \int_0^t \sigma^- a^\dagger e^{i(\omega_r-\omega_q)t}\right].$$

These integrals are straightforward,

$$\int_0^t \tilde{H}(t')dt' = g\left[\sigma^+ a^\dagger \frac{i - ie^{i(\omega_q+\omega_r)t}}{\omega_q+\omega_r} + \sigma^- a \frac{-i + ie^{i(\omega_q+\omega_r)t})}{\omega_q+\omega_r}\right.$$
$$\left. \sigma^+ a \frac{i - ie^{i(\omega_q-\omega_r)t}}{\omega_q-\omega_r} + \sigma^- a^\dagger \frac{-i + ie^{i(\omega_r-\omega_q)t}}{\omega_q-\omega_r}\right]. \tag{A.1}$$

There are four terms in eqn (A.1): the terms $\sigma^+ a^\dagger$ and $\sigma^- a$ are multiplied by a factor $g/(\omega_q+\omega_r)$, while the terms $\sigma^- a^\dagger$ and $\sigma^+ a$ are multiplied by $g/(\omega_q-\omega_r)$. Since we assumed $g \ll \omega_q, \omega_r$, the the major contributions to the time evolution operator come from the terms $\sigma^+ a$ and $\sigma^- a^\dagger$ and the other two can often be neglected.

Appendix B
Advanced quantum mechanics

B.1 Pure bipartite states and matrices

Pure bipartite states can be related in a natural way to matrices and vice versa. This correspondence is useful to simplify some calculations in quantum information theory. Let us start by considering a bipartite state $|A\rangle \in H \otimes H$ and its decomposition with respect to an orthonormal basis $|A\rangle = \sum_{ij} c_{ij}|i\rangle|j\rangle$. The coefficients c_{ij} can be used to construct an operator $\hat{A} \in \mathrm{B}(H)$ of the form

$$\boxed{\hat{A} = \sum_{ij} c_{ij}|i\rangle\langle j|.}$$

In a similar way, an operator $\hat{A} \in \mathrm{B}(H)$ defines a pure bipartite state. This follows from the fact that $\hat{A}$ can be decomposed as $\hat{A} = \sum_{ij} a_{ij}|i\rangle\langle j|$. The matrix elements a_{ij} define the bipartite state

$$\boxed{|A\rangle = \sum_{ij} a_{ij}|i\rangle|j\rangle.}$$

Note that this vector might not be normalized. In short, we have established the following relations:

bipartite state $\lvert A\rangle \in H \otimes H$		operator $\hat{A} \in \mathrm{B}(H)$
$\lvert A\rangle = \sum_{i,j} c_{ij}\lvert ij\rangle$	$\longmapsto$	$\hat{A} = \sum_{i,j} c_{ij}\lvert i\rangle\langle j\rvert$
operator $\hat{A} \in \mathrm{B}(H)$		**bipartite state $\lvert A\rangle \in H \otimes H$**
$\hat{A} = \sum_{i,j} a_{ij}\lvert i\rangle\langle j\rvert$	$\longmapsto$	$\lvert A\rangle = \sum_{i,j} a_{ij}\lvert i\rangle\lvert j\rangle.$

To better understand this correspondence, let us consider a concrete example. The Pauli operators $\hat{\sigma}_i$ define the two-qubit states

$$\hat{\sigma}_0 = \begin{bmatrix} 1 & 0 \\ 0 & 1 \end{bmatrix} \quad \leftrightarrow \quad |\sigma_0\rangle = |00\rangle + |11\rangle,$$
$$\hat{\sigma}_1 = \begin{bmatrix} 0 & 1 \\ 1 & 0 \end{bmatrix} \quad \leftrightarrow \quad |\sigma_1\rangle = |01\rangle + |10\rangle,$$
$$\hat{\sigma}_2 = \begin{bmatrix} 0 & -i \\ i & 0 \end{bmatrix} \quad \leftrightarrow \quad |\sigma_2\rangle = -i|01\rangle + i|10\rangle,$$
$$\hat{\sigma}_3 = \begin{bmatrix} 1 & 0 \\ 0 & -1 \end{bmatrix} \quad \leftrightarrow \quad |\sigma_3\rangle = |00\rangle - |11\rangle.$$

These two-qubit states are nothing but the Bell states (9.2, 9.3) up to a normalization factor.

The correspondence between bipartite states and operators can be promoted to an isomorphism between the Hilbert spaces $H \otimes H$ and $\mathrm{B}(H)$ by showing that the inner product $\langle A|B\rangle$ between two vectors $|A\rangle = \sum_{ij} a_{ij}|ij\rangle$ and $|B\rangle = \sum_{kl} b_{kl}|kl\rangle$ is equal to the Hilbert–Schmidt inner product between the corresponding operators, $\langle \hat{A}|\hat{B}\rangle_2 = \mathrm{Tr}[\hat{A}^\dagger \hat{B}]$. This is because[1]

$$\begin{aligned}
\langle A|B\rangle &= \sum_{ij}\sum_{kl} a_{ij}^* b_{kl} \langle ij|kl\rangle = \sum_{ij} a_{ij}^* b_{ij} \\
&= \sum_{ij} \langle i|\hat{A}^*|j\rangle\langle i|\hat{B}|j\rangle = \sum_{ij} \langle j|\hat{A}^\dagger|i\rangle\langle i|\hat{B}|j\rangle \\
&= \sum_{j} \langle j|\hat{A}^\dagger\hat{B}|j\rangle = \mathrm{Tr}[\hat{A}^\dagger\hat{B}] = \langle \hat{A}|\hat{B}\rangle_2.
\end{aligned}$$

In a similar way, one can verify that for two generic operators $\hat{A}, \hat{B} \in \mathrm{B}(H)$, we have $\langle \hat{A}|\hat{B}\rangle_2 = \mathrm{Tr}[\hat{A}^\dagger \hat{B}] = \langle A|B\rangle$.

The correspondence between bipartite states and operators is useful to calculate the partial trace of bipartite systems (see for example Section 9.5.1). Some useful properties to keep in mind are:

C1.1. $(\hat{A} \otimes \mathbb{1}_{\mathrm{B}})|\psi\rangle_{\mathrm{AB}} = |A\psi\rangle_{\mathrm{AB}}$,

C1.2. $(\mathbb{1}_{\mathrm{A}} \otimes \hat{B})|\psi\rangle_{\mathrm{AB}} = |\psi B^{\mathrm{T}}\rangle_{\mathrm{AB}}$,

C1.3. $\mathrm{Tr}_{\mathrm{B}}[|A\rangle\langle B|_{\mathrm{AB}}] = \hat{A}\hat{B}^\dagger$,

C1.4. $\mathrm{Tr}_{\mathrm{A}}[|A\rangle\langle B|_{\mathrm{AB}}] = \hat{A}^{\mathrm{T}}\hat{B}^*$,

where A^{T} is the transpose matrix, A^* is the conjugate matrix and $A^\dagger$ is the conjugate transpose. The proof of these properties is discussed below:

C1.1. A direct calculation leads to

[1] See Section 3.2.14 for more details about the Hilbert–Schmidt inner product.

$$
\begin{aligned}
(\hat{A} \otimes \mathbb{1}_{\text{B}})|\psi\rangle &= \Big[\Big(\sum_{ij} a_{ij}|i\rangle\langle j|\Big) \otimes \mathbb{1}_{\text{B}}\Big] \sum_{kl} \psi_{kl}|k\rangle|l\rangle \\
&= \sum_{ij}\sum_{kl} a_{ij}\psi_{kl}|i\rangle\langle j|k\rangle \otimes |l\rangle \\
&= \sum_{il}\Big(\sum_{j} a_{ij}\psi_{jl}\Big)|i\rangle \otimes |l\rangle = \sum_{il}(A\psi)_{il}|i\rangle \otimes |l\rangle \;=\; |A\psi\rangle,
\end{aligned}
$$

where A is the matrix associated with $\hat{A}$.

C1.2. A direct calculation leads to

$$
\begin{aligned}
(\mathbb{1}_{\text{A}} \otimes \hat{B})|\mathbf{x}\rangle &= \Big[\mathbb{1}_{\text{A}} \otimes \Big(\sum_{ij} b_{ij}|i\rangle\langle j|\Big)\Big] \sum_{kl} \psi_{kl}|k\rangle|l\rangle \\
&= \sum_{ij}\sum_{kl} b_{ij}\psi_{kl}|k\rangle \otimes |i\rangle\langle j|l\rangle \\
&= \sum_{ki}\Big(\sum_{l} \psi_{kl}b^{\text{T}}_{li}\Big)|k\rangle \otimes |i\rangle = \sum_{ki}(\psi B^{\text{T}})_{ki}|k\rangle \otimes |i\rangle \;=\; |\psi B^{\text{T}}\rangle,
\end{aligned}
$$

where B is the matrix associated with $\hat{B}$.

C1.3. A direct calculation leads to

$$
\begin{aligned}
\text{Tr}_{\text{B}}[|A\rangle\langle B|] &= \text{Tr}_{\text{B}}\Big[\sum_{ij} a_{ij}|i\rangle|j\rangle \sum_{ij} b^*_{kl}\langle k|\langle l|\Big] \\
&= \sum_{ij}\sum_{kl} a_{ij}b^*_{kl}|i\rangle\langle k| \;\text{Tr}_{\text{B}}[|j\rangle\langle l|] \\
&= \sum_{ij}\sum_{k} a_{ij}b^*_{kj}|i\rangle\langle k| \\
&= \sum_{ik}\Big(\sum_{j} a_{ij}b^{\dagger}_{jk}\Big)|i\rangle\langle k| = \sum_{ik}\left(AB^{\dagger}\right)_{ik}|i\rangle\langle k| = \hat{A}\hat{B}^{\dagger}.
\end{aligned}
$$

C1.4. A direct calculation leads to

$$
\begin{aligned}
\text{Tr}_{\text{A}}[|A\rangle\langle B|] &= \text{Tr}_{\text{A}}\Big[\sum_{kl} a_{ij}|i\rangle|j\rangle \sum_{kl} b^*_{kl}\langle k|\langle l|\Big] \\
&= \sum_{ij}\sum_{kl} a_{ij}b^*_{kl}\text{Tr}_{\text{A}}\left[|i\rangle\langle k|\right]|j\rangle\langle l| = \sum_{j}\sum_{kl} a_{kj}b^*_{kl}|j\rangle\langle l| \\
&= \sum_{jl}\Big(\sum_{k} a^{\text{T}}_{jk}b^*_{kl}\Big)|j\rangle\langle l| = \sum_{jl}\left(A^{\text{T}}B^*\right)_{jl}|j\rangle\langle l| = \hat{A}^{\text{T}}\hat{B}^*.
\end{aligned}
$$

B.2 Thermal density operators

Density operator formalism can be used to describe the equilibrium state of finite-temperature quantum systems. Consider a quantum system governed by a Hamiltonian

$$\hat{H} = \sum_n E_n |n\rangle\langle n|.$$

The system is in contact with a reservoir that keeps the energy of the system constant. The most general stationary state is a mixed state of the eigenstates of the Hamiltonian,

$$[\rho_{\text{th}}, \hat{H}] = 0 \qquad \Rightarrow \qquad \boxed{\rho_{\text{th}} = \sum_n \lambda_n |n\rangle\langle n|.}$$

The equilibrium state of the system is the one that maximizes the von Neumann entropy $S(\rho) = -k_{\text{B}}\text{Tr}[\rho \ln \rho]$ under two constraints:

1) The energy of the system must be a constant E,

$$\langle \hat{H} \rangle = \text{Tr}[\rho_{\text{th}}\hat{H}] = \sum_n \langle n|\rho_{\text{th}}\hat{H}|n\rangle = \sum_n \lambda_n E_n = E.$$

2) The equilibrium state must be normalized,

$$\text{Tr}[\rho_{\text{th}}] = \sum_n \lambda_n = 1.$$

To calculate the explicit expression of ρ_{th}, we will use the technique of Lagrange multipliers. The Lagrangian that we are trying to maximize is

$$\boxed{L(\rho) = S(\rho) - \alpha\text{Tr}[\rho] - \beta\text{Tr}[\rho\hat{H}],}$$

where α and β are two Lagrange multipliers. The term proportional to α ensures that the state is correctly normalized, while the term proportional to β ensures that the maximization is performed at a constant energy. The Lagrangian can be written in a more explicit form as

$$L = -k_{\text{B}} \sum_n \lambda_n \ln \lambda_n - \alpha \sum_n \lambda_n - \beta \sum_n \lambda_n E_n.$$

Let us calculate the stationary points of the Lagrangian,

$$0 = \frac{\partial L}{\partial \lambda_n} = \frac{\partial}{\partial \lambda_n}\Big(- k_{\mathrm{B}} \sum_j \lambda_j \ln \lambda_j - \alpha \sum_j \lambda_j - \beta \sum_j \lambda_j E_j \Big)$$
$$= -k_{\mathrm{B}} \sum_j (\delta_{jn} \ln \lambda_j + \delta_{jn}) - \alpha \sum_j \delta_{jn} - \beta \sum_j \delta_{jn} E_j$$
$$= -k_{\mathrm{B}}(\ln \lambda_n + 1) - \alpha - \beta E_n.$$

The solution of this equation is

$$\lambda_n = \frac{e^{-\beta E_n / k_{\mathrm{B}}}}{e^{1+\alpha/k_{\mathrm{B}}}}.$$

Since the state must be normalized to one, we have

$$1 = \mathrm{Tr}[\rho_{\mathrm{th}}] = \sum_n \lambda_n = \sum_n \frac{e^{-\beta E_n / k_{\mathrm{B}}}}{e^{1+\alpha/k_{\mathrm{B}}}} \qquad \Rightarrow \qquad e^{1+\alpha/k_{\mathrm{B}}} = \sum_n e^{-\beta E_n / k_{\mathrm{B}}},$$

and therefore

$$\lambda_n = \frac{e^{-\beta E_n / k_{\mathrm{B}}}}{\sum_n e^{-\beta E_n / k_{\mathrm{B}}}} = \frac{e^{-\beta E_n / k_{\mathrm{B}}}}{Z},$$

where in the last step we introduced the real number $Z = \sum_n e^{-\beta E_n / k_{\mathrm{B}}}$. In conclusion, the analytical expression of a thermal state is given by

$$\boxed{\rho_{\mathrm{th}} = \sum_n \lambda_n |n\rangle\langle n| = \sum_n \frac{e^{-\beta E_n / k_{\mathrm{B}}}}{Z} |n\rangle\langle n| = \frac{e^{-\beta \hat{H} / k_{\mathrm{B}}}}{Z}.}$$

Lastly, we need to clarify the role of β. At thermal equilibrium, the entropy is given by

$$S(\rho_{\mathrm{th}}) = -k_{\mathrm{B}} \sum_n \lambda_n \ln \lambda_n$$
$$= -k_{\mathrm{B}} \sum_n \lambda_n \ln \left(\frac{e^{-\beta E_n / k_{\mathrm{B}}}}{Z} \right) = -k_{\mathrm{B}} \sum_n \lambda_n \ln \left(e^{-\beta E_n / k_{\mathrm{B}}} \right) + k_{\mathrm{B}} \sum_n \lambda_n \ln Z$$
$$= \beta \sum_n \lambda_n E_n + \ln Z \sum_n \lambda_n = \beta E + k_{\mathrm{B}} \ln Z.$$

Therefore,

$$\frac{dS}{dE} = \beta.$$

Since the thermodynamic temperature T is defined by the expression,

$$\frac{dS}{dE} = \frac{1}{T},$$

we conclude that $\beta = 1/T$.

Appendix C
The quantum Fourier transform

C.1 Definition and circuit implementation

The quantum Fourier transform is defined as

$$\boxed{\hat{F}|j\rangle = \frac{1}{2^{n/2}} \sum_{k=0}^{2^n-1} e^{\frac{2\pi i}{2^n} jk} |k\rangle,} \tag{C.1}$$

where $|j\rangle$ is a basis vector of the Hilbert space $\mathbb{C}^n$. Let us investigate how the quantum Fourier transform operates on a generic n qubit state. To simplify the notation, we adopt the convention

$$\frac{j_0}{2} = 0.j_0, \qquad \frac{j_1}{2} + \frac{j_0}{2^2} = 0.j_1j_0, \qquad \frac{j_2}{2} + \frac{j_1}{2^2} + \frac{j_0}{2^3} = 0.j_2j_1j_0, \qquad \ldots$$

The QFT maps a basis vector $|j\rangle$ into a product state

$$\begin{aligned}
\hat{F}|j\rangle &= \frac{1}{2^{n/2}} \sum_{k=0}^{2^n-1} e^{\frac{2\pi i}{2^n} jk} |k\rangle = \frac{1}{2^{n/2}} \sum_{k_0=0}^{1} \cdots \sum_{k_{n-1}=0}^{1} e^{2\pi i j \sum_{l=1}^{n} \frac{k_{n-l}}{2^l}} |k_{n-1} \ldots k_0\rangle \\
&= \frac{1}{2^{n/2}} \sum_{k_0=0}^{1} \cdots \sum_{k_{n-1}=0}^{1} \bigotimes_{l=1}^{n} e^{\frac{2\pi i j}{2^l} k_{n-l}} |k_{n-l}\rangle \\
&= \frac{1}{2^{n/2}} \bigotimes_{l=1}^{n} \sum_{k_{n-l}=0}^{1} e^{\frac{2\pi i}{2^l} j k_{n-l}} |k_{n-l}\rangle \\
&= \frac{1}{2^{n/2}} \bigotimes_{l=1}^{n} \left[|0\rangle + e^{\frac{2\pi i}{2^l} j} |1\rangle \right] \\
&= \frac{1}{2^{n/2}} \left[|0\rangle + e^{2\pi i \frac{2^{n-1} j_{n-1} + \ldots + j_0}{2}} |1\rangle \right] \otimes \ldots \otimes \left[|0\rangle + e^{2\pi i \frac{2^{n-1} j_{n-1} + \ldots + j_0}{2^n}} |1\rangle \right] \\
&= \frac{1}{2^{n/2}} \left[|0\rangle + e^{2\pi i \frac{j_0}{2}} |1\rangle \right] \otimes \ldots \otimes \left[|0\rangle + e^{2\pi i \left(\frac{j_{n-1}}{2} + \ldots + \frac{j_0}{2^n} \right)} |1\rangle \right] \\
&= \frac{1}{2^{n/2}} \left[|0\rangle + e^{2\pi i 0.j_0} |1\rangle \right] \otimes \ldots \otimes \left[|0\rangle + e^{2\pi i 0.j_{n-1} \ldots j_0} |1\rangle \right].
\end{aligned} \tag{C.2}$$

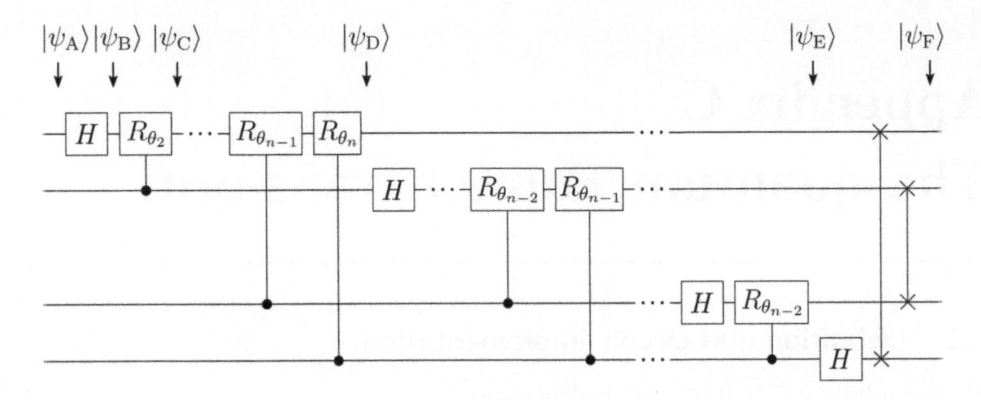

Fig. C.1 The quantum Fourier transform. The quantum circuit implementing the quantum Fourier transform. If the input of the circuit is the basis vector $|j\rangle$, the output is the product state (C.2).

This equation allows us to construct the quantum circuit of the QFT (see Fig. C.1). In this circuit, the single-qubit gate R_{θ_k} is defined as

$$R_{\theta_k} = \begin{bmatrix} 1 & 0 \\ 0 & e^{i\theta_k} \end{bmatrix} = \begin{bmatrix} 1 & 0 \\ 0 & e^{i2\pi/2^k} \end{bmatrix}.$$

Let us investigate how this circuit operates on a basis vector $|j\rangle = |j_{n-1} \cdots j_0\rangle$ where $j_i \in \mathbb{B}$. The qubits of the register are initialized in the state $|\psi_A\rangle = |j_{n-1} \cdots j_0\rangle$ by applying some $\hat{X}$ gates at the beginning of the circuit (not shown in the figure). The application of a Hadamard gate to the top qubit yields

$$\begin{aligned} |\psi_B\rangle = H_{n-1}|j_{n-1} \cdots j_0\rangle &= \frac{|0\rangle + e^{2\pi i \frac{j_{n-1}}{2}}|1\rangle}{\sqrt{2}}|j_{n-2} \cdots j_0\rangle \\ &= \frac{|0\rangle + e^{2\pi i\, 0.j_{n-1}}|1\rangle}{\sqrt{2}}|j_{n-2} \cdots j_0\rangle. \end{aligned}$$

A controlled unitary transforms this state into

$$\begin{aligned} |\psi_C\rangle = \mathsf{c}_{n-2}R_{\theta_2,n-1}|\psi_B\rangle &= \frac{|0\rangle + e^{2\pi i\, 0.j_{n-1}}e^{2\pi i\, 0.0j_{n-2}}|1\rangle}{\sqrt{2}}|j_{n-2} \cdots j_0\rangle \\ &= \frac{|0\rangle + e^{2\pi i\, 0.j_{n-1}j_{n-2}}|1\rangle}{\sqrt{2}}|j_{n-2} \cdots j_0\rangle. \end{aligned}$$

The next step is to apply a sequence of controlled operations to build up the correct relative phase for the top qubit,

$$|\psi_D\rangle = \frac{|0\rangle + e^{2\pi i 0.j_{n-1} \dots j_0}|1\rangle}{\sqrt{2}}|j_{n-2} \cdots j_0\rangle.$$

As shown in Fig. C.1, one can execute a similar procedure for the remaining qubits and obtain

$$|\psi_{\rm E}\rangle = \left[\frac{|0\rangle + e^{2\pi i 0.j_{n-1}j_{n-2}\ldots j_0}|1\rangle}{\sqrt{2}}\right]\left[\frac{|0\rangle + e^{2\pi i 0.j_{n-2}\ldots j_0}|1\rangle}{\sqrt{2}}\right]\ldots\left[\frac{|0\rangle + e^{2\pi i 0.j_0}|1\rangle}{\sqrt{2}}\right].$$

This state is very similar to eqn (C.2), the only difference being the reversed order of the qubits. To fix this issue, either the qubits are relabelled or a set of SWAP gates is applied at the end of the circuit, as illustrated in Fig. C.1. This procedure maps the state of the register into

$$|\psi_{\rm F}\rangle = \left[\frac{|0\rangle + e^{2\pi i 0.j_0}|1\rangle}{\sqrt{2}}\right]\ldots\left[\frac{|0\rangle + e^{2\pi i 0.j_{n-2}\ldots j_0}|1\rangle}{\sqrt{2}}\right]\left[\frac{|0\rangle + e^{2\pi i 0.j_{n-1}j_{n-2}\ldots j_0}|1\rangle}{\sqrt{2}}\right] = \\ = \frac{1}{2^{n/2}}\sum_{k=0}^{2^n-1} e^{\frac{2\pi i}{2^n}jk}|k\rangle \tag{C.3}$$

where in the last step we used eqn (C.2). This is the expected output of the quantum Fourier transform for the basis state $|j\rangle$. Note that the QFT is clearly a unitary operation since it is a composition of unitary gates.

How many gates are required to implement the quantum Fourier transform? We need n Hadamard gates and $n(n-1)/2$ controlled operations[1]. In addition, we need $\lfloor n/2 \rfloor$ SWAP gates at the end of the circuit. Therefore, the total number of elementary gates scales as $O(n^2)$. The best classical algorithm can compute the discrete Fourier transform of a vector with 2^n entries using $O(n2^n)$ gates. This algorithm is known as the fast Fourier algorithm (FFT) and was discovered by Cooley and Tukey in 1965 [698].

It is tempting to conclude that the quantum Fourier transform offers an exponential speed-up with respect to the FFT. There is an important caveat though: the output of the QFT might be a complex superposition state. To extract some useful information, the output state must be measured an exponential number of times to reconstruct the complex probability amplitudes. In addition, the Fourier transform of a generic input state $|\psi\rangle$ might require an exponential number of gates to prepare the state $|\psi\rangle$ itself. For these reasons, it is not straightforward to find useful applications of the quantum Fourier transform. The period finding and the quantum phase estimation algorithms presented in Sections 10.8 and 12.1 are the most important applications of the QFT.

[1] If all of the qubits are measured after the application of the QFT, then the controlled operations can be replaced by single-qubit operations. See Ref. [697] for more information.

Appendix D
The molecular Hamiltonian in second quantization

D.1 The one-body operator

Let us show that in second quantization the one-body operator

$$\hat{O}_1 = \sum_{j=1}^{N} \hat{h}(j) = \sum_{j=1}^{N} \left(-\frac{\nabla^2}{2} - \sum_{A=1}^{M} \frac{Z_A}{|\mathbf{r} - \mathbf{R}_A|} \right)$$

is given by $\sum_{p,q=1}^{K} h_{pq} a_p^\dagger a_q$ where $h_{pq} = \langle \chi_p | \hat{h} | \chi_q \rangle$ and $\{|\chi_j\rangle\}_{j=1}^{K}$ is an orthonormal basis of spin-orbitals.

To demonstrate this inequality, we show that the action of the operator $\sum_j \hat{h}(j)$ on the Slater determinant $|\chi_{p_1} \cdots \chi_{p_N}\rangle_-$ (where the indexes p_j are ordered in ascending order) is equal to the action of $\sum_{p,q=1}^{K} h_{pq} a_p^\dagger a_q$ on the corresponding Slater determinant $|\mathbf{n}\rangle$. For later convenience, we introduce a set S defined as $S = \{p_1, \ldots, p_N\}$. For example, if the system contains $N = 2$ electrons, the basis $B = \{\chi_1, \chi_2, \chi_3\}$ has three elements, and $S = \{p_1, p_2\} = \{1, 3\}$, then the Slater determinant $|\chi_{p_1}\chi_{p_2}\rangle_- = |\chi_1\chi_3\rangle_-$ is associated with $|\mathbf{n}\rangle = |101\rangle$.

Let us start from,

$$\begin{aligned}
\sum_{i,j=1}^{K} h_{ij} a_i^\dagger a_j |\mathbf{n}\rangle &= \sum_{i,j=1}^{K} h_{ij} \delta_{n_j,1} (-1)^{[\mathbf{n}]_j} a_i^\dagger |\mathbf{n} \oplus \mathbf{e}_j\rangle \\
&= \sum_{i,j=1}^{K} h_{ij} \delta_{n_j,1} (-1)^{[\mathbf{n}]_j} \delta_{(\mathbf{n}\oplus\mathbf{e}_j)_i,0} (-1)^{[\mathbf{n}\oplus\mathbf{e}_j]_i} |\mathbf{n} \oplus \mathbf{e}_j \oplus \mathbf{e}_i\rangle, \qquad \text{(D.1)}
\end{aligned}$$

where we used (12.75, 12.78). The first Kronecker delta can be expressed as

$$\delta_{n_j,1} = \sum_{\ell=1}^{N} \delta_{j,p_\ell}. \qquad \text{(D.2)}$$

Let us verify this equality with a simple example. Suppose that $|\chi_{p_1}\chi_{p_2}\rangle_- = |\chi_1\chi_3\rangle_-$, $S = \{1, 3\}$, and $|\mathbf{n}\rangle = |101\rangle$. Then, $\delta_{(101)_j,1}$ is given by

$$\begin{aligned} &\text{for } j=1: && \delta_{(101)_1,1}=\delta_{11}=1\\ &\text{for } j=2: && \delta_{(101)_2,1}=\delta_{01}=0\\ &\text{for } j=3: && \delta_{(101)_3,1}=\delta_{11}=1. \end{aligned}$$

This means that $\delta_{(101)_j,1}=\delta_{j,1}+\delta_{j,3}=\sum_{\ell=1}^{N}\delta_{j,p_\ell}$, which is consistent with eqn (D.2). Equation (D.1) becomes

$$\begin{aligned} \sum_{i,j=1}^{K} h_{ij}a_i^\dagger a_j|\mathbf{n}\rangle &= \sum_{i=1}^{K}\sum_{j=1}^{K}\sum_{\ell=1}^{N} h_{ij}\delta_{j,p_\ell}(-1)^{[\mathbf{n}]_j}\delta_{(\mathbf{n}\oplus\mathbf{e}_j)_i,0}(-1)^{[\mathbf{n}\oplus\mathbf{e}_j]_i}|\mathbf{n}\oplus\mathbf{e}_j\oplus\mathbf{e}_i\rangle\\ &= \sum_{i=1}^{K}\sum_{\ell=1}^{N} h_{ip_\ell}(-1)^{[\mathbf{n}]_{p_\ell}}\delta_{(\mathbf{n}\oplus\mathbf{e}_{p_\ell})_i,0}(-1)^{[\mathbf{n}\oplus\mathbf{e}_{p_\ell}]_i}|\mathbf{n}\oplus\mathbf{e}_{p_\ell}\oplus\mathbf{e}_i\rangle. \end{aligned} \quad \text{(D.3)}$$

The Kronecker delta appearing in this sum can be written as

$$\delta_{(\mathbf{n}\oplus\mathbf{e}_{p_\ell})_i,0}=\delta_{p_\ell,i}+\sum_{r\notin S}\delta_{i,r}. \quad \text{(D.4)}$$

Let us check that this equality is correct. Suppose that $|\chi_{p_1}\chi_{p_2}\rangle_-=|\chi_1\chi_3\rangle_-$, $S=\{1,3\}$, and $|\mathbf{n}\rangle=|101\rangle$. The quantity $\delta_{(\mathbf{n}\oplus\mathbf{e}_{p_\ell})_i,0}$ is given by

$$\begin{aligned} &\boxed{p_\ell=p_1, i=1}: && \delta_{(101\oplus\mathbf{e}_{p_1})_1,0}=\delta_{(101\oplus\mathbf{e}_1)_1,0}=\delta_{(001)_1,0}=\delta_{0,0}=1 && (p_\ell=i)\\ &\boxed{p_\ell=p_1, i=2}: && \delta_{(101\oplus\mathbf{e}_{p_1})_2,0}=\delta_{(101\oplus\mathbf{e}_1)_2,0}=\delta_{(001)_2,0}=\delta_{0,0}=1 && (i=r\notin S)\\ &p_\ell=p_1, i=3: && \delta_{(101\oplus\mathbf{e}_{p_1})_3,0}=\delta_{(101\oplus\mathbf{e}_1)_3,0}=\delta_{(001)_3,0}=\delta_{1,0}=0 &&\\ &p_\ell=p_2, i=1: && \delta_{(101\oplus\mathbf{e}_{p_2})_1,0}=\delta_{(101\oplus\mathbf{e}_3)_1,0}=\delta_{(100)_1,0}=\delta_{1,0}=0 &&\\ &\boxed{p_\ell=p_2, i=2}: && \delta_{(101\oplus\mathbf{e}_{p_2})_2,0}=\delta_{(101\oplus\mathbf{e}_3)_2,0}=\delta_{(100)_2,0}=\delta_{0,0}=1 && (i=r\notin S)\\ &\boxed{p_\ell=p_2, i=3}: && \delta_{(101\oplus\mathbf{e}_{p_2})_1,0}=\delta_{(101\oplus\mathbf{e}_3)_3,0}=\delta_{(100)_3,0}=\delta_{0,0}=1 && (p_\ell=i). \end{aligned}$$

These expressions are consistent with eqn (D.4). Therefore, eqn (D.3) can be written as

$$\begin{aligned} \sum_{i,j=1}^{K} h_{ij}a_i^\dagger a_j|\mathbf{n}\rangle &= \sum_{i=1}^{K}\sum_{\ell=1}^{N} h_{ip_\ell}(-1)^{[\mathbf{n}]_{p_\ell}}\left(\delta_{p_\ell,i}+\sum_{r\notin S}\delta_{i,r}\right)(-1)^{[\mathbf{n}\oplus\mathbf{e}_{p_\ell}]_i}|\mathbf{n}\oplus\mathbf{e}_{p_\ell}\oplus\mathbf{e}_i\rangle\\ &= \sum_{i=1}^{K}\sum_{\ell=1}^{N} h_{ip_\ell}(-1)^{[\mathbf{n}]_{p_\ell}}\delta_{p_\ell,i}(-1)^{[\mathbf{n}\oplus\mathbf{e}_{p_\ell}]_i}|\mathbf{n}\oplus\mathbf{e}_{p_\ell}\oplus\mathbf{e}_i\rangle\\ &\quad+\sum_{i=1}^{K}\sum_{\ell=1}^{N}\sum_{r\notin S} h_{ip_\ell}(-1)^{[\mathbf{n}]_{p_\ell}}\delta_{i,r}(-1)^{[\mathbf{n}\oplus\mathbf{e}_{p_\ell}]_i}|\mathbf{n}\oplus\mathbf{e}_{p_\ell}\oplus\mathbf{e}_i\rangle = \end{aligned}$$

$$= \sum_{\ell=1}^{N} h_{p_\ell p_\ell} (-1)^{[\mathbf{n}]_{p_\ell}} (-1)^{[\mathbf{n} \oplus \mathbf{e}_{p_\ell}]_{p_\ell}} |\mathbf{n}\rangle$$

$$+ \sum_{\ell=1}^{N} \sum_{r \notin S} h_{r p_\ell} (-1)^{[\mathbf{n}]_{p_\ell}} (-1)^{[\mathbf{n} \oplus \mathbf{e}_{p_\ell}]_r} |\mathbf{n} \oplus \mathbf{e}_{p_\ell} \oplus \mathbf{e}_r\rangle.$$

Since

$$(-1)^{[\mathbf{n}]_{p_\ell}} (-1)^{[\mathbf{n} \oplus \mathbf{e}_{p_\ell}]_{p_\ell}} = (-1)^{[\mathbf{n}]_{p_\ell}} (-1)^{[\mathbf{n}]_{p_\ell}} (-1)^{[\mathbf{e}_{p_\ell}]_{p_\ell}} = (-1)^{[\mathbf{e}_{p_\ell}]_{p_\ell}} = 1,$$

and $(-1)^{[\mathbf{n}]_{p_\ell}} = (-1)^{\ell-1}$, we finally arrive at

$$\boxed{\sum_{i,j=1}^{K} h_{ij} a_i^\dagger a_j |\mathbf{n}\rangle = \sum_{\ell=1}^{N} h_{p_\ell p_\ell} |\mathbf{n}\rangle + \sum_{\ell=1}^{N} \sum_{r \notin S} h_{r p_\ell} (-1)^{\ell-1} (-1)^{[\mathbf{n} \oplus \mathbf{e}_{p_\ell}]_r} |\mathbf{n} \oplus \mathbf{e}_{p_\ell} \oplus \mathbf{e}_r\rangle.} \tag{D.5}$$

Now we need to show that the action of the operator $\hat{O}_1 = \sum_{j=1}^{N} \hat{h}(j)$ onto the Slater determinant $|\chi_{p_1} \cdots \chi_{p_N}\rangle_-$ produces the same result. Let us start by calculating the action of $\hat{O}_1$ on a product state:

$$\hat{O}_1 |\chi_{p_1} \cdots \chi_{p_N}\rangle = \sum_{j=1}^{N} \hat{h}(j) |\chi_{p_1} \cdots \chi_{p_N}\rangle$$

$$= (\hat{h}|\chi_{p_1}\rangle)|\chi_{p_2} \cdots \chi_{p_N}\rangle + \ldots + |\chi_{p_1} \cdots \chi_{p_{N-1}}\rangle(\hat{h}|\chi_{p_N}\rangle). \tag{D.6}$$

To calculate the terms $\hat{h}|\chi_{p_j}\rangle$, we need to project the operator $\hat{h}$ onto the basis B,

$$\hat{h} = \sum_{r=1}^{K} \sum_{s=1}^{K} |\chi_r\rangle\langle\chi_r|\hat{h}|\chi_s\rangle\langle\chi_s| = \sum_{r=1}^{K} \sum_{s=1}^{K} h_{rs} |\chi_r\rangle\langle\chi_s|.$$

Thus,

$$\hat{h}|\chi_{p_j}\rangle = \sum_{r=1}^{K} \sum_{s=1}^{K} h_{rs} |\chi_r\rangle\langle\chi_s|\chi_{p_j}\rangle = \sum_{r=1}^{K} \sum_{s=1}^{K} h_{rs} \delta_{s p_j} |\chi_r\rangle = \sum_{r=1}^{K} h_{r p_j} |\chi_r\rangle.$$

Substituting this expression into (D.6), we obtain

$$\hat{O}_1|\chi_{p_1}\cdots\chi_{p_N}\rangle = \sum_{r=1}^{K} h_{rp_1}|\chi_r\rangle|\chi_{p_2}\cdots\chi_{p_N}\rangle + \ldots + \sum_{r=1}^{K} h_{rp_N}|\chi_{p_1}\cdots\chi_{p_{N-1}}\rangle|\chi_r\rangle$$

$$= \sum_{r=1}^{K}\sum_{\ell=1}^{N} h_{rp_\ell}|\chi_{p_1}\cdots\underset{\substack{\uparrow\\ \ell\text{-th position}}}{\chi_r}\cdots\chi_{p_N}\rangle.$$

This means that the action of $\hat{O}_1$ on a Slater determinant is given by

$$\hat{O}_1|\chi_{p_1}\cdots\chi_{p_N}\rangle_- = \hat{O}_1\hat{A}|\chi_{p_1}\cdots\chi_{p_N}\rangle = \hat{A}\hat{O}_1|\chi_{p_1}\cdots\chi_{p_N}\rangle$$
$$= \sum_{r=1}^{K}\sum_{\ell=1}^{N} h_{rp_\ell}\hat{A}|\chi_{p_1}\cdots\chi_r\cdots\chi_{p_N}\rangle = \sum_{r=1}^{K}\sum_{\ell=1}^{N} h_{rp_\ell}|\chi_{p_1}\cdots\chi_r\cdots\chi_{p_N}\rangle_-, \tag{D.7}$$

where we used the fact that $\hat{O}_1$ anti-commutes with the antisymmetrizer $\hat{A}$. To show that eqn (D.5) and (D.7) are identical, it is necessary to order the Slater determinants $|\chi_{p_1}\cdots\chi_r\cdots\chi_{p_N}\rangle_-$ in eqn (D.7) in ascending order. There are three possible cases:

1. If $r = p_\ell$, the vector $|\chi_{p_1}\cdots\chi_r\cdots\chi_{p_N}\rangle_-$ is already ordered.

2. If χ_r is equal to another spin-orbital χ_{p_j}, the vector $|\chi_{p_1}\cdots\chi_r\cdots\chi_{p_N}\rangle_-$ is zero.

3. If χ_r is different from all other spin-orbitals appearing in $|\chi_{p_1}\cdots\chi_r\cdots\chi_{p_N}\rangle_-$ (i.e. $r \notin S$), we need to order the sequence in ascending order. To this end, we can move the spin-orbital χ_r to the leftmost position and then move it to the right until the resulting sequence is ordered. Every time that χ_r moves to the left by one position, the Slater determinant must be multiplied by -1:

   $$|\chi_{p_1}\cdots\chi_{p_{\ell-1}}\chi_r\chi_{p_{\ell+1}}\cdots\chi_{p_N}\rangle_- = (-1)^{\ell-1}|\chi_r\chi_{p_1}\cdots\chi_{p_{\ell-1}}\chi_{p_{\ell+1}}\cdots\chi_{p_N}\rangle_-.$$

   Let us move χ_r to the right until we obtain the *ordered* vector $|\chi_{p_1}\cdots\chi_r\cdots\chi_{p_N}\rangle_-$. To calculate the number of times that the spin-orbital χ_r must shift to the right, it is useful to write the following relations:

   $$\begin{array}{rcl} |\chi_{p_1}\cdots\chi_{p_{\ell-1}}\chi_{p_\ell}\chi_{p_{\ell+1}}\cdots\chi_{p_N}\rangle_- & \leftrightarrow & |\mathbf{n}\rangle \\ |\chi_{p_1}\cdots\chi_{p_{\ell-1}}\chi_{p_{\ell+1}}\cdots\chi_{p_N}\rangle_- & \leftrightarrow & |\mathbf{n}\oplus\mathbf{e}_{p_\ell}\rangle \\ (\text{ordered})\rightarrow\ |\chi_{p_1}\cdots\chi_{p_{\ell-1}}\chi_{p_{\ell+1}}\cdots\chi_r\cdots\chi_{p_N}\rangle_- & \leftrightarrow & |\mathbf{n}\oplus\mathbf{e}_{p_\ell}\oplus\mathbf{e}_r\rangle. \end{array}$$

Therefore, χ_r must move to the right[1] a number of times equal to the number of electrons in the string $\mathbf{n} \oplus \mathbf{e}_{p_\ell}$ before the position r. This number is $(-1)^{[\mathbf{n}\oplus\mathbf{e}_{p_\ell}]_r}$.

In conclusion, equation (D.7) becomes

$$\sum_{r=1}^{K}\sum_{i=1}^{N} h_{rp_i}|\chi_{p_1}\cdots\chi_r\cdots\chi_{p_N}\rangle_- = \sum_{\ell=1}^{N} h_{p_\ell p_\ell}|\chi_{p_1}\cdots\chi_{p_N}\rangle_- + \sum_{r\notin S}^{K}\sum_{\ell=1}^{N}(-1)^{\ell-1}(-1)^{[\mathbf{n}\oplus\mathbf{e}_{p_\ell}]_r} h_{rp_\ell}|\chi_{p_1}\cdots\chi_r\cdots\chi_{p_N}\rangle_-.$$

This expression is equivalent to (D.5). Since Slater determinants form a basis of the Fock space, the operators $\sum_{j=1}^{N}\hat{h}(j)$ and $\sum_{p,q=1}^{K} h_{pq}a_p^\dagger a_q$ are equivalent. The calculations for the two-body operator are similar.

[1] For example, let us assume that $N = 3$, $K = 5$, and

$$|\chi_{p_1}\chi_{p_2}\chi_{p_3}\rangle_- = |\chi_1\chi_2\chi_4\rangle_- \qquad \leftrightarrow \qquad |\mathbf{n}\rangle = |11010\rangle.$$

Suppose that we need to order the sequence $|\chi_{p_1}\chi_r\chi_{p_3}\rangle_- = |\chi_1\chi_5\chi_4\rangle_-$ where $p_\ell = p_2$ and $\chi_r = \chi_5$. Then,

$$|\chi_1\chi_5\chi_4\rangle_- = (-1)^1|\chi_5\chi_1\chi_4\rangle_- = (-1)^{\ell-1}|\chi_5\chi_1\chi_4\rangle_-.$$

Now we need to move χ_5 to the right,

$$(-1)^{\ell-1}|\chi_5\chi_1\chi_4\rangle_- = (-1)^{\ell-1}(-1)^2|\chi_1\chi_4\chi_5\rangle_-.$$

Here, the number $(-1)^2$ corresponds to

$$(-1)^{[\mathbf{n}\oplus\mathbf{e}_{p_\ell}]_r} = (-1)^{[11010\oplus\mathbf{e}_{p_2}]_5} = (-1)^{[11010\oplus 01000]_5} = (-1)^{[10010]_5} = (-1)^2.$$

Appendix E
Oblivious amplitude amplification lemma

E.1 The lemma

In Section 11.5, we presented a method to simulate Hamiltonian dynamics with the Taylor series approach. In this method, the time evolution operator $\hat{U}_{\Delta t}$ is approximated with a linear combination of unitaries (LCU) of the form

$$\hat{V}_{\Delta t} = \sum_{j=0}^{M-1} c_j \hat{V}_j.$$

This operator satisfies $\|\hat{V}_{\Delta t} - \hat{U}_{\Delta t}\|_{\text{o}} = O(\varepsilon/m)$. As seen in Section 11.4, the LCU $\hat{V}_{\Delta t}$ can be implemented on a quantum computer by constructing a quantum circuit $\hat{W}$ such that

$$\hat{W}|\mathbf{0}\rangle \otimes |\psi\rangle = \sqrt{p(\mathbf{0})}|\mathbf{0}\rangle \otimes \frac{\hat{V}_{\Delta t}|\psi\rangle}{\|\hat{V}_{\Delta t}|\psi\rangle\|} + \sqrt{1 - p(\mathbf{0})}|\Phi\rangle, \tag{E.1}$$

where $|\mathbf{0}\rangle$ is the initial state of some ancilla qubits, $|\psi\rangle$ is the initial state of n qubits and $|\Phi\rangle$ is an arbitrary state of $n + p$ qubits orthogonal to $|\mathbf{0}\rangle$. At the end of the circuit, one obtains a quantum state proportional to $\hat{V}_{\Delta t}|\psi\rangle$ with success probability $p(\mathbf{0}) = \|\hat{V}_{\Delta t}|\psi\rangle\|^2/(\sum_j c_j)^2$. In the Taylor series method, the success probability is given by $p(\mathbf{0}) \approx 1/4 + \varepsilon^2/4m^2$ (see eqn 11.34). After multiple applications of the unitary $\hat{W}$, the success probability $p(\mathbf{0})^m$ decreases exponentially with m.

The oblivious amplitude amplification (OAA) method [349, 346] allows us to increase the success probability of each circuit to $p(\mathbf{0}) = 1 - O(\varepsilon/m)$. This method has some similarities to Grover's algorithm. The OAA method boils down to applying a quantum circuit $\hat{Y}$ to the initial state $|\mathbf{0}\rangle|\psi\rangle$ where

$$\hat{Y} = -\hat{W}(\mathbb{1} - 2\hat{P})\hat{W}^\dagger(\mathbb{1} - 2\hat{P})\hat{W},$$

$\hat{P} = |\mathbf{0}\rangle\langle\mathbf{0}| \otimes \mathbb{1}$ is a projector, and $\hat{W}$ is defined in eqn (E.1). Let us calculate the probability of measuring the ancilla qubits in $|\mathbf{0}\rangle$, i.e. $p(\mathbf{0})' = \text{Tr}[\hat{P}\hat{Y}|\mathbf{0}\psi\rangle\langle\mathbf{0}\psi|\hat{Y}^\dagger]$. Let us start from

$$\begin{aligned}\hat{P}\hat{Y}|\mathbf{0}\psi\rangle &= -\hat{P}\hat{W}(\mathbb{1}-2\hat{P})\hat{W}^\dagger(\mathbb{1}-2\hat{P})\hat{W}|\mathbf{0}\psi\rangle \\ &= -\hat{P}\hat{W}\hat{W}^\dagger\hat{W}|\mathbf{0}\psi\rangle + 2\hat{P}\left[\hat{W}\hat{P}\hat{W}^\dagger\hat{W} + \hat{W}\hat{W}^\dagger\hat{P}\hat{W}\right]|\mathbf{0}\psi\rangle - 4\hat{P}\hat{W}\hat{P}\hat{W}^\dagger\hat{P}\hat{W}|\mathbf{0}\psi\rangle \\ &= 3\hat{P}\hat{W}|\mathbf{0}\psi\rangle - 4\hat{P}\hat{W}\hat{P}\hat{W}^\dagger\hat{P}\hat{W}|\mathbf{0}\psi\rangle, \end{aligned} \tag{E.2}$$

where in the last step we used $\hat{W}\hat{W}^\dagger = \hat{W}^\dagger\hat{W} = \mathbb{1}$. Since

$$\hat{P}\hat{W}\hat{P} = |\mathbf{0}\rangle\langle\mathbf{0}| \otimes \frac{\hat{V}_{\Delta t}}{s},$$

eqn (E.2) becomes

$$\hat{P}\hat{Y}|\mathbf{0}\psi\rangle = 3\frac{\hat{V}_{\Delta t}}{s}|\mathbf{0}\psi\rangle - 4\frac{\hat{V}_{\Delta t}\hat{V}_{\Delta t}^\dagger\hat{V}_{\Delta t}}{s^3}|\mathbf{0}\psi\rangle.$$

Using the fact that $\hat{V}_{\Delta t}$ is well approximated by the unitary $\hat{U}_{\Delta t}$, one can write $\hat{V}_{\Delta t} = \hat{U}_{\Delta t} + \hat{R}$ where $\|\hat{R}\| = O(\varepsilon/m)$, and

$$\hat{V}_{\Delta t}^\dagger\hat{V}_{\Delta t} = \mathbb{1} + \underbrace{\hat{R}^\dagger\hat{U}_{\Delta t} + \hat{U}_{\Delta t}^\dagger\hat{R} + \hat{R}^\dagger\hat{R}}_{\hat{S}} = \mathbb{1} + \hat{S}, \qquad \text{with } \|\hat{S}\| = O(\varepsilon/m).$$

Therefore

$$\begin{aligned}\hat{P}\hat{Y}|\mathbf{0}\psi\rangle &= \left[\frac{3}{s} - \frac{4}{s^3}\right]\hat{U}_{\Delta t}|\mathbf{0}\psi\rangle + \underbrace{\left[\frac{3\hat{R}}{s} - \frac{4(\hat{U}_{\Delta t}\hat{S} + \hat{R} + \hat{R}\hat{S})}{s^3}\right]}_{\hat{X}}|\mathbf{0}\psi\rangle \\ &= \hat{U}_{\Delta t}|\mathbf{0}\psi\rangle + \hat{X}|\mathbf{0}\psi\rangle.\end{aligned}$$

In the last step, we used $s = 2$ and defined an operator $\hat{X}$ satisfying $\|\hat{X}\| = O(\varepsilon/m)$. The probability that the measurement collapses the state of the ancilla qubits onto $|\mathbf{0}\rangle$ is given by

$$p(\mathbf{0})' = \mathrm{Tr}\left[\hat{P}\hat{Y}|\mathbf{0}\psi\rangle\langle\mathbf{0}\psi|\hat{Y}^\dagger\right] = 1 + O(\varepsilon/m).$$

It is important to mention that the OAA procedure increases the success probability to $p(\mathbf{0})'_{\mathrm{s}} = 1 + O(\varepsilon/m)$ without producing a worse approximation of $\hat{U}_{\Delta t}|\psi\rangle$. Indeed,

$$\hat{P}\hat{Y}|\mathbf{0}\psi\rangle - |\mathbf{0}\rangle\hat{U}_{\Delta t}|\psi\rangle = |\mathbf{0}\rangle\hat{X}|\mathbf{0}\psi\rangle\ ,$$

and therefore

$$\|\hat{P}\hat{Y}|\mathbf{0}\psi\rangle - |\mathbf{0}\rangle\hat{U}_{\Delta t}|\psi\rangle\| = O(\varepsilon/m)\ .$$

Appendix F
Quantum signal processing

In this Appendix, we present some technical results about the quantum signal processing method. First, we prove an important theorem about the structure of the parametrized unitaries used in quantum signal processing.

Theorem F.1 For every set of parameters $\Phi \in [-\pi, \pi)^{r+1}$, the unitary operator

$$\hat{U}_\Phi = (\hat{A}^\dagger \otimes \mathbb{1}_n) \prod_{i=1}^{r} \left(\mathsf{Z}_{\phi_i} \mathsf{Z}_{-\frac{\pi}{4}} \hat{W} \mathsf{Z}_{\frac{\pi}{4}} \right) \mathsf{Z}_{\phi_0} (\hat{A} \otimes \mathbb{1}_n), \tag{F.1}$$

has the form

$$\hat{U}_\Phi = \bigoplus_{\lambda=1}^{2^n} \begin{bmatrix} p_\Phi(\lambda) & i\sqrt{1-\lambda^2} q_\Phi(\lambda) \\ i\sqrt{1-\lambda^2} q_\Phi^*(\lambda) & p_\Phi^*(\lambda) \end{bmatrix} \equiv \bigoplus_{\lambda=1}^{2^n} g_\Phi(\lambda) \tag{F.2}$$

where

1. $p_\Phi(\lambda)$ is a complex-valued polynomial of degree r and parity $r\%2$,
2. $q_\Phi(\lambda)$ is a complex-valued polynomial of degree at most $r-1$ and parity $(r-1)\%2$,
3. $|p_\Phi(\lambda)|^2 + (1-\lambda^2)|q_\Phi(\lambda)|^2 = 1$ over the interval $\lambda \in [-1, 1]$.

A consequence of this theorem is that $\hat{U}_\Phi$ is a block encoding of a complex-valued polynomial $p_\Phi(\lambda)$, with degree r, parity $r\%2$, and norm $\|p_\Phi\|_\infty \le 1$. Let us prove Theorem F.1.

Proof We prove this theorem by induction over r. For $r = 1$, the unitary operator in eqn (F.1) is given by

$$\hat{U}_\Phi = \mathsf{Z}_{\phi_1} \mathsf{Z}_{-\frac{\pi}{4}} \hat{W} \mathsf{Z}_{\frac{\pi}{4}} \mathsf{Z}_{\phi_0} = \bigoplus_{\lambda=1}^{2^n} \begin{bmatrix} e^{i(\phi_0+\phi_1)}\lambda & ie^{-i(\phi_0-\phi_1)}\sqrt{1-\lambda^2} \\ ie^{i(\phi_0-\phi_1)}\sqrt{1-\lambda^2} & e^{-i(\phi_0+\phi_1)}\lambda \end{bmatrix}, \tag{F.3}$$

and can be written as in eqn (F.2) by defining $p_\Phi(\lambda) = e^{i(\phi_0+\phi_1)}\lambda$ and $q_\Phi(\lambda) = e^{-i(\phi_0-\phi_1)}$. The polynomial p_Φ has degree $1 = r$ and parity $1 = r\%2$. The polynomial q_Φ has degree $0 = r-1$ and parity $0 = (r-1)\%2$. Furthermore,

$$|p_\Phi|^2 + (1-\lambda^2)|q_\Phi|^2 = \lambda^2 + (1-\lambda^2) = 1. \tag{F.4}$$

Therefore, eqn (F.2) holds for $r=1$. Let us assume it holds for some $r \geq 1$. Now define $\Phi' = (\phi_0 \dots \phi_r \phi_{r+1})$ and consider

$$\begin{aligned}\hat{U}_{\Phi'} &= \mathsf{Z}_{\phi_{r+1}} \mathsf{Z}_{-\frac{\pi}{4}} \hat{W} \mathsf{Z}_{\frac{\pi}{4}} \hat{U}_{\Phi} \\ &= \bigoplus_{\lambda=1}^{2^n} \begin{bmatrix} e^{i\phi_{r+1}}\lambda & ie^{i\phi_{r+1}}\sqrt{1-\lambda^2} \\ ie^{-i\phi_{r+1}}\sqrt{1-\lambda^2} & e^{-i\phi_{r+1}}\lambda \end{bmatrix} \begin{bmatrix} p_{\Phi}(\lambda) & i\sqrt{1-\lambda^2}q_{\Phi}(\lambda) \\ i\sqrt{1-\lambda^2}q_{\Phi}^*(\lambda) & p_{\Phi}^*(\lambda) \end{bmatrix} \\ &= \bigoplus_{\lambda=1}^{2^n} \begin{bmatrix} p_{\Phi'}(\lambda) & i\sqrt{1-\lambda^2}q_{\Phi'}(\lambda) \\ i\sqrt{1-\lambda^2}q_{\Phi'}^*(\lambda) & p_{\Phi'}^*(\lambda) \end{bmatrix}\end{aligned} \tag{F.5}$$

with

$$\begin{aligned} p_{\Phi'}(\lambda) &= e^{i\phi_{r+1}}\lambda p_{\Phi}(\lambda) - (1-\lambda^2)e^{i\phi_{r+1}}q_{\Phi}^*(\lambda)\,, \\ q_{\Phi'}(\lambda) &= e^{i\phi_{r+1}}\lambda q_{\Phi}(\lambda) + e^{i\phi_{r+1}}p_{\Phi}^*(\lambda)\,. \end{aligned} \tag{F.6}$$

From the induction hypothesis, $p_{\Phi'}(\lambda)$ is a complex-valued polynomial of degree $r+1$ and parity $(r+1)\%2$, and $q_{\Phi'}(\lambda)$ is a complex-valued polynomial of degree r and parity $r\%2$. Let us compute

$$\begin{aligned} |p_{\Phi'}|^2 + (1-\lambda)^2|q_{\Phi'}|^2 &= |\lambda p_{\Phi} - (1-\lambda^2)q_{\Phi}^*|^2 + (1-\lambda)^2|\lambda q_{\Phi} + p_{\Phi}^*|^2 \\ &= \lambda^2|p_{\Phi}|^2 + (1-\lambda^2)^2|q_{\Phi}|^2 - \lambda(1-\lambda^2)(p_{\Phi}^*q_{\Phi}^* + p_{\Phi}q_{\Phi}) \\ &+ (1-\lambda)^2\left(\lambda^2|q_{\Phi}|^2 + |p_{\Phi}|^2 + \lambda(p_{\Phi}^*q_{\Phi}^* + p_{\Phi}q_{\Phi})\right) \\ &= |p_{\Phi}|^2 + (1-\lambda)^2|q_{\Phi}|^2 = 1, \end{aligned} \tag{F.7}$$

where we used $|p_{\Phi}|^2 + (1-\lambda)^2|q_{\Phi}|^2 = 1$. We have just verified that if the induction hypothesis holds for some $r \geq 1$, it also holds for $r+1$. □

F.1 Numerical determination of parameters

Along with Theorem F.1, another important result holds: for any complex-valued polynomial p_r of degree r, parity $r\%2$, and norm $\|p_r\|_\infty \leq 1$, there exists a set of parameters $\tilde{\Phi} \in [-\pi,\pi)^{r+1}$ such that the unitary $\hat{U}_{\tilde{\Phi}}$ in eqn (F.1) is a block encoding of $p_r(\hat{H}')$. This result is very technical, and we defer the interested reader to Refs [352, 353, 28, 354, 355, 356] for its demonstration.

Here, we present a numerical least-squares minimization procedure [355] to determine the parameters $\tilde{\Phi}$ for a real-valued polynomial p_r using a classical computer.

- first, we define a mesh of points $\lambda_k = \cos\left(\frac{2k-1}{4\tilde{r}}\pi\right)$, with $k = 1\dots\tilde{r}$ e $\tilde{r} = \lceil\frac{r+1}{2}\rceil$,
- then, we require that the vector Φ is symmetric, i.e. $\Phi \in [-\pi,\pi)^{r+1}$ has the form

$$\Phi = (\phi_0, \phi_1, \phi_2, \dots, \phi_2, \phi_1, \phi_0), \tag{F.8}$$

- finally, we numerically minimize the quantity

$$\Delta(\Phi) = \sum_{k=1}^{\tilde{r}} \left| p_r(\lambda_k) - [g_\Phi(\lambda_k)]_{0,0} \right|^2 \tag{F.9}$$

starting from the initial point $\Phi_0 = (\pi/4, 0, 0, \ldots, 0, 0, \pi/4)$.

An implementation of this algorithm is reported at the end of this Appendix. Let us briefly describe the steps of the algorithm:

- the polynomial p_r has degree r and parity $r\%2$. Therefore, it has $\tilde{r}$ free parameters and is completely determined by the values it assumes on $\tilde{r}$ points. The chosen mesh is the set of nodes of the Chebyshev polynomial $T_{2r}(x)$, which are frequently used in polynomial interpolation because the resulting interpolating polynomial mitigates the Runge phenomenon.

- Why is the vector Φ symmetric? We know that the polynomial has degree r if and only if $\Phi \in [-\pi, \pi)^{r+1}$; on the other hand, p_r has $\tilde{r}$ free parameters, meaning that some elements of Φ must be redundant. The choice of symmetric phases was proposed in [355], and ensures that q_r is a real-valued polynomial like p_r.

- The initial condition $\Phi_0 = (\pi/4, 0, 0, \ldots, 0, 0, \pi/4)$ ensures that $[g_\Phi(\lambda_k)]_{0,0} \equiv 0$ and requires no prior knowledge about the target polynomial p_r. The function $\Delta(\Phi)$ is very intricated [354], and has multiple local and global minima. Hence, starting the optimization from a random point is prone to convergence issues. It was empirically observed that choosing Φ_0 as the initial point leads to a fast optimization that converges to the global minimum of $\Delta(\Phi)$ [355].

Example F.1 The function $f(\lambda) = \cos(t\lambda)$, with $t = 5.3$, can be approximated by a fourth order polynomial derived from a truncated Jacobi-Anger expansion,

$$A(\lambda) = J_0(-t)T_0(\lambda) - 2J_2(-t)T_2(\lambda) + 2J_4(-t)T_4(\lambda) = 0.613951 - 6.17501\lambda^2 + 6.394\lambda^4,$$

with norm $\|A\|_\infty = 0.876929$. The rescaled polynomial

$$p_4(\lambda) = \frac{A(\lambda)}{\|A\|_\infty} = 0.700115 - 7.04163\lambda^2 + 7.29136\lambda^4$$

satisfies the hypotheses of theorem F.1. Using the algorithm presented in this Appendix, one can compute parameters

$$\tilde{\Phi} = (0.059070, -0.278323, 0.120449, -0.278323, 0.059070)$$

such that $p_{\tilde{\Phi}}(\lambda) \simeq p_4(\lambda)$. The functions $f(\lambda)$, $p_4(\lambda)$, and $p_{\tilde{\Phi}}(\lambda)$ are shown below,

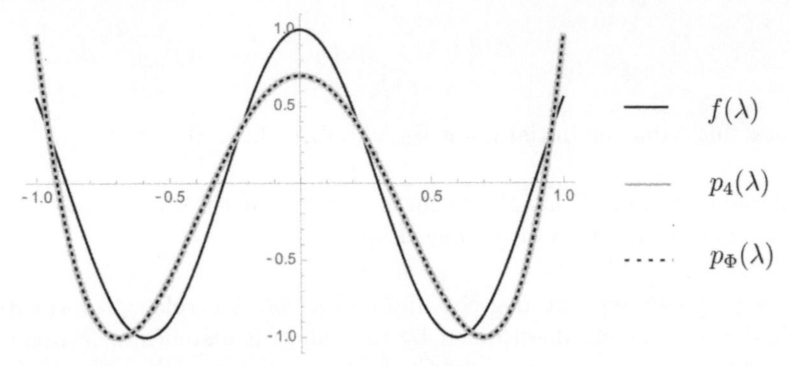

The code used to compute the parameters $\tilde{\Phi}$ is listed below.

```
import math
from matplotlib import pyplot as plt
from scipy.optimize import minimize

def poly(x,cf):
    pw = np.array([x**k for k in range(len(cf))])
    return np.dot(pw,cf)

def mesh(rtilde):
    x = np.array(range(1,rtilde+1))
    return np.cos((2*x-1)/(4.0*rtilde)*np.pi)

def rz(a):
    mat = np.eye(2,dtype=complex)
    mat[0,0] = np.exp( 1j*a)
    mat[1,1] = np.exp(-1j*a)
    return mat

def Wtilde(x):
    nx = len(x)
    mat = np.zeros((2,2,nx),dtype=complex)
    for j,xj in enumerate(x):
        mat[0,0,j] = mat[1,1,j] = xj
        mat[0,1,j] = mat[1,0,j] = 1j*np.sqrt(1-xj**2)
    return mat

def g(varphi,W_array):
    L = len(varphi)
    M = W_array.shape[2]
    Rz_array = np.zeros((2,2,L),dtype=complex)
    for j,xj in enumerate(varphi):
        Rz_array[:,:,j] = rz(xj)
    gvec = np.zeros(M,dtype=complex)
    for x in range(M):
        mat = Rz_array[:,:,0]
        for k in range(1,L):
            mat = np.dot(Rz_array[:,:,k],np.dot(W_array[:,:,x],mat))
        gvec[x] = mat[0,0]
    return gvec

def to_symmetric_phi(varphi,d):
    phi = np.zeros(d+1)
```

```
    for j,fj in enumerate(varphi):
        phi[j]   = fj
        phi[d-j] = fj
    return phi

def cost_function(varphi,d,fun,W_array):
    phi   = to_symmetric_phi(varphi,d)
    gvec  = g(phi,W_array)
    delta = 0.0
    for x in range(W_array.shape[2]):
        delta += np.abs(fun[x]-gvec[x].real)**2
    return delta
    return cst

cf = np.array([0.700115, 0, -7.04163, 0, 7.29136])
r       = len(cf)-1
rtilde  = math.ceil((r+1)/2.0)
print("coefficients of the polynomial = ",cf)
print("degree of the polynomial       = ",r)

x       = mesh(rtilde)
fun     = np.array([poly(xj,cf) for xj in x])
print("mesh      = ",x)
print("f(mesh)   = ",fun)

phi0    = np.zeros(rtilde)
phi0[0] = np.pi/4.0
res   = minimize(cost_function,x0=phi0,args=(r,fun,Wtilde(x)),method='CG',
                                     tol=1e-8,options={'maxiter':100})
gvec = g(to_symmetric_phi(res.x,r),Wtilde(x))
print("initial guess          ",to_symmetric_phi(phi0,r))
print("optimized parameters = ",to_symmetric_phi(res.x,r))
u  = np.arange(0,1,0.001)
hu = np.cos(5.3*u)
fu = np.array([poly(xj,cf) for xj in u])
gu = g(to_symmetric_phi(res.x,r),Wtilde(u))
plt.plot(u,hu,label=r'$f(\lambda)$',                lw=2,ls='-', c='black')
plt.plot(u,fu,label=r'$p_4(\lambda)$',              lw=3,ls='--',c='gray')
plt.plot(u,gu,label=r'$p_{\tilde \Phi}(\lambda)$',lw=1,ls='-', c='black')
plt.xlabel(r'$\lambda$'); plt.ylabel(r'$f(\lambda)$')
plt.legend()
plt.show()
```

References

[1] A. M. Turing, "On computable numbers, with an application to the Entscheidungsproblem," *Proc. London Math. Soc*, no. 1, pp. 230–265, 1937.

[2] A. Kitaev, A. Shen, and M. Vyalyi, *Classical and quantum computation.* American Mathematical Society, 2002.

[3] S. Aaronson, "How much structure is needed for huge quantum speedups?," *arXiv:2209.06930*, 2022.

[4] L. Szilard, "Über die Entropieverminderung in einem thermodynamischen System bei Eingriffen intelligenter Wesen," *Z. Phys*, vol. 53, no. 11–12, pp. 840–856, 1929.

[5] J. von Neumann and R. Beyer, *Mathematical foundations of quantum mechanics.* Princeton University Press, 1955.

[6] R. Landauer, "Irreversibility and heat generation in the computing process," *IBM J. Res. Dev*, vol. 5, pp. 183–191, 1961.

[7] C. H. Bennett, "Logical reversibility of computation," *IBM J. Res. Dev*, vol. 17, no. 6, pp. 525–532, 1973.

[8] C. H. Bennett, "The thermodynamics of computation: a review," *Int. J. Theor. Phys*, vol. 21, no. 12, pp. 905–940, 1982.

[9] E. Fredkin and T. Toffoli, "Conservative logic," *Int. J. Theor. Phys*, vol. 21, pp. 219–253, 1982.

[10] P. Benioff, "The computer as a physical system: A microscopic quantum mechanical Hamiltonian model of computers as represented by Turing machines," *J. Stat. Phys*, vol. 22, no. 5, pp. 563–591, 1980.

[11] R. P. Feynman, "Simulating physics with computers," *Int. J. Theor. Phys*, vol. 21, pp. 467–488, 1982.

[12] R. P. Feynman, "Quantum mechanical computers," *Opt. news*, vol. 11, no. 2, pp. 11–20, 1985.

[13] D. Deutsch, "Quantum theory, the Church-Turing principle and the universal quantum computer," *Proc. R. Soc. Lond*, vol. 400, pp. 97–117, 1985.

[14] D. E. Deutsch and R. Penrose, "Quantum computational networks," *Proc. R. Soc. Lond*, vol. 425, pp. 73–90, 1989.

[15] D. Deutsch and R. Jozsa, "Rapid solution of problems by quantum computation," *Proc. R. Soc. Lond*, vol. 439, no. 1907, pp. 553–558, 1992.

[16] E. Bernstein and U. Vazirani, "Quantum complexity theory," *SIAM J. Comput*, vol. 26, pp. 1411–1473, 1997.

[17] D. R. Simon, "On the power of quantum computation," *SIAM J. Comput*, vol. 26, no. 5, pp. 1474–1483, 1997.

[18] P. W. Shor, "Polynomial-time algorithms for prime factorization and discrete logarithms on a quantum computer," *SIAM J. Comput*, vol. 26, no. 5, pp. 1484–1509, 1997.

[19] L. K. Grover, "A fast quantum mechanical algorithm for database search," in *Proceedings of the 38th annual ACM symposium on theory of computing (STOC)*, pp. 212–219, 1996.
[20] F. Arute *et al.*, "Quantum supremacy using a programmable superconducting processor," *Nature*, vol. 574, pp. 505–510, 2019.
[21] H.-S. Zhong, H. Wang, Y.-H. Deng, M.-C. Chen, L.-C. Peng, Y.-H. Luo, J. Qin, D. Wu, X. Ding, Y. Hu, *et al.*, "Quantum computational advantage using photons," *Science*, vol. 370, no. 6523, pp. 1460–1463, 2020.
[22] E. Farhi, J. Goldstone, S. Gutmann, and M. Sipser, "Quantum computation by adiabatic evolution," *quant-ph/0001106*, 2000.
[23] E. Farhi, J. Goldstone, and S. Gutmann, "A quantum approximate optimization algorithm," *arXiv:1411.4028*, 2014.
[24] A. Peruzzo, J. McClean, P. Shadbolt, M.-H. Yung, X.-Q. Zhou, P. J. Love, A. Aspuru-Guzik, and J. L. O'Brien, "A variational eigenvalue solver on a photonic quantum processor," *Nat. Commun*, vol. 5, no. 1, p. 4213, 2014.
[25] S. Lloyd, "Universal quantum simulators," *Science*, vol. 400, pp. 1073–1078, 1996.
[26] D. S. Abrams and S. Lloyd, "Simulation of many-body Fermi systems on a universal quantum computer," *Phys. Rev. Lett*, vol. 79, pp. 2586–2589, 1997.
[27] D. W. Berry, A. M. Childs, R. Cleve, R. Kothari, and R. D. Somma, "Simulating Hamiltonian dynamics with a truncated Taylor series," *Phys. Rev. Lett*, vol. 114, no. 9, p. 090502, 2015.
[28] G. H. Low and I. L. Chuang, "Hamiltonian simulation by qubitization," *Quantum*, vol. 3, p. 163, 2019.
[29] M. Sipser, *Introduction to the Theory of Computation.* PWS Publishing, 2012.
[30] S. Arora and B. Barak, *Computational complexity: a modern approach.* Cambridge University Press, 2009.
[31] R. Sedgewick and K. Wayne, *Algorithms.* Addison-Wesley, 2011.
[32] D. E. Knuth, *The art of computer programming*, vol. 3. Pearson Education, 1997.
[33] A. V. Aho and J. E. Hopcroft, *The design and analysis of computer algorithms.* Pearson Education India, 1974.
[34] S. Baase, *Computer algorithms: introduction to design and analysis.* Pearson Education India, 2009.
[35] G. Brassard and P. Bratley, *Algorithms, theory and practice.* Prentice-Hall Englewood Cliffs, NJ, 1988.
[36] R. M. Karp, "On the computational complexity of combinatorial problems," *Networks*, vol. 5, no. 1, pp. 45–68, 1975.
[37] J. Van Leeuwen and J. Leeuwen, *Handbook of theoretical computer science: algorithms and complexity*, vol. 1. Elsevier, 1990.
[38] H. R. Lewis and C. H. Papadimitriou, *Elements of the theory of computation.* Prentice-Hall, 1997.
[39] P. G. H. Bachmann, *Analytische Zahlentheorie.* Teubner, 1894.
[40] E. Landau, *Handbuch der Lehre von der Verteilung der Primzahlen.* Teubner, 1909.
[41] G. Boole, *An investigation of the laws of thought: on which are founded the mathematical theories of logic and probabilities.* Walton and Maberly, 1854.

[42] C. E. Shannon, "A symbolic analysis of relay and switching circuits," *Electr. Eng*, vol. 57, no. 12, pp. 713–723, 1938.
[43] E. J. McCluskey, *Introduction to the theory of switching circuits.* McGraw-Hill, 1965.
[44] F. J. Hill and G. R. Peterson, *Introduction to switching theory and logical design.* John Wiley & Sons, 1981.
[45] E. A. Bender and S. Gill Williamson, *A short course in discrete mathematics.* Courier Dover Publications, 2005.
[46] K. Maruyama, F. Nori, and V. Vedral, "Colloquium: the physics of Maxwell's demon and information," *Rev. Mod. Phys*, vol. 81, pp. 1–23, 2009.
[47] H. S. Leff and A. F. Rex, *Maxwell's demon: entropy, information, computing.* Princeton University Press, 2014.
[48] N. Biggs, E. K. Lloyd, and R. J. Wilson, *Graph theory, 1736-1936.* Oxford University Press, 1986.
[49] J. A. Bondy and U. S. R. Murty, *Graph theory with applications.* Springer, 1976.
[50] G. Chartrand, *Introductory graph theory.* Courier Dover Publications, 1977.
[51] R. Trudeau, *Introduction to graph theory.* Courier Dover Publications, 1993.
[52] F. Harary, *Graph theory.* Addison-Wesley, 1969.
[53] J. E. Hopcroft, R. Motwani, and J. D. Ullman, *Introduction to automata theory, languages, and computation.* Addison-Wesley, 2000.
[54] S. C. Kleene, C. E. Shannon, and J. McCarthy, *Automata studies.* Princeton University Press, 1956.
[55] M. Agrawal, N. Kayal, and N. Saxena, "Primes is in P," *Ann. Math*, vol. 160, pp. 781–793, 2004.
[56] R. L. Rivest, A. Shamir, and L. Adleman, "A method for obtaining digital signatures and public-key cryptosystems," *Commun. ACM*, vol. 21, pp. 120–126, 1978.
[57] J. M. Pollard, "Theorems on factorization and primality testing," *Math. Proc. Camb. Philos. Soc*, vol. 76, no. 3, pp. 521–528, 1974.
[58] H. C. Williams, "A $p + 1$ method of factoring," *Math. Comput*, vol. 39, no. 159, pp. 225–234, 1982.
[59] H. W. Lenstra, "Factoring integers with elliptic curves," *Ann. Math*, vol. 126, no. 3, pp. 649–673, 1987.
[60] R. Crandall and C. Pomerance, *Prime numbers: a computational perspective.* Springer, 2006.
[61] C. Gidney and M. Ekerå, "How to factor 2048 bit RSA integers in 8 hours using 20 million noisy qubits," *Quantum*, vol. 5, p. 433, 2021.
[62] E. Gouzien and N. Sangouard, "Factoring 2048-bit RSA integers in 177 days with 13 436 qubits and a multimode memory," *Phys. Rev. Lett*, vol. 127, no. 14, p. 140503, 2021.
[63] M. Fleury, "Deux problèmes de géométrie de situation," *J. Math. Appl*, vol. 2, pp. 257–261, 1883.
[64] M. Garey, D. Johnson, and M. S. M. Collection, *Computers and intractability: a guide to the theory of NP-completeness.* W. H. Freeman, 1979.
[65] R. M. Karp, "Reducibility among combinatorial problems," in *Complexity of computer computations* (R. E. Miller and J. W. Thatcher, eds.), Plenum Press, 1972.

[66] J. E. Hopcroft and R. E. Tarjan, "A $|v| \log |v|$ algorithm for isomorphism of triconnected planar graphs," *J. Comput. Syst. Sci*, vol. 7, no. 3, pp. 323–331, 1973.
[67] J. E. Hopcroft and J. K. Wong, "Linear time algorithm for isomorphism of planar graphs (preliminary report)," in *Proceedings of the 6th annual ACM symposium on theory of computing (STOC)*, pp. 172–184, 1974.
[68] S. Plojak and Z. Tuza, "Maximum cuts and largest bipartite subgraphs," in *Combinatorial optimization* (W. Cook, L. Lováz, and S. Seymour, eds.), American Mathematical Society, 1995.
[69] P. Kaye, R. Laflamme, M. Mosca, *et al.*, *An introduction to quantum computing.* Oxford University Press, 2007.
[70] J. E. Savage, *Models of computation: exploring the power of computing.* Addison-Wesley, 1998.
[71] P. Ramond, *Group theory: a physicist's survey.* Cambridge University Press, 2010.
[72] J. F. Cornwell, *Group theory in physics: an introduction.* Academic press, 1997.
[73] M. Tinkham, *Group theory and quantum mechanics.* Courier Dover Publications, 2003.
[74] F. A. Cotton, *Chemical applications of group theory.* John Wiley & Sons, 2003.
[75] A. Kolmogorov, *Foundations of the theory of probability.* Courier Dover Publications, second english edition ed., 1956.
[76] W. Feller, *An introduction to probability theory and its applications.* John Wiley & Sons, 2008.
[77] J. Hartmanis and R. E. Stearns, "On the computational complexity of algorithms," *Trans. Am. Math. Soc*, vol. 117, pp. 285–306, 1965.
[78] A. A. Markov, "The theory of algorithms," *Am. Math. Soc. Transl*, vol. 15, pp. 1–14, 1960.
[79] M. L. Minsky, *Computation: finite and infinite machines.* Prentice-Hall, 1967.
[80] A. Schönhage, "Storage modification machines," *SIAM J. Comput*, vol. 9, no. 3, pp. 490–508, 1980.
[81] A. Church, "An unsolvable problem of elementary number theory," *Am. J. Math*, vol. 58, pp. 345–363, 1936.
[82] E. L. Post, "Finite combinatory processes: formulation 1," *J.Symb. Log*, vol. 1, no. 3, pp. 103–105, 1936.
[83] A. Vergis, K. Steiglitz, and B. Dickinson, "The complexity of analog computation," *Math. Comput. Simul*, vol. 28, no. 2, pp. 91–113, 1986.
[84] Y. Wu *et al.*, "Strong quantum computational advantage using a superconducting quantum processor," *Phys. Rev. Lett*, vol. 127, p. 180501, 2021.
[85] L. S. Madsen, F. Laudenbach, M. F. Askarani, F. Rortais, T. Vincent, J. F. Bulmer, F. M. Miatto, L. Neuhaus, L. G. Helt, M. J. Collins, *et al.*, "Quantum computational advantage with a programmable photonic processor," *Nature*, vol. 606, no. 7912, pp. 75–81, 2022.
[86] D. Harel and Y. A. Feldman, *Algorithmics: the spirit of computing.* Pearson Education, 2004.
[87] E. D. Reilly, *Concise encyclopedia of computer science.* John Wiley & Sons, 2004.
[88] S. A. Cook, "The complexity of theorem-proving procedures," in *Proceedings of the 3rd annual ACM symposium on theory of computing (STOC)*, pp. 151–158, 1971.

[89] R. E. Ladner, "On the structure of polynomial time reducibility," *J. ACM*, vol. 22, pp. 155–171, 1975.
[90] A. M. Turing, "Systems of logic based on ordinals," *Proc. London Math. Soc*, vol. 45, pp. 161–228, 1939.
[91] D. Van Melkebeek, *Randomness and completeness in computational complexity.* Springer, 2000.
[92] S. Aaronson, "BQP and the polynomial hierarchy," in *Proc. ACM*, pp. 141–150, 2010.
[93] A. Chi-Chih Yao, "Quantum circuit complexity," in *Proceedings of the 34th annual symposium on foundations of computer science (FOCS)*, pp. 352–361, 1993.
[94] A. Molina and J. Watrous, "Revisiting the simulation of quantum Turing machines by quantum circuits," *Proc. R. Soc. Lond*, vol. 475, p. 20180767, 2019.
[95] C. H. Bennett, E. Bernstein, G. Brassard, and U. Vazirani, "Strengths and weaknesses of quantum computing," *SIAM J. Comput*, vol. 26, pp. 1510–1523, 1997.
[96] I. L. Markov and Y. Shi, "Simulating quantum computation by contracting tensor networks," *SIAM J. Comput*, vol. 38, no. 3, pp. 963–981, 2008.
[97] S. Boixo, S. V. Isakov, *et al.*, "Simulation of low-depth quantum circuits as complex undirected graphical models," *arXiv:1712.05384*, 2017.
[98] A. J. McConnell, *Applications of tensor analysis.* Courier Dover Publications, 2014.
[99] I. L. Markov, A. Fatima, S. V. Isakov, and S. Boixo, "Quantum supremacy is both closer and farther than it appears," *arXiv:1807.10749*, 2018.
[100] F. Pan and P. Zhang, "Simulation of quantum circuits using the big-batch tensor network method," *Phys. Rev. Lett*, vol. 128, p. 030501, Jan 2022.
[101] J. Watrous, "Quantum computational complexity," *arXiv:0804.3401*, 2008.
[102] S. Aaronson, A. Bouland, and L. Schaeffer, "Lecture notes for the 28th McGill invitational workshop on computational complexity," 2016.
[103] A. Barenco, C. H. Bennett, R. Cleve, D. P. DiVincenzo, N. Margolus, P. Shor, T. Sleator, J. A. Smolin, and H. Weinfurter, "Elementary gates for quantum computation," *Phys. Rev. A*, vol. 52, pp. 3457–3467, 1995.
[104] J. Kempe, A. Kitaev, and O. Regev, "The complexity of the local Hamiltonian problem," *SIAM J. Comput*, vol. 35, no. 5, pp. 1070–1097, 2006.
[105] U. Vazirani, "A survey of quantum complexity theory," in *Proceedings of Symposia in Applied Mathematics*, vol. 58, pp. 193–220, 2002.
[106] Y. Choquet-Bruhat, C. DeWitt-Morette, and M. Dillard-Bleick, *Analysis, manifolds and physics.* Elsevier Science, 1982.
[107] I. M. Gelfand, *Lectures on linear algebra.* Courier Dover Publications, 1989.
[108] P. R. Halmos, *Finite-dimensional vector spaces.* Springer, 1993.
[109] A. Peres, *Quantum theory: concepts and methods.* Springer, 1995.
[110] G. Strang, *Introduction to linear algebra.* Wellesley-Cambridge Press, 2003.
[111] I. Bengtsson and K. Życzkowski, *Geometry of quantum states: an introduction to quantum entanglement.* Cambridge University Press, 2007.
[112] G. E. Shilov, *Linear algebra.* Courier Dover Publications, 2012.
[113] M. Reed and B. Simon, *Methods of modern mathematical physics: Fourier analysis, self-adjointness.* Elsevier, 1975.
[114] S. Lang, *Linear algebra.* Springer, 1987.

[115] J. Wilkinson, "The algebraic eigenvalue problem," in *Handbook for Automatic Computation, Volume II, Linear Algebra*, Springer-Verlag New York, 1971.
[116] W. Rossmann, *Lie groups: an introduction through linear groups.* Oxford University Press, 2002.
[117] R. R. Puri, *Mathematical methods of quantum optics.* Springer, 2001.
[118] J. W. Demmel, *Applied numerical linear algebra.* SIAM, 1997.
[119] L. N. Trefethen and D. Bau III, *Numerical linear algebra.* SIAM, 1997.
[120] B. N. Datta, *Numerical linear algebra and applications.* SIAM, 2010.
[121] J. Dongarra *et al.*, *LAPACK Users' guide.* SIAM, 1999.
[122] T. E. Oliphant, *A guide to NumPy.* Trelgol Publishing USA, 2006.
[123] P. Virtanen *et al.*, "SciPy 1.0: fundamental algorithms for scientific computing in Python," *Nat. Methods*, vol. 17, pp. 261–272, 2020.
[124] J. M. Jauch, *Foundations of quantum mechanics.* Addison-Wesley, 1968.
[125] P. Dirac, *The principles of quantum mechanics.* Oxford University Press, 1981.
[126] L. Landau and E. Lifshitz, *Quantum mechanics: non-relativistic theory.* Elsevier, 1981.
[127] A. Messiah, *Quantum mechanics.* Courier Dover Publications, 1999.
[128] J. J. Sakurai and J. Napolitano, *Modern quantum mechanics.* Addison-Wesley, 2011.
[129] A. Tonomura, J. Endo, T. Matsuda, T. Kawasaki, and H. Ezawa, "Demonstration of single-electron buildup of an interference pattern," *Am. J. Phys*, vol. 57, pp. 117–120, 1989.
[130] O. Nairz, M. Arndt, and A. Zeilinger, "Quantum interference experiments with large molecules," *Am. J. Phys*, vol. 71, pp. 319–325, 2003.
[131] S. Weinberg, *Lectures on quantum mechanics.* Cambridge University Press, 2015.
[132] W. K. Wootters and W. H. Zurek, "A single quantum cannot be cloned," *Nature*, vol. 299, pp. 802–803, 1982.
[133] R. Ghosh and L. Mandel, "Observation of nonclassical effects in the interference of two photons," *Phys. Rev. Lett*, vol. 59, pp. 1903–1905, 1987.
[134] B. Vlastakis, G. Kirchmair, Z. Leghtas, S. E. Nigg, L. Frunzio, S. M. Girvin, M. Mirrahimi, M. H. Devoret, and R. J. Schoelkopf, "Deterministically encoding quantum information using 100-photon Schrödinger cat states," *Science*, vol. 342, pp. 607–610, 2013.
[135] M. Mirrahimi, Z. Leghtas, V. V. Albert, S. Touzard, R. J. Schoelkopf, L. Jiang, and M. H. Devoret, "Dynamically protected cat-qubits: a new paradigm for universal quantum computation," *New J. Phys*, vol. 16, p. 045014, 2014.
[136] C. Wang, Y. Y. Gao, P. Reinhold, R. W. Heeres, N. Ofek, K. Chou, C. Axline, M. Reagor, J. Blumoff, *et al.*, "A Schrödinger cat living in two boxes," *Science*, vol. 352, pp. 1087–1091, 2016.
[137] A. Blais, A. L. Grimsmo, S. M. Girvin, and A. Wallraff, "Circuit quantum electrodynamics," *Rev. Mod. Phys*, vol. 93, p. 025005, May 2021.
[138] A. Wallraff, D. I. Schuster, A. Blais, L. Frunzio, R.-S. Huang, J. Majer, S. Kumar, S. M. Girvin, and R. J. Schoelkopf, "Strong coupling of a single photon to a superconducting qubit using circuit quantum electrodynamics," *Nature*, vol. 431, pp. 162–167, 2004.
[139] A. Blais, R.-S. Huang, A. Wallraff, S. M. Girvin, and R. J. Schoelkopf, "Cavity

quantum electrodynamics for superconducting electrical circuits: an architecture for quantum computation," *Phys. Rev. A*, vol. 69, p. 062320, 2004.

[140] Y. Chu, P. Kharel, W. H. Renninger, L. D. Burkhart, L. Frunzio, P. T. Rakich, and R. J. Schoelkopf, "Quantum acoustics with superconducting qubits," *Science*, vol. 358, pp. 199–202, 2017.

[141] R. Manenti, A. F. Kockum, A. Patterson, T. Behrle, J. Rahamim, G. Tancredi, F. Nori, and P. J. Leek, "Circuit quantum acoustodynamics with surface acoustic waves," *Nat. Commun*, vol. 8, pp. 1–6, 2017.

[142] Y. Tabuchi, S. Ishino, A. Noguchi, T. Ishikawa, R. Yamazaki, K. Usami, and Y. Nakamura, "Coherent coupling between a ferromagnetic magnon and a superconducting qubit," *Science*, vol. 349, pp. 405–408, 2015.

[143] J. J. Burnett, A. Bengtsson, M. Scigliuzzo, D. Niepce, M. Kudra, P. Delsing, and J. Bylander, "Decoherence benchmarking of superconducting qubits," *npj Quantum Inf*, vol. 5, pp. 1–8, 2019.

[144] J. Fink, M. Göppl, M. Baur, R. Bianchetti, P. Leek, A. Blais, and A. Wallraff, "Climbing the Jaynes-Cummings ladder and observing its nonlinearity in a cavity QED system," *Nature*, vol. 454, pp. 315–318, 2008.

[145] J. Gambetta, A. Blais, D. I. Schuster, A. Wallraff, L. Frunzio, J. Majer, M. H. Devoret, S. M. Girvin, and R. J. Schoelkopf, "Qubit-photon interactions in a cavity: measurement-induced dephasing and number splitting," *Phys. Rev. A*, vol. 74, p. 042318, 2006.

[146] D. I. Schuster, *Circuit quantum electrodynamics.* PhD thesis, Yale University, 2007.

[147] M. Alonso and E. J. Finn, *Quantum and statistical physics*, vol. 3. Addison-Wesley, 1968.

[148] G. W. Mackey, *Mathematical foundations of quantum mechanics.* Courier Dover Publications, 2013.

[149] A. S. Holevo, *Probabilistic and statistical aspects of quantum theory.* Springer, 2011.

[150] G. Ludwig, *Foundations of quantum mechanics.* Springer, 2012.

[151] R. P. Feynman, A. R. Hibbs, and D. F. Styer, *Quantum mechanics and path integrals.* Courier Dover Publications, 1964.

[152] C. Gerry and P. Knight, *Introductory quantum optics.* Cambridge University Press, 2005.

[153] M. Schuld, M. Fingerhuth, and F. Petruccione, "Implementing a distance-based classifier with a quantum interference circuit," *Europhys. Lett*, vol. 119, no. 6, p. 60002, 2017.

[154] M. Schuld and N. Killoran, "Quantum machine learning in feature Hilbert spaces," *Phys. Rev. Lett*, vol. 122, no. 4, p. 040504, 2019.

[155] F. Leymann and J. Barzen, "The bitter truth about gate-based quantum algorithms in the NISQ era," *Quantum Sci. Technol*, vol. 5, no. 4, p. 044007, 2020.

[156] M. Weigold, J. Barzen, F. Leymann, and M. Salm, "Data encoding patterns for quantum computing," in *Proceedings of the 27th conference on pattern languages of programs*, pp. 1–11, 2020.

[157] D. P. DiVincenzo, "The physical implementation of quantum computation," *Fortschr. Phys*, vol. 48, pp. 771–783, 2000.

[158] D. Maslov, "Advantages of using relative-phase toffoli gates with an application to

multiple control Toffoli optimization," *Phys. Rev. A*, vol. 93, no. 2, p. 022311, 2016.

[159] G. Craig, "Using quantum gates instead of ancilla bits." `https://algassert.com/circuits/2015/06/22/Using-Quantum-Gates-instead-of-Ancilla-Bits.html`.

[160] V. V. Shende and I. L. Markov, "On the cnot-cost of toffoli gates," *arXiv preprint arXiv:0803.2316*, 2008.

[161] E. Knill, "Approximation by quantum circuits," *quant-ph/9508006*, 1995.

[162] J. J. Vartiainen, M. Möttönen, and M. M. Salomaa, "Efficient decomposition of quantum gates," *Phys. Rev. Lett*, vol. 92, no. 17, p. 177902, 2004.

[163] M. Möttönen and J. J. Vartiainen, "Decompositions of general quantum gates," *quant-ph/0504100*, 2005.

[164] V. V. Shende, S. S. Bullock, and I. L. Markov, "Synthesis of quantum-logic circuits," *IEEE Trans. Comput.-Aided Des. Integr. Circuits Syst*, vol. 25, no. 6, pp. 1000–1010, 2006.

[165] A. Adedoyin, J. Ambrosiano, P. Anisimov, A. Bärtschi, W. Casper, G. Chennupati, C. Coffrin, H. Djidjev, D. Gunter, S. Karra, *et al.*, "Quantum algorithm implementations for beginners," *arXiv:1804.03719*, 2018.

[166] A. Y. Kitaev, "Quantum computations: algorithms and error correction," *Russ. Math. Surv*, vol. 52, p. 1191, 1997.

[167] V. Kliuchnikov, D. Maslov, and M. Mosca, "Fast and efficient exact synthesis of single qubit unitaries generated by Clifford and T gates," *Quantum Inf. Comput*, vol. 13, no. 7-8, pp. 607–630, 2013.

[168] V. Kliuchnikov, D. Maslov, and M. Mosca, "Asymptotically optimal approximation of single qubit unitaries by Clifford and T circuits using a constant number of ancillary qubits," *Phys. Rev. Lett*, vol. 110, no. 19, p. 190502, 2013.

[169] N. J. Ross and P. Selinger, "Optimal ancilla-free Clifford+T approximation of z-rotations," *Quantum Inf. Comput*, vol. 16, no. 11-12, pp. 901–953, 2016.

[170] V. V. Shende, S. S. Bullock, and I. L. Markov, "Synthesis of quantum logic circuits," in *Proceedings of the 2005 Asia and south Pacific design automation conference*, pp. 272–275, 2005.

[171] M. Plesch and Č. Brukner, "Quantum-state preparation with universal gate decompositions," *Phys. Rev. A*, vol. 83, no. 3, p. 032302, 2011.

[172] H. Buhrman, R. Cleve, J. Watrous, and R. De Wolf, "Quantum fingerprinting," *Phys. Rev. Lett*, vol. 87, p. 167902, 2001.

[173] L. Cincio, Y. Subaşı, A. T. Sornborger, and P. J. Coles, "Learning the quantum algorithm for state overlap," *New J. Phys*, vol. 20, no. 11, p. 113022, 2018.

[174] D. Aharonov, V. Jones, and Z. Landau, "A polynomial quantum algorithm for approximating the Jones polynomial," *Algorithmica*, vol. 55, pp. 395–421, 2009.

[175] D. Gottesman, *Stabilizer codes and quantum error correction.* PhD thesis, Caltech, 1997. arXiv:quant-ph/9705052.

[176] D. Gottesman, "The Heisenberg representation of quantum computers," *arXiv:quant-ph/9807006*, 1998.

[177] M. Nest, "Classical simulation of quantum computation, the gottesman-knill theorem, and slightly beyond," *arXiv preprint arXiv:0811.0898*, 2008.

[178] S. Aaronson and D. Gottesman, "Improved simulation of stabilizer circuits," *Physical Review A*, vol. 70, no. 5, p. 052328, 2004.

[179] M. Möttönen, J. J. Vartiainen, V. Bergholm, and M. M. Salomaa, "Quantum circuits for general multiqubit gates," *Phys. Rev. Lett*, vol. 93, no. 13, p. 130502, 2004.

[180] V. V. Shende and I. L. Markov, "Quantum circuits for incompletely specified two-qubit operators," *Quantum Inf. Comput*, vol. 5, pp. 49–57, 2005.

[181] M. Saeedi, M. Arabzadeh, M. S. Zamani, and M. Sedighi, "Block-based quantum-logic synthesis," *Quantum Inf. Comput*, vol. 11, pp. 262–277, 2010.

[182] Y. Nakajima, Y. Kawano, and H. Sekigawa, "A new algorithm for producing quantum circuits using KAK decompositions," *Quantum Inf. Comput*, vol. 6, pp. 67–80, 2006.

[183] R. Somma, G. Ortiz, E. Knill, and J. Gubernatis, "Quantum simulations of physics problems," *Int. J. Theor. Phys*, vol. 1, pp. 189–206, 2003.

[184] A. M. Gleason, "Measures on the closed subspaces of a Hilbert space," *J. Math. Mech*, pp. 885–893, 1957.

[185] M. G. A. Paris, "The modern tools of quantum mechanics," *Eur. Phys. J. Spec Top*, vol. 203, pp. 61–86, 2012.

[186] H.-P. Breuer and F. Petruccione, *The theory of open quantum systems*. Oxford University Press, 2002.

[187] J. Gemmer, M. Michel, and G. Mahler, *Quantum thermodynamics: emergence of thermodynamic behavior within composite quantum systems*. Springer, 2009.

[188] F. Schwabl, *Statistical mechanics*. Oxford University Press, 2002.

[189] U. Fano, "Description of states in quantum mechanics by density matrix and operator techniques," *Rev. Mod. Phys*, vol. 29, pp. 74–93, 1957.

[190] P. Busch, "Quantum states and generalized observables: a simple proof of Gleason's theorem," *Phys. Rev. Lett*, vol. 91, p. 120403, 2003.

[191] W. K. Wootters and B. D. Fields, "Optimal state-determination by mutually unbiased measurements," *Ann. Phys*, vol. 191, no. 2, pp. 363–381, 1989.

[192] T. Durt, B.-G. Englert, I. Bengtsson, and K. Życzkowski, "On mutually unbiased bases," *Int. J. Quant. Inf*, vol. 8, no. 04, pp. 535–640, 2010.

[193] E. Størmer, "Positive linear maps of operator algebras," *Acta Math*, vol. 110, pp. 233–278, 1963.

[194] M.-D. Choi, "Positive linear maps on C*-algebras," *Can. J. Math*, vol. 24, pp. 520–529, 1972.

[195] M.-D. Choi, "A Schwarz inequality for positive linear maps on C*-algebras," *Illinois J. Math*, vol. 18, pp. 565–574, 1974.

[196] E. B. Davies, *Quantum theory of open systems*. Academic Press, 1976.

[197] K. Kraus, A. Böhm, J. D. Dollard, and W. H. Wootters, *States, effects, and operations: fundamental notions of quantum theory*. Springer, 1983.

[198] A. Nayak and P. Sen, "Invertible quantum operations and perfect encryption of quantum states," *Quantum Inf. Comput*, vol. 7, pp. 103–110, 2007.

[199] A. Streltsov, H. Kampermann, S. Wölk, M. Gessner, and D. Bruß, "Maximal coherence and the resource theory of purity," *New J. Phys*, vol. 20, p. 053058, 2018.

[200] R. L. Frank and E. H. Lieb, "Monotonicity of a relative Rényi entropy," *J. Math. Phys*, vol. 54, p. 122201, 2013.

[201] G. Gour and M. M. Wilde, "Entropy of a quantum channel," *Phys. Rev. Res*, vol. 3, no. 2, p. 023096, 2021.

[202] S. Beigi, "Sandwiched Rényi divergence satisfies data processing inequality," *J. Math. Phys*, vol. 54, p. 122202, 2013.

[203] J. Johnson, C. Macklin, D. Slichter, R. Vijay, E. Weingarten, J. Clarke, and I. Siddiqi, "Heralded state preparation in a superconducting qubit," *Phys. Rev. Lett*, vol. 109, p. 050506, 2012.

[204] P. Magnard, P. Kurpiers, B. Royer, T. Walter, J.-C. Besse, S. Gasparinetti, M. Pechal, J. Heinsoo, S. Storz, *et al.*, "Fast and unconditional all-microwave reset of a superconducting qubit," *Phys. Rev. Lett*, vol. 121, p. 060502, 2018.

[205] M. Reed, *Entanglement and quantum error correction with superconducting qubits.* PhD thesis, Yale University, 2013. arXiv:1311.6759.

[206] S. Schlör, J. Lisenfeld, C. Müller, A. Bilmes, A. Schneider, D. P. Pappas, A. V. Ustinov, and M. Weides, "Correlating decoherence in transmon qubits: low frequency noise by single fluctuators," *Phys. Rev. Lett*, vol. 123, p. 190502, 2019.

[207] M. M. Wolf, *Partial transposition in quantum information theory.* PhD thesis, Technische Universität Braunschweig, 2003.

[208] M. M. Wolf, "Quantum channels & operations: guided tour." Lecture notes, url: `https://www-m5.ma.tum.de/foswiki/pub/M5/Allgemeines/MichaelWolf/QChannelLecture.pdf`, 2012.

[209] G. Benenti, G. Casati, and G. Strini, *Principles of quantum computation and information-volume I: basic concepts.* World scientific, 2004.

[210] W. F. Stinespring, "Positive functions on C*-algebras," *Proc. ACM*, vol. 6, pp. 211–211, 1955.

[211] M.-D. Choi, "Completely positive linear maps on complex matrices," *Linear Algebra Appl*, vol. 10, pp. 285–290, 1975.

[212] M. P. da Silva, O. Landon-Cardinal, and D. Poulin, "Practical characterization of quantum devices without tomography," *Phys. Rev. Lett*, vol. 107, no. 21, p. 210404, 2011.

[213] M. A. Nielsen, "A simple formula for the average gate fidelity of a quantum dynamical operation," *Physics Letters A*, vol. 303, no. 4, pp. 249 – 252, 2002.

[214] K.-E. Hellwig and K. Kraus, "Pure operations and measurements I," *Comm. Math. Phys*, vol. 11, no. 3, pp. 214–220, 1969.

[215] K.-E. Hellwig and K. Kraus, "Operations and measurements II," *Comm. Math. Phys*, vol. 16, no. 2, pp. 142–147, 1970.

[216] G. Lindblad, "On the generators of quantum dynamical semigroups," *Comm. Math. Phys*, vol. 48, no. 2, pp. 119–130, 1976.

[217] A. Royer, "Reduced dynamics with initial correlations, and time-dependent environment and Hamiltonians," *Phys. Rev. Lett*, vol. 77, no. 16, p. 3272, 1996.

[218] A. Shaji and E. C. G. Sudarshan, "Who's afraid of not completely positive maps?," *Phys. Lett. A*, vol. 341, no. 1-4, pp. 48–54, 2005.

[219] M. H. Stone, "On one-parameter unitary groups in Hilbert space," *Ann. Math*, vol. 33, no. 3, pp. 643–648, 1932.

[220] V. Gorini, A. Frigerio, M. Verri, A. Kossakowski, and E. Sudarshan, "Properties of quantum Markovian master equations," *Rev. Mod. Phys*, vol. 13, no. 2, pp. 149–

173, 1978.

[221] V. Gorini, A. Kossakowski, and E. C. G. Sudarshan, “Completely positive dynamical semigroups of n-level systems,” *J. Math. Phys*, vol. 17, no. 5, pp. 821–825, 1976.

[222] F. Haake, “Statistical treatment of open systems by generalized master equations,” in *Springer tracts in modern physics*, pp. 98–168, Springer, 1973.

[223] A. G. Redfield, “On the theory of relaxation processes,” *IBM J. Res. Dev*, vol. 1, no. 1, pp. 19–31, 1957.

[224] F. Bao, H. Deng, D. Ding, R. Gao, X. Gao, C. Huang, X. Jiang, H.-S. Ku, Z. Li, X. Ma, *et al.*, “Fluxonium: an alternative qubit platform for high-fidelity operations,” *Phys. Rev. Lett*, vol. 129, no. 1, p. 010502, 2022.

[225] H. Haken, *Handbuch der Physik.* Springer, 1970.

[226] W. H. Louisell, *Quantum statistical properties of radiation.* John Wiley & Sons, 1973.

[227] C. Cohen-Tannoudji, J. Dupont-Roc, and G. Grynberg, *Atom-photon interactions: basic processes and applications.* John Wiley & Sons, 1998.

[228] H. J. Carmichael, *Statistical methods in quantum optics 1: master equations and Fokker-Planck equations.* Springer, 2013.

[229] S. Gustavsson, F. Yan, G. Catelani, J. Bylander, A. Kamal, J. Birenbaum, D. Hover, D. Rosenberg, G. Samach, A. P. Sears, *et al.*, “Suppressing relaxation in superconducting qubits by quasiparticle pumping,” *Science*, vol. 354, no. 6319, pp. 1573–1577, 2016.

[230] A. A. Clerk, M. H. Devoret, S. M. Girvin, F. Marquardt, and R. J. Schoelkopf, “Introduction to quantum noise, measurement, and amplification,” *Rev. Mod. Phys*, vol. 82, no. 2, p. 1155, 2010.

[231] R. Schoelkopf, A. Clerk, S. Girvin, K. Lehnert, and M. Devoret, “Qubits as spectrometers of quantum noise,” in *Quantum noise in mesoscopic physics*, pp. 175–203, Springer, 2003.

[232] A. O. Caldeira and A. J. Leggett, “Quantum tunnelling in a dissipative system,” *Ann. Phys*, vol. 149, no. 2, pp. 374–456, 1983.

[233] D. Marion and K. Wüthrich, “Application of phase sensitive two-dimensional correlated spectroscopy (COSY) for measurements of ^{1}H-^{1}H spin-spin coupling constants in proteins,” in *NMR in structural biology: A collection of papers by Kurt Wüthrich*, pp. 114–121, World Scientific, 1995.

[234] M. H. Levitt, *Spin dynamics: basics of nuclear magnetic resonance.* John Wiley & Sons, 2013.

[235] J. Bylander, S. Gustavsson, F. Yan, F. Yoshihara, K. Harrabi, G. Fitch, D. G. Cory, Y. Nakamura, J.-S. Tsai, and W. D. Oliver, “Noise spectroscopy through dynamical decoupling with a superconducting flux qubit,” *Nat. Phys*, vol. 7, no. 7, pp. 565–570, 2011.

[236] F. Yan, S. Gustavsson, A. Kamal, J. Birenbaum, A. P. Sears, D. Hover, T. J. Gudmundsen, D. Rosenberg, G. Samach, *et al.*, “The flux qubit revisited to enhance coherence and reproducibility,” *Nat. Commun*, vol. 7, no. 1, pp. 1–9, 2016.

[237] M. J. Biercuk, H. Uys, A. P. VanDevender, N. Shiga, W. M. Itano, and J. J. Bollinger, “Optimized dynamical decoupling in a model quantum memory,” *Nature*, vol. 458, no. 7241, pp. 996–1000, 2009.

[238] R. Aguado and L. P. Kouwenhoven, "Double quantum dots as detectors of high-frequency quantum noise in mesoscopic conductors," *Phys. Rev. Lett*, vol. 84, no. 9, p. 1986, 2000.
[239] U. Gavish, Y. Levinson, and Y. Imry, "Detection of quantum noise," *Phys. Rev. B*, vol. 62, no. 16, p. R10637, 2000.
[240] W. Magnus, "On the exponential solution of differential equations for a linear operator," *Commun. Pure Appl. Math*, vol. 7, no. 4, pp. 649–673, 1954.
[241] E. Schrödinger, "The present status of quantum mechanics," *Naturwissenschaften*, vol. 23, pp. 1–26, 1935.
[242] A. Einstein, B. Podolsky, and N. Rosen, "Can quantum-mechanical description of physical reality be considered complete?," *Phys. Rev*, vol. 47, p. 777, 1935.
[243] J. S. Bell, "On the Einstein-Podolsky-Rosen paradox," *Phys. Phys. Fiz*, vol. 1, pp. 195–200, 1964.
[244] C. A. Kocher and E. D. Commins, "Polarization correlation of photons emitted in an atomic cascade," *Phys. Rev. Lett*, vol. 18, p. 575, 1967.
[245] S. J. Freedman and J. F. Clauser, "Experimental test of local hidden-variable theories," *Phys. Rev. Lett*, vol. 28, p. 938, 1972.
[246] A. Aspect, P. Grangier, and G. Roger, "Experimental tests of realistic local theories via Bell's theorem," *Phys. Rev. Lett*, vol. 47, p. 460, 1981.
[247] A. Aspect, P. Grangier, and G. Roger, "Experimental realization of the Einstein-Podolsky-Rosen-Bohm Gedankenexperiment: a new violation of Bell's inequalities," *Phys. Rev. Lett*, vol. 49, p. 91, 1982.
[248] F. A. Bovino, P. Varisco, A. M. Colla, G. Castagnoli, G. Di Giuseppe, and A. V. Sergienko, "Effective fiber-coupling of entangled photons for quantum communication," *Opt. Commun*, vol. 227, pp. 343–348, 2003.
[249] L. Steffen, Y. Salathe, M. Oppliger, P. Kurpiers, M. Baur, C. Lang, C. Eichler, G. Puebla-Hellmann, A. Fedorov, and A. Wallraff, "Deterministic quantum teleportation with feed-forward in a solid state system," *Nature*, vol. 500, pp. 319–322, 2013.
[250] J.-G. Ren, P. Xu, H.-L. Yong, L. Zhang, S.-K. Liao, J. Yin, W.-Y. Liu, W.-Q. Cai, M. Yang, L. Li, *et al.*, "Ground-to-satellite quantum teleportation," *Nature*, vol. 549, pp. 70–73, 2017.
[251] W. K. Wootters, "Entanglement of formation and concurrence," *Quantum Inf. Comput*, vol. 1, pp. 27–44, 2001.
[252] R. Horodecki, P. Horodecki, M. Horodecki, and K. Horodecki, "Quantum entanglement," *Rev. Mod. Phys*, vol. 81, p. 865, 2009.
[253] J. Eisert, M. Cramer, and M. B. Plenio, "Colloquium: Area laws for the entanglement entropy," *Rev. Mod. Phys*, vol. 82, no. 1, p. 277, 2010.
[254] M. B. Plenio and S. S. Virmani, "An introduction to entanglement theory," in *Quantum Inf. Coh*, pp. 173–209, Springer, 2014.
[255] K. Eckert, J. Schliemann, D. Bruß, and M. Lewenstein, "Quantum correlations in systems of indistinguishable particles," *Ann. Phys*, vol. 299, pp. 88–127, 2002.
[256] P. Zanardi, "Quantum entanglement in fermionic lattices," *Phys. Rev. A*, vol. 65, p. 042101, 2002.

[257] K. Edamatsu, "Entangled photons: generation, observation, and characterization," *Jpn. Appl. Phys. Lett*, vol. 46, p. 7175, 2007.

[258] Y. H. Shih and C. O. Alley, "New type of Einstein-Podolsky-Rosen-Bohm experiment using pairs of light quanta produced by optical parametric down conversion," *Phys. Rev. Lett*, vol. 61, p. 2921, 1988.

[259] Z. Y. Ou and L. Mandel, "Violation of Bell's inequality and classical probability in a two-photon correlation experiment," *Phys. Rev. Lett*, vol. 61, p. 50, 1988.

[260] P. G. Kwiat, K. Mattle, H. Weinfurter, A. Zeilinger, A. V. Sergienko, and Y. Shih, "New high-intensity source of polarization-entangled photon pairs," *Phys. Rev. Lett*, vol. 75, p. 4337, 1995.

[261] P. G. Kwiat, E. Waks, A. G. White, I. Appelbaum, and P. H. Eberhard, "Ultrabright source of polarization-entangled photons," *Phys. Rev. A*, vol. 60, p. R773, 1999.

[262] E. Schmidt, "Zur Theorie der linearen und nicht linearen Integralgleichungen Zweite Abhandlung," *Math. Ann*, vol. 64, pp. 161–174, 1907.

[263] C. H. Bennett and S. J. Wiesner, "Communication via one-and two-particle operators on Einstein-Podolsky-Rosen states," *Phys. Rev. Lett*, vol. 69, p. 2881, 1992.

[264] T. Schaetz, M. D. Barrett, D. Leibfried, J. Chiaverini, J. Britton, W. M. Itano, J. D. Jost, C. Langer, and D. J. Wineland, "Quantum dense coding with atomic qubits," *Phys. Rev. Lett*, vol. 93, p. 040505, 2004.

[265] B. P. Williams, R. J. Sadlier, and T. S. Humble, "Superdense coding over optical fiber links with complete Bell-state measurements," *Phys. Rev. Lett*, vol. 118, p. 050501, 2017.

[266] D. Wei, X. Yang, J. Luo, X. Sun, X. Zeng, and M. Liu, "NMR experimental implementation of three-parties quantum superdense coding," *Chi. Sci. Bull*, vol. 49, pp. 423–426, 2004.

[267] C. H. Bennett, G. Brassard, C. Crépeau, R. Jozsa, A. Peres, and W. K. Wootters, "Teleporting an unknown quantum state via dual classical and Einstein-Podolsky-Rosen channels," *Phys. Rev. Lett*, vol. 70, p. 1895, 1993.

[268] D. Bouwmeester, J.-W. Pan, K. Mattle, M. Eibl, H. Weinfurter, and A. Zeilinger, "Experimental quantum teleportation," *Nature*, vol. 390, pp. 575–579, 1997.

[269] D. Boschi, S. Branca, F. De Martini, L. Hardy, and S. Popescu, "Experimental realization of teleporting an unknown pure quantum state via dual classical and Einstein-Podolsky-Rosen channels," *Phys. Rev. Lett*, vol. 80, p. 1121, 1998.

[270] S. L. Braunstein and H. J. Kimble, "Teleportation of continuous quantum variables," *Phys. Rev. Lett*, vol. 80, p. 869, 1998.

[271] A. Furusawa, J. L. Sørensen, S. L. Braunstein, C. A. Fuchs, H. J. Kimble, and E. S. Polzik, "Unconditional quantum teleportation," *Science*, vol. 282, pp. 706–709, 1998.

[272] S. Massar and S. Popescu, "Optimal extraction of information from finite quantum ensembles," *Phys. Rev. Lett*, vol. 74, pp. 1259–1263, 1995.

[273] B. Yurke and D. Stoler, "Einstein-Podolsky-Rosen effects from independent particle sources," *Phys. Rev. Lett*, vol. 68, p. 1251, 1992.

[274] M. Żukowski, A. Zeilinger, M. A. Horne, and A. K. Ekert, "Event-ready-detectors Bell experiment via entanglement swapping," *Phys. Rev. Lett*, vol. 71, p. 4287, 1993.

[275] S. Bose, V. Vedral, and P. L. Knight, "Multiparticle generalization of entanglement swapping," *Phys. Rev. A*, vol. 57, p. 822, 1998.

[276] J.-W. Pan, D. Bouwmeester, H. Weinfurter, and A. Zeilinger, "Experimental entanglement swapping: entangling photons that never interacted," *Phys. Rev. Lett*, vol. 80, p. 3891, 1998.

[277] H. Bernien, B. Hensen, W. Pfaff, G. Koolstra, M. S. Blok, L. Robledo, T. Taminiau, M. Markham, D. J. Twitchen, L. Childress, *et al.*, "Heralded entanglement between solid-state qubits separated by 3 m," *Nature*, vol. 497, pp. 86–90, 2013.

[278] F. B. Basset, M. B. Rota, C. Schimpf, D. Tedeschi, K. D. Zeuner, S. F. C. da Silva, M. Reindl, V. Zwiller, K. D. Jöns, A. Rastelli, *et al.*, "Entanglement swapping with photons generated on demand by a quantum dot," *Phys. Rev. Lett*, vol. 123, p. 160501, 2019.

[279] J. F. Clauser, M. A. Horne, A. Shimony, and R. A. Holt, "Proposed experiment to test local hidden-variable theories," *Phys. Rev. Lett*, vol. 23, p. 880, 1969.

[280] W. Tittel, J. Brendel, H. Zbinden, and N. Gisin, "Violation of Bell inequalities by photons more than 10 km apart," *Phys. Rev. Lett*, vol. 81, p. 3563, 1998.

[281] R. Ursin, F. Tiefenbacher, T. Schmitt-Manderbach, H. Weier, T. Scheidl, M. Lindenthal, B. Blauensteiner, T. Jennewein, J. Perdigues, P. Trojek, B. Ömer, M. Fürst, M. Meyenburg, J. Rarity, Z. Sodnik, C. Barbieri, H. Weinfurter, and A. Zeilinger, "Entanglement-based quantum communication over 144 km," *Nat. Phys*, vol. 3, pp. 481–486, 2007.

[282] B. Hensen, H. Bernien, A. E. Dréau, A. Reiserer, N. Kalb, M. S. Blok, J. Ruitenberg, R. F. Vermeulen, R. N. Schouten, *et al.*, "Loophole-free Bell inequality violation using electron spins separated by 1.3 km," *Nature*, vol. 526, pp. 682–686, 2015.

[283] B. S. Cirel'son, "Quantum generalizations of Bell's inequality," *Lett. Math. Phys*, vol. 4, no. 2, pp. 93–100, 1980.

[284] N. Gisin and A. Peres, "Maximal violation of Bell's inequality for arbitrarily large spin," *Phys. Rev. A*, vol. 162, pp. 15–17, 1992.

[285] S. Popescu and D. Rohrlich, "Generic quantum nonlocality," *Phys. Lett. A*, vol. 166, pp. 293–297, 1992.

[286] R. F. Werner, "Quantum states with Einstein-Podolsky-Rosen correlations admitting a hidden-variable model," *Phys. Rev. A*, vol. 40, p. 4277, 1989.

[287] A. Peres, "Separability criterion for density matrices," *Phys. Rev. Lett*, vol. 77, p. 1413, 1996.

[288] G. Alber, T. Beth, M. Horodecki, P. Horodecki, R. Horodecki, M. Rötteler, H. Weinfurter, R. Werner, and A. Zeilinger, *Quantum information: an introduction to basic theoretical concepts and experiments*. Springer, 2003.

[289] S. L. Woronowicz, "Positive maps of low dimensional matrix algebras," *Rep. Math. Phys*, vol. 10, pp. 165–183, 1976.

[290] M. Horodecki, P. Horodecki, and R. Horodecki, "Separability of n-particle mixed states: necessary and sufficient conditions in terms of linear maps," *Phys. Lett. A*, vol. 283, pp. 1–7, 2001.

[291] R. Horodecki, P. Horodecki, and M. Horodecki, "Violating Bell's inequality by mixed spin-12 states: necessary and sufficient condition," *Phys. Lett. A*, vol. 200, pp. 340–344, 1995.

[292] M. J. Donald, M. Horodecki, and O. Rudolph, "The uniqueness theorem for entanglement measures," *J. Math. Phys*, vol. 43, pp. 4252–4272, 2002.

[293] C. H. Bennett, D. P. DiVincenzo, C. A. Fuchs, T. Mor, E. Rains, P. W. Shor, J. A. Smolin, and W. K. Wootters, "Quantum nonlocality without entanglement," *Phys. Rev. A*, vol. 59, p. 1070, 1999.

[294] L. Amico, R. Fazio, A. Osterloh, and V. Vedral, "Entanglement in many-body systems," *Rev. Mod. Phys*, vol. 80, no. 2, p. 517, 2008.

[295] E. H. Lieb and D. W. Robinson, "The finite group velocity of quantum spin systems," in *Statistical mechanics*, pp. 425–431, Springer, 1972.

[296] M. B. Hastings and T. Koma, "Spectral gap and exponential decay of correlations," *Comm. Math. Phys*, vol. 265, no. 3, pp. 781–804, 2006.

[297] M. B. Hastings, "An area law for one-dimensional quantum systems," *J. Stat. Mech*, vol. 2007, no. 08, p. P08024, 2007.

[298] D. Aharonov, I. Arad, Z. Landau, and U. Vazirani, "The 1D area law and the complexity of quantum states: a combinatorial approach," in *Proceedings of the 52nd annual symposium on foundations of computer science (FOCS)*, pp. 324–333, IEEE, 2011.

[299] F. G. Brandão and M. Horodecki, "Exponential decay of correlations implies area law," *Comm. Math. Phys*, vol. 333, no. 2, pp. 761–798, 2015.

[300] G. Vidal, "Efficient simulation of one-dimensional quantum many-body systems," *Phys. Rev. Lett*, vol. 93, no. 4, p. 040502, 2004.

[301] S. R. White, "Density matrix formulation for quantum renormalization groups," *Phys. Rev. Lett*, vol. 69, no. 19, p. 2863, 1992.

[302] G. K.-L. Chan and S. Sharma, "The density matrix renormalization group in quantum chemistry," *Annu. Rev. Phys. Chem*, vol. 62, pp. 465–481, 2011.

[303] U. Schollwöck, "The density-matrix renormalization group in the age of matrix product states," *Ann. Phys*, vol. 326, no. 1, pp. 96–192, 2011.

[304] R. Orús, "A practical introduction to tensor networks: matrix product states and projected entangled pair states," *Ann. Phys*, vol. 349, pp. 117–158, 2014.

[305] R. Orús, "Advances on tensor network theory: symmetries, fermions, entanglement, and holography," *Eur. Phys. J. B*, vol. 87, no. 11, p. 280, 2014.

[306] J. Biamonte and V. Bergholm, "Tensor networks in a nutshell," *arXiv:1708.00006*, 2017.

[307] P. Benioff, "Quantum mechanical Hamiltonian models of Turing machines," *J. Stat. Phys*, vol. 29, no. 3, pp. 515–546, 1982.

[308] Y. Manin, "Computable and uncomputable," *Sovetskoe Radio, Moscow*, vol. 128, 1980.

[309] Y. I. Manin, "Classical computing, quantum computing, and Shor's factoring algorithm," *Astérisque, société mathématique de France*, vol. 266, pp. 375–404, 2000.

[310] D. J. Bernstein and T. Lange, "Post-quantum cryptography," *Nature*, vol. 549, no. 7671, pp. 188–194, 2017.

[311] D. Moody *et al.*, *Status report on the second round of the NIST post-quantum cryptography standardization process.* NIST, 2019.

[312] M. A. Nielsen and I. L. Chuang, *Quantum Computation and Quantum Information.* Cambridge University Press, 2000.

[313] G. Brassard, P. Hoyer, M. Mosca, and A. Tapp, "Quantum amplitude amplification and estimation," in *Quantum Computation and Quantum Information* (S. J. Lomonaco, ed.), pp. 53–74, American Mathematical Society, 2002.

[314] T. Roy, L. Jiang, and D. I. Schuster, "Deterministic Grover search with a restricted oracle," *Phys. Rev. Res*, vol. 4, no. 2, p. L022013, 2022.

[315] S. Aaronson, "The limits of quantum computers," *Sci. Am*, vol. 298, no. 3, pp. 62–69, 2008.

[316] D. S. Abrams and S. Lloyd, "Nonlinear quantum mechanics implies polynomial-time solution for NP-complete and #P problems," *Phys. Rev. Lett*, vol. 81, no. 18, p. 3992, 1998.

[317] C. Zalka, "Grover's quantum searching algorithm is optimal," *Phys. Rev. A*, vol. 60, no. 4, p. 2746, 1999.

[318] P. W. Shor, "The early days of quantum computation," *arXiv:2208.09964*, 2022.

[319] D. Coppersmith, "An approximate Fourier transform useful in quantum factoring," *quant-ph/0201067*, 2002.

[320] G. H. Hardy, E. M. Wright, *et al.*, *An introduction to the theory of numbers.* Oxford University Press, 1979.

[321] C. Pomerance, "A tale of two sieves," in *Notices Amer. Math. Soc*, Am. Math. Soc, 1996.

[322] L. M. Vandersypen, M. Steffen, G. Breyta, C. S. Yannoni, M. H. Sherwood, and I. L. Chuang, "Experimental realization of Shor's quantum factoring algorithm using nuclear magnetic resonance," *Nature*, vol. 414, no. 6866, pp. 883–887, 2001.

[323] I. L. Markov and M. Saeedi, "Constant-optimized quantum circuits for modular multiplication and exponentiation," *Quantum Inf. Comput*, vol. 12, no. 5-6, pp. 361–394, 2012.

[324] E. G. Rieffel and W. H. Polak, *Quantum computing: A gentle introduction.* MIT Press, 2011.

[325] R. J. Lipton and K. W. Regan, *Quantum Algorithms via Linear Algebra: A Primer.* MIT Press, 2014.

[326] N. S. Yanofsky and M. A. Mannucci, *Quantum computing for computer scientists.* Cambridge University Press, 2008.

[327] M. Mosca, "Quantum algorithms," *arXiv:0808.0369*, 2008.

[328] A. Montanaro, "Quantum algorithms: an overview," *npj Quantum Inf*, vol. 2, no. 1, pp. 1–8, 2016.

[329] A. M. Childs and W. Van Dam, "Quantum algorithms for algebraic problems," *Rev. Mod. Phys*, vol. 82, no. 1, p. 1, 2010.

[330] M. Santha, "Quantum walk based search algorithms," in *International Conference on Theory and Applications of Models of Computation*, pp. 31–46, Springer, 2008.

[331] I. Kassal, S. P. Jordan, P. J. Love, M. Mohseni, and A. Aspuru-Guzik, "Polynomial-time quantum algorithm for the simulation of chemical dynamics," *Proc. Natl. Acad. Sci*, vol. 105, no. 48, pp. 18681–18686, 2008.

[332] G. Ortiz, J. E. Gubernatis, E. Knill, and R. Laflamme, "Quantum algorithms for fermionic simulations," *Phys. Rev. A*, vol. 64, no. 2, p. 022319, 2001.

[333] R. Somma, G. Ortiz, J. E. Gubernatis, E. Knill, and R. Laflamme, "Simulating physical phenomena by quantum networks," *Phys. Rev. A*, vol. 65, no. 4, p. 042323,

2002.

[334] A. M. Childs, *Quantum information processing in continuous time.* PhD thesis, Massachusetts Institute of Technology, 2004.

[335] A. M. Childs and Y. Su, "Nearly optimal lattice simulation by product formulas," *Phys. Rev. Lett*, vol. 123, no. 5, p. 050503, 2019.

[336] D. W. Berry, G. Ahokas, R. Cleve, and B. C. Sanders, "Efficient quantum algorithms for simulating sparse Hamiltonians," *Comm. Math. Phys*, vol. 270, no. 2, pp. 359–371, 2007.

[337] A. M. Childs and R. Kothari, "Limitations on the simulation of non-sparse Hamiltonians," *Quantum Inf. Comput*, vol. 10, 2009.

[338] J. M. Martyn, Z. M. Rossi, A. K. Tan, and I. L. Chuang, "Grand unification of quantum algorithms," *Phys. Rev. X Quantum*, vol. 2, no. 4, p. 040203, 2021.

[339] H. F. Trotter, "On the product of semigroups of operators," *Proc. Am. Math. Soc*, vol. 10, no. 4, pp. 545–551, 1959.

[340] M. Suzuki, "General theory of fractal path integrals with applications to many-body theories and statistical physics," *J. Math. Phys*, vol. 32, no. 2, pp. 400–407, 1991.

[341] A. M. Childs, A. Ostrander, and Y. Su, "Faster quantum simulation by randomization," *Quantum*, vol. 3, p. 182, 2019.

[342] A. M. Childs, Y. Su, M. C. Tran, N. Wiebe, and S. Zhu, "Theory of Trotter error with commutator scaling," *Phys. Rev. X*, vol. 11, no. 1, p. 011020, 2021.

[343] D. Aharonov and A. Ta-Shma, "Adiabatic quantum state generation and statistical zero knowledge," in *Proceedings of the 35th annual ACM symposium on theory of computing (STOC)*, pp. 20–29, 2003.

[344] I. Kassal, J. D. Whitfield, A. Perdomo-Ortiz, M.-H. Yung, and A. Aspuru-Guzik, "Simulating chemistry using quantum computers," *Annu. Rev. Phys. Chem*, vol. 62, pp. 185–207, 2011.

[345] A. Szabo and N. S. Ostlund, *Modern quantum chemistry: introduction to advanced electronic structure theory.* Courier Dover Publications, 2012.

[346] D. W. Berry, A. M. Childs, R. Cleve, R. Kothari, and R. D. Somma, "Exponential improvement in precision for simulating sparse Hamiltonians," in *Proceedings of the 46th annual ACM symposium on theory of computing (STOC)*, pp. 283–292, 2014.

[347] A. M. Childs, R. Cleve, E. Deotto, E. Farhi, S. Gutmann, and D. A. Spielman, "Exponential algorithmic speedup by a quantum walk," in *Proceedings of the 35th annual ACM symposium on theory of computing (STOC)*, pp. 59–68, 2003.

[348] A. M. Childs and N. Wiebe, "Hamiltonian simulation using linear combinations of unitary operations," *Quantum Inf. Comput*, vol. 12, no. 901, 2012.

[349] R. Kothari, *Efficient algorithms in quantum query complexity.* PhD thesis, University of Waterloo, 2014.

[350] A. Gilyén, Y. Su, G. H. Low, and N. Wiebe, "Quantum singular value transformation and beyond: exponential improvements for quantum matrix arithmetics," in *Proceedings of the 51st annual ACM SIGACT symposium on theory of computing*, pp. 193–204, 2019.

[351] A. M. Childs, D. Maslov, Y. Nam, N. J. Ross, and Y. Su, "Toward the first quantum simulation with quantum speedup," *Proc. Natl. Acad. Sci*, vol. 115, no. 38,

pp. 9456–9461, 2018.
[352] G. H. Low, T. J. Yoder, and I. L. Chuang, "Methodology of resonant equiangular composite quantum gates," *Phys. Rev. X*, vol. 6, no. 4, p. 041067, 2016.
[353] G. H. Low and I. L. Chuang, "Optimal Hamiltonian simulation by quantum signal processing," *Phys. Rev. Lett*, vol. 118, no. 1, p. 010501, 2017.
[354] J. Wang, Y. Dong, and L. Lin, "On the energy landscape of symmetric quantum signal processing," *Quantum*, vol. 6, p. 850, 2022.
[355] Y. Dong, X. Meng, K. B. Whaley, and L. Lin, "Efficient phase-factor evaluation in quantum signal processing," *Phys. Rev. A*, vol. 103, no. 4, p. 042419, 2021.
[356] L. Lin, "Lecture notes on quantum algorithms for scientific computation," *arXiv preprint arXiv:2201.08309*, 2022.
[357] I. M. Georgescu, S. Ashhab, and F. Nori, "Quantum simulation," *Rev. Mod. Phys*, vol. 86, no. 1, p. 153, 2014.
[358] Y. Cao, J. Romero, J. P. Olson, M. Degroote, P. D. Johnson, M. Kieferová, I. D. Kivlichan, T. Menke, B. Peropadre, N. P. Sawaya, *et al.*, "Quantum chemistry in the age of quantum computing," *Chem. Rev*, vol. 119, no. 19, pp. 10856–10915, 2019.
[359] B. Bauer, S. Bravyi, M. Motta, and G. K.-L. Chan, "Quantum algorithms for quantum chemistry and quantum materials science," *Chem. Rev*, vol. 120, no. 22, pp. 12685–12717, 2020.
[360] S. Wiesner, "Simulations of many-body quantum systems by a quantum computer," *quant-ph/9603028*, 1996.
[361] C. Zalka, "Efficient simulation of quantum systems by quantum computers," *Proc. R. Soc. Lond*, vol. 454, pp. 313–322, 1998.
[362] J. L. Dodd, M. A. Nielsen, M. J. Bremner, and R. T. Thew, "Universal quantum computation and simulation using any entangling Hamiltonian and local unitaries," *Phys. Rev. A*, vol. 65, no. 4, p. 040301, 2002.
[363] M. A. Nielsen, M. J. Bremner, J. L. Dodd, A. M. Childs, and C. M. Dawson, "Universal simulation of Hamiltonian dynamics for quantum systems with finite-dimensional state spaces," *Phys. Rev. A*, vol. 66, no. 2, p. 022317, 2002.
[364] N. Wiebe, D. Berry, P. Høyer, and B. C. Sanders, "Higher order decompositions of ordered operator exponentials," *J. Phys. A*, vol. 43, no. 6, p. 065203, 2010.
[365] J. Haah, M. B. Hastings, R. Kothari, and G. H. Low, "Quantum algorithm for simulating real time evolution of lattice Hamiltonians," *SIAM J. Comput*, pp. 250–284, 2021.
[366] M. Motta, E. Ye, J. R. McClean, Z. Li, A. J. Minnich, R. Babbush, and G. K. Chan, "Low rank representations for quantum simulation of electronic structure," *npj Quantum Inf*, vol. 7, no. 1, pp. 1–7, 2021.
[367] I. D. Kivlichan, J. McClean, N. Wiebe, C. Gidney, A. Aspuru-Guzik, G. K.-L. Chan, and R. Babbush, "Quantum simulation of electronic structure with linear depth and connectivity," *Phys. Rev. Lett*, vol. 120, no. 11, p. 110501, 2018.
[368] H. Yoshida, "Construction of higher-order symplectic integrators," *Phys. Lett. A*, vol. 150, no. 5-7, pp. 262–268, 1990.
[369] G. Benenti and G. Strini, "Quantum simulation of the single-particle Schrödinger equation," *Am. J. Phys*, vol. 76, no. 7, pp. 657–662, 2008.

[370] R. D. Somma, "Quantum simulations of one-dimensional quantum systems," *arXiv:1503.06319*, 2015.

[371] J. D. Whitfield, J. Biamonte, and A. Aspuru-Guzik, "Simulation of electronic structure Hamiltonians using quantum computers," *Mol. Phys*, vol. 109, no. 5, pp. 735–750, 2011.

[372] M. B. Hastings, D. Wecker, B. Bauer, and M. Troyer, "Improving quantum algorithms for quantum chemistry," *Quantum Inf. Comput*, vol. 15, no. 1-21, 2015.

[373] E. Campbell, "Random compiler for fast Hamiltonian simulation," *Phys. Rev. Lett*, vol. 123, no. 7, p. 070503, 2019.

[374] A. M. Childs, "On the relationship between continuous-and discrete-time quantum walk," *Comm. Math. Phys*, vol. 294, no. 2, pp. 581–603, 2010.

[375] D. W. Berry, A. M. Childs, and R. Kothari, "Hamiltonian simulation with nearly optimal dependence on all parameters," in *Proceedings of the 56th annual symposium on foundations of computer science (FOCS)*, pp. 792–809, IEEE, 2015.

[376] G. H. Low and N. Wiebe, "Hamiltonian simulation in the interaction picture," *arXiv:1805.00675*, 2018.

[377] D. Cremer, "Møller-plesset perturbation theory: from small molecule methods to methods for thousands of atoms," *WIREs Comput. Mol. Sci*, vol. 1, no. 4, pp. 509–530, 2011.

[378] R. J. Bartlett and M. Musiał, "Coupled-cluster theory in quantum chemistry," *Rev. Mod. Phys*, vol. 79, no. 1, p. 291, 2007.

[379] S. Wouters and D. Van Neck, "The density matrix renormalization group for *ab initio* quantum chemistry," *Eur. Phys. J. D*, vol. 68, no. 9, p. 272, 2014.

[380] M. Motta, D. M. Ceperley, G. K.-L. Chan, J. A. Gomez, E. Gull, S. Guo, C. A. Jiménez-Hoyos, T. N. Lan, *et al.*, "Towards the solution of the many-electron problem in real materials: equation of state of the hydrogen chain with state-of-the-art many-body methods," *Phys. Rev. X*, vol. 7, no. 3, p. 031059, 2017.

[381] A. Y. Kitaev, "Quantum measurements and the Abelian stabilizer problem," *quant-ph/9511026*, 1995.

[382] R. Cleve, A. Ekert, C. Macchiavello, and M. Mosca, "Quantum algorithms revisited," *Proc. R. Soc. Lond*, vol. 454, no. 1969, pp. 339–354, 1998.

[383] A. Aspuru-Guzik, A. D. Dutoi, P. J. Love, and M. Head-Gordon, "Simulated quantum computation of molecular energies," *Science*, vol. 309, no. 5741, pp. 1704–1707, 2005.

[384] S. Lee, J. Lee, H. Zhai, Y. Tong, A. M. Dalzell, A. Kumar, P. Helms, J. Gray, Z.-H. Cui, W. Liu, *et al.*, "Evaluating the evidence for exponential quantum advantage in ground-state quantum chemistry," *Nature Communications*, vol. 14, no. 1, p. 1952, 2023.

[385] M. Dobšíček, G. Johansson, V. Shumeiko, and G. Wendin, "Arbitrary accuracy iterative quantum phase estimation algorithm using a single ancillary qubit: a two-qubit benchmark," *Phys. Rev. A*, vol. 76, no. 3, p. 030306, 2007.

[386] S. P. Jordan, *Quantum computation beyond the circuit model.* PhD thesis, MIT, 2008. arXiv:0809.2307.

[387] M. Born and V. Fock, "Beweis des adiabatensatzes," *Zeitschrift für Physik*, vol. 51, no. 3, pp. 165–180, 1928.

[388] T. Kato, "On the adiabatic theorem of quantum mechanics," *J. Phys. Soc. Jpn*, vol. 5, no. 6, pp. 435–439, 1950.

[389] A. Messiah, *Quantum mechanics: volume II.* North-Holland Publishing Company Amsterdam, 1962.

[390] J. E. Avron and A. Elgart, "Adiabatic theorem without a gap condition: two-level system coupled to quantized radiation field," *Phys. Rev. A*, vol. 58, no. 6, p. 4300, 1998.

[391] S. Teufel, *Adiabatic perturbation theory in quantum dynamics.* Springer, 2003.

[392] E. Farhi, J. Goldstone, S. Gutmann, J. Lapan, A. Lundgren, and D. Preda, "A quantum adiabatic evolution algorithm applied to random instances of an NP-complete problem," *Science*, vol. 292, no. 5516, pp. 472–475, 2001.

[393] R. Babbush, P. J. Love, and A. Aspuru-Guzik, "Adiabatic quantum simulation of quantum chemistry," *Sci. Rep*, vol. 4, p. 6603, 2014.

[394] L. Veis and J. Pittner, "Adiabatic state preparation study of methylene," *J. Chem. Phys*, vol. 140, no. 21, p. 214111, 2014.

[395] D. Nagaj and S. Mozes, "New construction for a QMA-complete three-local Hamiltonian," *J. Math. Phys*, vol. 48, no. 7, p. 072104, 2007.

[396] D. Aharonov, W. Van Dam, J. Kempe, Z. Landau, S. Lloyd, and O. Regev, "Adiabatic quantum computation is equivalent to standard quantum computation," *SIAM review*, vol. 50, no. 4, pp. 755–787, 2008.

[397] W. Van Dam, M. Mosca, and U. Vazirani, "How powerful is adiabatic quantum computation?," in *Proceedings 42nd symposium on foundations of computer science (FOCS)*, pp. 279–287, IEEE, 2001.

[398] S. Jansen, M.-B. Ruskai, and R. Seiler, "Bounds for the adiabatic approximation with applications to quantum computation," *J. Math. Phys*, vol. 48, no. 10, p. 102111, 2007.

[399] T. Helgaker, P. Jorgensen, and J. Olsen, *Molecular electronic-structure theory.* John Wiley & Sons, 2014.

[400] R. D. Johnson III, "NIST 101, computational chemistry comparison and benchmark database," tech. rep., National Institute of Standards and Technology, 2019.

[401] R. Eisberg and R. Resnick, *Quantum physics of atoms, molecules, solids, nuclei, and particles.* John Wiley & Sons, 1985.

[402] D. A. McQuarrie, *Quantum chemistry.* University Science Books, 2008.

[403] C. W. Scherr and R. E. Knight, "Two-electron atoms III: A sixth-order perturbation study of the $^1S_1$ ground state," *Rev. Mod. Phys*, vol. 35, no. 3, p. 436, 1963.

[404] J. A. Pople, "Nobel lecture: Quantum chemical models," *Rev. Mod. Phys*, vol. 71, no. 5, p. 1267, 1999.

[405] W. Ritz, "Über eine neue Methode zur Lösung gewisser Variationsprobleme der mathematischen Physik," *J. fur Reine Angew. Math*, vol. 135, pp. 1–61, 1909.

[406] W. J. Hehre, R. F. Stewart, and J. A. Pople, "Self-consistent molecular-orbital methods. I. Use of Gaussian expansions of Slater-type atomic orbitals," *J. Chem. Phys*, vol. 51, no. 6, pp. 2657–2664, 1969.

[407] S. F. Boys, "Electronic wave functions. I. A general method of calculation for the stationary states of any molecular system," *Proc. R. Soc. Lond*, vol. 200, no. 1063, pp. 542–554, 1950.

[408] S. B. Bravyi and A. Y. Kitaev, "Fermionic quantum computation," *Ann. Phys*, vol. 298, no. 1, pp. 210–226, 2002.

[409] J. T. Seeley, M. J. Richard, and P. J. Love, "The Bravyi-Kitaev transformation for quantum computation of electronic structure," *J. Chem. Phys*, vol. 137, no. 22, p. 224109, 2012.

[410] S. McArdle, S. Endo, A. Aspuru-Guzik, S. C. Benjamin, and X. Yuan, "Quantum computational chemistry," *Rev. Mod. Phys*, vol. 92, no. 1, p. 015003, 2020.

[411] E. Wigner and P. Jordan, "Über das paulische äquivalenzverbot," *Z. Phys*, vol. 47, p. 631, 1928.

[412] P. J. O'Malley, R. Babbush, I. D. Kivlichan, J. Romero, J. R. McClean, R. Barends, J. Kelly, P. Roushan, A. Tranter, N. Ding, *et al.*, "Scalable quantum simulation of molecular energies," *Phys. Rev. X*, vol. 6, no. 3, p. 031007, 2016.

[413] V. E. Elfving, B. W. Broer, M. Webber, J. Gavartin, M. D. Halls, K. P. Lorton, and A. Bochevarov, "How will quantum computers provide an industrially relevant computational advantage in quantum chemistry?," *arXiv preprint arXiv:2009.12472*, 2020.

[414] M. Reiher, N. Wiebe, K. M. Svore, D. Wecker, and M. Troyer, "Elucidating reaction mechanisms on quantum computers," *Proc. Natl. Acad. Sci*, vol. 114, no. 29, pp. 7555–7560, 2017.

[415] D. W. Berry, C. Gidney, M. Motta, J. R. McClean, and R. Babbush, "Qubitization of arbitrary basis quantum chemistry leveraging sparsity and low rank factorization," *Quantum*, vol. 3, p. 208, 2019.

[416] J. J. Goings, A. White, J. Lee, C. S. Tautermann, M. Degroote, C. Gidney, T. Shiozaki, R. Babbush, and N. C. Rubin, "Reliably assessing the electronic structure of cytochrome P450 on today's classical computers and tomorrow's quantum computers," *Proc. Natl. Acad. Sci*, vol. 119, no. 38, p. e2203533119, 2022.

[417] J. R. McClean, J. Romero, R. Babbush, and A. Aspuru-Guzik, "The theory of variational hybrid quantum-classical algorithms," *New J. Phys*, vol. 18, no. 2, p. 023023, 2016.

[418] M. Cerezo, A. Arrasmith, R. Babbush, S. C. Benjamin, S. Endo, K. Fujii, J. R. McClean, K. Mitarai, X. Yuan, L. Cincio, *et al.*, "Variational quantum algorithms," *Nat. Rev. Phys*, pp. 1–20, 2021.

[419] M. Motta and J. E. Rice, "Emerging quantum computing algorithms for quantum chemistry," *WIREs Comput. Mol. Sci*, vol. 12, no. 3, p. e1580, 2022.

[420] A. Fedorov, N. Gisin, S. Beloussov, and A. Lvovsky, "Quantum computing at the quantum advantage threshold: a down-to-business review," *arXiv:2203.17181*, 2022.

[421] R. B. Griffiths and C.-S. Niu, "Semiclassical Fourier transform for quantum computation," *Phys. Rev. Lett*, vol. 76, no. 17, p. 3228, 1996.

[422] B. P. Lanyon, J. D. Whitfield, G. G. Gillett, M. E. Goggin, M. P. Almeida, I. Kassal, J. D. Biamonte, M. Mohseni, B. J. Powell, M. Barbieri, *et al.*, "Towards quantum chemistry on a quantum computer," *Nat. Chem*, vol. 2, no. 2, pp. 106–111, 2010.

[423] T. E. O'Brien, B. Tarasinski, and B. M. Terhal, "Quantum phase estimation of multiple eigenvalues for small-scale (noisy) experiments," *New J. Phys*, vol. 21, no. 2, p. 023022, 2019.

[424] P. M. Cruz, G. Catarina, R. Gautier, and J. Fernández-Rossier, "Optimizing quantum phase estimation for the simulation of Hamiltonian eigenstates," *Quantum Sci. Technol*, vol. 5, no. 4, p. 044005, 2020.

[425] C. Kang, N. P. Bauman, S. Krishnamoorthy, and K. Kowalski, "Optimized quantum phase estimation for simulating electronic states in various energy regimes," *J. Chem. Theory Comput*, vol. 18, no. 11, pp. 6567–6576, 2022.

[426] K. M. Svore, M. B. Hastings, and M. Freedman, "Faster phase estimation," *Quantum Inf. Comput*, vol. 14, 2013.

[427] R. D. Somma, "Quantum eigenvalue estimation via time series analysis," *New J. Phys*, vol. 21, no. 12, p. 123025, 2019.

[428] H. Mohammadbagherpoor, Y.-H. Oh, P. Dreher, A. Singh, X. Yu, and A. J. Rindos, "An improved implementation approach for quantum phase estimation on quantum computers," in *International conference on rebooting computing (ICRC)*, pp. 1–9, IEEE, 2019.

[429] K. Choi, D. Lee, J. Bonitati, Z. Qian, and J. Watkins, "Rodeo algorithm for quantum computing," *Phys. Rev. Lett*, vol. 127, no. 4, p. 040505, 2021.

[430] J. E. Avron and A. Elgart, "Adiabatic theorem without a gap condition," *Comm. Math. Phys*, vol. 203, no. 2, pp. 445–463, 1999.

[431] J. Du, N. Xu, X. Peng, P. Wang, S. Wu, and D. Lu, "NMR implementation of a molecular hydrogen quantum simulation with adiabatic state preparation," *Phys. Rev. Lett*, vol. 104, no. 3, p. 030502, 2010.

[432] Y.-H. Oh, H. Mohammadbagherpoor, P. Dreher, A. Singh, X. Yu, and A. J. Rindos, "Solving multi-coloring combinatorial optimization problems using hybrid quantum algorithms," *arXiv:1911.00595*, 2019.

[433] G. Pagano, A. Bapat, P. Becker, K. S. Collins, A. De, P. W. Hess, H. B. Kaplan, A. Kyprianidis, W. L. Tan, C. Baldwin, *et al.*, "Quantum approximate optimization of the long-range Ising model with a trapped-ion quantum simulator," *Proc. Natl. Acad. Sci*, vol. 117, no. 41, pp. 25396–25401, 2020.

[434] Z. Wang, S. Hadfield, Z. Jiang, and E. G. Rieffel, "Quantum approximate optimization algorithm for MAXCUT: A fermionic view," *Phys. Rev. A*, vol. 97, no. 2, p. 022304, 2018.

[435] L. Zhou, S.-T. Wang, S. Choi, H. Pichler, and M. D. Lukin, "Quantum approximate optimization algorithm: Performance, mechanism, and implementation on near-term devices," *Phys. Rev. X*, vol. 10, no. 2, p. 021067, 2020.

[436] G. E. Crooks, "Performance of the quantum approximate optimization algorithm on the maximum cut problem," *arXiv:1811.08419*, 2018.

[437] Z.-C. Yang, A. Rahmani, A. Shabani, H. Neven, and C. Chamon, "Optimizing variational quantum algorithms using Pontryagin's minimum principle," *Phys. Rev. X*, vol. 7, no. 2, p. 021027, 2017.

[438] Z. Jiang, E. G. Rieffel, and Z. Wang, "Near-optimal quantum circuit for Grover's unstructured search using a transverse field," *Phys. Rev. A*, vol. 95, no. 6, p. 062317, 2017.

[439] P. K. Barkoutsos, G. Nannicini, A. Robert, I. Tavernelli, and S. Woerner, "Improving variational quantum optimization using CVaR," *Quantum*, vol. 4, p. 256, 2020.

[440] J. S. Otterbach, R. Manenti, N. Alidoust, A. Bestwick, M. Block, B. Bloom, S. Caldwell, N. Didier, E. S. Fried, S. Hong, *et al.*, "Unsupervised machine learning on a hybrid quantum computer," *arXiv:1712.05771*, 2017.

[441] R. Shaydulin, I. Safro, and J. Larson, "Multistart methods for quantum approximate optimization," in *High-performance extreme computing conference (HPEC)*, pp. 1–8, IEEE, 2019.

[442] E. Farhi and A. W. Harrow, "Quantum supremacy through the quantum approximate optimization algorithm," *arXiv:1602.07674*, 2016.

[443] L. Zhu, H. L. Tang, G. S. Barron, F. Calderon-Vargas, N. J. Mayhall, E. Barnes, and S. E. Economou, "Adaptive quantum approximate optimization algorithm for solving combinatorial problems on a quantum computer," *Phys. Rev. Res*, vol. 4, no. 3, p. 033029, 2022.

[444] G. G. Guerreschi and A. Y. Matsuura, "QAOA for MAXCUT requires hundreds of qubits for quantum speed-up," *Sci. Rep*, vol. 9, no. 1, pp. 1–7, 2019.

[445] D. Wierichs, C. Gogolin, and M. Kastoryano, "Avoiding local minima in variational quantum eigensolvers with the natural gradient optimizer," *Phys. Rev. Res*, vol. 2, no. 4, p. 043246, 2020.

[446] S. Bravyi, A. Kliesch, R. Koenig, and E. Tang, "Obstacles to variational quantum optimization from symmetry protection," *Phys. Rev. Lett*, vol. 125, no. 26, p. 260505, 2020.

[447] R. J. Bartlett, S. A. Kucharski, and J. Noga, "Alternative coupled-cluster Ansätze II. The unitary coupled-cluster method," *Chem. Phys. Lett*, vol. 155, no. 1, pp. 133–140, 1989.

[448] N. Moll, P. Barkoutsos, L. S. Bishop, J. M. Chow, A. Cross, D. J. Egger, S. Filipp, A. Fuhrer, J. M. Gambetta, M. Ganzhorn, A. Kandala, A. Mezzacapo, P. Müller, W. Riess, G. Salis, J. Smolin, I. Tavernelli, and K. Temme, "Quantum optimization using variational algorithms on near-term quantum devices," *Quantum Sci. Technol*, vol. 3, pp. 030503–030520, jun 2018.

[449] J. Romero, R. Babbush, J. R. McClean, C. Hempel, P. J. Love, and A. Aspuru-Guzik, "Strategies for quantum computing molecular energies using the unitary coupled cluster Ansatz," *Quantum Sci. Technol*, vol. 4, no. 1, pp. 014008–014026, 2018.

[450] T. Albash and D. A. Lidar, "Adiabatic quantum computation," *Rev. Mod. Phys*, vol. 90, no. 1, pp. 015002–015066, 2018.

[451] F. A. Evangelista, G. K.-L. Chan, and G. E. Scuseria, "Exact parameterization of fermionic wave functions via unitary coupled cluster theory," *J. Chem. Phys*, vol. 151, no. 24, p. 244112, 2019.

[452] P. K. Barkoutsos, J. F. Gonthier, I. Sokolov, N. Moll, G. Salis, A. Fuhrer, M. Ganzhorn, D. J. Egger, M. Troyer, A. Mezzacapo, S. Filipp, and I. Tavernelli, "Quantum algorithms for electronic structure calculations: particle-hole Hamiltonian and optimized wave-function expansions," *Phys. Rev. A*, vol. 98, p. 022322, 2018.

[453] B. Cooper and P. J. Knowles, "Benchmark studies of variational, unitary and extended coupled cluster methods," *J. Chem. Phys*, vol. 133, no. 23, pp. 234102–234111, 2010.

[454] G. Harsha, T. Shiozaki, and G. E. Scuseria, "On the difference between variational and unitary coupled cluster theories," *J. Chem. Phys*, vol. 148, no. 4, pp. 044107–044113, 2018.
[455] A. L. Fetter and J. D. Walecka, *Quantum theory of many-particle systems.* Courier Dover Publications, 2012.
[456] R. M. Martin, *Electronic structure: basic theory and practical methods.* Cambridge University Press, 2020.
[457] R. A. Friesner, "*Ab initio* quantum chemistry: Methodology and applications," *Proc. Natl. Acad. Sci*, vol. 102, no. 19, pp. 6648–6653, 2005.
[458] T. Helgaker, S. Coriani, P. Jørgensen, K. Kristensen, J. Olsen, and K. Ruud, "Recent advances in wave function-based methods of molecular-property calculations," *Chem. Rev*, vol. 112, no. 1, pp. 543–631, 2012.
[459] F. Becca and S. Sorella, *Quantum Monte Carlo approaches for correlated systems.* Cambridge University Press, 2017.
[460] W. Foulkes, L. Mitas, R. Needs, and G. Rajagopal, "Quantum Monte Carlo simulations of solids," *Rev. Mod. Phys*, vol. 73, no. 1, p. 33, 2001.
[461] H. Paik, D. I. Schuster, L. S. Bishop, G. Kirchmair, G. Catelani, A. P. Sears, B. R. Johnson, M. J. Reagor, L. Frunzio, L. I. Glazman, S. M. Girvin, M. H. Devoret, and R. J. Schoelkopf, "Observation of high coherence in Josephson junction qubits measured in a three-dimensional circuit QED architecture," *Phys. Rev. Lett*, vol. 107, p. 240501, 2011.
[462] C. U. Lei, L. Krayzman, S. Ganjam, L. Frunzio, and R. J. Schoelkopf, "High coherence superconducting microwave cavities with indium bump bonding," *Appl. Phys. Lett*, vol. 116, no. 15, p. 154002, 2020.
[463] T. Brecht, W. Pfaff, C. Wang, Y. Chu, L. Frunzio, M. H. Devoret, and R. J. Schoelkopf, "Multilayer microwave integrated quantum circuits for scalable quantum computing," *npj Quantum Inf*, vol. 2, pp. 1–4, 2016.
[464] D. Rosenberg, S. Weber, D. Conway, D. Yost, J. Mallek, G. Calusine, R. Das, D. Kim, M. Schwartz, W. Woods, *et al.*, "3D integration and packaging for solid-state qubits," *IEEE Microw. Mag*, vol. 21, pp. 72–85, 2020.
[465] D. M. Pozar, *Microwave engineering, 3rd edition.* John Wiley & Sons, 2005.
[466] T. Brecht, M. Reagor, Y. Chu, W. Pfaff, C. Wang, L. Frunzio, M. H. Devoret, and R. J. Schoelkopf, "Demonstration of superconducting micromachined cavities," *Appl. Phys. Lett*, vol. 107, p. 192603, 2015.
[467] M. Reagor, H. Paik, G. Catelani, L. Sun, C. Axline, E. Holland, I. M. Pop, N. A. Masluk, T. Brecht, L. Frunzio, *et al.*, "Reaching 10 ms single photon lifetimes for superconducting aluminum cavities," *Appl. Phys. Lett*, vol. 102, p. 192604, 2013.
[468] M. Göppl, A. Fragner, M. Baur, R. Bianchetti, S. Filipp, J. M. Fink, P. J. Leek, G. Puebla, L. Steffen, and A. Wallraff, "Coplanar waveguide resonators for circuit quantum electrodynamics," *J. Appl. Phys*, vol. 104, no. 11, p. 113904, 2008.
[469] F. Valenti, F. Henriques, G. Catelani, N. Maleeva, L. Grünhaupt, U. von Lüpke, S. T. Skacel, P. Winkel, A. Bilmes, A. V. Ustinov, *et al.*, "Interplay between kinetic inductance, nonlinearity, and quasiparticle dynamics in granular aluminum microwave kinetic inductance detectors," *Phys. Rev. Appl*, vol. 11, p. 054087, 2019.
[470] J. Rahamim, T. Behrle, M. J. Peterer, A. Patterson, P. A. Spring, T. Tsunoda,

R. Manenti, G. Tancredi, and P. J. Leek, "Double-sided coaxial circuit QED with out-of-plane wiring," *Appl. Phys. Lett*, vol. 110, p. 222602, 2017.
[471] M. Reagor, C. B. Osborn, N. Tezak, A. Staley, G. Prawiroatmodjo, M. Scheer, N. Alidoust, E. A. Sete, N. Didier, M. P. da Silva, *et al.*, "Demonstration of universal parametric entangling gates on a multi-qubit lattice," *Sci. Adv*, vol. 4, no. eaao3603, 2018.
[472] J. M. Fink, *Quantum nonlinearities in strong coupling circuit QED.* PhD thesis, ETH Zurich, 2010.
[473] B. A. Mazin, "Microwave kinetic inductance detectors." Technical report, available at `https://web.physics.ucsb.edu/~bmazin/mkids.html`, 2005.
[474] R. Barends, *Photon-detecting superconducting resonators.* PhD thesis, Delft University of Technology, 2009.
[475] M. V. Göppl, *Engineering quantum electronic chips: realization and characterization of circuit quantum electrodynamics systems.* PhD thesis, ETH Zurich, 2009.
[476] S. Krinner, S. Storz, P. Kurpiers, P. Magnard, J. Heinsoo, R. Keller, J. Luetolf, C. Eichler, and A. Wallraff, "Engineering cryogenic setups for 100-qubit scale superconducting circuit systems," *EPJ Quantum Technol*, vol. 6, p. 2, 2019.
[477] P. Krantz, M. Kjaergaard, F. Yan, T. P. Orlando, S. Gustavsson, and W. D. Oliver, "A quantum engineer's guide to superconducting qubits," *Appl. Phys. Rev*, vol. 6, p. 021318, 2019.
[478] A. Megrant, C. Neill, R. Barends, B. Chiaro, Y. Chen, L. Feigl, J. Kelly, E. Lucero, M. Mariantoni, P. J. O Malley, *et al.*, "Planar superconducting resonators with internal quality factors above one million," *Appl. Phys. Lett*, vol. 100, p. 113510, 2012.
[479] A. Bruno, G. De Lange, S. Asaad, K. Van Der Enden, N. Langford, and L. DiCarlo, "Reducing intrinsic loss in superconducting resonators by surface treatment and deep etching of silicon substrates," *Appl. Phys. Lett*, vol. 106, p. 182601, 2015.
[480] A. Nersisyan, S. Poletto, N. Alidoust, R. Manenti, R. Renzas, C.-V. Bui, K. Vu, T. Whyland, Y. Mohan, E. A. Sete, *et al.*, "Manufacturing low dissipation superconducting quantum processors," in *International Electron Devices Meeting (IEDM)*, 2019.
[481] W. Woods, G. Calusine, A. Melville, A. Sevi, E. Golden, D. K. Kim, D. Rosenberg, J. L. Yoder, and W. D. Oliver, "Determining interface dielectric losses in superconducting coplanar-waveguide resonators," *Phys. Rev. Appl*, vol. 12, p. 014012, 2019.
[482] A. P. Place, L. V. Rodgers, P. Mundada, B. M. Smitham, M. Fitzpatrick, Z. Leng, A. Premkumar, J. Bryon, A. Vrajitoarea, S. Sussman, *et al.*, "New material platform for superconducting transmon qubits with coherence times exceeding 0.3 ms," *Nat. Commun*, vol. 12, no. 1, pp. 1–6, 2021.
[483] J. Wenner, R. Barends, R. Bialczak, Y. Chen, J. Kelly, E. Lucero, M. Mariantoni, A. Megrant, P. O Malley, D. Sank, *et al.*, "Surface loss simulations of superconducting coplanar waveguide resonators," *Appl. Phys. Lett*, vol. 99, p. 113513, 2011.
[484] R. Barends, N. Vercruyssen, A. Endo, P. J. De Visser, T. Zijlstra, T. M. Klapwijk, P. Diener, S. J. C. Yates, and J. J. A. Baselmans, "Minimal resonator loss for

circuit quantum electrodynamics," *Appl. Phys. Lett*, vol. 97, p. 023508, 2010.

[485] P. Macha, S. van Der Ploeg, G. Oelsner, E. Ilíichev, H.-G. Meyer, S. Wünsch, and M. Siegel, "Losses in coplanar waveguide resonators at mK temperatures," *Appl. Phys. Lett*, vol. 96, p. 062503, 2010.

[486] J. Gao, M. Daal, A. Vayonakis, S. Kumar, J. Zmuidzinas, B. Sadoulet, B. A. Mazin, P. K. Day, and H. G. Leduc, "Experimental evidence for a surface distribution of two-level systems in superconducting lithographed microwave resonators," *Appl. Phys. Lett*, vol. 92, p. 152505, 2008.

[487] T. Lindström, J. E. Healey, M. S. Colclough, C. M. Muirhead, and A. Y. Tzalenchuk, "Properties of superconducting planar resonators at mK temperatures," *Phys. Rev. B*, vol. 80, p. 132501, 2009.

[488] F. Henriques, F. Valenti, T. Charpentier, M. Lagoin, C. Gouriou, M. Martínez, L. Cardani, M. Vignati, L. Grünhaupt, D. Gusenkova, *et al.*, "Phonon traps reduce the quasiparticle density in superconducting circuits," *Appl. Phys. Lett*, vol. 115, no. 21, p. 212601, 2019.

[489] R. N. Simons, *Coplanar waveguide circuits, components, and systems.* John Wiley & Sons, 2004.

[490] N. M. Sundaresan, Y. Liu, D. Sadri, L. J. Szőcs, D. L. Underwood, M. Malekakhlagh, H. E. Türeci, and A. A. Houck, "Beyond strong coupling in a multimode cavity," *Phys. Rev. X*, vol. 5, p. 021035, 2015.

[491] Y. Sunada, S. Kono, J. Ilves, S. Tamate, T. Sugiyama, Y. Tabuchi, and Y. Nakamura, "Fast readout and reset of a superconducting qubit coupled to a resonator with an intrinsic Purcell filter," *Phys. Rev. Appl*, vol. 17, no. 4, p. 044016, 2022.

[492] C. R. H. McRae, H. Wang, J. Gao, M. R. Vissers, T. Brecht, A. Dunsworth, D. P. Pappas, and J. Mutus, "Materials loss measurements using superconducting microwave resonators," *Rev. Sci. Instrum*, vol. 91, no. 9, p. 091101, 2020.

[493] J. Verjauw, A. Potočnik, M. Mongillo, R. Acharya, F. Mohiyaddin, G. Simion, A. Pacco, T. Ivanov, D. Wan, A. Vanleenhove, *et al.*, "Investigation of microwave loss induced by oxide regrowth in high-Q niobium resonators," *Phys. Rev. Appl*, vol. 16, no. 1, p. 014018, 2021.

[494] M. V. P. Altoé *et al.*, "Localization and mitigation of loss in niobium superconducting circuits," *Phys. Rev. X Quantum*, vol. 3, p. 020312, 2022.

[495] A. Romanenko, R. Pilipenko, S. Zorzetti, D. Frolov, M. Awida, S. Belomestnykh, S. Posen, and A. Grassellino, "Three-dimensional superconducting resonators at $T < 20$ mK with photon lifetimes up to $\tau = 2$s," *Phys. Rev. Appl*, vol. 13, no. 3, p. 034032, 2020.

[496] J. Lee, Z. Sung, A. A. Murthy, M. Reagor, A. Grassellino, and A. Romanenko, "Discovery of Nb hydride precipitates in superconducting qubits," *arXiv:2108.10385*, 2021.

[497] G. Calusine, A. Melville, W. Woods, R. Das, C. Stull, V. Bolkhovsky, D. Braje, D. Hover, D. K. Kim, X. Miloshi, *et al.*, "Analysis and mitigation of interface losses in trenched superconducting coplanar waveguide resonators," *Appl. Phys. Lett*, vol. 112, p. 062601, 2018.

[498] A. P. Vepsäläinen, A. H. Karamlou, J. L. Orrell, A. S. Dogra, B. Loer, F. Vasconcelos, D. K. Kim, A. J. Melville, B. M. Niedzielski, J. L. Yoder, *et al.*, "Impact of

ionizing radiation on superconducting qubit coherence," *Nature*, vol. 584, no. 7822, pp. 551–556, 2020.

[499] L. Cardani, F. Valenti, N. Casali, G. Catelani, T. Charpentier, M. Clemenza, I. Colantoni, A. Cruciani, G. D'Imperio, L. Gironi, *et al.*, "Reducing the impact of radioactivity on quantum circuits in a deep-underground facility," *Nat. Commun*, vol. 12, no. 1, pp. 1–6, 2021.

[500] C. Wang, C. Axline, Y. Y. Gao, T. Brecht, Y. Chu, L. Frunzio, M. Devoret, and R. J. Schoelkopf, "Surface participation and dielectric loss in superconducting qubits," *Appl. Phys. Lett*, vol. 107, p. 162601, 2015.

[501] M. S. Khalil, M. J. A. Stoutimore, F. C. Wellstood, and K. D. Osborn, "An analysis method for asymmetric resonator transmission applied to superconducting devices," *J. Appl. Phys*, vol. 111, p. 054510, 2012.

[502] M. Jerger, S. Poletto, P. Macha, U. Hübner, E. Il'ichev, and A. V. Ustinov, "Frequency division multiplexing readout and simultaneous manipulation of an array of flux qubits," *Appl. Phys. Lett*, vol. 101, p. 042604, 2012.

[503] J. Kelly, R. Barends, A. G. Fowler, A. Megrant, E. Jeffrey, T. C. White, D. Sank, J. Y. Mutus, B. Campbell, Y. Chen, *et al.*, "State preservation by repetitive error detection in a superconducting quantum circuit," *Nature*, vol. 519, no. 7541, pp. 66–69, 2015.

[504] C. Song, K. Xu, W. Liu, C.-p. Yang, S.-B. Zheng, H. Deng, Q. Xie, K. Huang, Q. Guo, L. Zhang, *et al.*, "10-qubit entanglement and parallel logic operations with a superconducting circuit," *Phys. Rev. Lett*, vol. 119, p. 180511, 2017.

[505] K. Watanabe, K. Yoshida, T. Aoki, and S. Kohjiro, "Kinetic inductance of superconducting coplanar waveguides," *Jpn. Appl. Phys. Lett*, vol. 33, p. 5708, 1994.

[506] K. Yoshida, K. Watanabe, T. Kisu, and K. Enpuku, "Evaluation of magnetic penetration depth and surface resistance of superconducting thin films using coplanar waveguides," *IEEE Trans. Appl. Supercond*, vol. 5, no. 2, pp. 1979–1982, 1995.

[507] L. Frunzio, A. Wallraff, D. Schuster, J. Majer, and R. Schoelkopf, "Fabrication and characterization of superconducting circuit QED devices for quantum computation," *IEEE Trans. Appl. Supercond*, vol. 15, no. 2, pp. 860–863, 2005.

[508] W. A. Phillips, "Two-level states in glasses," *Rep. Prog. Phys*, vol. 50, p. 1657, 1987.

[509] C. Müller, J. H. Cole, and J. Lisenfeld, "Towards understanding two-level systems in amorphous solids: insights from quantum circuits," *Rep. Prog. Phys*, vol. 82, p. 124501, 2019.

[510] J. Gao, *The physics of superconducting microwave resonators.* PhD thesis, Caltech, 2008.

[511] R. Manenti, M. Peterer, A. Nersisyan, E. Magnusson, A. Patterson, and P. Leek, "Surface acoustic wave resonators in the quantum regime," *Phys. Rev. B*, vol. 93, no. 4, p. 041411, 2016.

[512] A. Blais, J. Gambetta, A. Wallraff, D. I. Schuster, S. M. Girvin, M. H. Devoret, and R. J. Schoelkopf, "Quantum information processing with circuit quantum electrodynamics," *Phys. Rev. A*, vol. 75, p. 032329, 2007.

[513] J. Majer, J. Chow, J. Gambetta, J. Koch, B. Johnson, J. Schreier, L. Frunzio, D. Schuster, A. A. Houck, A. Wallraff, *et al.*, "Coupling superconducting qubits via a cavity bus," *Nature*, vol. 449, no. 7161, pp. 443–447, 2007.

[514] R. W. Heeres, B. Vlastakis, E. Holland, S. Krastanov, V. V. Albert, L. Frunzio, L. Jiang, and R. J. Schoelkopf, "Cavity state manipulation using photon-number selective phase gates," *Phys. Rev. Lett*, vol. 115, no. 13, p. 137002, 2015.

[515] M. Brune, E. Hagley, J. Dreyer, X. Maitre, A. Maali, C. Wunderlich, J. Raimond, and S. Haroche, "Observing the progressive decoherence of the 'meter' in a quantum measurement," *Phys. Rev. Lett*, vol. 77, p. 4887, 1996.

[516] C. Monroe, D. Meekhof, B. King, and D. J. Wineland, "A 'Schrödinger cat' superposition state of an atom," *Science*, vol. 272, pp. 1131–1136, 1996.

[517] B. Johnson, M. Reed, A. Houck, D. Schuster, L. S. Bishop, E. Ginossar, J. Gambetta, L. DiCarlo, L. Frunzio, S. Girvin, *et al.*, "Quantum non-demolition detection of single microwave photons in a circuit," *Nat. Phys*, vol. 6, pp. 663–667, 2010.

[518] G. Kirchmair, B. Vlastakis, Z. Leghtas, S. E. Nigg, H. Paik, E. Ginossar, M. Mirrahimi, L. Frunzio, S. M. Girvin, and R. J. Schoelkopf, "Observation of quantum state collapse and revival due to the single-photon Kerr effect," *Nature*, vol. 495, pp. 205–209, 2013.

[519] C. Song, T. W. Heitmann, M. P. DeFeo, K. Yu, R. McDermott, M. Neeley, J. M. Martinis, and B. L. Plourde, "Microwave response of vortices in superconducting thin films of Re and Al," *Phys. Rev. B*, vol. 79, p. 174512, 2009.

[520] J. Kroll, F. Borsoi, K. van der Enden, W. Uilhoorn, D. de Jong, M. Quintero-Pérez, D. van Woerkom, A. Bruno, S. Plissard, D. Car, *et al.*, "Magnetic-field-resilient superconducting coplanar-waveguide resonators for hybrid circuit quantum electrodynamics experiments," *Phys. Rev. Appl*, vol. 11, p. 064053, 2019.

[521] B. Chiaro, A. Megrant, A. Dunsworth, Z. Chen, R. Barends, B. Campbell, Y. Chen, A. Fowler, I. Hoi, E. Jeffrey, *et al.*, "Dielectric surface loss in superconducting resonators with flux-trapping holes," *Supercond. Sci. Technol*, vol. 29, p. 104006, 2016.

[522] M. H. Devoret, "Quantum fluctuations in electrical circuits," *Les Houches, Session LXIII*, vol. 7, no. 8, pp. 133–135, 1995.

[523] S. M. Girvin, "Circuit QED: superconducting qubits coupled to microwave photons." Lecture notes, url: `http://www.capri-school.eu/capri16/lectureres/master_cqed_les_houches.pdf`, 2011.

[524] J. I. Cirac and P. Zoller, "Quantum computations with cold trapped ions," *Phys. Rev. Lett*, vol. 74, p. 4091, 1995.

[525] D. Leibfried, R. Blatt, C. Monroe, and D. Wineland, "Quantum dynamics of single trapped ions," *Rev. Mod. Phys*, vol. 75, p. 281, 2003.

[526] R. Blatt and D. Wineland, "Entangled states of trapped atomic ions," *Nature*, vol. 453, pp. 1008–1015, 2008.

[527] R. Blatt and C. F. Roos, "Quantum simulations with trapped ions," *Nat. Phys*, vol. 8, pp. 277–284, 2012.

[528] D. Jaksch and P. Zoller, "The cold atom Hubbard toolbox," *Ann. Phys*, vol. 315, pp. 52–79, 2005.

[529] M. Lewenstein, A. Sanpera, V. Ahufinger, B. Damski, A. Sen, and U. Sen, "Ultracold atomic gases in optical lattices: mimicking condensed matter physics and beyond," *Adv. Phys*, vol. 56, pp. 243–379, 2007.

[530] I. Bloch, J. Dalibard, and W. Zwerger, "Many-body physics with ultracold gases," *Rev. Mod. Phys*, vol. 80, p. 885, 2008.

[531] C. Gross and I. Bloch, "Quantum simulations with ultracold atoms in optical lattices," *Science*, vol. 357, pp. 995–1001, 2017.

[532] D. Loss and D. P. DiVincenzo, "Quantum computation with quantum dots," *Phys. Rev. A*, vol. 57, p. 120, 1998.

[533] B. E. Kane, "A silicon-based nuclear spin quantum computer," *Nature*, vol. 393, pp. 133–137, 1998.

[534] R. Vrijen, E. Yablonovitch, K. Wang, H. W. Jiang, A. Balandin, V. Roychowdhury, T. Mor, and D. DiVincenzo, "Electron-spin-resonance transistors for quantum computing in silicon-germanium heterostructures," *Phys. Rev. A*, vol. 62, p. 012306, 2000.

[535] R. de Sousa, J. D. Delgado, and S. D. Sarma, "Silicon quantum computation based on magnetic dipolar coupling," *Phys. Rev. A*, vol. 70, p. 052304, 2004.

[536] L. C. L. Hollenberg, A. D. Greentree, A. G. Fowler, and C. J. Wellard, "Two-dimensional architectures for donor-based quantum computing," *Phys. Rev. B*, vol. 74, p. 045311, 2006.

[537] A. Morello, J. J. Pla, F. A. Zwanenburg, K. W. Chan, K. Y. Tan, H. Huebl, M. Möttönen, C. D. Nugroho, C. Yang, J. A. van Donkelaar, *et al.*, "Single-shot readout of an electron spin in silicon," *Nature*, vol. 467, pp. 687–691, 2010.

[538] R. Hanson, O. Gywat, and D. D. Awschalom, "Room-temperature manipulation and decoherence of a single spin in diamond," *Phys. Rev. B*, vol. 74, p. 161203, 2006.

[539] M. G. Dutt, L. Childress, L. Jiang, E. Togan, J. Maze, F. Jelezko, A. Zibrov, P. Hemmer, and M. Lukin, "Quantum register based on individual electronic and nuclear spin qubits in diamond," *Science*, vol. 316, pp. 1312–1316, 2007.

[540] A. Imamog, D. D. Awschalom, G. Burkard, D. P. DiVincenzo, D. Loss, M. Sherwin, A. Small, *et al.*, "Quantum information processing using quantum dot spins and cavity QED," *Phys. Rev. Lett*, vol. 83, p. 4204, 1999.

[541] J. R. Petta, A. C. Johnson, J. M. Taylor, E. A. Laird, A. Yacoby, M. D. Lukin, C. M. Marcus, M. P. Hanson, and A. C. Gossard, "Coherent manipulation of coupled electron spins in semiconductor quantum dots," *Science*, vol. 309, no. 5744, pp. 2180–2184, 2005.

[542] D. Englund, D. Fattal, E. Waks, G. Solomon, B. Zhang, T. Nakaoka, Y. Arakawa, Y. Yamamoto, and J. Vučković, "Controlling the spontaneous emission rate of single quantum dots in a two-dimensional photonic crystal," *Phys. Rev. Lett*, vol. 95, p. 013904, 2005.

[543] R. Hanson, L. P. Kouwenhoven, J. R. Petta, S. Tarucha, and L. M. K. Vandersypen, "Spins in few-electron quantum dots," *Rev. Mod. Phys*, vol. 79, p. 1217, 2007.

[544] E. Knill, R. Laflamme, and G. J. Milburn, "A scheme for efficient quantum computation with linear optics," *Nature*, vol. 409, pp. 46–52, 2001.

[545] T. B. Pittman, B. C. Jacobs, and J. D. Franson, "Probabilistic quantum logic operations using polarizing beam splitters," *Phys. Rev. A*, vol. 64, p. 062311, 2001.

[546] J. D. Franson, M. Donegan, M. Fitch, B. Jacobs, and T. Pittman, "High-fidelity quantum logic operations using linear optical elements," *Phys. Rev. Lett*, vol. 89,

p. 137901, 2002.

[547] T. B. Pittman, M. J. Fitch, B. C. Jacobs, and J. D. Franson, "Experimental controlled-not logic gate for single photons in the coincidence basis," *Phys. Rev. A*, vol. 68, p. 032316, 2003.

[548] S. R. Ekanayake, T. Lehmann, A. S. Dzurak, R. G. Clark, and A. Brawley, "Characterization of SOS-CMOS FETs at low temperatures for the design of integrated circuits for quantum bit control and readout," *IEEE Trans. Electron Devices*, vol. 57, pp. 539–547, 2010.

[549] J. C. Bardin, E. Jeffrey, E. Lucero, T. Huang, O. Naaman, R. Barends, T. White, M. Giustina, D. Sank, P. Roushan, *et al.*, "A 28 *nm* bulk-CMOS 4-to-8GHz $<$ 2mW cryogenic pulse modulator for scalable quantum computing," in *Proceedings of the 2019 IEEE international solid-state circuits conference (ISSCC)*, pp. 456–458, 2019.

[550] E. Charbon, F. Sebastiano, A. Vladimirescu, H. Homulle, S. Visser, L. Song, and R. M. Incandela, "Cryo-CMOS for quantum computing," in *International Electron Devices Meeting (IEDM)*, pp. 13–5, 2016.

[551] B. Patra, R. M. Incandela, J. P. Van Dijk, H. A. Homulle, L. Song, M. Shahmohammadi, R. B. Staszewski, A. Vladimirescu, M. Babaie, F. Sebastiano, *et al.*, "Cryo-CMOS circuits and systems for quantum computing applications," *IEEE J. Solid-State Circuits*, vol. 53, pp. 309–321, 2017.

[552] B. Patra, X. Xue, J. van Dijk, N. Samkharadze, S. Subramanian, A. Corna, C. Jeon, F. Sheikh, E. Juarez-Hernandez, B. Esparza, *et al.*, "CMOS-based cryogenic control of silicon quantum circuits," *Bull. Am. Phys. Soc*, 2021.

[553] R. Acharya, S. Brebels, A. Grill, J. Verjauw, T. Ivanov, D. P. Lozano, D. Wan, J. van Damme, A. Vadiraj, M. Mongillo, *et al.*, "Overcoming I/O bottleneck in superconducting quantum computing: multiplexed qubit control with ultra-low-power, base-temperature cryo-CMOS multiplexer," *arXiv:2209.13060*, 2022.

[554] F. Lecocq, F. Quinlan, K. Cicak, J. Aumentado, S. Diddams, and J. Teufel, "Control and readout of a superconducting qubit using a photonic link," *Nature*, vol. 591, no. 7851, pp. 575–579, 2021.

[555] P. Magnard, S. Storz, P. Kurpiers, J. Schär, F. Marxer, J. Lütolf, T. Walter, J.-C. Besse, M. Gabureac, K. Reuer, *et al.*, "Microwave quantum link between superconducting circuits housed in spatially separated cryogenic systems," *Phys. Rev. Lett*, vol. 125, no. 26, p. 260502, 2020.

[556] M. H. Devoret and J. M. Martinis, "Implementing qubits with superconducting integrated circuits," in *Experimental aspects of quantum computing*, pp. 163–203, Springer, 2005.

[557] J. Q. You and F. Nori, "Superconducting circuits and quantum information," *Phys. Today*, vol. 58, p. 42, 2005.

[558] R. Schoelkopf and S. Girvin, "Wiring up quantum systems," *Nature*, vol. 451, pp. 664–669, 2008.

[559] J. Clarke and F. K. Wilhelm, "Superconducting quantum bits," *Nature*, vol. 453, pp. 1031–1042, 2008.

[560] M. Ansmann, *Benchmarking the superconducting Josephson phase qubit: the violation of Bell's inequality.* University of California, Santa Barbara, 2009.

[561] J. Q. You and F. Nori, "Atomic physics and quantum optics using superconducting circuits," *Nature*, vol. 474, pp. 589–597, 2011.

[562] W. D. Oliver and P. B. Welander, "Materials in superconducting quantum bits," *MRS Bull*, vol. 38, pp. 816–825, 2013.

[563] J. M. Gambetta, J. M. Chow, and M. Steffen, "Building logical qubits in a superconducting quantum computing system," *npj Quantum Inf*, vol. 3, pp. 1–7, 2017.

[564] G. Wendin, "Quantum information processing with superconducting circuits: a review," *Rep. Prog. Phys*, vol. 80, p. 106001, 2017.

[565] X. Gu, A. F. Kockum, A. Miranowicz, Y.-X. Liu, and F. Nori, "Microwave photonics with superconducting quantum circuits," *Phys. Rep*, vol. 718, pp. 1–102, 2017.

[566] A. Blais, S. M. Girvin, and W. D. Oliver, "Quantum information processing and quantum optics with circuit quantum electrodynamics," *Nat. Phys*, pp. 1–10, 2020.

[567] C. Rigetti, J. M. Gambetta, S. Poletto, B. L. Plourde, J. M. Chow, A. D. Córcoles, J. A. Smolin, S. T. Merkel, J. R. Rozen, G. A. Keefe, *et al.*, "Superconducting qubit in a waveguide cavity with a coherence time approaching 0.1 ms," *Phys. Rev. B*, vol. 86, no. 10, p. 100506, 2012.

[568] C. Wang, Y. Y. Gao, I. M. Pop, U. Vool, C. Axline, T. Brecht, R. W. Heeres, L. Frunzio, M. H. Devoret, G. Catelani, *et al.*, "Measurement and control of quasiparticle dynamics in a superconducting qubit," *Nat. Commun*, vol. 5, pp. 1–7, 2014.

[569] M. Hutchings, J. B. Hertzberg, Y. Liu, N. T. Bronn, G. A. Keefe, M. Brink, J. M. Chow, and B. Plourde, "Tunable superconducting qubits with flux-independent coherence," *Phys. Rev. Appl*, vol. 8, p. 044003, 2017.

[570] M. Takita, A. W. Cross, A. Córcoles, J. M. Chow, and J. M. Gambetta, "Experimental demonstration of fault-tolerant state preparation with superconducting qubits," *Phys. Rev. Lett*, vol. 119, p. 180501, 2017.

[571] D. Rosenberg, D. Kim, R. Das, D. Yost, S. Gustavsson, D. Hover, P. Krantz, A. Melville, L. Racz, G. O. Samach, *et al.*, "3D integrated superconducting qubits," *npj Quantum Inf*, vol. 3, pp. 1–5, 2017.

[572] M. Ware, B. R. Johnson, J. M. Gambetta, T. A. Ohki, J. M. Chow, and B. L. T. Plourde, "Cross-resonance interactions between superconducting qubits with variable detuning," *arXiv:1905.11480*, 2019.

[573] A. Premkumar, C. Weiland, S. Hwang, B. Jäck, A. P. Place, I. Waluyo, A. Hunt, V. Bisogni, J. Pelliciari, A. Barbour, *et al.*, "Microscopic relaxation channels in materials for superconducting qubits," *Nat. Commun*, vol. 2, no. 1, pp. 1–9, 2021.

[574] J. M. Chow, A. D. Córcoles, J. M. Gambetta, C. Rigetti, B. R. Johnson, J. A. Smolin, J. R. Rozen, G. A. Keefe, M. B. Rothwell, M. B. Ketchen, *et al.*, "Simple all-microwave entangling gate for fixed-frequency superconducting qubits," *Phys. Rev. Lett*, vol. 107, no. 8, p. 080502, 2011.

[575] J. M. Chow, J. M. Gambetta, A. D. Corcoles, S. T. Merkel, J. A. Smolin, C. Rigetti, S. Poletto, G. A. Keefe, M. B. Rothwell, J. R. Rozen, *et al.*, "Universal quantum gate set approaching fault-tolerant thresholds with superconducting qubits," *Phys. Rev. Lett*, vol. 109, no. 6, p. 060501, 2012.

[576] J. B. Chang, M. R. Vissers, A. D. Córcoles, M. Sandberg, J. Gao, D. W. Abraham, J. M. Chow, J. M. Gambetta, M. Beth Rothwell, G. A. Keefe, *et al.*, "Improved superconducting qubit coherence using titanium nitride," *Appl. Phys. Lett*, vol. 103, p. 012602, 2013.

[577] H. Deng, Z. Song, R. Gao, T. Xia, F. Bao, X. Jiang, H.-S. Ku, Z. Li, X. Ma, J. Qin, *et al.*, "Titanium nitride film on sapphire substrate with low dielectric loss for superconducting qubits," *Phys. Rev. Appl*, vol. 19, no. 2, p. 024013, 2023.

[578] A. Patterson, J. Rahamim, T. Tsunoda, P. Spring, S. Jebari, K. Ratter, M. Mergenthaler, G. Tancredi, B. Vlastakis, M. Esposito, *et al.*, "Calibration of a cross-resonance two-qubit gate between directly coupled transmons," *Phys. Rev. Appl*, vol. 12, p. 064013, 2019.

[579] J. Braumüller, M. Sandberg, M. R. Vissers, A. Schneider, S. Schlör, L. Grünhaupt, H. Rotzinger, M. Marthaler, A. Lukashenko, A. Dieter, *et al.*, "Concentric transmon qubit featuring fast tunability and an anisotropic magnetic dipole moment," *Appl. Phys. Lett*, vol. 108, p. 032601, 2016.

[580] R. Barends, J. Kelly, A. Megrant, D. Sank, E. Jeffrey, Y. Chen, Y. Yin, B. Chiaro, J. Mutus, C. Neill, *et al.*, "Coherent Josephson qubit suitable for scalable quantum integrated circuits," *Phys. Rev. Lett*, vol. 111, p. 080502, 2013.

[581] R. Barends, J. Kelly, A. Megrant, A. Veitia, D. Sank, E. Jeffrey, T. C. White, J. Mutus, A. G. Fowler, B. Campbell, *et al.*, "Superconducting quantum circuits at the surface code threshold for fault tolerance," *Nature*, vol. 508, no. 7497, pp. 500–503, 2014.

[582] A. Bengtsson, P. Vikstål, C. Warren, M. Svensson, X. Gu, A. F. Kockum, P. Krantz, C. Križan, D. Shiri, I.-M. Svensson, *et al.*, "Quantum approximate optimization of the exact-cover problem on a superconducting quantum processor," *arXiv:1912.10495*, 2019.

[583] M. Kjaergaard, M. E. Schwartz, A. Greene, G. O. Samach, A. Bengtsson, M. O'Keeffe, C. M. McNally, J. Braumüller, D. K. Kim, P. Krantz, *et al.*, "A quantum instruction set implemented on a superconducting quantum processor," *arXiv:2001.08838*, 2020.

[584] X. Y. Han, T. Q. Cai, X. G. Li, Y. K. Wu, Y. W. Ma, Y. L. Ma, J. H. Wang, H. Y. Zhang, Y. P. Song, and L. M. Duan, "Error analysis in suppression of unwanted qubit interactions for a parametric gate in a tunable superconducting circuit," *Phys. Rev. A*, vol. 102, p. 022619, 2020.

[585] B. Josephson, "Possible new effect in superconducting tunneling," *Phys. Lett*, vol. 1, pp. 251–253, 1962.

[586] R. P. Feynman, R. B. Leighton, and M. L. Sands, *The Feynman Lectures on Physics*, vol. 3. Addison-Wesley, 1963.

[587] J. M. Martinis and K. Osborne, "Superconducting qubits and the physics of Josephson junctions," *arXiv:cond-mat/0402415*, 2004.

[588] H. Ohta, "A self-consistent model of the Josephson junction," in *Superconducting Quantum Interference Devices and their Applications* (H. D. Hahlbonm and H. Lubbig, eds.), pp. 35–49, de Gruiter, Inc, 1977.

[589] R. De Luca, "Feynman's and Ohta's models of a Josephson junction," *Eur. J. Phys*, vol. 33, p. 1547, 2012.

[590] V. Ambegaokar and A. Baratoff, "Tunneling between superconductors," *Phys. Rev. Lett*, vol. 10, p. 486, 1963.

[591] A. Barone and G. Paterno, *Physics and applications of the Josephson effect*. Wiley, 1982.

[592] M. Büttiker, "Zero-current persistent potential drop across small-capacitance Josephson junctions," *Phys. Rev. B*, vol. 36, p. 3548, 1987.

[593] V. Bouchiat, D. Vion, P. Joyez, D. Esteve, and M. Devoret, "Quantum coherence with a single Cooper pair," *Phys. Scr*, vol. 1998, p. 165, 1998.

[594] Y. Nakamura, Y. A. Pashkin, and J. S. Tsai, "Coherent control of macroscopic quantum states in a single-Cooper-pair box," *Nature*, vol. 398, pp. 786–788, 1999.

[595] D. Vion, A. Aassime, A. Cottet, P. Joyez, H. Pothier, C. Urbina, D. Esteve, and M. H. Devoret, "Manipulating the quantum state of an electrical circuit," *Science*, vol. 296, no. 5569, pp. 886–889, 2002.

[596] A. Cottet, *Implementation of a quantum bit in a superconducting circuit*. PhD thesis, Université Paris 6, 2002.

[597] A. Zorin, F.-J. Ahlers, J. Niemeyer, T. Weimann, H. Wolf, V. Krupenin, and S. Lotkhov, "Background charge noise in metallic single-electron tunneling devices," *Phys. Rev. B*, vol. 53, no. 20, p. 13682, 1996.

[598] L. Geerligs, V. Anderegg, and J. Mooij, "Tunneling time and offset charging in small tunnel junctions," *Physica B Condens. Matter*, vol. 165, pp. 973–974, 1990.

[599] J. Koch, M. Y. Terri, J. Gambetta, A. A. Houck, D. Schuster, J. Majer, A. Blais, M. H. Devoret, S. M. Girvin, and R. J. Schoelkopf, "Charge-insensitive qubit design derived from the Cooper pair box," *Phys. Rev. A*, vol. 76, p. 042319, 2007.

[600] G. Ithier, E. Collin, P. Joyez, P. Meeson, D. Vion, D. Esteve, F. Chiarello, A. Shnirman, Y. Makhlin, J. Schriefl, *et al.*, "Decoherence in a superconducting quantum bit circuit," *Phys. Rev. B*, vol. 72, no. 13, p. 134519, 2005.

[601] Y. Makhlin and A. Shnirman, "Dephasing of solid-state qubits at optimal points," *Phys. Rev. Lett*, vol. 92, no. 17, p. 178301, 2004.

[602] R. Zhao, S. Park, T. Zhao, M. Bal, C. McRae, J. Long, and D. Pappas, "Merged-element transmon," *Phys. Rev. Appl*, vol. 14, p. 064006, 2020.

[603] N. Didier, E. A. Sete, M. P. da Silva, and C. Rigetti, "Analytical modeling of parametrically modulated transmon qubits," *Phys. Rev. A*, vol. 97, p. 022330, 2018.

[604] M. Malekakhlagh, E. Magesan, and D. C. McKay, "First-principles analysis of cross-resonance gate operation," *Phys. Rev. A*, vol. 102, no. 4, p. 042605, 2020.

[605] F. Motzoi, J. M. Gambetta, P. Rebentrost, and F. K. Wilhelm, "Simple pulses for elimination of leakage in weakly nonlinear qubits," *Phys. Rev. Lett*, vol. 103, p. 110501, 2009.

[606] I.-M. Pop, T. Fournier, T. Crozes, F. Lecocq, I. Matei, B. Pannetier, O. Buisson, and W. Guichard, "Fabrication of stable and reproducible submicron tunnel junctions," *J. Vac. Sci. Technol*, vol. 30, p. 010607, 2012.

[607] J. B. Hertzberg, E. J. Zhang, S. Rosenblatt, E. Magesan, J. A. Smolin, J.-B. Yau, V. P. Adiga, M. Sandberg, M. Brink, J. M. Chow, *et al.*, "Laser-annealing Josephson junctions for yielding scaled-up superconducting quantum processors," *npj Quantum Inf*, vol. 7, no. 1, pp. 1–8, 2021.

[608] M. Hutchings, J. B. Hertzberg, Y. Liu, N. T. Bronn, G. A. Keefe, M. Brink, J. M. Chow, and B. Plourde, "Tunable superconducting qubits with flux-independent coherence," *Phys. Rev. Appl*, vol. 8, no. 4, p. 044003, 2017.
[609] F. Yoshihara, K. Harrabi, A. Niskanen, Y. Nakamura, and J. S. Tsai, "Decoherence of flux qubits due to $1/f$ flux noise," *Phys. Rev. Lett*, vol. 97, no. 16, p. 167001, 2006.
[610] S. De Graaf, A. Adamyan, T. Lindström, D. Erts, S. Kubatkin, A. Y. Tzalenchuk, and A. Danilov, "Direct identification of dilute surface spins on Al_2O_3: Origin of flux noise in quantum circuits," *Phys. Rev. Lett*, vol. 118, no. 5, p. 057703, 2017.
[611] P. Kumar *et al.*, "Origin and reduction of $1/f$ magnetic flux noise in superconducting devices," *Phys. Rev. Appl*, vol. 6, no. 4, p. 041001, 2016.
[612] R. H. Koch, D. P. DiVincenzo, and J. Clarke, "Model for $1/f$ flux noise in SQUIDs and qubits," *Phys. Rev. Lett*, vol. 98, no. 26, p. 267003, 2007.
[613] F. Yan, J. Bylander, S. Gustavsson, F. Yoshihara, K. Harrabi, D. G. Cory, T. P. Orlando, Y. Nakamura, J.-S. Tsai, and W. D. Oliver, "Spectroscopy of low-frequency noise and its temperature dependence in a superconducting qubit," *Phys. Rev. B*, vol. 85, no. 17, p. 174521, 2012.
[614] J. Braumüller, L. Ding, A. P. Vepsäläinen, Y. Sung, M. Kjaergaard, T. Menke, R. Winik, D. Kim, B. M. Niedzielski, A. Melville, *et al.*, "Characterizing and optimizing qubit coherence based on SQUID geometry," *Phys. Rev. Appl*, vol. 13, no. 5, p. 054079, 2020.
[615] M. Kounalakis, C. Dickel, A. Bruno, N. K. Langford, and G. A. Steele, "Tuneable hopping and nonlinear cross-Kerr interactions in a high-coherence superconducting circuit," *npj Quantum Inf*, vol. 4, no. 1, pp. 1–7, 2018.
[616] C. Neill, P. Roushan, K. Kechedzhi, S. Boixo, S. V. Isakov, V. Smelyanskiy, A. Megrant, B. Chiaro, A. Dunsworth, K. Arya, *et al.*, "A blueprint for demonstrating quantum supremacy with superconducting qubits," *Science*, vol. 360, no. 6385, pp. 195–199, 2018.
[617] D. M. Abrams, N. Didier, S. A. Caldwell, B. R. Johnson, and C. A. Ryan, "Methods for measuring magnetic flux crosstalk between tunable transmons," *Phys. Rev. Appl*, vol. 12, no. 6, p. 064022, 2019.
[618] Z. Chen, J. Kelly, C. Quintana, R. Barends, B. Campbell, Y. Chen, B. Chiaro, A. Dunsworth, A. Fowler, E. Lucero, *et al.*, "Measuring and suppressing quantum state leakage in a superconducting qubit," *Phys. Rev. Lett*, vol. 116, no. 2, p. 020501, 2016.
[619] Z. Chen, *Metrology of quantum control and measurement in superconducting qubits.* PhD thesis, University of California, Santa Barbara, 2018.
[620] J. M. Gambetta, F. Motzoi, S. Merkel, and F. K. Wilhelm, "Analytic control methods for high-fidelity unitary operations in a weakly nonlinear oscillator," *Phys. Rev. A*, vol. 83, no. 1, p. 012308, 2011.
[621] F. Motzoi and F. K. Wilhelm, "Improving frequency selection of driven pulses using derivative-based transition suppression," *Phys. Rev. A*, vol. 88, no. 6, p. 062318, 2013.
[622] J. M. Chow, L. DiCarlo, J. M. Gambetta, F. Motzoi, L. Frunzio, S. M. Girvin, and R. J. Schoelkopf, "Optimized driving of superconducting artificial atoms for

improved single-qubit gates," *Phys. Rev. A*, vol. 82, no. 4, p. 040305, 2010.
[623] M. D. Reed *et al.*, "Fast reset and suppressing spontaneous emission of a superconducting qubit," *Appl. Phys. Lett*, vol. 96, no. 20, p. 203110, 2010.
[624] J. M. Chow, *Quantum information processing with superconducting qubits.* Yale University, 2010.
[625] M. Steffen, M. Ansmann, R. McDermott, N. Katz, R. C. Bialczak, E. Lucero, M. Neeley, E. M. Weig, A. N. Cleland, and J. M. Martinis, "State tomography of capacitively shunted phase qubits with high fidelity," *Phys. Rev. Lett*, vol. 97, no. 5, p. 050502, 2006.
[626] E. Lucero, J. Kelly, R. C. Bialczak, M. Lenander, M. Mariantoni, M. Neeley, A. O'Connell, D. Sank, H. Wang, M. Weides, *et al.*, "Reduced phase error through optimized control of a superconducting qubit," *Phys. Rev. A*, vol. 82, no. 4, p. 042339, 2010.
[627] E. Knill, R. Laflamme, R. Martinez, and C.-H. Tseng, "An algorithmic benchmark for quantum information processing," *Nature*, vol. 404, no. 6776, pp. 368–370, 2000.
[628] E. Knill, D. Leibfried, R. Reichle, J. Britton, R. B. Blakestad, J. D. Jost, C. Langer, R. Ozeri, S. Seidelin, and D. J. Wineland, "Randomized benchmarking of quantum gates," *Phys. Rev. A*, vol. 77, no. 1, p. 012307, 2008.
[629] B. R. Johnson, M. P. da Silva, C. A. Ryan, S. Kimmel, J. M. Chow, and T. A. Ohki, "Demonstration of robust quantum gate tomography via randomized benchmarking," *New J. Phys*, vol. 17, no. 11, p. 113019, 2015.
[630] D. C. McKay, C. J. Wood, S. Sheldon, J. M. Chow, and J. M. Gambetta, "Efficient Z gates for quantum computing," *Phys. Rev. A*, vol. 96, no. 2, p. 022330, 2017.
[631] D. T. Sank, *Fast, accurate state measurement in superconducting qubits.* PhD thesis, University of California, Santa Barbara, 2014.
[632] A. Wallraff, D. Schuster, A. Blais, L. Frunzio, J. Majer, M. Devoret, S. Girvin, and R. Schoelkopf, "Approaching unit visibility for control of a superconducting qubit with dispersive readout," *Phys. Rev. Lett*, vol. 95, no. 6, p. 060501, 2005.
[633] B. Yurke, L. Corruccini, P. Kaminsky, L. Rupp, A. Smith, A. Silver, R. Simon, and E. Whittaker, "Observation of parametric amplification and deamplification in a Josephson parametric amplifier," *Phys. Rev. A*, vol. 39, no. 5, p. 2519, 1989.
[634] I. Siddiqi, R. Vijay, F. Pierre, C. Wilson, M. Metcalfe, C. Rigetti, L. Frunzio, and M. Devoret, "RF-driven Josephson bifurcation amplifier for quantum measurement," *Phys. Rev. Lett*, vol. 93, no. 20, p. 207002, 2004.
[635] T. Yamamoto, K. Inomata, M. Watanabe, K. Matsuba, T. Miyazaki, W. D. Oliver, Y. Nakamura, and J. Tsai, "Flux-driven Josephson parametric amplifier," *Appl. Phys. Lett*, vol. 93, no. 4, p. 042510, 2008.
[636] C. Macklin, K. OBrien, D. Hover, M. Schwartz, V. Bolkhovsky, X. Zhang, W. Oliver, and I. Siddiqi, "A near-quantum-limited Josephson traveling-wave parametric amplifier," *Science*, vol. 350, no. 6258, pp. 307–310, 2015.
[637] D. I. Schuster, A. Wallraff, A. Blais, L. Frunzio, R.-S. Huang, J. Majer, S. Girvin, and R. Schoelkopf, "AC Stark shift and dephasing of a superconducting qubit strongly coupled to a cavity field," *Phys. Rev. Lett*, vol. 94, no. 12, p. 123602, 2005.

[638] E. M. Purcell, H. C. Torrey, and R. V. Pound, "Resonance absorption by nuclear magnetic moments in a solid," *Phys. Rev*, vol. 69, no. 1–2, p. 37, 1946.
[639] A. Houck, J. Schreier, B. Johnson, J. Chow, J. Koch, J. Gambetta, D. Schuster, L. Frunzio, M. Devoret, *et al.*, "Controlling the spontaneous emission of a superconducting transmon qubit," *Phys. Rev. Lett*, vol. 101, no. 8, p. 080502, 2008.
[640] E. Jeffrey, D. Sank, J. Mutus, T. White, J. Kelly, R. Barends, Y. Chen, Z. Chen, B. Chiaro, A. Dunsworth, *et al.*, "Fast accurate state measurement with superconducting qubits," *Phys. Rev. Lett*, vol. 112, no. 19, p. 190504, 2014.
[641] G. Zhang, Y. Liu, J. J. Raftery, and A. A. Houck, "Suppression of photon shot noise dephasing in a tunable coupling superconducting qubit," *npj Quantum Inf*, vol. 3, no. 1, pp. 1–4, 2017.
[642] F. Yan, D. Campbell, P. Krantz, M. Kjaergaard, D. Kim, J. L. Yoder, D. Hover, A. Sears, A. J. Kerman, T. P. Orlando, *et al.*, "Distinguishing coherent and thermal photon noise in a circuit quantum electrodynamical system," *Phys. Rev. Lett*, vol. 120, no. 26, p. 260504, 2018.
[643] A. Sears *et al.*, "Photon shot noise dephasing in the strong-dispersive limit of circuit QED," *Phys. Rev. B*, vol. 86, no. 18, p. 180504, 2012.
[644] J.-H. Yeh, J. LeFebvre, S. Premaratne, F. Wellstood, and B. Palmer, "Microwave attenuators for use with quantum devices below 100 mK," *J. Appl. Phys*, vol. 121, no. 22, p. 224501, 2017.
[645] A. Clerk and D. W. Utami, "Using a qubit to measure photon-number statistics of a driven thermal oscillator," *Phys. Rev. A*, vol. 75, no. 4, p. 042302, 2007.
[646] P. Bertet, C. Harmans, and J. Mooij, "Parametric coupling for superconducting qubits," *Phys. Rev. B*, vol. 73, no. 6, p. 064512, 2006.
[647] M. Neeley, M. Ansmann, R. C. Bialczak, M. Hofheinz, N. Katz, E. Lucero, A. O'Connell, H. Wang, A. N. Cleland, and J. M. Martinis, "Process tomography of quantum memory in a Josephson-phase qubit coupled to a two-level state," *Nat. Phys*, vol. 4, no. 7, pp. 523–526, 2008.
[648] L. DiCarlo, J. M. Chow, J. M. Gambetta, L. S. Bishop, B. R. Johnson, D. Schuster, J. Majer, A. Blais, L. Frunzio, S. Girvin, *et al.*, "Demonstration of two-qubit algorithms with a superconducting quantum processor," *Nature*, vol. 460, no. 7252, pp. 240–244, 2009.
[649] R. C. Bialczak, M. Ansmann, M. Hofheinz, E. Lucero, M. Neeley, A. D. O'Connell, D. Sank, H. Wang, J. Wenner, M. Steffen, *et al.*, "Quantum process tomography of a universal entangling gate implemented with Josephson phase qubits," *Nat. Phys*, vol. 6, no. 6, pp. 409–413, 2010.
[650] J. Chow, L. DiCarlo, J. Gambetta, A. Nunnenkamp, L. S. Bishop, L. Frunzio, M. Devoret, S. Girvin, and R. Schoelkopf, "Detecting highly entangled states with a joint qubit readout," *Phys. Rev. A*, vol. 81, no. 6, p. 062325, 2010.
[651] L. DiCarlo, M. D. Reed, L. Sun, B. R. Johnson, J. M. Chow, J. M. Gambetta, L. Frunzio, S. M. Girvin, M. H. Devoret, and R. J. Schoelkopf, "Preparation and measurement of three-qubit entanglement in a superconducting circuit," *Nature*, vol. 467, no. 7315, pp. 574–578, 2010.
[652] A. Dewes, F. Ong, V. Schmitt, R. Lauro, N. Boulant, P. Bertet, D. Vion, and D. Esteve, "Characterization of a two-transmon processor with individual single-

shot qubit readout," *Phys. Rev. Lett*, vol. 108, no. 5, p. 057002, 2012.

[653] F. Beaudoin, M. P. da Silva, Z. Dutton, and A. Blais, "First-order sidebands in circuit QED using qubit frequency modulation," *Phys. Rev. A*, vol. 86, no. 2, p. 022305, 2012.

[654] S. Caldwell, N. Didier, C. Ryan, E. Sete, A. Hudson, P. Karalekas, R. Manenti, M. da Silva, R. Sinclair, E. Acala, *et al.*, "Parametrically activated entangling gates using transmon qubits," *Phys. Rev. Appl*, vol. 10, no. 3, p. 034050, 2018.

[655] P. Leek, S. Filipp, P. Maurer, M. Baur, R. Bianchetti, J. Fink, M. Göppl, L. Steffen, and A. Wallraff, "Using sideband transitions for two-qubit operations in superconducting circuits," *Phys. Rev. B*, vol. 79, no. 18, p. 180511, 2009.

[656] P. De Groot, J. Lisenfeld, R. Schouten, S. Ashhab, A. Lupaşcu, C. Harmans, and J. Mooij, "Selective darkening of degenerate transitions demonstrated with two superconducting quantum bits," *Nat. Phys*, vol. 6, no. 10, pp. 763–766, 2010.

[657] S. Poletto, J. M. Gambetta, S. T. Merkel, J. A. Smolin, J. M. Chow, A. Córcoles, G. A. Keefe, M. B. Rothwell, J. Rozen, D. Abraham, *et al.*, "Entanglement of two superconducting qubits in a waveguide cavity via monochromatic two-photon excitation," *Phys. Rev. Lett*, vol. 109, no. 24, p. 240505, 2012.

[658] J. M. Chow, J. M. Gambetta, A. W. Cross, S. T. Merkel, C. Rigetti, and M. Steffen, "Microwave-activated conditional-phase gate for superconducting qubits," *New J. Phys*, vol. 15, no. 11, p. 115012, 2013.

[659] A. D. Córcoles, J. M. Gambetta, J. M. Chow, J. A. Smolin, M. Ware, J. Strand, B. L. Plourde, and M. Steffen, "Process verification of two-qubit quantum gates by randomized benchmarking," *Phys. Rev. A*, vol. 87, no. 3, p. 030301, 2013.

[660] T. Hime, P. Reichardt, B. Plourde, T. Robertson, C.-E. Wu, A. Ustinov, and J. Clarke, "Solid-state qubits with current-controlled coupling," *Science*, vol. 314, no. 5804, pp. 1427–1429, 2006.

[661] A. Niskanen, K. Harrabi, F. Yoshihara, Y. Nakamura, S. Lloyd, and J. S. Tsai, "Quantum coherent tunable coupling of superconducting qubits," *Science*, vol. 316, no. 5825, pp. 723–726, 2007.

[662] R. Harris, A. Berkley, M. Johnson, P. Bunyk, S. Govorkov, M. Thom, S. Uchaikin, A. Wilson, J. Chung, E. Holtham, *et al.*, "Sign-and magnitude-tunable coupler for superconducting flux qubits," *Phys. Rev. Lett*, vol. 98, no. 17, p. 177001, 2007.

[663] Y. Chen, C. Neill, P. Roushan, N. Leung, M. Fang, R. Barends, J. Kelly, B. Campbell, Z. Chen, B. Chiaro, *et al.*, "Qubit architecture with high coherence and fast tunable coupling," *Phys. Rev. Lett*, vol. 113, no. 22, p. 220502, 2014.

[664] D. C. McKay, S. Filipp, A. Mezzacapo, E. Magesan, J. M. Chow, and J. M. Gambetta, "Universal gate for fixed-frequency qubits via a tunable bus," *Phys. Rev. Appl*, vol. 6, no. 6, p. 064007, 2016.

[665] S. J. Weber, G. O. Samach, D. Hover, S. Gustavsson, D. K. Kim, A. Melville, D. Rosenberg, A. P. Sears, F. Yan, J. L. Yoder, *et al.*, "Coherent coupled qubits for quantum annealing," *Phys. Rev. Appl*, vol. 8, no. 1, p. 014004, 2017.

[666] F. Yan, P. Krantz, Y. Sung, M. Kjaergaard, D. L. Campbell, T. P. Orlando, S. Gustavsson, and W. D. Oliver, "Tunable coupling scheme for implementing high-fidelity two-qubit gates," *Phys. Rev. Appl*, vol. 10, no. 5, p. 054062, 2018.

[667] B. Foxen, C. Neill, A. Dunsworth, P. Roushan, B. Chiaro, A. Megrant, J. Kelly,

Z. Chen, K. Satzinger, R. Barends, *et al.*, "Demonstrating a continuous set of two-qubit gates for near-term quantum algorithms," *Phys. Rev. Lett*, vol. 125, no. 12, p. 120504, 2020.

[668] J. M. Gambetta, A. D. Córcoles, S. T. Merkel, B. R. Johnson, J. A. Smolin, J. M. Chow, C. A. Ryan, C. Rigetti, S. Poletto, T. A. Ohki, *et al.*, "Characterization of addressability by simultaneous randomized benchmarking," *Phys. Rev. Lett*, vol. 109, no. 24, p. 240504, 2012.

[669] P. Mundada, G. Zhang, T. Hazard, and A. Houck, "Suppression of qubit crosstalk in a tunable coupling superconducting circuit," *Phys. Rev. Appl*, vol. 12, no. 5, p. 054023, 2019.

[670] R. Sagastizabal, S. Premaratne, B. Klaver, M. Rol, V. Negîrneac, M. Moreira, X. Zou, S. Johri, N. Muthusubramanian, M. Beekman, *et al.*, "Variational preparation of finite-temperature states on a quantum computer," *npj Quantum Inf*, vol. 7, no. 1, pp. 1–7, 2021.

[671] J. Garbow, D. Weitekamp, and A. Pines, "Bilinear rotation decoupling of homonuclear scalar interactions," *Chem. Phys. Lett*, vol. 93, no. 5, pp. 504–509, 1982.

[672] K. Cooper, M. Steffen, R. McDermott, R. W. Simmonds, S. Oh, D. A. Hite, D. P. Pappas, and J. M. Martinis, "Observation of quantum oscillations between a Josephson phase qubit and a microscopic resonator using fast readout," *Phys. Rev. Lett*, vol. 93, no. 18, p. 180401, 2004.

[673] M. Steffen, M. Ansmann, R. C. Bialczak, N. Katz, E. Lucero, R. McDermott, M. Neeley, E. M. Weig, A. N. Cleland, and J. M. Martinis, "Measurement of the entanglement of two superconducting qubits via state tomography," *Science*, vol. 313, no. 5792, pp. 1423–1425, 2006.

[674] N. Schuch and J. Siewert, "Natural two-qubit gate for quantum computation using the XY interaction," *Phys. Rev. A*, vol. 67, no. 3, p. 032301, 2003.

[675] F. Arute, K. Arya, R. Babbush, D. Bacon, J. C. Bardin, R. Barends, A. Bengtsson, S. Boixo, M. Broughton, B. B. Buckley, *et al.*, "Observation of separated dynamics of charge and spin in the Fermi-Hubbard model," *arXiv preprint arXiv:2010.07965*, 2020.

[676] C. Huang, T. Wang, F. Wu, D. Ding, Q. Ye, L. Kong, F. Zhang, X. Ni, Z. Song, Y. Shi, *et al.*, "Quantum instruction set design for performance," *Phys. Rev. Lett*, vol. 130, no. 7, p. 070601, 2023.

[677] T. Abad, J. Fernández-Pendás, A. F. Kockum, and G. Johansson, "Universal fidelity reduction of quantum operations from weak dissipation," *Phys. Rev. Lett*, vol. 129, no. 15, p. 150504, 2022.

[678] E. A. Sete, N. Didier, A. Q. Chen, S. Kulshreshtha, R. Manenti, and S. Poletto, "Parametric-resonance entangling gates with a tunable coupler," *Phys. Rev. Appl*, vol. 16, no. 2, p. 024050, 2021.

[679] J. Li, K. Chalapat, and G. S. Paraoanu, "Entanglement of superconducting qubits via microwave fields: classical and quantum regimes," *Phys. Rev. B*, vol. 78, p. 064503, Aug 2008.

[680] C. Rigetti and M. Devoret, "Fully microwave-tunable universal gates in superconducting qubits with linear couplings and fixed transition frequencies," *Phys. Rev. B*, vol. 81, no. 13, p. 134507, 2010.

[681] D. P. DiVincenzo, ed., *Quantum Information Processing.* Lectures notes of the 44th IFF Spring School, Forschungszentrum Jülich, 2013.

[682] V. Tripathi, M. Khezri, and A. N. Korotkov, "Operation and intrinsic error budget of a two-qubit cross-resonance gate," *Phys. Rev. A*, vol. 100, p. 012301, 2019.

[683] E. Magesan and J. M. Gambetta, "Effective Hamiltonian models of the cross-resonance gate," *Phys. Rev. A*, vol. 101, no. 5, p. 052308, 2020.

[684] S. Sheldon, E. Magesan, J. M. Chow, and J. M. Gambetta, "Procedure for systematically tuning up cross-talk in the cross-resonance gate," *Phys. Rev. A*, vol. 93, no. 6, p. 060302, 2016.

[685] J. M. Martinis, "Superconducting phase qubits," *Quantum information processing*, vol. 8, no. 2, pp. 81–103, 2009.

[686] T. Orlando, J. Mooij, L. Tian, C. H. Van Der Wal, L. Levitov, S. Lloyd, and J. Mazo, "Superconducting persistent-current qubit," *Phys. Rev. B*, vol. 60, no. 22, p. 15398, 1999.

[687] J. Mooij, T. Orlando, L. Levitov, L. Tian, C. H. Van der Wal, and S. Lloyd, "Josephson persistent-current qubit," *Science*, vol. 285, no. 5430, pp. 1036–1039, 1999.

[688] I. Chiorescu, P. Bertet, K. Semba, Y. Nakamura, C. Harmans, and J. Mooij, "Coherent dynamics of a flux qubit coupled to a harmonic oscillator," *Nature*, vol. 431, no. 7005, pp. 159–162, 2004.

[689] P. Bertet, I. Chiorescu, G. Burkard, K. Semba, C. Harmans, D. P. DiVincenzo, and J. Mooij, "Dephasing of a superconducting qubit induced by photon noise," *Phys. Rev. Lett*, vol. 95, no. 25, p. 257002, 2005.

[690] V. E. Manucharyan, J. Koch, L. I. Glazman, and M. H. Devoret, "Fluxonium: single Cooper-pair circuit free of charge offsets," *Science*, vol. 326, pp. 113–116, 2009.

[691] I. M. Pop, K. Geerlings, G. Catelani, R. J. Schoelkopf, L. I. Glazman, and M. H. Devoret, "Coherent suppression of electromagnetic dissipation due to superconducting quasiparticles," *Nature*, vol. 508, no. 7496, pp. 369–372, 2014.

[692] L. Grünhaupt, M. Spiecker, D. Gusenkova, N. Maleeva, S. T. Skacel, I. Takmakov, F. Valenti, P. Winkel, H. Rotzinger, W. Wernsdorfer, *et al.*, "Granular aluminium as a superconducting material for high-impedance quantum circuits," *Nat. Mater*, vol. 18, no. 8, pp. 816–819, 2019.

[693] L. Ioffe and M. Feigel'man, "Possible realization of an ideal quantum computer in Josephson junction array," *Phys. Rev. B*, vol. 66, no. 22, p. 224503, 2002.

[694] P. Brooks, A. Kitaev, and J. Preskill, "Protected gates for superconducting qubits," *Phys. Rev. A*, vol. 87, no. 5, p. 052306, 2013.

[695] P. Groszkowski, A. Di Paolo, A. Grimsmo, A. Blais, D. Schuster, A. Houck, and J. Koch, "Coherence properties of the 0-π qubit," *New J. Phys*, vol. 20, no. 4, p. 043053, 2018.

[696] M. Carroll, S. Rosenblatt, P. Jurcevic, I. Lauer, and A. Kandala, "Dynamics of superconducting qubit relaxation times," *npj Quantum Inf*, vol. 8, p. 132, 2022.

[697] N. D. Mermin, *Quantum computer science: an introduction.* Cambridge University Press, 2007.

[698] J. W. Cooley and J. W. Tukey, "An algorithm for the machine calculation of complex Fourier series," *Math. Comput*, vol. 19, no. 90, pp. 297–301, 1965.

Index